Student Solutions Manual

D0129162

Intermediate Algebra

FIFTH EDITION

Alan S. Tussy
Citrus College

R. David Gustafson
Rock Valley College

Prepared by

Kristy Hill
Hinds Community College

BROOKS/COLE
CENGAGE Learning

Australia • Brazil • Japan • Korea • Mexico • Singapore • Spain • United Kingdom • United States

BROOKS/COLE
CENGAGE Learning

ISBN-13: 978-1-111-98758-9

ISBN-10: 1-111-98758-0

Brooks/Cole
20 Davis Drive
Belmont, CA 94002-3098
USA

Cengage Learning is a leading provider of customized learning solutions with office locations around the globe, including Singapore, the United Kingdom, Australia, Mexico, Brazil, and Japan. Locate your local office at **www.cengage.com/global**

Cengage Learning products are represented in Canada by Nelson Education, Ltd.

To learn more about Brooks/Cole, visit **www.cengage.com/brookscole**

Purchase any of our products at your local college store or at our preferred online store **www.cengagebrain.com**

Printed in the United States of America
2 3 4 5 6 20 19 18 17 16

TABLE OF CONTENTS

SECTION 1.1

VOCABULARY

1. A **variable** is a letter that stands for a number.

3. An **equation** is a mathematical sentence that contains an = sign.

5. Phrases such as *increased by* and *more than* are used to indicate the operation of **addition**. Phrases such as *decreased by* and *less than* are used to indicate the operation of **subtraction**.

CONCEPTS

7. a. $6x - 5$ is an **expression**
   b. $P = a + b + c$ is an **equation**
   c. $\dfrac{s+9t}{8}$ is an **expression**
   d. $\sqrt{2w^2}$ is an **expression**

NOTATION

9. a. $7d = h$ (Answers may vary.)
   b. $t = 2{,}500 - d$ (Answers may vary.)

GUIDED PRACTICE

11. $t - 7$

13. $0.54e$

15. $2p + 35$

17. $\frac{1}{2}(x+4)$

19. $\frac{3}{4}p$

21. $\frac{w}{p}$

23. $950 + 0.1v$

25. $w - 500$

27. $\frac{150}{m}$

29. $7(77 + h + 88)$

31. $3w$

33. $4d - 15$

35. $0.95(200 + t)$

37. $0.01d$ or $\frac{1}{100}d$

39. $c = 13u + 24$

41. $w = \frac{c}{75}$

43. $A = t + 15$

45. $c = 12b$

47. $h = \frac{t}{2} - 75$

49. $b = 300 - 6s$

51. $c = \frac{p}{12}$

Number of packages (p)	Cartons (c)
24	$c = \frac{\mathbf{24}}{12} = 2$
72	$c = \frac{\mathbf{72}}{12} = 6$
180	$c = \frac{\mathbf{180}}{12} = 15$

53. $n = 22.44 - K$

K	n
0	$n = 22.44 - \mathbf{0}$ $= 22.44$
1.01	$n = 22.44 - \mathbf{1.01}$ $= 21.43$
22.44	$n = 22.44 - \mathbf{22.44}$ $= 0$

TRY IT YOURSELF

55. a. $S - s$
    b. $s - S$

57. a. $|a - 2|$
    b. $|a| - 2$

59. a. $0.155a + 6$
    b. $0.155(a + 6)$

61. a. $(x - 14)^2$
    b. $x^2 - 14$

APPLICATIONS

63. PRODUCTION PLANNING
    $b = \mathbf{2}r,\ s = \mathbf{4}r$

65. REGISTERED DIETICIAN
    $C = I - (m + a)$
    Answers may vary due to the variables chosen.

67. $b = t - 10$

WRITING

69. Answers will vary.

CHALLENGE PROBLEMS

71. a. $(a - b) + (c - d)$ (answers may vary)
    b. $(a + b) - (c + d)$ (answers may vary)

SECTION 1.2

VOCABULARY

1. The set of **whole** numbers is {0, 1, 2, 3, 4, 5, …}, the set of **natural** numbers is {1, 2, 3, 4, 5, …}, and the set of **integers** is {…, −2, −1, 0, 1, 2, …}.

3. A **prime** number is a whole number greater than 1 that has only itself and 1 as factors. A **composite** number is a whole number greater than 1 that is not prime.

5. **Irrational** numbers are nonterminating, nonrepeating decimals.

7. $>, \geq, <,$ and $\leq$ are called **inequality** symbols.

CONCEPTS

9. $-2 + 6 = 4$ and $-2 - 6 = -8$, so **4 and −8** are both six units away from −2.

11. nonrepeating, irrational

13. repeating, rational

15.
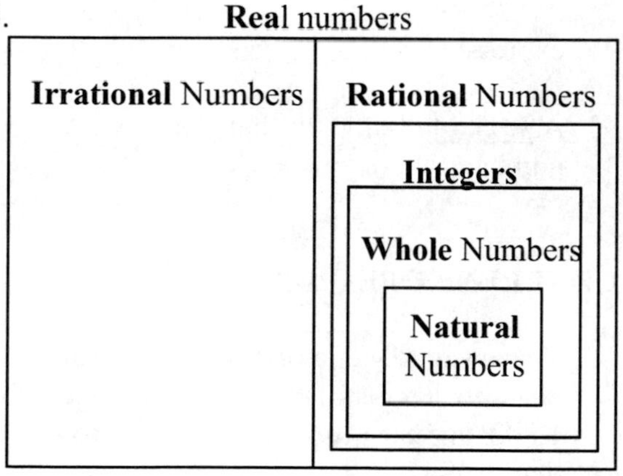

17. a. $12 < 19$
    b. $-5 \geq -6$

NOTATION

19. The symbol < means "**is less than**" and the symbol ≥ means "**is greater than or equal to**."

21. The symbols { } are called **braces**.

23. $\left\{\frac{a}{b} \,\middle|\, a \text{ and } b \text{ are integers, with } b \neq 0\right\}$

25. a. rational numbers
    b. irrational numbers
    c. real numbers

GUIDED PRACTICE

27. true

29. false; $-5 \in \mathbb{Z}$

31. false; $\mathbb{R} \not\subset W$ but $W \subseteq \mathbb{R}$

33. true

35. 1, 2, 9

37. −3, 0, 1, 2, 9

39. $\sqrt{3}, \pi$

41. 2

43. 2

45. 9

47
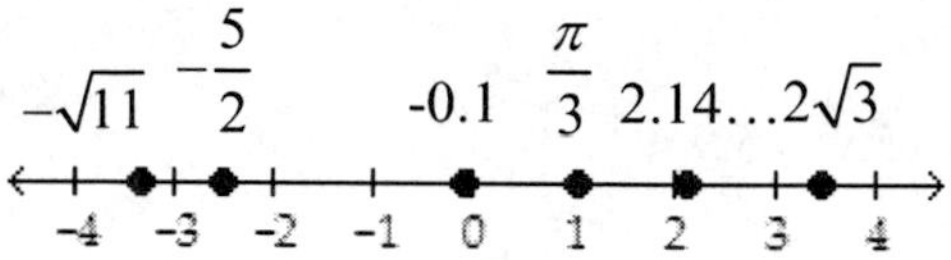

49.
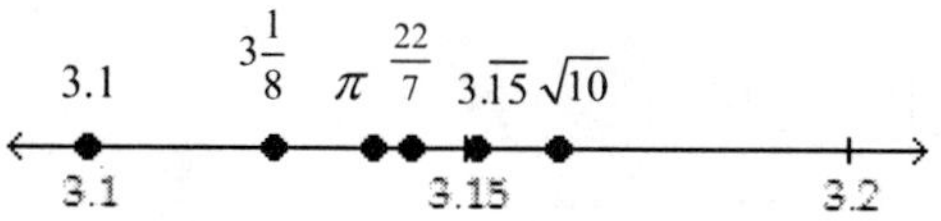

51. The set of prime numbers less than 8
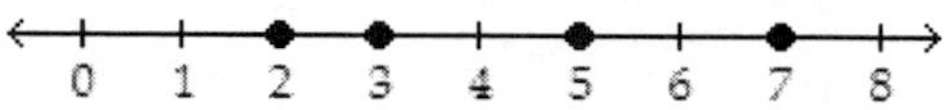

53. The set of odd integers between 10 and 18
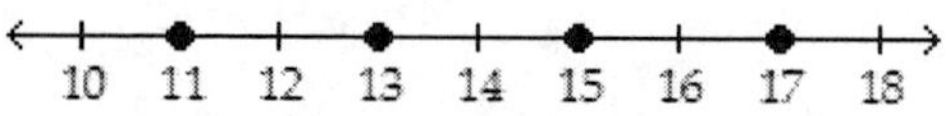

55. The set of positive odd integers less than 12

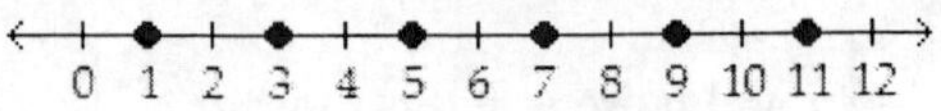

57. The set of even integers from −6 to 6.

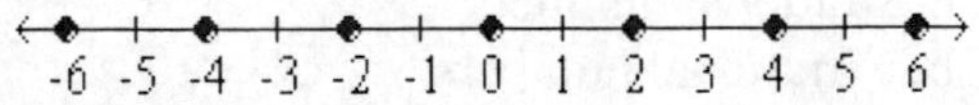

59. $-9 \boxed{<} -8$

61. $-(-5) \boxed{>} -10$ since $-(-5) = 5$ and $5 > -10$

63. $6.\overline{1} \boxed{>} -(-6)$
since $-(-6) = 6$

65. $-7.999 \boxed{<} -7.1$

67. $\frac{3}{5} \boxed{>} 0.06$

69. $-\frac{11}{15} \boxed{<} -0.73$

71. $\frac{27}{22} \boxed{<} 1.2\overline{28}$

73. $\frac{2}{125} \boxed{<} 0.\overline{016}$

75. $|20| = 20$

77. $|-5.9| = 5.9$

79. $-|-6| = -6$

81. $-\left|\frac{9}{4}\right| = -\frac{9}{4}$

APPLICATIONS

83. DRAFTING

$3\frac{2}{25} = 3.0800$
$\frac{77}{50} = 1.5400$
$\frac{15}{16} = 0.9375$
$2\frac{5}{8} = 2.6250$
$\frac{\pi}{4} \approx 0.7854$
$\sqrt{8} \approx 2.8284$

85. EROSION
Georgia had the greatest (1.3 m/yr).
Louisiana has the worst (−10.1 m/yr).

WRITING

87. Answers will vary.

89. Answers will vary.

REVIEW

91. expression

93. A **variable** is a letter that stands for a number.

CHALLENGE PROBLEMS

95. The absolute value of both positive and negative integers 1 through 999 are less than 1,000, and the absolute value of zero is less than 1,000.

999 (positive integers) + 999 (negative integers) + 1 (zero) = 1,999

There are 1,999 integers that have an absolute value that is less than 1,000.

97. Answers will vary.

SECTION 1.3

VOCABULARY

1. When we add two numbers, the result is called the **sum**. When we subtract two numbers, the result is called the **difference**.

3. The **reciprocal** of $\frac{5}{9}$ is $\frac{9}{5}$.

5. 6^2 can be read as "six **squared**," and 6^3 can be read as "six **cubed**."

7. In the expression 9 + 6[22 – (6 – 1)], the **parentheses** are the innermost grouping symbols, and the brackets are the **outermost** grouping symbols.

CONCEPTS

9. a. An exponent indicates repeated **multiplication**.
   b. Subtraction is the same as adding the **opposite** of the number being subtracted.

11. a. negative
    b. negative
    c. positive
    d. negative

NOTATION

13. a. $(-4)^2 = 16$
    b. $-4^2 = -16$

GUIDED PRACTICE

15. $-3 + (-5) = -8$

17. $-7.1 + 2.8 = -4.3$

19. $-9 + (-8) + 4 = -17 + 4$
    $= -13$

21.
$$\frac{1}{2}+\left(-\frac{1}{3}\right)=\frac{1}{2}\bullet\frac{3}{3}+\left(-\frac{1}{3}\right)\bullet\frac{2}{2}$$
$$=\frac{3}{6}+\left(-\frac{2}{6}\right)$$
$$=\frac{1}{6}$$

23. $-3 - 4 = -7$

25. $-3.3 - (-3.3) = 0$

27.
$$\frac{1}{2}-\left(-\frac{3}{5}\right)=\frac{1}{2}\bullet\frac{5}{5}-\left(-\frac{3}{5}\right)\bullet\frac{2}{2}$$
$$=\frac{5}{10}-\left(-\frac{6}{10}\right)$$
$$=\frac{5}{10}+\frac{6}{10}$$
$$=\frac{11}{10}$$

29. $-1 - 5 - (-4) = -6 + 4$
    $= -2$

31. $-2(6) = -12$

33. $-0.3(5) = -1.5$

35. $-5(6)(-2) = -30(-2)$
    $= 60$

37.
$$\left(-\frac{3}{5}\right)\left(\frac{10}{7}\right)=-\frac{3\bullet 10}{5\bullet 7}$$
$$=-\frac{30}{35}$$
$$=-\frac{6}{7}$$

39. $\frac{-8}{4} = -2$

41. $\frac{84}{-6} = -14$

43. $\frac{-10.8}{-1.2} = 9$

45.

$$-\frac{16}{5}\div\left(-\frac{10}{3}\right)=-\frac{16}{5}\bullet-\frac{3}{10}$$
$$=\frac{-16\bullet-3}{5\bullet10}$$
$$=\frac{48}{50}$$
$$=\frac{24}{25}$$

47.

$$6^4=6(6)(6)(6)$$
$$=1{,}296$$

49.

$$(-7.9)^2=(-7.9)(-7.9)$$
$$=62.41$$

51.

$$-5^2=-(5\cdot5)$$
$$=-25$$

53.

$$\left(-\frac{3}{5}\right)^3=\left(-\frac{3}{5}\right)\left(-\frac{3}{5}\right)\left(-\frac{3}{5}\right)$$
$$=-\frac{27}{125}$$

55.

$$\sqrt{64}=8$$

57.

$$-\sqrt{81}=-9$$

59.

$$-\sqrt{\frac{9}{16}}=-\frac{\sqrt{9}}{\sqrt{16}}$$
$$=-\frac{3}{4}$$

61.

$$\sqrt{0.04}=0.2$$

63.

$$3-5\cdot4=3-20$$
$$=-17$$

65.

$$4\cdot2^3=4\cdot8$$
$$=32$$

67.

$$-12\div3\cdot2=-4\cdot2$$
$$=-8$$

69.

$$7^2-(-9)^2=49-81$$
$$=-32$$

71.

$$(4+2\bullet3)^4=(4+6)^4$$
$$=(10)^4$$
$$=10{,}000$$

73.

$$\left(-3-\sqrt{25}\right)^2=(-3-5)^2$$
$$=(-8)^2$$
$$=(-8)(-8)$$
$$=64$$

75.

$$-2|4-8|=-2|-4|$$
$$=-2(4)$$
$$=-8$$

77.

$$2+3\left(\frac{25}{5}\right)+(-4)=2+3(5)+(-4)$$
$$=2+15+(-4)$$
$$=17+(-4)$$
$$=13$$

79.

$$30+6\left[-4-5(6-4)^2\right]=30+6\left[-4-5(2)^2\right]$$
$$=30+6\left[-4-5(4)\right]$$
$$=30+6[-4-20]$$
$$=30+6[-24]$$
$$=30+(-144)$$
$$=-114$$

81.

$$3-\left[3^3+(3-1)^3\right]=3-\left[3^3+(2)^3\right]$$
$$=3-[27+8]$$
$$=3-[35]$$
$$=3-35$$
$$=-32$$

83.

$$\frac{1}{3}\left(\frac{1}{6}\right)-\left(-\frac{1}{3}\right)^2=\frac{1}{3}\left(\frac{1}{6}\right)-\frac{1}{9}$$
$$=\frac{1}{18}-\frac{1}{9}$$
$$=\frac{1}{18}-\frac{1}{9}\left(\frac{2}{2}\right)$$
$$=\frac{1}{18}-\frac{2}{18}$$
$$=-\frac{1}{18}$$

85.

$$\frac{-2-5}{-7+(-7)}=\frac{-7}{-14}$$
$$=\frac{1}{2}$$

87.

$$\frac{|-25|-2(-5)}{2^4-9}=\frac{25-(-10)}{16-9}$$
$$=\frac{35}{7}$$
$$=5$$

89.

$$\frac{3[-9+2(7-3)]}{(8-5)(9-7)}=\frac{3[-9+2(4)]}{(3)(2)}$$
$$=\frac{3[-9+8]}{6}$$
$$=\frac{3[-1]}{6}$$
$$=\frac{-3}{6}$$
$$=-\frac{1}{2}$$

91.

$$-\frac{2}{3}a^2=-\frac{2}{3}(-\mathbf{6})^2$$
$$=-\frac{2}{3}(36)$$
$$=-\frac{2}{3}\left(\frac{36}{1}\right)$$
$$=-\frac{72}{3}$$
$$=-24$$

93.

$$\frac{y_2-y_1}{x_2-x_1}=\frac{\mathbf{-4-12}}{\mathbf{5-(-3)}}$$
$$=\frac{-16}{8}$$
$$=-2$$

95.

$$(x+y)(x^2-xy+y^2)=(-\mathbf{4}+\mathbf{5})((-\mathbf{4})^2-(-\mathbf{4})(\mathbf{5})+\mathbf{5}^2)$$
$$=(1)(16-(-20)+25)$$
$$=(1)(36+25)$$
$$=(1)(61)$$
$$=61$$

97.

$$\frac{-b+\sqrt{b^2-4ac}}{2a}=\frac{-\mathbf{2}+\sqrt{\mathbf{2}^2-4(\mathbf{1})(\mathbf{-3})}}{2(\mathbf{1})}$$
$$=\frac{-2+\sqrt{4-4(1)(-3)}}{2}$$
$$=\frac{-2+\sqrt{4-(-12)}}{2}$$
$$=\frac{-2+\sqrt{16}}{2}$$
$$=\frac{-2+4}{2}$$
$$=\frac{2}{2}$$
$$=1$$

99.

$$\sqrt{(x_2-x_1)^2+(y_2-y_1)^2}=\sqrt{(\mathbf{4-(-2)})^2+(\mathbf{-4-4})^2}$$
$$=\sqrt{(6)^2+(-8)^2}$$
$$=\sqrt{36+64}$$
$$=\sqrt{100}$$
$$=10$$

101.

$$-n\left(4n^2-27m^2\right)^3=-\frac{\mathbf{1}}{\mathbf{2}}\left[4\left(\frac{\mathbf{1}}{\mathbf{2}}\right)^2-27\left(\frac{\mathbf{1}}{\mathbf{3}}\right)^2\right]^3$$
$$=-\frac{1}{2}\left[4\left(\frac{1}{4}\right)-27\left(\frac{1}{9}\right)\right]^3$$
$$=-\frac{1}{2}[1-3]^3$$
$$=-\frac{1}{2}[-2]^3$$
$$=-\frac{1}{2}(-8)$$
$$=4$$

TRY IT YOURSELF

103. a.

$$100-20+5=80+5$$
$$=85$$

b.

$$100-(20+5)=100-25$$
$$=75$$

c.

$$100\div 20\cdot 5=5\cdot 5$$
$$=25$$

d.

$$100\div(20\cdot 5)=100\div 100$$
$$=1$$

105. a.

$$-11.2-(-3.9)=-11.2+3.9$$
$$=-7.3$$

b.

$$-3.9-(-11.2)=-3.9+11.2$$
$$=7.3$$

APPLICATIONS

107. THE STOCK MARKET
252+51-24+127+109+32+14-48-149-36
= 328
There was a net inflow of $328 billion.

109. TEMPERATURE EXTREMES
Find the range in temperatures for each city:
Atlanta: 105 – (–8) = 113
Boise: 111 – (–25) = 136
Helena: 105 – (–42) = 147
New York: 107 – (–3) = 110
Omaha: 114 – (–23) = 137

From smallest to the largest range:
New York, Atlanta, Boise, Omaha, and Helena

111. PEDIATRICS

$$\frac{6}{6+12}(27)=\frac{6}{18}(27)$$
$$=\frac{1}{3}(27)$$
$$=9$$

WRITING

113. Answers will vary.

REVIEW

115. $-2+5=3$
$-2-5=-7$
The numbers **3 and** –**7** are 5 units away from -2 on the number line.

117. {. . . , –2, –1, 0, 1, 2, . . . }

119. true

CHALLENGE PROBLEMS

121.

$$71-(1-2\cdot 5)^2+10=71-(1-10)^2+10$$
$$=71-(-9)^2+10$$
$$=71-81+10$$
$$=-10+10$$
$$=0$$

SECTION 1.4

VOCABULARY

1. A **term** is a product or quotient of numbers and/or variables, such as $6r$, $-t^3$, and $\frac{44}{m}$.

3. A term such as 9, that consists of a single number, is called a **constant** term.

5. The **commutative** properties of real numbers involve changing *order* and the **associative** property of real numbers involve changing *grouping*.

7. **Like** terms are terms with exactly the same variables raised to exactly the same powers.

CONCEPTS

9. a. $(x + y) + z = x + (y + z)$
   b. $xy = yx$
   c. $r(s + t) = rs + rt$

11. a. $a \cdot b = b \cdot \mathbf{a}$; commutative property of multiplication
    b $(ab)c = \mathbf{a(bc)}$; associative property of multiplication
    c. $0 \cdot a = \mathbf{0}$; multiplication property of 0
    d. $1 \cdot a = \mathbf{a}$; identity property of multiplication
    e. $a\left(\frac{1}{a}\right) = \mathbf{1}$; multiplicative inverse property

13. a. 0
    b. 1
    c. $-x$
    d. $\frac{1}{x}$

15. a. 5; $5 \div 5 = 1$
    b. $\frac{1}{5}$; $\frac{1}{5} \cdot 5 = 1$

NOTATION

17. multiplication by -1

GUIDED PRACTICE

19. terms: $3x^3, 11x^2, -x, 9$
    coefficients: $3, 11, -1, 9$

21. terms: $\frac{11}{12}a^4, -\frac{3}{4}b^2, 25b$
    coefficients: $\frac{11}{12}, -\frac{3}{4}, 25$

23. $3 + 7 = 7 + 3$

25. $3(2 + d) = 3 \cdot 2 + 3d$

27. $c + 0 = c$

29. $25 \cdot \frac{1}{25} = 1$

31. $8 + (7 + a) = (8 + 7) + a$

33. $(x + y)2 = 2(x + y)$

35. $9(8m) = 72m$

37. $5(-9q) = -45q$

39. $\frac{7}{8}x(-56) = \frac{7}{8}\left(-\frac{56}{1}\right)x$
    $= -49x$

41. $-4(8r)(-2y) = 64ry$

43. $9(9x + 2) = 9 \cdot 9x + 9 \cdot 2$
    $= 81x + 18$

45. $-4(-3t + 3) = -4(-3t) + -4(3)$
    $= 12t - 12$

47. $-(24 - d) = (-1)(24) - (-1)(d)$
    $= -24 + d$
    $= d - 24$

49. $\frac{2}{3}(3s^2 - 9) = \frac{2}{3} \cdot 3s^2 + \frac{2}{3} \cdot -9$
    $= 2s^2 - 6$

51. $0.7(m + 2n) = 0.7 \cdot m + 0.7 \cdot 2n$
    $= 0.7m + 1.4n$

53. $100(0.09x + 0.02y) = 100(0.09x) + 100(0.02y)$
    $= 9x + 2y$

55. $5(9t^2 - 12t - 3) = 5(9t^2) - 5(12t) - 5(3)$
$= 45t^2 - 60t - 15$

57. $3\left(\frac{4}{3}x - \frac{5}{3}y + \frac{1}{3}\right) = 3\left(\frac{4}{3}x\right) - 3\left(\frac{5}{3}y\right) + 3\left(\frac{1}{3}\right)$
$= 4x - 5y + 1$

59. $(16t + 24)\left(\frac{1}{8}\right) = 16t\left(\frac{1}{8}\right) + 24\left(\frac{1}{8}\right)$
$= 2t + 3$

61. $(y - 2)(-3) = -3 \cdot y - 3 \cdot -2$
$= -3y + 6$

63. $3x + 15x = 18x$

65. $0.7h - 3.8h = -3.1h$

67. $1.8x^2 - 5.1x^2 + 4.1x^2 = 0.8x^2$

69. $-8x + 5x - (-x) = -8x + 5x + x$
$= -2x$

71. $\frac{2}{5}ab - \left(-\frac{1}{2}ab\right) = \frac{2}{5}ab + \frac{1}{2}ab$
$= \frac{2}{5}ab \bullet \frac{2}{2} + \frac{1}{2}ab \bullet \frac{5}{5}$
$= \frac{4}{10}ab + \frac{5}{10}ab$
$= \frac{9}{10}ab$

73. $\frac{3}{5}t + \frac{1}{3}t = \frac{3}{5}t \bullet \frac{3}{3} + \frac{1}{3}t \bullet \frac{5}{5}$
$= \frac{9}{15}t + \frac{5}{15}t$
$= \frac{14}{15}t$

75. $-9a + 11ad - 35a + ad = (-9a - 35a) + (11ad + ad)$
$= -44a + 12ad$
$= 12ad - 44a$

77. $4m - t - (-2m) + 3t = 4m - t + 2m + 3t$
$= (4m + 2m) + (-t + 3t)$
$= 6m + 2t$

79. $2x^2 + 4(3x - x^2) + 3x = 2x^2 + 4(3x) + 4(-x^2) + 3x$
$= 2x^2 + 12x - 4x^2 + 3x$
$= -2x^2 + 15x$

81. $-3(p-2) + 2(p+3) - 5(p-1) = -3p + 6 + 2p + 6 - 5p + 5$
$= -6p + 17$

83. $36\left(\frac{2}{9}x - \frac{3}{4}\right) + 36\left(\frac{1}{2}\right) = 36\left(\frac{2}{9}x\right) + 36\left(-\frac{3}{4}\right) + 36\left(\frac{1}{2}\right)$
$= 8x - 27 + 18$
$= 8x - 9$

85. $24\left(\frac{5}{6}y - \frac{9}{8}\right) - 24\left(\frac{3}{24}y\right) = 24\left(\frac{5}{6}y\right) - 24\left(\frac{9}{8}\right) - 24\left(\frac{3}{24}y\right)$
$= 20y - 27 - 3y$
$= 17y - 27$

87.
$3[2(x+2)] - 5[3(x-5)] + 5x = 3[2x+4] - 5[3x-15] + 5x$
$= 6x + 12 - 15x + 75 + 5x$
$= -4x + 87$

89. $2\left[6\left(\frac{1}{3}a + 2b\right) - 8\left(\frac{1}{4}a - 2b\right) + 3\right]$
$= 2\left[6\left(\frac{1}{3}a\right) + 6(2b) - 8\left(\frac{1}{4}a\right) - 8(-2b) + 3\right]$
$= 2[2a + 12b - 2a + 16b + 3]$
$= 2[(2a - 2a) + (12b + 16b) + 3]$
$= 2[0a + 28b + 3]$
$= 2[28b + 3]$
$= 56b + 6$

TRY IT YOURSELF

91. $-(a + 2A + 1) - (a - A + 2) = -a - 2A - 1 - a + A - 2$
$= -2a - A - 3$

93. $8(2cd + 7c) - 2(cd - 3c)$
$= 8(2cd) + 8(7c) - 2(cd) - 2(-3c)$
$= 16cd + 56c - 2cd + 6c$
$= 14cd + 62c$

95. $6.4a^2 + 11.8a - 9.2a + 5.7$
$= 6.4a^2 + (11.8a - 9.2a) + 5.7$
$= 6.4a^2 + 2.6a + 5.7$

97. $-\frac{7}{16}x-\frac{3}{4}x=-\frac{7}{16}x-\frac{3}{4}\cdot\frac{4}{4}x$
$=-\frac{7}{16}x-\frac{12}{16}x$
$=-\frac{19}{16}x$

99. $-2[4(Z-9)-6(3Z-7)]-7(2Z-1)$
$=-2[4Z-36-18Z+42]-14Z+7$
$=-2[-14Z+6]-14Z+7$
$=28Z-12-14Z+7$
$=14Z-5$

101. $21\left(\frac{6}{7}h^2-\frac{15}{21}h\right)+21\left(\frac{1}{3}h\right)$
$=21\left(\frac{6}{7}h^2\right)+21\left(-\frac{15}{21}h\right)+21\left(\frac{1}{3}h\right)$
$=18h^2-15h+7h$
$=18h^2-8h$

103. $4.3(y+9)-8.1y=4.3(y)+4.3(9)-8.1y$
$=4.3y+38.7-8.1y$
$=(4.3y-8.1y)+38.7$
$=-3.8y+38.7$

105. $3x^2-(-2x^2)-5x^2=3x^2+2x^2-5x^2$
$=0x^2$
$=0$

107. $19a-\{-2[4a-2(a-16)]-3a\}$
$=19a-\{-2[4a-2a+32]-3a\}$
$=19a-\{-2[2a+32]-3a\}$
$=19a-\{-4a-64-3a\}$
$=19a-\{-7a-64\}$
$=19a+7a+64$
$=26a+64$

109. $\frac{1}{2}(4a-8)-6[2(5a-1)-a]$
$=\frac{1}{2}(4a)-\frac{1}{2}(8)-6[2(5a)+2(-1)-a]$
$=2a-4-6[10a-2-a]$
$=2a-4-6[9a-2]$
$=2a-4-6(9a)-6(-2)$
$=2a-4-54a+12$
$=-52a+8$

111. a. $12(8n)5=96n\cdot 5$
$=480n$
b. $12(8n+5)=96n+60$

113. a. $6a+6a+6a=18a$
b. $6a+6b+6c$ does not simplify

APPLICATIONS

115. PARKING AREAS
a. $20(x+6)\text{ m}^2$
b. $20\cdot 6+20\cdot x=120+20x$
$=(20x+120)\text{ m}^2$
c. $20(x+6)=20x+120$
distributive property

WRITING

117. Answers will vary.

119. Answers will vary.

REVIEW

121.
$\left(-\frac{3}{2}\right)\left(\frac{7}{12}\right)=-\frac{3\cdot 7}{2\cdot 12}$
$=-\frac{21}{24}$
$=-\frac{7}{8}$

123.
$-3|4-8|+(4+2\cdot 3)^3=-3|-4|+(4+6)^3$
$=-3(4)+(10)^3$
$=-12+1{,}000$
$=988$

CHALLENGE PROBLEMS

125.

$$\frac{x}{2}+\frac{x}{3}+\frac{x}{4}+\frac{x}{5}+\frac{x}{6}=\frac{x}{2}\bullet\frac{30}{30}+\frac{x}{3}\bullet\frac{20}{20}+\frac{x}{4}\bullet\frac{15}{15}+\frac{x}{5}\bullet\frac{12}{12}+\frac{x}{6}\bullet\frac{10}{10}$$

$$=\frac{30}{60}x+\frac{20}{60}x+\frac{15}{60}x+\frac{12}{60}x+\frac{10}{60}x$$

$$=\frac{87}{60}x$$

$$=\frac{29}{20}x$$

127. 1 and -1 are their own reciprocals

$$1=\frac{1}{1}$$

$$-1=\frac{-1}{1}=\frac{1}{-1}$$

SECTION 1.5

VOCABULARY

1. An **equation** is a statement that two expressions are equal.

3. A number that makes an equation true when substituted for the variable is called a **solution**.

5. An equation that is made true by any permissible replacement value for the variable is called an **identity**.

CONCEPTS

7. If $a = b$, then $a + c = b + \boxed{c}$, and $a - c = b - \boxed{c}$. **Adding** (or subtracting) the same number to (or from) **both** sides of an equation does not change the solution.

9. a. $\boxed{3} + 3 = 6$
   b. $\boxed{9} - 3 = 6$
   c. $3 \cdot \boxed{2} = 6$
   d. $\dfrac{\boxed{18}}{3} = 6$

11. a. all real numbers; $\mathbb{R}$
    b. no solution; $\varnothing$

NOTATION

13.

$$-2(x + 7) = 20$$
$$\boxed{-2x} - 14 = 20$$
$$-2x - 14 + \boxed{14} = 20 + \boxed{14}$$
$$-2x = 34$$
$$\frac{-2x}{\boxed{-2}} = \frac{34}{\boxed{-2}}$$
$$x = -17$$

$$-2(x + 7) = 20$$
$$-2(\boxed{\mathbf{-17}} + 7) \stackrel{?}{=} 20$$
$$-2(\boxed{-10}) \stackrel{?}{=} 20$$
$$\boxed{20} = 20$$

The solution is $\boxed{-17}$.

GUIDED PRACTICE

15. yes

$$3x + 2 = 17$$
$$3(\mathbf{5}) + 2 \stackrel{?}{=} 17$$
$$15 + 2 \stackrel{?}{=} 17$$
$$17 = 17$$

17. no

$$3(2m - 3) = 15$$
$$3(2 \cdot \mathbf{5} - 3) \stackrel{?}{=} 15$$
$$3(10 - 3) \stackrel{?}{=} 15$$
$$3(7) \stackrel{?}{=} 15$$
$$21 \neq 15$$

19.

$$2x - 12 = 0$$
$$2x - 12 + 12 = 0 + 12$$
$$2x = 12$$
$$\frac{2x}{2} = \frac{12}{2}$$
$$x = 6$$

21.

$$8k - 2 = 13$$
$$8k - 2 + 2 = 13 + 2$$
$$8k = 15$$
$$\frac{8k}{8} = \frac{15}{8}$$
$$k = \frac{15}{8}$$

23.

$$\frac{x}{4} - 6 = 1$$
$$4\left(\frac{x}{4} - 6\right) = 4(1)$$
$$4\left(\frac{x}{4}\right) - 4(6) = 4(1)$$
$$x - 24 = 4$$
$$x - 24 + 24 = 4 + 24$$
$$x = 28$$

25.

$$1.6a+(-4)=0.032$$
$$1.6a+(-4)+4=0.032+4$$
$$1.6a=4.032$$
$$\frac{1.6a}{1.6}=\frac{4.032}{1.6}$$
$$a=2.52$$

27.

$$0.7-4y=1.7$$
$$10(0.7-4y)=10(1.7)$$
$$7-40y=17$$
$$7-40y-7=17-7$$
$$-40y=10$$
$$\frac{-40y}{-40}=\frac{10}{-40}$$
$$y=-0.25$$

29.

$$-6-y=-13$$
$$-6-y+6=-13+6$$
$$-y=-7$$
$$\frac{-y}{-1}=\frac{-7}{-1}$$
$$y=7$$

31.

$$\frac{2}{3}c=10$$
$$\frac{3}{2}\left(\frac{2}{3}c\right)=\frac{3}{2}(10)$$
$$c=15$$

33.

$$-\frac{4}{5}s=2$$
$$-\frac{5}{4}\left(-\frac{4}{5}s\right)=-\frac{5}{4}(2)$$
$$s=-\frac{10}{4}$$
$$s=-\frac{5}{2}$$

35.

$$-\frac{7}{16}w-26=-19$$
$$-\frac{7}{16}w-26+26=-19+26$$
$$-\frac{7}{16}w=7$$
$$-\frac{16}{7}\left(-\frac{7}{16}w\right)=-\frac{16}{7}(7)$$
$$w=-16$$

37.

$$\frac{5}{6}k-7.5=7.5$$
$$\frac{5}{6}k-7.5+7.5=7.5+7.5$$
$$\frac{5}{6}k=15$$
$$\frac{6}{5}\left(\frac{5}{6}k\right)=\frac{6}{5}(15)$$
$$k=18$$

39.

$$8m+44=4m$$
$$8m+44-8m=4m-8m$$
$$44=-4m$$
$$\frac{44}{-4}=\frac{-4m}{-4}$$
$$-11=m$$
$$m=-11$$

41.

$$60t-50=15t-5$$
$$60t-50-15t=15t-5-15t$$
$$45t-50=-5$$
$$45t-50+50=-5+50$$
$$45t=45$$
$$\frac{45t}{45}=\frac{45}{45}$$
$$t=1$$

43.

$$9.8-16r=-15.7-r$$
$$9.8-16r+r=-15.7-r+r$$
$$9.8-15r=-15.7$$
$$10(9.8-15r)=10(-15.7)$$
$$98-150r=-157$$
$$98-150r-98=-157-98$$
$$-150r=-255$$
$$\frac{-150r}{-150}=\frac{-255}{-150}$$
$$r=1.7$$

45.

$$8b-2+b=5+5b+10$$
$$9b-2=5b+15$$
$$9b-2-5b=5b+15-5b$$
$$4b-2=15$$
$$4b-2+2=15+2$$
$$4b=17$$
$$\frac{4b}{4}=\frac{17}{4}$$
$$b=\frac{17}{4}$$

47.

$$3(k-4)=-36$$
$$3(k)-3(4)=-36$$
$$3k-12=-36$$
$$3k-12+12=-36+12$$
$$3k=-24$$
$$\frac{3k}{3}=\frac{-24}{3}$$
$$k=-8$$

49.

$$2(a-5)-(3a+1)=0$$
$$2a-10-3a-1=0$$
$$-a-11=0$$
$$-a-11+11=0+11$$
$$-a=11$$
$$\frac{-a}{-1}=\frac{11}{-1}$$
$$a=-11$$

51.

$$9(x+2)=-6(4-x)+18$$
$$9x+18=-24+6x+18$$
$$9x+18=-6+6x$$
$$9x+18-6x=-6+6x-6x$$
$$3x+18=-6$$
$$3x+18-18=-6-18$$
$$3x=-24$$
$$\frac{3x}{3}=\frac{-24}{3}$$
$$x=-8$$

53.

$$12+3(x-4)-21=5[5-4(4-x)]$$
$$12+3x-12-21=5[5-16+4x]$$
$$3x-21=5[-11+4x]$$
$$3x-21=-55+20x$$
$$3x-21-20x=-55+20x-20x$$
$$-17x-21=-55$$
$$-17x-21+21=-55+21$$
$$-17x=-34$$
$$\frac{-17x}{-17}=\frac{-34}{-17}$$
$$x=2$$

55.

$$\frac{1}{2}(a-2)=\frac{2}{3}a-6$$
$$6\left[\frac{1}{2}(a-2)\right]=6\left[\frac{2}{3}a-6\right]$$
$$3(a-2)=6\cdot\frac{2}{3}a-6\cdot 6$$
$$3a-6=4a-36$$
$$3a-6-4a=4a-36-4a$$
$$-a-6=-36$$
$$-a-6+6=-36+6$$
$$-a=-30$$
$$\frac{-a}{-1}=\frac{-30}{-1}$$
$$a=30$$

57.

$$\frac{1}{2}(3y+2)-\frac{5}{8}=\frac{3}{4}y$$
$$8\left[\frac{1}{2}(3y+2)-\frac{5}{8}\right]=8\left[\frac{3}{4}y\right]$$
$$4(3y+2)-5=6y$$
$$12y+8-5=6y$$
$$12y+3=6y$$
$$12y+3-12y=6y-12y$$
$$3=-6y$$
$$\frac{3}{-6}=\frac{-6y}{-6}$$
$$-\frac{1}{2}=y$$
$$y=-\frac{1}{2}$$

59.

$$\frac{a+1}{3}-\frac{a-1}{5}=\frac{8}{15}$$
$$15\left(\frac{a+1}{3}-\frac{a-1}{5}\right)=15\left(\frac{8}{15}\right)$$
$$5(a+1)-3(a-1)=8$$
$$5a+5-3a+3=8$$
$$2a+8=8$$
$$2a+8-8=8-8$$
$$2a=0$$
$$\frac{2a}{2}=\frac{0}{2}$$
$$a=0$$

61.

$$\frac{2z+3}{3}+\frac{3z-4}{6}=\frac{z-2}{2}$$
$$6\left(\frac{2z+3}{3}+\frac{3z-4}{6}\right)=6\left(\frac{z-2}{2}\right)$$
$$2(2z+3)+(3z-4)=3(z-2)$$
$$4z+6+3z-4=3z-6$$
$$7z+2=3z-6$$
$$7z+2-3z=3z-6-3z$$
$$4z+2=-6$$
$$4z+2-2=-6-2$$
$$4z=-8$$
$$\frac{4z}{4}=\frac{-8}{4}$$
$$z=-2$$

63.

$$0.45=16.95-0.25(75-3x)$$
$$100(0.45)=100[16.95-0.25(75-3x)]$$
$$45=1{,}695-25(75-3x)$$
$$45=1{,}695-1{,}875+75x$$
$$45=-180+75x$$
$$45+180=-180+75x+180$$
$$225=75x$$
$$\frac{225}{75}=\frac{75x}{75}$$
$$3=x$$
$$x=3$$

65.

$$0.04(12)+0.01t-0.02(12+t)=0$$
$$100[0.04(12)+0.01t-0.02(12+t)]=100(0)$$
$$4(12)+1t-2(12+t)=0$$
$$48+t-24-2t=0$$
$$24-t=0$$
$$24-t-24=0-24$$
$$-t=-24$$
$$\frac{-t}{-1}=\frac{-24}{-1}$$
$$t=24$$

67.

$$8x+3(2-x)=5x+6$$
$$8x+6-3x=5x+6$$
$$5x+6=5x+6$$
$$5x+6-5x=5x+6-5x$$
$$6=6$$

all real numbers; $\mathbb{R}$; identity

69.

$$2(x-3)=\frac{3}{2}(x-4)+\frac{x}{2}$$
$$2[2(x-3)]=2\left[\frac{3}{2}(x-4)+\frac{x}{2}\right]$$
$$2\bullet 2(x-3)=2\bullet\frac{3}{2}(x-4)+2\bullet\frac{x}{2}$$
$$4(x-3)=3(x-4)+x$$
$$4x-12=3x-12+x$$
$$4x-12=4x-12$$
$$4x-12-4x=4x-12-4x$$
$$-12=-12$$

all real numbers; $\mathbb{R}$; identity

71.

$$2x-6=-2x+4(x-2)$$
$$2x-6=-2x+4x-8$$
$$2x-6=2x-8$$
$$2x-6-2x=2x-8-2x$$
$$-6\neq-8$$

no solution; $\varnothing$; contradiction

73.

$$-3x=-2x+1-(5+x)$$
$$-3x=-2x+1-5-x$$
$$-3x=-3x-4$$
$$-3x+3x=-3x-4+3x$$
$$0\neq-4$$

no solution; $\varnothing$; contradiction

TRY IT YOURSELF

75.

$$\frac{3}{4}x-5=\frac{2}{3}x+\frac{1}{4}$$
$$12\left[\frac{3}{4}x-5\right]=12\left[\frac{2}{3}x+\frac{1}{4}\right]$$
$$12\cdot\frac{3}{4}x-12\cdot5=12\cdot\frac{2}{3}x+12\cdot\frac{1}{4}$$
$$9x-60=8x+3$$
$$9x-60-8x=8x+3-8x$$
$$x-60=3$$
$$x-60+60=3+60$$
$$x=63$$

77.

$$8x=x$$
$$8x-x=x-x$$
$$7x=0$$
$$\frac{7x}{7}=\frac{0}{7}$$
$$x=0$$

79.

$$-x+12=-17$$
$$-x+12-12=-17-12$$
$$-x=-29$$
$$\frac{-x}{-1}=\frac{-29}{-1}$$
$$x=29$$

81.

$$\frac{1}{2}b-\frac{19}{6}=\frac{1}{3}b+\frac{5}{6}$$
$$6\left[\frac{1}{2}b-\frac{19}{6}\right]=6\left[\frac{1}{3}b+\frac{5}{6}\right]$$
$$6\cdot\frac{1}{2}b-6\cdot\frac{19}{6}=6\cdot\frac{1}{3}b+6\cdot\frac{5}{6}$$
$$3b-19=2b+5$$
$$3b-19-2b=2b+5-2b$$
$$b-19=5$$
$$b-19+19=5+19$$
$$b=24$$

83.

$$a+18=5a-3+a$$
$$a+18=6a-3$$
$$a+18-6a=6a-3-6a$$
$$-5a+18=-3$$
$$-5a+18-18=-3-18$$
$$-5a=-21$$
$$\frac{-5a}{-5}=\frac{-21}{-5}$$
$$a=\frac{21}{5}$$

85.

$$-(2t-0.71)=0.9(1.4-t)$$
$$-2t+0.71=1.26-0.9t$$
$$-2t+0.71+0.9t=1.26-0.9t+0.9t$$
$$-1.1t+0.71=1.26$$
$$-1.1t+0.71-0.71=1.26-0.71$$
$$-1.1t=0.55$$
$$\frac{-1.1t}{-1.1}=\frac{0.55}{-1.1}$$
$$t=-0.5$$

87.

$$2(2x+1)=x+15+2x$$
$$4x+2=3x+15$$
$$4x+2-3x=3x+15-3x$$
$$x+2=15$$
$$x+2-2=15-2$$
$$x=13$$

89.

$$\frac{5}{2}a-12=\frac{1}{3}a+1$$
$$6\left[\frac{5}{2}a-12\right]=6\left[\frac{1}{3}a+1\right]$$
$$15a-72=2a+6$$
$$15a-72-2a=2a+6-2a$$
$$13a-72=6$$
$$13a-72+72=6+72$$
$$13a=78$$
$$\frac{13a}{13}=\frac{78}{13}$$
$$a=6$$

91.

$$\frac{4}{5}a=-12$$
$$\frac{5}{4}\cdot\frac{4}{5}a=\frac{5}{4}\cdot-12$$
$$a=-15$$

93.

$$0.06(a+200)+0.1a=172$$
$$100\left[0.06(a+200)+0.1a\right]=100[172]$$
$$6(a+200)+10a=17,200$$
$$6a+1,200+10a=17,200$$
$$16a+1,200=17,200$$
$$16a+1,200-1,200=17,200-1,200$$
$$16a=16,000$$
$$\frac{16a}{16}=\frac{16,000}{16}$$
$$a=1,000$$

95.

$$-4[p-(3-p)]=3(6p-2)$$
$$-4[p-3+p]=18p-6$$
$$-4[2p-3]=18p-6$$
$$-8p+12=18p-6$$
$$-8p+12-18p=18p-6-18p$$
$$-26p+12=-6$$
$$-26p+12-12=-6-12$$
$$-26p=-18$$
$$\frac{-26p}{-26}=\frac{-18}{-26}$$
$$p=\frac{9}{13}$$

97.

$$2(x-2)=\frac{2}{3}(3x+8)-2$$
$$3\left[2(x-2)\right]=3\left[\frac{2}{3}(3x+8)-2\right]$$
$$6(x-2)=2(3x+8)-6$$
$$6x-12=6x+16-6$$
$$6x-12=6x+10$$
$$6x-12-6x=6x+10-6x$$
$$-12\neq 10$$

no solution; $\varnothing$; contradiction

99.

$$13.5y+16.2=0$$
$$10[13.5y+16.2]=10[0]$$
$$135y+162=0$$
$$135y+162-162=0-162$$
$$135y=-162$$
$$\frac{135y}{135}=-\frac{162}{135}$$
$$y=-1.2$$

101.

$$\frac{4}{5}(x+5)=\frac{7}{8}(3x+23)-7$$
$$40\left[\frac{4}{5}(x+5)\right]=40\left[\frac{7}{8}(3x+23)-7\right]$$
$$32(x+5)=35(3x+23)-280$$
$$32x+160=105x+805-280$$
$$32x+160=105x+525$$
$$32x+160-105x=105x+525-105x$$
$$-73x+160=525$$
$$-73x+160-160=525-160$$
$$-73x=365$$
$$\frac{-73x}{-73}=\frac{365}{-73}$$
$$x=-5$$

103.

$$\frac{t-2}{5}+5t=\frac{7}{5}-\frac{t-2}{2}$$
$$10\left[\frac{t-2}{5}+5t\right]=10\left[\frac{7}{5}-\frac{t-2}{2}\right]$$
$$2(t-2)+50t=14-5(t-2)$$
$$2t-4+50t=14-5t+10$$
$$52t-4=24-5t$$
$$52t-4+5t=24-5t+5t$$
$$57t-4=24$$
$$57t-4+4=24+4$$
$$57t=28$$
$$\frac{57t}{57}=\frac{28}{57}$$
$$t=\frac{28}{57}$$

105.

$$6+4t-1=6-15t+12t-8$$
$$5+4t=-2-3t$$
$$5+4t+3t=-2-3t+3t$$
$$5+7t=-2$$
$$5+7t-5=-2-5$$
$$7t=-7$$
$$\frac{7t}{7}=\frac{-7}{7}$$
$$t=-1$$

LOOK ALIKES

107. a. $\frac{1}{2}(6x+8)-10-\frac{2}{3}(6x-9)$
$$=3x+4-10-4x+6$$
$$=(3x-4x)+(4-10+6)$$
$$=-x+0$$
$$=-x$$

b. $\frac{1}{2}(6x+8)-10=-\frac{2}{3}(6x-9)$
$$3x+4-10=-4x+6$$
$$3x-6=-4x+6$$
$$3x-6+4x=-4x+6+4x$$
$$7x-6=6$$
$$7x-6+6=6+6$$
$$7x=12$$
$$x=\frac{12}{7}$$

109. a. $-4\{6x-[3(7x-1)-x]\}+46x$
$$=-4\{6x-[21x-3-x]\}+46x$$
$$=-4\{6x-[20x-3]\}+46x$$
$$=-4\{6x-20x+3\}+46x$$
$$=-4\{-14x+3\}+46x$$
$$=56x-12+46x$$
$$=102x-12$$

b. $-4\{6x-[3(7x-1)-x]\}=46x$
$$-4\{6x-[21x-3-x]\}=46x$$
$$-4\{6x-[20x-3]\}=46x$$
$$-4\{6x-20x+3\}=46x$$
$$-4\{-14x+3\}=46x$$
$$56x-12=46x$$
$$56x-12-56x=46x-56x$$
$$-12=-10x$$
$$\frac{-12}{-10}=\frac{-10x}{-10}$$
$$\frac{6}{5}=x$$

WRITING

111. Answers will vary.

113. Answers will vary.

REVIEW

115. a. $a + b = b + a$
b. $(ab)c = a(bc)$
c. $a(b + c) = ab + ac$

117. a. $0 + a = a$
b. $1 \cdot a = a$

CHALLENGE PROBLEMS

119. Let $x = 4$ and solve the equation for k.

$$\begin{aligned} k + 3x - 6 &= 3kx - k + 16 \\ k + 3(4) - 6 &= 3k(4) - k + 16 \\ k + 12 - 6 &= 12k - k + 16 \\ k + 6 &= 11k + 16 \\ k + 6 - 6 &= 11k + 16 - 6 \\ k &= 11k + 10 \\ k - 11k &= 11k + 10 - 11k \\ -10k &= 10 \\ \frac{-10k}{-10} &= \frac{10}{-10} \\ k &= -1 \end{aligned}$$

SECTION 1.6

VOCABULARY

1. A **formula** is an equation that states a relationship between two or more variables.

3. The **volume** of a three-dimensional geometric solid is the amount of space it encloses.

CONCEPTS

5. a. area; ft.^2
   b. volume; ft.^3
   c. circumference; ft.
   d. perimeter; ft.

7. $$y = \frac{-5x+8}{2} = \frac{-5x}{\mathbf{2}} + \frac{8}{\mathbf{2}} = -\frac{5}{2}x + 4$$

NOTATION

9. $t = ad + bc$ for c

$$t - \boxed{ad} = ad + bc - \boxed{ad}$$
$$t - ad = \boxed{bc}$$
$$\frac{t-ad}{\boxed{b}} = \frac{bc}{\boxed{b}}$$
$$\frac{t-ad}{b} = \boxed{c}$$
$$c = \boxed{\frac{t-ad}{b}}$$

GUIDED PRACTICE

11. $P = 4s$
$= 4(2)$
$= 8$ yd

13. $P = 10 + 15 + 6 + 6$
$= 37$ in.

15. $$P = 2L + 2W$$
$$100 = 2(33.5) + 2W$$
$$100 = 67 + 2w$$
$$100 - 67 = 67 + 2w - 67$$
$$33 = 2w$$
$$\frac{33}{2} = \frac{2w}{2}$$
$$16.5 = w$$
The width is 16.5 in.

17. If the flower bed is in the shape of a square, then the length and the width are the same length. Let $x =$ the length of the length and the width.
$$P = 2L + 2W$$
$$26 = 2x + 2x$$
$$26 = 4x$$
$$\frac{26}{4} = \frac{4x}{4}$$
$$6.5 = x$$
Then length is 6.5 yards.

19. $$A = \frac{1}{2}bh = \frac{1}{2}(2.4)(8.5) = 10.2 \text{ ft}^2$$

21. $$A = s^2 = (17.2)^2 = 295.84 \text{ mi}^2$$

23. $$A = l \cdot w$$
$$30 = 12 \cdot w$$
$$30 = 12w$$
$$\frac{30}{12} = \frac{12w}{12}$$
$$2.5 = w$$

25. $A=\frac{1}{2}bh$

$42=\frac{1}{2}(7)(h)$

$42=\frac{7}{2}(h)$

$2(42)=2\left(\frac{7}{2}h\right)$

$84=7h$

$\frac{84}{7}=\frac{7h}{7}$

$12=h$

27. $C=\pi d$

$=7.5\pi$

$\approx(3.14159)(7.5)$

≈ 23.56 in.

29. $C=2\pi r$

$=2\cdot\pi\cdot 2\frac{1}{2}$

$\approx 2(3.14159)\left(\frac{5}{2}\right)$

≈ 15.71 ft.

31. $A=\pi r^2$

$=\pi\cdot(5.7)^2$

$=\pi(32.49)$

$\approx(3.14159)(32.49)$

$\approx 102.1\ \text{in.}^2$

33. $A=\pi r^2$

$=\pi\left(\frac{d}{2}\right)^2$

$=\pi\left(\frac{10\frac{1}{2}}{2}\right)^2$

$=\pi\left(5\frac{1}{4}\right)^2$

$=\pi(27.5625)$

$\approx(3.14159)(27.5625)$

$\approx 86.6\ \text{ft.}^2$

35. $V=lwh$

$=(2.51)(3.71)(10.21)$

$\approx 95.08\ \text{ft}^3$

37. $V=\frac{4}{3}\pi r^3$

$=\frac{4}{3}\pi(5.78)^3$

$\approx\frac{4}{3}(3.14159)(193.100552)$

$\approx 808.86\ \text{m}^3$

39. $d=rt$ for t

$\frac{d}{r}=\frac{rt}{r}$

$\frac{d}{r}=t$

$t=\frac{d}{r}$

41. $A=lwh$ for h

$\frac{A}{lw}=\frac{lwh}{lw}$

$\frac{A}{lw}=h$

$h=\frac{A}{lw}$

43. $V=\frac{1}{3}\pi r^2 h$ for h

$3\cdot V=3\cdot\frac{1}{3}\pi r^2 h$

$3V=\pi r^2 h$

$\frac{3V}{\pi r^2}=\frac{\pi r^2 h}{\pi r^2}$

$\frac{3V}{\pi r^2}=h$

$h=\frac{3V}{\pi r^2}$

45. $T=W+ma$ for W

$T-ma=W+ma-ma$

$T-ma=W$

$W=T-ma$

47.
$$h = 48t + \frac{1}{2}at^2 \quad \text{for } a$$
$$h = 48t + \frac{1}{2}at^2$$
$$h - 48t = 48t + \frac{1}{2}at^2 - 48t$$
$$h - 48t = \frac{1}{2}at^2$$
$$2(h - 48t) = 2\left(\frac{1}{2}at^2\right)$$
$$2h - 96t = at^2$$
$$\frac{2h - 96t}{t^2} = \frac{at^2}{t^2}$$
$$\frac{2h - 96t}{t^2} = a$$
$$a = \frac{2h - 96t}{t^2}$$

49.
$$A = \frac{1}{2}h(b_1 + b_2) \quad \text{for } b_2$$
$$2 \bullet A = 2 \bullet \frac{1}{2}h(b_1 + b_2)$$
$$2A = h(b_1 + b_2)$$
$$2A = b_1h + b_2h$$
$$2A - b_1h = b_1h + b_2h - b_1h$$
$$2A - b_1h = b_2h$$
$$\frac{2A - b_1h}{h} = \frac{b_2h}{h}$$
$$\frac{2A - b_1h}{h} = b_2$$
$$b_2 = \frac{2A - b_1h}{h} \quad \text{or}$$
$$b_2 = \frac{2A}{h} - b_1$$

51.
$$\ell = a + (n-1)d \quad \text{for } n$$
$$\ell = a + nd - d$$
$$\ell - a + d = a + nd - d - a + d$$
$$\ell - a + d = nd$$
$$\frac{\ell - a + d}{d} = \frac{nd}{d}$$
$$\frac{\ell - a + d}{d} = n$$
$$n = \frac{\ell - a + d}{d} \quad \text{or}$$
$$n = \frac{\ell - a}{d} + 1$$

53. $P = 2(w + h + l)$ for w
$$P = 2w + 2h + 2l$$
$$P - 2h - 2l = 2w + 2h + 2l - 2h - 2l$$
$$P - 2h - 2l = 2w$$
$$\frac{P - 2h - 2l}{2} = \frac{2w}{2}$$
$$\frac{P - 2h - 2l}{2} = w$$
$$w = \frac{P - 2h - 2l}{2} \quad \text{or}$$
$$w = \frac{P}{2} - h - l$$

55.
$$\lambda = A(x + B) \quad \text{for } A$$
$$\frac{\lambda}{x + B} = \frac{A(x + B)}{(x + B)}$$
$$\frac{\lambda}{x + B} = A$$
$$A = \frac{\lambda}{x + B}$$

57. $T_f = T_a(1 - F)$ for T_a
$$\frac{T_f}{1 - F} = \frac{T_a(1 - F)}{(1 - F)}$$
$$\frac{T_f}{1 - F} = T_a$$
$$T_a = \frac{T_f}{1 - F}$$

59. $l = a + (n-1)d$ for d

$$l - a = a + (n-1)d - a$$
$$l - a = (n-1)d$$
$$\frac{l-a}{(n-1)} = \frac{(n-1)d}{(n-1)}$$
$$\frac{l-a}{n-1} = d$$
$$d = \frac{l-a}{n-1}$$

61. $v = \frac{1}{t}(d_1 - d_2)$ for t

$$t[v] = t\left[\frac{1}{t}(d_1 - d_2)\right]$$
$$tv = 1(d_1 - d_2)$$
$$tv = d_1 - d_2$$
$$\frac{tv}{v} = \frac{d_1 - d_2}{v}$$
$$t = \frac{d_1 - d_2}{v}$$

63. $2x - 5y = 20$

$$2x - 5y - 2x = 20 - 2x$$
$$-5y = 20 - 2x$$
$$\frac{-5y}{-5} = \frac{20 - 2x}{-5}$$
$$y = \frac{20}{-5} + \frac{2x}{5}$$
$$y = \frac{2}{5}x - 4$$

65. $-4x = 12 + 3y$

$$-4x - 12 = 12 + 3y - 12$$
$$-4x - 12 = 3y$$
$$\frac{-4x - 12}{3} = \frac{3y}{3}$$
$$\frac{-4}{3}x - \frac{12}{3} = y$$
$$-\frac{4}{3}x - 4 = y$$
$$y = -\frac{4}{3}x - 4$$

TRY IT YOURSELF

67. $y = mx + b$ for x

$$y - b = mx + b - b$$
$$y - b = mx$$
$$\frac{y-b}{m} = \frac{mx}{m}$$
$$\frac{y-b}{m} = x$$
$$x = \frac{y-b}{m}$$

69. $L = 2d + 3.25(r + R)$ for R

$$L = 2d + 3.25r + 3.25R$$
$$L - 2d - 3.25r = 2d + 3.25r + 3.25R - 2d - 3.25r$$
$$L - 2d - 3.25r = 3.25R$$
$$\frac{L - 2d - 3.25r}{3.25} = \frac{3.25R}{3.25}$$
$$\frac{L - 2d - 3.25r}{3.25} = R$$
$$R = \frac{L - 2d - 3.25r}{3.25}$$

71. $s = \frac{1}{2}gt^2 + vt$ for g

$$s - vt = \frac{1}{2}gt^2 + vt - vt$$
$$s - vt = \frac{1}{2}gt^2$$
$$2(s - vt) = 2\left(\frac{1}{2}gt^2\right)$$
$$2(s - vt) = gt^2$$
$$\frac{2(s - vt)}{t^2} = \frac{gt^2}{t^2}$$
$$\frac{2(s - vt)}{t^2} = g$$
$$g = \frac{2(s - vt)}{t^2} \text{ or } g = \frac{2s - 2vt}{t^2}$$

73. $y - y_1 = m(x - x_1)$ for x

$$y - y_1 = mx - mx_1$$
$$y - y_1 + mx_1 = mx - mx_1 + mx_1$$
$$y - y_1 + mx_1 = mx$$
$$\frac{y - y_1 + mx_1}{m} = \frac{mx}{m}$$
$$\frac{y - y_1 + mx_1}{m} = x$$
$$x = \frac{y - y_1 + mx_1}{m}$$

75. $G = U - TS + pV$ for S

$$G - U - pV = U - TS + pV - U - pV$$
$$G - U - pV = -TS$$
$$\frac{G - U - pV}{-T} = \frac{-TS}{-T}$$
$$\frac{-G + U + pV}{T} = S$$
$$S = \frac{U + pV - G}{T}$$

77. $PV = nrt$ for r

$$\frac{PV}{nt} = \frac{nrt}{nt}$$
$$\frac{PV}{nt} = r$$
$$r = \frac{PV}{nt}$$

79. $E = IR + Ir$ for R

$$E - Ir = IR + Ir - Ir$$
$$E - Ir = IR$$
$$\frac{E - Ir}{I} = \frac{IR}{I}$$
$$\frac{E - Ir}{I} = R$$
$$R = \frac{E - Ir}{I}$$

81. $A = \frac{1}{3}(s_1 + s_2 + s_3)$ for s_3

$$3[A] = 3\left[\frac{1}{3}(s_1 + s_2 + s_3)\right]$$
$$3A = s_1 + s_2 + s_3$$
$$3A - s_1 - s_2 = s_1 + s_2 + s_3 - s_1 - s_2$$
$$3A - s_1 - s_2 = s_3$$
$$s_3 = 3A - s_1 - s_2$$

83. $S = \frac{n}{2}[2a + (n-1)d]$ for d

$$2[S] = 2\left[\frac{n}{2}[2a + (n-1)d]\right]$$
$$2S = n[2a + (n-1)d]$$
$$2S = 2an + n(n-1)d$$
$$2S - 2an = 2an + n(n-1)d - 2an$$
$$2S - 2an = n(n-1)d$$
$$\frac{2S - 2an}{n(n-1)} = \frac{n(n-1)d}{n(n-1)}$$
$$\frac{2S - 2an}{n(n-1)} = d$$
$$d = \frac{2S - 2an}{n(n-1)}$$

85. $d = \frac{4}{3}\pi h$ for h

$$3(d) = 3\left(\frac{4}{3}\pi h\right)$$
$$3d = 4\pi h$$
$$\frac{3d}{4\pi} = \frac{4\pi h}{4\pi}$$
$$\frac{3d}{4\pi} = h$$
$$h = \frac{3d}{4\pi}$$

APPLICATIONS

87. FLOOR MATS
Perimeter—add up all sides of the mat

46+40+6+10+6+46+6+10+6+40 = 216 in.

89. PAPER PRODUCTS

1st term: area of the bottom flap
2nd term: area of the left and right flaps
3rd term: area of the top flap
4th term: area of the face

$$A=\frac{1}{2}h_1(b_1+b_2)+b_3h_3+\frac{1}{2}b_1h_2+b_1b_3$$
$$=\frac{1}{2}(2)(6+2)+(3)(3)+\frac{1}{2}(6)(2.5)+(6)(3)$$
$$=\frac{1}{2}(2)(8)+(3)(3)+\frac{1}{2}(6)(2.5)+(6)(3)$$
$$=\frac{1}{2}(16)+9+\frac{1}{2}(15)+18$$
$$=8+9+7.5+18$$
$$=42.5\text{ in.}^2$$

91. PLANETS' TEMPERATURE RANGES

$$F=\frac{9}{5}C+32$$
$$F-32=\frac{9}{5}C+32-32$$
$$F-32=\frac{9}{5}C$$
$$\frac{5}{9}(F-32)=\frac{5}{9}\left(\frac{9}{5}C\right)$$
$$\frac{5}{9}(F-32)=C \quad\text{or}\quad C=\frac{5(F-32)}{9}$$

Mercury:

High °C
Let F = 810
$$C=\frac{5(810-32)}{9}=\frac{5(778)}{9}=\frac{3{,}890}{9}\approx 432$$

Low °C
Let F = –290
$$C=\frac{5(-290-32)}{9}=\frac{5(-322)}{9}=\frac{-1{,}610}{9}\approx -179$$

Earth:

High °C
Let F = 136
$$C=\frac{5(136-32)}{9}=\frac{5(104)}{9}=\frac{520}{9}\approx 58$$

Low °C
Let F = –129
$$C=\frac{5(-129-32)}{9}=\frac{5(-161)}{9}=\frac{-805}{9}\approx -89$$

Mars:

High °C
Let F = 63
$$C=\frac{5(63-32)}{9}=\frac{5(31)}{9}=\frac{155}{9}\approx 17$$

Low °C
Let F = –87
$$C=\frac{5(-87-32)}{9}=\frac{5(-119)}{9}=\frac{-595}{9}\approx -66$$

93. WIPER DESIGN

$$A=\frac{d\pi\left(r_1^2-r_2^2\right)}{360}$$
$$360(A)=360\left(\frac{d\pi\left(r_1^2-r_2^2\right)}{360}\right)$$
$$360A=d\pi\left(r_1^2-r_2^2\right)$$
$$\frac{360A}{\pi\left(r_1^2-r_2^2\right)}=\frac{d\pi\left(r_1^2-r_2^2\right)}{\pi\left(r_1^2-r_2^2\right)}$$
$$\frac{360A}{\pi\left(r_1^2-r_2^2\right)}=d$$

For $A = 513$ in.^{2}, $r_1 = 22$ and $r_2 = 8$

$$d=\frac{360A}{\pi\left(r_1^2-r_2^2\right)}=\frac{360(513)}{\pi\left(22^2-8^2\right)}=\frac{184{,}680}{\pi(484-64)}=\frac{184{,}680}{3.14(420)}=\frac{184{,}680}{1{,}318.8}\approx 140^\circ$$

For $A = 586$ in.^{2}, $r_1 = 22$ and $r_2 = 8$

$$d=\frac{360A}{\pi\left(r_1^2-r_2^2\right)}=\frac{360(586)}{\pi\left(22^2-8^2\right)}=\frac{210{,}960}{\pi(484-64)}=\frac{210{,}960}{3.14(420)}=\frac{210{,}960}{1{,}318.8}\approx 160^\circ$$

95. CHEMISTRY

$$PV = nR(T+273)$$

$$\frac{PV}{R(T+273)} = \frac{nR(T+273)}{R(T+273)}$$

$$\frac{PV}{R(T+273)} = n$$

Trial 1: Find n if $P = 0.900$ atmospheres, $V = 0.250$ liters, $T = 90°$ C, and $R = 0.082$.

$$n = \frac{(0.900)(0.250)}{0.082(90+273)}$$

$$= \frac{0.225}{0.082(363)}$$

$$= \frac{0.225}{29.766}$$

$$\approx 0.008$$

Trial 2: Find n if $P = 1.250$ atmospheres, $V = 1.560$ liters, $T = -10°$ C, and $R = 0.082$.

$$n = \frac{(1.250)(1.560)}{0.082(-10+273)}$$

$$= \frac{1.95}{0.082(263)}$$

$$= \frac{1.95}{21.566}$$

$$\approx 0.090$$

97. COST OF ELECTRICITY

$$C = 0.07n + 6.50$$

$$C - 6.50 = 0.07n + 6.50 - 6.50$$

$$C - 6.50 = 0.07n$$

$$\frac{C-6.50}{0.07} = \frac{0.07n}{0.07}$$

$$\frac{C-6.50}{0.07} = n$$

Find n if $C = \$49.97$

$$n = \frac{49.97 - 6.50}{0.07}$$

$$= \frac{43.47}{0.07}$$

$$= 621 \text{ kwh}$$

Find n if $C = \$76.50$

$$n = \frac{76.50 - 6.50}{0.07}$$

$$= \frac{70}{0.07}$$

$$= 1{,}000 \text{ kwh}$$

Find n if $C = \$125$

$$n = \frac{125 - 6.50}{0.07}$$

$$= \frac{118.5}{0.07}$$

$$\approx 1{,}692.9 \text{ kwh}$$

99. REGISTERED DIETICIAN

$$\text{BMI} = \frac{\text{weight (lb)} \cdot 703}{\text{height}^2 \text{ (in.}^2)}$$

$$(\text{height}^2)(\text{BMI}) = (\text{height}^2)\left(\frac{\text{weight (lb)} \cdot 703}{\text{height}^2 \text{ (in.}^2)}\right)$$

$$\text{BMI} \cdot \text{height}^2 \text{ (in.}^2) = \text{weight (lb)} \cdot 703$$

$$\frac{\text{BMI} \cdot \text{height}^2 \text{ (in.}^2)}{703} = \frac{\text{weight (lb)} \cdot 703}{703}$$

$$\frac{\text{BMI} \cdot \text{height}^2 \text{ (in.}^2)}{703} = \text{Weight (lb)}$$

WRITING

101. Answers will vary.

103. Answers will vary.

REVIEW

105. $(16b+8)\left(\frac{5}{4}\right)-8b = 16b\left(\frac{5}{4}\right)+8\left(\frac{5}{4}\right)-8b$

$= 20b+10-8b$

$= 12b+10$

107. $-5.7pt - p + 5.1pt + 12p = -0.6pt + 11p$

CHALLENGE PROBLEMS

109. To find the area of the shaded region subtract 2 times the area of the small circle from the area of the large circle.

Area of small circle:

$d = 8$, so $r = 8 \div 2 = 4$

$A = r^2\pi$

$A = 4^2\pi$

$A = 16\pi$

Area of large circle:

$r = 8$

$A = r^2\pi$

$A = 8^2\pi$

$A = 64\pi$

Area of shaded region:

$64\pi - 2(16\pi) = 64\pi - 32\pi$

$= 32\pi$

$= 100.5 \text{ ft}^2$

111. a. Chemistry; water and carbon dioxide molecules
b. Music; chords
c. Medicine; vitamins

SECTION 1.7

VOCABULARY

1. An **acute** angle has a measure of more than 0° and less than 90°.

3. If the sum of the measures of two angles equals 90°, the angles are called **complementary** angles.

5. If a triangle has a right angle, it is called a **right** triangle.

7. The sum of the measures of the **angles** of a triangle is 180°.

CONCEPTS

9.

	Decibels	Compared to conversation
Conversation	d	---
Vacuum cleaner	$d+15$	15 decibels more
Circular saw	$2d-10$	10 decibels less than twice
Whispering	$\frac{d}{2}-10$	10 decibels less than half

APPLICATIONS

11. AIRPLANES

Let x = number of seats on the B747.

Airplane	Number of seats
B747	x
B777	$x-125$ (125 less than the B747)
Combined	681

B747 seats + B777 seats=Combined Seats

$$x+(x-125)=681$$
$$2x-125=681$$
$$2x-125+125=681+125$$
$$2x=806$$
$$\frac{2x}{2}=\frac{806}{2}$$
$$x=403 \text{ seats}$$
$$x-125=403-125$$
$$=278 \text{ seats}$$

There are 403 seats on the B747 and 278 seats on the B777.

13. STATUE OF LIBERTY

Let x = height of the statute.

Statue	x
Base	$x+3$
Total height	305

Statue height + base height = total height

$$x+(x+3)=305$$
$$2x+3=305$$
$$2x+3-3=305-3$$
$$2x=302$$
$$\frac{2x}{2}=\frac{302}{2}$$
$$x=151$$
$$x+3=154$$

The height of the statue is 151 ft and the height of the base is 154 ft.

15. FLUTES

Let x = length of the first piece.

First Piece	x
Middle Piece	$2x-4$
Last piece	$\frac{2}{3}x$
Total length	29

1st piece+2nd piece+3rd piece = total length

$$x+(2x-4)+\frac{2}{3}x=29$$
$$3x-4+\frac{2}{3}x=29$$
$$3\left(3x-4+\frac{2}{3}x\right)=3(29)$$
$$9x-12+2x=87$$
$$11x-12=87$$
$$11x-12+12=87+12$$
$$11x=99$$
$$\frac{11x}{11}=\frac{99}{11}$$
$$x=9$$
$$2x-4=2(9)-4$$
$$=18-4$$
$$=14$$
$$\frac{2}{3}x=\frac{2}{3}(9)$$
$$=6$$

The first piece is 9 inches, the middle piece is 14 inches, and the last piece is 6 inches in length.

17. KITCHEN DRAWERS

front section	x in.
first partition	0.5 in.
middle section	$(x + 3)$ in.
second partition	0.5 in.
back section	$x + 3 + 3 = (x + 6)$ in.
Sum of lengths	28 in.

$$x + 0.5 + (x + 3) + 0.5 + (x + 6) = 28$$
$$3x + 10 = 28$$
$$3x + 10 - 10 = 28 - 10$$
$$3x = 18$$
$$\frac{3x}{3} = \frac{18}{3}$$
$$x = 6$$

The first partition should be 6 in. from the front.

19. TOURS

Let x = # of students supervised

Beginning Cost	\$1,810
Discount: \$15.50 · # of students	\$15.50x
Final Cost	\$1,500

Beginning Cost – Discount = Final Cost

$$1{,}810 - 15.50x = 1{,}500$$
$$1{,}810 - 15.50x - 1{,}810 = 1{,}500 - 1{,}810$$
$$-15.50x = -310$$
$$\frac{-15.50x}{-15.50} = \frac{-310}{-15.50}$$
$$x = 20$$

If they supervise 20 students, the cost for the chaperons will be \$1,500.

21. MOVING

Let x = # miles he can drive

Cost of Truck	\$41.50
Price per mile (35¢ · # of miles)	0.35x
Total cost	\$150

Cost of Truck + Price per mile = Total

$$41.50 + 0.35x = 150$$
$$41.50 + 0.35x - 41.50 = 150 - 41.50$$
$$0.35x = 108.50$$
$$\frac{0.35x}{0.35} = \frac{108.50}{0.35}$$
$$x = 310$$

He can drive 310 miles.

23. IRAS

Let x = # of shares of Safe Savings.
Then $2x$ = # of shares of Big Bank.

Share	price per share	number of shares	value of stock
Big Bank	\$115	$2x$	$115(2x)$
Safe Savings	\$97	x	$97(x)$

Value of BB + Value of SS = Total value

$$115(2x) + 97x = 49{,}050$$
$$230x + 97x = 49{,}050$$
$$327x = 49{,}050$$
$$\frac{327x}{327} = \frac{49{,}050}{327}$$
$$x = 150$$
$$2x = 2(150)$$
$$= 300$$

The couple owns 300 shares of Big Bank and 150 shares of Safe Savings.

25. PENSION FUNDS

Let x = # of shares of bond funds
Then $x - 2{,}000$ = # of shares of stock funds

Funds	price per share	number of shares	value of stock
Bonds	\$15	x	$15x$
Stocks	\$12	$x - 2{,}000$	$12(x–2{,}000)$

Bonds value + Stocks value = Total value

$$15x + 12(x - 2{,}000) = 165{,}000$$
$$15x + 12x - 24{,}000 = 165{,}000$$
$$27x - 24{,}000 = 165{,}000$$
$$27x - 24{,}000 + 24{,}000 = 165{,}000 + 24{,}000$$
$$27x = 189{,}000$$
$$\frac{27x}{27} = \frac{189{,}000}{27}$$
$$x = 7{,}000$$
$$x - 2{,}000 = 5{,}000$$

The pension fund owns 7,000 shares of bond funds and 5,000 shares of stock funds.

27. NURSING

The angles are supplementary (sum = 180°).

1st angle	x
2nd angle	$5x$
Sum of supplementary angles	180

$$1^{st}\text{ angle} + 2^{nd}\text{ angle} = 180$$
$$x + 5x = 180$$
$$6x = 180$$
$$\frac{6x}{6} = \frac{180}{6}$$
$$x = 30$$
$$5x = 150$$

The first angle is 30° and the second angle is 150°.

29. THE TOWER OF PISA

The angles are complementary (sum = 90°).

1st angle	x
2nd angle	$15x + 0.4$
Sum of complementary angles	90

$$1^{st}\text{ angle} + 2^{nd}\text{ angle} = 90$$
$$x + 15x + 0.4 = 90$$
$$16x + 0.4 = 90$$
$$16x + 0.4 - 0.4 = 90 - 0.4$$
$$\frac{16x}{16} = \frac{89.6}{16}$$
$$x = 5.6^{\circ}$$
$$15x + 0.4 = 84.4^{\circ}$$

a. The tower is leaning 5.6.°
b. 7 – 5.6 = 1.4°
   The tower is predicted to collapse when it leans 1.4° more.

31. STEPSTOOLS

Angle 1	x
Angle 2	$x + 10$
Angle 3	$(x + 10) + 10 = x + 20$
Sum of 3 angles of a triangle	180

Angle 1 + Angle 2 + Angle 3 = 180
$$x + (x + 10) + (x + 20) = 180$$
$$3x + 30 = 180$$
$$3x + 30 - 30 = 180 - 30$$
$$3x = 150$$
$$\frac{3x}{3} = \frac{150}{3}$$
$$x = 50$$

Angle 1 = 50°
Angle 2 = $x + 10$ = 60°
Angle 3 = $x + 20$ = 70°

33. SUPPLEMENTARY ANGLES AND PARALLEL LINES

The angles are supplementary.

Angle 1	$x + 50$
Angle 2	$2x - 20$
Sum of supplementary angles	180

$$1^{st}\text{ angle} + 2^{nd}\text{ angle} = 180$$
$$(x + 50) + (2x - 20) = 180$$
$$3x + 30 = 180$$
$$3x + 30 - 30 = 180 - 30$$
$$3x = 150$$
$$\frac{3x}{3} = \frac{150}{3}$$
$$x = 50$$

35. VERTICAL ANGLES

Vertical angles have the same measure.
$$3x + 10 = 5x - 10$$
$$3x + 10 - 10 = 5x - 10 - 10$$
$$3x = 5x - 20$$
$$3x - 5x = 5x - 20 - 5x$$
$$-2x = -20$$
$$\frac{-2x}{-2} = \frac{-20}{-2}$$
$$x = 10$$

37. e-READERS

Let x = width of the screen.
Then $x + 1.2$ = length/height of the screen.
Perimeter = 2(length) + 2(width)

$$P = 2L + 2W$$
$$16.8 = 2(x+1.2)+2(x)$$
$$16.8 = 2x+2.4+2x$$
$$16.8 = 4x+2.4$$
$$16.8-2.4 = 4x+2.4-2.4$$
$$14.4 = 4x$$
$$\frac{14.4}{4} = \frac{4x}{4}$$
$$3.6 = x$$
$$x+1.2 = 3.6+1.2$$
$$= 4.8$$

The width of the screen is 3.6 inches and the length of the screen is 4.8.

39. SWIMMING POOL

Let x = width of walkway.

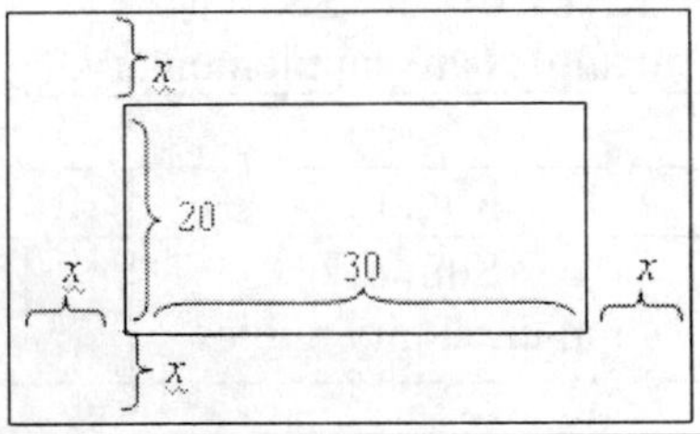

Width	$x + 20 + x = 2x + 20$
Length	$x + 30 + x = 2x + 30$
Perimeter	180

$$P = 2L + 2W$$
$$180 = 2(2x + 30) + 2(2x + 20)$$
$$180 = 4x + 60 + 4x + 40$$
$$180 = 8x + 100$$
$$180 - 100 = 8x + 100 - 100$$
$$80 = 8x$$
$$\frac{80}{8} = \frac{8x}{8}$$
$$10 = x$$

The walkway around the pool should be 10 feet wide.

41. RANCHING

width	x
length	$2x$
sum of three sides	624

$$x + 2x + x = 624$$
$$4x = 624$$
$$\frac{4x}{4} = \frac{624}{4}$$
$$x = 156$$

width = 156 ft
length = 2(156) = 312 ft

43. SOLAR HEATING

	height	width	Area
left rectangle	11	w	$11w$
right rectangle	8	$w + 3$	$8(w + 3)$

Area of left rectangle = Area of right rectangle

$$11w = 8(w+3)$$
$$11w = 8w+24$$
$$11w-8w = 8w+24-8w$$
$$3w = 24$$
$$\frac{3w}{3} = \frac{24}{3}$$
$$w = 8$$
$$w+3 = 8+3 = 11$$

The widths are 8 and 11 feet.

WRITING

45. Answers will vary.

REVIEW

47. Repeating

$$\frac{7}{9} = 9\overline{)7.000}^{\,0.777} = 0.\overline{7}$$

49. $\{\ldots, -4, -3, -2, -1, 0, 1, 2, 3, 4, \ldots\}$

CHALLENGE PROBLEMS

51. MOVING A STONE

force (weight) of stone	210 lb
distance to fulcrum	3 ft
force of woman	x lb
distance to fulcrum (length of bar is 10 – 3 ft that is on the other side of fulcrum)	7 ft

$$(\text{force})(\text{distance}) = (\text{force})(\text{distance})$$
$$210(3) = x(7)$$
$$630 = 7x$$
$$\frac{630}{7} = \frac{7x}{7}$$
$$90 = x$$

The woman must exert 90 lbs of force to move the stone.

53. BALANCING A LEVER

force 1	100
distance 1	4 + 8 = 12
force 2	70
distance 2	8
force 3	40
distance 3	x
force 4	200
distance 4	8

F1(D1) + F2(D2) = F3(D3) + F4(D4)

$$100(12) + 70(8) = 40(x) + 200(8)$$
$$1{,}200 + 560 = 40x + 1{,}600$$
$$1{,}760 = 40x + 1{,}600$$
$$1{,}760 - 1{,}600 = 40x + 1{,}600 - 1{,}600$$
$$160 = 40x$$
$$\frac{160}{40} = \frac{40x}{40}$$
$$4 = x$$

The distance of the smallest force from the fulcrum is 4 feet.

SECTION 1.8

VOCABULARY

1. **Percent** means parts per one hundred.

3. When the regular price of an item is reduced, the amount of reduction is called the **markdown**.

CONCEPTS

5. ☐ is ☐ % of ☐ ?

7. a. $I = \boxed{Prt}$
   b. $d = \boxed{rt}$
   c. Total value = amount · **price**
   d. Amount pure = amount · **strength**

9. $x + 20$

11. $0.055x + 0.07(10{,}850 - x) = 1{,}205$

Account	Principal ·	Rate ·	Time =	Interest
Bonds	x	0.055	1	$x(0.055)(1) = 0.055x$
Stocks	10,850-x	0.07	1	$(10{,}850-x)(0.07)(1) = 0.07(10{,}850 - x)$

13. $0.15x + 0.18x = 3{,}300$

	P ·	r ·	t =	I
Cattle futures	x	0.15	1	$0.15x$
Soybeans	x	0.18	1	$0.18x$

15. $7.45p + 8.25(50 - p) = 7.75(50)$

	Pounds ·	Price =	Value
M & M plain	p	7.45	$7.45p$
M & M peanut	$50 - p$	8.25	$8.25(50 - p)$
Mixture	50	7.75	7.75(50)

NOTATION

17. a. 0.025
    b. 6%

PRACTICE

19. $x = 0.05 \cdot 10.56$

21. $32.5 = 0.74x$

APPLICATIONS

23. ENERGY
103.3 is 20.2% of what?

$$103.3 = 0.202x$$
$$\frac{103.3}{0.202} = \frac{0.202x}{0.202}$$
$$511 \approx x$$

The world's energy consumption was about 511 quadrillion Btu.

25. BOATING ACCIDENTS
What is 6.5% of 4,730?

$$x = 0.065(4{,}730)$$
$$x \approx 307$$

Approximately 307 boating accidents were alcohol-related.

27. BUYING APPLIANCES
Let x = percent of markdown.

sale price	580.80
regular price	726
markdown	$x \cdot 726$

Sale price = Regular price – Markdown

$$580.80 = 726 - 726x$$
$$580.80 - 726 = 726 - 726x - 726$$
$$-145.2 = -726x$$
$$\frac{-145.2}{-726} = \frac{-726x}{-726}$$
$$0.2 = x$$
$$20\% = x$$

The markdown is 20% of the regular price.

29. FLEA MARKETS

Let x = wholesale price.

retail price	65
wholesale price	x
markup	$x \cdot 0.30$

Retail price = Wholesale price + Markup

$$65 = x + 0.30x$$
$$65 = 1.3x$$
$$\frac{65}{1.3} = \frac{1.3x}{1.3}$$
$$50 = x$$

She pays the manufacturer \$50 for each tool chest.

31. IMPROVING HORSEPOWER

Let x = percent of increase.

original horsepower	118
improved horsepower	129
% increase	$x \cdot 118$

Improved HP = Original HP + Increase

$$129 = 118 + 118x$$
$$129 - 118 = 118 + 118x - 118$$
$$11 = 118x$$
$$\frac{11}{118} = \frac{118x}{118}$$
$$0.093 = x$$
$$9.3\% = x$$

The increase is 9.3% of the original horsepower.

33. BROADWAY SHOWS

Season	Broadway attendance	% increase or decrease
2006-07	12.31 million	---
2007-08	12.27 million	$\frac{12.31 - 12.27}{12.27} \approx 0.003$ $\approx 0.3\%$
2008-09	12.15 million	$\frac{12.27 - 12.15}{12.15} \approx 0.0099$ $\approx 1.0\%$

35. HIGHEST RATES

Account	Principal	· Rate	· Time	= Interest
5-year CD	x	0.09	1	$0.09x$
Money Market	$12{,}000 - x$	0.08	1	$0.08(12{,}000 - x)$
Combined Interest	1,060			

CD Interest + MM Interest = Combined Interest

$$0.09x + 0.08(12{,}000 - x) = 1{,}060$$
$$0.09x + 960 - 0.08x = 1{,}060$$
$$0.01x + 960 = 1{,}060$$
$$0.01x + 960 - 960 = 1{,}060 - 960$$
$$0.01x = 100$$
$$\frac{0.01x}{0.01} = \frac{100}{0.01}$$
$$x = 10{,}000$$
$$12{,}000 - x = 2{,}000$$

She invested \$10,000 in the 5-year CD and \$2,000 in the Money Market.

37. INHERITANCES

Account	Principal	· Rate	· Time	= Interest
CD	x	0.07	1	$0.07x$
Biotech	$2x$	0.10	1	$0.10(2x)$
Combined Interest	4,050			

CD Int. + Biotech Int. = Combined Interest

$$0.07x + 0.10(2x) = 4{,}050$$
$$0.07x + 0.2x = 4{,}050$$
$$0.27x = 4{,}050$$
$$\frac{0.27x}{0.27} = \frac{4{,}050}{0.27}$$
$$x = 15{,}000$$
$$2x = 30{,}000$$

Total inheritance $= x + 2x$
$= 15{,}000 + 30{,}000$
$= 45{,}000$

a. She invested \$15,000 at 7% and \$30,000 at 10%.
b. She inherited \$45,000.

39. MONEY-LAUNDERING

Account	Principal	Rate	Time	Interest
Swiss bank	300,000	0.08	1	0.08(300,000)
Cayman Islands	x	0.05	1	$0.05x$
Total Interest	x + 300,000	0.0725	1	$0.0725(x+300{,}000)$

Swiss Int. + Cayman Int. = Total Interest

$$0.08(300{,}000)+0.05x=0.0725(x+300{,}000)$$
$$24{,}000+0.05x=0.0725x+21{,}750$$
$$24{,}000+0.05x-21{,}750=0.0725x+21{,}750-21{,}750$$
$$2{,}250+0.05x=0.0725x$$
$$2{,}250+0.05x-0.05x=0.0725x-0.05x$$
$$2{,}250=0.0225x$$
$$\frac{2{,}250}{0.0225}=\frac{0.0225x}{0.0225}$$
$$x=100{,}000$$

He deposited $100,000 in the Cayman Islands bank.

41. TRAVEL TIMES

	Rate	Time	Distance
Wife	35	t	$35t$
Husband	45	t	$45t$
Total Distance			20

Wife's D + Husb. D = Total D

$$35t+45t=20$$
$$80t=20$$
$$\frac{80t}{80}=\frac{20}{80}$$
$$t=\frac{1}{4}\text{ hour}$$
$$t=\frac{1}{4}\bullet 60\text{ min}$$
$$t=15\text{ min}$$

It will take them 15 minutes to meet.

43. CYCLING

	Rate	Time	Distance
cyclist	18	t	$18t$
support staff	45	$t-1$	$45(t-1)$

Since the staff will catch the cyclist, the distances for both will be the same (equal).

Cyclist Distance = Staff Distance

$$18t=45(t-1)$$
$$18t=45t-45$$
$$18t-45t=45t-45-45t$$
$$-27t=-45$$
$$\frac{-27t}{-27}=\frac{-45}{-27}$$
$$t=1.\overline{6}$$
$$t=1\frac{2}{3}$$
$$t-1=\frac{2}{3}$$

It will take the support staff $\frac{2}{3}$ hr to catch up with the cyclist.

45. RADIO COMMUNICATION

	Rate	Time	Distance
north	50	t	$50t$
south	40	t	$40t$
Total Distance Apart			135

North's Dis. + South's Dis. = Total Dis.

$$50t+40t=135$$
$$90t=135$$
$$\frac{90t}{90}=\frac{135}{90}$$
$$t=1.5\text{ hour}$$

If they start at 2:00 P.M, 1.5 hours later, or 3:30 P.M., they will lose radio contact.

47. JET SKIING

	Rate ·	Time =	Distance
upstream	12 – 4 = 8	3	24
downstream	12 + 4 = 16	t	$16t$

Distance upstream = Distance downstream

$$8(3) = 16t$$

$$24 = 16t$$

$$\frac{24}{16} = \frac{16t}{16t}$$

$$1.5 = t$$

It will take the rider 1.5 hours to return.

49. MIXING CANDY

	Pounds ·	Price =	Value
licorice	x	1.90	$1.90x$
gumdrops	5	2.20	2.20(5) = 11
Mixture	$x + 5$	2	$2(x + 5)$ $= 2x + 10$

Licorice value + gumdrop value = mix. value

$$1.90x + 11 = 2x + 10$$

$$1.90x + 11 - 2x = 2x + 10 - 2x$$

$$-0.1x + 11 = 10$$

$$-0.1x + 11 - 11 = 10 - 11$$

$$-0.1x = -1$$

$$\frac{-0.1x}{-0.1} = \frac{-1}{-0.1}$$

$$x = 10$$

He needs 10 pounds of licorice.

51. HEALTH FOODS

	Pounds ·	Price =	Value
pineapple	x	6.19	$6.19x$
banana	x	4.19	$4.19x$
raisins	2	2.39	2.39(2)
Mixture	$x + x + 2$	4.19	$4.19(2x + 2)$

Pine.value + Ban.value + Rai.value = mix. value

$$6.19x + 4.19x + 2.39(2) = 4.19(2x + 2)$$

$$6.19x + 4.19x + 4.78 = 8.38x + 8.38$$

$$10.38x + 4.78 = 8.38x + 8.38$$

$$10.38x + 4.78 - 8.38x = 8.38x + 8.38 - 8.38x$$

$$2x + 4.78 = 8.38$$

$$2x + 4.78 - 4.78 = 8.38 - 4.78$$

$$2x = 3.6$$

$$\frac{2x}{2} = \frac{3.6}{2}$$

$$x = 1.8$$

1.8 lb of each are needed.

53. GARDENING

	Pounds ·	Price =	Value
planting mix	x	1.57	$1.57x$
sawdust	$6{,}000 - x$	0.10	$0.10(6{,}000 - x)$
Mixture	6,000	1.08	1.08(6,000)

Planting mix.value + sawdust value = mix value

$$1.57x + 0.10(6{,}000 - x) = 1.08(6{,}000)$$

$$1.57x + 600 - 0.10x = 6{,}480$$

$$1.47x + 600 = 6{,}480$$

$$1.47x + 600 - 600 = 6{,}480 - 600$$

$$1.47x = 5{,}880$$

$$\frac{1.47x}{1.47} = \frac{5{,}880}{1.47}$$

$$x = 4{,}000$$

$$6{,}000 - x = 2{,}000$$

She needs 4,000 lbs of planting mix and 2,000 lbs of sawdust.

55. MIXING PAINT

	Gallons ·	% =	Value
Fruit Shake	x	0.035	$0.035x$
Watermelon	1	0.091	0.091
Mixture	$x + 1$	0.043	$0.043(x+1)$

Fruit Shake + Watermelon = mixture

$$0.035x + 0.091 = 0.043(x + 1)$$

$$0.035x + 0.091 = 0.043x + 0.043$$

$$0.035x + 0.091 - 0.043x = 0.043x + 0.043 - 0.043x$$

$$-0.008x + 0.091 = 0.043$$

$$-0.008x + 0.091 - 0.091 = 0.043 - 0.091$$

$$-0.008x = -0.048$$

$$x = 6$$

6 gallons of Fruit Shake are needed.

57. REGISTERED DIETICIAN

	Amount	· Strength	= Pure Concen.
super lean	x	0.12	$0.12x$
regular beef	8	0.30	0.30(8)
Mixture extra-lean	$x+8$	0.16	$0.16(x+8)$

Super-lean conc. + Reg. conc. = Mixture conc.

$$0.12x + 0.30(8) = 0.16(x+8)$$
$$0.12x + 2.4 = 0.16x + 1.28$$
$$100(0.12x + 2.4) = 100(0.16x + 1.28)$$
$$12x + 240 = 16x + 128$$
$$12x + 240 - 16x = 16x + 128 - 16x$$
$$-4x + 240 = 128$$
$$-4x + 240 - 240 = 128 - 240$$
$$-4x = -112$$
$$\frac{-4x}{-4} = \frac{-112}{-4}$$
$$x = 28$$

28 lb of extra-lean meat is needed.

59. DAIRY FOODS

	Amount	· Strength	= Pure concentrate
cream	x	0.22	$0.22x$
2% milk.	$20-x$	0.02	$0.02(20-x)$
Mixture	20	0.04	0.04(20)

Cream conc. + Milk conc. = Mixture conc.

$$0.22x + 0.02(20-x) = 0.04(20)$$
$$0.22x + 0.4 - 0.02x = 0.8$$
$$0.2x + 0.4 = 0.8$$
$$0.2x + 0.4 - 0.4 = 0.8 - 0.4$$
$$0.2x = 0.4$$
$$\frac{0.2x}{0.2} = \frac{0.4}{0.2}$$
$$x = 2$$

2 gal of cream is needed.

61. DILUTING SOLUTIONS

	Amount	· Strength	= Pure concentrate
water	x	0	$0x$
15% soln.	20	0.15	0.15(20)
Mixture	$x+20$	0.10	$0.10(x+20)$

Water conc. + Soln. conc. = Mixture conc.

$$0x + 0.15(20) = 0.10(x+20)$$
$$3 = 0.10x + 2$$
$$3 - 2 = 0.10x + 2 - 2$$
$$1 = 0.10x$$
$$\frac{1}{0.10} = \frac{0.10x}{0.10}$$
$$10 = x$$

10 oz of water is needed.

WRITING

63. Answers will vary.

65. Answers will vary.

67. Answers will vary.

REVIEW

69.
$$9x = 6x$$
$$9x - 6x = 6x - 6x$$
$$3x = 0$$
$$\frac{3x}{3} = \frac{0}{3}$$
$$x = 0$$

71. $$\frac{8(y-5)}{3}=2(y-4)$$
$$\frac{8y-40}{3}=2y-8$$
$$3\left(\frac{8y-40}{3}\right)=3(2y-8)$$
$$8y-40=6y-24$$
$$8y-40-6y=6y-24-6y$$
$$2y-40=-24$$
$$2y-40+40=-24+40$$
$$2y=16$$
$$\frac{2y}{2}=\frac{16}{2}$$
$$y=8$$

CHALLENGE PROBLEMS

73. Determine how many of problems he worked correctly on the original test.

What is 70% of 30?
$$x=0.70(30)$$
$$x=21$$

The total number of problems on the new assignment is 30 + 15 = 45.

What is 80% of 45?
$$x=0.8(45)$$
$$x=36$$

He must have a total of 36 problems correct on the new assignment to have a grade of 80%. So, the number of the new problems he must get correct would be
$$36 - 21 = 15.$$

CHAPTER 1 REVIEW

SECTION 1.1
The Language of Algebra

1. a. $C = 2t + 15$
   b. $l = \dfrac{25}{w}$
   c. $P = u - 3$

2.

p	$T = 30p$
6.0	$30(6.0) = 180$
6.5	$30(6.5) = 195$
7.0	$30(7.0) = 210$
7.5	$30(7.5) = 225$
8.0	$30(8.0) = 240$

3. $w - 43$

4. $\dfrac{1}{3}(x - 10)$

SECTION 1.2
The Real Number System

5. a. 7
   b. 0, 7

6. a. –5, 0, 7
   b. $-5, 0, 2.4, 7, -\dfrac{2}{3}, -3.\overline{6}, \dfrac{15}{4}$

7. a. $-\sqrt{3},\ \pi,\ 0.13242368...$
   b. all

8. a. $-5, -\sqrt{3}, -\dfrac{2}{3}, -3.\overline{6}$
   b. $2.4, 7, \pi, \dfrac{15}{4}, 0.13242368\ldots$

9. a. 7
   b. none

10. a. 0
    b. –5, 7

11.

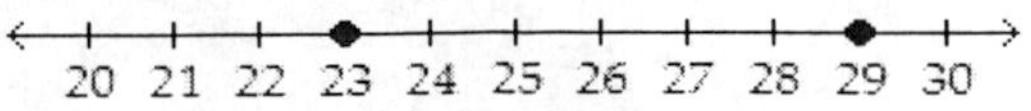

12.

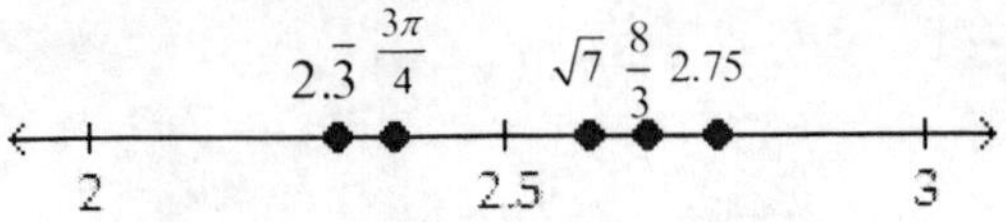

13. a. false
    b. true

14. a. false
    b. true

15. a. $-16 > -17$
    b. $-(-1.8) < 2\dfrac{1}{2}$

16. a. false
    b. true

17. $|-18| = 18$

18. $-|-6.26| = -6.26$

SECTION 1.3
Operations with Real Numbers

19. $-13 + (-14) = -27$

20. $-70.5 + 80.6 = 10.1$

21. $$-\frac{1}{2} - \frac{1}{4} = -\frac{1}{2}\cdot\frac{2}{2} - \frac{1}{4} = -\frac{2}{4} - \frac{1}{4} = -\frac{3}{4}$$

22. $-6 - (-8) = -6 + 8 = 2$

23. $(-4.2)(-3.0) = 12.6$

24. $$-\frac{1}{10}\cdot\frac{5}{16} = -\frac{5}{160} = -\frac{1}{32}$$

25. $$\frac{-2.2}{-11} = 0.2$$

26. $-\frac{9}{8} \div 21 = -\frac{9}{8} \cdot \frac{1}{21}$
$= -\frac{3}{56}$

27. $15 - 25 - 23 = -10 - 23$
$= -33$

28. $-3.5 + (-7.1) + 4.9 = -10.6 + 4.9$
$= -5.7$

29. $-3(-5)(-8) = 15(-8)$
$= -120$

30. $-1(-1)(-1)(-1) = 1(-1)(-1)$
$= -1(-1)$
$= 1$

31. $(-3)^5 = -243$

32. $\left(-\frac{2}{9}\right)^2 = \frac{(-2)^2}{9^2}$
$= \frac{4}{81}$

33. $(0.4)^3 = 0.064$

34. $-5^2 = -(5)(5) = -25$

35. $\sqrt{4} = 2$

36. $-\sqrt{100} = -10$

37. $\sqrt{\frac{9}{25}} = \frac{\sqrt{9}}{\sqrt{25}}$
$= \frac{3}{5}$

38. $\sqrt{0.64} = 0.8$

39. $-6 + 2(-5)^2 = -6 + 2(25)$
$= -6 + 50$
$= 44$

40. $\frac{-20}{4} - (-3)(-2)\left(-\sqrt{1}\right) = -5 - 6(-1)$
$= -5 + 6$
$= 1$

41. $4 - (5 - 9)^2 = 4 - (-4)^2$
$= 4 - 16$
$= -12$

42. $4 + 6[-1 - 5(25 - 3^3)] = 4 + 6[-1 - 5(25 - 27)]$
$= 4 + 6[-1 - 5(-2)]$
$= 4 + 6[-1 + 10]$
$= 4 + 6[9]$
$= 4 + 54$
$= 58$

43. $2|-1.3 + (-2.7)| = 2|-4|$
$= 2(4)$
$= 8$

44. $\frac{(7-6)^4 + 32}{36 - \left(\sqrt{16} + 1\right)^2} = \frac{(1)^4 + 32}{36 - (4 + 1)^2}$
$= \frac{1 + 32}{36 - (5)^2}$
$= \frac{33}{36 - 25}$
$= \frac{33}{11}$
$= 3$

45. $(-10)^3\left(\frac{-6}{-2}\right)(-1) = (-1{,}000)(3)(-1)$
$= -3{,}000(-1)$
$= 3{,}000$

46. $-(-2 \cdot 4)^2 \div 8 \cdot 2 = -(-8)^2 \div 8 \cdot 2$
$= -64 \div 8 \cdot 2$
$= -8 \cdot 2$
$= -16$

47. $(x + y)(x^2 - xy + y^2) = (-2 + 4)\left((-2)^2 - (-2)(4) + (4)^2\right)$
$= (2)(4 + 8 + 16)$
$= (2)(28)$
$= 56$

48. $\dfrac{-b-\sqrt{b^2-4ac}}{2a}=\dfrac{-(-3)-\sqrt{(-3)^2-4(2)(-2)}}{2(2)}$
$=\dfrac{3-\sqrt{9+16}}{4}$
$=\dfrac{3-\sqrt{25}}{4}$
$=\dfrac{3-5}{4}$
$=\dfrac{-2}{4}$
$=-\dfrac{1}{2}$

SECTION 1.4
Simplifying Algebraic Expressions Using Properties of Real Numbers

49. $3(x+7)=3x+21$

50. $t\cdot 5=5t$

51. $-x+x=0$

52. $(27+1)+99=27+(1+99)$

53. $\dfrac{1}{8}\cdot 8=1$

54. $0+m=m$

55. $1\cdot 9.87=9.87$

56. $5(-9)(0)(2{,}345)=0$

57. $(-3\cdot 5)2=-3(5\cdot 2)$

58. $(t+z)\cdot t=(z+t)\cdot t$

59. a. $\dfrac{102}{102}=1$
b. $\dfrac{-25}{1}=-25$

60. a. $\dfrac{0}{6}=0$
b. $\dfrac{5.88}{0}=\text{undefined}$

61. $8(9x+6)=72x+48$

62. $-(6y-2)=-6y+2$

63. $(3x-2y)1.2=3.6x-2.4y$

64. $\dfrac{3}{4}\left(8c^2-4c+1\right)=6c^2-3c+\dfrac{3}{4}$

65. $8(6k)=48k$

66. $(-7.5x)(-10y)=75xy$

67. $-9(-3p)(-7)=-189p$

68. $15a+7+30a+9=45a+16$

69. $3g^2+g^2-3g^2-g^2=0$

70. $-m+4(m-12n)-(-8n)=-m+4m-48n+8n$
$=3m-40n$

71. $\dfrac{7}{5}x-\dfrac{3}{4}x=\dfrac{7}{5}\cdot\dfrac{4}{4}x-\dfrac{3}{4}\cdot\dfrac{5}{5}x$
$=\dfrac{28}{20}x-\dfrac{15}{20}x$
$=\dfrac{13}{20}x$

72. $21.45l-45.99l=-24.54l$

73. $4\left[-2\left(a^3-a^2\right)-2\left(3a^2-6a^3\right)\right]$
$=4\left[-2a^3+2a^2-6a^2+12a^3\right]$
$=4\left[10a^3-4a^2\right]$
$=40a^3-16a^2$

74. $\dfrac{3}{4}(2h+9)-\dfrac{5}{4}(h-1)=\dfrac{6}{4}h+\dfrac{27}{4}-\dfrac{5}{4}h+\dfrac{5}{4}$
$=\dfrac{1}{4}h+\dfrac{32}{4}$
$=\dfrac{1}{4}h+8$

SECTION 1.5
Solving Linear Equations Using Properties of Equality

75. $6-x\stackrel{?}{=}2x+24$
$6-(-6)\stackrel{?}{=}2(-6)+24$
$6+6\stackrel{?}{=}-12+24$
$12=12$
Yes, –6 is a solution.

76. $\frac{5}{3}(x-3)=-12$

$$\frac{5}{3}(-\mathbf{6}-3)\overset{?}{=}-12$$

$$\frac{5}{3}(-9)\overset{?}{=}-12$$

$$-15 \neq -12$$

No, –6 is not a solution.

77.

$$\frac{x}{5}=-45$$

$$5\left(\frac{x}{5}\right)=5(-45)$$

$$x=-225$$

78.

$$t-3.67=4.23$$

$$t-3.67+3.67=4.23+3.67$$

$$t=7.9$$

79.

$$0.0035=0.25g$$

$$\frac{0.0035}{0.25}=\frac{0.25g}{0.25}$$

$$0.014=g$$

80.

$$0=x+4$$

$$0-4=x+4-4$$

$$-4=x$$

$$x=-4$$

81.

$$11-5x=-1$$

$$11-5x-11=-1-11$$

$$-5x=-12$$

$$\frac{-5x}{-5}=\frac{-12}{-5}$$

$$x=\frac{12}{5}$$

82.

$$-3x-7+x=6x+20-5x$$

$$-2x-7=x+20$$

$$-2x-7-x=x+20-x$$

$$-3x-7=20$$

$$-3x-7+7=20+7$$

$$-3x=27$$

$$\frac{-3x}{-3}=\frac{27}{-3}$$

$$x=-9$$

83.

$$-4(y-1)+(-3)=25$$

$$-4y+4+(-3)=25$$

$$-4y+1=25$$

$$-4y+1-1=25-1$$

$$-4y=24$$

$$\frac{-4y}{-4}=\frac{24}{-4}$$

$$y=-6$$

84.

$$5+3[2-13(x-1)]=17-18x$$

$$5+3[2-13x+13]=17-18x$$

$$5+3[-13x+15]=17-18x$$

$$5-39x+45=17-18x$$

$$-39x+50=17-18x$$

$$-39x+50+18x=17-18x+18x$$

$$-21x+50=17$$

$$-21x+50-50=17-50$$

$$-21x=-33$$

$$\frac{-21x}{-21}=\frac{-33}{-21}$$

$$x=\frac{11}{7}$$

85.

$$\frac{8}{3}(x-5)=\frac{2}{5}(x-4)$$
$$15\cdot\frac{8}{3}(x-5)=15\cdot\frac{2}{5}(x-4)$$
$$40(x-5)=6(x-4)$$
$$40x-200=6x-24$$
$$40x-200-6x=6x-24-6x$$
$$34x-200=-24$$
$$34x-200+200=-24+200$$
$$34x=176$$
$$\frac{34x}{34}=\frac{176}{34}$$
$$x=\frac{88}{17}$$

86.

$$\frac{3y}{4}-14=-\frac{y}{3}-1$$
$$12\left(\frac{3y}{4}-14\right)=12\left(-\frac{y}{3}-1\right)$$
$$9y-168=-4y-12$$
$$9y-168+4y=-4y-12+4y$$
$$13y-168=-12$$
$$13y-168+168=-12+168$$
$$13y=156$$
$$\frac{13y}{13}=\frac{156}{13}$$
$$y=12$$

87.

$$-k=-0.06$$
$$\frac{-k}{-1}=\frac{-0.06}{-1}$$
$$k=0.06$$

88.

$$\frac{5}{4}p=-10$$
$$\frac{4}{5}\cdot\frac{5}{4}p=\frac{4}{5}\cdot-10$$
$$p=-8$$

89.

$$\frac{4t+1}{3}-\frac{t+5}{6}=\frac{t-3}{6}$$
$$6\left(\frac{4t+1}{3}-\frac{t+5}{6}\right)=6\left(\frac{t-3}{6}\right)$$
$$2(4t+1)-(t+5)=t-3$$
$$8t+2-t-5=t-3$$
$$7t-3=t-3$$
$$7t-3-t=t-3-t$$
$$6t-3=-3$$
$$6t-3+3=-3+3$$
$$6t=0$$
$$\frac{6t}{6}=\frac{0}{6}$$
$$t=0$$

90.

$$33.9-0.5(75-3x)=0.9$$
$$33.9-37.5+1.5x=0.9$$
$$-3.6+1.5x=0.9$$
$$-3.6+1.5x+3.6=0.9+3.6$$
$$1.5x=4.5$$
$$\frac{1.5x}{1.5}=\frac{4.5}{1.5}$$
$$x=3$$

91.

$$2(x-6)=10+2x$$
$$2x-12=10+2x$$
$$2x-12-2x=10+2x-2x$$
$$-12\neq 10$$

no solution, $\varnothing$; contradiction

92.

$$-5x+2x-1=-(3x+1)$$
$$-5x+2x-1=-3x-1$$
$$-3x-1=-3x-1$$
$$-3x-1+3x=-3x-1+3x$$
$$-1=-1$$

all real numbers, $\mathbb{R}$; identity

SECTION 1.6
Solving Formula; Geometry

93. $10.5+12.5+3.5+4.5=31$ ft

94. $C = d\pi$

$$\approx 17(3.14159)$$
$$\approx 53.41 \text{ cm}$$
$$A = \pi r^2$$
$$= \pi\left(\frac{d}{2}\right)^2$$
$$= \pi\left(\frac{17}{2}\right)^2$$
$$= \pi(8.5)^2$$
$$= \pi(72.25)$$
$$\approx 3.14159(72.25)$$
$$\approx 226.98 \text{ cm}^2$$

95. $V = \frac{4}{3}\pi r^3$

$$= \frac{4}{3}\pi(7.5)^3$$
$$= \frac{4}{3}\pi(421.875)$$
$$\approx \frac{4}{3}(3.14159)(421.875)$$
$$\approx 1{,}767.15 \text{ m}^3$$

96. a. $A = s^2$

$$= 10^2$$
$$= 100 \text{ in.}^2$$

b. $V = \frac{1}{3}\pi r^2 h$

Find the diameter: 10 – 1 – 1 = 8 in.
So the radius is 4 in.

$$V = \frac{1}{3}\pi(4^2)(15)$$
$$= \frac{1}{3}\pi(16)(15)$$
$$= \frac{1}{3}(240)\pi$$
$$= 80\pi$$
$$\approx 251.3 \text{ in.}^3$$

97. $V = \frac{1}{3}\pi r^2 h$ for h

$$3 \cdot V = 3 \cdot \frac{1}{3}\pi r^2 h$$
$$3V = \pi r^2 h$$
$$\frac{3V}{\pi r^2} = \frac{\pi r^2 h}{\pi r^2}$$
$$\frac{3V}{\pi r^2} = h$$
$$h = \frac{3V}{\pi r^2}$$

98. $K = \frac{Mv_0^{\,2} + Iw^2}{2}$ for M

$$2 \cdot K = 2 \cdot \frac{Mv_0^{\,2} + Iw^2}{2}$$
$$2K = Mv_0^{\,2} + Iw^2$$
$$2K - Iw^2 = Mv_0^{\,2} + Iw^2 - Iw^2$$
$$2K - Iw^2 = Mv_0^{\,2}$$
$$\frac{2K - Iw^2}{v_0^{\,2}} = \frac{Mv_0^{\,2}}{v_0^{\,2}}$$
$$\frac{2K - Iw^2}{v_0^{\,2}} = M$$
$$M = \frac{2K - Iw^2}{v_0^{\,2}}$$

99. $\ell = a + (n-1)d$ for d

$$\ell = a + (n-1)d$$
$$\ell - a = a + (n-1)d - a$$
$$\ell - a = (n-1)d$$
$$\frac{\ell - a}{n-1} = \frac{(n-1)d}{n-1}$$
$$\frac{\ell - a}{n-1} = d$$
$$d = \frac{\ell - a}{n-1}$$

100. $9x-5y=35$ for y

$$9x-5y-9x=35-9x$$
$$-5y=35-9x$$
$$\frac{-5y}{-5}=\frac{35-9x}{-5}$$
$$y=-7+\frac{9}{5}x$$
$$y=\frac{9}{5}x-7$$

SECTION 1.7
Using Equations to Solve Problems

101.

O'Hare	x million
Atlanta	$(x+20.7)$ million
Total	159.3 million

$$x+(x+20.7)=159.3$$
$$2x+20.7=159.3$$
$$10(2x+20.7)=10(159.3)$$
$$20x+207=1{,}593$$
$$20x+207-207=1{,}593-207$$
$$20x=1{,}386$$
$$\frac{20x}{20}=\frac{1{,}386}{20}$$
$$x=69.3$$
$$x+20.7=90.0$$

O'Hare served 69.3 million passengers and Atlanta served 90.0 million.

102. $245-5c$

103.

Item stored	number sold	cost of storage for 1	total cost of storage
printers	x	\$1.50	$1.50x$
computers	$x+150$	\$2.50	$2.50(x+150)$

printer cost + computer cost=total storage cost

$$1.50x+2.50(x+150)=2{,}775$$
$$10[1.5x+2.5(x+150)]=10[2{,}775]$$
$$15x+25(x+150)=27{,}750$$
$$15x+25x+3{,}750=27{,}750$$
$$40x+3{,}750=27{,}750$$
$$40x+3{,}750-3{,}750=27{,}750-3{,}750$$
$$40x=24{,}000$$
$$\frac{40x}{40}=\frac{24{,}000}{40}$$
$$x=600$$

There are 600 printers in the warehouse.

104.

1st piece	x
2nd piece	$x+3$
3rd piece	$(x+3)+3=x+6$
4th piece	$(x+6)+3=x+9$
Total length	186

$$x+(x+3)+(x+6)+(x+9)=186$$
$$4x+18=186$$
$$4x+18-18=186-18$$
$$4x=168$$
$$\frac{4x}{4}=\frac{168}{4}$$
$$x=42$$
$$x+3=45$$
$$x+6=48$$
$$x+9=51$$

The lengths of the cable are 42 ft, 45 ft, 48 ft, and 51 ft.

105. The angles are supplementary (sum is 180°).

1st angle	x
2nd angle	$\frac{1}{2}x-15$
Sum	180

$$x+\left(\frac{1}{2}x-15\right)=180$$
$$\frac{3}{2}x-15=180$$
$$\frac{3}{2}x-15+15=180+15$$
$$\frac{3}{2}x=195$$
$$\frac{2}{3}\cdot\frac{3}{2}x=\frac{2}{3}\cdot 195$$
$$x=130$$
$$\frac{1}{2}x-15=50$$

The angle measures are 50° and 130°.

106. Let width = x.
Then the length = x + 13.

$$P=2L+2W$$
$$134=2(x+13)+2(x)$$
$$134=2x+26+2x$$
$$134=4x+26$$
$$134-26=4x+26-26$$
$$108=4x$$
$$\frac{108}{4}=\frac{4x}{4}$$
$$27=x$$
$$x=27$$
$$x+13=40$$

The dimensions are 27 in. by 40 in.

SECTION 1.8
More About Problem Solving

107. 99 is 80% of what?

$$99=0.80\cdot x$$
$$99=0.80x$$
$$\frac{99}{0.80}=\frac{0.80x}{0.80}$$
$$123.75=x$$
$$x\approx 124$$

It has been recorded about 124 days.

108. Let x = percent of markdown.

new fee	375
original fee	550
markdown	$x\cdot 550$

new fee = original fee – markdown

$$375=550-550x$$
$$375-550=550-550x-550$$
$$-175=-550x$$
$$\frac{-175}{-550}=\frac{-550x}{-550}$$
$$0.3\overline{18}=x$$
$$x\approx 0.32$$

The markdown is 32% of the original registration fee.

109. Let x = percent of decrease.

# sold in 2007	290,282
# sold in 2008	235,924
% decrease	$x\cdot 290{,}282$

sold 06 = # sold 05 – % Decrease

$$235{,}924=290{,}282-290{,}282x$$
$$235{,}924-290{,}282=290{,}282-290{,}282x-290{,}282$$
$$-54{,}358=-290{,}282x$$
$$\frac{-54{,}358}{-290{,}282}=\frac{-290{,}282x}{-290{,}282}$$
$$0.187=x$$
$$18.7\%=x$$

The percent decrease is 18.7%.

110. Let x = percent of increase.

# sold in 2007	193,900
# sold in 2008	198,309
% increase	$x\cdot 193{,}900$

sold 08 = # sold 07 + % Increase

$$198{,}309=193{,}900+193{,}900x$$
$$198{,}309-193{,}900=193{,}900+193{,}900x-193{,}900$$
$$4{,}409=193{,}900x$$
$$\frac{4{,}409}{193{,}900}=\frac{193{,}900x}{193{,}900}$$
$$0.023=x$$
$$2.3\%=x$$

The percent increase is 2.3%.

111.

Account	Principal ·	Rate ·	Time =	Interest
10%	x	0.10	1	$0.10x$
9%	$25{,}000-x$	0.09	1	$0.09(25{,}000-x)$
Combined Interest	2,430			

10% Int. + 9% Int. = Combined Int.

$$
\begin{aligned}
0.10x+0.09(25{,}000-x)&=2{,}430\\
0.10x+2{,}250-0.09x&=2{,}430\\
0.01x+2{,}250&=2{,}430\\
0.01x+2{,}250-2{,}250&=2{,}430-2{,}250\\
0.01x&=180\\
\frac{0.01x}{0.01}&=\frac{180}{0.01}\\
x&=18{,}000\\
25{,}000-x&=7{,}000
\end{aligned}
$$

She invested \$18,000 at 10% and \$7,000 at 9%.

112.

	Rate ·	Time =	Distance
celebrity	1	t	t
photographer	1.5	$t-1$	$1.5(t-1)$

Since the photographer will catch the celebrity, the distances for both will be the same (equal).

Celebrity Distance = Photographer Distance

$$
\begin{aligned}
t&=1.5(t-1)\\
t&=1.5t-1.5\\
t-1.5t&=1.5t-1.5-1.5t\\
-0.5t&=-1.5\\
\frac{-0.5t}{-0.5}&=\frac{-1.5}{-0.5}\\
t&=3\\
t-1&=2
\end{aligned}
$$

The photographer will catch the celebrity in 2 minutes.

113.

	Amount ·	Strength =	Pure concentrate
4% pesticide	x	0.04	$0.04x$
12% pesticide	20	0.12	$0.12(20)$
10% solution	$x+20$	0.10	$0.10(x+20)$

$$
\begin{aligned}
0.04x+0.12(20)&=0.10(x+20)\\
0.04x+2.4&=0.10x+2\\
100[0.04x+2.4]&=100[0.10x+2]\\
4x+240&=10x+200\\
4x+240-10x&=10x+200-10x\\
-6x+240&=200\\
-6x+240-240&=200-240\\
-6x&=-40\\
\frac{-6x}{-6}&=\frac{-40}{-6}\\
x&=\frac{20}{3}\\
x&=6\frac{2}{3}
\end{aligned}
$$

$6\frac{2}{3}$ gallons of the 4% solution are needed.

114.

Type of Coffee	Pounds ·	Price =	Value
Mild	x	7.50	$7.50x$
Robust	$90-x$	8.40	$8.40(90-x)$
Mixture	90	7.90	$7.90(90)$

mild value + robust value = mixture value

$$
\begin{aligned}
7.50x+8.40(90-x)&=7.90(90)\\
7.5x+756-8.4x&=711\\
10[7.5x+756-8.4x]&=10[711]\\
75x+7{,}560-84x&=7{,}110\\
-9x+7{,}560&=7{,}110\\
-9x+7{,}560-7{,}560&=7{,}110-7{,}560\\
-9x&=-450\\
\frac{-9x}{-9}&=\frac{-450}{-9}\\
x&=50\\
90-x&=90-50\\
&=40
\end{aligned}
$$

He needs 50 pounds of mild coffee and 40 pounds of robust coffee.

CHAPTER 1 TEST

1. a. For any nonzero real number a, $\frac{a}{0}$ is **undefined.**
   b. $>$, $\geq$, $<$, and $\leq$ are called **inequality** symbols.
   c. $9x^2$ and $7x^2$ are **like terms** because they have the same variable raised to exactly the same power.
   d. To **solve** an equation means to find all of the values of the variable that make the equation true.
   e. The **addition** property of **equality** says that adding the same number to both sides of an equation does not change the solution.

2. a. $s = T + 10$
   b. $A = \frac{1}{2}bh$

3. a. $0.64p$
   b. $2p + 15$
   c. $\frac{7}{8}x - 27$
   d. $3(x - 1)$

4. a. $-2, 0, 5$
   b. $-2, 0, -3\frac{3}{4}, 9.2, \frac{14}{5}, 5$
   c. $\pi, -\sqrt{7}$
   d. all

5. a. true
   b. false
   c. true
   d. true

6.

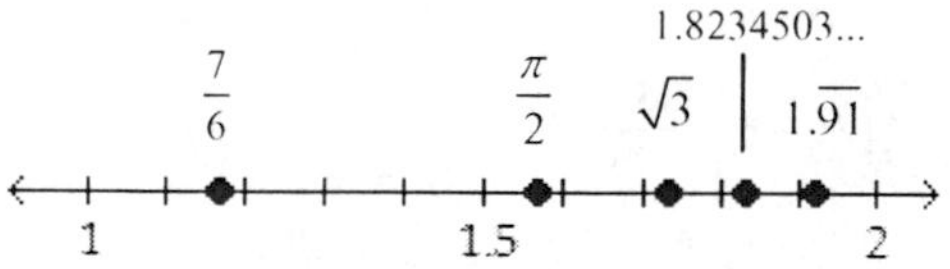

7.

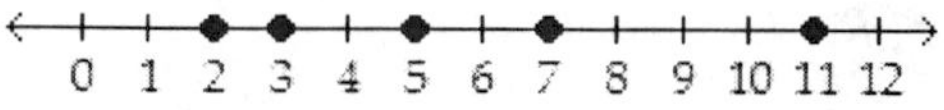

8. a. false
   b. false

9. $-\frac{5}{3}\left(-\frac{4}{25}\right) = \frac{4}{15}$

10. $$\begin{aligned}\frac{2|-4-2(3-1)|}{-3(\sqrt{9})(-2)} &= \frac{2|-4-2(2)|}{-3(3)(-2)} \\ &= \frac{2|-4-4|}{-9(-2)} \\ &= \frac{2|-8|}{18} \\ &= \frac{16}{18} \\ &= \frac{8}{9}\end{aligned}$$

11. $$\begin{aligned}10-3\left[5^2-6(-1-1)^3\right] &= 10-3\left[5^2-6(-2)^3\right] \\ &= 10-3\left[25-6(-8)\right] \\ &= 10-3[25+48] \\ &= 10-3[73] \\ &= 10-219 \\ &= -209\end{aligned}$$

12. $$\begin{aligned}\frac{(-3b+c)^2-17a}{-b+a^2bc} &= \frac{(-3(-3)+4)^2-17(2)}{-(-3)+(2)^2(-3)(4)} \\ &= \frac{(9+4)^2-17(2)}{3+(4)(-3)(4)} \\ &= \frac{(13)^2-34}{3-12(4)} \\ &= \frac{169-34}{3-48} \\ &= \frac{135}{-45} \\ &= -3\end{aligned}$$

13. $$\begin{aligned}\frac{8}{8+12}(250) &= \frac{8}{20}(250) \\ &= \frac{2}{5}(250) \\ &= 100 \text{ mg}\end{aligned}$$

14. a. commutative property of addition
    b. associative property of multiplication
    c. additive inverse property
    d. multiplicative identity property

15. $11.1n^2 - 7.3n + 15.1n - 9.8 = 11.1n^2 + 7.8n - 9.8$

16. $-5(9s)(-2t) = 90st$

17. $-7(c-4)-5[3(c-4)-2(c+2)]$
$$=-7c+28-5[3c-12-2c-4]$$
$$=-7c+28-5[c-16]$$
$$=-7c+28-5c+80$$
$$=-12c+108$$

18. $\frac{2}{9}(x+45y)-\frac{1}{4}(x-24y)$
$$=\frac{2}{9}x+10y-\frac{1}{4}x+6y$$
$$=\left(\frac{2}{9}-\frac{1}{4}\right)x+(10+6)y$$
$$=\left(\frac{2}{9}\cdot\frac{4}{4}-\frac{1}{4}\cdot\frac{9}{9}\right)x+16y$$
$$=\left(\frac{8}{36}-\frac{9}{36}\right)xy+16x$$
$$=-\frac{1}{36}x+16y$$

19. $9(x+4)+4-8x=4(x-5)+x$
$$9(x+4)+4-8x=4(x-5)+x$$
$$9x+36+4-8x=4x-20+x$$
$$x+40=5x-20$$
$$x+40-x=5x-20-x$$
$$40=4x-20$$
$$40+20=4x-20+20$$
$$60=4x$$
$$\frac{60}{4}=\frac{4x}{4}$$
$$x=15$$

20. $\frac{m-1}{5}=\frac{2m-3}{3}-2$
$$15\left[\frac{m-1}{5}\right]=15\left[\frac{2m-3}{3}-2\right]$$
$$3(m-1)=5(2m-3)-30$$
$$3m-3=10m-15-30$$
$$3m-3=10m-45$$
$$3m-3-10m=10m-45-10m$$
$$-7m-3=-45$$
$$-7m-3+3=-45+3$$
$$-7m=-42$$
$$\frac{-7m}{-7}=\frac{-42}{-7}$$
$$m=6$$

21. $6-(x-3)-5x=3[1-2(x+2)]$
$$6-x+3-5x=3[1-2x-4]$$
$$-6x+9=3[-2x-3]$$
$$-6x+9=-6x-9$$
$$-6x+9+6x=-6x-9+6x$$
$$9\neq-9$$

$\varnothing$ (no solution); contradiction

22. $\frac{1}{2}r-\frac{7}{6}=-\frac{1}{3}r+\frac{53}{6}$
$$6\left(\frac{1}{2}r-\frac{7}{6}\right)=6\left(-\frac{1}{3}r+\frac{53}{6}\right)$$
$$3r-7=-2r+53$$
$$3r-7+2r=-2r+53+2r$$
$$5r-7=53$$
$$5r-7+7=53+7$$
$$5r=60$$
$$\frac{5r}{5}=\frac{60}{5}$$
$$r=12$$

23. $1.6y+(-3)=y+1.02$
$$1.6(\mathbf{6.7})+(-3)\stackrel{?}{=}\mathbf{6.7}+1.02$$
$$10.72+(-3)\stackrel{?}{=}7.72$$
$$7.72=7.72$$

Yes, 6.7 is a solution.

24. $P=L+\frac{s}{f}i$ for i
$$P-L=L+\frac{s}{f}i-L$$
$$P-L=\frac{s}{f}i$$
$$\frac{f}{s}(P-L)=\frac{f}{s}\left(\frac{s}{f}i\right)$$
$$\frac{f(P-L)}{s}=i$$
$$i=\frac{f(P-L)}{s}$$

25. $y - y_1 = m(x - x_1)$ for x_1

$$y - y_1 = mx - mx_1$$
$$y - y_1 - mx = mx - mx_1 - mx$$
$$y - y_1 - mx = -mx_1$$
$$\frac{y - y_1 - mx}{-m} = \frac{-mx_1}{-m}$$
$$\frac{y_1 + mx - y}{m} = x_1$$
$$x_1 = \frac{y_1 + mx - y}{m}$$

26. $A = \pi r^2$

$$= \pi\left(\frac{d}{2}\right)^2$$
$$= \pi\left(\frac{36}{2}\right)^2$$
$$= \pi(18)^2$$
$$= 324\pi$$
$$\approx 1{,}018 \text{ m}^2$$

27. Let x = the number of passes of the sander.

$$0.9375 - 0.03125x = 0.6875$$
$$100{,}000(0.9375 - 0.03125x) = 100{,}000(0.6875)$$
$$93{,}750 - 3{,}125x = 68{,}750$$
$$93{,}750 - 3{,}125x - 93{,}750 = 68{,}750 - 93{,}750$$
$$-3{,}125x = -25{,}000$$
$$\frac{-3{,}125x}{-3{,}125} = \frac{-25{,}000}{-3{,}125}$$
$$x = 8$$

The craftsman needs to make 8 passes to obtain the desired thickness.

28. Let x = the number of units of 1-bedroom and 2-bedroom units.

Type of unit	number of units	monthly rent	monthly income
1-bedrm	x	\$950	$950x$
2-bedrm	x	\$1,200	$1{,}200x$

$$950x + 1{,}200x = 53{,}750$$
$$2{,}150x = 53{,}750$$
$$\frac{2{,}150x}{2{,}150} = \frac{53{,}750}{2{,}150}$$
$$x = 25$$

There were 25 1-bedroom units and 25 2-bedroom units.

29. Let x = the measure of the vertex angle. The two base angles in an isosceles triangle have equal measures, and the sum of all three angles is 180°.

vertex angle	x
base angles	$8x + 5$
sum of angles	180

$$x + (8x + 5) + (8x + 5) = 180$$
$$17x + 10 = 180$$
$$17x + 10 - 10 = 180 - 10$$
$$17x = 170$$
$$\frac{17x}{17} = \frac{170}{17}$$
$$x = 10$$
$$8x + 5 = 8(10) + 5$$
$$= 80 + 5$$
$$= 85$$

The vertex angle is 10° and the base angles are both 85°.

30.

width	x
length	$x + 5$
perimeter	26

$$P = 2L + 2W$$
$$26 = 2(x + 5) + 2(x)$$
$$26 = 2x + 10 + 2x$$
$$26 = 4x + 10$$
$$26 - 10 = 4x + 10 - 10$$
$$16 = 4x$$
$$\frac{16}{4} = \frac{4x}{4}$$
$$4 = x$$
$$9 = x + 5$$

The length is 9 cm and the width is 4 cm.

31. Let x = percent of decrease.

# sold in 2008	315,513
# sold in 2009	290,272
% decrease	$x \cdot 199{,}148$

$$\text{\# sold 09} = \text{\# sold 08} - \text{\% Decrease}$$
$$290{,}272 = 315{,}513 - 315{,}513x$$
$$290{,}272 - 315{,}513 = 315{,}513 - 315{,}513x - 315{,}513$$
$$-25{,}241 = -315{,}513x$$
$$\frac{-25{,}241}{-315{,}513} = \frac{-315{,}513x}{-315{,}513}$$
$$0.0799 = x$$
$$x = 8.0\%$$

32. Let t = time the planes travel.

	Rate	· Time	= Distance
Plane 1	450	t	$450t$
Plane 2	500	t	$500t$

Plane 1 distance + Plane 2 distance = 3,800

$$450t + 500t = 3{,}800$$
$$950t = 3{,}800$$
$$\frac{950t}{950} = \frac{3{,}800}{950}$$
$$t = 4$$

The planes will meet in 4 hours.

33.

Account	Principal	· Rate	· Time	= Interest
9%	x	0.09	1	$0.09x$
8%	$10{,}000 - x$	0.08	1	$0.08(10{,}000\text{-}x)$
Combined Interest	860			

$$0.09x + 0.08(10{,}000 - x) = 860$$
$$0.09x + 800 - 0.08x = 860$$
$$0.01x + 800 = 860$$
$$0.01x + 800 - 800 = 860 - 800$$
$$0.01x = 60$$
$$\frac{0.01x}{0.01} = \frac{60}{0.01}$$
$$x = 6{,}000$$
$$10{,}000 - x = 4{,}000$$

\$4,000 was invested at 8%.

34. Let x = the number of miles his policy will cover.

$$17(4) + 0.33x = 200$$
$$68 + 0.33x = 200$$
$$68 + 0.33x - 68 = 200 - 68$$
$$0.33x = 132$$
$$\frac{0.33x}{0.33} = \frac{132}{0.33}$$
$$x = 400$$

His policy will cover him driving up to 400 miles.

35.

	Amount ·	Strength =	Pure concentrate
40%	x	0.40	$0.40x$
10%	10	0.10	0.10(10)
Mixture	$x + 10$	0.25	$0.25(x + 10)$

$$
\begin{aligned}
0.40x + 0.10(10) &= 0.25(x + 10) \\
0.40x + 1 &= 0.25x + 2.5 \\
0.40x + 1 - 0.25x &= 0.25x + 2.5 - 0.25x \\
0.15x + 1 &= 2.5 \\
0.15x + 1 - 1 &= 2.5 - 1 \\
0.15x &= 1.5 \\
\frac{0.15x}{0.15} &= \frac{1.5}{0.15} \\
x &= 10
\end{aligned}
$$

10 ounces of the 40% gold alloy must be used.

36.

	Amount ·	Price =	Value
Skin Soother	x	2.40	$2.40x$
Cool Sport	$8 - x$	1.60	$1.60(8 - x)$
mixture	8	1.90	1.90(8)

Skin Soother + *Cool Sport* = mixture

$$
\begin{aligned}
2.40x + 1.60(8 - x) &= 1.90(8) \\
2.40x + 12.8 - 1.60x &= 15.2 \\
10(2.40x + 12.8 - 1.60x) &= 10(15.2) \\
24x + 128 - 16x &= 152 \\
8x + 128 &= 152 \\
8x + 128 - 128 &= 152 - 128 \\
8x &= 24 \\
\frac{8x}{8} &= \frac{24}{8} \\
x &= 3 \\
8 - x &= 8 - 3 \\
&= 5
\end{aligned}
$$

3 oz of *Skin Soother* and 5 oz of *Cool Sport* is needed.

SECTION 2.1

VOCABULARY

1. The pair of numbers (6, –2) is called an **ordered** pair.

3. The point (0, 0) is the **origin**.

5. Ordered pairs of numbers can be graphed on a **rectangular** coordinate system.

7. If a point is midway between two points *P* and *Q*, it is called the **midpoint** of segment *PQ*.

CONCEPTS

9. To plot (6, –3.5), we start at the **origin** and move 6 units to the **right** and then 3.5 units **down**.

11. Quadrant II

NOTATION

13. Yes.

$x-\text{coordinates: } 5.25 = 5\frac{1}{4} = \frac{21}{4}$

$y-\text{coordinates: } -\frac{3}{2} = -1.5 = -1\frac{1}{2}$

15. capital letters

17. $x^2 = x \cdot x$
x_2 represent the *x*-coordinate of a point.

PRACTICE

19 – 26. See graph below.

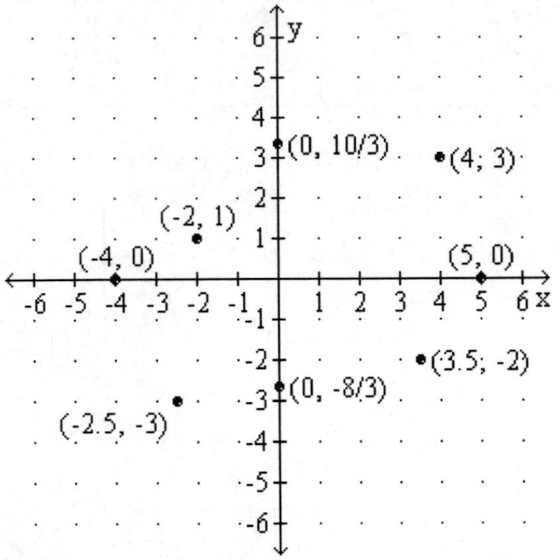

27. (2, 4)

29. (–2.5, –1.5)

31. (3, 0)

33. (0, 0)

35.

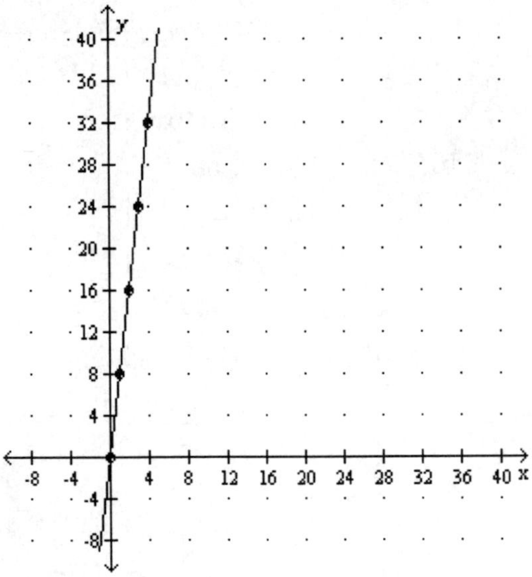

37. a. on the surface
b. diving deeper
c. 3 hours
d. 500 ft

39. a. \$2
b. \$9
c. 3 days

41.
$$\left(\frac{x_1+x_2}{2},\frac{y_1+y_2}{2}\right)=\left(\frac{0+6}{2},\frac{0+8}{2}\right)$$
$$=\left(\frac{6}{2},\frac{8}{2}\right)$$
$$=(3,4)$$

43.
$$\left(\frac{x_1+x_2}{2},\frac{y_1+y_2}{2}\right)=\left(\frac{6+12}{2},\frac{8+16}{2}\right)$$
$$=\left(\frac{18}{2},\frac{24}{2}\right)$$
$$=(9,12)$$

45.
$$\left(\frac{x_1+x_2}{2},\frac{y_1+y_2}{2}\right)=\left(\frac{-2+3}{2},\frac{-8+(-8)}{2}\right)$$
$$=\left(\frac{1}{2},\frac{-16}{2}\right)$$
$$=\left(\frac{1}{2},-8\right)$$

47.
$$\left(\frac{x_1+x_2}{2},\frac{y_1+y_2}{2}\right)=\left(\frac{7+(-10)}{2},\frac{1+4}{2}\right)$$
$$=\left(\frac{-3}{2},\frac{5}{2}\right)$$
$$=\left(-\frac{3}{2},\frac{5}{2}\right)$$

49. x – coordinate:
$$\frac{x_P+x_Q}{2}=x_{mid}$$
$$\frac{-8+x_Q}{2}=-2$$
$$2\left(\frac{-8+x_Q}{2}\right)=2(-2)$$
$$-8+x_Q=-4$$
$$-8+x_Q+8=-4+8$$
$$x_Q=4$$

y-coordinate:
$$\frac{y_P+y_Q}{2}=y_{mid}$$
$$\frac{5+x_Q}{2}=3$$
$$2\left(\frac{5+y_Q}{2}\right)=2(3)$$
$$5+y_Q=6$$
$$5+y_Q-5=6-5$$
$$y_Q=1$$

$Q\,(4, 1)$

51. x – coordinate:
$$\frac{x_P+x_Q}{2}=x_{mid}$$
$$\frac{x_P+6}{2}=-7$$
$$2\left(\frac{x_P+6}{2}\right)=2(-7)$$
$$x_P+6=-14$$
$$x_P+6-6=-14-6$$
$$x_P=-20$$

y-coordinate:
$$\frac{y_P+y_Q}{2}=y_{mid}$$
$$\frac{y_P+(-3)}{2}=-3$$
$$2\left(\frac{y_P-3}{2}\right)=2(-3)$$
$$y_P-3=-6$$
$$y_P-3+3=-6+3$$
$$y_P=-3$$

$Q\,(-20, -3)$

APPLICATIONS

53. ROAD MAPS
Jonesville (5, B)
Easley (1, B)
Hodges (2, E)
Union (6, C)

55. EARTHQUAKES
a. (2, –1)
b. no
c. yes

57. CERTIFIED FITNESS INSTRUCTOR
a. 50 beats/min
b. 10 minutes
c. 130 beats/min
d. 30 minutes
e. 6 minutes and 47 minutes

59. CAMPUS PARKING LOT
a. iii
b. iv
c. v
d. ii
e. i

61. U.S. POSTAGE
a. 78¢
b. 95¢
c. 3.5 oz

63. MULTICULTURAL STUDIES
a. T
b. R

WRITING

65. Answers will vary.

67. Answers will vary.

69. Answers will vary.

REVIEW

71.

$$\begin{aligned}-5^2-5-5(-5)&=-25-5-5(-5)\\&=-25-5+25\\&=-30+25\\&=-5\end{aligned}$$

73.

$$\begin{aligned}\frac{|-25|-2(-5)}{2^4-9}&=\frac{25-2(-5)}{16-9}\\&=\frac{25+10}{7}\\&=\frac{35}{7}\\&=5\end{aligned}$$

75.

$$\begin{aligned}P&=2L+2W\\P-2L&=2L+2W-2L\\P-2L&=2W\\\frac{P-2L}{2}&=\frac{2W}{2}\\\frac{P-2L}{2}&=W\\W&=\frac{P-2L}{2}\end{aligned}$$

CHALLENGE PROBLEMS

77.

$$\left(\frac{a+c}{2},\frac{b+d}{2}\right)$$

$$\begin{aligned}\left(\frac{\frac{a+c}{2}+a}{2},\frac{\frac{b+d}{2}+b}{2}\right)&=\left(\frac{2\left(\frac{a+c}{2}+a\right)}{2(2)},\frac{2\left(\frac{b+d}{2}+b\right)}{2(2)}\right)\\&=\left(\frac{a+c+2a}{4},\frac{b+d+2b}{4}\right)\\&=\left(\frac{3a+c}{4},\frac{3b+d}{4}\right)\end{aligned}$$

$$\begin{aligned}\left(\frac{\frac{a+c}{2}+c}{2},\frac{\frac{b+d}{2}+d}{2}\right)&=\left(\frac{2\left(\frac{a+c}{2}+c\right)}{2(2)},\frac{2\left(\frac{b+d}{2}+d\right)}{2(2)}\right)\\&=\left(\frac{a+c+2c}{4},\frac{b+d+2d}{4}\right)\\&=\left(\frac{a+3c}{4},\frac{b+3d}{4}\right)\end{aligned}$$

SECTION 2.2

VOCABULARY

1. A solution of an equation in two variables is an **ordered pair** of numbers that make a true statement when substituted into the equation.

3. The equation $y = -6x - 3$ is a **linear** equation in two variables.

5. The point where a graph intersects the y–axis is called the **y–intercept**, and the point where it intersects the x–axis is called the **x–intercept**.

CONCEPTS

7. (0, 3), (3, 1), (6, –1)

9. The exponent on each variable of a linear equation is an understood $\boxed{1}$. For example, $4x + 7y = 3$ can be thought of as $4x^{\boxed{1}} + 7y^{\boxed{1}} = 3$.

11. a. The x–intercept is the point the line crosses the x–axis; (–3 0).
The y–intercept is the point the line crosses the y–axis; (0, 4).
b. False

13. Pick an ordered pair such as (–10, 39). The only equation that ordered pair is a solution to is $\mathbf{y = -4x - 1}$.

NOTATION

15. a. the y–axis
b. the x–axis

GUIDED PRACTICE

17. a. yes

$$\mathbf{3} \stackrel{?}{=} -5(\mathbf{-1}) - 2$$
$$3 \stackrel{?}{=} 5 - 2$$
$$3 = 3$$

b. no

$$\mathbf{-13} \stackrel{?}{=} -5(\mathbf{3}) - 2$$
$$-13 \stackrel{?}{=} -15 - 2$$
$$-13 \neq -17$$

19. a. yes

$$\mathbf{11} - 6\left(\frac{\mathbf{1}}{\mathbf{6}}\right) \stackrel{?}{=} 10$$
$$11 - 1 \stackrel{?}{=} 10$$
$$10 = 10$$

b. no

$$\mathbf{-0.6} - 6(\mathbf{-2.1}) \stackrel{?}{=} 10$$
$$-0.6 + 12.6 \stackrel{?}{=} 10$$
$$12 \neq 10$$

21. $y = -x + 4$

x	$y = -x + 4$
–1	$-(\mathbf{-1}) + 4 = 1 + 4$ $= 5$
0	$-(\mathbf{0}) + 4 = 0 + 4$ $= 4$
2	$-(\mathbf{2}) + 4 = -2 + 4$ $= 2$

23. $y = -\frac{1}{3}x - 1$

x	$y = -\frac{1}{3}x - 1$
–3	$-\frac{1}{3}(\mathbf{-3}) - 1 = 1 - 1$ $= 0$
0	$-\frac{1}{3}(\mathbf{0}) - 1 = 0 - 1$ $= -1$
3	$-\frac{1}{3}(\mathbf{3}) - 1 = -1 - 1$ $= -2$

25. $y = \frac{x}{4} - 1$

Pick multiples of the denominator (4) for your x-values.

x	$y = \frac{x}{4} - 1$
4	$y = \frac{\mathbf{4}}{4} - 1$ $= 1 - 1$ $= 0$
0	$y = \frac{\mathbf{0}}{4} - 1$ $= 0 - 1$ $= -1$
–4	$y = \frac{\mathbf{-4}}{4} - 1$ $= -1 - 1$ $= -2$

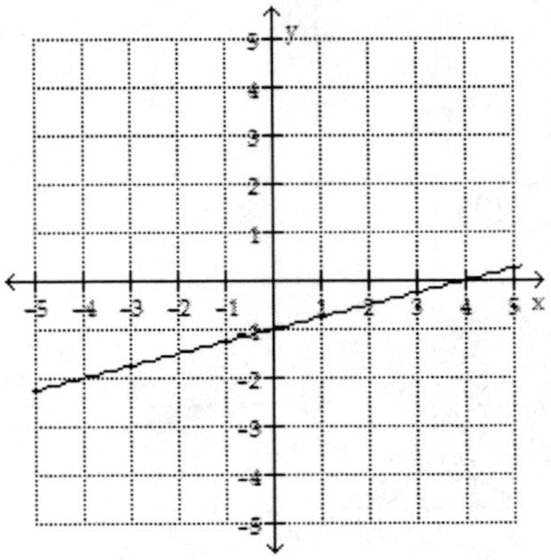

27. $y = -3x + 2$

x	$y = -3x + 2$
–1	$-3(\mathbf{-1}) + 2 = 5$
0	$-3(\mathbf{0}) + 2 = 2$
1	$-3(\mathbf{1}) + 2 = -1$

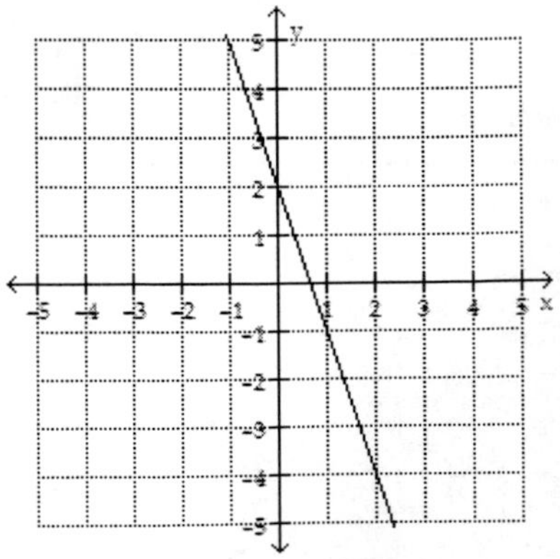

29. $y = 3 - x$

x	$Y = 3 - x$
2	$3 - (\mathbf{2}) = 1$
0	$3 - (\mathbf{0}) = 3$
1	$3 - (\mathbf{1}) = 2$

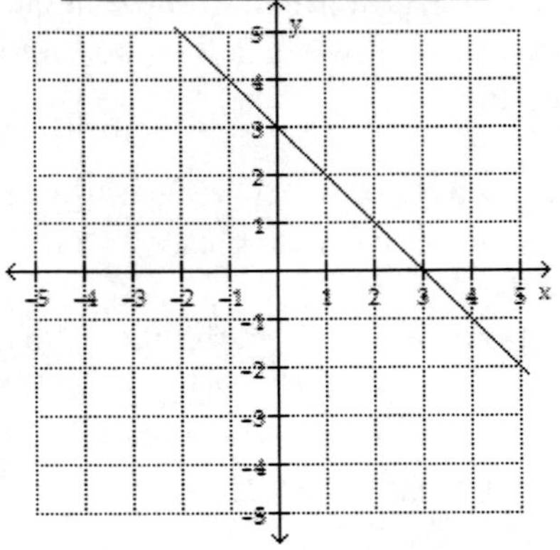

31. $y = x$

x	$y = x$
–2	**–2**
0	**0**
4	**4**

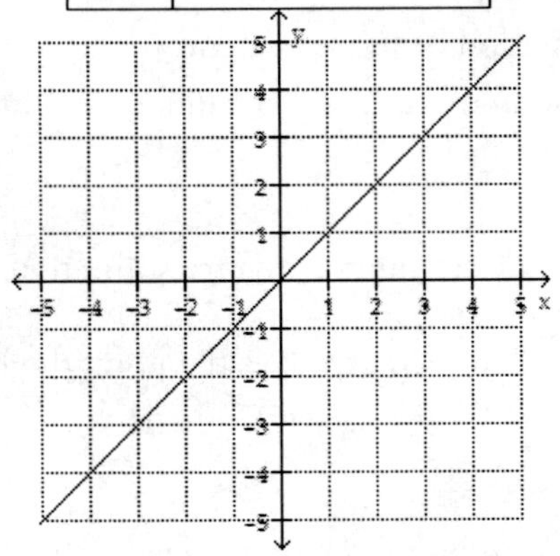

33. $3x + 4y = 12$

x–intercept:	y–intercept:
Let $y = 0$.	Let $x = 0$.
$3x + 4(0) = 12$	$3(0) + 4y = 12$
$3x = 12$	$4y = 12$
$x = 4$	$y = 3$
$(4,0)$	$(0,3)$

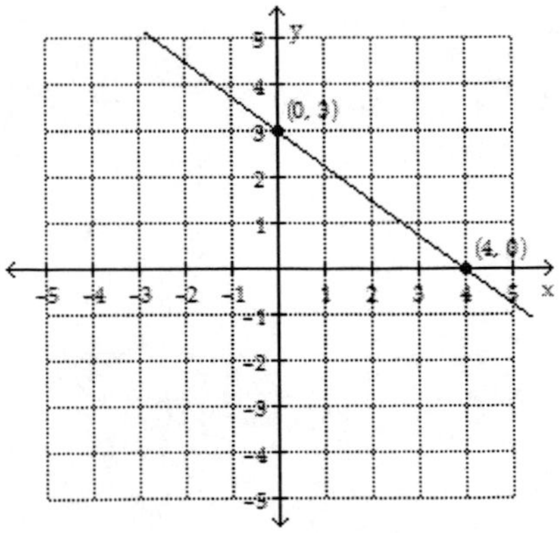

35. $3y = 6x - 9$

x–intercept: Let $y = 0$.

$$3(0) = 6x - 9$$
$$0 = 6x - 9$$
$$0 + 9 = 6x - 9 + 9$$
$$9 = 6x$$
$$\frac{9}{6} = x$$
$$\frac{3}{2} = x$$
$$\left(\frac{3}{2}, 0\right)$$

y–intercept: Let $x = 0$.

$$3y = 6(0) - 9$$
$$3y = -9$$
$$y = -3$$
$$(0, -3)$$

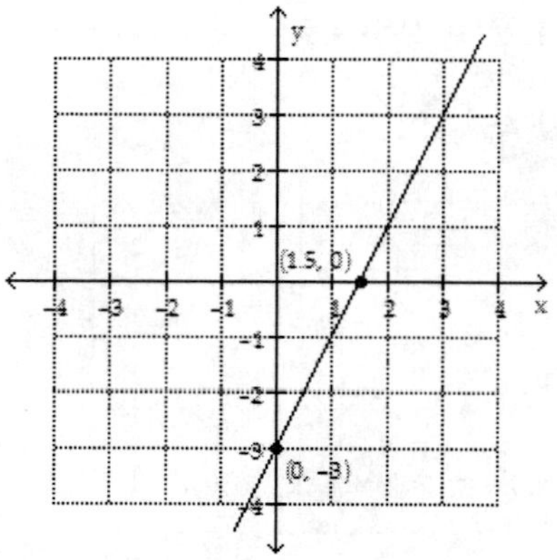

37. $3x + 4y - 8 = 0$

x–intercept: Let $y = 0$.

$$3x + 4(0) - 8 = 0$$
$$3x - 8 = 0$$
$$3x - 8 + 8 = 0 + 8$$
$$3x = 8$$
$$x = \frac{8}{3}$$
$$\left(\frac{8}{3}, 0\right)$$

y–intercept: Let $x = 0$.

$$3(0) + 4y - 8 = 0$$
$$4y - 8 = 0$$
$$4y - 8 + 8 = 0 + 8$$
$$4y = 8$$
$$y = 2$$
$$(0, 2)$$

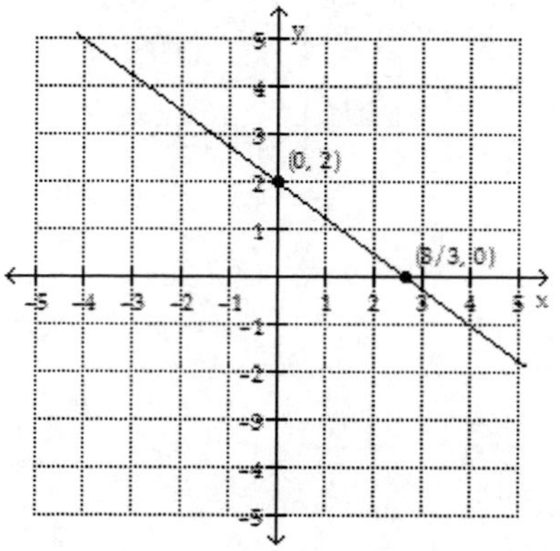

39. $3x = 4y - 11$

x–intercept: Let $y = 0$.

$$3x = 4(0) - 11$$
$$3x = -11$$
$$x = -\frac{11}{3}$$
$$\left(-\frac{11}{3}, 0\right)$$

y–intercept: Let $x = 0$.

$$3(0) = 4y - 11$$
$$0 = 4y - 11$$
$$0 + 11 = 4y - 11 + 11$$
$$11 = 4y$$
$$\frac{11}{4} = y$$
$$\left(0, \frac{11}{4}\right)$$

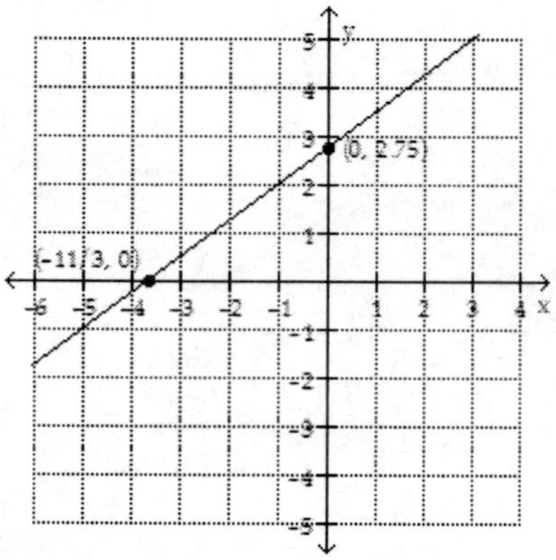

41. $x = 3$

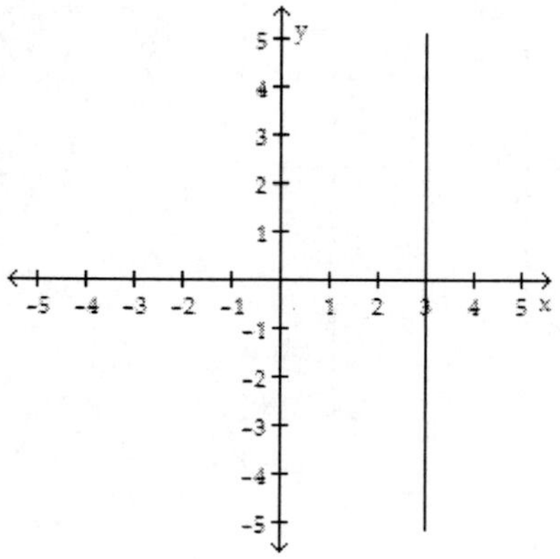

43. $y = -\frac{1}{2}$

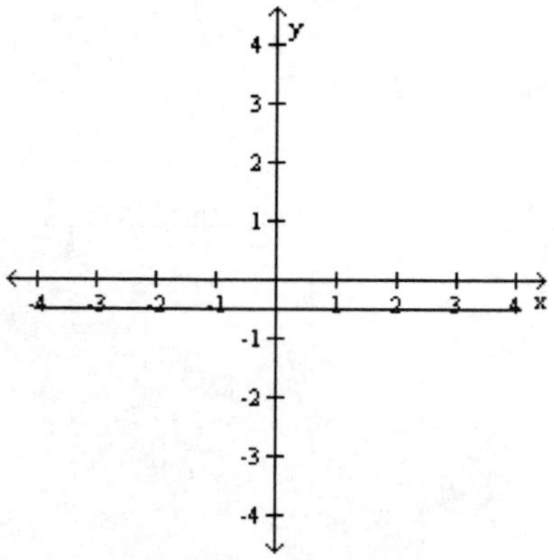

45.

$$y - 2 = 0$$
$$y - 2 + 2 = 0 + 2$$
$$y = 2$$

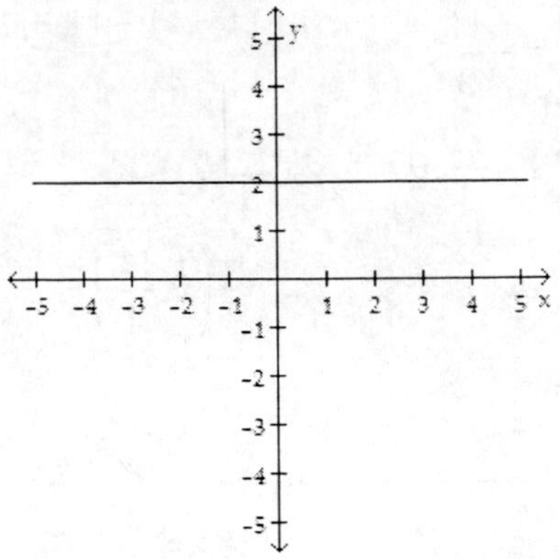

47.

$$-2x + 3 = 11$$
$$-2x + 3 - 3 = 11 - 3$$
$$-2x = 8$$
$$x = -4$$

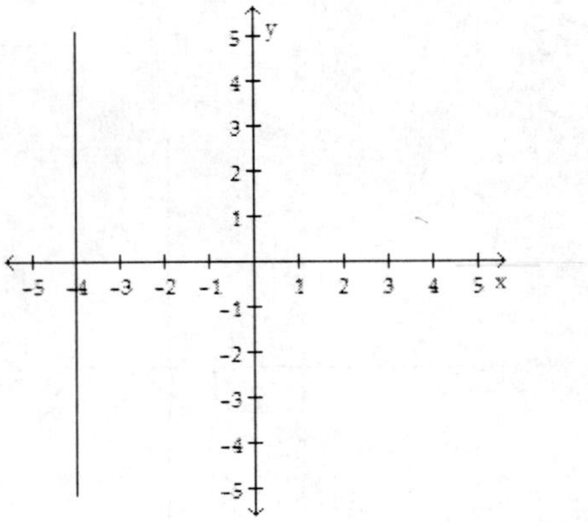

49. $y = 3.7x - 4.5$

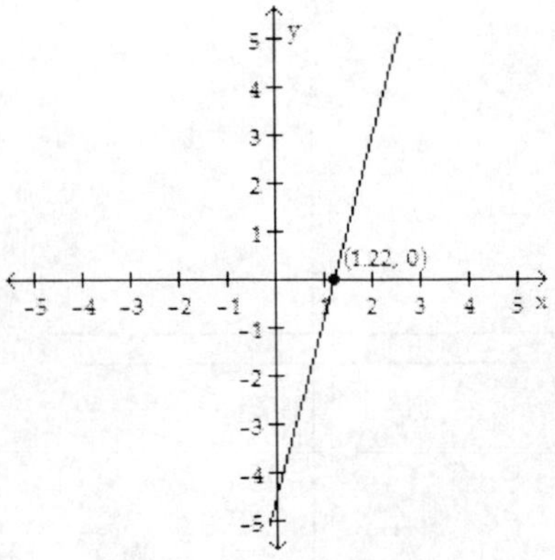

51. $1.5x - 3y = 7$

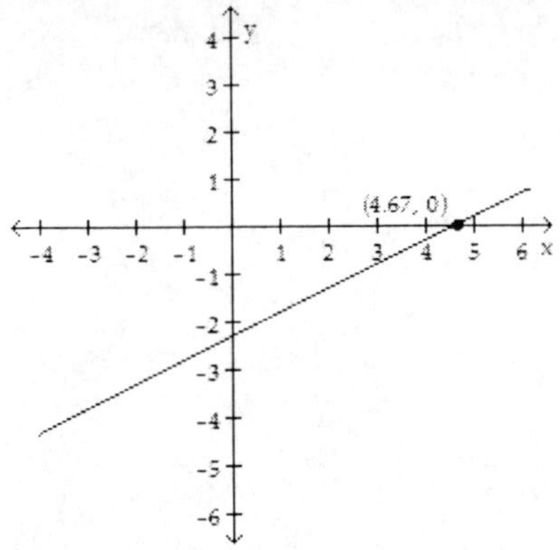

TRY IT YOURSELF

53. $5x - 4y = 13$

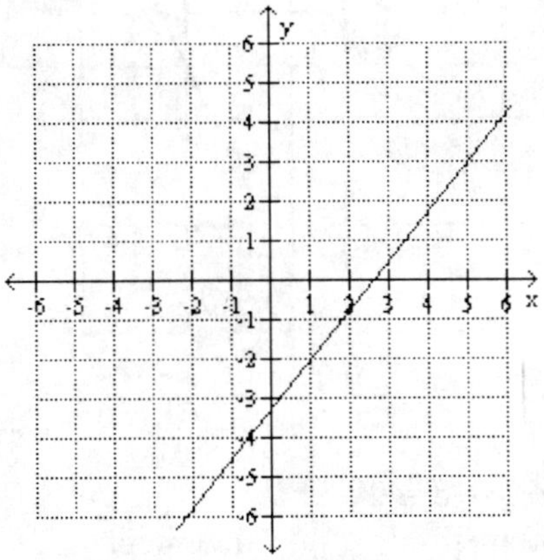

55. $y = \dfrac{5}{6}x - 5$

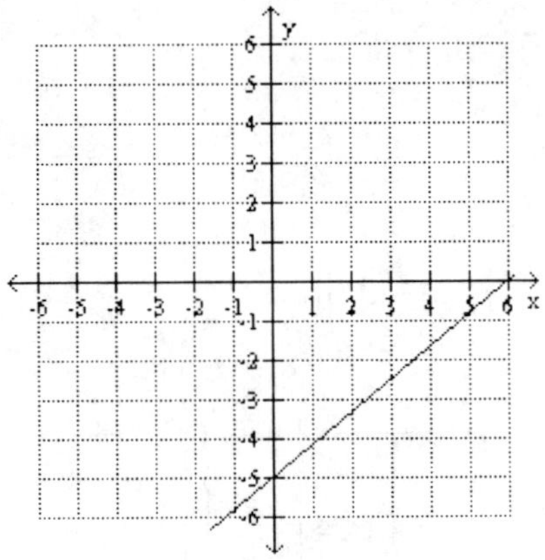

57. $x + 2y = -2$

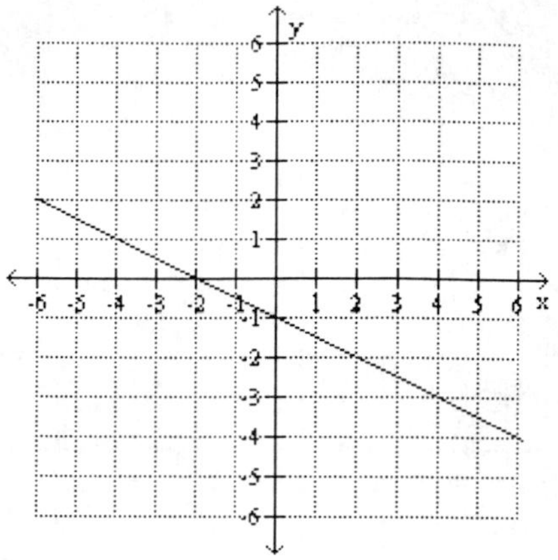

63. $x = 50 - 5y$

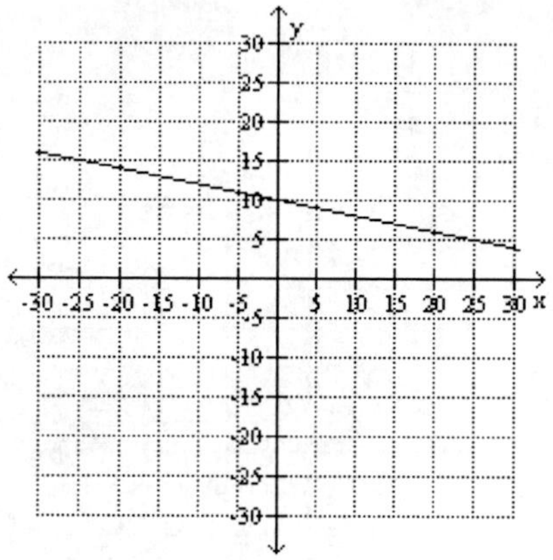

59. $y = \dfrac{5}{2}$

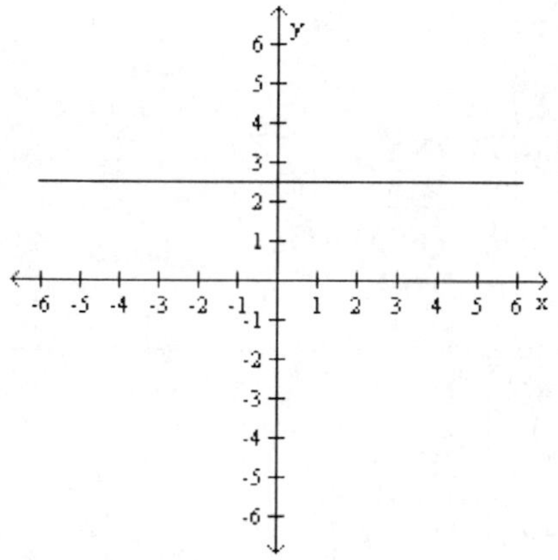

65. $y = \dfrac{1}{2}x$

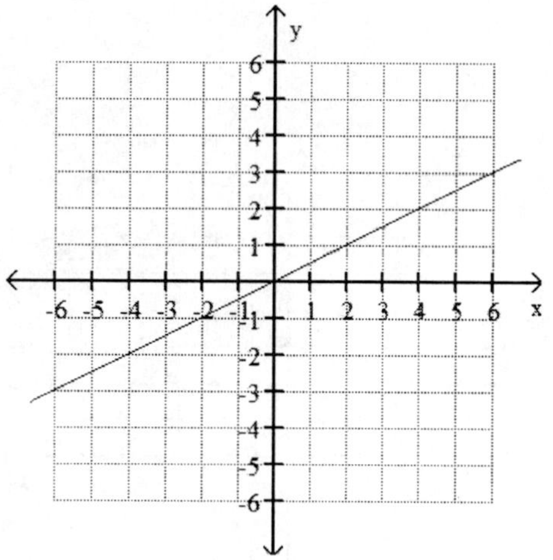

61. $y = 4x$

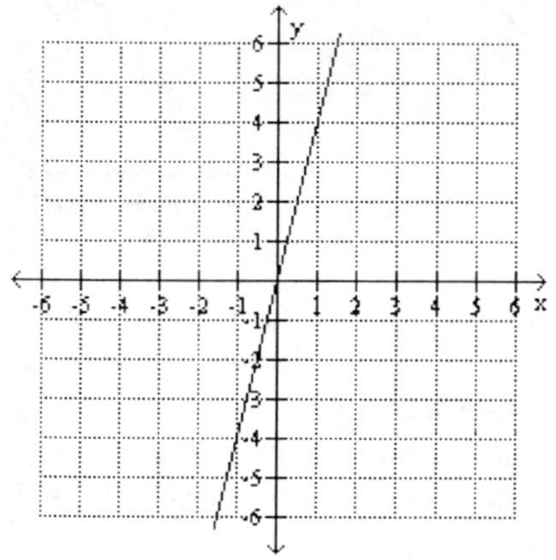

67. $y = 1.5x - 4$

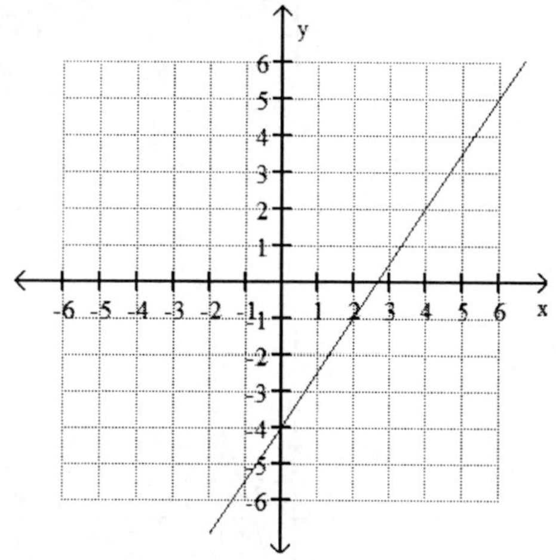

69. $3x + 5y = 0$

$$5y = -3x$$

$$y = -\frac{3}{5}x$$

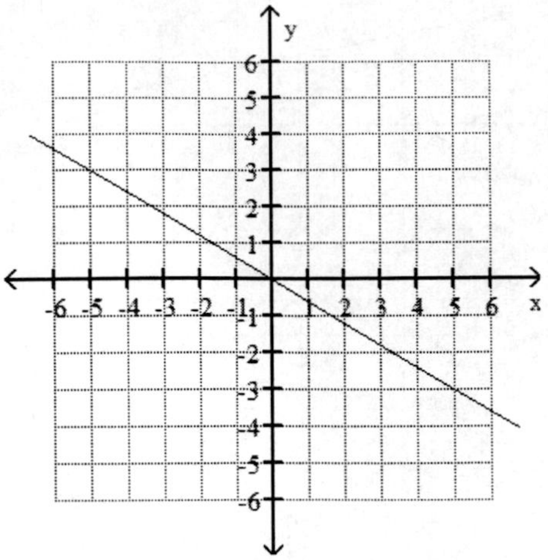

71. $x = -\frac{5}{3}$

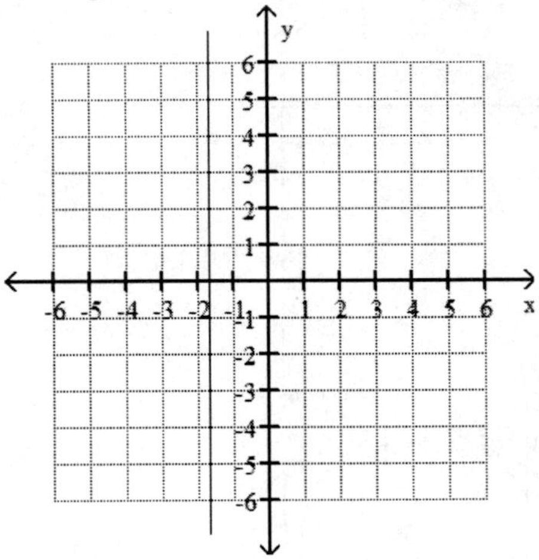

73. $y - 3x = -\frac{4}{3}$

$$y = 3x - \frac{4}{3}$$

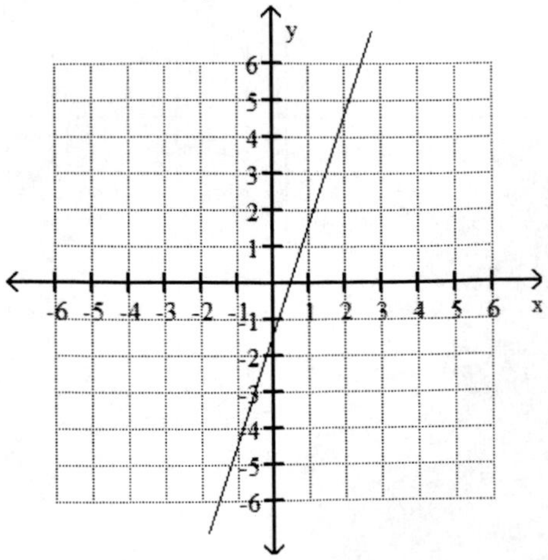

75. $4x - 5y = 0$

$$-5y = -4x$$

$$y = \frac{4}{5}x$$

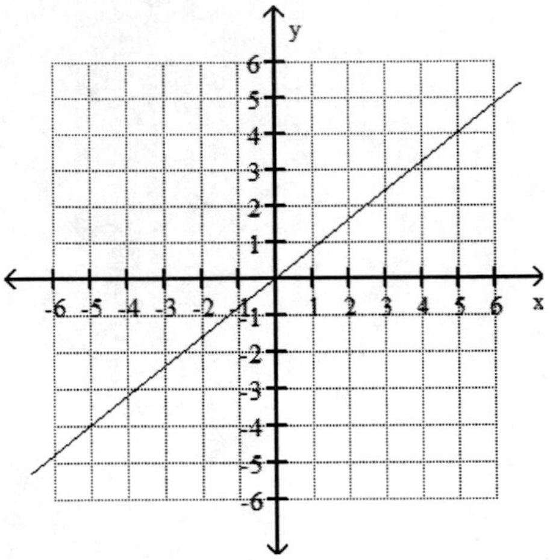

APPLICATIONS

77. SQUARE FOOTAGE

a. $s = 27t + 1{,}925$

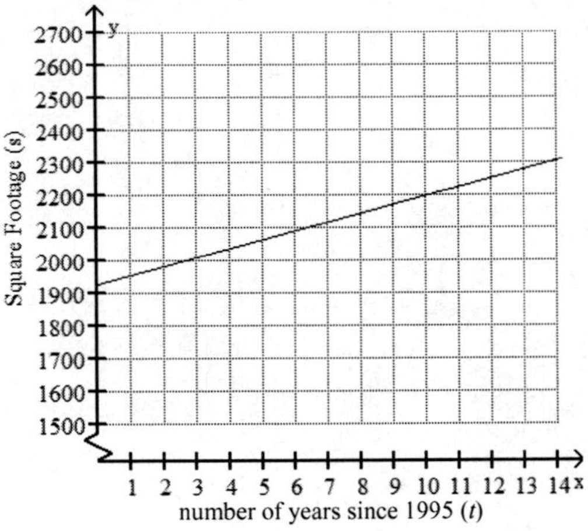

b. In 1995 ($t = 0$), the median number of square feet in a new single-family home was 1,925.

c. The median number of square feet in a new single-family home in 2008 would be approximately 2,275 because $t = 13$ since 2008 is 13 years after 1995.

79. UNION MEMBERSHIP

a. $p = -0.25t + 11.7$

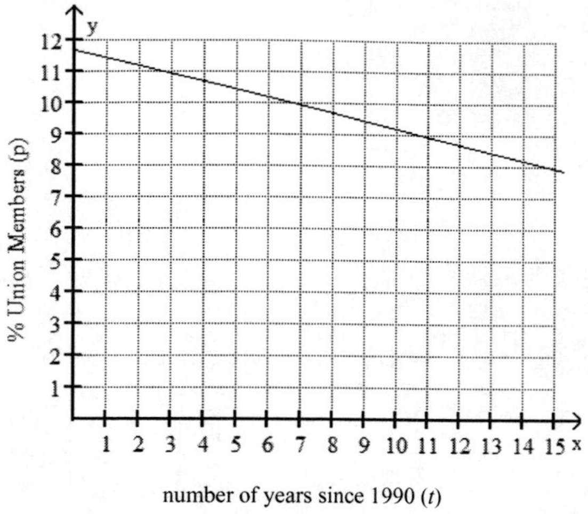

b. In 1990 ($t = 0$), 11.7% of the people working in the private sector were union members.

c. approximately 5.5%; $t = 25$ since 2015 is 25 years after 1990

81. LIVING LONGER

a. $a = 0.16t + 73.7$

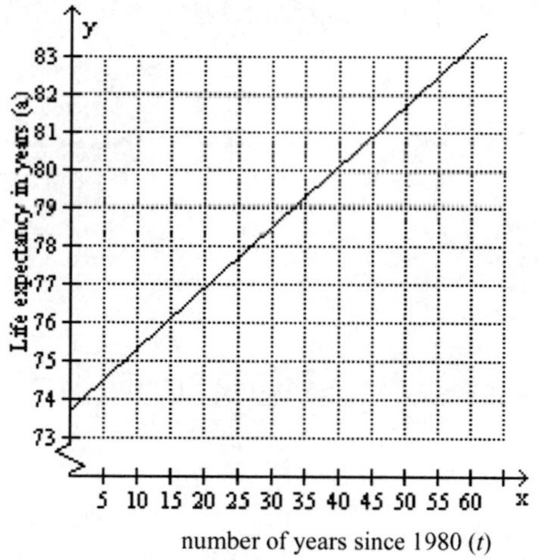

b. The average life expectancy in 1980 ($t = 0$) was 73.7 years.

c. For $t = 50$, the life expectancy is 81.7 years.

83. DEPRECIATION

The x–intercept indicates that in 8 years, the computer will have no value. The y–intercept indicates that when new, the computer was worth $3,000.

85. CAR DEPRECIATION

Let $y = 0$ and solve for x.

$$0 = -1{,}360x + 17{,}000$$
$$0 - 17{,}000 = -1{,}360x + 17{,}000 - 17{,}000$$
$$-17{,}000 = -1{,}360x$$
$$12.5 = x$$

In 12.5 years, the car will have no value.

87. DEMAND EQUATION

Let $p = 150$ and solve for q.

$$p = -\frac{1}{10}q + 170$$
$$150 = -\frac{1}{10}q + 170$$
$$10(150) = 10\left(-\frac{1}{10}q + 170\right)$$
$$1{,}500 = -q + 1{,}700$$
$$1{,}500 - 1{,}700 = -q + 1{,}700 - 1{,}700$$
$$-200 = -q$$
$$200 = q$$

200 Microwaves will be sold at a price of $150.

WRITING

89. Answers will vary.

91. Answers will vary.

REVIEW

93. {11, 13, 17, 19, 23, 29}

95. Quadrant III

97.

$$-4(-20s)(-6) = 80s(-6)$$
$$= -480s$$

99.

$$-4[-2(-3x - 8)] = -4[6x + 16]$$
$$= -24x - 64$$

CHALLENGE PROBLEMS

101.

$$\frac{1}{5}x = 6 - \frac{3}{10}y$$
$$10\left(\frac{1}{5}x\right) = 10\left(6 - \frac{3}{10}y\right)$$
$$2x = 60 - 3y$$
$$2x - 60 = 60 - 3y - 60$$
$$2x - 60 = -3y$$
$$-\frac{2}{3}x + 20 = y$$
$$y = -\frac{2}{3}x + 20$$

x	$y = -\frac{2}{3}x + 20$
3	$y = -\frac{2}{3}(3) + 20$ $= -2 + 20$ $= 18$
0	$y = -\frac{2}{3}(0) + 20$ $= 0 + 20$ $= 20$
−3	$y = -\frac{2}{3}(-3) + 20$ $= 2 + 20$ $= 22$

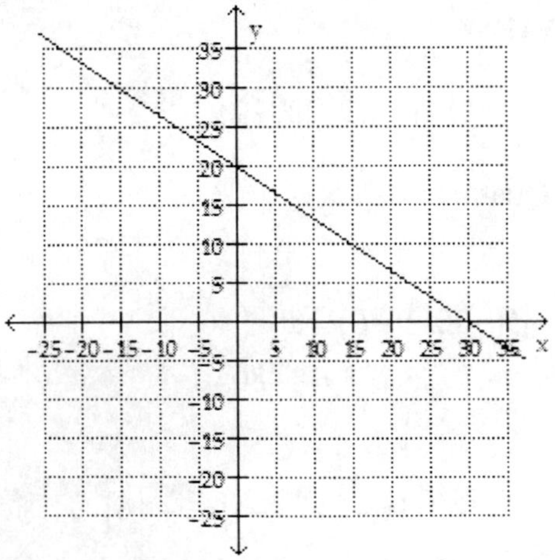

103. $y = \frac{x}{|x|}$

x	$y = \frac{x}{\lvert x\rvert}$
1	$y = \frac{1}{\lvert 1\rvert}$ $= \frac{1}{1}$ $= 1$
3	$y = \frac{3}{\lvert 3\rvert}$ $= \frac{3}{3}$ $= 1$
−3	$y = \frac{-3}{\lvert 3\rvert}$ $= \frac{-3}{3}$ $= -1$
−1	$y = \frac{-1}{\lvert 1\rvert}$ $= \frac{-1}{1}$ $= -1$
0	$y = \frac{0}{\lvert 0\rvert}$ $= \frac{0}{0}$ = undefined

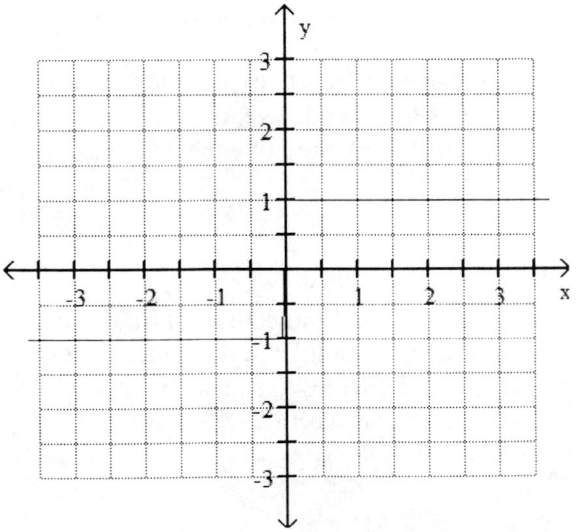

SECTION 2.3

VOCABULARY

1. A **ratio** is a comparison of two numbers using a quotient. In symbols, the **ratio** of a to b is $\frac{\boxed{a}}{\boxed{b}}$.

3. An average **rate** of **change** describes how much one quantity changes with respect to another.

5. The **change** in x (written Δx) is the horizontal run of the line between two points on the line and the change in y (written Δy) is the vertical **rise** of the line between two points on the line.

CONCEPTS

7. a. Number in 1990: **200**
   Number in 2010: **80**
   b. Change in the number of trick–or–treaters from 1990–2010: **–120**
   Time span: **20 years**
   c. $\text{Rate of change} = \dfrac{-120 \text{ trick-or-treaters}}{20 \text{ years}}$
   $= \mathbf{-6}$ trick-or-treaters/year

9. a. –6
   b. 2
   c. $\dfrac{-6}{2} = -3$

11. a. l_3; 0
    b. l_2; undefined
    c. l_1; 2
    d. l_4; –3

13. a. $\frac{5}{7}, \frac{4}{7}$; no
    b. $\frac{4}{7}, -\frac{7}{4}$; yes

NOTATION

15. The rate of change 10ft/year is read as "10 feet **per** year."

PRACTICE

17. Use the points (4, 0) and (–1, –7).

$$m = \frac{y_2 - y_1}{x_2 - x_1} = \frac{-7-0}{-1-4} = \frac{-7}{-5} = \frac{7}{5}$$

19. Use the points (–2, 1) and (1, –7).

$$m = \frac{y_2 - y_1}{x_2 - x_1} = \frac{-7-1}{1-(-2)} = \frac{-8}{3} = -\frac{8}{3}$$

21. Use the points (0, –5) and (–1, –2).

$$m = \frac{y_2 - y_1}{x_2 - x_1} = \frac{-2-(-5)}{-1-0} = \frac{3}{-1} = -3$$

23. Use the points (0, –5) and (2, –4).

$$m = \frac{y_2 - y_1}{x_2 - x_1} = \frac{-4-(-5)}{2-0} = \frac{1}{2}$$

25. (0, 0), (3, 9)

$$m = \frac{y_2 - y_1}{x_2 - x_1} = \frac{9-0}{3-0} = \frac{9}{3} = 3$$

27. $(-1, 8), (6, 1)$

$$m = \frac{y_2 - y_1}{x_2 - x_1} = \frac{1-8}{6-(-1)} = \frac{-7}{7} = -1$$

29. $(3, -1), (-6, 2)$

$$m = \frac{y_2 - y_1}{x_2 - x_1} = \frac{2-(-1)}{-6-3} = \frac{3}{-9} = -\frac{1}{3}$$

31. $(28, 50), (7, 17)$

$$m = \frac{y_2 - y_1}{x_2 - x_1} = \frac{17-50}{7-28} = \frac{-33}{-21} = \frac{11}{7}$$

33. $(7, 5), (-9, 5)$

$$m = \frac{y_2 - y_1}{x_2 - x_1} = \frac{5-5}{-9-7} = \frac{0}{-16} = 0$$

35. $(-7, -5), (-7, -2)$

$$m = \frac{y_2 - y_1}{x_2 - x_1} = \frac{-2-(-5)}{-7-(-7)} = \frac{3}{0} = \text{undefined}$$

37. $\left(\frac{1}{4}, \frac{9}{2}\right), \left(-\frac{3}{4}, 0\right)$

$$m = \frac{y_2 - y_1}{x_2 - x_1} = \frac{0-\frac{9}{2}}{-\frac{3}{4}-\frac{1}{4}} = \frac{-\frac{9}{2}}{-\frac{4}{4}} = \frac{-\frac{9}{2}}{-1} = \frac{9}{2}$$

39. $(0.7, -0.6), (-0.9, 0.2)$

$$m = \frac{y_2 - y_1}{x_2 - x_1} = \frac{0.2-(-0.6)}{-0.9-0.7} = \frac{0.8}{-1.6} = \frac{10}{10} \bullet \frac{0.8}{-1.6} = -\frac{8}{16} = -\frac{1}{2} = -0.5$$

41. parallel

$$m = \frac{y_2 - y_1}{x_2 - x_1} = \frac{2-4}{4-3} = \frac{-2}{1} = -2$$

43. perpendicular

$$m = \frac{y_2 - y_1}{x_2 - x_1} = \frac{5-1}{6-(-2)} = \frac{4}{8} = \frac{1}{2}$$

45. neither

$$m = \frac{y_2 - y_1}{x_2 - x_1} = \frac{6-4}{6-5} = \frac{2}{1} = 2$$

47. parallel

$$m = \frac{y_2 - y_1}{x_2 - x_1} = \frac{6.3-12.3}{6.2-3.2} = \frac{-6}{3} = -2$$

49. parallel; slopes are the same

(–3, –2) and (4, 5)

$$m = \frac{y_2 - y_1}{x_2 - x_1} = \frac{5-(-2)}{4-(-3)} = \frac{7}{7} = 1$$

(–1, 0) and (6, 7)

$$m = \frac{y_2 - y_1}{x_2 - x_1} = \frac{7-0}{6-(-1)} = \frac{7}{7} = 1$$

51. perpendicular; slopes are opposite reciprocals

(–2, –2) and (–5, 4)

$$m = \frac{y_2 - y_1}{x_2 - x_1} = \frac{4-(-2)}{-5-(-2)} = \frac{6}{-3} = -2$$

(–6, 2) and (–2, 4)

$$m = \frac{y_2 - y_1}{x_2 - x_1} = \frac{4-2}{-2-(-6)} = \frac{2}{4} = \frac{1}{2}$$

APPLICATIONS

53. U.S. MUSIC SALES

a. Use the points (1990, 285) and (2000, 940).

$$m = \frac{y_2 - y_1}{x_2 - x_1} = \frac{940-285}{2000-1990} = \frac{655}{10} = 65.5$$

An increase of 65.5 million units/yr

b. Use the points (2000, 940) and (2008, 380).

$$m = \frac{y_2 - y_1}{x_2 - x_1} = \frac{380-940}{2008-2000} = \frac{-560}{8} = -70$$

A decrease of 70 million units/yr

55. GLOBAL WARMING

Use the points (2100, 60.2) and (2000, 57.6).

$$m = \frac{y_2 - y_1}{x_2 - x_1} = \frac{60.2-57.6}{2100-2000} = \frac{2.6}{100} = 0.026^\circ \text{ F/yr}$$

57. FLYING

Part 1:

$$m = \frac{2,600 - 2,000}{28,000} = \frac{600}{28,000} = \frac{3}{140}$$

Part 2:

$$m = \frac{2,000 - 1,440}{8,400} = \frac{560}{8,400} = \frac{1}{15}$$

Part 3:

$$m = \frac{1,440 - 560}{17,600} = \frac{880}{17,600} = \frac{1}{20}$$

Part 2 is the steepest.

59. STEEP GRADES

Change 2.5 miles to feet by multiplying it by 5,280.

$$m = \frac{528 \text{ ft}}{2.5 \text{ mi}} = \frac{528}{2.5(5,280)} = \frac{528}{13,200} = \frac{1}{25} = 4\%$$

61. STAIRCASES

$$\frac{7\frac{1}{2}}{9} = \frac{\frac{15}{2}}{9} = \frac{15}{2} \div 9 = \frac{15}{2} \cdot \frac{1}{9} = \frac{15}{18} = \frac{5}{6}$$

63. BIKE RACING

$$\text{rate} = \frac{1,500 \text{ feet}}{6 \text{ minutes}} = 250 \text{ ft/min.}$$

65. SKIING

No; they are equally steep. The steepness of each course would be found by finding the slope of the same line.

WRITING

67. Answers will vary.

69. Answers will vary.

REVIEW

71. CANDY

	licorice	gumdrops	combined
# of pounds	x	$60 - x$	60
value	$1.90x$	$2.2(60-x)$	60(2)

$$\begin{aligned} 1.90x + 2.2(60 - x) &= 60(2) \\ 1.90x + 132 - 2.2x &= 120 \\ -0.3x + 132 &= 120 \\ -0.3x + 132 - 132 &= 120 - 132 \\ -0.3x &= -12 \\ x &= 40 \\ 60 - x = 60 - 40 &= 20 \end{aligned}$$

He needs 40 lbs of licorice and 20 lbs of gumdrops.

CHALLENGE PROBLEMS

73. Find the slope of the blue line.

$$\begin{aligned} m &= \frac{y_2 - y_1}{x_2 - x_1} \\ &= \frac{-2-4}{1-(-3)} \\ &= \frac{-6}{4} \\ &= -\frac{3}{2} \end{aligned}$$

The slope from (–2, 5) to (x, 0) must also be $-\frac{3}{2}$.

$$\begin{aligned} m &= \frac{y_2 - y_1}{x_2 - x_1} \\ \frac{-3}{2} &= \frac{0-5}{x-(-2)} \\ \frac{-3}{2} &= \frac{-5}{x+2} \\ -3(x+2) &= 2(-5) \\ -3x-6 &= -10 \\ -3x-6+6 &= -10+6 \\ -3x &= -4 \\ x &= \frac{4}{3} \end{aligned}$$

The slope from (–2, 5) to (3, y) must also be $-\frac{3}{2}$.

$$\begin{aligned} m &= \frac{y_2 - y_1}{x_2 - x_1} \\ \frac{-3}{2} &= \frac{y-5}{3-(-2)} \\ \frac{-3}{2} &= \frac{y-5}{5} \\ -3(5) &= 2(y-5) \\ -15 &= 2y-10 \\ -15+10 &= 2y-10+10 \\ -5 &= 2y \\ -\frac{5}{2} &= y \end{aligned}$$

75. (a, b), ($-b$, $-a$)

$$\begin{aligned} m &= \frac{y_2 - y_1}{x_2 - x_1} \\ &= \frac{-a-b}{-b-a} \\ &= \frac{-a-b}{-a-b} \\ &= 1 \end{aligned}$$

SECTION 2.4

VOCABULARY

1. The **slope-intercept** form of the equation of a line is $y = mx + b$.

CONCEPTS

3. $m = -\frac{2}{3}$; $b = (0, 1)$

5. a. (0, 0)
   b. none

7. a. $m = -\frac{2}{3}$
   b. (–1, 3)

9. The temperature lowers 5 degrees every 2 minutes, so the slope would be $-\frac{5}{2}$. Since the temperature starts at 75 degrees, the y-intercept is 75.

11. a. 8 because parallel lines have the same slope.
    b. The opposite reciprocal of $\frac{3}{4}$ is $-\frac{4}{3}$.

NOTATION

13.

$$y+2=\frac{1}{3}(x+3)$$
$$y+2=\boxed{\frac{1}{3}x}+1$$
$$y+2-\boxed{2}=\frac{1}{3}x+1-\boxed{2}$$
$$y=\frac{1}{3}x-\boxed{1}$$
$$m=\boxed{\frac{1}{3}},\quad b=\boxed{-1}$$

PRACTICE

15. $m = 3,\ b = 6$
$y = mx + b$
$y = 3x + 6$

17. $m = -\frac{2}{3},\ b = -\frac{7}{3}$
$y = mx + b$
$y = -\frac{2}{3}x - \frac{7}{3}$

19. $m = \frac{1}{2},\ b = 3$
$y = mx + b$
$y = \frac{1}{2}x + 3$

21. $m = -\frac{7}{5},\ b = -1$
$y = mx + b$
$y = -\frac{7}{5}x - 1$

23. $m = 7,\ (-7, 5)$
$$y - y_1 = m(x - x_1)$$
$$y-5=7(x-(-7))$$
$$y-5=7(x+7)$$
$$y-5=7x+49$$
$$y-5+5=7x+49+5$$
$$y=7x+54$$

25. $m = -9,\ (2, -4)$
$$y - y_1 = m(x - x_1)$$
$$y-(-4)=-9(x-2)$$
$$y+4=-9x+18$$
$$y+4-4=-9x+18-4$$
$$y=-9x+14$$

27. $m = 10,\ (1, 7)$
$$y - y_1 = m(x - x_1)$$
$$y-7=10(x-1)$$

29. $m = -\frac{2}{3},\ (-2, -4)$
$$y - y_1 = m(x - x_1)$$
$$y-(-4)=-\frac{2}{3}(x-(-2))$$
$$y+4=-\frac{2}{3}(x+2)$$

31. $m = 5,\ (4, -5)$

$$y - y_1 = m(x - x_1)$$
$$y - (-5) = 5(x - 4)$$
$$y + 5 = 5x - 20$$
$$y + 5 - 5 = 5x - 20 - 5$$
$$y = 5x - 25$$

33. $m = -9,\ (-3.5, 2.7)$

$$y - y_1 = m(x - x_1)$$
$$y - 2.7 = -9(x - (-3.5))$$
$$y - 2.7 = -9(x + 3.5)$$
$$y - 2.7 = -9x - 31.5$$
$$y - 2.7 + 2.7 = -9x - 31.5 + 2.7$$
$$y = -9x - 28.8$$

35. Find m.

$$m = \frac{y_2 - y_1}{x_2 - x_1}$$
$$= \frac{10 - 8}{2 - 6}$$
$$= \frac{2}{-4}$$
$$= -\frac{1}{2}$$

Use $m = -\frac{1}{2}$ and (6, 8)

$$y - y_1 = m(x - x_1)$$
$$y - 8 = -\frac{1}{2}(x - 6)$$
$$y - 8 = -\frac{1}{2}x + 3$$
$$y - 8 + 8 = -\frac{1}{2}x + 3 + 8$$
$$y = -\frac{1}{2}x + 11$$

37. Find m using the points (–1, 3) and (2, 5).

$$m = \frac{y_2 - y_1}{x_2 - x_1}$$
$$= \frac{5 - 3}{2 - (-1)}$$
$$= \frac{2}{3}$$

Use the point (2, 5) and $m = \frac{2}{3}$.

$$y - y_1 = m(x - x_1)$$
$$y - 5 = \frac{2}{3}(x - 2)$$
$$y - 5 = \frac{2}{3}x - \frac{4}{3}$$
$$y - 5 + 5 = \frac{2}{3}x - \frac{4}{3} + 5$$
$$y = \frac{2}{3}x + \frac{11}{3}$$

39. $y = x - 1$

$m = 1,\ (0, -1)$

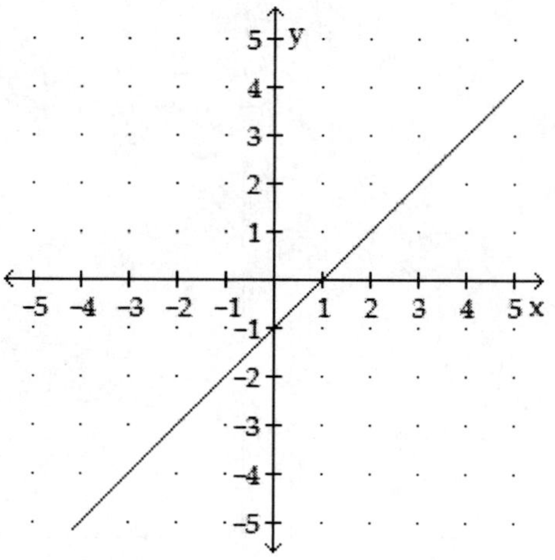

41. $y = -\frac{5}{4}x - 3$

$m = -\frac{5}{4}, \ (0, -3)$

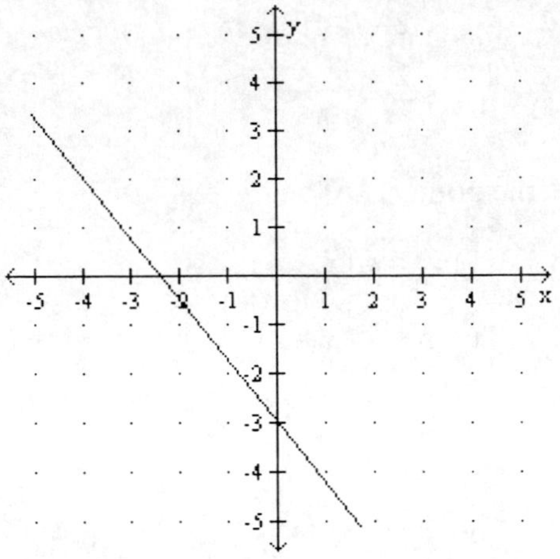

43.

$$5y - 8x - 30 = 0$$
$$5y - 8x - 30 + 8x + 30 = 0 + 8x + 30$$
$$5y = 8x + 30$$
$$y = \frac{8}{5}x + 6$$
$$m = \frac{8}{5}, \quad (0, 6)$$

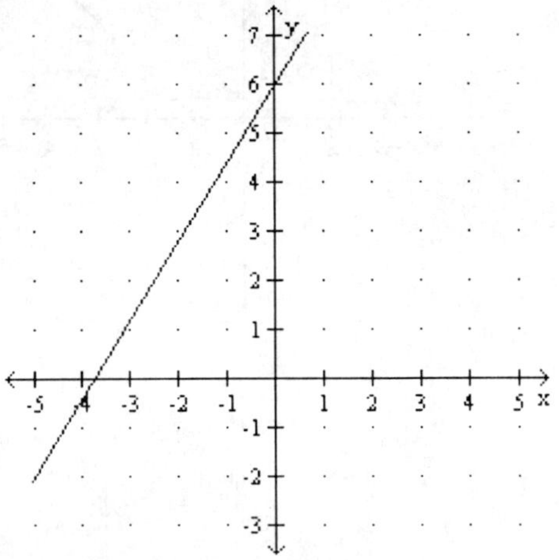

45.

$$4y - 3 = -3x - 11$$
$$4y - 3 + 3 = -3x - 11 + 3$$
$$4y = -3x - 8$$
$$y = -\frac{3}{4}x - 2$$
$$m = -\frac{3}{4}, \quad (0, -2)$$

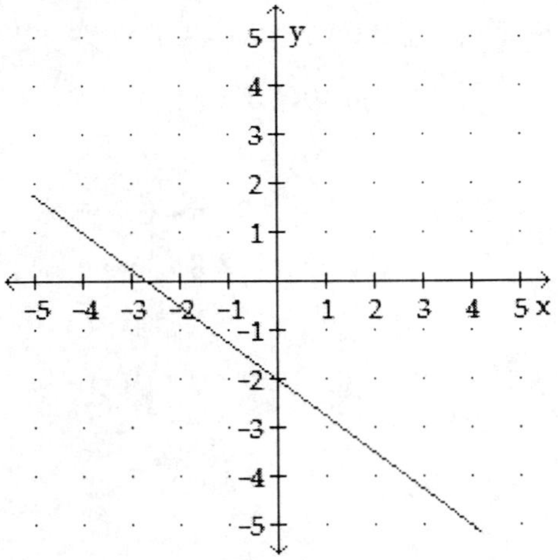

47.

$$3x - 2y = 8$$
$$3x - 2y - 3x = 8 - 3x$$
$$-2y = -3x + 8$$
$$\frac{-2y}{-2} = \frac{-3x}{-2} + \frac{8}{-2}$$
$$y = \frac{3}{2}x - 4$$
$$m = \frac{3}{2}, \quad (0, -4)$$

49.

$$-2(x + 3y) = 5$$
$$-2x - 6y = 5$$
$$-2x - 6y + 2x = 5 + 2x$$
$$-6y = 2x + 5$$
$$\frac{-6y}{-6} = \frac{2x}{-6} + \frac{5}{-6}$$
$$y = -\frac{1}{3}x - \frac{5}{6}$$
$$m = -\frac{1}{3}, \quad \left(0, -\frac{5}{6}\right)$$

51.

$y = 3x + 4$ $\quad$ $y = 3x - 7$

$m = 3$ $\quad$ $m = 3$

parallel lines

53.

$x + y = 2$ $\quad$ $y = x + 5$

$y = -x + 2$

$m = -1$ $\quad$ $m = 1$

perpendicular lines

55.

$3x + 6y = 1$ $\quad$ $y = \frac{1}{2}x$

$6y = -3x + 1$

$y = -\frac{1}{2}x + \frac{1}{6}$

$m = -\frac{1}{2}$ $\quad$ $m = \frac{1}{2}$

neither

57.

$y = 3$ $\quad$ $x = 4$

$m = 0$ $\quad$ $m = \text{undefined} = \frac{1}{0}$

perpendicular lines

59. Find the slope of the given line.

$y = 4x + 8$

$m = 4$

Use $m = 4$ and (2, 5) to find the equation of the line.

$$\begin{aligned} y - y_1 &= m(x - x_1) \\ y - 5 &= 4(x - 2) \\ y - 5 &= 4x - 8 \\ y - 5 + 5 &= 4x - 8 + 5 \\ y &= 4x - 3 \end{aligned}$$

61. Find the slope of the given line.

$y = 3x - 2$

$m = 3$

Use $m = 3$ and $\left(\frac{2}{3}, \frac{1}{4}\right)$ to find the equation of the line.

$$\begin{aligned} y - y_1 &= m(x - x_1) \\ y - \frac{1}{4} &= 3\left(x - \frac{2}{3}\right) \\ y - \frac{1}{4} &= 3x - 2 \\ y - \frac{1}{4} + \frac{1}{4} &= 3x - 2 + \frac{1}{4} \\ y &= 3x - \frac{7}{4} \end{aligned}$$

63. The slopes of perpendicular lines are opposite reciprocals. Since the slope of $y = 4x - 7$ is 4, the slope of the line perpendicular would be $-\frac{1}{4}$.

Use $m = -\frac{1}{4}$ and (–40, 10) to find the equation of the line.

$$\begin{aligned} y - y_1 &= m(x - x_1) \\ y - 10 &= -\frac{1}{4}(x - (-40)) \\ y - 10 &= -\frac{1}{4}x - 10 \\ y &= -\frac{1}{4}x - 0 \\ y &= -\frac{1}{4}x \end{aligned}$$

65. Find the slope of the given line.

$$7x = 9y - 2$$
$$7x - 9y - 7x = 9y - 2 - 9y - 7x$$
$$-9y = -7x - 2$$
$$\frac{-9y}{-9} = \frac{-7x}{-9} - \frac{2}{-9}$$
$$y = \frac{7}{9}x + \frac{2}{9}$$

The opposite reciprocal is $-\frac{9}{7}$.

Use $m = -\frac{9}{7}$ and (4, –2) to find the equation of the line.

$$y - y_1 = m(x - x_1)$$
$$y - (-2) = -\frac{9}{7}(x - 4)$$
$$y + 2 = -\frac{9}{7}x + \frac{36}{7}$$
$$y + 2 - 2 = -\frac{9}{7}x + \frac{36}{7} - \frac{14}{7}$$
$$y = -\frac{9}{7}x + \frac{22}{7}$$

TRY IT YOURSELF

67. First find m.

$$m = \frac{y_2 - y_1}{x_2 - x_1}$$
$$= \frac{-10 - 4}{-3 - 3}$$
$$= \frac{-14}{-6}$$
$$= \frac{7}{3}$$

Use the point (3, 4) and $m = \frac{7}{3}$.

$$y - y_1 = m(x - x_1)$$
$$y - 4 = \frac{7}{3}(x - 3)$$
$$y - 4 = \frac{7}{3}x - 7$$
$$y - 4 + 4 = \frac{7}{3}x - 7 + 4$$
$$y = \frac{7}{3}x - 3$$

69. $m = \frac{4}{3}$; (5, 9)

$$y - y_1 = m(x - x_1)$$
$$y - 9 = \frac{4}{3}(x - 5)$$
$$y - 9 = \frac{4}{3}x - \frac{20}{3}$$
$$y - 9 + 9 = \frac{4}{3}x - \frac{20}{3} + 9$$
$$y = \frac{4}{3}x + \frac{7}{3}$$

71. Find the slope of the given line.

$$4x - y = 7$$
$$4x - y - 4x = 7 - 4x$$
$$-y = -4x + 7$$
$$\frac{-y}{-1} = \frac{-4x}{-1} + \frac{7}{-1}$$
$$y = 4x - 7$$
$$m = 4$$

opposite reciprocal $= -\frac{1}{4}$

Use $m = -\frac{1}{4}$ and (2, 5) to find the equation of the line.

$$y - y_1 = m(x - x_1)$$
$$y - 5 = -\frac{1}{4}(x - 2)$$
$$y - 5 = -\frac{1}{4}x + \frac{1}{2}$$
$$y - 5 + 5 = -\frac{1}{4}x + \frac{1}{2} + 5$$
$$y = -\frac{1}{4}x + \frac{11}{2}$$

73. Use the points (4, 0) and (–1, –7) to find the slope.

$$\begin{aligned} m &= \frac{y_2 - y_1}{x_2 - x_1} \\ &= \frac{-7-0}{-1-4} \\ &= \frac{-7}{-5} \\ &= \frac{7}{5} \end{aligned}$$

Use the point (4, 0) and $m = \frac{7}{5}$.

$$\begin{aligned} y - y_1 &= m(x - x_1) \\ y - 0 &= \frac{7}{5}(x-4) \\ y &= \frac{7}{5}x - \frac{7}{5}(4) \\ y &= \frac{7}{5}x - \frac{28}{5} \end{aligned}$$

75. First find m.

$$\begin{aligned} m &= \frac{y_2 - y_1}{x_2 - x_1} \\ &= \frac{4-(-1)}{4-(-1)} \\ &= \frac{5}{5} \\ &= 1 \end{aligned}$$

Use the point (4, 4) and $m = 1$.

$$\begin{aligned} y - y_1 &= m(x - x_1) \\ y - 4 &= 1(x-4) \\ y - 4 &= 1x - 4 \\ y - 4 + 4 &= x - 4 + 4 \\ y &= x \end{aligned}$$

77. The slopes of parallel lines are the same. Since the slope of $y = 4x - 7$ is 4, the slope of the line parallel would also be 4.
Use $m = 4$ and (0, 0) to find the equation of the line.

$$\begin{aligned} y - y_1 &= m(x - x_1) \\ y - 0 &= 4(x-0) \\ y &= 4x - 0 \\ y &= 4x \end{aligned}$$

79. $m = -3$, (4, –6)

$$\begin{aligned} y - y_1 &= m(x - x_1) \\ y-(-6) &= -3(x-4) \\ y + 6 &= -3x + 12 \\ y + 6 - 6 &= -3x + 12 - 6 \\ y &= -3x + 6 \end{aligned}$$

81. Find the slope of the given line.

$$\begin{aligned} y &= 3x - 2 \\ m &= 3 \end{aligned}$$

opposite reciprocal $= -\frac{1}{3}$

Use $m = -\frac{1}{3}$ and $\left(\frac{2}{3}, \frac{1}{4}\right)$ to find the equation of the line.

$$\begin{aligned} y - y_1 &= m(x - x_1) \\ y - \frac{1}{4} &= -\frac{1}{3}\left(x - \frac{2}{3}\right) \\ y - \frac{1}{4} &= -\frac{1}{3}x + \frac{2}{9} \\ y - \frac{1}{4} + \frac{1}{4} &= -\frac{1}{3}x + \frac{2}{9} + \frac{1}{4} \\ y &= -\frac{1}{3}x + \frac{17}{36} \end{aligned}$$

83. Find m using the points (0, 4) and (–3, 0).

$$\begin{aligned} m &= \frac{y_2 - y_1}{x_2 - x_1} \\ &= \frac{0-4}{-3-0} \\ &= \frac{-4}{-3} \\ &= \frac{4}{3} \end{aligned}$$

Use the point (–3, 0) and $m = \frac{4}{3}$.

$$\begin{aligned} y - y_1 &= m(x - x_1) \\ y - 0 &= \frac{4}{3}(x-(-3)) \\ y &= \frac{4}{3}(x+3) \\ y &= \frac{4}{3}x + 4 \end{aligned}$$

85. Find the slope of the given line.

$$x = \frac{5}{4}y - 2$$
$$4(x) = 4\left(\frac{5}{4}y - 2\right)$$
$$4x = 5y - 8$$
$$4x + 8 = 5y - 8 + 8$$
$$4x + 8 = 5y$$
$$5y = 4x + 8$$
$$y = \frac{4}{5}x + \frac{8}{5}$$
$$m = \frac{4}{5}$$

Use $m = \frac{4}{5}$ and (4, –2) to find the equation of the line.

$$y - y_1 = m(x - x_1)$$
$$y - (-2) = \frac{4}{5}(x - 4)$$
$$y + 2 = \frac{4}{5}x - \frac{16}{5}$$
$$y + 2 - 2 = \frac{4}{5}x - \frac{16}{5} - 2$$
$$y = \frac{4}{5}x - \frac{26}{5}$$

APPLICATIONS

87. CERTIFIED FITNESS INSTRUCTOR

a. $c = 7.8m + 220$

b. Let $c = 300$.

$$c = 7.8m + 220$$
$$300 = 7.8m + 220$$
$$300 - 220 = 7.8m + 220 - 220$$
$$80 = 7.8m$$
$$\frac{80}{7.8} = \frac{7.8m}{7.8}$$
$$10.3 \approx m$$

The client must exercise about $10\frac{1}{4}$ minutes.

89. a. $y = -22{,}000{,}000t + 650{,}000{,}000$

b. The b-intercept indicates that the number of barrels produced in 1990 was 650,000,000. The slope indicates that the yearly decrease in production was about 22,000,000.

91. CRIMINOLOGY

a. Use (77,000, 575) and $m = \frac{1}{100}$

$$B - B_1 = m(p - p_1)$$
$$B - 575 = \frac{1}{100}(p - 77{,}000)$$
$$B - 575 = \frac{1}{100}p - 770$$
$$B - 575 + 575 = \frac{1}{100}p - 770 + 575$$
$$B = \frac{1}{100}p - 195$$

b. Find B if $p = 110{,}000$

$$B = \frac{1}{100}p - 195$$
$$= \frac{1}{100}(110{,}000) - 195$$
$$= 1{,}100 - 195$$
$$= 905$$

93. RUSSIA

a. Find t: 2000 – 1995 = 5

Use the point (5, 146,000,000) and $m = -505{,}000$.

$$y - y_1 = m(x - x_1)$$
$$y - 146{,}000{,}000 = -505{,}000(x - 5)$$
$$y - 146{,}000{,}000 = -505{,}000x + 2{,}525{,}000$$
$$y = -505{,}000x + 148{,}525{,}000$$
$$P = -505{,}000t + 148{,}525{,}000$$

b. Find t: 2015 – 1995 = 20.

$$P = -505{,}000(20) + 148{,}525{,}000$$
$$= -10{,}100{,}000 + 148{,}525{,}000$$
$$= 138{,}425{,}000$$

95. UNDERSEA DIVING

a. Write 2 ordered pairs in the form (depth, pressure). Find the slope of (0, 14.7) and (33, 29.4)

$$m=\frac{y_2-y_1}{x_2-x_1}$$
$$=\frac{29.4-14.7}{33-0}$$
$$=\frac{14.7}{33}$$
$$=\frac{14.7}{33}\cdot\frac{10}{10}$$
$$=\frac{147}{330}$$

Use the point (0, 14.7) and $m=\frac{147}{330}$.

$$p-p_1=m(d-d_1)$$
$$p-14.7=\frac{147}{330}(d-0)$$
$$p-14.7=\frac{147}{330}d-0$$
$$p=\frac{147}{330}d+14.7$$

b. Find p when $d=250$

$$p=\frac{147}{330}d+14.7$$
$$=\frac{147}{330}(250)+14.7$$
$$\approx 111.4+14.7$$
$$\approx 126.1\text{ lb/in.}^2$$

97. CRAIG'S LIST

Write ordered pairs in the form (age, value). Use (0, 1,750) for the purchase value and (3, 800) for the present value.

Find the rate of change (m).

$$m=\frac{y_2-y_1}{x_2-x_1}$$
$$=\frac{800-1,750}{3-0}$$
$$=\frac{-950}{3}$$

Use the point (3, 800) and $m=\frac{-950}{3}$.

$$y-y_1=m(x-x_1)$$
$$y-800=\frac{-950}{3}(x-3)$$
$$y-800=\frac{-950}{3}x+950$$
$$y-800+800=\frac{-950}{3}x+950+800$$
$$y=\frac{-950}{3}x+1,750$$

99. PAINTINGS

a. Use (1960, 100) and (2000, 600) to find the slope.

$$m=\frac{y_2-y_1}{x_2-x_1}$$
$$=\frac{600-100}{2000-1960}$$
$$=\frac{500}{40}$$
$$=12.5$$

If $m=12.5$ and $b=100$, then the equation would be:

$$y=mx+b$$
$$y=12.5x+100$$

b. Let $y=1,000$.
(1 billion = 1,000 million)

$$1,000=12.5x+100$$
$$900=12.5x$$
$$\frac{900}{12.5}=\frac{12.5x}{12.5}$$
$$72=x$$

$$1960+72=2032$$

101. REGRESSION EQUATIONS

a. Use (68, 155) and (73, 195) to find the slope.

$$m = \frac{y_2 - y_1}{x_2 - x_1} = \frac{195 - 155}{73 - 68} = \frac{40}{5} = 8$$

Use the point (68, 155) and $m = 8$.

$$y - y_1 = m(x - x_1)$$
$$y - 155 = 8(x - 68)$$
$$y - 155 = 8x - 544$$
$$y - 155 + 155 = 8x - 544 + 155$$
$$y = 8x - 389$$
$$w = 8h - 389$$

b. Let $h = 72$.

$$w = 8(72) - 389 = 576 - 389 = 187$$

103. CIGARETTE SMOKING

a. The y-intercept is (0m 632). It indicates that in 1980, about 632 billion cigarettes were consumed in the U.S.

b. The slope of the line is –10. This indicates that since 1980, the number of cigarettes consumed in the U.S. has decreased by about 10 billion each year.

WRITING

105. Answers will vary.

107. Answers will vary.

109. Answers will vary.

REVIEW

111. Let x = the amount invested at each %.

$$0.06x + 0.07x + 0.08x = 2,037$$
$$0.21x = 2,037$$
$$\frac{0.21x}{0.21} = \frac{2,037}{0.21}$$
$$x = \$9,700 \text{ in each account}$$
$$3(\$9,700) = \$29,100 \text{ in all}$$

CHALLENGE PROBLEMS

113. $m < 0$ and $b > 0$

SECTION 2.5

VOCABULARY

1. A set of ordered pairs is called a **relation**. The set of all first components of the ordered pairs is called the **domain** and the set of all second components is called the **range**.

3. Given a relation in x and y, if to each value of x in the domain there corresponds exactly one value of y in the range, y is said to be a **function** of x. We call x the independent **variable** and y the **dependent** variable.

5. We call $f(x) = 2x + 1$ a **linear** function because its graph is a straight line.

CONCEPTS

7. a. {(2000, 63), (2001, 56), (2002, 54), (2003, 50), (2004, 52), (2005, 51), (2006, 51)}
   b. D: {2000, 2001, 2002, 2003, 2004, 2005, 2006}
   R: {50, 51, 52, 54, 56, 63}
   c. **YEAR** **# COLLECTED**

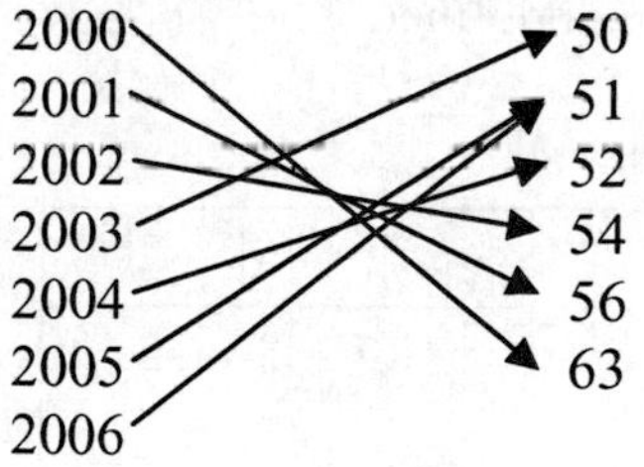

9. If $x = -4$, then the denominator of $\frac{1}{x+4}$ is 0 and undefined.

11. a. (0, –4)
    b. Let $y = 0$.

$$y = -6x - 4$$
$$0 = -6x - 4$$
$$4 = -6x$$
$$\frac{4}{-6} = x$$
$$-\frac{2}{3} = x$$
$$\left(-\frac{2}{3}, 0\right)$$

NOTATION

13. a. We read $f(x) = 5x - 6$ as "f $\boxed{\text{of}}$ x is equal to $5x$ minus 6."
    b. We read $g(t) = t + 9$ as "g $\boxed{\text{of}}$ t is equal to t plus 9."

15.
- If $y = 5x + 1$, find the value of y when $x = 8$.
- If $f(x) = 5x + 1$, find $\boxed{f(8)}$.

17. When graphing the function $f(x) = -x + 5$, the vertical axis of the rectangular coordinate system can be labeled $\boxed{f(x)}$ or $\boxed{y}$.

GUIDED PRACTICE

19. D:{–5, –1, 7, 8}; R:{–11, –6, –1, 3}

21. D:{–23, 0, 7}; R:{1, 35}

23. yes

25. no; (4, 2), (4, 4) and (4, 6)

27. no; (3, 4) and (3, –4) or (4, 3) and (4, –3)

29. yes

31. yes

33. no; (–1, 0) and (–1, 2)

35. yes

37. yes

39. no; (1, 1) and (1, –1)

41. yes

43. yes

45. no; (1, 1), (1, –1)

47. $f(x) = 3x$

$$f(3) = 3(3) \qquad f(-1) = 3(-1)$$
$$= 9 \qquad\qquad = -3$$

49. $f(x)=2x-3$

$f(3)=2(3)-3 = 6-3 = 3$

$f(-1)=2(-1)-3 = -2-3 = -5$

51. $g(x)=x^2-10$

$g(2)=\mathbf{2}^2-10 = 4-10 = -6$

$g(3)=\mathbf{3}^2-10 = 9-10 = -1$

53. $g(x)=-x^3+x$

$g(2)=-\mathbf{2}^3+\mathbf{2} = -8+2 = -6$

$g(3)=-\mathbf{3}^3+\mathbf{3} = -27+3 = -24$

55. $g(x)=(x+1)^2$

$g(2)=(\mathbf{2}+1)^2 = 3^2 = 9$

$g(3)=(\mathbf{3}+1)^2 = 4^2 = 16$

57. $g(x)=2x^2-x+1$

$g(2)=2(\mathbf{2})^2-\mathbf{2}+\mathbf{1} = 2(4)-2+1 = 8-2+1 = 7$

$g(3)=2(\mathbf{3})^2-\mathbf{3}+1 = 2(9)-3+1 = 18-3+1 = 16$

59. $h(x)=|x|+2$

$h(5)=|\mathbf{5}|+2 = 5+2 = 7$

$h(-2)=|\mathbf{-2}|+2 = 2+2 = 4$

61. $h(x)=\frac{1}{x+3}$

$h(5)=\frac{1}{\mathbf{5}+3} = \frac{1}{8}$

$h(-2)=\frac{1}{\mathbf{-2}+3} = \frac{1}{1} = 1$

63. $h(x)=\frac{x}{x-3}$

$h(2)=\frac{\mathbf{5}}{\mathbf{5}-3} = \frac{5}{2}$

$h(-2)=\frac{\mathbf{-2}}{\mathbf{-2}-3} = \frac{-2}{-5} = \frac{2}{5}$

65. $h(x)=\frac{x^2+2x-35}{x^2+5x+6}$

$h(5)=\frac{(\mathbf{5})^2+2(\mathbf{5})-35}{(\mathbf{5})^2+5(\mathbf{5})+6} = \frac{25+10-35}{25+25+6} = \frac{0}{56} = 0$

$h(-2)=\frac{(\mathbf{-2})^2+2(\mathbf{-2})-35}{(\mathbf{-2})^2+5(\mathbf{-2})+6} = \frac{4-4-35}{4-10+6} = \frac{-35}{0}$ = undefined

67. $f(t)=|t-2|$

t	$f(x)=\lvert t-2\rvert$
-1.7	$\lvert\mathbf{-1.7}-2\rvert=\lvert-3.7\rvert = 3.7$
0.9	$\lvert\mathbf{0.9}-2\rvert=\lvert-1.1\rvert = 1.1$
5.4	$\lvert\mathbf{5.4}-2\rvert=\lvert 3.4\rvert = 3.4$

69. $g(a) = a^3$

Input	Output
$-\frac{3}{4}$	$\left(-\frac{3}{4}\right)^3 = -\frac{3}{4}\left(-\frac{3}{4}\right)\left(-\frac{3}{4}\right)$ $= -\frac{27}{64}$
$\frac{1}{6}$	$\left(\frac{1}{6}\right)^3 = \frac{1}{6}\left(\frac{1}{6}\right)\left(\frac{1}{6}\right)$ $= \frac{1}{216}$
$\frac{5}{2}$	$\left(\frac{5}{2}\right)^3 = \frac{5}{2}\left(\frac{5}{2}\right)\left(\frac{5}{2}\right)$ $= \frac{125}{8}$

71. $g(x) = 2x$

$g(w) = 2(w) = 2w$ $\qquad$ $g(w+1) = 2(w+1) = 2w+2$

73. $g(x) = 3x - 5$

$g(w) = 3(w) - 5 = 3w - 5$ $\qquad$ $g(w+1) = 3(w+1) - 5 = 3w + 3 - 5 = 3x - 2$

75.

$$f(x) = -2x + 5$$
$$f(x) = 5$$
$$-2x + 5 = 5$$
$$-2x + 5 - 5 = 5 - 5$$
$$-2x = 0$$
$$x = 0$$

77.

$$f(x) = \frac{3}{2}x - 2$$
$$f(x) = -\frac{1}{2}$$
$$\frac{3}{2}x - 2 = -\frac{1}{2}$$
$$2\left(\frac{3}{2}x - 2\right) = 2\left(-\frac{1}{2}\right)$$
$$3x - 4 = -1$$
$$3x - 4 + 4 = -1 + 4$$
$$3x = 3$$
$$x = 1$$

79. a. The set of real numbers
b. The set of real numbers except 4

81. a. The set of real numbers
b. The set of real numbers except $-\frac{1}{2}$

83. $f(x) = 2x - 1$

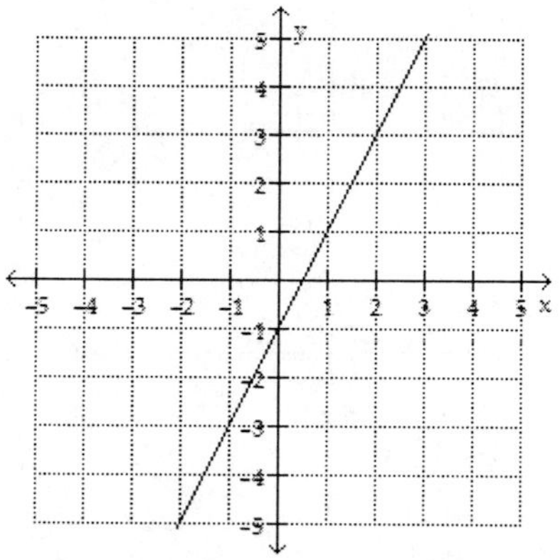

85. $f(x) = -\frac{3}{2}x - 3$

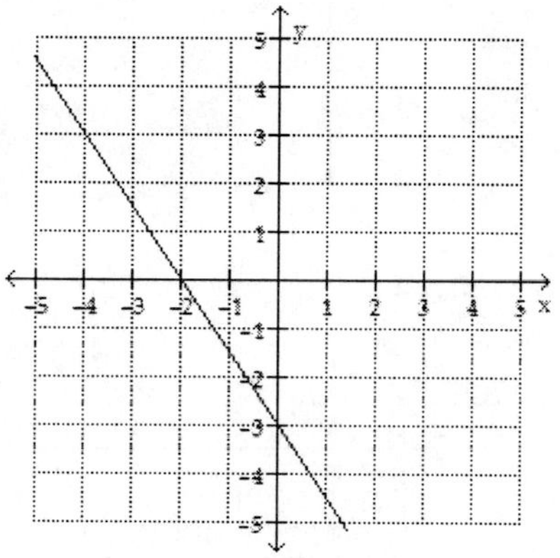

87. $f(x) = x$

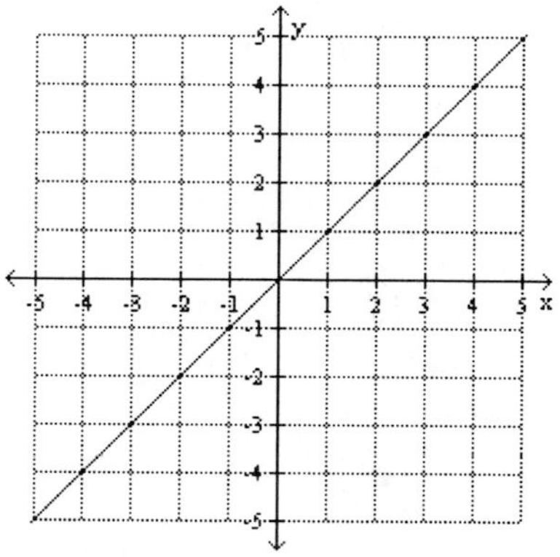

89. $f(x) = -4$

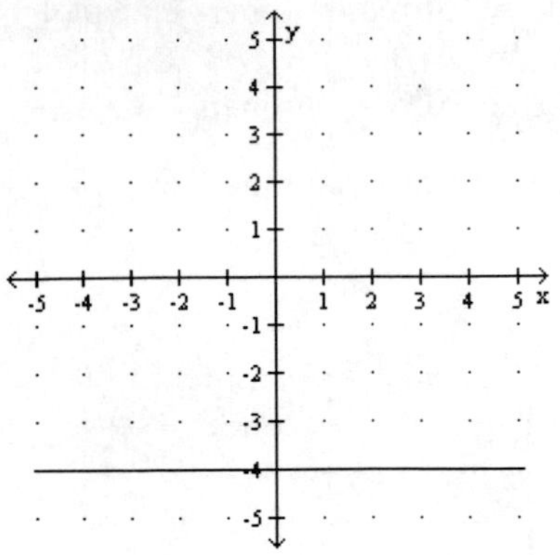

91. $g(x) = 0.75x$

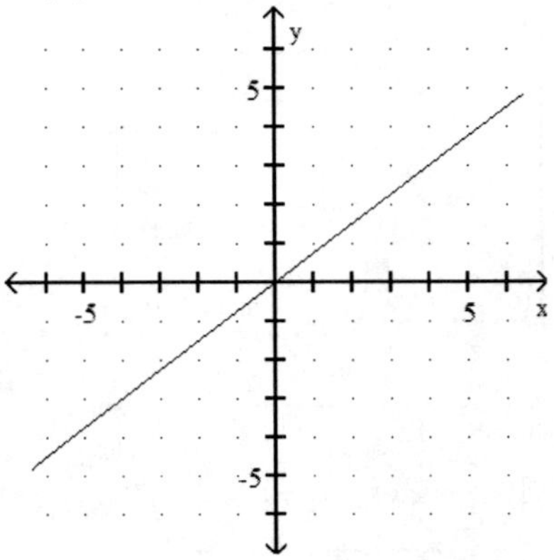

93. $g(x) = \frac{7}{8}x + 2$

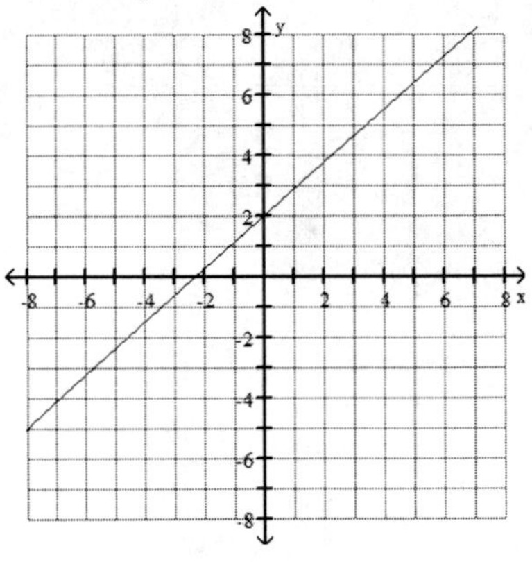

95. $m = 5, b = -3$

$$y = mx + b$$
$$y = 5x - 3$$
$$f(x) = 5x - 3$$

97. $m = \frac{1}{5}, x_1 = 10, y_1 = 1$

$$y - y_1 = m(x - x_1)$$
$$y - 1 = \frac{1}{5}(x - 10)$$
$$y - 1 = \frac{1}{5}x - 2$$
$$y - 1 + 1 = \frac{1}{5}x - 2 + 1$$
$$y = \frac{1}{5}x - 1$$
$$f(x) = \frac{1}{5}x - 1$$

99. First find the slope of the line:

$$m = \frac{y_2 - y_1}{x_2 - x_1}$$
$$= \frac{1-7}{-2-1}$$
$$= \frac{-6}{-3}$$
$$= 2$$

Use $m = 2, x_1 = 1, y_1 = 7$

$$y - y_1 = m(x - x_1)$$
$$y - 7 = 2(x - 1)$$
$$y - 7 = 2x - 2$$
$$y - 7 + 7 = 2x - 2 + 7$$
$$y = 2x + 5$$
$$f(x) = 2x + 5$$

101. The slope of the given line is $-\frac{2}{3}$. Since the lines are parallel, then use the same slope for the new line.

Use $m = -\frac{2}{3}, x_1 = 3\ y_1 = 0$

$$y - y_1 = m(x - x_1)$$
$$y - 0 = -\frac{2}{3}(x - 3)$$
$$y = -\frac{2}{3}x + 2$$
$$f(x) = -\frac{2}{3}x + 2$$

103. The slope of the given line is $-\frac{1}{6}$. The slope of the line perpendicular is the reciprocal with the opposite sign so slope would be 6.

Use $m = 6$, $x_1 = 1$ $y_1 = 2$

$$y - y_1 = m(x - x_1)$$
$$y - 2 = 6(x - 1)$$
$$y - 2 = 6x - 6$$
$$y - 2 + 2 = 6x - 6 + 2$$
$$y = 6x - 4$$
$$f(x) = 6x - 4$$

105. A horizontal line through the point (–8, 12) has the equation $f(x) = 12$.

APPLICATIONS

107. DECONGESTANTS

$$C(F) = \frac{5}{9}(F - 32)$$
$$C(68) = \frac{5}{9}(68 - 32)$$
$$= \frac{5}{9}(36)$$
$$= 20$$
$$C(77) = \frac{5}{9}(77 - 32)$$
$$= \frac{5}{9}(45)$$
$$= 25$$

between 20° C and 25° C

109. CONCESSIONAIRES

a. $p(b) = 4.75b - 125$

b. $p(b) = 4.75b - 125$

$$p(110) = 4.75(110) - 125$$
$$= 522.50 - 125$$
$$= \$397.50$$

111. NURSES

Write the two ordered pairs using the first coordinate as the years after 2000 and the second coordinate as the number of nurses: (5, 2,175,500) and (15, 2,586,500).

Find the slope of the line between those points:

$$m = \frac{y_2 - y_1}{x_2 - x_1}$$
$$= \frac{2,586,500 - 2,175,500}{15 - 5}$$
$$= \frac{411,000}{10}$$
$$= 41,100$$

a. Use $m = 41,100$, $x_1 = 5$, $y_1 = 2,175,500$

$$y - y_1 = m(x - x_1)$$
$$y - 2,175,500 = 41,100(x - 5)$$
$$y - 2,175,500 = 41,100x - 205,500$$
$$y = 41,100x + 1,970,000$$
$$N(t) = 41,100t + 1,970,000$$

b. Let $t = 25$ since 2025 is 25 years after 2000.

$$N(25) = 41,100(25) + 1,970,000$$
$$= 1,027,500 + 1,970,00$$
$$= 2,997,500$$

113. CERT. FITNESS INSTRUCTOR

Write the information as ordered pairs: (35, 90) and (55, 66).

a. Find the slope of the line between those points:

$$m = \frac{y_2 - y_1}{x_2 - x_1} = \frac{66-90}{55-35} = \frac{-24}{20} = -1.2$$

Use $m = -1.2$, $x_1 = 55$, $y_1 = 66$

$$y - y_1 = m(x - x_1)$$
$$y - 66 = -1.2(x - 55)$$
$$y - 66 = -1.2x + 66$$
$$y = -1.2x + 132$$
$$L(a) = -1.2a + 132$$

b. Let $a = 80$.

$$L(80) = -1.2(80) + 132 = -96 + 132 = 36\%$$

115. TAXES

a. $T(a) = 837.50 + 0.15(a - 8{,}375)$

$$T(25{,}000) = 837.50 + 0.15(25{,}000 - 8{,}375) = 837.50 + 0.15(16{,}625) = 837.50 + 2{,}493.75 = \$3{,}331.25$$

The tax on an adjusted gross income of $25,000 is $3,331.25.

b.

$T(a) = 4{,}681.25 + 0.25(a - 34{,}000)$

WRITING

117. Answers will vary.

119. Answers will vary.

REVIEW

121.

$$-2(t+4) + 5t + 1 = 3(t-4) + 7$$
$$-2t - 8 + 5t + 1 = 3t - 12 + 7$$
$$3t - 7 = 3t - 5$$
$$3t - 7 - 3t = 3t - 5 - 3t$$
$$-7 \neq -5$$
$$\varnothing$$

No Solution; contradiction

CHALLENGE PROBLEMS

123.

$$f(8) = 4(8) + 6 = 32 + 6 = 38$$
$$g(8) = 4$$
$$h(8) = 6$$

$$\frac{f(8) + g(8)}{h(8)} = \frac{38+4}{6} = \frac{42}{6} = 7$$

SECTION 2.6

VOCABULARY

1. Functions whose graphs are not straight lines are called **nonlinear** functions.

3. The set of **nonnegative** real numbers is the set of real numbers greater than or equal to 0.

CONCEPTS

5. a. the squaring function

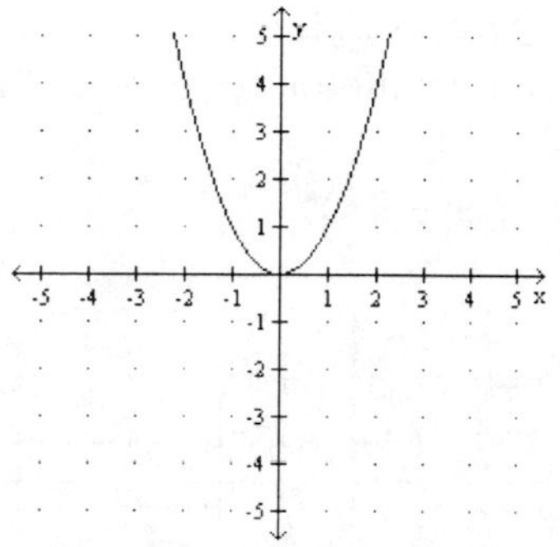

b. the cubing function

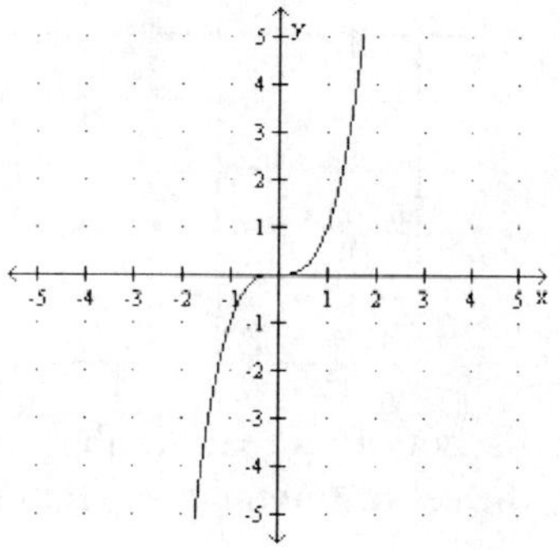

c. the absolute value function

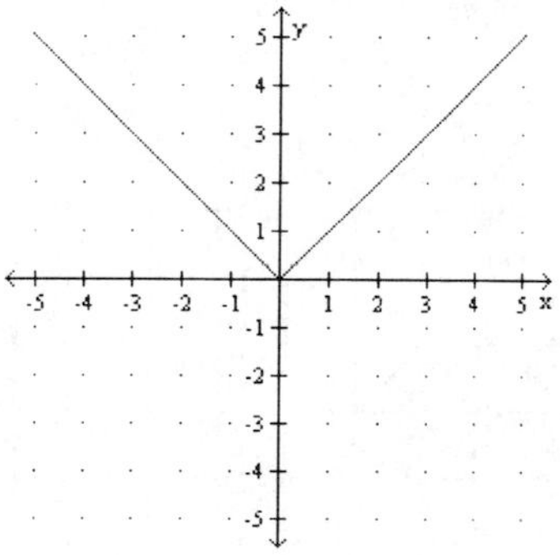

7. (5, 9)

9. a.

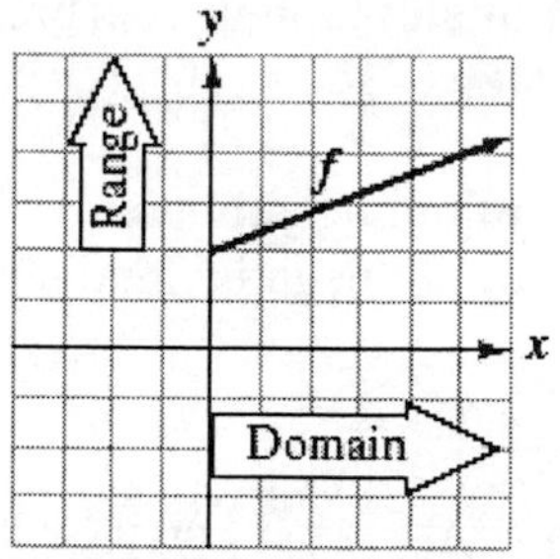

b. D: all nonnegative real numbers
R: all real numbers greater than or equal to 2

11. a. h is the x-value of the vertex: 4
b. h is the x-value of the vertex: 0
c. h is the x-value of the vertex: –2

13. a. (–2, 4) and (–2, –4)
b. No, since the x-value of –2 corresponds to more than one y-value, 4 and –4.

NOTATION

15. a. The graph of $f(x) = (x + 4)^3$ is the same as the graph of $f(x) = x^3$ except that it is shifted **4** units to the **left**.
b. The graph of $f(x) = x^3 + 4$ is the same as the graph of $f(x) = x^3$ except that it is shifted **4** units **up**.

GUIDED PRACTICE

17. a. –4
b. 0
c. 2
d. –1

19. a. 2
b. 4
c. –1 and 1
d. 2 and 4

21. D: the set of all real numbers
R: the set of all real numbers

23. D: the set of all real numbers
R: the set of all real numbers less than or equal to 5

25. D: the set of all real numbers
R: the set of all real numbers greater than or equal to –4

27. D: the set of nonnegative real numbers
R: the set of nonnegative real numbers

29. $f(x) = x^2 + 2$

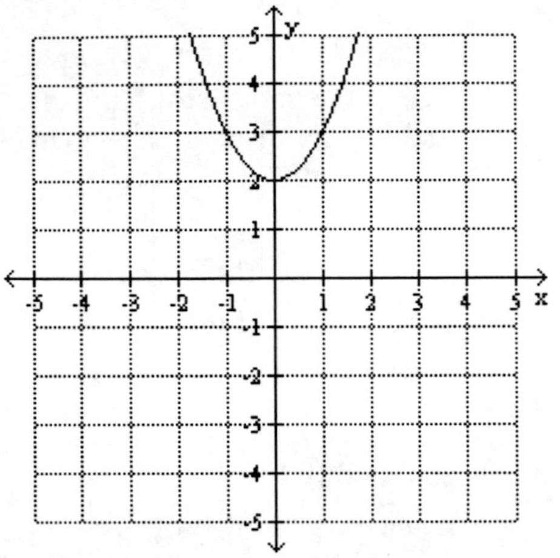

D: the set of all real numbers
R: the set of all real numbers greater than or equal to 2

31. $f(x) = x^3 - 3$

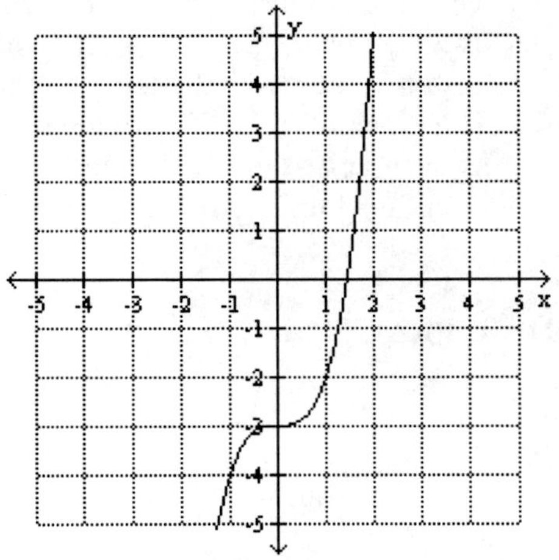

D: the set of all real numbers
R: the set of all real numbers

33. $f(x) = |x - 1|$

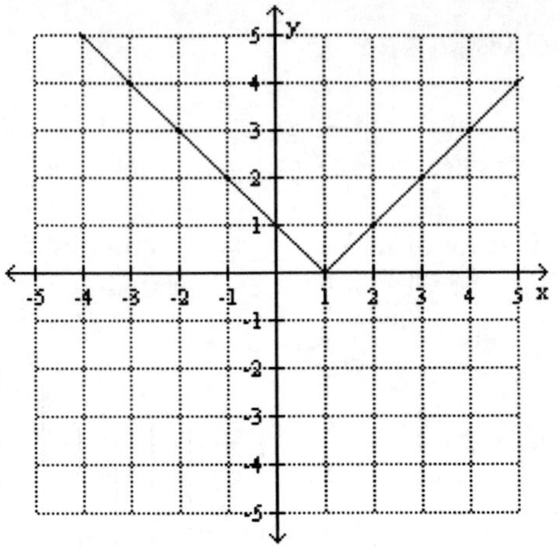

D: the set of all real numbers
R: the set of nonnegative real numbers

35. $f(x) = (x + 4)^2$

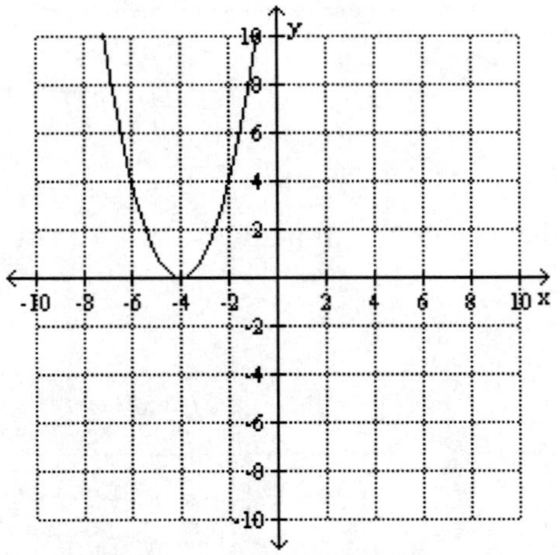

D: the set of all real numbers
R: the set of nonnegative real numbers

37. $g(x) = |x| - 2$

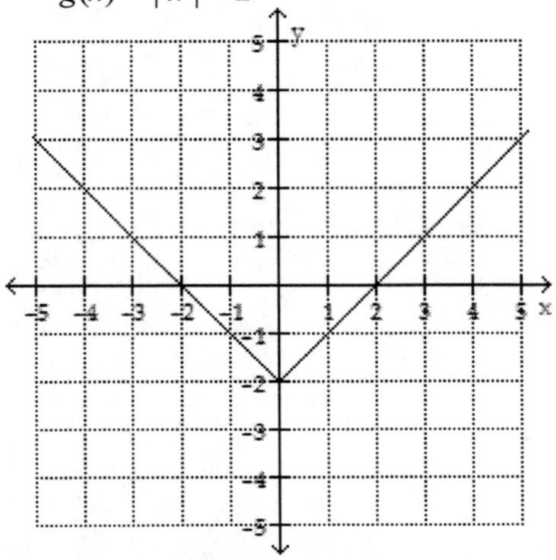

D: the set of real numbers
R: the set of real numbers greater than or equal to –2

39. $f(x) = (x + 1)^3$

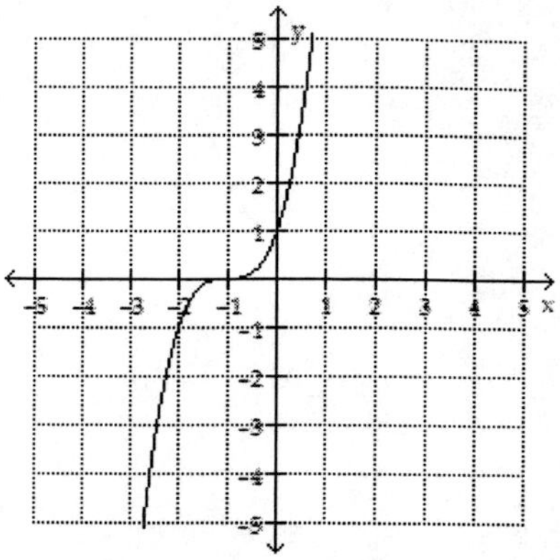

D: the set of real numbers
R: the set of real numbers

41. $g(x) = x^2 - 3$

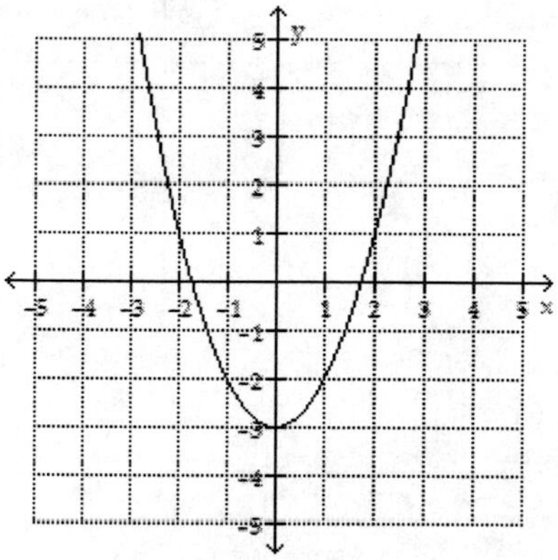

D: the set of real numbers
R: the set of real numbers greater than or equal to -3

43. $g(x) = (x - 4)^3$

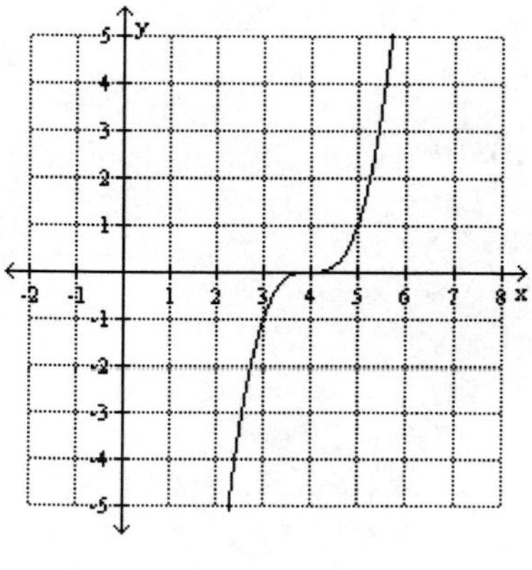

D: the set of real numbers
R: the set of real numbers

45. $g(x) = x^3 + 4$

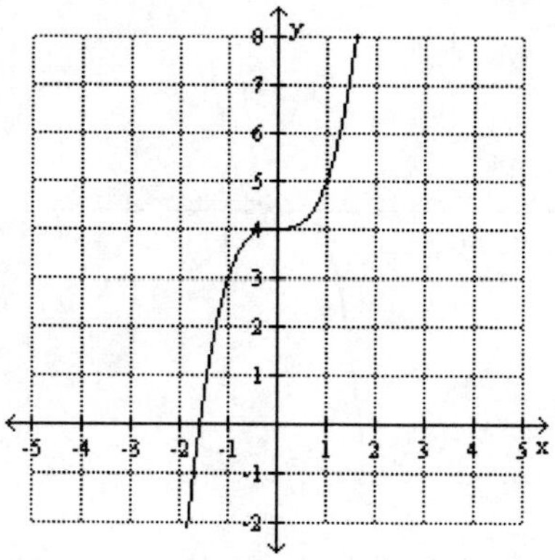

D: the set of real numbers
R: the set of real numbers

47. $g(x) = (x + 4)^2$

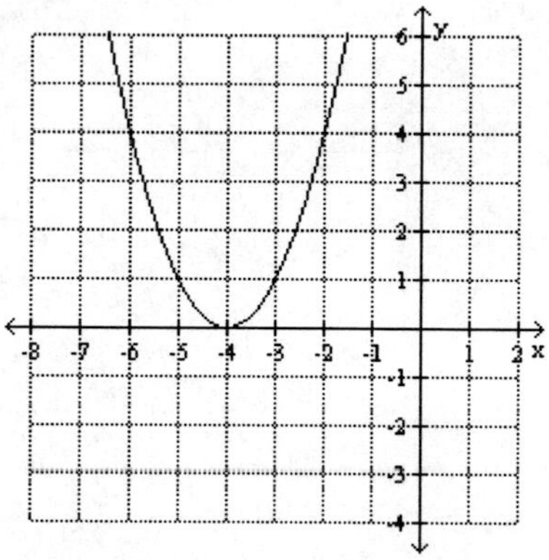

D: the set of real numbers
R: the set of nonnegative real numbers

49. $g(x) = |x - 2| - 1$

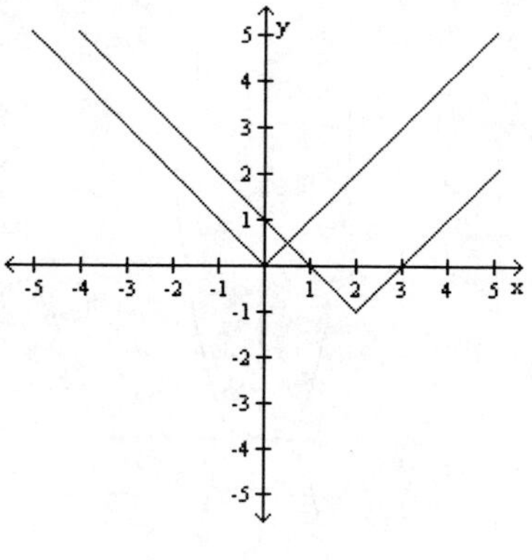

51. $g(x) = (x+1)^3 - 2$

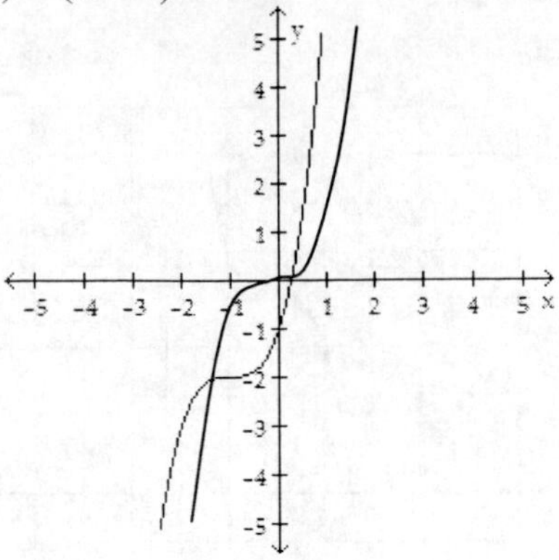

59. $g(x) = -x^2$

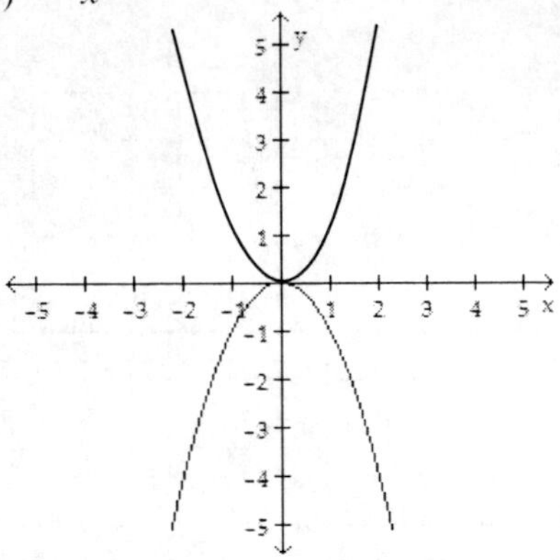

53. $g(x) = (x-2)^2 + 4$

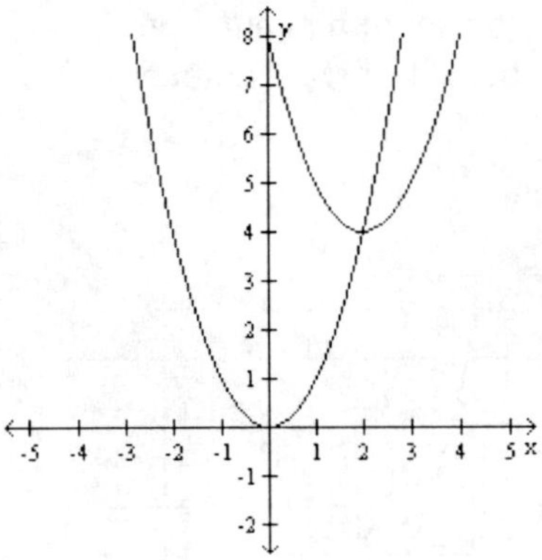

61. $g(x) = -|x+5|$

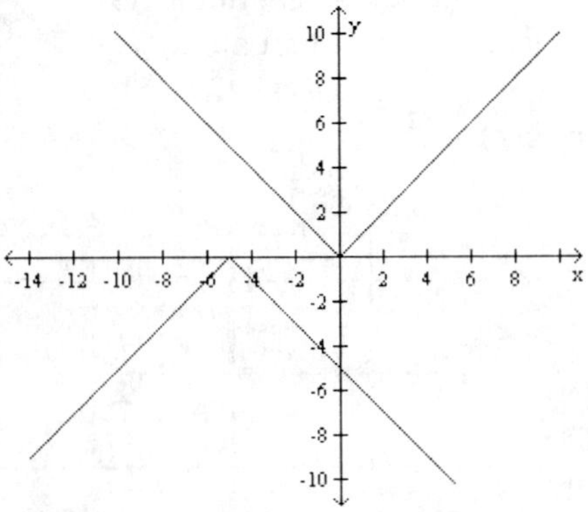

55. $g(x) = |x+3| + 5$

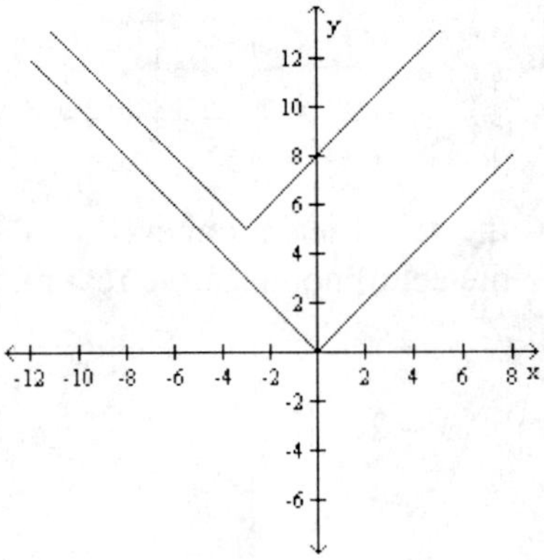

63. $g(x) = -x^2 + 3$

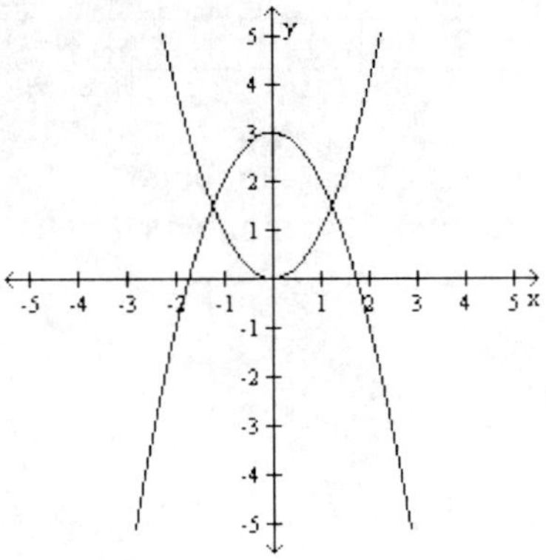

57. $g(x) = -x^3$

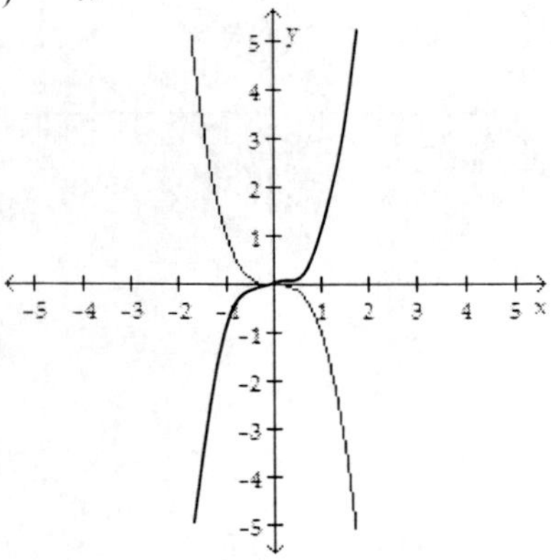

65. no; (0, 2), (0, –2)

67. yes

69. yes

71. no; (3, 0), (3, 1)

73. $f(x) = x^2 + 8$

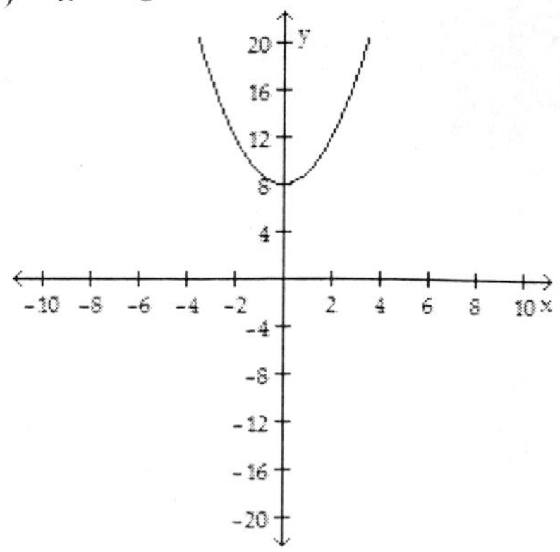

75. $f(x) = | x + 5 |$

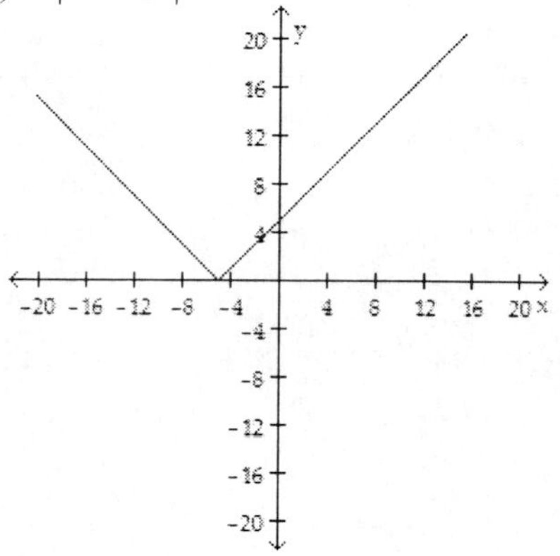

77. $f(x) = (x - 6)^2$

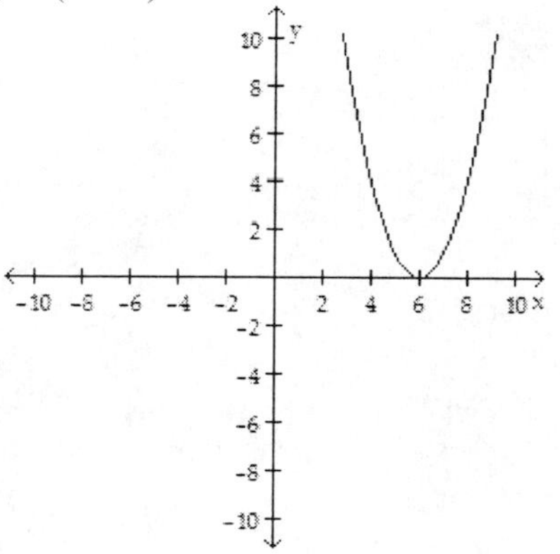

79. $f(x) = x^3 + 8$

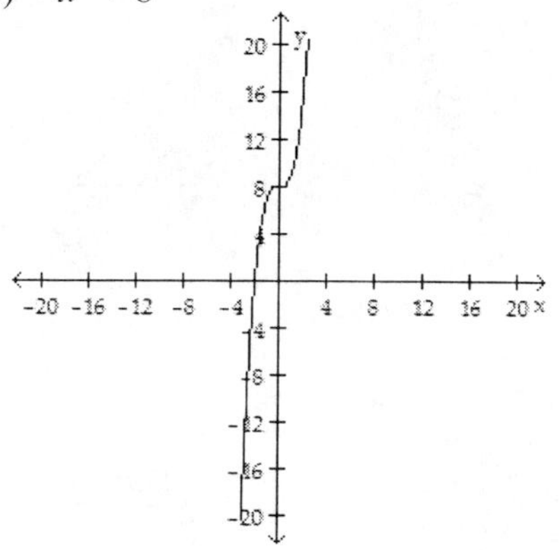

APPLICATIONS

81. OPTICS
$f(x) = | x |$

83. CENTER OF GRAVITY
a parabola

WRITING

85. Answers will vary.

87. Answers will vary.

89. Answers will very.

REVIEW

91.

$$\begin{aligned} T - W &= ma \\ T - W - T &= ma - T \\ -W &= ma - T \\ \frac{-W}{-1} &= \frac{ma - T}{-1} \\ W &= -ma + T \\ W &= T - ma \end{aligned}$$

93.

$$\begin{aligned} s &= \frac{1}{2}gt^2 + vt \\ s - vt &= \frac{1}{2}gt^2 + vt - vt \\ s - vt &= \frac{1}{2}gt^2 \\ 2(s - vt) &= 2\left(\frac{1}{2}gt^2\right) \\ 2(s - vt) &= gt^2 \\ \frac{2(s - vt)}{t^2} &= \frac{gt^2}{t^2} \\ \frac{2(s - vt)}{t^2} &= g \end{aligned}$$

CHALLENGE PROBLEMS

95. $f(x)=\begin{cases}|x| \text{ for } x \geq 0 \\ x^3 \text{ for } x < 0\end{cases}$

x	y
-3	$(\mathbf{-3})^3=-27$
-2	$(\mathbf{-2})^3=-8$
-1	$(\mathbf{-1})^3=-1$
0	$\lvert\mathbf{0}\rvert=0$
1	$\lvert\mathbf{1}\rvert=1$
2	$\lvert\mathbf{2}\rvert=2$
3	$\lvert\mathbf{3}\rvert=3$

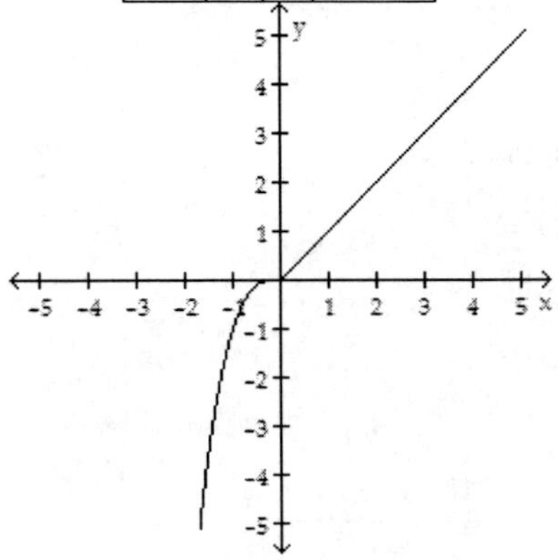

97. a. D: the set of all real numbers from –4 to 4
R: {–2, 1, 3}

b. D: the set of all real numbers except –3 and 1
R: the set of all real numbers

CHAPTER 2 REVIEW

SECTION 2.1
The Rectangular Coordinate System

1.

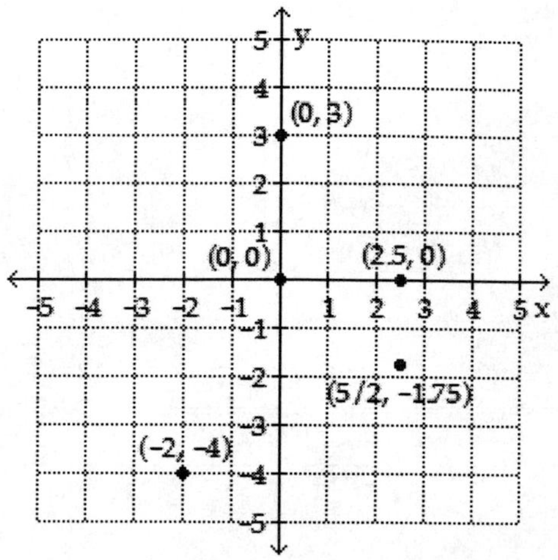

2. a. 1 foot below its normal level
   b. decreased by 3 feet (4 – 1 = 3)
   c. from day 3 to the beginning of day 4

3. a. $10 increments
   b. $800

4.

$$\left(\frac{x_1+x_2}{2},\frac{y_1+y_2}{2}\right)=\left(\frac{8+6}{2},\frac{-2+(-4)}{2}\right)$$
$$=\left(\frac{14}{2},\frac{-6}{2}\right)$$
$$=(7,-3)$$

SECTION 2.2
Graphing Linear Equations

5. Yes.

$$y=-5x+9$$
$$\mathbf{-6}\stackrel{?}{=}-5(\mathbf{3})+9$$
$$-6\stackrel{?}{=}-15+9$$
$$-6=-6$$

6. a. True, the point is a point on the line.
   b. False, the point is not a point on the line.

7. $y = -3x$

x	$y = -3x$
–3	$-3(\mathbf{-3}) = 9$
0	$-3(\mathbf{0}) = 0$
3	$-3(\mathbf{3}) = -9$

8. $y=\frac{1}{2}x-\frac{5}{2}$

x	$y=\frac{1}{2}x-\frac{5}{2}$
–3	$\frac{1}{2}(\mathbf{-3})-\frac{5}{2}=-\frac{3}{2}-\frac{5}{2}$ $=-\frac{8}{2}$ $=-4$
0	$\frac{1}{2}(\mathbf{0})-\frac{5}{2}=0-\frac{5}{2}$ $=-\frac{5}{2}$
3	$\frac{1}{2}(\mathbf{3})-\frac{5}{2}=\frac{3}{2}-\frac{5}{2}$ $=-\frac{2}{2}$ $=-1$

9. $y = 3x + 4$

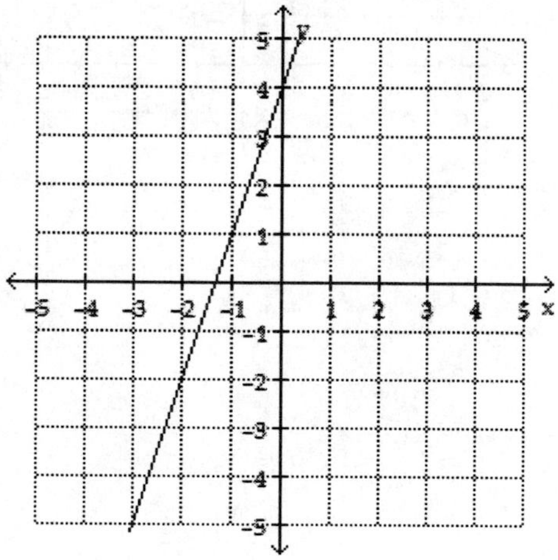

10. $y = -\frac{1}{3}x - 1$

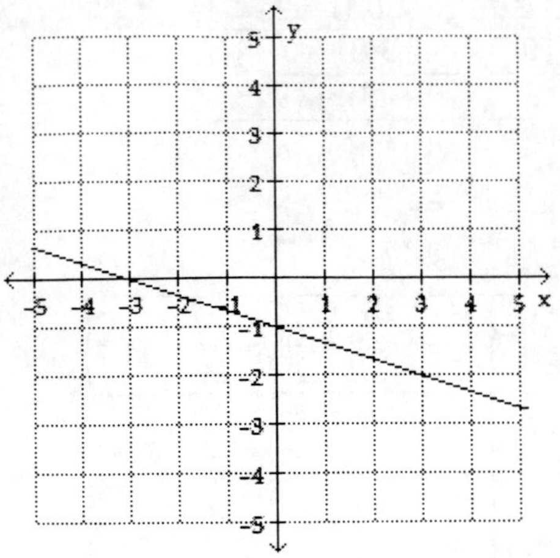

11. $2x + y = 4$

x-intercept:
Let $y = 0$.

$$2x + (0) = 4$$
$$2x = 4$$
$$x = 2$$
$$(2,0)$$

y-intercept:
Let $x = 0$.

$$2(0) + y = 4$$
$$y = 4$$
$$(0,4)$$

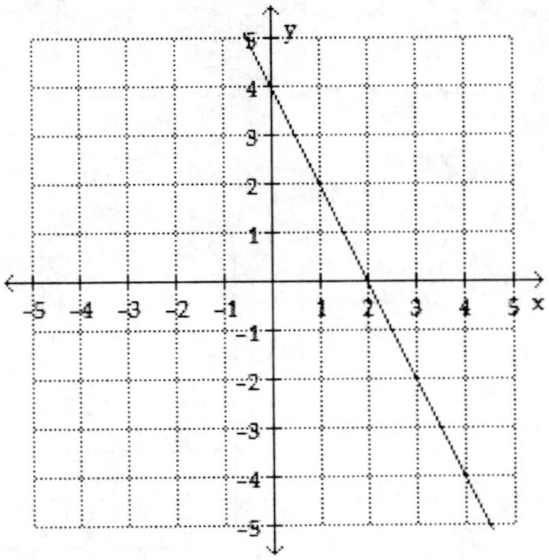

12. $3x - 4y - 8 = 0$

x-intercept:
Let $y = 0$.

$$3x - 4(0) - 8 = 0$$
$$3x - 8 = 0$$
$$3x = 8$$
$$x = 2.\overline{7}$$
$$(2.\overline{7},0)$$

y-intercept:
Let $x = 0$.

$$3(0) - 4y - 8 = 0$$
$$-4y - 8 = 0$$
$$-4y = 8$$
$$y = -2$$
$$(0,-2)$$

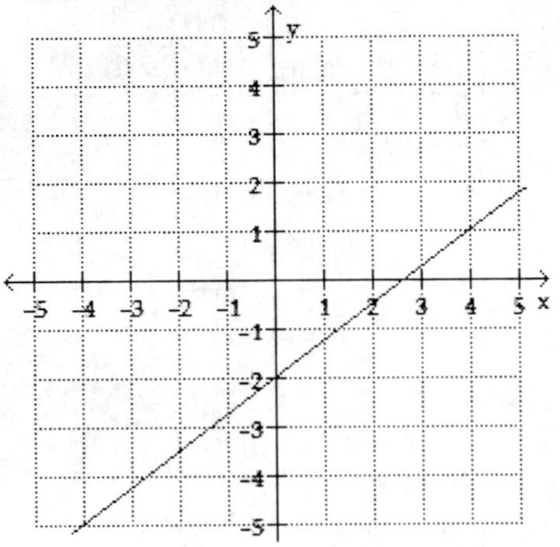

13. $y = 4$

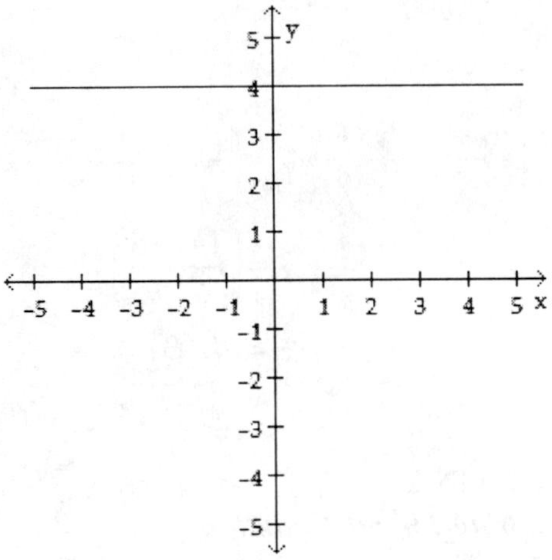

14. $x = -2$

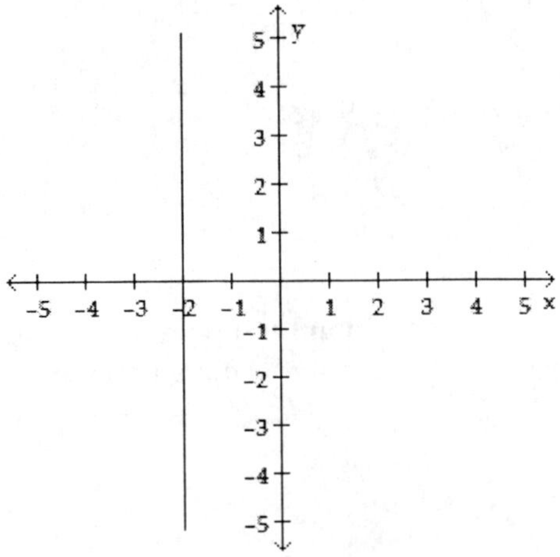

15. a. Let $x = 2$.

$$\begin{aligned} y &= -13{,}800x + 82{,}800 \\ &= -13{,}800(\mathbf{2}) + 82{,}800 \\ &= -27{,}600 + 82{,}800 \\ &= \$55{,}200 \end{aligned}$$

b. Let $y = 0$.

$$\begin{aligned} 0 &= -13{,}800x + 82{,}800 \\ 0 - 82{,}800 &= -13{,}800x + 82{,}800 - 82{,}800 \\ -82{,}800 &= -13{,}800x \\ \frac{-82{,}800}{-13{,}800} &= \frac{-13{,}800x}{-13{,}800} \\ 6 &= x \end{aligned}$$

It will have no value in 6 years.

16. $n = 0.36t + 24.81$

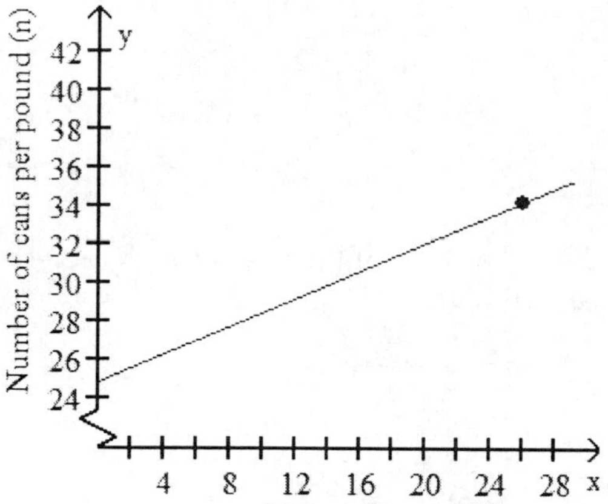

a. In 1980, it took about 25 cans to weigh one pound.

b. about 35

SECTION 2.3
Rate of Change and the Slope of a Line

17. For l_1, use (–3, –2) and (2, 2).

$$\begin{aligned} m &= \frac{y_2 - y_1}{x_2 - x_1} \\ &= \frac{2-(-2)}{2-(-3)} \\ &= \frac{4}{5} \end{aligned}$$

For l_2, use (0, 4) and (5, –4).

$$\begin{aligned} m &= \frac{y_2 - y_1}{x_2 - x_1} \\ &= \frac{-4-4}{5-0} \\ &= -\frac{8}{5} \end{aligned}$$

18. Use the points (2001, 3,520) and (2010, 8,047)

$$\begin{aligned} m &= \frac{y_2 - y_1}{x_2 - x_1} \\ &= \frac{8{,}047 - 3{,}520}{2010 - 2001} \\ &= \frac{4{,}527}{9} \\ &= 503 \text{ per year} \end{aligned}$$

19. (2, 5) and (5, 8)

$$\begin{aligned} m &= \frac{y_2 - y_1}{x_2 - x_1} \\ &= \frac{8-5}{5-2} \\ &= \frac{3}{3} \\ &= 1 \end{aligned}$$

20. (3, –2) and (–6, 12)

$$\begin{aligned} m &= \frac{y_2 - y_1}{x_2 - x_1} \\ &= \frac{12-(-2)}{-6-3} \\ &= \frac{14}{-9} \\ &= -\frac{14}{9} \end{aligned}$$

21. (–2, 4) and (8, 4)

$$\begin{aligned} m &= \frac{y_2 - y_1}{x_2 - x_1} \\ &= \frac{4-4}{8-(-2)} \\ &= \frac{0}{10} \\ &= 0 \end{aligned}$$

22. (–5, –4) and (–5, 8)

$$\begin{aligned} m &= \frac{y_2 - y_1}{x_2 - x_1} \\ &= \frac{8-(-4)}{-5-(-5)} \\ &= \frac{12}{0} \\ &= \text{undefined} \end{aligned}$$

23. Find the slope of the line containing the points (–2, 1) and (6, 5).

$$\begin{aligned} m &= \frac{y_2 - y_1}{x_2 - x_1} \\ &= \frac{5-1}{6-(-2)} \\ &= \frac{4}{8} \\ &= \frac{1}{2} \end{aligned}$$

Since $\frac{1}{2}$ and –2 are opposite reciprocals, the lines are perpendicular.

24. Find the slope of the line containing the points (7, 5) and (–1, 0).

$$\begin{aligned} m &= \frac{y_2 - y_1}{x_2 - x_1} \\ &= \frac{0-5}{-1-7} \\ &= \frac{-5}{-8} \\ &= \frac{5}{8} \end{aligned}$$

Find the slope of the line containing the points (2.2, 3.7) and (10.2, 8.7).

$$\begin{aligned} m &= \frac{y_2 - y_1}{x_2 - x_1} \\ &= \frac{8.7-3.7}{10.2-2.2} \\ &= \frac{5}{8} \end{aligned}$$

Since the slopes are the same, the lines parallel.

25.

$$\begin{aligned} \frac{\text{rise}}{\text{run}} = \frac{4}{1} &= \frac{24}{x} \\ 4x &= 24 \\ \frac{4x}{4} &= \frac{24}{4} \\ x &= 6 \end{aligned}$$

26. Find the slope.

$$\begin{aligned} m &= \frac{\text{rise}}{\text{run}} \\ &= \frac{63}{200} \\ &= 0.315 \\ &= 31.5\% \end{aligned}$$

SECTION 2.4
Writing Equations of Lines

27. $m = 3$, $(x_1, y_1) = (-8, 5)$

$$\begin{aligned} y - y_1 &= m(x - x_1) \\ y - 5 &= 3(x-(-8)) \\ y - 5 &= 3(x+8) \\ y - 5 &= 3x + 24 \\ y - 5 + 5 &= 3x + 24 + 5 \\ y &= 3x + 29 \end{aligned}$$

28. Find the slope of the line through (–2, 4) and (6, –9).

$$\begin{aligned} m &= \frac{y_2 - y_1}{x_2 - x_1} \\ &= \frac{-9-4}{6-(-2)} \\ &= \frac{-13}{8} \end{aligned}$$

$m = -\frac{13}{8}$, $(x_1, y_1) = (-2, 4)$

$$\begin{aligned} y - y_1 &= m(x - x_1) \\ y - 4 &= -\frac{13}{8}(x-(-2)) \\ y - 4 &= -\frac{13}{8}(x+2) \\ y - 4 &= -\frac{13}{8}x - \frac{13}{4} \\ y - 4 + 4 &= -\frac{13}{8}x - \frac{13}{4} + 4 \\ y &= -\frac{13}{8}x + \frac{3}{4} \end{aligned}$$

29. To find the slope of $3x - 2y = 7$, solve the equation for y.

$$3x - 2y = 7$$
$$-2y = -3x + 7$$
$$y = \frac{3}{2}x - \frac{7}{2}$$
$$m = \frac{3}{2}$$

Parallel lines have the same slope.

$m = \frac{3}{2}$, $(x_1, y_1) = (-3, -5)$

$$y - y_1 = m(x - x_1)$$
$$y - (-5) = \frac{3}{2}(x - (-3))$$
$$y + 5 = \frac{3}{2}(x + 3)$$
$$y + 5 = \frac{3}{2}x + \frac{9}{2}$$
$$y + 5 - 5 = \frac{3}{2}x + \frac{9}{2} - 5$$
$$y = \frac{3}{2}x - \frac{1}{2}$$

30. To find the slope of $3x - 2y = 7$, solve the equation for y.

$$3x - 2y = 7$$
$$-2y = -3x + 7$$
$$y = \frac{3}{2}x - \frac{7}{2}$$
$$m = \frac{3}{2}$$

Perpendicular lines have slopes that are opposite reciprocals. The opposite reciprocal of $\frac{3}{2}$ is $-\frac{2}{3}$.

$m = -\frac{2}{3}$, $(x_1, y_1) = (-3, -5)$

$$y - y_1 = m(x - x_1)$$
$$y - (-5) = -\frac{2}{3}(x - (-3))$$
$$y + 5 = -\frac{2}{3}(x + 3)$$
$$y + 5 = -\frac{2}{3}x - 2$$
$$y + 5 - 5 = -\frac{2}{3}x - 2 - 5$$
$$y = -\frac{2}{3}x - 7$$

31. Solve the equation for y.

$$3x + 4y = -12$$
$$3x + 4y - 3x = -12 - 3x$$
$$4y = -3x - 12$$
$$\frac{4y}{4} = \frac{-3x}{4} - \frac{12}{4}$$
$$y = -\frac{3}{4}x - 3$$
$$y = mx + b$$
$$m = -\frac{3}{4};\ (0, -3)$$

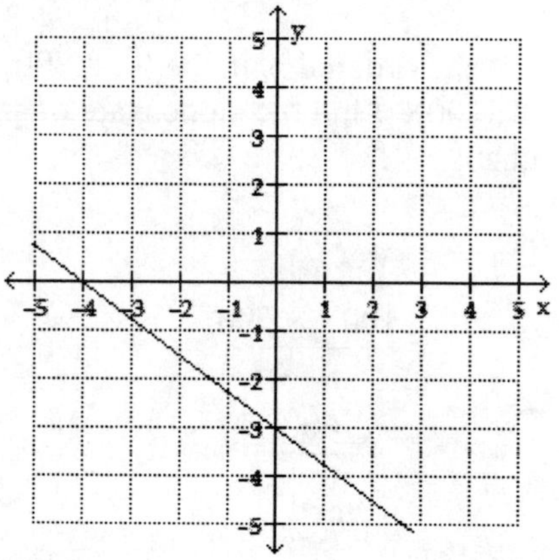

32. The y-intercept (b) is (0, 4) and the slope is $-\frac{4}{5}$.

Using $y = mx + b$, the equation of the line is $y = -\frac{4}{5}x + 4$.

33. a. $y = 0$
b. $x = 0$

34. Find the slope of both lines.

$$y = x + 15$$
$$m = 1$$

$$x + y = 4$$
$$y = -x + 4$$
$$m = -1$$

Since the slopes are opposite reciprocals, the lines are perpendicular.

35. $(p, L) = (21{,}000, 250)$ and $m = \dfrac{1}{150}$.

$$L - L_1 = m(p - p_1)$$
$$L - 250 = \frac{1}{150}(p - 21{,}000)$$
$$L - 250 = \frac{1}{150}p - 140$$
$$L - 250 + 250 = \frac{1}{150}p - 140 + 250$$
$$L = \frac{1}{150}p + 110$$

36. a. Let the ordered pairs be (0, 8,700) and (5, 100). Find the slope (rate of change).

$$m = \frac{y_2 - y_1}{x_2 - x_1} = \frac{100 - 8{,}700}{5 - 0} = \frac{-8{,}600}{5} = -1{,}720$$

Use (0, 8,700) and $m = -1{,}720$.

$$y - y_1 = m(x - x_1)$$
$$y - 8{,}700 = -1{,}720(x - 0)$$
$$y - 8{,}700 = -1{,}720x - 0$$
$$y - 8{,}700 + 8{,}700 = -1{,}720x - 0 + 8{,}700$$
$$y = -1{,}720x + 8{,}700$$

b. (0, 8,700) gives the value of the saw blade when it was new: \$8,700.

SECTION 2.5
An Introduction to Functions

37. D: {–4, –1, 2, 5}
R: {–2, 0, 16}

38. a. A **function** is a set of ordered pairs (a relation) in which to each first component there corresponds exactly one second component.
b. Given a relation in x and y, if to each value of x in the domain there corresponds exactly one value of y in the range, y is said to be a **function** of x. We call x the independent **variable** and y the **dependent** variable.

39. a. yes
b. no; (–1, 8), (–1, 9)
c. yes

40.
$$f(x) = -3x^3 - 7x^2 + 4$$
$$f(-2) = -3(\mathbf{-2})^3 - 7(\mathbf{-2})^2 + 4 = -3(-8) - 7(4) + 4 = 24 - 28 + 4 = 0$$

41. yes

41. yes

43. no; (25, 5) and (25, –5)

44. no; (3, 4) and (3, –4)

45.
$$f(x) = 3x + 2$$
$$f(-3) = 3(\mathbf{-3}) + 2 = -9 + 2 = -7$$

46.
$$g(a) = \frac{a^2 - 4a + 4}{2}$$
$$g(8) = \frac{(\mathbf{8})^2 - 4(\mathbf{8}) + 4}{2} = \frac{64 - 32 + 4}{2} = \frac{36}{2} = 18$$

47.
$$g(a) = \frac{a^2 - 4a + 4}{2}$$
$$g(-2) = \frac{(\mathbf{-2})^2 - 4(\mathbf{-2}) + 4}{2} = \frac{4 + 8 + 4}{2} = \frac{16}{2} = 8$$

48. $f(x) = 3x + 2$

$$\begin{aligned} f(t+2) &= 3(t+2) + 2 \\ &= 3t + 6 + 2 \\ &= 3t + 8 \end{aligned}$$

49. $f(x) = -8$

$$\begin{aligned} -5x + 7 &= -8 \\ -5x + 7 - 7 &= -8 - 7 \\ -5x &= -15 \\ x &= 3 \end{aligned}$$

50. $g(t) = \frac{3}{4}t - 1$

$$\begin{aligned} \frac{3}{4}x - 1 &= 0 \\ 4\left(\frac{3}{4}x - 1\right) &= 4(0) \\ 3x - 4 &= 0 \\ 3x - 4 + 4 &= 0 + 4 \\ 3x &= 4 \\ x &= \frac{4}{3} \end{aligned}$$

51. D: the set of real numbers

52. D: the set of real numbers

53. D: the set of real numbers except 2 (2 makes the denominator 0 and undefined)

54. D: the set of real numbers except –5

55. When written in function form, the coefficient of x is the slope and the constant is the y-intercept. So, for $g(x) = -2x - 16$, the slope is –2 and the y-intercept is (0, –16).

56. $f(x) = \frac{2}{3}x - 2$

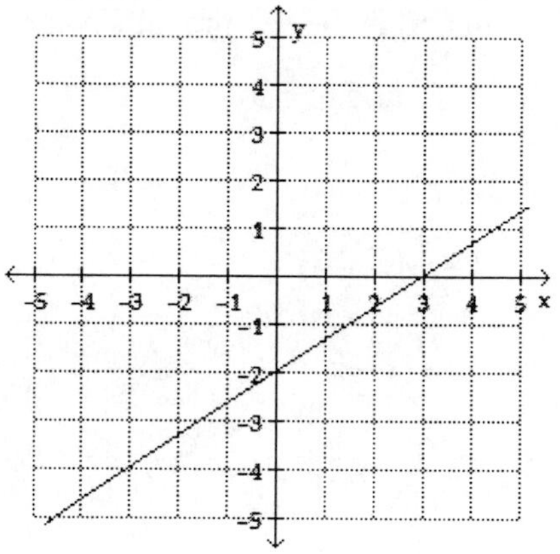

57. $m = \frac{9}{10}$ and $b = \frac{7}{8}$

$$\begin{aligned} y &= mx + b \\ y &= \frac{9}{10}x + \frac{7}{8} \end{aligned}$$

58. $m = \frac{1}{5}$, $x_1 = 10, y_1 = 1$

$$\begin{aligned} y - y_1 &= m(x - x_1) \\ y - 1 &= \frac{1}{5}(x - 10) \\ y - 1 &= \frac{1}{5}x - 2 \\ y - 1 + 1 &= \frac{1}{5}x - 2 + 1 \\ y &= \frac{1}{5}x - 1 \\ f(x) &= \frac{1}{5}x - 1 \end{aligned}$$

59. A horizontal line through (–8, 12) has an equation of $y = 12$, so $f(x) = 12$.

60. The slope of $g(x) = 4x - 70$ is 4. Parallel lines have identical slopes, so use $m = 4$, $x_1 = 2$, and $y_1 = 5$.

$$\begin{aligned} y - y_1 &= m(x - x_1) \\ y - 5 &= 4(x - 2) \\ y - 5 &= 4x - 8 \\ y - 5 + 5 &= 4x - 8 + 5 \\ y &= 4x - 3 \\ f(x) &= 4x - 3 \end{aligned}$$

61. The slope of $g(x) = -3x - 120$ is –3. The slope of a line perpendicular has an opposite sign and is the reciprocal, so use $m = \frac{1}{3}$, $x_1 = -6$, and $y_1 = 3$.

$$y - y_1 = m(x - x_1)$$
$$y - 3 = \frac{1}{3}(x + 6)$$
$$y - 3 = \frac{1}{3}x + 2$$
$$y - 3 + 3 = \frac{1}{3}x + 2 + 3$$
$$y = \frac{1}{3}x + 5$$
$$f(x) = \frac{1}{3}x + 5$$

62. a. Find the slope using (10, 5.25) and (30, 5.65).

$$m = \frac{y_2 - y_1}{x_2 - x_1} = \frac{5.65 - 5.25}{30 - 10} = \frac{0.4}{20} = 0.02$$

Use $m = 0.02$, and $(x_1, y_1) = (10, 5.25)$

$$y - y_1 = m(x - x_1)$$
$$y - 5.25 = 0.02(x - 10)$$
$$y - 5.25 = 0.02x - 0.2$$
$$y - 5.25 + 5.25 = 0.02x - 0.2 + 5.25$$
$$y = 0.02x + 5.05$$
$$R(t) = 0.02t + 5.05$$

b. $R(100) = 0.02(100) + 5.05$
$= 2 + 5.05$
$= 7.05$ milliohms

SECTION 2.6
Graphs of Functions

63. a. –4
b. 3
c. $x = 1$

64. a. 4
b. 1
c. $x = -4$ and $x = 2$

65. D: the set of real numbers
R: the set of real numbers

66. D: the set of real numbers
R: the set of real numbers greater than or equal to 1

67. $f(x) = |x + 2|$

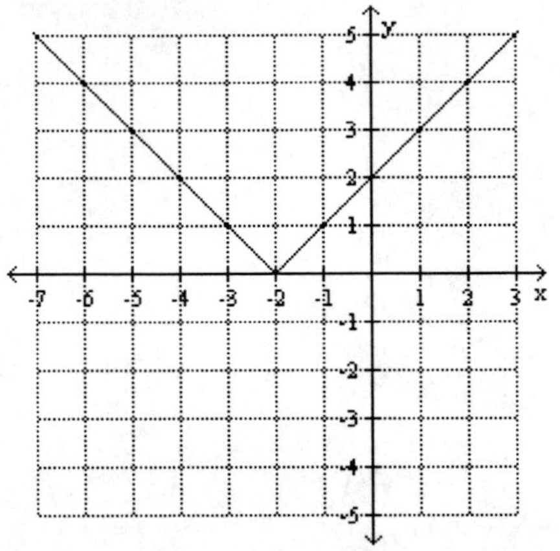

D: the set of all real numbers
R: the set of positive real numbers

68. a. The graph of $f(x) = x^2 + 6$ is the same as the graph of $f(x) = x^2$ except that it is shifted **6** units **up**.
b. The graph of $f(x) = (x + 6)^2$ is the same as the graph of $f(x) = x^2$ except that it is shifted **6** units **left**.

69. $g(x) = x^2 - 3$

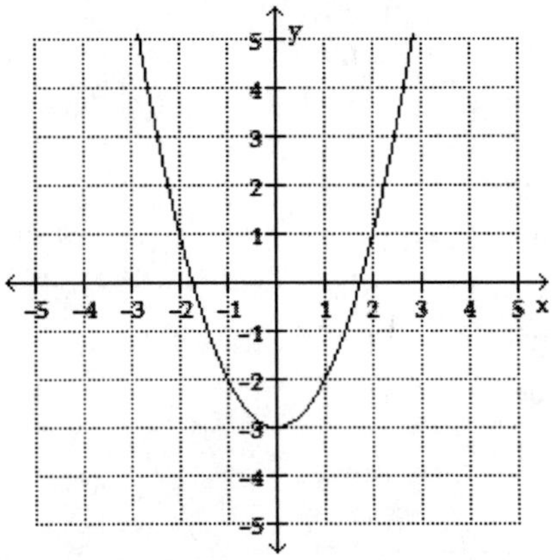

D: the set of all real numbers
R: the set of real numbers greater than or equal to –3

70. $g(x)=|x-4|$

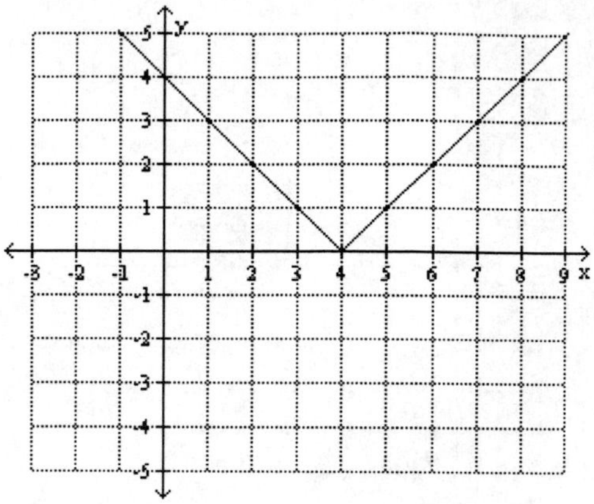

D: the set of all real numbers
R: the set of positive real numbers

71. $g(x)=(x-2)^3+1$

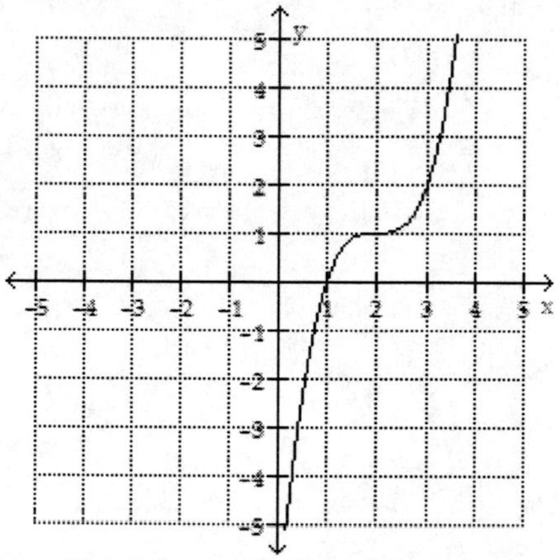

D: the set of all real numbers
R: the set of all real numbers

72. $g(x)=-x^3$

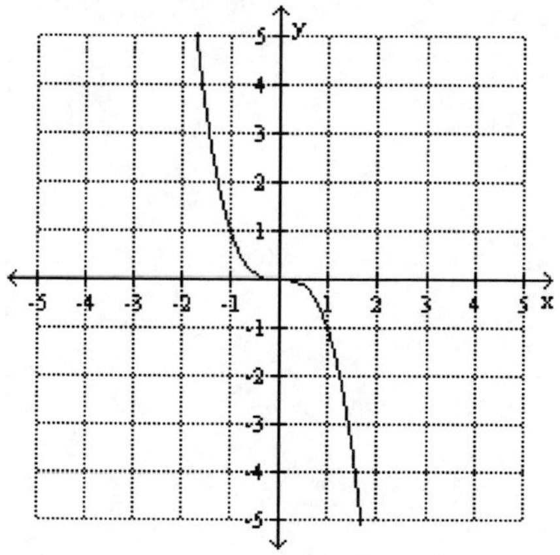

D: the set of all real numbers
R: the set of all real numbers

73. function

74. not a function; (0, 2) and (0, 3)

CHAPTER 2 TEST

1. a. Ordered pairs of numbers can be graphed on a **rectangular** coordinate system.
   b. A **function** is a set of ordered pairs (a relation) in which to each first component there corresponds exactly one second component.
   c. Given a relation in x and y, if to each value of x in the domain there corresponds exactly one value of y in the range, y is said to be a **function** of x. We call x the independent **variable** and y the **dependent** variable.
   d. For a function, the set of all possible values that can be used for the independent variable is called the **domain**. The set of all values of the dependent variable is called the **range**.
   e. A shift of the graph of a function to the left or right is called a horizontal **translation.**

2. a. 11 p.m.
   b. 10 a.m., 7 p.m.
   c. 4 hr
   d. 10 a.m. to 3 p.m.; 7 p.m. to 11 p.m.
   e. 3 p.m.

3. a. \$40
   b. \$80
   c. 3 days

4.

$$\left(\frac{x_1+x_2}{2},\frac{y_1+y_2}{2}\right)=\left(\frac{-2+7}{2},\frac{-5+-11}{2}\right)$$
$$=\left(\frac{5}{2},\frac{-16}{2}\right)$$
$$=\left(\frac{5}{2},-8\right)$$

5. $y=-\frac{2}{3}x-2$

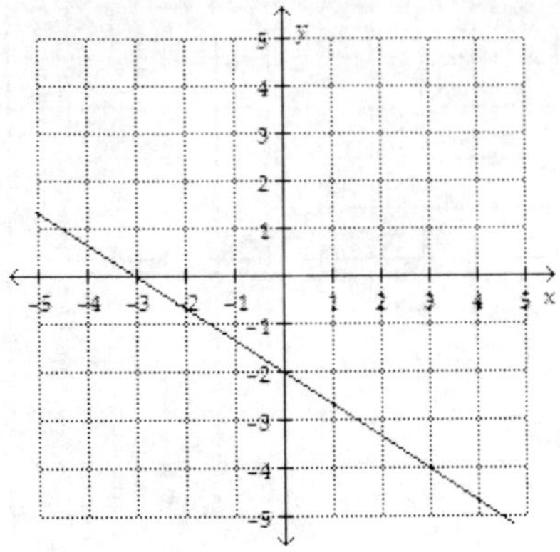

6. $2x-3y=10$

x-intercept:
Let $y=0$.
$$2x-3(0)=10$$
$$2x=10$$
$$x=5$$
$$(5,0)$$

y-intercept:
Let $x=0$.
$$2(0)-3y=10$$
$$-3y=10$$
$$y=-\frac{10}{3}$$
$$\left(0,-\frac{10}{3}\right)$$

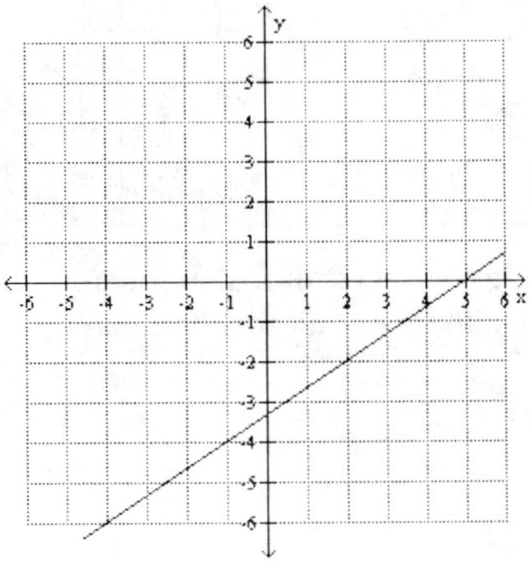

7. $y=-2$

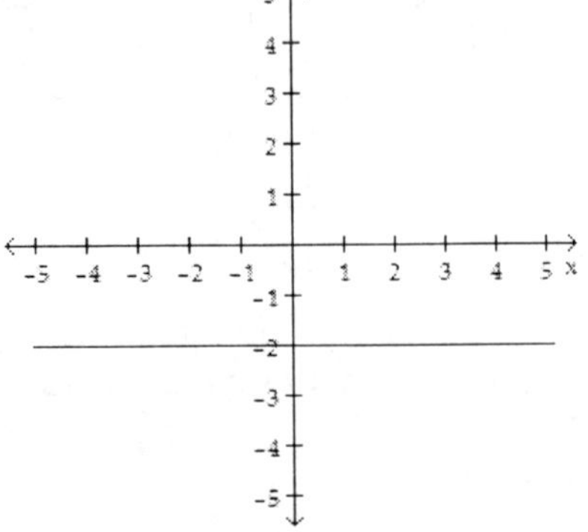

8. Use the ordered pairs (0, 35) and (10, 20).

$$m = \frac{y_2 - y_1}{x_2 - x_1}$$
$$= \frac{20-35}{10-0}$$
$$= \frac{-15}{10}$$
$$= -1.5 \text{ degrees per hour}$$

9. (–2, 4) and (6, 8)

$$m = \frac{y_2 - y_1}{x_2 - x_1}$$
$$= \frac{8-4}{6-(-2)}$$
$$= \frac{4}{8}$$
$$= \frac{1}{2}$$

10. Solve the equation for y.

$$2x - 3y = 8$$
$$2x - 3y - 2x = 8 - 2x$$
$$-3y = -2x + 8$$
$$\frac{-3y}{-3} = \frac{-2x}{-3} + \frac{8}{-3}$$
$$y = \frac{2}{3}x - \frac{8}{3}$$
$$y = mx + b$$
$$m = \frac{2}{3}$$

11. a. $x = 12$ is a vertical line. The slope of a vertical line is always **undefined**.

b. $y = 12$ is a horizontal line. The slope of a horizontal line is always **0**.

12. $m = 3$, $b = (0, 1)$

$y = mx + b$

$y = 3x + 1$

13. Find the slope.

$$m = \frac{y_2 - y_1}{x_2 - x_1}$$
$$= \frac{-10-6}{-4-(-2)}$$
$$= \frac{-16}{-2}$$
$$= 8$$

Use $m = 8$ and (–2, 6).

$$y - y_1 = m(x - x_1)$$
$$y - 6 = 8(x - (-2))$$
$$y - 6 = 8(x + 2)$$
$$y - 6 = 8x + 16$$
$$y - 6 + 6 = 8x + 16 + 6$$
$$y = 8x + 22$$

14. Solve the equation for y.

$$4x - 25 = -5y$$
$$\frac{4x}{-5} - \frac{25}{-5} = \frac{-5y}{-5}$$
$$-\frac{4}{5}x + 5 = y$$
$$y = -\frac{4}{5}x + 5$$
$$m = -\frac{4}{5}, b = (0, 5)$$

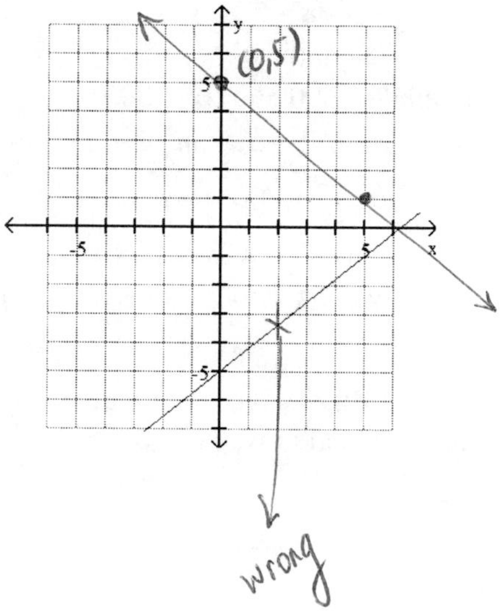

15. Find the slope of both equations.

$$y + 4x = 12 \qquad y = \frac{1}{4}x + 3$$
$$y = -4x + 12$$
$$m = -4 \qquad m = \frac{1}{4}$$

The slopes are opposite reciprocals, so the lines are perpendicular.

16. The slope of $y = -\frac{3}{2}x - 7$ is $-\frac{3}{2}$. The slope of the line parallel to that would also be $-\frac{3}{2}$. The y-intercept is 0 if it passes through the origin.

$$y = -\frac{3}{2}x + 0$$
$$y = -\frac{3}{2}x$$

17. a. Find the rate of change (m) using the ordered pairs (0, 4,000) and (6, 400).

$$m = \frac{y_2 - y_1}{x_2 - x_1}$$
$$= \frac{400 - 4{,}000}{6 - 0}$$
$$= \frac{-3{,}600}{6}$$
$$= -600$$

Use $m = -600$ and (0, 4,000)

$$y - y_1 = m(x - x_1)$$
$$y - 4{,}000 = -600(x - 0)$$
$$y - 4{,}000 = -600x$$
$$y - 4{,}000 + 4{,}000 = -600x + 4{,}000$$
$$y = -600x + 4{,}000$$

b. (0, 4,000); It gives the value of the copier when new, $4,000.

18. The rate of change, or the slope would be $\frac{11}{200}$. Use the ordered pair (20,000, 2,500) and the slope to find the linear equation.

$$y - y_1 = m(x - x_1)$$
$$y - 2{,}500 = \frac{11}{200}(x - 20{,}000)$$
$$y - 2{,}500 = \frac{11}{200}x - 100$$
$$y - 2{,}500 + 2{,}500 = \frac{11}{200}x - 1{,}100 + 2{,}500$$
$$y = \frac{11}{200}x + 1{,}400$$
$$n = \frac{11}{200}p + 1{,}400$$

19. a. no; (1, –8) and (1, 6)
b. yes
c. yes
d. no; (4, –4) and (4, 4)

20. the set of all real numbers except for 6

21.

$$-\frac{4}{5}x - 12 = 4$$
$$5\left(-\frac{4}{5}x - 12\right) = 5(4)$$
$$-4x - 60 = 20$$
$$-4x - 60 + 60 = 20 + 60$$
$$-4x = 80$$
$$x = -20$$

22. $f(x) = 8x - 9$ so $m = 8$ and y-intercept is (0, –9)

23. The slope of $g(x)=\frac{4}{5}x+\frac{1}{9}$ is $\frac{4}{5}$. The slope of a line perpendicular has slope that is opposite reciprocals, so the slope to use is $-\frac{5}{4}$.

Use $m=-\frac{5}{4}$ and (2, 0).

$$y-y_1=m(x-x_1)$$
$$y-0=-\frac{5}{4}(x-2)$$
$$y=-\frac{5}{4}x+\frac{5}{2}$$
$$f(x)=-\frac{5}{4}x+\frac{5}{2}$$

24. a. Use (1980, 14.5) and (2005, 10.5) to find the slope.

$$m=\frac{y_2-y_1}{x_2-x_1}$$
$$=\frac{10.5-14.5}{2005-1980}$$
$$=\frac{-4}{25}$$
$$=-0.16$$

Use $m=-0.16$ and (1980, 14.5)

$$y-y_1=m(x-x_1)$$
$$y-14.5=-0.16(x-1980)$$
$$y-14.5=-0.16x+316.8$$
$$y-14.5+14.5=-0.16x+316.8+14.5$$
$$y=-0.16x+331.3$$
$$T(m)=-0.16m+331.3$$

b. Let $m = 2015$.

$$T(2015)=-0.16(2015)+331.3$$
$$=-322.4+331.3$$
$$=8.9$$

25.

$$f(3)=3(3)+1$$
$$=9+1$$
$$=10$$

26.

$$g(-6)=(-6)^2-2(-6)+1$$
$$=36+12+1$$
$$=48+1$$
$$=49$$

27.

$$g\left(\frac{1}{4}\right)=\left(\frac{1}{4}\right)^2-2\left(\frac{1}{4}\right)+1$$
$$=\frac{1}{16}-\frac{1}{2}+1$$
$$=\frac{1}{16}-\frac{8}{16}+\frac{16}{16}$$
$$=\frac{9}{16}$$

28.

$$f(r+8)=3(r+8)+1$$
$$=3r+24+1$$
$$=3r+25$$

29. When $x = -2$, $\underline{\mathbf{y = -2}}$.

30. $\underline{\mathbf{x = 2}}$ when $y = 3$.

31. function

32. not a function; (2, 2) and (2, –2)

33. $f(x)=|x|+3$

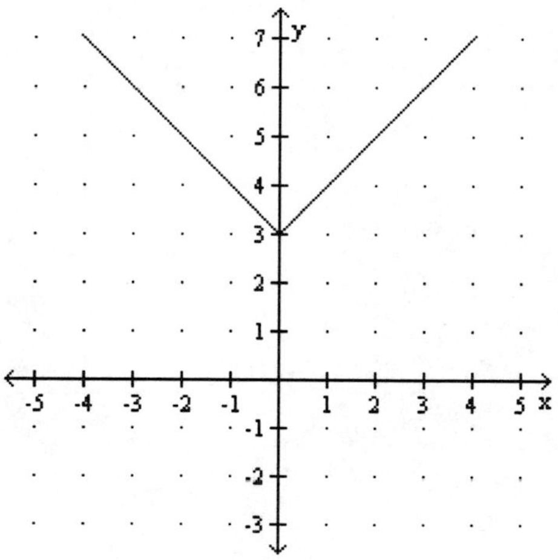

D: the set of real numbers
R: the set of real numbers greater than or equal to 3

34. $g(x)=(x-4)^3+1$

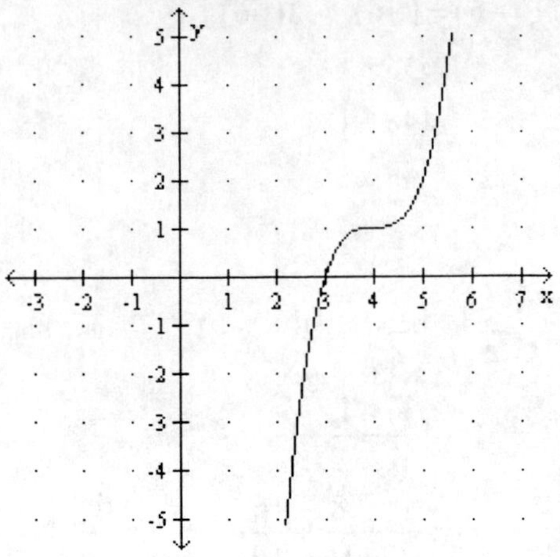

35. D: the set of real numbers
    R: the set of nonnegative real numbers

36. $f(x)=-x^2+1$

CHAPTER 2 CUMULATIVE REVIEW

1. a. 1, 2, 6, 7
   b. 0, 1, 2, 6, 7
   c. $-2, 0, 1, 2, \frac{13}{12}, 6, 7$
   d. $\sqrt{5}, \pi$
   e. -2
   f. $-2, 0, 1, 2, \frac{13}{12}, 6, 7, \sqrt{5}, \pi$
   g. 2, 7
   h. $-2, 0, 2, 6$

2. $-\frac{35}{8} \quad -\pi \quad -1\frac{1}{2} \quad -0.333... \quad \sqrt{2} \quad 3 \quad 4.25$

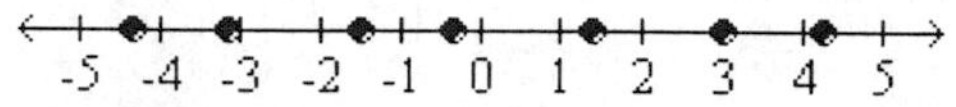

3.
$$\frac{|-5|+|-3|}{-|4|} = \frac{5+3}{-4}$$
$$= \frac{8}{-4}$$
$$= -2$$

4.
$$7-12\left[7^2-4(2-5)^2\right]$$
$$= 7-12\left[49-4(-3)^2\right]$$
$$= 7-12\left[49-4(9)\right]$$
$$= 7-12[49-36]$$
$$= 7-12[13]$$
$$= 7-156$$
$$= -149$$

5.
$$-\frac{16}{5} \div \left(-\frac{10}{3}\right) = -\frac{16}{5} \bullet \left(-\frac{3}{10}\right)$$
$$= \frac{48}{50}$$
$$= \frac{24}{25}$$

6.
$$\frac{(9-8)^4+21}{3^3-\left(\sqrt{16}\right)^2} = \frac{(1)^4+21}{3^3-(4)^2}$$
$$= \frac{1+21}{27-16}$$
$$= \frac{22}{11}$$
$$= 2$$

7.
$$-y-5xy = -(-3)-5(2)(-3)$$
$$= 3-10(-3)$$
$$= 3+30$$
$$= 33$$

8.
$$\frac{x^2-y^2}{2x+y} = \frac{(2)^2-(-3)^2}{2(2)+(-3)}$$
$$= \frac{4-9}{4-3}$$
$$= \frac{-5}{1}$$
$$= -5$$

9. a. associative property of addition
   b. distributive property
   c. commutative property of addition
   d. associative property of multiplication

10. a. -6
    b. -8

11. $-7s(-4t)(-1) = -28st$

12.
$$40\left(\frac{3}{8}y-\frac{1}{4}\right)+40\left(\frac{4}{5}\right) = 15y-10+32$$
$$= 15y+22$$

13.
$$-\frac{3}{4}s-\frac{1}{3}s = -\frac{9}{12}s-\frac{4}{12}s$$
$$= -\frac{13}{12}s$$

14. $-5\left[3(x-4)-2(x+2)\right]-7(x-3)$

$$
\begin{aligned}
&=-5[3x-12-2x-4]-7x+21\\
&=-5[x-16]-7x+21\\
&=-5x+80-7x+21\\
&=-12x+101
\end{aligned}
$$

15.

$$
\begin{aligned}
-\frac{9}{8}s&=3\\
-\frac{8}{9}\left(-\frac{9}{8}s\right)&=-\frac{8}{9}(3)\\
s&=-\frac{8}{3}
\end{aligned}
$$

16.

$$
\begin{aligned}
4(y-3)+4&=-3(y+5)\\
4y-12+4&=-3y-15\\
4y-8&=-3y-15\\
4y-8+3y&=-3y-15+3y\\
7y-8&=-15\\
7y-8+8&=-15+8\\
7y&=-7\\
y&=-1
\end{aligned}
$$

17.

$$
\begin{aligned}
2x-\frac{3(x-2)}{2}&=7-\frac{x-3}{3}\\
2x-\frac{3x-6}{2}&=7-\frac{x-3}{3}\\
6\left(2x-\frac{3x-6}{2}\right)&=6\left(7-\frac{x-3}{3}\right)\\
12x-3(3x-6)&=42-2(x-3)\\
12x-9x+18&=42-2x+6\\
3x+18&=-2x+48\\
3x+18+2x&=-2x+48+2x\\
5x+18&=48\\
5x+18-18&=48-18\\
5x&=30\\
x&=6
\end{aligned}
$$

18.

$$
\begin{aligned}
0.04(24)+0.02x&=0.04(12+x)\\
0.96+0.02x&=0.48+0.04x\\
100(0.96+0.02x)&=100(0.48+0.04x)\\
96+2x&=48+4x\\
96+2x-2x&=48+4x-2x\\
96&=48+2x\\
96-48&=48+2x-48\\
48&=2x\\
24&=x
\end{aligned}
$$

19.

$$
\begin{aligned}
-3x&=-2x+1-(5+x)\\
-3x&=-2x+1-5-x\\
-3x&=-3x-4\\
-3x+3x&=-3x-4+3x\\
0&=-4
\end{aligned}
$$

$\varnothing$

No Solution

20.

$$
\begin{aligned}
2\left[5(4-a)+2(a-1)\right]&=3-a\\
2[20-5a+2a-2]&=3-a\\
2[18-3a]&=3-a\\
36-6a&=3-a\\
36-6a+a&=3-a+a\\
36-5a&=3\\
36-5a-36&=3-36\\
-5a&=-33\\
\frac{-5a}{-5}&=\frac{-33}{-5}\\
a&=\frac{33}{5}
\end{aligned}
$$

21.

$$
\begin{aligned}
-Tx+3By&=c\\
-Tx+3By+Tx&=c+Tx\\
3By&=c+Tx\\
\frac{3By}{3y}&=\frac{c+Tx}{3y}\\
B&=\frac{c+Tx}{3y}
\end{aligned}
$$

22.

$$A = \frac{1}{2}h(b_1 + b_2)$$

$$2 \cdot A = 2 \cdot \frac{1}{2}h(b_1 + b_2)$$

$$2A = h(b_1 + b_2)$$

$$\frac{2A}{(b_1 + b_2)} = \frac{h(b_1 + b_2)}{(b_1 + b_2)}$$

$$\frac{2A}{(b_1 + b_2)} = h$$

23. Let x = the number of signatures she collects.

$$250 = 45 + 1.25x$$

$$250 - 45 = 45 + 1.25x - 45$$

$$205 = 1.25x$$

$$\frac{205}{1.25} = \frac{1.25x}{1.25}$$

$$164 = x$$

She needs 164 signatures.

24. Let x = the measure of the vertex angle. Then the measures each base angle is $2x - 10$. The sum of the measures of all three angles is 180 degrees.

$$(2x - 10) + (2x - 10) + x = 180$$

$$5x - 20 = 180$$

$$5x - 20 + 20 = 180 + 20$$

$$5x = 200$$

$$\frac{5x}{5} = \frac{200}{5}$$

$$x = 40$$

$$2x - 10 = 2(40) - 10$$

$$= 80 - 10$$

$$= 70$$

The measures of the angles are 70°, 70°, and 40°.

25.

Account	Principal ·	Rate ·	Time =	Interest
6%	x	0.06	1	$0.06x$
7%	$20{,}000 - x$	0.07	1	$0.07(20{,}000 - x)$
Combined Interest		1,260		

6% Interest + 7% Interest = Combined Interest

$$0.06x + 0.07(20{,}000 - x) = 1{,}260$$

$$0.06x + 1{,}400 - 0.07x = 1{,}260$$

$$-0.01x + 1{,}400 = 1{,}260$$

$$-0.01x + 1{,}400 - 1{,}400 = 1{,}260 - 1{,}400$$

$$-0.01x = -140$$

$$\frac{-0.01x}{-0.01} = \frac{-140}{-0.01}$$

$$x = 14{,}000$$

She invested \$14,000 at 6%.

26.

	Rate ·	Time =	Distance
going	x	5	$5x$
returning	$x + 26$	3	$3(x + 26)$

The distances for both trips are the same (equal).

Going = Returning

$$5x = 3(x + 26)$$

$$5x = 3x + 78$$

$$5x - 3x = 3x + 78 - 3x$$

$$2x = 78$$

$$x = 39$$

$$x + 26 = 39 + 26 = 65$$

39 mph going

65 mph returning

27.

$$\left(\frac{x_1 + x_2}{2}, \frac{y_1 + y_2}{2}\right) = \left(\frac{-5 + 7}{2}, \frac{-2 + 3}{2}\right)$$

$$= \left(\frac{2}{2}, \frac{1}{2}\right)$$

$$= \left(1, \frac{1}{2}\right)$$

28.

$$3y - 5x = 30$$
$$3\left(\frac{25}{3}\right) - 5(-1) \stackrel{?}{=} 30$$
$$25 + 5 \stackrel{?}{=} 30$$
$$30 = 30$$

Yes, the ordered pair is a solution to the equation.

29. $7x - 3y = 6$

$$-3y = -7x + 6$$
$$\frac{-3y}{-3} = \frac{-7x}{-3} + \frac{6}{-3}$$
$$y = \frac{7}{3}x - 2$$

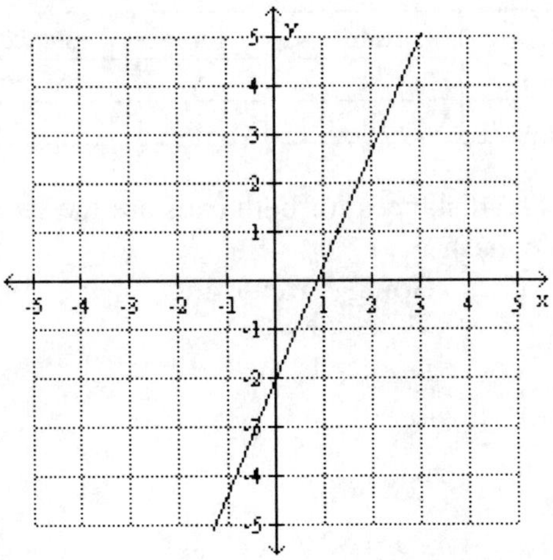

30. Use the ordered pairs (–2, 1) and (4, –4).

$$m = \frac{y_2 - y_1}{x_2 - x_1} = \frac{-4-1}{4-(-2)} = \frac{-5}{6} = -\frac{5}{6}$$

31. Find the slope.

$$m = \frac{y_2 - y_1}{x_2 - x_1} = \frac{-9-5}{8-(-2)} = \frac{-14}{10} = -\frac{7}{5}$$

Use $m = -\frac{7}{5}$ and (–2, 5)

$$y - y_1 = m(x - x_1)$$
$$y - 5 = -\frac{7}{5}(x - (-2))$$
$$y - 5 = -\frac{7}{5}(x + 2)$$
$$y - 5 = -\frac{7}{5}x - \frac{14}{5}$$
$$y - 5 + 5 = -\frac{7}{5}x - \frac{14}{5} + 5$$
$$y = -\frac{7}{5}x + \frac{11}{5}$$

32. Find the slope of $3x + y = 8$.

$$3x + y = 8$$
$$y = -3x + 8$$
$$m = -3$$

Use $m = -3$ and (–2, 3).

$$y - y_1 = m(x - x_1)$$
$$y - 3 = -3(x - (-2))$$
$$y - 3 = -3(x + 2)$$
$$y - 3 = -3x - 6$$
$$y - 3 + 3 = -3x - 6 + 3$$
$$y = -3x - 3$$

33 a. The rate of increase, or the slope would be $\frac{1}{1,000}$. Use the ordered pair (25,000, 50) and the slope to find the linear equation.

$$y - y_1 = m(x - x_1)$$
$$y - 50 = \frac{1}{1,000}(x - 25,000)$$
$$y - 50 = \frac{1}{1,000}x - 25$$
$$y - 50 + 50 = \frac{1}{1,000}x - 25 + 50$$
$$y = \frac{1}{1,000}x + 25$$
$$T = \frac{1}{1,000}p + 25$$

b. Let $p = 35,000$.

$$T = \frac{1}{1,000}p + 25$$
$$= \frac{1}{1,000}(35,000) + 25$$
$$= 35 + 25$$
$$= 60$$

34. Given a relation in x and y, if to each value of x in the domain there corresponds exactly one value of y in the range, then y is said to be a **function**.

35. a. When $x = 1$, **$y = 0$**.
b. When $y = 1$, **$x = 2$**.

36. D: the set of real numbers
R: the set of real numbers

37. no
D: {–6, 0, 1, 5}
R: {–12, 4, 7, 8}

38. no; (4, 2), (4, –2)

39.

$$-\frac{1}{5}x - 12 = 0$$
$$5\left(-\frac{1}{5}x - 12\right) = 5(0)$$
$$-x - 60 = 0$$
$$-x - 60 + 60 = 0 + 60$$
$$-x = 60$$
$$x = -60$$

40. yes

41.

$$f(-1) = 3(-1)^2 + 2$$
$$= 3(1) + 2$$
$$= 3 + 2$$
$$= 5$$

42.

$$g(0) = -2(0) - 1$$
$$= 0 - 1$$
$$= -1$$

43.

$$g(-2) = -2(-2) - 1$$
$$= 4 - 1$$
$$= 3$$

44.

$$f(-r) = 3(-r)^2 + 2$$
$$= 3r^2 + 2$$

45. the set of all real numbers except –1

46. $m = 6$ and $b = (0, 15)$

47. A horizontal line through the point $\left(\frac{3}{2}, -\frac{7}{8}\right)$ has the equation $f(x) = -\frac{7}{8}$.

48. a. Use (10, 5.76) and (25, 8.31) to find the slope.

$$m=\frac{y_2-y_1}{x_2-x_1}$$
$$=\frac{8.31-5.76}{25-10}$$
$$=\frac{2.55}{15}$$
$$=0.17$$

Use $m = 0.17$ and (10, 5.76) to find the linear function.

$$y-y_1=m(x-x_1)$$
$$y-5.76=0.17(x-10)$$
$$y-5.76=0.17x-1.7$$
$$y-5.76+5.76=0.17x-1.7+5.76$$
$$y=0.17x+4.06$$
$$M(t)=0.17t+4.06$$

b. Since 2020-1980 = 40, let $t = 40$.

$$M(40)=0.17(40)+4.06$$
$$=6.8+4.06$$
$$=10.86$$

Approximately 10.86 billion vehicle miles

49. $y = -x^2 + 1$

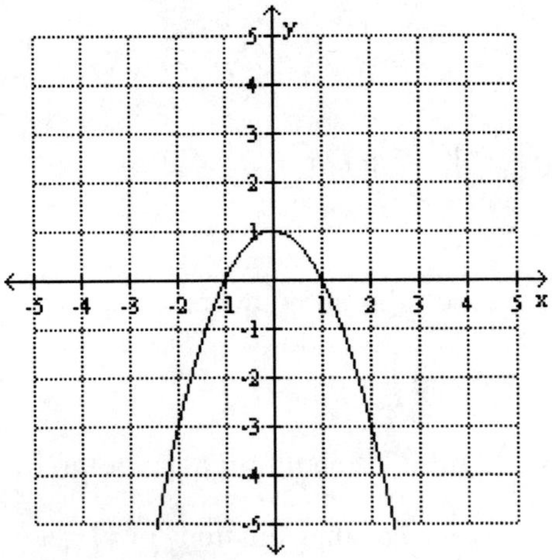

D: the set of real numbers
R: the set of real numbers less than or equal to 1

50. $g(x)=|x-3|-4$

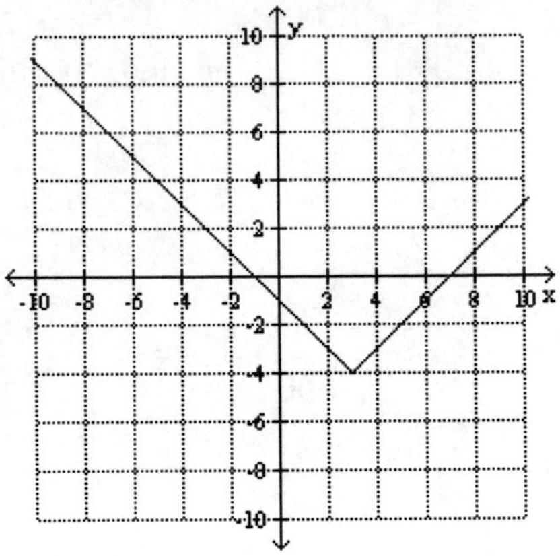

D: the set of real numbers
R: the set of real numbers greater than or equal to –4

SECTION 3.1

VOCABULARY

1. $\begin{cases} x+2y=4 \\ 2x-y=3 \end{cases}$ is called a **system** of linear equations.

3. If two equations have different graphs, they are called **independent** equations. Two equations with the same graph are called **dependent** equations.

CONCEPTS

5. a. true
   b. false
   c. true
   d. true

7. a. no solution; independent
   b. infinitely many solutions
   answers will vary: (–3, 0), (–2, –2), (0, –6); consistent

9. The lines intersect at ***x* = 1**.

NOTATION

11. The symbol { is called a **brace**. It is used when writing a system of equations.

GUIDED PRACTICE

13. Yes.

$$4x-y=-19 \qquad 3x+2y=-6$$
$$4(-4)-3\overset{?}{=}-19 \qquad 3(-4)+2(3)\overset{?}{=}-6$$
$$-16-3\overset{?}{=}-19 \qquad -12+6\overset{?}{=}-6$$
$$-19\overset{?}{=}-19 \qquad -6=-6$$

15. No.

$$y+2=\frac{1}{2}x \qquad 3x+2y=0$$
$$-3+2\overset{?}{=}\frac{1}{2}(2) \qquad 3(2)+2(-3)\overset{?}{=}0$$
$$-1\neq 1 \qquad 6-6\overset{?}{=}0$$
$$0=0$$

17. No.

$$2x+3y=2 \qquad 4x-9y=1$$
$$2\left(\frac{1}{2}\right)+3\left(\frac{1}{3}\right)\overset{?}{=}2 \qquad 4\left(\frac{1}{2}\right)-9\left(\frac{1}{3}\right)\overset{?}{=}1$$
$$1+1\overset{?}{=}2 \qquad 2-3\overset{?}{=}1$$
$$2=2 \qquad -1\neq 1$$

19. Yes.

$$2x+5y=2.1 \qquad 5x+y=-0.5$$
$$2(-0.2)+5(0.5)\overset{?}{=}2.1 \qquad 5(-0.2)+0.5\overset{?}{=}-0.5$$
$$-0.4+2.5\overset{?}{=}2.1 \qquad -1+0.5\overset{?}{=}-0.5$$
$$2.1=2.1 \qquad -0.5=-0.5$$

21. $\begin{cases} x+y=6 \\ x-y=2 \end{cases}$ (4, 2)

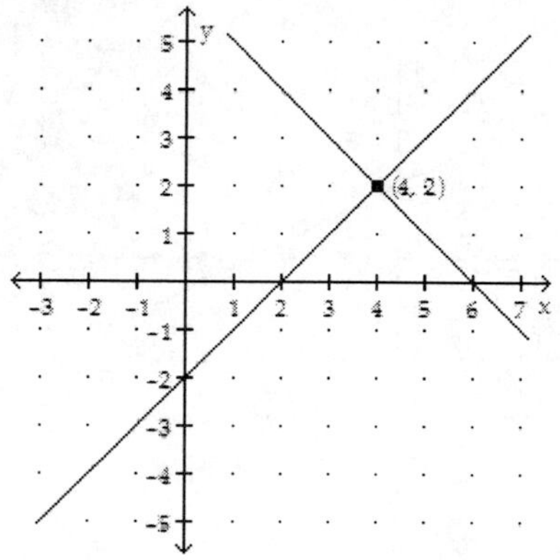

23. $\begin{cases} y=-2x+1 \\ x-2y=-7 \end{cases}$ (–1, 3)

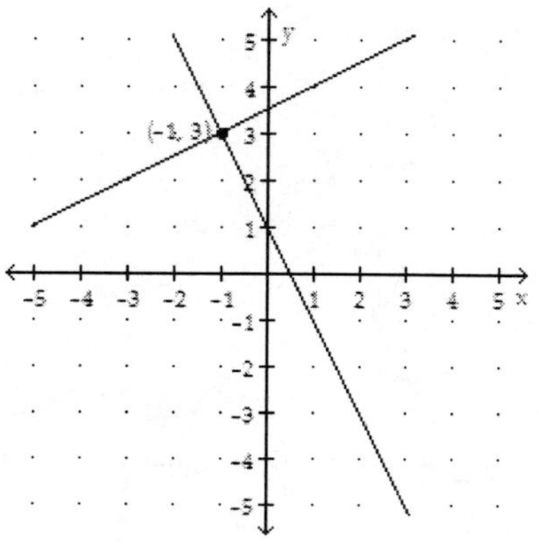

25. $\begin{cases} 3x - 3y = 4 \\ x - y = 4 \end{cases}$

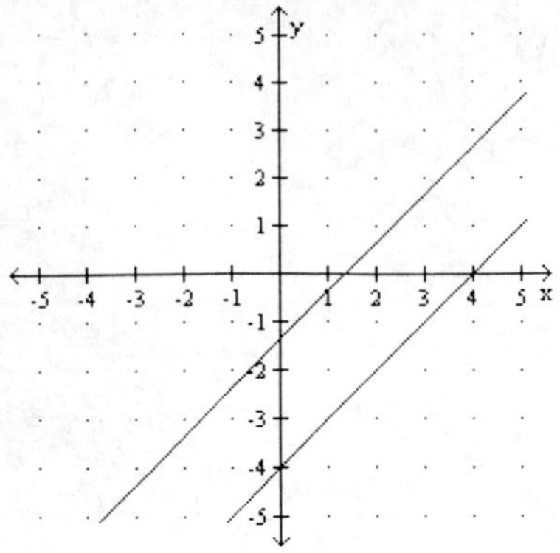

No solution, $\varnothing$, inconsistent system

27. $\begin{cases} x = 3 - 2y \\ 2x + 4y = 6 \end{cases}$

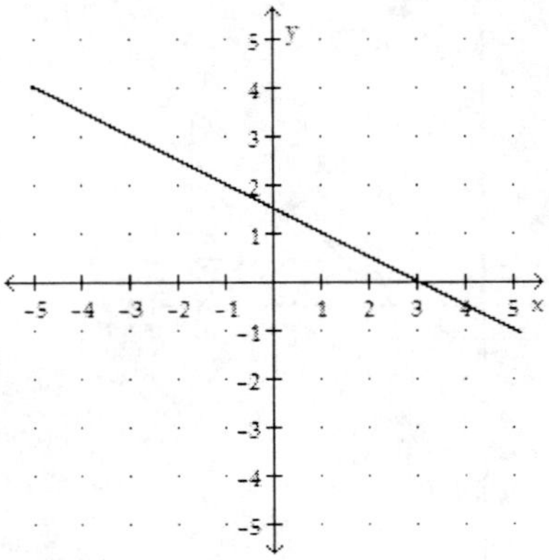

$\{(x, y)|\ x = 3 - 2y\}$ or $\{(x, y)|\ x + 2y = 3\}$

infinitely many solutions;
dependent equations

29. $\begin{cases} \dfrac{1}{6}x = \dfrac{1}{3}y + \dfrac{1}{2} \\ y = x \end{cases}$ $(-3, -3)$

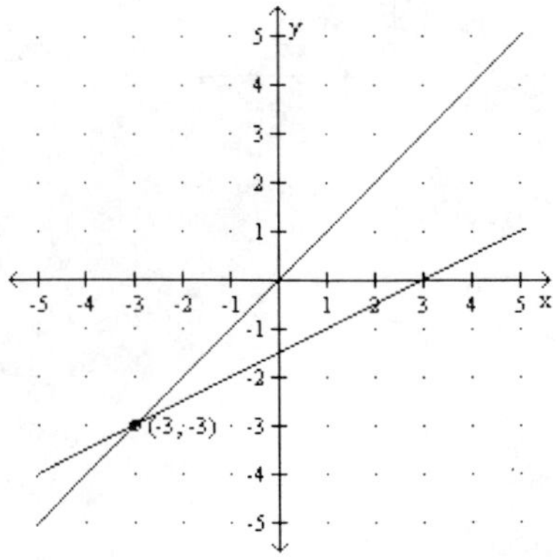

31. $\begin{cases} \dfrac{1}{3}x - \dfrac{7}{6}y = \dfrac{1}{2} \\ \dfrac{1}{5}y = \dfrac{1}{3}x + \dfrac{7}{15} \end{cases}$ $(-2, -1)$

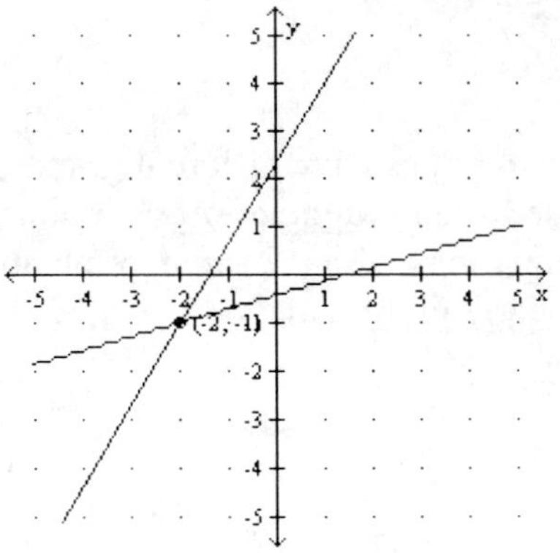

33. Rewrite it as $\begin{cases} y = 2x + 1 \\ y = 5 \end{cases}$, then $x = 2$.

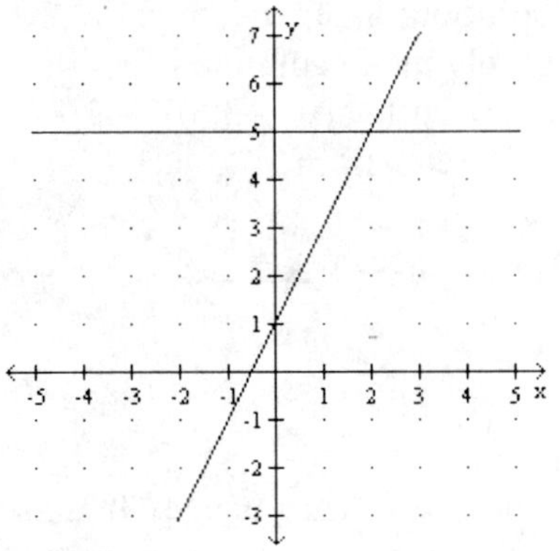

35. Rewrite it as $\begin{cases} y = -2x + 8 \\ y = 3x - 7 \end{cases}$, then $x = 3$.

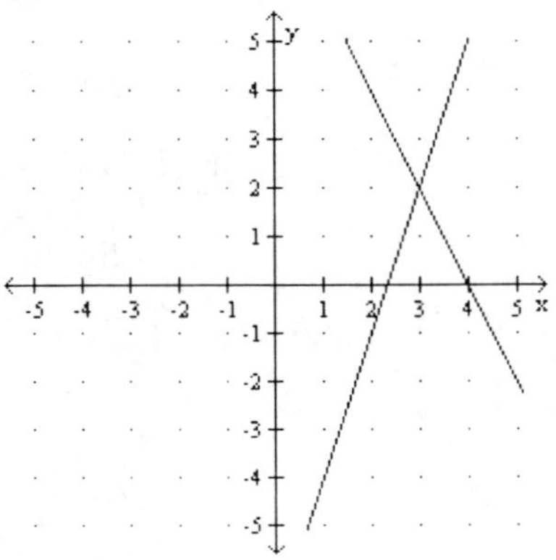

37. $\begin{cases} y = 3.2x - 1.5 \\ y = -2.7x - 3.7 \end{cases}$ $(-0.37, -2.69)$

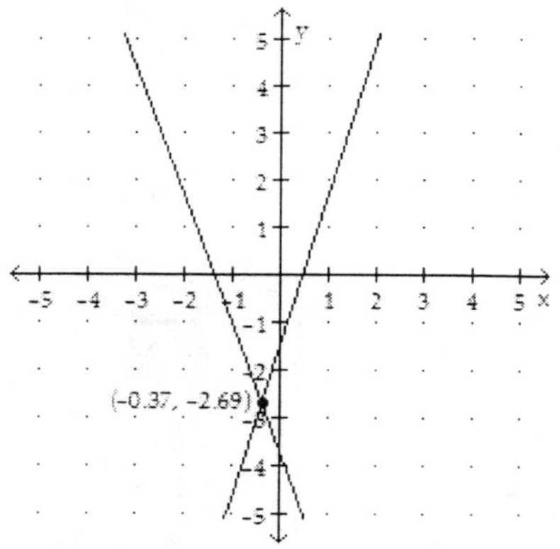

39. $\begin{cases} 1.7x + 2.3y = 3.2 \\ y = 0.25x + 8.95 \end{cases}$ (–7.64, 7.04)

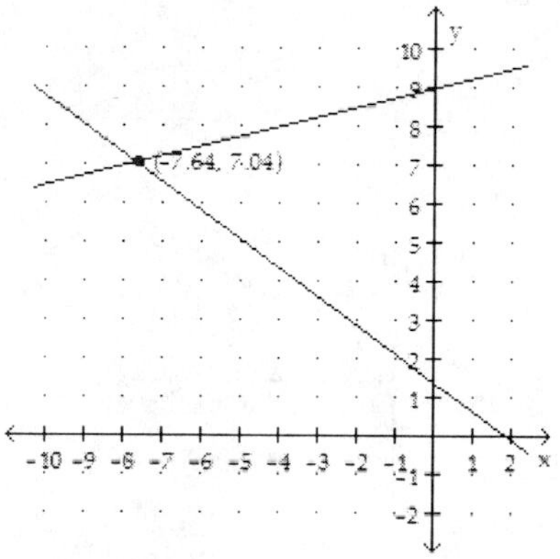

41. 4(*x* – 1) = 3*x*

Graph *y* = 4(*x* – 1) and *y* = 3*x* on the graphing calculator.
The *x*–coordinate of their intersection is **<u>x = 4</u>**.

43. 11*x* + 6(3 – *x*) = 3

Graph *y* = 11*x* + 6(3 – *x*) and *y* = 3 on the graphing calculator.
The *x*–coordinate of their intersection is **<u>x = –3</u>**.

TRY IT YOURSELF

45. $\begin{cases} y = -\dfrac{5}{2}x + \dfrac{1}{2} \\ 2x - \dfrac{3}{2}y = 5 \end{cases}$ (1, –2)

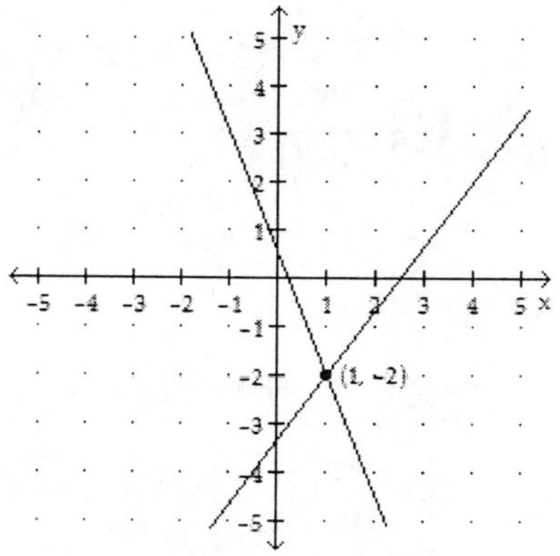

47. $\begin{cases} x + y = 0 \\ y = 2x - 6 \end{cases}$ (2, –2)

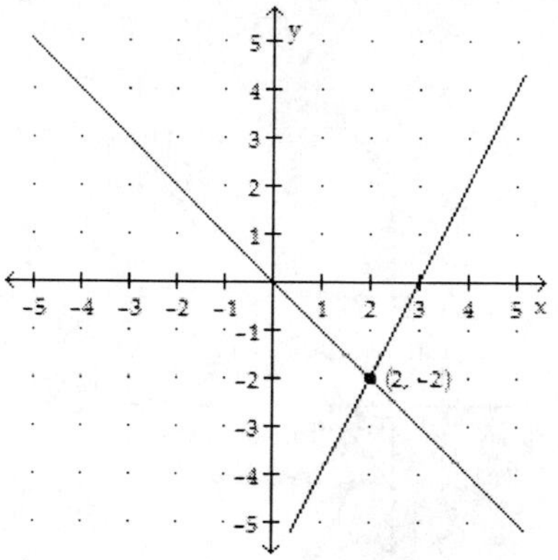

49. $\begin{cases} y = 3 \\ x = 2 \end{cases}$ (2, 3)

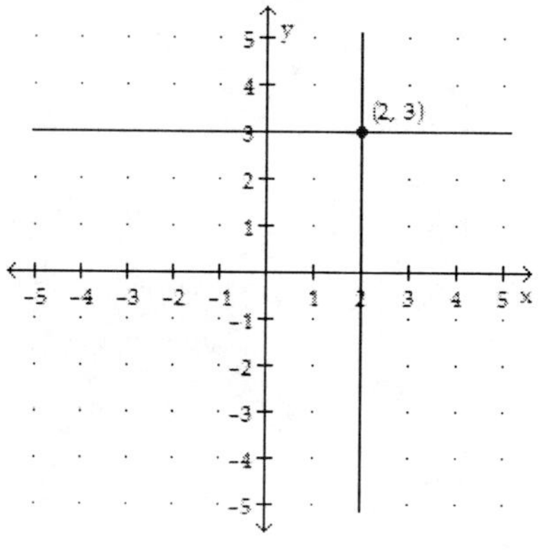

51. $\begin{cases} x = \dfrac{11-2y}{3} \\ y = \dfrac{11-6x}{4} \end{cases}$

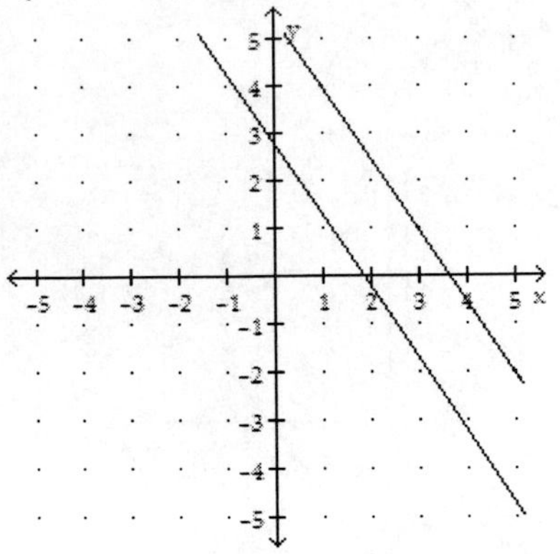

no solution;
inconsistent system

53. $\begin{cases} 4x - 3y = 5 \\ y = -2x \end{cases}$ $\left(\dfrac{1}{2}, -1\right)$

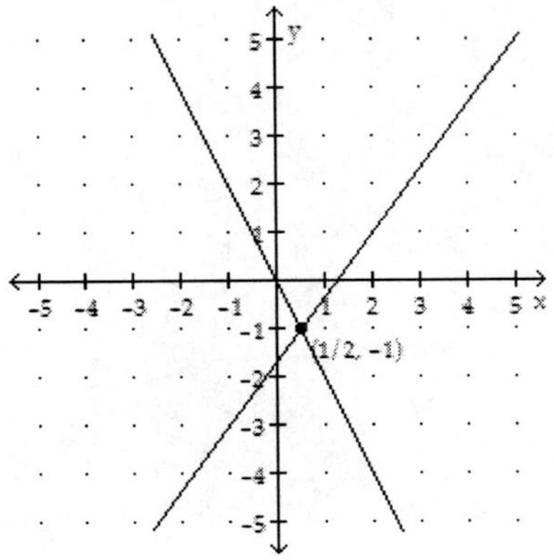

55. $\begin{cases} x = 13 - 4y \\ 3x = 4 + 2y \end{cases}$ $\left(3, \dfrac{5}{2}\right)$

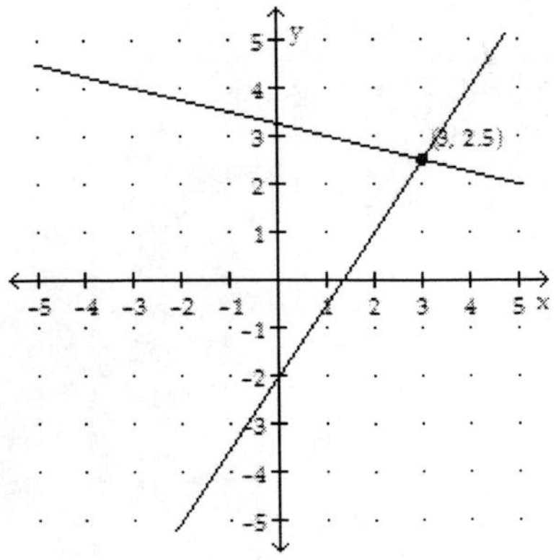

57. $\begin{cases} x = 2 \\ y = -\dfrac{1}{2}x + 2 \end{cases}$ (2, 1)

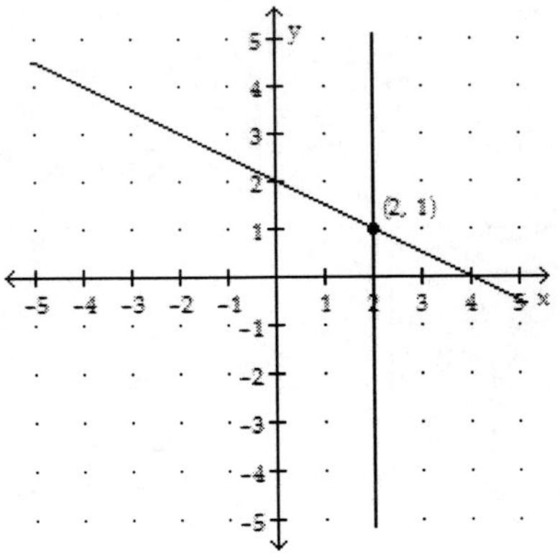

59. $\begin{cases} x = \dfrac{5}{2}y - 2 \\ x - \dfrac{5}{3}y + \dfrac{1}{3} = 0 \end{cases}$ (3, 2)

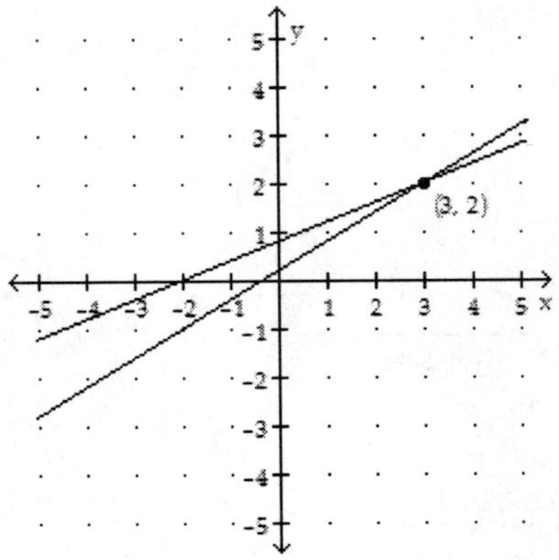

61. $\begin{cases} x + 3y = 6 \\ y = -\dfrac{1}{3}x + 2 \end{cases}$

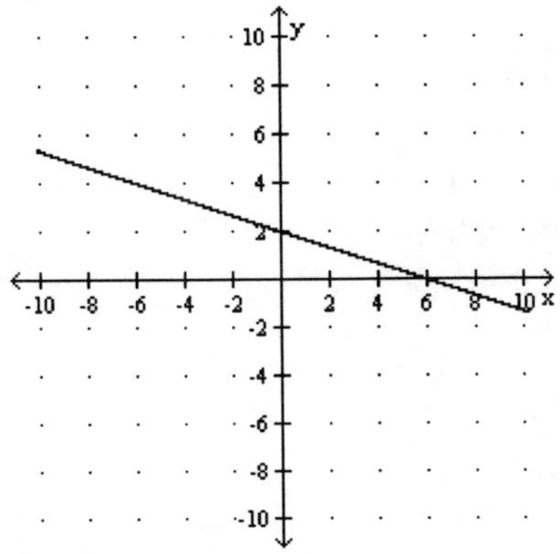

$\{(x, y) | x + 3y = 6\}$
infinitely many solutions;
dependent equation

63. $\begin{cases} x = -\dfrac{3}{2}y \\ 2x = 3y - 4 \end{cases}$ $\left(-1, \dfrac{2}{3}\right)$

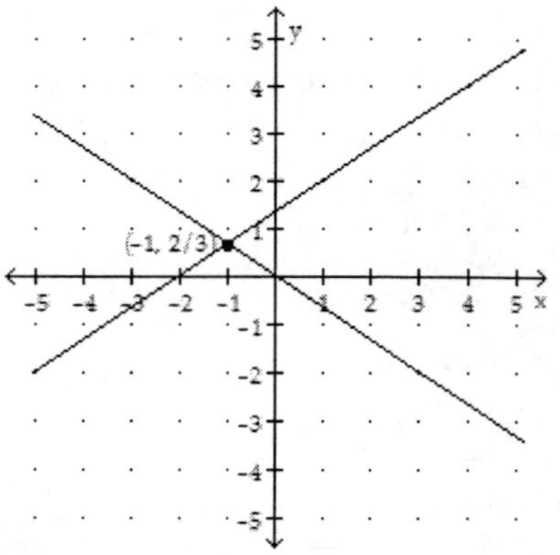

APPLICATIONS

65. FASHION DESIGNER
   a. (1978, 50%)
   b. In 1978, the percent share of the U.S. footwear market for shoes produced in United States and imports was the same: 50%.

67. HEARING TESTS
(2,000, 50)

69. LAW OF SUPPLY AND DEMAND
   a.

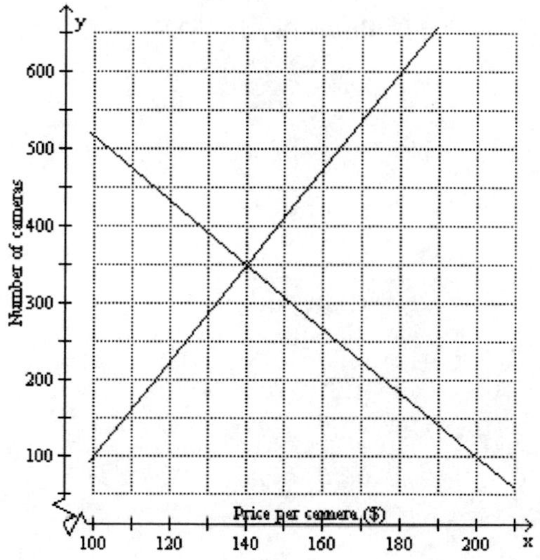

   b. \$140 – the x–value at the point of intersection
   c. As supply increases, demand decreases.

71. NAVIGATION

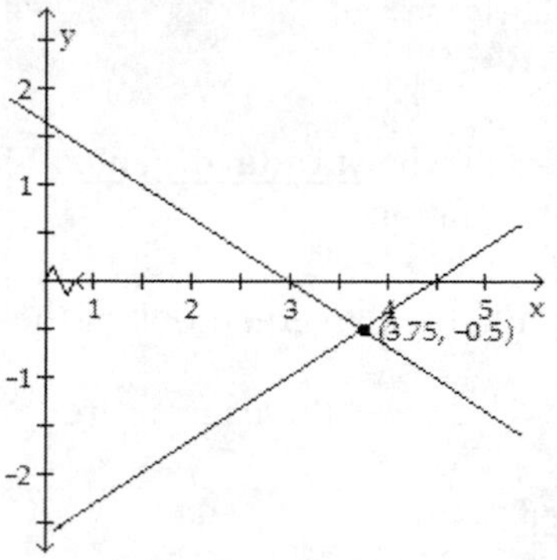

   a. yes
   b. No, the slopes would have to reach this point at the same time.

WRITING

73. Answers will vary.

75. Answers will vary.

REVIEW

77.
$$\begin{aligned} f(-1) &= -(-1)^3 + 2(-1) - 2 \\ &= -(-1) + 2(-1) - 2 \\ &= 1 - 2 - 2 \\ &= -3 \end{aligned}$$

79.
$$\begin{aligned} g(2) &= \frac{2-2}{9+2} \\ &= \frac{0}{11} \end{aligned}$$

81.
$$f(t) = -t^3 + 2t - 2$$

83 D: the set of all real numbers except –2

CHALLENGE PROBLEMS

85. Answers will vary.

One possible answer is $\begin{cases} x + y = -3 \\ x - y = -7 \end{cases}$.

SECTION 3.2

VOCABULARY

1. $Ax + By = C$ is the **standard/general** form of a linear equation.

3. When we add the two equations of the system $\begin{cases} x + y = 5 \\ x - y = -3 \end{cases}$, the y terms are **eliminated**.

CONCEPTS

5. y in the second equation

7. a. 3, –4 (answers will vary)
   b. 2; –3 (answers will vary)

9. elimination method

NOTATION

11. a.

$$\begin{cases} 4y = 8 - 7x \\ 2x = y - 3 \end{cases} \Rightarrow$$

$$\begin{cases} 4y + 7x = 8 - 7x + 7x \\ 2x - y = y - 3 - y \end{cases} \Rightarrow$$

$$\begin{cases} 7x + 4y = 8 \\ 2x - y = -3 \end{cases}$$

b.

$$\begin{cases} \frac{x}{5} + \frac{y}{10} = \frac{6}{5} \\ 0.3x - 0.9y = 17 \end{cases} \Rightarrow$$

$$\begin{cases} 10\left(\frac{x}{5} + \frac{y}{10}\right) = 10\left(\frac{6}{5}\right) \\ 10(0.3x - 0.9y) = 10(17) \end{cases} \Rightarrow$$

$$\begin{cases} 2x + y = 12 \\ 3x - 9y = 170 \end{cases}$$

GUIDED PRACTICE

13.

$$\begin{cases} y = 3x \\ x + y = 8 \end{cases}$$

$x + y = 8$ The 2nd equation

$x + (3x) = 8$

$4x = 8$

$\boxed{x = 2}$

$y = 3x$ The 1st equation

$y = 3(2)$

$\boxed{y = 6}$

The solution is (2, 6).

15.

$$\begin{cases} x = 2 + y \\ 2x + y = 13 \end{cases}$$

$2x + y = 13$ The 2nd equation

$2(2 + y) + y = 13$

$4 + 2y + y = 13$

$4 + 3y = 13$

$3y = 9$

$\boxed{y = 3}$

$x = 2 + y$ The 1st equation

$x = 2 + 3$

$\boxed{x = 5}$

The solution is (5, 3).

17.

$$\begin{cases} x+2y=6 \\ 3x-y=-10 \end{cases}$$

Solve the first equation for x.

$$\begin{cases} x=-2y+6 \\ 3x-y=-10 \end{cases}$$

$3x-y=-10$ The 2nd equation

$3(-2y+6)-y=-10$

$-6y+18-y=-10$

$-7y+18=-10$

$-7y=-28$

$\boxed{y=4}$

$x=-2y+6$ The 1st equation

$x=-2(4)+6$

$x=-8+6$

$\boxed{x=-2}$

The solution is $(-2, 4)$.

19.

$$\begin{cases} 0.3a+0.1b=0.5 \\ \frac{4}{3}a+\frac{1}{3}b=3 \end{cases}$$

Multiply the 1st equation by 10 to eliminate decimals and multiply the 2nd equation by 3 to eliminate fractions.

$$\begin{cases} 3a+b=5 \\ 4a+b=9 \end{cases}$$

Solve the 1st equation for b by subtracting $-3a$ from both sides.

$$\begin{cases} b=5-3a \\ 4a+b=9 \end{cases}$$

$4a+b=9$ The 2nd equation

$4a+(5-3a)=9$

$4a+5-3a=9$

$a+5=9$

$\boxed{a=4}$

$b=5-3a$ The 1st equation

$b=5-3(4)$

$b=5-12$

$\boxed{b=-7}$

The solution is $(4, -7)$.

21.

$$\begin{cases} x-y=7 \\ x+y=11 \end{cases}$$

eliminate the y

$$\begin{array}{r} x-y=7 \\ \underline{x+y=11} \\ 2x=18 \end{array}$$

$\boxed{x=9}$

$x-y=7$ 1st equation

$9-y=7$

$9-y-9=7-9$

$-y=-2$

$\frac{-y}{-1}=\frac{-2}{-1}$

$\boxed{y=2}$

The solution is (9, 2).

23.

$$\begin{cases} 2s+3t=-8 \\ 2s-3t=-8 \end{cases}$$

eliminate the t

$$\begin{array}{r} 2s+3t=-8 \\ \underline{2s-3t=-8} \\ 4s=-16 \end{array}$$

$\boxed{s=-4}$

$2s+3t=-8$ 1st equation

$2(-4)+3t=-8$

$-8+3t=-8$

$-8+3t+8=-8+8$

$3t=0$

$\boxed{t=0}$

The solution is $(-4, 0)$.

25.

$$\begin{cases} 5x+2y=11 \\ 7x+6y=9 \end{cases}$$

multiply 1st equation by -3

$$\begin{cases} -15x-6y=-33 \\ 7x+6y=9 \end{cases}$$

eliminate the y

$$\begin{aligned} -15x-6y&=-33 \\ \underline{7x+6y}&\underline{=9} \\ -8x&=-24 \end{aligned}$$

$$\boxed{x=3}$$

$5x+2y=11$ 1st equation

$$\begin{aligned} 5(3)+2y&=11 \\ 15+2y&=11 \\ 2y&=-4 \end{aligned}$$

$$\boxed{y=-2}$$

The solution is $(3, -2)$.

27.

$$\begin{cases} 5x+3y=72 \\ 3x+5y=56 \end{cases}$$

multiply 1st equation by 3

multiply 2nd equation by -5

$$\begin{cases} 15x+9y=216 \\ -15x-25y=-280 \end{cases}$$

eliminate the x

$$\begin{aligned} 15x+9y&=216 \\ \underline{-15x-25y}&\underline{=-280} \\ -16y&=-64 \end{aligned}$$

$$\boxed{y=4}$$

$5x+3y=72$ 1st equation

$$\begin{aligned} 5x+3(4)&=72 \\ 5x+12&=72 \\ 5x+12-12&=72-12 \\ 5x&=60 \end{aligned}$$

$$\boxed{x=12}$$

The solution is $(12, 4)$.

29.

$$\begin{cases} 2(a+b)=94 \\ 4(a-9)=3b-23 \end{cases}$$

Simplify each equation and write in general form.

$$\begin{cases} 2a+2b=94 \\ 4a-36=3b-23 \end{cases}$$

$$\begin{cases} 2a+2b=94 \\ 4a-3b=13 \end{cases}$$

Multiply 1st equation by -2.

$$\begin{cases} -4a-4b=-188 \\ 4a-3b=13 \end{cases}$$

eliminate the a

$$\begin{aligned} -4a-4b&=-188 \\ \underline{4a-3b}&\underline{=13} \\ -7b&=-175 \end{aligned}$$

$$\boxed{b=25}$$

$2a+2b=94$ 1st equation

$$\begin{aligned} 2a+2(25)&=94 \\ 2a+50&=94 \\ 2a+50-50&=94-50 \\ 2a&=44 \end{aligned}$$

$$\boxed{a=22}$$

The solution is $(22,25)$.

31.

$$\begin{cases} 2(x+y)+1=0 \\ 3x+4y=0 \end{cases}$$

Write the 1st equation in general form by distributing and subtracting 1 from both sides.

$$\begin{cases} 2x+2y=-1 \\ 3x+4y=0 \end{cases}$$

Multiply the 1st equation by – 2 and add both equations to eliminate *y*.

$$\begin{aligned} -4x-4y&=2 \\ 3x+4y&=0 \\ \hline -x\qquad&=2 \\ x&=-2 \end{aligned}$$

Substitute $x=-2$ into the 2nd equation and solve for *y*.

$$\begin{aligned} 3x+4y&=0 \\ 3(-2)+4y&=0 \\ -6+4y&=0 \\ 4y&=6 \\ y&=\frac{6}{4} \\ y&=\frac{3}{2} \end{aligned}$$

The solution is $\left(-2, \frac{3}{2}\right)$.

33.

$$\begin{cases} 2(a+b)=a+12 \\ a=14-2b \end{cases}$$

Write the 1st equation in general form.

$$\begin{cases} 2a+2b=a+12 \\ a=14-2b \end{cases}$$

$$\begin{cases} a+2b=12 \\ a=14-2b \end{cases}$$

Substitute $a=14-2b$ into the first equation.

$$a+2b=12 \quad \text{The 1st equation}$$

$$\begin{aligned} (14-2b)+2b&=12 \\ 14-2b+2b&=12 \\ 14&\neq 12 \end{aligned}$$

No solution; $\varnothing$

inconsistent system

35.

$$\begin{cases} 2x-\frac{5}{2}=y \\ 0.04x-0.02y=0.05 \end{cases}$$

Multiply the 2nd equation by 100 to eliminate decimals.

$$\begin{cases} 2x-\frac{5}{2}=y \\ 4x-2y=5 \end{cases}$$

$$4x-2y=5 \quad \text{The 2nd equation}$$

$$\begin{aligned} 4x-2\left(2x-\frac{5}{2}\right)&=5 \\ 4x-4x+5&=5 \\ 5&=5 \end{aligned}$$

$$\left\{(x,y)\,\middle|\,2x-\frac{5}{2}=y\right\}$$

infinitely many solutions

dependent equations

TRY IT YOURSELF

37.

$$\begin{cases} 3x-4y=9 \\ x+2y=8 \end{cases}$$

multiply 2nd equation by -3

$$\begin{cases} 3x-4y=9 \\ -3x-6y=-24 \end{cases}$$

eliminate the x

$$\begin{array}{r} 3x-4y=9 \\ \underline{-3x-6y=-24} \\ -10y=-15 \end{array}$$

$$y=\frac{-15}{-10}$$

$$\boxed{y=\frac{3}{2}}$$

$3x-4y=9$ 2nd equation

$$3x-4\left(\frac{3}{2}\right)=9$$

$$3x-6=9$$

$$3x-6+6=9+6$$

$$3x=15$$

$$\boxed{x=5}$$

The solution is $(5, \frac{3}{2})$.

39.

$$\begin{cases} 4(x-2)=-9y \\ 2(x-3y)=-3 \end{cases}$$

$$\begin{cases} 4x-8=-9y \\ 2x-6y=-3 \end{cases}$$

Write the 1st equation in general form.

$$\begin{cases} 4x+9y=8 \\ 2x-6y=-3 \end{cases}$$

Multiply 2nd equation by -2.

$$\begin{cases} 4x+9y=8 \\ -4x+12y=6 \end{cases}$$

eliminate the x

$$\begin{array}{r} 4x+9y=8 \\ \underline{-4x+12y=6} \\ 21y=14 \end{array}$$

$$y=\frac{14}{21}$$

$$\boxed{y=\frac{2}{3}}$$

$4x+9y=8$ 1st equation

$$4x+9\left(\frac{2}{3}\right)=8$$

$$4x+6=8$$

$$4x+6-6=8-6$$

$$4x=2$$

$$x=\frac{2}{4}$$

$$\boxed{x=\frac{1}{2}}$$

The solution is $\left(\frac{1}{2},\frac{2}{3}\right)$.

41.

$$\begin{cases} 0.16x - 0.08y = 0.32 \\ 2x - 4 = y \end{cases}$$

Multiply the 1st equation by 100 to eliminate decimals.

Write the 2nd equation in general form.

$$\begin{cases} 16x - 8y = 32 \\ 2x - y = 4 \end{cases}$$

Multiply the 2nd equation by -8 and add both equations to eliminate x.

$$\begin{aligned} 16x - 8y &= 32 \\ -16x + 8y &= -32 \\ \hline 0 &= 0 \end{aligned}$$

$$\{(x, y) 2x - 4 = y|\}$$

infinitely many solutions

dependent equations

43.

$$\begin{cases} x = \frac{2}{3}y \\ y = 4x + 50 \end{cases}$$

$$y = 4x + 50 \quad \text{The 2nd equation}$$

$$y = 4\left(\frac{2}{3}y\right) + 50$$

$$y = \frac{8}{3}y + 50$$

$$3(y) = 3\left(\frac{8}{3}y + 50\right)$$

$$3y = 8y + 150$$

$$3y - 8y = 8y + 150 - 8y$$

$$-5y = 150$$

$$y = -30$$

Substitute $y = -30$ in the 1st equation and solve for x.

$$x = \frac{2}{3}y \quad \text{The 1st equation}$$

$$x = \frac{2}{3}(-30)$$

$$x = -20$$

The solution is $(-20, -30)$.

45.

$$\begin{cases} \frac{m-n}{5} + \frac{m+n}{2} = 6 \\ \frac{m-n}{2} - \frac{m+n}{4} = 3 \end{cases}$$

Multiply the 1st equation by 10 and the 2nd equation by 4 to eliminate fractions.

$$\begin{cases} 2m - 2n + 5m + 5n = 60 \\ 2m - 2n - m - n = 12 \end{cases}$$

Combine like terms and write in general form.

$$\begin{cases} 7m + 3n = 60 \\ m - 3n = 12 \end{cases}$$

Add the equations to eliminate n.

$$\begin{aligned} 7m + 3n &= 60 \\ m - 3n &= 12 \\ \hline 8m \quad &= 72 \\ m &= 9 \end{aligned}$$

Substitute $m = 9$ into the 2nd equation and solve for n.

$$m - 3n = 12$$

$$9 - 3n = 12$$

$$-3n = 3$$

$$n = -1$$

The solution is $(9, -1)$.

47.

$$\begin{cases} 0.5x + 0.5y = 6 \\ \frac{x}{2} - \frac{y}{2} = -2 \end{cases}$$

Multiply the 1st equation by 10 to eliminate decimals and multiply the 2nd equation by 2 to eliminate fractions.

$$\begin{cases} 5x + 5y = 60 \\ x - y = -4 \end{cases}$$

Multiply the 2nd equation by 5 and add both equations to eliminate *y*.

$$\begin{aligned} 5x + 5y &= 60 \\ \underline{5x - 5y} &\underline{= -20} \\ 10x \quad &= 40 \\ x &= 4 \end{aligned}$$

Substitute $x = 4$ into the 1st equation and solve for *y*.

$$\begin{aligned} 5x + 5y &= 60 \\ 5(4) + 5y &= 60 \\ 20 + 5y &= 60 \\ 5y &= 40 \\ y &= 8 \end{aligned}$$

The solution is $(4, 8)$.

49.

$$\begin{cases} \frac{3}{4}x + \frac{2}{3}y = 7 \\ \frac{3}{5}x - \frac{1}{2}y = 18 \end{cases}$$

Multiply the 1st equation by 12 and the 2nd equation by 10 to eliminate fractions.

$$\begin{cases} 9x + 8y = 84 \\ 6x - 5y = 180 \end{cases}$$

Multiply the 1st equation by 5 and the 2nd equation by 8.

Add the equations to eliminate *y*.

$$\begin{aligned} 45x + 40y &= 420 \\ \underline{48x - 40y} &\underline{= 1,440} \\ 93x \quad &= 1,860 \\ x &= 20 \end{aligned}$$

Substitute $x = 20$ into the 1st equation and solve for *y*.

$$\begin{aligned} 9x + 8y &= 84 \\ 9(20) + 8y &= 84 \\ 180 + 8y &= 84 \\ 8y &= -96 \\ y &= -12 \end{aligned}$$

The solution is $(20, -12)$.

51.

$$\begin{cases} \frac{x}{4} = 1 + \frac{y}{5} \\ x = \frac{4}{5}(y+10) \end{cases}$$

Distribute $\frac{4}{5}$ on the 2nd equation.

$$\begin{cases} \frac{x}{4} = 1 + \frac{y}{5} \\ x = \frac{4}{5}y + 8 \end{cases}$$

Multiply the 1st equation by 20 to eliminate fractions.

$$\begin{cases} 5x = 20 + 4y \\ x = \frac{4}{5}y + 8 \end{cases}$$

Substitute $x = \frac{4}{5}y + 8$ into the 1st equation.

$$5x = 20 + 4y$$
$$5\left(\frac{4}{5}y + 8\right) = 20 + 4y$$
$$4y + 40 = 20 + 4y$$
$$4y + 40 - 4y = 20 + 4y - 4y$$
$$40 \neq 20$$

no solution; $\varnothing$

inconsistent system

53.

$$\begin{cases} \frac{3}{2}x - \frac{2}{3}y = 0 \\ \frac{3}{4}x + \frac{4}{3}y = \frac{5}{2} \end{cases}$$

Multiply the 1st equation by 6 and the 2nd equation by 12 to eliminate fractions.

$$\begin{cases} 9x - 4y = 0 \\ 9x + 16y = 30 \end{cases}$$

Multiply the 1st equation by -1 and add the equations to eliminate x.

$$-9x + 4y = 0$$
$$\underline{9x + 16y = 30}$$
$$20y = 30$$
$$y = \frac{30}{20}$$
$$y = \frac{3}{2}$$

Substitute $y = \frac{3}{2}$ into the 1st equation and solve for x.

$$9x - 4y = 0$$
$$9x - 4\left(\frac{3}{2}\right) = 0$$
$$9x - 6 = 0$$
$$9x = 6$$
$$x = \frac{6}{9}$$
$$x = \frac{2}{3}$$

The solution is $\left(\frac{2}{3}, \frac{3}{2}\right)$.

55.

$$\begin{cases} 12x - 5y - 21 = 0 \\ \frac{3}{4}x + \frac{2}{3}y = -\frac{13}{8} \end{cases}$$

Write the 1st equation in general form and multiply the 2nd equation by 24 to eliminate fractions.

$$\begin{cases} 12x - 5y = 21 \\ 18x + 16y = -39 \end{cases}$$

Multiply the 1st equation by 16 and the 2nd equation by 5 to create opposites. Add the equations to eliminate y.

$$\begin{aligned} 192x - 80y &= 336 \\ 90x + 80y &= -195 \\ \hline 282x \quad &= 141 \\ x &= \frac{141}{282} \\ x &= \frac{1}{2} \end{aligned}$$

Substitute $x = \frac{1}{2}$ into the 1st equation and solve for y.

$$\begin{aligned} 12x - 5y &= 21 \\ 12\left(\frac{1}{2}\right) - 5y &= 21 \\ 6 - 5y &= 21 \\ -5y &= 15 \\ y &= -3 \end{aligned}$$

The solution is $\left(\frac{1}{2}, -3\right)$.

57.

$$\begin{cases} y = -2.2x + 3.5 \\ y = -1.8x + 2.4 \end{cases}$$

Substitute $y = -1.8x + 2.4$ into the 1st equation for y.

$$\begin{aligned} y &= -2.2x + 3.5 \quad (1^{st}\text{ eq}) \\ -1.8x + 2.4 &= -2.2x + 3.5 \\ -1.8x + 2.4 + 2.2x &= -2.2x + 3.5 + 2.2x \\ 0.4x + 2.4 &= 3.5 \\ 0.4x + 2.4 - 2.4 &= 3.5 - 2.4 \\ 0.4x &= 1.1 \\ x &= 2.75 \end{aligned}$$

Substitute $x = 2.75$ into the 1st equation to solve for y.

$$\begin{aligned} y &= -2.2x + 3.5 \\ y &= -2.2(2.75) + 3.5 \\ y &= -6.05 + 3.5 \\ y &= -2.55 \end{aligned}$$

The solution is (2.75, –2.55).

59.

$$\begin{cases} 0.05a - 0.03b = 0.24 \\ 0.003a + 0.005b = 0.028 \end{cases}$$

Multiply the 1st equation by 100 and the 2nd equation by 1000 to eliminate decimals.

$$\begin{cases} 5a - 3b = 24 \\ 3a + 5b = 28 \end{cases}$$

Multiply the 1st equation by 5 and the 2nd equation by 3 to create opposites and then add to eliminate b.

$$\begin{aligned} 25a - 15b &= 120 \\ 9a + 15b &= 84 \\ \hline 34a \quad &= 204 \\ a &= 6 \end{aligned}$$

Substitute $a = 6$ into the 2nd equation to solve for b.

$$\begin{aligned} 3a + 5b &= 28 \\ 3(6) + 5b &= 28 \\ 18 + 5b &= 28 \\ 5b &= 10 \\ b &= 2 \end{aligned}$$

The solution is (6, 2).

61.

$$\begin{cases} \frac{1}{x}+\frac{1}{y}=\frac{5}{6} \\ \frac{1}{x}-\frac{1}{y}=\frac{1}{6} \end{cases}$$

Let $a=\frac{1}{x}$ and $b=\frac{1}{y}$.

$$\begin{cases} a+b=\frac{5}{6} \\ a-b=\frac{1}{6} \end{cases}$$

Add equations to eliminate b.

$$\begin{aligned} a+b&=\frac{5}{6} \\ a-b&=\frac{1}{6} \\ \hline 2a&=\frac{6}{6} \\ 2a&=1 \\ a&=\frac{1}{2} \end{aligned}$$

Substitute $a=\frac{1}{2}$ in the 1st equation and solve for b.

$$\begin{aligned} a+b&=\frac{5}{6} \\ \frac{1}{2}+b&=\frac{5}{6} \\ \frac{1}{2}+b-\frac{1}{2}&=\frac{5}{6}-\frac{1}{2} \\ b&=\frac{1}{3} \end{aligned}$$

$a=\frac{1}{2}=\frac{1}{x}$ so $x=2$

$b=\frac{1}{3}=\frac{1}{y}$ so $y=3$

The solution is $(2, 3)$.

63.

$$\begin{cases} \frac{1}{x}+\frac{2}{y}=-1 \\ \frac{2}{x}-\frac{1}{y}=-7 \end{cases}$$

Let $a=\frac{1}{x}$ and $b=\frac{1}{y}$.

$$\begin{cases} a+2b=-1 \\ 2a-b=-7 \end{cases}$$

Multiply the 2nd equation by 2 and add equations to eliminate b.

$$\begin{aligned} a+2b&=-1 \\ 4a-2b&=-14 \\ \hline 5a&=-15 \\ a&=-3 \end{aligned}$$

Substitute $a=-3$ in the 1st equation and solve for b.

$$\begin{aligned} a+2b&=-1 \\ -3+2b&=-1 \\ 2b&=2 \\ b&=1 \end{aligned}$$

$a=\frac{-3}{1}=\frac{1}{x}$ so $x=-\frac{1}{3}$

$b=\frac{1}{1}=\frac{1}{y}$ so $y=1$

The solution is $\left(-\frac{1}{3},1\right)$.

APPLICATIONS

65. AREA CODES

$$\begin{cases} A+H=1{,}715 \\ A-H=99 \end{cases}$$

Add the equations to eliminate H.

$$\begin{aligned} A+H &= 1{,}715 \\ A-H &= 99 \\ \hline 2A &= 1{,}814 \\ A &= 907 \end{aligned}$$

Substitute $A = 907$ into the 1st equation to solve for H.

$$\begin{aligned} A+H &= 1{,}715 \\ 907+H &= 1{,}715 \\ 907+H-907 &= 1{,}715-907 \\ H &= 808 \end{aligned}$$

The area code for Alaska is 907 and the area code for Hawaii is 808.

WRITING

67. Answers will vary.

69. Answers will vary.

71. Answers will vary.

REVIEW

73. Find two points on the line, such as (1, –1) and (–1, 4), and use the formula for slope.

$$\begin{aligned} m &= \frac{y_2 - y_1}{x_2 - x_1} \\ &= \frac{4-(-1)}{-1-1} \\ &= \frac{5}{-2} \\ &= -\frac{5}{2} \end{aligned}$$

75. Use the formula for slope.

$$\begin{aligned} m &= \frac{y_2 - y_1}{x_2 - x_1} \\ &= \frac{0-(-8)}{-5-0} \\ &= \frac{8}{-5} \\ &= -\frac{8}{5} \end{aligned}$$

77. Solve the equation for y.

$$\begin{aligned} 4x-3y &= -3 \\ 4x-3y-4x &= -3-4x \\ -3y &= -4x-3 \\ \frac{-3y}{-3} &= \frac{-4x}{-3} - \frac{3}{-3} \\ y &= \frac{4}{3}x+1 \\ m &= \frac{4}{3} \end{aligned}$$

CHALLENGE PROBLEMS

79.

$$\begin{cases} Ax+By=2 \\ Bx-Ay=-2 \end{cases}$$

Let $x=-3$ and $y=5$.

$$\begin{cases} A(-3)+B(5)=-2 \\ B(-3)-A(5)=-26 \end{cases}$$

$$\begin{cases} -3A+5B=-2 \\ -5A-3B=-26 \end{cases}$$

Multiply the first equation by 3 and the second by 5.

$$\begin{aligned} -9A+15B &= -6 \\ -25A-15B &= -130 \\ \hline -34A &= -136 \\ A &= 4 \end{aligned}$$

Substitute $A=4$ into the first equation and solve for B.

$$\begin{aligned} -3A+5B &= -2 \\ -3(4)+5B &= -2 \\ -12+5B &= -2 \\ 5B &= 10 \\ B &= 2 \end{aligned}$$

$A=4$ and $B=2$.

SECTION 3.3

VOCABULARY

1. $\begin{cases} 2x+y-3z=0 \\ 3x-y+4z=5 \\ 4x+2y-6z=0 \end{cases}$ is called a **system** of three linear equations in three variables. Each equation is written in **standard** $Ax + By + Cz = D$ form.

3. Solutions of a system of three equations in three variables, *x*, *y*, and *z* are written in the form (*x*, *y*, *z*) and are called ordered **triples**.

5. When three planes coincide, the equations of the system are **dependent**, and there are infinitely many solutions.

CONCEPTS

7. a. no solution
   b. no solution

NOTATION

9.
$$\begin{cases} x+y+4z=3 \\ 7x-2y+8z=15 \\ 3x+2y-z=4 \end{cases}$$

GUIDED PRACTICE

11. Yes.

$$x-y+z=2$$
$$2-1+1\stackrel{?}{=}2$$
$$2=2$$

$$2x+y-z=4$$
$$2(2)+1-1\stackrel{?}{=}4$$
$$4+1-1\stackrel{?}{=}4$$
$$4=4$$

$$2x-3y+z=2$$
$$2(2)-3(1)+1\stackrel{?}{=}2$$
$$4-3+1\stackrel{?}{=}2$$
$$2=2$$

13. No.

$$3x-2y-z=37$$
$$3(6)-2(-7)-(-5)\stackrel{?}{=}37$$
$$18+14+5\stackrel{?}{=}37$$
$$37=37$$

$$x-3y=27$$
$$(6)-3(-7)\stackrel{?}{=}27$$
$$6+21\stackrel{?}{=}27$$
$$27=27$$

$$2x+7y+2z=-48$$
$$2(6)+7(-7)+2(-5)\stackrel{?}{=}-48$$
$$12-49-10\stackrel{?}{=}-48$$
$$-47\neq-48$$

15.

$$\begin{cases} x+y+z=4 & (1) \\ 2x+y-z=1 & (2) \\ 2x-3y+z=1 & (3) \end{cases}$$

Add Equations 1 and 2 to eliminate z.

$$\begin{array}{rl} x+y+z=4 & (1) \\ \underline{2x+y-z=1} & (2) \\ 3x+2y \quad =5 & (4) \end{array}$$

Add Equations 3 and 2 to eliminate z.

$$\begin{array}{rl} 2x-3y+z=1 & (3) \\ \underline{2x+\ y-z=1} & (2) \\ 4x-2y \quad =2 & (5) \end{array}$$

Add Equations 4 and 5 to eliminate y.

$$\begin{array}{rl} 3x+2y=5 & (4) \\ \underline{4x-2y=2} & (5) \\ 7x \qquad =7 & \\ x=1 & \end{array}$$

Substitute $x=1$ into Equation 4 to find y.

$$\begin{array}{rl} 3x+2y=5 & (4) \\ 3(1)+2y=5 & \\ 3+2y=5 & \\ 2y=2 & \\ y=1 & \end{array}$$

Substitute $x=1$ and $y=1$ into

Equation 1 to find z.

$$\begin{array}{rl} x+y+z=4 & (1) \\ 1+1+z=4 & \\ 2+z=4 & \\ z=2 & \end{array}$$

The solution is (1, 1, 2).

17.

$$\begin{cases} 3x+2y-5z=3 & (1) \\ 4x-2y-3z=-10 & (2) \\ 5x-2y-2z=-11 & (3) \end{cases}$$

Add Equations 1 and 2 to eliminate y.

$$\begin{array}{rl} 3x+2y-5z=3 & (1) \\ \underline{4x-2y-3z=-10} & (2) \\ 7x-8z=-7 & (4) \end{array}$$

Add Equations 1 and 3 to eliminate y.

$$\begin{array}{rl} 3x+2y-5z=3 & (1) \\ \underline{5x-2y-2z=-11} & (3) \\ 8x-7z=-8 & (5) \end{array}$$

Multiply Equation 4 by 8 and Equation 5 by −7. Add to eliminate x.

$$\begin{array}{r} 56x-64z=-56 \\ \underline{-56x+49z=56} \\ -15z=0 \\ z=0 \end{array}$$

Substitute $z=0$ into Equation 4 and solve for x.

$$\begin{array}{rl} 7x-8z=-7 & (4) \\ 7x-8(0)=-7 & \\ 7x-0=-7 & \\ 7x=-7 & \\ x=-1 & \end{array}$$

Substitute $x=-1$ and $z=0$ into Equation 1 and solve for y.

$$\begin{array}{rl} 3x+2y-5z=3 & (1) \\ 3(-1)+2y-5(0)=3 & \\ -3+2y-0=3 & \\ -3+2y=3 & \\ 2y=6 & \\ y=3 & \end{array}$$

The solution is $(-1, 3, 0)$.

19.

$$\begin{cases} 2x+6y+3z=9 & (1) \\ 5x-3y-5z=3 & (2) \\ 4x+3y+2z=15 & (3) \end{cases}$$

Multiply Equation 2 by 2 and add Equations 1 and 2 to eliminate y.

$$\begin{aligned} 2x+6y+3z&=9 \quad (1) \\ \underline{10x-6y-10z}&\underline{=6} \quad (2) \\ 12x-7z&=15 \quad (4) \end{aligned}$$

Multiply Equation 3 by –2 and add Equations 1 and 3 to eliminate y.

$$\begin{aligned} 2x+6y+3z&=9 \quad (1) \\ \underline{-8x-6y-4z}&\underline{=-30} \quad (3) \\ -6x-z&=-21 \quad (5) \end{aligned}$$

Multiply Equation 5 by –7. Add Equations 4 and 5 to eliminate z.

$$\begin{aligned} 12x-7z&=15 \\ \underline{42x+7z}&\underline{=147} \\ 54x&=162 \\ x&=3 \end{aligned}$$

Substitute $x=3$ into Equation 4 and solve for z.

$$\begin{aligned} 12x-7z&=15 \quad (4) \\ 12(3)-7z&=15 \\ 36-7z&=15 \\ -7z&=-21 \\ z&=3 \end{aligned}$$

Substitute $x=3$ and $z=3$ into Equation 1 and solve for y.

$$\begin{aligned} 2x+6y+3z&=9 \quad (1) \\ 2(3)+6y+3(3)&=9 \\ 6+6y+9&=9 \\ 6y+15&=9 \\ 6y&=-6 \\ y&=-1 \end{aligned}$$

The solution is $(3,-1,\ 3)$.

21.

$$\begin{cases} 4x-5y-8z=-52 & (1) \\ 2x-3y-4z=-26 & (2) \\ 3x+7y+8z=31 & (3) \end{cases}$$

Multiply Equation 2 by –2 and add Equations 1 and 2 to eliminate x and z.

$$\begin{aligned} 4x-5y-8z&=-52 \quad (1) \\ \underline{-4x+6y+8z}&\underline{=52} \quad (2) \\ y&=0 \end{aligned}$$

Add Equations 1 and 3 to eliminate z.

$$\begin{aligned} 4x-5y-8z&=-52 \quad (1) \\ \underline{3x+7y+8z}&\underline{=31} \quad (3) \\ 7x+2y&=-21 \quad (4) \end{aligned}$$

Substitute $y=0$ into Equation 4 and solve for x.

$$\begin{aligned} 7x+2y&=-21 \quad (4) \\ 7x+2(0)&=-21 \\ 7x+0&=-21 \\ 7x&=-21 \\ x&=-3 \end{aligned}$$

Substitute $x=-3$ and $y=0$ into Equation 1 and solve for z.

$$\begin{aligned} 4x-5y-8z&=-52 \quad (1) \\ 4(-3)-5(0)-8z&=-52 \\ -12-0-8z&=-52 \\ -12-8z&=-52 \\ -8z&=-40 \\ z&=5 \end{aligned}$$

The solution is $(-3,\ 0,\ 5)$.

23.

$$\begin{cases} 3x+3z=6-4y \\ 7x-5z=46+2y \\ 4x=31-z \end{cases}$$

Write each equation in standard form.

$$\begin{cases} 3x+4y+3z=6 & (1) \\ 7x-2y-5z=46 & (2) \\ 4x+z=31 & (3) \end{cases}$$

Multiply Equation 2 by 2 and add Equations 1 and 2 to eliminate y.

$$\begin{array}{r} 3x+4y+3z=6 \\ \underline{14x-4y-10z=92} \\ 17x-7z=98 \quad (4) \end{array}$$

Multiply Equation 3 by 7 and add Equations 3 and 4 to eliminate z.

$$\begin{array}{r} 28x+7z=217 \\ \underline{17x-7z=98} \quad (4) \\ 45x=315 \\ x=7 \end{array}$$

Substitute $x = 7$ into Equation 4 and solve for z.

$$\begin{aligned} 17x-7z&=98 \quad (4) \\ 17(7)-7z&=98 \\ 119-7z&=98 \\ -7z&=-21 \\ z&=3 \end{aligned}$$

Substitute $x = 7$ and $z = 3$ into Equation 1 and solve for y.

$$\begin{aligned} 3x+4y+3z&=6 \quad (1) \\ 3(7)+4y+3(3)&=6 \\ 21+4y+9&=6 \\ 4y+30&=6 \\ 4y&=-24 \\ y&=-6 \end{aligned}$$

The solution is (7, –6, 3).

25.

$$\begin{cases} 2x+z=-2+y \\ 8x-3y=-2 \\ 6x-2y+3z=-4 \end{cases}$$

Write Equation 1 in standard form.

$$\begin{cases} 2x-y+z=-2 & (1) \\ 8x-3y=-2 & (2) \\ 6x-2y+3z=-4 & (3) \end{cases}$$

Multiply Equation 1 by –3 and add Equations 1 and 3 to eliminate x and z.

$$\begin{array}{r} -6x+3y-3z=6 \\ \underline{6x-2y+3z=-4} \\ y=2 \end{array}$$

Substitute $y = 2$ into Equation 2 and solve for x.

$$\begin{aligned} 8x-3y&=-2 \\ 8x-3(2)&=-2 \\ 8x-6&=-2 \\ 8x&=4 \\ x&=\frac{1}{2} \end{aligned}$$

Substitute $x = \frac{1}{2}$ and $y = 2$ into Equation 1 and solve for z.

$$\begin{aligned} 2x-y+z&=-2 \\ 2\left(\frac{1}{2}\right)-(2)+z&=-2 \\ 1-2+z&=-2 \\ -1+z&=-2 \\ z&=-1 \end{aligned}$$

The solution is $\left(\frac{1}{2},2,-1\right)$.

27.

$$\begin{cases} x+y+3z=35 & (1) \\ -x-3y=20 & (2) \\ 2y+z=-35 & (3) \end{cases}$$

Add Equations 1 and 2 to eliminate x.

$$\begin{array}{rl} x+y+3z=35 & (1) \\ \underline{-x-3y=20} & (2) \\ -2y+3z=55 & (4) \end{array}$$

Add Equation 3 and 4 to eliminate y.

$$\begin{array}{r} 2y+z=-35 \\ \underline{-2y+3z=55} \\ 4z=20 \\ z=5 \end{array}$$

Substitute $z = 5$ into Equation 3 and solve for y.

$$\begin{aligned} 2y+z&=-35 \\ 2y+5&=-35 \\ 2y&=-40 \\ y&=-20 \end{aligned}$$

Substitute $y = -20$ into Equation 2 and solve for x.

$$\begin{aligned} -x-3y&=20 \\ -x-3(-20)&=20 \\ -x+60&=20 \\ -x&=-40 \\ x&=40 \end{aligned}$$

The solution is (40, –20, 5).

29.

$$\begin{cases} 3x+2y-z=7 & (1) \\ 6x-3y=-2 & (2) \\ 3y-2z=8 & (3) \end{cases}$$

Multiply Equation 1 by –2 and add Equations 1 and 2 to eliminate x.

$$\begin{array}{rl} -6x-4y+2z=-14 & \\ \underline{6x-3y\qquad=-2} & \\ -7y+2z=-16 & (4) \end{array}$$

Add Equations 3 and 4 to eliminate z.

$$\begin{array}{rl} 3y-2z=8 & (3) \\ \underline{-7y+2z=-16} & (4) \\ -4y=-8 & \\ y=2 & \end{array}$$

Substitute $y = 2$ into Equation 3 and solve for z.

$$\begin{aligned} 3y-2z&=8 \quad (3) \\ 3(2)-2z&=8 \\ 6-2z&=8 \\ -2z&=2 \\ z&=-1 \end{aligned}$$

Substitute $y = 2$ into Equation 2 and solve for x.

$$\begin{aligned} 6x-3y&=-2 \quad (2) \\ 6x-3(2)&=-2 \\ 6x-6&=-2 \\ 6x&=4 \\ x&=\frac{4}{6} \\ x&=\frac{2}{3} \end{aligned}$$

The solution is $\left(\frac{2}{3},2,-1\right)$.

31.

$$\begin{cases} r+s-3t=21 & (1) \\ r+4s=9 & (2) \\ 5s+t=-4 & (3) \end{cases}$$

Multiply Equation 2 by –1 and add Equations 1 and 2 to eliminate r.

$$\begin{aligned} r+s-3t&=21 \\ \underline{-r-4s\quad}&\underline{=-9} \\ -3s-3t&=12 \quad (4) \end{aligned}$$

Multiply Equation 3 by 3 and add Equations 3 and 4 to eliminate t.

$$\begin{aligned} 15s+3t&=-12 \\ \underline{-3s-3t}&\underline{=12} \\ 12s&=0 \\ s&=0 \end{aligned}$$

Substitute $s=0$ into Equation 3 and solve for t.

$$\begin{aligned} 5s+t&=-4 \quad (3) \\ 5(0)+t&=-4 \\ 0+t&=-4 \\ t&=-4 \end{aligned}$$

Substitute $s=0$ into Equation 2 and solve for r.

$$\begin{aligned} r+4s&=9 \quad (2) \\ r+4(0)&=9 \\ r+0&=9 \\ r&=9 \end{aligned}$$

The solution is (9, 0, –4).

33.

$$\begin{cases} x-8z=-30 & (1) \\ 3x+y-4z=5 & (2) \\ y+7z=30 & (3) \end{cases}$$

Multiply Equation 3 by –1 and add Equations 2 and 3 to eliminate y.

$$\begin{aligned} 3x+y-4z&=5 \quad (2) \\ \underline{-y-7z}&\underline{=-30} \\ 3x-11z&=-25 \quad (4) \end{aligned}$$

Multiply Equation 1 by –3 and add Equations 1 and 4 to eliminate x.

$$\begin{aligned} -3x+24z&=90 \\ \underline{3x-11z}&\underline{=-25} \quad (4) \\ 13z&=65 \\ z&=5 \end{aligned}$$

Substitute $z=5$ into Equation 1 and solve for x.

$$\begin{aligned} x-8z&=-30 \quad (1) \\ x-8(5)&=-30 \\ x-40&=-30 \\ x&=10 \end{aligned}$$

Substitute $z=5$ into Equation 3 and solve for y.

$$\begin{aligned} y+7z&=30 \quad (3) \\ y+7(5)&=30 \\ y+35&=30 \\ y&=-5 \end{aligned}$$

The solution is (10, –5, 5).

35.

$$\begin{cases} 7a+9b-2c=-5 & (1) \\ 5a+14b-c=-11 & (2) \\ 2a-5b-c=3 & (3) \end{cases}$$

Multiply Equation 2 by –1 and add Equations 2 and 3 to eliminate c.

$$-5a-14b+c=11 \quad (2)$$
$$\underline{2a-5b-c=3} \quad (3)$$
$$-3a-19b=14 \quad (4)$$

Multiply Equation 2 by –2 and add Equations 1 and 2 to eliminate c.

$$7a+9b-2c=-5 \quad (1)$$
$$\underline{-10a-28b+2c=22}$$
$$-3a-19b=17 \quad (5)$$

Multiply Equation 4 by –1 and add Equations 4 and 5 to eliminate a.

$$3a+19b=-14 \quad (4)$$
$$\underline{-3a-19b=17} \quad (5)$$
$$0 \neq 3$$

No solution; $\varnothing$
Inconsistent system

37.

$$\begin{cases} 7x-y-z=10 & (1) \\ x-3y+z=2 & (2) \\ x+2y-z=1 & (3) \end{cases}$$

Add Equations 1 and 2 to eliminate z.

$$7x-y-z=10 \quad (1)$$
$$\underline{x-3y+z=2} \quad (2)$$
$$8x-4y=12 \quad (4)$$

Add Equations 2 and 3 to eliminate z.

$$x-3y+z=2 \quad (2)$$
$$\underline{x+2y-z=1} \quad (3)$$
$$2x-y=3 \quad (5)$$

Multiply Equation 5 by –4 and add equations 4 and 5 to eliminate x.

$$8x-4y=12 \quad (4)$$
$$\underline{-8x+4y=-12} \quad (5)$$
$$0=0$$

Infinitely many solutions;
dependent equations

TRY IT YOURSELF

39.

$$\begin{cases} 2a+3b-2c=18 & (1) \\ 5a-6b+c=21 & (2) \\ 4b-2c=6 & (3) \end{cases}$$

Multiply Equation 3 by –1 and add Equations 1 and 3 to eliminate c.

$$2a+3b-2c=18 \quad (1)$$
$$\underline{-4b+2c=-6} \quad (3)$$
$$2a-b=12 \quad (4)$$

Multiply Equation 2 by 2 and add Equations 1 and 2 to eliminate c.

$$2a+3b-2c=18 \quad (1)$$
$$\underline{10a-12b+2c=42} \quad (2)$$
$$12a-9b=60 \quad (5)$$

Multiply Equation 4 by –6 and add Equations 4 and 5 to eliminate a.

$$-12a+6b=-72 \quad (4)$$
$$\underline{12a-9b=60} \quad (5)$$
$$-3b=-12$$
$$b=4$$

Substitute $b = 4$ into Equation 3 and solve for c.

$$4b-2c=6 \quad (3)$$
$$4(4)-2c=6$$
$$16-2c=6$$
$$-2c=-10$$
$$c=5$$

Substitute $b = 4$ and $c = 5$ into Equation 1 and solve for a.

$$2a+3b-2c=18 \quad (1)$$
$$2a+3(4)-2(5)=18$$
$$2a+12-10=18$$
$$2a+2=18$$
$$2a=16$$
$$a=8$$

The solution is (8, 4, 5).

41.

$$\begin{cases} 2x+2y-z=2 & (1) \\ x+3z-24=0 & (2) \\ y=7-4z & (3) \end{cases}$$

Solve Equation 2 for x.

$$\begin{cases} 2x+2y-z=2 & (1) \\ x=-3z+24 & (2) \\ y=7-4z & (3) \end{cases}$$

Use substitution. Substitute $x = -3z + 24$ and $y = 7 - 4z$ into Equation 1 and solve for z.

$$\begin{aligned} 2x+2y-z&=2 \\ 2(-3z+24)+2(7-4z)-z&=2 \\ -6z+48+14-8z-z&=2 \\ -15z+62&=2 \\ -15z&=-60 \\ z&=4 \end{aligned}$$

Substitute $z = 4$ into Equation 3 and solve for y.

$$\begin{aligned} y&=7-4z \quad (3) \\ y&=7-4(4) \\ y&=7-16 \\ y&=-9 \end{aligned}$$

Substitute $z = 4$ into Equation 2 and solve for x.

$$\begin{aligned} x+3z-24&=0 \quad (2) \\ x+3(4)-24&=0 \\ x+12-24&=0 \\ x-12&=0 \\ x&=12 \end{aligned}$$

The solution is (12, –9, 4).

43.

$$\begin{cases} b+2c=7-a \\ a+c=2(4-b) \\ 2a+b+c=9 \end{cases}$$

Write all equations in general form.

$$\begin{cases} a+b+2c=7 & (1) \\ a+2b+c=8 & (2) \\ 2a+b+c=9 & (3) \end{cases}$$

Multiply Equation 2 by -2 and add Equations 1 and 2 to eliminate c.

$$\begin{aligned} a+\ \ b+2c&=7 \quad (1) \\ \underline{-2a-4b-2c}&\underline{=-16} \quad (2) \\ -a-3b\qquad&=-9 \quad (4) \end{aligned}$$

Multiply Equation 2 by -1 and add Equations 2 and 3 to eliminate c.

$$\begin{aligned} -a-2b-c&=-8 \quad (2) \\ \underline{2a+\ b+\ c}&\underline{=9} \quad (3) \\ a-b\qquad&=1 \quad (5) \end{aligned}$$

Add Equations 4 and 5 to eliminate a.

$$\begin{aligned} -a-3b&=-9 \quad (4) \\ \underline{a-\ b}&\underline{=1} \quad (5) \\ -4b&=-8 \\ b&=2 \end{aligned}$$

Substitute $b=2$ into Equation 5 and solve for a.

$$\begin{aligned} a-b&=1 \quad (5) \\ a-2&=1 \\ a&=3 \end{aligned}$$

Substitute $a=3$ and $b=2$ into Equation 1 and solve for c.

$$\begin{aligned} a+b+2c&=7 \quad (1) \\ 3+2+2c&=7 \\ 5+2c&=7 \\ 2c&=2 \\ c&=1 \end{aligned}$$

The solution is $(3, 2, 1)$.

45.

$$\begin{cases} 2x+y-z=1 & (1) \\ x+2y+2z=2 & (2) \\ 4x+5y+3z=3 & (3) \end{cases}$$

Multiply Equation 1 by 2 and add Equations 1 and 2 to eliminate z.

$$\begin{array}{rl} 4x+2y-2z=2 & (1) \\ \underline{x+2y+2z=2} & (2) \\ 5x+4y=4 & (4) \end{array}$$

Multiply Equation 1 by 3 and add Equations 1 and 3 to eliminate z.

$$\begin{array}{rl} 6x+3y-3z=3 & (1) \\ \underline{4x+5y+3z=3} & (3) \\ 10x+8y\quad\;=6 & (5) \end{array}$$

Multiply Equaiton 4 by -2 and add Equations 4 and 5 to eliminate x.

$$\begin{array}{r} -10x-8y=-8 \\ \underline{10x+8y=6} \\ 0\neq -2 \end{array}$$

Since $0\neq -2$, the system is inconsistent and has no solution.

47.

$$\begin{cases} 0.4x+0.3z=0.4 & (1) \\ 2y-6z=-1 & (2) \\ 4(2x+y)=9-3z & (3) \end{cases}$$

Multiply Equation 1 by 10 to clear decimals and write Equation 3 in standard form.

$$\begin{cases} 4x+3z=4 & (1) \\ 2y-6z=-1 & (2) \\ 8x+4y+3z=9 & (3) \end{cases}$$

Multiply Equation 1 by -2 and add Equations 1 and 3 to eliminate x.

$$\begin{array}{rl} -8x\qquad -6z=-8 & (1) \\ \underline{8x+4y+3z=9} & (3) \\ 4y-3z=1 & (4) \end{array}$$

Multiply Equation 4 by -2 and add Equations 4 and 2 to eliminate z.

$$\begin{array}{rl} -8y+6z=-2 & (4) \\ \underline{2y-6z=-1} & (2) \\ -6y=-3 & \\ y=\frac{1}{2} & \end{array}$$

Substitute $y=\frac{1}{2}$ into Equation 2 and solve for z.

$$\begin{array}{rl} 2y-6z=-1 & (2) \\ 2\left(\frac{1}{2}\right)-6z=-1 & \\ 1-6z=-1 & \\ -6z=-2 & \\ z=\frac{1}{3} & \end{array}$$

Substitute $z=\frac{1}{3}$ into Equation 1 and solve for x.

$$\begin{array}{rl} 4x+3z=4 & (1) \\ 4x+3\left(\frac{1}{3}\right)=4 & \\ 4x+1=4 & \\ 4x=3 & \\ x=\frac{3}{4} & \end{array}$$

The solution is $\left(\frac{3}{4},\frac{1}{2},\frac{1}{3}\right)$.

49.

$$\begin{cases} r+s+4t=3 & (1) \\ 3r+7t=0 & (2) \\ 3s+5t=0 & (3) \end{cases}$$

Multiply Equation 1 by –3 and add Equations 1 and 2 to eliminate r.

$$\begin{array}{rr} -3r-3s-12t=-9 & (1) \\ \underline{3r \qquad +7t=0} & (2) \\ -3s-5t=-9 & (4) \end{array}$$

Add Equations 3 and 4 to eliminate s.

$$\begin{array}{rr} 3s+5t=0 & (3) \\ \underline{-3s-5t=-9} & (4) \\ 0\neq -9 & \end{array}$$

No solution; $\varnothing$

Inconsistent System

51.

$$\begin{cases} 0.5a+0.3b=2.2 & (1) \\ 1.2c-8.5b=-24.4 & (2) \\ 3.3c+1.3a=29 & (3) \end{cases}$$

Multiply each equation by 10 to eliminate decimals and write in standard form.

$$\begin{cases} 5a+3b=22 & (1) \\ -85b+12c=-244 & (2) \\ 13a+33c=290 & (3) \end{cases}$$

Multiply Equation 1 by 85 and Equation 2 by 3. Add the equations.

$$\begin{array}{rr} 425a+255b \qquad =1,870 & (1) \\ \underline{-255b+36c=-732} & (2) \\ 425a+36c=1,138 & (4) \end{array}$$

Multiply Equation 3 by -36 and Equation 4 by 33. Add the equations.

$$\begin{array}{rr} -468a-1,188c=-10,440 & (3) \\ \underline{14,025a+1,188c=37,554} & (4) \\ 13,557a=27,114 & \\ a=2 & \end{array}$$

Substitute $a=2$ into Equation 1 and solve for b.

$$\begin{aligned} 5a+3b&=22 \quad (1) \\ 5(2)+3b&=22 \\ 10+3b&=22 \\ 3b&=12 \\ b&=4 \end{aligned}$$

Substitute $a=2$ into Equation 3 and solve for c.

$$\begin{aligned} 13a+33c&=290 \quad (3) \\ 13(2)+33c&=290 \\ 26+33c&=290 \\ 33c&=264 \\ c&=8 \end{aligned}$$

The solution is $(2, 4, 8)$.

53.

$$\begin{cases} 2x+3y=6-4z & (1) \\ 2x=3y+4z-4 & (2) \\ 4x+6y+8z=12 & (3) \end{cases}$$

Write each equation in standard form.

$$\begin{cases} 2x+3y+4z=6 & (1) \\ 2x-3y-4z=-4 & (2) \\ 4x+6y+8z=12 & (3) \end{cases}$$

Multiply Equation 1 by –2 and add Equations 1 and 3 to eliminate x.

$$\begin{array}{rl} -4x-6y-8z=-12 & (1) \\ \underline{4x+6y+8z=12} & (3) \\ 0=0 & \end{array}$$

Infinitely many solutions
dependent equations

55.

$$\begin{cases} a+b=2+c \\ a=3+b-c \\ -a+b+c-4=0 \end{cases}$$

Write each equation in standard form.

$$\begin{cases} a+b-c=2 & (1) \\ a-b+c=3 & (2) \\ a-b-c=-4 & (3) \end{cases}$$

Add Equations 1 and 2 to eliminate b and c.

$$\begin{array}{rl} a+b-c=2 & (1) \\ \underline{a-b+c=3} & (2) \\ 2a=5 & \\ a=2.5 & \end{array}$$

Multiply Equation 2 by –1 and add Equations 2 and 3 to eliminate a and b.

$$\begin{array}{rl} -a+b-c=-3 & (2) \\ \underline{a-b-c=-4} & (3) \\ -2c=-7 & \\ c=3.5 & \end{array}$$

Substitute $a = 2.5$ and $c = 3.5$ into Equation 1 and solve for b.

$$\begin{array}{rl} a+b-c=2 & (1) \\ 2.5+b-3.5=2 & \\ -1+b=2 & \\ b=3 & \end{array}$$

The solution is (2.5, 3, 3.5).

57.

$$\begin{cases} x+\frac{1}{3}y+z=13 \\ \frac{1}{2}x-y+\frac{1}{3}z=-2 \\ x+\frac{1}{2}y-\frac{1}{3}z=2 \end{cases}$$

Write each equation in standard form by muliplying by a LCD to eliminate fractions.

$$\begin{cases} 3x+y+3z=39 & (1) \\ 3x-6y+2z=-12 & (2) \\ 6x+3y-2z=12 & (3) \end{cases}$$

Multiply Equation 2 by -1 and add Equations 1 and 2.

$$\begin{array}{rl} -3x+6y-2z=12 & (2) \\ \underline{3x+\ \ y+3z=39} & (1) \\ 7y+z=51 & (4) \end{array}$$

Multiply Equation 1 by -2 and add Equations 1 and 3.

$$\begin{array}{rl} -6x-2y-6z=-78 & (1) \\ \underline{6x+3y-2z=12} & (3) \\ y-8z=-66 & (5) \end{array}$$

Multiply Equation 4 by 8 and add Equations 4 and 5.

$$\begin{array}{rl} 56y+8z=408 & (4) \\ \underline{y-8z=-66} & (5) \\ 57y=342 & \\ y=6 & \end{array}$$

Substitute $y=6$ into Equation 4.

$$\begin{array}{rl} 7y+z=51 & (4) \\ 7(6)+z=51 & \\ 42+z=51 & \\ z=9 & \end{array}$$

Substitute $y=6$ and $z=9$ into Equation 1.

$$\begin{array}{rl} 3x+y+3z=39 & (1) \\ 3x+6+3(9)=39 & \\ 3x+33=39 & \\ 3x=6 & \\ x=2 & \end{array}$$

The solution is $(2,\ 6,\ 9)$.

APPLICATIONS

59. a. infinitely many solutions, all lying on the line running down the binding
    b. 3 parallel planes (shelves); no solution
    c. each pair of planes (cards) intersects; no solution
    d. 3 planes (faces of die) intersect at a corner; 1 solution

61. NBA RECORDS

$$\begin{cases} x+y+z=259 & (1) \\ x-y=19 & (2) \\ x-z=22 & (3) \end{cases}$$

Add Equations 1 and 2 to eliminate y.

$$\begin{array}{ll} x+y+z=259 & (1) \\ \underline{x-y\quad\;\;=19} & (2) \\ 2x\quad\;+z=278 & (4) \end{array}$$

Add Equations 3 and 4 to eliminate z.

$$\begin{array}{ll} x-z=22 & (3) \\ \underline{2x+z=278} & (4) \\ 3x=300 & \\ x=100 & \end{array}$$

Substitute $x = 100$ into Equation 2 and solve for y.

$$\begin{array}{ll} x-y=19 & (2) \\ 100-y=19 & \\ -y=-81 & \\ y=81 & \end{array}$$

Substitute $x = 100$ into Equation 3 and solve for z.

$$\begin{array}{ll} x-z=22 & (3) \\ 100-z=22 & \\ -z=-78 & \\ z=78 & \end{array}$$

The solutions are 100, 81, and 78.

WRITING

63. Answers will vary.

65. Answers will vary.

REVIEW

67. $f(x)=|x|$

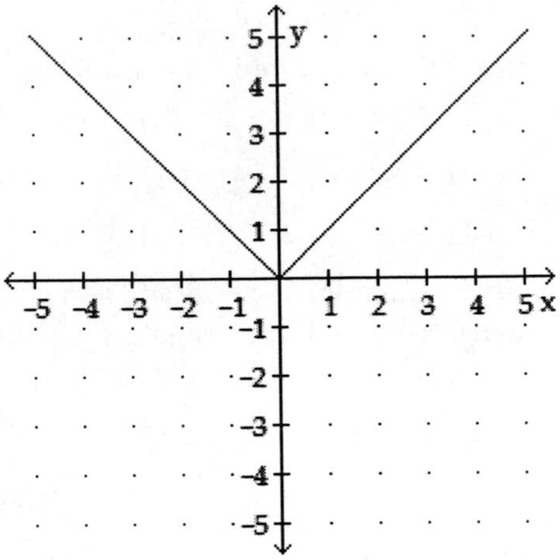

69. $h(x)=x^3$

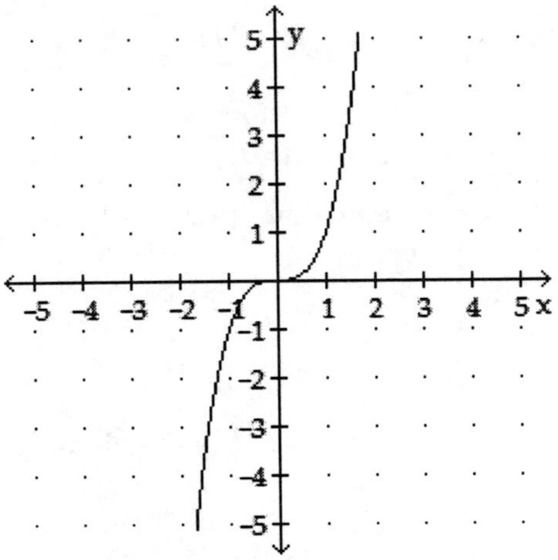

CHALLENGE PROBLEMS

71.

$$\begin{cases} w+x+y+z=3 & (1) \\ w-x+y+z=1 & (2) \\ w+x-y+z=1 & (3) \\ w+x+y-z=3 & (4) \end{cases}$$

Add Equations 3 and 4.

$$\begin{array}{rl} w+x-y+z=1 & (3) \\ \underline{w+x+y-z=3} & (4) \\ 2w+2x=4 & (5) \end{array}$$

Multiply Equation 1 by -1 and add Equations 1 and 2.

$$\begin{array}{rl} -w-x-y-z=-3 & (1) \\ \underline{w-x+y+z=1} & (2) \\ -2x=-2 & \\ x=1 & \end{array}$$

Substitute $x=1$ into Equation 5.

$$\begin{aligned} 2w+2x&=4 \quad (5) \\ 2w+2(1)&=4 \\ 2w+2&=4 \\ 2w&=2 \\ w&=1 \end{aligned}$$

Add Equations 2 and 3.

$$\begin{array}{rl} w-x+y+z=1 & (2) \\ \underline{w+x-y+z=1} & (3) \\ 2w+2z=2 & (6) \end{array}$$

Substitute $w=1$ into Equation 6.

$$\begin{aligned} 2w+2z&=2 \quad (6) \\ 2(1)+2z&=2 \\ 2+2z&=2 \\ 2z&=0 \\ z&=0 \end{aligned}$$

Substitute $x=1$, $w=1$, and $z=0$ into Equation 1.

$$\begin{aligned} w+x+y+z&=3 \quad (1) \\ 1+1+y+0&=3 \\ y+2&=3 \\ y&=1 \end{aligned}$$

The solution is $(1, 1, 1, 0)$.

73.

$$\begin{cases} 4a+b+2c-3d=-16 & \text{(Eq.1)} \\ 3a-3b+c-4d=-20 & \text{(Eq.2)} \\ a-2b-5c-d=4 & \text{(Eq.3)} \\ 5a+4b+3c-d=-10 & \text{(Eq.4)} \end{cases}$$

Multiply Equation 3 by –3 and add Equation 1 and 3 to get Equation 5.

$$\begin{array}{rl} 4a+\ b+\ 2c-3d=-16 & \text{(Eq.1)} \\ \underline{-3a+6b+15c+3d=-12} & \text{(Eq.3)} \\ a+7b+17c\qquad =-28 & \text{(Eq.5)} \end{array}$$

Multiply Equation 3 by –4 and add Equation 2 and 3 to get Equation 6.

$$\begin{array}{rl} 3a-3b+\quad c-4d=-20 & \text{(Eq.2)} \\ \underline{-4a+8b+20c+4d=-16} & \text{(Eq.3)} \\ -a+5b+21c\qquad =-36 & \text{(Eq.6)} \end{array}$$

Multiply Equation 3 by –1 and add Equation 4 and 3 to get Equation 7.

$$\begin{array}{rl} 5a+4b+3c-d=-10 & \text{(Eq.4)} \\ \underline{-a+2b+5c+d=-4} & \text{(Eq.3)} \\ 4a+6b+8c\qquad =-14 & \text{(Eq.7)} \end{array}$$

Add Equations 5 and 6 to get Equation 8.

$$\begin{array}{rl} a+7b+17c=-28 & \text{(Eq.5)} \\ \underline{-a+5b+21c=-36} & \text{(Eq.6)} \\ 12b+38c=-64 & \text{(Eq.8)} \end{array}$$

Multiply Equation 6 by 4 and add Equation 6 and 7 to get Equation 9.

$$\begin{array}{rl} -4a+20b+84c=-144 & \text{(Eq.6)} \\ \underline{4a+\ 6b+\ 8c=-14} & \text{(Eq.7)} \\ 26b+92c=-158 & \text{(Eq.9)} \end{array}$$

Multiply Equation 8 by 13 and multiply Equation 9 by –6. Add to solve for c.

$$\begin{array}{rl} 156b+494c=-832 & \text{(Eq.8)} \\ \underline{-156b-552c=948} & \text{(Eq.9)} \\ -58c=116 & \\ c=-2 & \end{array}$$

Substitute $c=-2$ into Equation 9 and solve for b.

$$\begin{aligned} 26b+92c&=-158 \\ 26b+92(-2)&=-158 \\ 26b-184&=-158 \\ 26b&=26 \\ b&=1 \end{aligned}$$

Substitute $c=-2$ and $b=1$ into Equation 5 and solve for a.

$$\begin{aligned} a+7b+17c&=-28 \\ a+7(1)+17(-2)&=-28 \\ a+7-34&=-28 \\ a-27&=-28 \\ a&=-1 \end{aligned}$$

Substitute $c=-2$, $b=1$ and $a=-1$ into Equation 3 and solve for d.

$$\begin{aligned} a-2b-5c-d&=4 \\ (-1)-2(1)-5(-2)-d&=4 \\ -1-2+10-d&=4 \\ 7-d&=4 \\ -d&=-3 \\ d&=3 \end{aligned}$$

The solution is (–1, 1, –2, 3).

SECTION 3.4

VOCABULARY

1. A **matrix** is a rectangular array of numbers written within brackets.

3. Of the order of a matrix is 3x4, it had 3 **rows** and 4 **columns**. We read 3x4 as "3 **by** 4."

5. Elementary **row** operations are used on the augmented matrix to produce a simpler matrix that gives to the solution of a system. This process is called **Gauss-Jordan**.

CONCEPTS

7. a. 2×3
   b. 3×4

9. a. $\begin{cases} x=-10 \\ y=6 \end{cases}$; The solution of the system is $\left(\boxed{10},\boxed{6}\right)$.

   b. $\begin{cases} x=-16 \\ y=8 \\ z=4 \end{cases}$; The solution of the system is $\left(\boxed{-16},\boxed{8},\boxed{4}\right)$.

NOTATION

11. a. interchange rows 1 and 2
    b. multiply row 1 by $\frac{1}{2}$
    c. add row 3 to 6 times row 2

GUIDED PRACTICE

13. $\left[\begin{array}{cc|c} 1 & 2 & 6 \\ 3 & -1 & -10 \end{array}\right]$

15. $\begin{cases} x+6y=7 \\ \quad\quad y=4 \end{cases}$

17. $\left[\begin{array}{cc|c} 1 & -4 & 4 \\ -3 & 1 & -6 \end{array}\right]$

19. $\left[\begin{array}{cc|c} 1 & -\frac{1}{3} & 2 \\ 1 & -4 & 4 \end{array}\right]$

21. $\left[\begin{array}{ccc|c} 3 & 6 & -9 & 0 \\ -2 & 2 & -2 & 5 \\ 1 & 5 & -2 & 1 \end{array}\right]$

23. $\left[\begin{array}{cccc} 1 & 2 & -3 & 0 \\ 0 & 3 & 1 & 1 \\ -2 & 2 & -2 & 5 \end{array}\right]$

25.

$\left[\begin{array}{cc|c} 1 & 1 & 2 \\ 1 & -1 & 0 \end{array}\right]$

$-R_1+R_2$

$\left[\begin{array}{cc|c} 1 & 1 & 2 \\ 0 & -2 & -2 \end{array}\right]$

$-\frac{1}{2}R_2$

$\left[\begin{array}{cc|c} 1 & 1 & 2 \\ 0 & 1 & 1 \end{array}\right]$

R_1-R_2

$\left[\begin{array}{cc|c} 1 & 0 & 1 \\ 0 & 1 & 1 \end{array}\right]$

This matrix represents the system

$\begin{cases} x=1 \\ y=1 \end{cases}$

The solution is $(1, 1)$.

27.

$$\left[\begin{array}{cc|c} 2 & 1 & 1 \\ 1 & 2 & -4 \end{array}\right]$$

$R_1 \leftrightarrow R_2$

$$\left[\begin{array}{cc|c} 1 & 2 & -4 \\ 2 & 1 & 1 \end{array}\right]$$

$-2R_1 + R_2$

$$\left[\begin{array}{cc|c} 1 & 2 & -4 \\ 0 & -3 & 9 \end{array}\right]$$

$-\frac{1}{3}R_2$

$$\left[\begin{array}{cc|c} 1 & 2 & -4 \\ 0 & 1 & -3 \end{array}\right]$$

$-2R_2 + R_1$

$$\left[\begin{array}{cc|c} 1 & 0 & 2 \\ 0 & 1 & -3 \end{array}\right]$$

This matrix represents the system

$$\begin{cases} x = -2 \\ y = -3 \end{cases}$$

The solution is $(2,\ -3)$.

29.

$$\left[\begin{array}{ccc|c} 1 & 1 & 1 & 6 \\ 1 & 2 & 1 & 8 \\ 1 & 1 & 2 & 7 \end{array}\right]$$

$-1R_1 + R_2$

$$\left[\begin{array}{ccc|c} 1 & 1 & 1 & 6 \\ 0 & 1 & 0 & 2 \\ 1 & 1 & 2 & 7 \end{array}\right]$$

$-1R_1 + R_3$

$$\left[\begin{array}{ccc|c} 1 & 1 & 1 & 6 \\ 0 & 1 & 0 & 2 \\ 0 & 0 & 1 & 1 \end{array}\right]$$

$-1R_2 + R_1$

$$\left[\begin{array}{ccc|c} 1 & 0 & 1 & 4 \\ 0 & 1 & 0 & 2 \\ 0 & 0 & 1 & 1 \end{array}\right]$$

$-1R_3 + R_1$

$$\left[\begin{array}{ccc|c} 1 & 0 & 0 & 3 \\ 0 & 1 & 0 & 2 \\ 0 & 0 & 1 & 1 \end{array}\right]$$

This matrix represents the system

$$\begin{cases} x = 3 \\ y = 2 \\ z = 1 \end{cases}$$

The solution is $(3,\ 2,\ 1)$.

31.

$$\left[\begin{array}{ccc|c} 3 & 1 & -3 & 5 \\ 1 & -2 & 4 & 10 \\ 1 & 1 & 1 & 13 \end{array}\right]$$

$R_1 \leftrightarrow R_3$

$$\left[\begin{array}{ccc|c} 1 & 1 & 1 & 13 \\ 1 & -2 & 4 & 10 \\ 3 & 1 & -3 & 5 \end{array}\right]$$

$-1R_1 + R_2$

$$\left[\begin{array}{ccc|c} 1 & 1 & 1 & 13 \\ 0 & -3 & 3 & -3 \\ 3 & 1 & -3 & 5 \end{array}\right]$$

$-\frac{1}{3}R_2$

$$\left[\begin{array}{ccc|c} 1 & 1 & 1 & 13 \\ 0 & 1 & -1 & 1 \\ 3 & 1 & -3 & 5 \end{array}\right]$$

$-3R_1 + R_3$

$$\left[\begin{array}{ccc|c} 1 & 1 & 1 & 13 \\ 0 & 1 & -1 & 1 \\ 0 & -2 & -6 & -34 \end{array}\right]$$

$2R_2 + R_3$

$$\left[\begin{array}{ccc|c} 1 & 1 & 1 & 13 \\ 0 & 1 & -1 & 1 \\ 0 & 0 & -8 & -32 \end{array}\right]$$

$-\frac{1}{8}R_3$

$$\left[\begin{array}{ccc|c} 1 & 1 & 1 & 13 \\ 0 & 1 & -1 & 1 \\ 0 & 0 & 1 & 4 \end{array}\right]$$

$-R_2 + R_1$

$$\left[\begin{array}{ccc|c} 1 & 0 & 2 & 12 \\ 0 & 1 & -1 & 1 \\ 0 & 0 & 1 & 4 \end{array}\right]$$

$-2R_3 + R_1$

$$\left[\begin{array}{ccc|c} 1 & 0 & 0 & 4 \\ 0 & 1 & -1 & 1 \\ 0 & 0 & 1 & 4 \end{array}\right]$$

$R_3 + R_2$

$$\left[\begin{array}{ccc|c} 1 & 0 & 0 & 4 \\ 0 & 1 & 0 & 5 \\ 0 & 0 & 1 & 4 \end{array}\right]$$

This matrix represents the system

$$\begin{cases} x = 4 \\ y = 5 \\ z = 4 \end{cases}$$

The solution is $(4, 5, 4)$.

33.

$$\left[\begin{array}{cc|c} 1 & -3 & 9 \\ -2 & 6 & 18 \end{array}\right]$$

$2R_1 + R_2$

$$\left[\begin{array}{cc|c} 1 & -3 & 9 \\ 0 & 0 & 36 \end{array}\right]$$

This matrix represents the system

$$\begin{cases} x - 3y = 9 \\ 0 + 0 = 36 \end{cases}$$

No solution. The system is inconsistent.

35.

$$\left[\begin{array}{cc|c} -4 & -4 & -12 \\ 1 & 1 & 3 \end{array}\right]$$

$\frac{1}{4}R_1$

$$\left[\begin{array}{cc|c} -1 & -1 & -3 \\ 1 & 1 & 3 \end{array}\right]$$

$R_1 \leftrightarrow R_2$

$$\left[\begin{array}{cc|c} 1 & 1 & 3 \\ -1 & -1 & -3 \end{array}\right]$$

$R_1 + R_2$

$$\left[\begin{array}{cc|c} 1 & 1 & 3 \\ 0 & 0 & 0 \end{array}\right]$$

This matrix represents the system

$$\begin{cases} x + y = 3 \\ 0 + 0 = 0 \end{cases}$$

$\{(x, y) | x + y = 3\}$

Infinitely many solutions.

The equations are dependent.

37.

$$\left[\begin{array}{ccc:c} 2 & 3 & -1 & -8 \\ 1 & -1 & -1 & -2 \\ -4 & 3 & 1 & 6 \end{array}\right]$$

$R_1 \leftrightarrow R_2$

$$\left[\begin{array}{ccc:c} 1 & -1 & -1 & -2 \\ 2 & 3 & -1 & -8 \\ -4 & 3 & 1 & 6 \end{array}\right]$$

$-2R_1 + R_2$

$$\left[\begin{array}{ccc:c} 1 & -1 & -1 & -2 \\ 0 & 5 & 1 & -4 \\ -4 & 3 & 1 & 6 \end{array}\right]$$

$4R_1 + R_3$

$$\left[\begin{array}{ccc:c} 1 & -1 & -1 & -2 \\ 0 & 5 & 1 & -4 \\ 0 & -1 & -3 & -2 \end{array}\right]$$

$-R_3 \leftrightarrow R_2$

$$\left[\begin{array}{ccc:c} 1 & -1 & -1 & -2 \\ 0 & 1 & 3 & 2 \\ 0 & 5 & 1 & -4 \end{array}\right]$$

$-5R_2 + R_3$

$$\left[\begin{array}{ccc:c} 1 & -1 & -1 & -2 \\ 0 & 1 & 3 & 2 \\ 0 & 0 & -14 & -14 \end{array}\right]$$

$-\frac{1}{14}R_3$

$$\left[\begin{array}{ccc:c} 1 & -1 & -1 & -2 \\ 0 & 1 & 3 & 2 \\ 0 & 0 & 1 & 1 \end{array}\right]$$

$R_2 + R_1$

$$\left[\begin{array}{ccc:c} 1 & 0 & 2 & 0 \\ 0 & 1 & 3 & 2 \\ 0 & 0 & 1 & 1 \end{array}\right]$$

$-2R_3 + R_1$

$$\left[\begin{array}{ccc:c} 1 & 0 & 0 & -2 \\ 0 & 1 & 3 & 2 \\ 0 & 0 & 1 & 1 \end{array}\right]$$

$-3R_3 + R_2$

$$\left[\begin{array}{ccc:c} 1 & 0 & 0 & -2 \\ 0 & 1 & 0 & -1 \\ 0 & 0 & 1 & 1 \end{array}\right]$$

The solution is $(-2, -1, 1)$.

39.

$$\left[\begin{array}{cc:c} 2 & -1 & -1 \\ 1 & -2 & 1 \end{array}\right]$$

$R_1 \leftrightarrow R_2$

$$\left[\begin{array}{cc:c} 1 & -2 & 1 \\ 2 & -1 & -1 \end{array}\right]$$

$-2R_1 + R_2$

$$\left[\begin{array}{cc:c} 1 & -2 & 1 \\ 0 & 3 & -3 \end{array}\right]$$

$\frac{1}{3}R_2$

$$\left[\begin{array}{cc:c} 1 & -2 & 1 \\ 0 & 1 & -1 \end{array}\right]$$

$2R_2 + R_1$

$$\left[\begin{array}{cc:c} 1 & 0 & -1 \\ 0 & 1 & -1 \end{array}\right]$$

The solution is $(-1, -1)$.

41.

$$\left[\begin{array}{cc:c} 3 & 4 & -12 \\ 9 & -2 & 6 \end{array}\right]$$

$\frac{1}{3}R_1$

$$\left[\begin{array}{cc:c} 1 & \frac{4}{3} & -4 \\ 9 & -2 & 6 \end{array}\right]$$

$-9R_1 + R_2$

$$\left[\begin{array}{cc:c} 1 & \frac{4}{3} & -4 \\ 0 & -14 & 42 \end{array}\right]$$

$-\frac{1}{14}R_2$

$$\left[\begin{array}{cc:c} 1 & \frac{4}{3} & -4 \\ 0 & 1 & -3 \end{array}\right]$$

$-\frac{4}{3}R_2 + R_1$

$$\left[\begin{array}{cc:c} 1 & 0 & 0 \\ 0 & 1 & -3 \end{array}\right]$$

The solution is $(0, -3)$.

43.

$$\left[\begin{array}{ccc|c} 2 & 1 & -1 & 1 \\ 1 & 2 & 2 & 2 \\ 4 & 5 & 3 & 3 \end{array}\right]$$

$R_1 \leftrightarrow R_2$

$$\left[\begin{array}{ccc|c} 1 & 2 & 2 & 2 \\ 2 & 1 & -1 & 1 \\ 4 & 5 & 3 & 3 \end{array}\right]$$

$-2R_1 + R_2$

$$\left[\begin{array}{ccc|c} 1 & 2 & 2 & 2 \\ 0 & -3 & -5 & -3 \\ 4 & 5 & 3 & 3 \end{array}\right]$$

$-\frac{1}{3}R_2$

$$\left[\begin{array}{ccc|c} 1 & 2 & 2 & 2 \\ 0 & 1 & \frac{5}{3} & 1 \\ 4 & 5 & 3 & 3 \end{array}\right]$$

$-4R_1 + R_3$

$$\left[\begin{array}{ccc|c} 1 & 2 & 2 & 2 \\ 0 & 1 & \frac{5}{3} & 1 \\ 0 & -3 & -5 & -5 \end{array}\right]$$

$3R_2 + R_3$

$$\left[\begin{array}{ccc|c} 1 & 2 & 2 & 2 \\ 0 & 1 & \frac{5}{3} & 1 \\ 0 & 0 & 0 & -2 \end{array}\right]$$

This system is inconsistent.

No solution.

45.

$$\left[\begin{array}{cc|c} 8 & -2 & 4 \\ 4 & -1 & 2 \end{array}\right]$$

$\frac{1}{8}R_1$

$$\left[\begin{array}{cc|c} 1 & -\frac{1}{4} & \frac{1}{2} \\ 4 & -1 & 2 \end{array}\right]$$

$-4R_1 + R_2$

$$\left[\begin{array}{cc|c} 1 & -\frac{1}{4} & \frac{1}{2} \\ 0 & 0 & 0 \end{array}\right]$$

Infinitely many solutions.

The equations are dependent.

47.

$$\left[\begin{array}{ccc|c} 2 & 1 & -2 & 6 \\ 4 & -1 & 1 & -1 \\ 6 & -2 & 3 & -5 \end{array}\right]$$

$\frac{1}{2}R_1$

$$\left[\begin{array}{ccc|c} 1 & \frac{1}{2} & -1 & 3 \\ 4 & -1 & 1 & -1 \\ 6 & -2 & 3 & -5 \end{array}\right]$$

$-4R_1 + R_2$

$$\left[\begin{array}{ccc|c} 1 & \frac{1}{2} & -1 & 3 \\ 0 & -3 & 5 & -13 \\ 6 & -2 & 3 & -5 \end{array}\right]$$

$-6R_1 + R_3$

$$\left[\begin{array}{ccc|c} 1 & \frac{1}{2} & -1 & 3 \\ 0 & -3 & 5 & -13 \\ 0 & -5 & 9 & -23 \end{array}\right]$$

$-\frac{1}{3}R_2$

$$\left[\begin{array}{ccc|c} 1 & \frac{1}{2} & -1 & 3 \\ 0 & 1 & -\frac{5}{3} & \frac{13}{3} \\ 0 & -5 & 9 & -23 \end{array}\right]$$

$5R_2 + R_3$

$$\left[\begin{array}{ccc|c} 1 & \frac{1}{2} & -1 & 3 \\ 0 & 1 & -\frac{5}{3} & \frac{13}{3} \\ 0 & 0 & \frac{2}{3} & -\frac{4}{3} \end{array}\right]$$

$\frac{3}{2}R_3$

$$\left[\begin{array}{ccc|c} 1 & \frac{1}{2} & -1 & 3 \\ 0 & 1 & -\frac{5}{3} & \frac{13}{3} \\ 0 & 0 & 1 & -2 \end{array}\right]$$

$-\frac{1}{2}R_2 + R_1$

$$\left[\begin{array}{ccc|c} 1 & 0 & -\frac{1}{6} & \frac{5}{6} \\ 0 & 1 & -\frac{5}{3} & \frac{13}{3} \\ 0 & 0 & 1 & -2 \end{array}\right]$$

$\frac{1}{6}R_3 + R_1$

$$\left[\begin{array}{ccc|c} 1 & 0 & 0 & \frac{1}{2} \\ 0 & 1 & -\frac{5}{3} & \frac{13}{3} \\ 0 & 0 & 1 & -2 \end{array}\right]$$

$\frac{5}{3}R_3 + R_2$

$$\left[\begin{array}{ccc|c} 1 & 0 & 0 & \frac{1}{2} \\ 0 & 1 & 0 & 1 \\ 0 & 0 & 1 & -2 \end{array}\right]$$

The solution is $\left(\frac{1}{2}, 1, -2\right)$.

49.

$$\left[\begin{array}{ccc|c} 6 & 1 & -1 & -2 \\ 1 & 2 & 1 & 5 \\ 0 & 5 & -1 & 2 \end{array}\right]$$

$R_1 \leftrightarrow R_2$

$$\left[\begin{array}{ccc|c} 1 & 2 & 1 & 5 \\ 6 & 1 & -1 & -2 \\ 0 & 5 & -1 & 2 \end{array}\right]$$

$-6R_1 + R_2$

$$\left[\begin{array}{ccc|c} 1 & 2 & 1 & 5 \\ 0 & -11 & -7 & -32 \\ 0 & 5 & -1 & 2 \end{array}\right]$$

$-\frac{1}{11}R_2$

$$\left[\begin{array}{ccc|c} 1 & 2 & 1 & 5 \\ 0 & 1 & \frac{7}{11} & \frac{32}{11} \\ 0 & 5 & -1 & 2 \end{array}\right]$$

$-5R_2 + R_3$

$$\left[\begin{array}{ccc|c} 1 & 2 & 1 & 5 \\ 0 & 1 & \frac{7}{11} & \frac{32}{11} \\ 0 & 0 & -\frac{46}{11} & -\frac{138}{11} \end{array}\right]$$

$-\frac{11}{46}R_3$

$$\left[\begin{array}{ccc|c} 1 & 2 & 1 & 5 \\ 0 & 1 & \frac{7}{11} & \frac{32}{11} \\ 0 & 0 & 1 & 3 \end{array}\right]$$

$-\frac{7}{11}R_3 + R_2$

$$\left[\begin{array}{ccc|c} 1 & 2 & 1 & 5 \\ 0 & 1 & 0 & 1 \\ 0 & 0 & 1 & 3 \end{array}\right]$$

$-2R_2 + R_1$

$$\left[\begin{array}{ccc|c} 1 & 0 & 1 & 3 \\ 0 & 1 & 0 & 1 \\ 0 & 0 & 1 & 3 \end{array}\right]$$

$-R_3 + R_1$

$$\left[\begin{array}{ccc|c} 1 & 0 & 0 & 0 \\ 0 & 1 & 0 & 1 \\ 0 & 0 & 1 & 3 \end{array}\right]$$

The solution is $(0, 1, 3)$.

51.

$$\left[\begin{array}{ccc|c} 5 & 3 & 0 & 4 \\ 0 & 3 & -4 & 4 \\ 1 & 0 & 1 & 1 \end{array}\right]$$

$\frac{1}{5}R_1$

$$\left[\begin{array}{ccc|c} 1 & \frac{3}{5} & 0 & \frac{4}{5} \\ 0 & 3 & -4 & 4 \\ 1 & 0 & 1 & 1 \end{array}\right]$$

$-R_1 + R_3$

$$\left[\begin{array}{ccc|c} 1 & \frac{3}{5} & 0 & \frac{4}{5} \\ 0 & 3 & -4 & 4 \\ 0 & -\frac{3}{5} & 1 & \frac{1}{5} \end{array}\right]$$

$\frac{1}{3}R_2$

$$\left[\begin{array}{ccc|c} 1 & \frac{3}{5} & 0 & \frac{4}{5} \\ 0 & 1 & -\frac{4}{3} & \frac{4}{3} \\ 0 & -\frac{3}{5} & 1 & \frac{1}{5} \end{array}\right]$$

$\frac{3}{5}R_2 + R_3$

$$\left[\begin{array}{ccc|c} 1 & \frac{3}{5} & 0 & \frac{4}{5} \\ 0 & 1 & -\frac{4}{3} & \frac{4}{3} \\ 0 & 0 & \frac{1}{5} & 1 \end{array}\right]$$

$5R_3$

$$\left[\begin{array}{ccc|c} 1 & \frac{3}{5} & 0 & \frac{4}{5} \\ 0 & 1 & -\frac{4}{3} & \frac{4}{3} \\ 0 & 0 & 1 & 5 \end{array}\right]$$

$\frac{4}{3}R_3 + R_2$

$$\left[\begin{array}{ccc|c} 1 & \frac{3}{5} & 0 & \frac{4}{5} \\ 0 & 1 & 0 & 8 \\ 0 & 0 & 1 & 5 \end{array}\right]$$

$-\frac{3}{5}R_2 + R_1$

$$\left[\begin{array}{ccc|c} 1 & 0 & 0 & -4 \\ 0 & 1 & 0 & 8 \\ 0 & 0 & 1 & 5 \end{array}\right]$$

The solution is $(-4, 8, 5)$.

APPLICATIONS

53. DIGITAL PHOTOGRAPHY
$512(512) = 262{,}144$

55. COMPLEMENTARY ANGLES

$$\begin{cases} x+y=90 \\ y=x+46 \end{cases}$$

Write the equations in standard form.

$$\begin{cases} x+y=90 \\ -x+y=46 \end{cases}$$

$$\left[\begin{array}{cc|c} 1 & 1 & 90 \\ -1 & 1 & 46 \end{array}\right]$$

$R_1 + R_2 = R_2$

$$\left[\begin{array}{cc|c} 1 & 1 & 90 \\ 0 & 2 & 136 \end{array}\right]$$

$\frac{1}{2}R_2$

$$\left[\begin{array}{cc|c} 1 & 1 & 90 \\ 0 & 1 & 68 \end{array}\right]$$

$-R_2 + R_1$

$$\left[\begin{array}{cc|c} 1 & 0 & 22 \\ 0 & 1 & 68 \end{array}\right]$$

The measures of the angles are 22° and 68°.

57. TRIANGLES

$$\begin{cases} A+B+C=180 \\ B=25+A \\ C=2A-5 \end{cases}$$

Write the equations in standard form.

$$\begin{cases} A+B+C=180 \\ A-B=-25 \\ 2A-C=5 \end{cases}$$

$$\left[\begin{array}{ccc|c} 1 & 1 & 1 & 180 \\ 1 & -1 & 0 & -25 \\ 2 & 0 & -1 & 5 \end{array}\right]$$

$-R_1 + R_2$

$$\left[\begin{array}{ccc|c} 1 & 1 & 1 & 180 \\ 0 & -2 & -1 & -205 \\ 2 & 0 & -1 & 5 \end{array}\right]$$

$-\frac{1}{2}R_2$

$$\left[\begin{array}{ccc|c} 1 & 1 & 1 & 180 \\ 0 & 1 & \frac{1}{2} & 102.5 \\ 2 & 0 & -1 & 5 \end{array}\right]$$

$-2R_1 + R_3$

$$\left[\begin{array}{ccc|c} 1 & 1 & 1 & 180 \\ 0 & 1 & \frac{1}{2} & 102.5 \\ 0 & -2 & -3 & -355 \end{array}\right]$$

$2R_2 + R_3$

$$\left[\begin{array}{ccc|c} 1 & 1 & 1 & 180 \\ 0 & 1 & \frac{1}{2} & 102.5 \\ 0 & 0 & -2 & -150 \end{array}\right]$$

$-\frac{1}{2}R_3$

$$\left[\begin{array}{ccc|c} 1 & 1 & 1 & 180 \\ 0 & 1 & \frac{1}{2} & 102.5 \\ 0 & 0 & 1 & 75 \end{array}\right]$$

$-\frac{1}{2}R_3 + R_2$

$$\left[\begin{array}{ccc|c} 1 & 1 & 1 & 180 \\ 0 & 1 & 0 & 65 \\ 0 & 0 & 1 & 75 \end{array}\right]$$

$-R_2 + R_1$

$$\left[\begin{array}{ccc|c} 1 & 0 & 1 & 115 \\ 0 & 1 & 0 & 65 \\ 0 & 0 & 1 & 75 \end{array}\right]$$

$-R_3 + R_1$

$$\left[\begin{array}{ccc|c} 1 & 0 & 0 & 40 \\ 0 & 1 & 0 & 65 \\ 0 & 0 & 1 & 75 \end{array}\right]$$

The measures of the angles are 40°, 65°, and 75°.

WRITING

59. Answers will vary.

61. Answers will vary.

REVIEW

63. $m = \dfrac{y_2 - y_1}{x_2 - x_1} \left(x_2 \neq x_1\right)$

CHALLENGE PROBLEMS

67. Let $a = x^2$, $b = y^2$, and $c = z^2$.

$$\left[\begin{array}{ccc|c} 1 & 1 & 1 & 14 \\ 2 & 3 & -2 & -7 \\ 1 & -5 & 1 & 8 \end{array}\right]$$

$-2R_1 + R_2$

$$\left[\begin{array}{ccc|c} 1 & 1 & 1 & 14 \\ 0 & 1 & -4 & -35 \\ 1 & -5 & 1 & 8 \end{array}\right]$$

$-R_1 + R_3$

$$\left[\begin{array}{ccc|c} 1 & 1 & 1 & 14 \\ 0 & 1 & -4 & -35 \\ 0 & -6 & 0 & -6 \end{array}\right]$$

$-\frac{1}{6}R_3$

$$\left[\begin{array}{ccc|c} 1 & 1 & 1 & 14 \\ 0 & 1 & -4 & -35 \\ 0 & 1 & 0 & 1 \end{array}\right]$$

$R_2 \leftrightarrow R_3$

$$\left[\begin{array}{ccc|c} 1 & 1 & 1 & 14 \\ 0 & 1 & 0 & 1 \\ 0 & 1 & -4 & -35 \end{array}\right]$$

$-R_2 + R_3$

$$\left[\begin{array}{ccc|c} 1 & 1 & 1 & 14 \\ 0 & 1 & 0 & 1 \\ 0 & 0 & -4 & -36 \end{array}\right]$$

$-\frac{1}{4}R_3$

$$\left[\begin{array}{ccc|c} 1 & 1 & 1 & 14 \\ 0 & 1 & 0 & 1 \\ 0 & 0 & 1 & 9 \end{array}\right]$$

$-R_2 + R_1$

$$\left[\begin{array}{ccc|c} 1 & 0 & 1 & 13 \\ 0 & 1 & 0 & 1 \\ 0 & 0 & 1 & 9 \end{array}\right]$$

$-R_3 + R_1$

$$\left[\begin{array}{ccc|c} 1 & 0 & 0 & 4 \\ 0 & 1 & 0 & 1 \\ 0 & 0 & 1 & 9 \end{array}\right]$$

Use substitution to solve for x, y, and z.

$$x^2 = a$$
$$x^2 = 4$$
$$x^2 = \pm 2$$

$$y^2 = b$$
$$y^2 = 1$$
$$y = \pm 1$$

$$z^2 = c$$
$$z^2 = 9$$
$$z = \pm 3$$

The solution is $(\pm 2, \pm 1, \pm 3)$.

SECTION 3.5

VOCABULARY

1. $\begin{vmatrix} 4 & 9 \\ -6 & 1 \end{vmatrix}$ is a 2 × 2 **determinant**. The numbers 4 and 1 lie along its main **diagonal**.

3. The **minor** of b_1 in $\begin{vmatrix} a_1 & b_1 & c_1 \\ a_2 & b_2 & c_2 \\ a_3 & b_3 & c_3 \end{vmatrix}$ is $\begin{vmatrix} a_2 & c_2 \\ a_3 & c_3 \end{vmatrix}$.

CONCEPTS

5. $\begin{vmatrix} a & b \\ c & d \end{vmatrix} = \boxed{ad} - \boxed{bc}$

7. It was expanded about the third column

9. $\begin{vmatrix} 1 & 2 & 0 \\ 3 & 1 & -1 \\ 8 & 4 & -1 \end{vmatrix}$

11. $\left(\frac{-28}{14}, \frac{-14}{14}, \frac{14}{14}\right) = (-2, -1, 1)$

NOTATION

13.

$$\begin{vmatrix} 5 & -2 \\ -2 & 6 \end{vmatrix} = 5(\boxed{6}) - (-2)(-2)$$
$$= \boxed{30} - 4$$
$$= 26$$

PRACTICE

15.

$$\begin{vmatrix} 2 & 3 \\ 2 & 5 \end{vmatrix} = 2(5) - 3(2)$$
$$= 10 - 6$$
$$= 4$$

17.

$$\begin{vmatrix} -9 & 7 \\ 4 & -2 \end{vmatrix} = -9(-2) - 7(4)$$
$$= 18 - 28$$
$$= -10$$

19.

$$\begin{vmatrix} 5 & 20 \\ 10 & 6 \end{vmatrix} = 5(6) - 20(10)$$
$$= 30 - 200$$
$$= -170$$

21.

$$\begin{vmatrix} -6 & -2 \\ 15 & 4 \end{vmatrix} = -6(4) - (-2)(15)$$
$$= -24 - (-30)$$
$$= 6$$

23.

$$\begin{vmatrix} -9 & -1 \\ -10 & -5 \end{vmatrix} = -9(-5) - (-1)(-10)$$
$$= 45 - 10$$
$$= 35$$

25.

$$\begin{vmatrix} 8 & 8 \\ -9 & -9 \end{vmatrix} = 8(-9) - 8(-9)$$
$$= -72 + 72$$
$$= 0$$

27.

$$\begin{vmatrix} 3 & 2 & 1 \\ 4 & 1 & 2 \\ 5 & 3 & 1 \end{vmatrix} = 3\begin{vmatrix} 1 & 2 \\ 3 & 1 \end{vmatrix} - 2\begin{vmatrix} 4 & 2 \\ 5 & 1 \end{vmatrix} + 1\begin{vmatrix} 4 & 1 \\ 5 & 3 \end{vmatrix}$$
$$= 3(1-6) - 2(4-10) + 1(12-5)$$
$$= 3(-5) - 2(-6) + 1(7)$$
$$= -15 + 12 + 7$$
$$= 4$$

29.

$$\begin{vmatrix} 1 & -2 & 3 \\ -2 & 1 & 1 \\ -3 & -2 & 1 \end{vmatrix} = 1\begin{vmatrix} 1 & 1 \\ -2 & 1 \end{vmatrix} - (-2)\begin{vmatrix} -2 & 1 \\ -3 & 1 \end{vmatrix} + 3\begin{vmatrix} -2 & 1 \\ -3 & -2 \end{vmatrix}$$

$$= 1(1+2) + 2(-2+3) + 3(4+3)$$
$$= 1(3) + 2(1) + 3(7)$$
$$= 3 + 2 + 21$$
$$= 26$$

31.

$$\begin{vmatrix} -2 & 5 & 1 \\ 0 & 3 & 4 \\ -1 & 2 & 6 \end{vmatrix} = -2\begin{vmatrix} 3 & 4 \\ 2 & 6 \end{vmatrix} - 5\begin{vmatrix} 0 & 4 \\ -1 & 6 \end{vmatrix} + 1\begin{vmatrix} 0 & 3 \\ -1 & 2 \end{vmatrix}$$

$$= -2(18-8) - 5(0+4) + 1(0+3)$$
$$= -2(10) - 5(4) + 1(3)$$
$$= -20 - 20 + 3$$
$$= -37$$

33.

$$\begin{vmatrix} 1 & -4 & 1 \\ 3 & 0 & -2 \\ 3 & 1 & -2 \end{vmatrix} = 1\begin{vmatrix} 0 & -2 \\ 1 & -2 \end{vmatrix} - (-4)\begin{vmatrix} 3 & -2 \\ 3 & -2 \end{vmatrix} + 1\begin{vmatrix} 3 & 0 \\ 3 & 1 \end{vmatrix}$$

$$= 1(0+2) + 4(-6+6) + 1(3-0)$$
$$= 1(2) + 4(0) + 1(3)$$
$$= 2 + 0 + 3$$
$$= 5$$

35.

$$\begin{vmatrix} 1 & 2 & 1 \\ -3 & 7 & 3 \\ -4 & 3 & -5 \end{vmatrix} = 1\begin{vmatrix} 7 & 3 \\ 3 & -5 \end{vmatrix} - 2\begin{vmatrix} -3 & 3 \\ -4 & -5 \end{vmatrix} + 1\begin{vmatrix} -3 & 7 \\ -4 & 3 \end{vmatrix}$$

$$= 1(-35-9) - 2(15+12) + 1(-9+28)$$
$$= 1(-44) - 2(27) + 1(19)$$
$$= -44 - 54 + 19$$
$$= -79$$

37.

$$\begin{vmatrix} 1 & 2 & 0 \\ 0 & 1 & 2 \\ 0 & 0 & 1 \end{vmatrix} = 1\begin{vmatrix} 1 & 2 \\ 0 & 1 \end{vmatrix} - 2\begin{vmatrix} 0 & 2 \\ 0 & 1 \end{vmatrix} + 0\begin{vmatrix} 0 & 1 \\ 0 & 0 \end{vmatrix}$$

$$= 1(1-0) - 2(0-0) + 0(0-0)$$
$$= 1(1) - 2(0) + 0(0)$$
$$= 1 - 0 + 0$$
$$= 1$$

39.

$$x = \frac{D_x}{D} = \frac{\begin{vmatrix} 6 & 1 \\ 2 & -1 \end{vmatrix}}{\begin{vmatrix} 1 & 1 \\ 1 & -1 \end{vmatrix}} = \frac{6(-1)-1(2)}{1(-1)-1(1)} = \frac{-6-2}{-1-1} = \frac{-8}{-2} = 4$$

$$y = \frac{D_y}{D} = \frac{\begin{vmatrix} 1 & 6 \\ 1 & 2 \end{vmatrix}}{\begin{vmatrix} 1 & 1 \\ 1 & -1 \end{vmatrix}} = \frac{1(2)-6(1)}{1(-1)-1(1)} = \frac{2-6}{-1-1} = \frac{-4}{-2} = 2$$

The solution is (4, 2).

41.

$$x = \frac{D_x}{D} = \frac{\begin{vmatrix} -21 & 2 \\ 11 & -2 \end{vmatrix}}{\begin{vmatrix} 1 & 2 \\ 1 & -2 \end{vmatrix}} = \frac{-21(-2)-2(11)}{1(-2)-2(1)} = \frac{42-22}{-2-2} = \frac{20}{-4} = -5$$

$$y = \frac{D_y}{D} = \frac{\begin{vmatrix} 1 & -21 \\ 1 & 11 \end{vmatrix}}{\begin{vmatrix} 1 & 2 \\ 1 & -2 \end{vmatrix}} = \frac{1(11)-(-21)(1)}{1(-2)-2(1)} = \frac{11+21}{-2-2} = \frac{32}{-4} = -8$$

The solution is (–5, –8).

43.

$$x=\frac{D_x}{D}=\frac{\begin{vmatrix}9 & -4\\ 8 & 2\end{vmatrix}}{\begin{vmatrix}3 & -4\\ 1 & 2\end{vmatrix}}=\frac{9(2)-(-4)(8)}{3(2)-(-4)(1)}=\frac{18+32}{6+4}=\frac{50}{10}=5$$

$$y=\frac{D_y}{D}=\frac{\begin{vmatrix}3 & 9\\ 1 & 8\end{vmatrix}}{\begin{vmatrix}3 & -4\\ 1 & 2\end{vmatrix}}=\frac{3(8)-9(1)}{3(2)-(-4)(1)}=\frac{24-9}{6+4}=\frac{15}{10}=\frac{3}{2}$$

The solution is $\left(5,\frac{3}{2}\right)$.

45.

$$x=\frac{D_x}{D}=\frac{\begin{vmatrix}31 & 3\\ 39 & 2\end{vmatrix}}{\begin{vmatrix}2 & 3\\ 3 & 2\end{vmatrix}}=\frac{31(2)-3(39)}{2(2)-3(3)}=\frac{62-117}{4-9}=\frac{-55}{-5}=11$$

$$y=\frac{D_y}{D}=\frac{\begin{vmatrix}2 & 31\\ 3 & 39\end{vmatrix}}{\begin{vmatrix}2 & 3\\ 3 & 2\end{vmatrix}}=\frac{2(39)-31(3)}{2(2)-3(3)}=\frac{78-93}{4-9}=\frac{-15}{-5}=3$$

The solution is (11, 3).

47.

$$x=\frac{D_x}{D}=\frac{\begin{vmatrix}11 & 2\\ 11 & 4\end{vmatrix}}{\begin{vmatrix}3 & 2\\ 6 & 4\end{vmatrix}}=\frac{11(4)-2(11)}{3(4)-2(6)}=\frac{44-22}{12-12}=\frac{22}{0}$$
undefined

$$y=\frac{D_y}{D}=\frac{\begin{vmatrix}3 & 11\\ 6 & 11\end{vmatrix}}{\begin{vmatrix}3 & 2\\ 6 & 4\end{vmatrix}}=\frac{3(11)-11(6)}{3(4)-2(6)}=\frac{33-66}{12-12}=\frac{-33}{0}$$
undefined

The system has no solution and is an inconsistent system.

49. Write the equations in standard form.

$$\begin{cases}5x+6y=12\\ 10x+12y=24\end{cases}$$

$$x=\frac{D_x}{D}=\frac{\begin{vmatrix}12 & 6\\ 24 & 12\end{vmatrix}}{\begin{vmatrix}5 & 6\\ 10 & 12\end{vmatrix}}=\frac{12(12)-6(24)}{5(12)-6(10)}=\frac{144-144}{60-60}=\frac{0}{0}$$

$$y=\frac{D_y}{D}=\frac{\begin{vmatrix}5 & 12\\ 10 & 24\end{vmatrix}}{\begin{vmatrix}5 & 6\\ 10 & 12\end{vmatrix}}=\frac{5(24)-12(10)}{5(12)-6(10)}=\frac{120-120}{60-60}=\frac{0}{0}$$

$\{(x,y)|5x+6y=12\}$

This system has infinitely many solutions and has dependent equations.

51.

$$x=\frac{D_x}{D}$$

$$=\frac{\begin{vmatrix}4 & 1 & 1\\ 0 & 1 & -1\\ 2 & -1 & 1\end{vmatrix}}{\begin{vmatrix}1 & 1 & 1\\ 1 & 1 & -1\\ 1 & -1 & 1\end{vmatrix}}$$

$$=\frac{4\begin{vmatrix}1 & -1\\ -1 & 1\end{vmatrix}-1\begin{vmatrix}0 & -1\\ 2 & 1\end{vmatrix}+1\begin{vmatrix}0 & 1\\ 2 & -1\end{vmatrix}}{1\begin{vmatrix}1 & -1\\ -1 & 1\end{vmatrix}-1\begin{vmatrix}1 & -1\\ 1 & 1\end{vmatrix}+1\begin{vmatrix}1 & 1\\ 1 & -1\end{vmatrix}}$$

$$=\frac{4(1-1)-1(0+2)+1(0-2)}{1(1-1)-1(1+1)+1(-1-1)}$$

$$=\frac{4(0)-1(2)+1(-2)}{1(0)-1(2)+1(-2)}$$

$$=\frac{0-2-2}{0-2-2}$$

$$=\frac{-4}{-4}$$

$$=1$$

$$y=\frac{D_y}{D}$$

$$=\frac{\begin{vmatrix}1 & 4 & 1\\ 1 & 0 & -1\\ 1 & 2 & 1\end{vmatrix}}{\begin{vmatrix}1 & 1 & 1\\ 1 & 1 & -1\\ 1 & -1 & 1\end{vmatrix}}$$

$$=\frac{1\begin{vmatrix}0 & -1\\ 2 & 1\end{vmatrix}-4\begin{vmatrix}1 & -1\\ 1 & 1\end{vmatrix}+1\begin{vmatrix}1 & 0\\ 1 & 2\end{vmatrix}}{1\begin{vmatrix}1 & -1\\ -1 & 1\end{vmatrix}-1\begin{vmatrix}1 & -1\\ 1 & 1\end{vmatrix}+1\begin{vmatrix}1 & 1\\ 1 & -1\end{vmatrix}}$$

$$=\frac{1(0+2)-4(1+1)+1(2-0)}{1(1-1)-1(1+1)+1(-1-1)}$$

$$=\frac{1(2)-4(2)+1(2)}{1(0)-1(2)+1(-2)}$$

$$=\frac{2-8+2}{0-2-2}$$

$$=\frac{-4}{-4}$$

$$=1$$

$$z=\frac{D_z}{D}$$

$$=\frac{\begin{vmatrix}1 & 1 & 4\\ 1 & 1 & 0\\ 1 & -1 & 2\end{vmatrix}}{\begin{vmatrix}1 & 1 & 1\\ 1 & 1 & -1\\ 1 & -1 & 1\end{vmatrix}}$$

$$=\frac{1\begin{vmatrix}1 & 0\\ -1 & 2\end{vmatrix}-1\begin{vmatrix}1 & 0\\ 1 & 2\end{vmatrix}+4\begin{vmatrix}1 & 1\\ 1 & -1\end{vmatrix}}{1\begin{vmatrix}1 & -1\\ -1 & 1\end{vmatrix}-1\begin{vmatrix}1 & -1\\ 1 & 1\end{vmatrix}+1\begin{vmatrix}1 & 1\\ 1 & -1\end{vmatrix}}$$

$$=\frac{1(2-0)-1(2-0)+4(-1-1)}{1(1-1)-1(1+1)+1(-1-1)}$$

$$=\frac{1(2)-1(2)+4(-2)}{1(0)-1(2)+1(-2)}$$

$$=\frac{2-2-8}{0-2-2}$$

$$=\frac{-8}{-4}$$

$$=2$$

The solution is (1, 1, 2).

53.

$$x=\frac{D_x}{D}$$

$$=\frac{\begin{vmatrix}-8 & 2 & -1\\ 10 & -1 & 7\\ -10 & 2 & -3\end{vmatrix}}{\begin{vmatrix}3 & 2 & -1\\ 2 & -1 & 7\\ 2 & 2 & -3\end{vmatrix}}$$

$$=\frac{-8\begin{vmatrix}-1 & 7\\ 2 & -3\end{vmatrix}-2\begin{vmatrix}10 & 7\\ -10 & -3\end{vmatrix}-1\begin{vmatrix}10 & -1\\ -10 & 2\end{vmatrix}}{3\begin{vmatrix}-1 & 7\\ 2 & -3\end{vmatrix}-2\begin{vmatrix}2 & 7\\ 2 & -3\end{vmatrix}-1\begin{vmatrix}2 & -1\\ 2 & 2\end{vmatrix}}$$

$$=\frac{-8(3-14)-2(-30+70)-1(20-10)}{3(3-14)-2(-6-14)-1(4+2)}$$

$$=\frac{-8(-11)-2(40)-1(10)}{3(-11)-2(-20)-1(6)}$$

$$=\frac{88-80-10}{-33+40-6}$$

$$=\frac{-2}{1}$$

$$=-2$$

$$y=\frac{D_y}{D}$$

$$=\frac{\begin{vmatrix}3 & -8 & -1\\ 2 & 10 & 7\\ 2 & -10 & -3\end{vmatrix}}{\begin{vmatrix}3 & 2 & -1\\ 2 & -1 & 7\\ 2 & 2 & -3\end{vmatrix}}$$

$$=\frac{3\begin{vmatrix}10 & 7\\ -10 & -3\end{vmatrix}-(-8)\begin{vmatrix}2 & 7\\ 2 & -3\end{vmatrix}-1\begin{vmatrix}2 & 10\\ 2 & -10\end{vmatrix}}{3\begin{vmatrix}-1 & 7\\ 2 & -3\end{vmatrix}-2\begin{vmatrix}2 & 7\\ 2 & -3\end{vmatrix}-1\begin{vmatrix}2 & -1\\ 2 & 2\end{vmatrix}}$$

$$=\frac{3(-30+70)+8(-6-14)-1(-20-20)}{3(3-14)-2(-6-14)-1(4+2)}$$

$$=\frac{3(40)+8(-20)-1(-40)}{3(-11)-2(-20)-1(6)}$$

$$=\frac{120-160+40}{-33+40-6}$$

$$=\frac{0}{1}$$

$$=0$$

$$z=\frac{D_z}{D}$$

$$=\frac{\begin{vmatrix}3 & 2 & -8\\ 2 & -1 & 10\\ 2 & 2 & -10\end{vmatrix}}{\begin{vmatrix}3 & 2 & -1\\ 2 & -1 & 7\\ 2 & 2 & -3\end{vmatrix}}$$

$$=\frac{3\begin{vmatrix}-1 & 10\\ 2 & -10\end{vmatrix}-2\begin{vmatrix}2 & 10\\ 2 & -10\end{vmatrix}-8\begin{vmatrix}2 & -1\\ 2 & 2\end{vmatrix}}{3\begin{vmatrix}-1 & 7\\ 2 & -3\end{vmatrix}-2\begin{vmatrix}2 & 7\\ 2 & -3\end{vmatrix}-1\begin{vmatrix}2 & -1\\ 2 & 2\end{vmatrix}}$$

$$=\frac{3(10-20)-2(-20-20)-8(4+2)}{3(3-14)-2(-6-14)-1(4+2)}$$

$$=\frac{3(-10)-2(-40)-8(6)}{3(-11)-2(-20)-1(6)}$$

$$=\frac{-30+80-48}{-33+40-6}$$

$$=\frac{2}{1}$$

$$=2$$

The solution is (–2, 0, 2).

TRY IT YOURSELF

55.

$$x=\frac{D_x}{D}=\frac{\begin{vmatrix}5&1&1\\10&-2&3\\-3&1&-4\end{vmatrix}}{\begin{vmatrix}2&1&1\\1&-2&3\\1&1&-4\end{vmatrix}}$$

$$=\frac{5\begin{vmatrix}-2&3\\1&-4\end{vmatrix}-1\begin{vmatrix}10&3\\-3&-4\end{vmatrix}+1\begin{vmatrix}10&-2\\-3&1\end{vmatrix}}{2\begin{vmatrix}-2&3\\1&-4\end{vmatrix}-1\begin{vmatrix}1&3\\1&-4\end{vmatrix}+1\begin{vmatrix}1&-2\\1&1\end{vmatrix}}$$

$$=\frac{5(8-3)-1(-40+9)+1(10-6)}{2(8-3)-1(-4-3)+1(1+2)}$$

$$=\frac{5(5)-1(-31)+1(4)}{2(5)-1(-7)+1(3)}$$

$$=\frac{25+31+4}{10+7+3}$$

$$=\frac{60}{20}$$

$$=3$$

$$y=\frac{D_y}{D}=\frac{\begin{vmatrix}2&5&1\\1&10&3\\1&-3&-4\end{vmatrix}}{\begin{vmatrix}2&1&1\\1&-2&3\\1&1&-4\end{vmatrix}}$$

$$=\frac{2\begin{vmatrix}10&3\\-3&-4\end{vmatrix}-5\begin{vmatrix}1&3\\1&-4\end{vmatrix}+1\begin{vmatrix}1&10\\1&-3\end{vmatrix}}{2\begin{vmatrix}-2&3\\1&-4\end{vmatrix}-1\begin{vmatrix}1&3\\1&-4\end{vmatrix}+1\begin{vmatrix}1&-2\\1&1\end{vmatrix}}$$

$$=\frac{2(-40+9)-5(-4-3)+1(-3-10)}{2(8-3)-1(-4-3)+1(1+2)}$$

$$=\frac{2(-31)-5(-7)+1(-13)}{2(5)-1(-7)+1(3)}$$

$$=\frac{-62+35-13}{10+7+3}$$

$$=\frac{-40}{20}$$

$$=-2$$

$$z=\frac{D_z}{D}=\frac{\begin{vmatrix}2&1&5\\1&-2&10\\1&1&-3\end{vmatrix}}{\begin{vmatrix}2&1&1\\1&-2&3\\1&1&-4\end{vmatrix}}$$

$$=\frac{2\begin{vmatrix}-2&10\\1&-3\end{vmatrix}-1\begin{vmatrix}1&10\\1&-3\end{vmatrix}+5\begin{vmatrix}1&-2\\1&1\end{vmatrix}}{2\begin{vmatrix}-2&3\\1&-4\end{vmatrix}-1\begin{vmatrix}1&3\\1&-4\end{vmatrix}+1\begin{vmatrix}1&-2\\1&1\end{vmatrix}}$$

$$=\frac{2(6-10)-1(-3-10)+5(1+2)}{2(8-3)-1(-4-3)+1(1+2)}$$

$$=\frac{2(-4)-1(-13)+5(3)}{2(5)-1(-7)+1(3)}$$

$$=\frac{-8+13+15}{10+7+3}$$

$$=\frac{20}{20}$$

$$=1$$

The solution is (3, –2, 1).

57. Write the first equation in standard form.

$$y=\frac{-2x+1}{3}$$

$$3y=-2x+1$$

$$2x+3y=1$$

$$\begin{cases}2x+3y=1\\3x-2y=8\end{cases}$$

$$x=\frac{D_x}{D}=\frac{\begin{vmatrix}1&3\\8&-2\end{vmatrix}}{\begin{vmatrix}2&3\\3&-2\end{vmatrix}}=\frac{1(-2)-3(8)}{2(-2)-3(3)}=\frac{-2-24}{-4-9}=\frac{-26}{-13}=2$$

$$y=\frac{D_y}{D}=\frac{\begin{vmatrix}2&1\\3&8\end{vmatrix}}{\begin{vmatrix}2&3\\3&-2\end{vmatrix}}=\frac{2(8)-1(3)}{2(-2)-3(3)}=\frac{16-3}{-4-9}=\frac{13}{-13}=-1$$

The solution is (2, –1).

59.

$$x=\frac{D_x}{D}=\frac{\begin{vmatrix}1&-3&0\\1&0&-8\\0&2&-4\end{vmatrix}}{\begin{vmatrix}4&-3&0\\6&0&-8\\0&2&-4\end{vmatrix}}=\frac{1\begin{vmatrix}0&-8\\2&-4\end{vmatrix}-(-3)\begin{vmatrix}1&-8\\0&-4\end{vmatrix}+0\begin{vmatrix}1&0\\0&2\end{vmatrix}}{4\begin{vmatrix}0&-8\\2&-4\end{vmatrix}-(-3)\begin{vmatrix}6&-8\\0&-4\end{vmatrix}+0\begin{vmatrix}6&0\\0&2\end{vmatrix}}$$

$$=\frac{1(0+16)+3(-4-0)+0(2-0)}{4(0+16)+3(-24-0)+0(12-0)}=\frac{1(16)+3(-4)+0(2)}{4(16)+3(-24)+0(12)}=\frac{16-12+0}{64-72+0}=\frac{4}{-8}=-\frac{1}{2}$$

$$y=\frac{D_y}{D}=\frac{\begin{vmatrix}4&1&0\\6&1&-8\\0&0&-4\end{vmatrix}}{\begin{vmatrix}4&-3&0\\6&0&-8\\0&2&-4\end{vmatrix}}=\frac{4\begin{vmatrix}1&-8\\0&-4\end{vmatrix}-1\begin{vmatrix}6&-8\\0&-4\end{vmatrix}+0\begin{vmatrix}6&1\\0&0\end{vmatrix}}{4\begin{vmatrix}0&-8\\2&-4\end{vmatrix}-(-3)\begin{vmatrix}6&-8\\0&-4\end{vmatrix}+0\begin{vmatrix}6&0\\0&2\end{vmatrix}}$$

$$=\frac{4(-4-0)-1(-24-0)+0(0-0)}{4(0+16)+3(-24-0)+0(12-0)}=\frac{4(-4)-1(-24)+0(0)}{4(16)+3(-24)+0(12)}=\frac{-16+24+0}{64-72+0}=\frac{8}{-8}=-1$$

$$z=\frac{D_z}{D}=\frac{\begin{vmatrix}4&-3&1\\6&0&1\\0&2&0\end{vmatrix}}{\begin{vmatrix}4&-3&0\\6&0&-8\\0&2&-4\end{vmatrix}}=\frac{4\begin{vmatrix}0&1\\2&0\end{vmatrix}-(-3)\begin{vmatrix}6&1\\0&0\end{vmatrix}+1\begin{vmatrix}6&0\\0&2\end{vmatrix}}{4\begin{vmatrix}0&-8\\2&-4\end{vmatrix}-(-3)\begin{vmatrix}6&-8\\0&-4\end{vmatrix}+0\begin{vmatrix}6&0\\0&2\end{vmatrix}}$$

$$=\frac{4(0-2)+3(0-0)+1(12-0)}{4(0+16)+3(-24-0)+0(12-0)}=\frac{4(-2)+3(0)+1(12)}{4(16)+3(-24)+0(12)}=\frac{-8+0+12}{64-72+0}=\frac{4}{-8}=-\frac{1}{2}$$

The solution is $\left(-\frac{1}{2},-1,-\frac{1}{2}\right)$.

61. Write the equations in standard form.

$$\begin{cases} 2x + y - z = 1 \\ x + 2y + 2z = 2 \\ 4x + 5y + 3z = 3 \end{cases}$$

$$x = \frac{D_x}{D}$$

$$= \frac{\begin{vmatrix} 1 & 1 & -1 \\ 2 & 2 & 2 \\ 3 & 5 & 3 \end{vmatrix}}{\begin{vmatrix} 2 & 1 & -1 \\ 1 & 2 & 2 \\ 4 & 5 & 3 \end{vmatrix}}$$

$$= \frac{1\begin{vmatrix} 2 & 2 \\ 5 & 3 \end{vmatrix} - 1\begin{vmatrix} 2 & 2 \\ 3 & 3 \end{vmatrix} - 1\begin{vmatrix} 2 & 2 \\ 3 & 5 \end{vmatrix}}{2\begin{vmatrix} 2 & 2 \\ 5 & 3 \end{vmatrix} - 1\begin{vmatrix} 1 & 2 \\ 4 & 3 \end{vmatrix} - 1\begin{vmatrix} 1 & 2 \\ 4 & 5 \end{vmatrix}}$$

$$= \frac{1(6-10) - 1(6-6) - 1(10-6)}{2(6-10) - 1(3-8) - 1(5-8)}$$

$$= \frac{1(-4) - 1(0) - 1(4)}{2(-4) - 1(-5) - 1(-3)}$$

$$= \frac{-4-0-4}{-8+5+3}$$

$$= \frac{-8}{0} \text{ is undefined}$$

Since the denominator determinant D is 0 and the numerator determinant D_x is not 0, the system is inconsistent and has no solutions.

63. Write the equations in standard form.

$$\begin{cases} 3x - 5y = 16 \\ -3x + 5y = 33 \end{cases}$$

$$x = \frac{D_x}{D} = \frac{\begin{vmatrix} 16 & -5 \\ 33 & 5 \end{vmatrix}}{\begin{vmatrix} 3 & -5 \\ -3 & 5 \end{vmatrix}} = \frac{16(5) - (-5)(33)}{3(5) - (-5)(-3)} = \frac{80+165}{15-15} = \frac{245}{0}$$

undefined

$$y = \frac{D_y}{D} = \frac{\begin{vmatrix} 3 & 16 \\ -3 & 33 \end{vmatrix}}{\begin{vmatrix} 3 & -5 \\ -3 & 5 \end{vmatrix}} = \frac{3(33) - 16(-3)}{3(5) - (-5)(-3)} = \frac{99+48}{15-15} = \frac{147}{0}$$

undefined

Since the denominator determinant D is 0 and the numerator determinant D_x is not 0, the system is inconsistent and has no solutions.

65. Write the equations in standard form. Multiply the second equation by 2 to eliminate fractions.

$$\begin{cases} x+y=1 \\ y+2z=5 \\ x-z=-3 \end{cases}$$

$$x=\frac{D_x}{D}=\frac{\begin{vmatrix}1 & 1 & 0\\ 5 & 1 & 2\\ -3 & 0 & -1\end{vmatrix}}{\begin{vmatrix}1 & 1 & 0\\ 0 & 1 & 2\\ 1 & 0 & -1\end{vmatrix}}$$

$$=\frac{1\begin{vmatrix}1 & 2\\ 0 & -1\end{vmatrix}-1\begin{vmatrix}5 & 2\\ -3 & -1\end{vmatrix}+0\begin{vmatrix}5 & 1\\ -3 & 0\end{vmatrix}}{1\begin{vmatrix}1 & 2\\ 0 & -1\end{vmatrix}-1\begin{vmatrix}0 & 2\\ 1 & -1\end{vmatrix}+0\begin{vmatrix}0 & 1\\ 1 & 0\end{vmatrix}}$$

$$=\frac{1(-1-0)-1(-5+6)+0(0+3)}{1(-1-0)-1(0-2)+0(0-1)}$$

$$=\frac{1(-1)-1(1)+0(3)}{1(-1)-1(-2)+0(-1)}$$

$$=\frac{-1-1+0}{-1+2+0}$$

$$=\frac{-2}{1}$$

$$=-2$$

$$y=\frac{D_y}{D}=\frac{\begin{vmatrix}1 & 1 & 0\\ 0 & 5 & 2\\ 1 & -3 & -1\end{vmatrix}}{\begin{vmatrix}1 & 1 & 0\\ 0 & 1 & 2\\ 1 & 0 & -1\end{vmatrix}}$$

$$=\frac{1\begin{vmatrix}5 & 2\\ -3 & -1\end{vmatrix}-1\begin{vmatrix}0 & 2\\ 1 & -1\end{vmatrix}+0\begin{vmatrix}0 & 5\\ 1 & -3\end{vmatrix}}{1\begin{vmatrix}1 & 2\\ 0 & -1\end{vmatrix}-1\begin{vmatrix}0 & 2\\ 1 & -1\end{vmatrix}+0\begin{vmatrix}0 & 1\\ 1 & 0\end{vmatrix}}$$

$$=\frac{1(-5+6)-1(0-2)+0(0-5)}{1(-1-0)-1(0-2)+0(0-1)}$$

$$=\frac{1(1)-1(-2)+0(-5)}{1(-1)-1(-2)+0(-1)}$$

$$=\frac{1+2+0}{-1+2+0}$$

$$=\frac{3}{1}$$

$$=3$$

$$z=\frac{D_z}{D}=\frac{\begin{vmatrix}1 & 1 & 1\\ 0 & 1 & 5\\ 1 & 0 & -3\end{vmatrix}}{\begin{vmatrix}1 & 1 & 0\\ 0 & 1 & 2\\ 1 & 0 & -1\end{vmatrix}}$$

$$=\frac{1\begin{vmatrix}1 & 5\\ 0 & -3\end{vmatrix}-1\begin{vmatrix}0 & 5\\ 1 & -3\end{vmatrix}+1\begin{vmatrix}0 & 1\\ 1 & 0\end{vmatrix}}{1\begin{vmatrix}1 & 2\\ 0 & -1\end{vmatrix}-1\begin{vmatrix}0 & 2\\ 1 & -1\end{vmatrix}+0\begin{vmatrix}0 & 1\\ 1 & 0\end{vmatrix}}$$

$$=\frac{1(-3-0)-1(0-5)+1(0-1)}{1(-1-0)-1(0-2)+0(0-1)}$$

$$=\frac{1(-3)-1(-5)+1(-1)}{1(-1)-1(-2)+0(-1)}$$

$$=\frac{-3+5-1}{-1+2+0}$$

$$=\frac{1}{1}$$

$$=1$$

The solution is (–2, 3, 1).

67.

$$x=\frac{D_x}{D}=\frac{\begin{vmatrix}0 & 3\\ -4 & -6\end{vmatrix}}{\begin{vmatrix}2 & 3\\ 4 & -6\end{vmatrix}}=\frac{0(-6)-3(-4)}{2(-6)-3(4)}=\frac{0+12}{-12-12}=\frac{12}{-24}=-\frac{1}{2}$$

$$y=\frac{D_y}{D}=\frac{\begin{vmatrix}2 & 0\\ 4 & -4\end{vmatrix}}{\begin{vmatrix}2 & 3\\ 4 & -6\end{vmatrix}}=\frac{2(-4)-0(4)}{2(-6)-3(4)}=\frac{-8-0}{-12-12}=\frac{-8}{-24}=\frac{1}{3}$$

The solution is $\left(-\frac{1}{2},\frac{1}{3}\right)$.

69.

$$x=\frac{D_x}{D}=\frac{\begin{vmatrix}6 & 3 & 4\\ -4 & -3 & -4\\ 12 & 6 & 8\end{vmatrix}}{\begin{vmatrix}2 & 3 & 4\\ 2 & -3 & -4\\ 4 & 6 & 8\end{vmatrix}}$$

$$=\frac{6\begin{vmatrix}-3 & -4\\ 6 & 8\end{vmatrix}-3\begin{vmatrix}-4 & -4\\ 12 & 8\end{vmatrix}+4\begin{vmatrix}-4 & -3\\ 12 & 6\end{vmatrix}}{2\begin{vmatrix}-3 & -4\\ 6 & 8\end{vmatrix}-3\begin{vmatrix}2 & -4\\ 4 & 8\end{vmatrix}+4\begin{vmatrix}2 & -3\\ 4 & 6\end{vmatrix}}$$

$$=\frac{6(-24+24)-3(-32+48)+4(-24+36)}{2(-24+24)-3(16+16)+4(12+12)}$$

$$=\frac{6(0)-3(16)+4(12)}{2(0)-3(32)+4(24)}=\frac{0-48+48}{0-96+96}=\frac{0}{0}$$

Since the denominator determinant D is 0 and the numerator determinant D_x is 0, the system is consistent, but the equations are dependent. There are infinitely many solutions.

71. –46,811

73 –60,527,941

APPLICATIONS

75.

$$\begin{cases}2x+y=180\\ y=30+x\end{cases}$$

$$\begin{cases}2x+y=180\\ -x+y=30\end{cases}$$

$$x=\frac{D_x}{D}=\frac{\begin{vmatrix}180 & 1\\ 30 & 1\end{vmatrix}}{\begin{vmatrix}2 & 1\\ -1 & 1\end{vmatrix}}=\frac{180(1)-1(30)}{2(1)-1(-1)}=\frac{180-30}{2+1}=\frac{150}{3}=50$$

$$y=\frac{D_y}{D}=\frac{\begin{vmatrix}2 & 180\\ -1 & 30\end{vmatrix}}{\begin{vmatrix}2 & 1\\ -1 & 1\end{vmatrix}}=\frac{2(30)-180(-1)}{2(1)-1(-1)}=\frac{60+180}{2+1}=\frac{240}{3}=80$$

$x = 50°$ and $y = 80°$.

WRITING

77. Answers will vary.

79. Answers will vary.

81. Answers will vary.

REVIEW

83. Find the slope of each using $y = mx + b$.

$$y = 2x - 7 \qquad m = 2$$

$$\begin{aligned} x - 2y &= 7 \\ -2y &= -x + 7 \\ y &= \frac{1}{2}x - \frac{7}{2} \\ m &= \frac{1}{2} \end{aligned}$$

No. Since the slopes are not opposite reciprocals, the lines are NOT perpendicular.

85. The graph of *g* is 2 units below the graph of *f*. *g*(*x*) is shifted down 2 units.

87. The *y*–intercept since the *x*–coordinate is 0.

89. The independent variable is *x* and the dependent variable is *y*.

CHALLENGE PROBLEMS

91.

$$\begin{vmatrix} x & y & 1 \\ -2 & 3 & 1 \\ 3 & 5 & 1 \end{vmatrix} = 0$$

$$\begin{aligned} x\begin{vmatrix} 3 & 1 \\ 5 & 1 \end{vmatrix} - y\begin{vmatrix} -2 & 1 \\ 3 & 1 \end{vmatrix} + 1\begin{vmatrix} -2 & 3 \\ 3 & 5 \end{vmatrix} &= 0 \\ x(3-5) - y(-2-3) + 1(-10-9) &= 0 \\ x(-2) - y(-5) + 1(-19) &= 0 \\ -2x + 5y - 19 &= 0 \\ 2x - 5y + 19 &= 0 \\ 2x - 5y &= -19 \end{aligned}$$

For (–2, 3):

$$\begin{aligned} 2(-2) - 5(3) &\stackrel{?}{=} -19 \\ -4 - 15 &\stackrel{?}{=} -19 \\ -19 &= -19 \end{aligned}$$

For (3, 5):

$$\begin{aligned} 2(3) - 5(5) &\stackrel{?}{=} -19 \\ 6 - 25 &\stackrel{?}{=} -19 \\ -19 &= -19 \end{aligned}$$

SECTION 3.6

VOCABULARY

1. A parallelogram is a four-sided figure with two pairs of **parallel** sides.

CONCEPTS

3. a. 70° and 75°
   b. alternate interior

5. a. $x + c$
   b $x - c$

NOTATION

7. a. 0.06
   b. 0.048
   c. 0.135

GUIDED PRACTICE

9. INVESTMENTS

	Principal ·	Rate ·	Time =	Interest
Township	x	0.05	1	$0.05x$
Ameritech	y	0.04	1	$0.04y$
Total	25,000			1,050

$$\begin{cases} x + y = 25{,}000 \\ 0.05x + 0.04y = 1{,}050 \end{cases}$$

11. CANDY

	Amount ·	Value =	Total Value
Dark chocolate	x	13.90	$13.90x$
White chocolate	y	5.10	$5.10y$
Mix	6	10.25	61.50

$$\begin{cases} x + y = 6 \\ 13.90x + 5.10y = 61.50 \end{cases}$$

APPLICATIONS

13. ELECTRONICS
Let R_1 and R_2 be the resistances.

$$\begin{cases} R_1 + R_2 = 1{,}375 \\ R_1 = 125 + R_2 \end{cases}$$

Use substitution.

$$R_1 + R_2 = 1{,}375$$
$$(125 + R_2) + R_2 = 1{,}375$$
$$125 + 2R_2 = 1{,}375$$
$$2R_2 = 1{,}250$$
$$R_2 = 625$$

Substitute $R_2 = 625$ in the 2nd equation and solve for R_1.

$$R_1 = 125 + R_2$$
$$R_1 = 125 + 625$$
$$R_1 = 750$$

The resistances were 750 ohms and 625 ohms.

15. AREA CODES

Let x = the area code for Montana and y = the area code for Idaho.

$$\begin{cases} x + y = 614 \\ x - y = 198 \end{cases}$$

Add the equations to eliminate y.

$$x + y = 614$$
$$\underline{x - y = 198}$$
$$2x = 812$$
$$x = 406$$

Substitute $x = 406$ into the first equation to solve for y.

$$x + y = 614$$
$$406 + y = 614$$
$$406 + y - 406 = 614 - 406$$
$$y = 208$$

The area code of Montana is 406 and the area code of Idaho is 208.

17. CAREER TO CAREER

Let x = the number of Gap stores in 2009 and y = the number of Aeropostale stores.

$$\begin{cases} x + y = 4,046 \\ x = 3\frac{1}{4}y \end{cases}$$

Use substitution.

$$x + y = 4,046$$
$$3\frac{1}{4}y + y = 4,046$$
$$4\frac{1}{4}y = 4,046$$
$$\frac{17}{4}y = 4,046$$
$$\frac{4}{17}\left(\frac{17}{4}y\right) = \frac{4}{17}(4,046)$$
$$y = 952$$

Substitute $y = 952$ into the second equation to find x.

$$x = 3\frac{1}{4}y$$
$$= 3\frac{1}{4}(952)$$
$$= 3,094$$

There were 3,094 Gap stores and 952 Aereopostale stores.

19. BRACING

In a parallelogram, opposite angles are equal ($x + y = 100$), and alternate interior angles are equal ($x - y = 50$).

$$\begin{cases} x + y = 100 \\ x - y = 50 \end{cases}$$

Add the equations to eliminate y.

$$x + y = 100$$
$$\underline{x - y = 50}$$
$$2x = 150$$
$$x = 75$$

Substitute $x = 75$ in the 1st equation and solve for y.

$$x + y = 100$$
$$75 + y = 100$$
$$y = 25$$

So, $x = 75^\circ$ and $y = 25^\circ$.

21. TRAFFIC SIGNALS

Angles x and y make a straight line, so their sum is 180°. Brace A and Brace B are perpendicular so their sum is 90°.

$$\begin{cases} x + y = 180 \\ \frac{2}{5}x + y = 90 \end{cases}$$

Multiply the 1st equation by -1.

$$-x - y = -180$$
$$\underline{\frac{2}{5}x + y = 90}$$
$$-\frac{3}{5}x = -90$$
$$-\frac{5}{3}\left(-\frac{3}{5}x\right) = -\frac{5}{3}(-90)$$
$$x = 150$$

Substitute $x = 150$ in the 1st equation and solve for y.

$$x + y = 180$$
$$150 + y = 180$$
$$y = 30$$
$$x = 150^\circ \text{ and } y = 30^\circ.$$

23. FENCING A FIELD

Let x = width of field and y = length of the field.

$$\begin{cases} 2x + 2y = 72 & \text{(perimeter)} \\ 3x + 2y = 88 & \text{(with partition)} \end{cases}$$

Multiply the 2nd equation by $--1$.

Add the equations to eliminate y.

$$2x + 2y = 72$$
$$\underline{-3x - 2y = -88}$$
$$-x = -16$$
$$x = 16$$

Substitute $x = 16$ in the 1st equation and solve for y.

$$2x + 2y = 72$$
$$2(16) + 2y = 72$$
$$32 + 2y = 72$$
$$2y = 40$$
$$y = 20$$

The width is 16 m and the length is 20 m.

25. ADVERTISING

Let x = 30 second spot and y = 15 second spot.

$$\begin{cases} 4x+6y=6{,}050 \\ 3x+5y=4{,}775 \end{cases}$$

Multiply the 1st equation by 3 and the 2nd equation by -4. Add to eliminate x.

$$\begin{aligned} 12x+18y&=18{,}150 \\ -12x-20y&=-19{,}100 \\ \hline -2y&=-950 \\ y&=475 \end{aligned}$$

Substitute $y=475$ in the 1st equation and solve for x.

$$\begin{aligned} 4x+6y&=6{,}050 \\ 4x+6(475)&=6{,}050 \\ 4x+2{,}850&=6{,}050 \\ 4x&=3{,}200 \\ x&=800 \end{aligned}$$

The 30-second spot costs \$800 and the 15-second spot costs \$475.

27. PRODUCTION PLANNING

Let x = number of racing bikes and y = number of mountain bikes.

	Racing	Mountain	Combined
Materials	$110x$	$140y$	26,150
Labor	$120x$	$180y$	31,800

$$\begin{cases} 110x+140y=26{,}150 \\ 120x+180y=31{,}800 \end{cases}$$

Multiply the 1st equation by -120 and the 2nd equation by 110. Add equations to eliminate x.

$$\begin{aligned} -13{,}200x-16{,}800y&=-3{,}138{,}000 \\ 13{,}200x+19{,}800y&=3{,}498{,}000 \\ \hline 3{,}000y&=360{,}000 \\ y&=120 \end{aligned}$$

Substitute $y=120$ in the 1st equation and solve for x.

$$\begin{aligned} 110x+140y&=26{,}150 \\ 110x+140(120)&=26{,}150 \\ 110x+16{,}800&=26{,}150 \\ 110x&=9{,}350 \\ x&=85 \end{aligned}$$

85 racing bikes and 120 mountain bikes can be built.

29. SUMMER CONCERTS

Let x = the cost of a Jonas Brothers ticket and y = the cost of an Elton John ticket.

$$\begin{cases} 2x+2y=560 \\ 4x+2y=806 \end{cases}$$

Multiply the second equation by –1 and add the equations to eliminate y.

$$\begin{aligned} 2x+2y&=560 \\ -4x-2y&=-806 \\ \hline -2x&=-246 \\ x&=123 \end{aligned}$$

$$\begin{aligned} 2x+2y&=560 \\ 2(123)+2y&=560 \\ 246+2y&=560 \\ 246+2y-246&=560-246 \\ 2y&=314 \\ y&=157 \end{aligned}$$

The cost of a Jonas Brothers ticket was \$123 and the cost of an Elton John ticket was \$157.

31. MAKING TIRES

a. Let x = number of tires made.

$$\begin{cases} C_1=15x+1{,}000 \\ C_2=10x+3{,}000 \end{cases}$$

Breakeven point:

$$\begin{aligned} C_1&=C_2 \\ 15x+1{,}000&=10x+3{,}000 \\ 5x+1{,}000&=3{,}000 \\ 5x&=2{,}000 \\ x&=400 \end{aligned}$$

The breakeven point is 400 tires.

b.

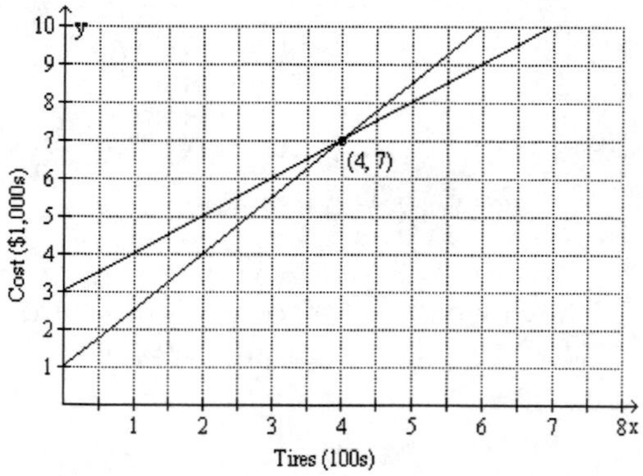

c. the second mold

33. PUBLISHING

a. Let x = number of books.

$$\begin{cases} C_1=5.98x+210 \\ C_2=5.95x+350 \end{cases}$$

Breakeven point:

$$\begin{aligned} C_1&=C_2 \\ 5.98x+210&=5.95x+350 \\ 0.03x+210&=350 \\ 0.03x&=140 \\ x&=4{,}666.\overline{6} \\ x&=4{,}666\tfrac{2}{3} \end{aligned}$$

The breakeven point is $4{,}666\frac{2}{3}$ books.

b. Let x = 5,100 in both equations.

$$\begin{aligned} C_1&=5.98x+210 \\ &=5.98(5{,}100)+210 \\ &=30{,}498+210 \\ &=30{,}708 \end{aligned}$$

$$\begin{aligned} C_2&=5.95x+350 \\ &=5.95(5{,}100)+350 \\ &=30{,}345+350 \\ &=30{,}695 \end{aligned}$$

The newer press costs less.

35. RECORDING COMPANIES

a. Let x = number of sets of CDs.

Retail price = production cost + setup cost

$$\begin{aligned} 45x&=28.20x+105{,}000 \\ 16.8x&=105{,}000 \\ x&\approx 6{,}250 \end{aligned}$$

b. At least 6,251 set must be sold to make a profit.

37. MANUFACTURING

a. Let x = number of water pumps sold.

$$\begin{cases} C_1 = 29x + 12{,}390 \\ C_2 = 50x \end{cases}$$

Breakeven point:

$$C_1 = C_2$$
$$29x + 12{,}390 = 50x$$
$$12{,}390 = 21x$$
$$590 = x$$

The breakeven point is 590 pumps.

b. Let x = number of water pumps sold.

$$\begin{cases} C_1 = 17x + 20{,}460 \\ C_2 = 50x \end{cases}$$

Breakeven point:

$$C_1 = C_2$$
$$17x + 20{,}460 = 50x$$
$$20{,}460 = 33x$$
$$620 = x$$

The breakeven point is 620 pumps

c. The company should use Process A for a smaller loss.

39. INVESTMENT CLUBS

Let x = amount at 10% and y = amount at 12%.

$$\begin{cases} x + y = 8{,}000 \\ 0.10x + 0.12y = 900 \end{cases}$$

Multiply the 1st equation by -10 and multiply the 2nd equation by 100.

$$-10x - 10y = -80{,}000$$
$$\underline{10x + 12y = 90{,}000}$$
$$2y = 10{,}000$$
$$y = 5{,}000$$

Substitute $y = 5{,}000$ in the 1st equation and solve for x.

$$x + y = 8{,}000$$
$$x + 5{,}000 = 8{,}000$$
$$x = 3{,}000$$

\$3,000 was invested at 10% and \$5,000 was invested at 12%.

41. INVESTING

Let x = amount at 5% and y = amount at 4.25%.

$$\begin{cases} y = 3x \\ 0.05x + 0.0425y = 1{,}420 \end{cases}$$

Multiply the 2nd equation by 10,000.

$$\begin{cases} y = 3x \\ 500x + 425y = 14{,}200{,}000 \end{cases}$$

Use substitution.

$$500x + 425y = 14{,}200{,}000$$
$$500x + 425(3x) = 14{,}200{,}000$$
$$500x + 1{,}275x = 14{,}200{,}000$$
$$1{,}775x = 14{,}200{,}000$$
$$x = \$8{,}000$$
$$y = 3x$$
$$y = 3(8{,}000)$$
$$y = \$24{,}000$$

She invested \$8,000 in the credit union and \$24,000 in the money market account.

43. CREDIT CARDS

Let x = amount owed on VISA card and y = amount owed on Robinson's May card.

$$\begin{cases} x + y = 16{,}500 \\ 0.015x + 0.0175y = 259.25 \end{cases}$$

Multiply the second equation by 10,000 to clear fractions.

$$\begin{cases} x + y = 16{,}500 \\ 150x + 175y = 2{,}592{,}500 \end{cases}$$

Multiply the first equation by -150 and add equations to eliminate x.

$$-150x - 150y = -2{,}475{,}000$$
$$\underline{150x + 175y = 2{,}592{,}500}$$
$$25y = 117{,}500$$
$$y = \$4{,}700$$
$$x + y = 16{,}500$$
$$x + 4{,}700 = 16{,}500$$
$$x = \$11{,}800$$

They owed \$11,800 on the VISA card and \$4,700 on the Robinson's May card.

45. AVIATION

Let s = the speed of the plane in still air and w = the speed of the wind. Then the speed of the plane flying with the wind is $s + w$ and the speed of the plane flying against the wind is $s - w$. Using the formula $\boldsymbol{d = rt}$, we find that $4.5(s + w)$ represents the distance traveled with the wind and $6(s - w)$ represents the distance traveled against the wind.

$$\begin{cases} 4.5(s+w) = 2{,}700 \\ 6(s-w) = 2{,}700 \end{cases}$$

Distribute.

$$\begin{cases} 4.5s + 4.5w = 2{,}700 \\ 6s - 6w = 2{,}700 \end{cases}$$

Multiply the 1st equation by 4 and the 2nd equation by 3 and add equations to eliminate the w.

$$\begin{aligned} 18s + 18w &= 10{,}800 \\ \underline{18s - 18w} &\underline{= 8{,}100} \\ 36s &= 18{,}900 \\ s &= 525 \\ 6s - 6w &= 2{,}700 \\ 6(525) - 6w &= 2{,}700 \\ 3{,}150 - 6w &= 2{,}700 \\ -6w &= -450 \\ w &= 75 \end{aligned}$$

The speed of the plane in still air is 525 mph and the speed of the wind is 75 mph.

47. AIRPORT WALKWAYS

Let s = the speed of the man walking and w = the speed of the walkway. Then the speed of the man walking on the walkway is $s + w$ and the speed of the man walking against the flow of the walkway is $s - w$. Using the formula $\boldsymbol{d = rt}$, we find that $40(s + w)$ represents the distance traveled on the walkway and $80(s - w)$ represents the distance going against the flow of the walkway.

$$\begin{cases} 40(s+w) = 320 \\ 80(s-w) = 320 \end{cases}$$

Distribute.

$$\begin{cases} 40s + 40w = 320 \\ 80s - 80w = 320 \end{cases}$$

Multiply the 1st equation by 2 and add equations to eliminate the w.

$$\begin{aligned} 80s + 80w &= 640 \\ \underline{80s - 80w} &\underline{= 320} \\ 160s &= 960 \\ s &= 6 \\ 40s + 40w &= 320 \\ 40(6) + 40w &= 320 \\ 240 + 40w &= 320 \\ 40w &= 80 \\ w &= 2 \end{aligned}$$

The rate of the man walking is 6 feet per second and the rate of the moving walkway is 2 feet per second.

49. JET SKIS

Let s = the speed of the jet ski in still water and w = the speed of the current. Then the speed of the jet ski going downstream (with the current) is $s + w$ and the speed of the jet ski upstream is $s - w$. Using the formula $\boldsymbol{d = rt}$, we find that $\frac{1}{3}(s + w)$ represents the distance traveled downstream and $\frac{1}{2}(s - w)$ represents the distance traveled upstream.

$$\begin{cases} \frac{1}{3}(s+w)=10 \\ \frac{1}{2}(s-w)=10 \end{cases}$$

Multiply the 1st equation by 3 and the 2nd equation by 2. Then add equations to eliminate the w.

$$\begin{aligned} s+w&=30 \\ \underline{s-w}&\underline{=20} \\ 2s&=50 \\ s&=25 \\ s+w&=30 \\ 25+w&=30 \\ w&=5 \end{aligned}$$

The speed of the jet ski in still water is 25 mph and the speed of the current is 5 mph.

51. MIXING CANDY

Let x = # of lb of Gummy Bears and y = # of lb of Jelly Beans

	Gummy Bears	Jelly Beans	Mixture
# of pounds	x	y	60
Value	$3.50x$	$5.50y$	4(60)

$$\begin{cases} x+y=60 \\ 3.50x+5.50y=4(60) \end{cases}$$

Multiply the 2nd equation by 100 to eliminate decimals.

$$\begin{cases} x+y=60 \\ 350x+550y=24{,}000 \end{cases}$$

Multiply the 1st equation by -350. Add equations to eliminate x.

$$\begin{aligned} -350x-350y&=-21{,}000 \\ \underline{350x+550y}&\underline{=24{,}000} \\ 200y&=3{,}000 \\ y&=15 \end{aligned}$$

Substitute $y = 15$ in the 1st equation and solve for x.

$$\begin{aligned} x+y&=60 \\ x+15&=60 \\ x&=45 \end{aligned}$$

45 lb of gummy bears and 15 lb of jelly beans

53. MIXING COFFEE

Let x = # of lb of regular coffee and y = # of lb of Kona coffee.

	Regular coffee	Kona coffee	Mixture
# of pounds	x	y	20
Value	$4x$	$11.50y$	6(20)

$$\begin{cases} x+y=20 \\ 4x+11.5y=6(20) \end{cases}$$

$$\begin{cases} x+y=20 \\ 4x+11.5y=120 \end{cases}$$

Multiply the 1st equation by -4.

Add equations to eliminate x.

$$\begin{aligned} -4x-\quad 4y&=-80 \\ 4x+11.5y&=120 \\ \hline 7.5y&=40 \\ y&=5\frac{1}{3} \\ x+y&=20 \\ x+5\frac{1}{3}&=20 \\ x&=14\frac{2}{3} \end{aligned}$$

$14\frac{2}{3}$ pounds of regular coffee and $5\frac{1}{3}$ pounds of Kona coffee are needed.

55. CONFETTI

Let x = # of lb of small flake confetti and y = # of lb of Mylar confetti stars. There are 2,000 lbs in 1 ton.

	small flake	Mylar stars	Mixture
# of pounds	x	y	2,000
Value	$14.50x$	$24.50y$	20(2,000)

$$\begin{cases} x+y=2{,}000 \\ 14.50x+24.50y=20(2{,}000) \end{cases}$$

Multiply the 2nd equation by 10 to clear decimals.

$$\begin{cases} x+y=2{,}000 \\ 145x+245y=400{,}000 \end{cases}$$

Multiply the 1st equation by -145.

Add equations to eliminate x.

$$\begin{aligned} -145x-145y&=-290{,}000 \\ 145x+245y&=400{,}000 \\ \hline 100y&=110{,}000 \\ y&=1{,}100 \\ x+y&=2{,}000 \\ x+1{,}100&=2{,}000 \\ x&=900 \end{aligned}$$

900 pounds of small flake confetti and 1,100 pounds of Mylar confetti stars are needed.

57. ANTIFREEZE

Let x = # of pints of 10% antifreeze and y = # of pints of 40% antifreeze.

	10% antifreeze	40% antifreeze	Mixture
# of pints	x	y	24
Value	$0.10x$	$0.40y$	0.30(24)

$$\begin{cases} x+y=24 \\ 0.10x+0.40y=0.30(24) \end{cases}$$

Multipy the 2nd equation by 10 to clear decimals.

$$\begin{cases} x+y=24 \\ x+4y=72 \end{cases}$$

Multiply the 1st equation by -1.

Add equations to eliminate x.

$$\begin{aligned} -x-y&=-24 \\ x+4y&=72 \\ \hline 3y&=48 \\ y&=16 \\ x+y&=24 \\ x+16&=24 \\ x&=8 \end{aligned}$$

8 pints of 10% antifreeze and 16 pints of 40% antifreeze are needed.

59. DERMATOLOGY

Let x = # of grams of 0.2% cream and y = # of grams of 0.7% cream

	0.2% solution	0.7% solution	0.3% mixture
# of grams	x	y	185
% Triclosan	$0.002x$	$0.007y$	0.003(185)

$$\begin{cases} x+y=185 \\ 0.002x+0.007y=0.003(185) \end{cases}$$

Multiply the 2nd equation by 1,000 to eliminate decimals.

$$\begin{cases} x+y=185 \\ 2x+7y=555 \end{cases}$$

Multiply the 1st equation by -2.

Add equations to eliminate x.

$$\begin{aligned} -2x-2y&=-370 \\ 2x+7y&=555 \\ \hline 5y&=185 \\ y&=37 \end{aligned}$$

Substitute $y=37$ in the 1st equation and solve for x.

$$\begin{aligned} x+y&=185 \\ x+37&=185 \\ x&=148 \end{aligned}$$

148 grams of 0.2% and 37 grams of 0.7%

WRITING

61. Answers will vary.

63. Answers will vary.

65. Answers will vary.

REVIEW

67. A **rational** number is any number that can be written as a fraction with an integer numerator and a nonzero integer denominator.

69. An equation that is true for all values of its variable is called an **identity**.

71. If a triangle has two sides with equal measures, it is called an **isosceles** triangle.

CHALLENGE PROBLEMS

73. MANAGING AN APARTMENT
Let x = the amount of rent the manager pays and let y = the amount of rent the paid by the other tenants.

$$\begin{cases} x + 5y = 5,520 \\ x = \frac{3}{4}y \end{cases}$$

Use substitution.

$$x + 5y = 5,520$$
$$\frac{3}{4}y + 5y = 5,520$$
$$\frac{23}{4}y = 5,520$$
$$y = 960$$
$$x = \frac{3}{4}y$$
$$x = \frac{3}{4}(960)$$
$$x = 720$$

The manager pays $720.

SECTION 3.7

VOCABULARY

1. If a point lies on the graph of an equation, it is a solution of the equation and the coordinates of the point **satisfy** the equation.

CONCEPTS

3. $\begin{cases} x+y+z=50 \\ 5x+6y+7z=295 \\ 2x+3y+4z=145 \end{cases}$

5. $y=ax^2+bx+c$

$-3=a(2)^2+b(2)+c$

$-3=4a+2b+c$

APPLICATIONS

7. MAKING STATUES

Let x = large, y = medium, and z = small type.

$$\begin{cases} x+y+z=180 & (1) \\ 5x+4y+3z=650 & (2) \\ 20x+12y+9z=2{,}100 & (3) \end{cases}$$

Multiply Equation 1 by -5 and add Equations 1 and 2.

$$\begin{array}{rl} -5x-5y-5z=-900 & (1) \\ \underline{5x+4y+3z=650} & (2) \\ -y-2z=-250 & (4) \end{array}$$

Multiply Equation 1 by -20 and add Equations 1 and 3.

$$\begin{array}{rl} -20x-20y-20z=-3{,}600 & (1) \\ \underline{20x+12y+\ 9z=2{,}100} & (3) \\ -8y-11z=-1{,}500 & (5) \end{array}$$

Multiply Equation 4 by -8 and add Equations 4 and 5.

$$\begin{array}{rl} 8y+16z=2{,}000 & (4) \\ \underline{-8y-11z=-1{,}500} & (5) \\ 5z=500 & \\ z=100 & \end{array}$$

Substitute $z=100$ into Equation 4.

$$\begin{array}{rl} -y-2z=-250 & (4) \\ -y-2(100)=-250 & \\ -y-200=-250 & \\ -y=-50 & \\ y=50 & \end{array}$$

Substitute $z=100$ and $y=50$ into Equation 1.

$$\begin{array}{rl} x+y+z=180 & (1) \\ x+50+100=180 & \\ x+150=180 & \\ x=30 & \end{array}$$

He must sell 30 large types, 50 medium types, and 100 small types.

9. NUTRITION

a.

Name of food	# of oz used	Oz. of fat	Oz. of carbs	Oz. of protein
A	a	$2a$	$3a$	$2a$
B	b	$3b$	$2b$	b
C	c	c	c	$2c$
Total		14	13	9

b. Write the 3 equations.

$$\begin{cases} 2a+3b+c=14 & (1) \\ 3a+2b+c=13 & (2) \\ 2a+b+2c=9 & (3) \end{cases}$$

Multiply Equation 1 by –2 and add Equations 1 and 3.

$$\begin{aligned} -4a-6b-2c&=-28 \quad (1) \\ 2a+\ b+2c&=\quad 9 \quad (3) \\ \hline -2a-5b \qquad &=-19 \quad (4) \end{aligned}$$

Multiply Equation 2 by –2 and add Equations 2 and 3.

$$\begin{aligned} -6a-4b-2c&=-26 \quad (2) \\ 2a+\ b+2c&=\quad 9 \quad (3) \\ \hline -4a-3b \qquad &=-17 \quad (5) \end{aligned}$$

Multiply Equation 4 by –-2 and add Equations 4 and 5.

$$\begin{aligned} 4a+10b&=38 \quad (4) \\ -4a-3b&=-17 \quad (5) \\ \hline 7b&=21 \\ b&=3 \end{aligned}$$

Substitute $b = 3$ into Equation 4 and solve for a.

$$\begin{aligned} -2a-5(3)&=-19 \\ -2a-15&=-19 \\ -2a&=-4 \\ a&=2 \end{aligned}$$

Substitute $a = 2$ and $b = 3$ into Equation 1 and solve for c.

$$\begin{aligned} 2(2)+3(3)+c&=14 \\ 4+9+c&=14 \\ 13+c&=14 \\ c&=1 \end{aligned}$$

She needs 2 oz of Food A, 3 oz of Food B, and 1 oz of Food C.

11. FASHION DESIGNER

Let x = # of coats, y = number of shirts, and z = # of slacks.

Change the time available from hours to minutes by multiplying by 60.

$$\begin{cases} 20x+15y+10z=6{,}900 & (1) \\ 60x+30y+24z=16{,}800 & (2) \\ 5x+12y+6z=3{,}900 & (3) \end{cases}$$

Divide Equation 1by 5 and Equation 2 by 6.

$$\begin{cases} 4x+3y+2z=1{,}380 & (1) \\ 10x+5y+4z=2{,}800 & (2) \\ 5x+12y+6z=3{,}900 & (3) \end{cases}$$

Multiply Equation 1 by -2 and add Equations 1 and 2.

$$\begin{aligned} -8x-6y-4z&=-2{,}760 \quad (1) \\ 10x+5y+4z&=2{,}800 \quad (2) \\ \hline 2x-y&=40 \quad (4) \end{aligned}$$

Multiply Equation 1 by -3 and add Equations 1 and 3.

$$\begin{aligned} -12x-\ 9y-6z&=-4{,}140 \quad (1) \\ 5x+12y+6z&=3{,}900 \quad (3) \\ \hline -7x+3y&=-240 \quad (5) \end{aligned}$$

Multiply Equation 4 by 3 and add Equations 4 and 5.

$$\begin{aligned} 6x-3y&=120 \quad (4) \\ -7x+3y&=-240 \quad (5) \\ \hline -x&=-120 \\ x&=120 \end{aligned}$$

Substitute $x=120$ into Equation 4.

$$\begin{aligned} 2x-y&=40 \quad (4) \\ 2(120)-y&=40 \\ 240-y&=40 \\ -y&=-200 \\ y&=200 \end{aligned}$$

Substitute $x=120$ and $y=200$ into Equation 1.

$$\begin{aligned} 4x+3y+2z&=1{,}380 \quad (1) \\ 4(120)+3(200)+2z&=1{,}380 \\ 1{,}080+2z&=1{,}380 \\ 2z&=300 \\ z&=150 \end{aligned}$$

120 coats, 200 shirts, and 150 slacks should be made.

13. NFL RECORDS

Let x = # of passes from Young, y = # of passes from Montana, and z = # of passes from Gannon.

$$\begin{cases} x = y + 30 \\ y = z + 39 \\ x + y + z = 156 \end{cases}$$

Solve the 2nd equation for z.

$$\begin{cases} x = y + 30 & (1) \\ z = y - 39 & (2) \\ x + y + z = 156 & (3) \end{cases}$$

Substitute $x = y + 30$ and $z = y - 39$ into Equation 3 and solve for y.

$$\begin{aligned} x + y + z &= 156 \quad (3) \\ (y+30)+(y)+(y-39) &= 156 \\ 3y - 9 &= 156 \\ 3y &= 165 \\ y &= 55 \end{aligned}$$

Substitute $y = 55$ into Equation 1.

$$\begin{aligned} x &= y + 30 \quad (1) \\ x &= 55 + 30 \\ x &= 85 \end{aligned}$$

Substitute $y = 55$ into Equation 2.

$$\begin{aligned} z &= y - 39 \quad (2) \\ z &= 55 - 39 \\ z &= 16 \end{aligned}$$

Rice caught 85 passes from Young, 55 passes from Montana, and 16 from Gannon.

15. EARTH'S ATMOSPHERE

Let x = % nitrogen, y = % oxygen, and z = % other gases.

$$\begin{cases} x = 12 + 3(y + z) & (1) \\ z = y - 20 & (2) \\ x + y + z = 100 & (3) \end{cases}$$

Write each equation in standard form.

$$\begin{cases} x - 3y - 3z = 12 & (1) \\ y - z = 20 & (2) \\ x + y + z = 100 & (3) \end{cases}$$

Multiply Equation 1 by -1 and add Equations 1 and 3.

$$\begin{aligned} -x + 3y + 3z &= -12 \quad (1) \\ \underline{x + y + z} &= \underline{100} \quad (2) \\ 4y + 4z &= 88 \quad (4) \end{aligned}$$

Multiply Equation 2 by 4 and add Equations 2 and 4.

$$\begin{aligned} 4y + 4z &= 88 \quad (4) \\ \underline{4y - 4z} &= \underline{80} \quad (2) \\ 8y &= 168 \\ y &= 21 \end{aligned}$$

Substitute $y = 21$ into Equation 2.

$$\begin{aligned} y - z &= 20 \quad (2) \\ 21 - z &= 20 \\ -z &= -1 \\ z &= 1 \end{aligned}$$

Substitute $z = 1$ and $y = 21$ into Equation 3.

$$\begin{aligned} x + y + z &= 100 \quad (3) \\ x + 21 + 1 &= 100 \\ x + 22 &= 100 \\ x &= 78 \end{aligned}$$

The Earth's atmosphere is 78% nitrogen, 21 % oxygen, and 1% other gases.

17. TRIANGLES

$$\begin{cases} A+B+C=180 \\ A=(B+C)-100 \\ C=2B-40 \end{cases}$$

Simplify.

$$\begin{cases} A+B+C=180 & (1) \\ A-B-C=-100 & (2) \\ -2B+C=-40 & (3) \end{cases}$$

Add Equations 1 and 2.

$$\begin{aligned} A+B+C&=180 \quad (1) \\ \underline{A-B-C}&\underline{=-100} \quad (2) \\ 2A&=80 \\ A&=40 \end{aligned}$$

Substitute $A=40$ into Equation 2.

$$\begin{aligned} A-B-C&=-100 \quad (2) \\ 40-B-C&=-100 \\ -B-C&=-140 \quad (4) \end{aligned}$$

Add Equations 3 and 4.

$$\begin{aligned} -2B+C&=-40 \quad (3) \\ \underline{-B-C}&\underline{=-140} \quad (4) \\ -3B&=-180 \\ B&=60 \end{aligned}$$

Substitute $B=60$ into Equation 3.

$$\begin{aligned} -2B+C&=-40 \quad (3) \\ -2(60)+C&=-40 \\ -120+C&=-40 \\ C&=80 \end{aligned}$$

The measure of the angles are $\angle A=40^\circ, \angle B=60^\circ$, and $\angle C=80^\circ$.

19. TV HISTORY

Let x= number of episodes of *X-Files*, y = number of episodes of *Will & Grace*, and z = number of episodes of *Seinfeld*.

$$\begin{cases} x+y+z=575 & (1) \\ x=21+z & (2) \\ y-z=14 & (3) \end{cases}$$

Solve Equation 3 for y.

$$\begin{cases} x+y+z=575 & (1) \\ x=21+z & (2) \\ y=14+z & (3) \end{cases}$$

Substitute Equations 2 and 3 into Equation 1 and solve for z.

$$\begin{aligned} x+y+z&=575 \quad (1) \\ (21+z)+(14+z)+z&=575 \\ 3z+35&=575 \\ 3z&=540 \\ z&=180 \end{aligned}$$

Substitute z = 180 into Equation 2.

$$\begin{aligned} x&=21+z \quad (2) \\ x&=21+180 \\ x&=201 \end{aligned}$$

Substitute z = 180 into Equation 3.

$$\begin{aligned} y&=14+z \quad (3) \\ y&=14+180 \\ y&=194 \end{aligned}$$

There were 201 episodes of *X-Files*, 194 episodes of *Will & Grace*, and 180 episodes of *Seinfeld*.

21. ICE SKATING

Let x = radius of left circle, y = radius of middle circle, and z = radius of right circle.

$$\begin{cases} x+y=10 & (1) \\ y+z=14 & (2) \\ x+z=18 & (3) \end{cases}$$

Multiply Equation 1 by –1 and add to Equation 2.

$$\begin{aligned} -x-y \quad &= -10 \quad (1) \\ \underline{y+z=14} \quad & \quad (2) \\ -x+z=4 \quad & \quad (4) \end{aligned}$$

Add Equations 3 and 4 to eliminate x.

$$\begin{aligned} x+z=18 \quad & (3) \\ \underline{-x+z=4} \quad & (4) \\ 2z=22 & \\ z=11 & \end{aligned}$$

Substitute z = 11 into Equation 2 to solve for y.

$$\begin{aligned} y+z=14 \quad & (2) \\ y+11=14 & \\ y=3 & \end{aligned}$$

Substitute z = 11 into Equation 3 to solve for x.

$$\begin{aligned} x+z=18 \quad & (3) \\ x+11=18 & \\ x=7 & \end{aligned}$$

The radius of the circle on the left is 7 yards, the radius of the circle in the middle is 3 yards, and the radius of the circle on the right is 11 yards.

23. POTPOURRI

Let x = rose petals, y = lavender, and z = buck–wheat hulls.

$$\begin{cases} x+y+z=10 & (1) \\ 6x+5y+4z=5.5(10) & (2) \\ x=2y & (3) \end{cases}$$

Multiply Equation 1 by -4 and add Equations 1 and 2.

$$\begin{aligned} -4x-4y-4z=-40 \quad & (1) \\ \underline{6x+5y+4z=55} \quad & (2) \\ 2x+y=15 \quad & (4) \end{aligned}$$

Substitute $x = 2y$ into Equation 4 and solve for y.

$$\begin{aligned} 2x+y=15 \quad & (4) \\ 2(2y)+y=15 & \\ 4y+y=15 & \\ 5y=15 & \\ y=3 & \end{aligned}$$

Substitute $y = 3$ into Equation 3 and solve for x.

$$\begin{aligned} x=2y \quad & (3) \\ x=2(3) & \\ x=6 & \end{aligned}$$

Substitute $y = 3$ and $x = 6$ into Equation 1 and solve for x.

$$\begin{aligned} x+y+z=10 \quad & (1) \\ 6+3+z=10 & \\ 9+z=10 & \\ z=1 & \end{aligned}$$

She should use 6 pounds of rose petals, 3 pounds of lavender, and 1 pound of buck–wheat hulls.

25. PIGGY BANKS

Let x = number of nickels, y = number of dimes and z = number of quarters.

$$\begin{cases} x+y+z=64 & (1) \\ 0.05x+0.10y+0.25z=6 & (2) \\ 0.10x+0.05y+0.25z=5 & (3) \end{cases}$$

Multiply Equations 2 and 3 by 100 to clear decimals.

$$\begin{cases} x+y+z=64 & (1) \\ 5x+10y+25z=600 & (2) \\ 10x+5y+25z=500 & (3) \end{cases}$$

Multiply Equation 1 by –5 and add Equations 1 and 2.

$$\begin{array}{rl} -5x-5y-5z=-320 & (1) \\ \underline{5x+10y+25z=600} & (2) \\ 5y+20z=280 & (4) \end{array}$$

Multiply Equation 1 by –10 and add Equations 1 and 3.

$$\begin{array}{rl} -10x-10y-10z=-640 & (1) \\ \underline{10x+5y+25z=500} & (3) \\ -5y+15z=-140 & (5) \end{array}$$

Add equations 4 and 5.

$$\begin{array}{rl} 5y+20z=280 & (4) \\ \underline{-5y+15z=-140} & (5) \\ 35z=140 & \\ z=4 & \end{array}$$

Substitute $z = 4$ into Equation 5.

$$\begin{aligned} -5y+15z&=-140 \quad (5) \\ -5y+15(4)&=-140 \\ -5y+60&=-140 \\ -5y&=-200 \\ y&=40 \end{aligned}$$

Substitute $z = 4$ and $y = 40$ into Equation 1.

$$\begin{aligned} x+y+z&=64 \quad (1) \\ x+40+4&=64 \\ x+44&=64 \\ x&=20 \end{aligned}$$

She had 20 nickels, 40 dimes, and 4 quarters in her piggy bank.

27. ASTRONOMY

Substitute the coordinates of (–2, 5), (2, –3), and (4, –1) into $y = ax^2 + bx + c$.

$$\begin{cases} a(-2)^2+b(-2)+c=5 \\ a(2)^2+b(2)+c=-3 \\ a(4)^2+b(4)+c=-1 \end{cases}$$

Simplify.

$$\begin{cases} 4a-2b+c=5 & (1) \\ 4a+2b+c=-3 & (2) \\ 16a+4b+c=-1 & (3) \end{cases}$$

Add Equations 1 and 2.

$$\begin{array}{rl} 4a-2b+c=5 & (1) \\ \underline{4a+2b+c=-3} & (2) \\ 8a+2c=2 & (4) \end{array}$$

Multiply Equation 1 by 2 and add Equations 1 and 3.

$$\begin{array}{rl} 8a-4b+2c=10 & (1) \\ \underline{16a+4b+c=-1} & (2) \\ 24a+3c=9 & (5) \end{array}$$

Multiply Equation 4 by -3 and add Equations 4 and 5.

$$\begin{array}{rl} -24a-6c=-6 & (4) \\ \underline{24a+3c=9} & (5) \\ -3c=3 & \\ c=-1 & \end{array}$$

Substitute $c = -1$ into Equation 4.

$$\begin{aligned} 8a+2c&=2 \quad (4) \\ 8a+2(-1)&=2 \\ 8a-2&=2 \\ 8a&=4 \\ a&=\frac{1}{2} \end{aligned}$$

Substitute $c = -1$ and $a = \frac{1}{2}$ into Equation 2.

$$\begin{aligned} 4a+2b+c&=-3 \quad (2) \\ 4\left(\frac{1}{2}\right)+2b+(-1)&=-3 \\ 2+2b-1&=-3 \\ 2b+1&=-3 \\ 2b&=-4 \\ b&=-2 \end{aligned}$$

The equation is $y=\frac{1}{2}x^2-2x-1$.

29. WALKWAYS

Substitute the coordinates of (1, 3), (3, 1), and (1, –1) into $x^2 + y^2 + Cx + Dy + E = 0$.

$$\begin{cases}(1)^2+(3)^2+C(1)+D(3)+E=0\\(3)^2+(1)^2+C(3)+D(1)+E=0\\(1)^2+(-1)^2+C(1)+D(-1)+E=0\end{cases}$$

Simplify.

$$\begin{cases}1+9+C+3D+E=0 & (1)\\9+1+3C+D+E=0 & (2)\\1+1+C-D+E=0 & (3)\end{cases}$$

$$\begin{cases}C+3D+E=-10 & (1)\\3C+D+E=-10 & (2)\\C-D+E=-2 & (3)\end{cases}$$

Multiply Equation 2 by -1 and add Equations 1 and 2.

$$\begin{array}{rl}C+3D+E=-10 & (1)\\ \underline{-3C-D-E=10} & (2)\\ -2C+2D=0 & (4)\end{array}$$

Multiply Equation 2 by -1 and add Equations 3 and 2.

$$\begin{array}{rl}C-D+E=-2 & (3)\\ \underline{-3C-D-E=10} & (2)\\ -2C-2D=8 & (5)\end{array}$$

Add Equations 4 and 5.

$$\begin{array}{rl}-2C+2D=0 & (4)\\ \underline{-2C-2D=8} & (5)\\ -4C=8 & \\ C=-2 & \end{array}$$

Substitute $C = -2$ into Equation 4.

$$\begin{array}{rl}-2C+2D=0 & (4)\\ -2(-2)+2D=0 & \\ 4+2D=0 & \\ 2D=-4 & \\ D=-2 & \end{array}$$

Substitute $C = -2$ and $D = -2$ into Equation 3.

$$\begin{array}{rl}C-D+E=-2 & (3)\\ -2-(-2)+E=-2 & \\ -2+2+E=-2 & \\ E=-2 & \end{array}$$

The equation is $x^2 + y^2 - 2x - 2y - 2 = 0$.

WRITING

31. Answers will vary.

REVIEW

33. yes

35. yes

37. no; (1, 2) and (1, –2)

39. no; (4, 2) and (4, –2)

CHALLENGE PROBLEMS

41. DIGITS PROBLEM

Let x = hundreds digit, y = tens digit, and z = ones digit.

$$\begin{cases}x+y+z=8\\2x+y=z\\100z+10y+x=82+2(100x+10y+z)\end{cases}$$

Write the equations in standard form.

$$\begin{cases}x+y+z=8 & (1)\\2x+y-z=0 & (2)\\199x+10y-98z=-82 & (3)\end{cases}$$

Add Equations 1 and 2.

$$\begin{array}{rl}x+y+z=8 & (1)\\ \underline{2x+y-z=0} & (2)\\ 3x+2y=8 & (4)\end{array}$$

Multiply Equation 1 by 98 and add Equations 1 and 3.

$$\begin{array}{rl}98x+98y+98z=784 & (1)\\ \underline{199x+10y-98z=-82} & (3)\\ 297x+108y=702 & (5)\end{array}$$

Multiply Equation 4 by –99 and add Equations 4 and 5.

$$\begin{array}{rl}-297x-198y=-792 & (4)\\ \underline{297x+108y=702} & (5)\\ -90y=-90 & \\ y=1 & \end{array}$$

Substitute $y = 1$ into Equation 4.

$$\begin{array}{rl}3x+2y=8 & (4)\\ 3x+2(1)=8 & \\ 3x+2=8 & \\ 3x=6 & \\ x=2 & \end{array}$$

Substitute $x = 2$ and $y = 1$ into Equation 1.

$$\begin{array}{rl}x+y+z=8 & (1)\\ 2+1+z=8 & \\ 3+z=8 & \\ z=5 & \end{array}$$

The number would be 215.

CHAPTER 3 REVIEW

SECTION 3.1
Solving Systems of Equations by Graphing

1. Yes.

$$x + 2y = 0 \qquad x + 4y = 1$$

$$-1 + 2\left(\frac{1}{2}\right) \stackrel{?}{=} 0 \qquad -1 + 4\left(\frac{1}{2}\right) \stackrel{?}{=} 1$$

$$-1 + 1 \stackrel{?}{=} 0 \qquad -1 + 2 \stackrel{?}{=} 1$$

$$0 = 0 \qquad 1 = 1$$

2. No.

$$3a - 2b + 7 = 0 \qquad -2a + b = -4$$

$$3(13) - 2(23) + 7 \stackrel{?}{=} 0 \qquad -2(13) + 23 \stackrel{?}{=} -4$$

$$39 - 46 + 7 \stackrel{?}{=} 0 \qquad -26 + 23 \stackrel{?}{=} -4$$

$$0 = 0 \qquad -3 \neq -4$$

3. a. Answers may vary.
   (1, 3), (2, 1), and (4, –3)
   b. Answers may vary.
   (0, –4), (2, –2), and (4, 0)
   c. The lines intersect at the point (3, –1).

4. The point of intersection is (2019, 15). In 2019, the weekly amount of time spent viewing live broadcast television, approximately 15 hours, will be the same as that spent viewing Internet video.

5. $\begin{cases} 2x + y = 11 \\ -x + 2y = 7 \end{cases}$

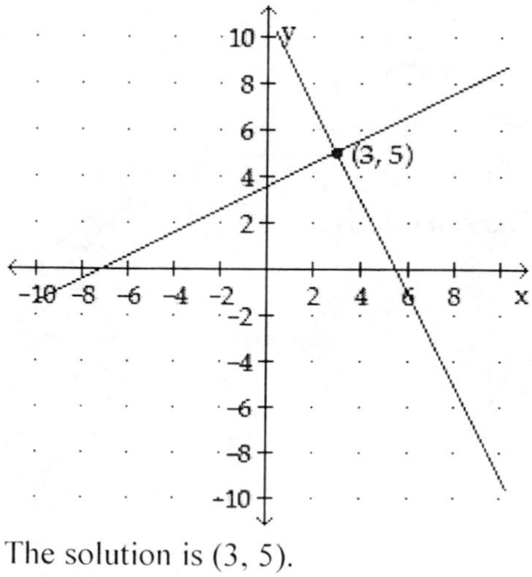

The solution is (3, 5).

6. $\begin{cases} y = -\dfrac{3}{2}x \\ 2x - 3y + 13 = 0 \end{cases}$

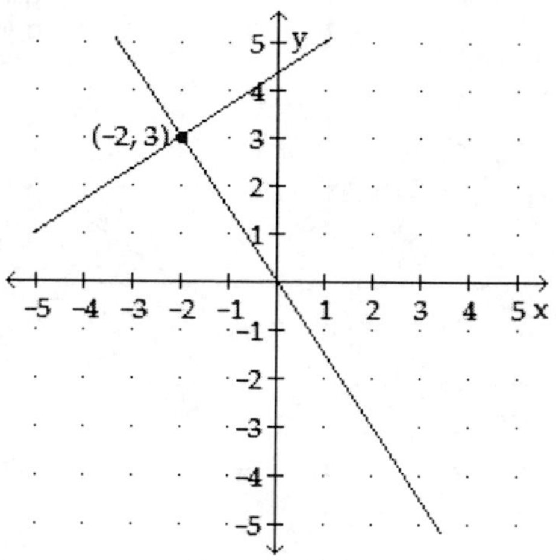

The solution is (–2, 3).

7. $\begin{cases} \dfrac{1}{2}x + \dfrac{1}{3}y = 2 \\ y = 6 - \dfrac{3}{2}x \end{cases}$

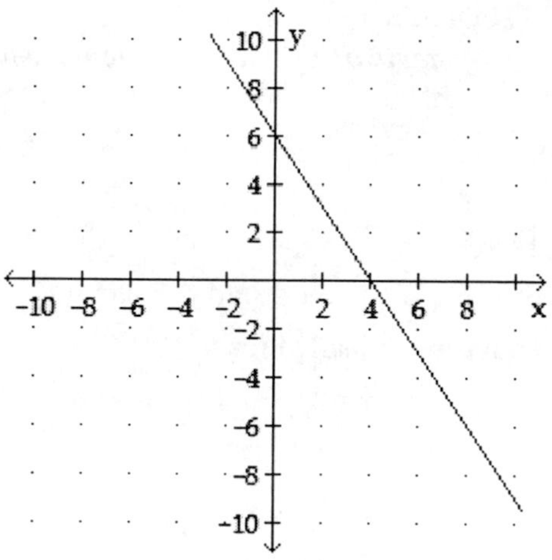

There are infinitely many solutions; the equations are dependent.

8. $\begin{cases} \frac{x}{3} - \frac{y}{2} = 1 \\ 6x - 9y = 3 \end{cases}$

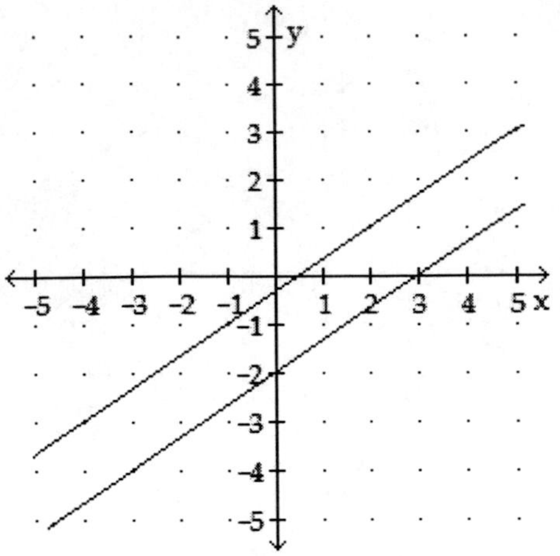

There are no solutions; the system is inconsistent.

9. The lines intersect at $x = 2$.

10. The lines intersect at $x = -1$.

SECTION 3.2
Solving Systems of Equations Algebraically

11.

$\begin{cases} x = y - 4 \\ 2x + 3y = 7 \end{cases}$

Substitute $x = y - 4$ into the second equation and solve for y.

$$2x + 3y = 7$$
$$2(y-4) + 3y = 7$$
$$2y - 8 + 3y = 7$$
$$5y - 8 = 7$$
$$5y = 15$$
$$y = 3$$

Subsitute $y = 3$ into the first equation and solve the x.

$$x = y - 4$$
$$x = 3 - 4$$
$$x = -1$$

The solution is $(-1, 3)$.

12.

$\begin{cases} y = 2x + 5 \\ 3x - 5y = -4 \end{cases}$

Substitute $y = 2x + 5$ into the second equation and solve for y.

$$3x - 5y = -4$$
$$3x - 5(2x+5) = -4$$
$$3x - 10x - 25 = -4$$
$$-7x - 25 = -4$$
$$-7x = 21$$
$$x = -3$$

Subsitute $x = -3$ into the first equation and solve for y.

$$y = 2x + 5$$
$$y = 2(-3) + 5$$
$$y = -6 + 5$$
$$y = -1$$

The solution is (-3, -1).

13.

$\begin{cases} x - 2y = 11 & (1) \\ x + 2y = -21 & (2) \end{cases}$

Add Equations 1 and 2.

$$x - 2y = 11 \quad (1)$$
$$\underline{x + 2y = -21} \quad (2)$$
$$2x = -10$$
$$x = -5$$

Substitute $x = -5$ into Equation 1.

$$x - 2y = 11$$
$$-5 - 2y = 11$$
$$-2y = 16$$
$$y = -8$$

The solution is (–5, –8).

14. $\begin{cases} 4a+5b=-9 \\ 6a=3b-3 \end{cases}$

Write the equations in standard form.

$$\begin{cases} 4a+5b=-9 & (1) \\ 6a-3b=-3 & (2) \end{cases}$$

Multiply Equation 1 by 3 and Equation 2 by 5.

$$\begin{aligned} 12a+15b&=-27 \\ \underline{30a-15b}&\underline{=-15} \\ 42a&=-42 \\ a&=-1 \end{aligned}$$

Substitute $a=-1$ into Equation 1.

$$\begin{aligned} 4a+5b&=-9 \\ 4(-1)+5b&=-9 \\ -4+5b&=-9 \\ 5b&=-5 \\ b&=-1 \end{aligned}$$

The solution is (–1, –1).

15. $\begin{cases} \frac{1}{2}a-\frac{3}{8}b=-9 \\ \frac{2}{3}a-\frac{1}{4}b=-8 \end{cases}$

Multiply the first equation by 8 and the second equation by 12 to eliminate fractions.

$$\begin{cases} 4a-3b=-72 & (1) \\ 8a-3b=-96 & (2) \end{cases}$$

Multiply Equation 1 by –1.

$$\begin{aligned} -4a+3b&=72 \quad (1) \\ \underline{8a-3b}&\underline{=-96} \quad (2) \\ 4a&=-24 \\ a&=-6 \end{aligned}$$

Substitute $a=-6$ into Equation 1.

$$\begin{aligned} 4a-3b&=-72 \quad (1) \\ 4(-6)-3b&=-72 \\ -24-3b&=-72 \\ -3b&=-48 \\ b&=16 \end{aligned}$$

The solution is (–6, 16).

16.

$$\begin{cases} y=\dfrac{x-3}{2} \\ x=\dfrac{2y+7}{2} \end{cases}$$

Multiply both equations by 2 to eliminate fractions

$$\begin{cases} 2y=x-3 \\ 2x=2y+7 \end{cases}$$

Write the equations in standard form.

$$\begin{cases} -x+2y=-3 \\ 2x-2y=7 \end{cases}$$

Add the equations to eliminate y.

$$\begin{aligned} -x+2y&=-3 \\ \underline{2x-2y}&\underline{=7} \\ x&=4 \end{aligned}$$

Substitute $x=4$ into the first equation and solve for y.

$$\begin{aligned} 2y&=x-3 \\ 2y&=4-3 \\ 2y&=1 \\ y&=\frac{1}{2} \end{aligned}$$

The solution is $\left(4, \frac{1}{2}\right)$.

17. $\begin{cases} 4x = 8y + 5 \\ 8x = 1 - 2y \end{cases}$

Write the equations in standard form.

$\begin{cases} 4x - 8y = 5 & (1) \\ 8x + 2y = 1 & (2) \end{cases}$

Multiply Equation 2 by 4.

$$\begin{aligned} 4x - 8y &= 5 \quad (1) \\ 32x + 8y &= 4 \quad (2) \\ \hline 36x &= 9 \\ x &= \frac{1}{4} \end{aligned}$$

Substitute $x = \frac{1}{4}$ into Equation 1.

$$\begin{aligned} 4x - 8y &= 5 \\ 4\left(\frac{1}{4}\right) - 8y &= 5 \\ 1 - 8y &= 5 \\ -8y &= 4 \\ y &= -\frac{1}{2} \end{aligned}$$

The solution is $\left(\frac{1}{4}, -\frac{1}{2}\right)$.

18. $\begin{cases} x + 3y = -2 \\ -2(x + 3y) = 4 \end{cases}$

Distribute –2 on Equation 2.

$\begin{cases} x + 3y = -2 & (1) \\ -2x - 6y = 4 & (2) \end{cases}$

Multiply Equation 1 by 2.

$$\begin{aligned} 2x + 6y &= -4 \quad (1) \\ -2x - 6y &= 4 \quad (2) \\ \hline 0 &= 0 \end{aligned}$$

$\{(x, y) | x + 3y = -2\}$

Infinitely many solutions

Dependent equations

19. $\begin{cases} 0.07x = 0.05 + 0.09y \\ 7x - 9y = 8 \end{cases}$

Multiply the first equation by 100 to eliminate decimals.

$\begin{cases} 7x = 5 + 9y & (1) \\ 7x - 9y = 8 & (2) \end{cases}$

Write the equations in standard form.

$$\begin{aligned} 7x - 9y &= 5 \quad (1) \\ 7x - 9y &= 8 \quad (2) \end{aligned}$$

Multiply Equation 1 by –1.

$$\begin{aligned} -7x + 9y &= -5 \quad (1) \\ 7x - 9y &= 8 \quad (2) \\ \hline 0 &\neq 3 \end{aligned}$$

No solution; $\varnothing$

Inconsistent system

20.

$\begin{cases} 0.1x + 0.2y = 1.1 \\ 2x - y = 2 \end{cases}$

Multiply the first equation by 10 to eliminate decimals.

$\begin{cases} x + 2y = 11 \\ 2x - y = 2 \end{cases}$

Solve the first equation for x.

$\begin{cases} x = 11 - 2y \\ 2x - y = 2 \end{cases}$

Substitute $x = 11 - 2y$ into the second equation and solve for y.

$$\begin{aligned} 2x - y &= 2 \\ 2(11 - 2y) - y &= 2 \\ 22 - 4y - y &= 2 \\ 22 - 5y &= 2 \\ -5y &= -20 \\ y &= 4 \end{aligned}$$

Subsitute $y = 4$ into the first equation and solve for x.

$$\begin{aligned} x &= 11 - 2y \\ x &= 11 - 2(4) \\ x &= 11 - 8 \\ x &= 3 \end{aligned}$$

The solution is $(3, 4)$.

21.

$$\begin{cases} y = -\frac{2}{3}x \\ 2x - 3y = -4 \end{cases}$$

Substitute $y = -\frac{2}{3}x$ into the second equation and solve for y.

$$2x - 3y = -4$$
$$2x - 3\left(-\frac{2}{3}x\right) = -4$$
$$2x + 2x = -4$$
$$4x = -4$$
$$x = -1$$

Subsitute $x = -1$ into the first equation and solve for y.

$$y = -\frac{2}{3}(-1)$$
$$y = \frac{2}{3}$$

The solution is $\left(-1, \frac{2}{3}\right)$.

22. Answers will vary.

SECTION 3.3
Solving Systems of Equations in Three Variables

23. No.

$$x - y + z = 4 \qquad x + 2y - z = -1$$
$$2 - (-1) + 1 \stackrel{?}{=} 4 \qquad 2 + 2(-1) - 1 \stackrel{?}{=} -1$$
$$2 + 1 + 1 \stackrel{?}{=} 4 \qquad 2 - 2 - 1 \stackrel{?}{=} -1$$
$$4 = 4 \qquad -1 = -1$$

$$x + y - 3z = -1$$
$$2 + (-1) - 3(1) \stackrel{?}{=} -1$$
$$2 - 1 - 3 \stackrel{?}{=} -1$$
$$-2 \neq -1$$

24. Yes; one solution

25.

$$\begin{cases} x - 2y + 3z = -7 & (1) \\ -x + 3y + 2z = -8 & (2) \\ 2x - y - z = 7 & (3) \end{cases}$$

Add Equations 1 and 2.

$$x - 2y + 3z = -7 \quad (1)$$
$$\underline{-x + 3y + 2z = -8} \quad (2)$$
$$y + 5z = -15 \quad (4)$$

Multiply Equation 2 by 2 and add Equation 3.

$$-2x + 6y + 4z = -16 \quad (2)$$
$$\underline{2x - y - z = 7} \quad (3)$$
$$5y + 3z = -9 \quad (5)$$

Multiply Equation 4 by –5 and add Equations 4 and 5.

$$-5y - 25z = 75 \quad (4)$$
$$\underline{5y + 3z = -9} \quad (5)$$
$$-22z = 66$$
$$z = -3$$

Substitute $z = -3$ into Equation 4.

$$y + 5z = -15 \quad (4)$$
$$y + 5(-3) = -15$$
$$y - 15 = -15$$
$$y = 0$$

Substitute $y = 0$ and $z = -3$ into Equation 1.

$$x - 2y + 3z = -7$$
$$x - 2(0) + 3(-3) = -7$$
$$x - 0 - 9 = -7$$
$$x - 9 = -7$$
$$x = 2$$

The solution is (2, 0, –3).

26. $\begin{cases} x+y+z=4 & (1) \\ x-2y-z=1 & (2) \\ 2x-y-2z=-1 & (3) \end{cases}$

Add Equations 1 and 2.

$$\begin{array}{rl} x+y+z=4 & (1) \\ \underline{x-2y-z=1} & (2) \\ 2x-y=5 & (4) \end{array}$$

Multiply Equation 1 by 2 and add Equations 1 and 3.

$$\begin{array}{rl} 2x+2y+2z=8 & (1) \\ \underline{2x-y-2z=-1} & (3) \\ 4x+y=7 & (5) \end{array}$$

Add Equations 4 and 5.

$$\begin{array}{rl} 2x-y=5 & (4) \\ \underline{4x+y=7} & (5) \\ 6x=12 & \\ x=2 & \end{array}$$

Substitute $x=2$ into Equation 4.

$$\begin{array}{rl} 2x-y=5 & (4) \\ 2(2)-y=5 & \\ 4-y=5 & \\ -y=1 & \\ y=-1 & \end{array}$$

Substitute $x=2$ and $y=-1$ into Equation 1.

$$\begin{aligned} x+y+z&=4 \\ 2+(-1)+z&=4 \\ 1+z&=4 \\ z&=3 \end{aligned}$$

The solution is (2, –1, 3).

27. $\begin{cases} x+y-z=-3 & (1) \\ x+z=2 & (2) \\ 2x-y+2z=3 & (3) \end{cases}$

Add Equations 1 and 3 to eliminate y.

$$\begin{array}{rl} x+y-z=-3 & (1) \\ \underline{2x-y+2z=3} & (3) \\ 3x+z=0 & (4) \end{array}$$

Multiply Equation 2 by -1 and add Equations 2 and 4 to eliminate z.

$$\begin{array}{rl} -x-z=-2 & (2) \\ \underline{3x+z=0} & (4) \\ 2x=-2 & \\ x=-1 & \end{array}$$

Substitute $x=-1$ into Equation 4 and solve for z.

$$\begin{array}{rl} 3x+z=0 & (4) \\ 3(-1)+z=0 & \\ -3+z=0 & \\ z=3 & \end{array}$$

Substitute $z=3$ and $x=-1$ into Equation 1 and solve for y.

$$\begin{array}{rl} x+y-z=-3 & (1) \\ -1+y-3=-3 & \\ y-4=-3 & \\ y=1 & \end{array}$$

The solution is $(-1, 1, 3)$.

28. $\begin{cases} b-4c=2 \\ a-b+2c=1 \\ 2a-2b=-2-5c \end{cases}$

Write Equation 3 in standard form.

$$\begin{cases} b-4c=2 & (1) \\ a-b+2c=1 & (2) \\ 2a-2b+5c=-2 & (3) \end{cases}$$

Multiply Equation 2 by –2 and add Equations 2 and 3.

$$\begin{array}{rl} -2a+2b-4c=-2 & (2) \\ \underline{2a-2b+5c=-2} & (3) \\ c=-4 & \end{array}$$

Substitute $c = -4$ into Equation 1.

$$\begin{aligned} b-4c&=2 \quad (1) \\ b-4(-4)&=2 \\ b+16&=2 \\ b&=-14 \end{aligned}$$

Substitute $b = -14$ and $c = -4$ into Equation 2.

$$\begin{aligned} a-b+2c&=1 \quad (2) \\ a-(-14)+2(-4)&=1 \\ a+14-8&=1 \\ a+6&=1 \\ a&=-5 \end{aligned}$$

The solution is (–5, –14, –4).

29. $\begin{cases} x+2z=10 & (1) \\ 3x+2y-3z=8 & (2) \\ y+4z=6 & (3) \end{cases}$

Multiply Equation 1 by –3 and add Equations 1 and 2.

$$\begin{array}{rl} -3x-6z=-30 & (1) \\ \underline{3x+2y-3z=8} & (2) \\ 2y-9z=-22 & (4) \end{array}$$

Multiply Equation 3 by –2 and add Equations 3 and 4.

$$\begin{array}{rl} -2y-8z=-12 & (3) \\ \underline{2y-9z=-22} & (4) \\ -17z=-34 & \\ z=2 & \end{array}$$

Substitute $z = 2$ into Equation 1.

$$\begin{aligned} x+2z&=10 \quad (1) \\ x+2(2)&=10 \\ x+4&=10 \\ x&=6 \end{aligned}$$

Substitute $z = 2$ into Equation 3.

$$\begin{aligned} y+4z&=6 \quad (3) \\ y+4(2)&=6 \\ y+8&=6 \\ y&=-2 \end{aligned}$$

The solution is (6, –2, 2).

30. $\begin{cases} x+3y+z=14 & (1) \\ x-5y=-19 & (2) \\ 3y+z=13 & (3) \end{cases}$

Multiply Equation 2 by –1 and add Equations 1 and 2.

$$\begin{aligned} x+3y+z&=14 \quad (1) \\ \underline{-x+5y \quad} &\underline{=19} \quad (2) \\ 8y+z&=33 \quad (4) \end{aligned}$$

Multiply Equation 4 by –1 and add Equations 3 and 4.

$$\begin{aligned} 3y+z&=13 \quad (3) \\ \underline{-8y-z}&\underline{=-33} \quad (4) \\ -5y&=-20 \\ y&=4 \end{aligned}$$

Substitute $y=4$ into Equation 3.

$$\begin{aligned} 3y+z&=13 \quad (3) \\ 3(4)+z&=13 \\ 12+z&=13 \\ z&=1 \end{aligned}$$

Substitute $y=4$ into Equation 2.

$$\begin{aligned} x-5y&=-19 \quad (2) \\ x-5(4)&=-19 \\ x-20&=-19 \\ x&=1 \end{aligned}$$

The solution is (1, 4, 1).

31. $\begin{cases} 2x+3y+z=-5 & (1) \\ -x+2y-z=-6 & (2) \\ 3x+y+2z=4 & (3) \end{cases}$

Multiply Equation 2 by 2 and add Equations 1and 2 to eliminate x.

$$\begin{aligned} 2x+3y+z&=-5 \quad (1) \\ \underline{-2x+4y-2z}&\underline{=-12} \quad (2) \\ 7y-z&=-17 \quad (4) \end{aligned}$$

Multiply Equation 2 by 3 and add Equations 2 and 3 to eliminate x.

$$\begin{aligned} -3x+6y-3z&=-18 \quad (2) \\ \underline{3x+y+2z}&\underline{=4} \quad (3) \\ 7y-z&=-14 \quad (5) \end{aligned}$$

Multiply Equation 4 by -1 and add Equations 4 and 5 to eliminate y and z.

$$\begin{aligned} -7y+z&=17 \quad (4) \\ \underline{7y-z}&\underline{=-14} \quad (5) \\ 0&\neq 3 \end{aligned}$$

No solution; Inconsistent system

32. $\begin{cases} 3x+3y+6z=-6 & (1) \\ -x-y-2z=2 & (2) \\ 2x+2y+4z=-4 & (3) \end{cases}$

Multiply Equation 2 by 3 and add Equations 1and 2 to eliminate x.

$$\begin{aligned} 3x+3y+6z&=-6 \quad (1) \\ \underline{-3x-3y-6z}&\underline{=6} \quad (2) \\ 0&=0 \end{aligned}$$

Infinitely many solutions;
Dependent equations

SECTION 3.4
Solving Systems of Equations Using Matrices

33. $\left[\begin{array}{cc|c} 5 & 4 & 3 \\ 1 & -1 & -3 \end{array}\right]$

34. $\left[\begin{array}{ccc|c} 1 & 2 & 3 & 6 \\ 1 & -3 & -1 & 4 \\ 6 & 1 & -2 & -1 \end{array}\right]$

35. a. $\begin{bmatrix} 1 & 3 & -2 \\ 6 & 12 & -6 \end{bmatrix}$

b. $\begin{bmatrix} 1 & 2 & -1 \\ 1 & 3 & -2 \end{bmatrix}$

c. $\begin{bmatrix} 0 & -6 & 6 \\ 1 & 3 & -2 \end{bmatrix}$

36. a. $\begin{bmatrix} 1 & 1 & 0 & -1 \\ 2 & -1 & 1 & 3 \\ 3 & -1 & -2 & 7 \end{bmatrix}$

b. $\begin{bmatrix} 2 & -1 & 1 & 3 \\ 3 & 3 & 0 & -3 \\ 3 & -1 & -2 & 7 \end{bmatrix}$

c. $\begin{bmatrix} 0 & -3 & 1 & 5 \\ 1 & 1 & 0 & -1 \\ 3 & -1 & -2 & 7 \end{bmatrix}$

37.

$\begin{bmatrix} 1 & -1 & 4 \\ 3 & 7 & -18 \end{bmatrix}$

$-3R_1 + R_2$

$\begin{bmatrix} 1 & -1 & 4 \\ 0 & 10 & -30 \end{bmatrix}$

$\frac{1}{10}R_2$

$\begin{bmatrix} 1 & -1 & 4 \\ 0 & 1 & -3 \end{bmatrix}$

$R_2 + R_1$

$\begin{bmatrix} 1 & 0 & 1 \\ 0 & 1 & -3 \end{bmatrix}$

The solution is $(1, -3)$.

38.

$\begin{bmatrix} 1 & 2 & -3 & 5 \\ 1 & 1 & 1 & 0 \\ 3 & 4 & 2 & -1 \end{bmatrix}$

$R_1 - R_2$

$\begin{bmatrix} 1 & 2 & -3 & 5 \\ 0 & 1 & -4 & 5 \\ 3 & 4 & 2 & -1 \end{bmatrix}$

$-3R_1 + R_3$

$\begin{bmatrix} 1 & 2 & -3 & 5 \\ 0 & 1 & -4 & 5 \\ 0 & -2 & 11 & -16 \end{bmatrix}$

$2R_2 + R_3$

$\begin{bmatrix} 1 & 2 & -3 & 5 \\ 0 & 1 & -4 & 5 \\ 0 & 0 & 3 & -6 \end{bmatrix}$

$\frac{1}{3}R_3$

$\begin{bmatrix} 1 & 2 & -3 & 5 \\ 0 & 1 & -4 & 5 \\ 0 & 0 & 1 & -2 \end{bmatrix}$

$4R_3 + R_2$

$\begin{bmatrix} 1 & 2 & -3 & 5 \\ 0 & 1 & 0 & -3 \\ 0 & 0 & 1 & -2 \end{bmatrix}$

$-2R_2 + R_1$

$\begin{bmatrix} 1 & 0 & -3 & 11 \\ 0 & 1 & 0 & -3 \\ 0 & 0 & 1 & -2 \end{bmatrix}$

$3R_3 + R_1$

$\begin{bmatrix} 1 & 0 & 0 & 5 \\ 0 & 1 & 0 & -3 \\ 0 & 0 & 1 & -2 \end{bmatrix}$

The solution is $(5, -3, -2)$.

39.

$$\begin{bmatrix} 16 & -8 & 32 \\ -2 & 1 & -4 \end{bmatrix}$$

$\frac{1}{16}R_1$

$$\begin{bmatrix} 1 & -\frac{1}{2} & 2 \\ -2 & 1 & -4 \end{bmatrix}$$

$2R_1 + R_2$

$$\begin{bmatrix} 1 & -\frac{1}{2} & 2 \\ 0 & 0 & 0 \end{bmatrix}$$

This matrix represents the system

$$\begin{cases} x - \frac{1}{2}y = 2 \\ 0 + 0 = 0 \end{cases}$$

Infinitely many solutions

Dependent equations

40.

$$\begin{bmatrix} 1 & 2 & -1 & 4 \\ 1 & 3 & 4 & 1 \\ 2 & 4 & -2 & 3 \end{bmatrix}$$

$-R_1 + R_3$

$$\begin{bmatrix} 1 & 2 & -1 & 4 \\ 0 & 1 & 5 & -3 \\ 2 & 4 & -2 & 3 \end{bmatrix}$$

$-2R_1 + R_2$

$$\begin{bmatrix} 1 & 2 & -1 & 4 \\ 0 & 1 & 5 & -3 \\ 0 & 0 & 0 & -5 \end{bmatrix}$$

This matrix represents the system

$$\begin{cases} x + 2y - z = 4 \\ y + 5z = -3 \\ 0 + 0 + 0 = -5 \end{cases}$$

No solution; $\varnothing$

Inconsistent system

SECTION 3.5
Solving Systems of Equations Using Determinants

41.

$$\begin{aligned} 2(3) - 3(-4) &= 6 + 12 \\ &= 18 \end{aligned}$$

42.

$$\begin{aligned} -3(-6) - (-4)(5) &= 18 + 20 \\ &= 38 \end{aligned}$$

43.

$$\begin{aligned} &-1\begin{vmatrix} -1 & 3 \\ -2 & 2 \end{vmatrix} - 2\begin{vmatrix} 2 & 3 \\ 1 & 2 \end{vmatrix} - 1\begin{vmatrix} 2 & -1 \\ 1 & -2 \end{vmatrix} \\ &= -1(-2+6) - 2(4-3) - 1(-4+1) \\ &= -1(4) - 2(1) - 1(-3) \\ &= -4 - 2 + 3 \\ &= -3 \end{aligned}$$

44.

$$\begin{aligned} &3\begin{vmatrix} -2 & -2 \\ 1 & -1 \end{vmatrix} + 2\begin{vmatrix} 1 & -2 \\ 2 & -1 \end{vmatrix} + 2\begin{vmatrix} 1 & -2 \\ 2 & 1 \end{vmatrix} \\ &= 3(2+2) + 2(-1+4) + 2(1+4) \\ &= 3(4) + 2(3) + 2(5) \\ &= 12 + 6 + 10 \\ &= 28 \end{aligned}$$

45.

$$x = \frac{D_x}{D} = \frac{\begin{vmatrix} 10 & 4 \\ 1 & -3 \end{vmatrix}}{\begin{vmatrix} 3 & 4 \\ 2 & -3 \end{vmatrix}} = \frac{10(-3) - 4(1)}{3(-3) - 4(2)} = \frac{-30 - 4}{-9 - 8} = \frac{-34}{-17} = 2$$

$$y = \frac{D_y}{D} = \frac{\begin{vmatrix} 3 & 10 \\ 2 & 1 \end{vmatrix}}{\begin{vmatrix} 3 & 4 \\ 2 & -3 \end{vmatrix}} = \frac{3(1) - 10(2)}{3(-3) - 4(2)} = \frac{3 - 20}{-9 - 8} = \frac{-17}{-17} = 1$$

The solution is (2, 1).

46.

$$x=\frac{D_x}{D}=\frac{\begin{vmatrix}-6 & -4\\ 5 & 2\end{vmatrix}}{\begin{vmatrix}-6 & -4\\ 3 & 2\end{vmatrix}}=\frac{-6(2)-(-4)(5)}{-6(2)-(-4)(3)}=\frac{-12+20}{-12+12}=\frac{8}{0}$$

No solution; inconsistent system

47.

$$x=\frac{D_x}{D}=\frac{\begin{vmatrix}0 & 2 & 1\\ 3 & 1 & 1\\ 5 & 1 & 2\end{vmatrix}}{\begin{vmatrix}1 & 2 & 1\\ 2 & 1 & 1\\ 1 & 1 & 2\end{vmatrix}}=\frac{0\begin{vmatrix}1 & 1\\ 1 & 2\end{vmatrix}-2\begin{vmatrix}3 & 1\\ 5 & 2\end{vmatrix}+1\begin{vmatrix}3 & 1\\ 5 & 1\end{vmatrix}}{1\begin{vmatrix}1 & 1\\ 1 & 2\end{vmatrix}-2\begin{vmatrix}2 & 1\\ 1 & 2\end{vmatrix}+1\begin{vmatrix}2 & 1\\ 1 & 1\end{vmatrix}}=\frac{0(2-1)-2(6-5)+1(3-5)}{1(2-1)-2(4-1)+1(2-1)}=\frac{0(1)-2(1)+1(-2)}{1(1)-2(3)+1(1)}=\frac{0-2-2}{1-6+1}=\frac{-4}{-4}=1$$

$$y=\frac{D_y}{D}=\frac{\begin{vmatrix}1 & 0 & 1\\ 2 & 3 & 1\\ 1 & 5 & 2\end{vmatrix}}{\begin{vmatrix}1 & 2 & 1\\ 2 & 1 & 1\\ 1 & 1 & 2\end{vmatrix}}=\frac{1\begin{vmatrix}3 & 1\\ 5 & 2\end{vmatrix}-0\begin{vmatrix}2 & 1\\ 1 & 2\end{vmatrix}+1\begin{vmatrix}2 & 3\\ 1 & 5\end{vmatrix}}{1\begin{vmatrix}1 & 1\\ 1 & 2\end{vmatrix}-2\begin{vmatrix}2 & 1\\ 1 & 2\end{vmatrix}+1\begin{vmatrix}2 & 1\\ 1 & 1\end{vmatrix}}=\frac{1(6-5)-0(4-1)+1(10-3)}{1(2-1)-2(4-1)+1(2-1)}=\frac{1(1)-0(3)+1(7)}{1(1)-2(3)+1(1)}=\frac{1-0+7}{1-6+1}=\frac{8}{-4}=-2$$

$$z=\frac{D_z}{D}=\frac{\begin{vmatrix}1 & 2 & 0\\ 2 & 1 & 3\\ 1 & 1 & 5\end{vmatrix}}{\begin{vmatrix}1 & 2 & 1\\ 2 & 1 & 1\\ 1 & 1 & 2\end{vmatrix}}=\frac{1\begin{vmatrix}1 & 3\\ 1 & 5\end{vmatrix}-2\begin{vmatrix}2 & 3\\ 1 & 5\end{vmatrix}+0\begin{vmatrix}2 & 1\\ 1 & 1\end{vmatrix}}{1\begin{vmatrix}1 & 1\\ 1 & 2\end{vmatrix}-2\begin{vmatrix}2 & 1\\ 1 & 2\end{vmatrix}+1\begin{vmatrix}2 & 1\\ 1 & 1\end{vmatrix}}=\frac{1(5-3)-2(10-3)+0(2-1)}{1(2-1)-2(4-1)+1(2-1)}=\frac{1(2)-2(7)+0(1)}{1(1)-2(3)+1(1)}=\frac{2-14+0}{1-6+1}=\frac{-12}{-4}=3$$

The solution is (1, –2, 3).

48.

$$x=\frac{D_x}{D}=\frac{\begin{vmatrix}2&3&1\\7&3&2\\-7&-1&-1\end{vmatrix}}{\begin{vmatrix}2&3&1\\1&3&2\\1&-1&-1\end{vmatrix}}$$

$$=\frac{2\begin{vmatrix}3&2\\-1&-1\end{vmatrix}-3\begin{vmatrix}7&2\\-7&-1\end{vmatrix}+1\begin{vmatrix}7&3\\-7&-1\end{vmatrix}}{2\begin{vmatrix}3&2\\-1&-1\end{vmatrix}-3\begin{vmatrix}1&2\\1&-1\end{vmatrix}+1\begin{vmatrix}1&3\\1&-1\end{vmatrix}}$$

$$=\frac{2(-3+2)-3(-7+14)+1(-7+21)}{2(-3+2)-3(-1-2)+1(-1-3)}$$

$$=\frac{2(-1)-3(7)+1(14)}{2(-1)-3(-3)+1(-4)}$$

$$=\frac{-2-21+14}{-2+9-4}$$

$$=\frac{-9}{3}$$

$$=-3$$

$$y=\frac{D_y}{D}=\frac{\begin{vmatrix}2&2&1\\1&7&2\\1&-7&-1\end{vmatrix}}{\begin{vmatrix}2&3&1\\1&3&2\\1&-1&-1\end{vmatrix}}$$

$$=\frac{2\begin{vmatrix}7&2\\-7&-1\end{vmatrix}-2\begin{vmatrix}1&2\\1&-1\end{vmatrix}+1\begin{vmatrix}1&7\\1&-7\end{vmatrix}}{2\begin{vmatrix}3&2\\-1&-1\end{vmatrix}-3\begin{vmatrix}1&2\\1&-1\end{vmatrix}+1\begin{vmatrix}1&3\\1&-1\end{vmatrix}}$$

$$=\frac{2(-7+14)-2(-1-2)+1(-7-7)}{2(-3+2)-3(-1-2)+1(-1-3)}$$

$$=\frac{2(7)-2(-3)+1(-14)}{2(-1)-3(-3)+1(-4)}$$

$$=\frac{14+6-14}{-2+9-4}$$

$$=\frac{6}{3}$$

$$=2$$

$$z=\frac{D_z}{D}=\frac{\begin{vmatrix}2&3&2\\1&3&7\\1&-1&-7\end{vmatrix}}{\begin{vmatrix}2&3&1\\1&3&2\\1&-1&-1\end{vmatrix}}$$

$$=\frac{2\begin{vmatrix}3&7\\-1&-7\end{vmatrix}-3\begin{vmatrix}1&7\\1&-7\end{vmatrix}+2\begin{vmatrix}1&3\\1&-1\end{vmatrix}}{2\begin{vmatrix}3&2\\-1&-1\end{vmatrix}-3\begin{vmatrix}1&2\\1&-1\end{vmatrix}+1\begin{vmatrix}1&3\\1&-1\end{vmatrix}}$$

$$=\frac{2(-21+7)-3(-7-7)+2(-1-3)}{2(-3+2)-3(-1-2)+1(-1-3)}$$

$$=\frac{2(-14)-3(-14)+2(-4)}{2(-1)-3(-3)+1(-4)}$$

$$=\frac{-28+42-8}{-2+9-4}$$

$$=\frac{6}{3}$$

$$=2$$

The solution is (–3, 2, 2).

SECTION 3.6
Problem Solving Using Systems of Two Equations

49. Let x = distance between Austin and Houston and y = distance Austin and San Antonio.

$$\begin{cases} x = 2y - 4 \\ x + y + 197 = 442 \end{cases}$$

Write the equations in standard form.

$$\begin{cases} x - 2y = -4 \\ x + y = 245 \end{cases}$$

Multiply the first equation by -1 and add the equations to eliminate x.

$$-x + 2y = 4$$
$$\underline{x + y = 245}$$
$$3y = 249$$
$$y = 83$$

Substitute $y = 83$ into the first equation and solve for x.

$$x = 2y - 4$$
$$x = 2(83) - 4$$
$$x = 166 - 4$$
$$x = 162$$

It is 162 miles from Austin to Houston and 83 miles from Austin to San Antonio.

50. Let b = boat's rate and c = current's rate.

	Rate · time = distance		
downstream	$b + c$	3	30
upstream	$b - c$	5	30

$$\begin{cases} 3(b + c) = 30 \\ 5(b - c) = 30 \end{cases}$$

Write equations in standard form.

$$\begin{cases} 3b + 3c = 30 \\ 5b - 5c = 30 \end{cases}$$

Multiply the first equation by -5 and the second equation by 3 to eliminate x.

$$-15b - 15c = -150$$
$$\underline{15b - 15c = 90}$$
$$-30c = -60$$
$$c = 2$$

Substitute $c = 2$ into the first equation and solve for b.

$$3b + 3c = 30$$
$$3b + 3(2) = 30$$
$$3b + 6 = 30$$
$$3b = 24$$
$$b = 8$$

The rate of the boat is 8 mph and the rate of the current is 2 mph.

51. Let x = oz of 6% solution and y = oz of 18% solution.

$$\begin{cases} x + y = 750 \\ 0.06x + 0.18y = 0.10(750) \end{cases}$$

Multiply the second equation by 100 to eliminate decimals.

$$\begin{cases} x + y = 750 & (1) \\ 6x + 18y = 7,500 & (2) \end{cases}$$

Multiply Equation 1 by –6.

$$-6x - 6y = -4,500 \quad (1)$$
$$\underline{6x + 18y = 7,500} \quad (2)$$
$$12y = 3,000$$
$$y = 250$$

Substitute $y = 250$ into Equation 1.

$$x + y = 750 \quad (1)$$
$$x + 250 = 750$$
$$x = 500$$

500 oz of the 6% solution and 250 oz of the 18% solution are needed.

52. Let x = amount invested at 6% and y = amount invested at 12%

$$\begin{cases} x + y = 10{,}000 \\ 0.06x + 0.12y = 960 \end{cases}$$

Multiply the second equation by 100 to eliminate decimals.

$$\begin{cases} x + y = 10{,}000 & (1) \\ 6x + 12y = 96{,}000 & (2) \end{cases}$$

Multiply Equation 1 by –6.

$$\begin{aligned} -6x - 6y &= -60{,}000 \quad (1) \\ \underline{6x + 12y} &\underline{= 96{,}000} \quad (2) \\ 6y &= 36{,}000 \\ y &= 6{,}000 \end{aligned}$$

Substitute y = 6,000 into Equation 1.

$$\begin{aligned} x + y &= 10{,}000 \quad (1) \\ x + 6{,}000 &= 10{,}000 \\ x &= 4{,}000 \end{aligned}$$

They invested \$4,000 at 6% and \$6,000 at 12%.

53. Let x = milliliters in one teaspoon and y = milliliters in one tablespoon.

$$\begin{cases} 2x + 5y = 85 & (1) \\ 5x + 2y = 55 & (2) \end{cases}$$

Multiply Equation 1 by –5 and Equation 2 by 2.

$$\begin{aligned} -10x - 25y &= -425 \quad (1) \\ \underline{10x + 4y} &\underline{= 110} \quad (2) \\ -21y &= -315 \\ y &= 15 \end{aligned}$$

Substitute y = 15 into Equation 1.

$$\begin{aligned} 2x + 5y &= 85 \quad (1) \\ 2x + 5(15) &= 85 \\ 2x + 75 &= 85 \\ 2x &= 10 \\ x &= 5 \end{aligned}$$

There are 5 mL in one teaspoon and 15 mL in one tablespoon.

54. Let x = number of bottles.

$$\begin{cases} C_1 = 0.04x + 250 \\ C_2 = 0.02x + 600 \end{cases}$$

Breakeven point:

$$\begin{aligned} C_1 &= C_2 \\ 0.04x + 250 &= 0.02x + 600 \\ 0.04x + 250 - 0.02x &= 0.02x + 600 - 0.02x \\ 0.02x + 250 &= 600 \\ 0.02x + 250 - 250 &= 600 - 250 \\ 0.02x &= 350 \\ x &= 17{,}500 \text{ bottles} \end{aligned}$$

The breakeven point is 17,500 bottles,

SECTION 3.7
Problem Solving Using Systems of Three Equations

55. Let x = the small bears, y = medium bears, and z = large bears.

$$\begin{cases} 3x+5y+10z=850 & (1) \\ 2x+3y+5z=480 & (2) \\ 6x+8y+12z=1,260 & (3) \end{cases}$$

Multiply Equation 1 by –2 and add Equations 1 and 3.

$$\begin{array}{rl} -6x-10y-20z=-1,700 & (1) \\ \underline{6x+8y+12z=1,260} & (3) \\ -2y-8z=-440 & (4) \end{array}$$

Multiply Equation 2 by –3 and add Equations 2 and 3.

$$\begin{array}{rl} -6x-9y-15z=-1,440 & (2) \\ \underline{6x+8y+12z=1,260} & (3) \\ -y-3z=-180 & (5) \end{array}$$

Multiply Equation 5 by –2 and add Equations 4 and 5.

$$\begin{array}{rl} -2y-8z=-440 & (4) \\ \underline{2y+6z=360} & (5) \\ -2z=-80 & \\ z=40 & \end{array}$$

Substitute $z = 40$ into Equation 4.

$$\begin{aligned} -2y-8z&=-440 \quad (4) \\ -2y-8(40)&=-440 \\ -2y-320&=-440 \\ -2y&=-120 \\ y&=60 \end{aligned}$$

Substitute $y = 60$ and $z = 40$ into Equation 1.

$$\begin{aligned} 3x+5y+10z&=850 \quad (1) \\ 3x+5(60)+10(40)&=850 \\ 3x+300+400&=850 \\ 3x+700&=850 \\ 3x&=150 \\ x&=50 \end{aligned}$$

The toy company produces 50 small bears, 60 medium bears, and 40 large bears.

56. Let x = amount invested at 5%, y = amount at 6%, and z = amount invested at 7%.

$$\begin{cases} x+y+z=22,000 \\ y=x+2,000 \\ 0.05x+0.06y+0.07z=1,370 \end{cases}$$

Multiply Equation 3 by 100 to clear decimals.

$$\begin{cases} x+y+z=22,000 & (1) \\ y=x+2,000 & (2) \\ 5x+6y+7z=137,000 & (3) \end{cases}$$

Multiply Equation 1 by –7 and add Equation 1 and 3.

$$\begin{array}{rl} -7x-7y-7z=-154,000 & (1) \\ \underline{5x+6y+7z=137,000} & (3) \\ -2x-y=-17,000 & (4) \end{array}$$

Substitute Equation 2 into Equation 4.

$$\begin{aligned} -2x-y&=-17,000 \quad (4) \\ -2x-(x+2,000)&=-17,000 \\ -2x-x-2,000&=-17,000 \\ -3x&=-15,000 \\ x&=5,000 \end{aligned}$$

Substitute $x = 5,000$ into Equation 2.

$$\begin{aligned} y&=x+2,000 \quad (2) \\ y&=5,000+2,000 \\ y&=7,000 \end{aligned}$$

Substitute $x = 5,000$ and $y = 7,000$ into Equation 1.

$$\begin{aligned} x+y+z&=22,000 \quad (1) \\ 5,000+7,000+z&=22,000 \\ 12,000+z&=22,000 \\ z&=10,000 \end{aligned}$$

She invested \$5,000 at 5%, \$7,000 at 6%, and \$10,000 at 7%.

57. Substitute the coordinates of (0, 0), (8, 12), and (12, 15) into $y = ax^2 + bx + c$.

$$\begin{cases} a(0)^2 + b(0) + c = 0 \\ a(8)^2 + b(8) + c = 12 \\ a(12)^2 + b(12) + c = 15 \end{cases}$$

Simplify.

$$\begin{cases} c = 0 & (1) \\ 64a + 8b + c = 12 & (2) \\ 144a + 12b + c = 15 & (3) \end{cases}$$

Multiply Equation 2 by 3 and Equation 4 by -2. Add Equations 2 and 3.

$$\begin{array}{rl} 192a + 24b + 3c = 36 & (1) \\ \underline{-288a - 24b - 2c = -30} & (2) \\ -96a + c = 6 & (4) \end{array}$$

Substitute $c = 0$ into Equation 4.

$$\begin{aligned} -96a + c &= 6 \quad (4) \\ -96a + 0 &= 6 \\ -96a &= 6 \\ a &= -\frac{1}{16} \end{aligned}$$

Substitute $c = 0$ and $a = -\frac{1}{16}$ into Equation 2.

$$\begin{aligned} 64a + 8b + c &= 12 \quad (2) \\ 64\left(-\frac{1}{16}\right) + 8b + 0 &= 12 \\ -4 + 8b &= 12 \\ 8b &= 16 \\ b &= 2 \end{aligned}$$

$$\left(-\frac{1}{16}, 2, 0\right)$$

58. Let x = number of cups from Mix A, y = cups from Mix B, and z = cups from Mix C.

$$\begin{cases} 5x + 6y + 8z = 24 & (1) \\ 2x + 3y + 3z = 10 & (2) \\ x + 2y + z = 5 & (3) \end{cases}$$

Multiply Equation 3 by –5 and add Equations 1 and 3.

$$\begin{array}{rl} 5x + 6y + 8z = 24 & (1) \\ \underline{-5x - 10y - 5z = -25} & (3) \\ -4y + 3z = -1 & (4) \end{array}$$

Multiply Equation 3 by –2 and add Equations 2 and 3.

$$\begin{array}{rl} 2x + 3y + 3z = 10 & (2) \\ \underline{-2x - 4y - 2z = -10} & (3) \\ -y + z = 0 & (5) \end{array}$$

Multiply Equation 5 by –4 and add Equations 4 and 5.

$$\begin{array}{rl} -4y + 3z = -1 & (4) \\ \underline{4y - 4z = 0} & (5) \\ -z = -1 & \\ z = 1 & \end{array}$$

Substitute $z = 1$ into Equation 5.

$$\begin{aligned} -y + z &= 0 \quad (5) \\ -y + 1 &= 0 \\ -y &= -1 \\ y &= 1 \end{aligned}$$

Substitute $y = 1$ and $z = 1$ into Equation 3.

$$\begin{aligned} x + 2y + z &= 5 \quad (3) \\ x + 2(1) + 1 &= 5 \\ x + 2 + 1 &= 5 \\ x + 3 &= 5 \\ x &= 2 \end{aligned}$$

You need 2 cups of Mix A, 1 cup of Mix B, and 1 cup of Mix C.

CHAPTER 3 TEST

1. a. $\begin{cases} 2x - 7y = 1 \\ 4x - y = -8 \end{cases}$ is called a **system** of linear equations.

   b. The matrix $\begin{bmatrix} -10 & 3 \\ 4 & 9 \end{bmatrix}$ has 2 **rows** and 2 **columns**.

   c. Solutions of a system of three equations in three variables *x*, *y*, and *z*, are written in the form (*x*, *y*, *z*) and are called ordered **triples**.

   d. The graph of the equation $2x + 3y + 4z = 5$ is a flat surface called a **plane**.

   e. A **matrix** is a rectangular array of numbers written with brackets.

2. $\begin{cases} 2x + y = 5 \\ y = 2x - 3 \end{cases}$

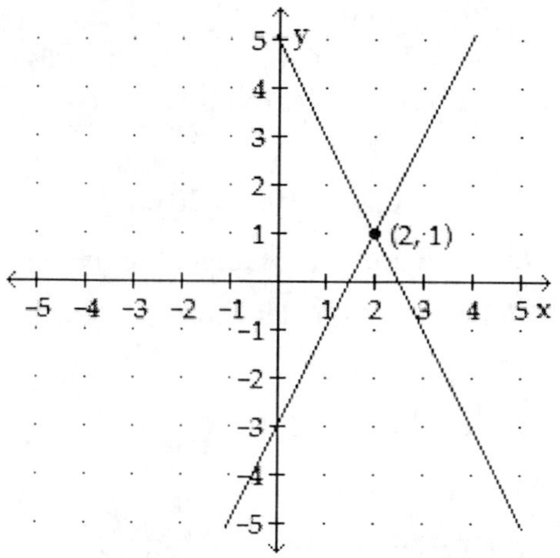

3. Yes, it is a solution.

$$10x - 12y = 3 \qquad\qquad 18x - 15y = 1$$

$$10\left(-\frac{1}{2}\right) - 12\left(-\frac{2}{3}\right) \stackrel{?}{=} 3 \qquad 18\left(-\frac{1}{2}\right) - 15\left(-\frac{2}{3}\right) \stackrel{?}{=} 1$$

$$-5 + 8 \stackrel{?}{=} 3 \qquad\qquad -9 + 10 \stackrel{?}{=} 1$$

$$3 = 3 \qquad\qquad 1 = 1$$

4. The point of intersection is (June 2006, 44%). That means that Governor Schwarzenegger's job approval and disapproval ratings were the same in June of 2006; approximately 44%.

5. a. inconsistent system; no solution, ∅

   b. dependent equations; infinitely many solutions

6. The graphs intersect at $x = 3$.

7.

$$\begin{cases} 2x - 4y = 14 \\ x + 2y = 7 \end{cases}$$

Solve the second equation for *x*.

$$\begin{cases} 2x - 4y = 14 \\ x = -2y + 7 \end{cases}$$

Substitute $x = -2y + 7$ into the first equation and solve for *y*.

$$2x + 4y = 14$$
$$2(-2y + 7) - 4y = 14$$
$$-4y + 14 - 4y = 14$$
$$-8y + 14 = 14$$
$$-8y = 0$$
$$y = 0$$

Substitute $y = 0$ into the second equation and solve for *x*.

$$x = -2y + 7$$
$$x = -2(0) + 7$$
$$x = 0 + 7$$
$$x = 7$$

The solution is $(7, 0)$.

8.

$$\begin{cases} 2c + 3d = -5 \\ 3c - 2d = 12 \end{cases}$$

Multiply the first equation by -3 and the second equation by 2 to eliminate *c*.

$$-6c - 9d = 15$$
$$\underline{6c - 4d = 24}$$
$$-13d = 39$$
$$d = -3$$

Substitute $d = -3$ into the first equation and solve for *c*.

$$2c + 3(-3) = -5$$
$$2c - 9 = -5$$
$$2c = 4$$
$$c = 2$$

The solution is $(2, -3)$.

9.

$$\begin{cases} 3(x+y) = x-3 \\ -y = \dfrac{2x+3}{3} \end{cases}$$

Write the equations in standard form.

$$\begin{cases} 3x+3y = x-3 \\ -3y = 2x+3 \end{cases}$$

$$\begin{cases} 2x+3y = -3 \\ 2x+3y = -3 \end{cases}$$

$$\{(x,y) | 2x+3y=-3\}$$

The equations are the same so they are dependent and there are infinitely many solutions.

10.

$$\begin{cases} 0.6x+0.5y = 1.2 \\ x - \dfrac{4}{9}y + \dfrac{5}{9} = 0 \end{cases}$$

Multiply the first equation by 10 to clear decimals and the second equation by 9 to eliminate fractions and write in standard form.

$$\begin{cases} 6x+5y = 12 & (1) \\ 9x-4y = -5 & (2) \end{cases}$$

Multiply Equation 1 by 4 and Equation 2 by 5.

$$\begin{aligned} 24x+20y &= 48 \quad (1) \\ 45x-20y &= -25 \quad (2) \\ \hline 69x &= 23 \\ x &= \frac{23}{69} \\ x &= \frac{1}{3} \end{aligned}$$

Substitute $x = \frac{1}{3}$ into Equation 1 and solve for y.

$$\begin{aligned} 6x+5y &= 12 \quad (1) \\ 6\left(\frac{1}{3}\right)+5y &= 12 \\ 2+5y &= 12 \\ 5y &= 10 \\ y &= 2 \end{aligned}$$

The solution is $\left(\frac{1}{3}, 2\right)$.

11. TRAFFIC SIGNS

The sum of the angles of a triangle is 180°.

$$\begin{cases} 2x+y = 180 \\ y = x+15 \end{cases}$$

Substitute $y = x+15$ into the first equation and solve for x.

$$\begin{aligned} 2x+(x+15) &= 180 \\ 3x+15 &= 180 \\ 3x &= 165 \\ x &= 55 \end{aligned}$$

Substitute $x = 55$ into the second equation and solve for y.

$$\begin{aligned} y &= 55+15 \\ y &= 70 \end{aligned}$$

The measure of the angles are 55° and 70°.

12. ANTIFREEZE

Let x = gallons of 40% solution and y = gallons of 80% solution.

$$\begin{cases} x+y = 20 \\ 0.40x+0.80y = 0.50(20) \end{cases}$$

Multiply the second equation by 100 and write in standard form.

$$\begin{cases} x+y = 20 \\ 40x+80y = 50(20) \end{cases}$$

$$\begin{cases} x+y = 20 \\ 40x+80y = 1{,}000 \end{cases}$$

Multiply the first equation by -40 and add the equations to eliminate x.

$$\begin{aligned} -40x-40y &= -800 \\ 40x+80y &= 1{,}000 \\ \hline 40y &= 200 \\ y &= 5 \end{aligned}$$

Substitute $y = 5$ into the first equation and solve for x.

$$\begin{aligned} x+5 &= 20 \\ x &= 15 \end{aligned}$$

15 gallons of 40% solution and 5 gallons of 80% solution are needed.

13. BREAK POINTS

Let x = number of impressions.

Total Cost = setup + cost per impression

$$\begin{cases} C_1 = 1{,}775 + 5.75x \\ C_2 = 3{,}975 + 4.15x \end{cases}$$

Breakpoint:

$$C_1 = C_2$$
$$1{,}775 + 5.75x = 3{,}975 + 4.15x$$
$$1{,}775 + 5.75x - 4.15x = 3{,}975 + 4.15x - 4.15x$$
$$1{,}775 + 1.6x = 3{,}975$$
$$1{,}775 + 1.6x - 1{,}775 = 3{,}975 - 1{,}775$$
$$1.6x = 2{,}200$$
$$x = 1.375 \text{ impressions}$$

14. PARALLELOGRAMS

Opposite angles in a parallelogram are equal and alternate interior angles are equal.

$$\begin{cases} x + y = 108 \\ x - y = 28 \end{cases}$$

Add the equations to eliminate y.

$$x + y = 108$$
$$\underline{x - y = 28}$$
$$2x = 136$$
$$x = 68$$

Substitute $x = 68$ into the first equation to solve for y.

$$x + y = 108$$
$$68 + y = 108$$
$$y = 40$$

15. STUDENT LOANS

Let x = amount of 2.5% loan and y = amount of the 4% loan.

$$\begin{cases} x + y = 8{,}500 & (1) \\ 0.025x + 0.04y = 265 & (2) \end{cases}$$

Multiply the Equation 2 by 1000 to eliminate decimals.

$$\begin{cases} x + y = 8{,}500 & (1) \\ 25x + 40y = 265{,}000 & (2) \end{cases}$$

Multiply the Equation 1 by –25 and add Equations 1 and 2 to solve for y.

$$-25x - 25y = -212{,}500$$
$$\underline{25x + 40y = \quad 265{,}000}$$
$$15y = 52{,}500$$
$$y = 3{,}500$$

Substitute $y = 3{,}500$ into Equation 1 and solve for x.

$$x + y = 8{,}500$$
$$x + 3{,}500 = 8{,}500$$
$$x = 5{,}000$$

The amount of the 2.5% loan was \$5,000 and the amount of the 4% loan was \$3,500.

16. GOURMET FRUIT

Let x = the cost of one pear and y = the cost of one apple.

$$\begin{cases} 6x + 4y = 25.50 & (1) \\ 4x + 10y = 31.30 & (2) \end{cases}$$

Multiply Equation 1 by 4 and Equation 2 by –6. Add the equations to solve for y.

$$24x + 16y = \quad 102$$
$$\underline{-24x - 60y = -187.8}$$
$$-44y = -85.8$$
$$y = 1.95$$

Substitute $y = 1.95$ into Equation 1 and solve for x.

$$6x + 4y = 22.5$$
$$6x + 4(1.95) = 25.5$$
$$6x + 7.8 = 25.5$$
$$6x = 17.7$$
$$x = 2.95$$

The price of one pear is \$2.95 and the price of one apple is \$1.95.

17. No.

Check Equation 1.

$$x - 2y + z = 5$$
$$-1 - 2\left(-\frac{1}{2}\right) + 5 \stackrel{?}{=} 5$$
$$-1 + 1 + 5 \stackrel{?}{=} 5$$
$$5 = 5$$

Check Equation 2.

$$2x + 4y = -4$$
$$2(-1) + 4\left(-\frac{1}{2}\right) \stackrel{?}{=} -4$$
$$-2 - 2 \stackrel{?}{=} -4$$
$$-4 = -4$$

Check Equation 3.

$$-6y + 4z = 22$$
$$-6\left(-\frac{1}{2}\right) + 4(5) \stackrel{?}{=} 22$$
$$3 + 20 \stackrel{?}{=} 22$$
$$23 \neq 22$$

18. It has no solutions since the three lines do not all intersect at the same point.

19.

$$\begin{cases} x + y + z = 4 & (1) \\ x + y - z = 6 & (2) \\ 2x - 3y + z = -1 & (3) \end{cases}$$

Multiply Equation 1 by -1 and add Equations 1 and 2.

$$\begin{array}{rl} -x - y - z = -4 & (1) \\ \underline{x + y - z = 6} & (2) \\ -2z = 2 & \\ z = -1 & \end{array}$$

Multiply Equation 1 by -2 and add Equations 1 and 3 to eliminate x.

$$\begin{array}{rl} -2x - 2y - 2z = -8 & (1) \\ \underline{2x - 3y + z = -1} & (2) \\ -5y - z = -9 & (4) \end{array}$$

Substitute $z = -1$ into Equation 4 and solve for y.

$$-5y - z = -9 \quad (4)$$
$$-5y - (-1) = -9$$
$$-5y + 1 = -9$$
$$-5y = -10$$
$$y = 2$$

Substitute $z = -1$ and $y = 2$ into Equation 1 and solve for x.

$$x + y + z = 4 \qquad (1)$$
$$x + 2 - 1 = 4$$
$$x + 1 = 4$$
$$x = 3$$

The solution is $(3, 2, -1)$.

20. Write the equations in standard form.

$$\begin{cases} -2y+z=1 & (1) \\ x+y+z=1 & (2) \\ x+5y=4 & (3) \end{cases}$$

Multiply Equation 1 by –1 and add Equations 1 and 2.

$$\begin{array}{rl} 2y-z=-1 & (1) \\ \underline{x+y+z=1} & (2) \\ x+3y=0 & (4) \end{array}$$

Multiply Equation 3 by –1 and add Equations 3 and 4.

$$\begin{array}{rl} -x-5y=-4 & (3) \\ \underline{x+3y=0} & (4) \\ -2y=-4 & \\ y=2 & \end{array}$$

Substitute $y=2$ into Equation 1.

$$\begin{aligned} -2y+z&=1 \quad (1) \\ -2(2)+z&=1 \\ -4+z&=1 \\ z&=5 \end{aligned}$$

Substitute $y=2$ into Equation 3.

$$\begin{aligned} x+5y&=4 \quad (3) \\ x+5(2)&=4 \\ x+10&=4 \\ x&=-6 \end{aligned}$$

The solution is (–6, 2, 5).

21. Let x = children's tickets sold, y = general admission tickets, and z = seniors tickets.

$$\begin{cases} x+y+z=100 \\ 3x+6y+5z=410 \\ x=2y \end{cases}$$

Write the third equation in standard form.

$$\begin{cases} x+y+z=100 & (1) \\ 3x+6y+5z=410 & (2) \\ x-2y=0 & (3) \end{cases}$$

Multiply Equation 1 by -5 to eliminate z and add Equations 1 and 2.

$$\begin{array}{rl} -5x-5y-5z=-500 & \\ \underline{3x+6y+5z=410} & \\ -2x+y=-90 & (4) \end{array}$$

Multiply Equation 3 by 2 and add Equations 3 and 4 to eliminate x.

$$\begin{array}{rl} 2x-4y=0 & (3) \\ \underline{-2x+y=-90} & (4) \\ -3y=-90 & \\ y=30 & \end{array}$$

Substitute $y=30$ into Equation 3 and solve for x.

$$\begin{aligned} x-2y&=0 \quad (3) \\ x-2(30)&=0 \\ x-60&=0 \\ x&=60 \end{aligned}$$

Substitute $y=30$ and $x=60$ into Equation 1 and solve for z.

$$\begin{aligned} x+y+z&=100 \quad (1) \\ 60+30+z&=100 \\ 90+z&=100 \\ z&=10 \end{aligned}$$

They sold 60 children's ticket, 30 general admission tickets, and 10 seniors' tickets.

22. Let x = the weight of the bar, y = the weight of the small plate, z = the weight of the large plate.

$$\begin{cases} x+2y+2z=155 & (1) \\ x+2y+4z=245 & (2) \\ x+6y+6z=375 & (3) \end{cases}$$

Multiply Equation 1 by –1 and add Equations 1 and 2.

$$\begin{aligned} -x-2y-2z&=-155 \\ x+2y+4z&=\ \ 245 \\ \hline 2z&=-90 \\ z&=45 \end{aligned}$$

Multiply Equation 3 by –1 and add Equations 1 and 3.

$$\begin{aligned} -x-6y-6z&=-375 \\ x+2y+4z&=\ \ 245 \\ \hline -4y-2z&=-130 \quad (4) \end{aligned}$$

Substitute $z = 45$ into Equation 4 and solve for y.

$$\begin{aligned} -4y-2z&=-130 \quad (4) \\ -4y-2(45)&=-130 \\ -4y-90&=-130 \\ -4y-90+90&=-130+90 \\ -4y&=-40 \\ y&=10 \end{aligned}$$

Substitute $y = 10$ and $z = 45$ into Equation 1 and solve for x.

$$\begin{aligned} x+2y+2z&=155 \\ x+2(10)+2(45)&=155 \\ x+20+90&=155 \\ x+110&=155 \\ x+110-110&=155-110 \\ x&=45 \end{aligned}$$

The bar weighs 45 lbs, the small plates weighs 10 lbs, and the large plate weights 45 lbs.

23. $\begin{bmatrix} 1 & 7 & -3 \\ 0 & -22 & 22 \end{bmatrix}$

24.

$$\begin{bmatrix} 1 & 1 & 4 \\ 2 & -1 & 2 \end{bmatrix}$$

$-2R_1 + R_2$

$$\begin{bmatrix} 1 & 1 & 4 \\ 0 & -3 & -6 \end{bmatrix}$$

$-\frac{1}{3}R_2$

$$\begin{bmatrix} 1 & 1 & 4 \\ 0 & 1 & 2 \end{bmatrix}$$

$-R_2 + R_1$

$$\begin{bmatrix} 1 & 0 & 2 \\ 0 & 1 & 2 \end{bmatrix}$$

The solution is $(2, 2)$.

25.

$$\begin{bmatrix} 1 & -3 & 2 & 1 \\ 1 & -2 & 3 & 5 \\ 2 & -6 & 4 & 3 \end{bmatrix}$$

$-1R_1 + R_2$

$$\begin{bmatrix} 1 & -3 & 2 & 1 \\ 0 & 1 & 1 & 4 \\ 2 & -6 & 4 & 3 \end{bmatrix}$$

$-2R_1 + R_2$

$$\begin{bmatrix} 1 & -3 & 2 & 1 \\ 0 & 1 & 1 & 4 \\ 0 & 0 & 0 & 1 \end{bmatrix}$$

This matrix represents the system

$$\begin{cases} x-3y+2z=1 \\ y+z=4 \\ 0=1 \end{cases}$$

There is No Solution, ∅.

Inconsistent system

26. $\begin{vmatrix} 2 & -3 \\ -4 & 5 \end{vmatrix} = 2(5)-(-3)(-4)$

$$\begin{aligned} &=10-12 \\ &=-2 \end{aligned}$$

27. $1\begin{vmatrix} 0 & 3 \\ -2 & 2 \end{vmatrix} - 2\begin{vmatrix} 2 & 3 \\ 1 & 2 \end{vmatrix} + 0\begin{vmatrix} 2 & 0 \\ 1 & -2 \end{vmatrix}$

$$= 1(0+6) - 2(4-3) + 0(-4-0)$$
$$= 1(6) - 2(1) + 0(-4)$$
$$= 6 - 2 + 0$$
$$= 4$$

28. The solution is (–3, 3).

$$x = \frac{D_x}{D} = \frac{\begin{vmatrix} -6 & -1 \\ -6 & 1 \end{vmatrix}}{\begin{vmatrix} 1 & -1 \\ 3 & 1 \end{vmatrix}} = \frac{-6(1) - (-1)(-6)}{1(1) - (-1)(3)} = \frac{-6-6}{1+3} = \frac{-12}{4} = -3$$

$$y = \frac{D_y}{D} = \frac{\begin{vmatrix} 1 & -6 \\ 3 & -6 \end{vmatrix}}{\begin{vmatrix} 1 & -1 \\ 3 & 1 \end{vmatrix}} = \frac{1(-6) - (-6)(3)}{1(1) - (-1)(3)} = \frac{-6+18}{1+3} = \frac{12}{4} = 3$$

CHAPTER 3 CUMULATIVE REVIEW

1. Real Numbers

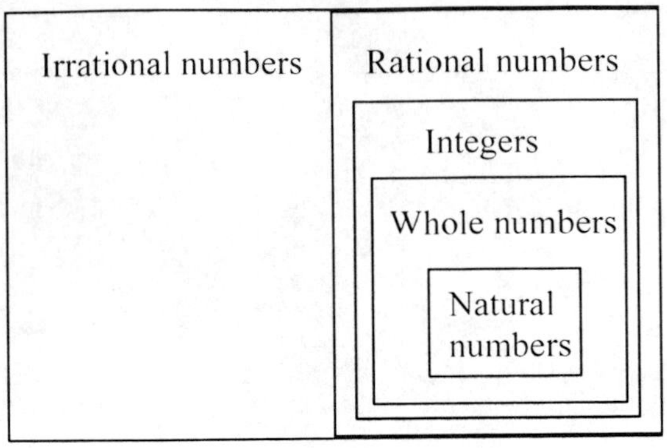

2. a. $-5.96 \boxed{<} -5.95$

   b. $-(-1) \;\square\; -|-18|$

   $1 \boxed{>} -18$

3. $a = -3,\ b = -5$

$$\begin{aligned}-|b| - ab^2 &= -|-5| - (-3)(-5)^2\\ &= -5 + 3(25)\\ &= -5 + 75\\ &= 70\end{aligned}$$

4.

$$\begin{aligned}\frac{14+2[2a-(b-a)]}{-b-2} &= \frac{14+2[2(-3)-(-5-(-3))]}{-(-5)-2}\\ &= \frac{14+2[-6-(-2)]}{5-2}\\ &= \frac{14+2[-6+2]}{3}\\ &= \frac{14+2[-4]}{3}\\ &= \frac{14-8}{3}\\ &= \frac{6}{3}\\ &= 2\end{aligned}$$

5.

$$\begin{aligned}40\left(\frac{3}{8}m - \frac{1}{4}\right) + 40\left(\frac{4}{5}\right) &= 15m - 10 + 32\\ &= 15m + 22\end{aligned}$$

6. $$\begin{aligned}&3[2(a+2)] - 5[3(a-5)] + 5a\\ &= 3[2a+4] - 5[3a-15] + 5a\\ &= 6a + 12 - 15a + 75 + 5a\\ &= -4a + 87\end{aligned}$$

7.

$$\begin{aligned}\frac{3}{4}x + 1.5 &= -19.5\\ \frac{3}{4}x + 1.5 - 1.5 &= -19.5 - 1.5\\ \frac{3}{4}x &= -21\\ \frac{4}{3}\left(\frac{3}{4}x\right) &= \frac{4}{3}(-21)\\ x &= -28\end{aligned}$$

8.

$$\begin{aligned}1 + 3[-2 + 6(4 - 2x)] &= -(x+3)\\ 1 + 3[-2 + 24 - 12x] &= -x - 3\\ 1 + 3[22 - 12x] &= -x - 3\\ 1 + 66 - 36x &= -x - 3\\ 67 - 36x &= -x - 3\\ 67 - 36x + x &= -x - 3 + x\\ 67 - 35x &= -3\\ 67 - 35x - 67 &= -3 - 67\\ -35x &= -70\\ \frac{-35x}{-35} &= \frac{-70}{-35}\\ x &= 2\end{aligned}$$

9.

$$\begin{aligned}\frac{x+7}{3} &= \frac{x-2}{5} - \frac{x}{15} + \frac{7}{3}\\ 15\left(\frac{x+7}{3}\right) &= 15\left(\frac{x-2}{5} - \frac{x}{15} + \frac{7}{3}\right)\\ 5(x+7) &= 3(x-2) - x + 5(7)\\ 5x + 35 &= 3x - 6 - x + 35\\ 5x + 35 &= 2x + 29\\ 3x + 35 &= 29\\ 3x &= -6\\ x &= -2\end{aligned}$$

10. $3p-6=4(p-2)+2-p$

$$3p-6=4p-8+2-p$$
$$3p-6=3p-6$$
$$-6=-6$$
$$0=0$$

$\Re$; identity

11.
$$\lambda = Ax + AB$$
$$\lambda - Ax = Ax + AB - Ax$$
$$\lambda - Ax = AB$$
$$\frac{\lambda - Ax}{A} = \frac{AB}{A}$$
$$\frac{\lambda - Ax}{A} = B$$

12.
$$v = \frac{d_1 - d_2}{t}$$
$$t(v) = t\left(\frac{d_1 - d_2}{t}\right)$$
$$tv = d_1 - d_2$$
$$tv - d_1 = d_1 - d_2 - d_1$$
$$tv - d_1 = -d_2$$
$$\frac{tv - d_1}{-1} = \frac{-d_2}{-1}$$
$$-tv + d_1 = d_2$$
$$d_1 - tv = d_2$$

13. Let x = number of weeks they will be in Las Vegas. Then, the number of weeks in LA = $x + 2$ and the weeks in Dallas = $2x - 1$.

LV + LA + Dallas = 17

$$x+(x+2)+(2x-1)=17$$
$$4x+1=17$$
$$4x+1-1=17-1$$
$$4x=16$$
$$x=4$$
$$x+2=4+2$$
$$=6$$
$$2x-1=2(4)-1$$
$$=8-1$$
$$=7$$

They spent 4 weeks in Las Vegas, 6 weeks in Los Angeles, and 7 weeks in Dallas.

14. Social Security: need new percentages
20% of \$3,518 = 0.20(3,518) = \$703.60

Medicare/Medicade:
19% of \$3,518 = 0.19(3,518) = \$668.42

Total:
703.60 + 668.42 = \$1,372.02 billion

15.

	Rate ·	Time	= Distance
driving	$x + 10$	0.25	$0.25(x+10)$
riding bus	x	0.5	$0.5x$

Since the distances are the same riding and driving, set the expressions equal to each other.

$$0.25(x+10)=0.5x$$
$$0.25x+2.5=0.5x$$
$$0.25x+2.5-0.25x=0.5x-0.25x$$
$$2.5=0.25x$$
$$10=x$$
$$20=x+10$$

Her average speed driving is 20 mph.

16.

	# of pounds ·	price =	value
Apples	x	4.60	$4.60x$
banana chips	$10 - x$	3.40	$3.40(10 - x)$

$$4.60x+3.40(10-x)=4(10)$$
$$4.60x+34-3.4x=40$$
$$1.2x+34=40$$
$$1.2x=6$$
$$x=5$$
$$10-x=5$$

5 pounds of apple slices and 5 pounds of banana chips should be used.

17. $3x = 4y - 11$

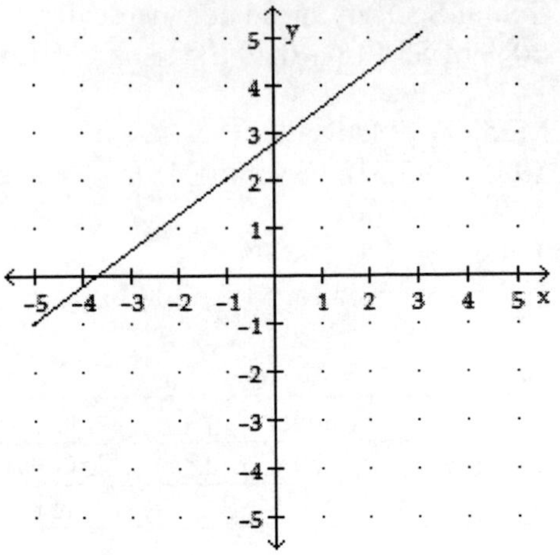

18. $y = -4$

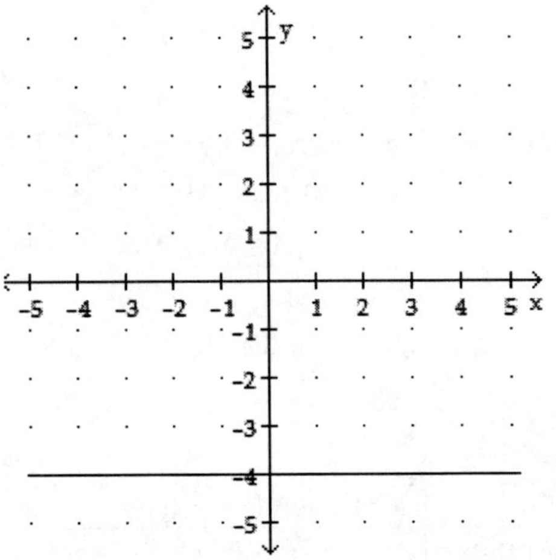

19. Use the order pairs (–5, 0) and (0, –4).

$$m = \frac{y_2 - y_1}{x_2 - x_1}$$
$$= \frac{-4 - 0}{0 - (-5)}$$
$$= \frac{-4}{5}$$
$$= -\frac{4}{5}$$

20. Compare the slopes.

(–3, –1) and (–3, 4)

$$m = \frac{y_2 - y_1}{x_2 - x_1}$$
$$= \frac{4 - (-1)}{-3 - (-3)}$$
$$= \frac{5}{0}$$
$$= \text{undefined}$$

$(-4, -2)$ and $(5, -2)$

$$m = \frac{y_2 - y_1}{x_2 - x_1}$$
$$= \frac{-2 - (-2)}{5 - (-4)}$$
$$= \frac{0}{9}$$
$$= 0$$

The lines are perpendicular.

21. Parallel lines have the same slope. Thus, the slope of the line parallel to $y = -3x$ would have a slope of $m = -3$.

$$y - y_1 = m(x - x_1)$$
$$y - 5 = -3(x - 4)$$
$$y - 5 = -3x + 12$$
$$y - 5 + 5 = -3x + 12 + 5$$
$$y = -3x + 17$$

22. Find the rate of change using the ordered pairs (0, 300) and (20, 650).

$$m = \frac{y_2 - y_1}{x_2 - x_1}$$
$$= \frac{650 - 300}{20 - 0}$$
$$= \frac{350}{20}$$
$$= 17.5$$

Let v = value of figurine and x = years after it is purchased.

$v = 17.5x + 300$

23.

$$f(10) = -(10)^2 - \frac{10}{2}$$
$$= -100 - 5$$
$$= -105$$

24.

$$f(r) = -(r)^2 - \frac{r}{2}$$
$$= -r^2 - \frac{r}{2}$$

25. Answers will vary.
$f(x) = x^3$

26. Yes. Each x–value, only corresponds with one y–value.

27. a. Find the slope using the points (5, 4,500) and (15, 2,800).
$$m = \frac{y_2 - y_1}{x_2 - x_1}$$
$$= \frac{2{,}800 - 4{,}500}{15 - 5}$$
$$= \frac{-1{,}700}{10}$$
$$= -170$$
Use $m = -170$ and (5, 4,500) to find the equation.
$$y - y_1 = m(x - x_1)$$
$$y - 4{,}500 = -170(x - 5)$$
$$y - 4{,}500 = -170x + 850$$
$$y = -170x + 5{,}350$$
$$A(t) = -170t + 5{,}350$$
b. Let $t = 25$.
$$A(25) = -170(25) + 5{,}350$$
$$= -4{,}250 + 5{,}350$$
$$= 1{,}100$$

28. a. $f(-1 = 1$
b. $x = 2$

29. No. It does not pass the vertical line test.

30. $f(x) = (x+4)^2$
D: the set of real numbers
R: the set of real numbers greater than or equal to 0

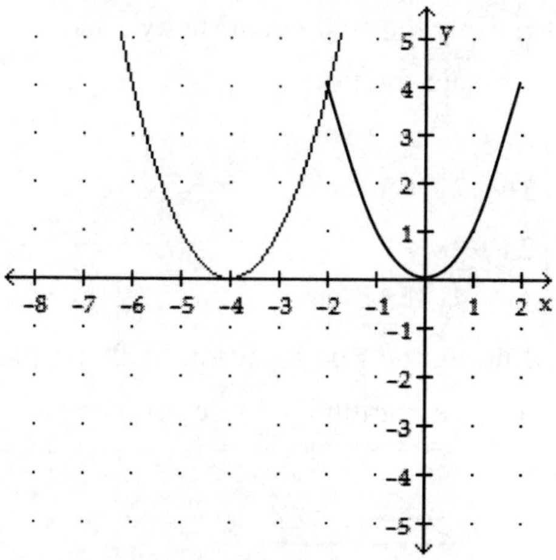

31. Since $g(x)$ is shifted 3 units up, the equation is $g(x) = |x| + 3$.

32. $\begin{cases} 3x - y = -3 \\ y = -2x - 7 \end{cases}$

The lines intersect at (–2, –3).

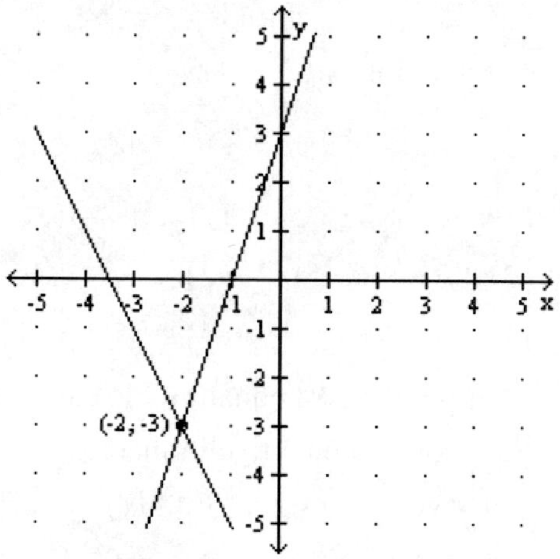

33. a. An inconsistent system has no solution.
b. Dependent equations have the same graph.

34.

$$\begin{cases} y = \dfrac{-2x+1}{3} \\ 3x - 2y = 8 \end{cases}$$

Multiply the first equation by 3 to eliminate fractions.

$$\begin{cases} 3y = -2x + 1 \\ 3x - 2y = 8 \end{cases}$$

$$\begin{cases} 2x + 3y = 1 \\ 3x - 2y = 8 \end{cases}$$

Multiply the first equation by 2 and the second equation by 3 to eliminate y.

$$\begin{aligned} 4x + 6y &= 2 \\ \underline{9x - 6y} &\underline{= 24} \\ 13x &= 26 \\ x &= 2 \end{aligned}$$

Substitute $x = 2$ into the second equation and solve for y.

$$\begin{aligned} 3x - 2y &= 8 \\ 3(2) - 2y &= 8 \\ 6 - 2y &= 8 \\ -2y &= 2 \\ y &= -1 \end{aligned}$$

The solution is $(2, -1)$.

35.

$$\begin{cases} -x + 3y + 2z = 5 & (1) \\ 3x + 2y + z = -1 & (2) \\ 2x - y + 3z = 4 & (3) \end{cases}$$

Multiply the first equation by 3 and add Equations 1 and 2 to eliminate x.

$$\begin{aligned} -3x + 9y + 6z &= 15 \quad (1) \\ \underline{3x + 2y + z} &\underline{= -1} \quad (2) \\ 11y + 7z &= 14 \quad (4) \end{aligned}$$

Multiply the first equation by 2 and add Equations 1 and 3 to eliminate x.

$$\begin{aligned} -2x + 6y + 4z &= 10 \quad (1) \\ \underline{2x - y + 3z} &\underline{= 4} \quad (3) \\ 5y + 7z &= 14 \quad (5) \end{aligned}$$

Multiply Equation 4 by -1 and add Equations 4 and 5 to eliminate z.

$$\begin{aligned} -11y - 7z &= -14 \quad (4) \\ \underline{5y + 7z} &\underline{= 14} \quad (5) \\ -6y &= 0 \\ y &= 0 \end{aligned}$$

Substitute $y = 0$ into Equation 5 and solve for z.

$$\begin{aligned} 5y + 7z &= 14 \quad (5) \\ 5(0) + 7z &= 14 \\ 0 + 7z &= 14 \\ 7z &= 14 \\ z &= 2 \end{aligned}$$

Substitute $y = 0$ and $z = 2$ into Equation 1 and solve for x.

$$\begin{aligned} -x + 3y + 2z &= 5 \quad (1) \\ -x + 3(0) + 2(2) &= 5 \\ -x + 0 + 4 &= 5 \\ -x + 4 &= 5 \\ -x &= 1 \\ x &= -1 \end{aligned}$$

The solution is $(-1, 0, 2)$.

36.

$$\begin{bmatrix} 1 & -1 & 5 \\ 2 & -5 & 1 \end{bmatrix}$$

$-2R_1 + R_2$

$$\begin{bmatrix} 1 & -1 & 5 \\ 0 & -3 & -9 \end{bmatrix}$$

$-\frac{1}{3}R_2$

$$\begin{bmatrix} 1 & -1 & 5 \\ 0 & 1 & 3 \end{bmatrix}$$

$R_2 + R_1$

$$\begin{bmatrix} 1 & 0 & 8 \\ 0 & 1 & 3 \end{bmatrix}$$

The solution is $(8, 3)$.

37. a. $\begin{vmatrix} 6 & 2 \\ -2 & 6 \end{vmatrix} = 6(6) - 2(-2)$

$= 36 + 4$

$= 40$

b.

$\begin{vmatrix} 2 & 1 & -3 \\ -2 & 2 & 4 \\ 1 & -2 & 2 \end{vmatrix} = 2\begin{vmatrix} 2 & 4 \\ -2 & 2 \end{vmatrix} - 1\begin{vmatrix} -2 & 4 \\ 1 & 2 \end{vmatrix} - 3\begin{vmatrix} -2 & 2 \\ 1 & -2 \end{vmatrix}$

$= 2(4+8) - 1(-4-4) - 3(4-2)$

$= 2(12) - 1(-8) - 3(2)$

$= 24 + 8 - 6$

$= 26$

38.

$x = \frac{D_x}{D} = \frac{\begin{vmatrix} 6 & 2 \\ 4 & -1 \end{vmatrix}}{\begin{vmatrix} 1 & 2 \\ 1 & -1 \end{vmatrix}} = \frac{6(-1) - 2(4)}{1(-1) - 2(1)} = \frac{-6-8}{-1-2} = \frac{-14}{-3} = \frac{14}{3}$

$y = \frac{D_y}{D} = \frac{\begin{vmatrix} 1 & 6 \\ 1 & 4 \end{vmatrix}}{\begin{vmatrix} 1 & 2 \\ 1 & -1 \end{vmatrix}} = \frac{1(4) - 6(1)}{1(-1) - 2(1)} = \frac{4-6}{-1-2} = \frac{-2}{-3} = \frac{2}{3}$

The solution is $\left(\frac{14}{3}, \frac{2}{3}\right)$.

39. Let x = number of \$67 phone and y = number of \$100 phone.

$$\begin{cases} x + y = 360 & (1) \\ 67x + 100y = 29,400 & (2) \end{cases}$$

Multiply Equation 1 by –67 and add Equations 1 and 2.

$$\begin{aligned} -67x - 67y &= -24,120 && (1) \\ 67x + 100y &= 29,400 && (2) \\ \hline 33y &= 5,280 \\ y &= 160 \end{aligned}$$

Substitute $y = 160$ into Equation 1 and solve for x.

$$\begin{aligned} x + y &= 360 && (1) \\ x + 160 &= 360 \\ x &= 200 \end{aligned}$$

The warehouse had 200 of the \$67 phones and 160 of the \$100 phones.

40. Let x = amount in 12-month CD, y = amount in 24-month CD, and z = amount in 36-month CD.

$$\begin{cases} x + y + z = 30,000 \\ 0.06x + 0.07y + 0.08z = 2,200 \\ z = 5x \end{cases}$$

Multiply the second equation by 100 to clear decimals.

$$\begin{cases} x + y + z = 30,000 & (1) \\ 6x + 7y + 8z = 220,000 & (2) \\ z = 5x & (3) \end{cases}$$

Multiply Equation 1 by –7 and add Equations 1 and 2.

$$\begin{aligned} -7x - 7y - 7z &= -210,000 && (1) \\ 6x + 7y + 8z &= 220,000 && (2) \\ \hline -x + z &= 10,000 && (4) \end{aligned}$$

Substitute $z = 5x$ into Equation 4 and solve for x.

$$\begin{aligned} -x + z &= 10,000 && (4) \\ -x + (5x) &= 10,000 \\ 4x &= 10,000 \\ x &= 2,500 \end{aligned}$$

Substitute $x = 2,500$ into Equation 3 and solve for z.

$$\begin{aligned} z &= 5x && (3) \\ z &= 5(2,500) \\ z &= 12,500 \end{aligned}$$

Substitute $x = 2,500$ and $z = 12,500$ into Equation 1 and solve for y.

$$\begin{aligned} x + y + z &= 30,000 && (1) \\ 2,500 + y + 12,500 &= 30,000 \\ 15,000 + y &= 30,000 \\ 15,000 + y - 15,000 &= 30,000 - 15,000 \\ y &= 15,000 \end{aligned}$$

She must invest \$2,500 in a 12-month CD, \$15,000 in a 24-month CD, and \$12,500 in a 36-month CD.

SECTION 4.1

VOCABULARY

1. <, >, ≤, and ≥ are **inequality** symbols.

3. The graph of a set of real numbers that is a portion of a number line is called an **interval**.

5. We read the **set–builder** notation $\{x/x<1\}$ as "the set of all real numbers x **such that** x is less than 1."

CONCEPTS

7. $7t-5>4$, $\frac{x}{2}\le -1$

9. $3x+6\le 6$

a. $x=0$

$$3(0)+6\overset{?}{\le}6$$
$$0+6\overset{?}{\le}6$$
$$6\le 6;\ \text{Yes}$$

b. $x=\frac{2}{3}$

$$3\left(\frac{2}{3}\right)+6\overset{?}{\le}6$$
$$2+6\overset{?}{\le}6$$
$$8\le 6;\ \text{No}$$

c. $x=-10$

$$3(-10)+6\overset{?}{\le}6$$
$$-30+6\overset{?}{\le}6$$
$$-24\le 6;\ \text{Yes}$$

d. $x=1.5$

$$3(1.5)+6\overset{?}{\le}6$$
$$4.5+6\overset{?}{\le}6$$
$$10.5\le 6;\ \text{No}$$

11. a. Since the inequality statement is true, the answer is all real numbers. Interval notation is $(-\infty,\infty)$ and the graph is

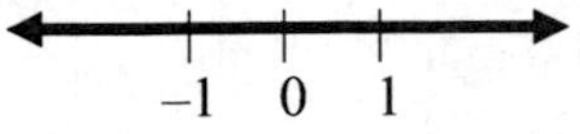

b. Since the inequality statement is false, the answer is no solution or ∅ and the graph is

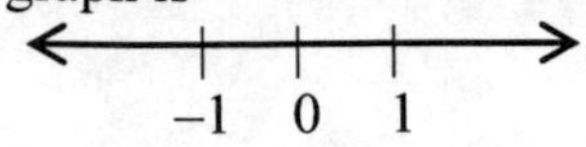

NOTATION

13. $-5x-1\ge -11$

$$-5x\ge \boxed{-10}$$
$$\frac{-5x}{-5}\boxed{\le}\frac{-10}{\boxed{-5}}$$
$$x\le \boxed{2}$$

The solution set is $\left(\boxed{-\infty},\ 2\right]$. Using set–builder notation it is $\left\{x\,\middle|\,\boxed{x\le 2}\right\}$.

15. If $-10>x$, then x $\boxed{<}$ -10.

GUIDED PRACTICE

17. $(-\infty,14)$; $\{x|x<14\}$

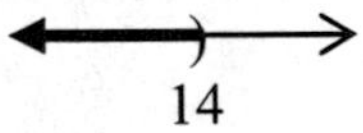

19. $[-2,\infty)$; $\{x|x\ge -2\}$

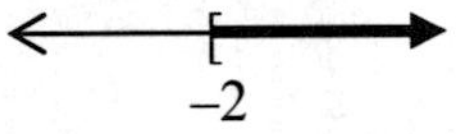

21.

$$x+4<5$$
$$x+4-4<5-4$$
$$x<1;\ (-\infty,\ 1)$$
$$\{x|x<1\}$$

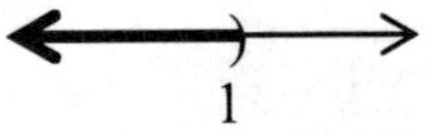

23.

$$3x>-9$$
$$\frac{3x}{3}>\frac{-9}{3}$$
$$x>-3\ ;\ (-3,\ \infty)$$
$$\{x|x>-3\}$$

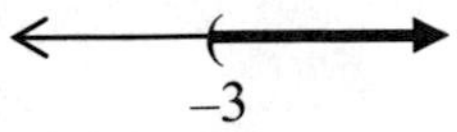

25.

$$2x-7\ge -29$$
$$2x-7+7\ge -29+7$$
$$2x\ge -22$$
$$\frac{2x}{2}\ge\frac{-22}{2}$$
$$x\ge -11;\ \left[-11,\infty\right)$$
$$\left\{x \mid x\ge -11\right\}$$

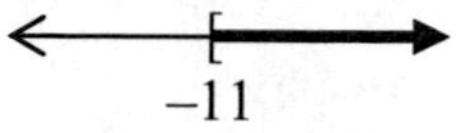

27.

$$9a+11\le 29$$
$$9a+11-11\le 29-11$$
$$9a\le 18$$
$$\frac{9a}{9}\le\frac{18}{9}$$
$$a\le 2;\ \left(-\infty,2\right]$$
$$\left\{a \mid a\le 2\right\}$$

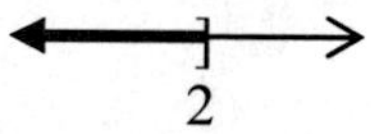

29.

$$2x+4+6x>2-3x+2$$
$$8x+4>4-3x$$
$$8x+4+3x>4-3x+3x$$
$$11x+4>4$$
$$11x+4-4>4-4$$
$$11x>0$$
$$\frac{11x}{11}>\frac{0}{11}$$
$$x>0;\ \left(0,\infty\right)$$
$$\left\{x \mid x>0\right\}$$

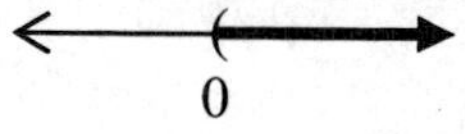

31.

$$t+1-3t\ge t-20$$
$$-2t+1\ge t-20$$
$$-2t+1-t\ge t-20-t$$
$$-3t+1\ge -20$$
$$-3t+1-1\ge -20-1$$
$$-3t\ge -21$$
$$\frac{-3t}{-3}\le\frac{-21}{-3}$$
$$t\le 7;\ \left(-\infty,7\right]$$
$$\left\{t \mid t\le 7\right\}$$

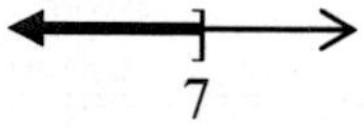

33.

$$4\le\frac{9}{10}x+1$$
$$4-1\le\frac{9}{10}x+1-1$$
$$3\le\frac{9}{10}x$$
$$\frac{10}{9}(3)\le\frac{10}{9}\left(\frac{9}{10}x\right)$$
$$\frac{10}{3}\le x$$
$$x\ge\frac{10}{3};\ \left[\frac{10}{3},\infty\right)$$
$$\left\{x \mid x\ge\frac{10}{3}\right\}$$

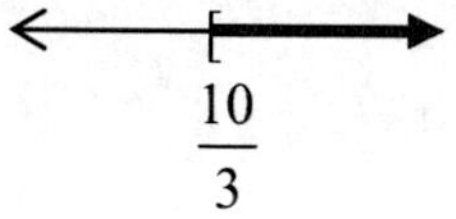

35.

$$-3 > \frac{7}{8}x - 1$$

$$-3 + 1 > \frac{7}{8}x - 1 + 1$$

$$-2 > \frac{7}{8}x$$

$$\frac{8}{7}(-2) > \frac{8}{7}\left(\frac{7}{8}x\right)$$

$$\frac{-16}{7} > x$$

$$x < -\frac{16}{7};\ \left(-\infty, -\frac{16}{7}\right)$$

$$\left\{x \middle| x < -\frac{16}{7}\right\}$$

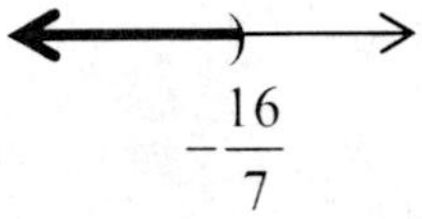

37.

$$\frac{3}{4}(x-3) < \frac{1}{3}(x-4)$$

$$12 \cdot \frac{3}{4}(x-3) < 12 \cdot \frac{1}{3}(x-4)$$

$$9(x-3) < 4(x-4)$$

$$9x - 27 < 4x - 16$$

$$9x - 27 - 4x < 4x - 16 - 4x$$

$$5x - 27 < -16$$

$$5x - 27 + 27 < -16 + 27$$

$$5x < 11$$

$$\frac{5x}{5} < \frac{11}{5}$$

$$x < \frac{11}{5};\ \left(-\infty, \frac{11}{5}\right)$$

$$\left\{x \middle| x < \frac{11}{5}\right\}$$

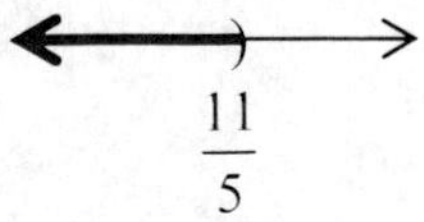

39.

$$\frac{2}{5}(3-2n) \geq \frac{3}{8}(2-3n)$$

$$40 \cdot \frac{2}{5}(3-2n) \geq 40 \cdot \frac{3}{8}(2-3n)$$

$$16(3-2n) \geq 15(2-3n)$$

$$48 - 32n \geq 30 - 45n$$

$$48 - 32n + 45n \geq 30 - 45n + 45n$$

$$48 + 13n \geq 30$$

$$48 + 13n - 48 \geq 30 - 48$$

$$13n \geq -18$$

$$\frac{13n}{13} \geq \frac{-18}{13}$$

$$n \geq -\frac{18}{13};\ \left[-\frac{18}{13}, \infty\right)$$

$$\left\{n \middle| n \geq -\frac{18}{13}\right\}$$

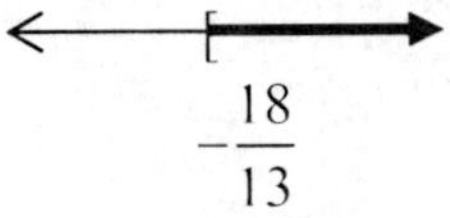

41.

$$2(5x-6) > 4x - 15 + 6x$$

$$10x - 12 > 10x - 15$$

$$10x - 12 - 10x > 10x - 15 - 10x$$

$$-12 > -15$$

$$(-\infty, \infty);\ \Re$$

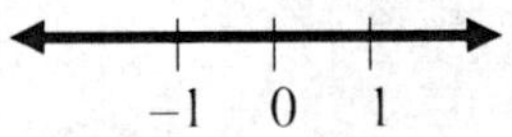

43.

$$\frac{5x+2}{-4} > \frac{5x+1}{-4}$$

$$-4\left(\frac{5x+2}{-4}\right) > -4\left(\frac{5x+1}{-4}\right)$$

$$5x + 2 < 5x + 1$$

$$5x + 2 - 5x < 5x + 1 - 5x$$

$$2 \not< 1$$

No solution; $\varnothing$

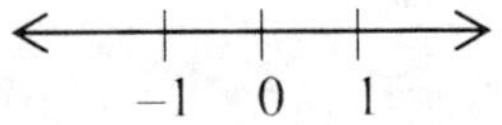

45.

$$-5t + 3 \le 5$$
$$-5t + 3 - 3 \le 5 - 3$$
$$-5t \le 2$$
$$\frac{-5t}{-5} \ge \frac{2}{-5}$$
$$t \ge -\frac{2}{5};\ \left[-\frac{2}{5}, \infty\right)$$
$$\left\{t \middle| t \ge -\frac{2}{5}\right\}$$

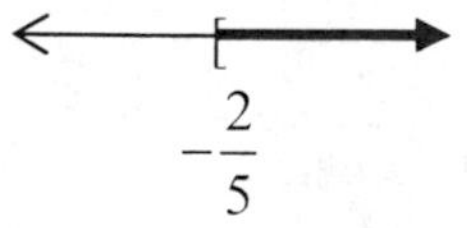

47.

$$\frac{2}{5} > \frac{4}{5}x$$
$$\frac{4}{5}x < \frac{2}{5}$$
$$\frac{5}{4} \cdot \frac{4}{5}x < \frac{5}{4} \cdot \frac{2}{5}$$
$$x < \frac{1}{2};\ \left(-\infty, \frac{1}{2}\right)$$
$$\left\{x \middle| x < \frac{1}{2}\right\}$$

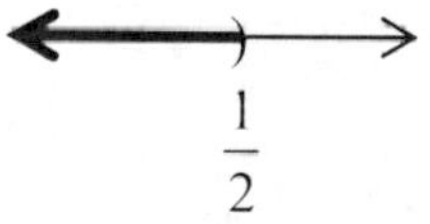

49.

$$-0.6x \le -36$$
$$\frac{-0.6x}{-0.6} \ge \frac{-36}{-0.6}$$
$$x \ge 60;\ [60, \infty)$$
$$\{x | x \ge 60\}$$

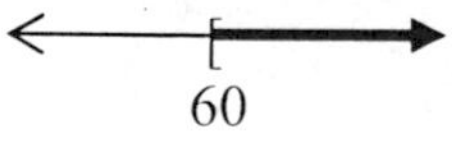

51.

$$7 < \frac{5}{3}a - 3$$
$$\frac{5}{3}a - 3 > 7$$
$$\frac{5}{3}a - 3 + 3 > 7 + 3$$
$$\frac{5}{3}a > 10$$
$$\frac{3}{5} \cdot \frac{5}{3}a > \frac{3}{5} \cdot 10$$
$$a > 6;\ (6, \infty)$$
$$\{a | a > 6\}$$

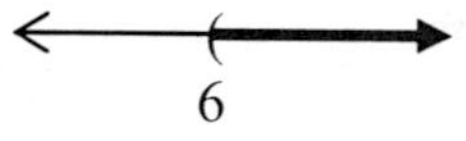

53.

$$-7y + 5 > -5y - 1$$
$$-7y + 5 - 5 > -5y - 1 - 5$$
$$-7y > -5y - 6$$
$$-7y + 5y > -5y - 6 + 5y$$
$$-2y > -6$$
$$\frac{-2y}{-2} < \frac{-6}{-2}$$
$$y < 3;\ (-\infty, 3)$$
$$\{y | y < 3\}$$

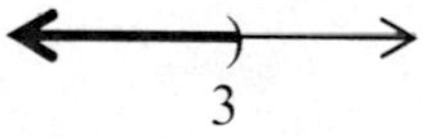

55.

$$\frac{6-d}{-2} \le -6$$
$$-2 \cdot \frac{6-d}{-2} \ge -2 \cdot -6$$
$$6 - d \ge 12$$
$$6 - d - 6 \ge 12 - 6$$
$$-d \ge 6$$
$$\frac{-d}{-1} \le \frac{6}{-1}$$
$$d \le -6;\ (-\infty, -6]$$
$$\{d | d \le -6\}$$

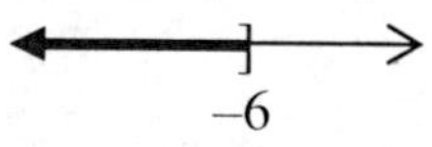

57.

$$0.4x+0.4 \le 0.1x+0.85$$
$$100(0.4x+0.4) \le 100(0.1x+0.85)$$
$$40x+40 \le 10x+85$$
$$40x+40-10x \le 10x+85-10x$$
$$30x+40 \le 85$$
$$30x+40-40 \le 85-40$$
$$30x \le 45$$
$$\frac{30x}{30} \le \frac{45}{30}$$
$$x \le 1.5 \text{ ; } (-\infty, 1.5]$$
$$\{x | x \le 1.5\}$$

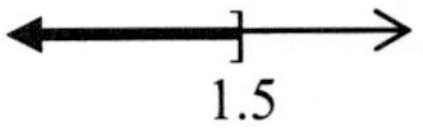

59.

$$3(z-2) \le 2(z+7)$$
$$3z-6 \le 2z+14$$
$$3z-6-2z \le 2z+14-2z$$
$$z-6 \le 14$$
$$z-6+6 \le 14+6$$
$$z \le 20 \text{ ; } (-\infty, 20]$$
$$\{z | z \le 20\}$$

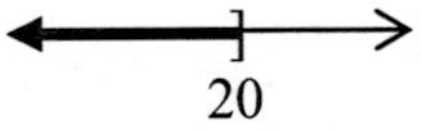

61.

$$\frac{3b+7}{3} \le \frac{2b-9}{2}$$
$$6 \cdot \frac{3b+7}{3} \le 6 \cdot \frac{2b-9}{2}$$
$$2(3b+7) \le 3(2b-9)$$
$$6b+14 \le 6b-27$$
$$6b+14-6b \le 6b-27-6b$$
$$14 \le -27$$

no solution; $\varnothing$

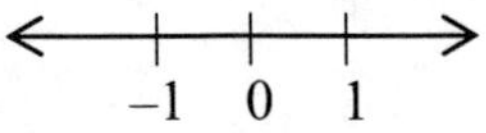

63.

$$\frac{x-7}{2} - \frac{x-1}{5} \ge -\frac{x}{4}$$
$$20\left(\frac{x-7}{2} - \frac{x-1}{5}\right) \ge 20\left(-\frac{x}{4}\right)$$
$$10(x-7) - 4(x-1) \ge 5(-x)$$
$$10x-70-4x+4 \ge -5x$$
$$6x-66 \ge -5x$$
$$6x-66-6x \ge -5x-6x$$
$$-66 \ge -11x$$
$$\frac{-66}{-11} \le \frac{-11x}{-11}$$
$$6 \le x$$
$$x \ge 6 \text{ ; } [6, \infty)$$
$$\{x | x \ge 6\}$$

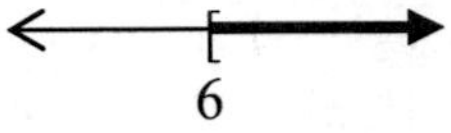

65.

$$\frac{1}{2}x+6 \ge 4+2x$$
$$2\left(\frac{1}{2}x+6\right) \ge 2(4+2x)$$
$$x+12 \ge 8+4x$$
$$x+12-4x \ge 8+4x-4x$$
$$-3x+12 \ge 8$$
$$-3x+12-12 \ge 8-12$$
$$-3x \ge -4$$
$$\frac{-3x}{-3} \le \frac{-4}{-3}$$
$$x \le \frac{4}{3} \text{ ; } \left(-\infty, \frac{4}{3}\right]$$
$$\left\{x \middle| x \le \frac{4}{3}\right\}$$

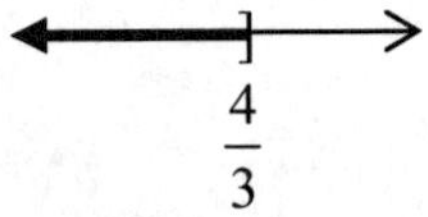

67.

$$5(2n+2) - n > 3n - 3(1-2n)$$
$$10n+10-n > 3n-3+6n$$
$$9n+10 > 9n-3$$
$$9n+10-9n > 9n-3-9n$$
$$10 > -3$$
$$(-\infty, \infty); \ \Re$$

69.

$$\frac{1}{2}y+2\geq\frac{1}{3}y-4$$
$$6\left(\frac{1}{2}y+2\right)\geq6\left(\frac{1}{3}y-4\right)$$
$$3y+12\geq2y-24$$
$$3y+12-2y\geq2y-24-2y$$
$$y+12\geq-24$$
$$y+12-12\geq-24-12$$
$$y\geq-36\ ;\ [-36,\infty)$$
$$\{y|y\geq-36\}$$

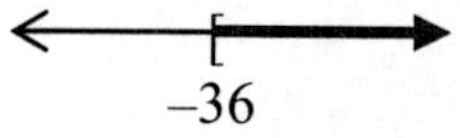

71.

$$-11(2-b)<4(2b+2)$$
$$-22+11b<8b+8$$
$$-22+11b-8b<8b+8-8b$$
$$-22+3b<8$$
$$-22+3b+22<8+22$$
$$3b<30$$
$$\frac{3b}{3}<\frac{30}{3}$$
$$b<10\ ;\ (-\infty,10)$$
$$\{b|b<10\}$$

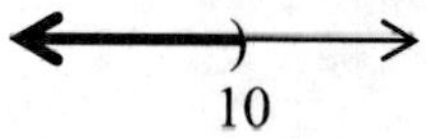

73.

$$\frac{2}{3}x+\frac{3}{2}(x-5)\leq x$$
$$6\left(\frac{2}{3}x+\frac{3}{2}(x-5)\right)\leq6(x)$$
$$4x+9(x-5)\leq6x$$
$$4x+9x-45\leq6x$$
$$13x-45\leq6x$$
$$13x-45-13x\leq6x-13x$$
$$-45\leq-7x$$
$$-7x\geq-45$$
$$\frac{-7x}{-7}\leq\frac{-45}{-7}$$
$$x\leq\frac{45}{7}\ ;\ \left(-\infty,\frac{45}{7}\right]$$
$$\left\{x|x\leq\frac{45}{7}\right\}$$

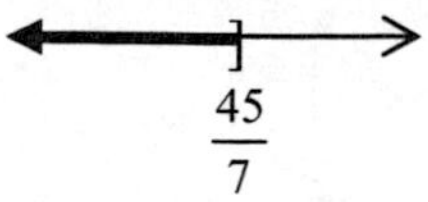

75.

$$5[3t-(t-4)]-11\leq-12(t-6)-(-t)$$
$$5[3t-t+4]-11\leq-12t+72+t$$
$$5[2t+4]-11\leq-11t+72$$
$$10t+20-11\leq-11t+72$$
$$10t+9\leq-11t+72$$
$$10t+9+11t\leq-11t+72+11t$$
$$21t+9\leq72$$
$$21t+9-9\leq72-9$$
$$21t\leq63$$
$$\frac{21t}{21}\leq\frac{63}{21}$$
$$t\leq3\ ;\ (-\infty,3]$$
$$\{t|t\leq3\}$$

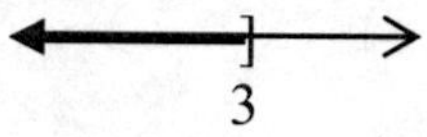

77.

$$f(x)<g(x)$$
$$\frac{1}{2}x-\frac{2}{3}<x+\frac{4}{3}$$
$$6\left(\frac{1}{2}x-\frac{2}{3}\right)<6\left(x+\frac{4}{3}\right)$$
$$3x-4<6x+8$$
$$3x-4-6x<6x+8-6x$$
$$-3x-4<8$$
$$-3x-4+4<8+4$$
$$-3x<12$$
$$\frac{-3x}{-3}<\frac{12}{-3}$$
$$x>-4;\ (-4,\infty)$$
$$\{x|x>-4\}$$

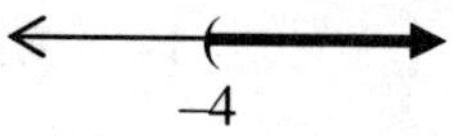

79.

$$y_2 > y_1$$
$$x + 0.3 > 0.7x - 0.15$$
$$100(x + 0.3) > 100(0.7x - 0.15)$$
$$100x + 30 > 70x - 15$$
$$100x + 30 - 70x > 70x - 15 - 70x$$
$$30x + 30 > -15$$
$$30x + 30 - 30 > -15 - 30$$
$$30x > -45$$
$$\frac{30x}{30} > \frac{-45}{30}$$
$$x > -1.5;\ (-1.5, \infty)$$
$$\{x \mid x > -1.5\}$$

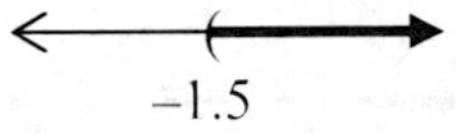

81. a.

$$12x - 33.16 \le 5.84$$
$$12x - 33.16 + 33.16 \le 5.84 + 33.16$$
$$12x \le 39$$
$$\frac{12x}{12} \le \frac{39}{12}$$
$$x \le 3.25; (-\infty, 3.25]$$
$$\{x \mid x \le 3.25\}$$

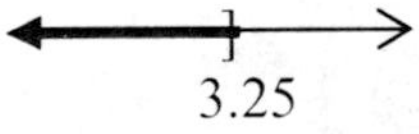

b.

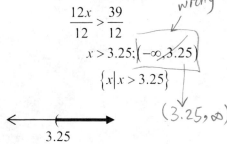

$$12x - 33.16 > 5.84$$
$$12x - 33.16 + 33.16 > 5.84 + 33.16$$
$$12x > 39$$
$$\frac{12x}{12} > \frac{39}{12}$$
$$x > 3.25; (-\infty, 3.25)$$
$$\{x \mid x > 3.25\}$$

wrong $(3.25, \infty)$

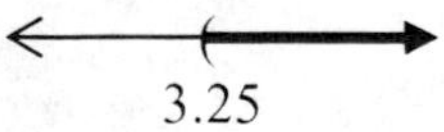

83. a.

$$3(2x + 2) > 5(x - 1) + 3x$$
$$6x + 6 > 5x - 5 + 3x$$
$$6x + 6 > 8x - 5$$
$$6x + 6 - 8x > 8x - 5 - 8x$$
$$-2x + 6 > -5$$
$$-2x + 6 - 6 > -5 - 6$$
$$-2x > -11$$
$$\frac{-2x}{-2} > \frac{-11}{-2}$$
$$x < \frac{11}{2}; \left(-\infty, \frac{11}{2}\right)$$
$$\left\{x \mid x < \frac{11}{2}\right\}$$

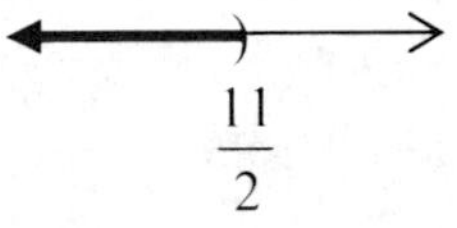

b.

$$3(2x + 2) < 5(x - 1) + 3x$$
$$6x + 6 < 5x - 5 + 3x$$
$$6x + 6 < 8x - 5$$
$$6x + 6 - 8x < 8x - 5 - 8x$$
$$-2x + 6 < -5$$
$$-2x + 6 - 6 < -5 - 6$$
$$-2x < -11$$
$$\frac{-2x}{-2} < \frac{-11}{-2}$$
$$x > \frac{11}{2}; \left(\frac{11}{2}, \infty\right)$$
$$\left\{x \mid x > \frac{11}{2}\right\}$$

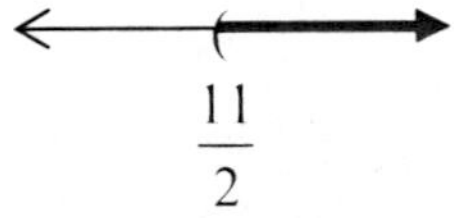

85. $(-\infty, 1)$

87. $[-4, \infty)$

APPLICATIONS

89. REAL ESTATE
   a. In the **South and Midwest**, the median sales price was less than the U.S. median price.
   b. In the **West and Northeast**, the median sales price was greater than or equal to the U.S. median price.

91. CAMPUS TO CAREERS

$$p(m) \geq 60$$
$$\frac{6}{5}m \geq 60$$
$$\frac{5}{6}\left(\frac{6}{5}m\right) \geq \frac{5}{6}(60)$$
$$m \geq 50$$

It will take 50 minutes or more to remove 60% of the air-borne particles.

93. NATIONAL PARKS

$$v(t) < 2{,}400{,}000$$
$$-100{,}000t + 3{,}600{,}000 < 2{,}400{,}000$$
$$-100{,}000t < -1{,}200{,}000$$
$$t > 12$$

If $t = 12$, the year would be 2002. So, the years after 2002 are when the number of visitors fell below 2,400,000.

95. MOVING DAY

Let x be the number of miles driving. Find when the cost of *Nationwide's* plan is less than the cost of *Valley Truck Rentals'* charge.

$$36.75 + 0.60x < 25.50 + 0.75x$$
$$36.75 + 0.60x - 0.75x < 25.50 + 0.75x - 0.75x$$
$$36.75 - 0.15x < 25.5$$
$$36.75 - 0.15x - 36.75 < 25.5 - 36.75$$
$$-0.15x < -11.25$$
$$\frac{-0.15x}{-0.15} < \frac{-11.25}{-0.15}$$
$$x > 75$$

When the number of miles is greater than 75, the *Nationwide* plan is cheaper.

97. BUSINESS LOSSES

Let x – the number of towels made and sold.

manufactoring cost > selling revenue

$$1{,}400 + 5x > 8.50x$$
$$1{,}400 + 5x - 5x > 8.50x - 5x$$
$$1{,}400 > 3.50x$$
$$\frac{1{,}400}{3.50} > \frac{3.50x}{3.50}$$
$$400 > x$$
$$x < 400$$

They could have sold as many as 399 towels to lose money.

99. AVERAGING GRADES

Let x = score on 4th exam

$$\frac{\text{sum of 4 exams}}{\text{number of exams}} \geq 80$$
$$\frac{70 + 77 + 85 + x}{4} \geq 80$$
$$\frac{232 + x}{4} \geq 80$$
$$4 \cdot \frac{232 + x}{4} \geq 4 \cdot 80$$
$$232 + x \geq 320$$
$$232 + x - 232 \geq 320 - 232$$
$$x \geq 88$$

She must make an **88 or higher** on her 4th exam.

101. WORK SCHEDULES

job	pay per hour	number of hr worked	amount earned
library	\$8	x	\$8x
construction	\$15	$25 - x$	$15(25 - x)$

$$8x + 15(25 - x) \geq 300$$
$$8x + 375 - 15x \geq 300$$
$$-7x + 375 \geq 300$$
$$-7x + 375 - 375 \geq 300 - 375$$
$$-7x \geq -75$$
$$\frac{-7x}{-7} \leq \frac{-75}{-7}$$
$$x \leq 10.7$$

He can work **10 hours** at the library.

103. FUNDRAISING

Let x = the number of hours rented in addition to the 3 original hours included in the $85.

$$19.50x + 85 \le 185$$
$$19.50x + 85 - 85 \le 185 - 85$$
$$19.50x \le 100$$
$$\frac{19.50x}{19.50} \le \frac{100}{19.50}$$
$$x \le 5.123$$
$$x \le 5$$

The number of additional hours (5) plus the original 3 hours equal the total number of hours rented.

5 + 3 = **8 hours**

WRITING

105. Answers will vary.

107. Answers will vary.

109. Answers will vary.

REVIEW

111. To find $f(-1)$, find the y–value of the graph when $x = -1$.

$f(-1) = 4$

To find $f(0)$, find the y–value of the graph when $x = 0$.

$f(0) = 5$

To find $f(2)$, find the y–value of the graph when $x = 2$.

$f(2) = 3$

CHALLENGE PROBLEMS

113. i and ii

115. MEDICAL PLANS

Let x = amount of hospital bill.
The "rest" would be the amount of the hospital bill (x) minus the amount of the deductible.

Plan 1:
The employee pays $100 + 30% of rest.

$$100 + 0.30(x - 100)$$

Plan 2:
The employee pays $200 + 20% rest

$$200 + 0.20(x - 200)$$

To be beneficial to the employee, he would want Plan 2 to be less than Plan 1.

$$\text{Plan 2} < \text{Plan 1}$$
$$200 + 0.20(x - 200) < 100 + 0.30(x - 100)$$
$$200 + 0.20x - 40 < 100 + 0.30x - 30$$
$$0.20x + 160 < 0.30x + 70$$
$$0.20x + 160 - 0.30x < 0.30x + 70 - 0.30x$$
$$-0.1x + 160 < 70$$
$$-0.1x + 160 - 160 < 70 - 160$$
$$-0.1x < -90$$
$$\frac{-0.1x}{-0.1} > \frac{-90}{-0.1}$$
$$x > 900$$

Plan 2 is better for the employee for any hospital bill **over $900**.

SECTION 4.2

VOCABULARY

1. The **<u>intersection</u>** of two sets is the set of elements that are common to both sets and the **<u>union</u>** of two sets is the set of elements that are in one set, or the other, or both.

3. $-6 < x + 1 \le 1$ is a **<u>double</u>** linear inequality.

CONCEPTS

5. a. The solution set of a compound inequality containing the word *and* includes all numbers that make **<u>both</u>** inequalities true.
   b. The solution set of a compound inequality containing the word *or* includes all numbers that make **<u>one</u>**, or the other, or **<u>both</u>** inequalities true.

7. a. When solving a compound inequality containing the word *and*, the solution set is the **<u>intersection</u>** of the solution sets of the inequalities.
   b. When solving a compound inequality containing the word *or*, the solution set is the **<u>union</u>** of the solution sets of the inequalities.

9. a. Let $x = -3$.

$$\frac{-3}{3}+1\overset{?}{\ge}0 \quad \text{and} \quad 2(-3)-3\overset{?}{<}-10$$

$$-1+1\overset{?}{\ge}0 \qquad -6-3\overset{?}{<}-10$$

$$0\overset{?}{\ge}0 \qquad -9\overset{?}{<}-10$$

true   false

No. Since both inequalities are not true, –3 is **not a solution** of the compound inequality.

   b. Let $x = -3$.

$$2(-3)\overset{?}{\le}0 \quad \text{or} \quad -3(-3)\overset{?}{<}-5$$

$$-6\overset{?}{\le}0 \qquad 9\overset{?}{<}-5$$

true   false

Yes. Since one inequality is true, –3 **is a solution** of the compound inequality.

11. a. $[-2, 1)$
    b. $[2, 2]$
    c. $\varnothing$

NOTATION

13. We read $\cup$ as **<u>union</u>** and $\cap$ as **<u>intersection</u>**.

15. all real numbers

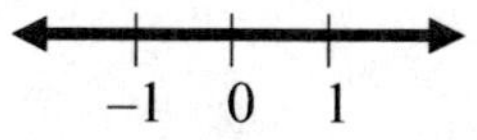

GUIDED PRACTICE

17. $\{4, 6\}$

19. $\{-3, 1, 2\}$

21. $\{-3, -1, 0, 1, 2, 4, 6, 8, 10\}$

23. $\{-3, 0, 1, 2, 3, 4, 5, 6, 8\}$

25. $x > -2$ and $x \le 5$
    $(-2, 5]$

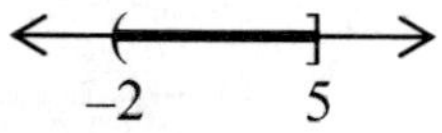

27. $2x - 1 > 3$ and $x + 8 \le 11$

$$2x-1+1>3+1 \qquad x+8-8\le 11-8$$

$$2x>4 \qquad x\le 3$$

$$\frac{2x}{2}>\frac{4}{2}$$

$$x>2$$

$(2, 3]$

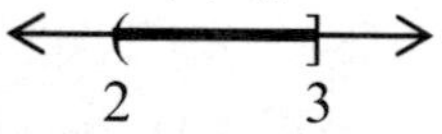

29. $6x+1<5x-3$ and $\frac{x}{2}+9\le 6$

$$6x+1-5x<5x-3-5x \qquad \frac{x}{2}+9-9\le 6-9$$

$$x+1<-3 \qquad \frac{x}{2}\le -3$$

$$x+1-1<-3-1 \qquad 2\left(\frac{x}{2}\right)\le 2(-3)$$

$$x<-4 \qquad x\le -6$$

$(-\infty, -6]$

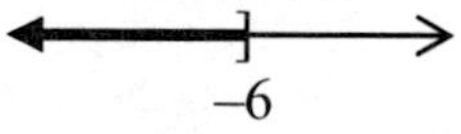

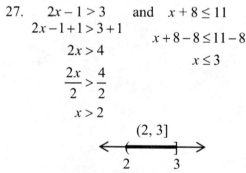

31. $x+2<-\frac{1}{3}x$ and $-6x<9x$

$$3(x+2)<3\left(-\frac{1}{3}x\right)$$

$$3x+6<-x$$

$$3x+6-3x<-x-3x$$

$$6<-4x$$

$$\frac{6}{-4}>\frac{-4x}{-4}$$

$$-\frac{3}{2}>x$$

$$x<-\frac{3}{2}$$

$$-6x-9x<9x-9x$$

$$-15x<0$$

$$\frac{-15x}{-15}>\frac{0}{-15}$$

$$x>0$$

No solution; ∅. There are no numbers that are less than $-\frac{3}{2}$ and greater than 0.

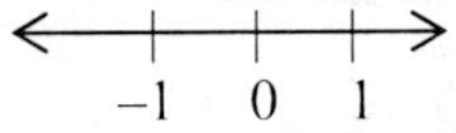

33.
$$4\le x+3\le 7$$

$$4-3\le x+3-3\le 7-3$$

$$1\le x\le 4$$

$$[1,\ 4]$$

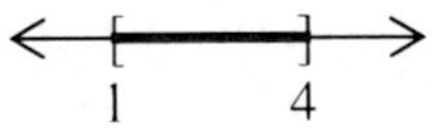

35.
$$0.9<2x-0.7<1.5$$

$$10(0.9)<10(2x-0.7)<10(1.5)$$

$$9<20x-7<15$$

$$9+7<20x-7+7<15+7$$

$$16<20x<22$$

$$\frac{16}{20}<\frac{20x}{20}<\frac{22}{20}$$

$$0.8<x<1.1$$

$$(0.8,\ 1.1)$$

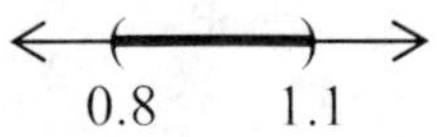

37. $x\le -2$ or $x>6$

$$(-\infty,-2]\cup(6,\infty)$$

39. $x-3<-4$ or $-x+2<0$

$$x-3+3<-4+3 \qquad -x+2-2<0-2$$

$$x<-1 \qquad -x<-2$$

$$\frac{-x}{-1}>\frac{-2}{-1}$$

$$x>2$$

$$(-\infty,-1)\cup(2,\infty)$$

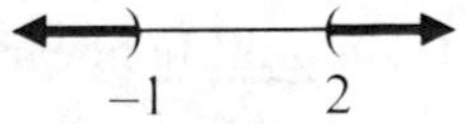

41. $3x+2<8$ or $2x-3>11$

$$3x+2-2<8-2 \qquad 2x-3+3>11+3$$

$$3x<6 \qquad 2x>14$$

$$x<2 \qquad x>7$$

$$(-\infty,2)\cup(7,\infty)$$

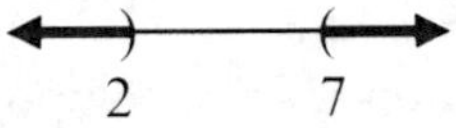

43. $2x>x+3$ or $\frac{x}{8}+1<\frac{13}{8}$

$$2x-x>x+3-x$$

$$x>3$$

$$8\left(\frac{x}{8}+1\right)<8\left(\frac{13}{8}\right)$$

$$x+8<13$$

$$x+8-8<13-8$$

$$x<5$$

$$(-\infty,\ \infty)$$

All real numbers are either greater than 3 or less than 5 or both.

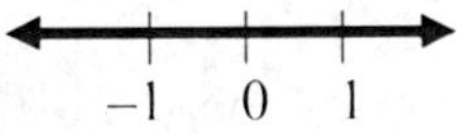

TRY IT YOURSELF

45.

$$-4(x+2)\ge 12 \quad \text{or} \quad 3x+8<11$$

$$-4x-8\ge 12 \qquad 3x+8-8<11-8$$

$$-4x-8+8\ge 12+8 \qquad 3x<3$$

$$-4x\ge 20 \qquad \frac{3x}{3}<\frac{3}{3}$$

$$\frac{-4x}{-4}\le\frac{20}{-4} \qquad x<1$$

$$x\le -5$$

$$(-\infty,1)$$

47.

$$2.2x < -19.8 \quad \text{and} \quad -4x < 40$$
$$\frac{2.2x}{2.2} < \frac{-19.8}{2.2} \qquad \frac{-4x}{-4} > \frac{40}{-4}$$
$$x < -9 \qquad x > -10$$
$$-10 < x < -9$$
$$(-10, -9)$$

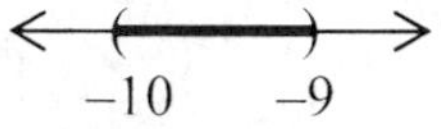

49.

$$-2 < -b + 3 < 5$$
$$-2 - 3 < -b + 3 - 3 < 5 - 3$$
$$-5 < -b < 2$$
$$\frac{-5}{-1} > \frac{-b}{-1} > \frac{2}{-1}$$
$$5 > b > -2$$
$$-2 < b < 5$$
$$(-2,\ 5)$$

51.

$$4.5x - 2 > 2.5 \quad \text{or} \quad \frac{1}{2}x \le 1$$
$$4.5x - 2 + 2 > 2.5 + 2 \qquad 2\left(\frac{1}{2}x\right) \le 2(1)$$
$$4.5x > 4.5 \qquad x \le 2$$
$$\frac{4.5x}{4.5} > \frac{4.5}{4.5}$$
$$x > 1$$
$$(-\infty, \infty)$$

All real numbers are either greater than 1 or less than 2 or both.

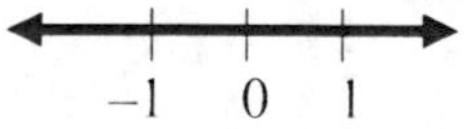

53.

$$5(x - 2) \ge 0 \quad \text{and} \quad -3x < 9$$
$$5x - 10 \ge 0 \qquad \frac{-3x}{-3} > \frac{9}{-3}$$
$$5x - 10 + 10 \ge 0 + 10 \qquad x > -3$$
$$5x \ge 10$$
$$x \ge 2$$
$$[2, \infty)$$

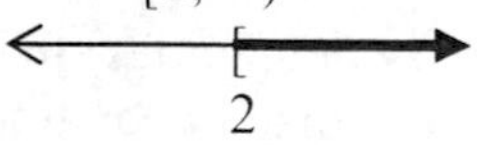

55.

$$-x < -2x \quad \text{and} \quad 3x > 2x$$
$$-x + 2x < -2x + 2x \qquad 3x - 2x > 2x - 2x$$
$$x < 0 \qquad x > 0$$

no solution; ∅

There are no numbers that are both less than and greater than 0.

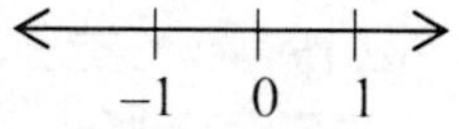

57.

$$-6 < -3(x - 4) \le 24$$
$$-6 < -3x + 12 \le 24$$
$$-6 - 12 < -3x + 12 - 12 \le 24 - 12$$
$$-18 < -3x \le 12$$
$$\frac{-18}{-3} > \frac{-3x}{-3} \ge \frac{12}{-3}$$
$$6 > x \ge -4$$
$$-4 \le x < 6$$
$$[-4,\ 6)$$

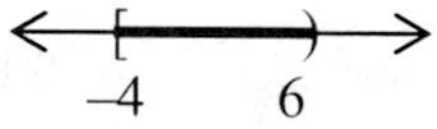

59.

$$2x + 1 \ge 5 \quad \text{and} \quad -3(x + 1) \ge -9$$
$$2x + 1 - 1 \ge 5 - 1 \qquad -3x - 3 \ge -9$$
$$2x \ge 4 \qquad -3x - 3 + 3 \ge -9 + 3$$
$$x \ge 2 \qquad -3x \ge -6$$
$$x \le 2$$
$$[2,\ 2]$$

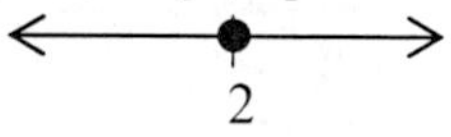

2 is the only number less than or equal to 2 AND greater than or equal to 2.

61.

$$\frac{4.5x-12}{2}<x \quad \text{or} \quad -15.3>-3(x-1.4)$$

$$2\left(\frac{4.5x-12}{2}\right)<2(x) \qquad -15.3>-3x+4.2$$

$$4.5x-12<2x \qquad -15.3-4.2>-3x+4.2-4.2$$

$$4.5x-12+12<2x+12 \qquad -19.5>-3x$$

$$4.5x<2x+12 \qquad \frac{-19.5}{-3}>\frac{-3x}{-3}$$

$$4.5x-2x<2x-2x+12 \qquad 6.5<x$$

$$2.5x<12 \qquad x>6.5$$

$$\frac{2.5x}{2.5}<\frac{12}{2.5}$$

$$x<4.8$$

$$(-\infty,4.8)\cup(6.5,\infty)$$

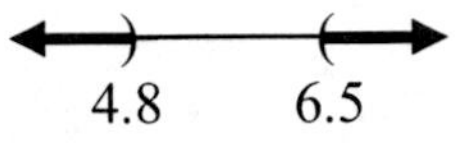

67.

$$\frac{x}{3}-\frac{x}{4}>\frac{1}{6} \quad \text{or} \quad \frac{x}{2}+\frac{2}{3}\le\frac{3}{4}$$

$$12\left(\frac{x}{3}-\frac{x}{4}\right)>12\left(\frac{1}{6}\right) \qquad 12\left(\frac{x}{2}+\frac{2}{3}\right)\le 12\left(\frac{3}{4}\right)$$

$$4x-3x>2 \qquad 6x+8\le 9$$

$$x>2 \qquad 6x+8-8\le 9-8$$

$$6x\le 1$$

$$\frac{6x}{6}\le\frac{1}{6}$$

$$x\le\frac{1}{6}$$

$$\left(-\infty,\frac{1}{6}\right]\cup(2,\infty)$$

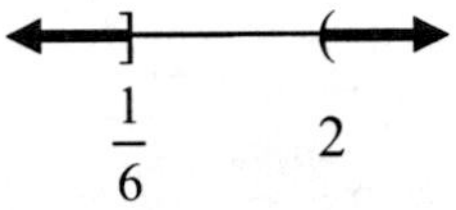

63.

$$\frac{x}{0.7}+5>4 \quad \text{and} \quad -4.8\le\frac{3x}{-0.125}$$

$$\frac{x}{0.7}+5-5>4-5 \qquad -0.125(-4.8)\ge -0.125\left(\frac{3x}{-0.125}\right)$$

$$\frac{x}{0.7}>-1 \qquad 0.6\ge 3x$$

$$0.7\left(\frac{x}{0.7}\right)>0.7(-1) \qquad \frac{0.6}{3}\ge\frac{3x}{3}$$

$$x>-0.7 \qquad 0.2\ge x$$

$$x\le 0.2$$

$$(-0.7,\ 0.2]$$

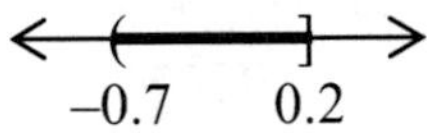

69.

$$0\le\frac{4-x}{3}\le 2$$

$$3(0)\le 3\left(\frac{4-x}{3}\right)\le 3(2)$$

$$0\le 4-x\le 6$$

$$0-4\le 4-x-4\le 6-4$$

$$-4\le -x\le 2$$

$$\frac{-4}{-1}\ge\frac{-x}{-1}\ge\frac{2}{-1}$$

$$4\ge x\ge -2$$

$$-2\le x\le 4$$

$$[-2,\ 4]$$

65.

$$-24<\frac{3}{2}x-6\le -15$$

$$-24+6<\frac{3}{2}x-6+6\le -15+6$$

$$-18<\frac{3}{2}x\le -9$$

$$2(-18)<2\left(\frac{3}{2}x\right)\le 2(-9)$$

$$-36<3x\le -18$$

$$-12<x\le -6$$

$$(-12,-6]$$

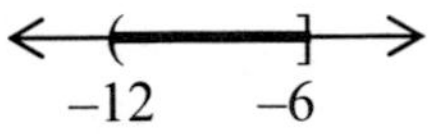

71.

$$x\le 6-\frac{1}{2}x \quad \text{and} \quad \frac{1}{2}x+1\ge 3$$

$$2(x)\le 2\left(6-\frac{1}{2}x\right) \qquad 2\left(\frac{1}{2}x+1\right)\ge 2(3)$$

$$2x\le 12-x \qquad x+2\ge 6$$

$$2x+x\le 12-x+x \qquad x+2-2\ge 6-2$$

$$3x\le 12 \qquad x\ge 4$$

$$x\le 4$$

$$[4,\ 4]$$

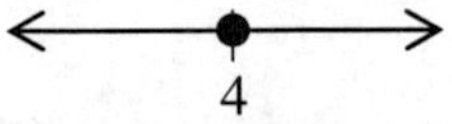

4 is the only number less than or equal to 4 AND greater than or equal to 4.

73.

$$-6 < f(x) \le 0$$
$$-6 < 3x - 9 \le 0$$
$$-6 + 9 < 3x - 9 + 9 \le 0 + 9$$
$$3 < 3x \le 9$$
$$\frac{3}{3} < \frac{3x}{3} \le \frac{9}{3}$$
$$1 < x \le 3$$
$$(1,3]$$

75.

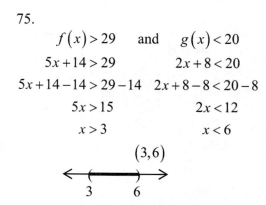

$$f(x) > 29 \quad \text{and} \quad g(x) < 20$$
$$5x + 14 > 29 \qquad 2x + 8 < 20$$
$$5x + 14 - 14 > 29 - 14 \qquad 2x + 8 - 8 < 20 - 8$$
$$5x > 15 \qquad 2x < 12$$
$$x > 3 \qquad x < 6$$
$$(3,6)$$

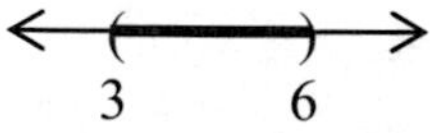

77. a.

$$3x - 2 \ge 4 \quad \text{and} \quad x + 6 \ge 12$$
$$3x - 2 + 2 \ge 4 + 2 \qquad x + 6 - 6 \ge 12 - 6$$
$$3x \ge 6 \qquad x \ge 6$$
$$x \ge 2$$
$$[6,\infty)$$

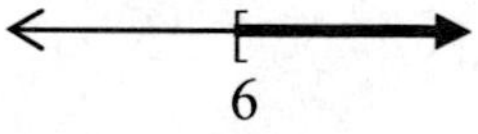

b.

$$3x - 2 \ge 4 \quad \text{or} \quad x + 6 \ge 12$$
$$3x - 2 + 2 \ge 4 + 2 \qquad x + 6 - 6 \ge 12 - 6$$
$$3x \ge 6 \qquad x \ge 6$$
$$x \ge 2$$
$$[2,\infty)$$

79. a.

$$2x + 1 \le 7 \quad \text{and} \quad 3x + 5 \ge 23$$
$$2x + 1 - 1 \le 7 - 1 \qquad 3x + 5 - 5 \ge 23 - 5$$
$$2x \le 6 \qquad 3x \ge 18$$
$$x \le 3 \qquad x \ge 6$$

no solution; $\varnothing$

There are no numbers that are both less than 3 and greater than 6.

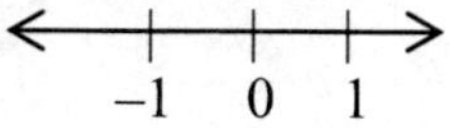

b.

$$2x + 1 \le 7 \quad \text{or} \quad 3x + 5 \ge 23$$
$$2x + 1 - 1 \le 7 - 1 \qquad 3x + 5 - 5 \ge 23 - 5$$
$$2x \le 6 \qquad 3x \ge 18$$
$$x \le 3 \qquad x \ge 6$$
$$(-\infty,3] \cup [6,\infty)$$

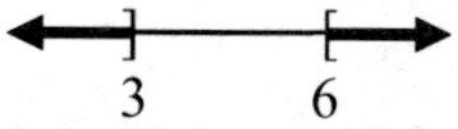

APPLICATIONS

81. BABY FURNITURE
   a. $128 \le 4s \le 192$
   b.

$$128 \le 4s \le 192$$
$$\frac{128}{4} \le \frac{4s}{4} \le \frac{192}{4}$$
$$32 \le s \le 48$$

83. from Campus to Careers

   a. (67, 77)
      5 degrees below is 72 – 5 = 67
      5 degrees above is 75 + 5 = 77
   b. (62, 82)
      10 degrees below is 72 – 10 = 62
      10 degrees above is 72 + 10 = 82

85. U.S. HEALTH CARE
   a. 2004
   b. 2004, 2005, 2004, 2007
   c. 2006 and 2007
   d. 2007

87. STREET INTERSECTIONS

a. intersection

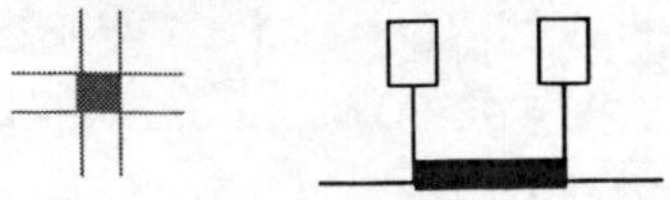

b. union

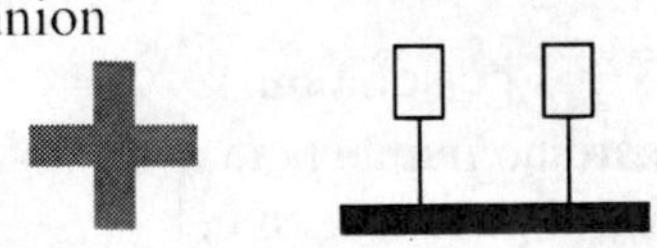

WRITING

89. Answers will vary.

91. Answers will vary.

93. Answers will vary.

REVIEW

95. Subtract 4,264 – 3,618 = 646 to find the increase.

646 is what percent of 3,618?

$$646 = x(3,618)$$
$$646 = 3,618x$$
$$\frac{646}{3,618} = \frac{3,618x}{3,618}$$
$$0.18 \approx x$$
$$x = 18\% \text{ increase}$$

CHALLENGE PROBLEMS

97.

$$-5 < \frac{x+2}{-2} < 0 \quad \text{or} \quad 2x + 10 \geq 30$$
$$-2(-5) > -2\left(\frac{x+2}{-2}\right) > -2(0) \quad 2x + 10 - 10 \geq 30 - 10$$
$$10 > x + 2 > 0 \quad 2x \geq 20$$
$$10 - 2 > x + 2 - 2 > 0 - 2 \quad \frac{2x}{2} \geq \frac{20}{2}$$
$$8 > x > -2 \quad x \geq 10$$
$$(-2, 8) \cup [10, \infty)$$

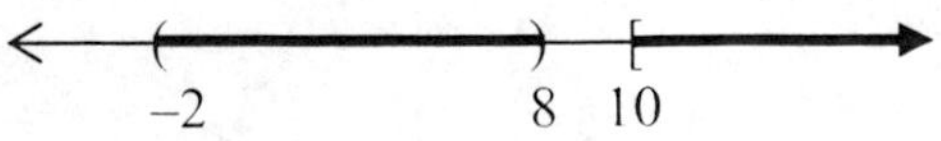

99.

$$x - 12 < 4x < 2x + 16$$
$$x - 12 < 4x \quad \text{and} \quad 4x < 2x + 16$$
$$-12 < 3x \quad 2x < 16$$
$$-4 < x \quad x < 8$$
$$x > -4$$
$$(-4, 8)$$

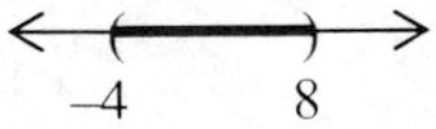

SECTION 4.3

VOCABULARY

1. The **absolute value** of a number is its distance from 0 on a number line.

3. To **isolate** the absolute value in $|3 - x| - 4 = 5$, we add 4 to both sides.

5. When two equations are joined by the word *or*, such as $x + 1 = 5$ *or* $x + 1 = -5$, we call the statement a **compound** equation.

CONCEPTS

7. To solve these absolute value equations and inequalities, we write and then solve equivalent **compound** equations and inequalities.

9. a. -2 and 2 are exactly 2 units from 0 on a number line.
   b. -1.99, -1, 0, 1, and 1.99 are less than 2 units from 0 on the number line.
   c. -3, -2.01, 2.01, and 3 are more than 2 units from 0 on the number line.

11. a.

$$|x-7| = 8$$

is equivalent to

$$x - 7 = \boxed{8} \text{ or } x - 7 = \boxed{-8}$$

b.

$$|x+10| = |x-3|$$

is equivalent to

$$x + 10 = \boxed{x-3} \text{ or } x + 10 = \boxed{-(x-3)}$$

13. a. $x = 8$ or $x = -8$
    b. $x \le -8$ or $x \ge 8$
    c. $-8 \le x \le 8$
    d. $5x - 1 = x + 3$ or $5x - 1 = -(x + 3)$

15. a. $|7x+6| = -8$ has no solution since the absolute value will never equal a negative number.
    b. $|7x+6| \le -8$ has no solution since the absolute value will never be less than or equal to a negative number.
    c. $|7x+6| \ge -8$ has solutions that include all real numbers since every absolute value will be greater than a negative number.

NOTATION

17. a. ii
    b. iii
    c. i

GUIDED PRACTICE

19.

$$|x| = 23$$

$$x = 23 \text{ or } x = -23$$

$$\boxed{x = 23, -23}$$

21.

$$|x-5| = 8$$

$$x - 5 = 8 \quad \text{or} \quad x - 5 = -8$$

$$x - 5 + 5 = 8 + 5 \qquad x - 5 + 5 = -8 + 5$$

$$x = 13 \qquad x = -3$$

$$\boxed{x = 13, -3}$$

23.

$$|3x+2| = 16$$

$$3x + 2 = 16 \quad \text{or} \quad 3x + 2 = -16$$

$$3x + 2 - 2 = 16 - 2 \qquad 3x + 2 - 2 = -16 - 2$$

$$3x = 14 \qquad 3x = -18$$

$$x = \frac{14}{3} \qquad x = -6$$

$$\boxed{x = \frac{14}{3}, -6}$$

25.

$$\left|\frac{x}{5}\right| = 10$$

$$\frac{x}{5} = 10 \quad \text{or} \quad \frac{x}{5} = -10$$

$$5\left(\frac{x}{5}\right) = 5(10) \qquad 5\left(\frac{x}{5}\right) = 5(-10)$$

$$x = 50 \qquad x = -50$$

$$\boxed{x = 50, -50}$$

27.

$$|2x+3.6|=9.8$$
$$2x+3.6=9.8 \quad \text{or} \quad 2x+3.6=-9.8$$
$$2x+3.6-3.6=9.8-3.6 \quad 2x+3.6-3.6=-9.8-3.6$$
$$2x=6.2 \quad 2x=-13.4$$
$$x=3.1 \quad x=-6.7$$
$$\boxed{x=3.1,-6.7}$$

29.

$$\left|\frac{7}{2}x+3\right|=-5$$

Since an absolute value can never be negative, there are no real numbers x that make $\left|\frac{7}{2}x+3\right|=-5$ true. The equation has no solution and the solution set is $\varnothing$.

31.

$$|x-3|-19=3$$
$$|x-3|-19+19=3+19$$
$$|x-3|=22$$
$$x-3=22 \quad \text{or} \quad x-3=-22$$
$$x-3+3=22+3 \quad x-3+3=-22+3$$
$$x=25 \quad x=-19$$
$$\boxed{x=25,-19}$$

33.

$$|3x-7|+8=22$$
$$|3x-7|+8-8=22-8$$
$$|3x-7|=14$$
$$3x-7=14 \quad \text{or} \quad 3x-7=-14$$
$$3x-7+7=14+7 \quad 3x-7+7=-14+7$$
$$3x=21 \quad 3x=-7$$
$$x=7 \quad x=-\frac{7}{3}$$
$$\boxed{x=7,-\frac{7}{3}}$$

35.

$$|3-4x|+1=6$$
$$|3-4x|+1-1=6-1$$
$$|3-4x|=5$$
$$3-4x=5 \quad \text{or} \quad 3-4x=-5$$
$$3-4x-3=5-3 \quad 3-4x-3=-5-3$$
$$-4x=2 \quad -4x=-8$$
$$x=-\frac{1}{2} \quad x=2$$
$$\boxed{x=-\frac{1}{2},2}$$

37.

$$\left|\frac{7}{8}x+5\right|-2=7$$
$$\left|\frac{7}{8}x+5\right|-2+2=7+2$$
$$\left|\frac{7}{8}x+5\right|=9$$
$$\frac{7}{8}x+5=9 \quad \text{or} \quad \frac{7}{8}x+5=-9$$
$$\frac{7}{8}x+5-5=9-5 \quad \frac{7}{8}x+5-5=-9-5$$
$$\frac{7}{8}x=4 \quad \frac{7}{8}x=-14$$
$$\frac{8}{7}\left(\frac{7}{8}x\right)=\frac{8}{7}(4) \quad \frac{8}{7}\left(\frac{7}{8}x\right)=\frac{8}{7}(-14)$$
$$x=\frac{32}{7} \quad x=-16$$
$$\boxed{x=\frac{32}{7},-16}$$

39.

$$\left|\frac{1}{5}x+2\right|-8=-8$$
$$\left|\frac{1}{5}x+2\right|-8+8=-8+8$$
$$\left|\frac{1}{5}x+2\right|=0$$
$$\frac{1}{5}x+2=0$$
$$\frac{1}{5}x+2-2=0-2$$
$$\frac{1}{5}x=-2$$
$$5\left(\frac{1}{5}x\right)=5(-2)$$
$$\boxed{x=-10}$$

41.

$$2|3x+24|=0$$
$$\frac{2|3x+24|}{2}=\frac{0}{2}$$
$$|3x+24|=0$$
$$3x+24=0$$
$$3x+24-24=0-24$$
$$3x=-24$$
$$\boxed{x=-8}$$

43.

$$-5|2x-9|+14=14$$
$$-5|2x-9|+14-14=14-14$$
$$-5|2x-9|=0$$
$$\frac{-5|2x-9|}{-5}=\frac{0}{-5}$$
$$|2x-9|=0$$
$$2x-9=0$$
$$2x-9+9=0+9$$
$$2x=9$$
$$\boxed{x=\frac{9}{2}}$$

45.

$$6-3|10x+5|=6$$
$$6-3|10x+5|-6=6-6$$
$$-3|10x+5|=0$$
$$\frac{-3|10x+5|}{-3}=\frac{0}{-3}$$
$$|10x+5|=0$$
$$10x+5=0$$
$$10x+5-5=0-5$$
$$10x=-5$$
$$x=-\frac{5}{10}$$
$$\boxed{x=-\frac{1}{2}}$$

47.

$$|5x-12|=|4x-16|$$

$5x-12=4x-16$	or	$5x-12=-(4x-16)$
$5x-12=4x-16$		$5x-12=-4x+16$
$5x-12-4x=4x-16-4x$		$5x-12+4x=-4x+16+4x$
$x-12=-16$		$9x-12=16$
$x-12+12=-16+12$		$9x-12+12=16+12$
$x=-4$		$9x=28$
$x=-4$		$x=\frac{28}{9}$

$$\boxed{x=-4,\ \frac{28}{9}}$$

49.

$$|10x|=|x-18|$$

$10x=x-18$	or	$10x=-(x-18)$
$10x=x-18$		$10x=-x+18$
$10x-x=x-18-x$		$10x+x=-x+18+x$
$9x=-18$		$11x=18$
$x=-2$		$x=\frac{18}{11}$

$$\boxed{x=-2,\ \frac{18}{11}}$$

51.

$$|2-x|=|3x+2|$$

$$2-x=3x+2 \quad \text{or} \quad 2-x=-(3x+2)$$

$$2-x=3x+2 \qquad 2-x=-3x-2$$

$$2-x-3x=3x+2-3x \qquad 2-x+3x=-3x-2+3x$$

$$2-4x=2 \qquad 2+2x=-2$$

$$2-4x-2=2-2 \qquad 2+2x-2=-2-2$$

$$-4x=0 \qquad 2x=-4$$

$$x=0 \qquad x=-2$$

$$\boxed{x=0,-2}$$

53.

$$|5x-7|=|4(x+1)|$$

$$5x-7=4(x+1) \quad \text{or} \quad 5x-7=-4(x+1)$$

$$5x-7=4x+4 \qquad 5x-7=-4x-4$$

$$5x-7+7=4x+4+7 \qquad 5x-7+7=-4x-4+7$$

$$5x=4x+11 \qquad 5x=-4x+3$$

$$5x-4x=4x+11-4x \qquad 5x+4x=-4x+3+4x$$

$$x=11 \qquad 9x=3$$

$$x=11 \qquad x=\frac{1}{3}$$

$$\boxed{x=11,\frac{1}{3}}$$

55.

$$|x|<4$$

$$-4<x<4$$

$$(-4,4)$$

57.

$$|x+9|\le 12$$

$$-12\le x+9\le 12$$

$$-12-9\le x+9-9\le 12-9$$

$$-21\le x\le 3$$

$$[-21,3]$$

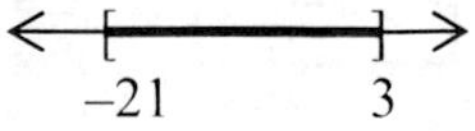

59.

$$|3x-2|<10$$

$$-10<3x-2<10$$

$$-10+2<3x-2+2<10+2$$

$$-8<3x<12$$

$$-\frac{8}{3}<x<4$$

$$\left(-\frac{8}{3},4\right)$$

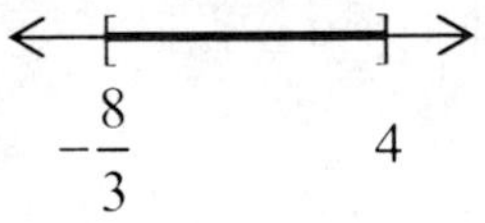

61.

$$|5x-12|<-5$$

No solution; $\varnothing$. Since $|5x-12|$ can never be negative, there are no real numbers x that can make the equation true.

63.

$$|x|>3$$

$$x<-3 \quad \text{or} \quad x>3$$

$$(-\infty,-3)\cup(3,\infty)$$

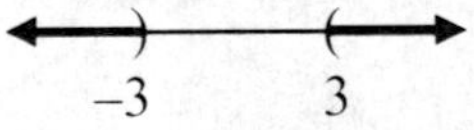

65.

$$|x-12|>24$$

$$x-12<-24 \quad \text{or} \quad x-12>24$$

$$x-12+12<-24+12 \qquad x-12+12>24+12$$

$$x<-12 \qquad x>36$$

$$(-\infty,-12)\cup(36,\infty)$$

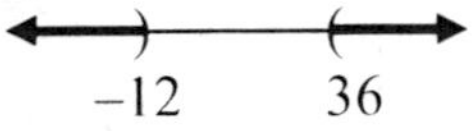

67.

$$|5x-1|-2\ge 0$$
$$|5x-1|-2+2\ge 0+2$$
$$|5x-1|\ge 2$$
$$5x-1\le -2 \quad \text{or} \quad 5x-1\ge 2$$
$$5x-1+1\le -2+1 \qquad 5x-1+1\ge 2+1$$
$$5x\le -1 \qquad 5x\ge 3$$
$$x\le -\frac{1}{5} \qquad x\ge \frac{3}{5}$$
$$\left(-\infty,-\frac{1}{5}\right]\cup\left[\frac{3}{5},\infty\right)$$

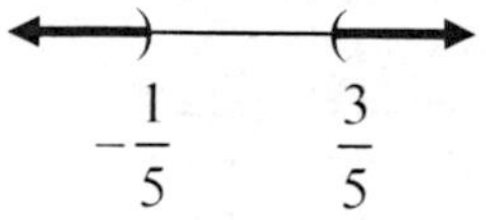

69.

$$|4x+3|>-5$$
$$(-\infty,\infty)$$

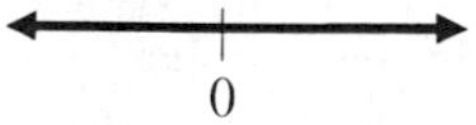

Since $|4x+3|$ is always greater than or equal to 0 for any real number x, then this absolute value inequality is true for all real numbers, $\mathbb{R}$.

71. $f(x)=|x+3|$

$$3=|x+3|$$
$$3=x+3 \quad \text{or} \quad -3=x+3$$
$$3-3=x+3-3 \qquad -3-3=x+3-3$$
$$0=x \qquad -6=x$$
$$\boxed{x=0,-6}$$

73. $f(x)=|2(x-1)+4|$

$$|2(x-1)+4|<4$$
$$|2x-2+4|<4$$
$$|2x+2|<4$$
$$-4<2x+2<4$$
$$-4-2<2x+2-2<4-2$$
$$-6<2x<2$$
$$\frac{-6}{2}<\frac{2x}{2}<\frac{2}{2}$$
$$-3<x<1$$
$$(-3,1)$$

TRY IT YOURSELF

75.

$$|3x+2|+1>15$$
$$|3x+2|+1-1>15-1$$
$$|3x+2|>14$$
$$3x+2<-14 \quad \text{or} \quad 3x+2>14$$
$$3x+2-2<-14-2 \qquad 3x+2-2>14-2$$
$$3x<-16 \qquad 3x>12$$
$$x<-\frac{16}{3} \qquad x>4$$
$$\left(-\infty,-\frac{16}{3}\right)\cup(4,\infty)$$

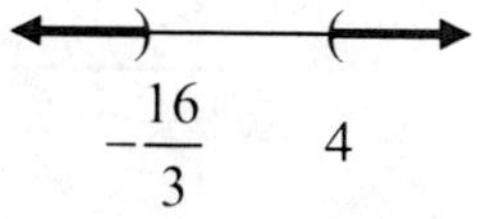

77.

$$6\left|\frac{x-2}{3}\right|\le 24$$
$$\frac{6\left|\frac{x-2}{3}\right|}{6}\le\frac{24}{6}$$
$$\left|\frac{x-2}{3}\right|\le 4$$
$$-4\le\frac{x-2}{3}\le 4$$
$$3(-4)\le 3\left(\frac{x-2}{3}\right)\le 3(4)$$
$$-12\le x-2\le 12$$
$$-12+2\le x-2+2\le 12+2$$
$$-10\le x\le 14$$
$$[-10,14]$$

79.

$$-7 = 2 - |0.3x - 3|$$
$$-7 - 2 = 2 - |0.3x - 3| - 2$$
$$-9 = -|0.3x - 3|$$
$$\frac{-9}{-1} = \frac{-|0.3x - 3|}{-1}$$
$$9 = |0.3x - 3|$$
$$|0.3x - 3| = 9$$
$$0.3x - 3 = 9 \quad \text{or} \quad 0.3x - 3 = -9$$
$$0.3x - 3 + 3 = 9 + 3 \quad 0.3x - 3 + 3 = -9 + 3$$
$$0.3x = 12 \quad 0.3x = -6$$
$$x = 40 \quad x = -20$$
$$\boxed{x = 40, -20}$$

81.

$$|2 - 3x| \geq -8$$
$$(-\infty, \infty)$$

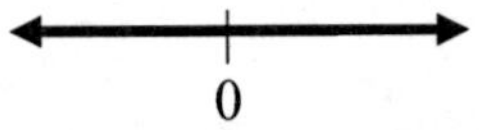

Since $|2 - 3x|$ is always greater than –8 for any real number x, then this absolute value inequality is true for all real numbers, $\mathbb{R}$.

83.

$$|7x + 12| = |x - 6|$$
$$7x + 12 = x - 6 \quad \text{or} \quad 7x + 12 = -(x - 6)$$
$$7x + 12 = x - 6 \quad 7x + 12 = -x + 6$$
$$7x + 12 - 12 = x - 6 - 12 \quad 7x + 12 - 12 = -x + 6 - 12$$
$$7x = x - 18 \quad 7x = -x - 6$$
$$7x - x = x - 18 - x \quad 7x + x = -x - 6 + x$$
$$6x = -18 \quad 8x = -6$$
$$x = -3 \quad x = -\frac{3}{4}$$
$$\boxed{x = -3, -\frac{3}{4}}$$

85.

$$3|2 - 3x| + 2 \leq 2$$
$$3|2 - 3x| + 2 - 2 \leq 2 - 2$$
$$3|2 - 3x| \leq 0$$
$$3|2 - 3x| \leq 0$$
$$\frac{3|2 - 3x|}{3} \leq \frac{0}{3}$$
$$|2 - 3x| \leq 0$$
$$2 - 3x = 0$$
$$2 - 3x - 2 = 0 - 2$$
$$-3x = -2$$
$$\frac{-3x}{-3} = \frac{-2}{-3}$$
$$x = \frac{2}{3}$$

87. $-14 = |x - 3|$

Since an absolute value can never be negative, there are no real numbers x that make $-14 = |x - 3|$ true. The equation has no solution and the solution set is $\varnothing$.

89.

$$\frac{6}{5} = \left|\frac{3x}{5} + \frac{x}{2}\right|$$
$$\frac{3x}{5} + \frac{x}{2} = \frac{6}{5} \quad \text{or} \quad \frac{3x}{5} + \frac{x}{2} = -\frac{6}{5}$$
$$10\left(\frac{3x}{5} + \frac{x}{2}\right) = 10\left(\frac{6}{5}\right) \quad 10\left(\frac{3x}{5} + \frac{x}{2}\right) = 10\left(-\frac{6}{5}\right)$$
$$6x + 5x = 12 \quad 6x + 5x = -12$$
$$11x = 12 \quad 11x = -12$$
$$x = \frac{12}{11} \quad x = -\frac{12}{11}$$
$$\boxed{x = \frac{12}{11}, -\frac{12}{11}}$$

91.

$$-|2x-3| < -7$$
$$\frac{-|2x-3|}{-1} > \frac{-7}{-1}$$
$$|2x-3| > 7$$
$$2x-3 < -7 \quad \text{or} \quad 2x-3 > 7$$
$$2x-3+3 < -7+3 \qquad 2x-3+3 > 7+3$$
$$2x < -4 \qquad 2x > 10$$
$$x < -2 \qquad x > 5$$
$$(-\infty,-2)\cup(5,\infty)$$

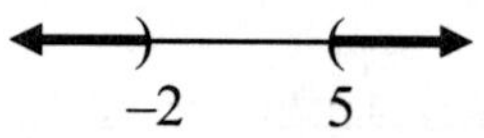

93.

$$|0.5x+1| < -23$$

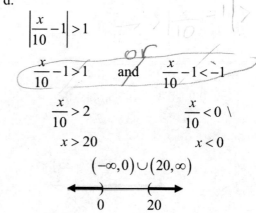

No solution; $\varnothing$. Since $|0.5x+1|$ can never be negative, there are no real numbers x that can make the equation true.

LOOK ALIKES

95. a.

$$\frac{x}{10}-1=1$$
$$\frac{x}{10}-1+1=1+1$$
$$\frac{x}{10}=2$$
$$10\left(\frac{x}{10}\right)=10(2)$$
$$x=20$$

b.

$$\left|\frac{x}{10}-1\right|=1$$
$$\frac{x}{10}-1=1 \quad \text{and} \quad \frac{x}{10}-1=-1$$
$$\frac{x}{10}=2 \qquad \frac{x}{10}=0$$
$$x=20 \qquad x=0$$
$$x=20,0$$

c.

$$\frac{x}{10}-1>1$$
$$\frac{x}{10}>2$$
$$x>20$$
$$(20,\infty)$$

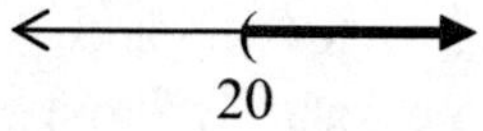

d.

$$\left|\frac{x}{10}-1\right|>1$$
$$\frac{x}{10}-1>1 \quad \text{and} \quad \frac{x}{10}-1<-1$$
$$\frac{x}{10}>2 \qquad \frac{x}{10}<0$$
$$x>20 \qquad x<0$$
$$(-\infty,0)\cup(20,\infty)$$

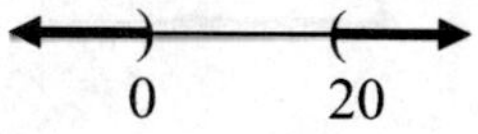

97. a.

$$0.9-0.3x=8.4$$
$$0.9-0.3x-0.9=8.4-0.9$$
$$-0.3x=7.5$$
$$x=-25$$

b.

$$|0.9-0.3x|=8.4$$

$0.9-0.3x=8.4$ and $0.9-0.3x=-8.4$

$0.9-0.3x-0.9=8.4-0.9$ $\quad 0.9-0.3x-0.9=-8.4-0.9$

$-0.3x=7.5$ $\quad -0.3x=-9.3$

$x=-25$ $\quad x=31$

$$x=-25,31$$

c.

$$0.9-0.3x>8.4$$
$$0.9-0.3x-0.9>8.4-0.9$$
$$-0.3x>7.5$$
$$x<-25$$
$$(-\infty,-25)$$

−25

d.

$$|0.9-0.3x|>8.4$$

$0.9-0.3x>8.4$ and $0.9-0.3x<-8.4$

$0.9-0.3x-0.9>8.4-0.9$ $\quad 0.9-0.3x-0.9<-8.4-0.9$

$-0.3x>7.5$ $\quad -0.3x<-9.3$

$x<-25$ $\quad x>31$

$$(-\infty,-25)\cup(31,\infty)$$

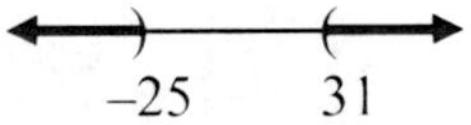

−25 31

99. a.

$$|8x-40|\le 16$$
$$-16\le 8x-40\le 16$$
$$-16+40\le 8x-40+40\le 16+40$$
$$24\le 8x\le 56$$
$$3\le x\le 7$$
$$[3,7]$$

3 7

b.

$$|8x-40|\ge 16$$

$8x-40\ge 16$ or $8x-40\le -16$

$8x\ge 56$ $\quad 8x\le 24$

$x\ge 7$ $\quad x\le 3$

$$(-\infty,3]\cup[7,\infty)$$

3 7

101. a.

$$\left|\frac{4x-4}{3}\right|-1>11$$
$$\left|\frac{4x-4}{3}\right|-1+1>11+1$$
$$\left|\frac{4x-4}{3}\right|>12$$

$\frac{4x-4}{3}>12$ or $\frac{4x-4}{3}<-12$

$3\left(\frac{4x-4}{3}\right)>3(12)$ $\quad 3\left(\frac{4x-4}{3}\right)<3(-12)$

$4x-4>36$ $\quad 4x-4<-36$

$4x>40$ $\quad 4x<-32$

$x>10$ $\quad x<-8$

$$(-\infty,-8)\cup(10,\infty)$$

−8 10

b.

$$\left|\frac{4x-4}{3}\right|-1\le 11$$
$$\left|\frac{4x-4}{3}\right|-1+1\le 11+1$$
$$\left|\frac{4x-4}{3}\right|\le 12$$
$$-12\le\frac{4x-4}{3}\le 12$$
$$3(-12)\le 3\left(\frac{4x-4}{3}\right)\le 3(12)$$
$$-36\le 4x-4\le 36$$
$$-36+4\le 4x-4+4\le 36+4$$
$$-32\le 4x\le 40$$
$$-8\le x\le 10$$
$$[-8,10]$$

−8 10

APPLICATIONS

103. TEMPERATURE RANGES

$$|t-78| \le 8$$
$$-8 \le t-78 \le 8$$
$$-8+78 \le t-78+78 \le 8+78$$
$$70^\circ \le t \le 86^\circ$$

105. AUTO MECHANICS

a. $|c-0.6^\circ| \le 0.5^\circ$

b. $|c-0.6^\circ| \le 0.5^\circ$
$$-0.5 \le c-0.6 \le 0.5$$
$$-0.5+0.6 \le c-0.6+0.6 \le 0.5+0.6$$
$$0.1 \le c \le 1.1$$
$$[0.1^\circ, 1.1^\circ]$$

107. ERROR ANALYSIS

Trial 1: $p = 22.91\%$
$$|\mathbf{22.91} - 25.46| \overset{?}{\le} 1.00$$
$$|-2.55| \overset{?}{\le} 1.00$$
$$2.55 \not\le 1.00$$
no

Trial 2: $p = 26.45\%$
$$|\mathbf{26.45} - 25.46| \overset{?}{\le} 1.00$$
$$|0.99| \overset{?}{\le} 1.00$$
$$0.99 \le 1.00$$
yes

Trial 3: $p = 26.49\%$
$$|\mathbf{26.49} - 25.46| \overset{?}{\le} 1.00$$
$$|1.03| \overset{?}{\le} 1.00$$
$$1.03 \not\le 1.00$$
no

Trial 4: $p = 24.76\%$
$$|\mathbf{24.76} - 25.46| \overset{?}{\le} 1.00$$
$$|-0.70| \overset{?}{\le} 1.00$$
$$0.70 \le 1.00$$
yes

26.45% and 24.76% satisfy the inequality.

WRITING

109. Answers will vary.

111. Answers will vary.

REVIEW

113. The angles x and y are supplementary (their sum is 180°).

$$\begin{cases} x+y=180 \\ y=30+2x \end{cases}$$
$$x+y=180$$
$$x+(30+2x)=180$$
$$3x+30=180$$
$$3x+30-30=180-30$$
$$3x=150$$
$$x=50^\circ$$
$$y=30+2(50)$$
$$y=130^\circ$$

CHALLENGE PROBLEMS

115. a. $k<0$
b. $k=0$

SECTION 4.4

VOCABULARY

1. $4x - 2y \geq -8$ is an example of a **linear** inequality in **two** variables.

3. The graph of a linear inequality in two variables is a region of the coordinate plane on one side of a **boundary** line.

CONCEPTS

5. a. Yes, (3, 1) is a solution.

$$3(\mathbf{3}) - 2(\mathbf{1}) \overset{?}{\geq} 5$$
$$9 - 2 \overset{?}{\geq} 5$$
$$7 \geq 5$$

b. No, (0, 3) is not a solution.

$$3(\mathbf{0}) - 2(\mathbf{3}) \overset{?}{\geq} 5$$
$$0 - 6 \overset{?}{\geq} 5$$
$$-6 \not\geq 5$$

c. Yes, (–1, –4) is a solution.

$$3(\mathbf{-1}) - 2(\mathbf{-4}) \overset{?}{\geq} 5$$
$$-3 + 8 \overset{?}{\geq} 5$$
$$5 \geq 5$$

d. No, $\left(1, \frac{1}{2}\right)$ is not a solution.

$$3(\mathbf{1}) - 2\left(\frac{\mathbf{1}}{\mathbf{2}}\right) \overset{?}{\geq} 5$$
$$3 - 1 \overset{?}{\geq} 5$$
$$2 \not\geq 5$$

7. a. Use the formula $y = mx + b$ to find the slope (m) and y-intercept (b).

$$y = mx + b$$
$$y = 3x - 1$$

$m = 3$ and $b = (0, -1)$

b. To find the x-intercept, let $y = 0$.

$$2x + 3(\mathbf{0}) = -6$$
$$2x = -6$$
$$x = -3$$

x- intercept: (–3, 0)

To find the y-intercept, let $x = 0$.

$$2(\mathbf{0}) + 3y = -6$$
$$3y = -6$$
$$y = -2$$

y-intercept: (0, –2)

NOTATION

9. a. no, dashed
   b. yes, solid
   c. yes, solid
   d. no, dashed

GUIDED PRACTICE

11. $y > x + 1$

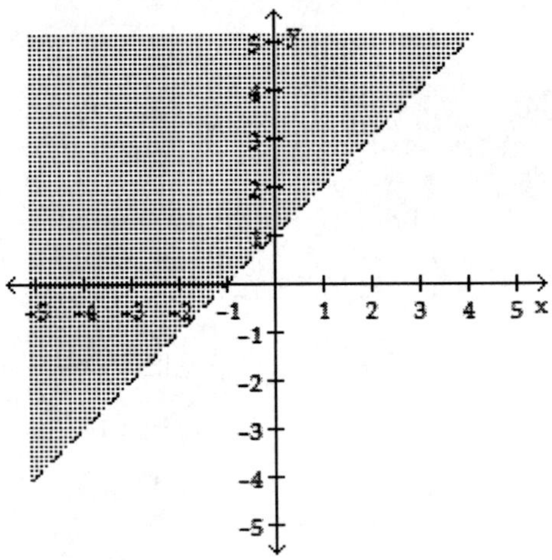

13. $y \geq -\frac{3}{2}x + 1$

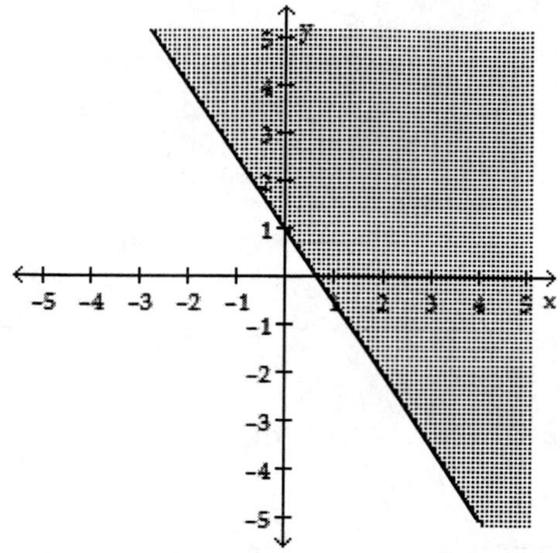

15. $2x + y \leq 6$

$y \leq -2x + 6$

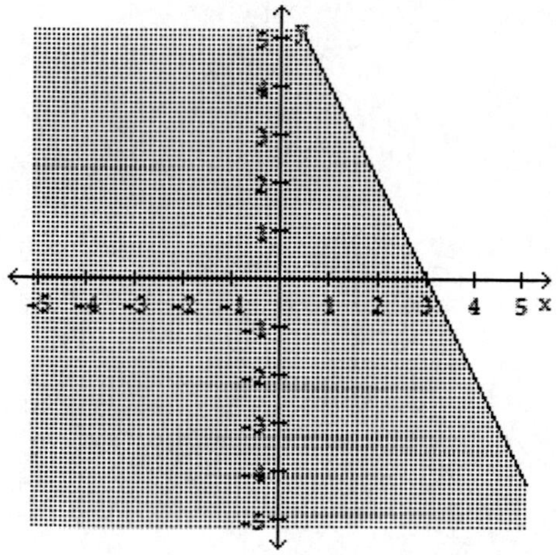

17. $3x + 5y > -9$

$5y > -3x - 9$

$y > -\frac{3}{5}x - \frac{9}{5}$

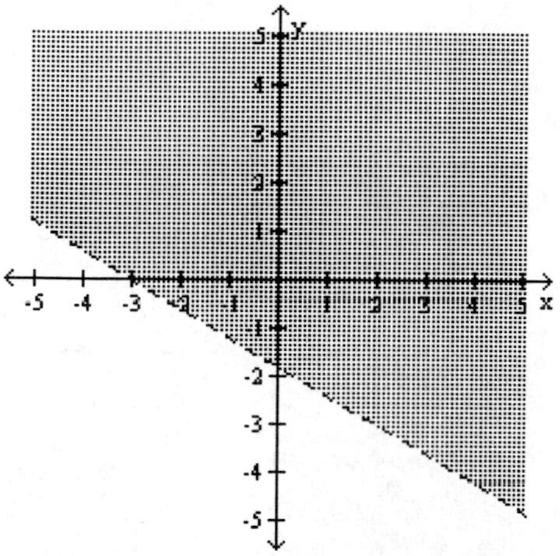

19. $y \geq x$

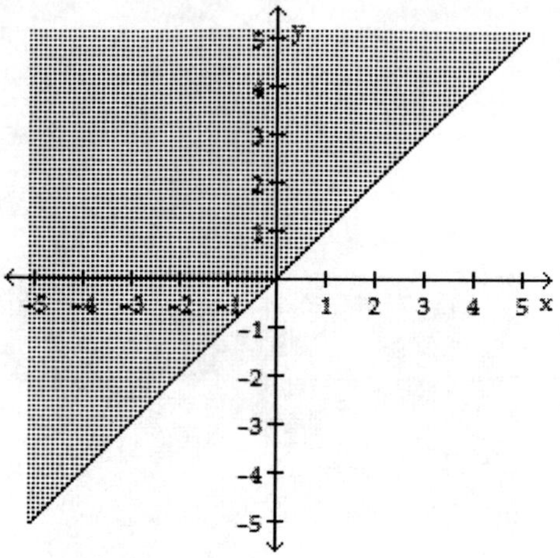

21. $y < -\frac{x}{2}$

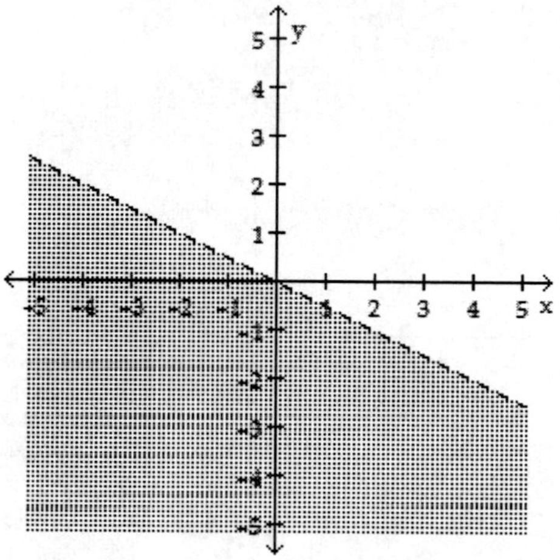

23. $x < 4$

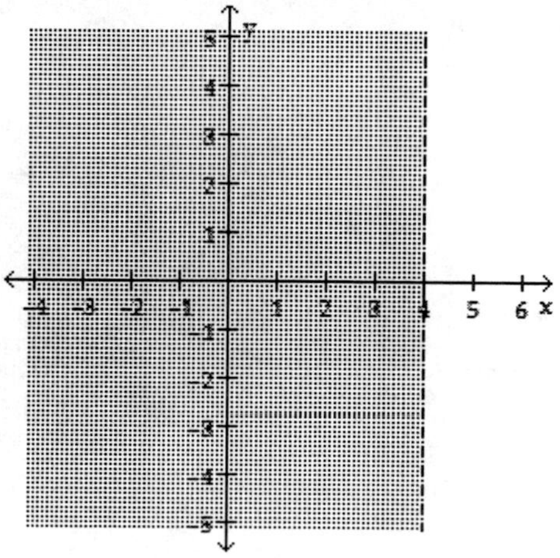

25. $y < 0$

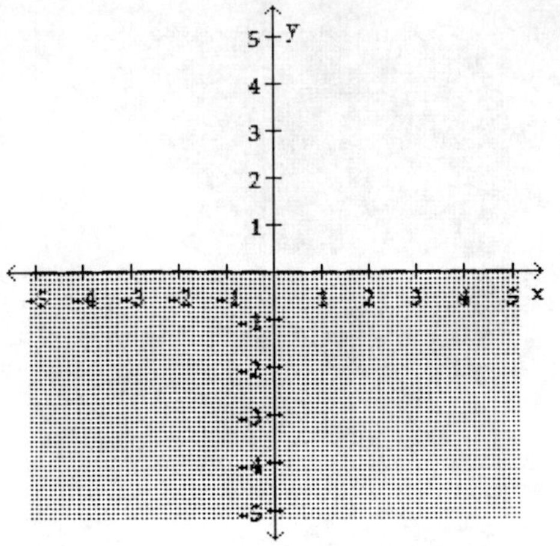

TRY IT YOURSELF

27. $3x \geq -y + 3$

$y \geq -3x + 3$

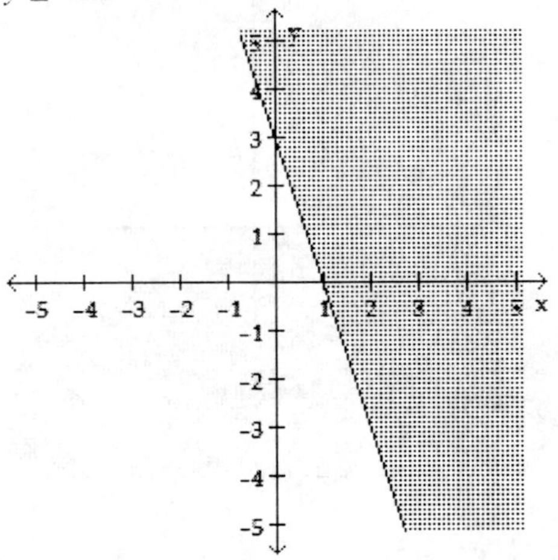

29. $3x + y > 2 + x$

$y > -2x + 2$

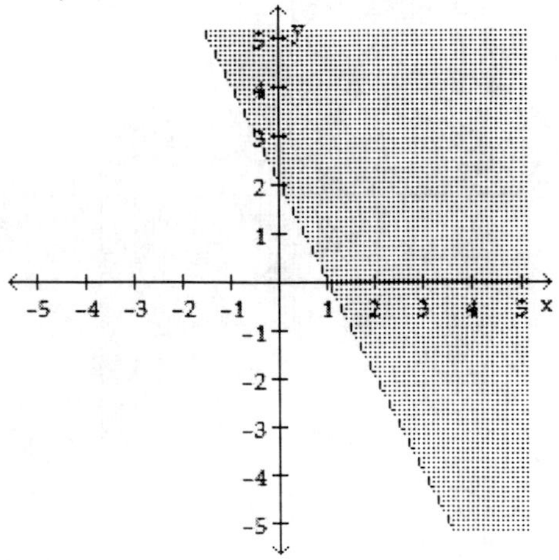

31. $y \geq \dfrac{8}{3}$

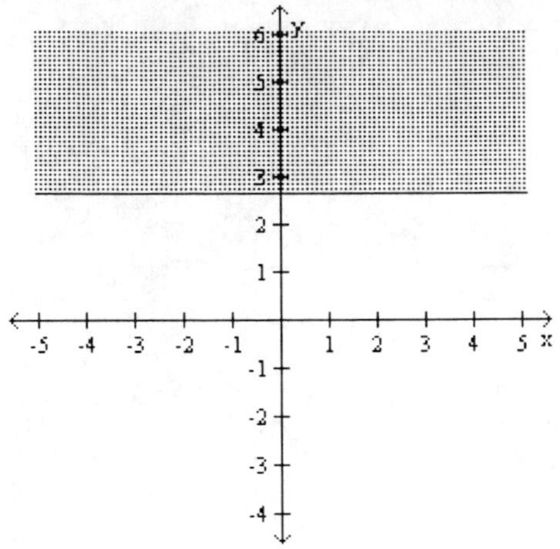

33. $y + 4x \geq 0$

$y \geq -4x$

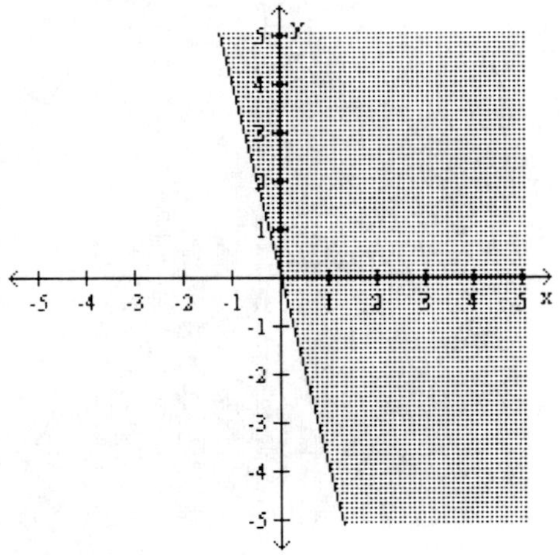

35. $\dfrac{x}{2} + \dfrac{y}{2} \leq 2$

$2\left(\dfrac{x}{2} + \dfrac{y}{2}\right) \leq 2(2)$

$x + y \leq 4$

$y \leq -x + 4$

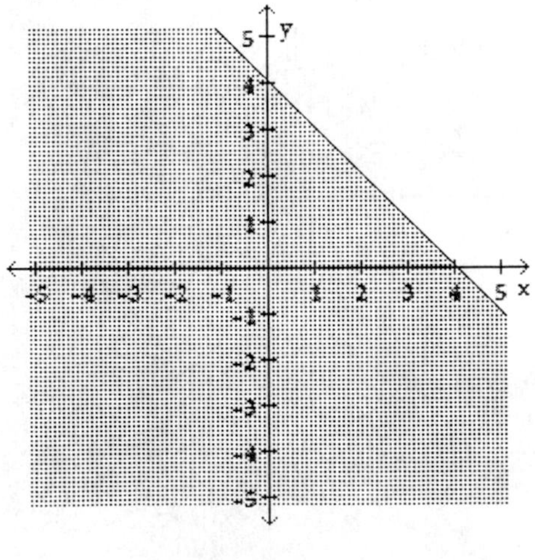

37. $y - 4.5 < 0$

$y < 4.5$

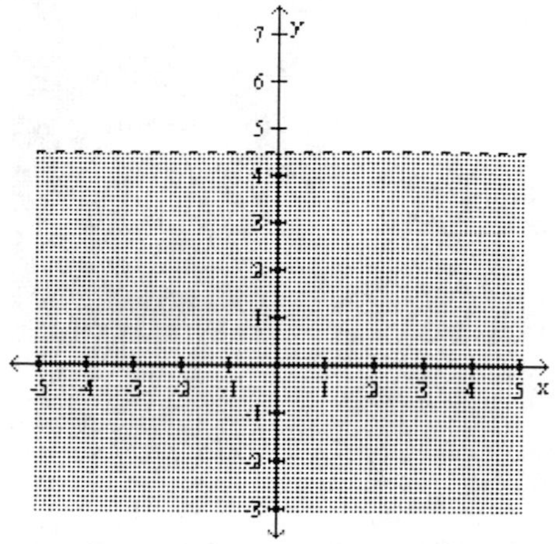

39. $x < -\frac{1}{2}y$

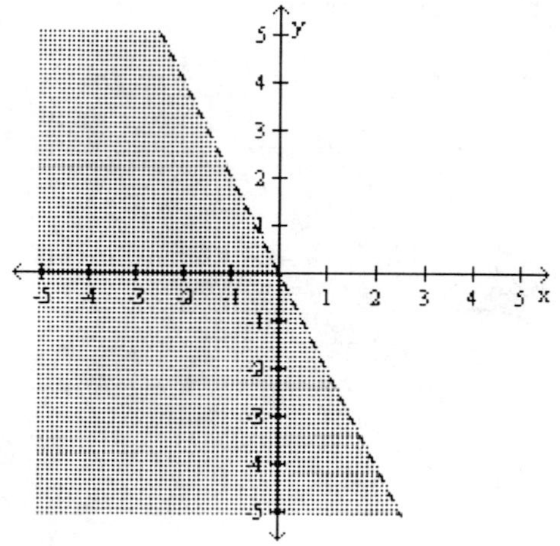

41. $0.3x + 0.4y \geq -1.2$

$10(0.3x + 0.4y) \geq 10(-1.2)$

$3x + 4y \geq -12$

$4y \geq -3x - 12$

$y \geq -\frac{3}{4}x - 3$

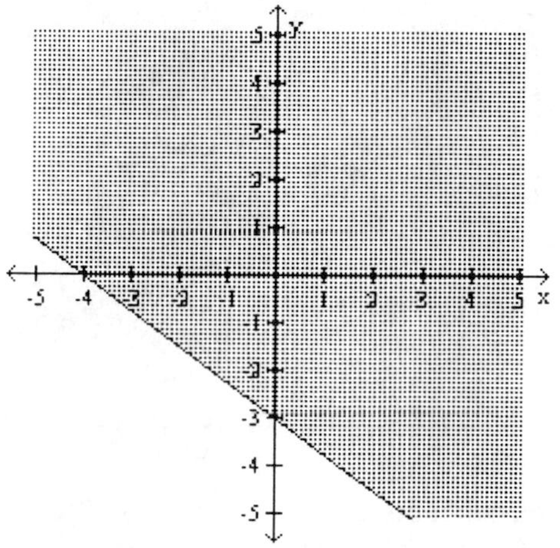

43. a. $5x - 3y < -15$

$-3y < -5x - 15$

$y > \frac{5}{3}x + 5$

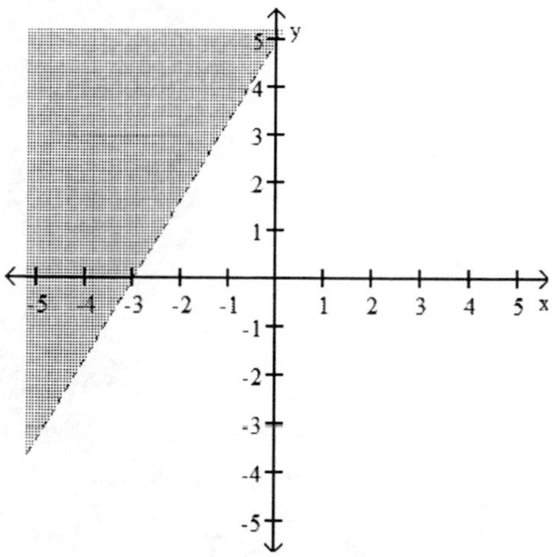

b. $5x - 3y \geq -15$

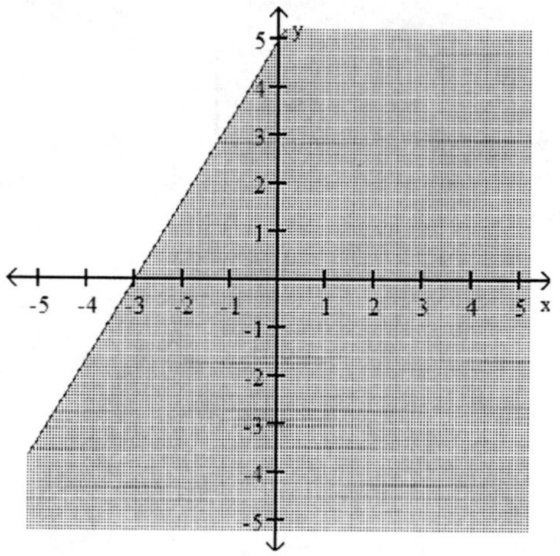

45. a. $y + 2x \geq 0$

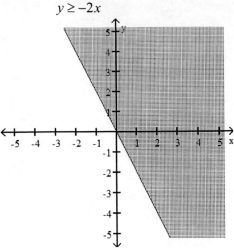

$y \geq -2x$

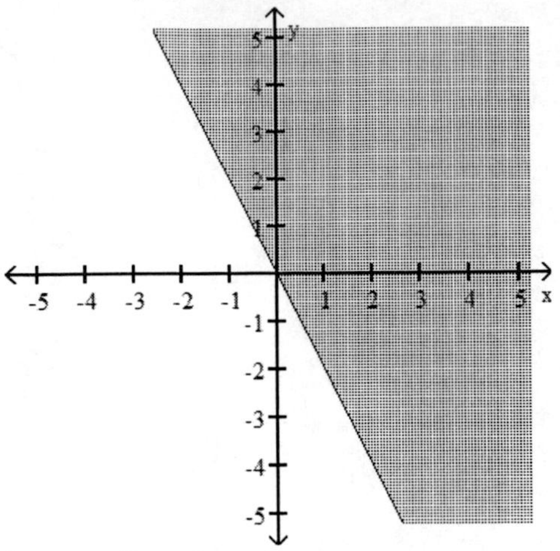

b. $y + 2x < 0$

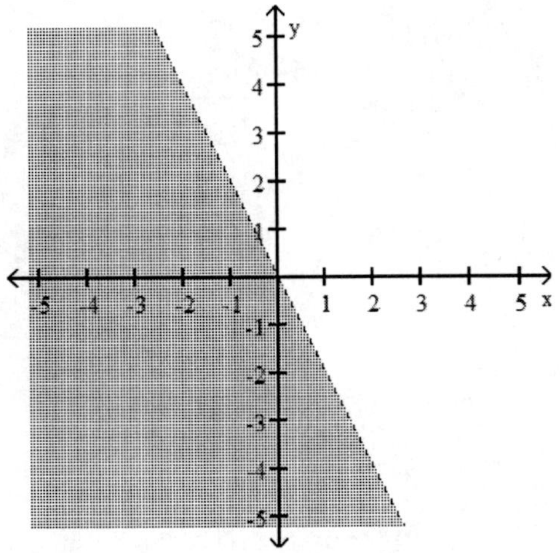

47. $y < 0.27x - 1$

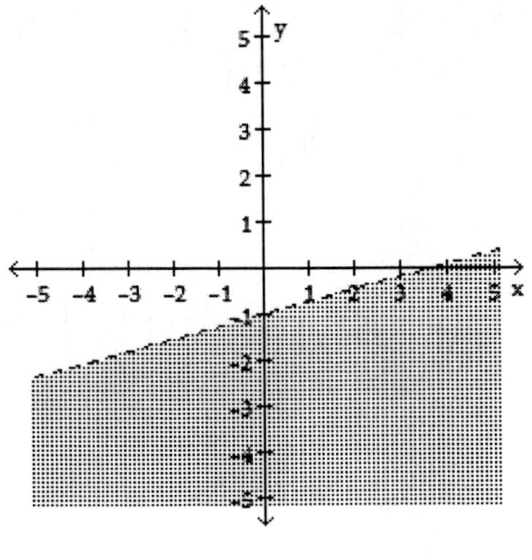

49. $y \geq -2.37x + 1.5$

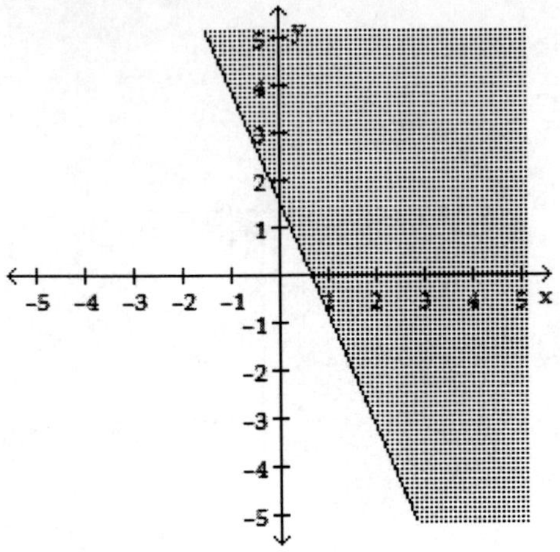

APPLICATIONS

51. THE KOREAN WAR

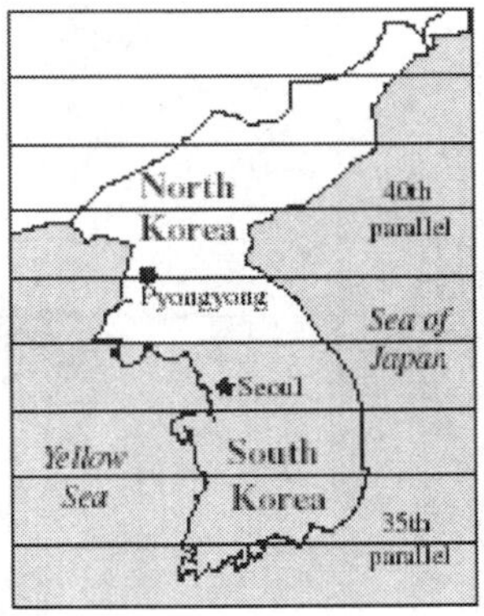

53. RESTAURANT SEATING
$4x + 6y \leq 120$
Possible combinations are (5, 15), (15, 10), and (20, 5).

55. SPORTING GOODS
$10x + 15y \geq 1{,}200$

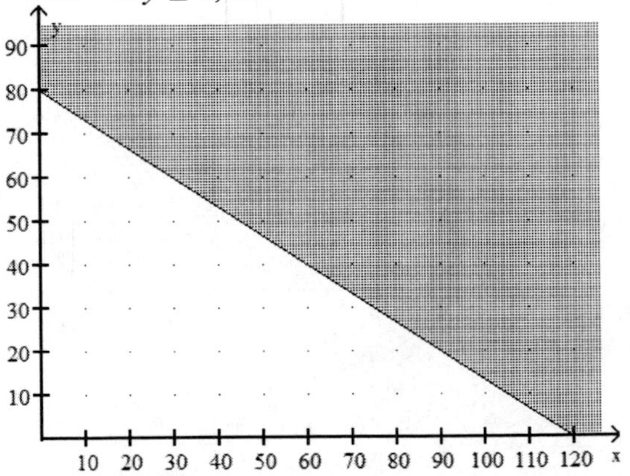

Possible combinations are (40, 80), (80, 80), and (120, 0).

WRITING

57. Answers will vary.

39. Answers will vary.

REVIEW

61.

$$\begin{aligned} 4x - y &= -19 \\ 4(\mathbf{-4}) - \mathbf{3} &\stackrel{?}{=} -19 \\ -16 - 3 &\stackrel{?}{=} -19 \\ -19 &= -19 \end{aligned} \qquad \begin{aligned} 3x + 2y &= -6 \\ 3(\mathbf{-4}) + 2(\mathbf{3}) &\stackrel{?}{=} -6 \\ -12 + 6 &\stackrel{?}{=} -6 \\ -6 &= -6 \end{aligned}$$

Yes, (–4, 3) is a solution.

63.

$$\begin{cases} x = \frac{2}{3}y \\ y = 4x + 5 \end{cases}$$

$$y = 4\left(\frac{2}{3}y\right) + 5$$

$$y = \frac{8}{3}y + 5$$

$$3(y) = 3\left(\frac{8}{3}y + 5\right)$$

$$3y = 8y + 15$$

$$3y - 8y = 8y + 15 - 8y$$

$$-5y = 15$$

$$y = -3$$

$$x = \frac{2}{3}(-3) = -2$$

$$(-2, -3)$$

CHALLENGE PROBLEMS

65. Answers will vary.

67. a. $3x + 2y > 6$
    b. $x \leq 3$

SECTION 4.5

VOCABULARY

1. $\begin{cases} x + y \le 2 \\ x - 3y > 10 \end{cases}$ is a system of linear **inequalities** in two variables.

3. To determine which half-plane to shade when graphing a linear inequality, we see whether the coordinates of a test **point** satisfy the inequality.

CONCEPTS

5. a. Yes.

$$\mathbf{2} + (\mathbf{-3}) \overset{?}{\le} 2 \qquad \mathbf{2} - 3(\mathbf{-3}) \overset{?}{>} 10$$
$$-1 \le 2 \qquad 2 + 9 \overset{?}{>} 10$$
$$11 > 10$$

b. No.

$$\mathbf{12} + (\mathbf{-1}) \overset{?}{\le} 2 \qquad \mathbf{12} - 3(\mathbf{-1}) \overset{?}{>} 10$$
$$11 \not\le 2 \qquad 12 + 3 \overset{?}{>} 10$$
$$15 > 10$$

c. No.

$$\mathbf{0} + (\mathbf{-3}) \overset{?}{\le} 2 \qquad \mathbf{0} - 3(\mathbf{-3}) \overset{?}{>} 10$$
$$-3 \le 2 \qquad 0 + 9 \overset{?}{>} 10$$
$$9 \not> 10$$

d. Yes.

$$\mathbf{-0.5} + (\mathbf{-5}) \overset{?}{\le} 2 \qquad \mathbf{-0.5} - 3(\mathbf{-5}) \overset{?}{>} 10$$
$$-5.5 \le 2 \qquad -0.5 + 15 \overset{?}{>} 10$$
$$14.5 > 10$$

7. a. False
   b. True
   c. True
   d. False
   e. True
   f. True

GUIDED PRACTICE

9. $\begin{cases} y < 3x + 2 \\ 2x + y < 3 \end{cases} \Rightarrow \begin{cases} y < 3x + 2 \\ y < -2x + 3 \end{cases}$

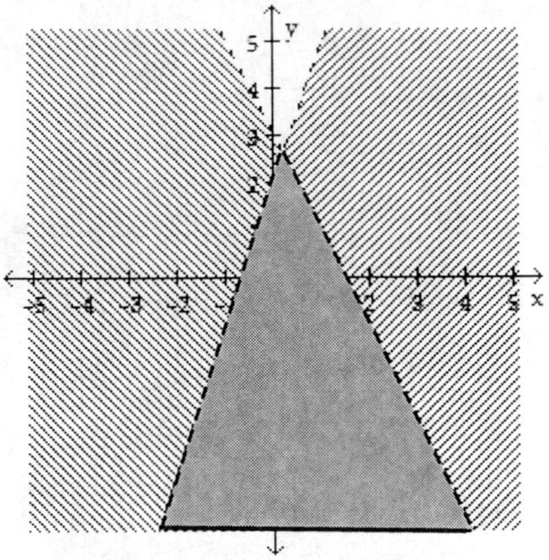

11. $\begin{cases} 3x + y \le 1 \\ -x + 2y \ge 6 \end{cases}$

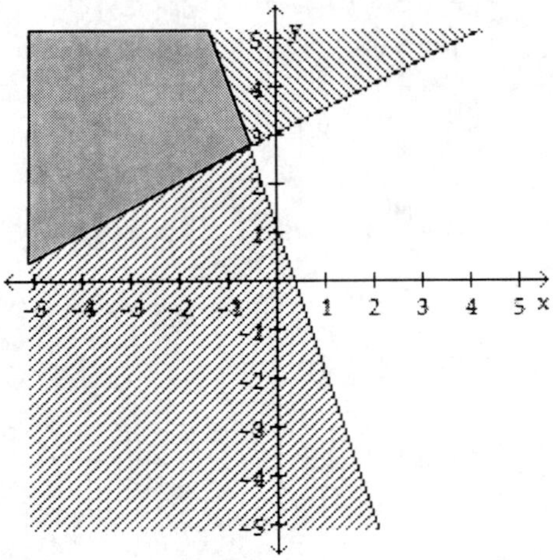

13. $\begin{cases} y - x \le 2 \\ y > -2 \\ x < 2 \end{cases} \Rightarrow \begin{cases} y \le x + 2 \\ y > -2 \\ x < 2 \end{cases}$

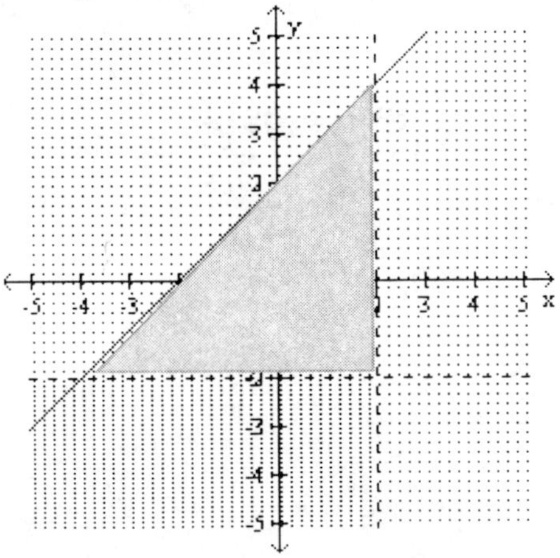

15. $\begin{cases} 2x+3y \le 6 \\ 3x+y \le 1 \\ x \le 0 \end{cases}$

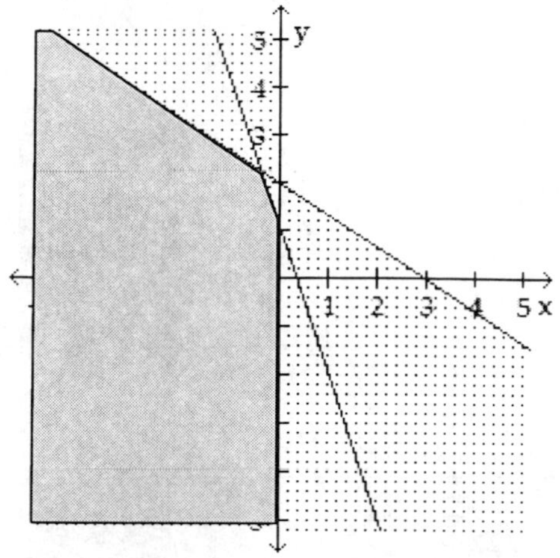

17. $-2 \le x < 0$

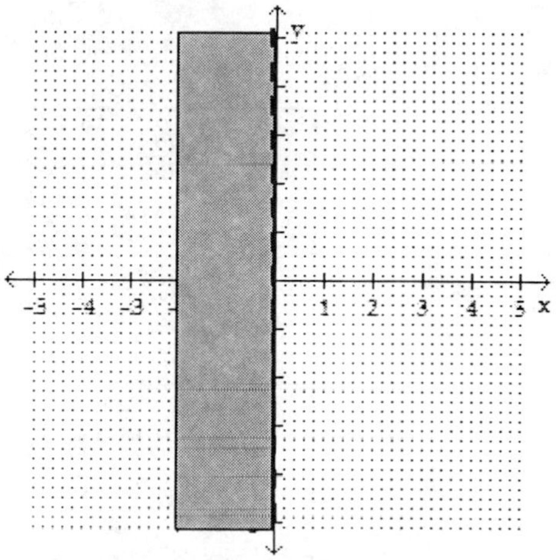

19. $y < -2$ or $y > 3$

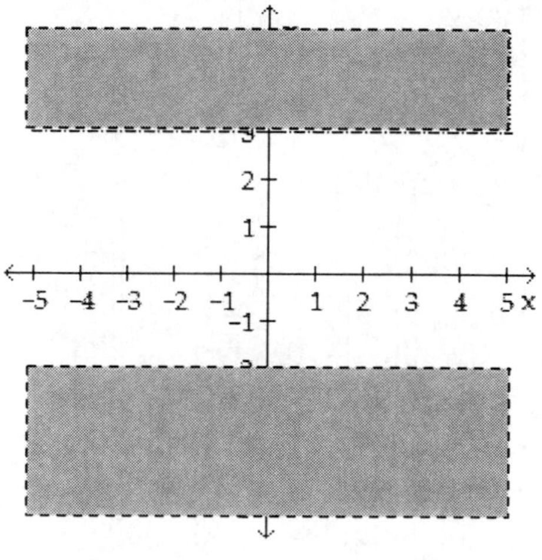

TRY IT YOURSELF

21. $\begin{cases} 2x < 3y \\ 2x+3y \ge 12 \end{cases} \Rightarrow \begin{cases} y > \frac{2}{3}x \\ y \ge -\frac{2}{3}x+4 \end{cases}$

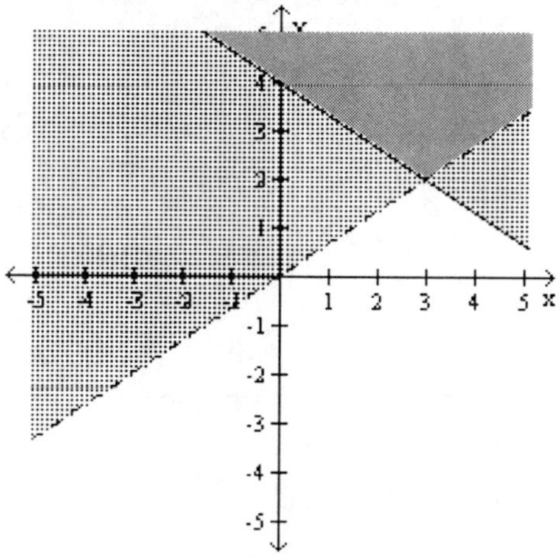

23. $\begin{cases} x > 0 \\ y > 0 \end{cases}$

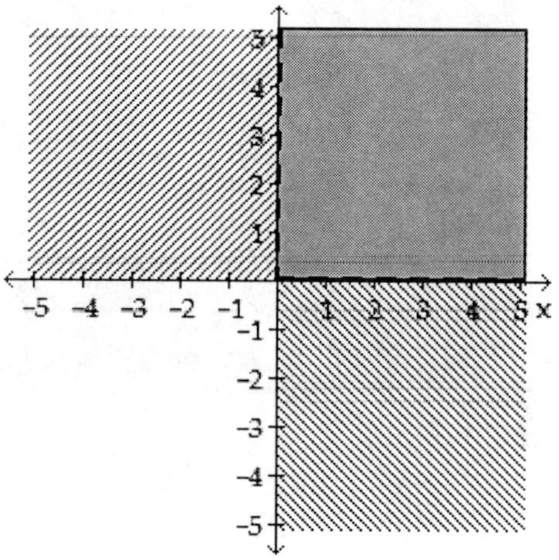

25. $\begin{cases} y \ge x \\ y \le \frac{1}{3}x+1 \\ x > -3 \end{cases}$

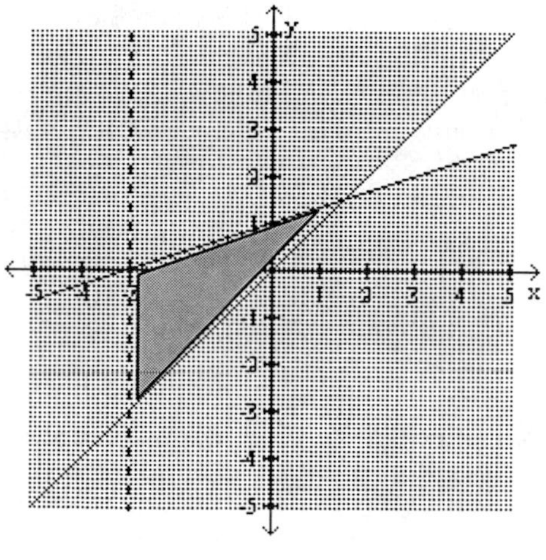

27. $5 > y \geq 2 \quad \Rightarrow \quad 2 \leq y < 5$

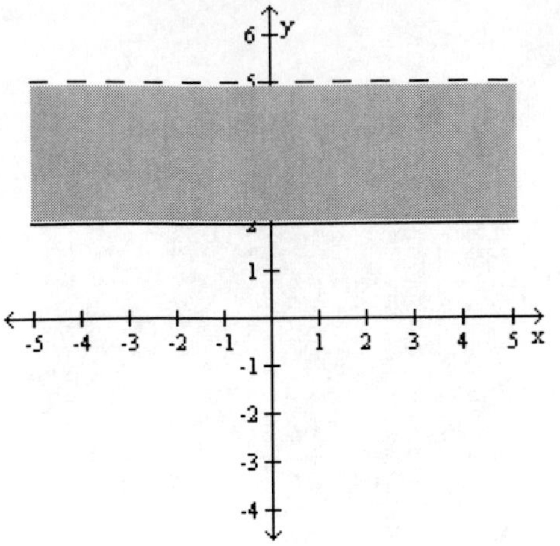

29. $\begin{cases} x + y < 2 \\ x + y \leq 1 \end{cases}$

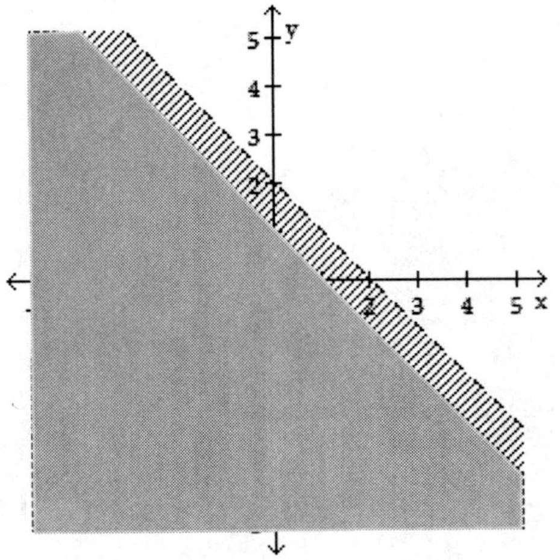

31. $\begin{cases} y < -\dfrac{3}{2}x - 3 \\ 3x + 2y \geq 2 \end{cases} \quad \Rightarrow \quad \begin{cases} y < -\dfrac{3}{2}x - 3 \\ y \geq -\dfrac{3}{2}x + 1 \end{cases}$

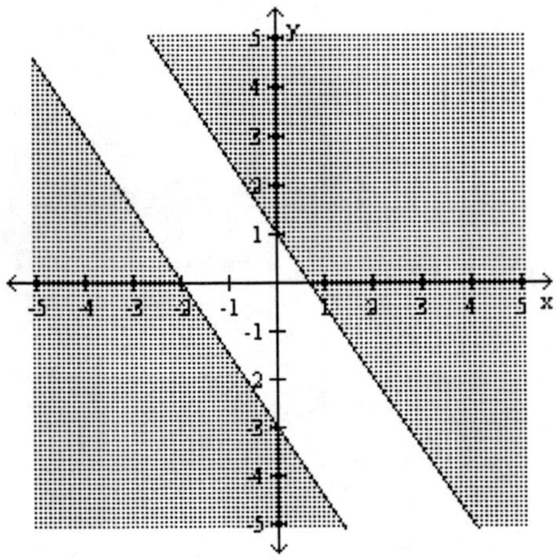

No solution; $\varnothing$. The graphs do not intersect.

33. $\begin{cases} 3y - 5x < 0 \\ 5x - 3y \geq -12 \end{cases} \quad \Rightarrow \quad \begin{cases} y < \dfrac{5}{3}x \\ y \leq \dfrac{5}{3}x + 4 \end{cases}$

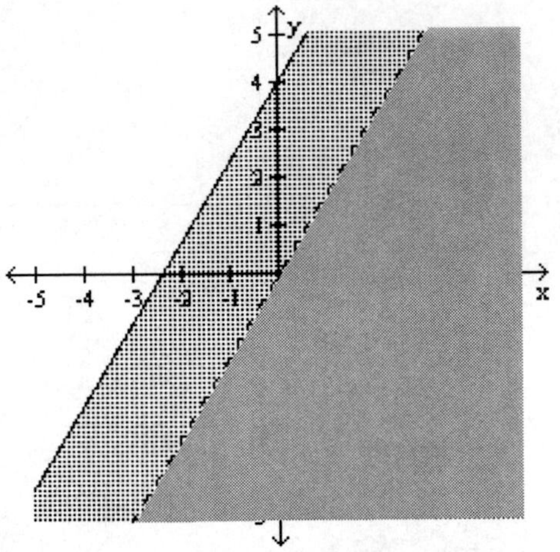

35. $-x \leq 1$ or $x \geq 2$
$x \geq -1$ or $x \geq 2$

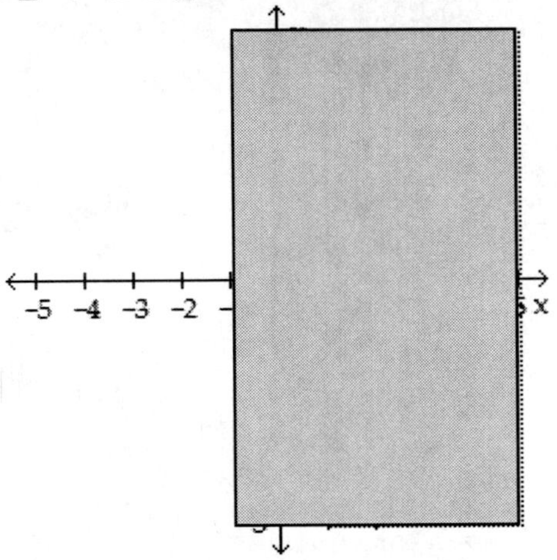

37. $\begin{cases} x < 1 \\ x > -1 \\ x - y + 4 \geq 0 \\ y - x \geq -4 \end{cases} \quad \Rightarrow \quad \begin{cases} x < 1 \\ x > -1 \\ y \leq x + 4 \\ y \geq x - 4 \end{cases}$

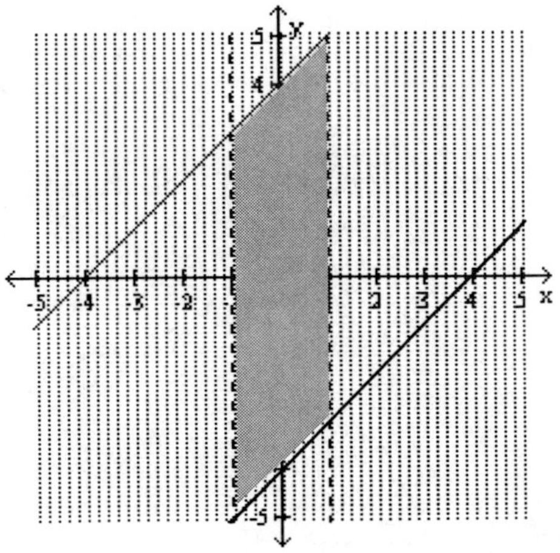

39. $\begin{cases} 2x-3y\le 3 \\ 3y\le 2x-3 \end{cases} \Rightarrow \begin{cases} y\ge \frac{2}{3}x-1 \\ y\le \frac{2}{3}x-1 \end{cases}$

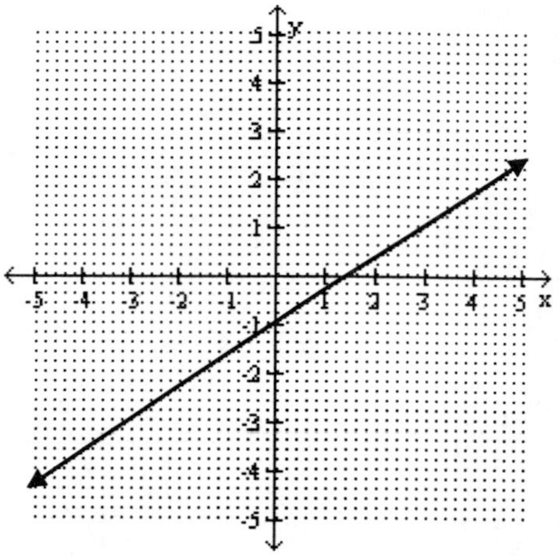

Only the line $y=\frac{2}{3}x-1$ is shaded on both.

41. $\begin{cases} y<3x+2 \\ y<-2x+3 \end{cases}$

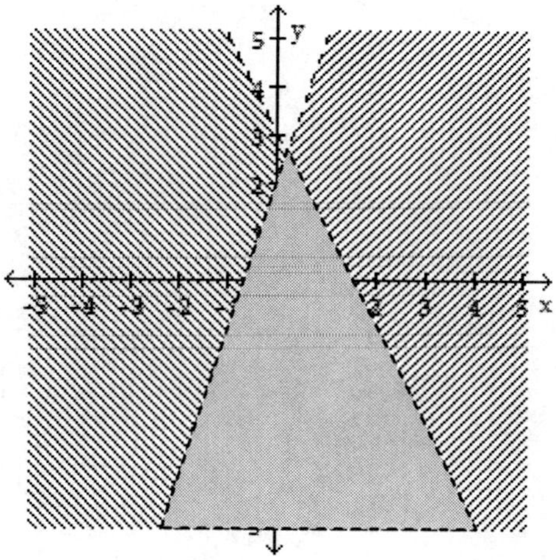

43. $\begin{cases} 2x+y\ge 6 \\ y\le 2(2x-3) \end{cases}$

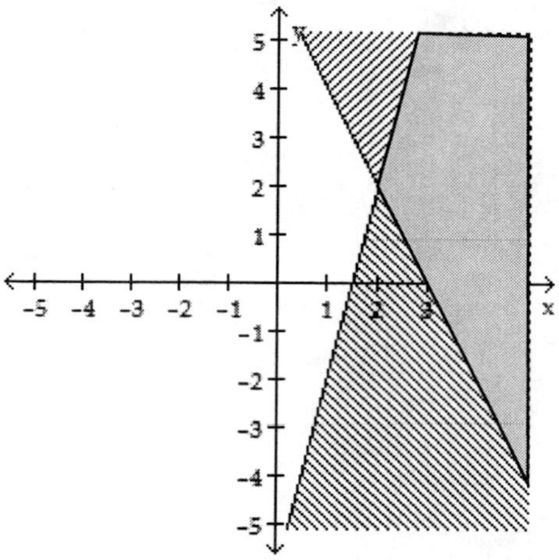

APPLICATIONS

45. FOOTBALL

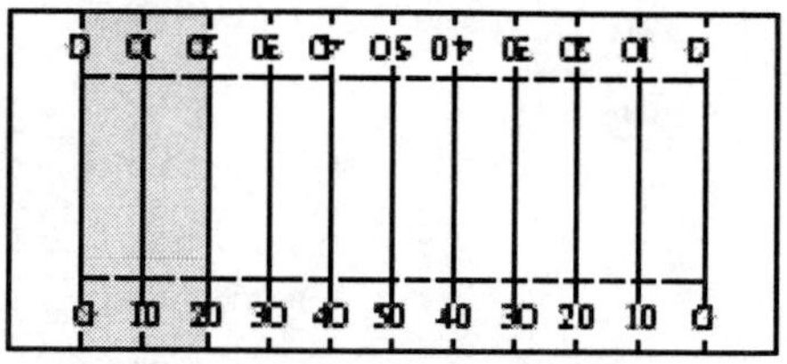

47. NO-FLY ZONES

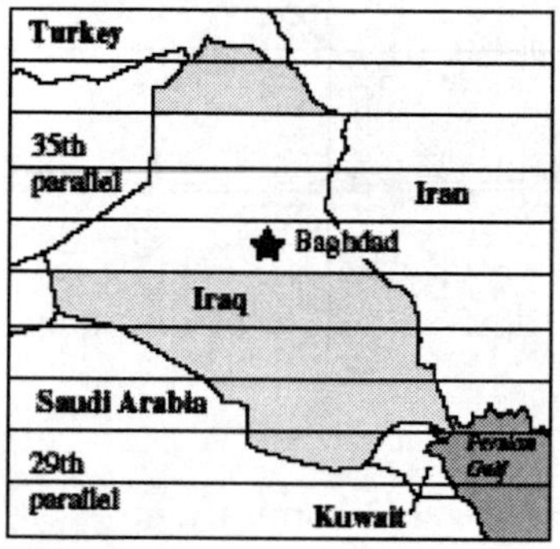

49. COMPACT DISCS

$\begin{cases} 10x+15y\ge 30 \\ 10x+15y\le 60 \\ x\ge 0 \\ y\ge 0 \end{cases}$

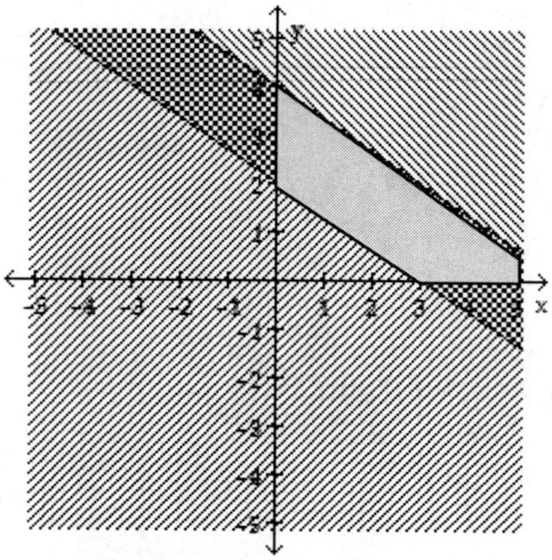

1 \$10 CD and 2 \$15 CDs
4 \$10 CDs and 1 \$15 CD

51. FURNACE EQUIPMENT

$$\begin{cases} 500x + 200y \le 2{,}000 \\ y > x \\ x \ge 0 \\ y \ge 0 \end{cases}$$

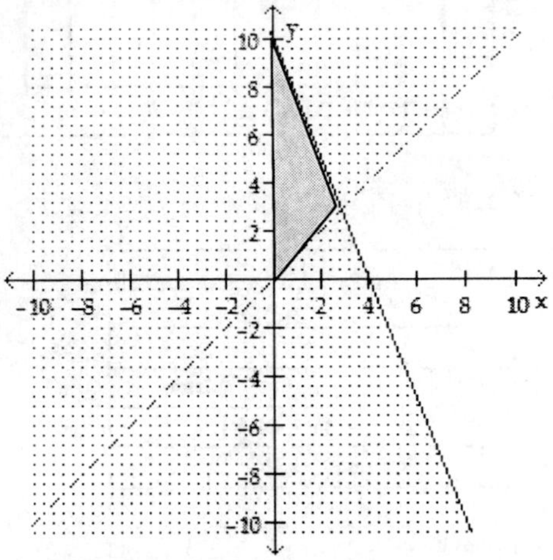

1 air cleaner and 2 humidifiers
2 air cleaners and 3 humidifiers

WRITING

53. Answers will vary.

55. Answers will vary.

REVIEW

57. IV

59. II

CHALLENGE PROBLEMS

61. $x < -2$ or $y \ge 3$

63. a. No
    b. Yes

CHAPTER 4 REVIEW

SECTION 4.1
Solving Linear Inequalities in One Variable

1. a. No.

$$3x-6<x-10$$
$$3(-2)-6\overset{?}{<}(-2)-10$$
$$-6-6\overset{?}{<}-2-10$$
$$-12\not<-12$$

b. Yes.

$$\frac{x}{2}-3\geq 4(x+1)$$
$$\frac{-2}{2}-3\overset{?}{\geq}4(-2+1)$$
$$-1-3\overset{?}{\geq}4(-1)$$
$$-4\geq -4$$

2. $[-5,\infty)$, $\{x|x\geq -5\}$

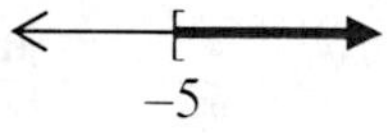

3.

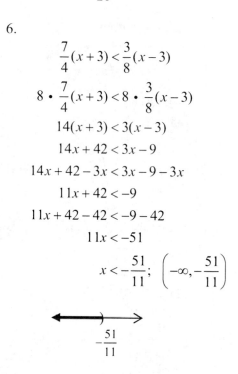

$$5(x-2)\leq 5$$
$$5x-10\leq 5$$
$$5x-10+10\leq 5+10$$
$$5x\leq 15$$
$$x\leq 3;\quad (-\infty,3]$$

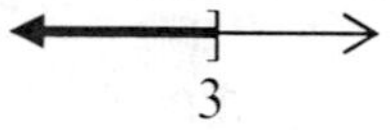

4.

$$0.3x-0.4\geq 1.2-0.1x$$
$$10(0.3x-0.4)\geq 10(1.2-0.1x)$$
$$3x-4\geq 12-x$$
$$3x-4+x\geq 12-x+x$$
$$4x-4\geq 12$$
$$4x-4+4\geq 12+4$$
$$4x\geq 16$$
$$x\geq 4;\quad [4,\infty)$$

5.

$$-16<-\frac{4}{5}t$$
$$-\frac{5}{4}(-16)>-\frac{5}{4}\left(-\frac{4}{5}t\right)$$
$$20>t$$
$$t<20;\quad (-\infty,\ 20)$$

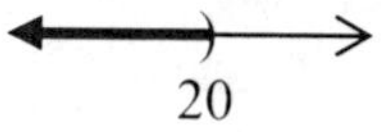

6.

$$\frac{7}{4}(x+3)<\frac{3}{8}(x-3)$$
$$8\cdot\frac{7}{4}(x+3)<8\cdot\frac{3}{8}(x-3)$$
$$14(x+3)<3(x-3)$$
$$14x+42<3x-9$$
$$14x+42-3x<3x-9-3x$$
$$11x+42<-9$$
$$11x+42-42<-9-42$$
$$11x<-51$$
$$x<-\frac{51}{11};\quad \left(-\infty,-\frac{51}{11}\right)$$

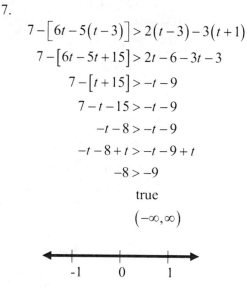

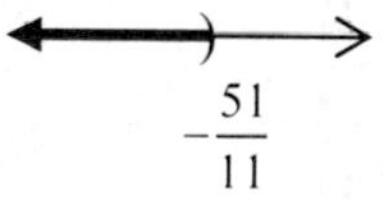

7.

$$7-[6t-5(t-3)]>2(t-3)-3(t+1)$$
$$7-[6t-5t+15]>2t-6-3t-3$$
$$7-[t+15]>-t-9$$
$$7-t-15>-t-9$$
$$-t-8>-t-9$$
$$-t-8+t>-t-9+t$$
$$-8>-9$$

true

$$(-\infty,\infty)$$

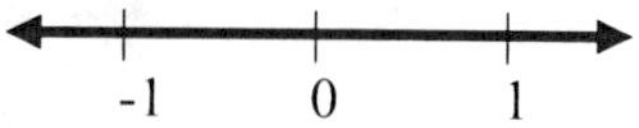

8.

$$\frac{2b+7}{2} \le \frac{3b-1}{3}$$
$$6\left(\frac{2b+7}{2}\right) \le 6\left(\frac{3b-1}{3}\right)$$
$$3(2b+7) \le 2(3b-1)$$
$$6b+21 \le 6b-2$$
$$6b+21-6b \le 6b-2-6b$$
$$21 \le -2$$
false

$\varnothing$; no solution

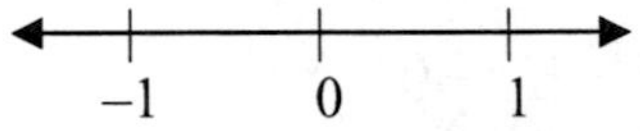

9.

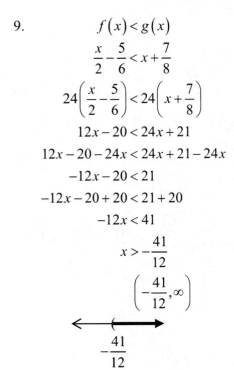

$$f(x) < g(x)$$
$$\frac{x}{2}-\frac{5}{6} < x+\frac{7}{8}$$
$$24\left(\frac{x}{2}-\frac{5}{6}\right) < 24\left(x+\frac{7}{8}\right)$$
$$12x-20 < 24x+21$$
$$12x-20-24x < 24x+21-24x$$
$$-12x-20 < 21$$
$$-12x-20+20 < 21+20$$
$$-12x < 41$$
$$x > -\frac{41}{12}$$
$$\left(-\frac{41}{12}, \infty\right)$$

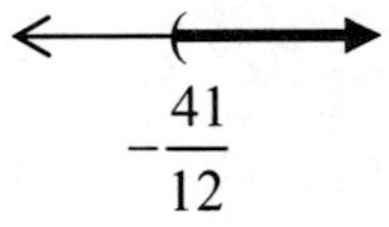

10. $f(x) \le 6$

$$6(4x-1)-11(2x-1) \le 6$$
$$24x-6-22x+11 \le 6$$
$$2x+5 \le 6$$
$$2x+5-5 \le 6-5$$
$$2x \le 1$$
$$x \le \frac{1}{2}$$

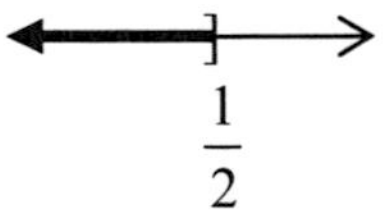

11.

$$s(g) > 21{,}250$$
$$1{,}700g > 21{,}250$$
$$\frac{1{,}700g}{1{,}700} > \frac{21{,}250}{1{,}700}$$
$$g > 12.5$$

More than 12.5 gallons of paint per week

12. Let x = number of cards printed.
Speedy Cleaners = $45 + 0.02x$
Vista Printing = $24 + 0.03x$
Speedy Cleaners < Vista Printing

$$45+0.02x < 24+0.03x$$
$$45+0.02x-0.03x < 24+0.03x-0.03x$$
$$45-0.01x < 24$$
$$45-0.01x-45 < 24-45$$
$$-0.01x < -21$$
$$x > 2{,}100$$

More than 2,100 cards

13.

Account	Principal	· Rate	· Time	= Interest
6%	10,000	0.06	1	0.06(10,000)
7%	x	0.07	1	0.07(x)
Combined Interest	2,000			

6% Interest + 7% Interest ≥ Combined Interest

$$0.06(10{,}000)+0.07x \ge 2{,}000$$
$$600+0.07x \ge 2{,}000$$
$$600+0.07x-600 \ge 2{,}000-600$$
$$0.07x \ge 1{,}400$$
$$\frac{0.07x}{0.07} \ge \frac{1{,}400}{0.07}$$
$$x \ge 20{,}000$$

At least **$20,000** or more at 7%.

14. Let x = score from the 6th judge.

$$\frac{5.3+4.8+4.7+4.9+5.1+x}{6} > 5.0$$
$$\frac{24.8+x}{6} > 5.0$$
$$6\left(\frac{24.8+x}{6}\right) > 6(5.0)$$
$$24.8+x > 30$$
$$24.8+x-24.8 > 30-24.8$$
$$x > 5.2$$

She needs to receive a score that is greater than 5.2.

15. Let x = number of hours on the phone.

$$200x+300(15-x)\ge 4{,}000$$
$$200x+4{,}500-300x\ge 4{,}000$$
$$-100x+4{,}500\ge 4{,}000$$
$$-100x+4{,}500-4{,}500\ge 4{,}000-4{,}500$$
$$-100x\ge -500$$
$$\frac{-100x}{-100}\le\frac{-500}{-100}$$
$$x\le 5$$

The most she can spend on the phone and earn $4,000 is 5 hours.

16. Answers will vary.

SECTION 4.2
Solving Compound Inequalities

17. $\{-3,3\}$

18. $\{-6,-5,-3,0,3,6,8\}$

19. Yes.
$-4<0$ and $-4>-5$

20. No.

$-4+3\overset{?}{<}-3(-4)-1$ and $4(-4)-3\overset{?}{>}3(-4)$

$-1\overset{?}{<}12-1$ $\quad -16-3\overset{?}{>}-12$

$-1<11$ $\quad -19\not>-12$

21. $(-3, 3)$: 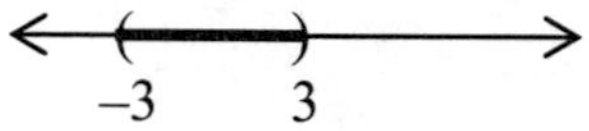

$[1, 6]$: 

$(-3,3)\cup[1,6]$: 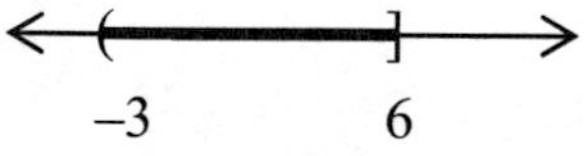

22. $(-\infty, 2]$: 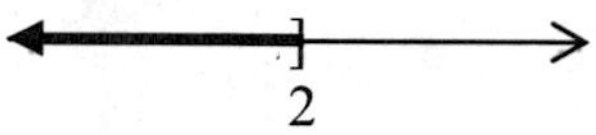

$[1, 4)$: 

$(-\infty,2]\cap[1,4)$: 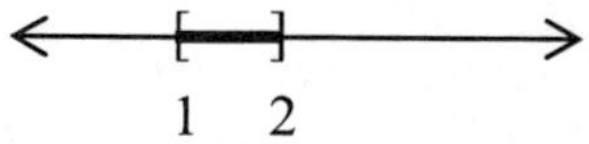

23.

$-2x>8$ and $x+4\ge -6$

$\frac{-2x}{-2}<\frac{8}{-2}$ $\quad x+4-4\ge -6-4$

$x<-4$ $\quad x\ge -10$

$$-10\le x<-4$$
$$[-10,-4)$$

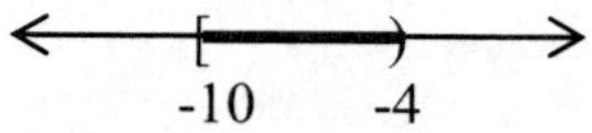

24.

$5(x+2)\le 4(x+1)$ and $11+x<0$

$5x+10\le 4x+4$ $\quad 11+x-11<0-11$

$5x+10-4x\le 4x+4-4x$ $\quad x<-11$

$x+10\le 4$

$x+10-10\le 4-10$

$x\le -6$

$$(-\infty,-11)$$

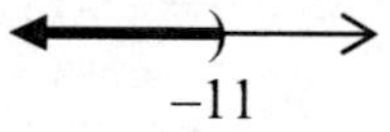

25.

$\frac{2}{5}x-2<-\frac{4}{5}$ and $\frac{x}{-3}<-1$

$5\left(\frac{2}{5}x-2\right)<5\left(-\frac{4}{5}\right)$ $\quad -3\left(-\frac{x}{3}\right)>-3(-1)$

$2x-10<-4$ $\quad x>3$

$2x-10+10<-4+10$

$2x<6$

$x<3$

no solution; $\varnothing$

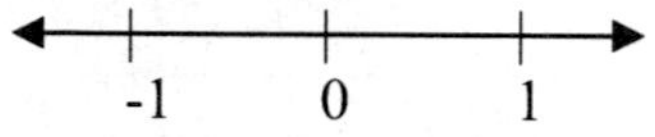

26.

$4\left(x-\frac{1}{4}\right)\le 3x-1$ and $x\ge 0$

$4x-1\le 3x-1$

$4x-1-3x\le 3x-1-3x$

$x-1\le -1$

$x-1+1\le -1+1$

$x\le 0$

$$[0,0]$$

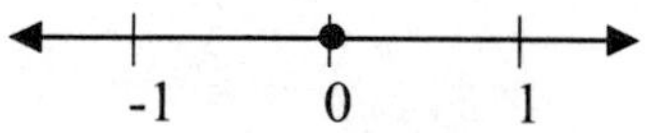

27.

$$3 < 3x+4 < 10$$
$$3-4 < 3x+4-4 < 10-4$$
$$-1 < 3x < 6$$
$$\frac{-1}{3} < \frac{3x}{3} < \frac{6}{3}$$
$$-\frac{1}{3} < x < 2$$
$$\left(-\frac{1}{3}, 2\right)$$

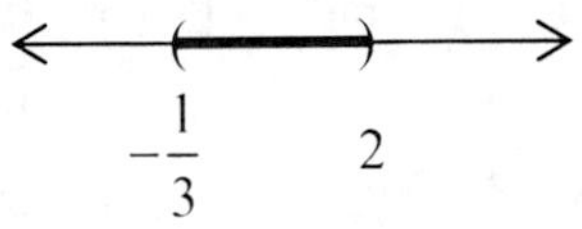

28.

$$-2 \le \frac{5-x}{2} \le 2$$
$$2(-2) \le 2\left(\frac{5-x}{2}\right) \le 2(2)$$
$$-4 \le 5-x \le 4$$
$$-4-5 \le 5-x-5 \le 4-5$$
$$-9 \le -x \le -1$$
$$\frac{-9}{-1} \ge \frac{-x}{-1} \ge \frac{-1}{-1}$$
$$9 \ge x \ge 1$$
$$1 \le x \le 9$$
$$[1,9]$$

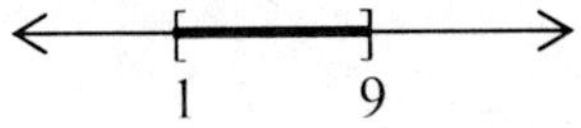

29. Yes.

$-4 < 1.6$ or $-4 \not> -3.9$

true false

30. No.

$-4+1 \stackrel{?}{<} 2(-4)-1$ or $4(-4)-3 \stackrel{?}{>} 3(-4)$

$-3 \not< -9$ $-19 \not> -12$

false false

31.

$x+1 < -4$ or $x-4 > 0$

$x+1-1 < -4-1$ $x-4+4 > 0+4$

$x < -5$ $x > 4$

$$(-\infty - 5) \cup (4, \infty)$$

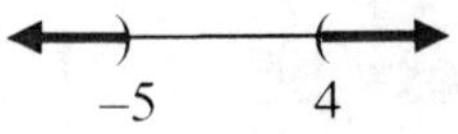

32.

$\frac{x}{2}+3 > -2$ or $4-x > 4$

$2\left(\frac{x}{2}+3\right) > 2(-2)$ $4-x-4 > 4-4$

$x+6 > -4$ $-x > 0$

$x+6-6 > -4-6$ $\frac{-x}{-1} < \frac{0}{-1}$

$x > -10$ $x < 0$

$$(-\infty, \infty)$$

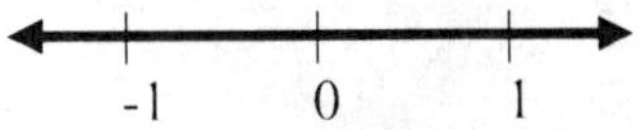

33. Area = length · width

$= x \cdot 4$

$= 4x$

$17 \le 4x \le 25$

$\frac{17}{4} \le \frac{4x}{4} \le \frac{25}{4}$

$4.25 \le x \le 6.25$

34. a. ii and iv

b. i and iii

35.

$f(x) < -48$ or $f(x) \ge 32$

$\frac{5}{4}x - 140 < -48$ $\frac{5}{4}x - 140 \ge 32$

$\frac{5}{4}x < 92$ $\frac{5}{4}x \ge 172$

$4\left(\frac{5}{4}x\right) < 4(92)$ $4\left(\frac{5}{4}x\right) \ge 4(172)$

$5x < 368$ $5x \ge 688$

$x < 73.6$ $x \ge 137.6$

$$(-\infty, 73.6) \cup [137.6, \infty)$$

36.

$$-4 \geq f(x) > -12$$
$$-4 \geq 3x - 5 > -12$$
$$-4 + 5 \geq 3x - 5 + 5 > -12 + 5$$
$$1 \geq 3x > -7$$
$$\frac{1}{3} \geq x > -\frac{7}{3}$$
$$-\frac{7}{3} < x \leq \frac{1}{3}$$
$$\left(-\frac{7}{3}, \frac{1}{3}\right]$$

SECTION 4.3
Solving Absolute Value Equations and Inequalities

37.

$$|4x| = 8$$
$$4x = 8 \quad \text{or} \quad 4x = -8$$
$$x = 2 \qquad x = -2$$
$$\boxed{x = 2, -2}$$

38.

$$2|3x+1| - 1 = 19$$
$$2|3x+1| - 1 + 1 = 19 + 1$$
$$2|3x+1| = 20$$
$$\frac{2|3x+1|}{2} = \frac{20}{2}$$
$$|3x+1| = 10$$
$$3x + 1 = 10 \quad \text{or} \quad 3x + 1 = -10$$
$$3x + 1 - 1 = 10 - 1 \quad 3x + 1 - 1 = -10 - 1$$
$$3x = 9 \qquad 3x = -11$$
$$x = 3 \qquad x = -\frac{11}{3}$$
$$\boxed{x = 3, -\frac{11}{3}}$$

39.

$$\left|\frac{3}{2}x - 4\right| - 10 = -1$$
$$\left|\frac{3}{2}x - 4\right| - 10 + 10 = -1 + 10$$
$$\left|\frac{3}{2}x - 4\right| = 9$$
$$\frac{3}{2}x - 4 = 9 \quad \text{or} \quad \frac{3}{2}x - 4 = -9$$
$$2\left(\frac{3}{2}x - 4\right) = 2(9) \quad 2\left(\frac{3}{2}x - 4\right) = 2(-9)$$
$$3x - 8 = 18 \qquad 3x - 8 = -18$$
$$3x - 8 + 8 = 18 + 8 \quad 3x - 8 + 8 = -18 + 8$$
$$3x = 26 \qquad 3x = -10$$
$$x = \frac{26}{3} \qquad x = -\frac{10}{3}$$
$$\boxed{x = \frac{26}{3}, -\frac{10}{3}}$$

40.

$$\left|\frac{2-x}{3}\right| = -4$$

No solution. Since an absolute value can never be negative, there are no real numbers x that can make the equation true.

41.

$$|-4(2x-6)| = 0$$
$$-4(2x-6) = 0$$
$$-8x + 24 = 0$$
$$-8x + 24 - 24 = 0 - 24$$
$$-8x = -24$$
$$\frac{-8x}{-8} = \frac{-24}{-8}$$
$$x = 3$$
$$\boxed{x = 3}$$

42.

$$\left|\frac{3}{8}+\frac{x}{3}\right|=\frac{5}{12}$$

$$\frac{3}{8}+\frac{x}{3}=\frac{5}{12} \quad \text{or} \quad \frac{3}{8}+\frac{x}{3}=-\frac{5}{12}$$

$$24\left(\frac{3}{8}+\frac{x}{3}\right)=24\left(\frac{5}{12}\right) \quad 24\left(\frac{3}{8}+\frac{x}{3}\right)=24\left(-\frac{5}{12}\right)$$

$$9+8x=10 \quad 9+8x=-10$$

$$9+8x-9=10-9 \quad 9+8x-9=-10-9$$

$$8x=1 \quad 8x=-19$$

$$x=\frac{1}{8} \quad x=-\frac{19}{8}$$

$$\boxed{x=\frac{1}{8},-\frac{19}{8}}$$

43.

$$|3x+2|=|2x-3|$$

$$3x+2=2x-3 \quad \text{or} \quad 3x+2=-(2x-3)$$

$$3x+2=2x-3 \quad 3x+2=-2x+3$$

$$3x+2-2x=2x-3-2x \quad 3x+2+2x=-2x+3+2x$$

$$x+2=-3 \quad 5x+2=3$$

$$x=-5 \quad 5x=1$$

$$x=-5 \quad x=\frac{1}{5}$$

$$\boxed{x=-5,\frac{1}{5}}$$

44.

$$\left|\frac{2(1-x)+1}{2}\right|=\left|\frac{3x-2}{3}\right|$$

$$\left|\frac{2-2x+1}{2}\right|=\left|\frac{3x-2}{3}\right|$$

$$\left|\frac{-2x+3}{2}\right|=\left|\frac{3x-2}{3}\right|$$

$$\frac{-2x+3}{2}=\frac{3x-2}{3} \quad \text{or} \quad \frac{-2x+3}{2}=-\left(\frac{3x-2}{3}\right)$$

$$6\left(\frac{-2x+3}{2}\right)=6\left(\frac{3x-2}{3}\right) \quad 6\left(\frac{-2x+3}{2}\right)=-6\left(\frac{3x-2}{3}\right)$$

$$3(-2x+3)=2(3x-2) \quad 3(-2x+3)=-2(3x-2)$$

$$-6x+9=6x-4 \quad -6x+9=-6x+4$$

$$-6x+9-6x=6x-4-6x \quad -6x+9+6x=-6x+4+6x$$

$$-12x+9=-4 \quad 9\neq 4$$

$$-12x+9-9=-4-9$$

$$-12x=-13$$

$$\boxed{x=\frac{13}{12}}$$

45.

$$|x|\leq 3$$

$$-3\leq x\leq 3$$

$$[-3,3]$$

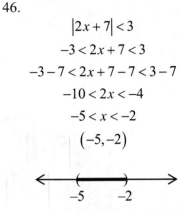

46.

$$|2x+7|<3$$

$$-3<2x+7<3$$

$$-3-7<2x+7-7<3-7$$

$$-10<2x<-4$$

$$-5<x<-2$$

$$(-5,-2)$$

47.

$$2|5-3x|\leq 28$$

$$\frac{2|5-3x|}{2}\leq\frac{28}{2}$$

$$|5-3x|\leq 14$$

$$-14\leq 5-3x\leq 14$$

$$-14-5\leq 5-3x-5\leq 14-5$$

$$-19\leq -3x\leq 9$$

$$\frac{-19}{-3}\geq\frac{-3x}{-3}\geq\frac{9}{-3}$$

$$\frac{19}{3}\geq x\geq -3$$

$$-3\leq x\leq\frac{19}{3}$$

$$\left[-3,\frac{19}{3}\right]$$

48.

$$\left|\frac{2}{3}x+14\right|+6<6$$

$$\left|\frac{2}{3}x+14\right|+6-6<6-6$$

$$\left|\frac{2}{3}x+14\right|<0$$

No solution. Since an absolute value can never be negative (less than zero), there are no real numbers x that can make the equation true.

49.

$$|x|>1$$

$$x<-1 \quad \text{or} \quad x>1$$

$$(-\infty,-1)\cup(1,\infty)$$

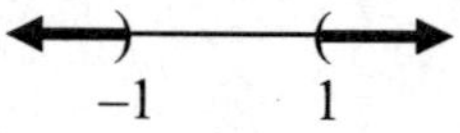

50.

$$\left|\frac{1-5x}{3}\right|\geq 7$$

$$\frac{1-5x}{3}\leq -7 \quad \text{or} \quad \frac{1-5x}{3}\geq 7$$

$$3\left(\frac{1-5x}{3}\right)\leq 3(-7) \qquad 3\left(\frac{1-5x}{3}\right)\geq 3(7)$$

$$1-5x\leq -21 \qquad 1-5x\geq 21$$

$$1-5x-1\leq -21-1 \qquad 1-5x-1\geq 21-1$$

$$-5x\leq -22 \qquad -5x\geq 20$$

$$x\geq \frac{22}{5} \qquad x\leq -4$$

$$(-\infty,-4]\cup\left[\frac{22}{5},\infty\right)$$

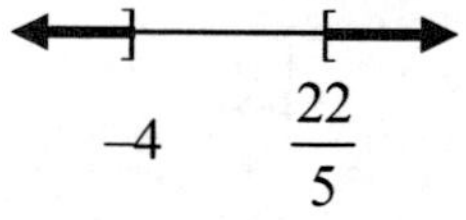

51.

$$|3x-8|-4>0$$

$$|3x-8|-4+4>0+4$$

$$|3x-8|>4$$

$$3x-8<-4 \quad \text{or} \quad 3x-8>4$$

$$3x-8+8<-4+8 \qquad 3x-8+8>4+8$$

$$3x<4 \qquad 3x>12$$

$$x<\frac{4}{3} \qquad x>4$$

$$\left(-\infty,\frac{4}{3}\right)\cup(4,\infty)$$

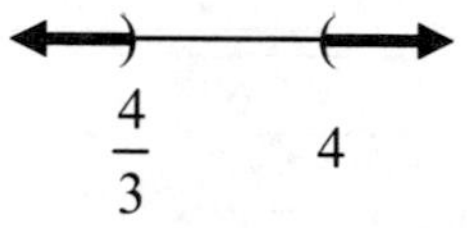

52.

$$\left|\frac{3}{2}x-14\right|\geq 0$$

$$(-\infty,\ \infty)$$

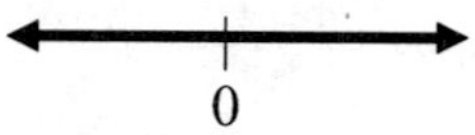

Since $\left|\frac{3}{2}x-14\right|$ is always greater than or equal to 0 for any real number x, then this absolute value inequality is true for all real numbers, $\mathbb{R}$.

53. a. $|w-8|\leq 2$

b.

$$|w-8|\leq 2$$

$$-2\leq w-8\leq 2$$

$$-2+8\leq w-8+8\leq 2+8$$

$$6\leq w\leq 10$$

$$[6,10]$$

54.
$$f(x)=\frac{1}{3}|6x|-1$$
$$\frac{1}{3}|6x|-1=5$$
$$\frac{1}{3}|6x|-1+1=5+1$$
$$\frac{1}{3}|6x|=6$$
$$3\cdot\frac{1}{3}|6x|=3\cdot 6$$
$$|6x|=18$$
$$6x=18 \quad \text{or} \quad 6x=-18$$
$$x=3 \qquad x=-3$$
$$\boxed{x=3,-3}$$

55.
$$f(x)=25.5$$
$$2|3(x+4)|+1.5=25.5$$
$$2|3(x+4)|+1.5-1.5=25.5-1.5$$
$$2|3(x+4)|=24$$
$$\frac{2|3(x+4)|}{2}=\frac{24}{2}$$
$$|3(x+4)|=12$$
$$|3x+12|=12$$
$$3x+12=12 \quad \text{and} \quad 3x+12=-12$$
$$3x+12-12=12-12 \quad 3x+12-12=-12-12$$
$$3x=0 \qquad 3x=-24$$
$$x=0 \qquad x=-8$$
$$\boxed{x=0,-8}$$

56.
$$f(x)<5$$
$$|7-x|<5$$
$$-5<7-x<5$$
$$-5-7<7-x-7<5-7$$
$$-12<-x<-2$$
$$\frac{-12}{-1}<\frac{-x}{-1}<\frac{-2}{-1}$$
$$12>x>2$$
$$2<x<12$$
$$(2,12)$$

57. Since $|0.04x-8.8|$ is always greater than or equal to 0 for any real number x, this absolute value inequality has no solution.

58. Since $\left|\frac{3x}{50}+\frac{1}{45}\right|$ is always greater than or equal to 0 for any real number x, this absolute value inequality is true for all real numbers, $\mathbb{R}$.

SECTION 4.4
Linear Inequalities in Two Variables

59. No.
$$6x-4y\geq 15$$
$$6(-1)-4(-4)\overset{?}{\geq}15$$
$$-6+16\overset{?}{\geq}15$$
$$10\not\geq 15$$

60. Yes. The $\geq$ sign includes the boundary line.

61. $2x+3y>6$
$$y>-\frac{2}{3}x+2$$

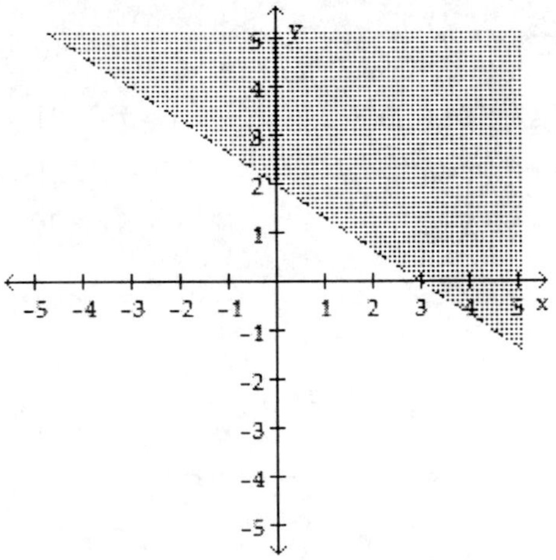

62. $y \le 4 - x$

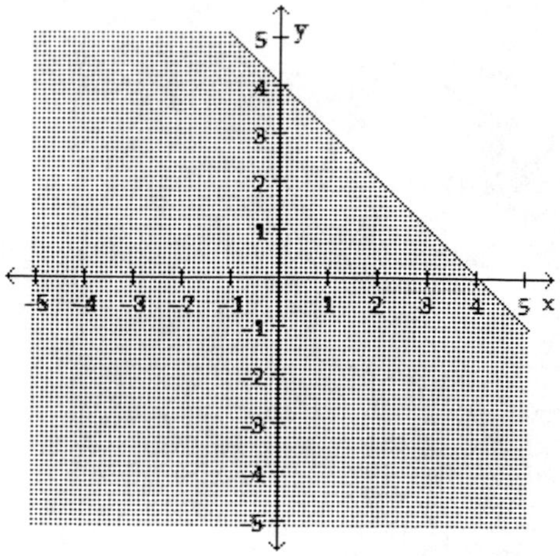

63. $y < \frac{1}{2}x$

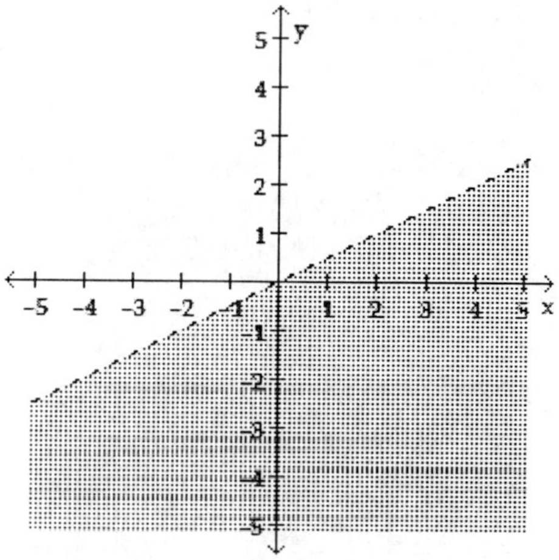

64. $x \ge -\frac{3}{2}$

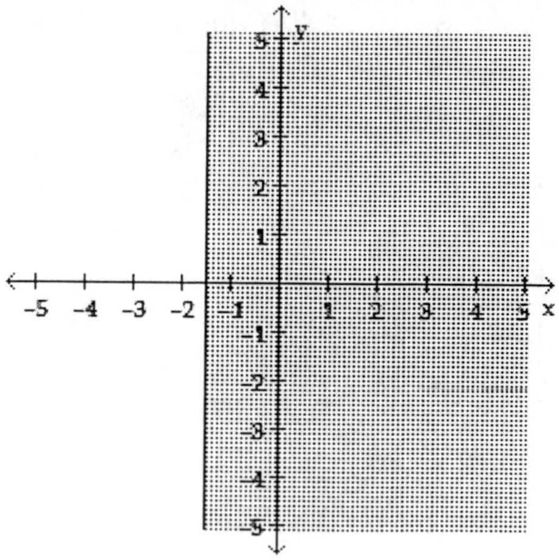

65. $m = \frac{3}{4}$, y-intercept = (0, –3)

$$y = \frac{3}{4}x - 3$$
$$4(y) = 4\left(\frac{3}{4}x - 3\right)$$
$$4y = 3x - 12$$
$$-3x + 4y = -12$$
$$3x - 4y = 12$$
$$3x - 4y > 12$$

666 $6x + 4y \ge 10{,}200$

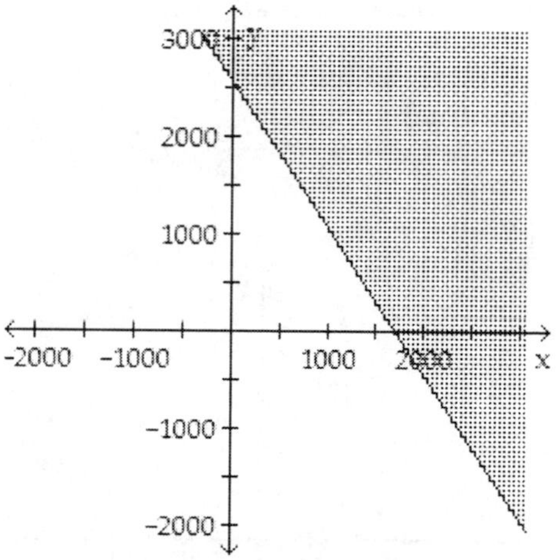

Answers will vary. Three possible answers are (1,800, 0), (1,000, 1,500), and (2,000, 2,000).

SECTION 4.5
Systems of Linear Inequalities

67. Yes.

$$y \le -x + 1$$
$$-2 \overset{?}{\le} -1 + 1$$
$$-2 \le 0$$

$$2x - y > 2$$
$$2(1) - (-2) \overset{?}{>} 2$$
$$2 + 2 \overset{?}{>} 2$$
$$4 > 2$$

68. a. True
    b. False
    c. True
    d. False
    e. True
    f. True

69. $\begin{cases} y \ge x+1 \\ 3x+2y<6 \end{cases}$

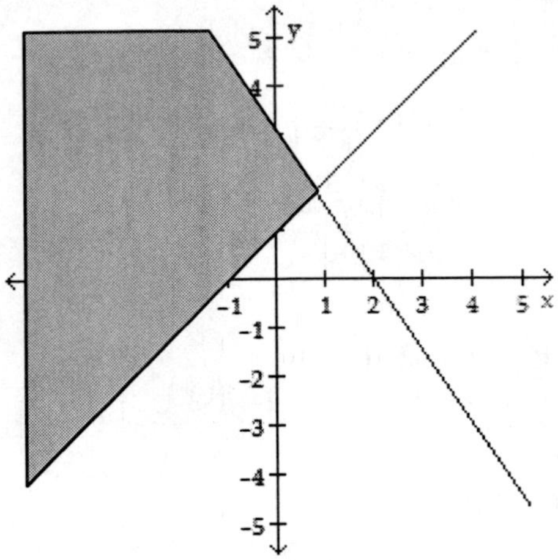

70. $\begin{cases} x-y<3 \\ y \le 0 \\ x \ge 0 \end{cases}$

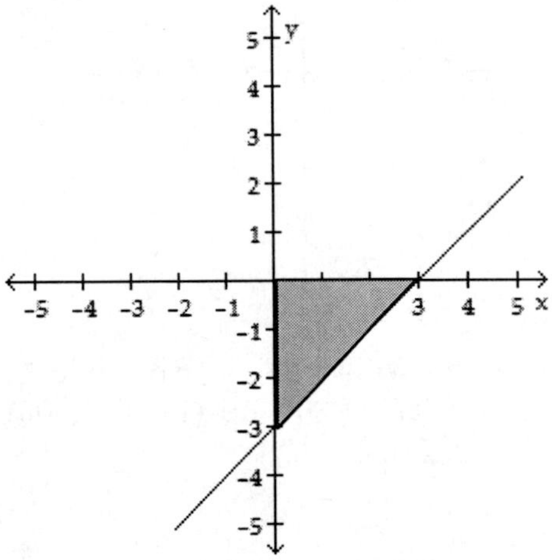

71. $-2<x<4$

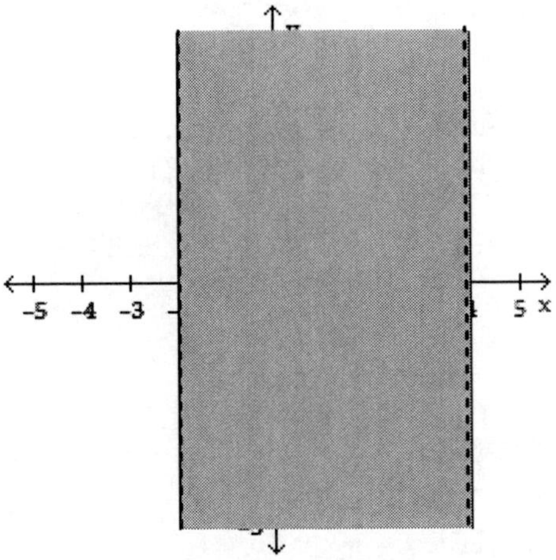

72. $y \le -2$ or $y>1$

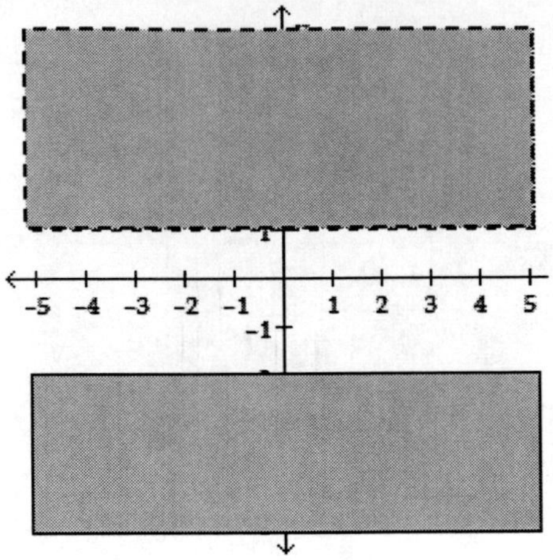

73. $\begin{cases} 20x+30y \ge 300 \\ 20x+30y \le 600 \end{cases}$

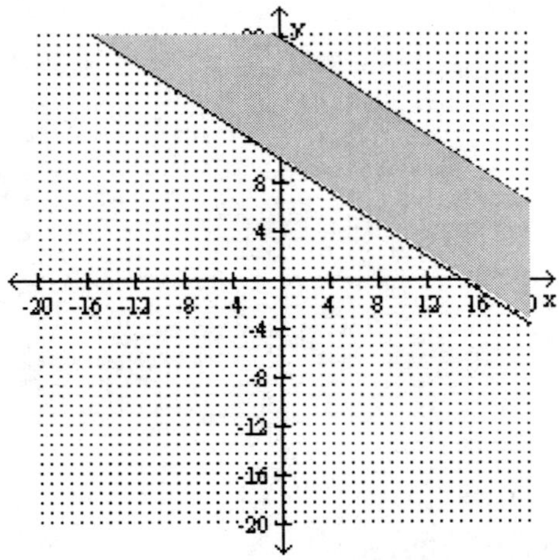

Possible solutions are 5 \$20 shirts and 15 \$30 shirts or 15 \$20 shirts and 10 \$30 shirts.

74.

$\begin{cases} x \boxed{\ge} 35 \\ x \boxed{\le} 130 \\ y \boxed{\ge} -56x+280 \\ y \boxed{\le} -18x+90 \end{cases}$

CHAPTER 4 TEST

1. a. $<, >, \leq$, and $\geq$ are **inequality** symbols.
   b. ∞ is a symbol representing **infinity**.
   c. $x + 1 > 2$ or $2x - 3 \leq 8$ is a **compound** inequality.
   d. We read $\cup$ as **union** and $\cap$as **intersection**.
   e. $\begin{cases} x + y > 10 \\ 3x - 2y \geq 4 \end{cases}$ is a **system** of linear equations in **two** variables.

2. Yes.

$$3(-2-2) \overset{?}{\leq} 2(-2+7)$$
$$3(-4) \overset{?}{\leq} 2(5)$$
$$-12 \leq 10$$

3.

$$\frac{2}{3}t - 1 > 7$$
$$3\left(\frac{2}{3}t - 1\right) > 3(7)$$
$$2t - 3 > 21$$
$$2t - 3 + 3 > 21 + 3$$
$$2t > 24$$
$$t > 12;\ (12, \infty)$$

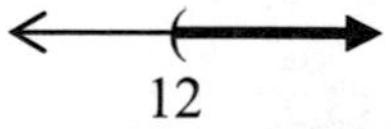

4.

$$-2(2x+3) \geq 14$$
$$-4x - 6 \geq 14$$
$$-4x - 6 + 6 \geq 14 + 6$$
$$-4x \geq 20$$
$$x \leq -5;\ (-\infty, -5]$$

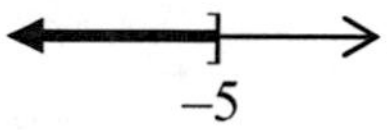

5.

$$\frac{x}{4} - \frac{1}{3} > \frac{5}{6} + \frac{x}{3}$$
$$12\left(\frac{x}{4} - \frac{1}{3}\right) > 12\left(\frac{5}{6} + \frac{x}{3}\right)$$
$$3x - 4 > 10 + 4x$$
$$3x - 4 - 3x > 10 + 4x - 3x$$
$$-4 > 10 + x$$
$$-4 - 10 > 10 + x - 10$$
$$-14 > x$$
$$x < -14;\ (-\infty, 14)$$

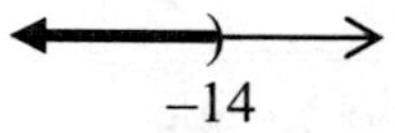

6.

$$4 - 4[3t - 2(3 - t)] \leq -15t - (5t - 28)$$
$$4 - 4[3t - 6 + 2t] \leq -15t - 5t + 28$$
$$4 - 4[5t - 6] \leq -20t + 28$$
$$4 - 20t + 24 \leq -20t + 28$$
$$28 - 20t \leq -20t + 28$$
$$28 - 20t + 20t \leq -20t + 28 + 20t$$
$$28 \leq 28$$
$$\text{true};\ (-\infty, \infty)$$

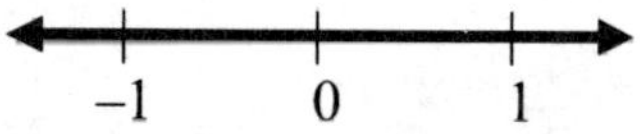

7. Let x = grade on Exam 5.

$$\frac{70 + 79 + 85 + 88 + x}{5} > 80$$
$$\frac{322 + x}{5} > 80$$
$$5\left(\frac{322 + x}{5}\right) > 5(80)$$
$$322 + x > 400$$
$$322 + x - 322 > 400 - 322$$
$$x > 78$$

She must make higher than 78.

8. Let x = number of hours working.

$$175+80x \le 1{,}000$$
$$175+80x-175 \le 1{,}000-175$$
$$80x \le 825$$
$$x \le 10.31.25$$

$x \le 10.3$ plus the first hour means that the crew can only work up to 11 hours.

9.

$$w(t) \le 6{,}550$$
$$150t+5{,}200 \le 6{,}550$$
$$150t+5{,}200-5{,}200 \le 6{,}550-5{,}200$$
$$150t \le 1{,}350$$
$$t \le 9$$

The years from 1990 to 1999.

10. Let x = number of hours worked.
Corner Bakery: $15.50x - 135$
Main Street Bakery: $12.80x$

$$15.50x-135 \ge 12.80x$$
$$15.50x-135-15.50x \ge 12.80x-15.50x$$
$$-135 \ge -2.70x$$
$$50 \le x$$
$$x \ge 50$$

50 hours or more at Corner Bakery would be more profitable than Main Street.

11. No.

$$x+6 \ge 10 \qquad 3x-8>4$$
$$4+6 \overset{?}{\ge} 10 \qquad 3(4)-8 \overset{?}{>} 4$$
$$10 \ge 10 \qquad 12-8 \overset{?}{>} 4$$
$$4 \not> 4$$

12. $\{-4, 0, 11\}$

13. $\{-5, -4, 0, 7, 8, 9, 10, 11\}$

14. a. $(-3,6)\cup[5,\infty)$

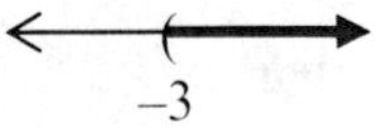

b. $[-2,7]\cap(-\infty,1)$

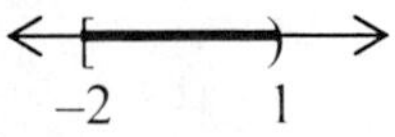

15.

$$3x \ge -2x+5 \quad \text{and} \quad 7 \ge 4x-2$$
$$3x+2x \ge -2x+5+2x \quad 7+2 \ge 4x-2+2$$
$$5x \ge 5 \qquad 9 \ge 4x$$
$$x \ge 1 \qquad \frac{9}{4} \ge x$$
$$\left[1,\frac{9}{4}\right]$$

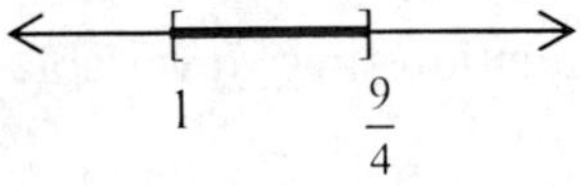

16.

$$3x<-9 \quad \text{or} \quad -\frac{x}{4}<-2$$
$$\frac{3x}{3}<\frac{-9}{3} \qquad -4\left(-\frac{x}{4}\right)>-4(-2)$$
$$x<-3 \qquad x>8$$
$$(-\infty,-3)\cup(8,\infty)$$

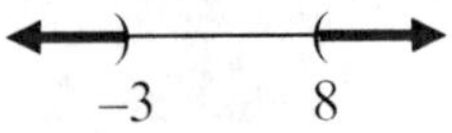

17.

$$-2<\frac{x-4}{3}<4$$
$$3(-2)<3\left(\frac{x-4}{3}\right)<3(4)$$
$$-6<x-4<12$$
$$-6+4<x-4+4<12+4$$
$$-2<x<16$$
$$(-2,16)$$

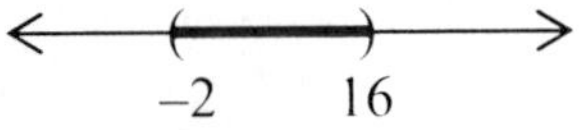

18.

$$\frac{4}{5}(x+1)>1 \quad \text{and} \quad -(0.3x+1.5)>2.9-0.2x$$

$$5\cdot\frac{4}{5}(x+1)>5\cdot 1 \qquad -0.3x-1.5>2.9-0.2x$$

$$4(x+1)>5 \qquad 10(-0.3x-1.5)>10(2.9-0.2x)$$

$$4x+4>5 \qquad -3x-15>29-2x$$

$$4x+4-4>5-4 \qquad -3x-15+15>29-2x+15$$

$$4x>1 \qquad -3x>44-2x$$

$$x>\frac{1}{4} \qquad -3x+2x>44-2x+2x$$

$$x>\frac{1}{4} \qquad -x>44$$

$$x>\frac{1}{4} \qquad x<-44$$

no solution; $\varnothing$

There are no real numbers greater than $\frac{1}{4}$ and less than -44.

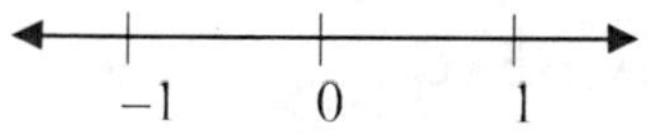

19.

$$f(x)>24.8 \quad \text{and} \quad -72.8\le f(x)$$

$$0.08x+6.48>24.8 \qquad -72.8\le 0.08x+6.48$$

$$0.08x>18.32 \qquad -79.28\le 0.08x$$

$$x>229 \qquad -991\le x$$

$$(229,\infty)$$

20. a. iii.
b. i.
c. iv.
d. ii.

21.

$$|4-3x|=19$$

$$4-3x=19 \quad \text{or} \quad 4-3x=-19$$

$$4-3x-4=19-4 \qquad 4-3x-4=-19-4$$

$$-3x=15 \qquad -3x=-23$$

$$x=-5 \qquad x=\frac{23}{3}$$

$$\boxed{x=-5,\frac{23}{3}}$$

22.

$$|3x+4|=|x+12|$$

$$3x+4=x+12 \quad \text{or} \quad 3x+4=-(x+12)$$

$$3x+4=x+12 \qquad 3x+4=-x-12$$

$$3x+4-4=x+12-4 \qquad 3x+4-4=-x-12-4$$

$$3x=x+8 \qquad 3x=-x-16$$

$$3x-x=x+8-x \qquad 3x+x=-x-16+x$$

$$2x=8 \qquad 4x=-16$$

$$x=4 \qquad x=-4$$

$$\boxed{x=4,-4}$$

23.

$$10=4\left|\frac{3x}{8}-\frac{3x}{2}\right|+6$$

$$10-6=4\left|\frac{3x}{8}-\frac{3x}{2}\right|+6-6$$

$$4=4\left|\frac{3x}{8}-\frac{3x}{2}\right|$$

$$\frac{4}{4}=\frac{4\left|\frac{3x}{8}-\frac{3x}{2}\right|}{4}$$

$$1=\left|\frac{3x}{8}-\frac{3x}{2}\right|$$

$$\frac{3x}{8}-\frac{3x}{2}=1 \quad \text{or} \quad \frac{3x}{8}-\frac{3x}{2}=-1$$

$$8\left(\frac{3x}{8}-\frac{3x}{2}\right)=8(1) \qquad 8\left(\frac{3x}{8}-\frac{3x}{2}\right)=8(-1)$$

$$3x-12x=8 \qquad 3x-12x=-8$$

$$-9x=8 \qquad -9x=-8$$

$$x=-\frac{8}{9} \qquad x=\frac{8}{9}$$

$$\boxed{x=-\frac{8}{9},\frac{8}{9}}$$

24.

$$|16x|=-16$$

No solution. Since an absolute value can never be negative, there are no real numbers x that can make the equation true.

25.

$$5|20-2x|=0$$
$$\frac{5|20-2x|}{5}=\frac{0}{5}$$
$$|20-2x|=0$$
$$20-2x=0$$
$$20-2x-20=0-20$$
$$-2x=-20$$
$$x=10$$
$$\boxed{x=10}$$

26.

$$|x-0.0625|\le 0.0015$$
$$-0.0015\le x-0.0625\le 0.0015$$
$$10000(-0.0015)\le 10000(x-0.0625)\le 10000(0.0015)$$
$$-15\le 10{,}000x-625\le 15$$
$$-15+625\le 10{,}000x-625+625\le 15+625$$
$$610\le 10{,}000x\le 640$$
$$\frac{610}{10{,}000}\le\frac{10{,}000x}{10{,}000}\le\frac{640}{10{,}000}$$
$$0.0610\le x\le 0.0640$$
$$[0.0610,\ 0.0640]$$

27.

$$|x+3|\le 4$$
$$-4\le x+3\le 4$$
$$-4-3\le x+3-3\le 4-3$$
$$-7\le x\le 1$$
$$[-7,1]$$

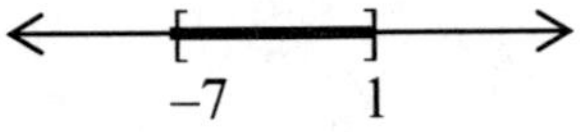

28.

$$\left|\frac{x-2}{2}\right|>5.5$$
$$\frac{x-2}{2}<-5.5 \quad \text{or} \quad \frac{x-2}{2}>5.5$$
$$2\left(\frac{x-2}{2}\right)<2(-5.5) \quad 2\left(\frac{x-2}{2}\right)>2(5.5)$$
$$x-2<-11 \qquad x-2>11$$
$$x-2+2<-11+2 \qquad x-2+2>11+2$$
$$x<-9 \qquad x>13$$
$$(-\infty,-9)\cup(13,\infty)$$

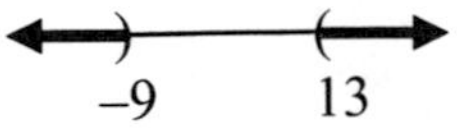

29.

$$|4-2x|+48>50$$
$$|4-2x|+48-48>50-48$$
$$|4-2x|>2$$
$$4-2x<-2 \quad \text{or} \quad 4-2x>2$$
$$4-2x-4<-2-4 \qquad 4-2x-4>2-4$$
$$-2x<-6 \qquad -2x>-2$$
$$x>3 \qquad x<1$$
$$(-\infty,1)\cup(3,\infty)$$

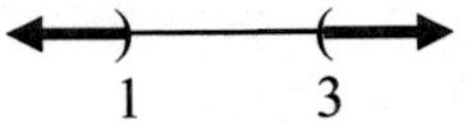

30.

$$2|3(x-2)|\le 4$$
$$2|3x-6|\le 4$$
$$\frac{2|3x-6|}{2}\le\frac{4}{2}$$
$$|3x-6|\le 2$$
$$-2\le 3x-6\le 2$$
$$-2+6\le 3x-6+6\le 2+6$$
$$4\le 3x\le 8$$
$$\frac{4}{3}\le x\le\frac{8}{3}$$
$$\left[\frac{4}{3},\frac{8}{3}\right]$$

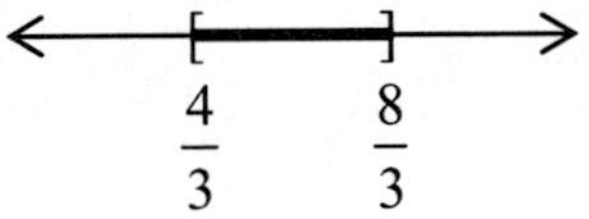

31.

$$|4.5x-0.9|\ge -0.7$$

Since $|4.5x-0.9|$ is always greater than or equal to 0 for any real number x, this absolute value inequality is true for all real numbers, $\mathbb{R}$. $(-\infty, \infty)$

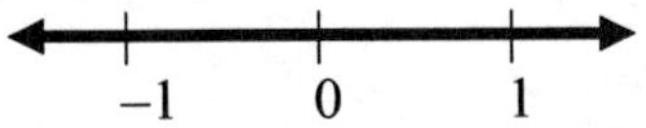

32. $-160\ge|4x|$

Since the absolute value is always positive, -160 will never be greater than or equal to $|4x|$ so there is No Solution, $\varnothing$.

33.

$$f(x) < 3$$
$$|2x + 9| < 3$$
$$-3 < 2x + 9 < 3$$
$$-3 - 9 < 2x + 9 - 9 < 3 - 9$$
$$-12 < 2x < -6$$
$$-6 < x < -3$$
$$(-6, -3)$$

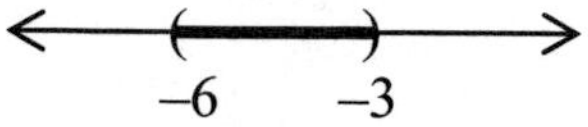

34.

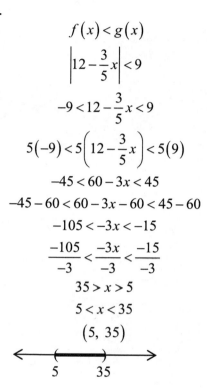

$$f(x) < g(x)$$
$$\left|12 - \frac{3}{5}x\right| < 9$$
$$-9 < 12 - \frac{3}{5}x < 9$$
$$5(-9) < 5\left(12 - \frac{3}{5}x\right) < 5(9)$$
$$-45 < 60 - 3x < 45$$
$$-45 - 60 < 60 - 3x - 60 < 45 - 60$$
$$-105 < -3x < -15$$
$$\frac{-105}{-3} < \frac{-3x}{-3} < \frac{-15}{-3}$$
$$35 > x > 5$$
$$5 < x < 35$$
$$(5,\ 35)$$

35. $5x + 3y \le 10$

$$y \le -\frac{5}{3}x + \frac{10}{3}$$

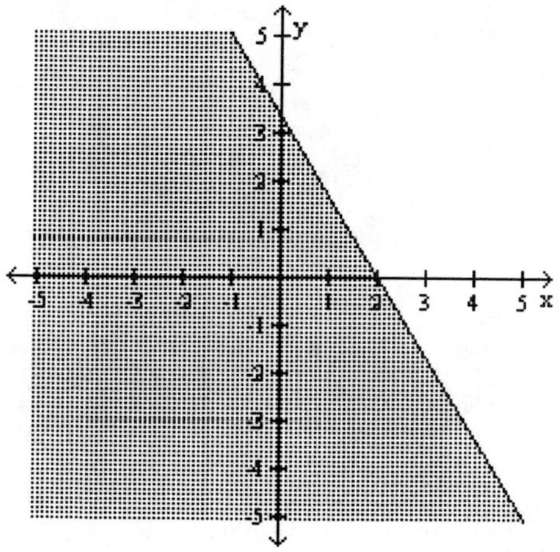

36. $y < x$

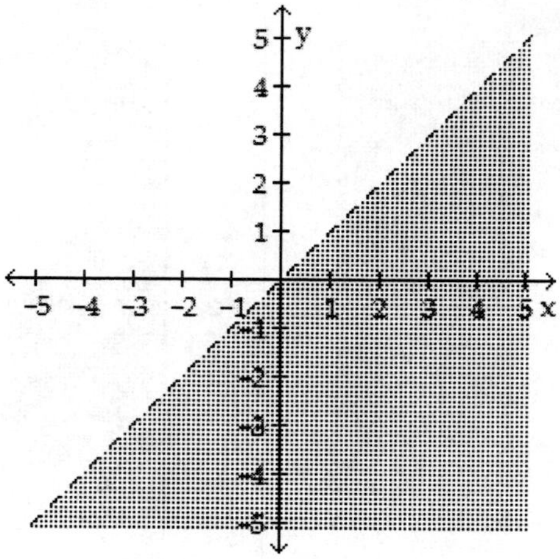

37. $\begin{cases} 2x - 3y \ge 6 \\ y < -x + 1 \end{cases}$

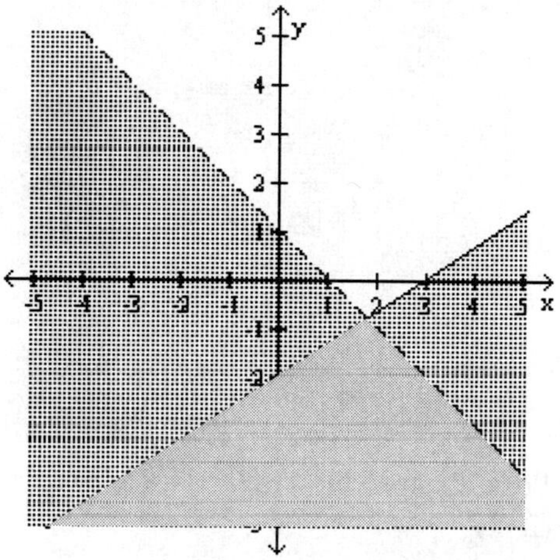

38. $-2 \le y < 5$

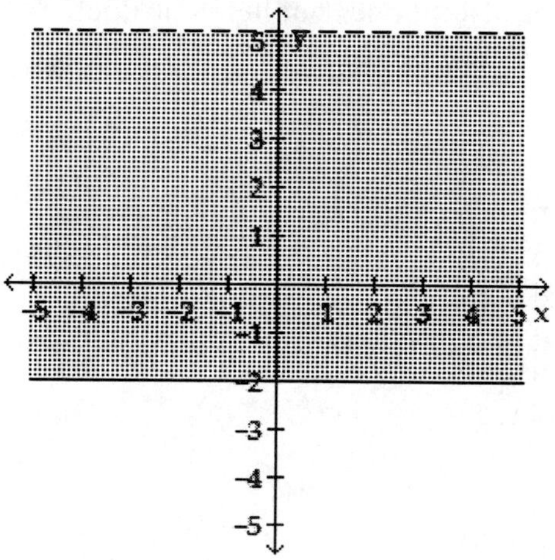

39. $x < -3$ or $x \geq 4$

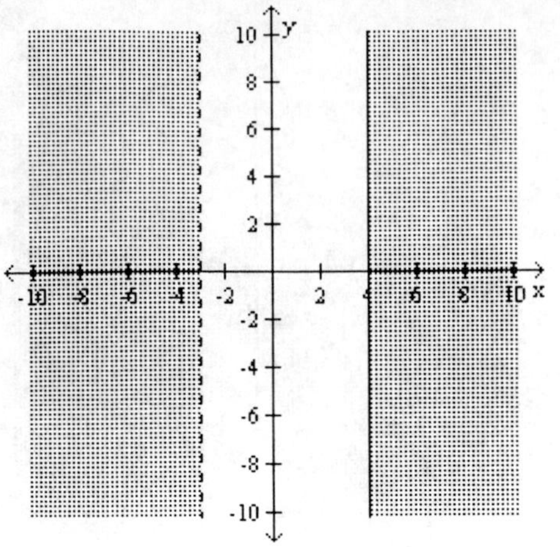

40. $x + 3y < 9$

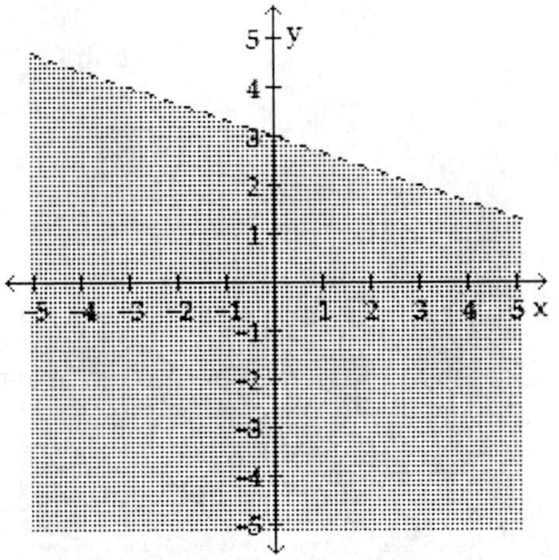

Possible answers are (1, 1), (2, 1) and (2, 2).

41. a. (3, –4) is a solution of inequality 2.
    b. No; It does not lie in the double shaded region.

42.

$$\begin{cases} y \boxed{\leq} 60 \\ y \boxed{\geq} 27 \\ y \boxed{\geq} -11x + 852 \\ y \boxed{\leq} -5x + 445 \end{cases}$$

43. a. ii
    b. i
    c. iv
    d. iii

44. No. $10 \leq x < 3$ implies that $10 \leq 3$ which is untrue.

CHAPTER 4 CUMULATIVE REVIEW

1. The set of **rational** numbers together with the set of **irrational** numbers form the set of real numbers.

2. $p = 0.125$

$$\frac{p}{2} = \frac{0.125}{2} = 0.0625$$

$$\frac{p}{4} = \frac{0.125}{4} = 0.03125$$

$$0.433p = 0.433(0.125) = 0.054125$$

3.
$$\begin{aligned}|x| - xy &= |2| - 2(-4)\\ &= 2 + 8\\ &= 10\end{aligned}$$

4.
$$\begin{aligned}\frac{x^2 - y^2}{3x + y} &= \frac{(2)^2 - (-4)^2}{3(2) + (-4)}\\ &= \frac{4-16}{6-4}\\ &= \frac{-12}{2}\\ &= -6\end{aligned}$$

5. a. $9 + x = x + 9$
   Commutative property of addition
   b. $6(10n) = (6 \cdot 10)n$
   Associative property of multiplication

6. a. $-x$
   b. $\frac{1}{x}$

7.
$$\begin{aligned}-(a+2)-(a-b) &= -a - 2 - a + b\\ &= -2a + b - 2\end{aligned}$$

8.
$$\begin{aligned}36\left(\frac{2}{9}t - \frac{3}{4}\right) + 36\left(\frac{1}{2}\right) &= 8t - 27 + 18\\ &= 8t - 9\end{aligned}$$

9.
$$\begin{aligned}6(x-1) &= 2(x+3)\\ 6x - 6 &= 2x + 6\\ 6x - 6 - 2x &= 2x + 6 - 2x\\ 4x - 6 &= 6\\ 4x - 6 + 6 &= 6 + 6\\ 4x &= 12\\ x &= 3\end{aligned}$$

10.
$$\begin{aligned}\frac{5b}{2} - 10 &= \frac{b}{3} + 3\\ 6\left(\frac{5b}{2} - 10\right) &= 6\left(\frac{b}{3} + 3\right)\\ 15b - 60 &= 2b + 18\\ 15b - 60 - 2b &= 2b + 18 - 2b\\ 13b - 60 &= 18\\ 13b - 60 + 60 &= 18 + 60\\ 13b &= 78\\ b &= 6\end{aligned}$$

11.
$$\begin{aligned}2a - 5 &= -2a + 4(a-2) + 1\\ 2a - 5 &= -2a + 4a - 8 + 1\\ 2a - 5 &= 2a - 7\\ 2a - 5 - 2a &= 2a - 7 - 2a\\ -5 &= -7\end{aligned}$$

no solution; $\varnothing$

12.
$$\begin{aligned}\frac{2z+3}{3} + \frac{3z-4}{6} &= \frac{z-2}{2}\\ 6\left(\frac{2z+3}{3} + \frac{3z-4}{6}\right) &= 6\left(\frac{z-2}{2}\right)\\ 2(2z+3) + (3z-4) &= 3(z-2)\\ 4z + 6 + 3z - 4 &= 3z - 6\\ 7z + 2 &= 3z - 6\\ 7z + 2 - 3z &= 3z - 6 - 3z\\ 4z + 2 &= -6\\ 4z + 2 - 2 &= -6 - 2\\ 4z &= -8\\ z &= -2\end{aligned}$$

13.

$$l = a + (n-1)d$$
$$l - a = a + (n-1)d - a$$
$$l - a = (n-1)d$$
$$\frac{l-a}{n-1} = \frac{(n-1)d}{n-1}$$
$$\frac{l-a}{n-1} = d$$

14.

Change $11\frac{3}{4}$ inches to feet by dividing by 12.

$$11\frac{3}{4} \div 12 = 0.9791\overline{6}$$

Area = length · width
≈ 205 · 0.979
≈ 201 ft^2

15.

$$I = Prt$$
$$1,775 = x(0.08875)(1)$$
$$1,775 = 0.08875x$$
$$\frac{1,775}{0.08875} = \frac{0.08875x}{0.08875}$$
$$\$20,000 = x$$

16.

	Rate ·	Time =	Distance
Runner 1	12	t	$12t$
Runner 2	10	t	$10t$
Total Distance Apart			0.5

Runner 1 Dis.–Runner 2 Dis =Total Dis.Apart

$$12t - 10t = 0.5$$
$$2t = 0.5$$
$$\frac{2t}{2} = \frac{0.5}{2}$$
$$t = \frac{\frac{1}{2}}{2}$$
$$t = \frac{1}{2} \div 2$$
$$t = \frac{1}{2} \bullet \frac{1}{2}$$
$$t = \frac{1}{4} \text{ hour}$$

It will take $\frac{1}{4}$ hr before they are one-quarter of a mile apart.

17.

$$m = \frac{y_2 - y_1}{x_2 - x_1}$$
$$= \frac{-8-0}{0-(-5)}$$
$$= -\frac{8}{5}$$

18. a. Find the rate of change using the points (2000, 1,316,000) and (2005, 1,448,000).

$$m = \frac{y_2 - y_1}{x_2 - x_1}$$
$$= \frac{1,448,000 - 1,316,000}{2005 - 2000}$$
$$= \frac{132,000}{5}$$
$$= 26,400 \text{ prisoners per year}$$

b. The greatest rate of change was from **1990-1995**.
Find the rate of change using the points (1990, 743,000) and (1995, 1,079,000).

$$m = \frac{y_2 - y_1}{x_2 - x_1}$$
$$= \frac{1,079,000 - 743,000}{1995 - 1990}$$
$$= \frac{336,000}{5}$$
$$= 67,200 \text{ prisoners per year}$$

19. Find the slope of both lines.

$3x = y + 4$	$y = 3(x - 4) - 1$
$3x - 4 = y$	$y = 3x - 12 - 1$
$y = 3x - 4$	$y = 3x - 13$
$m = 3$	$m = 3$

Since the slopes are the same, the lines are **parallel**.

20. Find the slope.
$3x + y = 8$
$y = -3x + 8$
$m = -3$
Perpendicular lines have slopes that are opposite reciprocals. The opposite reciprocal of –3 is $\frac{1}{3}$.

Use point-slope formula with $m = \frac{1}{3}$ and the point (–2, 3).

$$y - y_1 = m(x - x_1)$$
$$y - 3 = \frac{1}{3}(x - (-2))$$
$$y - 3 = \frac{1}{3}(x + 2)$$
$$y - 3 = \frac{1}{3}x + \frac{2}{3}$$
$$y - 3 + 3 = \frac{1}{3}x + \frac{2}{3} + 3$$
$$y = \frac{1}{3}x + \frac{11}{3}$$

21. D: {–12, –6, 5, 8}
R: {–6, 4, 6}

22. No. (1, 1) and (1, –1)

23.
$$f(-2) = 3(\mathbf{-2})^2 - (\mathbf{-2})$$
$$= 3(4) + 2$$
$$= 12 + 2$$
$$= 14$$

24.
$$f(t) = 3(t)^2 - (t)$$
$$= 3t^2 - t$$

25. a. Use points (20, 7) and (70, 4)
$$m = \frac{y_2 - y_1}{x_2 - x_1}$$
$$= \frac{7-4}{20-70}$$
$$= \frac{3}{-50}$$
$$= -0.06$$
Use $m = -0.06$ and (70, 4)
$$y - y_1 = m(x - x_1)$$
$$y - 4 = -0.06(x - 70)$$
$$y - 4 = -0.06x + 4.2$$
$$y = -0.06x + 8.2$$
$$f(x) = -0.06x + 8.2$$
b. $f(90) = -0.06(90) + 8.2$
$= -5.4 + 8.2$
$= 2.8$ L/min

26. a. 1
b. 2 and –2
c. $D: (-\infty, \infty)$ $R: [0, \infty)$

27. Not a function: (3, 4) and (3, –8)

28. It defines a function since every value of x yields only one value of y.

29. The graph of $g(x) = |x| - 2$ is the same graph as $g(x) = |x|$, but shifted 2 units down.

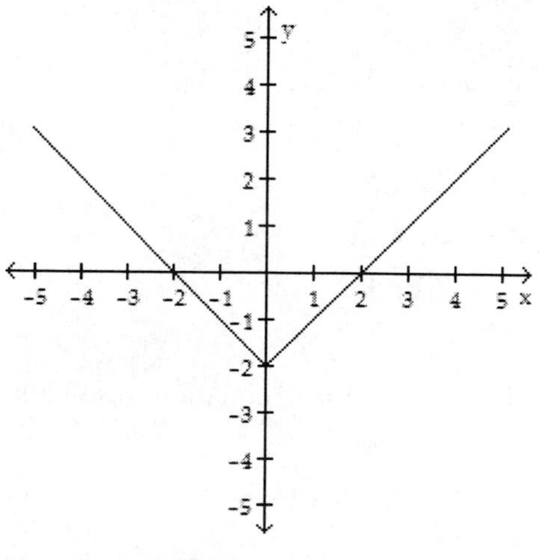

D: the set of real numbers, $\mathbb{R}$
R: the set of real numbers greater than or equal to –2

30. Yes.

31. (2, 1)

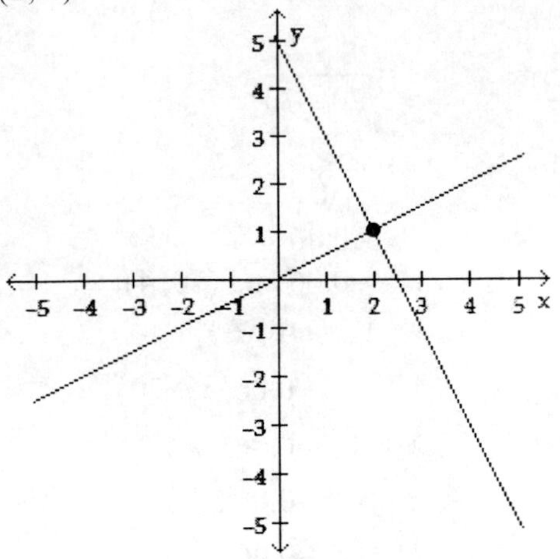

32. The point of intersection is (82, 50). In 1982, 50% of the degrees awarded went to men and 50% of the degrees went to women.

33.

$$\begin{cases} \frac{x}{10}+\frac{y}{5}=\frac{1}{2} \\ \frac{x}{2}-\frac{y}{5}=\frac{13}{10} \end{cases}$$

Multiply both equation by 10 to eliminate fractions.

$$\begin{cases} 10\left(\frac{x}{10}+\frac{y}{5}\right)=10\left(\frac{1}{2}\right) \\ 10\left(\frac{x}{2}-\frac{y}{5}\right)=10\left(\frac{13}{10}\right) \end{cases}$$

$$\begin{cases} x+2y=5 \\ 5x-2y=13 \end{cases}$$

$$\begin{aligned} x+2y&=5 \\ 5x-2y&=13 \\ \hline 6x&=18 \\ x&=3 \end{aligned}$$

Let $x=3$ in the 1st equation to solve for y.

$$\begin{aligned} 3+2y&=5 \\ 3+2y-3&=5-3 \\ 2y&=2 \\ y&=1 \end{aligned}$$

$(3,1)$

34.

$$\begin{cases} 3x=4-y \\ 4x-3y=-1+2x \end{cases}$$

Solve the first equation for y.

$$\begin{aligned} 3x-4&=4-y-4 \\ 3x-4&=-y \\ -3x+4&=y \end{aligned}$$

Substitute $-3x+4$ for y into the second equation.

$$\begin{aligned} 4x-3(-3x+4)&=-1+2x \\ 4x+9x-12&=-1+2x \\ 13x-12&=-1+2x \\ 13x-12-2x&=-1+2x-2x \\ 11x-12&=-1 \\ 11x-12+12&=-1+12 \\ 11x&=11 \\ x&=1 \end{aligned}$$

Substitute $x=1$ into the first equation.

$$\begin{aligned} -3x+4&=y \\ -3(1)+4&=y \\ -3+4&=y \\ 1&=y \end{aligned}$$

$(1,1)$

35.

$$\begin{cases} x+y+z=1 & \text{Equation (1)} \\ 2x-y-z=-4 & \text{Equation (2)} \\ x-2y+z=4 & \text{Equation (3)} \end{cases}$$

Use elimination method with (1) and (2).

$$\begin{aligned} x+y+z &= 1 \\ 2x-y-z &= -4 \\ \hline 3x &= -3 \\ x &= -1 \end{aligned}$$

Use $x=-1$ and elimination in (2) and (3).

$$\begin{aligned} -2-\ y-z &= -4 \\ -1-2y+z &= 4 \\ \hline -3-3y &= 0 \\ -3-3y+3 &= 0+3 \\ -3y &= 3 \\ y &= -1 \end{aligned}$$

Substitute $x=-1$ and $y=-1$ into (1) and solve for z.

$$\begin{aligned} -1+(-1)+z &= 1 \\ -2+z &= 1 \\ -2+z+2 &= 1+2 \\ z &= 3 \end{aligned}$$

$(-1,\ -1,\ 3)$

36.

$$\left(\begin{array}{cc|c} 4 & -3 & -1 \\ 3 & 4 & -7 \end{array}\right)$$

$R_1 - R_2$

$$\left(\begin{array}{cc|c} 1 & -7 & 6 \\ 3 & 4 & -7 \end{array}\right)$$

$-3R_1 + R_2$

$$\left(\begin{array}{cc|c} 1 & -7 & 6 \\ 0 & 25 & -25 \end{array}\right)$$

$\frac{1}{25}R_2$

$$\left(\begin{array}{cc|c} 1 & -7 & 6 \\ 0 & 1 & -1 \end{array}\right)$$

This matrix represents the system

$$\begin{cases} x-7y=6 \\ y=-1 \end{cases}$$

Use substitution.

$$\begin{aligned} x-7y &= 6 \\ x-7(-1) &= 6 \\ x+7 &= 6 \\ x &= -1 \end{aligned}$$

The solution is $(-1,-1)$.

37.

$$\begin{aligned} \begin{vmatrix} -9 & 7 \\ 4 & -2 \end{vmatrix} &= -9(-2)-7(4) \\ &= 18-28 \\ &= -10 \end{aligned}$$

38.

$$x = \frac{D_x}{D} = \frac{\begin{vmatrix} 11 & 2 \\ 9 & 6 \end{vmatrix}}{\begin{vmatrix} 5 & 2 \\ 7 & 6 \end{vmatrix}} = \frac{11(6)-2(9)}{5(6)-2(7)} = \frac{66-18}{30-14} = \frac{48}{16} = 3$$

$$y = \frac{D_y}{D} = \frac{\begin{vmatrix} 5 & 11 \\ 7 & 9 \end{vmatrix}}{\begin{vmatrix} 5 & 2 \\ 7 & 6 \end{vmatrix}} = \frac{5(9)-11(7)}{5(6)-2(7)} = \frac{45-77}{30-14} = \frac{-32}{16} = -2$$

The solution is (3, –2).

39. Let x = number of pieces of software.
Earnings = $29.95x$
Expenses = $5.45x$
Investment = 18,375

Earnings – Expenses = Investment

$$29.95x - 5.45x = 18,375$$
$$24.5x = 18,375$$
$$\frac{24.5x}{24.5} = \frac{18,375}{24.5}$$
$$x = 750$$

40. Let x = # of \$5 tickets, y = # of \$3 tickets, and z = # of \$2 tickets.

$$\begin{cases} x + y + z = 750 & (1) \\ 5x + 3y + 2z = 2,625 & (2) \\ x = 2z & (3) \end{cases}$$

Multiply Equation 1 by -3 and add equations 1 and 2 to eliminate y.

$$\begin{array}{l} -3x - 3y - 3z = -2,250 \\ \underline{5x + 3y + 2z = 2,625} \\ 2x \qquad - \; z = 375 \quad (4) \end{array}$$

Multiply Equation 3 by -2 and write it in standard form. Add Equations 3 and 4 to eliminate x.

$$\begin{array}{ll} -2x + 4z = 0 & (3) \\ \underline{2x - z = 375} & (4) \\ 3z = 375 & \\ z = 125 & \end{array}$$

Substitute $z = 125$ into Equation 4 and solve the equation for x.

$$2x - z = 375 \quad (4)$$
$$2x - (125) = 375$$
$$2x = 500$$
$$x = 250$$

Substitute $z = 125$ and $x = 250$ into Equation 1 and solve the equation for y.

$$x + y + z = 750 \quad (1)$$
$$250 + y + 125 = 750$$
$$y + 375 = 750$$
$$y = 375$$

There are 250 \$5 tickets, 375 \$3 tickets and 125 \$2 tickets.

41.

$$-3(x-4) \ge x - 32$$
$$-3x + 12 \ge x - 32$$
$$-3x + 12 - x \ge x - 32 - x$$
$$-4x + 12 \ge -32$$
$$-4x + 12 - 12 \ge -32 - 12$$
$$-4x \ge -44$$
$$\frac{-4x}{-4} \le \frac{-44}{-4}$$
$$x \le 11; \quad (-\infty, 11]$$

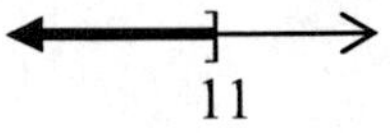

42.

$$-8 < -3x + 1 < 10$$
$$-8 - 1 < -3x + 1 - 1 < 10 - 1$$
$$-9 < -3x < 9$$
$$\frac{-9}{-3} > \frac{-3x}{-3} > \frac{9}{-3}$$
$$3 > x > -3$$
$$-3 < x < 3$$
$$(-3, 3)$$

43.

$$\begin{array}{lcl} 3x + 2 < 8 & \text{or} & 2x - 3 > 11 \\ 3x + 2 - 2 < 8 - 2 & & 2x - 3 + 3 > 11 + 3 \\ 3x < 6 & & 2x > 14 \\ x < 2 & & x > 7 \end{array}$$
$$(-\infty, 2) \cup (7, \infty)$$

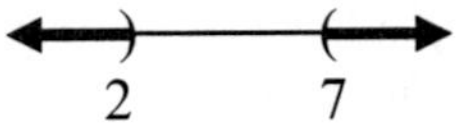

44.

$$\begin{array}{lcl} 5x - 3 \ge 2 & \text{and} & 6 \ge 4x - 3 \\ 5x - 3 + 3 \ge 2 + 3 & & 6 + 3 \ge 4x - 3 + 3 \\ 5x \ge 5 & & 9 \ge 4x \\ x \ge 1 & & \frac{9}{4} \ge x \\ & & x \le \frac{9}{4} \end{array}$$
$$\left[1, \frac{9}{4}\right]$$

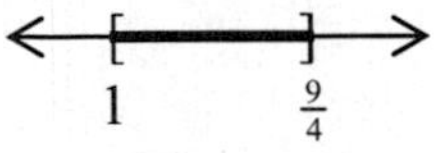

45.

$$2|4x-3|+1=19$$
$$2|4x-3|+1-1=19-1$$
$$2|4x-3|=18$$
$$\frac{2|4x-3|}{2}=\frac{18}{2}$$
$$|4x-3|=9$$
$$4x-3=9 \quad \text{or} \quad 4x-3=-9$$
$$4x-3+3=9+3 \quad 4x-3+3=-9+3$$
$$4x=12 \quad 4x=-6$$
$$x=3 \quad x=-\frac{3}{2}$$

46. $f(x)=g(x)$

$$|2x-1|=|3x+4|$$
$$2x-1=3x+4 \quad \text{or} \quad 2x-1=-(3x+4)$$
$$2x-1=3x+4 \quad 2x-1=-3x-4$$
$$2x-1-2x=3x+4-2x \quad 2x-1+3x=-3x-4+3x$$
$$-1=x+4 \quad 5x-1=-4$$
$$-1-4=x+4-4 \quad 5x-1+1=-4+1$$
$$-5=x \quad 5x=-3$$
$$x=-5 \quad x=-\frac{3}{5}$$

47.

$$|3x-2|\le 4$$
$$-4\le 3x-2\le 4$$
$$-4+2\le 3x-2+2\le 4+2$$
$$-2\le 3x\le 6$$
$$-\frac{2}{3}\le x\le 2$$
$$\left[-\frac{2}{3},2\right]$$

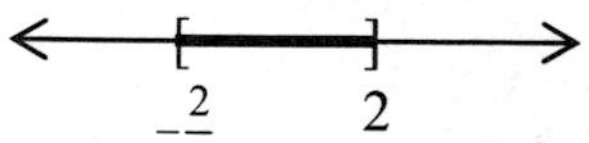

48.

$$f(x)>4$$
$$|2x+3|-1>4$$
$$|2x+3|-1+1>4+1$$
$$|2x+3|>5$$
$$2x+3<-5 \quad \text{or} \quad 2x+3>5$$
$$2x+3-3<-5-3 \quad 2x+3-3>5-3$$
$$2x<-8 \quad 2x>2$$
$$x<-4 \quad x>1$$
$$(-\infty,-4)\cup(1,\infty)$$

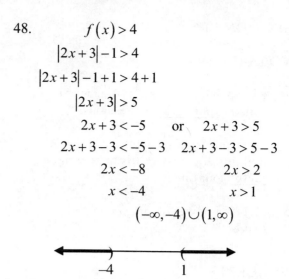

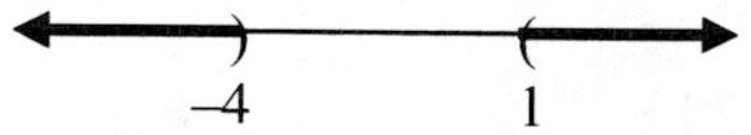

49 $2x-3y\le 12$

$$y\ge\frac{2}{3}x-4$$

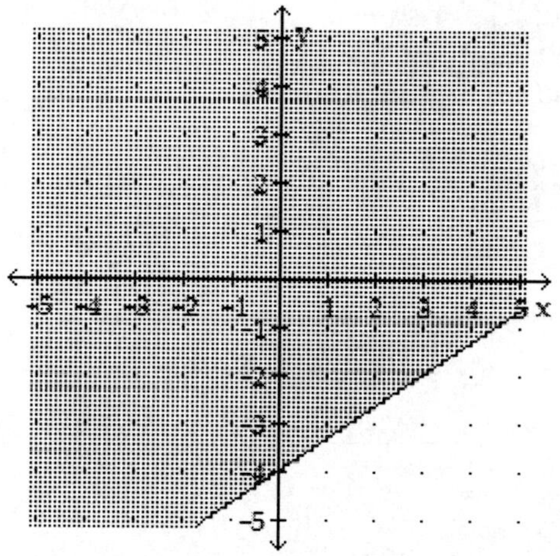

50. $\begin{cases} y<x+2 \\ 3x+y\le 6 \end{cases}$

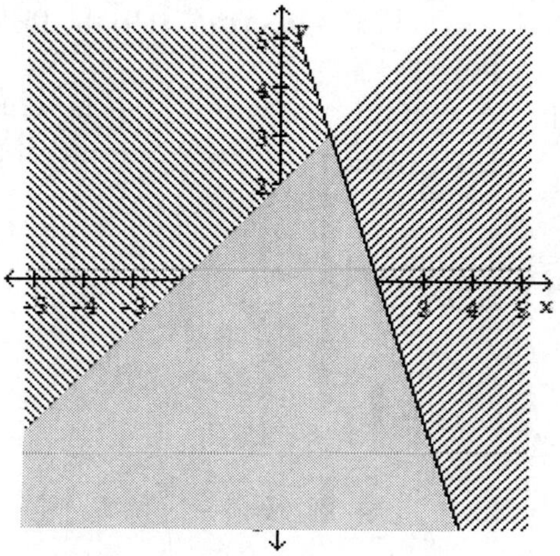

SECTION 5.1

VOCABULARY

1. Expressions such as x^4, 10^3, and $(5t)^2$ are called **exponential** expressions.

3. The expression x^4 represents a repeated multiplication where x is to be written as a **factor** four times.

5. $(h^3)^7$ is a **power** of an exponential expression.

CONCEPTS

7. a. $x^m x^n = x^{m+n}$
   b. $(x^m)^n = x^{mn}$
   c. $(xy)^n = x^n y^n$
   d. $\left(\frac{x}{y}\right)^n = \frac{x^n}{y^n}$
   e. $x^0 = 1$
   f. $x^{-n} = \frac{1}{x^n}$
   g. $\frac{x^m}{x^n} = x^{m-n}$
   h. $\left(\frac{x}{y}\right)^{-n} = \left(\frac{y}{x}\right)^n$
   i. $\frac{x^{-m}}{y^{-n}} = \frac{y^n}{x^m}$

9. To raise an exponential expression to a power, such as $\left(x^9\right)^4$, keep the base x and **multiply** the exponents.

11. a. Any nonzero base raised to the 0 power is **1**.
    b. Another way to write x^{-n} is to write its **reciprocal** and change the sign of the exponent.

NOTATION

13.
$$\frac{x^5x^4}{x^{-2}} = \frac{x^{\boxed{9}}}{x^{-2}}$$
$$= x^{9-\boxed{(-2)}}$$
$$= x^{\boxed{11}}$$

GUIDED PRACTICE

15. a. base: $6x$; exponent 3
    b. base: x; exponent 5
    c. base: b; exponent 6

17. a. base: $\frac{n}{4}$; exponent 3
    b. base: $m - 8$; exponent 6
    c. base: 10; exponent 0

19.
$$x^2x^3 = x^{2+3}$$
$$= x^5$$

21.
$$y^3y^7y^2 = y^{3+7+2}$$
$$= y^{12}$$

23.
$$-6t^8\left(t^{15}\right) = -6t^{8+15}$$
$$= -6t^{23}$$

25.
$$aba^3b^4 = a^{1+3}b^{1+4}$$
$$= a^4b^5$$

27.
$$\left(2^3\right)^2 = 2^{3\cdot 2}$$
$$= 2^6$$
$$= 64$$

29.
$$\left(x^4\right)^7 = x^{4(7)}$$
$$= x^{28}$$

31.
$$\left(r^8r^3\right)^5 = \left(r^{8+3}\right)^5$$
$$= \left(r^{11}\right)^5$$
$$= r^{55}$$

33.

$$\left(g^4\right)^5\left(g^2\right)^6 = \left(g^{4\cdot5}\right)\left(g^{2\cdot6}\right)$$
$$= g^{20}g^{12}$$
$$= g^{20+12}$$
$$= g^{32}$$

35.

$$\left(x^5y\right)^4 = x^{5\cdot4}y^4$$
$$= x^{20}y^4$$

37.

$$\left(4m^7\right)^3 = 4^3m^{7\cdot3}$$
$$= 64m^{21}$$

39.

$$\left(\frac{m^{10}}{n}\right)^8 = \frac{m^{10\cdot8}}{n^{1\cdot8}}$$
$$= \frac{m^{80}}{n^8}$$

41.

$$\left(\frac{3a^{16}}{7n^{11}}\right)^2 = \frac{3^2a^{16\cdot2}}{7^2n^{11\cdot2}}$$
$$= \frac{9a^{32}}{49n^{22}}$$

43.

$$8^0 = 1$$

45.

$$\left(-6t\right)^0 = 1$$

47.

$$60h^0 = 60(1)$$
$$= 60$$

49.

$$-3s^0t = -3(1)(t)$$
$$= -3t$$

51.

$$5^{-2} = \frac{1}{5^2}$$
$$= \frac{1}{25}$$

53.

$$(-3)^{-3} = \frac{1}{(-3)^3}$$
$$= \frac{1}{-27}$$
$$= -\frac{1}{27}$$

55.

$$8x^{-9} = 8\left(\frac{1}{x^9}\right)$$
$$= \frac{8}{x^9}$$

57.

$$-h^{-1} = -\frac{1}{h}$$

59.

$$m^{-4}m^{-6} = m^{-4+(-6)}$$
$$= m^{-10}$$
$$= \frac{1}{m^{10}}$$

61.

$$\left(s^2\right)^{-3} = s^{2\cdot-3}$$
$$= s^{-6}$$
$$= \frac{1}{s^6}$$

63.

$$\frac{1}{a^{-5}} = a^5$$

65.

$$\frac{1}{7^{-2}} = 7^2$$
$$= 49$$

67.

$$\frac{2^{-4}}{1^{-10}} = \frac{1^{10}}{2^4}$$
$$= \frac{1}{16}$$

69.

$$-\frac{t^{-6}}{12p^{-8}}=-\frac{p^8}{12t^6}$$

71.

$$\begin{aligned}\frac{m^{15}}{m^3}&=m^{15-3}\\&=m^{12}\end{aligned}$$

73.

$$\begin{aligned}\frac{33y^{-2}}{y^{10}}&=\frac{33}{y^{10}y^2}\\&=\frac{33}{y^{10+2}}\\&=\frac{33}{y^{12}}\end{aligned}$$

75.

$$\begin{aligned}\frac{t^4t^9}{t^{-1}}&=t^4t^9t^1\\&=t^{4+9+1}\\&=t^{14}\end{aligned}$$

77.

$$\begin{aligned}\frac{m^3n^2}{6mn^{11}}&=\frac{1}{6}m^{3-1}n^{2-11}\\&=\frac{1}{6}m^2n^{-9}\\&=\frac{m^2}{6n^9}\end{aligned}$$

79.

$$\begin{aligned}\left(\frac{2}{3}\right)^{-2}&=\left(\frac{3}{2}\right)^2\\&=\frac{3^2}{2^2}\\&=\frac{9}{4}\end{aligned}$$

81.

$$\begin{aligned}\left(\frac{g^{20}}{t^{30}}\right)^{-4}&=\left(\frac{t^{30}}{g^{20}}\right)^4\\&=\frac{t^{30\cdot4}}{g^{20\cdot4}}\\&=\frac{t^{120}}{g^{80}}\end{aligned}$$

TRY IT YOURSELF

83.

$$\begin{aligned}\left(2a^2a^3b^0\right)^4&=\left(2a^{2+3}b^0\right)^4\\&=\left(2a^5\cdot1\right)^4\\&=\left(2a^5\right)^4\\&=2^4a^{5(4)}\\&=16a^{20}\end{aligned}$$

85.

$$\begin{aligned}\left(\frac{4a^{-2}b}{3ab^{-3}}\right)^3&=\left(\frac{4}{3}a^{-2-1}b^{1-(-3)}\right)^3\\&=\left(\frac{4}{3}a^{-3}b^4\right)^3\\&=\left(\frac{4b^4}{3a^3}\right)^3\\&=\frac{4^3b^{4(3)}}{3^3a^{3(3)}}\\&=\frac{64b^{12}}{27a^9}\end{aligned}$$

87.

$$\begin{aligned}-\frac{8t^{-3}t^{-11}}{t^{-14}}&=-\frac{8t^{-3+(-11)}}{t^{-14}}\\&=-\frac{8t^{-14}}{t^{-14}}\\&=-8t^{-14-(-14)}\\&=-8t^0\\&=-8(1)\\&=-8\end{aligned}$$

89.

$$\begin{aligned}\left(-x^8\right)^2y^4x^3x^0&=x^{8\cdot2}y^4x^3x^0\\&=x^{16}y^4x^3x^0\\&=x^{16+3+0}y^4\\&=x^{19}y^4\end{aligned}$$

91.

$$\left(\frac{x^{-5}}{x^2}\right)^{-4}\left(\frac{x^7}{x^{-8}}\right)^3=\left(\frac{x^2}{x^{-5}}\right)^4\left(\frac{x^7}{x^{-8}}\right)^3$$
$$=\left(x^2x^5\right)^4\left(x^7x^8\right)^3$$
$$=\left(x^{2+5}\right)^4\left(x^{7+8}\right)^3$$
$$=\left(x^7\right)^4\left(x^{15}\right)^3$$
$$=\left(x^{7\cdot 4}\right)\left(x^{15\cdot 3}\right)$$
$$=x^{28}x^{45}$$
$$=x^{28+45}$$
$$=x^{73}$$

93.

$$5^2r^{-5}\left(r^6\right)^3=25r^{-5}\left(r^{6(3)}\right)$$
$$=25r^{-5}r^{18}$$
$$=25r^{-5+18}$$
$$=25r^{13}$$

95.

$$m^{-4}m^2m^{-8}=m^{-4+2+(-8)}$$
$$=m^{-10}$$
$$=\frac{1}{m^{10}}$$

97.

$$\left(\frac{4a^2b^3z^{-4}}{3a^{-2}b^{-1}z^3}\right)^{-3}=\left(\frac{4}{3}a^{2-(-2)}b^{3-(-1)}z^{-4-3}\right)^{-3}$$
$$=\left(\frac{4}{3}a^4b^4z^{-7}\right)^{-3}$$
$$=\left(\frac{4a^4b^4}{3z^7}\right)^{-3}$$
$$=\left(\frac{3z^7}{4a^4b^4}\right)^3$$
$$=\frac{3^3z^{7(3)}}{4^3a^{4(3)}b^{4(3)}}$$
$$=\frac{27z^{21}}{64a^{12}b^{12}}$$

99.

$$\frac{\left(4c^{-3}d\right)^0}{25(d+3)^0}=\frac{1}{25\cdot 1}$$
$$=\frac{1}{25}$$

101.

$$\frac{\left(3x^2\right)^{-2}}{x^3x^{-4}x^0}=\frac{3^{-2}x^{2(-2)}}{x^{3+(-4)+0}}$$
$$=\frac{x^{-4}}{9x^{-1}}$$
$$=\frac{1}{9}x^{-4-(-1)}$$
$$=\frac{1}{9}x^{-3}$$
$$=\frac{1}{9x^3}$$

103.

$$\left(\frac{3\left(d^{-1}\right)^{-5}}{8\left(d^{-4}\right)^{-2}}\right)^{-2}=\left(\frac{3d^{-1(-5)}}{8d^{-4(-2)}}\right)^{-2}$$
$$=\left(\frac{3d^5}{8d^8}\right)^{-2}$$
$$=\left(-\frac{3}{8}d^{5-8}\right)^{-2}$$
$$=\left(\frac{3}{8}d^{-3}\right)^{-2}$$
$$=\left(\frac{3}{8d^3}\right)^{-2}$$
$$=\left(\frac{8d^3}{3}\right)^2$$
$$=\frac{8^2d^{3\cdot 2}}{3^2}$$
$$=\frac{64d^6}{9}$$

105.

$$\frac{(-3cd^2)^3(c^{-1}d^{-3})^3}{(c^3d)^5} = \frac{((-3)^3c^3d^{2\cdot3})(c^{-1\cdot3}d^{-3\cdot3})}{(c^{3\cdot5}d^{1\cdot5})}$$
$$= \frac{(-27c^3d^6)(c^{-3}d^{-9})}{c^{15}d^5}$$
$$= \frac{-27c^{3+(-3)}d^{6+(-9)}}{c^{15}d^5}$$
$$= \frac{-27c^0d^{-3}}{c^{15}d^5}$$
$$= -27c^{0-15}d^{-3-5}$$
$$= -27c^{-15}d^{-8}$$
$$= -\frac{27}{c^{15}d^8}$$

LOOK ALIKES

107. a.

$$x^4 \cdot x^4 = x^{4+4}$$
$$= x^8$$

b.

$$(x^4)^4 = x^{4(4)}$$
$$= x^{16}$$

c.

$$x^4 + x^4 = 2x^4$$

109. a.

$$(-2t^3t)^5 = (-2t^{3+1})^5$$
$$= (-2t^4)^5$$
$$= (-2)^5(t^4)^5$$
$$= -32t^{4(5)}$$
$$= -32t^{20}$$

b.

$$-2t^3t^5 = -2t^{3+5}$$
$$= -2t^8$$

c.

$$(2t^3t)^{-5} = (2t^{3+1})^{-5}$$
$$= (2t^4)^{-5}$$
$$= \frac{1}{(2t^4)^5}$$
$$= \frac{1}{(2)^5(t^4)^5}$$
$$= \frac{1}{32t^{4(5)}}$$
$$= \frac{1}{32t^{20}}$$

111. a.

$$6^{-1} = \frac{1}{6}$$

b.

$$-6^{-1} = -\frac{1}{6}$$

c.

$$-(-6)^{-1} = \frac{-1}{-6}$$
$$= \frac{1}{6}$$

113. a.

$$8xy^{-2} = \frac{8x}{y^2}$$

b.

$$(8xy)^{-2} = \frac{1}{(8xy)^2}$$
$$= \frac{1}{(8)^2x^2y^2}$$
$$= \frac{1}{64x^2y^2}$$

c.

$$-8^{-2}xy = -\frac{xy}{8^2}$$
$$= -\frac{xy}{64}$$

115.

$$(3.68)^0 = 1$$
$$1 = 1$$

117.

$$\left(\frac{5.4}{2.7}\right)^{-4} = \left(\frac{2.7}{5.4}\right)^{4}$$
$$(2)^{-4} = (0.5)^{4}$$
$$0.0625 = 0.0625$$

APPLICATIONS

119. LICENSE PLATES

$$26 \cdot 26 \cdot 26 \cdot 10 \cdot 10 \cdot 10 \cdot 10 \cdot = 26^3 \cdot 10^4$$
$$= 175{,}760{,}000$$

121. BIOLOGY

Bacterium: $\dfrac{1}{1{,}000{,}000} = \dfrac{1}{10^6} = 10^{-6}$ m

Molecule: $\dfrac{1}{1{,}000{,}000{,}000} = \dfrac{1}{10^9} = 10^{-9}$ m

123. LANDSCAPE ARCHITECT

$V =$ volume of cube - volume of hole

$$= \left(3x^4\right)^3 - \left(x^4\right)^3$$
$$= 27x^{12} - x^{12}$$
$$= 26x^{12} \text{ in.}^3$$

WRITING

125. Answers will vary.

127. Answers will vary.

REVIEW

129.

$$-9x + 5 \geq 15$$
$$-9x + 5 - 5 \geq 15 - 5$$
$$-9x \geq 10$$
$$\frac{-9x}{-9} \leq \frac{10}{-9}$$
$$x \leq -\frac{10}{9}$$
$$\left(-\infty, -\frac{10}{9}\right]$$

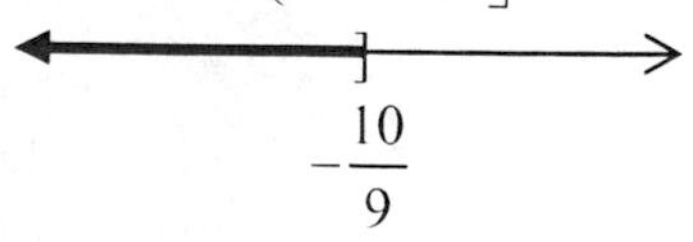

CHALLENGE PROBLEMS

131.

$$\left(2^{-1} + 3^{-1} - 4^{-1}\right)^{-1} = \left(\frac{1}{2} + \frac{1}{3} - \frac{1}{4}\right)^{-1}$$
$$= \left(\frac{1}{2} \cdot \frac{6}{6} + \frac{1}{3} \cdot \frac{4}{4} - \frac{1}{4} \cdot \frac{3}{3}\right)^{-1}$$
$$= \left(\frac{6}{12} + \frac{4}{12} - \frac{3}{12}\right)^{-1}$$
$$= \left(\frac{7}{12}\right)^{-1}$$
$$= \frac{12}{7}$$

133.

$$\frac{8^{5a}\left(8^{6a}\right)^5}{8^{-2a} \cdot 8^{a} \cdot 8^{4a}} = \frac{8^{5a}8^{6a(5)}}{8^{-2a+a+4a}}$$
$$= \frac{8^{5a}8^{30a}}{8^{3a}}$$
$$= \frac{8^{35a}}{8^{3a}}$$
$$= 8^{35a-3a}$$
$$= 8^{32a}$$

SECTION 5.2

VOCABULARY

1. 7.4×10^{10} is written in **scientific** notation and 7,400,000 is written in **standard** notation.

CONCEPTS

3. A positive number is written in scientific notation when it is written in the form $N \times$ **10^n**, where $1 \le N < 10$ and n is an **integer**.

5. a. N = 2.316; n = 54
   b. N = 1.07; n = –21

NOTATION

7. a. 60.22 is not between 1 and 10.
   b. 0.6022 is not between 1 and 10.

GUIDED PRACTICE

9. 3.9×10^3; move the decimal 3 places to the left

11. 7.8×10^{-3}; move the decimal 3 places to the right

13. 1.73×10^{14}; move the decimal 14 places to the left

15. 9.6×10^{-6}; move the decimal 6 places to the right

17. 2.03×10^{-9}; move the decimal 9 places to the right

19. 5.016×10^{16}; move the decimal 16 places to the left

21. $23.65 \times 10^6 = 2.365 \times 10^7$

23. $90.09 \times 10^{-11} = 9.009 \times 10^{-10}$

25. $0.0317 \times 10^{-2} = 3.17 \times 10^{-4}$

27. $0.0527 \times 10^5 = 5.27 \times 10^3$

29. $323 \times 10^5 = 3.23 \times 10^7$

31. $6{,}000 \times 10^{-7} = 6.0 \times 10^{-4}$

33. 27,000; move the decimal 4 places to the right

35. 0.00323; move the decimal 3 places to the left

37. 796,000,000; move the decimal 8 places to the right

39. 0.00000035; move the decimal 7 places to the left

41. 5.23; the decimal does not move

43. 80,000,000,000,000; move the decimal 13 places to the right

45.
$$\begin{aligned}(1.3\times10^4)(2.0\times10^5) &= (1.3\times2.0)(10^{4+5})\\ &= 2.6\times10^9\end{aligned}$$

47.
$$\begin{aligned}(7.9\times10^5)(2.3\times10^6) &= (7.9\times2.3)(10^{5+6})\\ &= 18.17\times10^{11}\\ &= 1.817\times10^{12}\end{aligned}$$

49.
$$\begin{aligned}(9.1\times10^{-5})(5.5\times10^{12}) &= (9.1\times5.5)(10^{-5+12})\\ &= 50.05\times10^7\\ &= 5.005\times10^8\end{aligned}$$

51.
$$\begin{aligned}(9.0\times10^{-1})(8.0\times10^{-6}) &= (9.0\times8.0)(10^{-1+(-6)})\\ &= 72.0\times10^{-7}\\ &= 7.2\times10^{-6}\end{aligned}$$

53.
$$\begin{aligned}\frac{8.6\times10^{15}}{2.0\times10^6} &= \frac{8.6}{2.0}\times10^{15-6}\\ &= 4.3\times10^9\end{aligned}$$

55.

$$\frac{2.193\times10^{32}}{4.3\times10^{20}}=\frac{2.193}{4.3}\times10^{32-20}$$
$$=0.51\times10^{12}$$
$$=5.1\times10^{11}$$

57.

$$\frac{2.686\times10^{10}}{7.9\times10^{-7}}=\frac{2.686}{7.9}\times10^{10-(-7)}$$
$$=0.34\times10^{17}$$
$$=3.4\times10^{16}$$

59.

$$\frac{4.2\times10^{-12}}{8.4\times10^{-5}}=\frac{4.2}{8.4}\times10^{-12-(-5)}$$
$$=0.5\times10^{-7}$$
$$=5.0\times10^{-8}$$

61.

$$\frac{4,500,000,000,000}{0.0002}=\frac{4.5\times10^{12}}{2\times10^{-4}}$$
$$=\frac{4.5}{2}\times10^{12-(-4)}$$
$$=2.25\times10^{16}$$
$$=22,500,000,000,000,000$$

63.

$$\frac{0.00000128}{0.0004}=\frac{1.28\times10^{-6}}{4.0\times10^{-4}}$$
$$=\frac{1.28}{4.0}\times10^{-6-(-4)}$$
$$=0.32\times10^{-2}$$
$$=3.2\times10^{-3}$$
$$=0.0032$$

65.

$$(89,000,000,000)(4,500,000,000)$$
$$=(8.9\times10^{10})(4.5\times10^{9})$$
$$=40.05\times10^{10+9}$$
$$=40.05\times10^{19}$$
$$=4.005\times10^{20}$$
$$=400,500,000,000,000,000,000$$

67.

$$\frac{(640,000)(2,700,000)}{120,000}=\frac{(6.4\times10^{5})(2.7\times10^{6})}{1.2\times10^{5}}$$
$$=\frac{(6.4\times2.7)(10^{5+6})}{1.2\times10^{5}}$$
$$=\frac{17.28\times10^{11}}{1.2\times10^{5}}$$
$$=\frac{17.28}{1.2}\times10^{11-5}$$
$$=14.4\times10^{6}$$
$$=1.44\times10^{7}$$
$$=14,400,000$$

69.

$$\frac{(15,000,000)(7,000,000,000)}{25,000,000}=\frac{(1.5\times10^{7})(7.0\times10^{9})}{2.5\times10^{7}}$$
$$=\frac{(1.5\times7.0)(10^{7+9})}{2.5\times10^{7}}$$
$$=\frac{10.5\times10^{16}}{2.5\times10^{7}}$$
$$=\frac{10.5}{2.5}\times10^{16-7}$$
$$=4.2\times10^{9}$$
$$=4,200,000,000$$

71.

$$\frac{(0.0000000039)(0.00095)}{(0.0195)(4,000)}=\frac{(3.9\times10^{-9})(9.5\times10^{-4})}{(1.95\times10^{-2})(4.0\times10^{3})}$$
$$=\frac{(3.9\times9.5)(10^{-9+(-4)})}{(1.95\times4.0)(10^{-2+3})}$$
$$=\frac{37.05\times10^{-13}}{7.8\times10^{1}}$$
$$=\frac{37.05}{7.8}\times10^{-13-1}$$
$$=4.75\times10^{-14}$$
$$=0.0000000000000475$$

APPLICATIONS

73. FIVE-CARD POKER
2,600,000 TO 1

75. ATOMS

$1{,}000{,}000{,}000{,}000{,}000{,}000{,}000 = 1.0\times10^{21}$

77. NATIONAL DEBT

$$\begin{aligned}\frac{\$13{,}950{,}000{,}000{,}000}{310{,}000{,}000} &= \frac{1.395\times10^{13}}{3.1\times10^{8}}\\ &= \frac{1.395}{3.1}\times10^{13-8}\\ &= 0.45\times10^{5}\\ &= 4.5\times10^{4}\\ &= \$45{,}000\end{aligned}$$

79. ATOMS

$$\begin{aligned}m &= \frac{1}{2{,}000} \text{ of } 1.7\times10^{-24}\\ &= \frac{1}{2{,}000}\left(1.7\times10^{-24}\right)\\ &= \frac{1.7\times10^{-24}}{2\times10^{3}}\\ &= \frac{1.7}{2}\times10^{-24-3}\\ &= 0.85\times10^{-27}\\ &= 8.5\times10^{-28}\text{ g}\end{aligned}$$

81. LIGHT YEAR

Find the number of seconds in a year.

$$\begin{aligned}1\text{ min} &= 60\text{ sec}\\ 60\text{ min} &= 1\text{ hour}\\ 24\text{ hours} &= 1\text{ day}\\ 365\text{ days} &= 1\text{ year}\end{aligned}$$

$$(60)(60)(24)(365) = 31{,}536{,}000 \text{ sec per year}$$

Find the number of meters in one light year.

$$\begin{aligned}31{,}536{,}000 &= 3.1536\times10^{7}\\ 300{,}000{,}000 &= 3\times10^{8}\\ \left(3.1536\times10^{7}\right)\left(3\times10^{8}\right) &= (3.1536\times3)\left(10^{7+8}\right)\\ &= 9.4608\times10^{15}\\ &\approx 9.5\times10^{15}\text{ m}\end{aligned}$$

83. THE BIG DIPPER

a.

$$\begin{aligned}t &= \frac{d}{r}\\ &= \frac{4.65\times10^{14}}{1.86\times10^{5}}\\ &= \frac{4.65}{1.86}\times10^{14-5}\\ &= 2.5\times10^{9}\\ &= 2{,}500{,}000{,}000\text{ sec}\end{aligned}$$

b.

$$\begin{aligned}\frac{2.5\times10^{9}\text{ sec}}{1}\cdot\frac{1\text{min}}{60\text{sec}}\cdot\frac{1\text{ hr}}{60\text{ min}}\cdot\frac{1\text{ day}}{24\text{ hr}}\cdot\frac{1\text{ year}}{365\text{ day}} &= \frac{2.5\times10^{9}}{3.1536\times10^{7}}\\ &= 0.7927\times10^{2}\\ &\approx 79\text{ years}\end{aligned}$$

85. COMETS

$$\begin{aligned}1.3\left(9.3\times10^{7}\right) &= (1.3\times9.3)\left(10^{7}\right)\\ &= 12.09\times10^{7}\\ &= 1.209\times10^{8}\text{ mi}\end{aligned}$$

WRITING

87. Answers will vary.

89. Answers will vary.

91. Answers will vary.

REVIEW

93.

$$\begin{array}{lcl} 4x \geq -x+5 & \text{and} & 6 \geq 4x-3\\ 4x+x \geq -x+5+x & & 6+3 \geq 4x-3+3\\ 5x \geq 5 & & 9 \geq 4x\\ x \geq 1 & & \frac{9}{4} \geq x \end{array}$$

$$1 \leq x \leq \frac{9}{4};\quad \left[1, \frac{9}{4}\right]$$

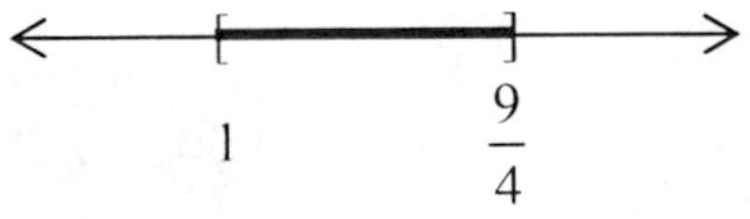

95.

$$
\begin{aligned}
3x+2<8 \quad &\text{or} \quad 2x-3>11\\
3x+2-2<8-2 \quad &\quad 2x-3+3>11+3\\
3x<6 \quad &\quad 2x>14\\
x<2 \quad &\quad x>7
\end{aligned}
$$

$$(-\infty,2)\cup(7,\infty)$$

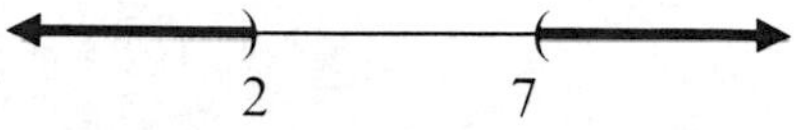

CHALLENGE PROBLEMS

97.

$$
\begin{aligned}
-\frac{1}{2.5\times 10^{-24}} &= -\frac{1}{2.5}\times 10^{24}\\
&= -0.4\times 10^{24}\\
&= -4.0\times 10^{23}
\end{aligned}
$$

SECTION 5.3

VOCABULARY

1. A **polynomial** is the sum of one or more algebraic terms whose variables have whole–number exponents.

3. For the polynomial $7x^2 - 5x - 12$, the **leading** term is $7x^2$, and the leading **coefficient** is 7. The **constant** term is –12.

5 The **degree** of the term $6x^5$ is 5 because x appears as a factor 5 times: $6 \cdot x \cdot x \cdot x \cdot x \cdot x$.

7. Terms having the same variables with the same exponents are called **like** terms.

CONCEPTS

9. a. no
   b. yes
   c. no
   d. yes

11. a. $-15x^{\boxed{6}}$
    b. $\frac{9}{8}ab^{\boxed{5}}$

13. a. $f(1) = 15$
    b. $f(-3) = 5$
    c. $x = 0, -2, -4$
    d. $D = (-\infty, \infty)$; $R = (-\infty, \infty)$

15. a. $g(1) = 2$
    b. $g(-4) = -5$
    c. $x = -2, 2$
    d. $D: = (-\infty, \infty)$; $R = (-\infty, 4]$

NOTATION

17.

$$h(t) = -t^3 - t^2 + 2t + 1$$
$$h(\boxed{3}) = -(\boxed{3})^3 - (\boxed{3})^2 + 2(3) + 1$$
$$= \boxed{-27} - 9 + 6 + 1$$
$$= \boxed{-29}$$

GUIDED PRACTICE

19. Terms: $25x^2, -x, 4$
    Coefficients: 25, –1, 4
    Degrees: 2, 1, 0
    Degree of the polynomial: 2
    Type of polynomial: Trinomial

21. Terms: $5a^7b^4, -33a^2b$
    Coefficients: 5, –33
    Degrees: 11, 3
    Degree of the polynomial: 11
    Type of polynomial: binomial

23. $f(x) = 5x^4 - 2x^2 + 9x + 5$

a.

$$f(-1) = 5(-1)^4 - 2(-1)^2 + 9(-1) + 5$$
$$= 5(1) - 2(1) + 9(-1) + 5$$
$$= 5 - 2 - 9 + 5$$
$$= -1$$

b.

$$f(2) = 5(2)^4 - 2(2)^2 + 9(2) + 5$$
$$= 5(16) - 2(4) + 9(2) + 5$$
$$= 80 - 8 + 18 + 5$$
$$= 95$$

25. $h(t) = \frac{1}{4}t^2 - \frac{5}{8}t$

a.

$$h(-4) = \frac{1}{4}(-4)^2 - \frac{5}{8}(-4)$$
$$= \frac{1}{4}(16) - \frac{5}{8}(-4)$$
$$= 4 + 2.5$$
$$= 6.5$$

b.

$$h(8) = \frac{1}{4}(8)^2 - \frac{5}{8}(8)$$
$$= \frac{1}{4}(64) - \frac{5}{8}(8)$$
$$= 16 - 5$$
$$= 11$$

27. $f(x) = 2.75x^2 - 4.7x + 1.5$

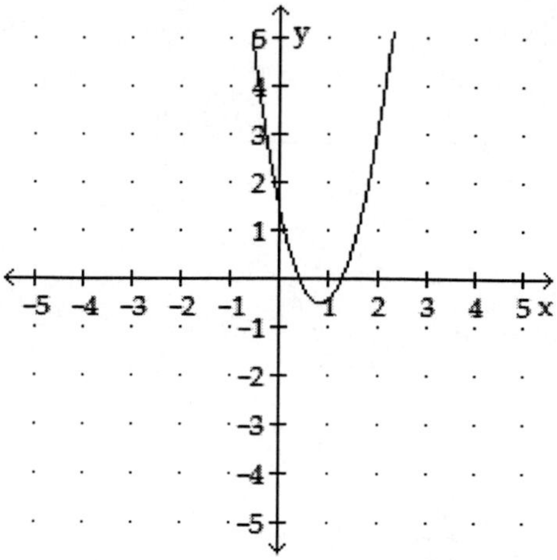

29. $15x^2 + 4x - 5x + 5x^2 + 9 = 20x^2 - x + 9$

31. $-7y^2 - 11y^3 - 4 + y^3 - y^2 = -10y^3 - 8y^2 - 4$

33. $1.4ab^2 - 4.8ab - 0.2a + 5.4ab + 3.7ab^2$
$= 5.1ab^2 + 0.6ab - 0.2a$

35. $\frac{9}{4}rst^2 - \frac{5}{3}rst - \frac{1}{2}rst^2 + \frac{5}{6}rst$
$= \frac{7}{4}rst^2 - \frac{5}{6}rst$

37. $(3x^2 + 2x + 1) + (2x^2 - 7x + 5)$
$= 5x^2 - 5x + 6$

39. $(9p^2q^2 + p - q) + (-p^2q^2 - p - q + 8)$
$= 8p^2q^2 - 2q + 8$

41. $\left(\frac{1}{5}h^6 + \frac{3}{4}h^2\right) + \left(\frac{2}{3}h^6 - \frac{1}{12}h^2\right)$
$= \frac{13}{15}h^6 + \frac{2}{3}h^2$

43.

$$\begin{array}{r} 6a^4 + 9a^2 + a \\ +\ \underline{2a^4 - 13a^2 + a} \\ 8a^4 - 4a^2 + 2a \end{array}$$

45. $(6x^3 + 3x - 2) - (2x^3 + 3x^2 + 5)$
$= 6x^3 + 3x - 2 - 2x^3 - 3x^2 - 5$
$= 4x^3 - 3x^2 + 3x - 7$

47. $(2m^2n^2 + 2m - n) - (-2m^2n^2 - 2m + n)$
$= 2m^2n^2 + 2m - n + 2m^2n^2 + 2m - n$
$= 4m^2n^2 + 4m - 2n$

49. $(7.1y^3 + 4.9y^2 + 0.1y) - (-8.4y^3 - 0.1y)$
$= 7.1y^3 + 4.9y^2 + 0.1y + 8.4y^3 + 0.1y$
$= 15.5y^3 + 4.9y^2 + 0.2y$

51. $\left(\frac{5}{2}w^3 + \frac{1}{4}w^2 + \frac{3}{5}\right) - \left(\frac{1}{3}w^3 + \frac{1}{2}w^2 - \frac{1}{5}\right)$
$= \frac{5}{2}w^3 + \frac{1}{4}w^2 + \frac{3}{5} - \frac{1}{3}w^3 - \frac{1}{2}w^2 + \frac{1}{5}$
$= \frac{13}{6}w^3 - \frac{1}{4}w^2 + \frac{4}{5}$

53.

$$\begin{array}{rr} 3x^2 - 4x + 17 & 3x^2 - 4x + 17 \\ -\ \underline{(2x^2 + 4x - 5)} & +\ \underline{-2x^2 - 4x + 5} \\ & x^2 - 8x + 22 \end{array}$$

55.

$$\begin{array}{rr} 4x^3 \quad +6a & 4x^3 \quad +6a \\ -\ \underline{(4x^3 - 2x^2 - a)} & +\ \underline{-4x^3 + 2x^2 + a} \\ & 2x^2 + 7a \end{array}$$

57.
$\left[\left(-2x^2y^3 - xy^2 + 7x^2\right) + \left(5x^2y^3 + 3xy^2 - x^2\right)\right]$
$-\left(3x^2y^3 + 4xy^2 - 3x^2\right)$
$= \left[3x^2y^3 + 2xy^2 + 6x^2\right] - \left(3x^2y^3 + 4xy^2 - 3x^2\right)$
$= 3x^2y^3 + 2xy^2 + 6x^2 - 3x^2y^3 - 4xy^2 + 3x^2$
$= -2xy^2 + 9x^2$

59.
$\left[\left(2x^2 - 4x + 3\right) - \left(8x^2 + 5x - 3\right)\right] + \left(-2x^2 + 7x - 4\right)$
$= \left[2x^2 - 4x + 3 - 8x^2 - 5x + 3\right] + \left(-2x^2 + 7x - 4\right)$
$= \left[-6x^2 - 9x + 6\right] + \left(-2x^2 + 7x - 4\right)$
$= -8x^2 - 2x + 2$

61. $(2.8b^2 + 1.2bc + 4.2c^2) + (5.1b^2 - 7.6bc - 3.9c^2)$
$= 2.8b^2 + 1.2bc + 4.2c^2 + 5.1b^2 - 7.6bc - 3.9c^2$
$= 7.9b^2 - 6.4bc + 0.3c^2$

63. $(3x^2 + 4x - 3) + (2x^2 - 3x - 1) - (x^2 + x + 7)$
$= 3x^2 + 4x - 3 + 2x^2 - 3x - 1 - x^2 - x - 7$
$= 4x^2 - 11$

65.

$$\begin{array}{r} 3x^3-2x^2+4x-3 \\ + \quad -2x^3+3x^2+3x-2 \\ \underline{5x^3-7x^2+7x-12} \\ 6x^3-6x^2+14x-17 \end{array}$$

67.
$[(3ay^3+ay^2)+(-ay^3+6ay-3a)]-(ay^3-2ay^2+2a)$
$= [2ay^3+ay^2+6ay-3a]-(ay^3-2ay^2+2a)$
$= 2ay^3+ay^2+6ay-3a-ay^3+2ay^2-2a$
$= ay^3+3ay^2+6ay-5a$

69. $(0.2xy^7+0.8xy^5)-(0.5xy^7-0.6xy^5+0.2xy)$
$= 0.2xy^7+0.8xy^5-0.5xy^7+0.6xy^5-0.2xy$
$= -0.3xy^7+1.4xy^5-0.2xy$

71.

$$\begin{array}{r} -5y^3+4y^2-11y+3 \\ \underline{-(-2y^3-14y^2+17y-32)} \end{array}$$

$$\begin{array}{r} -5y^3+4y^2-11y+3 \\ \underline{2y^3+14y^2-17y+32} \\ -3y^3+18y^2-28y+35 \end{array}$$

73. $(1-2x-x^2+4x^3)+(x^3-5x^2+x+8)$
$= 1-2x-x^2+4x^3+x^3-5x^2+x+8$
$= 5x^3-6x^2-x+9$

75. $(6k^4+2k^2-16)-(k^3-5k)$
$= 6k^4+2k^2-16-k^3+5k$
$= 6k^4-k^3+2k^2+5k-16$

77. $f(x)-g(x)=(x^2+2x)-(4x^2-2x-1)$
$=x^2+2x-4x^2+2x+1$
$=-3x^2+4x+1$

79. $f(x)+g(x)$

$$\left(\frac{1}{3}x^6-\frac{1}{6}x^4-\frac{4}{3}x^2\right)+\left(-\frac{1}{6}x^6-\frac{1}{2}x^4+\frac{5}{6}x^2\right)$$
$$=\left(\frac{1}{3}-\frac{1}{6}\right)x^6+\left(-\frac{1}{6}-\frac{1}{2}\right)x^4+\left(-\frac{4}{3}+\frac{5}{6}\right)x^2$$
$$=\left(\frac{2}{6}-\frac{1}{6}\right)x^6+\left(-\frac{1}{6}-\frac{3}{6}\right)x^4+\left(-\frac{8}{6}+\frac{5}{6}\right)x^2$$
$$=\frac{1}{6}x^6-\frac{4}{6}x^4-\frac{3}{6}x^2$$
$$=\frac{1}{6}x^6-\frac{2}{3}x^4-\frac{1}{2}x^2$$

LOOK ALIKES

81. a.

$$\begin{array}{r} 3d^3-4d^2-3d+5 \\ \underline{+11d^3 \qquad\quad -8d-2} \\ 14d^3-4d^2-11d+3 \end{array}$$

b.

$$\begin{array}{r} 3d^3-4d^2-3d+5 \\ \underline{-(11d^3 \qquad\quad -8d-2)} \end{array}$$

$$\begin{array}{r} 3d^3-4d^2-3d+5 \\ \underline{-11d^3 \qquad\quad +8d+2} \\ -8d^3-4d^2+5d+7 \end{array}$$

83. a. $(-8m^2-3m)+(-11m^2+6m+10)$
$=-8m^2-3m-11m^2+6m+10$
$=-19m^2+3m+10$

b. $(-8m^2-3m)-(-11m^2+6m+10)$
$=-8m^2-3m+11m^2-6m-10$
$=3m^2-9m-10$

APPLICATIONS

85. JUGGLING
$f(t)=-16t^2+22t+4$
$f(1)=-16(1)^2+22(1)+4$
$=-16(1)+22(1)+4$
$=-16+22+4$
$=10$ ft

87. STORAGE TANKS
$V(r)=4.2r^3+37.7r^2$
$V(4)=4.2(4)^3+37.7(4)^2$
$=4.2(64)+37.7(16)$
$=268.8+603.2$
$=872$ ft^3

89. PACKAGING

$$f(x) = 4x^3 - 44x^2 + 120x$$

$$\begin{aligned} f(3) &= 4(3)^3 - 44(3)^2 + 120(3) \\ &= 4(27) - 44(9) + 120(3) \\ &= 108 - 396 + 360 \\ &= 72 \text{ in.}^3 \end{aligned}$$

91. LANDSCAPE ARCHITECT

Pillar vol. = Cone vol. + Cylinder vol.

$$\begin{aligned} \text{Volume} &= \frac{1}{3}\pi r^2 h + \pi r^2 h \\ &= \left(\frac{1}{3} + 1\right)\pi r^2 h \\ &= \left(\frac{1}{3} + \frac{3}{3}\right)\pi r^2 h \\ &= \frac{4}{3}\pi r^2 h \end{aligned}$$

93. LABOR STATISTICS

a. $J(9) = 12$. In 2009, there were about 12 million manufacturing jobs in the U.S.

b. $x = 3$. In 2003, there were about 14.5 million manufacturing jobs in the U.S.

95. CALCULUS

a. 5 terms

b. degree of 4

c. $\frac{1}{6}$

d. $f(1) = 1 + (1) + \frac{(1)^2}{2} + \frac{(1)^3}{6} + \frac{(1)^4}{24}$

$$\begin{aligned} &= 1 + 1 + \frac{1}{2} + \frac{1}{6} + \frac{1}{24} \\ &= \frac{24}{24} + \frac{24}{24} + \frac{12}{24} + \frac{4}{24} + \frac{1}{24} \\ &= \frac{65}{24} \end{aligned}$$

WRITING

97. Answers will vary.

99. Answers will vary.

101. Answers will vary.

REVIEW

103.
$$\begin{aligned} |x| &\le 5 \\ -5 \le x &\le 5 \\ &[-5, 5] \end{aligned}$$

105.
$$\begin{aligned} |x - 4| &< 5 \\ -5 < x - 4 &< 5 \\ -5 + 4 < x - 4 + 4 &< 5 + 4 \\ -1 < x &< 9 \\ &(-1, 9) \end{aligned}$$

CHALLENGE PROBLEMS

107.
$$\begin{aligned} &(5x^3y - 5xy + 2x) - (8x^3y - 7xy + 11x) \\ &= 5x^3y - 5xy + 2x - 8x^3y + 7xy - 11x \\ &= -3x^3y + 2xy - 9x \end{aligned}$$

109. $f(x) = 2x^3 - 3x^2 - 11x + 6$

x	$f(x)$
-3	$2(-3)^3 - 3(-3)^2 - 11(-3) + 6$ $= 2(-27) - 3(9) + 33 + 6$ $= -42$
-2	$2(-2)^3 - 3(-2)^2 - 11(-2) + 6$ $= 2(-8) - 3(4) + 22 + 6$ $= 0$
-1	$2(-1)^3 - 3(-1)^2 - 11(-1) + 6$ $= 2(-1) - 3(1) + 11 + 6$ $= 12$
0	$2(0)^3 - 3(0)^2 - 11(0) + 6$ $= 2(0) - 3(0) - 0 + 6$ $= 6$
1	$2(1)^3 - 3(1)^2 - 11(1) + 6$ $= 2(1) - 3(1) - 11 + 6$ $= -6$
2	$2(2)^3 - 3(2)^2 - 11(2) + 6$ $= 2(8) - 3(4) - 22 + 6$ $= -12$
3	$2(3)^3 - 3(3)^2 - 11(3) + 6$ $= 2(27) - 3(9) - 33 + 6$ $= 0$
4	$2(4)^3 - 3(4)^2 - 11(4) + 6$ $= 2(64) - 3(16) - 44 + 6$ $= 42$

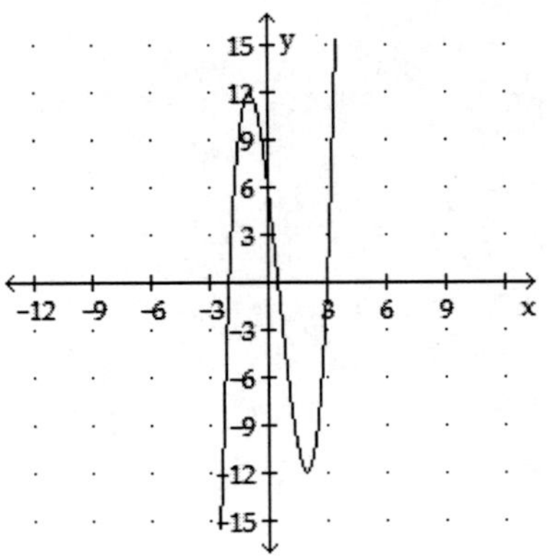

D: $(-\infty, \infty)$; R: $(-\infty, \infty)$

111. $f(x) = 2x^2 - 4x + 2$

x	$f(x)$
-1	$2(-1)^2 - 4(-1) + 2 = 2(1) + 4 + 2$ $= 2 + 4 + 2$ $= 8$
0	$2(0)^2 - 4(0) + 2 = 2(0) - 0 + 2$ $= 0 - 0 + 2$ $= 2$
1	$2(1)^2 - 4(1) + 2 = 2(1) - 4 + 2$ $= 2 - 4 + 2$ $= 0$
2	$2(2)^2 - 4(2) + 2 = 2(4) - 8 + 2$ $= 8 - 8 + 2$ $= 2$
3	$2(3)^2 - 4(3) + 2 = 2(9) - 12 + 2$ $= 18 - 12 + 2$ $= 8$

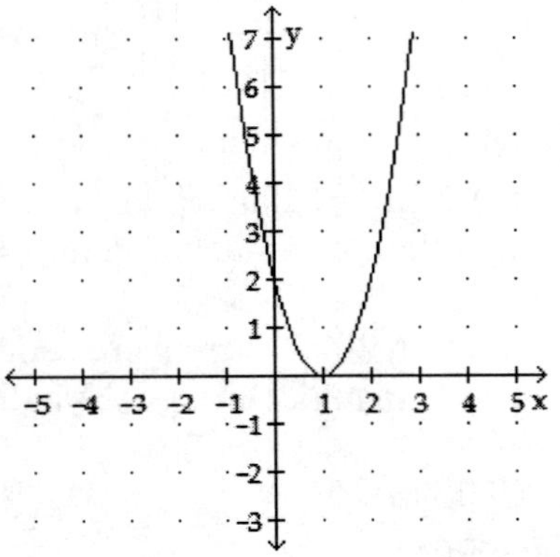

D: $(-\infty, \infty)$; R: $[0, \infty)$

SECTION 5.4

VOCABULARY

1. The expression $(2x^3)(3x^4)$ is the product of two **monomials** and the expression $(x + 4)(x - 5)$ is the product of two **binomials**.

3. FOIL is an acronym for **First** terms, **Outer** terms, **Inner** terms, and **Last** terms.

CONCEPTS

5. a. To multiply a monomial by a monomial, we multiply the numerical **factors** and then multiply the variable factors.
   b. To multiply a polynomial by a monomial, we multiply each **term** of the polynomial by the monomial.
   c. To multiply a polynomial by a polynomial, we multiply each **term** of one polynomial by each term of the other polynomial.

7. The product of the sum and difference of the same two terms is the **square** of the first term minus the **square** of the second term.
   $(x+y)(x-y)=x^2-y^2$

9. a. $(4b-1)+(2b-1)=6b-2$
   b. $(4b-1) - (2b-1)=4b-1-2b+1 = 2b$
   c. $(4b-1)(2b-1)=8b^2-4b-2b+1 = 8b^2-6b+1$

GUIDED PRACTICE

11. $(2a^2)(3a^5)=6a^7$

13.
$$\left(\frac{1}{6}g^4h^5\right)(36g^7h^{11})=\left(\frac{1}{6}\bullet 36\right)g^{4+7}h^{5+11} = 6g^{11}h^{16}$$

15.
$$10x^5(3x^5+2x^2)=10x^5(3x^5)+10x^5(2x^2) = 30x^{10}+20x^7$$

17.
$$-2d^4(3d^3-3d^2+2d) = -2d^4(3d^3)-2d^4(-3d^2)-2d^4(2d) = -6d^7+6d^6-4d^5$$

19.
$$7rs^3t(r^2+s^2-t^2) = 7rs^3t(r^2)+7rs^3t(s^2)+7rs^3t(-t^2) = 7r^3s^3t+7rs^5t-7rs^3t^3$$

21.
$$(7x^6y^3z-4x^2yz^4)4xy^6z^3 = (7x^6y^3z)4xy^6z^3-(4x^2yz^4)4xy^6z^3 = 28x^{6+1}y^{3+6}z^{1+3}-16x^{2+1}y^{1+6}z^{4+3} = 28x^7y^9z^4-16x^3y^7z^7$$

23.
$$(7t-2)(2t+3)=14t^2+21t-4t-6 = 14t^2+17t-6$$

25.
$$(3y^3-4)(2y^3-1)=6y^6-3y^3-8y^3+4 = 6y^6-11y^3+4$$

27.
$$\left(4b+\frac{1}{2}\right)\left(8b-\frac{3}{4}\right)=32b^2-3b+4b-\frac{3}{8} = 32b^2+b-\frac{3}{8}$$

29.
$$(9b^3c-c)(3b^2-bc)=27b^5c-9b^4c^2-3b^2c+bc^2$$

31.
$$(3y+1)(2y^2+3y+2) = 3y(2y^2+3y+2)+1(2y^2+3y+2) = 6y^3+9y^2+6y+2y^2+3y+2 = 6y^3+11y^2+9y+2$$

33.

$$\begin{aligned}&(x-y)(x^2+xy+y^2)\\&=x(x^2+xy+y^2)-y(x^2+xy+y^2)\\&=x^3+x^2y+xy^2-x^2y-xy^2-y^3\\&=x^3-y^3\end{aligned}$$

35.

$$\begin{array}{r}3a^2+4a-2\\ \underline{2a+3}\\ 9a^2+12a-6\\ \underline{6a^3+8a^2-4a+0}\\ 6a^3+17a^2+8a-6\end{array}$$

37.

$$\begin{array}{r}2x^2-10x+14\\ \underline{x^2+\ 2x\ -\ 3}\\ -6x^2+30x-42\\ 4x^3-20x^2+28x+\ 0\\ \underline{2x^4-10x^3+14x^2+\ 0+\ 0}\\ 2x^4-6x^3-12x^2+58x-42\end{array}$$

39.

$$\begin{aligned}6p^2(3p-4)(p+3)&=6p^2(3p^2+9p-4p-12)\\&=6p^2(3p^2+5p-12)\\&=18p^4+30p^3-72p^2\end{aligned}$$

41.

$$\begin{aligned}(4my)(3m-y)(2m-y)&=(12m^2y-4my^2)(2m-y)\\&=24m^3y-12m^2y^2-8m^2y^2+4my^3\\&=24m^3y-20m^2y^2+4my^3\end{aligned}$$

43.

$$\begin{aligned}(2a+b)^2&=(2a)^2+2(2a)(b)+(b)^2\\&=4a^2+4ab+b^2\end{aligned}$$

45.

$$\begin{aligned}(5r^2+6)^2&=(5r^2)^2+2(5r^2)(6)+(6)^2\\&=25r^4+60r^2+36\end{aligned}$$

47.

$$\begin{aligned}\left(\frac{1}{4}b-2\right)^2&=\left(\frac{1}{4}b\right)^2-2\left(\frac{1}{4}b\right)(2)+(-2)^2\\&=\frac{1}{16}b^2-b+4\end{aligned}$$

49.

$$\begin{aligned}(9ab^2-4)^2&=(9ab^2)^2-2(9ab^2)(4)+(-4)^2\\&=81a^2b^4-72ab^2+16\end{aligned}$$

51.

$$\begin{aligned}(5y+2.4)(5y-2.4)&=(5y)^2-(2.4)^2\\&=25y^2-5.76\end{aligned}$$

53.

$$\begin{aligned}\left(x^2y-\frac{6}{5}\right)\left(x^2y+\frac{6}{5}\right)&=(x^2y)^2-\left(\frac{6}{5}\right)^2\\&=x^4y^2-\frac{36}{25}\end{aligned}$$

55.

$$\begin{aligned}[(6a+b)+4]^2&=(6a+b)^2+2(6a+b)(4)+(4)^2\\&=\left[(6a)^2+2(6a)(b)+(b)^2\right]+8(6a+b)+16\\&=36a^2+12ab+b^2+48a+8b+16\end{aligned}$$

57.

$$\begin{aligned}[7-(x-y)]^2&=(7)^2+2(7)(x-y)+(x-y)^2\\&=49-14(x-y)+\left[(x)^2+2(x)(-y)+(-y)^2\right]\\&=49-14x+14y+x^2-2xy+y^2\end{aligned}$$

59.

$$\begin{aligned}[5+(2n+p)][5-(2n+p)]&=(5)^2-(2n+p)^2\\&=25-\left[(2n)^2+2(2n)(p)+(p)^2\right]\\&=25-\left[4n^2+4np+p^2\right]\\&=25-4n^2-4np-p^2\end{aligned}$$

61.

$$(3x-2)^3 = (3x-2)(3x-2)^2$$
$$= (3x-2)\left[(3x)^2 - 2(3x)(2) + (2)^2\right]$$
$$= (3x-2)(9x^2 - 12x + 4)$$
$$= 27x^3 - 36x^2 + 12x - 18x^2 + 24x - 8$$
$$= 27x^3 - 54x^2 + 36x - 8$$

63.

$$f(x) = x^2 - 8x + 2$$
$$f(b+1) = (b+1)^2 - 8(b+1) + 2$$
$$= \left[(b)^2 + 2(b)(1) + (1)^2\right] - 8b - 8 + 2$$
$$= b^2 + 2b + 1 - 8b - 8 + 2$$
$$= b^2 - 6b - 5$$

65.

$$f(x) = x^2 + 4x - 9$$
$$f(a-6) = (a-6)^2 + 4(a-6) - 9$$
$$= \left[(a)^2 - 2(a)(6) + (6)^2\right] + 4(a-6) - 9$$
$$= a^2 - 12a + 36 + 4a - 24 - 9$$
$$= a^2 - 8a + 3$$

67.

$$(7x-1)^2 - (x-4)(x+3)$$
$$= \left[(7x)^2 - 2(7x)(1) + (1)^2\right] - \left[x^2 + 3x - 4x - 12\right]$$
$$= 49x^2 - 14x + 1 - \left[x^2 - x - 12\right]$$
$$= 49x^2 - 14x + 1 - x^2 + x + 12$$
$$= 48x^2 - 13x + 13$$

69.

$$(3x-4)^2 - (2x+3)^2$$
$$= (9x^2 - 24x + 16) - (4x^2 + 12x + 9)$$
$$= 9x^2 - 24x + 16 - 4x^2 - 12x - 9$$
$$= 5x^2 - 36x + 7$$

71.

$$(3.21x - 7.85)(2.87x + 4.59)$$
$$= 9.2127x^2 + 14.7339x - 22.5295x - 36.0315$$
$$= 9.2127x^2 - 7.7956x - 36.0315$$

73.

$$(-17.3y + 4.35)^2$$
$$= (-17.3y)^2 + 2(-17.3y)(4.35) + (4.35)^2$$
$$= 299.29y^2 - 150.51y + 18.9225$$

75.

$$(2a-b)(4a^2 + 2ab + b^2)$$
$$= 2a(4a^2 + 2ab + b^2) - b(4a^2 + 2ab + b^2)$$
$$= 8a^3 + 4a^2b + 2ab^2 - 4a^2b - 2ab^2 - b^3$$
$$= 8a^3 - b^3$$

77.

$$(2b+3)(5b-1) - (b+2)^2$$
$$= \left[10b^2 - 2b + 15b - 3\right] - \left[b^2 + 2(b)(2) + 2^2\right]$$
$$= \left[10b^2 + 13b - 3\right] - \left[b^2 + 4b + 4\right]$$
$$= 10b^2 + 13b - 3 - b^2 - 4b - 4$$
$$= 9b^2 + 9b - 7$$

79.

$$(11m^2 + 3n^3)(5m + 2n^2)$$
$$= 55m^3 + 22m^2n^2 + 15mn^3 + 6n^5$$

81.

$$(-5s^2tu)(-3s^4t^4u) = 15s^{2+4}t^{1+4}u^{1+1}$$
$$= 15s^6t^5u^2$$

83.

$$(a+b+c)(2a-b-2c)$$
$$= a(2a-b-2c) + b(2a-b-2c) + c(2a-b-2c)$$
$$= 2a^2 - ab - 2ac + 2ab - b^2 - 2bc + 2ac - bc - 2c^2$$
$$= 2a^2 + ab - b^2 - 3bc - 2c^2$$

85.

$$(a+b)(a-b)(a-3b) = (a^2 - b^2)(a-3b)$$
$$= a^3 - 3a^2b - ab^2 + 3b^3$$

87.

$$(4k-13)^2 = (4k)^2 - 2(4k)(13) + (-13)^2$$
$$= 16k^2 - 104k + 169$$

89.

$$\left(\frac{1}{2}x-4\right)\left(\frac{1}{2}x+16\right)=\frac{1}{4}x^2+8x-2x-64$$
$$=\frac{1}{4}x^2+6x-64$$

91. $\left(-3ab^2\right)\left(5ab\right)=-15a^2b^3$

93.

$$\left(y^3+2\right)\left(y^3-2\right)=\left(y^3\right)^2-2^2$$
$$=y^6-4$$

95.

$$\left(3a^3+4c\right)^2=\left(3a^3\right)^2+2\left(3a^3\right)\left(4c\right)+\left(4c\right)^2$$
$$=9a^6+24a^3c+16c^2$$

97.

$$\frac{1}{2}m^2n\left(m^2+n^3\right)\left(-12mn\right)$$
$$=\frac{1}{2}m^2n\left(-12mn\right)\left(m^2+n^3\right)$$
$$=-6m^3n^2\left(m^2+n^3\right)$$
$$=-6m^5n^2-6m^3n^5$$

99.

$$\left(2x^2y^3z^5\right)\left(4xy^5z\right)\left(-5y^6z^6\right)$$
$$=\left(2\cdot4\cdot-5\right)x^{2+1}y^{3+5+6}z^{5+1+6}$$
$$=-40x^3y^{14}z^{12}$$

101.

$$\left(5s-t^3\right)^3$$
$$=\left(5s-t^3\right)\left(5s-t^3\right)^2$$
$$=\left(5s-t^3\right)\left[\left(5s\right)^2-2\left(5s\right)\left(t^3\right)+\left(t^3\right)^2\right]$$
$$=\left(5s-t^3\right)\left(25s^2-10st^3+t^6\right)$$
$$=125s^3-50s^2t^3+5st^6-25s^2t^3+10st^6-t^9$$
$$=125s^3-75s^2t^3+15st^6-t^9$$

103.

$$\left[\left(3d+f\right)+2\right]^2$$
$$=\left(3d+f\right)^2+2\left(3d+f\right)\left(2\right)+\left(2\right)^2$$
$$=\left[\left(3d\right)^2+2\left(3d\right)\left(f\right)+f^2\right]+4\left(3d+f\right)+4$$
$$=9d^2+6df+f^2+12d+4f+4$$

105.

$$f\left(t\right)\cdot g\left(t\right)=\left(0.4t-3\right)\left(0.5t-3\right)$$
$$=0.2t^2-1.2t-1.5t+9$$
$$=0.2t^2-2.7t+9$$

107.

$$s\left(x\right)\cdot t\left(x\right)=\left(5x^5+6\right)\left(5x^2-6\right)$$
$$=\left(5x^5\right)^2-\left(6\right)^2$$
$$=25x^{10}-36$$

LOOK ALIKES

109. a. $\left(-3ab^2\right)\left(5ab\right)=-15a^2b^3$

b. $\left(-3a+b^2\right)\left(5a+b\right)$

$$=-15a^2-3ab+5ab^2+b^3$$

111. a. $\left(2x^2+5\right)+\left(3x^2+2x+1\right)$

$$=5x^2+2x+6$$

b. $\left(2x^2+5\right)\left(3x^2+2x+1\right)$

$$=6x^4+4x^3+2x^2+15x^2+10x+5$$
$$=6x^4+4x^3+17x^2+10x+5$$

APPLICATIONS

113. GEOMETRY

a. Length = $(x + 4)$; Width = $(x - 2)$

$$\begin{aligned}\text{Area} &= (\text{length})(\text{width})\\ &= (x+4)(x-2)\\ &= x^2 - 2x + 4x - 8\\ &= x^2 + 2x - 8\end{aligned}$$

b. Base = $(b + 5)$; Height = $(b - 2)$

$$\begin{aligned}\text{Area} &= \frac{1}{2}(\text{base})(\text{height})\\ &= \frac{1}{2}(b+5)(b-2)\\ &= \frac{1}{2}\left(b^2 - 2b + 5b - 10\right)\\ &= \frac{1}{2}\left(b^2 + 3b - 10\right)\\ &= \frac{1}{2}b^2 + \frac{3}{2}b - 5\end{aligned}$$

115. THE YELLOW PAGES

a. Area = (length)(width)

$$= (x-y)(x+y)$$

b. Area = (length)(width)

$$\begin{aligned}&= x(x-y)\\ &= x^2 - xy\end{aligned}$$

c. Area = (length)(width)

$$\begin{aligned}&= y(x-y)\\ &= xy - y^2\end{aligned}$$

d. They represent the same area.

$$(x-y)(x+y) \stackrel{?}{=} \left(x^2 - xy\right) + \left(xy - y^2\right)$$

$$x^2 - y^2 = x^2 - y^2$$

117. QUILTS

$$\begin{aligned}(36-2x)(46-2x) &= 1{,}656 - 72x - 92x + 4x^2\\ &= 4x^2 - 164x + 1{,}656 \text{ in.}^2\end{aligned}$$

WRITING

119. Answers will vary.

121. Answers will vary.

REVIEW

123. $2x + y \le 2$

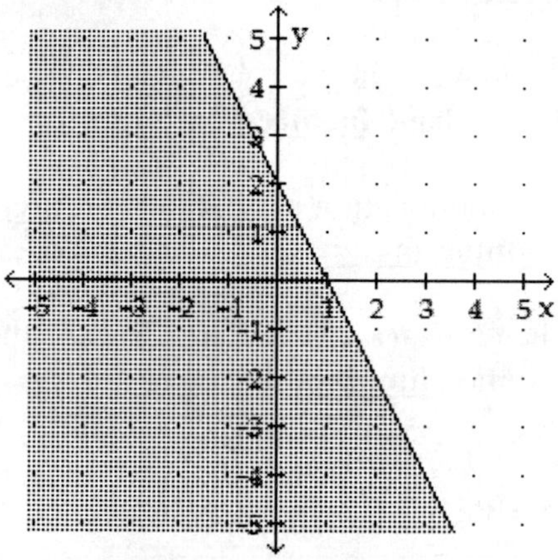

125. $\begin{cases} y - 2 < 3x \\ y + 2x < 3 \end{cases}$

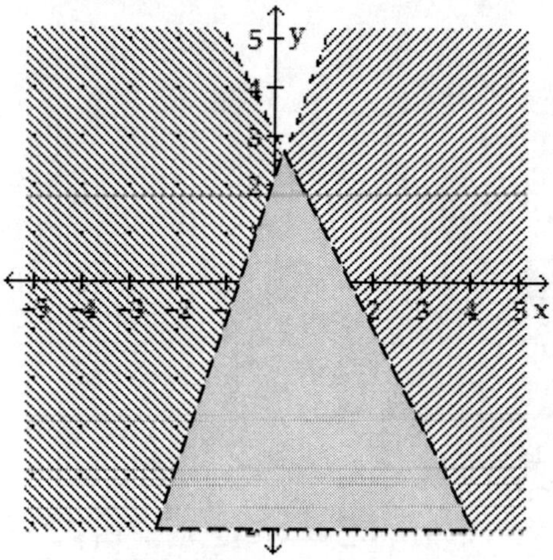

CHALLENGE PROBLEMS

127.

$$\begin{aligned}&ab^{-2}c^{-3}\left(a^{-4}bc^3 + a^{-3}b^4c^3\right)\\ &= a^{-3}b^{-1}c^0 + a^{-2}b^2c^0\\ &= \frac{1}{a^3b} + \frac{b^2}{a^2}\end{aligned}$$

129.

$$a^{2n}\left(a^n + a^{2n}\right) = a^{3n} + a^{4n}$$

131.

$$f(x) = x^2 - 4x - 7$$

$$\begin{aligned}f(a+h) - f(a) &= \left[(a+h)^2 - 4(a+h) - 7\right] - \left(a^2 - 4a - 7\right)\\ &= a^2 + 2ah + h^2 - 4a - 4h - 7 - a^2 + 4a + 7\\ &= 2ah + h^2 - 4h\end{aligned}$$

SECTION 5.5

VOCABULARY

1. When we write $2x + 4$ as $2(x + 2)$, we say that we have **factored** $2x + 4$.

3. The abbreviation GCF stands for **greatest common factor**.

5. The terms $t(t - 5)$ and $9(t - 5)$ have the common **binomial** factor $t - 5$.

CONCEPTS

7. $6xy^2$

9. a. 4
   b. no
   c. 5; s

GUIDED PRACTICE

11. $14x = 2 \cdot 7 \cdot x$
    $24x^2 = 2 \cdot 2 \cdot 2 \cdot 3 \cdot x \cdot x$
    $\text{GCF} = 2x$

13. $45a^5b = 3 \cdot 3 \cdot 5 \cdot a \cdot a \cdot a \cdot a \cdot a \cdot b$
    $30a^4 = 2 \cdot 3 \cdot 5 \cdot a \cdot a \cdot a \cdot a$
    $\text{GCF} = 3 \cdot 5 \cdot a \cdot a \cdot a \cdot a = 15a^4$

15. $16y^4 = 2 \cdot 2 \cdot 2 \cdot 2 \cdot y \cdot y \cdot y \cdot y$
    $40y^2 = 2 \cdot 2 \cdot 2 \cdot 5 \cdot y \cdot y$
    $24y^3 = 2 \cdot 2 \cdot 2 \cdot 3 \cdot y \cdot y \cdot y$
    $\text{GCF} = 2 \cdot 2 \cdot 2 \cdot y \cdot y = 8y^2$

17. $24r^3s^5 = 2 \cdot 2 \cdot 2 \cdot 3 \cdot r \cdot r \cdot r \cdot s \cdot s \cdot s \cdot s \cdot s$
    $36r^8s^4 = 2 \cdot 2 \cdot 3 \cdot 3 \cdot r \cdot r \cdot r \cdot r \cdot r \cdot r \cdot r \cdot r \cdot s \cdot s \cdot s \cdot s$
    $48r^4s^4 = 2 \cdot 2 \cdot 2 \cdot 2 \cdot 3 \cdot r \cdot r \cdot r \cdot r \cdot s \cdot s \cdot s \cdot s$
    $\text{GCF} = 2 \cdot 2 \cdot 3 \cdot r \cdot r \cdot r \cdot s \cdot s \cdot s \cdot s$
    $= 12r^3s^4$

19. $18x^4y^3z^9 = 2 \cdot 3 \cdot 3 \cdot x \cdot x \cdot x \cdot x \cdot y \cdot y \cdot y \cdot z \cdot z \cdot z \cdot z \cdot z \cdot z \cdot z \cdot z \cdot z$
    $54xy^2z^6 = 2 \cdot 3 \cdot 3 \cdot 3 \cdot x \cdot y \cdot y \cdot z \cdot z \cdot z \cdot z \cdot z \cdot z$
    $36xy^5z^8 = 2 \cdot 2 \cdot 3 \cdot 3 \cdot x \cdot y \cdot y \cdot y \cdot y \cdot y \cdot z \cdot z \cdot z \cdot z \cdot z \cdot z \cdot z \cdot z$
    $\text{GCF} = 2 \cdot 3 \cdot 3 \cdot x \cdot y \cdot y \cdot z \cdot z \cdot z \cdot z \cdot z \cdot z$
    $= 18xy^2z^6$

21. $10(c - d) = 2 \cdot 5 \cdot (c - d)$
    $c(c - d) = c \cdot (c - d)$
    $\text{GCF} = (c - d)$

23. $24x + 16 = 8(3x + 2)$

25. $3y^3 + 5y^2 - 9y = y(3y^2 + 5y - 9)$

27. $45a^2 - 9a = 9a(5a - 1)$

29. $15x^2y - 10x^2y^2 = 5x^2y(3 - 2y)$

31. $14r + 15$ is Prime

33. $27a^9b^4c^5 - 12a^7b^4c^2 + 30a^6b^4c^3$
    $= 3a^6b^4c^2(9a^3c^3 - 4a + 10c)$

35. $-a - b = -(a + b)$

37. $-5xy + y - 4 = -(5xy - y + 4)$

39. $-8a - 16 = -8(a + 2)$

41. $-6x^2 - 3xy = -3x(2x + y)$

43. $-18a^2b + 12ab^2 = -6ab(3a - 2b)$

45. $-8a^4c^8 + 28a^3c^8 - 20a^2c^9$
    $= -4a^2c^8(2a^2 - 7a + 5c)$

47. $4(x + y) + t(x + y) = (x + y)(4 + t)$

49. $(a - b)r - (a - b)s + (a - b)t$
    $= (a - b)(r - s + t)$

51. $3(m + n + p) + x(m + n + p)$
    $= (m + n + p)(3 + x)$

53. $3x(x + 7)^2 - 2(x + 7)^2$
    $= (x + 7)^2(3x - 2)$

55.
$$ax + 8x + ay + 8y = x(a + 8) + y(a + 8)$$
$$= (a + 8)(x + y)$$

57.
$$3c - c^2 + 3d - cd = c(3 - c) + d(3 - c)$$
$$= (c + d)(3 - c)$$

59.
$$a^2 + ab - 4a - 4b = a(a + b) - 4(a + b)$$
$$= (a + b)(a - 4)$$

61.
$$t^3-3t^2-7t+21=t^2(t-3)-7(t-3)$$
$$=(t-3)(t^2-7)$$

63.
$$x^2+yx+x+y=x(x+y)+1(x+y)$$
$$=(x+y)(x+1)$$

65.
$$1-n-m+mn=1(1-n)-m(1-n)$$
$$=(1-m)(1-n)$$

67.
$$y^3-12+3y-4y^2=y^3-4y^2+3y-12$$
$$=y^2(y-4)+3(y-4)$$
$$=(y-4)(y^2+3)$$

69.
$$st+rv+sv+rt=st+sv+rt+rv$$
$$=s(t+v)+r(t+v)$$
$$=(t+v)(s+r)$$

71.
$$28a^3b^3c+14a^3c-4b^3c-2c$$
$$=2c(14a^3b^3+7a^3-2b^3-1)$$
$$=2c\left[7a^3(2b^3+1)-1(2b^3+1)\right]$$
$$=2c(2b^3+1)(7a^3-1)$$

73.
$$mpx+mqx+npx+nqx$$
$$=x[mp+mq+np+nq]$$
$$=x\left[m(p+q)+n(p+q)\right]$$
$$=x(p+q)(m+n)$$

75.
$$d_1d_2=fd_2+fd_1$$
$$d_1d_2=f(d_2+d_1)$$
$$\frac{d_1d_2}{d_2+d_1}=\frac{f(d_2+d_1)}{d_2+d_1}$$
$$\frac{d_1d_2}{d_2+d_1}=f$$
$$f=\frac{d_1d_2}{d_2+d_1}$$

77.
$$b^2x^2+a^2y^2=a^2b^2$$
$$b^2x^2+a^2y^2-a^2y^2=a^2b^2-a^2y^2$$
$$b^2x^2=a^2b^2-a^2y^2$$
$$b^2x^2=a^2(b^2-y^2)$$
$$\frac{b^2x^2}{b^2-y^2}=\frac{a^2(b^2-y^2)}{b^2-y^2}$$
$$\frac{b^2x^2}{b^2-y^2}=a^2$$
$$a^2=\frac{b^2x^2}{b^2-y^2}$$

79.
$$rx-ty=by$$
$$rx-ty+ty=by+ty$$
$$rx=by+ty$$
$$rx=y(b+t)$$
$$\frac{rx}{b+t}=\frac{y(b+t)}{b+t}$$
$$\frac{rx}{b+t}=y$$

81.
$$S(1-r)=a-lr$$
$$S-Sr=a-lr$$
$$S-Sr+lr=a-lr+lr$$
$$S-Sr+lr=a$$
$$S-Sr+lr-S=a-S$$
$$lr-Sr=a-S$$
$$r(l-S)=a-S$$
$$\frac{r(l-S)}{l-S}=\frac{a-S}{l-S}$$
$$r=\frac{a-S}{l-S}\bullet\frac{-1}{-1}$$
$$r=\frac{S-a}{S-l}$$

TRY IT YOURSELF

83. $-63u^3+28u^2=-7u^2(9u-4)$

85.
$$a^3b^2-3+a^3-3b^2=a^3b^2-3b^2+a^3-3$$
$$=b^2(a^3-3)+1(a^3-3)$$
$$=(a^3-3)(b^2+1)$$

87.
$$45x^{10}y^3-63x^7y^7+81x^{10}y^{10}$$
$$=9x^7y^3(5x^3-7y^4+9x^3y^7)$$

89.
$$ar^2t-br^2t+as^2t-bs^2t$$
$$=t(ar^2-br^2+as^2-bs^2)$$
$$=t[r^2(a-b)+s^2(a-b)]$$
$$=t(a-b)(r^2+s^2)$$

91.
$16a^3b^3-27c^3d^3$ is prime

93.
$$\frac{3}{5}ax^4+\frac{1}{5}bx^2-\frac{4}{5}ax^3=\frac{1}{5}x^2(3ax^2+b-4ax)$$

95.
$$24x^{50}-18x^{40}-42x^{30}=6x^{30}(4x^{20}-3x^{10}-7)$$

97.
$$a(2x+y)-b(2x+y)+c(2x+y)$$
$$=(2x+y)(a-b+c)$$

99.
$$45y^{12}+30y^{10}+25y^8-5y^6$$
$$=5y^6(9y^6+6y^4+5y^2-1)$$

101.
$$a^2x+bx-a^2-b=x(a^2+b)-1(a^2+b)$$
$$=(a^2+b)(x-1)$$

103.
$$ab+b^2+bc+ac+bc+c^2$$
$$=(ab+b^2+bc)+(ac+bc+c^2)$$
$$=b(a+b+c)+c(a+b+c)$$
$$=(a+b+c)(b+c)$$

105. a. $5a^3+6a^2+15a+18$
$$=a^2(5a+6)+3(5a+6)$$
$$=(5a+6)(a^2+3)$$

b. $3a^3+6a^2+15a+18$
$$=3(a^3+2a^2+5a+6)$$

APPLICATIONS

107. GEOMETRIC FORMULAS
$$A=\frac{1}{2}b_1h+\frac{1}{2}b_2h$$
$$=\frac{1}{2}h(b_1+b_2)$$
It gives the formula for the area of a trapezoid.

109. LANDSCAPING
$$4r^2-\pi r^2=r^2(4-\pi)$$

WRITING

111. Answers will vary.

113. Answers will vary.

115. Answers will vary.

REVIEW

117. INVESTMENTS
Let x = the amount invested in each account.

Account	Principal	· Rate	· Time	= Interest
7%	x	0.07	1	$0.07x$
8%	x	0.08	1	$0.08x$
10.5%	x	0.105	1	$0.105x$
Combined Interest	$1,249.50			

7% int. + 8% int. + 10.5% int.= Combined int.

$$0.07x + 0.08x + 0.105x = 1,249.50$$
$$0.255x = 1,249.50$$
$$\frac{0.255x}{0.255} = \frac{1,249.50}{0.255}$$
$$x = \$4,900$$

$4,900 is invested in each account.

CHALLENGE PROBLEMS

119.
$$\begin{aligned}&2n^4p - 2n^2 - n^3p^2 + np + 2mn^3p - 2mn\\&= n\left[2n^3p - 2n - n^2p^2 + p + 2mn^2p - 2m\right]\\&= n\left[2n^3p - n^2p^2 + 2mn^2p - 2n + p - 2m\right]\\&= n\left[\left(2n^3p - n^2p^2 + 2mn^2p\right) + \left(-2n + p - 2m\right)\right]\\&= n\left[n^2p(2n - p + 2m) - 1(2n - p + 2m)\right]\\&= n(2n - p + 2m)\left(n^2p - 1\right)\end{aligned}$$

121.
$$\begin{aligned}&6(x^3 - 7x + 1)^2 - 3(x^3 - 7x + 1)^3\\&= 3(x^3 - 7x + 1)^2[2 - (x^3 - 7x + 1)]\\&= 3(x^3 - 7x + 1)^2(2 - x^3 + 7x - 1)\\&= 3(x^3 - 7x + 1)^2(-x^3 + 7x + 1)\end{aligned}$$

123.
$$\begin{aligned}x^2(a-b) + 83(b-a) &= x^2(a-b) - 83(a-b)\\&= (a-b)(x^2 - 83)\end{aligned}$$

125.
$$\begin{aligned}x^{n+2} + x^{n+3} &= x^2\left(x^{n+2-2} + x^{n+3-2}\right)\\&= x^2\left(x^n + x^{n+1}\right)\end{aligned}$$

127.
$$\begin{aligned}t^5 + 4t^{-6} &= t^{-3}\left(t^{5-(-3)} + 4t^{-6-(-3)}\right)\\&= t^{-3}\left(t^8 + 4t^{-3}\right)\end{aligned}$$

SECTION 5.6

VOCABULARY

1. Since $y^2 + 2y + 1 = (y + 1)^2$, we call $y^2 + 2y + 1$ a **perfect** square trinomial.

3. The **leading** coefficient of the trinomial $x^2 - 3x + 2$ is 1, the **coefficient** of the middle term is –3, and the last term is **2**. The trinomial is written in **descending** powers of x.

CONCEPTS

5. a. positive
   b. negative
   c. positive

7.

Factors of 8	Sum of the factors of 8
1(8) = 8	1 + 8 = 9
2(4) = 8	2 + 4 = 6
–1(–8) = 8	–1 + (–8) = –9
–2(–4) = 8	–2 + (–4) = –6

The numbers –1 and 8 are two integers whose **product** is 8 and whose **sum** is –9.

9. No.

$$(3t-1)(5t-6) = 15t^2 - 18t - 5t + 6$$
$$= 15t^2 - 23t + 6$$

11. $x^2 - 6x + 8 = (x - 4)(x - 2)$

13. $2a^2 + 9a + 4 = (2a + 1)(a + 4)$

NOTATION

15. $a = 4,\ b = -4,\ c = 1$

GUIDED PRACTICE

17. $a^2 + 18a + 81 = (a + 9)^2$

19. $4y^2 + 28y + 49 = (2y + 7)^2$

21. $y^4 - 10y^2 + 25 = (y^2 - 5)^2$

23. $9b^4 - 12b^2c^2 + 4c^4 = (3b^2 - 2c^2)^2$

25. $x^2 + 5x + 6 = (x + 3)(x + 2)$

27. $t^2 + 14t + 48 = (t + 6)(t + 8)$

29. $x^2 - 16x + 55 = (x - 5)(x - 11)$

31. $y^2 - 17y + 72 = (y - 8)(y - 9)$

33. $-13y^2 + 30 + y^4 = y^4 - 13y^2 + 30$
$= (y^2 - 10)(y^2 - 3)$

35. $50 + 15x^2 + x^4 = x^4 + 15x^2 + 50$
$= (x^2 + 10)(x^2 + 5)$

37. $3x^2 + 12x - 63 = 3(x^2 + 4x - 21)$
$= 3(x + 7)(x - 3)$

39. $15x^2 + 45x + 30 = 15(x^2 + 3x + 2)$
$= 15(x + 1)(x + 2)$

41. $2x^2y - 12xy^2 - 14y^3 = 2y(x^2 - 6xy - 7y^2)$
$= 2y(x + y)(x - 7y)$

43. $3s^2t^2 + 18st^3 - 48t^4 = 3t^2(s^2 + 6st - 16t^2)$
$= 3t^2(s + 8t)(s - 2t)$

45. $5x^2 + 13x + 6 = (5x + 3)(x + 2)$

47. $7a^2 + 12a + 5 = (7a + 5)(a + 1)$

49. $3r^2 + 13r - 10 = (3r - 2)(r + 5)$

51. $11y^2 + 32y - 3 = (11y - 1)(y + 3)$

53. $2t^2 + 9t - 6$ is prime

55. $13a^2 + 2a + 1$ is prime

57. $8a^2 - 18ay + 7y^2 = (4a - 7y)(2a - y)$

59. $7g^2 - 12gh - 4h^2 = (7g + 2h)(g - 2h)$

61. $-18p^2 + 14pq + 4q^2 = -2(9p^2 - 7pq - 2q^2)$
$= -2(9p + 2q)(p - q)$

63. $-30x^2 + 25xy + 20y^2 = -5(6x^2 - 5xy - 4y^2)$
$= -5(2x + y)(3x - 4y)$

65. $9a^2b^4 + 15a^2b^2 + 4a^2 = a^2(9b^4 + 15b^2 + 4)$
$= a^2(3b^2 + 4)(3b^2 + 1)$

67. $8b^4c^8 + 14b^2c^8 - 15c^8 = c^8(8b^4 + 14b^2 - 15)$
$= c^8(2b^2 + 5)(4b^2 - 3)$

69. $10b^6 - 19b^4 + 6b^2 = b^2(10b^4 - 19b^2 + 6)$
$= b^2(5b^2 - 2)(2b^2 - 3)$

71. $9m^7 + 6m^5 - 8m^3 = m^3(9m^4 + 6m^2 - 8)$
$= m^3(3m^2 + 4)(3m^2 - 2)$

73. $(x + a)^2 + 2(x + a) + 1$
Let $y = (x + a)$
$y^2 + 2y + 1 = (y + 1)(y + 1)$
$= (y + 1)^2$
Substitute $(x + a)$ for y.
$= (x + a + 1)^2$

75. $(a + b)^2 - 2(a + b) - 24$
Let $x = (a + b)$
$x^2 - 2x - 24 = (x - 6)(x + 4)$
Substitute $(a + b)$ for x.
$= (a + b - 6)(a + b + 4)$

77. $14(q - r)^2 - 17(q - r) - 6$
Let $x = (q - r)$.
$14x^2 - 17x - 6 = (2x - 3)(7x + 2)$
Substitute $(q - r)$ for x.
$= [2(q - r) - 3][7(q - r) + 2]$
$= (2q - 2r - 3)(7q - 7r + 2)$

79. $16(s + t)^2 - 6(s + t) - 27$
Let $x = (s + t)$.
$16x^2 - 6x - 27 = (8x + 9)(2x - 3)$
Substitute $(s + t)$ for x.
$= [8(s + t) + 9][2(s + t) - 3]$
$= (8s + 8t + 9)(2s + 2t - 3)$

TRY IT YOURSELF

81. $32 - a^2 + 4a = -a^2 + 4a + 32$
$= -(a^2 - 4a - 32)$
$= -(a - 8)(a + 4)$

83. $6z^2 + 17z + 12 = (2z + 3)(3z + 4)$

85. $25y^2 - 10y + 1 = (5y - 1)^2$

87. $64h^6 + 24h^5 - 4h^4 = 4h^4(16h^2 + 6h - 1)$
$= 4h^4(8h - 1)(2h + 1)$

89. $5x^2 + 4x + 1$ is prime

91. $b^4x^2 - 12b^2x^2 + 35x^2 = x^2(b^4 - 12b^2 + 35)$
$= x^2(b^2 - 5)(b^2 - 7)$

93. $-3a^2 + ab + 2b^2 = -(3a^2 - ab - 2b^2)$
$= -(3a + 2b)(a - b)$

95. $56a^2 + 42a - 70 = 14(4a^2 + 3a - 5)$

97. $6(t + w)^2 + 11(t + w) - 10$
Let $x = (t + w)$.
$6x^2 + 11x - 10 = (3x - 2)(2x + 5)$
Substitute $(t + w)$ for x.
$= [3(t + w) - 2][2(t + w) + 5]$
$= (3t + 3w - 2)(2t + 2w + 5)$

99. $12y^6 + 23y^3 + 10 = (3y^3 + 2)(4y^3 + 5)$

101. $6a^2(m + n) + 13a(m + n) - 15(m + n)$
$= (m + n)(6a^2 + 13a - 15)$
$= (m + n)(6a - 5)(a + 3)$

103. $20a^2 - 60ab + 45b^2 = 5(4a^2 - 12ab + 9b^2)$
$= 5(2a - 3b)(2a - 3b)$
$= 5(2a - 3b)^2$

105. $-3a^2x^2 + 15a^2x - 18a^2 = -3a^2(x^2 - 5x + 6)$
$= -3a^2(x - 2)(x - 3)$

107. $25m^8 - 60m^4n + 36n^2 = (5m^4 - 6n)^2$

LOOK ALIKE

109. a. $x^2 - 5x + 6 = (x - 2)(x - 3)$
b. $x^4 - 5x^2 + 6 = (x^2 - 2)(x^2 - 3)$
c. $x^6 - 5x^3 + 6 = (x^3 - 2)(x^3 - 3)$

111. a. $3x^2 - 11x + 8 = (3x - 8)(x - 1)$
b. $3x^2 - 11xy + 8y^2 = (3x - 8y)(x - y)$

APPLICATIONS

113. CHECKERS
Area of square $= s^2$
$s^2 = 25x^2 - 40x + 16$
$= (5x - 4)(5x - 4)$
$= (5x - 4)^2$

The length of the side is $5x - 4$.

WRITING

115. Answers will vary.

117. Answers will vary.

REVIEW

119.

$$
\begin{aligned}
f(x) &= |2x-1| \\
f(-2) &= |2(-2)-1| \\
&= |-4-1| \\
&= |-5| \\
&= 5
\end{aligned}
$$

121.

$$
\begin{aligned}
f(x) &= -2x+5 \\
-7 &= -2x+5 \\
-7-5 &= -2x+5-5 \\
-12 &= -2x \\
6 &= x
\end{aligned}
$$

CHALLENGE PROBLEMS

123. b is the sum of the factors of 9(–1) = –9.

9 + (–1) = 8
–9 + 1 = –8
3 + (–3) = 0
–3 + 3 = 0

b can be 8, 0, or –8.

125.

$$
\begin{aligned}
x^{2n}+2x^{n}+1 &= (x^{n}+1)(x^{n}+1) \\
&= (x^{n}+1)^{2}
\end{aligned}
$$

127.

$$
\begin{aligned}
x^{4n}+2x^{2n}y^{2n}+y^{4n} &= (x^{2n}+y^{2n})(x^{2n}+y^{2n}) \\
&= (x^{2n}+y^{2n})^{2}
\end{aligned}
$$

SECTION 5.7

VOCABULARY

1. When the polynomial $4x^2 - 25$ is written as $(2x)^2 - (5)^2$, we see that it is the difference of two **squares**.

CONCEPTS

3. a. 1, 4, 9, 16, 25, 36, 49, 64, 81, 100
   b. 1, 8, 27, 64, 125, 216, 343, 512, 729, 1,000

5. a. $F^2 - L^2 = (F+L)(F-L)$
   b. $F^3 + L^3 = (F+L)(F^2 - FL + L^2)$
   c. $F^3 - L^3 = (F-L)(F^2 + FL + L^2)$

NOTATION

7. Answers will vary.
   a. $x^2 - 4$
   b. $(x-4)^2$
   c. $x^2 + 4$
   d. $x^3 + 8$
   e. $(x+8)^3$

GUIDED PRACTICE

9.
$$\begin{aligned} x^2 - 16 &= (x)^2 - (4)^2 \\ &= (x+4)(x-4) \end{aligned}$$

11.
$$\begin{aligned} 9y^2 - 64 &= (3y)^2 - (8)^2 \\ &= (3y+8)(3y-8) \end{aligned}$$

13.
$$\begin{aligned} 144 - c^2 &= (12)^2 - (c)^2 \\ &= (12+c)(12-c) \end{aligned}$$

15.
$$\begin{aligned} 100m^2 - 1 &= (10m)^2 - (1)^2 \\ &= (10m+1)(10m-1) \end{aligned}$$

17.
$$\begin{aligned} 81a^2 - 49b^2 &= (9a)^2 - (7b)^2 \\ &= (9a-7b)(9a+7b) \end{aligned}$$

19.
$x^2 + 25$ is prime

21.
$$\begin{aligned} 9r^4 - 121s^2 &= (3r^2)^2 - (11s)^2 \\ &= (3r^2 - 11s)(3r^2 + 11s) \end{aligned}$$

23.
$$\begin{aligned} 16t^2 - 25w^4 &= (4t)^2 - (5w^2)^2 \\ &= (4t + 5w^2)(4t - 5w^2) \end{aligned}$$

25.
$$\begin{aligned} 100r^2s^4 - t^4 &= (10rs^2)^2 - (t^2)^2 \\ &= (10rs^2 + t^2)(10rs^2 - t^2) \end{aligned}$$

27.
$$\begin{aligned} 36x^4y^2 - 49z^6 &= (6x^2y)^2 - (7z^3)^2 \\ &= (6x^2y - 7z^3)(6x^2y + 7z^3) \end{aligned}$$

29.
$$\begin{aligned} x^4 - y^4 &= (x^2)^2 - (y^2)^2 \\ &= (x^2 + y^2)(x^2 - y^2) \\ &= (x^2 + y^2)(x+y)(x-y) \end{aligned}$$

31.
$$\begin{aligned} 16a^4 - 81b^4 &= (4a^2)^2 - (9b^2)^2 \\ &= (4a^2 + 9b^2)(4a^2 - 9b^2) \\ &= (4a^2 + 9b^2)(2a+3b)(2a-3b) \end{aligned}$$

33.
$$(x+y)^2 - z^2 = (x+y+z)(x+y-z)$$

35.
$$(r-s)^2 - t^4 = (r-s+t^2)(r-s-t^2)$$

37.

$$\begin{aligned}2x^2-288&=2(x^2-144)\\&=2(x+12)(x-12)\end{aligned}$$

39.

$$\begin{aligned}3x^3-243x&=3x(x^2-81)\\&=3x(x+9)(x-9)\end{aligned}$$

41.

$$\begin{aligned}5ab^4-5a&=5a(b^4-1)\\&=5a(b^2+1)(b^2-1)\\&=5a(b^2+1)(b+1)(b-1)\end{aligned}$$

43.

$$\begin{aligned}64b-4b^5&=4b(16-b^4)\\&=4b(4+b^2)(4-b^2)\\&=4b(4+b^2)(2+b)(2-b)\end{aligned}$$

45.

$$\begin{aligned}c+d+c^2-d^2&=(c+d)+(c^2-d^2)\\&=1(c+d)+(c-d)(c+d)\\&=(c+d)(1+c-d)\end{aligned}$$

47.

$$\begin{aligned}a^2-b^2+2a-2b&=(a^2-b^2)+(2a-2b)\\&=(a+b)(a-b)+2(a-b)\\&=(a-b)(a+b+2)\end{aligned}$$

49.

$$\begin{aligned}x^2+12x+36-y^2&=(x+6)^2-y^2\\&=(x+6-y)(x+6+y)\end{aligned}$$

51.

$$\begin{aligned}x^2-2x+1-9z^2&=(x^2-2x+1)-9z^2\\&=(x-1)^2-9z^2\\&=(x-1-3z)(x-1+3z)\end{aligned}$$

53.

$$\begin{aligned}a^3+125&=(a+5)(a^2-5a+5^2)\\&=(a+5)(a^2-5a+25)\end{aligned}$$

55.

$$\begin{aligned}8r^3+s^3&=(2r+s)((2r)^2-2rs+s^2)\\&=(2r+s)(4r^2-2rs+s^2)\end{aligned}$$

57.

$$\begin{aligned}64t^6-27v^3&=(4t^2-3v)((4t^2)^2+(4t^2)(3v)+(3v)^2)\\&=(4t^2-3v)(16t^4+12t^2v+9v^2)\end{aligned}$$

59.

$$\begin{aligned}x^3-216y^6&=(x-6y^2)(x^2+6xy^2+(6y^2)^2)\\&=(x-6y^2)(x^2+6xy^2+36y^4)\end{aligned}$$

61.

$$\begin{aligned}&(a-b)^3+27\\&=(a-b)^3+(3)^3\\&=((a-b)+3)((a-b)^2-3(a-b)+9)\\&=(a-b+3)(a^2-2ab+b^2-3a+3b+9)\end{aligned}$$

63.

$$\begin{aligned}&64-(a+b)^3\\&=(4)^3-(a+b)^3\\&=(4-(a+b))(4^2+4(a+b)+(a+b)^2)\\&=(4-a-b)(16+4a+4b+a^2+2ab+b^2)\end{aligned}$$

65.

$$\begin{aligned}x^6-1&=(x^3+1)(x^3-1)\\&=(x+1)(x^2-x+1)(x-1)(x^2+x+1)\end{aligned}$$

67.

$$\begin{aligned}&x^{12}-y^6\\&=(x^6-y^3)(x^6+y^3)\\&=(x^2-y)(x^4+x^2y+y^2)(x^2+y)(x^4-x^2y+y^2)\end{aligned}$$

69.

$$\begin{aligned}5x^3+625&=5(x^3+125)\\&=5(x+5)(x^2-5x+25)\end{aligned}$$

71.

$$4x^5 - 256x^2 = 4x^2\left(x^3 - 64\right)$$
$$= 4x^2\left(x - 4\right)\left(x^2 + 4x + 16\right)$$

TRY IT YOURSELF

73.

$$64a^3 - 125b^6$$
$$= (4a)^3 - \left(5b^2\right)^3$$
$$= \left(4a - 5b^2\right)\left((4a)^2 + (4a)\left(5b^2\right) + \left(5b^2\right)^2\right)$$
$$= \left(4a - 5b^2\right)\left(16a^2 + 20ab^2 + 25b^4\right)$$

75.

$$288b^2 - 2b^6 = 2b^2\left(144 - b^4\right)$$
$$= 2b^2\left(12 + b^2\right)\left(12 - b^2\right)$$

77.

$$x^2 - y^2 + 8x + 8y = \left(x^2 - y^2\right) + (8x + 8y)$$
$$= (x - y)(x + y) + 8(x + y)$$
$$= (x + y)(x - y + 8)$$

79.

$$x^9 + y^9 = \left(x^3\right)^3 + \left(y^3\right)^3$$
$$= \left(x^3 + y^3\right)\left(x^6 - x^3y^3 + y^6\right)$$
$$= (x + y)\left(x^2 - xy + y^2\right)\left(x^6 - x^3y^3 + y^6\right)$$

81.

$$144a^2t^2 - 169b^6 = (12at)^2 - \left(13b^3\right)^2$$
$$= \left(12at + 13b^3\right)\left(12at - 13b^3\right)$$

83.

$100a^2 + 9b^2$ is prime

85.

$$81c^4d^4 - 16t^4 = \left(9c^2d^2\right)^2 - \left(4t^2\right)$$
$$= \left(9c^2d^2 + 4t^2\right)\left(9c^2d^2 - 4t^2\right)$$
$$= \left(9c^2d^2 + 4t^2\right)(3cd + 2t)(3cd - 2t)$$

87.

$$128u^2v^3 - 2t^3u^2 = 2u^2\left(64v^3 - t^3\right)$$
$$= 2u^2(4v - t)\left(16v^2 + 4tv + t^2\right)$$

89.

$$y^2 - (2x - t)^2 = (y + (2x - t))(y - (2x - t))$$
$$= (y + 2x - t)(y - 2x + t)$$

91.

$$x^2 + 20x + 100 - 9z^2 = (x + 10)^2 - 9z^2$$
$$= (x + 10 - 3z)(x + 10 + 3z)$$

93.

$$(c - d)^3 + 216$$
$$= (c - d)^3 + 6^3$$
$$= (c - d + 6)\left((c - d)^2 - 6(c - d) + 6^2\right)$$
$$= (c - d + 6)\left(c^2 - 2cd + d^2 - 6c + 6d + 36\right)$$

95.

$$\frac{1}{36} - y^4 = \left(\frac{1}{6}\right)^2 - \left(y^2\right)^2$$
$$= \left(\frac{1}{6} + y^2\right)\left(\frac{1}{6} - y^2\right)$$

97.

$$m^6 - 64$$
$$= \left(m^3\right)^2 - 8^2$$
$$= \left(m^3 + 8\right)\left(m^3 - 8\right)$$
$$= (m + 2)\left(m^2 - 2m + 4\right)(m - 2)\left(m^2 + 2m + 4\right)$$

99.

$$(a + b)x^3 + 27(a + b)$$
$$= (a + b)\left(x^3 + 27\right)$$
$$= (a + b)(x + 3)\left(x^2 - 3x + 9\right)$$

101.

$$x^9 - y^{12}z^{15} = \left(x^3\right)^3 - \left(y^4z^5\right)^3$$
$$= \left(x^3 - y^4z^5\right)\left(x^6 + x^3y^4z^5 + y^8z^{10}\right)$$

LOOK ALIKES

103. a. $q^2-64=(q+8)(q-8)$

b. $q^3-64=(q-4)(q^2+4q+16)$

105. a. $d^2-25=(d-5)(d+5)$

b. $d^3-125=(d-5)(d^2+5d+25)$

107. a. $a^6-b^3=(a^2-b)(a^4+a^2b+b^2)$

b. $a^6+b^3=(a^2+b)(a^4-a^2b+b^2)$

109.

a. $125m^3+8n^3=(5m+2n)(25m^2-10mn+4n^2)$

b. $125m^3-8n^3=(5m-2n)(25m^2+10mn+4n^2)$

APPLICATIONS

111. CANDY

$$V=\frac{4}{3}\pi r_1^3-\frac{4}{3}\pi r_2^3$$
$$=\frac{4}{3}\pi(r_1^3-r_2^3)$$
$$=\frac{4}{3}\pi(r_1-r_2)(r_1^2+r_1r_2+r_2^2)$$

WRITING

113. Answers will vary.

115. Answers will vary.

REVIEW

117. (–2, –1); $m=-\frac{2}{3}$

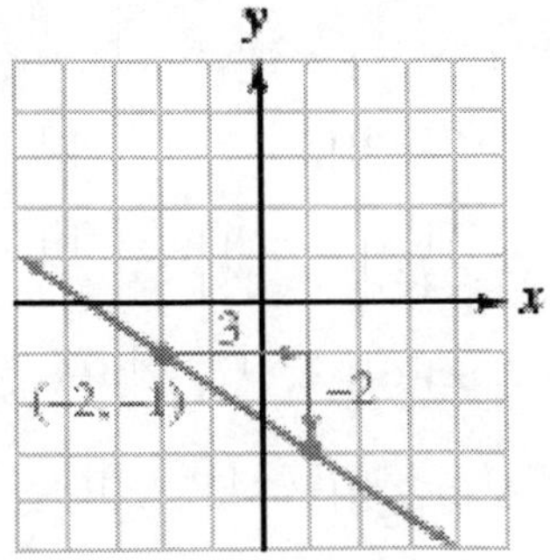

119. $y=-4$

CHALLENGE PROBLEMS

121. $4x^{2n}-9y^{2n}=(2x^n)^2-(3y^n)^2$
$=(2x^n-3y^n)(2x^n+3y^n)$

123. $a^{3b}-c^{3b}=(a^b)^3-(c^b)^3$
$=(a^b-c^b)(a^{2b}+a^bc^b+c^{2b})$

125.

$x^{32}-y^{32}$

$=(x^{16}+y^{16})(x^{16}-y^{16})$

$=(x^{16}+y^{16})(x^8+y^8)(x^8-y^8)$

$=(x^{16}+y^{16})(x^8+y^8)(x^4+y^4)(x^4-y^4)$

$=(x^{16}+y^{16})(x^8+y^8)(x^4+y^4)(x^2+y^2)(x^2-y^2)$

$=(x^{16}+y^{16})(x^8+y^8)(x^4+y^4)(x^2+y^2)(x+y)(x-y)$

SECTION 5.8

VOCABULARY

1. Each factor of a completely factored expression will be **prime**.

3. $x^3 + y^3$ is called a sum of two **cubes** and $x^3 - y^3$ is called a difference of two **cubes**.

CONCEPTS

5. In any factoring problem, always factor out any **common** factors first.

7. If a polynomial has three terms, try to factor it as a **trinomial**.

9. Multiply the factors of $y^2z^3(x + 6)(x + 1)$ to see if the product is $x^2y^2z^3 + 7xy^2z^3 + 6y^2z^3$.

NOTATION

11.
$$18a^3b + 3a^2b^2 - 6ab^3 = \boxed{3ab}\left(6a^2 + ab - 2b^2\right)$$
$$= 3ab\left(3a + \boxed{2b}\right)\left(\boxed{2a} - b\right)$$

TRY IT YOURSELF

13.
$$4a^2bc + 4abc - 120bc$$
$$= 4bc\left(a^2 + a - 30\right)$$
$$= 4bc(a + 6)(a - 5)$$

15.
$$-3x^2y - 6xy^2 + 12xy = -3xy(x + 2y - 4)$$

17.
$$y^3\left(y^2 - 1\right) - 27\left(y^2 - 1\right)$$
$$= \left(y^2 - 1\right)\left(y^3 - 27\right)$$
$$= (y + 1)(y - 1)(y - 3)\left(y^2 + 3y + 9\right)$$

19.
$$36x^4 - 36 = 36\left(x^4 - 1\right)$$
$$= 36\left(x^2 + 1\right)\left(x^2 - 1\right)$$
$$= 36\left(x^2 + 1\right)(x + 1)(x - 1)$$

21.
$16c^2g^2 + h^4$ is prime

23.
$$-14x + 8 + 6x^2 = 6x^2 - 14x + 8$$
$$= 2\left(3x^2 - 7x + 4\right)$$
$$= 2(3x - 4)(x - 1)$$

25.
$$4x^2y^2 + 4xy^2 + y^2 = y^2\left(4x^2 + 4x + 1\right)$$
$$= y^2(2x + 1)(2x + 1)$$
$$= y^2(2x + 1)^2$$

27.
$$4x^2y^2z^2 - 26x^2y^2z^3 = 2x^2y^2z^2(2 - 13z)$$

29.
$$9a^6 - 48a^4 + 64a^2 = a^2\left(9a^4 - 48a^2 + 64\right)$$
$$= a^2\left(3a^2 - 8\right)^2$$

31.
$$6a^5 - 6a^3b^2 - 6a^2b^3 + 6b^5$$
$$= 6\left(a^5 - a^3b^2 - a^2b^3 + b^5\right)$$
$$= 6\left[a^3\left(a^2 - b^2\right) - b^3\left(a^2 - b^2\right)\right]$$
$$= 6\left(a^2 - b^2\right)\left(a^3 - b^3\right)$$
$$= 6(a - b)(a + b)(a - b)\left(a^2 + ab + b^2\right)$$
$$= 6(a - b)^2(a + b)\left(a^2 + ab + b^2\right)$$

33.
$$(x - y)^3 + 125$$
$$= \left[(x - y) + 5\right]\left[(x - y)^2 - 5(x - y) + 5^2\right]$$
$$= (x - y + 5)\left[(x - y)^2 - 5(x - y) + 25\right]$$

35.

$$2(a-b)^2+5(a-b)+3$$
$$=(2(a-b)+3)((a-b)+1)$$
$$=(2a-2b+3)(a-b+1)$$

37.

$$6x^2-63-13x=6x^2-13x-63$$
$$=(2x-9)(3x+7)$$

39.

$$-17x^2+16+x^4=x^4-17x^2+16$$
$$=(x^2-1)(x^2-16)$$
$$=(x+1)(x-1)(x+4)(x-4)$$

41.

$$x^2+10x+25-y^8=(x^2+10x+25)-y^8$$
$$=(x+5)^2-y^8$$
$$=(x+5-y^4)(x+5+y^4)$$

43.

$$9x^2-6x+1-25y^2=(9x^2-6x+1)-25y^2$$
$$=(3x-1)^2-25y^2$$
$$=(3x-1-5y)(3x-1+5y)$$

45.

$$a^2-b^2+4b-4=a^2-(b^2-4b+4)$$
$$=a^2-(b-2)^2$$
$$=(a+(b-2))(a-(b-2))$$
$$=(a+b-2)(a-b+2)$$

47.

$$ax^2-2axy+ay^2-x^2+2xy-y^2$$
$$=(ax^2-2axy+ay^2)-(x^2-2xy+y^2)$$
$$=a(x^2-2xy+y^2)-(x^2-2xy+y^2)$$
$$=a(x-y)^2-1(x-y)^2$$
$$=(x-y)^2(a-1)$$

49.

$$\frac{81}{16}x^4-y^{40}=\left(\frac{9}{4}x^2+y^{20}\right)\left(\frac{9}{4}x^2-y^{20}\right)$$
$$=\left(\frac{9}{4}x^2+y^{20}\right)\left(\frac{3}{2}x+y^{10}\right)\left(\frac{3}{2}x-y^{10}\right)$$

51.

$$16m^{16}-16$$
$$=16(m^{16}-1)$$
$$=16(m^8+1)(m^8-1)$$
$$=16(m^8+1)(m^4+1)(m^4-1)$$
$$=16(m^8+1)(m^4+1)(m^2+1)(m^2-1)$$
$$=16(m^8+1)(m^4+1)(m^2+1)(m+1)(m-1)$$

53.

$$9y^5+6y^4+y^3+3y+1$$
$$=(9y^5+6y^4+y^3)+(3y+1)$$
$$=y^3(9y^2+6y+1)+(3y+1)$$
$$=y^3(3y+1)^2+1(3y+1)$$
$$=(3y+1)(y^3(3y+1)+1)$$
$$=(3y+1)(3y^4+y^3+1)$$

55.

$$x^3-xy^2-4x^2+4y^2$$
$$=(x^3-xy^2)-(4x^2-4y^2)$$
$$=x(x^2-y^2)-4(x^2-y^2)$$
$$=(x^2-y^2)(x-4)$$
$$=(x-4)(x+y)(x-y)$$

57.

$$c^3-4a^2c+4abc-b^2c$$
$$=c(c^2-4a^2+4ab-b^2)$$
$$=c\left[c^2-(4a^2-4ab+b^2)\right]$$
$$=c\left[c^2-(2a-b)^2\right]$$
$$=c(c+2a-b)(c-(2a-b))$$
$$=c(c+2a-b)(c-2a+b)$$

59.

$$\begin{aligned}&9x^4+6x^3+x^2+3x+1\\&=\left(9x^4+6x^3+x^2\right)+(3x+1)\\&=x^2\left(9x^2+6x+1\right)+1(3x+1)\\&=x^2(3x+1)^2+1(3x+1)\\&=(3x+1)\left(x^2(3x+1)+1\right)\\&=(3x+1)\left(3x^3+x^2+1\right)\end{aligned}$$

61.

$$\begin{aligned}(2x-1)^2+4(2x-1)+4&=\left[(2x-1)+2\right]^2\\&=(2x-1+2)^2\\&=(2x+1)^2\end{aligned}$$

63.

a^2+b^2+25 is prime

APPLICATIONS

65. GRAPHIC DESIGN
$1-16x^2=(1-4x)(1+4x)$ in.^{2}

WRITING

67. Answers will vary.

REVIEW

69.

$$\begin{aligned}\begin{vmatrix}-6 & -2\\ 15 & 4\end{vmatrix}&=-6(4)-(-2)(15)\\&=-24+30\\&=6\end{aligned}$$

71.

$$\begin{aligned}\begin{vmatrix}-1 & 2 & 1\\ 2 & 1 & -3\\ 1 & 1 & 1\end{vmatrix}&=-1\begin{vmatrix}1 & -3\\ 1 & 1\end{vmatrix}-2\begin{vmatrix}2 & -3\\ 1 & 1\end{vmatrix}+1\begin{vmatrix}2 & 1\\ 1 & 1\end{vmatrix}\\&=-1(1+3)-2(2+3)+1(2-1)\\&=-1(4)-2(5)+1(1)\\&=-4-10+1\\&=-13\end{aligned}$$

CHALLENGE PROBLEMS

73.

$$\begin{aligned}x^4+x^2+1&=x^4+x^2+1+x^2-x^2\\&=x^4+2x^2+1-x^2\\&=\left(x^4+2x^2+1\right)-x^2\\&=\left(x^2+1\right)^2-x^2\\&=\left(x^2+1-x\right)\left(x^2+1+x\right)\\&=\left(x^2-x+1\right)\left(x^2+x+1\right)\end{aligned}$$

75.

$$\begin{aligned}2a^{2n}+2a^n-24&=2\left(a^{2n}+a^n-12\right)\\&=2\left(a^n+4\right)\left(a^n-3\right)\end{aligned}$$

77.

$$\begin{aligned}&54a^{3n}+16b^{3n}\\&=2\left(27a^{3n}+8b^{3n}\right)\\&=2\left(3a^n+2b^n\right)\left(9a^{2n}-6a^nb^n+4b^{2n}\right)\end{aligned}$$

79.

$$\begin{aligned}12m^{4n}+10m^{2n}+2&=2\left(6m^{4n}+5m^{2n}+1\right)\\&=2\left(3m^{2n}+1\right)\left(2m^{2n}+1\right)\end{aligned}$$

SECTION 5.9

VOCABULARY

1. A **quadratic** equation is any equation that can be written in the form $ax^2 + bx + c = 0$, where $a \neq 0$.

3. To write the quadratic equation $x^2 - 3x = 15$ in **standard** form, we subtract 15 from both sides.

CONCEPTS

5. a. yes
   b. no
   c. yes
   d. no

7. a. At least one is 0.
   b. By the **zero**-factor property, if $ab = 0$, then $a = \boxed{0}$ or $b = \boxed{0}$.

9.

$$a = -5$$
$$a^2 - 9a + 20 = 0$$
$$(-5)^2 - 9(-5) + 20 \stackrel{?}{=} 0$$
$$25 + 45 + 20 \stackrel{?}{=} 0$$
$$90 \neq 0$$

No, -5 is not a solution.

$$a = 4$$
$$a^2 - 9a + 20 = 0$$
$$(4)^2 - 9(4) + 20 \stackrel{?}{=} 0$$
$$16 - 36 + 20 \stackrel{?}{=} 0$$
$$0 = 0$$

Yes, 4 is a solution.

11. a. $10 + x + x = 10 + 2x$
    b. $20 + x + x = 20 + 2x$

NOTATION

13.

$$y^2 - 2y - 8 = 0$$
$$(y-4)\left(\boxed{y+2}\right) = 0$$
$$\boxed{y-4} = 0 \quad \text{or} \quad y + 2 = 0$$
$$y = 4 \qquad\qquad y = \boxed{-2}$$

GUIDED PRACTICE

15.

$$z^2 + 8z + 15 = 0$$
$$(z+5)(z+3) = 0$$
$$z + 5 = 0 \quad \text{or} \quad z + 3 = 0$$
$$z = -5 \qquad\qquad z = -3$$

17.

$$x^2 + 6x + 8 = 0$$
$$(x+4)(x+2) = 0$$
$$x + 4 = 0 \quad \text{or} \quad x + 2 = 0$$
$$x = -4 \qquad\qquad x = -2$$

19.

$$2x^2 - 3x + 1 = 0$$
$$(2x-1)(x-1) = 0$$
$$2x - 1 = 0 \quad \text{or} \quad x - 1 = 0$$
$$2x = 1 \qquad\qquad x = 1$$
$$x = \frac{1}{2}$$

21.

$$3m^2 + 10m + 3 = 0$$
$$(3m+1)(m+3) = 0$$
$$3m + 1 = 0 \quad \text{or} \quad m + 3 = 0$$
$$3m = -1 \qquad\qquad m = -3$$
$$m = -\frac{1}{3}$$

23.

$$x^2 + x = 0$$
$$x(x+1) = 0$$
$$x = 0 \quad \text{or} \quad x + 1 = 0$$
$$x = 0 \qquad\qquad x = -1$$

25.

$$4x^2 = 8x$$
$$4x^2 - 8x = 0$$
$$4x(x-2) = 0$$
$$4x = 0 \quad \text{or} \quad x - 2 = 0$$
$$x = 0 \qquad x = 2$$

27.

$$y^2 - 16 = 0$$
$$(y+4)(y-4) = 0$$
$$y + 4 = 0 \quad \text{or} \quad y - 4 = 0$$
$$y = -4 \qquad y = 4$$

29.

$$16y^2 = 9$$
$$16y^2 - 9 = 0$$
$$(4y+3)(4y-3) = 0$$
$$4y + 3 = 0 \quad \text{or} \quad 4y - 3 = 0$$
$$4y = -3 \qquad 4y = 3$$
$$y = -\frac{3}{4} \qquad y = \frac{3}{4}$$

31.

$$\frac{3a^2}{2} = \frac{1}{2} - a$$
$$2\left(\frac{3a^2}{2}\right) = 2\left(\frac{1}{2} - a\right)$$
$$3a^2 = 1 - 2a$$
$$3a^2 + 2a - 1 = 0$$
$$(3a-1)(a+1) = 0$$
$$3a - 1 = 0 \quad \text{or} \quad a + 1 = 0$$
$$3a = 1 \qquad a = -1$$
$$a = \frac{1}{3}$$

33.

$$x^2 - \frac{2}{5} = -\frac{9}{5}x$$
$$5\left(x^2 - \frac{2}{5}\right) = 5\left(-\frac{9}{5}x\right)$$
$$5x^2 - 2 = -9x$$
$$5x^2 + 9x - 2 = 0$$
$$(5x-1)(x+2) = 0$$
$$5x - 1 = 0 \quad \text{or} \quad x + 2 = 0$$
$$5x = 1 \qquad x = -2$$
$$x = \frac{1}{5}$$

35.

$$\frac{8}{3}a^2 = 1 - \frac{10}{3}a$$
$$3\left(\frac{8}{3}a^2\right) = 3\left(1 - \frac{10}{3}a\right)$$
$$8a^2 = 3 - 10a$$
$$8a^2 + 10a - 3 = 0$$
$$(2a+3)(4a-1) = 0$$
$$2a + 3 = 0 \quad \text{or} \quad 4a - 1 = 0$$
$$2a = -3 \qquad 4a = 1$$
$$a = -\frac{3}{2} \qquad a = \frac{1}{4}$$

37.

$$\frac{3}{16}m^2 - \frac{27}{16} = 0$$
$$16\left(\frac{3}{16}m^2 - \frac{27}{16}\right) = 16(0)$$
$$3m^2 - 27 = 0$$
$$3\left(m^2 - 9\right) = 0$$
$$3(m+3)(m-3) = 0$$
$$m + 3 = 0 \quad \text{or} \quad m - 3 = 0$$
$$m = -3 \qquad m = 3$$

39.

$$(x+7)^2=-2(x+7)-1$$
$$x^2+7x+7x+49=-2x-14-1$$
$$x^2+14x+49=-2x-15$$
$$x^2+16x+64=0$$
$$(x+8)(x+8)=0$$
$$x+8=0 \quad \text{or} \quad x+8=0$$
$$x=-8 \qquad x=-8$$

41.

$$(m+4)(2m+3)-22=10m$$
$$2m^2+3m+8m+12-22=10m$$
$$2m^2+11m-10=10m$$
$$2m^2+m-10=0$$
$$(2m+5)(m-2)=0$$
$$2m+5=0 \quad \text{or} \quad m-2=0$$
$$2m=-5 \qquad m=2$$
$$m=-\frac{5}{2}$$

43.

$$x^2-3x+3=1$$
$$x^2-3x+2=0$$
$$(x-2)(x-1)=0$$
$$x-2=0 \quad \text{or} \quad x-1=0$$
$$x=2 \qquad x=1$$

45.

$$x^3-6x^2+8x+2=2$$
$$x^3-6x^2+8x=0$$
$$x(x^2-6x+8)=0$$
$$x(x-2)(x-4)=0$$
$$x=0,\ x-2=0,\ \text{or}\ x-4=0$$
$$x=2 \qquad x=4$$

47.

$$x^3-4x^2=21x$$
$$x^3-4x^2-21x=0$$
$$x(x^2-4x-21)=0$$
$$x(x-7)(x+3)=0$$
$$x=0,\ x-7=0,\ \text{or}\ x+3=0$$
$$x=7 \qquad x=-3$$

49.

$$y^3-49y=0$$
$$y(y^2-49)=0$$
$$y(y+7)(y-7)=0$$
$$y=0,\ y+7=0,\ \text{or}\ y-7=0$$
$$y=-7 \qquad y=7$$

51.

$$-z^4+37z^2-36=0$$
$$-(z^4-37z^2+36)=0$$
$$-(z^2-1)(z^2-36)=0$$
$$-(z+1)(z-1)(z+6)(z-6)=0$$
$$z+1=0,\ z-1=0,\ z+6=0,\ \text{or}\ z-6=0$$
$$z=-1 \qquad z=1 \qquad z=-6 \qquad z=6$$

53.

$$x^3-3x^2-4x+12=0$$
$$(x^3-3x^2)-(4x-12)=0$$
$$x^2(x-3)-4(x-3)=0$$
$$(x-3)(x^2-4)=0$$
$$(x-3)(x+2)(x-2)=0$$
$$x-3=0 \qquad x+2=0 \qquad x-2=0$$
$$x=3 \qquad x=-2 \qquad x=2$$

55. $2x^2-7x+4=0$
The solutions are the x-intercepts.
$x=2.78$ or $x=0.72$

57. $-3x^3-2x^2+5=0$
The solutions are the x-intercepts.
$x=1$

59.

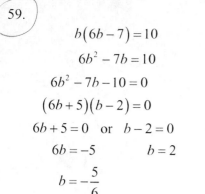

$$b(6b-7)=10$$
$$6b^2-7b=10$$
$$6b^2-7b-10=0$$
$$(6b+5)(b-2)=0$$
$$6b+5=0 \quad \text{or} \quad b-2=0$$
$$6b=-5 \qquad b=2$$
$$b=-\frac{5}{6}$$

61.

$$x^3+x^2=0$$
$$x^2(x+1)=0$$
$$x=0, \quad x=0, \quad \text{or} \quad x+1=0$$
$$x=-1$$

63.

$$\frac{x^2}{5}-\frac{4}{5}=-\frac{3}{5}x$$
$$5\left(\frac{x^2}{5}-\frac{4}{5}\right)=5\left(-\frac{3}{5}x\right)$$
$$x^2-4=-3x$$
$$x^2+3x-4=0$$
$$(x+4)(x-1)=0$$
$$x+4=0 \quad \text{or} \quad x-1=0$$
$$x=-4 \qquad x=1$$

65.

$$a^3-2a^2-16a+32=0$$
$$a^2(a-2)-16(a-2)=0$$
$$(a-2)(a^2-16)=0$$
$$(a-2)(a-4)(a+4)=0$$
$$a-2=0, \quad a-4=0, \quad a+4=0$$
$$a=2 \qquad a=4 \qquad a=-4$$

67.

$$6y^2=25y$$
$$6y^2-25y=0$$
$$y(6y-25)=0$$
$$y=0 \quad \text{or} \quad 6y-25=0$$
$$y=0 \qquad 6y=25$$
$$y=\frac{25}{6}$$

69.

$$a^3-18a^2=-81a$$
$$a^3-18a^2+81a=0$$
$$a(a^2-18a+81)=0$$
$$a(a-9)^2=0$$
$$a=0 \quad \text{or} \quad a-9=0$$
$$a=9$$

9 is a repeated solution

71.

$$7t^2-2t-5=0$$
$$(7t+5)(t-1)=0$$
$$7t+5=0 \quad \text{or} \quad t-1=0$$
$$7t=-5 \qquad t=1$$
$$t=-\frac{5}{7}$$

73.

$$3x^3+3x^2=12(x+1)$$
$$3x^3+3x^2=12x+12$$
$$3x^3+3x^2-12x-12=0$$
$$\frac{3x^3}{3}+\frac{3x^2}{3}-\frac{12x}{3}-\frac{12}{3}=\frac{0}{3}$$
$$x^3+x^2-4x-4=0$$
$$(x^3+x^2)-(4x+4)=0$$
$$x^2(x+1)-4(x+1)=0$$
$$(x+1)(x^2-4)=0$$
$$(x+1)(x+2)(x-2)=0$$
$$x+1=0, \quad x+2=0, \quad \text{or} \quad x-2=0$$
$$x=-1 \qquad x=-2 \qquad x=2$$

75.

$$64t^2-81=0$$
$$(8t-9)(8t+9)=0$$
$$8t-9=0 \quad \text{or} \quad 8t+9=0$$
$$8t=9 \qquad 8t=-9$$
$$t=\frac{9}{8} \qquad t=-\frac{9}{8}$$

77.

$$\frac{x^2(6x+37)}{35}=x$$
$$35\cdot\frac{x^2(6x+37)}{35}=35\cdot x$$
$$x^2(6x+37)=35x$$
$$6x^3+37x^2=35x$$
$$6x^3+37x^2-35x=0$$
$$x(6x^2+37x-35)=0$$
$$x(6x-5)(x+7)=0$$
$$x=0,\quad 6x-5=0,\quad \text{or}\quad x+7=0$$
$$6x=5 \qquad x=-7$$
$$x=\frac{5}{6}$$

79.

$$0=d^2-5d-66$$
$$0=(d-11)(d+6)$$
$$d-11=0 \quad \text{or} \quad d+6=0$$
$$d=11 \qquad d=-6$$

81.

$$n(3n-4)=(n-6)^2+11n-1$$
$$n(3n-4)=n^2-12n+36+11n-1$$
$$3n^2-4n=n^2-n+35$$
$$2n^2-3n-35=0$$
$$(2n+7)(n-5)=0$$
$$2n+7=0 \quad \text{or} \quad n-5=0$$
$$2n=-7 \qquad n=5$$
$$n=-\frac{7}{2}$$

83.

$$(x-5.5)(x+3)=0$$
$$x-5.5=0 \quad \text{or} \quad x+3=0$$
$$x=5.5 \qquad x=-3$$

85.

$$-x^4+34x^2-225=0$$
$$-(x^4-34x^2+225)=0$$
$$-(x^2-25)(x^2-9)=0$$
$$-(x+5)(x-5)(x+3)(x-3)=0$$
$$x+5=0,\quad x-5=0,\quad x+3=0,\quad \text{or}\quad x-3=0$$
$$x=-5 \qquad x=5 \qquad x=-3 \qquad x=3$$

87.

$$f(t)=g(t)$$
$$3t^2-2t+2=8t^2+t$$
$$0=5t^2+3t-2$$
$$0=(5t-2)(t+1)$$
$$5t-2=0 \qquad t+1=0$$
$$5t=2 \qquad t=-1$$
$$t=\frac{2}{5}$$

89.

$$s(x)=t(x)$$
$$(x+11)(x+3)=(x+5)(2x+1)$$
$$x^2+3x+11x+33=2x^2+x+10x+5$$
$$x^2+14x+33=2x^2+11x+5$$
$$0=x^2-3x-28$$
$$0=(x-7)(x+4)$$
$$x-7=0 \qquad x+4=0$$
$$x=7 \qquad x=-4$$

APPLICATIONS

91. INTEGER PROBLEM
Let x = first integer
Then $x + 2$ = second integer
Product indicates multiplication.

$$x(x+2)=288$$
$$x^2+2x=288$$
$$x^2+2x-288=0$$
$$(x+18)(x-16)=0$$
$$x+18=0 \quad \text{or} \quad x-16=0$$
$$x=-18 \qquad x=16$$
$$x+2=-16 \qquad x+2=18$$

The positive integers are 16 and 18.

93. KITCHENS

Let the height = x and then the base = $2x$.

The area of a triangle is $A = \frac{1}{2}bh$.

$$A = \frac{1}{2}bh$$
$$16 = \frac{1}{2}(2x)(x)$$
$$16 = x^2$$
$$0 = x^2 - 16$$
$$0 = (x+4)(x-4)$$
$$x + 4 = 0 \quad \text{or} \quad x - 4 = 0$$
$$x = -4 \qquad x = 4$$

Since the height cannot be negative, $h = 4$ feet. So, $b = 2(4) = 8$ feet.

95. COOKING

$$(\text{length})(\text{width}) = \text{Area}$$
$$w(w+6) = 160$$
$$w^2 + 6w = 160$$
$$w^2 + 6w - 160 = 0$$
$$(w+16)(w-10) = 0$$
$$w + 16 = 0 \quad \text{or} \quad w - 10 = 0$$
$$w = -16 \qquad w = 10$$
$$w + 6 = -10 \qquad w + 6 = 16$$

Since length and width cannot be negative integers, the width is 10 inches and the length is 16 inches.

97. LANDSCAPE ARCHITECT

Let x = the width of the walkway.

$$(\text{length})(\text{width}) = \text{Area of pool}$$
$$(2x+6)(2x+8) - 48 = 120$$
$$4x^2 + 28x + 48 - 48 = 120$$
$$4x^2 + 28x - 120 = 0$$
$$4(x^2 + 7x - 30) = 0$$
$$4(x+10)(x-3) = 0$$
$$x + 10 = 0 \quad \text{or} \quad x - 3 = 0$$
$$x = -10 \qquad x = 3$$

Since the width cannot be negative, the width of the walkway is 3 feet.

99. FRAMES

Let x = the width of the frame. There are 2 parts of the frame on each side of the picture so the width of the picture is $(8 - 2x)$ and the length is $(10 - 2x)$.

Area of picture = length(width)

$$15 = (10 - 2x)(8 - 2x)$$
$$15 = 80 - 20x - 16x + 4x^2$$
$$0 = 4x^2 - 36x + 65$$
$$0 = (2x - 13)(2x - 5)$$
$$2x - 13 = 0 \qquad 2x - 5 = 0$$
$$2x = 13 \qquad 2x = 5$$
$$x = 6.5 \qquad x = 2.5$$

Since 6.5 would be too large for each end of the frame, the width of the frame would have to be 2.5 cm.

101. ARCHITECTURE

Let x = width of big room

Then $2x$ = length of big room

Area of big room = $x(2x) = 2x^2$

Area of larger room = 560

Area of small room = 12(width) = $12x$

A of big = A of larger + A of smaller

$$2x^2 = 560 + 12x$$
$$2x^2 - 12x - 560 = 0$$
$$\frac{2x^2}{2} - \frac{12x}{2} - \frac{560}{2} = \frac{0}{2}$$
$$x^2 - 6x - 280 = 0$$
$$(x - 20)(x + 14) = 0$$
$$x - 20 = 0 \quad \text{or} \quad x + 14 = 0$$
$$x = 20 \qquad x = -14$$
$$2x = 40$$

The dimensions are 20 ft by 40 ft.

84.

103. BURP GUNS

Let $h = 0$.

$$h = -16t^2 + 63t + 4$$
$$0 = -16t^2 + 63t + 4$$
$$16t^2 - 63t - 4 = 0$$
$$(t-4)(16t+1) = 0$$
$$t - 4 = 0 \qquad 16t + 1 = 0$$
$$t = 4 \qquad t = -\frac{1}{16}$$

Since the time cannot be negative, $t = 4$ seconds.

105. BUNGEE JUMPING

$$h = -16t^2 + 212$$
$$148 = -16t^2 + 212$$
$$148 + 16t^2 - 212 = -16t^2 + 212 + 16t^2 - 212$$
$$16t^2 - 64 = 0$$
$$\frac{16t^2}{16} - \frac{64}{16} = \frac{0}{16}$$
$$t^2 - 4 = 0$$
$$(t+2)(t-2) = 0$$
$$t + 2 = 0 \text{ or } t - 2 = 0$$
$$t = -2 \qquad t = 2$$

It will take 2 seconds.

107. FORENSIC MEDICINE

$$E = \frac{1}{2}mv^2$$
$$54 = \frac{1}{2}(3)v^2$$
$$54 = \frac{3}{2}v^2$$
$$\frac{2}{3}(54) = \frac{2}{3}\left(\frac{3}{2}v^2\right)$$
$$36 = v^2$$
$$0 = v^2 - 36$$
$$0 = (v+6)(v-6)$$
$$v + 6 = 0 \text{ or } v - 6 = 0$$
$$v = -6 \qquad v = 6$$

The velocity is 6 m/sec.

109. BREAK–EVEN POINT

$$C(x) = R(x)$$
$$\frac{1}{8}x^2 - x + 6 = \frac{1}{4}x^2$$
$$8\left(\frac{1}{8}x^2 - x + 6\right) = 8\left(\frac{1}{4}x^2\right)$$
$$x^2 - 8x + 48 = 2x^2$$
$$0 = x^2 + 8x - 48$$
$$0 = (x+12)(x-4)$$
$$x + 12 = 0 \text{ or } x - 4 = 0$$
$$x = -12 \qquad x = 4$$

He must sell 4 guitars.

WRITING

111. Answers will vary.

113. Answers will vary.

115. Answers will vary.

REVIEW

117. ALUMINUM FOIL

Change inches to feet by dividing by 12. Change yards to feet by multiplying by 3.

$$12 \text{ inches} = \frac{12}{12}$$
$$= 1 \text{ foot}$$
$$8\frac{1}{3} \text{ yd} = 8\frac{1}{3}(3)$$
$$= \frac{25}{3}\left(\frac{3}{1}\right)$$
$$= 25 \text{ ft}$$
$$A = LW$$
$$= 25(1)$$
$$= 25 \text{ ft}^2$$

CHALLENGE PROBLEMS

119. $x=\frac{1}{4}$ or $x=-\frac{4}{3}$

$$\left(x-\frac{1}{4}\right)\left(x+\frac{4}{3}\right)=0$$

$$x^2+\frac{4}{3}x-\frac{1}{4}x-\frac{1}{3}=0$$

$$x^2+\frac{13}{12}x-\frac{1}{3}=0$$

$$12\left(x^2+\frac{13}{12}x-\frac{1}{3}\right)=12(0)$$

$$12x^2+13x-4=0$$

121.

$$\frac{r^2}{105}+\frac{r}{140}=\frac{1}{42}$$

$$420\left(\frac{r^2}{105}+\frac{r}{140}\right)=420\left(\frac{1}{42}\right)$$

$$4r^2+3r=10$$

$$4r^2+3r-10=0$$

$$(4r-5)(r+2)=0$$

$$4r-5=0 \quad \text{or} \quad r+2=0$$

$$4r=5 \qquad r=-2$$

$$r=\frac{5}{4}$$

123

$$5x^2+3x+2=2x(x+1)+3x^2$$

$$5x^2+3x+2=2x^2+2x+3x^2$$

$$5x^2+3x+2=2x+5x^2$$

$$5x^2+3x+2-2x-5x^2-=2x+5x^2-2x-5x^2$$

$$x+2=0$$

The polynomial is not a quadratic equation because after simplifying, we obtain $x+2=0$, which is a linear equation.

.

CHAPTER 5 REVIEW

SECTION 5.1
Exponents

1. $3^5 = 3 \cdot 3 \cdot 3 \cdot 3 \cdot 3$
   $= 243$

2. $-2^5 = -(2 \cdot 2 \cdot 2 \cdot 2 \cdot 2)$
   $= -32$

3. $(-4)^3 = (-4) \cdot (-4) \cdot (-4)$
   $= -64$

4. $\left(\frac{2}{3}\right)^2 = \left(\frac{2}{3}\right)\left(\frac{2}{3}\right)$
   $= \frac{4}{9}$

5.
$$x^4 \cdot x^2 = x^{4+2}$$
$$= x^6$$

6.
$$m^{-3}n^{-4}m^6n^{-1} = m^{-3+6}n^{-4+(-1)}$$
$$= m^3n^{-5}$$
$$= \frac{m^3}{n^5}$$

7.
$$\frac{\left(4m^6\right)^3}{m^2} = \frac{4^3m^{6(3)}}{m^2}$$
$$= \frac{64m^{18}}{m^2}$$
$$= 64m^{18-2}$$
$$= 64m^{16}$$

8.
$$\left(-t^2\right)^2\left(t^3\right)^3 = \left(t^4\right)\left(t^9\right)$$
$$= t^{4+9}$$
$$= t^{13}$$

9.
$$\left(3x^2y^3\right)^2 = 3^2 \cdot x^{2(2)} \cdot y^{3(2)}$$
$$= 9x^4y^6$$

10.
$$\left(\frac{x^4}{b}\right)^4 = \frac{x^{4(4)}}{b^{1(4)}}$$
$$= \frac{x^{16}}{b^4}$$

11.
$$-10x^0 = -10(1)$$
$$= -10$$

12.
$$\frac{1}{5h^{-12}} = \frac{h^{12}}{5}$$

13.
$$\left(\frac{2a}{b}\right)^{-1} = \left(\frac{b}{2a}\right)^1$$
$$= \frac{b}{2a}$$

14.
$$\frac{x}{5x^{-4}} = \frac{x \cdot x^4}{5}$$
$$= \frac{x^5}{5}$$

15.
$$\left(3x^{-3}\right)^{-2} = 3^{-2}x^{-3(-2)}$$
$$= 3^{-2}x^6$$
$$= \frac{x^6}{3^2}$$
$$= \frac{x^6}{9}$$

16.
$$\frac{2x^{-4}x^3}{9} = \frac{2x^{-4+3}}{9}$$
$$= \frac{2x^{-1}}{9}$$
$$= \frac{2}{9x}$$

17.

$$-\left(\frac{c^{-3}}{c^{-5}}\right)^5 = -\left(c^{-3-(-5)}\right)^5$$
$$= -\left(c^2\right)^5$$
$$= -c^{2(5)}$$
$$= -c^{10}$$

18.

$$\left(\frac{4}{5}\right)^{-2} = \left(\frac{5}{4}\right)^2$$
$$= \frac{25}{16}$$

19.

$$\frac{3^{-3}}{4^{-2}} = \frac{4^2}{3^3}$$
$$= \frac{16}{27}$$

20.

$$\frac{1}{4^{-3}} = \frac{4^3}{1}$$
$$= 64$$

21.

$$\left(\frac{s^7}{s^{-8}}\right)^3\left(\frac{s^{-5}}{s^2}\right)^{-4} = \left(s^{7-(-8)}\right)^3\left(s^{-5-2}\right)^{-4}$$
$$= \left(s^{15}\right)^3\left(s^{-7}\right)^{-4}$$
$$= s^{15(3)}s^{-7(-4)}$$
$$= s^{45}s^{28}$$
$$= s^{45+28}$$
$$= s^{73}$$

22.

$$\left(\frac{-2a^4bc^{-6}}{a^{-3}b^2c^{-5}}\right)^{-3} = \left(-2a^{4-(-3)}b^{1-2}c^{-6-(-5)}\right)^{-3}$$
$$= \left(-2a^7b^{-1}c^{-1}\right)^{-3}$$
$$= \left(-\frac{2a^7}{bc}\right)^{-3}$$
$$= \left(-\frac{bc}{2a^7}\right)^3$$
$$= -\frac{b^3c^3}{2^3a^{7(3)}}$$
$$= -\frac{b^3c^3}{8a^{21}}$$

SECTION 5.2
Scientific Notation

23. $19{,}300{,}000{,}000 = 1.93 \times 10^{10}$

24. $0.00000002735 = 2.735 \times 10^{-8}$

25. $7.2777 \times 10^7 = 72{,}777{,}000$

26. $8.3 \times 10^{-9} = 0.0000000083$

27.

$$t = \frac{d}{r}$$
$$= \frac{228{,}000{,}000}{300{,}000}$$
$$= \frac{2.28\times10^8}{3\times10^5}$$
$$= \frac{2.28}{3}\times\frac{10^8}{10^5}$$
$$= 0.76\times10^3$$
$$= 7.6\times10^2 \text{ sec.}$$

28.

$$\left(1.67248\times10^{-24}\right)\left(1\times10^6\right)$$
$$= \left(1.67248\times1\right)\times\left(10^{-24+6}\right)$$
$$= 1.67248\times10^{-18}\text{ grams}$$

29.

$$\frac{(616{,}000{,}000)(0.000009)}{0.00066}$$
$$=\frac{(6.16\times10^{8})(9\times10^{-6})}{6.6\times10^{-4}}$$
$$=\frac{55.44\times10^{2}}{6.6\times10^{-4}}$$
$$=\frac{55.44}{6.6}\times10^{2-(-4)}$$
$$=8.4\times10^{6}$$

30.

$$\frac{0.0000000495}{(33{,}000)(800{,}000{,}000)}=\frac{4.95\times10^{-8}}{(3.3\times10^{4})(8.0\times10^{8})}$$
$$=\frac{4.95\times10^{-8}}{(3.3\times8.0)(10^{4}\times10^{8})}$$
$$=\frac{4.95\times10^{-8}}{26.4\times10^{12}}$$
$$=\frac{4.95}{26.4}\times10^{-8-12}$$
$$=0.1875\times10^{-20}$$
$$=1.875\times10^{-21}$$

SECTION 5.3
Polynomials and Polynomial Functions

31 No, it is not a polynomial.

32. Yes, it is a polynomial.

33. Yes, it is a polynomial.

34. No, it is not a polynomial.

35. binomial
degree 2

36. monomial
degree: 3 + 1 = 4

37. none of these
degree: 4

38. trinomial
degree: 4 + 4 = 8

39. $f(x)=x^3-5x^2-2x-5$
$$f(3)=3^3-5(3)^2-2(3)-5$$
$$=27-5(9)-2(3)-5$$
$$=27-45-6-5$$
$$=-29$$

40. $V(r)=4.19r^3+25.13r^2$
$$V(2)=4.19(2)^3+25.13(2)^2$$
$$=4.19(8)+25.13(4)$$
$$=33.52+100.52$$
$$=134.04$$
$$\approx134\text{ in.}^3$$

41.

$$9t^3-5t^2-5t+3t^2-7t^3$$
$$=(9t^3-7t^3)+(-5t^2+3t^2)+(-5t)$$
$$=2t^3-2t^2-5t$$

42.

$$\frac{5}{4}ab^2c-\frac{5}{3}abc-\frac{1}{2}ab^2c+\frac{5}{6}abc$$
$$=\left(\frac{5}{4}ab^2c-\frac{1}{2}ab^2c\right)+\left(-\frac{5}{3}abc+\frac{5}{6}abc\right)$$
$$=\left(\frac{5}{4}ab^2c-\frac{2}{4}ab^2c\right)+\left(-\frac{10}{6}abc+\frac{5}{6}abc\right)$$
$$=\frac{3}{4}ab^2c-\frac{5}{6}abc$$

43.

$$(2x^2y^3-5x^2y+9y)+(x^2y^3-3x^2y-y)$$
$$=3x^2y^3-8x^2y+8y$$

44.

$$(8m^2-7m)-(-11m^2+6m+9)$$
$$=8m^2-7m+11m^2-6m-9$$
$$=19m^2-13m-9$$

45.

$$\left(\frac{2}{3}s^6-\frac{1}{6}s^4\right)+\left(-\frac{1}{6}s^6-\frac{3}{2}s^4\right)$$
$$=\left(\frac{2}{3}s^6-\frac{1}{6}s^6\right)+\left(-\frac{1}{6}s^4-\frac{3}{2}s^4\right)$$
$$=\left(\frac{4}{6}s^6-\frac{1}{6}s^6\right)+\left(-\frac{1}{6}s^4-\frac{9}{6}s^4\right)$$
$$=\frac{3}{6}s^6-\frac{10}{6}s^4$$
$$=\frac{1}{2}s^6-\frac{5}{3}s^4$$

46.

$$\begin{array}{rr} & -10k^4-4k^3+5k^2-k+1 \\ - & \underline{-16k^4+2k^3-4k^2-k+3} \end{array}$$

$$\begin{array}{rr} & -10k^4-4k^3+5k^2-k+1 \\ + & \underline{16k^4-2k^3+4k^2+k-3} \\ & 6k^4-6k^3+9k^2-2 \end{array}$$

47.

$$\left[\left(-c^2d^2+5c^2d-10cd^2\right)+\left(11c^2d^2-c^2d+9cd^2\right)\right]$$
$$-\left(6c^2d^2+4c^2d-5cd^2\right)$$
$$=\left[10c^2d^2+4c^2d-cd^2\right]-\left(6c^2d^2+4c^2d-5cd^2\right)$$
$$=10c^2d^2+4c^2d-cd^2-6c^2d^2-4c^2d+5cd^2$$
$$=4c^2d^2+4cd^2$$

48.

$$f(x)+g(x)=\left(x^5+3x^2+2x\right)+\left(4x^5+2x^2-5x\right)$$
$$=x^5+3x^2+2x+4x^5+2x^2-5x$$
$$=5x^5+5x^2-3x$$

49.

$$s(t)-g(t)$$
$$=\left(1.7t^3-0.5t^2+0.4t-0.3\right)-\left(0.5t^3-0.9t^2-0.8t+1.1\right)$$
$$=1.7t^3-0.5t^2+0.4t-0.3-0.5t^3+0.9t^2+0.8t-1.1$$
$$=1.2t^3+0.4t^2+1.2t-1.4$$

50. a. $f(0)=-1$
b. $x=-2, -1, 1$
c. $D=(-\infty, \infty)$
$R=[-1, \infty)$

SECTION 5.4
Multiplying Polynomials

51.

$$\left(8a^2\right)\left(-\frac{1}{2}a\right)=-4a^3$$

52.

$$\left(-3xy^2z\right)\left(-2xz^3\right)(xz)=6x^{1+1+1}y^2z^{1+3+1}$$
$$=6x^3y^2z^5$$

53.

$$2xy^2\left(x^3y-4xy^5\right)=2x^4y^3-8x^2y^7$$

54.

$$-a^2b\left(-a^2-2ab+b^2\right)=a^4b+2a^3b^2-a^2b^3$$

55.

$$\left(3x^2+2\right)(2x-4)=6x^3-12x^2+4x-8$$

56.

$$(5at-6)^2=(5at)^2-2(5at)(6)+(-6)^2$$
$$=25a^2t^2-60at+36$$

57.

$$\left(7c^4d^3-d\right)\left(7c^4d^3+d\right)=\left(7c^4d^3\right)^2-(d)^2$$
$$=49c^8d^6-d^2$$

58.

$$\left(5x^2-4x\right)\left(3x^2-2x+10\right)$$
$$=5x^2\left(3x^2-2x+10\right)-4x\left(3x^2-2x+10\right)$$
$$=15x^4-10x^3+50x^2-12x^3+8x^2-40x$$
$$=15x^4-22x^3+58x^2-40x$$

59.

$$(r+s)(r-s)(r-3s)=\left(r^2-s^2\right)(r-3s)$$
$$=r^3-3r^2s-rs^2+3s^3$$

60.

$$\left(3c-\frac{3}{4}\right)^2=(3c)^2-2(3c)\left(\frac{3}{4}\right)+\left(-\frac{3}{4}\right)^2$$
$$=9c^2-\frac{9}{2}c+\frac{9}{16}$$

61.

$$[5-(a-b)]^2 = 5^2 - 2(5)(a-b) + (a-b)^2$$
$$= 25 - 10(a-b) + (a^2 - 2ab + b^2)$$
$$= 25 - 10a + 10b + a^2 - 2ab + b^2$$

62.

$$(2x-y-2z)(x+y+z)$$
$$= 2x(x+y+z) - y(x+y+z) - 2z(x+y+z)$$
$$= 2x^2 + 2xy + 2xz - xy - y^2 - yz - 2xz - 2yz - 2z^2$$
$$= 2x^2 + xy - y^2 - 3yz - 2z^2$$

63.

$$(4a+1)^2 + (5a-2)^2$$
$$= (16a^2 + 8a + 1) + (25a^2 - 20a + 4)$$
$$= 41a^2 - 12a + 5$$

64.

$$f(x) = x^2 + 5x - 1$$
$$f(b-3) = (b-3)^2 + 5(b-3) - 1$$
$$= (b^2 - 6b + 9) + (5b - 15) - 1$$
$$= b^2 - b - 7$$

65.

$$f(x) \cdot g(x) = (0.5x - 0.4)(0.1x + 2.1)$$
$$= 0.05x^2 + 1.05x - 0.04x - 0.84$$
$$= 0.05x^2 + 1.01x - 0.84$$

66.

$$s(x) \cdot h(x) = (x+3)(x^2 - 3x + 9)$$
$$= x^3 - 3x^2 + 9x + 3x^2 - 9x + 27$$
$$= x^3 + 27$$

67. a. $P = 2L + 2W$
$$P = 2(2x) + 2(4x-1)$$
$$= 4x + 8x - 2$$
$$= (12x - 2) \text{ in.}$$

b. $A = LW$
$$A = 2x(4x-1)$$
$$= (8x^2 - 2x) \text{ in.}^2$$

c. $V = LWH$
$$V = 2x(4x-1)(2x+2)$$
$$= (8x^2 - 2x)(2x+2)$$
$$= 16x^3 + 16x^2 - 4x^2 - 4x$$
$$= (16x^3 + 12x^2 - 4x) \text{ in.}^3$$

68. a. Let x = length, $x + 1$ = width, and $x + 2$ = height.
$$f(x) = x(x+1)(x+2)$$
$$f(x) = x(x^2 + 2x + 1x + 2)$$
$$f(x) = x(x^2 + 3x + 2)$$
$$f(x) = x^3 + 3x^2 + 2x$$

b. $f(5) = (5)^3 + 3(5)^2 + 2(5)$
$$= 125 + 3(25) + 2(5)$$
$$= 125 + 75 + 10$$
$$= 210 \text{ in.}^3$$

SECTION 5.5
The Greatest Common Factor and Factoring by Grouping

69. $42 = 2 \cdot 3 \cdot 7$
$36 = 2 \cdot 2 \cdot 3 \cdot 3$
$54 = 2 \cdot 3 \cdot 3 \cdot 3$
GCF $= 2 \cdot 3 =$ **6**

70. $6x^2y^5 = 2 \cdot 3 \cdot x \cdot x \cdot y \cdot y \cdot y \cdot y \cdot y$
$15xy^3 = 3 \cdot 5 \cdot x \cdot y \cdot y \cdot y$
GCF $= 3 \cdot x \cdot y \cdot y \cdot y =$ $\mathbf{3xy^3}$

71. $4x^4 + 8 = 4(x^4 + 2)$

72. $\frac{3x^3}{5} - \frac{6x^2}{5} + \frac{x}{5} = \frac{x}{5}(3x^2 - 6x + 1)$

73. $6x - 11$ is prime

74. $7a^4b^2 + 49a^3b = 7a^3b(ab+7)$

75.

$$\begin{aligned} &5x^2(x+y) - 15x^3(x+y) \\ &= 5x^2[1(x+y) - 3x(x+y)] \\ &= 5x^2(x+y)(1-3x) \end{aligned}$$

76.

$$\begin{aligned} &27x^3y^3z^3 + 81x^4y^5z^2 - 90x^2y^3z^7 \\ &= 9x^2y^3z^2(3xz + 9x^2y^2 - 10z^5) \end{aligned}$$

77.

$$-x - 9 = -(x+9)$$

78.

$$\begin{aligned} 4r - 7 &= -(-4r+7) \\ &= -(7-4r) \end{aligned}$$

79.

$$-7b^3 + 14c = -7(b^3 - 2c)$$

80.

$$\begin{aligned} &-49a^3b^2(a-b)^4 + 63a^2b^4(a-b)^3 \\ &= -7a^2b^2(a-b)^3(7a(a-b) - 9b^2) \\ &= -7a^2b^2(a-b)^3(7a^2 - 7ab - 9b^2) \end{aligned}$$

81.

$$\begin{aligned} xy + 2y + 4x + 8 &= (xy + 2y) + (4x + 8) \\ &= y(x+2) + 4(x+2) \\ &= (x+2)(y+4) \end{aligned}$$

82.

$$\begin{aligned} &r^2y - ar - ry + a + r - 1 \\ &= r^2y - ar + r - ry + a - 1 \\ &= (r^2y - ar + r) + (-ry + a - 1) \\ &= r(ry - a + 1) - 1(ry - a + 1) \\ &= (ry - a + 1)(r - 1) \end{aligned}$$

83.

$$\begin{aligned} t^3 - 9 + t - 9t^2 &= t^3 - 9t^2 + t - 9 \\ &= t^2(t-9) + 1(t-9) \\ &= (t-9)(t^2+1) \end{aligned}$$

84.

$$\begin{aligned} 1 - x - 3z + 3xz &= 1(1-x) - 3z(1-x) \\ &= (1-x)(1-3z) \end{aligned}$$

85.

$$\begin{aligned} m_1m_2 &= mm_2 + mm_1 \\ m_1m_2 - mm_1 &= mm_2 + mm_1 - mm_1 \\ m_1m_2 - mm_1 &= mm_2 \\ m_1(m_2 - m) &= mm_2 \\ \frac{m_1(m_2 - m)}{m_2 - m} &= \frac{mm_2}{m_2 - m} \\ m_1 &= \frac{mm_2}{m_2 - m} \end{aligned}$$

86.

$$\begin{aligned} A &= 2\pi r^2 + 2\pi rh \\ A &= 2\pi r(r+h) \end{aligned}$$

SECTION 5.6
Factoring Trinomials

87. $x^2 + 10x + 25 = (x+5)^2$

88. $49a^6 + 84a^3b^2 + 36b^4 = (7a^3 + 6b^2)^2$

89. $y^2 - 21y + 20 = (y-20)(y-1)$

90. $z^2 + 30 - 11z = z^2 - 11z + 30 = (z-5)(z-6)$

91. $-x^2 - 3x + 28 = -(x^2 + 3x - 28) = -(x+7)(x-4)$

92. $a^2 - 24b^2 - 5ab = a^2 - 5ab - 24b^2 = (a-8b)(a+3b)$

93. $4a^2 - 5a + 1 = (4a-1)(a-1)$

94. $3b^2 + 2b + 1$
prime

95. $y^8 + y^7 - 2y^6 = y^6(y^2 + y - 2)$
$= y^6(y + 2)(y - 1)$

96. $27r^2st + 90rst - 72st = 9st(3r^2 + 10r - 8)$
$= 9st(3r - 2)(r + 4)$

97. $6t^2(r + s) + 13t(r + s) - 15(r + s)$
$= (r + s)(6t^2 + 13t - 15)$
$= (r + s)(t + 3)(6t - 5)$

98. $v^4 - 13v^2 + 42 = (v^2 - 6)(v^2 - 7)$

99. $w^8 - w^4 - 90 = (w^4 - 10)(w^4 + 9)$

100. $(s + t)^2 - 2(s + t) + 1$
Let $x = (s + t)$
$x^2 - 2x + 1 = (x - 1)^2$
Substitute $(s + t)$ for x in the answer.
$(x - 1)^2 = (s + t - 1)^2$

SECTION 5.7
The Difference of Two Squares; The Sum and Difference of Two Cubes

101. $z^2 - 16 = (z + 4)(z - 4)$

102. $x^2y^4 - 64z^6 = (xy^2 + 8z^3)(xy^2 - 8z^3)$

103. $a^2b^2 + c^2$
prime

104. $c^2 - (a + b)^2 = (c + a + b)(c - a - b)$

105.
$$10m^6 - 160m^2 = 10m^2(m^4 - 16)$$
$$= 10m^2(m^2 + 4)(m^2 - 4)$$
$$= 10m^2(m^2 + 4)(m + 2)(m - 2)$$

106.
$$m^2 - n^2 + m + n = (m^2 - n^2) + (m + n)$$
$$= (m + n)(m - n) + 1(m + n)$$
$$= (m + n)(m - n + 1)$$

107.
$$32a^4c - 162b^4c$$
$$= 2c(16a^4 - 81b^4)$$
$$= 2c(4a^2 + 9b^2)(4a^2 - 9b^2)$$
$$= 2c(4a^2 + 9b^2)(2a + 3b)(2a - 3b)$$

108.
$$k^2 + 2k + 1 - 9m^2 = (k^2 + 2k + 1) - 9m^2$$
$$= (k + 1)^2 - 9m^2$$
$$= (k + 1 + 3m)(k + 1 - 3m)$$

109.
$$t^3 + 64 = (t + 4)(t^2 - 4t + 16)$$

110.
$$8a^3 - 125b^9$$
$$= (2a - 5b^3)((2a)^2 + (2a)(5b^3) + (5b^3)^2)$$
$$= (2a - 5b^3)(4a^2 + 10ab^3 + 25b^6)$$

111.
$$4d^7 + 4d^4 = 4d^4(d^3 + 1)$$
$$= 4d^4(d + 1)(d^2 - d + 1)$$

112.
$$(b + c)^3 + 27$$
$$= ((b + c) + 3)((b + c)^2 - 3(b + c) + 3^2)$$
$$= (b + c + 3)(b^2 + 2bc + c^2 - 3b - 3c + 9)$$

SECTION 5.8
Summary of Factoring Techniques

113.
$$4q^2rs + 4qrst - 120rst^2 = 4rs(q^2 + qt - 30t^2)$$
$$= 4rs(q + 6t)(q - 5t)$$

114.

$$\begin{aligned} &2(m+n)^2+(m+n)-3 \\ &\quad=(2(m+n)+3)((m+n)-1) \\ &\quad=(2m+2n+3)(m+n-1) \end{aligned}$$

115.

$$\begin{aligned} z^2-4+zx-2x &= (z^2-4)+(zx-2x) \\ &= (z+2)(z-2)+x(z-2) \\ &= (z-2)(z+2+x) \end{aligned}$$

116.

m^4+16n^2 is prime

117.

$$\begin{aligned} x^2+4x+4-4p^4 &= (x^2+4x+4)-4p^4 \\ &= (x+2)^2-4p^4 \\ &= (x+2-2p^2)(x+2+2p^2) \end{aligned}$$

118.

$$\begin{aligned} &y^2+3y+2+2x+xy \\ &\quad=(y^2+3y+2)+(2x+xy) \\ &\quad=(y+2)(y+1)+x(2+y) \\ &\quad=(y+2)(y+1+x) \end{aligned}$$

119.

$$\begin{aligned} &4a^3b^3c^2+256c^2 \\ &\quad=4c^2(a^3b^3+64) \\ &\quad=4c^2(ab+4)(a^2b^2-4ab+16) \end{aligned}$$

120.

$$\begin{aligned} -13a^2+36+a^4 &= a^4-13a^2+36 \\ &= (a^2-4)(a^2-9) \\ &= (a+2)(a-2)(a+3)(a-3) \end{aligned}$$

121.

$$\begin{aligned} &4x^4+12x^3+9x^2+2x+3 \\ &\quad=(4x^4+12x^3+9x^2)+(2x+3) \\ &\quad=x^2(4x^2+12x+9)+(2x+3) \\ &\quad=x^2(2x+3)(2x+3)+1(2x+3) \\ &\quad=(2x+3)(x^2(2x+3)+1) \\ &\quad=(2x+3)(2x^3+3x^2+1) \end{aligned}$$

122.

$$\begin{aligned} V &= \frac{\pi}{2}r_1^2h-\frac{\pi}{2}r_2^2h \\ &= \frac{\pi}{2}h(r_1^2-r_2^2) \\ &= \frac{\pi}{2}h(r_1-r_2)(r_1+r_2) \end{aligned}$$

SECTION 5.9
Solving Equations by Factoring

123.

$$\begin{aligned} 4x^2-3x&=0 \\ x(4x-3)&=0 \\ x=0 \quad \text{or} \quad 4x-3&=0 \\ 4x&=3 \\ x&=\frac{3}{4} \end{aligned}$$

124.

$$\begin{aligned} x^2&=36 \\ x^2-36&=0 \\ (x+6)(x-6)&=0 \\ x+6=0 \quad &\text{or} \quad x-6=0 \\ x=-6 \quad &\quad\quad\;\; x=6 \end{aligned}$$

125.

$$\begin{aligned} 12x^2&=5-4x \\ 12x^2+4x-5&=0 \\ (2x-1)(6x+5)&=0 \\ 2x-1=0 \quad &\text{or} \quad 6x+5=0 \\ 2x=1 \quad &\quad\quad\;\; 6x=-5 \\ x=\frac{1}{2} \quad &\quad\quad\;\; x=-\frac{5}{6} \end{aligned}$$

126.

$$\begin{aligned} -d^4+10d^2-9&=0 \\ -(d^4-10d^2+9)&=0 \\ -(d^2-1)(d^2-9)&=0 \\ -(d+1)(d-1)(d+3)(d-3)&=0 \\ d+1=0, \quad d-1=0, \quad d+3&=0, \quad d-3=0 \\ d=-1 \quad\;\; d=1 \quad\;\; d&=-3 \quad\;\; d=3 \end{aligned}$$

127.

$$t^2(15t-2)=8t$$
$$15t^3-2t^2-8t=0$$
$$t(15t^2-2t-8)=0$$
$$t(3t+2)(5t-4)=0$$
$$t=0,\quad 3t+2=0,\quad \text{or}\quad 5t-4=0$$
$$3t=-2 \qquad 5t=4$$
$$t=-\frac{2}{3} \qquad t=\frac{4}{5}$$

128.

$$u^3=\frac{1}{3}u(19u+14)$$
$$3\cdot u^3=3\cdot\frac{1}{3}u(19u+14)$$
$$3u^3=u(19u+14)$$
$$3u^3=19u^2+14u$$
$$3u^3-19u^2-14u=0$$
$$u(3u^2-19u-14)=0$$
$$u(3u+2)(u-7)=0$$
$$u=0,\quad 3u+2=0,\quad \text{or}\quad u-7=0$$
$$3u=-2 \qquad u=7$$
$$u=-\frac{2}{3}$$

129.

$$(y+7)^2+8=-2(y+7)+7$$
$$(y+7)(y+7)+8=-2(y+7)+7$$
$$y^2+7y+7y+49+8=-2y-14+7$$
$$y^2+14y+57=-2y-7$$
$$y^2+14y+57+2y+7=0$$
$$y^2+16y+64=0$$
$$(y+8)(y+8)=0$$
$$y+8=0\quad \text{or}\quad y+8=0$$
$$y=-8 \qquad y=-8$$

130.

$$x^3+7x^2-x-7=0$$
$$x^2(x+7)-1(x+7)=0$$
$$(x+7)(x^2-1)=0$$
$$(x+7)(x+1)(x-1)=0$$
$$x+7=0,\quad x+1=0,\quad x-1=0$$
$$x=-7 \qquad x=-1 \qquad x=1$$

131. $f(x)=x^2-4x+2$

$$x^2-4x+2=-1$$
$$x^2-4x+3=0$$
$$(x-1)(x-3)=0$$
$$x-1=0\quad \text{or}\quad x-3=0$$
$$x=1\quad \text{or}\quad x=3$$

132. $f(x)=g(x)$

$$9x^2-14x=x+6$$
$$9x^2-15x-6=0$$
$$3(3x^2-5x-2)=0$$
$$3(3x+1)(x-2)=0$$
$$3x+1=0\quad \text{and}\quad x-2=0$$
$$3x=-1 \qquad x=2$$
$$x=-\frac{1}{3}$$

133. $h(t)=-16t^2+64t$

$$0=-16t^2+64t$$
$$16t^2-64t=0$$
$$16t(t-4)=0$$
$$16t(t-4)=0$$
$$16t=0\quad \text{and}\quad t-4=0$$
$$t=0 \qquad t=4$$

134. Let x = one side of the base and $x + 3$ = the other side of the base. The area of the base $(B) = x(x + 3)$.

$$V = \frac{1}{3}Bh$$
$$210 = \frac{1}{3}x(x+3)(9)$$
$$210 = \frac{1}{3}(9)(x)(x+3)$$
$$210 = 3x(x+3)$$
$$210 = 3x^2 + 9x$$
$$0 = 3x^2 + 9x - 210$$
$$0 = 3(x^2 + 3x - 70)$$
$$0 = 3(x+10)(x-7)$$
$$x+10 = 0 \quad \text{or} \quad x-7 = 0$$
$$x = -10 \qquad x = 7$$

Since you cannot have negative lengths, the dimensions of the base are 7 and $(7 + 3) = 10$.
7 m by 10 m

135. Let x = the width of the border and 200 is the area of the poster.

$$(2x+10)(2x+20) - 200 = 400$$
$$4x^2 + 20x + 40x + 200 - 200 = 400$$
$$4x^2 + 60x = 400$$
$$4x^2 + 60x - 400 = 0$$
$$4(x^2 + 15x - 100) = 0$$
$$4(x+20)(x-5) = 0$$
$$x+20 = 0 \quad \text{or} \quad x-5 = 0$$
$$x = -20 \qquad x = 5$$

Since the width cannot be negative, the width must be 5 feet.

136. The x – intercepts are $x = -\frac{1}{2}$ and $x = 1$.

CHAPTER 5 TEST

1. a. Since $x^2 + 16x + 64$ is the square of $x + 8$, it is called a **perfect**-square trinomial.
   b. When we factor a polynomial, we write a sum of terms as a **product** of factors.
   c. The statement $x^2 - x - 12 = (x - 4)(x + 3)$ shows that the trinomial $x^2 - x - 12$ factors as the product of two **binomials**.
   d. A **quadratic** equation is any equation that can be written in the form $ax^2 + bx + c = 0$, where $a \neq 0$.
   e. $x^2 - y^2$ is called a **difference** of two squares and $x^3 + y^3$ is called a sum of two **cubes**.
   f. The **greatest common** factor of $12x^3$ and $8x^5$ is $4x^3$.

2. a. $x^5 \cdot x^5 = x^{5+5} = x^{10}$
   b. $x^5 \cdot x^5 \cdot x^5 = x^{5+5+5} = x^{15}$

3.

$$\begin{aligned} 9^{-1}a^{-5}m^3\left(a^5m^{-4}\right)^2 &= 9^{-1}a^{-5}m^3a^{5\cdot2}m^{-4\cdot2} \\ &= 9^{-1}a^{-5}m^3a^{10}m^{-8} \\ &= 9^{-1}a^{-5+10}m^{3-8} \\ &= 9^{-1}a^5m^{-5} \\ &= \frac{a^5}{9m^5} \end{aligned}$$

4.

$$\begin{aligned} \left(\frac{-2x^2y^3}{5}\right)^3 &= \frac{(-2)^3x^{2(3)}y^{3(3)}}{5^3} \\ &= -\frac{8x^6y^9}{125} \end{aligned}$$

5.

$$\begin{aligned} \frac{\left(-4b^2\right)^3\left(b^{-3}\right)^3}{b^5} &= \frac{(-4)^3b^{2\cdot3}b^{-3\cdot3}}{b^5} \\ &= \frac{-64b^6b^{-9}}{b^5} \\ &= -\frac{64b^{6-9}}{b^5} \\ &= -\frac{64b^{-3}}{b^5} \\ &= -\frac{64}{b^5b^3} \\ &= -\frac{64}{b^8} \end{aligned}$$

6.

$$\begin{aligned} \left(\frac{3m^2n^3}{m^4n^{-2}}\right)^{-2} &= \left(3m^{2-4}n^{3-(-2)}\right)^{-2} \\ &= \left(3m^{-2}n^5\right)^{-2} \\ &= \left(\frac{3n^5}{m^2}\right)^{-2} \\ &= \left(\frac{m^2}{3n^5}\right)^2 \\ &= \frac{m^{2(2)}}{3^2n^{5(2)}} \\ &= \frac{m^4}{9n^{10}} \end{aligned}$$

7.

$$\begin{aligned} \frac{3.19\times10^{12}}{\left(2.2\times10^{-4}\right)\left(5.0\times10^9\right)} &= \frac{3.19\times10^{12}}{(2.2\times5.0)\left(10^{-4}\times10^9\right)} \\ &= \frac{3.19\times10^{12}}{11\times10^5} \\ &= \frac{3.19}{11}\times10^{12-5} \\ &= 0.29\times10^7 \\ &= 2.9\times10^6 \\ &= 2{,}900{,}000 \end{aligned}$$

8. Multiply 1.86×10^5 by 60 since there are 60 seconds in 1 minute.

$$(1.86\times10^5)(60) = (1.86\times10^5)(6\times10^1)$$
$$= (1.86\times6)\times10^{5+1}$$
$$= 11.16\times10^6$$
$$= 1.116\times10^7 \text{ mi/min}$$

9. a. The coefficients are 3, –4, –3, and $-\frac{5}{3}$. The degree is 5.

b. The coefficients are 8, –1, 1, –6, and 4. The degree is 13 (9 + 4 = 13).

10. $h(t) = -16t^2 + 80t + 10$

$$h(2.5) = -16(2.5)^2 + 80(2.5) + 10$$
$$= -16(6.25) + 80(2.5) + 10$$
$$= -100 + 200 + 10$$
$$= 110 \text{ ft}$$

11. a. $f(4) = 0$

b. $x = 2$ and $x = 6$

c. D: $(-\infty, \infty)$

R: $(-\infty, 2]$

12. $f(x) - g(x)$

$$= (2.39x^3 + 4.17x + 9.89) - (1.24x^3 - 1.33x - 2.11)$$
$$= 2.39x^3 + 4.17x + 9.89 - 1.24x^3 + 1.33x + 2.11$$
$$= 1.15x^3 + 5.5x + 12$$

13.

$$(-2x^2y^3 + 6xy + 5y^2) - (-4x^2y^3 - 7xy + 2y^2)$$
$$= -2x^2y^3 + 6xy + 5y^2 + 4x^2y^3 + 7xy - 2y^2$$
$$= 2x^2y^3 + 13xy + 3y^2$$

14.

$$\left(\frac{1}{2}x^5 + \frac{1}{3}x^2\right) + \left(\frac{3}{2}x^5 - \frac{1}{5}x^2\right)$$
$$= \left(\frac{1}{2}x^5 + \frac{3}{2}x^5\right) + \left(\frac{1}{3}x^2 - \frac{1}{5}x^2\right)$$
$$= \left(\frac{1}{2}x^5 + \frac{3}{2}x^5\right) + \left(\frac{5}{15}x^2 - \frac{3}{15}x^2\right)$$
$$= \frac{4}{2}x^5 + \frac{2}{15}x^2$$
$$= 2x^5 + \frac{2}{15}x^2$$

15.

$$(a^2yz^4)(2ay^5z)(-6ay^6z^7)$$
$$= -12a^{2+1+1}y^{1+5+6}z^{4+1+7}$$
$$= -12a^4y^{12}z^{12}$$

16.

$$-5a^2b(3ab^3 - 2ab^4) = -15a^3b^4 + 10a^3b^5$$

17.

$$(3y^5 + 1)(2y^2 + 3y + 2)$$
$$= 3y^5(2y^2 + 3y + 2) + 1(2y^2 + 3y + 2)$$
$$= 6y^7 + 9y^6 + 6y^5 + 2y^2 + 3y + 2$$

18.

$$(0.6d - 2)(0.1d + 3) = 0.06d^2 + 1.8d - 0.2d - 6$$
$$= 0.06d^2 + 1.6d - 6$$

19.

$$[6 + (m - n)]^2 = 6^2 + 2(6)(m - n) + (m - n)^2$$
$$= 36 + 12(m - n) + (m^2 - 2mn + n^2)$$
$$= 36 + 12m - 12n + m^2 - 2mn + n^2$$

20.

$$2s(4s + 5t)(4s - 5t) = 2s((4s)^2 - (5t)^2)$$
$$= 2s(16s^2 - 25t^2)$$
$$= 32s^3 - 50st^2$$

21.

$$(4t-3)^2-(t+1)(t-4)$$
$$=(4t)^2-2(4t)(3)+(3)^2-\left[t^2-4t+t-4\right]$$
$$=16t^2-24t+9-t^2+4t-t+4$$
$$=15t^2-21t+13$$

22.

$$f(x)=x^2-3x+6$$
$$f(c+1)=(c+1)^2-3(c+1)+6$$
$$=c^2+2c+1-3c-3+6$$
$$=c^2-c+4$$

23.

$$f(x)\cdot g(x)=(x^2+7x+49)(x-7)$$
$$=x^2(x-7)+7x(x-7)+49(x-7)$$
$$=x^3-7x^2+7x^2-49x+49x-343$$
$$=x^3-343$$

24. a. $(4a+3b)+(4a-b)=8a+2b$

b. $(4a+3b)(4a-b)$

$$=16a^2-4ab+12ab-3b^2$$
$$=16a^2+8ab-3b^2$$

25.

$$12a^3b^2c-3a^2b^2c^2+6abc^3$$
$$=3abc(4a^2b-abc+2c^2)$$

26.

$$hk+bz+hz+bk=hk+hz+bk+bz$$
$$=h(k+z)+b(k+z)$$
$$=(k+z)(h+b)$$

27.

$$x^2-x-30=(x-6)(x+5)$$

28.

$$y^4-81=(y^2+9)(y^2-9)$$
$$=(y^2+9)(y+3)(y-3)$$

29.

$$-7x^4+13x^3+2x^2=-x^2(7x^2-13x-2)$$
$$=-x^2(7x+1)(x-2)$$

30.

$$s^4-13s^2+36=(s^2-4)(s^4-9)$$
$$=(s+2)(s-2)(s+3)(s-3)$$

31.

$$25m^2-40mn+16n^2=(5m-4n)^2$$

32.

$16b^2+25$ is prime

33.

$$5x^3+625=5(x^3+125)$$
$$=5(x+5)(x^2-5x+25)$$

34.

$$64a^3-125b^6$$
$$=(4a-5b^2)\left[(4a)^2+(4a)(5b^2)+(5b^2)^2\right]$$
$$=(4a-5b^2)(16a^2+20ab^2+25b^4)$$

35.

$$(x-y)^2+3(x-y)-10$$
$$=\left[(x-y)+5\right]\left[(x-y)-2\right]$$
$$=(x-y+5)(x-y-2)$$

36.

$$6b^2+bc-2c^2=(3b+2c)(2b-c)$$

37.

$$a^2-b^2+a+b=(a+b)(a-b)+1(a+b)$$
$$=(a+b)(a-b+1)$$

38.

$$n^2-6n+9-4m^2=(n^2-6n+9)-4m^2$$
$$=(n-3)^2-4m^2$$
$$=(n-3+2m)(n-3-2m)$$

39.

$$a^6-1=(a^3+1)(a^3-1)$$
$$=(a+1)(a^2-a+1)(a-1)(a^2+a+1)$$

40.

$$v_1v_3-v_3v=v_1v$$
$$v_1v_3-v_3v+v_3v=v_1v+v_3v$$
$$v_1v_3=v_1v+v_3v$$
$$v_1v_3=v(v_1+v_3)$$
$$\frac{v_1v_3}{v_1+v_3}=\frac{v(v_1+v_3)}{v_1+v_3}$$
$$\frac{v_1v_3}{v_1+v_3}=v$$

41.

$$5m^2=25m$$
$$5m^2-25m=0$$
$$5m(m-5)=0$$
$$5m=0 \quad \text{or} \quad m-5=0$$
$$m=0 \qquad m=5$$

42.

$$\frac{2}{5}x^2+\frac{3}{5}x-7=0$$
$$5\left(\frac{2}{5}x^2+\frac{3}{5}x-7\right)=5(0)$$
$$2x^2+3x-35=0$$
$$(2x-7)(x+5)=0$$
$$2x-7=0 \quad \text{or} \quad x+5=0$$
$$2x=7 \qquad x=-5$$
$$x=\frac{7}{2}$$

43.

$$(3x+1)^2+2x=x^2+2(x+5)$$
$$9x^2+6x+1+2x=x^2+2x+10$$
$$9x^2+8x+1=x^2+2x+10$$
$$8x^2+6x-9=0$$
$$(4x-3)(2x+3)=0$$
$$4x-3=0 \quad \text{or} \quad 2x+3=0$$
$$4x=3 \qquad 2x=-3$$
$$x=\frac{3}{4} \qquad x=-\frac{3}{2}$$

44.

$$x^3+8x^2-9x=0$$
$$x(x^2+8x-9)=0$$
$$x(x+9)(x-1)=0$$
$$x=0, \quad x+9=0, \quad \text{or} \quad x-1=0$$
$$x=-9 \qquad x=1$$

45.

$$x^3-16x=16-x^2$$
$$x^3+x^2-16x-16=0$$
$$x^2(x+1)-16(x+1)=0$$
$$(x^2-16)(x+1)=0$$
$$(x+4)(x-4)(x+1)=0$$
$$x+4=0, \quad x-4=0, \quad x+1=0$$
$$x=-4 \qquad x=4 \qquad x=-1$$

46. Let b = the length of the base. Then the height, h, would be $2b-1$.

$$A=\frac{1}{2}bh$$
$$14=\frac{1}{2}(b)(2b-1)$$
$$2(14)=2\left(\frac{1}{2}(b)(2b-1)\right)$$
$$28=b(2b-1)$$
$$28=2b^2-b$$
$$0=2b^2-b-28$$
$$0=(2b+7)(b-4)$$
$$2b+7=0 \quad \text{and} \quad b-4=0$$
$$2b=-7 \qquad b=4$$
$$b=-\frac{7}{2} \text{ (crossed out)}$$
$$2b-1=2(4)-1$$
$$=7$$

`The base = 4 m and the height = 7 m.

47.

$$h(t) = -16t^2 + 16t + 96$$
$$= -16(t^2 - t - 6)$$
$$= -16(t-3)(t+2)$$

$t - 3 = 0$ or $t + 2 = 0$

$t = 3$ $\quad t = -2$

Since the time cannot be negative, $t = 3$ seconds.

48.

$$(2w+5)(2w+10) - 50 = 54$$
$$4w^2 + 10w + 20w + 50 - 50 = 54$$
$$4w^2 + 30w = 54$$
$$4w^2 + 30w - 54 = 0$$
$$2(2w^2 + 15w - 27) = 0$$
$$2(2w-3)(w+9) = 0$$

$2w - 3 = 0$ or $w + 9 = 0$

$2w = 3$ $\quad w = -9$

$w = \frac{3}{2}$

Since the width cannot be negative, then $w = 1.5$ ft.

CHAPTER 5 CUMULATIVE REVIEW

1. a. true
   b. false
   c. false
   d. true
   e. true

2.

$$\begin{aligned}-3\left|(3^2\cdot 5-2^3\cdot 6)^2\right| &= -3\left|(9\cdot 5-8\cdot 6)^2\right|\\ &= -3\left|(45-48)^2\right|\\ &= -3\left|(-3)^2\right|\\ &= -3|9|\\ &= -3(9)\\ &= -27\end{aligned}$$

3.

$$\begin{aligned}-\frac{7}{16}x-\frac{3}{4}x &= -\frac{7}{16}x-\frac{3}{4}\left(\frac{4}{4}\right)x\\ &= -\frac{7}{16}x-\frac{12}{16}x\\ &= -\frac{19}{16}x\end{aligned}$$

4.

$$\begin{aligned}\frac{x+2}{5}-4x &= \frac{8}{5}-\frac{x+9}{2}\\ 10\left(\frac{x+2}{5}-4x\right) &= 10\left(\frac{8}{5}-\frac{x+9}{2}\right)\\ 2(x+2)-10(4x) &= 2(8)-5(x+9)\\ 2x+4-40x &= 16-5x-45\\ 4-38x &= -29-5x\\ 4-38x+38x &= -29-5x+38x\\ 4 &= -29+33x\\ 4+29 &= -29+33x+29\\ 33 &= 33x\\ 1 &= x\end{aligned}$$

5.

$$\begin{aligned}F &= \frac{9}{5}C+32\\ F-32 &= \frac{9}{5}C+32-32\\ F-32 &= \frac{9}{5}C\\ \frac{5}{9}(F-32) &= \frac{5}{9}\left(\frac{9}{5}C\right)\\ \frac{5}{9}(F-32) &= C\end{aligned}$$

6. Volume of prism – Volume of cylinder

$$\begin{aligned}lwh-\pi r^2h &= 12(12)(18)-\pi(4)^2(18)\\ &= 2{,}592-288\pi\\ &= 1{,}687.22\text{ cm}^3\end{aligned}$$

7. Let x = travel time of each plane.
   The distance of plane 1 is $500x$ and the distance of plane 2 is $550x$.

$$\begin{aligned}500x+550x &= 3{,}675\\ 1050x &= 3{,}675\\ x &= 3.5\text{ hrs}\end{aligned}$$

8.

$$\begin{aligned}24(0.15)+x(0.0) &= 0.10(x+24)\\ 3.6 &= 0.10x+2.4\\ 3.6-2.4 &= 0.10x+2.4-2.4\\ 1.2 &= 0.10x\\ \frac{1.2}{0.10} &= \frac{0.10x}{0.10}\\ 12 &= x\end{aligned}$$

12 oz of water are needed.

9. Yes, it is a solution.

$$\begin{aligned}y-6x &= 10\\ 11-6\left(\frac{1}{6}\right) &\overset{?}{=} 10\\ 11-1 &\overset{?}{=} 10\\ 10 &= 10\end{aligned}$$

10. x-intercept, let $y = 0$: y-intercept, let $x = 0$

$$\begin{aligned} 2x-5y&=10 \\ 2x-5(0)&=10 \\ 2x&=10 \\ x&=5 \\ &(5,0) \end{aligned} \qquad \begin{aligned} 2x-5y&=10 \\ 2(0)-5y&=10 \\ -5y&=10 \\ y&=-2 \\ &(0,-2) \end{aligned}$$

11. Use the points (2011, 563) and (2019, 943) to find the slope.

$$\begin{aligned} m&=\frac{y_2-y_1}{x_2-x_1} \\ &=\frac{943-563}{2019-2011} \\ &=\frac{380}{8} \\ &=47.5 \end{aligned}$$

An increase of \$47.5 billion per year.

12. The slope of $y = 7$ is 0.

$$\begin{aligned} y&=mx+b \\ y&=7 \\ y&=0x+7 \\ m&=0 \end{aligned}$$

13. The line that is perpendicular to a line with $m=-\frac{1}{8}$ has a slope of $m=8$ because perpendicular lines have slopes that are opposite reciprocals.

Use $m=8$ and $(-2,5)$ for (x_1,y_1).

$$\begin{aligned} y-y_1&=m(x-x_1) \\ y-5&=8(x-(-2)) \\ y-5&=8(x+2) \\ y-5&=+8x+16 \\ y-5+5&=+8x+16+5 \\ y&=+8x+21 \end{aligned}$$

14. Find the slope of the line going through the points (–5, 4) and (8, –6).

$$\begin{aligned} m&=\frac{y_2-y_1}{x_2-x_1} \\ &=\frac{-6-4}{8-(-5)} \\ &=-\frac{10}{13} \end{aligned}$$

Use $m=-\frac{10}{13}$ and (–5, 4) for (x_1,y_1).

$$\begin{aligned} y-y_1&=m(x-x_1) \\ y-4&=-\frac{10}{13}(x-(-5)) \\ y-4&=-\frac{10}{13}x-\frac{50}{13} \\ y-4+4&=-\frac{10}{13}x-\frac{50}{13}+4 \\ y&=-\frac{10}{13}x+\frac{2}{13} \end{aligned}$$

15. $h(x)=x^4-x^3-x^2-x-1$

$$\begin{aligned} h(1)&=(1)^4-(1)^3-(1)^2-(1)-1 \\ &=1-1-1-1-1 \\ &=-3 \end{aligned}$$

16. $2\frac{1}{2}¢=0.025$

$f(x)=0.025x+95$

17.

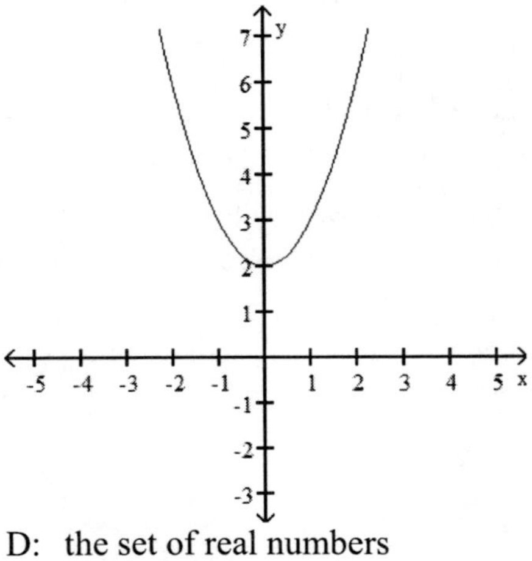

D: the set of real numbers

R: the set of real numbers greater than or equal to 2

18. a. $h(5) = 4$
    b. $h(0) =$ ~~0~~ 1
    c. $h(x) = 3$ when $x = -2$ and $x = 4$

19. No, it is not a function. Answers may vary. Examples are (1, 2) and (1, –2).

20. $\begin{cases} y = -2x + 1 \\ x - 2y = -7 \end{cases}$

Solve by equations for y and graph using slope and y-intercepts.

$\begin{cases} y = -2x + 1 \\ y = \dfrac{1}{2}x + \dfrac{7}{2} \end{cases}$

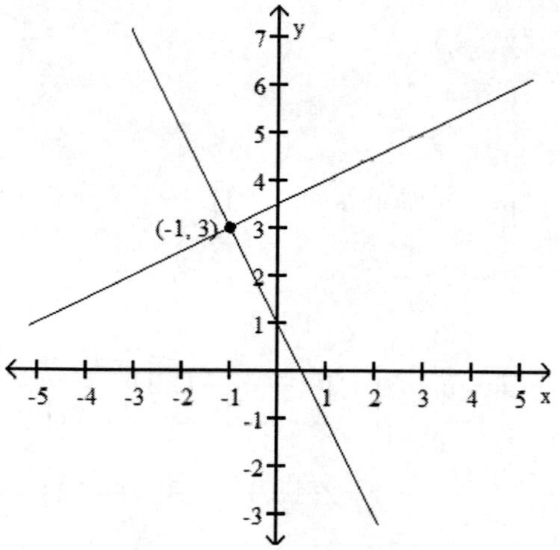

The graphs intersect at (–1, 3).

21. $\begin{cases} 4x + 6y = 5 \\ 8x = 3(1 + 3y) \end{cases}$

$\begin{cases} 4x + 6y = 5 \\ 8x = 3 + 9y \end{cases}$

Write the equations in general form.

$\begin{cases} 4x + 6y = 5 \\ 8x - 9y = 3 \end{cases}$

Multiply the first equation by –2 and add the equations to eliminate x.

$$\begin{aligned} -8x - 12y &= -10 \\ 8x - 9y &= 3 \\ \hline -21y &= -7 \\ y &= \frac{7}{21} \\ y &= \frac{1}{3} \end{aligned}$$

Substitute $y = \dfrac{1}{3}$ into the first equation and solve for x.

$$\begin{aligned} 4x + 6y &= 5 \\ 4x + 6\left(\frac{1}{3}\right) &= 5 \\ 4x + 2 &= 5 \\ 4x &= 3 \\ x &= \frac{3}{4} \end{aligned}$$

The solution is $\left(\dfrac{3}{4}, \dfrac{1}{3}\right)$.

22. $\begin{cases} 3x+2y-z=-8 & (1) \\ 2x-y+7z=10 & (2) \\ 2x+2y-3z=-10 & (3) \end{cases}$

Multiply Equation 2 by 2 and add Equations 1 and 2 to obtain Equation 4.

$$\begin{aligned} 3x+2y-\ \ z&=-8 \\ \underline{4x-2y+14z}&\underline{=20} \\ 7x+13z&=12 \end{aligned}$$

Multiply Equation 2 by 2 and add Equations 3 and 2 to obtain Equation 5.

$$\begin{aligned} 2x+2y-\ 3z&=-10 \\ \underline{4x-2y+14z}&\underline{=20} \\ 6x+11z&=10 \end{aligned}$$

Multiply Equation 4 by 6 and Equation 5 by –7 and add to find z.

$$\begin{aligned} 42x+78z&=72 \\ \underline{-42x-77z}&\underline{=-70} \\ z&=\ 2 \end{aligned}$$

Substitute $z = 2$ into Equation 4 and solve for x.

$$\begin{aligned} 7x+13z&=12 \\ 7x+13(2)&=12 \\ 7x+26&=12 \\ 7x&=-14 \\ x&=-2 \end{aligned}$$

Substitute $x = -2$ and $z = 2$ into Equation 1 and solve for y.

$$\begin{aligned} 3(-2)+2y-(2)&=-8 \\ -6+2y-2&=-8 \\ 2y-8&=-8 \\ 2y&=0 \\ y&=0 \end{aligned}$$

The solution is (–2, 0, 2).

23.

$$\left[\begin{array}{cc|c} 2 & 1 & 1 \\ 1 & 2 & -4 \end{array}\right]$$

$R_1 \leftrightarrow R_2$

$$\left[\begin{array}{cc|c} 1 & 2 & -4 \\ 2 & 1 & 1 \end{array}\right]$$

$-2R_1 + R_2$

$$\left[\begin{array}{cc|c} 1 & 2 & -4 \\ 0 & -3 & 9 \end{array}\right]$$

$-\frac{1}{3}R_2$

$$\left[\begin{array}{cc|c} 1 & 2 & -4 \\ 0 & 1 & -3 \end{array}\right]$$

This matrix represents the system

$$\begin{cases} x+2y=-4 \\ y=-3 \end{cases}$$

$$\begin{aligned} x-6&=-4 \\ x&=2 \end{aligned}$$

The solution is $(2,\ -3)$.

24.

$$\begin{vmatrix} 2 & -3 & 4 \\ -1 & 2 & 4 \\ 3 & -3 & 1 \end{vmatrix} = 2\begin{vmatrix} 2 & 4 \\ -3 & 1 \end{vmatrix} - (-3)\begin{vmatrix} -1 & 4 \\ 3 & 1 \end{vmatrix} + 4\begin{vmatrix} -1 & 2 \\ 3 & -3 \end{vmatrix}$$

$$\begin{aligned} &= 2(2+12)+3(-1-12)+4(3-6) \\ &= 2(14)+3(-13)+4(-3) \\ &= 28-39-12 \\ &= -23 \end{aligned}$$

25. Let x = amount invested at 7.5% and y = amount at 6%.

$$\begin{cases} x+y=50{,}000 \\ 0.075x+0.06y=3{,}240 \end{cases}$$

Multiply Equation 1 by –0.075 and add Equations 1 and 2.

$$\begin{aligned} -0.075x-0.075y&=-3{,}750 \\ \underline{0.075x+0.06y}&\underline{=\ \ 3{,}240} \\ -0.015y&=\ -510 \\ y&=34{,}000 \end{aligned}$$

Substitute y = 34,000 in Equation 1 and solve for x.

$$\begin{aligned} x+34{,}000&=50{,}000 \\ x&=16{,}000 \end{aligned}$$

$16,000 at 7.5% and $34,000 at 6%

26. Let x = measure of first angle, y = measure of second angle, and z = measure of third angle.

$$\begin{cases} x + 2y = 72 + z \\ z - 26 = y \\ x + y + z = 180 \end{cases}$$

Write each equation in standard form.

$$\begin{cases} x + 2y - z = 72 \\ y - z = -26 \\ x + y + z = 180 \end{cases}$$

Multiply Equation 1 by −1 and add Equations 1 and 3 to get Equation 4.

$$\begin{array}{r} -x - 2y + z = -72 \\ \underline{x + y + z = 180} \\ -y + 2z = 108 \end{array}$$

Add Equations 2 and 4.

$$\begin{array}{r} y - z = -26 \\ \underline{-y + 2z = 108} \\ z = 82 \end{array}$$

Substitute z = 82 into Equation 2.

$$y - 82 = -26$$
$$y = 56$$

Substitute y = 56 and z = 82 into Equation 3.

$$x + 56 + 82 = 180$$
$$x + 138 = 180$$
$$x = 42$$

The angles have measures of 42°, 56°, and 82°.

27.

$$-9(t-3) + 2t \le 8(4-t)$$
$$-9t + 27 + 2t \le 32 - 8t$$
$$-7t + 27 \le 32 - 8t$$
$$t + 27 \le 32$$
$$t \le 5$$
$$(-\infty, 5]$$

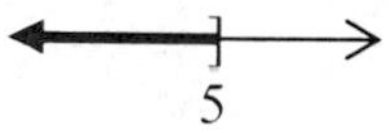

28.

$$-6 \le \frac{1}{3}h + 1 < 0$$
$$3(-6) \le 3\left(\frac{1}{3}h + 1\right) < 3(0)$$
$$-18 \le h + 3 < 0$$
$$-18 - 3 \le h + 3 - 3 < 0 - 3$$
$$-21 \le h < -3$$
$$[-21, -3)$$

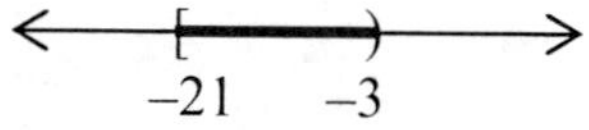

29.

$$|m + 5| \ge 7$$
$$m + 5 \ge 7 \quad \text{or} \quad m + 5 \le -7$$
$$m \ge 2 \qquad m \le -12$$
$$(-\infty, -12] \cup [2, \infty)$$

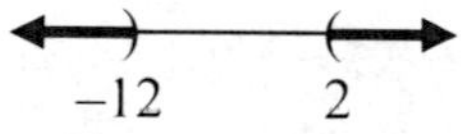

30.

$$4.5x - 1 < -10 \quad \text{or} \quad 6 - 2x \ge 12$$
$$4.5x < -9 \qquad -2x \ge 6$$
$$x < -2 \qquad x \le -3$$
$$(-\infty, -2)$$

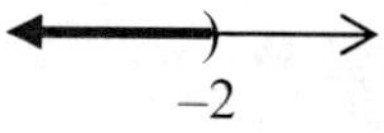

31. iii. $10 \le x \le 15$

32. $\begin{cases} x - y < 4 \\ y \le 0 \\ x \ge 0 \end{cases}$

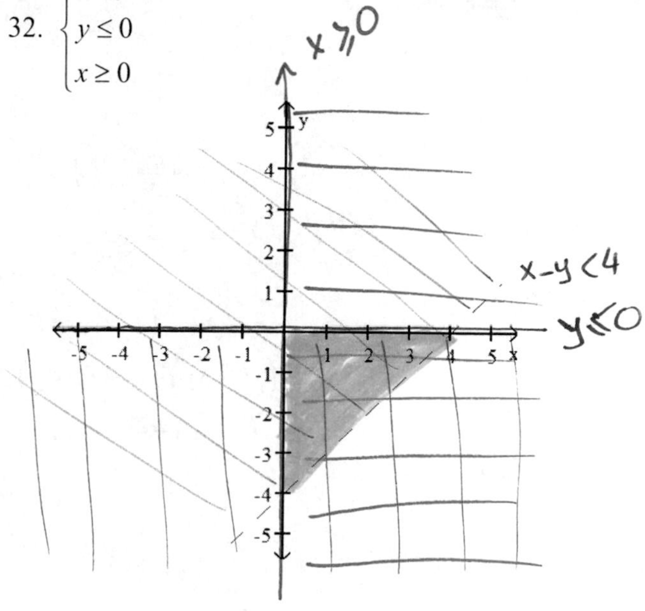

33.

$$\left(\frac{-3a^4b^2}{-9a^5b^{-2}}\right)^{-2}=\left(\frac{-9a^5b^{-2}}{-3a^4b^2}\right)^2$$
$$=\left(\frac{-9}{-3}a^{5-4}b^{-2-2}\right)^2$$
$$=\left(3ab^{-4}\right)^2$$
$$=\left(\frac{3a}{b^4}\right)^2$$
$$=\frac{9a^2}{b^8}$$

34. $9.0895\times10^{-8}=0.00000090895$

0.000000090895

35.

$$f(x)-g(x)=\left(0.4x^3+0.6x\right)-\left(-0.3x^3+0.2x\right)$$
$$=0.4x^3+0.6x+0.3x^3-0.2x$$
$$=0.7x^3+0.4x$$

36.

$$\left(4a^2b-3c^3\right)\left(9a^2b-2c^3\right)$$
$$=36a^4b^2-8a^2bc^3-27a^2bc^3+6c^6$$
$$=36a^4b^2-35a^2bc^3+6c^6$$

37.

$$(2a-b)\left(4a^2+2ab+b^2\right)$$
$$=2a\left(4a^2+2ab+b^2\right)-b\left(4a^2+2ab+b^2\right)$$
$$=8a^3+4a^2b+2ab^2-4a^2b-2ab^2-b^3$$
$$=8a^3-b^3$$

38.

$$(3k+1)^2+(2k-4)^2$$
$$=\left(9k^2+6k+1\right)+\left(4k^2-16k+16\right)$$
$$=13k^2-22k+17$$

$-10k$

39.

$$x^2+4y-xy-4x=x^2-xy-4x+4y$$
$$=\left(x^2-xy\right)+\left(-4x+4y\right)$$
$$=x(x-y)-4(x-y)$$
$$=(x-y)(x-4)$$

40.

$$6s^4-216s^2=6s^2\left(s^2-36\right)$$
$$=6s^2(s+6)(s-6)$$

41.

$$8x^6+125y^3=\left(2x^2\right)^3+(5y)^3$$
$$=\left(2x^2+5y\right)\left(4x^4-10x^2y+25y^2\right)$$

42.

$$-3a^2+ab+2b^2=-\left(3a^2-ab-2b^2\right)$$
$$=-(3a+2b)(a-1)$$

43.

$$a^3+5a^2+6a+a^2b+5ab+6b$$
$$=a\left(a^2+5a+6\right)+b\left(a^2+5a+6\right)$$
$$=(a+b)\left(a^2+5a+6\right)$$
$$=(a+b)(a+2)(a+3)$$

44.

$$x^2=\frac{1}{2}(x+1)$$
$$x^2=\frac{1}{2}x+\frac{1}{2}$$
$$x^2-\frac{1}{2}x-\frac{1}{2}=0$$
$$2\left(x^2-\frac{1}{2}x-\frac{1}{2}\right)=2(0)$$
$$2x^2-x-1=0$$
$$(2x+1)(x-1)=0$$
$$2x+1=0 \quad\text{and}\quad x-1=0$$
$$2x=-1 \qquad x=1$$
$$x=-\frac{1}{2}$$

45.

$$m_1 m_2 = m_2 g + m_1 g$$
$$m_1 m_2 - m_1 g = m_2 g + m_1 g - m_1 g$$
$$m_1 m_2 - m_1 g = m_2 g$$
$$m_1 (m_2 - g) = m_2 g$$
$$\frac{m_1 (m_2 - g)}{m_2 - g} = \frac{m_2 g}{m_2 - g}$$
$$m_1 = \frac{m_2 g}{m_2 - g}$$

46.

$$C(n) = \frac{1}{2}(n^2 - n)$$
$$66 = \frac{1}{2}(n^2 - n)$$
$$2 \cdot 66 = 2 \cdot \frac{1}{2}(n^2 - n)$$
$$132 = n^2 - n$$
$$0 = n^2 - n - 132$$
$$0 = (n - 12)(n + 11)$$
$$n - 12 = 0 \qquad n + 11 = 00$$
$$n = 12 \qquad n = -11$$

12 telephones are need to make 66 connections.

SECTION 6.1

VOCABULARY

1. A quotient of two polynomials, such as $\frac{x^2+x}{x^2-3x}$, is called a **rational** expression.

3. In the rational expression $\frac{(x+2)(3x-1)}{(x+2)(4x+2)}$, the binomial $x+2$ is a common **factor** of the numerator and the denominator.

5. Because of the division by 0, the expression $\frac{8}{0}$ is **undefined**.

7. The **domain** of a function is the set of all permissible input values for the variable.

CONCEPTS

9. a. $f(1) = -1$
   b. $f(4) = 2$
   c. $x = 2$
   d. $x = 5$

11. a. $\frac{3\cdot\cancel{5}\cdot\cancel{x}\cdot\cancel{y}\cdot y}{\cancel{5}\cdot 7\cdot\cancel{x}\cdot x\cdot x\cdot\cancel{y}} = \frac{3y}{7x^2}$

    b. $\frac{\cancel{(x+8)}(x-3)}{(x+2)\cancel{(x+8)}} = \frac{x-3}{x+2}$

    c. $\frac{a^3(a-9)}{(9-a)(9+a)} = \frac{a^3\cancel{(a-9)}}{-1\cancel{(a-9)}(9+a)} = -\frac{a^3}{a+9}$

13. a. iii.
    b. v.
    c. i.
    d. vi.
    e. iv.
    f. ii.

15. a. $f(0) = \frac{2(0)+1}{(0)^2+3(0)-4}$
$= \frac{0+1}{0+0-4}$
$= -\frac{1}{4}$

    b. $f(2) = \frac{2(2)+1}{(2)^2+3(2)-4}$
$= \frac{4+1}{4+6-4}$
$= \frac{5}{6}$

    c. $f(1) = \frac{2(1)+1}{(1)^2+3(1)-4}$
$= \frac{2+1}{1+3-4}$
$= \frac{3}{0}$; undefined

NOTATION

17. No. All the answers given are equivalent to the answer in the book except for $\frac{4-x}{x-4}$.

GUIDED PRACTICE

19.
$f(x) = \frac{2}{x}$
$x \neq 0$
$(-\infty,0)\cup(0,\infty)$

21.
$f(x) = \frac{2x}{x+2}$
$x+2 \neq 0$
$x \neq -2$
$(-\infty,-2)\cup(-2,\infty)$

23.

$$f(x)=\frac{3x-1}{x-x^2}$$

$$x-x^2\neq 0$$

$$x(1-x)\neq 0$$

$$x\neq 0 \quad \text{and} \quad 1-x\neq 0$$

$$x\neq 0 \qquad\qquad 1\neq x$$

$$(-\infty,0)\cup(0,1)\cup(1,\infty)$$

25.

$$f(x)=\frac{x^2+3x+2}{x^2-x-56}$$

$$x^2-x-56\neq 0$$

$$(x-8)(x+7)\neq 0$$

$$x-8\neq 0 \quad \text{and} \quad x+7\neq 0$$

$$x\neq 8 \qquad\qquad x\neq -7$$

$$(-\infty,-7)\cup(-7,8)\cup(8,\infty)$$

27.

$$\frac{12a^3}{18a}=\frac{\cancel{2}\cdot 2\cdot\cancel{3}\cdot\cancel{a}\cdot a\cdot a}{\cancel{2}\cdot\cancel{3}\cdot 3\cdot\cancel{a}}$$

$$=\frac{2a^2}{3}$$

29.

$$\frac{15a^2}{25a^8}=\frac{3\cdot\cancel{5}\cdot\cancel{a}\cdot\cancel{a}}{5\cdot\cancel{5}\cdot\cancel{a}\cdot\cancel{a}\cdot a\cdot a\cdot a\cdot a\cdot a\cdot a}$$

$$=\frac{3}{5a^6}$$

31.

$$\frac{27st}{36st^2}=\frac{\cancel{3}\cdot\cancel{3}\cdot 3\cdot\cancel{s}\cdot\cancel{t}}{2\cdot 2\cdot\cancel{3}\cdot\cancel{3}\cdot\cancel{s}\cdot\cancel{t}\cdot t}$$

$$=\frac{3}{4t}$$

33.

$$\frac{24x^3y^{10}}{18x^4y^3}=\frac{\cancel{2}\cdot 2\cdot 2\cdot\cancel{3}\cdot\cancel{x}\cdot\cancel{x}\cdot\cancel{x}\cdot\cancel{y}\cdot\cancel{y}\cdot\cancel{y}\cdot y\cdot y\cdot y\cdot y\cdot y\cdot y\cdot y}{\cancel{2}\cdot\cancel{3}\cdot 3\cdot\cancel{x}\cdot\cancel{x}\cdot\cancel{x}\cdot x\cdot\cancel{y}\cdot\cancel{y}\cdot\cancel{y}}$$

$$=\frac{4y^7}{3x}$$

35.

$$\frac{4x^2}{2x^3-12x^2}=\frac{4x^2}{2x^2(x-6)}$$

$$=\frac{2\cdot\cancel{2}\cdot\cancel{x}\cdot\cancel{x}}{\cancel{2}\cdot\cancel{x}\cdot\cancel{x}(x-6)}$$

$$=\frac{2}{x-6}$$

37.

$$\frac{24n^4}{16n^4+24n^3}=\frac{24n^4}{8n^3(2n+3)}$$

$$=\frac{\cancel{2}\cdot\cancel{2}\cdot\cancel{2}\cdot 3\cdot\cancel{n}\cdot\cancel{n}\cdot\cancel{n}\cdot n}{\cancel{2}\cdot\cancel{2}\cdot\cancel{2}\cdot\cancel{n}\cdot\cancel{n}\cdot\cancel{n}(2n+3)}$$

$$=\frac{3n}{2n+3}$$

39.

$$\frac{2x+18}{x^2-81}=\frac{2\cancel{(x+9)}}{\cancel{(x+9)}(x-9)}$$

$$=\frac{2}{x-9}$$

41.

$$\frac{4a^2-25}{20a-50}=\frac{(2a+5)\cancel{(2a-5)}}{10\cancel{(2a-5)}}$$

$$=\frac{2a+5}{10}$$

43.

$$\frac{5x^2-10x}{x^2-4x+4}=\frac{5x\cancel{(x-2)}}{(x-2)\cancel{(x-2)}}$$

$$=\frac{5x}{x-2}$$

45.

$$\frac{x^2+2x+1}{x^2+4x+3}=\frac{\cancel{(x+1)}(x+1)}{\cancel{(x+1)}(x+3)}$$

$$=\frac{x+1}{x+3}$$

47.

$$\frac{3d^2+13d+4}{3d^2+7d+2}=\frac{\cancel{(3d+1)}(d+4)}{\cancel{(3d+1)}(d+2)}$$
$$=\frac{d+4}{d+2}$$

49.

$$\frac{2h^2+9h-5}{4h^2-4h+1}=\frac{\cancel{(2h-1)}(h+5)}{\cancel{(2h-1)}(2h-1)}$$
$$=\frac{h+5}{2h-1}$$

51.

$$f(x)=\frac{x^2-1}{x^2+5x-6}$$
$$=\frac{(x+1)\cancel{(x-1)}}{(x+6)\cancel{(x-1)}}$$
$$=\frac{x+1}{x+6}; x\neq -6 \text{ and } x\neq 1$$

53.

$$f(x)=\frac{9x^2+81x}{x^3+9x^2}$$
$$=\frac{9x\cancel{(x+9)}}{x^2\cancel{(x+9)}}$$
$$=\frac{9}{x}; x\neq 0 \text{ and } x\neq -9$$

55.

$$g(x)=\frac{x^3+27}{x^3+3x^2+4x+12}$$
$$=\frac{(x)^3+(3)^3}{x^2(x+3)+4(x+3)}$$
$$=\frac{\cancel{(x+3)}(x^2-3x+9)}{\cancel{(x+3)}(x^2+4)}$$
$$=\frac{x^2-3x+9}{x^2+4};\quad x\neq -3$$

57.

$$s(a)=\frac{a^3-a^2-6a+6}{a^3-1}$$
$$=\frac{a^2(a-1)-6(a-1)}{(a)^3-(1)^3}$$
$$=\frac{\cancel{(a-1)}(a^2-6)}{\cancel{(a-1)}(a^2+a+1)}$$
$$=\frac{a^2-6}{a^2+a+1};\quad a\neq 1$$

59.

$$\frac{3m^2-2mn-n^2}{mn-m^2}=\frac{(3m+n)(m-n)}{m(n-m)}$$
$$=-\frac{(3m+n)\cancel{(m-n)}}{m\cancel{(m-n)}}$$
$$=-\frac{3m+n}{m} \text{ or } \frac{-(3m+n)}{m}$$

61.

$$\frac{b^2-a^2}{a-b}=\frac{(b-a)(b+a)}{(a-b)}$$
$$=-\frac{\cancel{(b-a)}(b+a)}{\cancel{(b-a)}}$$
$$=-(b+a)$$
$$=-b-a$$

63.

$$\frac{4-x^2}{x^2-x-2}=\frac{(2-x)(2+x)}{(x-2)(x+1)}$$
$$=-\frac{\cancel{(2-x)}(x+2)}{\cancel{(2-x)}(x+1)}$$
$$=-\frac{x+2}{x+1} \text{ or } \frac{-(x+2)}{x+1}$$

65.

$$\frac{20x^3-20x^4}{x^2-2x+1}=\frac{20x^3(1-x)}{(x-1)(x-1)}$$
$$=-\frac{20x^3\cancel{(x-1)}}{\cancel{(x-1)}(x-1)}$$
$$=-\frac{20x^3}{x-1}$$

67. $f(x)=\dfrac{x}{x-2}$

D: $(-\infty, 2)\cup(2,\infty)$; R: $(-\infty, 1)\cup(1,\infty)$

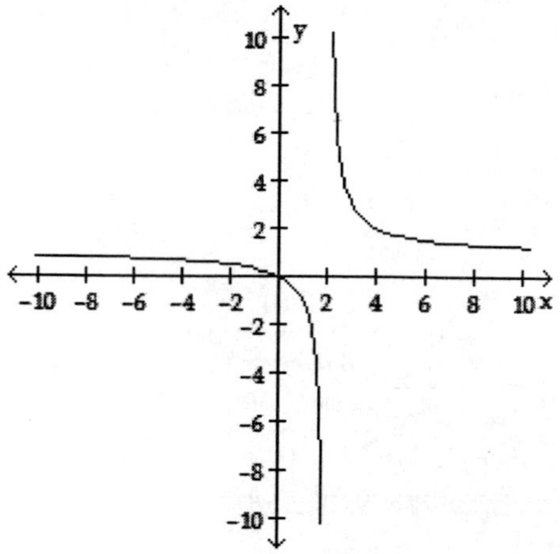

69. $f(x)=\dfrac{x+1}{x^2-4}=\dfrac{x+1}{(x+2)(x-2)}$

D: $(-\infty,-2)\cup(-2,2)\cup(2,\infty)$; R: $(-\infty,\infty)$

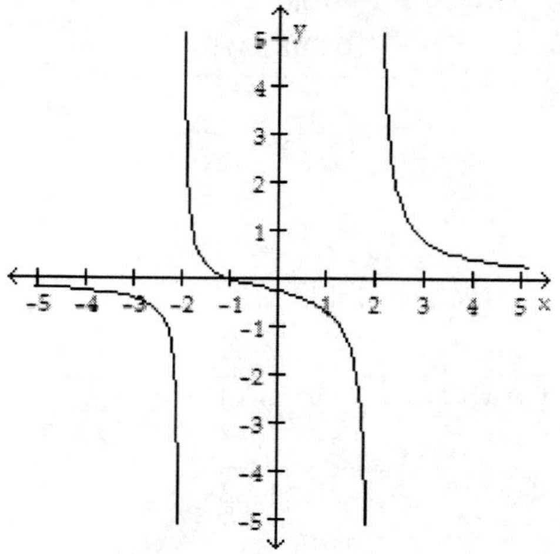

TRY IT YOURSELF

71.

$$\frac{x^2+x-30}{3x^2-3x-60}=\frac{(x+6)(x-5)}{3(x^2-x-20)}$$

$$=\frac{(x+6)\cancel{(x-5)}}{3\cancel{(x-5)}(x+4)}$$

$$=\frac{x+6}{3(x+4)}$$

73.

$$\frac{a^2-4}{a^3-8}=\frac{(a+2)\cancel{(a-2)}}{\cancel{(a-2)}(a^2+2a+4)}$$

$$=\frac{a+2}{a^2+2a+4}$$

75.

$$\frac{m^3-mn^2}{mn^2+m^2n-2m^3}=\frac{m(m^2-n^2)}{m(n^2+mn-2m^2)}$$

$$=\frac{m(m+n)(m-n)}{m(n-m)(n+2m)}$$

$$=\frac{-1\cancel{m}(m+n)\cancel{(n-m)}}{\cancel{m}\cancel{(n-m)}(n+2m)}$$

$$=-\frac{m+n}{n+2m}\ \text{or}\ \frac{-m-n}{n+2m}$$

77.

$$\frac{sx+4s-3x-12}{sx+4s+6x+24}=\frac{(sx+4s)+(-3x-12)}{(sx+4s)+(6x+24)}$$

$$=\frac{s(x+4)-3(x+4)}{s(x+4)+6(x+4)}$$

$$=\frac{\cancel{(x+4)}(s-3)}{\cancel{(x+4)}(s+6)}$$

$$=\frac{s-3}{s+6}$$

79.

$$\frac{2x^2-3x-9}{2x^2+3x-9}=\frac{(2x+3)(x-3)}{(2x-3)(x+3)}$$

Does not simplify

81.

$$\frac{3x+6y}{x+2y}=\frac{3\cancel{(x+2y)}}{\cancel{x+2y}}$$

$$=3$$

83.

$$\frac{x^4+3x^3+9x^2}{x^3-27}=\frac{x^2\cancel{(x^2+3x+9)}}{(x-3)\cancel{(x^2+3x+9)}}$$

$$=\frac{x^2}{x-3}$$

85.

$$\frac{2x^2+2x-12}{x^3+3x^2-4x-12}=\frac{2\left(x^2+x-6\right)}{x^2(x+3)-4(x+3)}$$
$$=\frac{2(x+3)(x-2)}{(x+3)\left(x^2-4\right)}$$
$$=\frac{2\cancel{(x+3)}\cancel{(x-2)}}{\cancel{(x+3)}\cancel{(x-2)}(x+2)}$$
$$=\frac{2}{(x+2)}$$

87.

$$\frac{4x^2+8x+3}{6+x-2x^2}=\frac{\cancel{(2x+3)}(2x+1)}{\cancel{(3+2x)}(2-x)}$$
$$=\frac{2x+1}{2-x}\quad\text{or}\quad-\frac{2x+1}{x-2}$$

89.

$$\frac{x^2-6x+9}{81-x^4}=\frac{(x-3)(x-3)}{\left(9+x^2\right)\left(9-x^2\right)}$$
$$=\frac{(x-3)(x-3)}{\left(9+x^2\right)(3-x)(3+x)}$$
$$=-\frac{\cancel{(x-3)}(x-3)}{\left(9+x^2\right)\cancel{(x-3)}(3+x)}$$
$$=-\frac{x-3}{\left(9+x^2\right)(3+x)}\quad\text{or}\quad-\frac{x-3}{\left(9+x^2\right)(x+3)}$$

91.

$$\frac{16p^3q^2}{24pq^8}=\frac{\cancel{2}\cdot\cancel{2}\cdot\cancel{2}\cdot2\cdot\cancel{p}\cdot p\cdot p\cdot\cancel{q}\cdot\cancel{q}}{\cancel{2}\cdot\cancel{2}\cdot\cancel{2}\cdot3\cdot\cancel{p}\cdot\cancel{q}\cdot\cancel{q}\cdot q\cdot q\cdot q\cdot q\cdot q\cdot q}$$
$$=\frac{2p^2}{3q^6}$$

93.

$$\frac{t^3-5t^2+6t}{9t-t^3}=\frac{t\left(t^2-5t+6\right)}{t\left(9-t^2\right)}$$
$$=\frac{t(t-3)(t-2)}{t(3-t)(3+t)}$$
$$=-\frac{\cancel{t}\cancel{(t-3)}(t-2)}{\cancel{t}\cancel{(t-3)}(3+t)}$$
$$=-\frac{t-2}{3+t}\quad\text{or}\quad-\frac{t-2}{t+3}$$

APPLICATIONS

95. ENVIRONMENTAL CLEANUP

a.
$$f(50)=\frac{50{,}000(50)}{100-50}$$
$$=\frac{2{,}500{,}000}{50}$$
$$=\$50{,}000$$

b.
$$f(80)=\frac{50{,}000(80)}{100-80}$$
$$=\frac{4{,}000{,}000}{20}$$
$$=\$200{,}000$$

97. UTILITY COSTS

a. $c(n)=0.09n+7.50$

b. $c(n)=\dfrac{0.09n+7.50}{n}$

c.
$$c(775)=\frac{0.09(775)+7.50}{775}\quad\widehat{ABC}$$
$$=\frac{69.75+7.50}{775}$$
$$=\frac{77.25}{775}$$
$$\approx\$0.10$$

99. CAMPUS TO CAREERS

a.
$$f(15)=\frac{15^2+2(15)}{2(15)+2}$$
$$=\frac{225+30}{30+2}$$
$$=\frac{255}{32}$$
$$\approx 7.968$$
$$\approx 8\text{ days}$$

b. $t+2=20$

$t=18$

$$f(18)=\frac{18^2+2(18)}{2(18)+2}$$
$$=\frac{324+36}{36+2}$$
$$=\frac{360}{38}$$
$$\approx 9.473$$
$$\approx 9.5\text{ days}$$

WRITING

101. Answers will vary.

103. Answers will vary.

REVIEW

105.

$$\left(a^2-4a-3\right)(a-2)$$
$$=a^2(a-2)-4a(a-2)-3(a-2)$$
$$=a^3-2a^2-4a^2+8a-3a+6$$
$$=a^3-6a^2+5a+6$$

107.

$$-3mn^2\left(m^3-7mn-2m^2\right)$$
$$=-3m^4n^2+21m^2n^3+6m^3n^2$$

CHALLENGE PROBLEMS

109.

$$\frac{x^{32}-1}{x^{16}-1}=\frac{\left(x^{16}+1\right)\cancel{\left(x^{16}-1\right)}}{\cancel{x^{16}-1}}$$
$$=x^{16}+1$$

111.

$$\frac{a^6-64}{\left(a^2+2a+4\right)\left(a^2-2a+4\right)}$$
$$=\frac{\left(a^3+8\right)\left(a^3-8\right)}{\left(a^2+2a+4\right)\left(a^2-2a+4\right)}$$
$$=\frac{(a+2)\cancel{\left(a^2-2a+4\right)}(a-2)\cancel{\left(a^2+2a+4\right)}}{\cancel{\left(a^2+2a+4\right)}\cancel{\left(a^2-2a+4\right)}}$$
$$=(a+2)(a-2)$$

113. $f(x)=\dfrac{1}{x-1}$

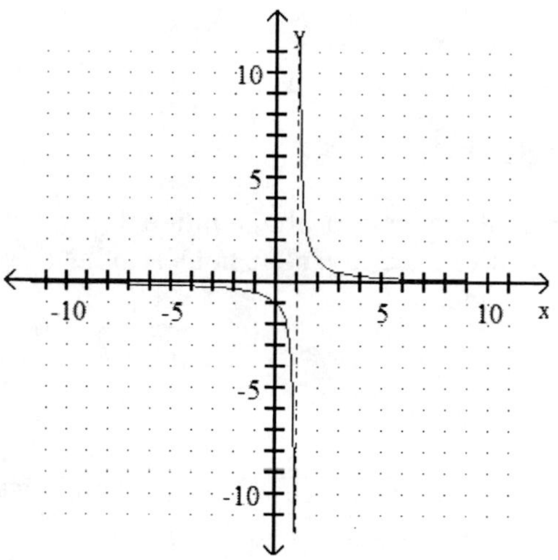

SECTION 6.2

VOCABULARY

1. $\dfrac{a^2-9}{a^2-49} \bullet \dfrac{a-7}{a+3}$ is the product of two **rational** expressions.

3. To find the reciprocal of a rational expression, we **invert** its numerator and denominator.

CONCEPTS

5. To multiply rational expressions, multiply their **numerators** and multiply their **denominators**. In symbols,

$$\frac{A}{B} \bullet \frac{C}{D} = \boxed{\frac{AC}{BC}}$$

NOTATION

7.

$$\begin{aligned}\frac{x^2+3x}{5x-25} \bullet \frac{x-5}{x+3} &= \frac{(x^2+3x)\boxed{(x-5)}}{(5x-25)\boxed{(x+3)}} \\ &= \frac{\boxed{x}(x+3)(x-5)}{\boxed{5}(x-5)(x+3)} \\ &= \frac{\boxed{x}}{5}\end{aligned}$$

9. $\dfrac{x-3}{x+3} \neq \dfrac{3-x}{3+x}$

The other two answers are equivalent.

GUIDED PRACTICE

11.

$$\begin{aligned}\frac{3}{4} \bullet \frac{11}{3} &= \frac{\cancel{3} \bullet 11}{4 \bullet \cancel{3}} \\ &= \frac{11}{4}\end{aligned}$$

13.

$$\begin{aligned}\frac{15}{24} \bullet \frac{16}{25} &= \frac{15 \bullet 16}{24 \bullet 25} \\ &= \frac{\cancel{3} \bullet \cancel{5} \bullet \cancel{2} \bullet \cancel{2} \bullet \cancel{2} \bullet 2}{\cancel{2} \bullet \cancel{2} \bullet \cancel{2} \bullet \cancel{3} \bullet \cancel{5} \bullet 5} \\ &= \frac{2}{5}\end{aligned}$$

15.

$$\begin{aligned}\frac{3a}{10} \bullet \frac{2}{15a^4} &= \frac{3a \bullet 2}{10 \bullet 15a^4} \\ &= \frac{\cancel{3} \bullet \cancel{a} \bullet \cancel{2}}{\cancel{2} \bullet 5 \bullet 5 \bullet \cancel{3} \bullet \cancel{a} \bullet a \bullet a \bullet a} \\ &= \frac{1}{25a^3}\end{aligned}$$

17.

$$\begin{aligned}\frac{12x^{61}}{7y^{15}} \bullet \frac{y}{8x^{27}} &= \frac{12x^{61} \bullet y}{7y^{15} \bullet 8x^{27}} \\ &= \frac{\cancel{2} \bullet \cancel{2} \bullet 3 \bullet \cancel{x^{27}} \bullet x^{34} \bullet \cancel{y}}{7 \bullet \cancel{y} \bullet y^{14} \bullet \cancel{2} \bullet \cancel{2} \bullet 2 \bullet \cancel{x^{27}}} \\ &= \frac{3x^{34}}{14y^{14}}\end{aligned}$$

19.

$$\begin{aligned}\frac{y^2+6y+9}{15y} \bullet \frac{3y^2}{2y+6} &= \frac{(y^2+6y+9)3y^2}{15y(2y+6)} \\ &= \frac{(y+3)\cancel{(y+3)}\cancel{(3)}\cancel{(y)}(y)}{\cancel{(3)}(5)\cancel{(y)}(2)\cancel{(y+3)}} \\ &= \frac{y(y+3)}{10}\end{aligned}$$

21.

$$\begin{aligned}\frac{x^2+x-6}{5x} \bullet \frac{5x-10}{x+3} &= \frac{(x^2+x-6)(5x-10)}{5x(x+3)} \\ &= \frac{\cancel{(x+3)}(x-2)\cancel{(5)}(x-2)}{\cancel{5}x\cancel{(x+3)}} \\ &= \frac{(x-2)^2}{x}\end{aligned}$$

23.

$$\frac{x^2+2x+1}{9x^3}\bullet\frac{2x^2-2x}{2x^2-2}=\frac{(x^2+2x+1)(2x^2-2x)}{9x^3(2x^2-2)}$$

$$=\frac{(x+1)(x+1)2x(x-1)}{9x(x)(x)(2)(x^2-1)}$$

$$=\frac{\cancel{(x+1)}(x+1)\cancel{(2)}\cancel{(x)}\cancel{(x-1)}}{9\cancel{x}(x)(x)\cancel{(2)}\cancel{(x-1)}\cancel{(x+1)}}$$

$$=\frac{x+1}{9x^2}$$

25.

$$\frac{x^3+3x^2-3x-9}{x}\bullet\frac{1}{x^2+3x}=\frac{(x^3+3x^2-3x-9)(1)}{(x)(x^2+3x)}$$

$$=\frac{(x^2-3)\cancel{(x+3)}}{x\cancel{(x+3)}}$$

$$=\frac{x^2-3}{x}$$

27.

$$\frac{2x^2-x-3}{x^2-1}\bullet\frac{x^2+x-2}{2x^2+x-6}$$

$$=\frac{(2x^2-x-3)(x^2+x-2)}{(x^2-1)(2x^2+x-6)}$$

$$=\frac{\cancel{(2x-3)}\cancel{(x+1)}\cancel{(x+2)}\cancel{(x-1)}}{\cancel{(x+1)}\cancel{(x-1)}\cancel{(2x-3)}\cancel{(x+2)}}$$

$$=1$$

29.

$$\frac{3t^2-t-2}{6t^2-5t-6}\bullet\frac{4t^2-9}{2t^2+5t+3}$$

$$=\frac{(3t^2-t-2)(4t^2-9)}{(6t^2-5t-6)(2t^2+5t+3)}$$

$$=\frac{\cancel{(3t+2)}(t-1)\cancel{(2t+3)}\cancel{(2t-3)}}{\cancel{(3t+2)}\cancel{(2t-3)}\cancel{(2t+3)}(t+1)}$$

$$=\frac{t-1}{t+1}$$

31.

$$\frac{x^2+4xy+4y^2}{2x^2+4xy}\bullet\frac{3x-6y}{x^2-4y^2}$$

$$=\frac{(x^2+4xy+4y^2)(3x-6y)}{(2x^2+4xy)(x^2-4y^2)}$$

$$=\frac{\cancel{(x+2y)}\cancel{(x+2y)}(3)\cancel{(x-2y)}}{2x\cancel{(x+2y)}\cancel{(x+2y)}\cancel{(x-2y)}}$$

$$=\frac{3}{2x}$$

33.

$$\frac{3a^2+7ab+2b^2}{a^2+2ab}\bullet\frac{a^2-ab}{3a^2+ab}$$

$$=\frac{(3a^2+7ab+2b^2)(a^2-ab)}{(a^2+2ab)(3a^2+ab)}$$

$$=\frac{\cancel{(3a+b)}\cancel{(a+2b)}\cancel{(a)}(a-b)}{a\cancel{(a+2b)}\cancel{(a)}\cancel{(3a+b)}}$$

$$=\frac{a-b}{a}$$

35.

$$15x\left(\frac{x+1}{15x}\right)=\frac{\cancel{15x}(x+1)}{\cancel{15x}}$$

$$=x+1$$

37.

$$12y\left(\frac{y+8}{6y}\right)=\frac{2\bullet\cancel{6y}(y+8)}{\cancel{6y}}$$

$$=2(y+8)$$

$$=2y+16$$

39.

$$\left(6a-a^2\right)\bullet\frac{a^3}{2a^3-12a^2+6a-36}$$
$$=\frac{\left(6a-a^2\right)\left(a^3\right)}{2a^3-12a^2+6a-36}$$
$$=\frac{a(6-a)\left(a^3\right)}{2\left(a^3-6a^2+3a-18\right)}$$
$$=\frac{a(6-a)\left(a^3\right)}{2\left(a^2+3\right)(a-6)}$$
$$=-\frac{a\cancel{(a-6)}\left(a^3\right)}{2\left(a^2+3\right)\cancel{(a-6)}}$$
$$=-\frac{a^4}{2\left(a^2+3\right)}$$

41.

$$\left(x^2+x-2cx-2c\right)\bullet\frac{x^2+3x+2}{4c^2-x^2}$$
$$=\frac{\left(x^2+x-2cx-2c\right)\left(x^2+3x+2\right)}{4c^2-x^2}$$
$$=\frac{\left(x(x+1)-2c(x+1)\right)(x+2)(x+1)}{(2c+x)(2c-x)}$$
$$=-\frac{(x+1)\cancel{(x-2c)}(x+2)(x+1)}{(x+2c)\cancel{(x-2c)}}$$
$$=-\frac{(x+1)^2(x+2)}{(x+2c)}$$

43.

$$\left(\frac{x-3}{x^3+4}\right)^2=\left(\frac{x-3}{x^3+4}\right)\left(\frac{x-3}{x^3+4}\right)$$
$$=\frac{(x-3)(x-3)}{\left(x^3+4\right)\left(x^3+4\right)}$$
$$=\frac{x^2-3x-3x+9}{x^6+4x^3+4x^3+16}$$
$$=\frac{x^2-6x+9}{x^6+8x^3+16}$$

45.

$$\left(\frac{2m^2-m-3}{x^2-1}\right)^2$$
$$=\left(\frac{2m^2-m-3}{x^2-1}\right)\left(\frac{2m^2-m-3}{x^2-1}\right)$$
$$=\frac{\left(2m^2-m-3\right)\left(2m^2-m-3\right)}{\left(x^2-1\right)\left(x^2-1\right)}$$
$$=\frac{4m^4-2m^3-6m^2-2m^3+m^2+3m-6m^2+3m+9}{x^4-x^2-x^2+1}$$
$$=\frac{4m^4-4m^3-11m^2+6m+9}{x^4-2x^2+1}$$

47.

$$\frac{6}{11}\div\frac{36}{55}=\frac{6}{11}\bullet\frac{55}{36}$$
$$=\frac{6\bullet 55}{11\bullet 36}$$
$$=\frac{\cancel{6}\bullet 5\bullet\cancel{11}}{\cancel{11}\bullet\cancel{6}\bullet 6}$$
$$=\frac{5}{6}$$

49.

$$\frac{12}{5}\div\frac{24}{45}=\frac{12}{5}\bullet\frac{45}{24}$$
$$=\frac{12\bullet 45}{5\bullet 24}$$
$$=\frac{\cancel{2}\bullet\cancel{2}\bullet\cancel{3}\bullet 3\bullet 3\bullet\cancel{5}}{\cancel{5}\bullet\cancel{2}\bullet\cancel{2}\bullet 2\bullet\cancel{3}}$$
$$=\frac{9}{2}$$

51.

$$\frac{22x^3}{y^2}\div\frac{33x^9}{y^7}=\frac{22x^3}{y^2}\bullet\frac{y^7}{33x^9}$$
$$=\frac{22x^3\bullet y^7}{y^2\bullet 33x^9}$$
$$=\frac{2\bullet\cancel{11}\bullet\cancel{x}\bullet\cancel{x}\bullet\cancel{x}\bullet\cancel{y}\bullet\cancel{y}\bullet y\bullet y\bullet y\bullet y\bullet y}{\cancel{y}\bullet\cancel{y}\bullet 3\bullet\cancel{11}\bullet\cancel{x}\bullet\cancel{x}\bullet\cancel{x}\bullet x\bullet x\bullet x\bullet x\bullet x\bullet x}$$
$$=\frac{2y^5}{3x^6}$$

53.

$$\frac{pq^{29}}{50} \div \frac{p^{10}q^{38}}{15} = \frac{pq^{29}}{50} \bullet \frac{15}{p^{10}q^{38}}$$
$$= \frac{pq^{29}(15)}{50p^{10}q^{38}}$$
$$= \frac{\cancel{p} \bullet \cancel{q^{29}} \bullet 3 \bullet \cancel{5}}{2 \bullet \cancel{5} \bullet 5 \bullet \cancel{p} \bullet p^9 \bullet \cancel{q^{29}} \bullet q^9}$$
$$= \frac{3}{10p^9q^9}$$

55.

$$\frac{x^{12}}{x^3 - 8} \div \frac{x^2}{x^2 - 2x}$$
$$= \frac{x^{12}}{x^3 - 8} \bullet \frac{x^2 - 2x}{x^2}$$
$$= \frac{x^{12}(x^2 - 2x)}{(x^3 - 8)(x^2)}$$
$$= \frac{\cancel{x} \bullet \cancel{x} \bullet x \bullet x \bullet x \bullet x \bullet x \bullet x \bullet x \bullet x \bullet x \bullet x \bullet x\,\cancel{(x-2)}}{\cancel{x} \bullet \cancel{x}\,\cancel{(x-2)}(x^2 + 2x + 4)}$$
$$= \frac{x^{11}}{x^2 + 2x + 4}$$

57.

$$\frac{a^2 - 18a + 81}{a^{40}} \div \frac{(a-9)^3}{a^{37}} = \frac{a^2 - 18a + 81}{a^{40}} \bullet \frac{a^{37}}{(a-9)^3}$$
$$= \frac{(a-9)^2}{a^{40}} \bullet \frac{a^{37}}{(a-9)^3}$$
$$= \frac{\cancel{a^{37}}\,\cancel{(a-9)^2}}{\cancel{a^{37}} \bullet a^3\,\cancel{(a-9)^2}\,(a-9)}$$
$$= \frac{1}{a^3(a-9)}$$

59.

$$\frac{3n^2 + 5n - 2}{12n^2 - 13n + 3} \div \frac{n^2 + 3n + 2}{4n^2 + 5n - 6}$$
$$= \frac{3n^2 + 5n - 2}{12n^2 - 13n + 3} \bullet \frac{4n^2 + 5n - 6}{n^2 + 3n + 2}$$
$$= \frac{(3n^2 + 5n - 2)(4n^2 + 5n - 6)}{(12n^2 - 13n + 3)(n^2 + 3n + 2)}$$
$$= \frac{\cancel{(3n-1)}\,\cancel{(n+2)}\,\cancel{(4n-3)}\,(n+2)}{\cancel{(4n-3)}\,\cancel{(3n-1)}\,\cancel{(n+2)}\,(n+1)}$$
$$= \frac{n+2}{n+1}$$

61.

$$\frac{5cd + d^2}{6d^2} \div \frac{125c^3 + d^3}{6c + 6d}$$
$$= \frac{5cd + d^2}{6d^2} \bullet \frac{6c + 6d}{125c^3 + d^3}$$
$$= \frac{(5cd + d^2)(6c + 6d)}{6d^2(125c^3 + d^3)}$$
$$= \frac{\cancel{d}\,\cancel{(5c+d)}\,\cancel{(6)}\,(c+d)}{\cancel{6}\,\cancel{d}\,(d)\,\cancel{(5c+d)}\,(25c^2 - 5cd + d^2)}$$
$$= \frac{c+d}{d(25c^2 - 5cd + d^2)}$$

63.

$$\frac{y^3 - 9y}{y + 2} \div (y - 3) = \frac{y^3 - 9y}{y + 2} \bullet \frac{1}{y - 3}$$
$$= \frac{(y^3 - 9y)}{(y+2)(y-3)}$$
$$= \frac{y(y^2 - 9)}{(y+2)(y-3)}$$
$$= \frac{y\,\cancel{(y-3)}\,(y+3)}{(y+2)\,\cancel{(y-3)}}$$
$$= \frac{y(y+3)}{y+2}$$

65.

$$(x+1)\div\frac{x^2+2x+1}{2}=(x+1)\bullet\frac{2}{x^2+2x+1}$$
$$=\frac{2(x+1)}{x^2+2x+1}$$
$$=\frac{2\cancel{(x+1)}}{\cancel{(x+1)}(x+1)}$$
$$=\frac{2}{x+1}$$

67.

$$\frac{6a^2-7a-3}{a^2-1}\div\frac{4a^2-12a+9}{a^2-1}\bullet\frac{2a^2-a-3}{3a^2-2a-1}$$
$$=\frac{6a^2-7a-3}{a^2-1}\bullet\frac{a^2-1}{4a^2-12a+9}\bullet\frac{2a^2-a-3}{3a^2-2a-1}$$
$$=\frac{\cancel{(2a-3)}\cancel{(3a+1)}\cancel{(a+1)}\cancel{(a-1)}\cancel{(2a-3)}(a+1)}{\cancel{(a+1)}\cancel{(a-1)}\cancel{(2a-3)}\cancel{(2a-3)}\cancel{(3a+1)}(a-1)}$$
$$=\frac{a+1}{a-1}$$

69.

$$\frac{2x^2-2x-4}{x^2+2x-8}\bullet\frac{3x^2+15x}{x+1}\div\frac{4x^2-100}{x^2-x-20}$$
$$=\frac{2x^2-2x-4}{x^2+2x-8}\bullet\frac{3x^2+15x}{x+1}\bullet\frac{x^2-x-20}{4x^2-100}$$
$$=\frac{(2x^2-2x-4)(3x^2+15x)(x^2-x-20)}{(x^2+2x-8)(x+1)(4x^2-100)}$$
$$=\frac{2(x^2-x-2)(3x)(x+5)(x-5)(x+4)}{(x+4)(x-2)(x+1)(4)(x^2-25)}$$
$$=\frac{\cancel{2}\cancel{(x-2)}\cancel{(x+1)}(3x)\cancel{(x+5)}\cancel{(x-5)}\cancel{(x+4)}}{\cancel{(x+4)}\cancel{(x-2)}\cancel{(x+1)}(\cancel{2}\bullet 2)\cancel{(x+5)}\cancel{(x-5)}}$$
$$=\frac{3x}{2}$$

71.

$$\frac{x^2-6x+9}{4-x^2}\div\frac{x^2-9}{x^2-8x+12}$$
$$=\frac{x^2-6x+9}{4-x^2}\bullet\frac{x^2-8x+12}{x^2-9}$$
$$=\frac{(x^2-6x+9)(x^2-8x+12)}{(4-x^2)(x^2-9)}$$
$$=\frac{(x-3)(x-3)(x-6)(x-2)}{(2-x)(2+x)(x+3)(x-3)}$$
$$=\frac{-\cancel{(x-3)}(x-3)(x-6)\cancel{(2-x)}}{\cancel{(2-x)}(2+x)(x+3)\cancel{(x-3)}}$$
$$=-\frac{(x-3)(x-6)}{(2+x)(x+3)}$$

73.

$$\frac{2x^2-2x-12}{x^2-4}\bullet\frac{x^2-x-2}{x^3-9x}$$
$$=\frac{(2x^2-2x-12)(x^2-x-2)}{(x^2-4)(x^3-9x)}$$
$$=\frac{2(x-3)(x+2)(x-2)(x+1)}{(x+2)(x-2)(x)(x-3)(x+3)}$$
$$=\frac{2\cancel{(x-3)}\cancel{(x+2)}\cancel{(x-2)}(x+1)}{\cancel{(x+2)}\cancel{(x-2)}(x)\cancel{(x-3)}(x+3)}$$
$$=\frac{2(x+1)}{x(x+3)}$$

75.

$$\frac{p^3-q^3}{p^2-q^2}\bullet\frac{q^2+pq}{p^3+p^2q+pq^2}$$
$$=\frac{(p^3-q^3)(q^2+pq)}{(p^2-q^2)(p^3+p^2q+pq^2)}$$
$$=\frac{\cancel{(p-q)}\cancel{(p^2+pq+q^2)}(q)\cancel{(q+p)}}{\cancel{(p+q)}\cancel{(p-q)}(p)\cancel{(p^2+pq+q^2)}}$$
$$=\frac{q}{p}$$

77.

$$\frac{10r^2s}{6rs^2}\bullet\frac{3r^3}{2rs}=\frac{10r^2s\bullet 3r^3}{6rs^2\bullet 2rs}$$
$$=\frac{\cancel{2}\bullet 5\bullet \cancel{r}\bullet \cancel{r}\bullet \cancel{s}\bullet \cancel{3}\bullet r\bullet r\bullet r}{\cancel{2}\bullet \cancel{3}\bullet \cancel{r}\bullet \cancel{s}\bullet s\bullet 2\bullet \cancel{r}\bullet s}$$
$$=\frac{5r^3}{2s^2}$$

79.

$$10(h-9)\left(\frac{h-3}{9-h}\right)=\frac{10(h-9)(h-3)}{9-h}$$
$$=-\frac{10\cancel{(h-9)}(h-3)}{\cancel{(h-9)}}$$
$$=-10(h-3)$$
$$=-10h+30$$

81.

$$\frac{2x^2+5xy+3y^2}{3x^2-5xy+2y^2}\div\frac{2x^2+xy-3y^2}{3x^2-5xy+2y^2}$$
$$=\frac{2x^2+5xy+3y^2}{3x^2-5xy+2y^2}\bullet\frac{3x^2-5xy+2y^2}{2x^2+xy-3y^2}$$
$$=\frac{(2x^2+5xy+3y^2)(3x^2-5xy+2y^2)}{(3x^2-5xy+2y^2)(2x^2+xy-3y^2)}$$
$$=\frac{\cancel{(2x+3y)}(x+y)\cancel{(3x-2y)}\cancel{(x-y)}}{\cancel{(3x-2y)}(x-y)\cancel{(2x+3y)}\cancel{(x-y)}}$$
$$=\frac{x+y}{x-y}$$

83.

$$(4x^2-9)\div\frac{2x^2+5x+3}{x+2}\bullet\frac{1}{2x-3}$$
$$=\frac{4x^2-9}{1}\bullet\frac{x+2}{2x^2+5x+3}\bullet\frac{1}{2x-3}$$
$$=\frac{(4x^2-9)(x+2)}{(2x^2+5x+3)(2x-3)}$$
$$=\frac{\cancel{(2x+3)}\cancel{(2x-3)}(x+2)}{\cancel{(2x+3)}(x+1)\cancel{(2x-3)}}$$
$$=\frac{x+2}{x+1}$$

85.

$$\frac{x^3-3x^2-25x+75}{x^3-27}\bullet\frac{2x^3+6x^2+18x}{x^2+10x+25}$$
$$=\frac{(x^3-3x^2-25x+75)(2x^3+6x^2+18x)}{(x^3-27)(x^2+10x+25)}$$
$$=\frac{(x^2-25)(x-3)(2x)(x^2+3x+9)}{(x-3)(x^2+3x+9)(x+5)(x+5)}$$
$$=\frac{(x-5)\cancel{(x+5)}\cancel{(x-3)}(2x)\cancel{(x^2+3x+9)}}{\cancel{(x-3)}\cancel{(x^2+3x+9)}\cancel{(x+5)}(x+5)}$$
$$=\frac{2x(x-5)}{x+5}$$

87.

$$f(x)\bullet g(x)=\frac{x^2+x-6}{x^2-6x+9}\bullet\frac{x^2-9}{x^2-4}$$
$$=\frac{(x+3)\cancel{(x-2)}(x+3)\cancel{(x-3)}}{\cancel{(x-3)}(x-3)(x+2)\cancel{(x-2)}}$$
$$=\frac{(x+3)^2}{(x-3)(x+2)}$$

89.

$$s(x)\bullet f(x)=\frac{x^2-16}{x^2-25}\div\frac{5x+20}{10x^2-50x}$$
$$=\frac{x^2-16}{x^2-25}\bullet\frac{10x^2-50x}{5x+20}$$
$$=\frac{\cancel{(x+4)}(x-4)(\cancel{5}\bullet 2x)\cancel{(x-5)}}{\cancel{(x-5)}(x+5)\cancel{(5)}\cancel{(x+4)}}$$
$$=\frac{2x(x-4)}{x+5}$$

91. a.

$$\frac{t^2+9t+20}{9t+36}\bullet\frac{9t+45}{t+4}=\frac{(t^2+9t+20)(9t+45)}{(9t+36)(t+4)}$$
$$=\frac{\cancel{(t+4)}(t+5)\cancel{(9)}(t+5)}{\cancel{(9)}\cancel{(t+4)}(t+4)}$$
$$=\frac{(t+5)^2}{(t+4)}$$

b.

$$\frac{t^2+9t+20}{9t+36}\div\frac{9t+45}{t+4}=\frac{t^2+9t+20}{9t+36}\bullet\frac{t+4}{9t+45}$$
$$=\frac{(t^2+9t+20)(t+4)}{(9t+36)(9t+45)}$$
$$=\frac{\cancel{(t+4)}\cancel{(t+5)}(t+4)}{(9)\cancel{(t+4)}(9)\cancel{(t+5)}}$$
$$=\frac{t+4}{81}$$

93. a.

$$\frac{4r-8}{5r}\div\frac{5r-10}{4r^2}=\frac{4r-8}{5r}\bullet\frac{4r^2}{5r-10}$$
$$=\frac{(4r-8)(4r^2)}{5r(5r-10)}$$
$$=\frac{(4)\cancel{(r-2)}(4\cancel{r}r)}{(5\cancel{r})(5)\cancel{(r-2)}}$$
$$=\frac{16r}{25}$$

b.

$$\frac{4r-8}{5r}\bullet\frac{5r-10}{4r^2}=\frac{(4r-8)(5r-10)}{5r(4r^2)}$$
$$=\frac{(\cancel{4})(r-2)(\cancel{5})(r-2)}{(\cancel{5}r)(\cancel{4}r^2)}$$
$$=\frac{(r-2)^2}{r^3}$$

APPLICATIONS

95. PHYSICS EXPERIMENTS

Trial 1:

$$D=rt$$
$$=\left(\frac{k_1^2+3k_1+2}{k_1-3}\right)\left(\frac{k_1^2-3k_1}{k_1+1}\right)$$
$$=\frac{\cancel{(k_1+1)}(k_1+2)(k_1)\cancel{(k_1-3)}}{\cancel{(k_1-3)}\cancel{(k_1+1)}}$$
$$=k_1(k_1+2)$$

Trial 2:

$$t=D\div r$$
$$=(k_2^2+11k_2+30)\div\left(\frac{k_2^2+6k_2+5}{k_2+1}\right)$$
$$=\frac{k_2^2+11k_2+30}{1}\bullet\frac{k_2+1}{k_2^2+6k_2+5}$$
$$=\frac{\cancel{(k_2+5)}(k_2+6)\cancel{(k_2+1)}}{\cancel{(k_2+5)}\cancel{(k_2+1)}}$$
$$=k_2+6$$

WRITING

97. Answers will vary.

99. Answers will vary.

REVIEW

101. a. $x^m x^n = x^{m+n}$

b. $\left(x^m\right)^n = x^{mn}$

c. $(xy)^n = x^n y^n$

d. $\left(\frac{x}{y}\right)^n = \frac{x^n}{y^n}$

e. $x^0 = 1$

f. $x^{-n} = \frac{1}{x^n}$

CHALLENGE PROBLEMS

103. $\frac{x^2}{y}\boxed{\div}\frac{x}{y^2}\boxed{\bullet}\frac{x^2}{y^2}=\frac{x^3}{y}$

SECTION 6.3

VOCABULARY

1. The rational expressions $\frac{7}{6n}$ and $\frac{n+1}{6n}$ have a common **denominator** of $6n$.

3. To **build** a rational expression, we multiply it by a form of 1. For example, $\frac{2}{n^2}\bullet\frac{8}{8}=\frac{16}{8n^2}$.

CONCEPTS

5. To add or subtract rational expressions that have the same denominator, add or subtract the **numerators**, and write the sum or difference over the common **denominators**. In symbols, if $\frac{A}{D}$ and $\frac{B}{D}$ are rational expressions, $\frac{A}{D}+\frac{B}{D}=\frac{\boxed{A+B}}{D}$ or $\frac{A}{D}-\frac{B}{D}=\frac{\boxed{A-B}}{D}$.

7. When a number is multiplied by **1**, its value does not change.

9. a. ii; adding or subtracting rational expressions
   b. simplifying a rational expression

11. a. once
    b. twice
    c. $3(x-2)^2$

13. a. $40x^2 = 2\cdot 2\cdot 2\cdot 5\cdot x\cdot x$
    b. $2x^2 - 6x = 2x(x-3)$
    c. $n^2 - 64 = (n-8)(n+8)$

NOTATION

15.
$$\frac{6x-1}{3x-1}+\frac{3x-2}{3x-1}=\frac{6x-1+\boxed{3x-2}}{3x-1}$$
$$=\frac{9x-\boxed{3}}{3x-1}$$
$$=\frac{3\left(\boxed{3x-1}\right)}{3x-1}$$
$$=3$$

17.
$$\frac{8}{3v}-\frac{1}{4v^2}=\frac{8}{3v}\bullet\frac{\boxed{4v}}{\boxed{4v}}-\frac{1}{4v^2}\bullet\frac{\boxed{3}}{\boxed{3}}$$
$$=\frac{\boxed{32v}}{12v^2}-\frac{3}{\boxed{12v^2}}$$
$$=\frac{32v-3}{\boxed{12v^2}}$$

GUIDED PRACTICE

19.
$$\frac{8}{3x}+\frac{5}{3x}=\frac{8+5}{3x}$$
$$=\frac{13}{3x}$$

21.
$$\frac{t}{4r}+\frac{t}{4r}=\frac{t+t}{4r}$$
$$=\frac{2t}{4r}$$
$$=\frac{t}{2r}$$

23.
$$\frac{4y}{y-4}-\frac{16}{y-4}=\frac{4y-16}{y-4}$$
$$=\frac{4\cancel{(y-4)}}{\cancel{y-4}}$$
$$=4$$

25.

$$\frac{3x}{x^2-9}-\frac{9}{x^2-9}=\frac{3x-9}{x^2-9}$$

$$=\frac{3\cancel{(x-3)}}{\cancel{(x-3)}(x+3)}$$

$$=\frac{3}{x+3}$$

27. $12xy = 2\cdot 2\cdot 3\cdot x\cdot y$
$18x^2y = 2\cdot 3\cdot 3\cdot x\cdot x\cdot y$
$\text{LCD} = 2\cdot 2\cdot 3\cdot 3\cdot x\cdot x\cdot y = 36x^2y$

29. $x^2 + 3x = x(x + 3)$
$x^2 - 9 = (x + 3)(x - 3)$
$\text{LCD} = x(x + 3)(x - 3)$

31. $x^3 + 27 = (x + 3)(x^2 + 3x + 9)$
$x^2 + 6x + 9 = (x + 3)(x + 3)$
$\text{LCD} = (x + 3)^2(x^2 + 3x + 9)$

33. $2x^2 + 5x + 3 = (2x + 3)(x + 1)$
$4x^2 + 12x + 9 = (2x + 3)(2x + 3)$
$x^2 + 2x + 1 = (x + 1)(x + 1)$
$\text{LCD} = (x + 1)^2(2x + 3)^2$

35.

$$\frac{11}{5m}-\frac{5}{6m}=\frac{11}{5m}\left(\frac{6}{6}\right)-\frac{5}{6m}\left(\frac{5}{5}\right)$$

$$=\frac{66}{30m}-\frac{25}{30m}$$

$$=\frac{66-25}{30m}$$

$$=\frac{41}{30m}$$

37.

$$\frac{3}{4ab^2}-\frac{5}{2a^2b}=\frac{3}{4ab^2}\left(\frac{a}{a}\right)-\frac{5}{2a^2b}\left(\frac{2b}{2b}\right)$$

$$=\frac{3a}{4a^2b^2}-\frac{10b}{4a^2b^2}$$

$$=\frac{3a-10b}{4a^2b^2}$$

39.

$$\frac{3}{x+2}+\frac{5}{x-4}=\frac{3}{x+2}\left(\frac{x-4}{x-4}\right)+\frac{5}{x-4}\left(\frac{x+2}{x+2}\right)$$

$$=\frac{3(x-4)}{(x+2)(x-4)}+\frac{5(x+2)}{(x+2)(x-4)}$$

$$=\frac{3x-12}{(x+2)(x-4)}+\frac{5x+10}{(x+2)(x-4)}$$

$$=\frac{3x-12+5x+10}{(x+2)(x-4)}$$

$$=\frac{8x-2}{(x+2)(x-4)} \text{ or } \frac{2(4x-1)}{(x+2)(x-4)}$$

41.

$$\frac{6x}{x+3}-\frac{4x}{x-3}=\frac{6x}{x+3}\left(\frac{x-3}{x-3}\right)-\frac{4x}{x-3}\left(\frac{x+3}{x+3}\right)$$

$$=\frac{6x^2-18x}{(x+3)(x-3)}-\frac{4x^2+12x}{(x+3)(x-3)}$$

$$=\frac{6x^2-18x-\left(4x^2+12x\right)}{(x+3)(x-3)}$$

$$=\frac{6x^2-18x-4x^2-12x}{(x+3)(x-3)}$$

$$=\frac{2x^2-30x}{(x+3)(x-3)} \text{ or } \frac{2x(x-15)}{(x+3)(x-3)}$$

43.

$$\frac{a-1}{a+4}+\frac{a+3}{a-5}=\frac{a-1}{a+4}\left(\frac{a-5}{a-5}\right)+\frac{a+3}{a-5}\left(\frac{a+4}{a+4}\right)$$

$$=\frac{a^2-5a-a+5}{(a+4)(a-5)}+\frac{a^2+4a+3a+12}{(a+4)(a-5)}$$

$$=\frac{a^2-5a-a+5+a^2+4a+3a+12}{(a+4)(a-5)}$$

$$=\frac{2a^2+a+17}{(a+4)(a-5)}$$

45.

$$\frac{n+2}{n-4}-\frac{n+5}{n+4}=\frac{n+2}{n-4}\left(\frac{n+4}{n+4}\right)-\frac{n+5}{n+4}\left(\frac{n-4}{n-4}\right)$$
$$=\frac{n^2+4n+2n+8}{(n+4)(n-4)}-\frac{n^2-4n+5n-20}{(n+4)(n-4)}$$
$$=\frac{n^2+4n+2n+8-\left(n^2-4n+5n-20\right)}{(n+4)(n-4)}$$
$$=\frac{n^2+4n+2n+8-n^2+4n-5n+20}{(n+4)(n-4)}$$
$$=\frac{5n+28}{(n+4)(n-4)}$$

47.

$$4+\frac{1}{x-2}=4\left(\frac{x-2}{x-2}\right)+\frac{1}{x-2}$$
$$=\frac{4x-8}{x-2}+\frac{1}{x-2}$$
$$=\frac{4x-8+1}{x-2}$$
$$=\frac{4x-7}{x-2}$$

49.

$$x+\frac{4x}{7x-3}=x\left(\frac{7x-3}{7x-3}\right)+\frac{4x}{7x-3}$$
$$=\frac{7x^2-3x}{7x-3}+\frac{4x}{7x-3}$$
$$=\frac{7x^2-3x+4x}{7x-3}$$
$$=\frac{7x^2+x}{7x-3}\quad\text{or}\quad\frac{x(7x+1)}{7x-3}$$

51.

$$\frac{1}{x+3}+\frac{2}{x^2+4x+3}=\frac{1}{x+3}+\frac{2}{(x+3)(x+1)}$$
$$=\frac{1}{x+3}\left(\frac{x+1}{x+1}\right)+\frac{2}{(x+3)(x+1)}$$
$$=\frac{x+1}{(x+3)(x+1)}+\frac{2}{(x+3)(x+1)}$$
$$=\frac{x+1+2}{(x+3)(x+1)}$$
$$=\frac{\cancel{x+3}}{\cancel{(x+3)}(x+1)}$$
$$=\frac{1}{x+1}$$

53.

$$\frac{m}{m^2+9m+20}-\frac{4}{m^2+7m+12}$$
$$=\frac{m}{(m+4)(m+5)}-\frac{4}{(m+4)(m+3)}$$
$$=\frac{m}{(m+4)(m+5)}\left(\frac{m+3}{m+3}\right)-\frac{4}{(m+4)(m+3)}\left(\frac{m+5}{m+5}\right)$$
$$=\frac{m^2+3m}{(m+3)(m+4)(m+5)}-\frac{4m+20}{(m+3)(m+4)(m+5)}$$
$$=\frac{m^2+3m-(4m+20)}{(m+3)(m+4)(m+5)}$$
$$=\frac{m^2+3m-4m-20}{(m+3)(m+4)(m+5)}$$
$$=\frac{m^2-m-20}{(m+3)(m+4)(m+5)}$$
$$=\frac{(m-5)\cancel{(m+4)}}{(m+3)\cancel{(m+4)}(m+5)}$$
$$=\frac{m-5}{(m+3)(m+5)}$$

55.

$$\frac{x}{x^2+5x+6}+\frac{x}{x^2-4}$$
$$=\frac{x}{(x+2)(x+3)}+\frac{x}{(x+2)(x-2)}$$
$$=\frac{x}{(x+2)(x+3)}\left(\frac{x-2}{x-2}\right)+\frac{x}{(x+2)(x-2)}\left(\frac{x+3}{x+3}\right)$$
$$=\frac{x^2-2x}{(x+2)(x-2)(x+3)}+\frac{x^2+3x}{(x+2)(x-2)(x+3)}$$
$$=\frac{x^2-2x+x^2+3x}{(x+2)(x-2)(x+3)}$$
$$=\frac{2x^2+x}{(x+2)(x-2)(x+3)}$$
$$=\frac{x(2x+1)}{(x+2)(x-2)(x+3)}$$

57.

$$\frac{x+2}{6x-42}-\frac{x-3}{5x-35}=\frac{x+2}{6(x-7)}-\frac{x-3}{5(x-7)}$$
$$=\frac{x+2}{6(x-7)}\left(\frac{5}{5}\right)-\frac{x-3}{5(x-7)}\left(\frac{6}{6}\right)$$
$$=\frac{5x+10}{30(x-7)}-\frac{6x-18}{30(x-7)}$$
$$=\frac{5x+10-(6x-18)}{30(x-7)}$$
$$=\frac{5x+10-6x+18}{30(x-7)}$$
$$=\frac{-x+28}{30(x-7)}$$

59.

$$\frac{5x}{x-3}+\frac{4x}{3-x}=\frac{5x}{x-3}-\frac{4x}{x-3}$$
$$=\frac{5x-4x}{x-3}$$
$$=\frac{x}{x-3}$$

61.

$$\frac{9m}{m-n}-\frac{2}{n-m}=\frac{9m}{m-n}+\frac{2}{m-n}$$
$$=\frac{9m+2}{m-n}$$

63.

$$\frac{5x}{x+1}+\frac{3}{x+1}-\frac{2x}{x+1}=\frac{5x+3-2x}{x+1}$$
$$=\frac{3x+3}{x+1}$$
$$=\frac{3\cancel{(x+1)}}{\cancel{x+1}}$$
$$=3$$

65.

$$\frac{1}{x+y}-\frac{1}{x-y}+\frac{2y}{x^2-y^2}$$
$$=\frac{1}{x+y}-\frac{1}{x-y}+\frac{2y}{(x+y)(x-y)}$$
$$=\frac{1}{x+y}\left(\frac{x-y}{x-y}\right)-\frac{1}{x-y}\left(\frac{x+y}{x+y}\right)+\frac{2y}{(x-y)(y+x)}$$
$$=\frac{x-y}{(x+y)(x-y)}-\frac{x+y}{(x+y)(x-y)}+\frac{2y}{(x-y)(y+x)}$$
$$=\frac{x-y-(x+y)+2y}{(x+y)(x-y)}$$
$$=\frac{x-y-x-y+2y}{(x+y)(x-y)}$$
$$=\frac{0}{(x+y)(x-y)}$$
$$=0$$

67.

$$\frac{3x}{2x-1}+\frac{x+1}{3x+2}-\frac{2}{6x^2+x-2}$$
$$=\frac{3x}{2x-1}+\frac{x+1}{3x+2}-\frac{2}{(2x-1)(3x+2)}$$
$$=\frac{3x}{2x-1}\left(\frac{3x+2}{3x+2}\right)+\frac{x+1}{3x+2}\left(\frac{2x-1}{2x-1}\right)-\frac{2}{(2x-1)(3x+2)}$$
$$=\frac{3x(3x+2)}{(2x-1)(3x+2)}+\frac{(x+1)(2x-1)}{(2x-1)(3x+2)}-\frac{2}{(2x-1)(3x+2)}$$
$$=\frac{9x^2+6x}{(2x-1)(3x+2)}+\frac{(2x^2-x+2x-1)}{(2x-1)(3x+2)}-\frac{2}{(2x-1)(3x+2)}$$
$$=\frac{9x^2+6x}{x(2x-1)(3x+2)}+\frac{2x^2+x-1}{x(2x-1)(3x+2)}-\frac{2}{(2x-1)(3x+2)}$$
$$=\frac{9x^2+6x+2x^2+x-1-2}{(2x-1)(3x+2)}$$
$$=\frac{11x^2+7x-3}{(2x-1)(3x+2)}$$

69.

$$\frac{8}{x^2-9}+\frac{2}{x-3}-\frac{6}{x}$$
$$=\frac{8}{(x+3)(x-3)}+\frac{2}{x-3}-\frac{6}{x}$$
$$=\frac{8}{(x+3)(x-3)}\left(\frac{x}{x}\right)+\frac{2}{x-3}\left(\frac{x(x+3)}{x(x+3)}\right)-\frac{6}{x}\left(\frac{(x+3)(x-3)}{(x+3)(x-3)}\right)$$
$$=\frac{8x}{x(x+3)(x-3)}+\frac{2x(x+3)}{x(x+3)(x-3)}-\frac{6(x^2-9)}{x(x+3)(x-3)}$$
$$=\frac{8x}{x(x+3)(x-3)}+\frac{2x^2+6x}{x(x+3)(x-3)}-\frac{6x^2-54}{x(x+3)(x-3)}$$
$$=\frac{8x+2x^2+6x-(6x^2-54)}{x(x+3)(x-3)}$$
$$=\frac{8x+2x^2+6x-6x^2+54}{x(x+3)(x-3)}$$
$$=\frac{-4x^2+14x+54}{x(x+3)(x-3)}$$
$$=\frac{-2(2x^2-7x-27)}{x(x+3)(x-3)}$$

71.

$$\frac{7}{2b}-\frac{11}{3a}=\frac{7}{2b}\left(\frac{3a}{3a}\right)-\frac{11}{3a}\left(\frac{2b}{2b}\right)$$
$$=\frac{21a}{6ab}-\frac{22b}{6ab}$$
$$=\frac{21a-22b}{6ab}$$

73.

$$\frac{s+7}{s+3}-\frac{s-3}{s+7}=\frac{s+7}{s+3}\left(\frac{s+7}{s+7}\right)-\frac{s-3}{s+7}\left(\frac{s+3}{s+3}\right)$$
$$=\frac{(s+7)(s+7)}{(s+3)(s+7)}-\frac{(s-3)(s+3)}{(s+3)(s+7)}$$
$$=\frac{s^2+7s+7s+49}{(s+3)(s+7)}-\frac{s^2-9}{(s+3)(s+7)}$$
$$=\frac{s^2+7s+7s+49-(s^2-9)}{(s+3)(s+7)}$$
$$=\frac{s^2+7s+7s+49-s^2+9}{(s+3)(s+7)}$$
$$=\frac{14s+58}{(s+3)(s+7)}$$

75.

$$\frac{x-y}{2}+\frac{x+y}{3}=\frac{x-y}{2}\left(\frac{3}{3}\right)+\frac{x+y}{3}\left(\frac{2}{2}\right)$$
$$=\frac{3x-3y}{6}+\frac{2x+2y}{6}$$
$$=\frac{3x-3y+2x+2y}{6}$$
$$=\frac{5x-y}{6}$$

77.

$$\frac{3x^2+3x}{x^2-5x+6}-\frac{3x^2-3x+12}{x^2-5x+6}$$
$$=\frac{3x^2+3x-(3x^2-3x+12)}{x^2-5x+6}$$
$$=\frac{3x^2+3x-3x^2+3x-12}{x^2-5x+6}$$
$$=\frac{6x-12}{x^2-5x+6}$$
$$=\frac{6\cancel{(x-2)}}{\cancel{(x-2)}(x-3)}$$
$$=\frac{6}{x-3}$$

79.

$$\frac{a^2+ab}{a^3-b^3}-\frac{b^2}{b^3-a^3}=\frac{a^2+ab}{a^3-b^3}+\frac{b^2}{a^3-b^3}$$
$$=\frac{a^2+ab+b^2}{a^3-b^3}$$
$$=\frac{\cancel{a^2+ab+b^2}}{(a-b)\cancel{(a^2+ab+b^2)}}$$
$$=\frac{1}{a-b}$$

81.

$$2x+3+\frac{1}{x+1}=2x\left(\frac{x+1}{x+1}\right)+3\left(\frac{x+1}{x+1}\right)+\frac{1}{x+1}$$
$$=\frac{2x^2+2x}{x+1}+\frac{3x+3}{x+1}+\frac{1}{x+1}$$
$$=\frac{2x^2+2x+3x+3+1}{x+1}$$
$$=\frac{2x^2+5x+4}{x+1}$$

83.

$$\frac{d}{d^2+11d+30}-\frac{5}{d^2+9d+20}$$
$$=\frac{d}{(d+5)(d+6)}-\frac{5}{(d+4)(d+5)}$$
$$=\frac{d}{(d+5)(d+6)}\left(\frac{d+4}{d+4}\right)-\frac{5}{(d+4)(d+5)}\left(\frac{d+6}{d+6}\right)$$
$$=\frac{d^2+4d}{(d+5)(d+6)(d+4)}-\frac{5d+30}{(d+5)(d+6)(d+4)}$$
$$=\frac{d^2+4d-(5d+30)}{(d+5)(d+6)(d+4)}$$
$$=\frac{d^2+4d-5d-30}{(d+5)(d+6)(d+4)}$$
$$=\frac{d^2-d-30}{(d+5)(d+6)(d+4)}$$
$$=\frac{(d-6)\cancel{(d+5)}}{\cancel{(d+5)}(d+6)(d+4)}$$
$$=\frac{d-6}{(d+6)(d+4)}$$

85.

$$\frac{3}{x+1}-\frac{2}{x-1}+\frac{x+3}{x^2-1}$$
$$=\frac{3}{x+1}-\frac{2}{x-1}+\frac{x+3}{(x-1)(x+1)}$$
$$=\frac{3}{x+1}\left(\frac{x-1}{x-1}\right)-\frac{2}{x-1}\left(\frac{x+1}{x+1}\right)+\frac{x+3}{(x-1)(x+1)}$$
$$=\frac{3(x-1)}{(x+1)(x-1)}-\frac{2(x+1)}{(x+1)(x-1)}+\frac{x+3}{(x+1)(x-1)}$$
$$=\frac{3x-3}{(x+1)(x-1)}-\frac{2x+2}{(x+1)(x-1)}+\frac{x+3}{(x+1)(x-1)}$$
$$=\frac{3x-3-(2x+2)+x+3}{(x+1)(x-1)}$$
$$=\frac{3x-3-2x-2+x+3}{(x+1)(x-1)}$$
$$=\frac{2x-2}{(x+1)(x-1)}$$
$$=\frac{2\cancel{(x-1)}}{(x+1)\cancel{(x-1)}}$$
$$=\frac{2}{x+1}$$

87.

$$\frac{8}{9y^2}+\frac{1}{6y^4}=\frac{8}{9y^2}\left(\frac{2y^2}{2y^2}\right)+\frac{1}{6y^4}\left(\frac{3}{3}\right)$$
$$=\frac{16y^2}{18y^4}+\frac{3}{18y^4}$$
$$=\frac{16y^2+3}{18y^4}$$

89.

$$\frac{6}{b^2-9}\bullet\frac{b+3}{2b+4}=\frac{6(b+3)}{(b^2-9)(2b+4)}$$
$$=\frac{\cancel{2}(3)\cancel{(b+3)}}{\cancel{(b+3)}(b-3)\cancel{(2)}(b+2)}$$
$$=\frac{3}{(b-3)(b+2)}$$

91.

$$\frac{4a}{a-5}+a=\frac{4a}{a-5}+a\left(\frac{a-5}{a-5}\right)$$
$$=\frac{4a}{a-5}+\frac{a^2-5a}{a-5}$$
$$=\frac{4a+a^2-5a}{a-5}$$
$$=\frac{a^2-a}{a-5}\text{ or }\frac{a(a-1)}{a-5}$$

93.

$$\frac{2a+1}{3a-2}-\frac{a-4}{2-3a}=\frac{2a+1}{3a-2}+\frac{a-4}{3a-2}$$
$$=\frac{2a+1+a-4}{3a-2}$$
$$=\frac{3a-3}{3a-2}$$
$$=\frac{3(a-1)}{3a-2}$$

95.

$$\frac{x^2+x}{3x-15} \div \frac{(x+1)^2}{6x-30} = \frac{x^2+x}{3x-15} \bullet \frac{6x-30}{(x+1)^2}$$

$$= \frac{(x^2+x)(6x-30)}{(3x-15)(x+1)^2}$$

$$= \frac{x\cancel{(x+1)}(2)\cancel{(3)}\cancel{(x-5)}}{\cancel{3}\cancel{(x-5)}\cancel{(x+1)}(x+1)}$$

$$= \frac{2x}{x+1}$$

97.

$$\frac{z^2-9}{z^2+4z+3} \div \frac{z^2-3z}{(z+1)^2} = \frac{z^2-9}{z^2+4z+3} \bullet \frac{(z+1)^2}{z^2-3z}$$

$$= \frac{(z^2-9)(z+1)^2}{(z^2+4z+3)(z^2-3z)}$$

$$= \frac{\cancel{(z+3)}\cancel{(z-3)}\cancel{(z+1)}(z+1)}{\cancel{(z+1)}\cancel{(z+3)}(z)\cancel{(z-3)}}$$

$$= \frac{z+1}{z}$$

99.

$$\frac{27p^4}{35q} \div \frac{9p}{21q} = \frac{27p^4}{35q} \bullet \frac{21q}{9p}$$

$$= \frac{27p^4(21q)}{35q(9p)}$$

$$= \frac{\cancel{3}\bullet\cancel{3}\bullet 3\bullet\cancel{p}\bullet p\bullet p\bullet p\bullet 3\bullet\cancel{7}\bullet\cancel{q}}{\cancel{7}\bullet 5\bullet\cancel{q}\bullet\cancel{3}\bullet\cancel{3}\bullet\cancel{p}}$$

$$= \frac{9p^3}{5}$$

101.

$$\frac{6}{5d^2-5d} - \frac{3}{5d-5} = \frac{6}{5d(d-1)} - \frac{3}{5(d-1)}$$

$$= \frac{6}{5d(d-1)} - \frac{3}{5(d-1)}\left(\frac{d}{d}\right)$$

$$= \frac{6}{5d(d-1)} - \frac{3d}{5d(d-1)}$$

$$= \frac{6-3d}{5d(d-1)} \text{ or } \frac{3(2-d)}{5d(d-1)}$$

103.

$$\frac{s^3t}{4s^2-9t^2} \bullet \frac{4s^2-12st+9t^2}{s^3t^2}$$

$$= \frac{s^3t(4s^2-12st+9t^2)}{(4s^2-9t^2)(s^3t^2)}$$

$$= \frac{\cancel{s}\bullet\cancel{s}\bullet\cancel{s}\bullet\cancel{t}\,\cancel{(2s-3t)}(2s-3t)}{\cancel{(2s-3t)}(2s+3t)(\cancel{s}\bullet\cancel{s}\bullet\cancel{s}\bullet\cancel{t}\bullet t)}$$

$$= \frac{2s-3t}{t(2s+3t)}$$

105.

$$\frac{4}{x^2-2x-3} - \frac{x}{3x^2-7x-6}$$

$$= \frac{4}{(x-3)(x+1)} - \frac{x}{(3x+2)(x-3)}$$

$$= \frac{4}{(x-3)(x+1)}\left(\frac{3x+2}{3x+2}\right) - \frac{x}{(3x+2)(x-3)}\left(\frac{x+1}{x+1}\right)$$

$$= \frac{12x+8}{(x-3)(x+1)(3x+2)} - \frac{x^2+x}{(3x+2)(x-3)(x+1)}$$

$$= \frac{12x+8-(x^2+x)}{(x-3)(x+1)(3x+2)}$$

$$= \frac{12x+8-x^2-x}{(x-3)(x+1)(3x+2)}$$

$$= \frac{-x^2+11x+8}{(x-3)(x+1)(3x+2)}$$

107.

$$f(x)+g(x) = \frac{3x}{x^2-25} + \frac{4}{x+5}$$

$$= \frac{3x}{(x+5)(x-5)} + \frac{4}{x+5}$$

$$= \frac{3x}{(x+5)(x-5)} + \frac{4}{x+5}\left(\frac{x-5}{x-5}\right)$$

$$= \frac{3x}{(x+5)(x-5)} + \frac{4x-20}{(x+5)(x-5)}$$

$$= \frac{3x+4x-20}{(x+5)(x-5)}$$

$$= \frac{7x-20}{(x+5)(x-5)}$$

109. a.

$$\begin{aligned}\frac{x-1}{x+2}+\frac{x+2}{x-1}&=\frac{x-1}{x+2}\left(\frac{x-1}{x-1}\right)+\frac{x+2}{x-1}\left(\frac{x+2}{x+2}\right)\\&=\frac{x^2-x-x+1}{(x+2)(x-1)}+\frac{x^2+2x+2x+4}{(x+2)(x-1)}\\&=\frac{x^2-2x+1}{(x+2)(x-1)}+\frac{x^2+4x+4}{(x+2)(x-1)}\\&=\frac{x^2-2x+1+x^2+4x+4}{(x+2)(x-1)}\\&=\frac{2x^2+2x+5}{(x+2)(x-1)}\end{aligned}$$

b.

$$\begin{aligned}\frac{x-1}{x+2}\bullet\frac{x+2}{x-1}&=\frac{\cancel{(x-1)}\,\cancel{(x+2)}}{\cancel{(x+2)}\,\cancel{(x-1)}}\\&=1\end{aligned}$$

APPLICATIONS

111. DRAFTING

45°–45°–90° triangle:

$$\begin{aligned}\frac{10}{r}+\frac{10}{r}+10&=\frac{10}{r}+\frac{10}{r}+10\left(\frac{r}{r}\right)\\&=\frac{10}{r}+\frac{10}{r}+\frac{10r}{r}\\&=\frac{10+10+10r}{r}\\&=\frac{10r+20}{r}\end{aligned}$$

30°–60°–90° triangle:

$$\begin{aligned}\frac{1}{2}\bullet\frac{6}{t}+3+\frac{6}{t}&=\frac{3}{t}+3+\frac{6}{t}\\&=\frac{3}{t}+3\left(\frac{t}{t}\right)+\frac{6}{t}\\&=\frac{3}{t}+\frac{3t}{t}+\frac{6}{t}\\&=\frac{3+3t+6}{t}\\&=\frac{3t+9}{t}\end{aligned}$$

WRITING

113. Answers will vary.

115. Answers will vary.

REVIEW

117.

$$\begin{aligned}a(a-6)&=-9\\a^2-6a&=-9\\a^2-6a+9&=0\\(a-3)(a-3)&=0\end{aligned}$$

$a-3=0$ or $a-3=0$

$a=3$ $\qquad a=3$

119.

$$\begin{aligned}y^3+y^2&=0\\y^2(y+1)&=0\end{aligned}$$

$y+1=0$ or $y=0$ or $y=0$

$y=-1$

CHALLENGE PROBLEMS

121. Answers will vary. Possible answers are $\frac{x}{x^2+5x+6}$ and $\frac{3}{x^2+5x+6}$.

123.

$$\left(\frac{3}{x-3}-\frac{1}{x}\right)\div\frac{12x+18}{x^3-9x}$$

$$=\left(\frac{3}{x-3}\bullet\frac{x}{x}-\frac{1}{x}\bullet\frac{x-3}{x-3}\right)\div\frac{12x+18}{x^3-9x}$$

$$=\left(\frac{3x}{x(x-3)}-\frac{x-3}{x(x-3)}\right)\div\frac{12x+18}{x^3-9x}$$

$$=\left(\frac{3x-x+3}{x(x-3)}\right)\div\frac{12x+18}{x^3-9x}$$

$$=\frac{2x+3}{x(x-3)}\div\frac{12x+18}{x^3-9x}$$

$$=\frac{2x+3}{x(x-3)}\bullet\frac{x^3-9x}{12x+18}$$

$$=\frac{(2x+3)(x^3-9x)}{x(x-3)(12x+18)}$$

$$=\frac{\cancel{(2x+3)}\cancel{(x)}\cancel{(x-3)}(x+3)}{\cancel{x}\cancel{(x-3)}(6)\cancel{(2x+3)}}$$

$$=\frac{x+3}{6}$$

125.

$$\left(\frac{3x}{x+1}-\frac{6}{x^2-1}+\frac{4}{x-1}\right)\left(\frac{x^3-1}{9x^2-4}\right)$$

$$=\left(\frac{3x}{x+1}\bullet\frac{x-1}{x-1}-\frac{6}{(x-1)(x+1)}+\frac{4}{x-1}\bullet\frac{x+1}{x+1}\right)\left(\frac{x^3-1}{9x^2-4}\right)$$

$$=\left(\frac{3x^2-3x}{(x-1)(x+1)}-\frac{6}{(x-1)(x+1)}+\frac{4x+4}{(x-1)(x+1)}\right)\left(\frac{x^3-1}{9x^2-4}\right)$$

$$=\left(\frac{3x^2-3x-6+4x+4}{(x-1)(x+1)}\right)\left(\frac{x^3-1}{9x^2-4}\right)$$

$$=\left(\frac{3x^2+x-2}{(x-1)(x+1)}\right)\left(\frac{x^3-1}{9x^2-4}\right)$$

$$=\frac{(3x^2+x-2)(x^3-1)}{(x-1)(x+1)(9x^2-4)}$$

$$=\frac{\cancel{(3x-2)}\cancel{(x+1)}\cancel{(x-1)}(x^2+x+1)}{\cancel{(x-1)}\cancel{(x+1)}\cancel{(3x-2)}(3x+2)}$$

$$=\frac{x^2+x+1}{3x+2}$$

SECTION 6.4

VOCABULARY

1. $\dfrac{\frac{x}{y}+\frac{1}{x}}{\frac{1}{y}+\frac{2}{x}}$ and $\dfrac{\frac{5a^2}{b}}{\frac{b}{2a^3}}$ are examples of complex **rational** expressions, or more simply, **complex** fractions.

CONCEPTS

3. $\dfrac{t^2}{t^2}$

5. Method 1: To simplify a complex fraction, write its numerator and denominator as **single** rational expressions. Then perform the indicated **division** by multiplying the numerator of the complex fraction by the **reciprocal** of the denominator.

NOTATION

7.

$$\frac{\frac{5m^2}{6}}{\frac{25m}{3}}=\frac{5m^2}{6}\boxed{\div}\frac{25m}{3}$$

$$=\frac{5m^2}{6}\bullet\boxed{\frac{3}{25m}}$$

$$=\frac{5\bullet\boxed{m}\bullet m\bullet\boxed{3}}{2\bullet\boxed{3}\bullet 5\bullet 5\bullet\boxed{m}}$$

$$=\frac{m}{\boxed{10}}$$

9. a. The fraction $\dfrac{\frac{a}{b}}{\frac{c}{d}}$ is equivalent to $\dfrac{a}{b}\boxed{\div}\dfrac{c}{d}$.

b. numerator: $6-k-\dfrac{5}{k}$

denominator: k^2-9

GUIDED PRACTICE

11.

$$\frac{\frac{a^6}{2}}{\frac{3a}{4}}=\frac{a^6}{2}\div\frac{3a}{4}$$

$$=\frac{a^6}{2}\bullet\frac{4}{3a}$$

$$=\frac{\cancel{a}\bullet a^5\bullet\cancel{2}\bullet 2}{\cancel{2}\bullet 3\cancel{a}}$$

$$=\frac{2a^5}{3}$$

13.

$$\frac{\frac{20x}{y}}{\frac{36x}{y^2}}=\frac{20x}{y}\div\frac{36x}{y^2}$$

$$=\frac{20x}{y}\bullet\frac{y^2}{36x}$$

$$=\frac{\cancel{2}\bullet\cancel{2}\bullet 5\bullet\cancel{x}\bullet\cancel{y}\bullet y}{\cancel{y}\bullet 3\bullet 3\bullet\cancel{2}\bullet\cancel{2}\bullet\cancel{x}}$$

$$=\frac{5y}{9}$$

15.

$$\frac{\frac{18x^5}{35y^2}}{\frac{2x^8}{21y^5}}=\frac{18x^5}{35y^2}\div\frac{2x^8}{21y^5}$$

$$=\frac{18x^5}{35y^2}\bullet\frac{21y^5}{2x^8}$$

$$=\frac{\cancel{2}\bullet 3\bullet 3\bullet\cancel{x}\bullet\cancel{x}\bullet\cancel{x}\bullet\cancel{x}\bullet\cancel{x}\bullet 3\bullet\cancel{7}\bullet\cancel{y}\bullet\cancel{y}\bullet y\bullet y\bullet y}{5\bullet\cancel{7}\bullet\cancel{y}\bullet\cancel{y}\bullet\cancel{2}\bullet x\bullet x\bullet x\bullet\cancel{x}\bullet\cancel{x}\bullet\cancel{x}\bullet\cancel{x}\bullet\cancel{x}}$$

$$=\frac{27y^3}{5x^3}$$

17.

$$\frac{\frac{16c}{77d^4}}{\frac{28c^7}{55d^0}} = \frac{16c}{77d^4} \div \frac{28c^7}{55d^0}$$

$$= \frac{16c}{77d^4} \bullet \frac{55d^0}{28c^7}$$

$$= \frac{2 \cdot 2 \cdot \cancel{2} \cdot \cancel{2} \cdot \cancel{c} \cdot 5 \cdot \cancel{11} \cdot 1}{7 \cdot \cancel{11} \cdot d \cdot d \cdot d \cdot d \cdot \cancel{2} \cdot \cancel{2} \cdot 7 \cdot c \cdot c \cdot c \cdot c \cdot c \cdot c \cdot \cancel{c}}$$

$$= \frac{20}{49c^6 d^4}$$

19.

$$\frac{\frac{3}{a} - 2}{a - 3} = \frac{\frac{3}{a} - 2}{a - 3} \bullet \frac{a}{a}$$

$$= \frac{3 - 2a}{a^2 - 3a}$$

21.

$$\frac{4p - \frac{4}{p}}{12 - \frac{4}{p}} = \frac{4p - \frac{4}{p}}{12 - \frac{4}{p}} \bullet \frac{p}{p}$$

$$= \frac{4p^2 - 4}{12p - 4}$$

$$= \frac{\cancel{4}(p^2 - 1)}{\cancel{4}(3p - 1)}$$

$$= \frac{p^2 - 1}{3p - 1}$$

23.

$$\frac{\frac{y}{x} - \frac{x}{y}}{\frac{1}{x} + \frac{1}{y}} = \frac{\frac{y}{x} - \frac{x}{y}}{\frac{1}{x} + \frac{1}{y}} \bullet \frac{xy}{xy}$$

$$= \frac{\frac{xy^2}{x} - \frac{x^2 y}{y}}{\frac{xy}{x} + \frac{xy}{y}}$$

$$= \frac{y^2 - x^2}{y + x}$$

$$= \frac{(y - x)\cancel{(y + x)}}{\cancel{y + x}}$$

$$= y - x$$

25.

$$\frac{\frac{1}{a} - \frac{1}{b}}{\frac{a}{b} - \frac{b}{a}} = \frac{\frac{1}{a} - \frac{1}{b}}{\frac{a}{b} - \frac{b}{a}} \bullet \frac{ab}{ab}$$

$$= \frac{\frac{ab}{a} - \frac{ab}{b}}{\frac{a^2 b}{b} - \frac{ab^2}{a}}$$

$$= \frac{b - a}{a^2 - b^2}$$

$$= \frac{b - a}{(a - b)(a + b)}$$

$$= -\frac{\cancel{a - b}}{\cancel{(a - b)}(a + b)}$$

$$= -\frac{1}{a + b}$$

27.

$$\frac{\frac{2}{a^2} + \frac{1}{a}}{\frac{2}{a} + \frac{1}{a^2}} = \frac{\frac{2}{a^2} + \frac{1}{a}}{\frac{2}{a} + \frac{1}{a^2}} \bullet \frac{a^2}{a^2}$$

$$= \frac{\frac{2a^2}{a^2} + \frac{a^2}{a}}{\frac{2a^2}{a} + \frac{a^2}{a^2}}$$

$$= \frac{2 + a}{2a + 1}$$

29.

$$\frac{\frac{3}{b^2} - \frac{4}{b} + 1}{1 - \frac{1}{b} - \frac{6}{b^2}} = \frac{\frac{3}{b^2} - \frac{4}{b} + 1}{1 - \frac{1}{b} - \frac{6}{b^2}} \bullet \frac{b^2}{b^2}$$

$$= \frac{\frac{3b^2}{b^2} - \frac{4b^2}{b} + b^2}{b^2 - \frac{b^2}{b} - \frac{6b^2}{b^2}}$$

$$= \frac{3 - 4b + b^2}{b^2 - b - 6}$$

$$= \frac{b^2 - 4b + 3}{b^2 - b - 6}$$

$$= \frac{\cancel{(b - 3)}(b - 1)}{\cancel{(b - 3)}(b + 2)}$$

$$= \frac{b - 1}{b + 2}$$

31.

$$\frac{x^{-2}-y^{-2}}{x^{-1}-y^{-1}}=\frac{\frac{1}{x^2}-\frac{1}{y^2}}{\frac{1}{x}-\frac{1}{y}}$$

$$=\frac{\frac{1}{x^2}-\frac{1}{y^2}}{\frac{1}{x}-\frac{1}{y}}\bullet\frac{x^2y^2}{x^2y^2}$$

$$=\frac{\frac{x^2y^2}{x^2}-\frac{x^2y^2}{y^2}}{\frac{x^2y^2}{x}-\frac{x^2y^2}{y}}$$

$$=\frac{y^2-x^2}{xy^2-x^2y}$$

$$=\frac{\cancel{(y-x)}(y+x)}{xy\cancel{(y-x)}}$$

$$=\frac{y+x}{xy}$$

33.

$$\frac{a-b^{-2}}{b-a^{-2}}=\frac{a-\frac{1}{b^2}}{b-\frac{1}{a^2}}$$

$$=\frac{a-\frac{1}{b^2}}{b-\frac{1}{a^2}}\bullet\frac{a^2b^2}{a^2b^2}$$

$$=\frac{a^3b^2-\frac{a^2b^2}{b^2}}{a^2b^3-\frac{a^2b^2}{a^2}}$$

$$=\frac{a^3b^2-a^2}{a^2b^3-b^2}$$

$$=\frac{a^2(ab^2-1)}{b^2(a^2b-1)}$$

35.

$$\frac{\frac{3}{z-3}+\frac{2}{z-2}}{\frac{5z}{z^2-5z+6}}=\frac{\frac{3}{z-3}+\frac{2}{z-2}}{\frac{5z}{(z-3)(z-2)}}$$

$$=\frac{\frac{3}{z-3}+\frac{2}{z-2}}{\frac{5z}{(z-3)(z-2)}}\bullet\frac{(z-3)(z-2)}{(z-3)(z-2)}$$

$$=\frac{\frac{3(z-3)(z-2)}{z-3}+\frac{2(z-3)(z-2)}{z-2}}{\frac{5z(z-3)(z-2)}{(z-3)(z-2)}}$$

$$=\frac{3(z-2)+2(z-3)}{5z}$$

$$=\frac{3z-6+2z-6}{5z}$$

$$=\frac{5z-12}{5z}$$

37.

$$\frac{\frac{2}{x+3}-\frac{1}{x-3}}{\frac{3}{x^2-9}}$$

$$=\frac{\frac{2}{x+3}-\frac{1}{x-3}}{\frac{3}{(x+3)(x-3)}}$$

$$=\frac{\frac{2}{x+3}-\frac{1}{x-3}}{\frac{3}{(x+3)(x-3)}}\bullet\frac{(x+3)(x-3)}{(x+3)(x-3)}$$

$$=\frac{\frac{2(x+3)(x-3)}{x+3}-\frac{1(x+3)(x-3)}{x-3}}{\frac{3(x+3)(x-3)}{(x+3)(x-3)}}$$

$$=\frac{2(x-3)-(x+3)}{3}$$

$$=\frac{2x-6-x-3}{3}$$

$$=\frac{x-9}{3}$$

39.

$$\frac{1+\frac{x}{y}}{1-\frac{x}{y}}=\frac{1+\frac{x}{y}}{1-\frac{x}{y}}\bullet\frac{y}{y}$$

$$=\frac{y+\frac{xy}{y}}{y-\frac{xy}{y}}$$

$$=\frac{y+x}{y-x}$$

41.

$$\frac{\frac{x^2+5x+6}{3xy}}{\frac{9-x^2}{6xy}}=\frac{x^2+5x+6}{3xy}\div\frac{9-x^2}{6xy}$$

$$=\frac{x^2+5x+6}{3xy}\bullet\frac{6xy}{9-x^2}$$

$$=\frac{(x+2)(x+3)}{3xy}\bullet\frac{6xy}{(3+x)(3-x)}$$

$$=\frac{2\bullet\cancel{3xy}\,(x+2)\,\cancel{(x+3)}}{\cancel{3xy}\,\cancel{(x+3)}\,(3-x)}$$

$$=\frac{2(x+2)}{3-x}$$

43.

$$\frac{1+\frac{6}{x}+\frac{8}{x^2}}{1+\frac{1}{x}-\frac{12}{x^2}}=\frac{1+\frac{6}{x}+\frac{8}{x^2}}{1+\frac{1}{x}-\frac{12}{x^2}}\bullet\frac{x^2}{x^2}$$

$$=\frac{x^2+\frac{6x^2}{x}+\frac{8x^2}{x^2}}{x^2+\frac{x^2}{x}-\frac{12x^2}{x^2}}$$

$$=\frac{x^2+6x+8}{x^2+x-12}$$

$$=\frac{(x+2)\,\cancel{(x+4)}}{(x-3)\,\cancel{(x+4)}}$$

$$=\frac{x+2}{x-3}$$

45.

$$\frac{\frac{ac-ad-c+d}{a^3-1}}{\frac{c^2-2cd+d^2}{a^2+a+1}}$$

$$=\frac{ac-ad-c+d}{a^3-1}\div\frac{c^2-2cd+d^2}{a^2+a+1}$$

$$=\frac{ac-ad-c+d}{a^3-1}\bullet\frac{a^2+a+1}{c^2-2cd+d^2}$$

$$=\frac{a(c-d)-1(c-d)}{(a-1)(a^2+a+1)}\bullet\frac{(a^2+a+1)}{(c-d)(c-d)}$$

$$=\frac{\cancel{(a-1)}\,\cancel{(c-d)}\,\cancel{(a^2+a+1)}}{\cancel{(a-1)}\,\cancel{(a^2+a+1)}\,\cancel{(c-d)}\,(c-d)}$$

$$=\frac{1}{c-d}$$

47.

$$\frac{\frac{8}{x}+\frac{x}{8}}{\frac{x}{8}-\frac{8}{x}}=\frac{\frac{8}{x}+\frac{x}{8}}{\frac{x}{8}-\frac{8}{x}}\bullet\frac{8x}{8x}$$

$$=\frac{64+x^2}{x^2-64}$$

$$=\frac{64+x^2}{(x+8)(x-8)}$$

49.

$$-\frac{a}{\frac{1}{a}+\frac{1}{b}+\frac{1}{c}}=-\frac{a}{\frac{1}{a}+\frac{1}{b}+\frac{1}{c}}\bullet\frac{abc}{abc}$$

$$=-\frac{a^2bc}{bc+ac+ab}$$

51.

$$\frac{\frac{1}{a+1}+1}{\frac{3}{a-1}+1}=\frac{\frac{1}{a+1}+1}{\frac{3}{a-1}+1}\bullet\frac{(a+1)(a-1)}{(a+1)(a-1)}$$

$$=\frac{\frac{(a+1)(a-1)}{a+1}+(a+1)(a-1)}{\frac{3(a+1)(a-1)}{a-1}+(a+1)(a-1)}$$

$$=\frac{a-1+a^2-1}{3(a+1)+a^2-1}$$

$$=\frac{a^2+a-2}{3a+3+a^2-1}$$

$$=\frac{a^2+a-2}{a^2+3a+2}$$

$$=\frac{\cancel{(a+2)}(a-1)}{\cancel{(a+2)}(a+1)}$$

$$=\frac{a-1}{a+1}$$

53.

$$\frac{5ab^2}{\frac{ab}{25}}=\frac{5ab^2}{1}\div\frac{ab}{25}$$

$$=\frac{5ab^2}{1}\bullet\frac{25}{ab}$$

$$=\frac{5\bullet\cancel{a}\bullet\cancel{b}\bullet b\bullet 5\bullet 5}{\cancel{a}\bullet\cancel{b}}$$

$$=125b$$

55.

$$\frac{a-4+\frac{1}{a}}{-\frac{1}{a}-a+4}=\frac{a-4+\frac{1}{a}}{-\frac{1}{a}-a+4}\bullet\frac{a}{a}$$

$$=\frac{a^2-4a+\frac{a}{a}}{-\frac{a}{a}-a^2+4a}$$

$$=\frac{a^2-4a+1}{-1-a^2+4a}$$

$$=\frac{a^2-4a+1}{-a^2+4a-1}$$

$$=\frac{\cancel{a^2-4a+1}}{-\cancel{(a^2-4a+1)}}$$

$$=-1$$

57.

$$\frac{\frac{2}{a+1}+\frac{1}{a+3}}{\frac{2a}{a^2+4a+3}}=\frac{\frac{2}{a+1}+\frac{1}{a+3}}{\frac{2a}{(a+1)(a+3)}}$$

$$=\frac{\frac{2}{a+1}+\frac{1}{a+3}}{\frac{2a}{(a+1)(a+3)}}\bullet\frac{(a+1)(a+3)}{(a+1)(a+3)}$$

$$=\frac{2(a+3)+(a+1)}{2a}$$

$$=\frac{2a+6+a+1}{2a}$$

$$=\frac{3a+7}{2a}$$

59.

$$\frac{y}{x^{-1}-y^{-1}}=\frac{y}{\frac{1}{x}-\frac{1}{y}}$$

$$=\frac{y}{\frac{1}{x}-\frac{1}{y}}\bullet\frac{xy}{xy}$$

$$=\frac{xy^2}{\frac{xy}{x}-\frac{xy}{y}}$$

$$=\frac{xy^2}{y-x}$$

61.

$$\frac{\frac{1}{x^2}-\frac{3}{xy}+\frac{2}{y^2}}{\frac{2}{x^2}-\frac{1}{xy}-\frac{1}{y^2}}=\frac{\frac{1}{x^2}-\frac{3}{xy}+\frac{2}{y^2}}{\frac{2}{x^2}-\frac{1}{xy}-\frac{1}{y^2}}\bullet\frac{x^2y^2}{x^2y^2}$$

$$=\frac{\frac{x^2y^2}{x^2}-\frac{3x^2y^2}{xy}+\frac{2x^2y^2}{y^2}}{\frac{2x^2y^2}{x^2}-\frac{x^2y^2}{xy}-\frac{x^2y^2}{y^2}}$$

$$=\frac{y^2-3xy+2x^2}{2y^2-xy-x^2}$$

$$=\frac{\cancel{(y-x)}(y-2x)}{(2y+x)\cancel{(y-x)}}$$

$$=\frac{y-2x}{2y+x}$$

63.

$$\frac{5xy}{1+\frac{1}{xy}}=\frac{5xy}{1+\frac{1}{xy}}\bullet\frac{xy}{xy}$$

$$=\frac{5x^2y^2}{xy+\frac{xy}{xy}}$$

$$=\frac{5x^2y^2}{xy+1}$$

65.

$$\frac{\frac{2}{y-1}-\frac{2}{y}}{\frac{3}{y-1}-\frac{1}{1-y}}=\frac{\frac{2}{y-1}-\frac{2}{y}}{\frac{3}{y-1}+\frac{1}{y-1}}$$

$$=\frac{\frac{2}{y-1}-\frac{2}{y}}{\frac{3}{y-1}+\frac{1}{y-1}}\bullet\frac{y(y-1)}{y(y-1)}$$

$$=\frac{\frac{2y(y-1)}{y-1}-\frac{2y(y-1)}{y}}{\frac{3y(y-1)}{y-1}+\frac{y(y-1)}{y-1}}$$

$$=\frac{2y-2(y-1)}{3y+y}$$

$$=\frac{2y-2y+2}{4y}$$

$$=\frac{2}{4y}$$

$$=\frac{1}{2y}$$

67.

$$\frac{\frac{t}{x^2-y^2}}{\frac{t}{x+y}}=\frac{\frac{t}{(x+y)(x-y)}}{\frac{t}{x+y}}$$

$$=\frac{\frac{t}{(x+y)(x-y)}}{\frac{t}{x+y}}\bullet\frac{(x+y)(x-y)}{(x+y)(x-y)}$$

$$=\frac{\frac{t(x+y)(x-y)}{(x+y)(x-y)}}{\frac{t(x+y)(x-y)}{x+y}}$$

$$=\frac{\not{t}}{\not{t}(x-y)}$$

$$=\frac{1}{x-y}$$

69.

$$\frac{\frac{4}{cd}}{c^{-1}+d^{-1}}=\frac{\frac{4}{cd}}{\frac{1}{c}+\frac{1}{d}}$$

$$=\frac{\frac{4}{cd}}{\frac{1}{c}+\frac{1}{d}}\bullet\frac{cd}{cd}$$

$$=\frac{\frac{4cd}{cd}}{\frac{cd}{c}+\frac{cd}{d}}$$

$$=\frac{4}{d+c}$$

APPLICATIONS

71. ENGINEERING

$$k=\frac{1}{\frac{1}{k_1}+\frac{1}{k_2}}$$

$$=\frac{1}{\frac{1}{k_1}+\frac{1}{k_2}}\bullet\frac{k_1k_2}{k_1k_2}$$

$$=\frac{k_1k_2}{\frac{k_1k_2}{k_1}+\frac{k_1k_2}{k_2}}$$

$$=\frac{k_1k_2}{k_2+k_1}$$

73. KITCHEN UTENSILS

$$\frac{8-\frac{2}{d}}{6-\frac{2}{d}}=\frac{8-\frac{2}{d}}{6-\frac{2}{d}}\bullet\frac{d}{d}$$

$$=\frac{8d-\frac{2d}{d}}{6d-\frac{2d}{d}}$$

$$=\frac{8d-2}{6d-2}$$

$$=\frac{\not{2}(4d-1)}{\not{2}(3d-1)}$$

$$=\frac{4d-1}{3d-1}$$

WRITING

75. Answers will vary.

77. Answers will vary.

REVIEW

79.

$$\frac{8(a-5)}{3}=2(a-4)$$

$$\frac{8a-40}{3}=2a-8$$

$$3\left(\frac{8a-40}{3}\right)=3(2a-8)$$

$$8a-40=6a-24$$

$$2a-40=-24$$

$$2a=16$$

$$a=8$$

81.

$$a^4-13a^2+36=0$$

$$(a^2-9)(a^2-4)=0$$

$$(a+3)(a-3)(a+2)(a-2)=0$$

$$a+3=0,\quad a-3=0,\quad a+2=0,\quad a-2=0$$

$$a=-3\qquad a=3\qquad a=-2\qquad a=2$$

CHALLENGE PROBLEMS

83.

$$a+\frac{a}{1+\frac{a}{a+1}}=a+\left(\frac{a}{1+\frac{a}{a+1}}\bullet\frac{a+1}{a+1}\right)$$

$$=a+\left(\frac{a(a+1)}{1(a+1)+\frac{a}{a+1}(a+1)}\right)$$

$$=a+\left(\frac{a^2+a}{a+1+a}\right)$$

$$=a+\frac{a^2+a}{2a+1}$$

$$=\frac{a}{1}\left(\frac{2a+1}{2a+1}\right)+\frac{a^2+a}{2a+1}$$

$$=\frac{2a^2+a}{2a+1}+\frac{a^2+a}{2a+1}$$

$$=\frac{2a^2+a+a^2+a}{2a+1}$$

$$=\frac{3a^2+2a}{2a+1}$$

85.

$$\frac{x-\dfrac{1}{1-\dfrac{x}{2}}}{\dfrac{3}{x+\dfrac{2}{3}}+x}=\frac{x-\dfrac{1}{1\bullet\dfrac{2}{2}-\dfrac{x}{2}}}{\dfrac{3}{x\bullet\dfrac{3}{3}+\dfrac{2}{3}}+x}$$

$$=\frac{x-\dfrac{1}{\dfrac{2}{2}-\dfrac{x}{2}}}{\dfrac{3}{\dfrac{3x}{3}+\dfrac{2}{3}}+x}$$

$$=\frac{x-\dfrac{1}{\dfrac{2-x}{2}}}{\dfrac{3}{\dfrac{3x+2}{3}}+x}$$

$$=\frac{x-\left(1\div\dfrac{2-x}{2}\right)}{\left(3\div\dfrac{3x+2}{3}\right)+x}$$

$$=\frac{x-\left(\dfrac{1}{1}\bullet\dfrac{2}{2-x}\right)}{\left(\dfrac{3}{1}\bullet\dfrac{3}{3x+2}\right)+x}$$

$$=\frac{x-\dfrac{2}{2-x}}{\dfrac{9}{3x+2}+x}$$

$$=\frac{\dfrac{x}{1}\bullet\dfrac{2-x}{2-x}-\dfrac{2}{2-x}}{\dfrac{9}{3x+2}+\dfrac{x}{1}\bullet\dfrac{3x+2}{3x+2}}$$

$$=\frac{\dfrac{2x-x^2-2}{2-x}}{\dfrac{9+3x^2+2x}{3x+2}}$$

$$=\frac{2x-x^2-2}{2-x}\div\frac{9+3x^2+2x}{3x+2}$$

$$=\frac{-x^2+2x-2}{2-x}\bullet\frac{3x+2}{3x^2+2x+9}$$

$$=\frac{\left(-x^2+2x-2\right)(3x+2)}{(2-x)\left(3x^2+2x+9\right)}$$

87.

$$\left(x^{-1}y^{-1}\right)\left(x^{-1}+y^{-1}\right)^{-1}=\left(\frac{1}{xy}\right)\left(\frac{1}{\left(x^{-1}+y^{-1}\right)}\right)$$

$$=\left(\frac{1}{xy}\right)\left(\frac{1}{\dfrac{1}{x}+\dfrac{1}{y}}\right)$$

$$=\left(\frac{1}{xy}\right)\left(\frac{1}{\dfrac{1}{x}+\dfrac{1}{y}}\bullet\frac{xy}{xy}\right)$$

$$=\left(\frac{1}{xy}\right)\left(\frac{xy}{\dfrac{xy}{x}+\dfrac{xy}{y}}\right)$$

$$=\left(\frac{1}{xy}\right)\left(\frac{xy}{y+x}\right)$$

$$=\frac{\cancel{xy}}{\cancel{xy}(y+x)}$$

$$=\frac{1}{y+x}$$

89.

$$\left(\frac{\dfrac{x}{y}-\dfrac{y}{x}}{\dfrac{x+y}{x}}\right)\div\left(\frac{\dfrac{x^2}{y}-y}{\dfrac{y^2}{x}-x}\right)$$

$$=\left(\frac{\dfrac{x}{y}-\dfrac{y}{x}}{\dfrac{x+y}{x}}\bullet\frac{xy}{xy}\right)\div\left(\frac{\dfrac{x^2}{y}-y}{\dfrac{y^2}{x}-x}\bullet\frac{xy}{xy}\right)$$

$$=\left(\frac{x^2-y^2}{xy+y^2}\right)\div\left(\frac{x^3-xy^2}{y^3-x^2y}\right)$$

$$=\left[\frac{\cancel{(x+y)}(x-y)}{y\cancel{(x+y)}}\right]\div\left[-\frac{x\cancel{(x+y)}\cancel{(x-y)}}{y\cancel{(y+x)}\cancel{(x-y)}}\right]$$

$$=\left(\frac{x-y}{y}\right)\div\left(-\frac{x}{y}\right)$$

$$=\frac{x-y}{y}\bullet-\frac{y}{x}$$

$$=-\frac{x-y}{x}$$

$$=\frac{y-x}{x}$$

SECTION 6.5

VOCABULARY

1. The expression $\dfrac{18x^7}{9x^4}$ is a monomial divided by a **monomial**. The expression $\dfrac{6x^3y-4x^2y^2+8xy^3-2y^4}{2x^4}$ is a **polynomial** divided by a monomial. The expression $\dfrac{x^2-8x+12}{x-6}$ is a trinomial divided by a **binomial**.

3. Divisor, Quotient, Dividend, Remainder

$$\begin{array}{r}
x-2 \\
\text{(Divisor)}\; x-6\,\overline{)\,x^2-8x-4} \;\text{(Dividend)} \\
\underline{x^2-6x} \\
-2x-4 \\
\underline{-2x+12} \\
-16 \;\text{(Remainder)}
\end{array}$$

(Quotient: $x-2$)

CONCEPTS

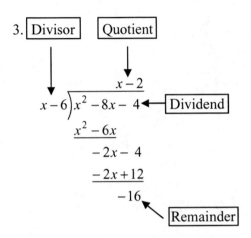

5. a. To divide a polynomial by a monomial, divide each **term** of the polynomial by the monomial.

b. $\dfrac{18x+9}{9}=\dfrac{18x}{\boxed{9}}+\dfrac{9}{\boxed{9}}$

c. $\dfrac{30x^2+12x-24}{6}=\dfrac{30x^2}{\boxed{6}}+\dfrac{12x}{\boxed{6}}-\dfrac{24}{\boxed{6}}$

7.
$$\begin{aligned}
&(2x-1)(x^2+3x-4) \\
&=2x(x^2+3x-4)-1(x^2+3x-4) \\
&=2x^3+6x^2-8x-x^2-3x+4 \\
&=2x^3+5x^2-11x+4
\end{aligned}$$

NOTATION

9.
$$\begin{array}{r}
x+7 \\
x+4\,\overline{)\,x^2+11x+28} \\
\underline{-(\boxed{x^2}+4x)} \\
\boxed{7x}+28 \\
\underline{-(7x+\boxed{28})} \\
0
\end{array}$$

11. $3a^2+5+\dfrac{6}{3a-2}$

13. $\dfrac{x^2-x-12}{x-4}$, $x-4\overline{)x^2-x-12}$, or $(x^2-x-12)\div(x-4)$

GUIDED PRACTICE

15. $\dfrac{4x^2y^3}{8x^5y^2}=\dfrac{y}{2x^3}$

17. $\dfrac{33a^2b^2}{44a^4b^2}=\dfrac{3}{4a^2}$

19.
$$\begin{aligned}
\frac{4x^4+6x}{2}&=\frac{4x^4}{2}+\frac{6x}{2} \\
&=2x^4+3x
\end{aligned}$$

21.
$$\begin{aligned}
\frac{4x^2-x^3}{6x}&=\frac{4x^2}{6x}-\frac{x^3}{6x} \\
&=\frac{2x}{3}-\frac{x^2}{6}
\end{aligned}$$

23.
$$\begin{aligned}
\frac{54a^3y^2-18a^4y^3}{27a^2y^2}&=\frac{54a^3y^2}{27a^2y^2}-\frac{18a^4y^3}{27a^2y^2} \\
&=2a-\frac{2a^2y}{3}
\end{aligned}$$

25.

$$\frac{24x^6y^7 - 12x^5y^{12} + 36xy}{-48x^2y^3}$$
$$= \frac{24x^6y^7}{-48x^2y^3} - \frac{12x^5y^{12}}{-48x^2y^3} + \frac{36xy}{-48x^2y^3}$$
$$= -\frac{x^4y^4}{2} + \frac{x^3y^9}{4} - \frac{3}{4xy^2}$$

27.

$$\begin{array}{r} x+2 \\ x+3\overline{)x^2+5x+6} \\ \underline{x^2+3x} \\ 2x+6 \\ \underline{2x+6} \\ 0 \end{array}$$

29.

$$\begin{array}{r} x-3 \\ x-7\overline{)x^2-10x+21} \\ \underline{x^2-\ 7x} \\ -3x+21 \\ -\underline{3x+21} \\ 0 \end{array}$$

31.

$$\begin{array}{r} 4x-5 \\ 4x+1\overline{)16x^2-16x-5} \\ \underline{16x^2+\ 4x} \\ -20x-5 \\ \underline{-20x-5} \\ 0 \end{array}$$

33.

$$\begin{array}{r} 3x^2+4x+3 \\ 2x-3\overline{)6x^3-x^2-6x-9} \\ \underline{6x^3-9x^2} \\ 8x^2-6x \\ \underline{8x^2-12x} \\ 6x-9 \\ \underline{6x-9} \\ 0 \end{array}$$

35.

$$\begin{array}{r} t^2+2t+1+\frac{3}{t+6} \\ t+6\overline{)t^3+8t^2+13t+9} \\ \underline{t^3+6t^2} \\ 2t^2+13t \\ \underline{2t^2+12t} \\ t+9 \\ \underline{t+6} \\ 3 \end{array}$$

37.

$$\begin{array}{r} 2x^2+5x-3+\frac{-8}{3x-2} \\ 3x-2\overline{)\ 6x^3+11x^2-19x-2} \\ \underline{6x^3-4x^2} \\ 15x^2-19x \\ \underline{15x^2-10x} \\ -9x-2 \\ \underline{-9x+6} \\ -8 \end{array}$$

39. Write the dividend in descending order.

$$\begin{array}{r} a+1 \\ a+1\overline{)a^2+2a+1} \\ \underline{a^2+\ a} \\ a+1 \\ \underline{a+1} \\ 0 \end{array}$$

41. Write the dividend in descending order.

$$\begin{array}{r} 2y+2 \\ 5y-2\overline{)10y^2+6y-4} \\ \underline{10y^2-4y} \\ 10y-4 \\ \underline{10y-4} \\ 0 \end{array}$$

43. Write the dividend in descending order.

$$\begin{array}{r} 3x^2 - x + 2 \\ 3x+2\overline{)9x^3 + 3x^2 + 4x + 4} \\ \underline{9x^3 + 6x^2} \\ -3x^2 + 4x \\ \underline{-3x^2 - 2x} \\ 6x + 4 \\ \underline{6x + 4} \\ 0 \end{array}$$

45. Write the dividend in descending order and insert $0x^3$ for the missing term.

$$\begin{array}{r} 4x^3 - 3x^2 + 3x + 1 \\ 4x+3\overline{)16x^4 + 0x^3 + 3x^2 + 13x + 3} \\ \underline{16x^4 + 12x^3} \\ -12x^3 + 3x^2 \\ \underline{-12x^3 - 9x^2} \\ 12x^2 + 13x \\ \underline{12x^2 + 9x} \\ 4x + 3 \\ \underline{4x + 3} \\ 0 \end{array}$$

47. Insert $0a^2$ and $0a$ for the missing terms.

$$\begin{array}{r} 4a^2 - 2a + 1 \\ 2a+1\overline{)8a^3 + 0a^2 + 0a + 1} \\ \underline{8a^3 + 4a^2} \\ -4a^2 + 0a \\ \underline{-4a^2 - 2a} \\ 2a + 1 \\ \underline{2a + 1} \\ 0 \end{array}$$

49. Insert $0a$ for the missing term.

$$\begin{array}{r} 5a^2 - 3a - 4 \\ 3a-4\overline{)15a^3 - 29a^2 + 0a + 16} \\ \underline{15a^3 - 20a^2} \\ -9a^2 + 0a \\ \underline{-9a^2 + 12a} \\ -12a + 16 \\ \underline{-12a + 16} \\ 0 \end{array}$$

51. Write the dividend in descending order.

$$\begin{array}{r} x^2 + 3x + 4 \\ x^2+2x-3\overline{)x^4 + 5x^3 + 7x^2 - x - 12} \\ \underline{x^4 + 2x^3 - 3x^2} \\ 3x^3 + 10x^2 - x \\ \underline{3x^3 + 6x^2 - 9x} \\ 4x^2 + 8x - 12 \\ \underline{4x^2 + 8x - 12} \\ 0 \end{array}$$

53. Write the dividend in descending order.

$$\begin{array}{r} 2x + 3 + \dfrac{20x - 13}{3x^2 - 7x + 4} \\ 3x^2-7x+4\overline{)6x^3 - 5x^2 + 7x - 1} \\ \underline{6x^3 - 14x^2 + 8x} \\ 9x^2 - x - 1 \\ \underline{9x^2 - 21x + 12} \\ 20x - 13 \end{array}$$

TRY IT YOURSELF

55. Write the dividend in descending order.

$$\begin{array}{r} 6y - 12 \\ y-2\overline{)6y^2 - 24y + 24} \\ \underline{6y^2 - 12y} \\ -12y + 24 \\ \underline{-12y + 24} \\ 0 \end{array}$$

57.

$$\begin{array}{r} 4a^2 - 3a + \dfrac{7}{a+1} \\ a+1\overline{)4a^3 + a^2 - 3a + 7} \\ \underline{4a^3 + 4a^2} \\ -3a^2 - 3a \\ \underline{-3a^2 - 3a} \\ 0+7 \end{array}$$

59.

$$\begin{array}{r} x^4 + x^2 + 4 \\ x^2 - 2\overline{)x^6 - x^4 + 2x^2 - 8} \\ \underline{x^6 - 2x^4} \\ x^4 + 2x^2 \\ \underline{x^4 - 2x^2} \\ 4x^2 - 8 \\ \underline{4x^2 - 8} \\ 0 \end{array}$$

61.

$$\frac{5a^5 - 10a}{25a^3} = \frac{5a^5}{25a^3} - \frac{10a}{25a^3}$$
$$= \frac{a^2}{5} - \frac{2}{5a^2}$$

63.

$$\begin{array}{r} s + 5 + \frac{-10}{2s+3} \\ 2s + 3\overline{)2s^2 + 13s + 5} \\ \underline{2s^2 + 3s} \\ 10s + 5 \\ \underline{10s + 15} \\ -10 \end{array}$$

65.

$$\frac{40m^{17}n^{20}}{35m^{15}n^{30}} = \frac{8m^2}{7n^{10}}$$

67. Insert $0m$ into the divisor for the missing m term.

$$\begin{array}{r} m - 4 + \frac{m+3}{m^2+1} \\ m^2 + 0m + 1\overline{)m^3 - 4m^2 + 2m - 1} \\ \underline{m^3 + 0m^2 + m} \\ -4m^2 + m - 1 \\ \underline{-4m^2 + 0m - 4} \\ m + 3 \end{array}$$

69. Insert missing terms in the dividend.

$$\begin{array}{r} y^2 + 4y + 16 \\ y - 4\overline{)y^3 + 0y^2 + 0y - 64} \\ \underline{y^3 - 4y^2} \\ 4y^2 + 0y \\ \underline{4y^2 - 16y} \\ 16y - 64 \\ \underline{16y - 64} \\ 0 \end{array}$$

71.

$$\begin{array}{r} a^4 - a^2 + 1 \\ a^4 + 2a^2 - 3\overline{)a^8 + a^6 - 4a^4 + 5a^2 - 3} \\ \underline{a^8 + 2a^6 - 3a^4} \\ -a^6 - a^4 + 5a^2 \\ \underline{-a^6 - 2a^4 + 3a^2} \\ a^4 + 2a^2 - 3 \\ \underline{a^4 + 2a^2 - 3} \\ 0 \end{array}$$

73.

$$\frac{40x^3z^2 - 8x^2z - 4z}{4xz} = \frac{40x^3z^2}{4xz} - \frac{8x^2z}{4xz} - \frac{4z}{4xz}$$
$$= 10x^2z - 2x - \frac{1}{x}$$

75. Insert missing terms in the dividend and the divisor.

$$\begin{array}{r} x^2 - 2 + \frac{-x^2 + 7x + 4}{x^3 + 2x + 1} \\ x^3 + 0x^2 + 2x + 1\overline{)x^5 + 0x^4 + 0x^3 + 0x^2 + 3x + 2} \\ \underline{x^5 + 0x^4 + 2x^3 + x^2} \\ -2x^3 - x^2 + 3x + 2 \\ \underline{-2x^3 + 0x^2 - 4x - 2} \\ -x^2 + 7x + 4 \end{array}$$

77. Write the divisor in descending order.

$$\begin{array}{r} 2x^2 - x + 1 \\ 4x^2 - x + 3 \overline{\smash{\big)} 8x^4 - 6x^3 + 11x^2 - 4x + 3} \\ \underline{8x^4 - 2x^3 + 6x^2} \\ -4x^3 + 5x^2 - 4x \\ \underline{-4x^3 + x^2 - 3x} \\ 4x^2 - x + 3 \\ \underline{4x^2 - x + 3} \\ 0 \end{array}$$

79.

$$\frac{15x^2 + 9x - 3}{27} = \frac{15x^2}{27} + \frac{9x}{27} - \frac{3}{27}$$
$$= \frac{5x^2}{9} + \frac{x}{3} - \frac{1}{9}$$

81. Write the dividend in descending order and insert $0x$ into the divisor for the missing x term.

$$\begin{array}{r} x^4 - x^2 - 3 \\ x^2 + 3 \overline{\smash{\big)} x^6 + 2x^4 - 6x^2 - 9} \\ \underline{x^6 + 3x^4} \\ -x^4 - 6x^2 \\ \underline{-x^4 - 3x^2} \\ -3x^2 - 9 \\ \underline{-3x^2 - 9} \\ 0 \end{array}$$

83. $\dfrac{f(x)}{g(x)} = \dfrac{4x^4 + 20x^3 - x^2 - 2x + 15}{x + 5}$

$$\begin{array}{r} 4x^3 - x + 3 \\ x + 5 \overline{\smash{\big)} 4x^4 + 20x^3 - x^2 - 2x + 15} \\ \underline{4x^4 + 20x^3} \\ -x^2 - 2x \\ \underline{-x^2 - 5x} \\ 3x + 15 \\ \underline{3x + 15} \\ 0 \end{array}$$

85.

$$\begin{array}{r} 4x^2 + 6x + 10 \\ x - \frac{1}{2} \overline{\smash{\big)} 4x^3 + 4x^2 + 7x - 5} \\ \underline{4x^3 - 2x^2} \\ 6x^2 + 7x \\ \underline{6x^2 - 3x} \\ 10x - 5 \\ \underline{10x + 5} \\ 0 \end{array}$$

87.

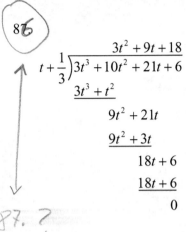

$$\begin{array}{r} 3t^2 + 9t + 18 \\ t + \frac{1}{3} \overline{\smash{\big)} 3t^3 + 10t^2 + 21t + 6} \\ \underline{3t^3 + t^2} \\ 9t^2 + 21t \\ \underline{9t^2 + 3t} \\ 18t + 6 \\ \underline{18t + 6} \\ 0 \end{array}$$

APPLICATIONS

89. ADVERTISING

$$L = \frac{A}{W}$$
$$= \frac{x^3 - 4x^2 + x + 6}{x + 1}$$

$$\begin{array}{r} x^2 - 5x + 6 \\ x + 1 \overline{\smash{\big)} x^3 - 4x^2 + x + 6} \\ \underline{x^3 + x^2} \\ -5x^2 + x \\ \underline{-5x^2 - 5x} \\ 6x + 6 \\ \underline{6x + 6} \\ 0 \end{array}$$

The length is $x^2 - 5x + 6$.

91. WINTER TRAVEL

Dog Sled:

$$r=\frac{d}{t}$$

$$=\frac{12x^2+13x-14}{4x+7}$$

$$\begin{array}{r} 3x-2 \\ 4x+7\overline{)12x^2+13x-14} \\ \underline{12x^2+21x} \\ -8x-14 \\ \underline{-8x-14} \\ 0 \end{array}$$

$$r=3x-2$$

Snowshoes:

$$t=\frac{d}{r}$$

$$=\frac{3x^2+19x+20}{3x+4}$$

$$\begin{array}{r} x+5 \\ 3x+4\overline{)3x^2+19x+20} \\ \underline{3x^2+\ 4x} \\ 15x+20 \\ \underline{15x+20} \\ 0 \end{array}$$

$$t=x+5$$

WRITING

93. Answers will vary.

REVIEW

95.

$$2\left(x^2+4x-1\right)+3\left(2x^2-2x+2\right)$$
$$=2x^2+8x-2+6x^2-6x+6$$
$$=8x^2+2x+4$$

97.

$$-2\left(3y^3-2y+7\right)-\left(y^2+2y-4\right)+4\left(y^3+2y-1\right)$$
$$=-6y^3+4y-14-y^2-2y+4+4y^3+8y-4$$
$$=-2y^3-y^2+10y-14$$

CHALLENGE PROBLEMS

99.

$$\begin{array}{r} \frac{3}{4}c-1 \\ 4c+3\overline{)3c^2-\frac{7}{4}c-3} \\ \underline{3c^2+\frac{9}{4}c} \\ -4c-3 \\ \underline{-4c-3} \\ 0 \end{array}$$

101.

$$\begin{array}{r} c^2-d^2+4-\frac{1}{c^2+6} \\ c^2+6\overline{)c^4-c^2d^2+10c^2-6d^2+23} \\ \underline{c^4\qquad\quad +6c^2} \\ -c^2d^2+4c^2-6d^2 \\ \underline{-c^2d^2\qquad -6d^2} \\ 4c^2\qquad +23 \\ \underline{4c^2\qquad +24} \\ -1 \end{array}$$

103.

$$\begin{array}{r} 9.8x+16.4-\frac{36.5}{x-2} \\ x-2\overline{)9.8x^2-3.2x-69.3} \\ \underline{9.8x^2-19.6x} \\ 16.4x-69.3 \\ \underline{16.4x-32.8} \\ -36.5 \end{array}$$

SECTION 6.6

VOCABULARY

1. The method of dividing $x^2 + 2x - 9$ by $x - 4$ shown below is called **synthetic** division.

3. In Exercise 1, the synthetic **divisor** is 4.

5. The factor **theorem** tells us how to find one factor of a polynomial if the remainder of a certain division is 0.

CONCEPTS

7. a. $(5x^3 + x - 3) \div (x + 2)$

   b. $5x^2 - 10x + 21 - \dfrac{45}{x+2}$

9. Rather than substituting 8 for x in $P(x) = 6x^3 - x^2 - 17x + 9$, we can divide the polynomial **$6x^3 - x^2 - 17x + 9$** by **$x - 8$** to find $P(8)$.

NOTATION

11.

$\boxed{2}$	6	$\boxed{1}$	−23	$\boxed{2}$
		$\boxed{12}$	$\boxed{26}$	6
	$\boxed{6}$	13	3	$\boxed{8}$

GUIDED PRACTICE

13.

1	2	1	−3
		2	3
	2	3	0

$$\frac{2x^2 + x - 3}{x - 1} = 2x + 3$$

15.

5	5	−27	10
		25	−10
	5	−2	0

$$\frac{5x^2 - 27x + 10}{x - 5} = 5x - 2$$

17.

3	3	−13	12
		9	−12
	3	−4	0

$$\frac{3x^2 - 13x + 12}{x - 3} = 3x - 4$$

19.

6	5	−24	−36
		30	36
	5	6	0

$$\frac{5x^2 - 24x - 36}{x - 6} = 5x + 6$$

21. Write the numerator in descending order and insert 0 for any missing terms.

2	1	−3	0	4
		2	−2	−4
	1	−1	−2	0

$$\frac{a^3 - 3a^2 + 4}{a - 2} = a^2 - a - 2$$

23. Write the numerator in descending order and insert 0 for any missing terms.

4	3	0	−47	−4
		12	48	4
	3	12	1	0

$$\frac{a^3 - 47a - 4}{a - 4} = 3a^2 + 12a + 1$$

25. Write the numerator in descending order and insert 0 for any missing terms.

$$\begin{array}{r|rrrr} 3 & 3 & 0 & -31 & 13 \\ & & 9 & 27 & -12 \\ \hline & 3 & 9 & -4 & 1 \end{array}$$

$$\frac{3b^3-31b+13}{b-3}=3b^2+9b-4+\frac{1}{b-3}$$

27. Write the numerator in descending order and insert 0 for any missing terms.

$$\begin{array}{r|rrrr} 2 & 4 & 0 & -1 & -18 \\ & & 8 & 16 & 30 \\ \hline & 4 & 8 & 15 & 12 \end{array}$$

$$\frac{4t^3-t-18}{t-2}=4t^2+8t+15+\frac{12}{t-2}$$

29. Write each polynomial in descending order.

$$\begin{array}{r|rrrr} -1 & 1 & -4 & 1 & 6 \\ & & -1 & 5 & -6 \\ \hline & 1 & -5 & 6 & 0 \end{array}$$

$$\frac{x^3-4x^2+x+6}{x+1}=x^2-5x+6$$

31. Write each polynomial in descending order.

$$\begin{array}{r|rrrr} -8 & 3 & 20 & -36 & -42 \\ & & -24 & 32 & 32 \\ \hline & 3 & -4 & -4 & -10 \end{array}$$

$$\frac{3x^3+20x^2-36x-42}{x+8}=3x^2-4x-4+\frac{-10}{x+8}$$

33. Write each polynomial in descending order.

$$\begin{array}{r|rrrr} -5 & 2 & 7 & -3 & 8 \\ & & -10 & 15 & -60 \\ \hline & 2 & -3 & 12 & -52 \end{array}$$

$$\frac{2x^3+7x^2-3x+8}{x+5}=2x^2-3x+12+\frac{-52}{x+5}$$

35. Write each polynomial in descending order.

$$\begin{array}{r|rrrr} -10 & 1 & 8 & -17 & 27 \\ & & -10 & 20 & -30 \\ \hline & 1 & -2 & 3 & -3 \end{array}$$

$$\frac{x^3+8x^2-17x+27}{x+10}=x^2-2x+3+\frac{-3}{x+10}$$

37.

$$\begin{array}{r|rrr} 0.2 & 7.2 & -2.1 & 0.5 \\ & & 1.44 & -0.132 \\ \hline & 7.2 & -0.66 & 0.368 \end{array}$$

$$\frac{7.2x^2-2.1x+0.5}{x-0.2}=7.2x-0.66+\frac{0.368}{x-0.2}$$

39.

$$\begin{array}{r|rrrr} -57 & 9 & 0 & 0 & -25 \\ & & -513 & 29,241 & -1,666,737 \\ \hline & 9 & -513 & 29,241 & -1,666,762 \end{array}$$

$$\frac{9x^3-25}{x+57}=9x^2-513x+29,241-\frac{1,666,762}{x+57}$$

41.

$$\begin{aligned} P(x)&=2x^3-4x^2+2x-1 \\ P(1)&=2(1)^3-4(1)^2+2(1)-1 \\ &=2(1)-4(1)+2(1)-1 \\ &=2-4+2-1 \\ &=-1 \end{aligned}$$

$$\begin{array}{r|rrrr} 1 & 2 & -4 & 2 & -1 \\ & & 2 & -2 & 0 \\ \hline & 2 & -2 & 0 & -1 \end{array}$$

The remainder, -1, equals $P(1)$.

43.

$$P(x)=2x^3-4x^2+2x-1$$
$$\begin{aligned}P(-2)&=2(-2)^3-4(-2)^2+2(-2)-1\\&=2(-8)-4(4)+2(-2)-1\\&=-16-16-4-1\\&=-37\end{aligned}$$

$$\begin{array}{r|rrrr} -2 & 2 & -4 & 2 & -1 \\ & & -4 & 16 & -36 \\ \hline & 2 & -8 & 18 & -37 \end{array}$$

The remainder, -37, equals $P(-2)$.

45.

$$P(x)=2x^3-4x^2+2x-1$$
$$\begin{aligned}P(3)&=2(3)^3-4(3)^2+2(3)-1\\&=2(27)-4(9)+2(3)-1\\&=54-36+6-1\\&=23\end{aligned}$$

$$\begin{array}{r|rrrr} 3 & 2 & -4 & 2 & -1 \\ & & 6 & 6 & 24 \\ \hline & 2 & 2 & 8 & 23 \end{array}$$

The remainder, 23, equals $P(3)$.

47.

$$P(x)=2x^3-4x^2+2x-1$$
$$\begin{aligned}P(0)&=2(0)^3-4(0)^2+2(0)-1\\&=2(0)-4(0)+2(0)-1\\&=0-0+0-1\\&=-1\end{aligned}$$

$$\begin{array}{r|rrrr} 0 & 2 & -4 & 2 & -1 \\ & & 0 & 0 & 0 \\ \hline & 2 & -4 & 2 & -1 \end{array}$$

The remainder, -1, equals $P(0)$.

49.

$$Q(x)=x^4-3x^3+2x^2+x-3$$
$$\begin{aligned}Q(-1)&=(-1)^4-3(-1)^3+2(-1)^2+(-1)-3\\&=1-3(-1)+2(1)-1-3\\&=1+3+2-1-3\\&=2\end{aligned}$$

$$\begin{array}{r|rrrrr} -1 & 1 & -3 & 2 & 1 & -3 \\ & & -1 & 4 & -6 & 5 \\ \hline & 1 & -4 & 6 & -5 & 2 \end{array}$$

The remainder, 2, equals $Q(-1)$.

51.

$$Q(x)=x^4-3x^3+2x^2+x-3$$
$$\begin{aligned}Q(2)&=(2)^4-3(2)^3+2(2)^2+(2)-3\\&=16-3(8)+2(4)+2-3\\&=16-24+8+2-3\\&=-1\end{aligned}$$

$$\begin{array}{r|rrrrr} 2 & 1 & -3 & 2 & 1 & -3 \\ & & 2 & -2 & 0 & 2 \\ \hline & 1 & -1 & 0 & 1 & -1 \end{array}$$

The remainder, -1, equals $Q(2)$.

53.

$$Q(x)=x^4-3x^3+2x^2+x-3$$
$$\begin{aligned}Q(3)&=(3)^4-3(3)^3+2(3)^2+(3)-3\\&=81-3(27)+2(9)+3-3\\&=81-81+18+3-3\\&=18\end{aligned}$$

$$\begin{array}{r|rrrrr} 3 & 1 & -3 & 2 & 1 & -3 \\ & & 3 & 0 & 6 & 21 \\ \hline & 1 & 0 & 2 & 7 & 18 \end{array}$$

The remainder, 18, equals $Q(3)$.

55.

$$Q(x) = x^4 - 3x^3 + 2x^2 + x - 3$$
$$Q(-3) = (-3)^4 - 3(-3)^3 + 2(-3)^2 + (-3) - 3$$
$$= 81 - 3(-27) + 2(9) - 3 - 3$$
$$= 81 + 81 + 18 - 3 - 3$$
$$= 174$$

$$\begin{array}{r|rrrrr} -3 & 1 & -3 & 2 & 1 & -3 \\ & & -3 & 18 & -60 & 177 \\ \hline & 1 & -6 & 20 & -59 & 174 \end{array}$$

The remainder, 174, equals $Q(-3)$.

57.

$$\begin{array}{r|rrrr} 2 & 1 & -4 & 1 & -2 \\ & & 2 & -4 & -6 \\ \hline & 1 & -2 & -3 & -8 \end{array}$$

The remainder is -8 so $P(2) = -8$.

59.

$$\begin{array}{r|rrrr} 3 & 2 & 0 & 1 & 2 \\ & & 6 & 18 & 57 \\ \hline & 2 & 6 & 19 & 59 \end{array}$$

The remainder is 59 so $P(3) = 59$.

61.

$$\begin{array}{r|rrrrr} -2 & 1 & -2 & 1 & -3 & 2 \\ & & -2 & 8 & -18 & 42 \\ \hline & 1 & -4 & 9 & -21 & 44 \end{array}$$

The remainder is 44 so $P(-2) = 44$.

63.

$$\begin{array}{r|rrrrrr} -\frac{1}{2} & 3 & 0 & 0 & 0 & 0 & 1 \\ & & -\frac{3}{2} & \frac{3}{4} & -\frac{3}{8} & \frac{3}{16} & -\frac{3}{32} \\ \hline & 3 & -\frac{3}{2} & \frac{3}{4} & -\frac{3}{8} & \frac{3}{16} & \frac{29}{32} \end{array}$$

The remainder is $\frac{29}{32}$ so $P\left(-\frac{1}{2}\right) = \frac{29}{32}$.

65.

$$\begin{array}{r|rrrr} 3 & 1 & -3 & 5 & -15 \\ & & 3 & 0 & 15 \\ \hline & 1 & 0 & 5 & 0 \end{array}$$

The remainder is 0, so $x - 3$ is a factor.

67.

$$\begin{array}{r|rrr} -2 & 3 & -7 & 4 \\ & & -6 & 26 \\ \hline & 3 & -13 & 30 \end{array}$$

The remainder is 30,
so $x + 2$ is NOT a factor.

69. Write the numerator in descending order and insert 0 for any missing terms.

$$\begin{array}{r|rrrr} -1 & 6 & 5 & 0 & 4 \\ & & -6 & 1 & -1 \\ \hline & 6 & -1 & 1 & 3 \end{array}$$

$$\frac{6x^3 + 5x^2 + 4}{x + 1} = 6x^2 - x + 1 + \frac{3}{x + 1}$$

71.

$$\begin{array}{r|rrr} -2 & 1 & -5 & 14 \\ & & -2 & 14 \\ \hline & 1 & -7 & 28 \end{array}$$

$$\frac{x^2 - 5x + 14}{x + 2} = x - 7 + \frac{28}{x + 2}$$

73.

$$\begin{array}{r|rrrrrr} 1 & 1 & 0 & 0 & 0 & 0 & -1 \\ & & 1 & 1 & 1 & 1 & 1 \\ \hline & 1 & 1 & 1 & 1 & 1 & 0 \end{array}$$

$$\frac{a^5 - 1}{a - 1} = a^4 + a^3 + a^2 + a + 1$$

75.

$$\begin{array}{r|rrrrrr} 4 & -6 & 14 & 38 & 4 & 25 & -36 \\ & & -24 & -40 & -8 & -16 & 36 \\ \hline & -6 & -10 & -2 & -4 & 9 & 0 \end{array}$$

$$\frac{-6c^5+14c^4+38c^3+4c^2+7c+36}{c-4} = -6c^4-10c^3-2c^2-4c+9$$

77.

$$\begin{array}{r|rrrr} -\frac{1}{3} & 9 & 3 & -21 & -7 \\ & & -3 & 0 & 7 \\ \hline & 9 & 0 & -21 & 0 \end{array}$$

$$\frac{9a^3+3a^2-21a-7}{a+\frac{1}{3}} = 9a^2-21$$

79.

$$\begin{array}{r|rrrrr} -3 & 4 & 12 & -1 & -1 & 12 \\ & & -12 & 0 & 3 & -6 \\ \hline & 4 & 0 & -1 & 2 & 6 \end{array}$$

$$\frac{4x^4+12x^3-x^2-x+12}{x+3} = 4x^3-x+2+\frac{6}{x+3}$$

81.

$$\begin{array}{r|rrrr} 8 & 3 & -25 & 10 & -16 \\ & & 24 & -8 & 16 \\ \hline & 3 & -1 & 2 & 0 \end{array}$$

$$\frac{3x^3-25x^2+10x-16}{x-8} = 3x^2-x+2$$

83.

$$\begin{array}{r|rrrr} 10 & 2 & -16 & -35 & -50 \\ & & 20 & 40 & 50 \\ \hline & 2 & 4 & 5 & 0 \end{array}$$

$$\frac{2x^3-16x^2-35x-50}{x-10} = 2x^2+4x+5$$

85. Write the numerator in descending order and insert 0 for any missing terms.

$$\begin{array}{r|rrrr} -2 & 4 & 5 & 0 & -1 \\ & & -8 & 6 & -12 \\ \hline & 4 & -3 & 6 & -13 \end{array}$$

$$\frac{4x^3+5x^2-1}{x+2} = 4x^2-3x+6-\frac{13}{x+2}$$

87.

$$\begin{array}{r|rrrr} -\frac{3}{4} & 8 & -10 & -32 & -15 \\ & & -6 & 12 & 15 \\ \hline & 8 & -16 & -20 & 0 \end{array}$$

$$\frac{8a^3-10a^2-32a-15}{a+\frac{3}{4}} = 8a^2-16a-20$$

WRITING

89. Answers will vary.

91. Answers will vary.

REVIEW

93.

$$|3x-7|+8=22$$
$$|3x-7|+8-8=22-8$$
$$|3x-7|=14$$
$$3x-7=14 \quad \text{and} \quad 3x-7=-14$$
$$3x=21 \qquad 3x=-7$$
$$x=7 \qquad x=-\frac{7}{3}$$

95. $6-3|10x+5|=6$

$$6-3|10x+5|-6=6-6$$
$$-3|10x+5|=0$$
$$\frac{-3|10x+5|}{-3}=\frac{0}{-3}$$
$$|10x+5|=0$$
$$10x+5=0$$
$$10x=-5$$
$$x=-\frac{1}{2}$$

CHALLENGE PROBLEMS

97. During the synthetic division, every column would add to be zero except for the last, which would stay 1. Thus, the remainder would be 1.

99.

$$\begin{array}{r|rrrrrrr} 2 & 1 & 0 & 0 & 0 & 0 & 0 & 0 \\ & & 2 & 4 & 8 & 16 & 32 & 64 \\ \hline & & 2 & 4 & 8 & 16 & 32 & 64 \end{array}$$

The remainder is 64 so $P(2)=64$.

SECTION 6.7

VOCABULARY

1. Equations that contain one or more rational expressions, such as $\frac{x}{x+2} = 4 + \frac{10}{x+2}$, are called **rational** equations.

CONCEPTS

3. a. $\frac{7}{5x} - \frac{1}{2} = \frac{5}{6x} + \frac{1}{3}$ is an equation.
   b. $\frac{4}{x^2-4} - \frac{5}{x-2}$ is an expression.
   c. $\frac{27p^4}{35q} \div \frac{9p}{21q}$ is an expression.
   d. $\frac{4}{t+3} + \frac{8}{t^2-9} = \frac{2}{t-3}$ is an equation.
   e. $\dfrac{\frac{y}{x} - \frac{x}{y}}{\frac{1}{y} - \frac{1}{x}}$ is an expression.
   f. $\frac{t^2+t-6}{t^2-6t+9} \bullet \frac{t^2-9}{t^2-4}$ is an expression.

5. a. 3, 0
   b. 3, 0
   c. 3, 0

NOTATION

7.

$$\frac{10}{3y} - \frac{7}{30} = \frac{9}{2y}$$
$$\boxed{30y}\left(\frac{10}{3y} - \frac{7}{30}\right) = 30y\left(\boxed{\frac{9}{2y}}\right)$$
$$\boxed{30y}\left(\frac{10}{3y}\right) - \boxed{30y}\left(\frac{7}{30}\right) = \boxed{30y}\left(\frac{9}{2y}\right)$$
$$100 - \boxed{7y} = 135$$
$$-7y = \boxed{35}$$
$$y = -5$$

GUIDED PRACTICE

9.

$$\frac{1}{4} + \frac{9}{x} = 1$$
$$4x\left(\frac{1}{4} + \frac{9}{x}\right) = 4x(1)$$
$$x(1) + 4(9) = 4x$$
$$x + 36 = 4x$$
$$36 = 3x$$
$$12 = x$$

11.

$$\frac{1}{a} = \frac{1}{3} - \frac{2}{3a}$$
$$3a\left(\frac{1}{a}\right) = 3a\left(\frac{1}{3} - \frac{2}{3a}\right)$$
$$3(1) = a(1) - 1(2)$$
$$3 = a - 2$$
$$3 + 2 = a - 2 + 2$$
$$5 = a$$

13.

$$\frac{18}{y+1} + \frac{2}{5} = 4$$
$$5(y+1)\left(\frac{18}{y+1} + \frac{2}{5}\right) = 5(y+1)(4)$$
$$5(18) + 2(y+1) = 20(y+1)$$
$$90 + 2y + 2 = 20y + 20$$
$$2y + 92 = 20y + 20$$
$$2y + 92 - 20y = 20y + 20 - 20y$$
$$-18y + 92 = 20$$
$$-18y + 92 - 92 = 20 - 92$$
$$-18y = -72$$
$$\frac{-18y}{-18} = \frac{-72}{-18}$$
$$y = 4$$

15.

$$\frac{1}{2}+\frac{x}{x-1}=3$$
$$2(x-1)\left(\frac{1}{2}+\frac{x}{x-1}\right)=2(x-1)(3)$$
$$1(x-1)+2(x)=6(x-1)$$
$$x-1+2x=6x-6$$
$$3x-1=6x-6$$
$$3x-1-3x=6x-6-3x$$
$$-1=3x-6$$
$$-1+6=3x-6+6$$
$$5=3x$$
$$\frac{5}{3}=x$$

17.

$$\frac{4}{t+3}+\frac{8}{t^2-9}=\frac{2}{t-3}$$
$$\frac{4}{t+3}+\frac{8}{(t+3)(t-3)}=\frac{2}{t-3}$$
$$(t+3)(t-3)\left(\frac{4}{t+3}+\frac{8}{(t+3)(t-3)}\right)=(t+3)(t-3)\left(\frac{2}{t-3}\right)$$
$$4(t-3)+8=2(t+3)$$
$$4t-12+8=2t+6$$
$$4t-4=2t+6$$
$$2t-4=6$$
$$2t=10$$
$$t=5$$

19.

$$\frac{4}{x^2-4}-\frac{5}{x-2}=\frac{1}{x+2}$$
$$\frac{4}{(x+2)(x-2)}-\frac{5}{x-2}=\frac{1}{x+2}$$
$$(x+2)(x-2)\left(\frac{4}{(x+2)(x-2)}-\frac{5}{x-2}\right)=(x+2)(x-2)\left(\frac{1}{x+2}\right)$$
$$4-5(x+2)=1(x-2)$$
$$4-5x-10=x-2$$
$$-5x-6=x-2$$
$$-6x-6=-2$$
$$-6x=4$$
$$x=-\frac{4}{6}$$
$$x=-\frac{2}{3}$$

21.

$$\frac{2}{x-2}+\frac{10}{x+5}=\frac{2x}{x^2+3x-10}$$
$$\frac{2}{x-2}+\frac{10}{x+5}=\frac{2x}{(x+5)(x-2)}$$
$$(x+5)(x-2)\left(\frac{2}{x-2}+\frac{10}{x+5}\right)=(x+5)(x-2)\left(\frac{2x}{(x+5)(x-2)}\right)$$
$$2(x+5)+10(x-2)=2x$$
$$2x+10+10x-20=2x$$
$$12x-10=2x$$
$$-10=-10x$$
$$1=x$$

23.

$$\frac{1}{n+2}-\frac{2}{n-3}=\frac{-2n}{n^2-n-6}$$
$$\frac{1}{n+2}-\frac{2}{n-3}=\frac{-2n}{(n-3)(n+2)}$$
$$(n-3)(n+2)\left(\frac{1}{n+2}-\frac{2}{n-3}\right)=(n-3)(n+2)\left(\frac{-2n}{(n-3)(n+2)}\right)$$
$$n-3-2(n+2)=-2n$$
$$n-3-2n-4=-2n$$
$$-n-7=-2n$$
$$-7=-n$$
$$7=n$$

25.

$$\frac{x}{8}=\frac{x-12}{3x-27}-\frac{1}{3}$$
$$\frac{x}{8}=\frac{x-12}{3(x-9)}-\frac{1}{3}$$
$$24(x-9)\left(\frac{x}{8}\right)=24(x-9)\left(\frac{x-12}{3(x-9)}-\frac{1}{3}\right)$$
$$3x(x-9)=8(x-12)-8(x-9)$$
$$3x^2-27x=8x-96-8x+72$$
$$3x^2-27x=-24$$
$$3x^2-27x+24=0$$
$$3(x^2-9x+8)=0$$
$$3(x-1)(x-8)=0$$
$$x-1=0 \quad \text{and} \quad x-8=0$$
$$x=1 \qquad\qquad x=8$$

27.

$$\frac{p-1}{2}+1=\frac{3}{p}$$
$$2p\left(\frac{p-1}{2}+1\right)=2p\left(\frac{3}{p}\right)$$
$$p(p-1)+2p=2(3)$$
$$p^2-p+2p=6$$
$$p^2+p-6=0$$
$$(p+3)(p-2)=0$$
$$p+3=0 \text{ or } p-2=0$$
$$p=-3 \qquad p=2$$

29.

$$\frac{16}{t+3}+\frac{7}{t-2}=3$$
$$(t+3)(t-2)\left(\frac{16}{t+3}+\frac{7}{t-2}\right)=(t+3)(t-2)(3)$$
$$16(t-2)+7(t+3)=3(t^2+t-6)$$
$$16t-32+7t+21=3t^2+3t-18$$
$$23t-11=3t^2+3t-18$$
$$0=3t^2-20t-7$$
$$0=(3t+1)(t-7)$$
$$3t+1=0 \text{ or } t-7=0$$
$$3t=-1 \qquad t=7$$
$$t=-\frac{1}{3}$$

31.

$$\frac{2}{5x-5}+\frac{x-2}{15}=\frac{4}{5x-5}$$
$$\frac{2}{5(x-1)}+\frac{x-2}{15}=\frac{4}{5(x-1)}$$
$$15(x-1)\left(\frac{2}{5(x-1)}+\frac{x-2}{15}\right)=15(x-1)\left(\frac{4}{5(x-1)}\right)$$
$$3(2)+(x-1)(x-2)=3(4)$$
$$6+x^2-2x-x+2=12$$
$$x^2-3x+8=12$$
$$x^2-3x-4=0$$
$$(x-4)(x+1)=0$$
$$x-4=0 \text{ or } x+1=0$$
$$x=4 \qquad x=-1$$

33.

$$\frac{1}{3x-18}+\frac{5}{6-x}=\frac{1}{3}$$
$$\frac{1}{3(x-6)}-\frac{5}{x-6}=\frac{1}{3}$$
$$3(x-6)\left(\frac{1}{3(x-6)}-\frac{5}{x-6}\right)=3(x-6)\left(\frac{1}{3}\right)$$
$$1-3(5)=x-6$$
$$1-15=x-6$$
$$-14=x-6$$
$$-8=x$$

35.

$$\frac{7}{3x-9}+\frac{1}{3-x}=\frac{4}{9}$$
$$\frac{7}{3(x-3)}-\frac{1}{x-3}=\frac{4}{9}$$
$$9(x-3)\left(\frac{7}{3(x-3)}-\frac{1}{x-3}\right)=9(x-3)\left(\frac{4}{9}\right)$$
$$3(7)-9=4(x-3)$$
$$21-9=4x-12$$
$$12=4x-12$$
$$24=4x$$
$$6=x$$

37.

$$4-\frac{3x}{x-9}=\frac{5x-72}{x-9}$$
$$(x-9)\left(4-\frac{3x}{x-9}\right)=(x-9)\left(\frac{5x-72}{x-9}\right)$$
$$4(x-9)-3x=5x-72$$
$$4x-36-3x=5x-72$$
$$x-36=5x-72$$
$$-36=4x-72$$
$$36=4x$$
$$9=x$$

No solution; 9 is extraneous

39.

$$\frac{6}{x+3}+\frac{48}{x^2-2x-15}-\frac{7}{x-5}=0$$
$$\frac{6}{x+3}+\frac{48}{(x-5)(x+3)}-\frac{7}{x-5}=0$$
$$(x-5)(x+3)\left(\frac{6}{x+3}+\frac{48}{(x-5)(x+3)}-\frac{7}{x-5}\right)=(x-5)(x+3)(0)$$
$$6(x-5)+48-7(x+3)=0$$
$$6x-30+48-7x-21=0$$
$$-x-3=0$$
$$-x=3$$
$$x=-3$$

No solution; –3 is extraneous

41.

$$Q=\frac{A-I}{L}$$
$$L(Q)=L\left(\frac{A-I}{L}\right)$$
$$LQ=A-I$$
$$LQ+I=A-I+I$$
$$LQ+I=A$$
$$A=LQ+I$$

43.

$$I=\frac{E}{R_L+r}$$
$$(R_L+r)(I)=(R_L+r)\left(\frac{E}{R_L+r}\right)$$
$$IR_L+Ir=E$$
$$IR_L+Ir-IR_L=E-IR_L$$
$$Ir=E-IR_L$$
$$\frac{Ir}{I}=\frac{E-IR_L}{I}$$
$$r=\frac{E-IR_L}{I}$$

45.

$$\mu_R=\frac{n_1(n_1+n_2+1)}{2}$$
$$\mu_R=\frac{n_1^2+n_1n_2+n_1}{2}$$
$$2(\mu_R)=2\left(\frac{n_1^2+n_1n_2+n_1}{2}\right)$$
$$2\mu_R=n_1^2+n_1n_2+n_1$$
$$2\mu_R-n_1^2-n_1=n_1^2+n_1n_2+n_1-n_1^2-n_1$$
$$2\mu_R-n_1^2-n_1=n_1n_2$$
$$\frac{2\mu_R-n_1^2-n_1}{n_1}=\frac{n_1n_2}{n_1}$$
$$\frac{2\mu_R-n_1^2-n_1}{n_1}=n_2$$

47.

$$P=\frac{Q_1}{Q_2-Q_1}$$
$$(Q_2-Q_1)(P)=(Q_2-Q_1)\left(\frac{Q_1}{Q_2-Q_1}\right)$$
$$PQ_2-PQ_1=Q_1$$
$$PQ_2-PQ_1+PQ_1=Q_1+PQ_1$$
$$PQ_2=Q_1+PQ_1$$
$$PQ_2=Q_1(1+P)$$
$$\frac{PQ_2}{1+P}=\frac{Q_1(1+P)}{1+P}$$
$$\frac{PQ_2}{1+P}=Q_1$$

49.

$$\frac{1}{R}=\frac{1}{R_1}+\frac{1}{R_2}+\frac{1}{R_3}$$
$$RR_1R_2R_3\left(\frac{1}{R}\right)=RR_1R_2R_3\left(\frac{1}{R_1}+\frac{1}{R_2}+\frac{1}{R_3}\right)$$
$$R_1R_2R_3=RR_2R_3+RR_1R_3+RR_1R_2$$
$$R_1R_2R_3=R(R_2R_3+R_1R_3+R_1R_2)$$
$$\frac{R_1R_2R_3}{R_2R_3+R_1R_3+R_1R_2}=\frac{R(R_2R_3+R_1R_3+R_1R_2)}{R_2R_3+R_1R_3+R_1R_2}$$
$$\frac{R_1R_2R_3}{R_2R_3+R_1R_3+R_1R_2}=R$$

51.

$$\frac{E}{e}=\frac{R+r}{r}$$
$$er\left(\frac{E}{e}\right)=er\left(\frac{R+r}{r}\right)$$
$$Er=eR+er$$
$$Er-er=eR$$
$$r(E-e)=eR$$
$$r=\frac{eR}{E-e}$$

53.

$$\frac{x+2}{x+3}-1=\frac{-1}{x^2+2x-3}$$
$$\frac{x+2}{x+3}-1=\frac{-1}{(x+3)(x-1)}$$
$$(x+3)(x-1)\left(\frac{x+2}{x+3}-1\right)=(x+3)(x-1)\left(\frac{-1}{(x+3)(x-1)}\right)$$
$$(x-1)(x+2)-(x+3)(x-1)=-1$$
$$(x^2+2x-x-2)-(x^2-x+3x-3)=-1$$
$$(x^2+x-2)-(x^2+2x-3)=-1$$
$$x^2+x-2-x^2-2x+3=-1$$
$$-x+1=-1$$
$$-x=-2$$
$$x=2$$

55.

$$\frac{3}{y}+\frac{7}{2y}=13$$
$$2y\left(\frac{3}{y}+\frac{7}{2y}\right)=2y(13)$$
$$2(3)+7=26y$$
$$6+7=26y$$
$$13=26y$$
$$\frac{13}{26}=y$$
$$\frac{1}{2}=y$$

57.

$$\frac{3}{r}+\frac{12}{r^2-4r}=-\frac{7}{r-4}$$
$$\frac{3}{r}+\frac{12}{r(r-4)}=-\frac{7}{r-4}$$
$$r(r-4)\left(\frac{3}{r}+\frac{12}{r(r-4)}\right)=r(r-4)\left(-\frac{7}{r-4}\right)$$
$$3(r-4)+12=-7r$$
$$3r-12+12=-7r$$
$$3r=-7r$$
$$10r=0$$
$$r=0$$

No solution; 0 is extraneous

59.

$$\frac{x+4}{2x+14}-\frac{x}{2x+6}=\frac{3}{16}$$
$$\frac{x+4}{2(x+7)}-\frac{x}{2(x+3)}=\frac{3}{16}$$
$$16(x+7)(x+3)\left(\frac{x+4}{2(x+7)}-\frac{x}{2(x+3)}\right)=16(x+7)(x+3)\left(\frac{3}{16}\right)$$
$$8(x+3)(x+4)-8x(x+7)=3(x+7)(x+3)$$
$$8(x^2+3x+4x+12)-8x^2-56x=3(x^2+3x+7x+21)$$
$$8(x^2+7x+12)-8x^2-56x=3(x^2+10x+21)$$
$$8x^2+56x+96-8x^2-56x=3x^2+30x+63$$
$$96=3x^2+30x+63$$
$$96-96=3x^2+30x+63-96$$
$$0=3x^2+30x-33$$
$$0=\frac{3x^2}{3}+\frac{30x}{3}-\frac{33}{3}$$
$$0=x^2+10x-11$$
$$0=(x+11)(x-1)$$
$$x+11=0 \quad \text{or} \quad x-1=0$$
$$x=-11 \qquad x=1$$

61.

$$\frac{-10}{t+3}=1-\frac{11}{t-3}$$
$$(t+3)(t-3)\left(\frac{-10}{t+3}\right)=(t+3)(t-3)\left(1-\frac{11}{t-3}\right)$$
$$-10(t-3)=(t+3)(t-3)-11(t+3)$$
$$-10t+30=t^2-9-11t-33$$
$$-10t+30=t^2-11t-42$$
$$0=t^2-t-72$$
$$0=(t+8)(t-9)$$
$$t+8=0 \quad \text{or} \quad t-9=0$$
$$t=-8 \qquad t=9$$

63.

$$\frac{x+2}{2x-6}+\frac{3}{3-x}=\frac{x}{2}$$
$$\frac{x+2}{2(x-3)}-\frac{3}{x-3}=\frac{x}{2}$$
$$2(x-3)\left(\frac{x+2}{2(x-3)}-\frac{3}{x-3}\right)=2(x-3)\left(\frac{x}{2}\right)$$
$$x+2-6=x(x-3)$$
$$x-4=x^2-3x$$
$$0=x^2-4x+4$$
$$0=(x-2)(x-2)$$
$$x-2=0 \quad \text{or} \quad x-2=0$$
$$x=2 \qquad x=2$$

A repeated solution of 2

65.

$$\frac{2}{x}+\frac{1}{2}=\frac{9}{4x}-\frac{1}{2x}$$
$$4x\left(\frac{2}{x}+\frac{1}{2}\right)=4x\left(\frac{9}{4x}-\frac{1}{2x}\right)$$
$$4(2)+2x(1)=9-2(1)$$
$$8+2x=9-2$$
$$8+2x=7$$
$$2x=-1$$
$$x=-\frac{1}{2}$$

67.

$$\frac{3-5y}{2+y}=\frac{-5y-3}{y-2}$$
$$(2+y)(y-2)\left(\frac{3-5y}{2+y}\right)=(2+y)(y-2)\left(\frac{-5y-3}{y-2}\right)$$
$$(y-2)(3-5y)=(2+y)(-5y-3)$$
$$3y-5y^2-6+10y=-10y-6-5y^2-3y$$
$$-5y^2+13y-6=-5y^2-13y-6$$
$$13y-6=-13y-6$$
$$26y-6=-6$$
$$26y=0$$
$$y=0$$

69.

$$\frac{21}{x^2-4}-\frac{14}{x+2}=\frac{3}{2-x}$$
$$\frac{21}{(x-2)(x+2)}-\frac{14}{x+2}=-\frac{3}{x-2}$$
$$(x+2)(x-2)\left(\frac{21}{(x-2)(x+2)}-\frac{14}{x+2}\right)=(x+2)(x-2)\left(-\frac{3}{x-2}\right)$$
$$21-14(x-2)=-3(x+2)$$
$$21-14x+28=-3x-6$$
$$-14x+49=-3x-6$$
$$-11x+49=-6$$
$$-11x=-55$$
$$x=5$$

71.

$$\frac{x-4}{x-3}-\frac{x-2}{3-x}=x-3$$
$$\frac{x-4}{x-3}+\frac{x-2}{x-3}=x-3$$
$$(x-3)\left(\frac{x-4}{x-3}+\frac{x-2}{x-3}\right)=(x-3)(x-3)$$
$$x-4+x-2=x^2-6x+9$$
$$2x-6=x^2-6x+9$$
$$0=x^2-8x+15$$
$$0=(x-3)(x-5)$$
$$x-3=0 \quad \text{or} \quad x-5=0$$
$$x=3 \qquad x=5$$

3 is extraneous

73.

$$\frac{a+2}{a+1}=\frac{a-4}{a-3}$$

$$(a+1)(a-3)\left(\frac{a+2}{a+1}\right)=(a+1)(a-3)\left(\frac{a-4}{a-3}\right)$$

$$(a-3)(a+2)=(a+1)(a-4)$$

$$a^2+2a-3a-6=a^2-4a+a-4$$

$$a^2-a-6=a^2-3a-4$$

$$-a-6=-3a-4$$

$$2a-6=-4$$

$$2a=2$$

$$a=1$$

75.

$$x^{-1}-3=4x^{-1}$$

$$\frac{1}{x}-3=\frac{4}{x}$$

$$x\left(\frac{1}{x}-3\right)=x\left(\frac{4}{x}\right)$$

$$1-3x=4$$

$$-3x=3$$

$$x=-1$$

77.

$$\frac{5}{y-1}+\frac{3}{y-3}=\frac{8}{y-2}$$

$$(y-1)(y-3)(y-2)\left(\frac{5}{y-1}+\frac{3}{y-3}\right)=(y-1)(y-3)(y-2)\left(\frac{8}{y-2}\right)$$

$$5(y-3)(y-2)+3(y-1)(y-2)=8(y-1)(y-3)$$

$$5(y^2-5y+6)+3(y^2-3y+2)=8(y^2-4y+3)$$

$$5y^2-25y+30+3y^2-9y+6=8y^2-32y+24$$

$$8y^2-34y+36=8y^2-32y+24$$

$$-34y+36=-32y+24$$

$$-2y+36=24$$

$$-2y=-12$$

$$y=6$$

79.

$$\frac{3}{s-2}+\frac{s-14}{2s^2-3s-2}-\frac{4}{2s+1}=0$$

$$\frac{3}{s-2}+\frac{s-14}{(s-2)(2s+1)}-\frac{4}{2s+1}=0$$

Multiply each term by the LCD of $(s-2)(2s+1)$.

$$\frac{3}{s-2}+\frac{s-14}{(s-2)(2s+1)}-\frac{4}{2s+1}=0$$

$$3(2s+1)+(s-14)-4(s-2)=0$$

$$6s+3+s-14-4s+8=0$$

$$3s-3=0$$

$$3s=3$$

$$s=1$$

81.

$$\frac{x}{x+2}=1-\frac{3x+2}{x^2+4x+4}$$

$$\frac{x}{x+2}=1-\frac{3x+2}{(x+2)(x+2)}$$

$$(x+2)^2\left(\frac{x}{x+2}\right)=(x+2)^2\left(1-\frac{3x+2}{(x+2)(x+2)}\right)$$

$$x(x+2)=(x+2)^2-(3x+2)$$

$$x^2+2x=x^2+4x+4-3x-2$$

$$x^2+2x=x^2+x+2$$

$$x^2+2x-x^2=x^2+x+2-x^2$$

$$2x=x+2$$

$$x=2$$

83.

$$3x^{-2}-4x^{-1}+1=0$$

$$\frac{3}{x^2}-\frac{4}{x}+1=0$$

$$x^2\left(\frac{3}{x^2}-\frac{4}{x}+1\right)=x^2(0)$$

$$3-4x+x^2=0$$

$$x^2-4x+3=0$$

$$(x-1)(x-3)=0$$

$$x-1=0 \quad \text{and} \quad x-3=0$$

$$x=1 \qquad\qquad x=3$$

85.

$$\frac{5}{2z^2+z-3}-\frac{2}{2z+3}=\frac{z+1}{z-1}-1$$

$$\frac{5}{(2z+3)(z-1)}-\frac{2}{2z+3}=\frac{z+1}{z-1}-1$$

$$(2z+3)(z-1)\left(\frac{5}{(2z+3)(z-1)}-\frac{2}{2z+3}\right)=(2z+3)(z-1)\left(\frac{z+1}{z-1}-1\right)$$

$$5-2(z-1)=(2z+3)(z+1)-(2z+3)(z-1)$$

$$5-2z+2=2z^2+2z+3z+3-(2z^2-2z+3z-3)$$

$$-2z+7=(2z^2+5z+3)-(2z^2+z-3)$$

$$-2z+7=2z^2+5z+3-2z^2-z+3$$

$$-2z+7=4z+6$$

$$7=6z+6$$

$$1=6z$$

$$\frac{1}{6}=z$$

87.

$$\frac{5}{3x+12}-\frac{1}{9}=\frac{x-1}{3x}$$

$$\frac{5}{3(x+4)}-\frac{1}{9}=\frac{x-1}{3x}$$

$$9x(x+4)\left(\frac{5}{3(x+4)}-\frac{1}{9}\right)=9x(x+4)\left(\frac{x-1}{3x}\right)$$

$$3x(5)-x(x+4)=3(x+4)(x-1)$$

$$15x-x^2-4x=3(x^2-x+4x-4)$$

$$-x^2+11x=3x^2-3x+12x-12$$

$$-x^2+11x+x^2-11x=3x^2-3x+12x-12+x^2-11x$$

$$0=4x^2-2x-12$$

$$\frac{0}{2}=\frac{4x^2}{2}-\frac{2x}{2}-\frac{12}{2}$$

$$0=2x^2-x-6$$

$$0=(2x+3)(x-2)$$

$$2x+3=0 \quad \text{or} \quad x-2=0$$

$$2x=-3 \qquad x=2$$

$$x=-\frac{3}{2}$$

89. a.

$$\frac{11}{12}-\frac{3}{2x}+\frac{4}{x}=\frac{11}{12}\left(\frac{x}{x}\right)-\frac{3}{2x}\left(\frac{6}{6}\right)+\frac{4}{x}\left(\frac{12}{12}\right)$$

$$=\frac{11x}{12x}-\frac{18}{12x}+\frac{48}{12x}$$

$$=\frac{11x-18+48}{12x}$$

$$=\frac{11x+30}{12x}$$

b.

$$\frac{11}{12}-\frac{3}{2x}=\frac{4}{x}$$

$$12x\left(\frac{11}{12}-\frac{3}{2x}\right)=12x\left(\frac{4}{x}\right)$$

$$11x-18=48$$

$$11x=66$$

$$x=6$$

91. a.

$$\frac{m}{m-2}-\frac{1}{m-3}=\frac{m}{m-2}\left(\frac{m-3}{m-3}\right)-\frac{1}{m-3}\left(\frac{m-2}{m-2}\right)$$

$$=\frac{m(m-3)}{(m-2)(m-3)}-\frac{m-2}{(m-2)(m-3)}$$

$$=\frac{m^2-3m-m+2}{(m-2)(m-3)}$$

$$=\frac{m^2-4m+2}{(m-2)(m-3)}$$

b.

$$\frac{m}{m-2}-\frac{1}{m-3}=1$$

$$(m-2)(m-3)\left(\frac{m}{m-2}-\frac{1}{m-3}\right)=(m-2)(m-3)(1)$$

$$m(m-3)-(m-2)=(m-2)(m-3)$$

$$m^2-3m-m+2=m^2-5m+6$$

$$-4m+2=-5m+6$$

$$m+2=6$$

$$m=4$$

APPLICATIONS

93. PHOTOGRAPHY

a.

$$\frac{1}{f}=\frac{1}{s_1}+\frac{1}{s_2}$$

$$fs_1s_2\left(\frac{1}{f}\right)=fs_1s_2\left(\frac{1}{s_1}+\frac{1}{s_2}\right)$$

$$s_1s_2=fs_2+fs_1$$

$$s_1s_2=f\left(s_2+s_1\right)$$

$$\frac{s_1s_2}{s_2+s_1}=f$$

$$f=\frac{s_1s_2}{s_2+s_1}$$

b. Since there are 12 inches in 1 foot, convert 5 ft to inches by multiplying: 5(12) = 60 inches
Let $s_1 = 5$ and $s_2 = 60$.

$$f=\frac{s_1s_2}{s_2+s_1}$$
$$=\frac{5(60)}{60+5}$$
$$=\frac{300}{65}$$
$$=\frac{60}{13}$$
$$=4\frac{8}{13}\text{ in.}$$

95. ACCOUNTING

a.

$$V=C-\left(\frac{C-S}{L}\right)N$$

$$L(V)=L\left(C-\left(\frac{C-S}{L}\right)N\right)$$

$$LV=CL-(C-S)N$$

$$LV=CL-CN+SN$$

$$LV-CL=CL-CN+SN-CL$$

$$LV-CL=SN-CN$$

$$L(V-C)=SN-CN$$

$$\frac{L(V-C)}{V-C}=\frac{SN-CN}{V-C}$$

$$L=\frac{SN-CN}{V-C}$$

b. Find L, if C = \$25,000, N = 4, V = \$13,000 and S = 1,000.

$$L=\frac{SN-CN}{V-C}$$
$$=\frac{(1{,}000)(4)-(25{,}000)(4)}{13{,}000-25{,}000}$$
$$=\frac{4{,}000-100{,}000}{-12{,}000}$$
$$=\frac{-96{,}000}{-12{,}000}$$
$$=8\text{ years}$$

WRITING

97. Answers will vary.

99. Answers will vary.

REVIEW

101. 9.0×10^{9}

103. 4.4×10^{-22}

CHALLENGE PROBLEMS

105.

$$\left(\frac{1}{2}\right)^{-1}=\frac{5b^{-1}}{2}+2b\left(b+1\right)^{-1}$$

$$2=\frac{5}{2b}+\frac{2b}{b+1}$$

$$2b(b+1)(2)=2b(b+1)\left(\frac{5}{2b}+\frac{2b}{b+1}\right)$$

$$4b(b+1)=5(b+1)+2b(2b)$$

$$4b^2+4b=5b+5+4b^2$$

$$4b^2+4b-4b^2=5b+5+4b^2-4b^2$$

$$4b=5b+5$$

$$-b=5$$

$$b=-5$$

107. $f(x)=\dfrac{x^3-3x^2+12}{x}$

$$4=\frac{x^3-3x^2+12}{x}$$

$$x\cdot 4=x\cdot\frac{x^3-3x^2+12}{x}$$

$$4x=x^3-3x^2+12$$

$$0=x^3-3x^2-4x+12$$

$$0=x^2(x-3)-4(x-3)$$

$$0=(x-3)(x^2-4)$$

$$0=(x-3)(x+2)(x-2)$$

$x-3=0 \quad x+2=0 \quad x-2=0$

$x=3 \quad x=-2 \quad x=2$

109. $f(x)=\dfrac{2x^3+x^2}{98x+49}$

$$1=\frac{2x^3+x^2}{98x+49}$$

$$(98x+49)(1)=(98x+49)\left(\frac{2x^3+x^2}{98x+49}\right)$$

$$98x+49=2x^3+x^2$$

$$0=2x^3+x^2-98x-49$$

$$0=x^2(2x+1)-49(2x+1)$$

$$0=(2x+1)(x^2-49)$$

$$0=(2x+1)(x-7)(x+7)$$

$2x+1=0 \quad x-7=0 \quad x+7=0$

$2x=-1 \quad x=7 \quad x=-7$

$x=-\dfrac{1}{2}$

SECTION 6.8

VOCABULARY

1. In this section, we call problems that involve:
   - people or machines completing jobs, shared-**work** problems.
   - moving vehicles, uniform **motion** problems.

CONCEPTS

3. If a job can be completed in x hours, then the rate of work can be expressed as $\frac{1}{x}$ of the job is completed per hour.

5.

	rate	· time =	work completed
1st crew	$\frac{1}{15}$	x	$\frac{x}{15}$
2nd crew	$\frac{1}{8}$	x	$\frac{x}{8}$

7.

	r	$\cdot\ t$	$= d$
Running	x	$\frac{12}{x}$	12
Bicycling	$x + 15$	$\frac{12}{x+15}$	12

NOTATION

9. $\frac{41}{9} = 4\frac{5}{9}$ hr

APPLICATIONS

11. ROOFING HOUSES

Let x = the number of days it takes the homeowner and the professional to roof the house working together.

Homeowner's work in 1 day $= \frac{1}{7}$

Professional's work in 1 day $= \frac{1}{4}$

Amount of work together in 1 day $= \frac{1}{x}$

$$\frac{1}{7} + \frac{1}{4} = \frac{1}{x}$$

$$28x\left(\frac{1}{7} + \frac{1}{4}\right) = 28x\left(\frac{1}{x}\right)$$

$$4x + 7x = 28$$

$$11x = 28$$

$$x = \frac{28}{11} = 2\frac{6}{11} \text{ days}$$

13. HOUSEPAINTING

a. Let x = the number of days it takes both painters to paint the house working together.

Santos' work in 1 day $= \frac{1}{3}$

Mays' work in 1 day $= \frac{1}{5}$

Work done together in 1 day $= \frac{1}{x}$

Santos' work + Mays' work = work together

$$\frac{1}{3} + \frac{1}{5} = \frac{1}{x}$$

$$15x\left(\frac{1}{3} + \frac{1}{5}\right) = 15x\left(\frac{1}{x}\right)$$

$$5x + 3x = 15$$

$$8x = 15$$

$$x = \frac{15}{8}$$

$$x = 1\frac{7}{8} \text{ days}$$

It will take them $1\frac{7}{8}$ day to paint the house.

b. Santos:

$$1\frac{7}{8}(220) = \$412.50$$

Mays:

$$1\frac{7}{8}(200) = \$375.00$$

15. FARMING

Let x = the number of minutes it takes the belts to move the corn working together.

Conveyor belt's work in 1 minute = $\frac{1}{10}$

Smaller belt's work in 1 minute = $\frac{1}{14}$

Amount of work together in 1 minute = $\frac{1}{x}$

$$\frac{1}{10}+\frac{1}{14}=\frac{1}{x}$$
$$70x\left(\frac{1}{10}+\frac{1}{14}\right)=70x\left(\frac{1}{x}\right)$$
$$7x+5x=70$$
$$12x=70$$
$$x=\frac{70}{12}$$
$$x=5\frac{10}{12}$$
$$x=5\frac{5}{6}\text{ mins}$$

17. THRILL RIDES

Let x = the number of seconds it take the three pies to fill the pool working together.

1st pipe's work in 1 second = $\frac{1}{10}$

2nd pipe's work in 1 second = $\frac{1}{15}$

3rd pipe's work in 1 second = $\frac{1}{20}$

Amount of work together in 1 second = $\frac{1}{x}$

$$\frac{1}{10}+\frac{1}{15}+\frac{1}{20}=\frac{1}{x}$$
$$60x\left(\frac{1}{10}+\frac{1}{15}+\frac{1}{20}\right)=60x\left(\frac{1}{x}\right)$$
$$6x+4x+3x=60$$
$$13x=60$$
$$x=\frac{60}{13}=4\frac{8}{13}\text{ seconds}$$

19. FILLING PONDS

Let x be the number of weeks it takes both pipes and the seepage/removal to fill the pool working together. Since the seepage is EMPTYING the pool, subtract the work of the seepage from the sum of the work of the pipes.

Amount of work 1st pipe in 1 week = $\frac{1}{3}$

Amount of work 2nd pipe in 1 week = $\frac{1}{5}$

Amount of work removal in 1 week = $\frac{1}{10}$

Amount of work together in 1 week = $\frac{1}{x}$

$$\frac{1}{3}+\frac{1}{5}-\frac{1}{10}=\frac{1}{x}$$
$$30x\left(\frac{1}{3}+\frac{1}{5}-\frac{1}{10}\right)=30x\left(\frac{1}{x}\right)$$
$$10x+6x-3x=30$$
$$13x=30$$
$$x=\frac{30}{13}$$
$$x=2\frac{4}{13}\text{ weeks}$$

21. FINE DINING

Let x = the waiter's time. Then the bus boy's time is $x + 5$. Their time together is 6 minutes.

Waiter's work in 1 minute = $\frac{1}{x}$

Bus boy's work in 1 minute = $\frac{1}{x+5}$

Amount of work together in 1 minute = $\frac{1}{6}$

$$\frac{1}{x}+\frac{1}{x+5}=\frac{1}{6}$$
$$6x(x+5)\left(\frac{1}{x}+\frac{1}{x+5}\right)=6x(x+5)\left(\frac{1}{6}\right)$$
$$6(x+5)+6x=x(x+5)$$
$$6x+30+6x=x^2+5x$$
$$12x+30=x^2+5x$$
$$0=x^2-7x-30$$
$$0=(x-10)(x+3)$$
$$x-10=0 \quad\text{or}\quad x+3=0$$
$$x=10 \qquad x=\cancel{-3}$$

Waiter's time = 10 minutes

Bus boy's time = 10 + 5 = 15 minutes

23. FUND-RAISING LETTERS

Let x = number of hours it takes the faster worker and $x + 6$ is the time it takes the slower worker.

Faster worker's work in 1 hour = $\frac{1}{x}$

Slower worker's work in 1 hour = $\frac{1}{x+6}$

Amount of work together in 1 minute = $\frac{1}{4}$

$$\frac{1}{x}+\frac{1}{x+6}=\frac{1}{4}$$
$$4x(x+6)\left(\frac{1}{x}+\frac{1}{x+6}\right)=4x(x+6)\left(\frac{1}{4}\right)$$
$$4(x+6)+4x=x(x+6)$$
$$4x+24+4x=x^2+6x$$
$$8x+24=x^2+6x$$
$$0=x^2-2x-24$$
$$0=(x-6)(x+4)$$
$$x-6=0 \quad \text{or} \quad x+4=0$$
$$x=6 \qquad x=\cancel{-4}$$

Faster worker's time is 6 hours.
Slower worker's time is 6 + 6 = 12 hours.

25. WEBMASTER

Let x = the time it takes for Dell PowerEdge to send emails. Then the time for Cisco Systems would be $x - 1.5$.

Dell's work in 1 hour = $\frac{1}{x}$

Cisco's work in 1 hour = $\frac{1}{x-1.5}$

Amount of work together in 1 hour = $\frac{1}{1.8}$

$$\frac{1}{x}+\frac{1}{x-1.5}=\frac{1}{1.8}$$
$$1.8x(x-1.5)\left(\frac{1}{x}+\frac{1}{x-1.5}\right)=1.8x(x-1.5)\left(\frac{1}{1.8}\right)$$
$$1.8(x-1.5)+1.8x=x(x-1.5)$$
$$1.8x-2.7+1.8x=x^2-1.5x$$
$$0=x^2-5.1x+2.7$$
$$0=10x^2-51x+27$$
$$0=(2x-9)(5x-3)$$
$$2x-9=0 \quad \text{and} \quad 5x-3=0$$
$$2x=9 \qquad 5x=3$$
$$x=4.5 \qquad x=\cancel{0.6}$$

Dell's time is 4.5 hours.
Cisco's time is 4.5 – 1.5 = 3 hours.

27. DETAILING A CAR

Let x be the number of hours it takes the son to wash and wax the car.

Dad's work in 1 hour = $\frac{1}{3}$

Son's work in 1 hour = $\frac{1}{x}$

Amount of work together in 1 hour = $\frac{1}{1}=1$

$$\frac{1}{3}+\frac{1}{x}=1$$
$$3x\left(\frac{1}{3}+\frac{1}{x}\right)=3x(1)$$
$$x+3=3x$$
$$3=2x$$
$$\frac{3}{2}=x$$
$$x=1\frac{1}{2}\text{ hours}$$

29. OYSTERS

Change 8½ minutes to seconds by multiply by 60 since there are 60 seconds in one minute.

$$\left(8\frac{1}{2}\right)(60)=510\text{ seconds}$$

Let x = number of seconds they can open 100 oysters together.

Racz's time in one second = $\frac{1}{140}$

Novice's time in one second = $\frac{1}{510}$

Time together is one second = $\frac{1}{x}$

$$\frac{1}{140}+\frac{1}{510}=\frac{1}{x}$$
$$7{,}140x\left(\frac{1}{140}+\frac{1}{510}\right)=7{,}140x\left(\frac{1}{x}\right)$$
$$51x+14x=7{,}140$$
$$65x=7{,}140$$
$$x=\frac{7{,}140}{65}$$
$$x\approx 110\text{ seconds}$$

31. TRUCK DELIVERIES

Let x = the rate going.
Then $x - 10$ = rate coming back.

	distance	rate	time
Going	75	x	$t_1 = \frac{d}{r} = \frac{75}{x}$
Coming	75	$x - 10$	$t_1 = \frac{d}{r} = \frac{75}{x-10}$

The time for the trip going + 2 hour equals the time for the return trip.

$$t_1 + 1 = t_2$$

$$\frac{75}{x} + 2 = \frac{75}{x-10}$$

$$x(x-10)\left(\frac{75}{x} + 2\right) = x(x-10)\left(\frac{75}{x-10}\right)$$

$$75(x-10) + 2x(x-10) = 75x$$

$$75x - 750 + 2x^2 - 20x = 75x$$

$$2x^2 + 55x - 750 = 75x$$

$$2x^2 - 20x - 750 = 0$$

$$x^2 - 10x - 375 = 0$$

$$(x-25)(x+15) = 0$$

$$x - 25 = 0 \quad \text{or} \quad x + 15 = 0$$

$$x = 25 \qquad x = \cancel{-15}$$

$$x - 10 = 15$$

Since the rates cannot be negative, the rate going must be 25 mph and the rate on the return trip is 15 mph.

33. TRAIN TRAVEL

Let x = the rate of the empty train.
Then $x - 25$ = rate of the full train.

	distance	rate	time
Empty Train	60	x	$t_1 = \frac{d}{r} = \frac{60}{x}$
Full Train	60	$x - 25$	$t_1 = \frac{d}{r} = \frac{60}{x-25}$

Since the total time for the trip is 5.5, add the times and set the sum equal to 5.5.

$$t_1 + t_2 = 5.5$$

$$\frac{60}{x} + \frac{60}{x-25} = 5.5$$

$$x(x-25)\left(\frac{60}{x} + \frac{60}{x-25}\right) = x(x-25)(5.5)$$

$$60(x-25) + 60x = 5.5x(x-25)$$

$$60x - 1{,}500 + 60x = 5.5x^2 - 137.5x$$

$$120x - 1{,}500 = 5.5x^2 - 137.5x$$

$$0 = 5.5x^2 - 257.5x + 1{,}500$$

$$0 = 10\left(5.5x^2 - 257.5x + 1{,}500\right)$$

$$0 = 55x^2 - 2575x + 15{,}000$$

$$\frac{0}{5} = \frac{55x^2}{5} - \frac{2575x}{5} + \frac{15{,}000}{5}$$

$$0 = 11x^2 - 515x + 3{,}000$$

$$0 = (11x - 75)(x - 40)$$

$$11x - 75 = 0 \quad \text{or} \quad x - 40 = 0$$

$$x \approx 6.82 \qquad x = 40$$

$$x - 25 = -\cancel{18.18} \qquad x - 25 = 15$$

Since the rates cannot be negative, the rates for the train must be 40 mph when empty and 15 mph when full.

35. RATES OF SPEED

Let x = the rate of the first train.
Then $x + 10$ = rate of the second train.

$$r = \frac{d}{t}$$

	distance	rate	time
Train 1	315	x	$t_1 + 2$
Train 2	315	$x + 10$	t_2

Solve for t for both trains:

Train 1:

$$t_1 + 2 = \frac{315}{x}$$

$$t_1 = \frac{315}{x} - 2$$

Train 2:

$$t_2 = \frac{315}{x+10}$$

Set the times equal and solve for x.

$$\frac{315}{x+10} = \frac{315}{x} - 2$$

$$x(x+10)\left(\frac{315}{x+10}\right) = x(x+10)\left(\frac{315}{x} - 2\right)$$

$$315x = 315(x+10) - 2x(x+10)$$

$$315x = 315x + 3{,}150 - 2x^2 - 20x$$

$$315x - 315x = 315x + 3{,}150 - 2x^2 - 20x - 315x$$

$$0 = -2x^2 - 20x + 3{,}150$$

$$2x^2 + 20x - 3{,}150 = 0$$

$$\frac{2x^2}{2} + \frac{20x}{2} - \frac{3{,}150}{2} = \frac{0}{2}$$

$$x^2 + 10x - 1{,}575 = 0$$

$$(x-35)(x+45) = 0$$

$$x - 35 = 0 \quad \text{or} \quad x + 45 = 0$$

$$x = 35 \qquad\qquad x = -45$$

$$x + 10 = 45$$

The first train traveled 35 mph and the second train traveled 45 mph.

37. COMPARING TRAVEL

Let x = the rate of the car.
Then $x + 90$ = rate of the plane.

	distance	rate	time
plane	600	$x + 90$	$t = \frac{d}{r} = \frac{600}{x+90}$
car	240	x	$t = \frac{d}{r} = \frac{240}{x}$

Set the times equal and solve for x.

$$\frac{600}{x+90} = \frac{240}{x}$$

$$x(x+90)\left(\frac{600}{x+90}\right) = x(x+90)\left(\frac{240}{x}\right)$$

$$600x = 240(x+90)$$

$$600x = 240x + 21{,}600$$

$$360x = 21{,}600$$

$$x = 60$$

$$x + 90 = 150$$

The rate of the car is 60 mph and the rate of the plane is 150 mph.

39. BOATING

Going downstream, the rate is the speed of the boat in still water, 6, plus the speed of the current, c. Going upstream, the rate is the speed of the boat in still water, 6, minus the speed of the current, c.

	distance	rate	time
downstream	16	$6+c$	$t=\frac{d}{r}=\frac{16}{6+c}$
upstream	16	$6-c$	$t=\frac{d}{r}=\frac{16}{6-c}$

The time going downstream plus the time going upstream equals the total time of 6 hours.

$$\frac{16}{6+c}+\frac{16}{6-c}=6$$

$$(6+c)(6-c)\left(\frac{16}{6+c}+\frac{16}{6-c}\right)=6(6+c)(6-c)$$

$$16(6-c)+16(6+c)=6(36-c^2)$$

$$96-16c+96+16c=216-6c^2$$

$$192=216-6c^2$$

$$6c^2-24=0$$

$$6(c^2-4)=0$$

$$6(c+2)(c-2)=0$$

$$c+2=0 \text{ or } c-2=0$$

$$c=\not{-2} \qquad c=2$$

The rate of the current is 2 mph.

41. BOATING

Going downstream, the rate is the speed of the boat in still water, 12, plus the speed of the current, c. Going upstream, the rate is the speed of the boat in still water, 12, minus the speed of the current, c.

	distance	rate	time
downstream	45	$12+c$	$t=\frac{d}{r}$ $=\frac{45}{12+c}$
upstream	27	$12-c$	$t=\frac{d}{r}$ $=\frac{27}{12-c}$

Since the time is the same for both trips, set the time for both trips equal and solve for the rate of the current, c.

$$\frac{45}{12+c}=\frac{27}{12-c}$$

$$(12+c)(12-c)\left(\frac{45}{12+c}\right)=(12+c)(12-c)\left(\frac{27}{12-c}\right)$$

$$45(12-c)=27(12+c)$$

$$540-45c=324+27c$$

$$540=324+72c$$

$$216=72c$$

$$3=c$$

The rate of the current is 3 mph.

43. AIRPORT WALKWAYS

With the walkway, the rate is the speed of the walkway, 1.5, plus the man's normal walking rate, x. Going against the walkway, the rate is the speed of the walkway, 1.5, minus the man's normal walking rate, x.

	distance	rate	time
On the walkway	65	$x+1.5$	$t=\frac{d}{r}=\frac{65}{x+1.5}$
Against the Walkway	35	$x-1.5$	$t=\frac{d}{r}=\frac{35}{x-1.5}$

Since the time is the same for both, set the time for both trips equal and solve for the rate of the wind, x.

$$\frac{65}{x+1.5}=\frac{35}{x-1.5}$$
$$(x+1.5)(x-1.5)\left(\frac{65}{x+1.5}\right)=(x+1.5)(x-1.5)\left(\frac{35}{x-1.5}\right)$$
$$65(x-1.5)=35(x+1.5)$$
$$65x-97.5=35x+52.5$$
$$65x-97.5-35x=35x+52.5-35x$$
$$30x-97.5=52.5$$
$$30x-97.5+97.5=52.5+97.5$$
$$30x=150$$
$$x=5$$

The man's rate of walking is 5 mph.

WRITING

45. Answers will vary.

REVIEW

47.

$$\left(\frac{m^{10}}{n}\right)^8=\frac{m^{10\cdot 8}}{n^{1\cdot 8}}$$
$$=\frac{m^{80}}{n^8}$$

49.

$$-w^{-2}=-\frac{1}{w^2}$$

51.

$$-\frac{4x^{-9}\cdot x^{-3}}{x^{-12}}=-\frac{4x^{-9+(-3)}}{x^{-12}}$$
$$=-\frac{4x^{-12}}{x^{-12}}$$
$$=-4x^{-12-(-12)}$$
$$=-4x^0$$
$$=-4(1)$$
$$=-4$$

53.

$$\left(-x^2\right)^5y^7y^3x^{-2}y^0=-x^{10}y^7y^3x^{-2}y^0$$
$$=-x^{10+(-2)}y^{7+3+0}$$
$$=-x^8y^{10}$$

CHALLENGE PROBLEMS

55. FIREPLACES

Mason's work in 1 hour = $\frac{1}{18}$

Assistant's work in 1 hour = $\frac{1}{x}$

$$6\left(\frac{1}{18}+\frac{1}{x}\right)+10\left(\frac{1}{x}\right)=1$$
$$\frac{6}{18}+\frac{6}{x}+\frac{10}{x}=1$$
$$\frac{1}{3}+\frac{6}{x}+\frac{10}{x}=1$$
$$3x\left(\frac{1}{3}+\frac{6}{x}+\frac{10}{x}\right)=3x(1)$$
$$x+18+30=3x$$
$$x+48=3x$$
$$48=2x$$
$$24=x$$

It takes the assistant 24 hours to construct a fireplace working alone.

VOCABULARY

1. A **ratio** is the quotient of two numbers or two quantities with the same units.

3. In $\frac{50}{3}=\frac{x}{9}$, the terms 50 and 9 are called the **extremes** and the terms 3 and x are called the **means**. In a proportion, the product of the **extremes** is equal to the product of the **means**.

5. If two angles of one triangle have the same measure as two angles of a second triangle, the triangles are **similar**.

7. The equation $y=\frac{k}{x}$ defines **inverse** variation: As x increases, y **decreases**.

CONCEPTS

9. a. direct
   b. inverse

11. direct

13. inverse

15. direct

17. inverse

NOTATION

19.
$$\frac{7}{6}=\frac{x+3}{12}$$
$$\boxed{7}(12)=\boxed{6}(x+3)$$
$$84=6x+\boxed{18}$$
$$\boxed{66}=6x$$
$$\boxed{11}=x$$

GUIDED PRACTICE

21.
$$\frac{x}{5}=\frac{15}{25}$$
$$x(25)=15(5)$$
$$25x=75$$
$$x=3$$

23.
$$\frac{r-2}{3}=\frac{r}{5}$$
$$5(r-2)=3(r)$$
$$5r-10=3r$$
$$-10=-2r$$
$$5=r$$

25.
$$\frac{5}{5z+3}=\frac{3}{2z+6}$$
$$5(2z+6)=3(5z+3)$$
$$10z+30=15z+9$$
$$21=5z$$
$$\frac{21}{5}=z$$

27.
$$\frac{x-2}{x}=\frac{x+1}{x+2}$$
$$(x+2)(x-2)=x(x+1)$$
$$x^2-4=x^2+x$$
$$x^2-4-x^2=x^2+x-x^2$$
$$-4=x$$

29.
$$\frac{2}{3x}=\frac{6x}{36}$$
$$2(36)=6x(3x)$$
$$72=18x^2$$
$$0=18x^2-72$$
$$0=18(x^2-4)$$
$$0=18(x+2)(x-2)$$
$$x+2=0 \quad \text{or} \quad x-2=0$$
$$x=-2 \qquad x=2$$

31.

$$\frac{2}{c}=\frac{c-3}{2}$$
$$2(2)=c(c-3)$$
$$4=c^2-3c$$
$$0=c^2-3c-4$$
$$0=(c-4)(c+1)$$
$$c-4=0 \quad \text{or} \quad c+1=0$$
$$c=4 \qquad c=-1$$

33.

$$\frac{1}{x+3}=\frac{-2x}{x+5}$$
$$1(x+5)=-2x(x+3)$$
$$x+5=-2x^2-6x$$
$$2x^2+6x+x+5=0$$
$$2x^2+7x+5=0$$
$$(2x+5)(x+1)=0$$
$$2x+5=0 \quad \text{or} \quad x+1=0$$
$$2x=-5 \qquad x=-1$$
$$x=-\frac{5}{2}$$

35.

$$\frac{2b}{b+5}=\frac{-b}{3b+8}$$
$$2b(3b+8)=-b(b+5)$$
$$6b^2+16b=-b^2-5b$$
$$7b^2+21b=0$$
$$7b(b+3)=0$$
$$7b=0 \quad \text{or} \quad b+3=0$$
$$b=0 \qquad b=-3$$

37. $A=kp^2$

39. $z=\frac{k}{t^3}$

41. $C=kxyz$

43. $P=\frac{ka^2}{j^3}$

45. r varies directly as t

47. b varies inversely as h

49. U varies jointly as r, the square of s, and t.

51. P varies directly as m and inversely as n.

TRY IT YOURSELF

53.

$$\frac{b+4}{5}=\frac{3b-6}{3}$$
$$3(b+4)=5(3b-6)$$
$$3b+12=15b-30$$
$$12=12b-30$$
$$42=12b$$
$$\frac{42}{12}=\frac{12b}{12}$$
$$\frac{7}{2}=b$$

55.

$$\frac{5}{b+3}=\frac{b}{2}$$
$$5(2)=b(b+3)$$
$$10=b^2+3b$$
$$0=b^2+3b-10$$
$$0=(b+5)(b-2)$$
$$b+5=0 \quad \text{or} \quad b-2=0$$
$$b=-5 \qquad b=2$$

57.

$$\frac{9z+6}{z^2+3z}=\frac{7}{z+3}$$
$$(9z+6)(z+3)=7(z^2+3z)$$
$$9z^2+27z+6z+18=7z^2+21z$$
$$9z^2+33z+18=7z^2+21z$$
$$9z^2+33z+18-7z^2-21z=0$$
$$2z^2+12z+18=0$$
$$2(z^2+6z+9)=0$$
$$2(z+3)(z+3)=0$$
$$z+3=0 \quad \text{or} \quad z+3=0$$
$$z=-3 \qquad z=-3$$

Since $z=-3$ makes the denominator 0, there is No Solution to this problem.

59.

$$\frac{h^2}{5} = \frac{h}{2h-9}$$
$$h^2(2h-9) = 5(h)$$
$$2h^3 - 9h^2 = 5h$$
$$2h^3 - 9h^2 - 5h = 0$$
$$h(2h^2 - 9h - 5) = 0$$
$$h(2h+1)(h-5) = 0$$
$$h = 0, \quad 2h+1 = 0, \quad \text{or} \quad h - 5 = 0$$
$$2h = -1 \qquad h = 5$$
$$h = -\frac{1}{2}$$

61.

$$\frac{x}{x+2} = \frac{6}{x+2}$$
$$x(x+2) = 6(x+2)$$
$$x^2 + 2x = 6x + 12$$
$$x^2 + 2x - 6x - 12 = 0$$
$$x^2 - 4x - 12 = 0$$
$$(x-6)(x+2) = 0$$
$$x - 6 = 0 \quad \text{or} \quad x + 2 = 0$$
$$x = 6 \qquad x = -2$$

63.

$$\frac{t^2-1}{5} = \frac{1-t^2}{2t}$$
$$2t(t^2-1) = 5(1-t^2)$$
$$2t^3 - 2t = 5 - 5t^2$$
$$2t^3 - 2t - 5 + 5t^2 = 0$$
$$2t^3 + 5t^2 - 2t - 5 = 0$$
$$(2t^3 + 5t^2) + (-2t - 5) = 0$$
$$t^2(2t+5) - 1(2t+5) = 0$$
$$(2t+5)(t^2-1) = 0$$
$$(2t+5)(t+1)(t-1) = 0$$
$$2t + 5 = 0, \quad t + 1 = 0, \quad \text{or} \quad t - 1 = 0$$
$$2t = -5 \qquad t = -1 \qquad t = 1$$
$$t = -\frac{5}{2}$$

65.

$$\frac{2.5x+1}{2} = \frac{4.5}{12}$$
$$12(2.5x+1) = 2(4.5)$$
$$30x + 12 = 9$$
$$30x = -3$$
$$x = -0.1$$

67.

$$\frac{t}{10} = \frac{10}{t}$$
$$t(t) = 10(10)$$
$$t^2 = 100$$
$$t^2 - 100 = 0$$
$$(t-10)(t+10) = 0$$
$$t - 10 = 0 \quad \text{or} \quad t + 10 = 0$$
$$t = 10 \qquad t = -10$$

APPLICATIONS

69. CAFFEINE

Mountain Dew:

$$\frac{44 \text{ oz}}{12 \text{ oz}} = \frac{x \text{ mg caffeine}}{55 \text{ mg caffeine}}$$
$$44(55) = x(12)$$
$$2{,}420 = 12x$$
$$202 \approx x$$

Pepsi:

$$\frac{44 \text{ oz}}{12 \text{ oz}} = \frac{x \text{ mg caffeine}}{38 \text{ mg caffeine}}$$
$$44(38) = x(12)$$
$$1{,}672 = 12x$$
$$139 \approx x$$

Coca-Cola Classic:

$$\frac{44 \text{ oz}}{12 \text{ oz}} = \frac{x \text{ mg caffeine}}{34 \text{ mg caffeine}}$$
$$44(34) = x(12)$$
$$1{,}496 = 12x$$
$$125 \approx x$$

71. WALLPAPERING

$$\frac{x \text{ gal. of adhesive}}{500 \text{ sq. ft}} = \frac{0.5 \text{ gal. of adhesive}}{140 \text{ sq. ft}}$$

$$140x = 0.5(500)$$
$$140x = 250$$
$$x = 1.786$$
$$x \approx 2$$

He will need about 2 gallons

73. ERGONOMICS

Change 5 feet and 11 inches to inches

5(12) + 11 = 71 inches

Eye:

$$\frac{48.5}{69} = \frac{x}{71}$$
$$48.5(71) = x(69)$$
$$3443.5 = 69x$$
$$49.9 = x$$

Seat:

$$\frac{17.1}{69} = \frac{x}{71}$$
$$17.1(71) = x(69)$$
$$1,214.1 = 69x$$
$$17.6 = x$$

Elbow:

$$\frac{27.0}{69} = \frac{x}{71}$$
$$27.0(71) = x(69)$$
$$1,917 = 69x$$
$$27.8 = x$$

The heights should be 49.9 inches for the eye, 17.6 inches for the seat, and 27.8 inches for the elbow.

75. DRAWING

$$\frac{3}{5} = \frac{7.5}{x}$$
$$3(x) = 7.5(5)$$
$$3x = 37.5$$
$$x = 12.5$$

The length is 12.5 in.

77. FLAGPOLES

$$\frac{5}{6} = \frac{h}{30}$$
$$5(30) = h(6)$$
$$150 = 6h$$
$$25 = h$$

The flagpole is 25 ft. tall.

79. TOWERS

Change the length of the shadows to inches.

$$\frac{6 \text{ ft}}{3 \text{ ft } 9 \text{ in.}} = \frac{h}{75 \text{ ft}}$$
$$\frac{6 \text{ ft}}{45 \text{ in.}} = \frac{h}{900 \text{ in.}}$$
$$6(900) = 45h$$
$$5,400 = 45h$$
$$120 = h$$

The tower is 120 ft. tall.

81. FLIGHT PATHS

$$\frac{1,000}{150} = \frac{5,280}{x}$$
$$1,000(x) = 5,280(150)$$
$$1,000x = 792,000$$
$$x = 792 \text{ ft}$$

It will gain 792 ft.

83. WEBMASTER

a. false

b. false

c. true

85. GRAVITY

Let m = mass and F = force.

Find k if $m = 5$ and $F = 49$.

$$F = km$$
$$49 = k(5)$$
$$49 = 5k$$
$$9.8 = k$$

Find F if $k = 9.8$ and $t = 12$.

$$F = km$$
$$F = (9.8)(12)$$
$$F = 117.6 \text{ newtons}$$

The force is 117.6 newtons.

87. FINDING DISTANCE

Let d = distance driven and g = gallons of gasoline.

Find k if d = 288 and g = 12.

$$d = kg$$
$$288 = k(12)$$
$$288 = 12k$$
$$24 = k$$

Find d if k = 24 and g = 18.

$$d = kg$$
$$d = 24(18)$$
$$d = 432$$

The car can go 432 miles.

89. FARMING

Let d = days feed will last and a = number of animals.

Find k if d = 10 and a = 25.

$$d = \frac{k}{a}$$
$$10 = \frac{k}{25}$$
$$25(10) = 25\left(\frac{k}{25}\right)$$
$$250 = k$$

Find d if k = 250 and a = 10.

$$d = \frac{k}{a}$$
$$d = \frac{250}{10}$$
$$d = 25$$

The feed will last 25 days.

91. GAS PRESSURE

Let v = volume of gas and p = pressure.

Find k when v = 20 and p = 6.

$$v = \frac{k}{p}$$
$$20 = \frac{k}{6}$$
$$6(20) = 6\left(\frac{k}{6}\right)$$
$$120 = k$$

Find v when k = 120 and p = 10.

$$v = \frac{k}{p}$$
$$v = \frac{120}{10}$$
$$v = 12 \text{ in.}^3$$

The pressure would be 12 in.^3

93. TRUCKING COSTS

Let c = costs, t = # of trucks, and h = # of hours.

Find k if c = 1,800, t = 4 and h = 6.

$$c = thk$$
$$1{,}800 = (4)(6)k$$
$$1{,}800 = 24k$$
$$75 = k$$

Find c if k = 75, t = 10, and h = 12.

$$c = thk$$
$$c = (10)(12)(75)$$
$$c = 120(75)$$
$$c = 9{,}000$$

The cost is $9,000.

95. ELECTRONICS

Let v = voltage, r = resistance, and c = current.

$$v = rc$$
$$6 = r \bullet 2$$
$$6 = 2r$$
$$3 = r$$
$$r = 3 \text{ ohms}$$

97. STRUCTURAL ENGINEERING

Let F = force of deflection, w = width, d = depth, and k = constant. Find k using the given information that $F = 1.1$, $w = 4$, and $d = 4$.

$$F = \frac{k}{wd^3}$$

$$1.1 = \frac{k}{4(4)^3}$$

$$1.1 = \frac{k}{4(64)}$$

$$1.1 = \frac{k}{256}$$

$$256(1.1) = 256\left(\frac{k}{256}\right)$$

$$281.6 = k$$

Let $k = 281.6$, $w = 2$, and $d = 8$ to find the force of deflection.

$$F = \frac{k}{wd^3}$$

$$= \frac{281.6}{2(8)^3}$$

$$= \frac{281.6}{2(512)}$$

$$= \frac{281.6}{1{,}024}$$

$$= 0.275 \text{ in.}$$

99. ELECTRONICS

Let R = resistance, l = length, and d = diameter of the wire, and k = constant. Find k using the given information that $R = 11.2$, $l = 80$, and $d = 0.01$.

$$R = \frac{kl}{d^2}$$

$$11.2 = \frac{80k}{(0.01)^2}$$

$$11.2 = \frac{80k}{0.0001}$$

$$0.00112 = 80k$$

$$0.000014 = k$$

Let $k = 0.000014$, $l = 160$, and $d = 0.04$ to find the resistance.

$$R = \frac{kl}{d^2}$$

$$R = \frac{160(0.000014)}{(0.04)^2}$$

$$R = \frac{0.00224}{0.0016}$$

$$R = 1.4$$

101. TENSION IN A STRING

Let T = tension of the string, s = speed of the ball, and r = radius of the circle, and k = constant. Find k.

$$T = \frac{ks^2}{r}$$

$$6 = \frac{k \cdot (6)^2}{3}$$

$$6 = \frac{k \cdot 36}{3}$$

$$6 = 12k$$

$$0.5 = k$$

Find T given $k = 0.5$, $s = 8$ and $r = 2.5$.

$$T = \frac{ks^2}{r}$$

$$= \frac{0.5 \cdot (8)^2}{2.5}$$

$$= \frac{0.5(64)}{2.5}$$

$$= \frac{32}{2.5}$$

$$= 12.8 \text{ lb}$$

WRITING

103. Answers will vary.

REVIEW

105.

$$\left(\frac{5}{2}w^3+\frac{1}{4}w^2+\frac{3}{5}\right)-\left(\frac{1}{3}w^3+\frac{1}{2}w^2-\frac{1}{5}\right)$$
$$=\frac{5}{2}w^3+\frac{1}{4}w^2+\frac{3}{5}-\frac{1}{3}w^3-\frac{1}{2}w^2+\frac{1}{5}$$
$$=\frac{13}{6}w^3-\frac{1}{4}w^2+\frac{4}{5}$$

107.

$$(3y+1)(2y^2+3y+2)$$
$$=3y(2y^2+3y+2)+1(2y^2+3y+2)$$
$$=6y^3+9y^2+6y+2y^2+3y+2$$
$$=6y^3+11y^2+9y+2$$

CHALLENGE PROBLEMS

109. Answers will vary.

CHAPTER 6 REVIEW

SECTION 6.1
Rational Functions and Simplifying Rational Expressions

1. a. $f(2) = 1$

   b. $f(-1) = -\frac{1}{2}$

   c. $x = 0$

2. To find the domain, set the denominator equal to 0 to determine what values make the rational expression undefined.

$$x^2 + 2x - 24 = 0$$
$$(x+6)(x-4) = 0$$
$$x+6=0 \quad \text{or} \quad x-4=0$$
$$x=-6 \qquad x=4$$

Since $x \neq -6$ and $x \neq 4$, the domain is $(-\infty,-6)\cup(-6,4)\cup(4,\infty)$.

3. $n(t) = \frac{28t}{t^2+1}$

$$n(3) = \frac{28(3)}{(3)^2+1} = \frac{84}{9+1} = \frac{84}{10} = 8.4$$

Three hours after the injection, the concentration of pain medication in the patient's bloodstream was 8.4 milligrams per liter.

4. $y = 3$ and $x = 0$ are the asymptotes.
D = $(-\infty,0)\cup(0,\infty)$; R = $(-\infty,3)\cup(3,\infty)$

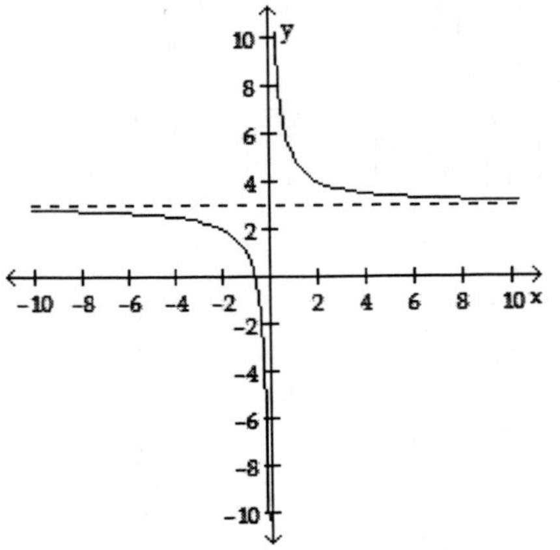

5.

$$\frac{48x^2y}{76xy^8} = \frac{\cancel{2}\cdot\cancel{2}\cdot 2\cdot 2\cdot 3\cdot \cancel{x}\cdot x\cdot \cancel{y}}{\cancel{2}\cdot\cancel{2}\cdot 19\cdot \cancel{x}\cdot \cancel{y}\cdot y\cdot y\cdot y\cdot y\cdot y\cdot y\cdot y} = \frac{12x}{19y^7}$$

6.

$$\frac{x^2-49}{x^2+14x+49} = \frac{\cancel{(x+7)}(x-7)}{\cancel{(x+7)}(x+7)} = \frac{x-7}{x+7}$$

7.

$$\frac{x^2-2x+4}{2x^5+16x^2} = \frac{x^2-2x+4}{2x^2(x^3+8)} = \frac{\cancel{x^2-2x+4}}{2x^2(x+2)\cancel{(x^2-2x+4)}} = \frac{1}{2x^2(x+2)}$$

8.

$$\frac{x^2+6x+36}{x^7-216x^4} = \frac{\cancel{x^2+6x+36}}{x^4(x-6)\cancel{(x^2+6x+36)}} = \frac{1}{x^4(x-6)}$$

9.

$$\frac{5ac-5ad+5bc-5bd}{5d^2-5c^2} = \frac{5(ac-ad+bc-bd)}{5(d^2-c^2)}$$
$$= \frac{5[a(c-d)+b(c-d)]}{5(d-c)(d+c)}$$
$$= \frac{5(c-d)(a+b)}{5(d-c)(d+c)}$$
$$= -\frac{\cancel{5}\cancel{(c-d)}(a+b)}{\cancel{5}\cancel{(c-d)}(d+c)}$$
$$= -\frac{a+b}{d+c}$$
$$= \frac{-a-b}{d+c}$$

10.

$$\frac{m^3+m^2n-2mn^2}{2m^3-mn^2-m^2n}=\frac{m\left(m^2+mn-2n^2\right)}{m\left(2m^2-n^2-mn\right)}$$
$$=\frac{m\left(m^2+mn-2n^2\right)}{m\left(2m^2-mn-n^2\right)}$$
$$=\frac{\cancel{m}\,(m+2n)\,\cancel{(m-n)}}{\cancel{m}\,(2m+n)\,\cancel{(m-n)}}$$
$$=\frac{m+2n}{2m+n}$$

11.

$$\frac{6x^2-5x-4}{9x^2-24x+16}=\frac{\cancel{(3x-4)}\,(2x+1)}{\cancel{(3x-4)}\,(3x-4)}$$
$$=\frac{2x+1}{3x-4}$$

12.

$$\frac{2m-2n}{n-m}=\frac{2\cancel{(m-n)}}{-1\cancel{(m-n)}}$$
$$=-2$$

13.

$$\frac{s^2+t^2}{s-t}$$

14.

$$\frac{3m^2-10m+8}{6-m-m^2}=-\frac{3m^2-10m+8}{m^2+m-6}$$
$$=-\frac{(3m-4)\,\cancel{(m-2)}}{(m+3)\,\cancel{(m-2)}}$$
$$=-\frac{3m-4}{m+3}\quad\text{or}\quad\frac{4-3m}{m+3}$$

SECTION 6.2

Multiplying and Dividing Rational Expressions

15.

$$\frac{3x^3y^4}{35}\bullet\frac{10}{21x^5y^4}=\frac{\cancel{3}\cancel{x}\cancel{x}\cancel{x}\cancel{y}\cancel{y}\cancel{y}\cancel{y}\bullet 2\bullet\cancel{5}}{\cancel{5}\bullet 7\bullet\cancel{3}\bullet 7\cancel{x}\cancel{x}\cancel{x}xx\cancel{y}\cancel{y}\cancel{y}\cancel{y}}$$
$$=\frac{2}{49x^2}$$

16

$$f(x)\bullet g(x)=\frac{x^3+4x^2+4x}{x^2-x-6}\bullet\frac{9-x^2}{x^2+5x+6}$$
$$=\frac{x(x+2)(x+2)}{(x+2)(x-3)}\bullet\frac{(3-x)(3+x)}{(x+3)(x+2)}$$
$$=\frac{x\cancel{(x+2)}\,\cancel{(x+2)}}{\cancel{(x+2)}\,\cancel{(x-3)}}\bullet\frac{-\cancel{(x-3)}\,\cancel{(3+x)}}{\cancel{(x+3)}\,\cancel{(x+2)}}$$
$$=-x$$

17.

$$\frac{2a^2-5a-3}{4a^3-36a}\div\frac{2a^2+5a+2}{2a^2+5a-3}$$
$$=\frac{2a^2-5a-3}{4a\left(a^2-9\right)}\bullet\frac{2a^2+5a-3}{2a^2+5a+2}$$
$$=\frac{\cancel{(2a+1)}\,\cancel{(a-3)}}{4a\cancel{(a-3)}\,\cancel{(a+3)}}\bullet\frac{(2a-1)\,\cancel{(a+3)}}{\cancel{(2a+1)}\,(a+2)}$$
$$=\frac{2a-1}{4a(a+2)}$$

18.

$$\frac{t^4-4t^2}{t}\div\left(t^3+2t^2\right)$$
$$=\frac{t^4-4t^2}{t}\bullet\frac{1}{t^3+2t^2}$$
$$=\frac{t^2\left(t^2-4\right)}{t}\bullet\frac{1}{t^2(t+2)}$$
$$=\frac{\cancel{t^2}\,(t-2)\,\cancel{(t+2)}}{t}\bullet\frac{1}{\cancel{t^2}\,\cancel{(t+2)}}$$
$$=\frac{t-2}{t}$$

19.

$$\left(\frac{h-2}{h^3+4}\right)^2=\frac{(h-2)^2}{\left(h^3+4\right)^2}$$
$$=\frac{(h-2)(h-2)}{\left(h^3+4\right)\left(h^3+4\right)}$$
$$=\frac{h^2-2h-2h+4}{h^6+4h^3+4h^3+16}$$
$$=\frac{h^2-4h+4}{h^6+8h^3+16}$$

20.

$$\frac{m^2+3m+9}{m^2+4m+mr+4r} \div \frac{m^3-27}{am+ar+6m+6r}$$
$$=\frac{m^2+3m+9}{m^2+4m+mr+4r} \bullet \frac{am+ar+6m+6r}{m^3-27}$$
$$=\frac{m^2+3m+9}{m(m+4)+r(m+4)} \bullet \frac{a(m+r)+6(m+r)}{(m-3)(m^2+3m+9)}$$
$$=\frac{\cancel{m^2+3m+9}}{\cancel{(m+r)}(m+4)} \bullet \frac{(a+6)\cancel{(m+r)}}{(m-3)\cancel{(m^2+3m+9)}}$$
$$=\frac{a+6}{(m+4)(m-3)}$$

21.

$$\frac{8m^2+6mn-9n^2}{2m^2+5mn+3n^2} \bullet \frac{6m^2+5mn-4n^2}{12m^2+7mn-12n^2}$$
$$=\frac{\cancel{(2m+3n)}\cancel{(4m-3n)}}{\cancel{(2m+3n)}(m+n)} \bullet \frac{\cancel{(3m+4n)}(2m-n)}{\cancel{(3m+4n)}\cancel{(4m-3n)}}$$
$$=\frac{2m-n}{m+n}$$

22.

$$\frac{x^3+3x^2+2x}{2x^2-2x-12} \div \frac{x^3-3x^2-4x}{3x^2-3x} \bullet \frac{2x^2-4x-16}{x^2+3x+2}$$
$$=\frac{x^3+3x^2+2x}{2x^2-2x-12} \bullet \frac{3x^2-3x}{x^3-3x^2-4x} \bullet \frac{2x^2-4x-16}{x^2+3x+2}$$
$$=\frac{x(x^2+3x+2)}{2(x^2-x-6)} \bullet \frac{3x(x-1)}{x(x^2-3x-4)} \bullet \frac{2(x^2-2x-8)}{x^2+3x+2}$$
$$=\frac{\cancel{x}\cancel{(x+1)}\cancel{(x+2)}}{\cancel{2}(x-3)\cancel{(x+2)}} \bullet \frac{3x(x-1)}{\cancel{x}\cancel{(x-4)}\cancel{(x+1)}} \bullet \frac{\cancel{2}\cancel{(x-4)}\cancel{(x+2)}}{(x+1)\cancel{(x+2)}}$$
$$=\frac{3x(x-1)}{(x-3)(x+1)}$$

SECTION 6.3
Adding and Subtracting Rational Expressions

23.

$$\frac{5y}{x-y}-\frac{3}{x-y}=\frac{5y-3}{x-y}$$

24.

$$\frac{d^2}{c^3-d^3}+\frac{c^2+cd}{c^3-d^3}=\frac{d^2+c^2+cd}{c^3-d^3}$$
$$=\frac{\cancel{c^2+cd+d^2}}{(c-d)\cancel{(c^2+cd+d^2)}}$$
$$=\frac{1}{c-d}$$

25.

$$\frac{4}{t-3}+\frac{6}{3-t}=\frac{4}{t-3}-\frac{6}{t-3}$$
$$=\frac{4-6}{t-3}$$
$$=\frac{-2}{t-3}$$
$$=-\frac{2}{t-3}$$

26.

$$\frac{p+3}{p^2+13p+12}-\frac{2p+4}{p^2+13p+12}$$
$$=\frac{(p+3)-(2p+4)}{p^2+13p+12}$$
$$=\frac{p+3-2p-4}{(p+1)(p+12)}$$
$$=\frac{-p-1}{(p+1)(p+12)}$$
$$=-\frac{\cancel{p+1}}{\cancel{(p+1)}(p+12)}$$
$$=-\frac{1}{p+12}$$

27. $15a^2h = 3 \cdot 5 \cdot a \cdot a \cdot h$
$20ah^3 = 2 \cdot 2 \cdot 5 \cdot a \cdot h \cdot h \cdot h$
$\text{LCD} = 2 \cdot 2 \cdot 3 \cdot 5 \cdot a \cdot a \cdot h \cdot h \cdot h$
$= 60a^2h^3$

28. $ab^2 - ab = ab(b-1)$
$ab^2 = a \cdot b \cdot b$
$b^2 - b = b(b-1)$
$\text{LCD} = ab^2(b-1)$

29. $x^2 - 4x - 5 = (x-5)(x+1)$
$x^2 - 25 = (x-5)(x+5)$
$\text{LCD} = (x-5)(x+1)(x+5)$

30. $m^2 - 4m + 4 = (m-2)(m-2)$
$m^3 - 8 = (m-2)(m^2 + 2m + 4)$
$\text{LCD} = (m-2)^2(m^2 + 2m + 4)$

31.

$$9 - \frac{1}{a+1} = \frac{9}{1}\left(\frac{a+1}{a+1}\right) - \frac{1}{a+1}$$
$$= \frac{9a+9}{a+1} - \frac{1}{a+1}$$
$$= \frac{9a+9-1}{a+1}$$
$$= \frac{9a+8}{a+1}$$

32.

$$\frac{5x}{14z^2} + \frac{y^2}{16z} = \frac{5x}{14z^2}\left(\frac{8}{8}\right) + \frac{y^2}{16z}\left(\frac{7z}{7z}\right)$$
$$= \frac{40x}{112z^2} + \frac{7y^2z}{112z^2}$$
$$= \frac{40x + 7y^2z}{112z^2}$$

33.

$$\frac{4x}{x-4} - \frac{3}{x+3} = \frac{4x}{x-4}\left(\frac{x+3}{x+3}\right) - \frac{3}{x+3}\left(\frac{x-4}{x-4}\right)$$
$$= \frac{4x^2+12x}{(x-4)(x+3)} - \frac{3x-12}{(x-4)(x+3)}$$
$$= \frac{4x^2+12x-(3x-12)}{(x-4)(x+3)}$$
$$= \frac{4x^2+12x-3x+12}{(x-4)(x+3)}$$
$$= \frac{4x^2+9x+12}{(x-4)(x+3)}$$

34.

$$\frac{2a+4}{3} - \frac{9}{a+2} = \frac{2a+4}{3}\left(\frac{a+2}{a+2}\right) - \frac{9}{a+2}\left(\frac{3}{3}\right)$$
$$= \frac{2a^2+4a+4a+8}{3(a+2)} - \frac{27}{3(a+2)}$$
$$= \frac{2a^2+4a+4a+8-27}{3(a+2)}$$
$$= \frac{2a^2+8a-19}{3(a+2)}$$

35.

$$\frac{6}{a^2-9} - \frac{5}{a^2-a-6}$$
$$= \frac{6}{(a+3)(a-3)} - \frac{5}{(a-3)(a+2)}$$
$$= \frac{6}{(a+3)(a-3)}\left(\frac{a+2}{a+2}\right) - \frac{5}{(a-3)(a+2)}\left(\frac{a+3}{a+3}\right)$$
$$= \frac{6a+12}{(a+3)(a-3)(a+2)} - \frac{5a+15}{(a+3)(a-3)(a+2)}$$
$$= \frac{6a+12-(5a+15)}{(a+3)(a-3)(a+2)}$$
$$= \frac{6a+12-5a-15}{(a+3)(a-3)(a+2)}$$
$$= \frac{\cancel{a-3}}{(a+3)\cancel{(a-3)}(a+2)}$$
$$= \frac{1}{(a+3)(a+2)}$$

36.

$$\frac{4}{3xy-6y} - \frac{4}{10-5x} = \frac{4}{3y(x-2)} - \frac{4}{5(2-x)}$$
$$= \frac{4}{3y(x-2)} + \frac{4}{5(x-2)}$$
$$= \frac{4}{3y(x-2)}\left(\frac{5}{5}\right) + \frac{4}{5(x-2)}\left(\frac{3y}{3y}\right)$$
$$= \frac{20}{15y(x-2)} + \frac{12y}{15y(x-2)}$$
$$= \frac{20+12y}{15y(x-2)}$$

37.

$$\frac{a}{a+2}-\frac{3}{a^2+2a}+\frac{1}{2a+4}$$
$$=\frac{a}{a+2}-\frac{3}{a(a+2)}+\frac{1}{2(a+2)}$$
$$=\frac{a}{a+2}\left(\frac{2a}{2a}\right)-\frac{3}{a(a+2)}\left(\frac{2}{2}\right)+\frac{1}{2(a+2)}\left(\frac{a}{a}\right)$$
$$=\frac{2a^2}{2a(a+2)}-\frac{6}{2a(a+2)}+\frac{a}{2a(a+2)}$$
$$=\frac{2a^2-6+a}{2a(a+2)}$$
$$=\frac{2a^2+a-6}{2a(a+2)}$$
$$=\frac{(2a-3)\cancel{(a+2)}}{2a\cancel{(a+2)}}$$
$$=\frac{2a-3}{2a}$$

38.

$$n(x)-s(x)=\frac{x+7}{x+3}-\frac{x-3}{x+7}$$
$$=\frac{x+7}{x+3}\left(\frac{x+7}{x+7}\right)-\frac{x-3}{x+7}\left(\frac{x+3}{x+3}\right)$$
$$=\frac{x^2+7x+7x+49}{(x+3)(x+7)}-\frac{x^2-3x+3x-9}{(x+3)(x+7)}$$
$$=\frac{x^2+14x+49}{(x+3)(x+7)}+\frac{-x^2+9}{(x+3)(x+7)}$$
$$=\frac{14x+58}{(x+3)(x+7)}$$

SECTION 6.4

Simplifying Complex Fractions

39.

$$\frac{\frac{4a^3b^2}{9c}}{\frac{14a^3b}{9c^4}}=\frac{\frac{4a^3b^2}{9c}}{\frac{14a^3b}{9c^4}}\bullet\frac{9c^4}{9c^4}$$
$$=\frac{\left(9c^4\right)\left(\frac{4a^3b^2}{9c}\right)}{\left(9c^4\right)\left(\frac{14a^3b}{9c^4}\right)}$$
$$=\frac{c^3\left(4a^3b^2\right)}{14a^3b}$$
$$=\frac{4a^3b^2c^3}{14a^3b}$$
$$=\frac{2bc^3}{7}$$

40.

$$\frac{\frac{p^2-9}{6pt}}{\frac{p^2+5p+6}{3pt}}=\frac{\frac{p^2-9}{6pt}}{\frac{p^2+5p+6}{3pt}}\bullet\frac{6pt}{6pt}$$
$$=\frac{6pt\left(\frac{p^2-9}{6pt}\right)}{6pt\left(\frac{p^2+5p+6}{3pt}\right)}$$
$$=\frac{p^2-9}{2\left(p^2+5p+6\right)}$$
$$=\frac{\cancel{(p+3)}(p-3)}{2(p+2)\cancel{(p+3)}}$$
$$=\frac{p-3}{2(p+2)}$$

41.

$$\frac{\frac{1}{a}+\frac{2}{b}}{\frac{2}{a}-\frac{1}{b}}=\frac{\frac{1}{a}+\frac{2}{b}}{\frac{2}{a}-\frac{1}{b}}\bullet\frac{ab}{ab}$$

$$=\frac{ab\left(\frac{1}{a}+\frac{2}{b}\right)}{ab\left(\frac{2}{a}-\frac{1}{b}\right)}$$

$$=\frac{b+2a}{2b-a}$$

42.

$$\frac{1-\frac{1}{x}-\frac{2}{x^2}}{1+\frac{4}{x}+\frac{3}{x^2}}=\frac{1-\frac{1}{x}-\frac{2}{x^2}}{1+\frac{4}{x}+\frac{3}{x^2}}\bullet\frac{x^2}{x^2}$$

$$=\frac{x^2\left(1-\frac{1}{x}-\frac{2}{x^2}\right)}{x^2\left(1+\frac{4}{x}+\frac{3}{x^2}\right)}$$

$$=\frac{x^2-x-2}{x^2+4x+3}$$

$$=\frac{(x-2)\cancel{(x+1)}}{\cancel{(x+1)}(x+3)}$$

$$=\frac{x-2}{x+3}$$

43.

$$\frac{(x-y)^{-2}}{x^{-2}-y^{-2}}=\frac{\frac{1}{(x-y)^2}}{\frac{1}{x^2}-\frac{1}{y^2}}$$

$$=\frac{\frac{1}{(x-y)^2}}{\frac{1}{x^2}-\frac{1}{y^2}}\bullet\frac{x^2y^2(x-y)^2}{x^2y^2(x-y)^2}$$

$$=\frac{x^2y^2(x-y)^2\left(\frac{1}{(x-y)^2}\right)}{x^2y^2(x-y)^2\left(\frac{1}{x^2}-\frac{1}{y^2}\right)}$$

$$=\frac{x^2y^2}{y^2(x-y)^2-x^2(x-y)^2}$$

$$=\frac{x^2y^2}{(x-y)^2\left(y^2-x^2\right)}$$

44.

$$\frac{1+\frac{1}{b+d}}{\frac{1}{b+d}-1}=\frac{1+\frac{1}{b+d}}{\frac{1}{b+d}-1}\bullet\frac{b+d}{b+d}$$

$$=\frac{b+d+1}{1-(b+d)}$$

$$=\frac{b+d+1}{1-b-d}$$

45.

$$\frac{\frac{2b}{b-1}-\frac{3}{b}}{\frac{1}{b-1}+\frac{2}{b}}=\frac{\frac{2b}{b-1}-\frac{3}{b}}{\frac{1}{b-1}+\frac{2}{b}}\bullet\frac{b(b-1)}{b(b-1)}$$

$$=\frac{2b(b)-3(b-1)}{b+2(b-1)}$$

$$=\frac{2b^2-3b+3}{b+2b-2}$$

$$=\frac{2b^2-3b+3}{3b-2}$$

46.

$$\frac{\frac{8}{r+3}}{\frac{4}{r-2}-\frac{2}{r^2+r-6}}=\frac{\frac{8}{r+3}}{\frac{4}{r-2}-\frac{2}{(r+3)(r-2)}}$$

$$=\frac{\frac{8}{r+3}}{\frac{4}{r-2}-\frac{2}{(r+3)(r-2)}}\bullet\frac{(r+3)(r-2)}{(r+3)(r-2)}$$

$$=\frac{8(r-2)}{4(r+3)-2}$$

$$=\frac{8r-16}{4r+12-2}$$

$$=\frac{8r-16}{4r+10}$$

$$=\frac{\overset{4}{\cancel{8}}(r-2)}{\cancel{2}(2r+5)}$$

$$=\frac{4(r-2)}{2r+5}\text{ or }\frac{4r-8}{2r+5}$$

SECTION 6.5
Dividing Polynomials

47.

$$\frac{25h^4k^7}{55hk^9}=\frac{5}{11}h^{4-1}k^{7-9}$$
$$=\frac{5}{11}h^3k^{-2}$$
$$=\frac{5h^3}{11k^2}$$

48.

$$\left(5x^3y^3z^{10}\right)\div\left(10x^3y^6z^{20}\right)=\frac{5x^3y^3z^{10}}{10x^3y^6z^{20}}$$
$$=\frac{1}{2}x^{3-3}y^{3-6}z^{10-20}$$
$$=\frac{1}{2}x^0y^{-3}z^{-10}$$
$$=\frac{1}{2y^3z^{10}}$$

49.

$$\frac{36a+32}{6}=\frac{36a}{6}+\frac{32}{6}$$
$$=6a+\frac{16}{3}$$

50.

$$\frac{30x^3y^2-15x^2y-10xy^2}{-10xy}$$
$$=\frac{30x^3y^2}{-10xy}-\frac{15x^2y}{-10xy}-\frac{10xy^2}{-10xy}$$
$$=-3x^2y+\frac{3x}{2}+y$$

51.

$$\begin{array}{r} b+4 \\ b+5\overline{\smash{)}b^2+9b+20} \\ \underline{b^2+5b} \\ 4b+20 \\ \underline{4b+20} \\ 0 \end{array}$$

52.

$$\begin{array}{r} v^2-3v-10 \\ 3v+1\overline{\smash{)}3v^3-8v^2-33v-10} \\ \underline{3v^3+v^2} \\ -9v^2-33v \\ \underline{-9v^2-3v} \\ -30v-10 \\ \underline{-30v-10} \\ 0 \end{array}$$

53. $f(x)\div g(x)=\left(x^3+8\right)\div\left(x+2\right)$

$$\begin{array}{r} x^2-2x+4 \\ x+2\overline{\smash{)}x^3+0x^2+0x+8} \\ \underline{x^3+2x^2} \\ -2x^2+0x \\ \underline{-2x^2-4x} \\ 4x+8 \\ \underline{4x+8} \\ 0 \end{array}$$

54.

$$\begin{array}{r} 2m-5+\dfrac{-4}{4m+1} \\ 4m+1\overline{\smash{)}\;8m^2-18m-9} \\ \underline{8m^2+2m} \\ -20m-9 \\ \underline{-20m-5} \\ -4 \end{array}$$

55.

$$\begin{array}{r} 3a-2+\dfrac{-15a+2}{a^2+5} \\ a^2+0a+5\overline{\smash{)}3a^3-2a^2+0a-8} \\ \underline{3a^3+0a^2+15a} \\ -2a^2-15a-8 \\ \underline{-2a^2+0a-10} \\ -15a+2 \end{array}$$

56.

$$m^4 - m^2 + 1 + \frac{2}{m^4 + 2m^2 - 3}$$

$$m^4 + 2m^2 - 3 \overline{\big) \; m^8 + m^6 - 4m^4 + 5m^2 - 1}$$

$$\underline{m^8 + 2m^6 - 3m^4}$$

$$-m^6 - m^4 + 5m^2$$

$$\underline{-m^6 - 2m^4 + 3m^2}$$

$$m^4 + 2m^2 - 1$$

$$\underline{m^4 + 2m^2 - 3}$$

$$2$$

SECTION 6.6
Solving Rational Expressions

57.

$$\left(x^2 - 13x + 42\right) \div \left(x - 6\right) = x - 7$$

6	1	−13	42
		6	−42
	1	−7	0

58.

$$\frac{m^3 - 6m^2 + 11m - 6}{m - 3} = m^2 - 3m + 2$$

3	1	−6	11	−6
		3	−9	6
	1	−3	2	0

59.

$$\frac{-3n^5 + 10n^4 + 7n^3 + 2n^2 + 9n - 4}{n - 4}$$

$$= -3n^4 - 2n^3 - n^2 - 2n + 1$$

4	−3	10	7	2	9	−4
		−12	−8	−4	−8	4
	−3	−2	−1	−2	1	0

60.

$$\frac{4x^3 + 5x^2 - 1}{x + 2} = 4x^2 - 3x + 6 + \frac{-13}{x + 2}$$

−2	4	5	0	−1
		−8	6	−12
	4	−3	6	−13

61.

$$\frac{3a^4 + 3a^3 - a^2 + 3a + 10}{a + 1} = 3a^3 - a + 4 + \frac{6}{a + 1}$$

−1	3	3	−1	3	10
		−3	0	1	−4
	3	0	−1	4	6

62.

$$\frac{x^4 + 1}{x - 3} = x^3 + 3x^2 + 9x + 27 + \frac{82}{x - 3}$$

3	1	0	0	0	1
		3	9	27	81
	1	3	9	27	82

63.

$$P\left(x\right) = x^4 - 2x^3 + x^2 - 3x + 12$$

-2	1	-2	1	-3	12
		-2	8	-18	42
	1	-4	9	-21	54

The remainder is 54.

$P\left(-2\right) = 54$

64.

$$P\left(x\right) = x^3 - 13x^2 - 27$$

5	1	−13	0	−27
		5	−40	−200
	1	−8	−40	−227

The remainder is -227.

$P\left(5\right) = -227$

65. Yes.

$$\begin{array}{r|rrrr} 5 & 1 & -3 & -8 & -10 \\ & & 5 & 10 & 10 \\ \hline & 1 & 2 & 2 & 0 \end{array}$$

The remainder is 0, so $(x-5)$ is a factor.

66. No.

$$\begin{array}{r|rrrr} -5 & 1 & 4 & -5 & 5 \\ & & -5 & 5 & 0 \\ \hline & 1 & -1 & 0 & 5 \end{array}$$

The remainder is not 0,

so $(x+5)$ is not a factor.

SECTION 6.7
Solving Rational Equations

67.

$$\frac{4}{x}-\frac{1}{10}=\frac{7}{2x}$$
$$10x\left(\frac{4}{x}-\frac{1}{10}\right)=10x\left(\frac{7}{2x}\right)$$
$$40-x=35$$
$$40-x-40=35-40$$
$$-x=-5$$
$$x=5$$

68.

$$\frac{11}{t}=\frac{6}{t-7}$$
$$t(t-7)\left(\frac{11}{t}\right)=t(t-7)\left(\frac{6}{t-7}\right)$$
$$11(t-7)=6t$$
$$11t-77=6t$$
$$-77=-5t$$
$$\frac{77}{5}=t$$

69.

$$\frac{3}{y}-\frac{2}{y+1}=\frac{1}{2}$$
$$2y(y+1)\left(\frac{3}{y}-\frac{2}{y+1}\right)=2y(y+1)\left(\frac{1}{2}\right)$$
$$3(2)(y+1)-2(2y)=y(y+1)$$
$$6y+6-4y=y^2+y$$
$$2y+6=y^2+y$$
$$0=y^2-y-6$$
$$0=(y-3)(y+2)$$
$$y-3=0 \quad \text{or} \quad y+2=0$$
$$y=3 \qquad\qquad y=-2$$

70.

$$\frac{2}{3x+15}-\frac{1}{18}=\frac{1}{3x+12}$$
$$\frac{2}{3(x+5)}-\frac{1}{18}=\frac{1}{3(x+4)}$$
$$18(x+5)(x+4)\left(\frac{2}{3(x+5)}-\frac{1}{18}\right)=18(x+5)(x+4)\left(\frac{1}{3(x+4)}\right)$$
$$6(x+4)2-(x+5)(x+4)=6(x+5)$$
$$12(x+4)-(x^2+4x+5x+20)=6x+30$$
$$12x+48-x^2-4x-5x-20=6x+30$$
$$-x^2+3x+28=6x+30$$
$$-x^2+3x+28+x^2-3x-28=6x+30+x^2-3x-28$$
$$0=x^2+3x+2$$
$$0=(x+1)(x+2)$$
$$x+1=0 \quad \text{or} \quad x+2=0$$
$$x=-1 \qquad\qquad x=-2$$

71.

$$\frac{3}{x+2}=\frac{1}{2-x}+\frac{2}{x^2-4}$$
$$\frac{3}{x+2}=\frac{-1}{x-2}+\frac{2}{(x-2)(x+2)}$$
$$(x-2)(x+2)\left(\frac{3}{x+2}\right)=(x-2)(x+2)\left(\frac{-1}{x-2}+\frac{2}{(x-2)(x+2)}\right)$$
$$3(x-2)=-(x+2)+2$$
$$3x-6=-x-2+2$$
$$3x-6=-x$$
$$-6=-4x$$
$$\frac{3}{2}=x$$

72.

$$\frac{x+3}{x-5}+\frac{2x^2+6}{x^2-7x+10}=\frac{3x}{x-2}$$
$$\frac{x+3}{x-5}+\frac{2x^2+6}{(x-2)(x-5)}=\frac{3x}{x-2}$$
$$(x-5)(x-2)\left(\frac{x+3}{x-5}+\frac{2x^2+6}{(x-2)(x-5)}\right)=(x-2)(x-5)\left(\frac{3x}{x-2}\right)$$
$$(x-2)(x+3)+2x^2+6=3x(x-5)$$
$$x^2+3x-2x-6+2x^2+6=3x^2-15x$$
$$3x^2+x=3x^2-15x$$
$$3x^2+x-3x^2=3x^2-15x-3x^2$$
$$x=-15x$$
$$x+15x=-15x+15x$$
$$16x=0$$
$$x=0$$

73.

$$\frac{5a}{a-3}-7=\frac{15}{a-3}$$
$$(a-3)\left(\frac{5a}{a-3}-7\right)=(a-3)\left(\frac{15}{a-3}\right)$$
$$5a-7(a-3)=15$$
$$5a-7a+21=15$$
$$-2a+21=15$$
$$-2a=-6$$
$$a=\not{3};\ \text{No solution}$$

3 is extraneous

74. a.

$$\frac{10}{x^2-4x}-\frac{4}{x}+\frac{5}{x-4}$$
$$=\frac{10}{x(x-4)}-\frac{4}{x}+\frac{5}{x-4}$$
$$=\frac{10}{x(x-4)}-\frac{4}{x}\left(\frac{x-4}{x-4}\right)+\frac{5}{x-4}\left(\frac{x}{x}\right)$$
$$=\frac{10}{x(x-4)}-\frac{4x-16}{x(x-4)}+\frac{5x}{x(x-4)}$$
$$=\frac{10-(4x-16)+5x}{x(x-4)}$$
$$=\frac{10-4x+16+5x}{x(x-4)}$$
$$=\frac{x+26}{x(x-4)}$$

b.

$$\frac{10}{x^2-4x}-\frac{4}{x}=\frac{5}{x-4}$$
$$x(x-4)\left(\frac{10}{x(x-4)}-\frac{4}{x}\right)=x(x-4)\left(\frac{5}{x-4}\right)$$
$$10-4(x-4)=5x$$
$$10-4x+16=5x$$
$$26-4x=5x$$
$$26=9x$$
$$\frac{26}{9}=x$$

75.

$$H=\frac{2ab}{a+b}$$
$$(a+b)H=(a+b)\left(\frac{2ab}{a+b}\right)$$
$$Ha+Hb=2ab$$
$$Ha+Hb-Hb=2ab-Hb$$
$$Ha=2ab-Hb$$
$$Ha=b(2a-H)$$
$$\frac{Ha}{2a-H}=\frac{b(2a-H)}{2a-H}$$
$$\frac{Ha}{2a-H}=b$$

76.

$$\frac{x^2}{a^2}-\frac{y^2}{b^2}=1$$
$$a^2b^2\left(\frac{x^2}{a^2}-\frac{y^2}{b^2}\right)=a^2b^2(1)$$
$$x^2b^2-a^2y^2=a^2b^2$$
$$b^2x^2-a^2y^2-b^2x^2=a^2b^2-b^2x^2$$
$$-a^2y^2=a^2b^2-b^2x^2$$
$$\frac{-a^2y^2}{-a^2}=\frac{a^2b^2-b^2x^2}{-a^2}$$
$$y^2=\frac{b^2x^2-a^2b^2}{a^2}$$

77.

$$\frac{1}{R}=\frac{1}{R_1}+\frac{1}{R_2}$$
$$RR_1R_2\left(\frac{1}{R}\right)=RR_1R_2\left(\frac{1}{R_1}+\frac{1}{R_2}\right)$$
$$R_1R_2=RR_2+RR_1$$
$$R_1R_2=R(R_2+R_1)$$
$$\frac{R_1R_2}{R_2+R_1}=R$$
$$R=\frac{R_1R_2}{R_2+R_1}$$

78.

$$k=\frac{ma}{F}$$
$$F(k)=F\left(\frac{ma}{F}\right)$$
$$Fk=ma$$
$$F=\frac{ma}{k}$$

SECTION 6.8

Problem Solving Using Rational Equations

79. a. a. $\frac{1}{10}$ of the job per hour

b. $\frac{x}{10}$ of the job is completed

80. Let x be the number of hours it takes both pipes to drain the pipe working together.

Amount of work 1st pipe in 1 hour = $\frac{1}{24}$

Amount of work 2nd pipe in 1 hour = $\frac{1}{36}$

Amount of work together in 1 hour = $\frac{1}{x}$

$$\frac{1}{24}+\frac{1}{36}=\frac{1}{x}$$
$$72x\left(\frac{1}{24}+\frac{1}{36}\right)=72x\left(\frac{1}{x}\right)$$
$$3x+2x=72$$
$$5x=72$$
$$x=\frac{72}{5}$$
$$x=14\frac{2}{5}\text{ hours}$$

It takes $14\frac{2}{5}$ hours for the pipes to drain the tank working together.

81. Let x be the number of days it takes an experienced electrician to wire the house. Then the apprentice's time is $x + 5$.

Work of experienced man in 1 day $= \dfrac{1}{x}$

Work of apprentice in 1 day $= \dfrac{1}{x+5}$

Amount of work together in 1 day $= \dfrac{1}{6}$

$$\frac{1}{x}+\frac{1}{x+5}=\frac{1}{6}$$
$$6x(x+5)\left(\frac{1}{x}+\frac{1}{x+5}\right)=6x(x+5)\left(\frac{1}{6}\right)$$
$$6(x+5)+6x=x(x+5)$$
$$6x+30+6x=x^2+5x$$
$$12x+30=x^2+5x$$
$$0=x^2-7x-30$$
$$0=(x-10)(x+3)$$
$$x-10=0 \quad \text{or} \quad x+3=0$$
$$x=10 \qquad x=\not{-3}$$

It takes the experienced electrician 10 days to wire the house and 10 + 5 = 15 days for the apprentice.

82. Let x be the time it takes to fill the sink if the drain is open and the faucet is on.

Amount of work of faucet in 1 min. $= \dfrac{1}{2}$

Amount of work of drain in 1 min. $= \dfrac{1}{3}$

Amount of work together in 1 hour $= \dfrac{1}{x}$

$$\frac{1}{2}-\frac{1}{3}=\frac{1}{x}$$
$$6x\left(\frac{1}{2}-\frac{1}{3}\right)=6x\left(\frac{1}{x}\right)$$
$$3x-2x=6$$
$$x=6$$

It takes 6 minutes for the sink to fill.

83.

Going downwind, the rate would be the speed of the helicopter in still wind (75) plus the speed of the wind, x. Going upwind, the rate would be the speed of the helicopter in still air (75) minus the speed of the wind, x.

	distance	rate	time
downwind	40	$75+x$	$t=\frac{d}{r}=\frac{40}{75+x}$
upwind	35	$75-x$	$t=\frac{d}{r}=\frac{35}{75-x}$

Since the time is the same for both trips, set the time for both trips equal and solve for the rate of the wind, x.

$$\frac{40}{75+x}=\frac{35}{75-x}$$
$$(75+x)(75-x)\left(\frac{40}{75+x}\right)=(75+x)(75-x)\left(\frac{35}{75-x}\right)$$
$$40(75-x)=35(75+x)$$
$$3{,}000-40x=2{,}625+35x$$
$$3{,}000=2{,}625+75x$$
$$375=75x$$
$$5=x$$

The rate of the wind is 5 mph.

84.

	distance	rate	time
usual trip	200	x	$t = \dfrac{200}{x}$
with traffic	200	$x - 10$	$t + 1 = \dfrac{200}{x-10}$ $t = \dfrac{200}{x-10} - 1$

Since the time is the same for both trips, set the time for both trips equal and solve for x.

$$\frac{200}{x} = \frac{200}{x-10} - 1$$
$$x(x-10)\left(\frac{200}{x}\right) = x(x-10)\left(\frac{200}{x-10} - 1\right)$$
$$200(x-10) = 200x - x(x-10)$$
$$200x - 2{,}000 = 200x - x^2 + 10x$$
$$200x - 2{,}000 - 200x = 200x - x^2 + 10x - 200x$$
$$-2{,}000 = -x^2 + 10x$$
$$x^2 - 10x - 2{,}000 = 0$$
$$(x-50)(x+40) = 0$$
$$x - 50 = 0 \quad \text{or} \quad x + 40 = 0$$
$$x = 50 \qquad\qquad x = -40$$

Since the rate cannot be negative, her usual speed is 50 mph.

SECTION 6.9
Proportion and Variation

85.

$$\frac{x+1}{8} = \frac{4x-2}{24}$$
$$24(x+1) = 8(4x-2)$$
$$24x + 24 = 32x - 16$$
$$-8x + 24 = -16$$
$$-8x = -40$$
$$x = 5$$

86.

$$\frac{1}{x+6} = \frac{x+10}{12}$$
$$12(x+6)\left(\frac{1}{x+6}\right) = 12(x+6)\left(\frac{x+10}{12}\right)$$
$$12 = (x+6)(x+10)$$
$$12 = x^2 + 10x + 6x + 60$$
$$12 = x^2 + 16x + 60$$
$$0 = x^2 + 16x + 48$$
$$0 = (x+4)(x+12)$$
$$x + 4 = 0 \quad \text{or} \quad x + 12 = 0$$
$$x = -4 \qquad\qquad x = -12$$

87.

$$\frac{-r}{r+2} = \frac{3r}{r-2}$$
$$(r+2)(r-2)\frac{-r}{r+2} = (r+2)(r-2)\frac{3r}{r-2}$$
$$-r(r-2) = 3r(r+2)$$
$$-r^2 + 2r = 3r^2 + 6r$$
$$0 = 4r^2 + 4r$$
$$0 = 4r(r+1)$$
$$4r = 0 \quad \text{or} \quad r + 1 = 0$$
$$r = 0 \qquad\qquad r = -1$$

88.

$$\frac{18}{t^2} = \frac{3t-3}{t}$$
$$t^2\left(\frac{18}{t^2}\right) = t^2\left(\frac{3t-3}{t}\right)$$
$$18 = t(3t-3)$$
$$18 = 3t^2 - 3t$$
$$0 = 3t^2 - 3t - 18$$
$$0 = 3(t^2 - t - 6)$$
$$0 = 3(t-3)(t+2)$$
$$t - 3 = 0 \quad \text{or} \quad t + 2 = 0$$
$$t = 3 \qquad\qquad t = -2$$

89.

$$\frac{\text{actual height}}{\text{height of shadow}}$$

$$\frac{x}{44}=\frac{4}{2.5}$$
$$2.5x=4(44)$$
$$2.5x=176$$
$$x=70.4\text{ ft.}$$

90.

$$\frac{4\text{ bottles}}{2\text{ gallons}}=\frac{x\text{ bottles}}{10\text{ gallons}}$$
$$\frac{4}{2}=\frac{x}{10}$$
$$4(10)=x(2)$$
$$40=2x$$
$$20=x$$
$$x=20\text{ bottles}$$

91.

$$\frac{1}{12}=\frac{5.5}{x}$$
$$1(x)=5.5(12)$$
$$x=66\text{ in.}$$

92. Let t = property tax, k = constant, and a = assessed valuation.

$$t=ka$$
$$1,575=k(90,000)$$
$$1,575=90,000k$$
$$0.0175=k$$

Find t if a = 312,000 and k – 0.0175.

$$t=0.0175(312,000)$$
$$t=\$5,460$$

93. Let c = current, k = constant, and r = resistance.

$$c=\frac{k}{r}$$
$$2.5=\frac{k}{150}$$
$$150(2.5)=150\left(\frac{k}{150}\right)$$
$$375=k$$

Find c if k = 375 and r = 2(150) = 300.

$$c=\frac{k}{r}$$
$$=\frac{375}{300}$$
$$=1.25\text{ amps}$$

94. Let f = force of wind, k = constant, A = area of sign, and v = velocity of wind.

$$f=kAv^2$$
$$1.98=k(3\cdot 1.5)(10)^2$$
$$1.98=k(4.5)(100)$$
$$1.98=450k$$
$$0.0044=k$$

Find f if k = 0.0044 and v = 80.

$$f=kAv^2$$
$$f=0.0044(1.5\cdot 3)(80)^2$$
$$f=0.0044(4.5)(6,400)$$
$$f=126.72\text{ lb}$$

95. inverse variation

96.

$$x_1=\frac{kt^3}{x_2}$$
$$1.6=\frac{k(8)^3}{64}$$
$$1.6=\frac{512k}{64}$$
$$1.6=8k$$
$$0.2=k$$

CHAPTER 6 TEST

1. a. A quotient of two polynomials, such as $\frac{x-8}{x^2-2x-3}$, is called a **rational** expression.

   b. The **reciprocal** of $\frac{x+1}{x-7}$ is $\frac{x-7}{x+1}$.

   c. $\dfrac{\frac{x}{4}+\frac{1}{x}}{\frac{1}{8}+\frac{1}{x}}$ is an example of a **complex** fraction.

   d. When solving a rational equation, if we obtain a number that does not satisfy the original equation, the number is called an **extraneous** solution.

   e. An equation that states that two ratios are equal, such as $\frac{1}{2}=\frac{3}{6}$, is called a **proportion**.

2. Answers may vary. $(x + 2)$ is not a factor of the entire numerator, and therefore, cannot be removed.

3.
$$\frac{12x^2y^3z^2}{18x^3y^6z^2}=\frac{\cancel{2}\cdot 2\cdot\cancel{3}\cdot\cancel{x}\cdot\cancel{x}\cdot\cancel{y}\cdot\cancel{y}\cdot\cancel{y}\cdot\cancel{z}\cdot\cancel{z}}{\cancel{2}\cdot\cancel{3}\cdot 3\cdot\cancel{x}\cdot\cancel{x}\cdot x\cdot\cancel{y}\cdot\cancel{y}\cdot\cancel{y}\cdot y\cdot y\cdot y\cdot\cancel{z}\cdot\cancel{z}}$$
$$=\frac{2}{3xy^3}$$

4.
$$\frac{2x+4}{x^2-4}=\frac{2\cancel{(x+2)}}{\cancel{(x+2)}(x-2)}$$
$$=\frac{2}{x-2}$$

5.
$$\frac{3y-6z}{2z-y}=\frac{3(y-2z)}{2z-y}$$
$$=\frac{3\cancel{(y-2z)}}{-1\cancel{(y-2z)}}$$
$$=-3$$

6.
$$\frac{2x^2+7xy+3y^2}{4xy+12y^2}=\frac{(2x+y)\cancel{(x+3y)}}{4y\cancel{(x+3y)}}$$
$$=\frac{2x+y}{4y}$$

7. a. $f(5)=1$

   b. $f(-1)=-\frac{1}{2}$

   c. $f(1)=-1$

   d. $x=2$

8.
$$p(t)=\frac{200t}{t+1}$$
$$p(7)=\frac{200(7)}{(7)+1}$$
$$=\frac{1,400}{8}$$
$$=175$$

9. No. $\frac{x-4}{x^2+x-18}\neq\frac{x+4}{x^2+x-18}$

10. Set the denominator equal to zero.
$$x-x^2=0$$
$$x(1-x)=0$$
$$x=0 \quad\text{or}\quad 1-x=0$$
$$x=0 \quad\text{or}\quad 1=x$$
$$(-\infty,0)\cup(0,1)\cup(1,\infty)$$
The set of all real numbers except 0 and 1

11.
$$\frac{x^2}{x^3z^2y^2}\cdot\frac{x^2z^4}{y^2z}=\frac{x^4z^4}{x^3y^4z^3}$$
$$=\frac{xz}{y^4}$$

12.
$$\frac{a^2+12a+20}{a^2-4}\cdot\frac{a^2-12a+20}{a^2-100}$$
$$=\frac{\cancel{(a+2)}\cancel{(a+10)}}{\cancel{(a+2)}\cancel{(a-2)}}\cdot\frac{\cancel{(a-2)}\cancel{(a-10)}}{\cancel{(a+10)}\cancel{(a-10)}}$$
$$=1$$

13.
$$\frac{xu+2u+5x+10}{u^2+2u-15}\cdot\frac{13u-39}{x^2+3x+2}$$
$$=\frac{u(x+2)+5(x+2)}{(u-3)(u+5)}\cdot\frac{13(u-3)}{(x+1)(x+2)}$$
$$=\frac{\cancel{(u+5)}\cancel{(x+2)}}{\cancel{(u-3)}\cancel{(u+5)}}\cdot\frac{13\cancel{(u-3)}}{(x+1)\cancel{(x+2)}}$$
$$=\frac{13}{x+1}$$

14.

$$\frac{x^3+y^3}{16x^2}\div\frac{x^2-xy+y^2}{8x^2+8xy}=\frac{x^3+y^3}{16x^2}\cdot\frac{8x^2+8xy}{x^2-xy+y^2}$$

$$=\frac{(x+y)\cancel{(x^2-xy+y^2)}}{\cancel{8}\cdot 2\cdot \cancel{x}\cdot x}\cdot\frac{\cancel{8}\cancel{x}(x+y)}{\cancel{x^2-xy+y^2}}$$

$$=\frac{(x+y)^2}{2x}$$

15. $f(a)\div g(a)=\frac{a^2+7a+12}{a+3}\div\frac{16-a^2}{a-4}$

$$=\frac{a^2+7a+12}{a+3}\cdot\frac{a-4}{16-a^2}$$

$$=\frac{(a+3)(a+4)}{(a+3)}\cdot\frac{a-4}{(4-a)(4+a)}$$

$$=\frac{\cancel{(a+3)}\cancel{(a+4)}}{\cancel{(a+3)}}\cdot\frac{\cancel{a-4}}{-\cancel{(a-4)}\cancel{(4+a)}}$$

$$=-1$$

16.

$$\frac{(2x-3)^3}{x^2-2x+1}\div\frac{3x^2+7x+2}{3x^2-2x-1}\cdot\frac{x^2+x-2}{2x^7-3x^6}$$

$$=\frac{(2x-3)^3}{x^2-2x+1}\cdot\frac{3x^2-2x-1}{3x^2+7x+2}\cdot\frac{x^2+x-2}{2x^7-3x^6}$$

$$=\frac{\cancel{(2x-3)}(2x-3)(2x-3)}{\cancel{(x-1)}\cancel{(x-1)}}\cdot\frac{\cancel{(3x+1)}\cancel{(x-1)}}{\cancel{(3x+1)}\cancel{(x+2)}}\cdot\frac{\cancel{(x+2)}\cancel{(x-1)}}{x^6\cancel{(2x-3)}}$$

$$=\frac{(2x-3)^2}{x^6}$$

17.

$$\frac{-3t+4}{t^2+t-20}+\frac{6+5t}{t^2+t-20}=\frac{-3t+4+6+5t}{t^2+t-20}$$

$$=\frac{2t+10}{t^2+t-20}$$

$$=\frac{2\cancel{(t+5)}}{\cancel{(t+5)}(t-4)}$$

$$=\frac{2}{t-4}$$

18.

$$\frac{3wx}{wx-5}+\frac{wx+10}{5-wx}=\frac{3wx}{wx-5}-\frac{wx+10}{wx-5}$$

$$=\frac{3wx-(wx+10)}{wx-5}$$

$$=\frac{3wx-wx-10}{wx-5}$$

$$=\frac{2wx-10}{wx-5}$$

$$=\frac{2\cancel{(wx-5)}}{\cancel{(wx-5)}}$$

$$=2$$

19. $f(x)+g(x)=(8x-5)+\left(\frac{5x+4}{3x+1}\right)$

$$=\frac{8x}{1}\left(\frac{3x+1}{3x+1}\right)-\frac{5}{1}\left(\frac{3x+1}{3x+1}\right)+\frac{5x+4}{3x+1}$$

$$=\frac{24x^2+8x}{3x+1}-\frac{15x+5}{3x+1}+\frac{5x+4}{3x+1}$$

$$=\frac{24x^2+8x-(15x+5)+5x+4}{3x+1}$$

$$=\frac{24x^2+8x-15x-5+5x+4}{3x+1}$$

$$=\frac{24x^2-2x-1}{3x+1}$$

20.

$$\frac{a+3}{a^2-a-2}-\frac{a-4}{a^2-2a-3}$$

$$=\frac{a+3}{(a-2)(a+1)}-\frac{a-4}{(a-3)(a+1)}$$

$$=\frac{a+3}{(a-2)(a+1)}\left(\frac{a-3}{a-3}\right)-\frac{a-4}{(a-3)(a+1)}\left(\frac{a-2}{a-2}\right)$$

$$=\frac{a^2-9}{(a-2)(a+1)(a-3)}-\frac{a^2-6a+8}{(a-2)(a+1)(a-3)}$$

$$=\frac{a^2-9-(a^2-6a+8)}{(a-2)(a+1)(a-3)}$$

$$=\frac{a^2-9-a^2+6a-8}{(a-2)(a+1)(a-3)}$$

$$=\frac{6a-17}{(a-2)(a+1)(a-3)}$$

21.

$$\frac{\dfrac{2u^2w^3}{v^2}}{\dfrac{4uw^4}{uv}} = \frac{\dfrac{2u^2w^3}{v^2}}{\dfrac{4uw^4}{uv}} \bullet \frac{uv^2}{uv^2}$$

$$= \frac{\left(\dfrac{uv^2}{1}\right)\left(\dfrac{2u^2w^3}{v^2}\right)}{\left(\dfrac{uv^2}{1}\right)\left(\dfrac{4uw^4}{uv}\right)}$$

$$= \frac{2u^3w^3}{4uvw^4}$$

$$= \frac{u^2}{2vw}$$

22.

$$\frac{\dfrac{4}{3k}+\dfrac{k}{k+1}}{\dfrac{k}{k+1}-\dfrac{3}{k}} = \frac{\dfrac{4}{3k}+\dfrac{k}{k+1}}{\dfrac{k}{k+1}-\dfrac{3}{k}} \bullet \frac{3k(k+1)}{3k(k+1)}$$

$$= \frac{3k(k+1)\left(\dfrac{4}{3k}\right)+3k(k+1)\left(\dfrac{k}{k+1}\right)}{3k(k+1)\left(\dfrac{k}{k+1}\right)-3k(k+1)\left(\dfrac{3}{k}\right)}$$

$$= \frac{4(k+1)+3k(k)}{3k(k)-9(k+1)}$$

$$= \frac{4k+4+3k^2}{3k^2-9k-9}$$

$$= \frac{3k^2+4k+4}{3k^2-9k-9}$$

23.

$$\frac{18x^2y^3-12x^3y^2+9xy}{-3xy^4}$$

$$= \frac{18x^2y^3}{-3xy^4} - \frac{12x^3y^2}{-3xy^4} + \frac{9xy}{-3xy^4}$$

$$= \frac{-6x}{y} + \frac{4x^2}{y^2} - \frac{3}{y^3}$$

24.

$$y^2-2y+4-\frac{56}{y+2}$$

$$y+2\,\overline{)\,y^3+0y^2+0y-48}$$

$$\underline{y^3+2y^2}$$

$$-2y^2+0y$$

$$\underline{-2y^2-4y}$$

$$4y-48$$

$$\underline{4y+8}$$

$$-56$$

25.

2	4	3	2	−7
		8	22	48
	4	11	24	41

The remainder is 41, so $P(2)=41$.

26.

-3	1	3	-16	-27	63
		-3	0	48	-63
	1	0	-16	21	0

The remainder is 0 so $x+3$ is a factor of $P(x)$.

27.

$$\frac{34}{x^2}+\frac{13}{20x}=\frac{3}{2x}$$

$$20x^2\left(\frac{34}{x^2}+\frac{13}{20x}\right)=20x^2\left(\frac{3}{2x}\right)$$

$$20(34)+x(13)=10x(3)$$

$$680+13x=30x$$

$$680=17x$$

$$40=x$$

28.

$$\frac{u-2}{u-3}+3=u+\frac{u-4}{3-u}$$
$$\frac{u-2}{u-3}+3=u-\frac{u-4}{u-3}$$
$$u-3\left(\frac{u-2}{u-3}+3\right)=u-3\left(u-\frac{u-4}{u-3}\right)$$
$$u-2+3(u-3)=u(u-3)-(u-4)$$
$$u-2+3u-9=u^2-3u-u+4$$
$$4u-11=u^2-4u+4$$
$$4u-11-4u+11=u^2-4u+4-4u+11$$
$$0=u^2-8u+15$$
$$0=(u-3)(u-5)$$
$$u-3=0 \quad \text{or} \quad u-5=0$$
$$u=3 \qquad u=5$$
$u=5$; 3 is extraneous.

29.

$$\frac{3}{x-2}=\frac{x+3}{2x}$$
$$2x(x-2)\left(\frac{3}{x-2}\right)=2x(x-2)\left(\frac{x+3}{2x}\right)$$
$$2x(3)=(x-2)(x+3)$$
$$6x=x^2+3x-2x-6$$
$$6x=x^2+x-6$$
$$6x-6x=x^2+x-6-6x$$
$$0=x^2-5x-6$$
$$0=(x+1)(x-6)$$
$$x+1=0 \quad \text{or} \quad x-6=0$$
$$x=-1 \qquad x=6$$

30.

$$\frac{4}{m^2-9}+\frac{5}{m^2-m-12}=\frac{7}{m^2-7m+12}$$
$$\frac{4}{(m+3)(m-3)}+\frac{5}{(m-4)(m+3)}=\frac{7}{(m-3)(m-4)}$$
Multiply by LCD of $(m+3)(m-3)(m-4)$
$$4(m-4)+5(m-3)=7(m+3)$$
$$4m-16+5m-15=7m+21$$
$$9m-31=7m+21$$
$$2m-31=21$$
$$2m=52$$
$$m=26$$

31.

$$\frac{AB}{C}=\frac{DE}{F}$$
$$CF\left(\frac{AB}{C}\right)=CF\left(\frac{DE}{F}\right)$$
$$ABF=CDE$$
$$\frac{ABF}{AB}=\frac{CDE}{AB}$$
$$F=\frac{CDE}{AB}$$

32.

$$\frac{1}{r}=\frac{1}{r_1}+\frac{1}{r_2}$$
$$rr_1r_2\left(\frac{1}{r}\right)=rr_1r_2\left(\frac{1}{r_1}+\frac{1}{r_2}\right)$$
$$r_1r_2=rr_2+rr_1$$
$$r_1r_2-rr_2=rr_2+rr_1-rr_2$$
$$r_2(r_1-r)=rr_1$$
$$\frac{r_2(r_1-r)}{r_1-r}=\frac{rr_1}{r_1-r}$$
$$r_2=\frac{rr_1}{r_1-r}$$

33. Let x = the number of hours it takes both crews to roof the house working together.

1st crew's work in 1 hour = $\frac{1}{12}$

2nd crew's work in 1 hour = $\frac{1}{10}$

Amount of work together in 1 hour = $\frac{1}{x}$

$$\frac{1}{12}+\frac{1}{10}=\frac{1}{x}$$
$$60x\left(\frac{1}{12}+\frac{1}{10}\right)=60x\left(\frac{1}{x}\right)$$
$$5x+6x=60$$
$$11x=60$$
$$x=\frac{60}{11}$$
$$x=5\frac{5}{11} \text{ hours}$$

No, they will not be able to finish in 5 hours. They will have to work in the rain $\frac{5}{11}$ of an hour.

34. Let x = the number of hours it the supervisor to clean the room.

Supervisor's work in 1 minute = $\frac{1}{x}$

Technician's work in 1 minute = $\frac{1}{x+45}$

Work together in 1 minute = $\frac{1}{30}$

$$\frac{1}{x}+\frac{1}{x+45}=\frac{1}{30}$$
$$30x(x+45)\left(\frac{1}{x}+\frac{1}{x+45}\right)=30x(x+45)\left(\frac{1}{30}\right)$$
$$30(x+45)+30x=x(x+45)$$
$$30x+1,350+30x=x^2+45x$$
$$60x+1,350=x^2+45x$$
$$0=x^2-15x-1,350$$
$$0=(x-45)(x+30)$$
$$x-45=0 \quad \text{or} \quad x+30=0$$
$$x=45 \qquad x=\cancel{-30}$$

The time cannot be negative. Thus, the supervisor's time is 45 minutes and the technician's time is 45 + 45 = 90 minutes.

35. Let x = the rate walking.

	distance	rate	time
walking	24	x	$t=\frac{d}{r}=\frac{24}{x}$
biking	24	$x+5$	$t=\frac{d}{r}=\frac{24}{x+5}$

Time biking + time walking = total time
$$\frac{24}{x}+\frac{24}{x+5}=11$$
$$x(x+5)\left(\frac{24}{x}+\frac{24}{x+5}\right)=x(x+5)(11)$$
$$24(x+5)+24(x)=11x(x+5)$$
$$24x+120+24x=11x^2+55x$$
$$48x+120=11x^2+55x$$
$$48x+120-48x-120=11x^2+55x-48x-120$$
$$0=11x^2+7x-120$$
$$0=(11x+40)(x-3)$$
$$11x+40=0 \quad \text{or} \quad x-3=0$$
$$11x=-40 \qquad x=3$$
$$x=-\frac{40}{11}$$

The rate cannot be negative. Thus, the rate walking is 3 mph.

36. Let x = the rate of the current.

	distance	rate	time
upstream	8	$6-x$	$t=\frac{d}{r}=\frac{8}{6-x}$
downstream	8	$6+x$	$t=\frac{d}{r}=\frac{8}{6+x}$

Time up + time down = total time

$$\frac{8}{6+x}+\frac{8}{6-x}=3$$
$$(6+x)(6-x)\left(\frac{8}{6+x}+\frac{8}{6-x}\right)=3(6+x)(6-x)$$
$$8(6-x)+8(6+x)=3(36-x^2)$$
$$48-8x+48+8x=108-3x^2$$
$$96=108-3x^2$$
$$3x^2-12=0$$
$$3(x^2-4)=0$$
$$3(x-2)(x+2)=0$$
$$x-2=0 \quad \text{or} \quad x+2=0$$
$$x=2 \qquad x=\cancel{-2}$$

The rate cannot be negative. Thus, the rate of the current is 2 mph.

37.
$$\frac{5}{1.5}=\frac{x}{24}$$
$$5(24)=1.5x$$
$$120=1.5x$$
$$80=x$$
$$x=80 \text{ ft}$$

38.
$$\frac{12}{57.99}=\frac{16}{x}$$
$$12x=16(57.99)$$
$$12x=927.84$$
$$x=\$77.32$$

39. $r=\frac{kx}{yz}$

Find k.
$$1.5=\frac{k(1{,}000)}{(50)(10)}$$
$$1.5=\frac{1{,}000k}{500}$$
$$1.5=2k$$
$$0.75=k$$
Find r.
$$r=\frac{(0.75)(0.8)}{(0.02)(0.1)}$$
$$=\frac{0.6}{0.002}$$
$$=300$$

40. Answers will vary.
Example of a direct variation graph of the amount of money someone will earn if they earn \$5 per hour.

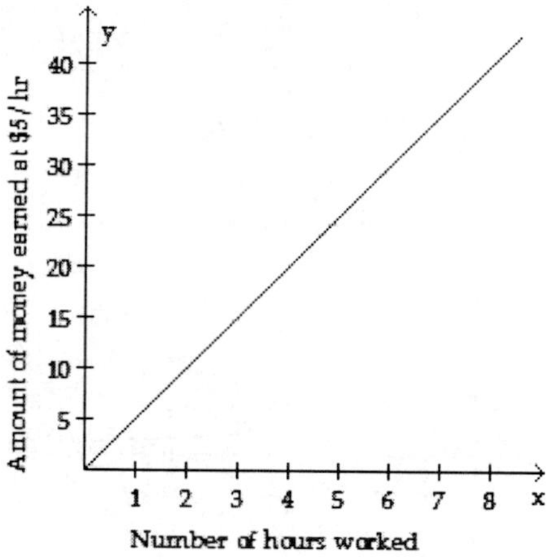

41. Let L = loudness and d = distance.

$$L = \frac{k}{d^2}$$
$$100 = \frac{k}{30^2}$$
$$100 = \frac{k}{900}$$
$$90{,}000 = k$$

Find L if $k = 90{,}000$ and $d = 60$.

$$L = \frac{k}{d^2} = \frac{90{,}000}{60^2} = \frac{90{,}000}{3{,}600} = 25 \text{ decibels}$$

42. $f(x) = \frac{2}{x}$ for $x > 0$

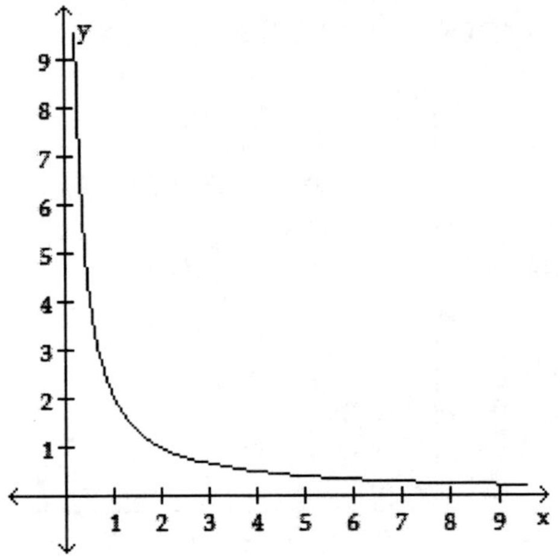

The horizontal asymptote is $y = 0$, the x-axis and the vertical asymptote is $x = 0$, the y-axis.

CHAPTER 6 CUMULATIVE REVIEW

1.

$$\begin{aligned} &12-6\left[\left(130-4^3\right)-2\right] \\ &=12-6\left[\left(130-64\right)-2\right] \\ &=12-6\left[66-2\right] \\ &=12-6\left[64\right] \\ &=12-384 \\ &=-372 \end{aligned}$$

2.

$$\begin{aligned} &9\left(a^3+3a\right)-5\left(3a-a^3\right)-8\left(-a-a^3\right) \\ &=9a^3+27a-15a+5a^3+8a+8a^3 \\ &=22a^3+20a \end{aligned}$$

3.

$$\begin{aligned} \frac{3x-4}{6}-\frac{x-2}{2}&=\frac{-2x-3}{3} \\ 6\left(\frac{3x-4}{6}-\frac{x-2}{2}\right)&=6\left(\frac{-2x-3}{3}\right) \\ 3x-4-3\left(x-2\right)&=2\left(-2x-3\right) \\ 3x-4-3x+6&=-4x-6 \\ 2&=-4x-6 \\ 8&=-4x \\ -2&=x \end{aligned}$$

4.

$$\begin{aligned} l&=a+\left(n-1\right)d \\ l&=a+dn-d \\ l-a+d&=dn \\ \frac{l-a+d}{d}&=n \end{aligned}$$

5.

$$\begin{aligned} r&=\frac{d}{2} \\ r&=\frac{25}{2} \\ A&=r^2\pi \\ &=\left(\frac{25}{2}\right)^2\pi \\ &=156.25\pi \\ &\approx 490.9\text{ in.}^2 \end{aligned}$$

6. Let <B = x°. Then <A = (x + 10)° and <C = (x + 10) + 10 = x + 20°. The sum of all three angles is 180°.

$$\begin{aligned} \angle A+\angle B+\angle C&=180 \\ \left(x+10\right)+\left(x\right)+\left(x+20\right)&=180 \\ 3x+30&=180 \\ 3x&=150 \\ x&=50 \\ \angle B&=50^\circ \\ \angle A&=50+10=60^\circ \\ \angle C&=50+20=70^\circ \end{aligned}$$

7. $5{,}530{,}000-6{,}231{,}000=-701{,}000$

$-701{,}000$ is what percent of $6{,}231{,}000$?

$$\begin{aligned} -701{,}000&=x\left(6{,}231{,}000\right) \\ \frac{-701{,}000}{6{,}231{,}000}&=\frac{x\left(6{,}231{,}000\right)}{6{,}231{,}000} \\ -0.113&\approx x \\ x&\approx 11.3\%\text{ decrease} \end{aligned}$$

8. (–5, 2) and (13, –3)

$$\begin{aligned} \left(\frac{x_1+x_2}{2},\frac{y_1+y_2}{2}\right)&=\left(\frac{-5+13}{2},\frac{2+\left(-3\right)}{2}\right) \\ &=\left(\frac{8}{2},\frac{-1}{2}\right) \\ &=\left(4,-\frac{1}{2}\right) \end{aligned}$$

9.

$$\begin{aligned} \frac{\text{change in }y}{\text{change in }x}&=\frac{83.8-78.8}{2050-2000} \\ &=\frac{5}{50} \\ &=\frac{1}{10} \\ &=0.1 \end{aligned}$$

Life expectancy will increase 0.1 each year during this period.

10. x – intercept: Let $y = 0$.

$$7x-10(0)=-7$$
$$7x=-7$$
$$x=-1$$
$$(-1,0)$$

y – intercept: Let $x = 0$.

$$7(0)-10y=-7$$
$$-10y=-7$$
$$y=\frac{7}{10}$$
$$\left(0,\frac{7}{10}\right)$$

11.

$$y-y_1=m(x-x_1)$$
$$y-5=-7(x-7)$$
$$y-5=-7x+49$$
$$y=-7x+54$$

12.

$$m=\frac{y_2-y_1}{x_2-x_1}$$
$$=\frac{-6-5}{2-(-4)}$$
$$=\frac{-11}{6}$$
$$y-y_1=m(x-x_1)$$
$$y-(-6)=-\frac{11}{6}(x-2)$$
$$y+6=-\frac{11}{6}x+\frac{11}{3}$$
$$y+6-6=-\frac{11}{6}x+\frac{11}{3}-6$$
$$y=-\frac{11}{6}x+\frac{11}{3}-\frac{18}{3}$$
$$y=-\frac{11}{6}x-\frac{7}{3}$$

13. a. $-\frac{7}{3}$

b. 1

c. $y=-\frac{7}{3}x-7$

d. Yes.

$$y=-\frac{7}{3}x-7$$
$$-42\stackrel{?}{=}-\frac{7}{3}(15)-7$$
$$-42\stackrel{?}{=}-35-7$$
$$-42=-42$$

14. $g(x)=-3x^3+x-4$

$$g(-2)=-3(-2)^3+(-2)-4$$
$$=-3(-8)-2-4$$
$$=24-2-4$$
$$=18$$

15. a. No. Answers may vary. Examples of ordered pairs that proves it is not a functions are (1, –1) and (1, 1).

b. No. Answers may vary. Examples of ordered pairs that proves it is not a functions are (0, 2) and (0, –2).

16. a. Find m using (8, 365) and (18, 384).

$$m=\frac{y_2-y_1}{x_2-x_1}$$
$$=\frac{384-365}{18-8}$$
$$=\frac{19}{10}$$
$$=1.9$$
$$y-y_1=m(x-x_1)$$
$$y-365=1.9(x-8)$$
$$y-365=1.9x-15.2$$
$$y-365+365=1.9x-15.2+365$$
$$y=1.9x+349.8$$
$$c(t)=1.9t+349.8$$

b. Let $t = 30$; 2020 is 30 years after 1990.

$$c(30)=1.9(30)+349.8$$
$$=57+349.8$$
$$=406.8$$

17. $f(x) = |x + 4|$

D: $(-\infty, \infty)$; R: $[0, \infty)$

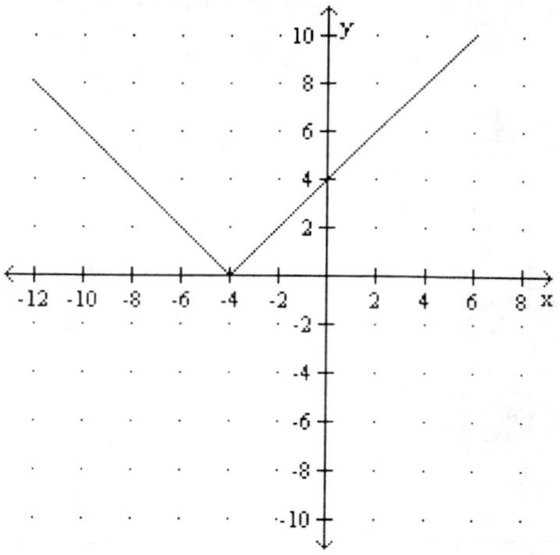

18. $\begin{cases} x = y + 3 \\ \frac{1}{4}x - \frac{1}{6}y = \frac{1}{3} \end{cases} = \begin{cases} x = y + 3 \\ 3x - 2y = 4 \end{cases}$

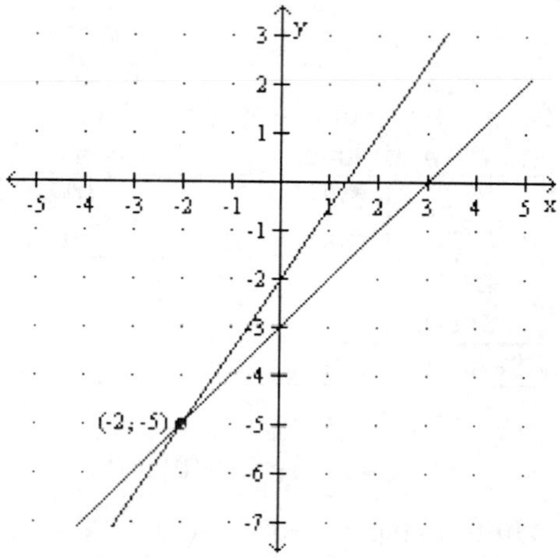

19.

$\begin{cases} 0.6T_1 - 0.8T_2 = 0 \\ 0.8T_1 + 0.6T_2 = 100 \end{cases}$

Multiply each equation by 10 to eliminate decimals.

$\begin{cases} 6T_1 - 8T_2 = 0 \\ 8T_1 + 6T_2 = 1{,}000 \end{cases}$

Multiply the first equation by 3 and the second equation by 4. Add the equations to eliminate T_2.

$$\begin{aligned} 18T_1 - 24T_2 &= 0 \\ \underline{32T_1 + 24T_2} &\underline{= 4{,}000} \\ 50T_1 \qquad &= 4{,}000 \\ T_1 &= 80 \end{aligned}$$

Substitute $T_1 = 80$ into the first equation and solve for T_2.

$$\begin{aligned} 6T_1 - 8T_2 &= 0 \\ 6(80) - 8T_2 &= 0 \\ 480 - 8T_2 &= 0 \\ -8T_2 &= -480 \\ T_2 &= 60 \end{aligned}$$

20.

$$\begin{cases} x+2y+3z=11 & (1) \\ 5x-y-z=8 & (2) \\ 2x+y-3z=-14 & (3) \end{cases}$$

Add Equations 1 and 3 to eliminate z.

$$\begin{array}{rl} x+2y+3z=11 & (1) \\ \underline{2x+y-3z=-14} & (3) \\ 3x+3y=-3 & (4) \end{array}$$

Multiply Equation 2 by 3 and add Equations 1 and 2 to eliminate z.

$$\begin{array}{rl} x+2y+3z=11 & (1) \\ \underline{15x-3y-3z=24} & (2) \\ 16x-y=35 & (5) \end{array}$$

Multiply Equation 5 by 3 and add Equations 4 and 5 to eliminate y.

$$\begin{array}{rl} 3x+3y=-3 & (4) \\ \underline{48x-3y=105} & (5) \\ 51x=102 & \\ x=2 & \end{array}$$

Substitute $x = 2$ into Equation 4 and solve for y.

$$\begin{array}{rl} 3x+3y=-3 & (4) \\ 3(2)+3y=-3 & \\ 6+3y=-3 & \\ 3y=-9 & \\ y=-3 & \end{array}$$

Substitute $x = 2$ and $y = -3$ into Equation 1 and solve for z.

$$\begin{array}{rl} x+2y+3z=11 & (1) \\ (2)+2(-3)+3z=11 & \\ 2-6+3z=11 & \\ -4+3z=11 & \\ 3z=15 & \\ z=5 & \end{array}$$

The solution is (2, –3, 5).

21.

$$x=\frac{D_x}{D}=\frac{\begin{vmatrix} 10 & -4 \\ 2 & -7 \end{vmatrix}}{\begin{vmatrix} 5 & -4 \\ 1 & -7 \end{vmatrix}}=\frac{-70-(-8)}{-35-(-4)}=\frac{-62}{-31}$$

$$x=2$$

$$y=\frac{D_y}{D}=\frac{\begin{vmatrix} 5 & 10 \\ 1 & 2 \end{vmatrix}}{\begin{vmatrix} 5 & -4 \\ 1 & -7 \end{vmatrix}}=\frac{10-10}{-35-(-4)}=\frac{0}{-31}$$

$$y=0$$

22.

$$\begin{vmatrix} 3 & -2 \\ -2 & 4 \end{vmatrix}=3(4)-(-2)(-2)=12-4=8$$

23. Let x = # of ounces of 50% silver and y = # of ounces of 25% silver

	50% silver	25% silver	40% mixture
# of ounces	x	y	20
% silver	$0.50x$	$0.25y$	0.40(20)

$$\begin{cases} x+y=20 \\ 0.50x+0.25y=0.40(20) \end{cases}$$

Multiply the 2nd equation by 100 to eliminate decimals.

$$\begin{cases} x+y=20 \\ 50x+25y=800 \end{cases}$$

Multiply the 1st equation by -50.
Add equations to eliminate x.

$$\begin{array}{r} -50x-50y=-1{,}000 \\ \underline{50x+25y=800} \\ -25y=-200 \\ y=8 \end{array}$$

Substitute $y = 8$ in the 1st equation and solve for x.

$$\begin{array}{r} x+y=20 \\ x+8=20 \\ x=12 \end{array}$$

12 ounces of 50% and 8 ounces of 25%

24. Let x = Pizza Hut, y = Domino's, and z = Papa John's.

$$\begin{cases} x+y+z=28 & (1) \\ x=y+z & (2) \\ z=y-2 & (3) \end{cases}$$

Substitute Equation 3 into Equation 2.

$$x=y+(y-2)$$
$$x=2y-2 \qquad (4)$$

Substitute Equations 3 and 4 into Equation 1 and solve for y.

$$x+y+z=28$$
$$(2y-2)+y+(y-2)=28$$
$$4y-4=28$$
$$4y=32$$
$$y=8$$

Substitute $y = 8$ into Equation 3 and solve for z.

$$z=y-2 \quad (3)$$
$$z=8-2$$
$$z=6$$

Substitute $y = 8$ and $z = 6$ into Equation 2 and solve for x.

$$x=y+z \quad (2)$$
$$x=8+6$$
$$x=14$$

Pizza Hut has 14%, Domino's has 8%, and Papa John's has 6%.

25.

$$\frac{1}{2}x+6\ge 4+2x$$
$$2\left(\frac{1}{2}x+6\right)\ge 2(4+2x)$$
$$x+12\ge 8+4x$$
$$-3x+12\ge 8$$
$$-3x\ge -4$$
$$x\le \frac{4}{3};\ \left(-\infty,\frac{4}{3}\right]$$

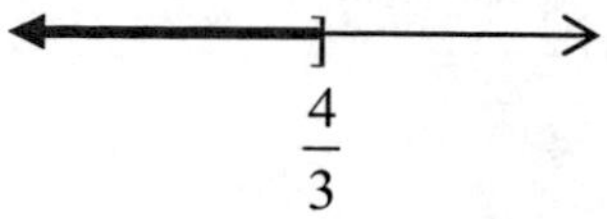

26.

$$5(x+2)\le 4(x+1) \quad \text{and} \quad 11+x<0$$
$$5x+10\le 4x+4 \qquad 11+x-11<0-11$$
$$5x+10-10\le 4x+4-10 \qquad x<-11$$
$$5x\le 4x-6$$
$$5x-4x\le 4x-6-4x$$
$$x\le -6$$
$$(-\infty,-11)$$

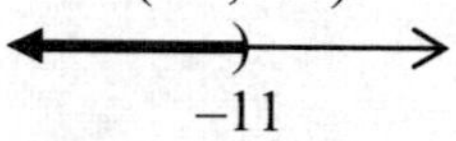

27.

$$-4(x+2)\ge 12 \quad \text{or} \quad 3x+8<11$$
$$-4x-8\ge 12 \qquad 3x+8-8<11-8$$
$$-4x-8+8\ge 12+8 \qquad 3x<3$$
$$-4x\ge 20 \qquad \frac{3x}{3}<\frac{3}{3}$$
$$\frac{-4x}{-4}\le\frac{20}{-4} \qquad x<1$$
$$x\le -5$$
$$(-\infty,1)$$

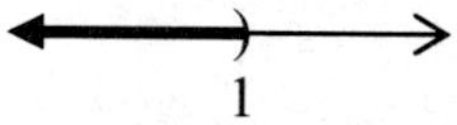

28.

$$\left|\frac{3a}{5}-2\right|+1\ge\frac{6}{5}$$
$$\left|\frac{3a}{5}-2\right|+1-1\ge\frac{6}{5}-1$$
$$\left|\frac{3a}{5}-2\right|\ge\frac{1}{5}$$
$$\frac{3a}{5}-2\ge\frac{1}{5} \quad \text{or} \quad \frac{3a}{5}-2\le-\frac{1}{5}$$
$$5\left(\frac{3a}{5}-2\right)\ge 5\left(\frac{1}{5}\right) \qquad 5\left(\frac{3a}{5}-2\right)\le 5\left(-\frac{1}{5}\right)$$
$$3a-10\ge 1 \qquad 3a-10\le -1$$
$$3a\ge 11 \qquad 3a\le 9$$
$$a\ge\frac{11}{3} \qquad a\le 3$$
$$(-\infty,3]\cup\left[\frac{11}{3},\infty\right)$$

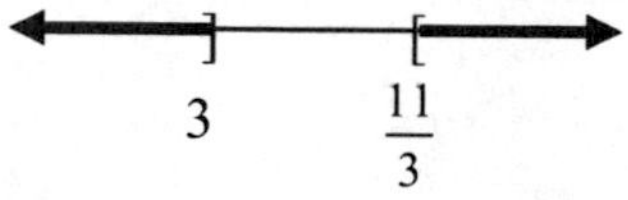

29. $y < 4 - x$

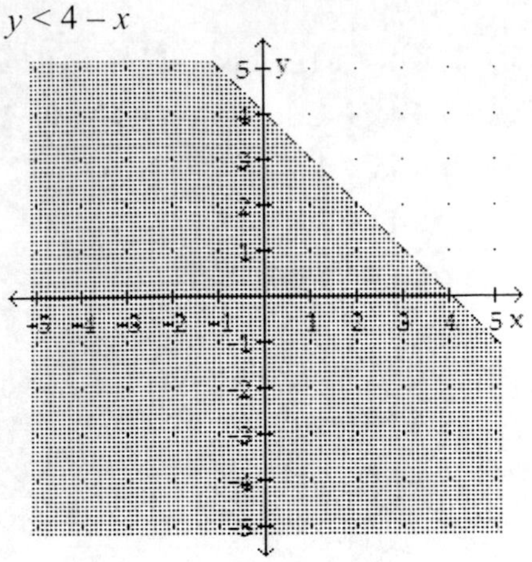

30. $\begin{cases} 3x + 2y > 6 \\ x + y \le 2 \end{cases}$

$$3x + 2y > 6 \qquad x + y \le 2$$

$$3(4) + 2(-2) \overset{?}{>} 6 \qquad 4 + (-2) \overset{?}{\le} 2$$

$$12 - 4 \overset{?}{>} 6 \qquad 2 \le 2$$

$$8 > 6$$

Yes, (4, –2) is a solution.

31.

$$\left(3bb^2b^3c^0\right)^4 = \left(3b^{1+2+3} \cdot 1\right)^4$$
$$= \left(3b^6\right)^4$$
$$= 3^4 b^{6\cdot 4}$$
$$= 81b^{24}$$

32.

$$9^2 d^{-8}\left(d^9\right)^2 = 81d^{-8}d^{18}$$
$$= 81d^{-8+18}$$
$$= 81d^{10}$$

33.

$$-\frac{x^{-9}}{20k^{-7}} = -\frac{k^7}{20x^9}$$

34.

$$\left(\frac{2x^{-2}y^3}{x^2x^3y^4}\right)^{-3} = \left(\frac{2x^{-2}y^3}{x^5y^4}\right)^{-3}$$
$$= \left(2x^{-2-5}y^{3-4}\right)^{-3}$$
$$= \left(2x^{-7}y^{-1}\right)^{-3}$$
$$= \left(\frac{2}{x^7y}\right)^{-3}$$
$$= \left(\frac{x^7y}{2}\right)^{3}$$
$$= \frac{x^{7(3)}y^{1(3)}}{2^3}$$
$$= \frac{x^{21}y^3}{8}$$

35. penny: $\$10^{-2}$
dime: $\$10^{-1}$
one dollar bill: $\$10^0$
one hundred thousand dollar bill: $\$10^5$

36.

$$\frac{(0.00024)(96,000,000)}{640,000,000} = \frac{\left(2.4\times10^{-4}\right)\left(9.6\times10^7\right)}{6.4\times10^8}$$
$$= \frac{23.04\times10^3}{6.4\times10^8}$$
$$= \frac{2.304\times10^4}{6.4\times10^8}$$
$$= 0.36\times10^{-4}$$
$$= 3.6\times10^{-5}$$
$$= 0.000036$$

37. a. $f(-1) = 4$
b. $f(2) = 5$
c. $x = -2$ and $x = 1$
d. D: $(-\infty, \infty)$; R: $(-\infty, \infty)$

38. Let $t = 3$.

$$h(t) = -16t^2 + 144t$$
$$h(3) = -16(3)^2 + 144(3)$$
$$= -16(9) + 432$$
$$= -144 + 432$$
$$= 288$$

The height is 288 feet 3 seconds after firing.

39. The degree is 7.
(3 + 4 = 7)

40. $\frac{9}{4}c^2 - \frac{5}{3}c - \frac{1}{2}c^2 + \frac{5}{6}c$
$= \frac{7}{4}c^2 - \frac{5}{6}c$

41. $f(x) + g(x)$
$(x^3 + 3x^2 - 2x + 7) + (x^3 - 2x^2 + 2x + 5)$
$= 2x^3 + x^2 + 12$

42. $(2m^2n^2 + 2m - n) - (-2m^2n^2 - 2m + n)$
$= 2m^2n^2 + 2m - n + 2m^2n^2 + 2m - n$
$= 4m^2n^2 + 4m - 2n$

43.
$$\left(\frac{1}{16}r^9s^{10}\right)(32r^2s^{10}) = \left(\frac{1}{16}\cdot 32\right)r^{9+2}s^{10+10}$$
$$= 2r^{11}s^{20}$$

44.
$$-3a^8(4a^4 + 3a^3 - 4a^2) = -12a^{12} - 9a^{11} + 12a^{10}$$

45.
$$(2x^3 - 1)^2 = (2x^3 - 1)(2x^3 - 1)$$
$$= 4x^6 - 2x^3 - 2x^3 + 1$$
$$= 4x^6 - 4x^3 + 1$$

46.
$(a + b + c)(2a - b - 2c)$
$= a(2a - b - 2c) + b(2a - b - 2c) + c(2a - b - 2c)$
$= 2a^2 - ab - 2ac + 2ab - b^2 - 2bc + 2ac - bc - 2c^2$
$= 2a^2 + ab - b^2 - 3bc - 2c^2$

47. $3r^2s^3 - 6rs^4 = 3rs^3(r - 2s)$

48. $5(x - y) - a(x - y) = (x - y)(5 - a)$

49.
$$xu + yv + xv + yu = xu + yu + xv + yv$$
$$= u(x + y) + v(x + y)$$
$$= (x + y)(u + v)$$

50.
$$3 - 10x^2 + 8x^4 = 8x^4 - 10x^2 + 3$$
$$= (2x^2 - 1)(4x^4 - 3)$$

51.
$$(x - y)^2 + 3(x - y) - 10 = (x - y + 5)(x - y - 2)$$

52.
$$81x^4 - 16y^4$$
$$= (9x^2 + 4y^2)(9x^2 - 4y^2)$$
$$= (9x^2 + 4y^2)(3x + 2y)(3x - 2y)$$

53.
$$8x^3 - 27y^6 = (2x - 3y^2)(4x^2 + 6xy^2 + 9y^4)$$

54.
$$x^2 + 10x + 25 - 16z^2 = (x^2 + 10x + 25) - 16z^2$$
$$= (x + 5)^2 - 16z^2$$
$$= (x + 5 - 4z)(x + 5 + 4z)$$

55.
$$b^2x^2 + a^2y^2 = a^2b^2$$
$$b^2x^2 + a^2y^2 - b^2x^2 = a^2b^2 - b^2x^2$$
$$a^2y^2 = a^2b^2 - b^2x^2$$
$$a^2y^2 = b^2(a^2 - x^2)$$
$$\frac{a^2y^2}{a^2 - x^2} = \frac{b^2(a^2 - x^2)}{a^2 - x^2}$$
$$\frac{a^2y^2}{a^2 - x^2} = b^2$$

56.
$$6x^2 + 7 = -23x$$
$$6x^2 + 23x + 7 = 0$$
$$(2x + 7)(3x + 1) = 0$$
$2x + 7 = 0$ or $3x + 1 = 0$
$2x = -7 \qquad 3x = -1$
$x = -\frac{7}{2} \qquad x = -\frac{1}{3}$

57.

$$x^3 - 4x = 0$$
$$x(x^2 - 4) = 0$$
$$x(x+2)(x-2) = 0$$
$$x = 0,\ x+2 = 0,\ \text{or}\ x - 2 = 0$$
$$x = -2 \qquad x = 2$$

58. Let the width = x and the length = x + 3.

$$A = LW$$
$$108 = x(x+3)$$
$$108 = x^2 + 3x$$
$$0 = x^2 + 3x - 108$$
$$0 = (x+12)(x-9)$$
$$x + 12 = 0 \ \text{or}\ x - 9 = 0$$
$$x = -12 \qquad x = 9$$

Since the width cannot be negative, the width = 9 and the length = 9 + 3 = 12.
9 in. by 12 in.

59.

$$\frac{2x^2y + xy - 6y}{3x^2y + 5xy - 2y} = \frac{y(2x^2 + x - 6)}{y(3x^2 + 5x - 2)}$$
$$= \frac{\cancel{y}(2x-3)\cancel{(x+2)}}{\cancel{y}(3x-1)\cancel{(x+2)}}$$
$$= \frac{2x-3}{3x-1}$$

60. Set the denominator equal to 0.

$$x^2 - 2x = 0$$
$$x(x-2) = 0$$
$$x = 0 \ \text{or}\ x - 2 = 0$$
$$x = 2$$

The domain is all real numbers except 0 and 2. $D:(-\infty,0)\cup(0,2)\cup(2,\infty)$

61.

$$(10n - n^2)\bullet\frac{n^6}{n^4 - 10n^3 - 2n^2 + 20n}$$
$$= \frac{(10n - n^2)(n^6)}{n^4 - 10n^3 - 2n^2 + 20n}$$
$$= \frac{n(10-n)(n^6)}{n(n^2-2)(n-10)}$$
$$= -\frac{\cancel{n}\cancel{(n-10)}(n^6)}{\cancel{n}(n^2-2)\cancel{(n-10)}}$$
$$= -\frac{n^6}{n^2 - 2}$$

62.

$$\frac{p^3 - q^3}{q^2 - p^2} \div \frac{p^3 + p^2q + pq^2}{q^2 + pq}$$
$$= \frac{p^3 - q^3}{q^2 - p^2}\bullet\frac{q^2 + pq}{p^3 + p^2q + pq^2}$$
$$= \frac{(p-q)(p^2 + pq + q^2)}{(q+p)(q-p)}\bullet\frac{q(q+p)}{p(p^2 + pq + q^2)}$$
$$= \frac{\cancel{(p-q)}\cancel{(p^2 + pq + q^2)}}{-\cancel{(q+p)}\cancel{(p-q)}}\bullet\frac{q\cancel{(q+p)}}{p\cancel{(p^2 + pq + q^2)}}$$
$$= -\frac{q}{p}$$

63.

$$\frac{2}{x+y}+\frac{3}{x-y}-\frac{x-3y}{x^2-y^2}$$
$$=\frac{2}{x+y}+\frac{3}{x-y}-\frac{x-3y}{(x+y)(x-y)}$$
$$=\frac{2}{x+y}\left(\frac{x-y}{x-y}\right)+\frac{3}{x-y}\left(\frac{x+y}{x+y}\right)-\frac{x-3y}{(x+y)(x-y)}$$
$$=\frac{2x-2y}{(x+y)(x-y)}+\frac{3x+3y}{(x+y)(x-y)}-\frac{x-3y}{(x+y)(x-y)}$$
$$=\frac{2x-2y+3x+3y-(x-3y)}{(x+y)(x-y)}$$
$$=\frac{2x-2y+3x+3y-x+3y}{(x+y)(x-y)}$$
$$=\frac{4x+4y}{(x+y)(x-y)}$$
$$=\frac{4(x+y)}{(x+y)(x-y)}$$
$$=\frac{4}{x-y}$$

64.

$$\frac{\frac{18}{c^2}+\frac{11}{c}+1}{1-\frac{3}{c}-\frac{10}{c^2}}=\frac{\frac{18}{c^2}+\frac{11}{c}+1}{1-\frac{3}{c}-\frac{10}{c^2}}\bullet\frac{c^2}{c^2}$$
$$=\frac{\frac{18c^2}{c^2}+\frac{11c^2}{c}+c^2}{c^2-\frac{3c^2}{c}-\frac{10c^2}{c^2}}$$
$$=\frac{18+11c+c^2}{c^2-3c-10}$$
$$=\frac{\cancel{(c+2)}(c+9)}{\cancel{(c+2)}(c-5)}$$
$$=\frac{c+9}{c-5}$$

65.

$$\frac{5y^4+45y^3}{15y^2}=\frac{5y^4}{15y^2}+\frac{45y^3}{15y^2}$$
$$=\frac{y^2}{3}+3y$$

66.

$$\begin{array}{r} 4x^2-x-1 \\ 4x+5\overline{)16x^3+16x^2-9x-5} \\ \underline{16x^3+20x^2} \\ -4x^2-9x \\ \underline{-4x^2-5x} \\ -4x-5 \\ \underline{-4x-5} \\ 0 \end{array}$$

67.

$$\frac{3}{x-2}+\frac{x^2}{(x+3)(x-2)}=\frac{x+4}{x+3}$$
$$(x+3)(x-2)\left(\frac{3}{x-2}+\frac{x^2}{(x+3)(x-2)}\right)=(x+3)(x-2)\left(\frac{x+4}{x+3}\right)$$
$$3(x+3)+x^2=(x-2)(x+4)$$
$$3x+9+x^2=x^2+4x-2x-8$$
$$x^2+3x+9=x^2+2x-8$$
$$x^2+3x+9-x^2=x^2+2x-8-x^2$$
$$3x+9=2x-8$$
$$x+9=-8$$
$$x=-17$$

68.

$$\frac{5x-3}{x+2}=\frac{5x+3}{x-2}$$
$$(x+2)(x-2)\left(\frac{5x-3}{x+2}\right)=(x+2)(x-2)\left(\frac{5x+3}{x-2}\right)$$
$$(x-2)(5x-3)=(x+2)(5x+3)$$
$$5x^2-3x-10x+6=5x^2+3x+10x+6$$
$$5x^2-13x+6=5x^2+13x+6$$
$$5x^2-13x+6-5x^2=5x^2+13x+6-5x^2$$
$$-13x+6=13x+6$$
$$-26x+6=6$$
$$-26x=0$$
$$x=0$$

69. $\dfrac{x^4-9x^3+x^2-7x-20}{x-9}$

$$\begin{array}{r|rrrrr} 9 & 1 & -9 & 1 & -7 & -20 \\ & & 9 & 0 & 9 & 18 \\ \hline & 1 & 0 & 1 & 2 & -2 \end{array}$$

$$x^3+x+2+\frac{-2}{x-9}$$

70. Let x = time when both doors are opened.

$$\frac{1}{6}+\frac{1}{10}=\frac{1}{x}$$

$$60x\left(\frac{1}{6}+\frac{1}{10}\right)=60x\left(\frac{1}{x}\right)$$

$$10x+6x=60$$

$$16x=60$$

$$x=\frac{60}{16}$$

$$x=3\frac{3}{4}$$

It takes $3\frac{3}{4}$ minutes when both doors are opened.

71. It will rise sharply.

72.

$$\frac{1.5\text{ c flour}}{15\text{ brownies}}=\frac{x\text{ c flour}}{130\text{ brownies}}$$

$$1.5(130)=x(15)$$

$$195=15x$$

$$13=x$$

13 cups of flour are needed.

73. Let d = days feed will last and a = number of animals.

Find k if $d = 4$ and $a = 300$.

$$d=\frac{k}{a}$$

$$4=\frac{k}{300}$$

$$300(4)=300\left(\frac{k}{300}\right)$$

$$1,200=k$$

Find d if $k = 1,200$ and $a = 1,200$.

$$d=\frac{k}{a}$$

$$d=\frac{1,200}{1,200}$$

$$d=1$$

The feed will last 1 day.

74. a. iii.
b. i.
c. iv.
d. ii.

SECTION 7.1

VOCABULARY

1. $5x^2$ is the **square** root of $25x^4$ because $(5x^2)^2 = 25x^4$. The **cube** root of 216 is 6 because $6^3 = 216$.

3. The radical symbol $\sqrt{\ }$ represents the **positive** or principle square root of a number.

5. The number 100 has two square roots. The positive or **principal** square root of 100 is 10.

7. When we write $\sqrt{b^4} = b^2$, we say that we have **simplified** the radical expression.

9. $f(x) = \sqrt{x}$ and $g(x) = \sqrt[3]{x}$ are **radical** functions.

CONCEPTS

11 b is a square root of a if $b^2 = \boxed{a}$.

13. $\sqrt{-4}$ is not a real number, because no real number **squared** equals –4.

15. $\sqrt{x^2} = \boxed{|x|}$ and $\sqrt[3]{x^3} = \boxed{x}$.

17. a. $f(11) = 3$
 b. $f(2) = 0$
 c. $f(-1)$ is undefined
 d. $x = 6$
 e. none
 f. D: $[2, \infty)$ and R: $[0, \infty)$

19. a. $f(-8) = -5$
 b. $f(0) = -3$
 c. $x = 1$
 d. D: $(-\infty, \infty)$ and R: $(-\infty, \infty)$

NOTATION

21. a. $\sqrt{x^2} = |x|$
 b. $\sqrt[3]{x^3} = x$

GUIDED PRACTICE

23.
$$\sqrt{100} = 10$$

25.
$$-\sqrt{64} = -1(8) = -8$$

27.
$$\sqrt{\frac{1}{9}} = \frac{\sqrt{1}}{\sqrt{9}} = \frac{1}{3}$$

29.
$$\sqrt{0.25} = 0.5$$

31.
$\sqrt{-81}$ is not real

33.
$$\sqrt{121} = 11$$

35.
$$\sqrt{12} = 3.4641$$

37.
$$\sqrt{679.25} = 26.0624$$

39.
$$\sqrt{4x^2} = \sqrt{(2x)^2} = |2x| = 2|x|$$

41.
$$\sqrt{81h^4} = \sqrt{(9h^2)^2} = |9h^2| = 9h^2$$

43.
$$\sqrt{36s^6} = \sqrt{(6s^3)^2} = |6s^3| = 6|s^3|$$

45.

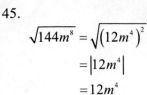

$$\sqrt{144m^8} = \sqrt{(12m^4)^2} = |12m^4| = 12m^4$$

47.

$$\sqrt{y^2 - 2y + 1} = \sqrt{(y-1)^2} = |y-1|$$

49.

$$\sqrt{a^4 + 6a^2 + 9} = \sqrt{(a^2+3)^2} = a^2 + 3$$

51. $f(x) = -\sqrt{x}$

x	y
0	$-\sqrt{0} = 0$
1	$-\sqrt{1} = -1$
4	$-\sqrt{4} = -2$
9	$-\sqrt{9} = -3$
16	$-\sqrt{16} = -4$

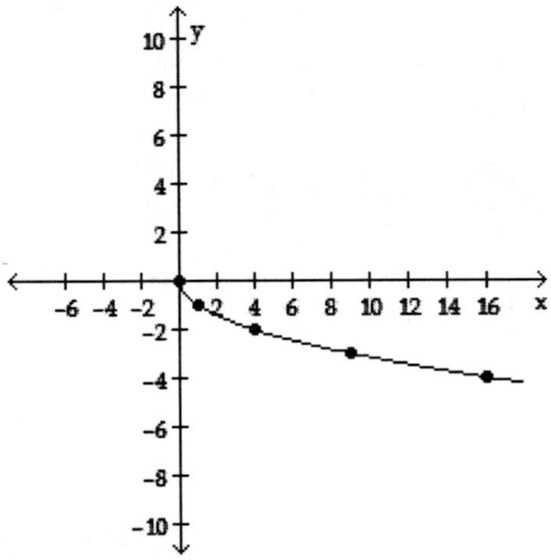

D: $[0, \infty)$ and R: $(-\infty, 0]$

53. $f(x) = \sqrt{x+4}$

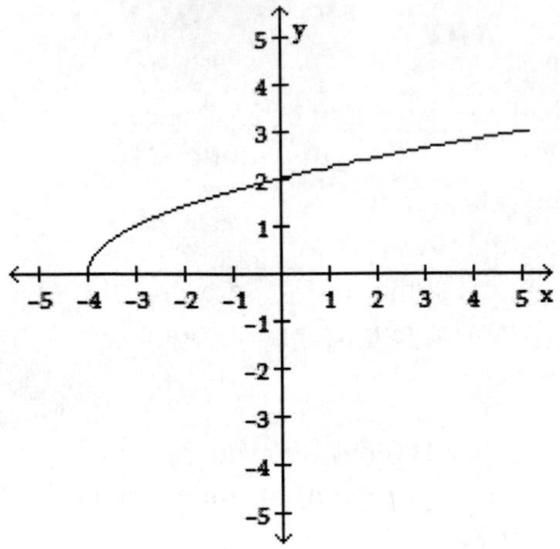

$D: [-4, \infty)$; $R: [0, \infty)$

55. $f(x) = \sqrt{x+6}$

$x + 6 \geq 0$

$x \geq -6$

$D: [-6, \infty)$

57. $g(x) = \sqrt{8 - 2x}$

$8 - 2x \geq 0$

$-2x \geq -8$

$x \leq 4$

$D: (-\infty, 4]$

59. $s(t) = \sqrt{9t - 4}$

$9t - 4 \geq 0$

$9t \geq 4$

$x \geq \frac{4}{9}$

$D: \left[\frac{4}{9}, \infty\right)$

61. $c(x) = \sqrt{0.5x - 20}$

$0.5x - 20 \geq 0$

$0.5x \geq 20$

$x \geq 40$

$D: [40, \infty)$

63. $f(x)=\sqrt{3x+1}$

a.

$$\begin{aligned} f(8) &= \sqrt{3(8)+1} \\ &= \sqrt{24+1} \\ &= \sqrt{25} \\ &= 5 \end{aligned}$$

b.

$$\begin{aligned} f(-2) &= \sqrt{3(-2)+1} \\ &= \sqrt{-6+1} \\ &= \sqrt{-5} \\ &= \text{undefined} \end{aligned}$$

65. $g(x)=\sqrt[3]{x-4}$

a.

$$\begin{aligned} g(12) &= \sqrt[3]{12-4} \\ &= \sqrt[3]{8} \\ &= 2 \end{aligned}$$

b.

$$\begin{aligned} g(-23) &= \sqrt[3]{-23-4} \\ &= \sqrt[3]{-27} \\ &= -3 \end{aligned}$$

67. $f(x)=\sqrt{x^2+1}$

a.

$$\begin{aligned} f(4) &= \sqrt{4^2+1} \\ &= \sqrt{16+1} \\ &= \sqrt{17} \\ &= 4.1231 \end{aligned}$$

b.

$$\begin{aligned} f(2.35) &= \sqrt{(2.35)^2+1} \\ &= \sqrt{5.5225+1} \\ &= \sqrt{6.5225} \\ &= 2.5539 \end{aligned}$$

69. $g(x)=\sqrt[3]{x^2+1}$

a.

$$\begin{aligned} g(6) &= \sqrt[3]{6^2+1} \\ &= \sqrt[3]{36+1} \\ &= \sqrt[3]{37} \\ &= 3.3322 \end{aligned}$$

b.

$$\begin{aligned} g(21.57) &= \sqrt[3]{(21.57)^2+1} \\ &= \sqrt[3]{465.2649+1} \\ &= \sqrt[3]{466.2649} \\ &= 7.7543 \end{aligned}$$

71.

$$\sqrt[3]{1}=1$$

73.

$$\sqrt[3]{-125}=-5$$

75.

$$\begin{aligned} \sqrt[3]{\frac{8}{27}} &= \frac{\sqrt[3]{8}}{\sqrt[3]{27}} \\ &= \frac{2}{3} \end{aligned}$$

77.

$$\sqrt[3]{64}=4$$

79.

$$\sqrt[3]{-216a^3}=-6a$$

81.

$$\sqrt[3]{-1,000p^6q^3}=-10p^2q$$

83. $f(x)=\sqrt[3]{x}-3$

x	y
-8	$\sqrt[3]{-8}-3=-2-3$ $=-5$
-1	$\sqrt[3]{-1}-3=-1-3$ $=-4$
0	$\sqrt[3]{0}-3=0-3$ $=-3$
1	$\sqrt[3]{1}-3=1-3$ $=-2$
8	$\sqrt[3]{8}-3=2-3$ $=-1$

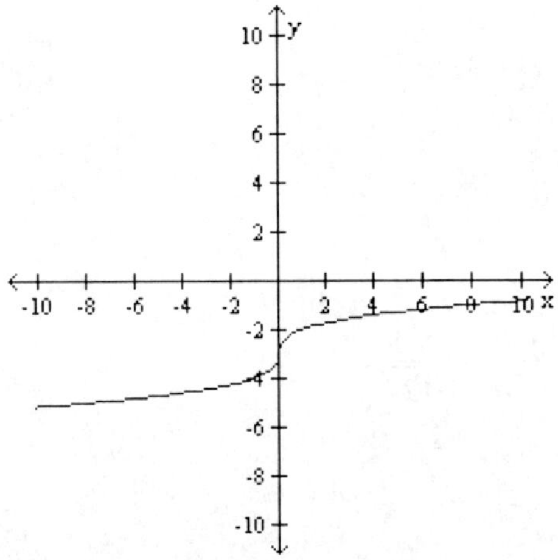

D: $(-\infty, \infty)$ and R: $(-\infty, \infty)$

85. $f(x)=\sqrt[3]{x-3}$

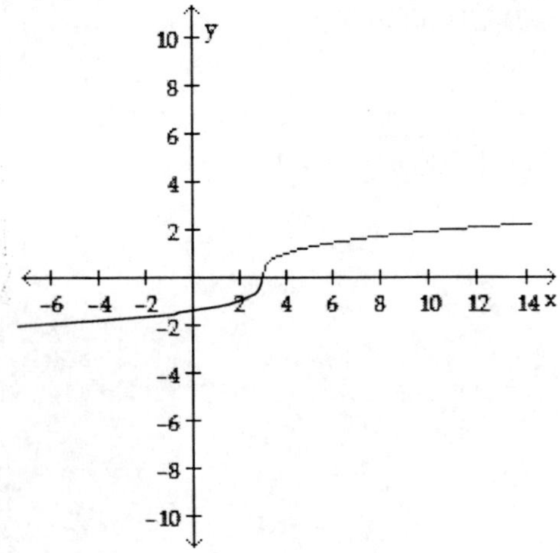

$D:(-\infty,\infty);\ \ R:(-\infty,\infty)$

87.

$$\sqrt[4]{81}=\sqrt[4]{(3)^4}$$
$$=3$$

89.

$$-\sqrt[5]{243}=-\sqrt[5]{(3)^5}$$
$$=-1(3)$$
$$=-3$$

91.

$\sqrt[4]{-256}$ is not a real number

93.

$$\sqrt[5]{-\frac{1}{32}}=\sqrt[5]{\frac{(1)^5}{(-2)^5}}$$
$$=-\frac{1}{2}$$

95.

$$\sqrt[5]{32a^5}=\sqrt[5]{(2a)^5}$$
$$=2a$$

97.

$$\sqrt[4]{81a^4}=(3a)^4$$
$$=3|a|$$

99.

$$\sqrt[6]{k^{12}}=(k^2)^6$$
$$=k^2$$

101.

$$\sqrt[4]{(m+4)^8}=\sqrt[4]{\left[(m+4)^2\right]^4}$$
$$=(m+4)^2$$

TRY IT YOURSELF

103.

$$\sqrt[3]{64s^9t^6}=\sqrt[3]{(4s^3t^2)^3}$$
$$=4s^3t^2$$

105.

$$-\sqrt{49b^8}=-7b^4$$

107.

$$-\sqrt[5]{-\frac{1}{32}} = -\sqrt[5]{-\frac{(1)^5}{(2)^5}}$$
$$= -1\left(-\frac{1}{2}\right)$$
$$= \frac{1}{2}$$

109.

$$\sqrt[3]{-125m^6} = \sqrt[3]{\left(-5m^2\right)^3}$$
$$= -5m^2$$

111.

$$\sqrt{400m^{16}n^2} = \left|20m^8n\right|$$
$$= 20m^8\left|n\right|$$

113.

$$\sqrt[6]{64a^6b^6} = \sqrt[6]{(2ab)^6}$$
$$= \left|2ab\right|$$
$$= 2\left|ab\right|$$

115.

$\sqrt[4]{-81}$ is not a real number

117.

$$\sqrt{n^2 + 12n + 36} = \sqrt{(n+6)^2}$$
$$= \left|n + 6\right|$$

119. a. $\sqrt{64} =$ ~~4~~ 8

b. $\sqrt[3]{64} = 4$

121. a. $\sqrt{81} = 9$

b. $\sqrt[4]{81} = 3$

APPLICATIONS

123. EMBROIDERY

$$r = \sqrt{\frac{A}{\pi}}$$
$$= \sqrt{\frac{38.5}{\pi}}$$
$$\approx \sqrt{12.3}$$
$$\approx 3.5$$

$$d = 2r$$
$$= 2(3.5)$$
$$= 7.0 \text{ in.}$$

125. SHOELACES

$$S = 2\left[H + L + (p-1)\sqrt{H^2 + V^2}\right]$$
$$= 2\left[50 + 250 + (6-1)\sqrt{50^2 + 20^2}\right]$$
$$= 2\left[50 + 250 + (5)\sqrt{2{,}500 + 400}\right]$$
$$= 2\left[50 + 250 + (5)\sqrt{2{,}900}\right]$$
$$= 2\left[50 + 250 + (5)(53.85164807)\right]$$
$$= 2\left[50 + 250 + 269.26\right]$$
$$= 2\left[569.26\right]$$
$$= 1{,}138.5 \text{ mm}$$

127. PULSE RATES
Change his height from feet to inches by multiplying by 12 since there are 12 inches in 1 foot.

$$8 \text{ ft } 5.5 \text{ in.} = 8(12) + 5.5 \text{ inches}$$
$$= 96 + 5.5 \text{ inches}$$
$$= 101.5 \text{ inches}$$

$$p(t) = \frac{590}{\sqrt{t}}$$
$$p(101.5) = \frac{590}{\sqrt{101.5}}$$
$$\approx 58.6 \text{ beats/min.}$$

129. BIOLOGY

The volume required for each rat is 125 ft^3, so to find the volume for 5 rats, multiply 125 times 5.

$$\begin{aligned} d(V) &= \sqrt[3]{12\left(\frac{V}{\pi}\right)} \\ &= \sqrt[3]{12\left(\frac{5\cdot 125}{\pi}\right)} \\ &= \sqrt[3]{12\left(\frac{625}{\pi}\right)} \\ &= \sqrt[3]{12(198.9)} \\ &= \sqrt[3]{2{,}386.8} \\ &= 13.4\text{ ft} \end{aligned}$$

131. COLLECTIBLES

$$\begin{aligned} r &= \sqrt[n]{\frac{A}{P}} - 1 \\ &= \sqrt[5]{\frac{950}{800}} - 1 \\ &= \sqrt[5]{1.1875} - 1 \\ &= 1.035 - 1 \\ &= 0.035 \\ &= 3.5\% \end{aligned}$$

WRITING

133. Answers will vary.

135. Answers will vary.

REVIEW

137.

$$\frac{x^2-3xy-4y^2}{x^2+cx-2yx-2cy} \div \frac{x^2-2xy-3y^2}{x^2+cx-4yx-4cy}$$

$$= \frac{x^2-3xy-4y^2}{x^2+cx-2yx-2cy} \cdot \frac{x^2+cx-4yx-4cy}{x^2-2xy-3y^2}$$

$$= \frac{(x-4y)(x+y)}{x(x+c)-2y(x+c)} \cdot \frac{x(x+c)-4y(x+c)}{(x-3y)(x+y)}$$

$$= \frac{(x-4y)\cancel{(x+y)}}{(x-2y)\cancel{(x+c)}} \cdot \frac{(x-4y)\cancel{(x+c)}}{(x-3y)\cancel{(x+y)}}$$

$$= \frac{(x-4y)^2}{(x-2y)(x-3y)}$$

CHALLENGE PROBLEMS

139. $f(x) = -\sqrt{x-2}+3$

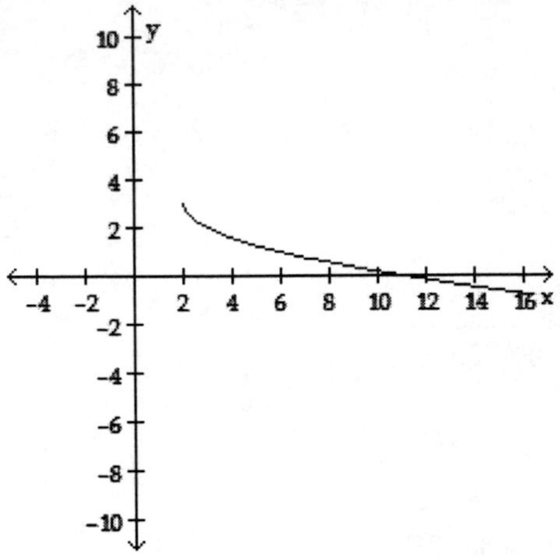

D: $[2, \infty)$ and R: $(-\infty, 3]$

SECTION 7.2

VOCABULARY

1. The expressions $4^{1/2}$ and $(-8)^{-2/3}$ have **rational (or fractional)** exponents.

3. We read $27^{-1/3}$ as "27 to the **negative** one-third power."

5. In the radical expression $\sqrt[4]{16x^8}$, 4 is the **index**, and $16x^8$ is the **radicand**.

CONCEPTS

7.

Radical form	Exponential form	Base	Exponent
$\sqrt[5]{25}$	$25^{1/5}$	25	$\frac{1}{5}$
$\left(\sqrt[3]{-27}\right)^2$	$(-27)^{2/3}$	−27	$\frac{2}{3}$
$\left(\sqrt[4]{16}\right)^{-3}$	$16^{-3/4}$	16	$-\frac{3}{4}$
$\left(\sqrt{81}\right)^3$	$81^{3/2}$	81	$\frac{3}{2}$
$-\sqrt{\frac{9}{64}}$	$-\left(\frac{9}{64}\right)^{1/2}$	$\frac{9}{64}$	$\frac{1}{2}$

9. Simplify each number.

$$8^{2/3} = 4$$
$$(-125)^{1/3} = -5$$
$$-16^{-1/4} = -\frac{1}{2}$$
$$4^{3/2} = 8$$
$$-\left(\frac{9}{100}\right)^{-1/2} = -\frac{10}{3}$$

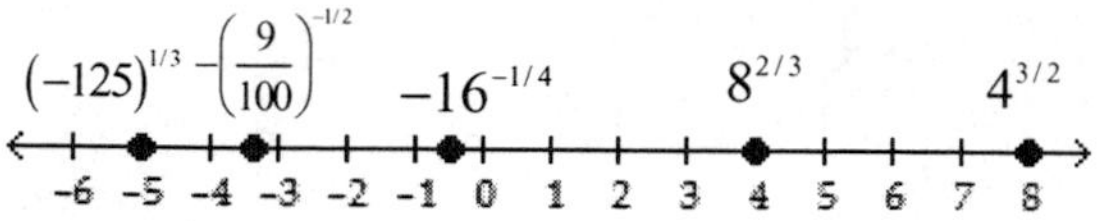

11. $x^{1/n} = \boxed{\sqrt[n]{x}}$

13. $x^{-m/n} = \boxed{\dfrac{1}{x^{m/n}}}$

NOTATION

15.

$$\left(100a^4\right)^{3/2} = \left(\sqrt{\boxed{100a^4}}\right)^3$$
$$= \left(\boxed{10a^2}\right)^3$$
$$= 1{,}000a^6$$

GUIDED PRACTICE

17.

$$125^{1/3} = \sqrt[3]{125}$$
$$= 5$$

19.

$$81^{1/4} = \sqrt[4]{81}$$
$$= 3$$

21.

$$32^{1/5} = \sqrt[5]{32}$$
$$= 2$$

23.

$$(-216)^{1/3} = \sqrt[3]{-216}$$
$$= -6$$

25.

$$-16^{1/4} = -\sqrt[4]{16}$$
$$= -2$$

27.

$$\left(\frac{1}{4}\right)^{1/2} = \sqrt{\frac{1}{4}}$$
$$= \frac{\sqrt{1}}{\sqrt{4}}$$
$$= \frac{1}{2}$$

29.

$$\left(4x^4\right)^{1/2} = \sqrt{4x^4}$$
$$= 2x^2$$

31.

$$\left(x^2\right)^{1/2} = \sqrt{x^2}$$
$$= |x|$$

33.
$$(-64p^8)^{1/2} = \sqrt{-64p^8}$$
is not a real number

35.
$$(-27n^9)^{1/3} = \sqrt[3]{-27n^9}$$
$$= -3n^3$$

37.
$$(-64x^8)^{1/8} = \sqrt[8]{-64x^8}$$
is not a real number

39.
$$\left[(x+1)^6\right]^{1/6} = \sqrt[6]{(x+1)^6}$$
$$= |x+1|$$

41.
$$36^{3/2} = \left(\sqrt{36}\right)^3$$
$$= (6)^3$$
$$= 216$$

43.
$$16^{3/4} = \left(\sqrt[4]{16}\right)^3$$
$$= (2)^3$$
$$= 8$$

45.
$$\left(-\frac{1}{216}\right)^{2/3} = \left(\sqrt[3]{-\frac{1}{216}}\right)^2$$
$$= \left(-\frac{1}{6}\right)^2$$
$$= \frac{1}{36}$$

47.
$$-4^{5/2} = -\left(\sqrt{4}\right)^5$$
$$= -2^5$$
$$= -32$$

49.
$$(25x^4)^{3/2} = \left(\sqrt{25x^4}\right)^3$$
$$= (5x^2)^3$$
$$= 125x^6$$

51.
$$(-8x^6y^3)^{2/3} = \left(\sqrt[3]{-8x^6y^3}\right)^2$$
$$= (-2x^2y)^2$$
$$= 4x^4y^2$$

53.
$$(81x^4y^8)^{3/4} = \left(\sqrt[4]{81x^4y^8}\right)^3$$
$$= (3xy^2)^3$$
$$= 27x^3y^6$$

55.
$$-\left(\frac{x^5}{32}\right)^{4/5} = -\left(\sqrt[5]{\frac{x^5}{32}}\right)^4$$
$$= -\left(\frac{x}{2}\right)^4$$
$$= -\frac{x^4}{16}$$

57. $\sqrt[5]{8abc} = (8abc)^{1/5}$

59. $\sqrt[3]{a^2 - b^2} = (a^2 - b^2)^{1/3}$

61. $(6x^3y)^{1/4} = \sqrt[4]{6x^3y}$

63. $(2s^2 - t^2)^{1/2} = \sqrt{2s^2 - t^2}$

65.
$$4^{-1/2} = \frac{1}{4^{1/2}}$$
$$= \frac{1}{\sqrt{4}}$$
$$= \frac{1}{2}$$

67.

$$125^{-1/3} = \frac{1}{125^{1/3}} = \frac{1}{\sqrt[3]{125}} = \frac{1}{5}$$

69.

$$-\left(1{,}000y^3\right)^{-2/3} = -\frac{1}{\left(1{,}000y^3\right)^{2/3}}$$
$$= -\frac{1}{\left(\sqrt[3]{1{,}000y^3}\right)^2}$$
$$= -\frac{1}{(10y)^2}$$
$$= -\frac{1}{100y^2}$$

71.

$$\left(-\frac{27}{8}\right)^{-4/3} = \left(-\frac{8}{27}\right)^{4/3} = \left(\sqrt[3]{-\frac{8}{27}}\right)^4 = \left(-\frac{2}{3}\right)^4 = \frac{16}{81}$$

73.

$$\left(\frac{16}{81y^4}\right)^{-3/4} = \left(\frac{81y^4}{16}\right)^{3/4} = \left(\sqrt[4]{\frac{81y^4}{16}}\right)^3 = \left(\frac{3y}{2}\right)^3 = \frac{27y^3}{8}$$

75.

$$\frac{1}{9^{-5/2}} = 9^{5/2} = \left(\sqrt{9}\right)^5 = 3^5 = 243$$

77.

$$9^{3/7}9^{2/7} = 9^{3/7+2/7} = 9^{5/7}$$

79.

$$6^{-2/3}6^{-4/3} = 6^{-6/3} = 6^{-2} = \frac{1}{6^2} = \frac{1}{36}$$

81.

$$\left(m^{2/3}m^{1/3}\right)^6 = \left(m^{2/3+1/3}\right)^6 = \left(m^{3/3}\right)^6 = \left(m^1\right)^6 = m^6$$

83.

$$\left(a^{1/2}b^{1/3}\right)^{3/2} = a^{(1/2)(3/2)}b^{(1/3)(3/2)} = a^{3/4}b^{1/2}$$

85.

$$\frac{3^{4/3}3^{1/3}}{3^{2/3}} = \frac{3^{4/3+1/3}}{3^{2/3}} = \frac{3^{5/3}}{3^{2/3}} = 3^{5/3-2/3} = 3^{3/3} = 3^1 = 3$$

87.

$$\frac{a^{3/4}a^{3/4}}{a^{1/2}} = \frac{a^{3/4+3/4}}{a^{1/2}}$$
$$= \frac{a^{6/4}}{a^{1/2}}$$
$$= \frac{a^{3/2}}{a^{1/2}}$$
$$= a^{3/2-1/2}$$
$$= a^{2/2}$$
$$= a$$

89.

$$y^{1/3}\left(y^{2/3} + y^{5/3}\right) = y^{1/3+2/3} + y^{1/3+5/3}$$
$$= y^{3/3} + y^{6/3}$$
$$= y + y^2$$

91.

$$x^{3/5}\left(x^{7/5} - x^{-3/5} + 1\right) = x^{3/5+7/5} - x^{3/5+(-3/5)} + x^{3/5}$$
$$= x^{10/5} - x^0 + x^{3/5}$$
$$= x^2 - 1 + x^{3/5}$$

93.

$$\sqrt[4]{5^2} = 5^{2/4}$$
$$= 5^{1/2}$$
$$= \sqrt{5}$$

95.

$$\sqrt[9]{11^3} = 11^{3/9}$$
$$= 11^{1/3}$$
$$= \sqrt[3]{11}$$

97.

$$\sqrt[6]{p^3} = p^{3/6}$$
$$= p^{1/2}$$
$$= \sqrt{p}$$

99.

$$\sqrt[10]{x^2y^2} = \left(x^2y^2\right)^{1/10}$$
$$= x^{2(1/10)}y^{2(1/10)}$$
$$= x^{2/10}y^{2/10}$$
$$= x^{1/5}y^{1/5}$$
$$= \sqrt[5]{xy}$$

101

$$\sqrt[9]{\sqrt{c}} = \left(c^{1/2}\right)^{1/9}$$
$$= c^{(1/2)(1/9)}$$
$$= c^{1/18}$$
$$= \sqrt[18]{c}$$

103.

$$\sqrt[5]{\sqrt[3]{7m}} = \left[(7m)^{1/3}\right]^{1/5}$$
$$= (7m)^{(1/3)(1/5)}$$
$$= (7m)^{1/15}$$
$$= \sqrt[15]{7m}$$

105.

$$15^{1/3} = \sqrt[3]{15}$$
$$= 2.47$$

107.

$$(1.045)^{2/5} = \left(\sqrt[5]{1.045}\right)^2$$
$$= (1.01)^2$$
$$= 1.02$$

TRY IT YOURSELF

109.

$$\left(25y^2\right)^{1/2} = \sqrt{25y^2}$$
$$= 5y$$

111.

$$-\left(\frac{a^4}{81}\right)^{3/4} = -\left(\sqrt[4]{\frac{a^4}{81}}\right)^3$$
$$= -\left(\frac{a}{3}\right)^3$$
$$= -\frac{a^3}{27}$$

113.

$$16^{-3/2} = \frac{1}{16^{3/2}}$$
$$= \frac{1}{\left(\sqrt{16}\right)^3}$$
$$= \frac{1}{4^3}$$
$$= \frac{1}{64}$$

115.

$$\frac{p^{8/5}p^{7/5}}{p^2} = \frac{p^{8/5+7/5}}{p^2}$$
$$= \frac{p^{15/5}}{p^2}$$
$$= \frac{p^3}{p^2}$$
$$= p^{3-2}$$
$$= p$$

117.

$$\left(-27x^6\right)^{-1/3} =$$
$$= \left(-\frac{1}{27x^6}\right)^{1/3}$$
$$= \sqrt[3]{-\frac{1}{27x^6}}$$
$$= -\frac{1}{3x^2}$$
$$= -\frac{1}{3}x^2$$

119.

$$\frac{1}{32^{-1/5}} = 32^{1/5}$$
$$= \sqrt[5]{32}$$
$$= 2$$

121.

$$n^{1/5}\left(n^{2/5} - n^{-1/5}\right) = n^{1/5+2/5} - n^{1/5+(-1/5)}$$
$$= n^{3/5} - n^0$$
$$= n^{3/5} - 1$$

123. Wrong

$$\frac{1}{9^{-5/2}} = 9^{5/2}$$ ($4^{-5/2}$ handwritten, circled)
$$= \left(\sqrt{9}\right)^5$$
$$= 3^5$$
$$= 243$$

(handwritten correction:)
$$\frac{1}{4^{-5/2}} = 4^{5/2}$$
$$= \left(\sqrt[2]{4}\right)^5$$
$$= (2)^5$$
$$= 32$$

125.

$$\left(m^4\right)^{1/2} = \sqrt{m^4}$$
$$= m^2$$

127.

$$\sqrt[4]{25b^2} = \left(5^2b^2\right)^{1/4}$$
$$= 5^{2(1/4)}b^{2(1/4)}$$
$$= 5^{2/4}b^{2/4}$$
$$= 5^{1/2}b^{1/2}$$
$$= \sqrt{5b}$$

129

$$\left(16x^4\right)^{1/4} = \sqrt[4]{16x^4}$$
$$= |2x|$$
$$= 2|x| = \boxed{2x}$$

← we assume all variables represent positive real numbers.

131.

$$-\left(8a^3b^6\right)^{-2/3} = -\frac{1}{\left(8a^3b^6\right)^{2/3}}$$
$$= -\frac{1}{\left(\sqrt[3]{8a^3b^6}\right)^2}$$
$$= -\frac{1}{\left(2ab^2\right)^2}$$
$$= -\frac{1}{4a^2b^4}$$

133. a.

$$-125^{2/3} = -\left(\sqrt[3]{125}\right)^2$$
$$= -(5)^2$$
$$= -25$$

b.

$$(-125)^{2/3} = \left(\sqrt[3]{-125}\right)^2$$
$$= (-5)^2$$
$$= 25$$

c.

$$(-125)^{-2/3} = \frac{1}{(-125)^{2/3}}$$
$$= \frac{1}{\left(\sqrt[3]{-125}\right)^2}$$
$$= \frac{1}{(-5)^2}$$
$$= \frac{1}{25}$$

$-125^{-2/3}$

d.

$$\frac{1}{(-125)^{-2/3}} = (-125)^{2/3}$$
$$= \left(\sqrt[3]{-125}\right)^2$$
$$= (-5)^2$$
$$= 25$$

135. a.

$$\left(64a^4\right)^{1/2} = \sqrt{64a^4}$$
$$= 8a^2$$

b.

$$\left(64a^4\right)^{-1/2} = \frac{1}{\left(64a^4\right)^{1/2}}$$
$$= \frac{1}{\sqrt{64a^4}}$$
$$= \frac{1}{8a^2}$$

c.

$$-\left(64a^4\right)^{1/2} = -\sqrt{64a^4}$$
$$= -8a^2$$

d.

$$\frac{1}{\left(64a^4\right)^{1/2}} = \frac{1}{\sqrt{64a^4}}$$
$$= \frac{1}{8a^2}$$

APPLICATIONS

137. BALLISTIC PENDULUMS

Let $M = 6.0$, $m = 0.0625$, $g = 32$, and $h = 0.9$.

$$v = \frac{m+M}{m}(2gh)^{1/2}$$
$$= \frac{0.0625 + 6.0}{0.0625}(2 \cdot 32 \cdot 0.9)^{1/2}$$
$$= \frac{6.0625}{0.0625}(57.6)^{1/2}$$
$$= 97\left(\sqrt{57.6}\right)$$
$$= 97(7.589)$$
$$= 736 \text{ ft/sec}$$

139. RELATIVITY

Let $c = 186{,}000$, $v = 160{,}000$, and $m_0 = 1$.

$$m = m_0\left(1 - \frac{v^2}{c^2}\right)^{-1/2}$$
$$= 1\left(1 - \frac{160{,}000^2}{186{,}000^2}\right)^{-1/2}$$
$$= 1\left(1 - \frac{2.56 \times 10^{10}}{3.4596 \times 10^{10}}\right)^{-1/2}$$
$$= 1(1 - 0.74)^{-1/2}$$
$$= 1(0.26)^{-1/2}$$
$$= 1\left(\frac{1}{0.26}\right)^{1/2}$$
$$= 1\left(\sqrt{\frac{1}{0.26}}\right)$$
$$= 1(1.96)$$
$$= 1.96 \text{ units}$$

141. GENERAL CONTRACTOR

Let $a = 40$ and $b = 64$.

$$L = \left(a^{2/3} + b^{2/3}\right)^{3/2}$$
$$= \left(40^{2/3} + 64^{2/3}\right)^{3/2}$$
$$= \left[\left(\sqrt[3]{40}\right)^2 + \left(\sqrt[3]{64}\right)^2\right]^{3/2}$$
$$= \left[(3.42)^2 + (4)^2\right]^{3/2}$$
$$= [11.6964 + 16]^{3/2}$$
$$= (27.6964)^{3/2}$$
$$= \left(\sqrt{27.6964}\right)^3$$
$$= (5.26)^3$$
$$= 145.5 \text{ in.}$$

Divide by 12 since there are 12 inches in 1 foot.

$$L = \frac{145.5}{12}$$
$$= 12.1 \text{ ft}$$

WRITING

143. Answers will vary.

REVIEW

145. COMMUTING TIME

$$t = \frac{k}{r}$$
$$3 = \frac{k}{50}$$
$$50(3) = 50\left(\frac{k}{50}\right)$$
$$150 = k$$

$$t = \frac{150}{60}$$
$$t = 2.5 \text{ hours}$$

CHALLENGE PROBLEMS

147. Yes.

$$16^{2/4} = \left(\sqrt[4]{16}\right)^2 \qquad 16^{1/2} = \sqrt{16}$$
$$= 2^2 \qquad = 4$$
$$= 4$$

SECTION 7.3

VOCABULARY

1. Radical expressions such as $\sqrt[3]{4}$ and $6\sqrt[3]{4}$ with the same index and the same radicand are called **like** radicals.

3. The largest perfect square **factor** of 27 is 9. The largest **perfect**-cube factor of 16 is 8.

CONCEPTS

5. The product rule for radicals: $\sqrt[n]{ab} = \sqrt[n]{a}\sqrt[n]{b}$. In words, the *n*th root of the **product** of two numbers is equal to the product of their *n*th **roots**.

7. a. $\sqrt{4 \cdot 5}$
   b. $\sqrt{4}\sqrt{5}$
   c. $\sqrt{4 \cdot 5} = \sqrt{4}\sqrt{5}$

9. a. Answers will vary. Possible answers are $\sqrt{5}$ and $\sqrt[3]{5}$. No, the expressions cannot be added.
   b. Answers will vary. Possible answers are $\sqrt{5}$ and $\sqrt{6}$. No, the expressions cannot be added.

NOTATION

11.
$$\begin{aligned}\sqrt[3]{32k^4} &= \sqrt[3]{\boxed{8k^3} \cdot 4k} \\ &= \sqrt[3]{\boxed{8k^3}}\sqrt[3]{4k} \\ &= 2k\sqrt[3]{\boxed{4k}}\end{aligned}$$

GUIDED PRACTICE

13.
$$\begin{aligned}\sqrt{50} &= \sqrt{25 \cdot 2} \\ &= \sqrt{25}\sqrt{2} \\ &= 5\sqrt{2}\end{aligned}$$

15.
$$\begin{aligned}8\sqrt{45} &= 8\left(\sqrt{9 \cdot 5}\right) \\ &= 8\left(\sqrt{9}\sqrt{5}\right) \\ &= 8\left(3\sqrt{5}\right) \\ &= 24\sqrt{5}\end{aligned}$$

17.
$$\begin{aligned}\sqrt[3]{32} &= \sqrt[3]{8 \cdot 4} \\ &= \sqrt[3]{8}\sqrt[3]{4} \\ &= 2\sqrt[3]{4}\end{aligned}$$

19.
$$\begin{aligned}\sqrt[4]{48} &= \sqrt[4]{16 \cdot 3} \\ &= \sqrt[4]{16}\sqrt[4]{3} \\ &= 2\sqrt[4]{3}\end{aligned}$$

21.
$$\begin{aligned}\sqrt{75a^2} &= \sqrt{25a^2}\sqrt{3} \\ &= 5a\sqrt{3}\end{aligned}$$

23.
$$\begin{aligned}\sqrt{128a^3b^5} &= \sqrt{64a^2b^4}\sqrt{2ab} \\ &= 8ab^2\sqrt{2ab}\end{aligned}$$

25.
$$\begin{aligned}2\sqrt[3]{-54x^6} &= 2\sqrt[3]{-27x^6}\sqrt[3]{2} \\ &= 2\left(-3x^2\sqrt[3]{2}\right) \\ &= -6x^2\sqrt[3]{2}\end{aligned}$$

27.
$$\begin{aligned}\sqrt[4]{32x^{12}y^4} &= \sqrt[4]{16x^{12}y^4}\sqrt[4]{2} \\ &= 2x^3y\sqrt[4]{2}\end{aligned}$$

29.
$$\begin{aligned}\sqrt{242} &= \sqrt{121}\sqrt{2} \\ &= 11\sqrt{2}\end{aligned}$$

31.
$$\begin{aligned}\sqrt{112a^3} &= \sqrt{16a^2}\sqrt{7a} \\ &= 4a\sqrt{7a}\end{aligned}$$

33.

$$-\sqrt[5]{96a^4} = -\sqrt[5]{32}\sqrt[5]{3a^4} = -2\sqrt[5]{3a^4}$$

35.

$$\sqrt[3]{405x^{12}y^4} = \sqrt[3]{27x^{12}y^3}\sqrt[3]{15y} = 3x^4y\sqrt[3]{15y}$$

37.

$$\sqrt{\frac{11}{9}} = \frac{\sqrt{11}}{\sqrt{9}} = \frac{\sqrt{11}}{3}$$

39.

$$\sqrt[4]{\frac{3}{625}} = \frac{\sqrt[4]{3}}{\sqrt[4]{625}} = \frac{\sqrt[4]{3}}{5}$$

41.

$$\sqrt[5]{\frac{3x^{10}}{32}} = \frac{\sqrt[5]{3x^{10}}}{\sqrt[5]{32}} = \frac{x^2\sqrt[5]{3}}{2}$$

43.

$$\sqrt{\frac{z^2}{16x^2}} = \frac{\sqrt{z^2}}{\sqrt{16x^2}} = \frac{z}{4x}$$

45.

$$\frac{\sqrt{500}}{\sqrt{5}} = \sqrt{\frac{500}{5}} = \sqrt{100} = 10$$

47.

$$\frac{\sqrt{98x^3}}{\sqrt{2x}} = \sqrt{\frac{98x^3}{2x}} = \sqrt{49x^2} = 7x$$

49.

$$\frac{\sqrt[3]{48x^7}}{\sqrt[3]{6x}} = \sqrt[3]{\frac{48x^7}{6x}} = \sqrt[3]{8x^6} = 2x^2$$

51.

$$\frac{\sqrt[3]{189a^5}}{\sqrt[3]{7a}} = \sqrt[3]{\frac{189a^5}{7a}} = \sqrt[3]{27a^4} = \sqrt[3]{27a^3}\sqrt[3]{a} = 3a\sqrt[3]{a}$$

53.

$$5\sqrt{7} + 3\sqrt{7} = (5+3)\sqrt{7} = 8\sqrt{7}$$

55.

$$20\sqrt[3]{4} - 15\sqrt[3]{4} = (20-15)\sqrt[3]{4} = 5\sqrt[3]{4}$$

57.

$$4 + \sqrt{8} + \sqrt{2} + 8 = 4 + 2\sqrt{2} + \sqrt{2} + 8 = (4+8) + (2+1)\sqrt{2} = 12 + 3\sqrt{2}$$

59.

$$\sqrt{98} - \sqrt{50} - \sqrt{72} = \sqrt{49}\sqrt{2} - \sqrt{25}\sqrt{2} - \sqrt{36}\sqrt{2} = 7\sqrt{2} - 5\sqrt{2} - 6\sqrt{2} = -4\sqrt{2}$$

61.

$$8 + \sqrt[3]{32} - \sqrt[3]{108} - 7 = 8 + \sqrt[3]{8}\sqrt[3]{4} - \sqrt[3]{27}\sqrt[3]{4} - 7 = 8 + 2\sqrt[3]{4} - 3\sqrt[3]{4} - 7 = 1 - \sqrt[3]{4}$$

63.

$$14\sqrt[4]{32} - 15\sqrt[4]{2} = 14\sqrt[4]{16}\sqrt[4]{2} - 15\sqrt[4]{2} = 14\left(2\sqrt[4]{2}\right) - 15\left(\sqrt[4]{2}\right) = 28\sqrt[4]{2} - 15\sqrt[4]{2} = 13\sqrt[4]{2}$$

65.

$$4\sqrt{2x}+6\sqrt{2x}=10\sqrt{2x}$$

67.

$$\begin{aligned}&\sqrt{18t}+\sqrt{300t}-\sqrt{243t}\\&=\sqrt{9}\sqrt{2t}+\sqrt{100}\sqrt{3t}-\sqrt{81}\sqrt{3t}\\&=3\sqrt{2t}+10\sqrt{3t}-9\sqrt{3t}\\&=3\sqrt{2t}+\sqrt{3t}\end{aligned}$$

69.

$$\begin{aligned}&2\sqrt[3]{16}-\sqrt[3]{54}-3\sqrt[3]{128}\\&=2\sqrt[3]{8}\sqrt[3]{2}-\sqrt[3]{27}\sqrt[3]{2}-3\sqrt[3]{64}\sqrt[3]{2}\\&=2\left(2\sqrt[3]{2}\right)-3\sqrt[3]{2}-3\left(4\sqrt[3]{2}\right)\\&=4\sqrt[3]{2}-3\sqrt[3]{2}-12\sqrt[3]{2}\\&=-11\sqrt[3]{2}\end{aligned}$$

71.

$$\begin{aligned}\sqrt[4]{64}+5\sqrt[4]{4}-\sqrt[4]{324}&=\sqrt[4]{16}\sqrt[4]{4}+5\sqrt[4]{4}-\sqrt[4]{81}\sqrt[4]{4}\\&=2\sqrt[4]{4}+5\sqrt[4]{4}-3\sqrt[4]{4}\\&=4\sqrt[4]{4}\end{aligned}$$

TRY IT YOURSELF

73.

$$\begin{aligned}\sqrt[6]{m^{11}}&=\sqrt[6]{m^6}\sqrt[6]{m^5}\\&=m\sqrt[6]{m^5}\end{aligned}$$

75.

$$\begin{aligned}2\sqrt[3]{64a}+2\sqrt[3]{8a}&=2\sqrt[3]{64}\sqrt[3]{a}+2\sqrt[3]{8}\sqrt[3]{a}\\&=2\left(4\sqrt[3]{a}\right)+2\left(2\sqrt[3]{a}\right)\\&=8\sqrt[3]{a}+4\sqrt[3]{a}\\&=12\sqrt[3]{a}\end{aligned}$$

77.

$$\begin{aligned}&\sqrt{8y^7}+\sqrt{32y^7}-\sqrt{2y^7}\\&=\sqrt{4y^6}\sqrt{2y}+\sqrt{16y^6}\sqrt{2y}-\sqrt{y^6}\sqrt{2y}\\&=2y^3\sqrt{2y}+4y^3\sqrt{2y}-y^3\sqrt{2y}\\&=5y^3\sqrt{2y}\end{aligned}$$

79.

$$\begin{aligned}\sqrt{32b}&=\sqrt{16}\sqrt{2b}\\&=4\sqrt{2b}\end{aligned}$$

81.

$$\begin{aligned}\sqrt{\frac{125n^5}{64n}}&=\sqrt{\frac{125n^4}{64}}\\&=\frac{\sqrt{125n^4}}{\sqrt{64}}\\&=\frac{5n^2\sqrt{5}}{8}\end{aligned}$$

83.

$$\begin{aligned}2\sqrt[3]{125}-5\sqrt[3]{64}&=2(5)-5(4)\\&=10-20\\&=-10\end{aligned}$$

85.

$$\begin{aligned}\sqrt{300xy}&=\sqrt{100}\sqrt{3xy}\\&=10\sqrt{3xy}\end{aligned}$$

87.

$$\begin{aligned}\sqrt[4]{\frac{5x}{16z^4}}&=\frac{\sqrt[4]{5x}}{\sqrt[4]{16z^4}}\\&=\frac{\sqrt[4]{5x}}{2z}\end{aligned}$$

89.

$$8\sqrt[5]{7a^2}-7\sqrt[5]{7a^2}=\sqrt[5]{7a^2}$$

91.

$$\begin{aligned}&\sqrt[5]{x^6y^2}+\sqrt[5]{32x^6y^2}+\sqrt[5]{x^6y^2}\\&=\sqrt[5]{x^5}\sqrt[5]{xy^2}+\sqrt[5]{32x^5}\sqrt[5]{xy^2}+\sqrt[5]{x^5}\sqrt[5]{xy^2}\\&=x\sqrt[5]{xy^2}+2x\sqrt[5]{xy^2}+x\sqrt[5]{xy^2}\\&=4x\sqrt[5]{xy^2}\end{aligned}$$

93.

$$\begin{aligned}\sqrt[4]{208m^4n}&=\sqrt[4]{16m^4}\sqrt[4]{13n}\\&=2m\sqrt[4]{13n}\end{aligned}$$

95.

$$\sqrt[3]{\frac{a^7}{64a}} = \sqrt[3]{\frac{a^6}{64}}$$
$$= \frac{\sqrt[3]{a^6}}{\sqrt[3]{64}}$$
$$= \frac{a^2}{4}$$

97.

$$\sqrt[3]{\frac{7}{64}} = \frac{\sqrt[3]{7}}{\sqrt[3]{64}}$$
$$= \frac{\sqrt[3]{7}}{4}$$

99.

$$\sqrt{80} + \sqrt{45} - \sqrt{27} = \sqrt{16}\sqrt{5} + \sqrt{9}\sqrt{5} - \sqrt{9}\sqrt{3}$$
$$= 4\sqrt{5} + 3\sqrt{5} - 3\sqrt{3}$$
$$= (4+3)\sqrt{5} - 3\sqrt{3}$$
$$= 7\sqrt{5} - 3\sqrt{3}$$

101.

$$\sqrt[5]{64t^{11}} = \sqrt[5]{32t^{10}}\sqrt[5]{2t}$$
$$= 2t^2\sqrt[5]{2t}$$

103.

$$\sqrt[3]{24x} + \sqrt[3]{3x} = \sqrt[3]{8}\sqrt[3]{3x} + \sqrt[3]{3x}$$
$$= 2\sqrt[3]{3x} + \sqrt[3]{3x}$$
$$= 3\sqrt[3]{3x}$$

105. a.

$$\sqrt{20} + \sqrt{20} = \sqrt{4}\sqrt{5} + \sqrt{4}\sqrt{5}$$
$$= 2\sqrt{5} + 2\sqrt{5}$$
$$= 4\sqrt{5}$$

b.

$$\sqrt{21} + \sqrt{21} = 2\sqrt{21}$$

107. a.

$$\sqrt{9x^2} - \sqrt{25x^2} + \sqrt{16x^2} = 3x - 5x + 4x$$
$$= 2x$$

b.

$$\sqrt{9x^3} - \sqrt{25x^3} + \sqrt{16x^3}$$
$$= 3x\sqrt{x} - 5x\sqrt{x} + 4x\sqrt{x}$$
$$= 2x\sqrt{x}$$

109. a.

$$3\sqrt{16} + \sqrt{54} = 3(4) + \sqrt{9}\sqrt{6}$$
$$= 12 + 3\sqrt{6}$$

b.

$$3\sqrt[3]{16} + \sqrt[3]{54} = 3\sqrt[3]{8}\sqrt[3]{2} + \sqrt[3]{27}\sqrt[3]{2}$$
$$= 6\sqrt[3]{2} + 3\sqrt[3]{2}$$
$$= 9\sqrt[3]{2}$$

111. a.

$$24\sqrt[5]{6x} + 16\sqrt[5]{6x} = 40\sqrt[5]{6x}$$

b.

$$24\sqrt[4]{6x} + 16\sqrt[4]{6x} = 40\sqrt[4]{6x}$$

APPLICATIONS

113. GENERAL CONTRACTORS
Let $a = 20$, $b = 16$, and $c = 14$.

$$L = \sqrt{\frac{b^2}{2} + \frac{c^2}{2} - \frac{a^2}{2}}$$
$$= \sqrt{\frac{16^2}{2} + \frac{14^2}{2} - \frac{20^2}{4}}$$
$$= \sqrt{\frac{256}{2} + \frac{196}{2} - \frac{400}{4}}$$
$$= \sqrt{128 + 98 - 100}$$
$$= \sqrt{126}$$
$$= \sqrt{9}\sqrt{14}$$
$$= 3\sqrt{14} \text{ ft}$$

$$L \approx 11.2 \text{ ft}$$

115. BLOW DRYERS
Let $P = 1{,}200$ and $R = 16$.

$$I = \sqrt{\frac{P}{R}}$$
$$= \sqrt{\frac{1{,}200}{16}}$$
$$= \sqrt{75}$$
$$= \sqrt{25}\sqrt{3}$$
$$= 5\sqrt{3} \text{ amps}$$

$$I \approx 8.7 \text{ amps}$$

117. DUCTWORK
Add up all of the sides.

$$\begin{aligned}&4\sqrt{80}+2\sqrt{20}+2\sqrt{45}+2\sqrt{75}\\&=4\sqrt{16}\sqrt{5}+2\sqrt{4}\sqrt{5}+2\sqrt{9}\sqrt{5}+2\sqrt{25}\sqrt{3}\\&=4\left(4\sqrt{5}\right)+2\left(2\sqrt{5}\right)+2\left(3\sqrt{5}\right)+2\left(5\sqrt{3}\right)\\&=16\sqrt{5}+4\sqrt{5}+6\sqrt{5}+10\sqrt{3}\\&=\left(26\sqrt{5}+10\sqrt{3}\right)\text{ in.}\\&\approx 75.5\text{ in.}\end{aligned}$$

WRITING

119. Answers will vary.

121. Answers will vary.

REVIEW

123.

$$\begin{aligned}3x^2y^3\left(-5x^3y^{-4}\right)&=-15x^5y^{-1}\\&=-\frac{15x^5}{y}\end{aligned}$$

125.

$$\begin{array}{r}3p+4-\dfrac{5}{2p-5}\\2p-5\overline{\smash{\big)}\,6p^2-7p-25}\\\underline{6p^2-15p}\quad\;\;\\8p-25\\\underline{8p-20}\\-5\end{array}$$

CHALLENGE PROBLEMS

127.

$$\begin{aligned}\frac{\sqrt{24}}{3}+\frac{\sqrt{6}}{5}&=\frac{2\sqrt{6}}{3}+\frac{\sqrt{6}}{5}\\&=\frac{2\sqrt{6}}{3}\left(\frac{5}{5}\right)+\frac{\sqrt{6}}{5}\left(\frac{3}{3}\right)\\&=\frac{10\sqrt{6}}{15}+\frac{3\sqrt{6}}{15}\\&=\frac{13\sqrt{6}}{15}\end{aligned}$$

129.

$$\begin{aligned}\sqrt[3]{\frac{3b}{8}}-9\sqrt[3]{3b}&=\frac{\sqrt[3]{3b}}{\sqrt[3]{8}}-9\sqrt[3]{3b}\\&=\frac{\sqrt[3]{3b}}{2}-9\sqrt[3]{3b}\\&=\frac{\sqrt[3]{3b}}{2}-\frac{18\sqrt[3]{3b}}{2}\\&=-\frac{17\sqrt[3]{3b}}{2}\end{aligned}$$

131.

$$\begin{aligned}\sqrt{25x+25}-\sqrt{x+1}&=\sqrt{25(x+1)}-\sqrt{x+1}\\&=\sqrt{25}\sqrt{x+1}-\sqrt{x+1}\\&=5\sqrt{x+1}-\sqrt{x+1}\\&=4\sqrt{x+1}\end{aligned}$$

SECTION 7.4

VOCABULARY

1. In this section, we used the **product** rule for radicals in reverse: $\sqrt[n]{a}\cdot\sqrt[n]{b}=\sqrt[n]{ab}$.

3. To **rationalize** the denominator of $\frac{4}{\sqrt{5}}$, we multiply the fraction by $\frac{\sqrt{5}}{\sqrt{5}}$.

5. To obtain a **perfect** cube radicand in the denominator of $\frac{\sqrt[3]{7}}{\sqrt[3]{5n}}$, we multiply the fraction by $\frac{\sqrt[3]{25n^2}}{\sqrt[3]{25n^2}}$.

CONCEPTS

7.

	Why isn't it in simplified form?	Simp. Form	Approx.
$\frac{3}{\sqrt{2}}$	A radical appears on the denominator.	$\frac{3\sqrt{2}}{2}$	2.121320344
$\frac{\sqrt{18}}{2}$	There is a perfect square factor in the radicand: 9.	$\frac{3\sqrt{2}}{2}$	2.121320344
$\sqrt{\frac{9}{2}}$	The radicand contains a fraction.	$\frac{3\sqrt{2}}{2}$	2.121320344

9.
$$\begin{aligned}(5-\sqrt{x})^2&=(5)^2-2(5)(\sqrt{x})+(\sqrt{x})^2\\&=25-10\sqrt{x}+x\end{aligned}$$

11. a. $4\sqrt{6}+2\sqrt{6}=6\sqrt{6}$

b.
$$\begin{aligned}4\sqrt{6}\left(2\sqrt{6}\right)&=8\sqrt{36}\\&=8(6)\\&=48\end{aligned}$$

c. cannot be simplified

d. $3\sqrt{2}\left(-2\sqrt{3}\right)=-6\sqrt{6}$

NOTATION

13.
$$\begin{aligned}5\sqrt{8}\cdot7\sqrt{6}&=5(7)\sqrt{8}\boxed{\sqrt{6}}\\&=35\sqrt{\boxed{48}}\\&=35\sqrt{\boxed{16}\cdot3}\\&=35\left(\boxed{4}\right)\sqrt{3}\\&=140\sqrt{3}\end{aligned}$$

GUIDED PRACTICE

15.
$$\begin{aligned}\sqrt{3}\sqrt{15}&=\sqrt{45}\\&=\sqrt{9}\sqrt{5}\\&=3\sqrt{5}\end{aligned}$$

17.
$$\begin{aligned}2\sqrt{3}\sqrt{6}&=2\sqrt{18}\\&=2\sqrt{9}\sqrt{2}\\&=2\left(3\sqrt{2}\right)\\&=6\sqrt{2}\end{aligned}$$

19.
$$\begin{aligned}\left(3\sqrt[3]{9}\right)\left(2\sqrt[3]{3}\right)&=3(2)\sqrt[3]{9}\sqrt[3]{3}\\&=6\sqrt[3]{27}\\&=6(3)\\&=18\end{aligned}$$

21.
$$\begin{aligned}\sqrt[3]{2}\cdot\sqrt[3]{12}&=\sqrt[3]{24}\\&=\sqrt[3]{8}\sqrt[3]{3}\\&=2\sqrt[3]{3}\end{aligned}$$

23.
$$\begin{aligned}6\sqrt{ab^3}\left(8\sqrt{ab}\right)&=48\sqrt{a^2b^4}\\&=48ab^2\end{aligned}$$

25.
$$\begin{aligned}\sqrt[4]{5a^3}\sqrt[4]{125a^2}&=\sqrt[4]{625a^5}\\&=\sqrt[4]{625a^4}\sqrt[4]{a}\\&=5a\sqrt[4]{a}\end{aligned}$$

27.

$$3\sqrt{5}\left(4-\sqrt{5}\right)=3\sqrt{5}(4)-3\sqrt{5}\left(\sqrt{5}\right)$$
$$=3(4)\sqrt{5}-3(5)$$
$$=12\sqrt{5}-15$$

29.

$$\sqrt{2}\left(4\sqrt{6}+2\sqrt{7}\right)=4\left(\sqrt{2}\sqrt{6}\right)+2\left(\sqrt{2}\sqrt{7}\right)$$
$$=4\sqrt{12}+2\sqrt{14}$$
$$=4\sqrt{4}\sqrt{3}+2\sqrt{14}$$
$$=4\left(2\sqrt{3}\right)+2\sqrt{14}$$
$$=8\sqrt{3}+2\sqrt{14}$$

31.

$$-2\sqrt{5x}\left(4\sqrt{2x}-3\sqrt{3}\right)$$
$$=-2(4)\sqrt{5x}\sqrt{2x}+2(3)\sqrt{5x}\sqrt{3}$$
$$=-8\sqrt{10x^2}+6\sqrt{15x}$$
$$=-8\sqrt{x^2}\sqrt{10}+6\sqrt{15x}$$
$$=-8x\sqrt{10}+6\sqrt{15x}$$

33.

$$\sqrt[3]{2}\left(4\sqrt[3]{4}+\sqrt[3]{12}\right)=\sqrt[3]{2}\left(4\sqrt[3]{4}\right)+\sqrt[3]{2}\left(\sqrt[3]{12}\right)$$
$$=4\sqrt[3]{2\cdot 4}+\sqrt[3]{2\cdot 12}$$
$$=4\sqrt[3]{8}+\sqrt[3]{24}$$
$$=4(2)+\sqrt[3]{8}\sqrt[3]{3}$$
$$=8+2\sqrt[3]{3}$$

35.

$$\left(\sqrt{2}+1\right)\left(\sqrt{2}-3\right)$$
$$=\sqrt{2}\sqrt{2}+\sqrt{2}(-3)+1\left(\sqrt{2}\right)+1(-3)$$
$$=\sqrt{4}-3\sqrt{2}+\sqrt{2}-3$$
$$=2-2\sqrt{2}-3$$
$$=-1-2\sqrt{2}$$

37.

$$\left(\sqrt{3x}-\sqrt{2y}\right)\left(\sqrt{3x}+\sqrt{2y}\right)$$
$$=\sqrt{3x}\sqrt{3x}+\sqrt{3x}\sqrt{2y}-\sqrt{2y}\sqrt{3x}-\sqrt{2y}\sqrt{2y}$$
$$=\sqrt{9x^2}+\sqrt{6xy}-\sqrt{6xy}-\sqrt{4y^2}$$
$$=3x-2y$$

39.

$$\left(2\sqrt[3]{4}-3\sqrt[3]{2}\right)\left(3\sqrt[3]{4}+2\sqrt[3]{10}\right)$$
$$=2\sqrt[3]{4}\left(3\sqrt[3]{4}\right)+2\sqrt[3]{4}\left(2\sqrt[3]{10}\right)-3\sqrt[3]{2}\left(3\sqrt[3]{4}\right)-3\sqrt[3]{2}\left(2\sqrt[3]{10}\right)$$
$$=6\sqrt[3]{16}+4\sqrt[3]{40}-9\sqrt[3]{8}-6\sqrt[3]{20}$$
$$=6\sqrt[3]{8}\sqrt[3]{2}+4\sqrt[3]{8}\sqrt[3]{5}-9\sqrt[3]{8}-6\sqrt[3]{20}$$
$$=6(2)\sqrt[3]{2}+4(2)\sqrt[3]{5}-9(2)-6\sqrt[3]{20}$$
$$=12\sqrt[3]{2}+8\sqrt[3]{5}-18-6\sqrt[3]{20}$$

41.

$$\left(\sqrt[3]{5z}+\sqrt[3]{3}\right)\left(\sqrt[3]{5z}+2\sqrt[3]{3}\right)$$
$$=\sqrt[3]{5z}\sqrt[3]{5z}+\sqrt[3]{5z}\left(2\sqrt[3]{3}\right)+\sqrt[3]{3}\left(\sqrt[3]{5z}\right)+\sqrt[3]{3}\left(2\sqrt[3]{3}\right)$$
$$=\sqrt[3]{25z^2}+2\sqrt[3]{15z}+\sqrt[3]{15z}+2\sqrt[3]{9}$$
$$=\sqrt[3]{25z^2}+3\sqrt[3]{15z}+2\sqrt[3]{9}$$

43.

$$\left(\sqrt{7}\right)^2=\sqrt{7}\sqrt{7}$$
$$=\sqrt{49}$$
$$=7$$

45.

$$\left(\sqrt[3]{12}\right)^3=\sqrt[3]{12}\sqrt[3]{12}\sqrt[3]{12}$$
$$=\sqrt[3]{1,728}$$
$$=12$$

47.

$$\left(3\sqrt{2}\right)^2=3\sqrt{2}\cdot 3\sqrt{2}$$
$$=3(3)\sqrt{2}\sqrt{2}$$
$$=9\sqrt{4}$$
$$=9(2)$$
$$=18$$

49.

$$\left(-2\sqrt[3]{2x^2}\right)^3=\left(-2\sqrt[3]{2x^2}\right)\left(-2\sqrt[3]{2x^2}\right)\left(-2\sqrt[3]{2x^2}\right)$$
$$=-8\sqrt[3]{8x^6}$$
$$=-8\left(2x^2\right)$$
$$=-16x^2$$

51.

$$\left(6-\sqrt{3}\right)^2=(6)^2-2(6)\left(\sqrt{3}\right)+\left(\sqrt{3}\right)^2$$
$$=36-12\sqrt{3}+3$$
$$=39-12\sqrt{3}$$

53.

$$\left(\sqrt{3x}+\sqrt{3}\right)^2$$
$$=\left(\sqrt{3x}+\sqrt{3}\right)\left(\sqrt{3x}+\sqrt{3}\right)$$
$$=\sqrt{3x}\sqrt{3x}+\sqrt{3x}\sqrt{3}+\sqrt{3}\sqrt{3x}+\sqrt{3}\sqrt{3}$$
$$=\sqrt{9x^2}+\sqrt{9x}+\sqrt{9x}+\sqrt{9}$$
$$=3x+\sqrt{9}\sqrt{x}+\sqrt{9}\sqrt{x}+3$$
$$=3x+3\sqrt{x}+3\sqrt{x}+3$$
$$=3x+6\sqrt{x}+3$$

55.

$$\sqrt{\frac{2}{7}}=\frac{\sqrt{2}}{\sqrt{7}}$$
$$=\frac{\sqrt{2}}{\sqrt{7}}\cdot\frac{\sqrt{7}}{\sqrt{7}}$$
$$=\frac{\sqrt{14}}{\sqrt{49}}$$
$$=\frac{\sqrt{14}}{7}$$

57.

$$\sqrt{\frac{8}{3}}=\frac{\sqrt{8}}{\sqrt{3}}$$
$$=\frac{\sqrt{8}}{\sqrt{3}}\cdot\frac{\sqrt{3}}{\sqrt{3}}$$
$$=\frac{\sqrt{24}}{\sqrt{9}}$$
$$=\frac{\sqrt{4}\sqrt{6}}{3}$$
$$=\frac{2\sqrt{6}}{3}$$

59.

$$\frac{4}{\sqrt{6}}=\frac{4}{\sqrt{6}}\cdot\frac{\sqrt{6}}{\sqrt{6}}$$
$$=\frac{4\sqrt{6}}{\sqrt{36}}$$
$$=\frac{4\sqrt{6}}{6}$$
$$=\frac{2\sqrt{6}}{3}$$

61.

$$\frac{1}{\sqrt[3]{2}}=\frac{1}{\sqrt[3]{2}}\cdot\frac{\sqrt[3]{4}}{\sqrt[3]{4}}$$
$$=\frac{\sqrt[3]{4}}{\sqrt[3]{8}}$$
$$=\frac{\sqrt[3]{4}}{2}$$

63.

$$\frac{3}{\sqrt[3]{9}}=\frac{3}{\sqrt[3]{9}}\cdot\frac{\sqrt[3]{3}}{\sqrt[3]{3}}$$
$$=\frac{3\sqrt[3]{3}}{\sqrt[3]{27}}$$
$$=\frac{3\sqrt[3]{3}}{3}$$
$$=\sqrt[3]{3}$$

65.

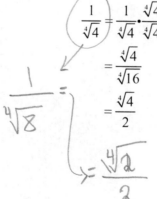

$$\frac{1}{\sqrt[4]{4}}=\frac{1}{\sqrt[4]{4}}\cdot\frac{\sqrt[4]{4}}{\sqrt[4]{4}}$$
$$=\frac{\sqrt[4]{4}}{\sqrt[4]{16}}$$
$$=\frac{\sqrt[4]{4}}{2}$$

$$\frac{1}{\sqrt[4]{8}}=\frac{\sqrt[4]{2}}{2}$$

67.

$$\frac{\sqrt{10y^2}}{\sqrt{2y^3}}=\sqrt{\frac{10y^2}{2y^3}}$$
$$=\sqrt{\frac{5}{y}}$$
$$=\frac{\sqrt{5}}{\sqrt{y}}$$
$$=\frac{\sqrt{5}}{\sqrt{y}}\bullet\frac{\sqrt{y}}{\sqrt{y}}$$
$$=\frac{\sqrt{5y}}{y}$$

69.

$$\frac{\sqrt{48x^2}}{\sqrt{8x^2y}}=\sqrt{\frac{48x^2}{8x^2y}}$$
$$=\sqrt{\frac{6}{y}}$$
$$=\frac{\sqrt{6}}{\sqrt{y}}\bullet\frac{\sqrt{y}}{\sqrt{y}}$$
$$=\frac{\sqrt{6y}}{y}$$

71.

$$\frac{\sqrt[3]{12t^3}}{\sqrt[3]{54t^2}}=\sqrt[3]{\frac{12t^3}{54t^2}}$$
$$=\sqrt[3]{\frac{2t}{9}}$$
$$=\frac{\sqrt[3]{2t}}{\sqrt[3]{9}}\bullet\frac{\sqrt[3]{3}}{\sqrt[3]{3}}$$
$$=\frac{\sqrt[3]{6t}}{\sqrt[3]{27}}$$
$$=\frac{\sqrt[3]{6t}}{3}$$

73.

$$\frac{\sqrt[3]{4a^6}}{\sqrt[3]{2a^5b}}=\sqrt[3]{\frac{4a^6}{2a^5b}}$$
$$=\sqrt[3]{\frac{2a}{b}}$$
$$=\frac{\sqrt[3]{2a}}{\sqrt[3]{b}}\bullet\frac{\sqrt[3]{b^2}}{\sqrt[3]{b^2}}$$
$$=\frac{\sqrt[3]{2ab^2}}{\sqrt[3]{b^3}}$$
$$=\frac{\sqrt[3]{2ab^2}}{b}$$

75.

$$\frac{23}{\sqrt{50p^5}}=\frac{23}{\sqrt{25p^4}\sqrt{2p}}$$
$$=\frac{23}{5p^2\sqrt{2p}}$$
$$=\frac{23}{5p^2\sqrt{2p}}\bullet\frac{\sqrt{2p}}{\sqrt{2p}}$$
$$=\frac{23\sqrt{2p}}{5p^2(2p)}$$
$$=\frac{23\sqrt{2p}}{10p^3}$$

77.

$$\frac{7}{\sqrt{24b^3}}=\frac{7}{\sqrt{4b^2}\sqrt{6b}}$$
$$=\frac{7}{2b\sqrt{6b}}\bullet\frac{\sqrt{6b}}{\sqrt{6b}}$$
$$=\frac{7\sqrt{6b}}{2b(6b)}$$
$$=\frac{7\sqrt{6b}}{12b^2}$$

79.

$$\sqrt[3]{\frac{5}{16}}=\frac{\sqrt[3]{5}}{\sqrt[3]{16}}$$
$$=\frac{\sqrt[3]{5}}{\sqrt[3]{8}\sqrt[3]{2}}$$
$$=\frac{\sqrt[3]{5}}{2\sqrt[3]{2}}\bullet\frac{\sqrt[3]{4}}{\sqrt[3]{4}}$$
$$=\frac{\sqrt[3]{20}}{2(2)}$$
$$=\frac{\sqrt[3]{20}}{4}$$

81.

$$\sqrt[3]{\frac{4}{81}}=\frac{\sqrt[3]{4}}{\sqrt[3]{81}}$$
$$=\frac{\sqrt[3]{4}}{\sqrt[3]{27}\sqrt[3]{3}}$$
$$=\frac{\sqrt[3]{4}}{3\sqrt[3]{3}}\bullet\frac{\sqrt[3]{9}}{\sqrt[3]{9}}$$
$$=\frac{\sqrt[3]{36}}{3\sqrt[3]{27}}$$
$$=\frac{\sqrt[3]{36}}{3(3)}$$
$$=\frac{\sqrt[3]{36}}{9}$$

83.

$$\frac{19}{\sqrt[3]{5c^2}}=\frac{19}{\sqrt[3]{5c^2}}\bullet\frac{\sqrt[3]{25c}}{\sqrt[3]{25c}}$$
$$=\frac{19\sqrt[3]{25c}}{\sqrt[3]{125c^3}}$$
$$=\frac{19\sqrt[3]{25c}}{5c}$$

85.

$$\frac{\sqrt[3]{3}}{\sqrt[3]{2r}}=\frac{\sqrt[3]{3}}{\sqrt[3]{2r}}\bullet\frac{\sqrt[3]{4r^2}}{\sqrt[3]{4r^2}}$$
$$=\frac{\sqrt[3]{12r^2}}{\sqrt[3]{8r^3}}$$
$$=\frac{\sqrt[3]{12r^2}}{2r}$$

87.

$$\frac{\sqrt[4]{2}}{\sqrt[4]{3t^2}}=\frac{\sqrt[4]{2}}{\sqrt[4]{3t^2}}\bullet\frac{\sqrt[4]{27t^2}}{\sqrt[4]{27t^2}}$$
$$=\frac{\sqrt[4]{54t^2}}{\sqrt[4]{81t^4}}$$
$$=\frac{\sqrt[4]{54t^2}}{3t}$$

89.

$$\frac{25}{\sqrt[4]{8a}}=\frac{25}{\sqrt[4]{8a}}\bullet\frac{\sqrt[4]{2a^3}}{\sqrt[4]{2a^3}}$$
$$=\frac{25\sqrt[4]{2a^3}}{\sqrt[4]{16a^4}}$$
$$=\frac{25\sqrt[4]{2a^3}}{2a}$$

91.

$$\frac{\sqrt{2}}{\sqrt{5}+3}=\frac{\sqrt{2}}{\sqrt{5}+3}\bullet\frac{\sqrt{5}-3}{\sqrt{5}-3}$$
$$=\frac{\sqrt{2}\left(\sqrt{5}-3\right)}{\left(\sqrt{5}+3\right)\left(\sqrt{5}-3\right)}$$
$$=\frac{\sqrt{10}-3\sqrt{2}}{\sqrt{25}-9}$$
$$=\frac{\sqrt{10}-3\sqrt{2}}{5-9}$$
$$=\frac{\sqrt{10}-3\sqrt{2}}{-4}$$
$$=\frac{-\sqrt{10}+3\sqrt{2}}{4}$$
$$=\frac{3\sqrt{2}-\sqrt{10}}{4}$$

93.

$$\frac{2}{\sqrt{x}+1}=\frac{2}{\sqrt{x}+1}\bullet\frac{\sqrt{x}-1}{\sqrt{x}-1}$$
$$=\frac{2\left(\sqrt{x}-1\right)}{\left(\sqrt{x}+1\right)\sqrt{x}-1}$$
$$=\frac{2\sqrt{x}-2}{\sqrt{x^2}-1}$$
$$=\frac{2\left(\sqrt{x}-1\right)}{x-1}$$

95.

$$\frac{\sqrt{7}-\sqrt{2}}{\sqrt{2}+\sqrt{7}} = \frac{\sqrt{7}-\sqrt{2}}{\sqrt{2}+\sqrt{7}} \cdot \frac{\sqrt{2}-\sqrt{7}}{\sqrt{2}-\sqrt{7}}$$
$$= \frac{(\sqrt{7}-\sqrt{2})(\sqrt{2}-\sqrt{7})}{(\sqrt{2}+\sqrt{7})(\sqrt{2}-\sqrt{7})}$$
$$= \frac{\sqrt{7}\sqrt{2}+\sqrt{7}(-\sqrt{7})-\sqrt{2}\sqrt{2}-\sqrt{2}(-\sqrt{7})}{\sqrt{2}\sqrt{2}-\sqrt{7}\sqrt{7}}$$
$$= \frac{\sqrt{14}-\sqrt{49}-\sqrt{4}+\sqrt{14}}{\sqrt{4}-\sqrt{49}}$$
$$= \frac{\sqrt{14}-7-2+\sqrt{14}}{2-7}$$
$$= \frac{-9+2\sqrt{14}}{-5}$$
$$= \frac{-(9-2\sqrt{14})}{-5}$$
$$= \frac{9-2\sqrt{14}}{5}$$

97.

$$\frac{\sqrt{x}-\sqrt{y}}{\sqrt{x}+\sqrt{y}} = \frac{\sqrt{x}-\sqrt{y}}{\sqrt{x}+\sqrt{y}} \cdot \frac{\sqrt{x}-\sqrt{y}}{\sqrt{x}-\sqrt{y}}$$
$$= \frac{(\sqrt{x}-\sqrt{y})(\sqrt{x}-\sqrt{y})}{(\sqrt{x}+\sqrt{y})(\sqrt{x}-\sqrt{y})}$$
$$= \frac{\sqrt{x}\sqrt{x}-\sqrt{x}\sqrt{y}-\sqrt{y}\sqrt{x}+\sqrt{y}\sqrt{y}}{\sqrt{x}\sqrt{x}-\sqrt{y}\sqrt{y}}$$
$$= \frac{\sqrt{x^2}-\sqrt{xy}-\sqrt{xy}+\sqrt{y^2}}{\sqrt{x^2}-\sqrt{y^2}}$$
$$= \frac{x-2\sqrt{xy}+y}{x-y}$$

99.

$$\frac{\sqrt{x}+3}{x} = \frac{\sqrt{x}+3}{x} \cdot \frac{\sqrt{x}-3}{\sqrt{x}-3}$$
$$= \frac{(\sqrt{x}+3)(\sqrt{x}-3)}{x(\sqrt{x}-3)}$$
$$= \frac{\sqrt{x}\sqrt{x}-3(3)}{x\sqrt{x}-3x}$$
$$= \frac{\sqrt{x^2}-9}{x\sqrt{x}-3x}$$
$$= \frac{x-9}{x\sqrt{x}-3x}$$
$$= \frac{x-9}{x(\sqrt{x}-3)}$$

101.

$$\frac{\sqrt{x}+\sqrt{y}}{\sqrt{x}} = \frac{\sqrt{x}+\sqrt{y}}{\sqrt{x}} \cdot \frac{\sqrt{x}-\sqrt{y}}{\sqrt{x}-\sqrt{y}}$$
$$= \frac{(\sqrt{x}+\sqrt{y})(\sqrt{x}-\sqrt{y})}{\sqrt{x}(\sqrt{x}-\sqrt{y})}$$
$$= \frac{\sqrt{x}\sqrt{x}-\sqrt{y}\sqrt{y}}{\sqrt{x}(\sqrt{x}-\sqrt{y})}$$
$$= \frac{\sqrt{x^2}-\sqrt{y^2}}{\sqrt{x}(\sqrt{x}-\sqrt{y})}$$
$$= \frac{x-y}{\sqrt{x}(\sqrt{x}-\sqrt{y})}$$

TRY IT YOURSELF

103.

$$\sqrt{x}\left(\sqrt{14x}+\sqrt{2}\right) = \sqrt{14x^2}+\sqrt{2x}$$
$$= x\sqrt{14}+\sqrt{2x}$$

105.

$$\frac{3\sqrt{2}-5\sqrt{3}}{2\sqrt{3}-3\sqrt{2}}$$

$$=\frac{3\sqrt{2}-5\sqrt{3}}{2\sqrt{3}-3\sqrt{2}}\bullet\frac{2\sqrt{3}+3\sqrt{2}}{2\sqrt{3}+3\sqrt{2}}$$

$$=\frac{\left(3\sqrt{2}-5\sqrt{3}\right)\left(2\sqrt{3}+3\sqrt{2}\right)}{\left(2\sqrt{3}-3\sqrt{2}\right)\left(2\sqrt{3}+3\sqrt{2}\right)}$$

$$=\frac{3\sqrt{2}\left(2\sqrt{3}\right)+3\sqrt{2}\left(3\sqrt{2}\right)-5\sqrt{3}\left(2\sqrt{3}\right)-5\sqrt{3}\left(3\sqrt{2}\right)}{2\sqrt{3}\left(2\sqrt{3}\right)-3\sqrt{2}\left(3\sqrt{2}\right)}$$

$$=\frac{6\sqrt{6}+9\sqrt{4}-10\sqrt{9}-15\sqrt{6}}{4\sqrt{9}-9\sqrt{4}}$$

$$=\frac{6\sqrt{6}+9(2)-10(3)-15\sqrt{6}}{4(3)-9(2)}$$

$$=\frac{6\sqrt{6}+18-30-15\sqrt{6}}{12-18}$$

$$=\frac{-12-9\sqrt{6}}{-6}$$

$$=\frac{-\not{3}\left(4+3\sqrt{6}\right)}{-\not{3}(2)}$$

$$=\frac{4+3\sqrt{6}}{2}$$

$$=\frac{3\sqrt{6}+4}{2}$$

107.

$$\left(10\sqrt[3]{2x}\right)^3=\left(10\sqrt[3]{2x}\right)\left(10\sqrt[3]{2x}\right)\left(10\sqrt[3]{2x}\right)$$
$$=1{,}000\sqrt[3]{8x^3}$$
$$=1{,}000(2x)$$
$$=2{,}000x$$

109.

$$-4\sqrt[3]{5r^2s}\left(5\sqrt[3]{2r}\right)=-20\sqrt[3]{10r^3s}$$
$$=-20\sqrt[3]{r^3}\sqrt[3]{10s}$$
$$=-20r\sqrt[3]{10s}$$

111.

$$\left(3p+\sqrt{5}\right)^2=(3p)^2+2(3p)\left(\sqrt{5}\right)+\left(\sqrt{5}\right)^2$$
$$=9p^2+6p\sqrt{5}+5$$

113.

$$\sqrt{\frac{72m^8}{25m^3}}=\sqrt{\frac{72m^5}{25}}$$
$$=\frac{\sqrt{72m^5}}{\sqrt{25}}$$
$$=\frac{\sqrt{36m^4}\sqrt{2m}}{5}$$
$$=\frac{6m^2\sqrt{2m}}{5}$$

115.

$$\sqrt[4]{3n^2}\sqrt[4]{27n^3}=\sqrt[4]{81n^5}$$
$$=\sqrt[4]{81n^4}\sqrt[4]{n}$$
$$=3n\sqrt[4]{n}$$

117.

$$\frac{\sqrt[3]{x}}{\sqrt[3]{9}}=\frac{\sqrt[3]{x}}{\sqrt[3]{9}}\bullet\frac{\sqrt[3]{3}}{\sqrt[3]{3}}$$
$$=\frac{\sqrt[3]{3x}}{\sqrt[3]{27}}$$
$$=\frac{\sqrt[3]{3x}}{3}$$

119.

$$\left(3\sqrt{2r}-2\right)^2$$
$$=\left(3\sqrt{2r}-2\right)\left(3\sqrt{2r}-2\right)$$
$$=3\sqrt{2r}\left(3\sqrt{2r}\right)+3\sqrt{2r}(-2)-2\left(3\sqrt{2r}\right)-2(-2)$$
$$=9\sqrt{4r^2}-6\sqrt{2r}-6\sqrt{2r}+4$$
$$=9(2r)-12\sqrt{2r}+4$$
$$=18r-12\sqrt{2r}+4$$

121.

$$\sqrt{x(x+3)}\sqrt{x^3(x+3)}=\sqrt{x^4(x+3)^2}$$
$$=x^2(x+3)$$

123.

$$\frac{2z-1}{\sqrt{2z}-1}=\frac{2z-1}{\sqrt{2z}-1}\bullet\frac{\sqrt{2z}+1}{\sqrt{2z}+1}$$
$$=\frac{(2z-1)(\sqrt{2z}+1)}{(\sqrt{2z}-1)(\sqrt{2z}+1)}$$
$$=\frac{(2z-1)(\sqrt{2z}+1)}{\sqrt{2z}\sqrt{2z}-1(1)}$$
$$=\frac{(2z-1)(\sqrt{2z}+1)}{\sqrt{4z^2}-1}$$
$$=\frac{(2z-1)(\sqrt{2z}+1)}{2z-1}$$
$$=\frac{(\sqrt{2z}+1)\cancel{(2z-1)}}{\cancel{2z-1}}$$
$$=\sqrt{2z}+1$$

125.a.

$$(3\sqrt{a})^2=(3\sqrt{a})(3\sqrt{a})$$
$$=9\sqrt{a^2}$$
$$=9a$$

b.

$$(3+\sqrt{a})^2=(3)^2+2(3)(\sqrt{a})+(\sqrt{a})^2$$
$$=9+6\sqrt{a}+a$$

127.a.

$$(\sqrt{m-6})^2=(\sqrt{m-6})(\sqrt{m-6})$$
$$=m-6$$

b.

$$(\sqrt{m}-6)^2=(\sqrt{m})^2-2(\sqrt{m})(6)+(-6)^2$$
$$=m-12\sqrt{m}+36$$

APPLICATIONS

129. STATISTICS

$$\frac{1}{\sigma\sqrt{2\pi}}=\frac{1}{\sigma\sqrt{2\pi}}\bullet\frac{\sqrt{2\pi}}{\sqrt{2\pi}}$$
$$=\frac{\sqrt{2\pi}}{\sigma\sqrt{4\pi^2}}$$
$$=\frac{\sqrt{2\pi}}{\sigma(2\pi)}$$
$$=\frac{\sqrt{2\pi}}{2\pi\sigma}$$

131. TRIGONOMETRY

$$\frac{\text{length of side } AC}{\text{length of side } AB}=\frac{1}{\sqrt{2}}$$
$$=\frac{1}{\sqrt{2}}\bullet\frac{\sqrt{2}}{\sqrt{2}}$$
$$=\frac{\sqrt{2}}{\sqrt{4}}$$
$$=\frac{\sqrt{2}}{2}$$

WRITING

133. Answers will vary.

135. Answers will vary.

137. Answers will vary.

REVIEW

139.

$$\frac{8}{b-2}+\frac{3}{2-b}=-\frac{1}{b}$$
$$\frac{8}{b-2}-\frac{3}{b-2}=-\frac{1}{b}$$
$$b(b-2)\left(\frac{8}{b-2}-\frac{3}{b-2}\right)=b(b-2)\left(-\frac{1}{b}\right)$$
$$8b-3b=-(b-2)$$
$$5b=-b+2$$
$$6b=2$$
$$b=\frac{2}{6}$$
$$b=\frac{1}{3}$$

CHALLENGE PROBLEMS

141.

$$\sqrt{2}\bullet\sqrt[3]{2}=2^{1/2}\bullet 2^{1/3}$$
$$=2^{(1/2)+(1/3)}$$
$$=2^{3/6+2/6}$$
$$=2^{5/6}$$
$$=\sqrt[6]{2^5}$$
$$=\sqrt[6]{32}$$

SECTION 7.5

VOCABULARY

1. Equations such as $\sqrt{x+4}-4=5$ and $\sqrt[3]{x+1}=12$ are called **radical** equations.

3. When we square both sides of a radical equation, we say we are **raising** both sides to the second power.

5. Proposed solutions of a radical equation that do not satisfy it are called **extraneous** solutions.

CONCEPTS

7. a. The power rule for solving radical equations states that if *x*, *y*, and *n* are real numbers and $x = y$, then $x^{\boxed{n}} = y^{\boxed{n}}$.

   b. $\sqrt[n]{a^n} = \boxed{a}$

9. a. square both sides
   b. subtract 3 from both sides
   c. add $\sqrt{2x+9}$ to both sides

11.

$$\begin{aligned}\left(\sqrt{x}-3\right)^2 &= \left(\sqrt{x}\right)^2 - 2(3)\left(\sqrt{x}\right) + (-3)^2 \\ &= x - 6\sqrt{x} + 9\end{aligned}$$

NOTATION

13.

$$\begin{aligned}\sqrt{3x+3}-1&=5\\ \sqrt{3x+3}&=\boxed{6}\\ \left(\sqrt{3x+3}\right)^{\boxed{2}}&=(6)^{\boxed{2}}\\ \boxed{3x+3}&=36\\ 3x&=\boxed{33}\\ x&=\boxed{11}\end{aligned}$$

$\boxed{\text{Yes}}$ it checks.

GUIDED PRACTICE

15.

$$\begin{aligned}\sqrt{a-3}&=1\\ \left(\sqrt{a-3}\right)^2&=(1)^2\\ a-3&=1\\ a&=4\end{aligned}$$

17.

$$\begin{aligned}\sqrt{4x+5}&=5\\ \left(\sqrt{4x+5}\right)^2&=5^2\\ 4x+5&=25\\ 4x&=20\\ x&=5\end{aligned}$$

19.

$$\begin{aligned}\sqrt{6x+13}&=7\\ \left(\sqrt{6x+13}\right)^2&=(7)^2\\ 6x+13&=49\\ 6x&=36\\ x&=6\end{aligned}$$

21.

$$\begin{aligned}\sqrt{\frac{1}{3}x-2}&=8\\ \left(\sqrt{\frac{1}{3}x-2}\right)^2&=8^2\\ \frac{1}{3}x-2&=64\\ \frac{1}{3}x&=66\\ 3\left(\frac{1}{3}x\right)&=3(66)\\ x&=198\end{aligned}$$

23.

$$\sqrt{2x+11}+2=x$$
$$\sqrt{2x+11}=x-2$$
$$\left(\sqrt{2x+11}\right)^2=(x-2)^2$$
$$2x+11=x^2-4x+4$$
$$0=x^2-6x-7$$
$$0=(x-7)(x+1)$$
$$x-7=0 \quad \text{or} \quad x+1=0$$
$$x=7 \qquad x=\cancel{-1}$$

25.

$$\sqrt{2r-3}+9=r$$
$$\sqrt{2r-3}=r-9$$
$$\left(\sqrt{2r-3}\right)^2=(r-9)^2$$
$$2r-3=r^2-9r-9r+81$$
$$2r-3=r^2-18r+81$$
$$0=r^2-20r+84$$
$$0=(r-14)(r-6)$$
$$r-14=0 \quad \text{or} \quad r-6=0$$
$$r=14 \qquad r=\cancel{6}$$

27.

$$\sqrt{3t+7}-t=1$$
$$\sqrt{3t+7}=t+1$$
$$\left(\sqrt{3t+7}\right)^2=(t+1)^2$$
$$3t+7=t^2+2t+1$$
$$0=t^2-t-6$$
$$0=(t-3)(t+2)$$
$$t-3=0 \quad \text{or} \quad t+2=0$$
$$t=3 \qquad t=\cancel{-2}$$

29.

$$\sqrt{9-a}-a=3$$
$$\sqrt{9-a}=a+3$$
$$\left(\sqrt{9-a}\right)^2=(a+3)^2$$
$$9-a=a^2+6a+9$$
$$0=a^2+7a$$
$$0=a(a+7)$$
$$a=0 \quad \text{or} \quad a+7=0$$
$$a=0 \qquad a=\cancel{-7}$$

31.

$$\sqrt{5x}+10=8$$
$$\sqrt{5x}=-2$$
$$\left(\sqrt{5x}\right)=(-2)^2$$
$$5x=4$$
$$x=\cancel{\frac{4}{5}}$$

No Solution

33.

$$\sqrt{5-x}+10=9$$
$$\sqrt{5-x}=-1$$
$$\left(\sqrt{5-x}\right)^2=(-1)^2$$
$$5-x=1$$
$$-x=-4$$
$$x=\cancel{4}$$

No Solution

35.

$$\sqrt[3]{7n-1}=3$$
$$\left(\sqrt[3]{7n-1}\right)^3=(3)^3$$
$$7n-1=27$$
$$7n=28$$
$$n=4$$

37.

$$\sqrt[3]{x^3-7}=x-1$$
$$\left(\sqrt[3]{x^3-7}\right)^3=(x-1)^3$$
$$x^3-7=(x-1)(x-1)(x-1)$$
$$x^3-7=(x-1)(x^2-x-x+1)$$
$$x^3-7=(x-1)(x^2-2x+1)$$
$$x^3-7=x(x^2-2x+1)-1(x^2-2x+1)$$
$$x^3-7=x^3-2x^2+x-x^2+2x-1$$
$$x^3-7=x^3-3x^2+3x-1$$
$$x^3-7-x^3=x^3-3x^2+3x-1-x^3$$
$$-7=-3x^2+3x-1$$
$$3x^2-3x+1-7=0$$
$$3x^2-3x-6=0$$
$$\frac{3x^2}{3}-\frac{3x}{3}-\frac{6}{3}=\frac{0}{3}$$
$$x^3-x-2=0$$
$$(x-2)(x+1)=0$$
$$x-2=0 \quad \text{or} \quad x+1=0$$
$$x=2 \qquad x=-1$$

39.

$$\left(m^3+26\right)^{1/3}=m+2$$
$$\sqrt[3]{m^3+26}=m+2$$
$$\left(\sqrt[3]{m^3+26}\right)^3=(m+2)^3$$
$$m^3+26=(m+2)(m+2)(m+2)$$
$$m^3+26=(m+2)(m^2+4m+4)$$
$$m^3+26=m^3+4m^2+4m+2m^2+8m+8$$
$$m^3+26=m^3+6m^2+12m+8$$
$$26=6m^2+12m+8$$
$$0=6m^2+12m-18$$
$$0=6(m^2+2m-3)$$
$$0=6(m+3)(m-1)$$
$$m+3=0 \quad \text{or} \quad m-1=0$$
$$m=-3 \qquad m=1$$

41.

$$(5r+14)^{1/3}=4$$
$$\sqrt[3]{5r+14}=4$$
$$\left(\sqrt[3]{5r+14}\right)^3=4^3$$
$$5r+14=64$$
$$5r=50$$
$$r=10$$

43. $f(x)=\sqrt[4]{3x+1}$

$$\sqrt[4]{3x+1}=4$$
$$\left(\sqrt[4]{3x+1}\right)^4=4^4$$
$$3x+1=256$$
$$3x=255$$
$$x=85$$

45. $f(x)=\sqrt[3]{3x-6}$

$$\sqrt[3]{3x-6}=-3$$
$$\left(\sqrt[3]{3x-6}\right)^3=(-3)^3$$
$$3x-6=-27$$
$$3x=-21$$
$$x=-7$$

47.

$$\sqrt{3x+12}=\sqrt{5x-12}$$
$$\left(\sqrt{3x+12}\right)^2=\left(\sqrt{5x-12}\right)^2$$
$$3x+12=5x-12$$
$$-2x+12=-12$$
$$-2x=-24$$
$$x=12$$

49.

$$2\sqrt{4x+1}=\sqrt{x+4}$$
$$\left(2\sqrt{4x+1}\right)^2=\left(\sqrt{x+4}\right)^2$$
$$4(4x+1)=x+4$$
$$16x+4=x+4$$
$$15x+4=4$$
$$15x=0$$
$$x=0$$

51.

$$\sqrt{6t+9}=3\sqrt{t}$$
$$\left(\sqrt{6t+9}\right)^2=\left(3\sqrt{t}\right)^2$$
$$6t+9=9t$$
$$9=3t$$
$$3=t$$

53.

$$(34x+26)^{1/3}=4(x-1)^{1/3}$$
$$\sqrt[3]{34x+26}=4\sqrt[3]{x-1}$$
$$\left(\sqrt[3]{34x+26}\right)^3=\left(4\sqrt[3]{x-1}\right)^3$$
$$34x+26=64(x-1)$$
$$34x+26=64x-64$$
$$-30x+26=-64$$
$$-30x=-90$$
$$x=3$$

55.

$$\sqrt{x-5}+\sqrt{x}=5$$
$$\sqrt{x-5}=5-\sqrt{x}$$
$$\left(\sqrt{x-5}\right)^2=\left(5-\sqrt{x}\right)^2$$
$$x-5=25-5\sqrt{x}-5\sqrt{x}+x$$
$$x-5=25-10\sqrt{x}+x$$
$$x-5-25-x=25-10\sqrt{x}+x-25-x$$
$$-30=-10\sqrt{x}$$
$$\frac{-30}{-10}=\frac{-10\sqrt{x}}{-10}$$
$$3=\sqrt{x}$$
$$3^2=\left(\sqrt{x}\right)^2$$
$$9=x$$

57.

$$\sqrt{z+3}-\sqrt{z}=1$$
$$\sqrt{z+3}=1+\sqrt{z}$$
$$\left(\sqrt{z+3}\right)^2=\left(1+\sqrt{z}\right)^2$$
$$z+3=1+\sqrt{z}+\sqrt{z}+\sqrt{z^2}$$
$$z+3=1+2\sqrt{z}+z$$
$$z+3-z-1=1+2\sqrt{z}+z-z-1$$
$$2=2\sqrt{z}$$
$$\frac{2}{2}=\frac{2\sqrt{z}}{2}$$
$$1=\sqrt{z}$$
$$(1)^2=\left(\sqrt{z}\right)^2$$
$$1=z$$

59.

$$3=\sqrt{y+4}-\sqrt{y+7}$$
$$3+\sqrt{y+7}=\sqrt{y+4}$$
$$\left(3+\sqrt{y+7}\right)^2=\left(\sqrt{y+4}\right)^2$$
$$9+3\sqrt{y+7}+3\sqrt{y+7}+y+7=y+4$$
$$9+6\sqrt{y+7}+y+7=y+4$$
$$16+y+6\sqrt{y+7}=y+4$$
$$6\sqrt{y+7}=-12$$
$$\frac{6\sqrt{y+7}}{6}=\frac{-12}{6}$$
$$\sqrt{y+7}=-2$$
$$\left(\sqrt{y+7}\right)^2=(-2)^2$$
$$y+7=4$$
$$y=\cancel{-3}$$

No Solution

61.

$$2=\sqrt{2u+7}-\sqrt{u}$$
$$2+\sqrt{u}=\sqrt{2u+7}$$
$$\left(2+\sqrt{u}\right)^2=\left(\sqrt{2u+7}\right)^2$$
$$4+2\sqrt{u}+2\sqrt{u}+\sqrt{u^2}=2u+7$$
$$4+4\sqrt{u}+u=2u+7$$
$$4+4\sqrt{u}+u-4-u=2u+7-4-u$$
$$4\sqrt{u}=u+3$$
$$\left(4\sqrt{u}\right)^2=\left(u+3\right)^2$$
$$16u=u^2+3u+3u+9$$
$$16u=u^2+6u+9$$
$$16u-16u=u^2+6u+9-16u$$
$$0=u^2-10u+9$$
$$0=(u-1)(u-9)$$
$$u-1=0 \quad \text{or} \quad u-9=0$$
$$u=1 \qquad\qquad u=9$$

63.

$$v=\sqrt{2gh}$$
$$v^2=\left(\sqrt{2gh}\right)^2$$
$$v^2=2gh$$
$$\frac{v^2}{2g}=\frac{2gh}{2g}$$
$$\frac{v^2}{2g}=h$$

65.

$$T=2\pi\sqrt{\frac{\ell}{32}}$$
$$(T)^2=\left(2\pi\sqrt{\frac{\ell}{32}}\right)^2$$
$$T^2=4\pi^2\bullet\frac{\ell}{32}$$
$$T^2=\frac{4\pi^2\ell}{32}$$
$$T^2=\frac{\pi^2\ell}{8}$$
$$8\left(T^2\right)=8\left(\frac{\pi^2\ell}{8}\right)$$
$$8T^2=\pi^2\ell$$
$$\frac{8T^2}{\pi^2}=\frac{\pi^2\ell}{\pi^2}$$
$$\frac{8T^2}{\pi^2}=\ell$$

67.

$$r=\sqrt[3]{\frac{A}{P}}-1$$
$$r+1=\sqrt[3]{\frac{A}{P}}$$
$$(r+1)^3=\left(\sqrt[3]{\frac{A}{P}}\right)^3$$
$$(r+1)^3=\frac{A}{P}$$
$$P(r+1)^3=P\left(\frac{A}{P}\right)$$
$$P(r+1)^3=A$$

69.

$$L_A = L_B\sqrt{1-\frac{v^2}{c^2}}$$

$$\frac{L_A}{L_B} = \frac{L_B\sqrt{1-\frac{v^2}{c^2}}}{L_B}$$

$$\frac{L_A}{L_B} = \sqrt{1-\frac{v^2}{c^2}}$$

$$\left(\frac{L_A}{L_B}\right)^2 = \left(\sqrt{1-\frac{v^2}{c^2}}\right)^2$$

$$\frac{L_A^{\ 2}}{L_B^{\ 2}} = 1-\frac{v^2}{c^2}$$

$$\frac{L_A^{\ 2}}{L_B^{\ 2}} - 1 = -\frac{v^2}{c^2}$$

$$-c^2\left(\frac{L_A^{\ 2}}{L_B^{\ 2}} - 1\right) = -c^2\left(-\frac{v^2}{c^2}\right)$$

$$-c^2\left(\frac{L_A^{\ 2}}{L_B^{\ 2}} - 1\right) = v^2$$

$$c^2\left(-\frac{L_A^{\ 2}}{L_B^{\ 2}} + 1\right) = v^2$$

$$c^2\left(1-\frac{L_A^{\ 2}}{L_B^{\ 2}}\right) = v^2$$

TRY IT YOURSELF

71.

$$2\sqrt{x} = \sqrt{5x-16}$$

$$\left(2\sqrt{x}\right)^2 = \left(\sqrt{5x-16}\right)^2$$

$$4x = 5x-16$$

$$-x = -16$$

$$x = 16$$

73.

$$\sqrt{x+5} + \sqrt{x-3} = 4$$

$$\sqrt{x+5} = 4 - \sqrt{x-3}$$

$$\left(\sqrt{x+5}\right)^2 = \left(4-\sqrt{x-3}\right)^2$$

$$x+5 = 16 - 4\sqrt{x-3} - 4\sqrt{x-3} + x - 3$$

$$x+5 = 13 - 8\sqrt{x-3} + x$$

$$x+5-x-13 = 13 - 8\sqrt{x-3} + x - x - 13$$

$$-8 = -8\sqrt{x-3}$$

$$\frac{-8}{-8} = \frac{-8\sqrt{x-3}}{-8}$$

$$1 = \sqrt{x-3}$$

$$1^2 = \left(\sqrt{x-3}\right)^2$$

$$1 = x-3$$

$$4 = x$$

75.

$$n = \left(n^3+n^2-1\right)^{1/3}$$

$$n = \sqrt[3]{n^3+n^2-1}$$

$$(n)^3 = \left(\sqrt[3]{n^3+n^2-1}\right)^3$$

$$n^3 = n^3+n^2-1$$

$$n^3 - n^3 = n^3 + n^2 - 1 - n^3$$

$$0 = n^2 - 1$$

$$0 = (n+1)(n+1)$$

$$n+1 = 0 \quad \text{or} \quad n-1=0$$

$$n = -1 \qquad\qquad n = 1$$

77.

$$\sqrt{y+2} + y = 4$$

$$\sqrt{y+2} = 4 - y$$

$$\left(\sqrt{y+2}\right)^2 = (4-y)^2$$

$$y+2 = 16 - 4y - 4y + y^2$$

$$y+2 = 16 - 8y + y^2$$

$$y+2-y-2 = 16 - 8y + y^2 - y - 2$$

$$0 = y^2 - 9y + 14$$

$$0 = (y-2)(y-7)$$

$$y-2 = 0 \quad \text{or} \quad y-7=0$$

$$y = 2 \qquad\qquad y = 7$$

79.

$$\sqrt[3]{x+8}=-2$$
$$\left(\sqrt[3]{x+8}\right)^3=(-2)^3$$
$$x+8=-8$$
$$x=-16$$

81.

$$2=\sqrt{x+5}-\sqrt{x}+1$$
$$2+\sqrt{x}-1=\sqrt{x+5}-\sqrt{x}+1+\sqrt{x}-1$$
$$1+\sqrt{x}=\sqrt{x+5}$$
$$\left(1+\sqrt{x}\right)^2=\left(\sqrt{x+5}\right)^2$$
$$1+\sqrt{x}+\sqrt{x}+x=x+5$$
$$1+2\sqrt{x}+x=x+5$$
$$1+2\sqrt{x}+x-x-1=x+5-x-1$$
$$2\sqrt{x}=4$$
$$\frac{2\sqrt{x}}{2}=\frac{4}{2}$$
$$\sqrt{x}=2$$
$$\left(\sqrt{x}\right)^2=2^2$$
$$x=4$$

83.

$$x=\frac{\sqrt{12x-5}}{2}$$
$$2(x)=2\left(\frac{\sqrt{12x-5}}{2}\right)$$
$$2x=\sqrt{12x-5}$$
$$(2x)^2=\left(\sqrt{12x-5}\right)^2$$
$$4x^2=12x-5$$
$$4x^2-12x+5=0$$
$$(2x-5)(2x-1)=0$$
$$2x-5=0 \quad \text{or} \quad 2x-1=0$$
$$2x=5 \qquad 2x=1$$
$$x=\frac{5}{2} \qquad x=\frac{1}{2}$$

85.

$$\left(n^2+6n+3\right)^{1/2}=\left(n^2-6n-3\right)^{1/2}$$
$$\sqrt{n^2+6n+3}=\sqrt{n^2-6n-3}$$
$$\left(\sqrt{n^2+6n+3}\right)^2=\left(\sqrt{n^2-6n-3}\right)^2$$
$$n^2+6n+3=n^2-6n-3$$
$$n^2+6n+3-n^2=n^2-6n-3-n^2$$
$$6n+3=-6n-3$$
$$12n+3=-3$$
$$12n=-6$$
$$n=-\frac{6}{12}$$
$$n=-\frac{1}{2}$$

87.

$$\sqrt{x-5}-\sqrt{x+3}=4$$
$$\sqrt{x-5}=4+\sqrt{x+3}$$
$$\left(\sqrt{x-5}\right)^2=\left(4+\sqrt{x+3}\right)^2$$
$$x-5=16+4\sqrt{x+3}+4\sqrt{x+3}+x+3$$
$$x-5=19+8\sqrt{x+3}+x$$
$$x-5-x-19=19+8\sqrt{x+3}+x-x-19$$
$$-24=8\sqrt{x+3}$$
$$\frac{-24}{8}=\frac{8\sqrt{x+3}}{8}$$
$$-3=\sqrt{x+3}$$
$$(-3)^2=\left(\sqrt{x+3}\right)^2$$
$$9=x+3$$
$$\not{6}=x$$

No Solution

89.

$$\sqrt[4]{10y+6}=2\sqrt[4]{y}$$
$$\left(\sqrt[4]{10y+6}\right)^4=\left(2\sqrt[4]{y}\right)^4$$
$$10y+6=16y$$
$$6=6y$$
$$1=y$$
$$y=1$$

91.

$$\sqrt{-5x+24}=6-x$$
$$\left(\sqrt{-5x+24}\right)^2=(6-x)^2$$
$$-5x+24=36-6x-6x+x^2$$
$$-5x+24=36-12x+x^2$$
$$-5x+24+5x-24=36-12x+x^2+5x-24$$
$$0=x^2-7x+12$$
$$0=(x-3)(x-4)$$
$$x-3=0 \quad \text{or} \quad x-4=0$$
$$x=3 \qquad\qquad x=4$$

93.

$$\sqrt{2x}+5=1$$
$$\sqrt{2x}=-4$$
$$\left(\sqrt{2x}\right)^2=(-4)^2$$
$$2x=16$$
$$x=\not{8}$$

No Solution

95.

$$\sqrt{6x+2}-\sqrt{5x+3}=0$$
$$\sqrt{6x+2}=\sqrt{5x+3}$$
$$\left(\sqrt{6x+2}\right)^2=\left(\sqrt{5x+3}\right)^2$$
$$6x+2=5x+3$$
$$x+2=3$$
$$x=1$$

97.

$$f(x)=g(x)$$
$$\sqrt{x+16}=7-\sqrt{x+9}$$
$$\left(\sqrt{x+16}\right)^2=\left(7-\sqrt{x+9}\right)^2$$
$$x+16=49-7\sqrt{x+9}-7\sqrt{x+9}+x+9$$
$$x+16=49-14\sqrt{x+9}+x+9$$
$$x+16=58+x-14\sqrt{x+9}$$
$$-42=-14\sqrt{x+9}$$
$$\frac{-42}{-14}=\frac{-14\sqrt{x+9}}{-14}$$
$$3=\sqrt{x+9}$$
$$3^2=\left(\sqrt{x+9}\right)^2$$
$$9=x+9$$
$$0=x$$

99.

$$f(x)=0$$
$$\sqrt[4]{x+8}-\sqrt[4]{2x}=0$$
$$\sqrt[4]{x+8}=\sqrt[4]{2x}$$
$$\left(\sqrt[4]{x+8}\right)^4=\left(\sqrt[4]{2x}\right)^4$$
$$x+8=2x$$
$$x+8-x=2x-x$$
$$8=x$$

101.a.

$$3\sqrt{5n-9}=\sqrt{5n}$$
$$\left(3\sqrt{5n-9}\right)^2=\left(\sqrt{5n}\right)^2$$
$$9(5n-9)=5n$$
$$45n-81=5n$$
$$-81=-40n$$
$$\frac{81}{40}=n$$

b.

$$3+\sqrt{5n-9}=\sqrt{5n}$$
$$\left(3+\sqrt{5n-9}\right)^2=\left(\sqrt{5n}\right)^2$$
$$(3)^2+2(3)\left(\sqrt{5n-9}\right)+\left(\sqrt{5n-9}\right)^2=5n$$
$$9+6\sqrt{5n-9}+5n-9=5n$$
$$6\sqrt{5n-9}+5n=5n$$
$$6\sqrt{5n-9}=0$$
$$\frac{6\sqrt{5n-9}}{6}=\frac{0}{6}$$
$$\sqrt{5n-9}=0$$
$$\left(\sqrt{5n-9}\right)^2=(0)^2$$
$$5n-9=0$$
$$5n=9$$
$$n=\frac{9}{5}$$

103.a.

$$\sqrt{2x}-10=0$$
$$\sqrt{2x}=10$$
$$\left(\sqrt{2x}\right)^2=(10)^2$$
$$2x=100$$
$$x=50$$

b.

$$\sqrt{2x}+10=0$$
$$\sqrt{2x}=-10$$
$$\left(\sqrt{2x}\right)^2=(-10)^2$$
$$2x=100$$
$$x=\cancel{50}\text{; no solution}$$

APPLICATIONS

105. HIGHWAY DESIGN

$$s=3\sqrt{r}$$
$$40=3\sqrt{r}$$
$$(40)^2=\left(3\sqrt{r}\right)^2$$
$$1{,}600=9r$$
$$\frac{1{,}600}{9}=\frac{9r}{9}$$
$$178=r$$
$$r=178\text{ ft}$$

107. WIND POWER

$$v=\sqrt[3]{\frac{P}{0.02}}$$
$$29=\sqrt[3]{\frac{P}{0.02}}$$
$$(29)^3=\left(\sqrt[3]{\frac{P}{0.02}}\right)^3$$
$$24{,}389=\frac{P}{0.02}$$
$$0.02(24{,}389)=0.02\left(\frac{P}{0.02}\right)$$
$$488=P$$
$$P=488\text{ watts}$$

109. GENERAL CONTRACTOR

$$\ell=\sqrt{f^2+h^2}$$
$$10=\sqrt{f^2+6^2}$$
$$10=\sqrt{f^2+36}$$
$$(10)^2=\left(\sqrt{f^2+36}\right)^2$$
$$100=f^2+36$$
$$100-100=f^2+36-100$$
$$0=f^2-64$$
$$0=(f+8)(f-8)$$
$$f+8=0 \quad \text{or} \quad f-8=0$$
$$f=\cancel{-8} \qquad f=8\text{ ft}$$

111. SUPPLY AND DEMAND

$$\sqrt{5x} = \sqrt{100 - 3x^2}$$
$$\left(\sqrt{5x}\right)^2 = \left(\sqrt{100 - 3x^2}\right)^2$$
$$5x = 100 - 3x^2$$
$$3x^2 + 5x - 100 = 0$$
$$(3x + 20)(x - 5) = 0$$
$$3x + 20 = 0 \quad \text{or} \quad x - 5 = 0$$
$$3x = -20 \qquad x = 5$$
$$x = -\frac{20}{3}$$

The equilibrium price is \$5.

WRITING

113. Answers will vary.

115. Answers will vary.

117. Answers will vary.

119. Answers will vary.

REVIEW

121. Let I = intensity and d = distance.

$$I = \frac{k}{d^2}$$
$$40 = \frac{k}{5^2}$$
$$40 = \frac{k}{25}$$
$$25(40) = 25\left(\frac{k}{25}\right)$$
$$1{,}000 = k$$
$$I = \frac{k}{d^2}$$
$$I = \frac{1{,}000}{20^2}$$
$$= \frac{1{,}000}{400}$$
$$= 2.5 \text{ foot-candles}$$

123.

$$\frac{12}{0.166044} = \frac{30}{x}$$
$$12x = 30(0.166044)$$
$$12x = 4.98132$$
$$x = 0.41511 \text{ in.}$$

CHALLENGE PROBLEMS

125.

$$\sqrt[4]{x} = \sqrt{\frac{x}{4}}$$
$$\left(\sqrt[4]{x}\right)^2 = \left(\sqrt{\frac{x}{4}}\right)^2$$
$$\sqrt[4]{x^2} = \frac{x}{4}$$
$$\left(\sqrt[4]{x^2}\right)^2 = \left(\frac{x}{4}\right)^2$$
$$\sqrt[4]{x^4} = \frac{x^2}{16}$$
$$x = \frac{x^2}{16}$$
$$16(x) = 16\left(\frac{x^2}{16}\right)$$
$$16x = x^2$$
$$0 = x^2 - 16x$$
$$0 = x(x - 16)$$
$$x = 0 \qquad x - 16 = 0$$
$$x = 0 \qquad x = 16$$

127.

$$\sqrt{x+2}+\sqrt{2x}=\sqrt{18-x}$$
$$\left(\sqrt{x+2}+\sqrt{2x}\right)^2=\left(\sqrt{18-x}\right)^2$$
$$x+2+\sqrt{x+2}\sqrt{2x}+\sqrt{x+2}\sqrt{2x}+2x=18-x$$
$$3x+2+2\sqrt{x+2}\sqrt{2x}=18-x$$
$$3x+2+2\sqrt{x+2}\sqrt{2x}-3x-2=18-x-3x-2$$
$$2\sqrt{x+2}\sqrt{2x}=-4x+16$$
$$\frac{2\sqrt{x+2}\sqrt{2x}}{2}=\frac{-4x}{2}+\frac{16}{2}$$
$$\sqrt{x+2}\sqrt{2x}=-2x+8$$
$$\left(\sqrt{x+2}\sqrt{2x}\right)^2=\left(-2x+8\right)^2$$
$$2x(x+2)=4x^2-16x-16x+64$$
$$2x^2+4x=4x^2-32x+64$$
$$2x^2+4x-2x^2-4x=4x^2-32x+64-2x^2-4x$$
$$0=2x^2-36x+64$$
$$\frac{0}{2}=\frac{2x^2}{2}-\frac{36x}{2}+\frac{64}{2}$$
$$0=x^2-18x+32$$
$$0=(x-2)(x-16)$$
$$x-2=0 \quad \text{or} \quad x-16=0$$
$$x=2 \qquad x=\not{16}$$

129.

$$\sqrt{2\sqrt{x+1}}=\sqrt{16-4x}$$
$$\left(\sqrt{2\sqrt{x+1}}\right)^2=\left(\sqrt{16-4x}\right)^2$$
$$2\sqrt{x+1}=16-4x$$
$$\frac{2\sqrt{x+1}}{2}=\frac{16-4x}{2}$$
$$\sqrt{x+1}=8-2x$$
$$\left(\sqrt{x+1}\right)^2=(8-2x)^2$$
$$x+1=64-32x+4x^2$$
$$0=4x^2-33x+63$$
$$0=(4x-21)(x-3)$$
$$4x-21=0 \quad x-3=0$$
$$4x=21 \qquad x=3$$
$$x=\not{\frac{21}{4}}$$

VOCABULARY

1. In a right triangle, the side opposite the 90° angle is called the **hypotenuse**.

3. The **Pythagorean** Theorem states that in a right triangle, the sum of the squares of the lengths of the two legs is equal to the square of the hypotenuse.

CONCEPTS

5. If a and b are the lengths of the legs of a right triangle and c is the length of the hypotenuse, then $a^2+b^2=c^2$. This is called the **Pythagorean** equation.

7. In an isosceles right triangle, the length of the hypotenuse is $\sqrt{2}$ times the length of one leg.

9. The length of the longer leg of a 30°–60°–90° triangle is $\sqrt{3}$ times the length of the shorter leg.

11. The formula to find the distance between two points (x_1,y_1) and (x_2,y_2) is $d=\sqrt{(x_2-x_1)^2+(y_2-y_1)^2}$.

NOTATION

13.

$$\begin{aligned}\sqrt{(-1-3)^2+[2-(-4)]^2}&=\sqrt{(-4)^2+[\boxed{6}]^2}\\&=\sqrt{\boxed{52}}\\&=\sqrt{\boxed{4}\cdot 13}\\&=\boxed{2}\sqrt{13}\\&\approx 7.21\end{aligned}$$

GUIDED PRACTICE

15.

$$\begin{aligned}a^2+b^2&=c^2\\6^2+8^2&=c^2\\36+64&=c^2\\100&=c^2\\\sqrt{100}&=c\\10&=c\\c&=10\text{ ft}\end{aligned}$$

17.

$$\begin{aligned}a^2+b^2&=c^2\\8^2+15^2&=c^2\\64+225&=c^2\\289&=c^2\\\sqrt{289}&=c\\17&=c\\c&=17\text{ ft}\end{aligned}$$

19.

$$\begin{aligned}a^2+b^2&=c^2\\a^2+9^2&=41^2\\a^2+81&=1{,}681\\a^2&=1{,}600\\a&=\sqrt{1{,}600}\\a&=40\text{ ft}\end{aligned}$$

21.

$$\begin{aligned}a^2+b^2&=c^2\\10^2+b^2&=26^2\\100+b^2&=676\\b^2&=576\\b&=\sqrt{576}\\b&=24\text{ cm}\end{aligned}$$

23. In an isosceles right triangle, the length of the hypotenuse is the length of one leg times $\sqrt{2}$ and the lengths of the legs are equal.

$$x = 2$$

$$\begin{aligned} h &= x\sqrt{2} \\ &= 2\sqrt{2} \\ &\approx 2.83 \end{aligned}$$

25. In an isosceles right triangle, the length of the hypotenuse is the length of one leg times $\sqrt{2}$ and the lengths of the legs are equal.

$$x = 3.2$$

$$\begin{aligned} h &= x\sqrt{2} \\ &= 3.2\sqrt{2} \\ &\approx 4.53 \text{ ft} \end{aligned}$$

27. In an isosceles right triangle, the length of the hypotenuse is the length of one leg times $\sqrt{2}$ and the lengths of the legs are equal.

$$3 = x\sqrt{2}$$

$$\frac{3}{\sqrt{2}} = \frac{x\sqrt{2}}{\sqrt{2}}$$

$$\frac{3}{\sqrt{2}} \bullet \frac{\sqrt{2}}{\sqrt{2}} = x$$

$$\frac{3\sqrt{2}}{\sqrt{4}} = x$$

$$\frac{3\sqrt{2}}{2} = x$$

$$2.12 \approx x$$

$$x = y$$

$$y = \frac{3\sqrt{2}}{2}$$

$$y \approx 2.12$$

29. The diagonal of the square is the hypotenuse of the right triangle.

$$10 = x\sqrt{2}$$

$$\frac{10}{\sqrt{2}} = \frac{x\sqrt{2}}{\sqrt{2}}$$

$$\frac{10}{\sqrt{2}} \bullet \frac{\sqrt{2}}{\sqrt{2}} = x$$

$$\frac{10\sqrt{2}}{\sqrt{4}} = x$$

$$\frac{10\sqrt{2}}{2} = x$$

$$5\sqrt{2} = x$$

$$7.07 \approx x$$

$$x = y$$

$$y = 5\sqrt{2}$$

$$y \approx 7.07 \text{ in.}$$

31. The shorter leg of a 30°–60°–90° triangle is half as long as the hypotenuse. The longer leg of a 30°–60°–90° triangle is the length of the shorter leg times $\sqrt{3}$.

$$x = 5\sqrt{3} \qquad h = 2(5)$$

$$x \approx 8.66 \qquad h = 10$$

33. The shorter leg of a 30°–60°–90° triangle is half as long as the hypotenuse. The longer leg of a 30°–60°–90° triangle is the length of the shorter leg times $\sqrt{3}$.

$$x = 75\sqrt{3} \qquad h = 2(75)$$

$$x \approx 129.90 \text{ cm} \qquad h = 150 \text{ cm}$$

35.

$$x\sqrt{3} = 40$$
$$\frac{x\sqrt{3}}{\sqrt{3}} = \frac{40}{\sqrt{3}}$$
$$x = \frac{40}{\sqrt{3}} \cdot \frac{\sqrt{3}}{\sqrt{3}}$$
$$x = \frac{40\sqrt{3}}{3}$$
$$\approx 23.09$$

$$h = 2x$$
$$= 2\left(\frac{40\sqrt{3}}{3}\right)$$
$$= \frac{80\sqrt{3}}{3}$$
$$\approx 46.19$$

37.

$$x\sqrt{3} = 55$$
$$\frac{x\sqrt{3}}{\sqrt{3}} = \frac{55}{\sqrt{3}}$$
$$x = \frac{55}{\sqrt{3}} \cdot \frac{\sqrt{3}}{\sqrt{3}}$$
$$x = \frac{55\sqrt{3}}{3}$$
$$\approx 31.75 \text{ mm}$$

$$h = 2x$$
$$= 2\left(\frac{55\sqrt{3}}{3}\right)$$
$$= \frac{110\sqrt{3}}{3}$$
$$\approx 63.51 \text{ mm}$$

39.

$$2x = 100$$
$$x = 50$$

$$y = x\sqrt{3}$$
$$= 50\sqrt{3}$$
$$\approx 86.60$$

41.

$$2x = 1.5$$
$$x = 0.75$$

$$y = x\sqrt{3}$$
$$= 0.75\sqrt{3}$$
$$\approx 1.30 \text{ ft}$$

43.

$$d = \sqrt{(x_2 - x_1)^2 + (y_2 - y_1)^2}$$
$$= \sqrt{(3-0)^2 + (-4-0)^2}$$
$$= \sqrt{(3)^2 + (-4)^2}$$
$$= \sqrt{9+16}$$
$$= \sqrt{25}$$
$$= 5$$

45.

$$d = \sqrt{(x_2 - x_1)^2 + (y_2 - y_1)^2}$$
$$= \sqrt{[3-(-2)]^2 + [4-(-8)]^2}$$
$$= \sqrt{(5)^2 + (12)^2}$$
$$= \sqrt{25+144}$$
$$= \sqrt{169}$$
$$= 13$$

47.

$$d = \sqrt{(x_2 - x_1)^2 + (y_2 - y_1)^2}$$
$$= \sqrt{(12-6)^2 + (16-8)^2}$$
$$= \sqrt{(6)^2 + (8)^2}$$
$$= \sqrt{36+64}$$
$$= \sqrt{100}$$
$$= 10$$

49.

$$d = \sqrt{(x_2 - x_1)^2 + (y_2 - y_1)^2}$$
$$= \sqrt{[3-(-2)]^2 + (4-1)^2}$$
$$= \sqrt{(5)^2 + (3)^2}$$
$$= \sqrt{25+9}$$
$$= \sqrt{34}$$

51.

$$d = \sqrt{(x_2 - x_1)^2 + (y_2 - y_1)^2}$$
$$= \sqrt{[3-(-1)]^2 + [-4-(-6)]^2}$$
$$= \sqrt{(4)^2 + (2)^2}$$
$$= \sqrt{16+4}$$
$$= \sqrt{20}$$
$$= \sqrt{4 \cdot 5}$$
$$= 2\sqrt{5}$$

53.

$$d = \sqrt{(x_2 - x_1)^2 + (y_2 - y_1)^2}$$
$$= \sqrt{[-5-(-2)]^2 + [8-(-1)]^2}$$
$$= \sqrt{(-3)^2 + (9)^2}$$
$$= \sqrt{9+81}$$
$$= \sqrt{90}$$
$$= \sqrt{9 \cdot 10}$$
$$= 3\sqrt{10}$$

APPLICATIONS

55. SOCCER

a. The diagonal would be the hypotenuse of the right triangle with equal sides of 64 and 100 m.

$$a^2 + b^2 = c^2$$
$$64^2 + 100^2 = c^2$$
$$4{,}096 + 10{,}000 = c^2$$
$$14{,}096 = c^2$$
$$\sqrt{14{,}096} = c$$
$$118.73 \text{ meters} \approx c$$

b. The diagonal would be the hypotenuse of the right triangle with equal sides of 75 and 110 m.

$$a^2 + b^2 = c^2$$
$$75^2 + 110^2 = c^2$$
$$5{,}625 + 12{,}100 = c^2$$
$$17{,}725 = c^2$$
$$\sqrt{17{,}725} = c$$
$$133.14 \text{ meters} \approx c$$

57. CUBES

Find the length of the blue diagonal. It would be the hypotenuse of a 45°–45°–90° triangle whose legs (x) are 7.

$$d = x\sqrt{2}$$
$$= 7\sqrt{2} \text{ cm}$$

The diagonal of the cube is the hypotenuse of the right triangle that has one leg (side of the cube) whose length is 7 cm and the other leg (blue diagonal) whose length is $7\sqrt{2}$ cm. Use the Pythagorean Theorem to find the length of the green diagonal.

$$(7)^2 + \left(7\sqrt{2}\right)^2 = c^2$$
$$49 + 49(2) = c^2$$
$$147 = c^2$$
$$c = \sqrt{147}$$
$$c = \sqrt{49 \cdot 3}$$
$$c = 7\sqrt{3} \text{ cm.}$$

59. WASHINGTON, D.C.
The x– and y–axes divide the square into 4 isosceles triangles (45°–45°–90°). If the area of the square is 100, then the length of the sides would be 10. Thus, the hypotenuse is 10. Find the length of the legs and that would be the distance of a corner from the origin.

$$x\sqrt{2} = 10$$
$$\frac{x\sqrt{2}}{\sqrt{2}} = \frac{10}{\sqrt{2}}$$
$$x = \frac{10}{\sqrt{2}} \cdot \frac{\sqrt{2}}{\sqrt{2}}$$
$$x = \frac{10\sqrt{2}}{2}$$
$$x = 5\sqrt{2}$$

The coordinates would all be $5\sqrt{2}$ units from the origin in all four directions:

$$\left(5\sqrt{2}, 0\right) = (7.07, 0)$$
$$\left(-5\sqrt{2}, 0\right) = (-7.07, 0)$$
$$\left(0, 5\sqrt{2}\right) = (0, 7.07)$$
$$\left(0, -5\sqrt{2}\right) = (0, -7.07)$$

61. HARDWARE
If the sides of the nut are 10 mm, then half of the side is 5 mm, which would be the length of the side opposite the 30° angle. The height is one–half the side opposite the 60° angle. To find the total height, multiply 2 times the length of the side opposite the 60° angle.

$$h = 2\left(x\sqrt{3}\right)$$
$$= 2\left(5\sqrt{3}\right)$$
$$= 10\sqrt{3} \text{ mm}$$
$$\approx 17.32 \text{ mm}$$

63. BASEBALL
Find the distance from 3rd base to 1st base.

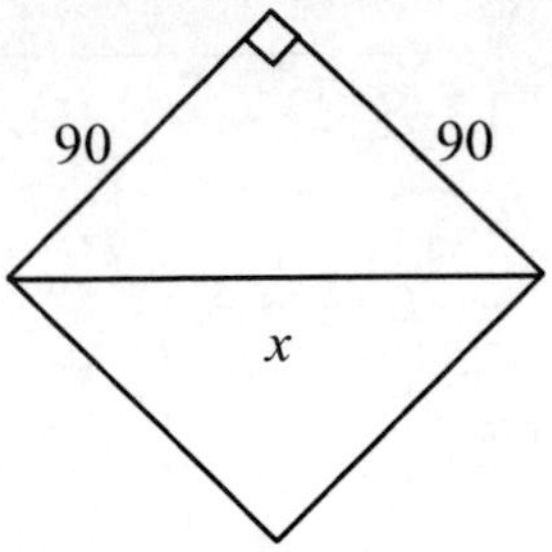

Use the Pythagorean Theorem to find x.

$$90^2 + 90^2 = x^2$$
$$8{,}100 + 8{,}100 = x^2$$
$$16{,}200 = x^2$$
$$\sqrt{16{,}200} = x$$
$$\sqrt{8{,}100 \cdot 2} = x$$
$$90\sqrt{2} = x$$
$$x = 90\sqrt{2} \text{ ft}$$
$$x \approx 127.3 \text{ ft}$$

If the ball lands 10 feet behind 3rd base, use the following triangle to find how far he must throw the ball.

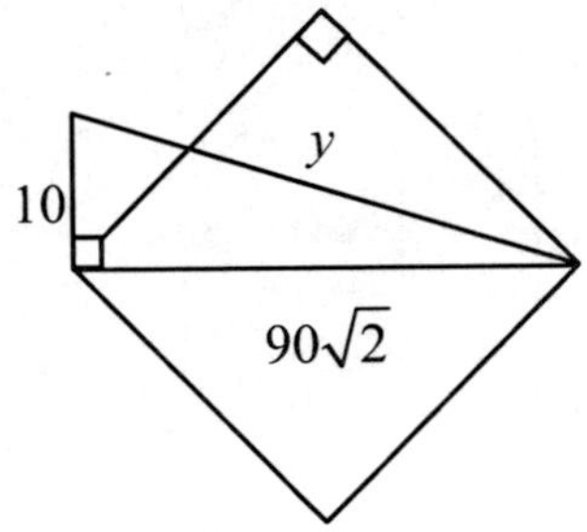

$$10^2 + \left(90\sqrt{2}\right)^2 = y^2$$
$$100 + 8{,}100(2) = y^2$$
$$16{,}300 = y^2$$
$$\sqrt{16{,}300} = y$$
$$\sqrt{100 \cdot 163} = y$$
$$10\sqrt{163} = y$$
$$y = 10\sqrt{163} \text{ ft}$$
$$y \approx 127.7 \text{ ft}$$

65. CLOTHESLINES

Find one–half of the amount stretched by using the Pythagorean Theorem. The blue line (amount stretched) is the hypotenuse (c). Let $a = 1$, and $b = \frac{15}{2} = 7.5$.

$$\begin{aligned} a^2 + b^2 &= c^2 \\ 1^2 + (7.5)^2 &= c^2 \\ 1 + 56.25 &= c^2 \\ 57.25 &= c^2 \\ \sqrt{57.25} &= c \\ c &\approx 7.566 \\ 2c &\approx 15.132 \end{aligned}$$

When stretched, the line is 15.132 ft. To find the amount the line is stretched, find the difference in the original length of the line and the length when stretched:

$$\begin{aligned} 15.132 - 15 &= 0.132 \\ &\approx 0.13 \text{ ft.} \end{aligned}$$

67. ART HISTORY

a. Use the ordered pairs (5, 0) and (8, 21)

$$\begin{aligned} d &= \sqrt{(x_2 - x_1)^2 + (y_2 - y_1)^2} \\ &= \sqrt{(8-5)^2 + (21-0)^2} \\ &= \sqrt{(3)^2 + (21)^2} \\ &= \sqrt{9 + 441} \\ &= \sqrt{450} \\ &= 21.21 \text{ units} \end{aligned}$$

b. Use the ordered pairs (10, 13) and (2, 11).

$$\begin{aligned} d &= \sqrt{(x_2 - x_1)^2 + (y_2 - y_1)^2} \\ &= \sqrt{(2-10)^2 + (11-13)^2} \\ &= \sqrt{(-8)^2 + (-2)^2} \\ &= \sqrt{64 + 4} \\ &= \sqrt{68} \\ &= 8.25 \text{ units} \end{aligned}$$

c. Use the ordered pairs (7, 19) and (12, 7).

$$\begin{aligned} d &= \sqrt{(x_2 - x_1)^2 + (y_2 - y_1)^2} \\ &= \sqrt{(12-7)^2 + (7-19)^2} \\ &= \sqrt{(5)^2 + (-12)^2} \\ &= \sqrt{25 + 144} \\ &= \sqrt{169} \\ &= 13.00 \text{ units} \end{aligned}$$

69. PACKAGING

Let $a = 24$, $b = 24$, and $c = 4$.

$$\begin{aligned} d &= \sqrt{a^2 + b^2 + c^2} \\ &= \sqrt{24^2 + 24^2 + 4^2} \\ &= \sqrt{576 + 576 + 16} \\ &= \sqrt{1{,}168} \\ &\approx 34.2 \text{ in.} \end{aligned}$$

Yes, the bone will fit in the box because the length of the diagonal (34.2 in.) is longer than the length of the bone (34 in.).

WRITING

71. Answers will vary.

73. Answers will vary.

REVIEW

75. DISCOUNT BUYING
Let c = cost of each unit and let x = number of units purchased.
Unit cost · number = total cost

$$\text{unit cost } (c) = \frac{\text{total cost}}{\text{number of units}(x)}$$

	unit cost	number	total cost
first purchase	$c = \dfrac{224}{x}$	x	224
second purchase	$c - 4 = \dfrac{224}{x+1}$	$x + 1$	224

Solve the second unit cost for c.

$$c - 4 = \frac{224}{x+1}$$
$$c = \frac{224}{x+1} + 4$$

Set the costs equal and solve the equation for x.

$$\frac{224}{x} = \frac{224}{x+1} + 4$$
$$x(x+1)\left(\frac{224}{x}\right) = x(x+1)\left(\frac{224}{x+1} + 4\right)$$
$$224(x+1) = 224(x) + 4x(x+1)$$
$$224x + 224 = 224x + 4x^2 + 4x$$
$$224x + 224 - 224x = 224x + 4x^2 + 4x - 224x$$
$$224 = 4x^2 + 4x$$
$$224 - 224 = 4x^2 + 4x - 224$$
$$0 = 4x^2 + 4x - 224$$
$$\frac{0}{4} = \frac{4x^2}{4} + \frac{4x}{4} - \frac{224}{4}$$
$$0 = x^2 + x - 56$$
$$0 = (x-7)(x+8)$$
$$x - 7 = 0 \quad \text{or} \quad x + 8 = 0$$
$$x = 7 \qquad\qquad x = -8$$

Since the answer cannot be negative, he originally bought 7 motors.

CHALLENGE PROBLEMS

77. Using a 45°–45°–90° triangle on the face of the cube, you can draw a diagonal whose measure would be $a\sqrt{2}$ in. Now you have a triangle involving the following sides: base of the cubes whose measure is a, diagonal of a face whose length is $a\sqrt{2}$, and the diagonal of the cube (d). Use the Pythagorean Theorem to find the length of the diagonal (d).

$$(a)^2 + \left(a\sqrt{2}\right)^2 = d^2$$
$$a^2 + a^2(2) = d^2$$
$$a^2 + 2a^2 = d^2$$
$$3a^2 = d^2$$
$$d = \sqrt{3a^2}$$
$$d = a\sqrt{3} \text{ in.}$$

79.

$$d = \sqrt{(x_2 - x_1)^2 + (y_2 - y_1)^2}$$
$$= \sqrt{\left(\sqrt{12} - \sqrt{48}\right)^2 + \left(\sqrt{24} - \sqrt{150}\right)^2}$$
$$= \sqrt{\left(2\sqrt{3} - 4\sqrt{3}\right)^2 + \left(2\sqrt{6} - 5\sqrt{6}\right)^2}$$
$$= \sqrt{\left(-2\sqrt{3}\right)^2 + \left(-3\sqrt{6}\right)^2}$$
$$= \sqrt{12 + 54}$$
$$= \sqrt{66}$$

SECTION 7.7

VOCABULARY

1. The **imaginary** number i is defined as $i = \sqrt{-1}$. We call i^{25} a **power** of i.

3. For the complex number 2 + 5i, we call 2 the **real** part and 5 the **imaginary** part.

CONCEPTS

5. a. $i = \boxed{\sqrt{-1}}$
   b. $i^2 = \boxed{-1}$
   c. $i^3 = \boxed{-i}$
   d. $i^4 = \boxed{1}$
   e. four

7. a. To add (or subtract) complex numbers, add (or subtract) their **real** parts and add (or subtract) their **imaginary** parts.
   b. To multiply two complex numbers, such as (2 + 3i)(3 + 5i), we can use the **FOIL** method.

9. a. $2 + 3i$
   b. $2 - 0i$
   c. $0 + 3i$

11.

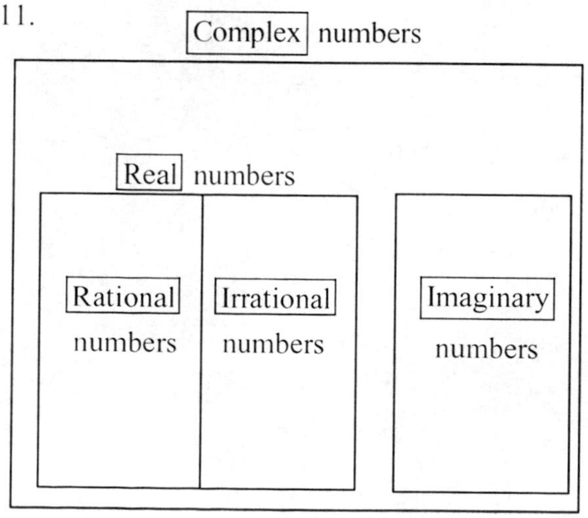

NOTATION

13.

$$(3+2i)(3-i) = \boxed{9} - 3i + \boxed{6i} - 2i^2$$
$$= 9 + 3i + \boxed{2}$$
$$= \boxed{11} + 3i$$

15. a. true
    b. false
    c. false
    d. false

GUIDED PRACTICE

17.

$$\sqrt{-9} = \sqrt{-1 \cdot 9}$$
$$= 3i$$

19.

$$\sqrt{-7} = \sqrt{-1 \cdot 7}$$
$$= i\sqrt{7} \text{ or } \sqrt{7}i$$

21.

$$\sqrt{-24} = \sqrt{-1}\sqrt{4}\sqrt{6}$$
$$= 2i\sqrt{6}$$
$$= 2\sqrt{6}i$$

23.

$$-\sqrt{-72} = -\sqrt{-1}\sqrt{36}\sqrt{2}$$
$$= -6i\sqrt{2}$$
$$= -6\sqrt{2}i$$

25.

$$5\sqrt{-81} = 5\sqrt{-1}\sqrt{81}$$
$$= 5(9i)$$
$$= 45i$$

27.

$$\sqrt{-\frac{25}{9}} = \frac{\sqrt{-1}\sqrt{25}}{\sqrt{9}}$$
$$= \frac{5i}{3}$$
$$= \frac{5}{3}i$$

29. a. $5 + 0i$
b. $0 + 7i$

31. a. $1 + 5i$
b. $-3 + 2i\sqrt{2}$

33. a. $76 - 3i\sqrt{6}$
b. $-7 + i\sqrt{19}$

35. a. $-6 - 3i$
b. $3 + i\sqrt{6}$

37.
$$\begin{aligned}(3+4i)+(5-6i)&=(3+5)+(4-6)i\\&=8-2i\end{aligned}$$

39.
$$\begin{aligned}(6-i)+(9+3i)&=(6+9)+(-1+3)i\\&=15+2i\end{aligned}$$

41.
$$\begin{aligned}(7-3i)-(4+2i)&=(7-3i)+(-4-2i)\\&=(7-4)+(-3-2)i\\&=3-5i\end{aligned}$$

43.
$$\begin{aligned}\left(8+\sqrt{-25}\right)-\left(7+\sqrt{-4}\right)&=(8+5i)-(7+2i)\\&=(8-7)+(5-2)i\\&=1+3i\end{aligned}$$

45.
$$\begin{aligned}\sqrt{-1}\sqrt{-36}&=(i)(6i)\\&=6i^2\\&=-6\end{aligned}$$

47.
$$\begin{aligned}\sqrt{-2}\sqrt{-12}&=\left(i\sqrt{2}\right)\left(i\sqrt{12}\right)\\&=i^2\sqrt{24}\\&=i^2\sqrt{4}\sqrt{6}\\&=2i^2\sqrt{6}\\&=-2\sqrt{6}\end{aligned}$$

49.
$$3(2-9i)=6-27i$$

51.
$$7(5-4i)=35-28i$$

53.
$$\begin{aligned}2i(7-3i)&=14i-6i^2\\&=14i-6(-1)\\&=14i+6\\&=6+14i\end{aligned}$$

55.
$$\begin{aligned}-5i(5-5i)&=-25i+25i^2\\&=-25i-25\\&=-25-25i\end{aligned}$$

57.
$$\begin{aligned}(2+i)(3-i)&=6-2i+3i-i^2\\&=6+i-(-1)\\&=6+i+1\\&=7+i\end{aligned}$$

59.
$$\begin{aligned}(3-2i)(2+3i)&=6+9i-4i-6i^2\\&=6+5i-6(-1)\\&=6+5i+6\\&=12+5i\end{aligned}$$

61.
$$\begin{aligned}(4+i)(3-i)&=12-4i+3i-i^2\\&=12-i-(-1)\\&=12-i+1\\&=13-i\end{aligned}$$

63.
$$\begin{aligned}(2+i)^2&=(2+i)(2+i)\\&=4+2i+2i+i^2\\&=4+4i-1\\&=3+4i\end{aligned}$$

65.
$$\begin{aligned}(2+6i)(2-6i)&=4-36i^2\\&=4-36(-1)\\&=4+36\\&=40\end{aligned}$$

67.

$$\begin{aligned}(-4-7i)(-4+7i) &= 16-49i^2 \\ &= 16-49(-1) \\ &= 16+49 \\ &= 65\end{aligned}$$

69.

$$\begin{aligned}\frac{9}{5+i} &= \frac{9}{5+i}\bullet\frac{(5-i)}{(5-i)} \\ &= \frac{9(5-i)}{(5+i)(5-i)} \\ &= \frac{45-9i}{25-i^2} \\ &= \frac{45-9i}{25-(-1)} \\ &= \frac{45-9i}{26} \\ &= \frac{45}{26}-\frac{9}{26}i\end{aligned}$$

71.

$$\begin{aligned}\frac{11i}{4-7i} &= \frac{11i}{4-7i}\bullet\frac{4+7i}{4+7i} \\ &= \frac{11i(4+7i)}{(4-7i)(4+7i)} \\ &= \frac{44i+77i^2}{16-49i^2} \\ &= \frac{44i+77(-1)}{16-49(-1)} \\ &= \frac{44i-77}{16+49} \\ &= \frac{-77+44i}{65} \\ &= -\frac{77}{65}+\frac{44}{65}i\end{aligned}$$

73.

$$\begin{aligned}\frac{3-2i}{4-i} &= \frac{3-2i}{4-i}\bullet\frac{4+i}{4+i} \\ &= \frac{(3-2i)(4+i)}{(4-i)(4+i)} \\ &= \frac{12+3i-8i-2i^2}{16-i^2} \\ &= \frac{12-5i-2(-1)}{16-(-1)} \\ &= \frac{12-5i+2}{16-(-1)} \\ &= \frac{14-5i}{17} \\ &= \frac{14}{17}-\frac{5}{17}i\end{aligned}$$

75.

$$\begin{aligned}\frac{7+4i}{2-5i} &= \frac{7+4i}{2-5i}\bullet\frac{2+5i}{2+5i} \\ &= \frac{(7+4i)(2+5i)}{(2-5i)(2+5i)} \\ &= \frac{14+35i+8i+20i^2}{4-25i^2} \\ &= \frac{14+43i+20(-1)}{4-25(-1)} \\ &= \frac{14+43i-20}{4+25} \\ &= \frac{-6+43i}{29} \\ &= -\frac{6}{29}+\frac{43}{29}i\end{aligned}$$

77.

$$\frac{7+3i}{4-2i}=\frac{7+3i}{4-2i}\bullet\frac{(4+2i)}{(4+2i)}$$
$$=\frac{(7+3i)(4+2i)}{(4-2i)(4+2i)}$$
$$=\frac{28+14i+12i+6i^2}{16-4i^2}$$
$$=\frac{28+26i+6(-1)}{16-4(-1)}$$
$$=\frac{28+26i-6}{16+4}$$
$$=\frac{22+26i}{20}$$
$$=\frac{22}{20}+\frac{26}{20}i$$
$$=\frac{11}{10}+\frac{13}{10}i$$

79.

$$\frac{1-3i}{3+i}=\frac{1-3i}{3+i}\bullet\frac{3-i}{3-i}$$
$$=\frac{(1-3i)(3-i)}{(3+i)(3-i)}$$
$$=\frac{3-i-9i+3i^2}{9-i^2}$$
$$=\frac{3-10i+3(-1)}{9-(-1)}$$
$$=\frac{3-10i-3}{10}$$
$$=\frac{-10i}{10}$$
$$=-i$$
$$=0-i$$

81.

$$\frac{8+\sqrt{-144}}{2+\sqrt{-9}}=\frac{8+12i}{2+3i}$$
$$=\frac{8+12i}{2+3i}\bullet\frac{2-3i}{2-3i}$$
$$=\frac{(8+12i)(2-3i)}{(2+3i)(2-3i)}$$
$$=\frac{16-24i+24i-36i^2}{4-9i^2}$$
$$=\frac{16-36(-1)}{4-9(-1)}$$
$$=\frac{16+36}{4+9}$$
$$=\frac{52}{13}$$
$$=4+0i$$

83.

$$\frac{-4-\sqrt{-4}}{2+\sqrt{-1}}=\frac{-4-2i}{2+i}$$
$$=\frac{-4-2i}{2+i}\bullet\frac{2-i}{2-i}$$
$$=\frac{(-4-2i)(2-i)}{(2+i)(2-i)}$$
$$=\frac{-8+4i-4i+2i^2}{4-i^2}$$
$$=\frac{-8+2(-1)}{4-(-1)}$$
$$=\frac{-8-2}{4+1}$$
$$=\frac{-10}{5}$$
$$=-2+0i$$

85.

$$\frac{5}{3i}=\frac{5}{3i}\bullet\frac{i}{i}$$
$$=\frac{5i}{3i^2}$$
$$=\frac{5i}{3(-1)}$$
$$=-\frac{5i}{3}$$
$$=0-\frac{5}{3}i$$

87.

$$\begin{aligned}-\frac{2}{7i}&=-\frac{2}{7i}\bullet\frac{i}{i}\\&=-\frac{2i}{7i^2}\\&=-\frac{2i}{7(-1)}\\&=\frac{2i}{7}\\&=0+\frac{2}{7}i\end{aligned}$$

89.

$$\begin{aligned}i^{21}&=i^{4\bullet5+1}\\&=\left(i^4\right)^5\bullet i^1\\&=(1)^5\bullet i\\&=1\bullet i\\&=i\end{aligned}$$

91.

$$\begin{aligned}i^{27}&=i^{4\bullet6+3}\\&=\left(i^4\right)^6\bullet i^3\\&=(1)^6\bullet i^3\\&=1\bullet i^3\\&=i^3\\&=-i\end{aligned}$$

93.

$$\begin{aligned}i^{100}&=i^{4\bullet25}\\&=\left(i^4\right)^{25}\\&=(1)^{25}\\&=1\end{aligned}$$

95.

$$\begin{aligned}i^{42}&=i^{4\bullet10+2}\\&=\left(i^4\right)^{10}\bullet i^2\\&=(1)^{10}\bullet i^2\\&=1\bullet i^2\\&=i^2\\&=-1\end{aligned}$$

TRY IT YOURSELF

97.

$$\begin{aligned}(3-i)-(-1+10i)&=3-i+1-10i\\&=4-11i\end{aligned}$$

99.

$$\begin{aligned}&\left(2-\sqrt{-16}\right)\left(3+\sqrt{-4}\right)\\&=(2-4i)(3+2i)\\&=6+4i-12i-8i^2\\&=6-8i-8(-1)\\&=6-8i+8\\&=14-8i\end{aligned}$$

101.

$$\begin{aligned}(-6-9i)+(4+3i)&=-6-9i+4+3i\\&=-2-6i\end{aligned}$$

103.

$$\begin{aligned}\frac{-2i}{3+2i}&=\frac{-2i}{3+2i}\bullet\frac{3-2i}{3-2i}\\&=\frac{-2i(3-2i)}{(3+2i)(3-2i)}\\&=\frac{-6i+4i^2}{9-4i^2}\\&=\frac{-6i+4(-1)}{9-4(-1)}\\&=\frac{-6i-4}{9+4}\\&=\frac{-4-6i}{13}\\&=-\frac{4}{13}-\frac{6}{13}i\end{aligned}$$

105.

$$\begin{aligned}6i(2-3i)&=12i-18i^2\\&=12i-18(-1)\\&=12i+18\\&=18+12i\end{aligned}$$

107.

$$\begin{aligned}\frac{4}{5i^{35}}&=\frac{4}{5i^{35}}\bullet\frac{i}{i}\\&=\frac{4i}{5i^{36}}\\&=\frac{4i}{5(1)}\\&=\frac{4i}{5}\\&=0+\frac{4}{5}i\end{aligned}$$

109.

$$\begin{aligned}(2+i\sqrt{2})(3-i\sqrt{2}) &= 6-2i\sqrt{2}+3i\sqrt{2}-2i^2\\ &= 6+i\sqrt{2}-2(-1)\\ &= 6+i\sqrt{2}+2\\ &= 8+i\sqrt{2}\end{aligned}$$

111.

$$\begin{aligned}\frac{5+9i}{1-i} &= \frac{5+9i}{1-i}\bullet\frac{1+i}{1+i}\\ &= \frac{(5+9i)(1+i)}{(1-i)(1+i)}\\ &= \frac{5+5i+9i+9i^2}{1-i^2}\\ &= \frac{5+14i+9(-1)}{1-(-1)}\\ &= \frac{5+14i-9}{2}\\ &= \frac{-4+14i}{2}\\ &= -\frac{4}{2}+\frac{14}{2}i\\ &= -2+7i\end{aligned}$$

113.

$$\begin{aligned}(4-8i)^2 &= 4^2-2(4)(8i)+(-8i)^2\\ &= 16-64i+64i^2\\ &= 16-64i+64(-1)\\ &= 16-64i-64\\ &= -48-64i\end{aligned}$$

115.

$$\begin{aligned}\frac{\sqrt{5}-\sqrt{3}i}{\sqrt{5}+\sqrt{3}i} &= \frac{(\sqrt{5}-\sqrt{3}i)(\sqrt{5}-\sqrt{3}i)}{(\sqrt{5}+\sqrt{3}i)(\sqrt{5}-\sqrt{3}i)}\\ &= \frac{5-\sqrt{15}i-\sqrt{15}i+3i^2}{5-3i^2}\\ &= \frac{5-2\sqrt{15}i+3(-1)}{5-3(-1)}\\ &= \frac{5-2\sqrt{15}i-3}{5+3}\\ &= \frac{2-2\sqrt{15}i}{8}\\ &= \frac{2}{8}-\frac{2\sqrt{15}}{8}i\\ &= \frac{1}{4}-\frac{\sqrt{15}}{4}i\end{aligned}$$

117. a.

$$\begin{aligned}\sqrt{-8} &= \sqrt{-1}\sqrt{4}\sqrt{2}\\ &= 2i\sqrt{2}\end{aligned}$$

b.

$$\begin{aligned}\sqrt[3]{-8} &= \sqrt[3]{-1}\sqrt[3]{8}\\ &= -2\end{aligned}$$

119. a.

$$\begin{aligned}(2i)^2 &= (2i)(2i)\\ &= 4i^2\\ &= 4(-1)\\ &= -4\end{aligned}$$

b.

$$\begin{aligned}(2+i)^2 &= (2)^2+2(2)(i)+(i)^2\\ &= 4+4i-1\\ &= 3+4i\end{aligned}$$

APPLICATIONS

121. FRACTALS

Step 1: i^2+i

Step 2 $$\begin{aligned}(i^2+i)^2 &= (i^2+i)(i^2+i)+i\\ &= i^4+i^3+i^3+i^2+i\\ &= i^4+2i^3+i^2+i\\ &= 1+2(-i)+(-1)+i\\ &= 1-2i-1+i\\ &= -i\end{aligned}$$

Step 3 $$\begin{aligned}(-i)^2+i &= i^2+i\\ &= -1+i\end{aligned}$$

WRITING

123. Answers will vary.

125. Answers will vary.

REVIEW

127. WIND SPEEDS
Let x = the rate of the wind.

	distance	rate	time
with a tail wind	330	$200+x$	$t_1=\frac{d}{r}$ $=\frac{330}{200+x}$
against the wind	330	$200-x$	$t_1=\frac{d}{r}$ $=\frac{330}{200-x}$

Since the total time for the trip is $3\frac{1}{3}$ hours, add the times and set the sum equal to $3\frac{1}{3}$.

$$t_1+t_2=3\frac{1}{3}$$
$$\frac{330}{200+x}+\frac{330}{200-x}=\frac{10}{3}$$
$$3(200+x)(200-x)\left(\frac{330}{200+x}+\frac{330}{200-x}\right)=3(200+x)(200-x)\left(\frac{10}{3}\right)$$
$$3\cdot 330(200-x)+3\cdot 330(200+x)=10(200+x)(200-x)$$
$$990(200-x)+990(200+x)=10\left(40{,}000-x^2\right)$$
$$198{,}000-990x+198{,}000+990x=400{,}000-10x^2$$
$$396{,}000=400{,}000-10x^2$$
$$-4{,}000=-10x^2$$
$$\frac{-4{,}000}{-10}=\frac{-10x^2}{-10}$$
$$400=x^2$$
$$0=x^2-400$$
$$0=(x+20)(x-20)$$
$$x+20=0 \text{ or } x-20=0$$
$$x=-20 \qquad x=20$$

Since the rates cannot be negative, the rates of the wind is 20 mph.

CHALLENGE PROBLEMS

129.

$$\left(i^{349}\right)^{-i^{456}}=\left(i^{349}\right)^{-i^{4}}=\left(i^{349}\right)^{-1}$$
$$=i^{-349}$$
$$=\frac{1}{i^{349}}$$
$$=\frac{1}{i^{1}}$$
$$=\frac{1}{i}$$
$$=\frac{1}{i}\bullet\frac{i}{i}$$
$$=\frac{i}{i^2}$$
$$=\frac{i}{-1}$$
$$=-i$$

CHAPTER 7 REVIEW

SECTION 7.1
Radical Expressions and Radical Functions

1. a. $f(1)=2$
   b. $f(-3)=0$
   c. $x=6$
   d. D: $[-3,\infty)$; R: $[0,\infty)$

2. a. $\sqrt{100a^2}=10|a|$
   b. $\sqrt{100a^2}=10a$

3.
$$\sqrt{49}=7$$

4.
$$-\sqrt{121}=-11$$

5.
$$\sqrt{\frac{225}{49}}=\frac{\sqrt{225}}{\sqrt{49}}$$
$$=\frac{15}{7}$$

6.
$\sqrt{-4}$ is not real

7.
$$\sqrt{100a^{12}}=10a^6$$

8.
$$\sqrt{25x^2}=|5x|$$
$$=5|x|$$

9.
$$\sqrt{x^8}=\sqrt{(x^4)^2}$$
$$=x^4$$

10.
$$\sqrt{x^2+4x+4}=\sqrt{(x+2)^2}$$
$$=|x+2|$$

11.
$$\sqrt[3]{-27}=-3$$

12.
$$-\sqrt[3]{216}=-6$$

13.
$$\sqrt[3]{64x^6y^3}=4x^2y$$

14.
$$\sqrt[3]{\frac{x^9}{125}}=\frac{\sqrt[3]{x^9}}{\sqrt[3]{125}}$$
$$=\frac{x^3}{5}$$

15.
$$\sqrt[6]{64}=2$$

16.
$$\sqrt[5]{-32}=-2$$

17.
$$\sqrt[4]{256x^8y^4}=|4x^2y|$$
$$=4x^2|y|$$

18.
$$\sqrt[15]{(x+1)^{15}}=x+1$$

19.
$$-\sqrt[4]{\frac{1}{16}}=-\frac{\sqrt[4]{1}}{\sqrt[4]{16}}$$
$$=-\frac{1}{2}$$

20.
$\sqrt[4]{-16}$ is not real

21.
$\sqrt[6]{-1}$ is not real

22.
$$\sqrt[3]{0}=0$$

23.
$$3x+15>0$$
$$3x>-15$$
$$x>-5$$
$$D:[-5,\infty)$$

24. Find the length of each side.

$$s(A)=\sqrt{A}$$
$$s(169)=\sqrt{169}$$
$$=13 \text{ ft}$$

25.

$$A(V)=6\sqrt[3]{V^2}$$
$$A(8)=6\sqrt[3]{8^2}$$
$$=6\sqrt[3]{64}$$
$$=6(4)$$
$$=24 \text{ cm}^2$$

26.

$$g(-1.9)=\sqrt[3]{(-1.9)^2+9}$$
$$\approx 2.3276$$

27. $f(x)=\sqrt{x}$

D: $[0, \infty)$ $\quad R$: $[0, \infty)$

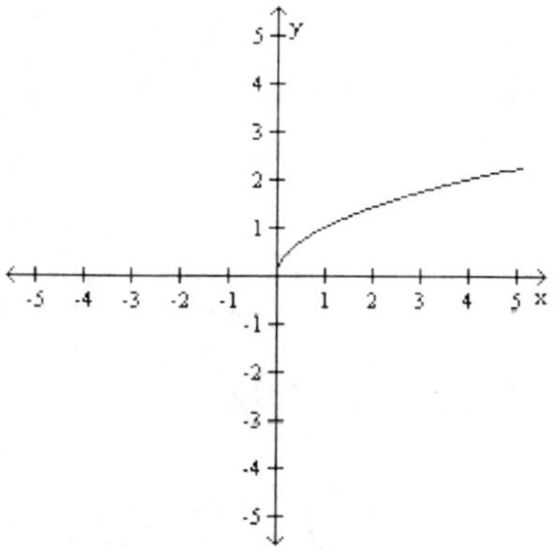

28. $f(x)=\sqrt[3]{x}$

D: $(-\infty, \infty)$ $\quad R$: $(-\infty, \infty)$

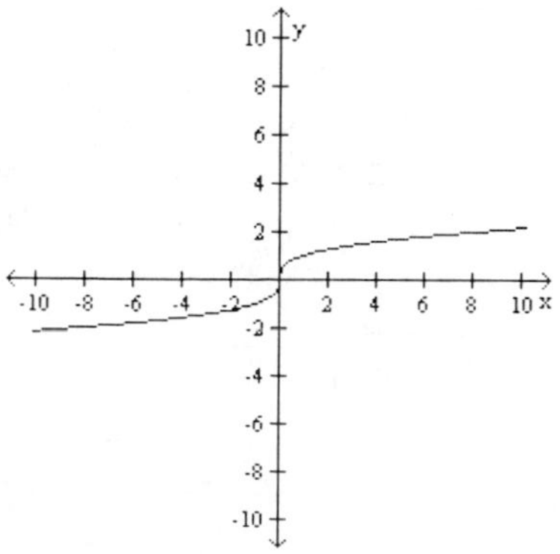

29. $f(x)=\sqrt{x+2}$

D: $[-2, \infty)$ $\quad R$: $[0, \infty)$

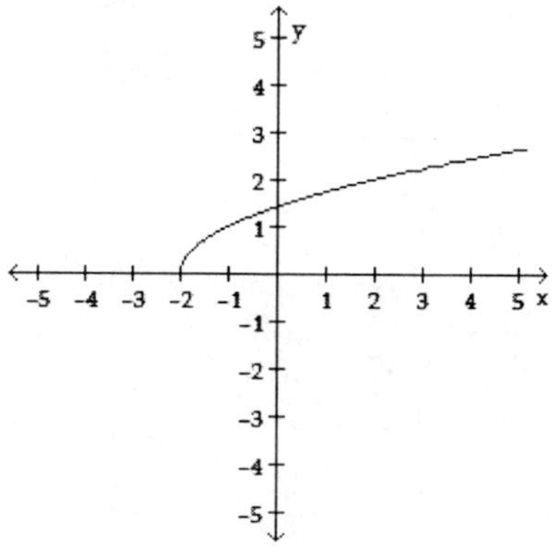

30. $f(x)=-\sqrt[3]{x}+3$

D: $(-\infty, \infty)$ $\quad R$: $(-\infty, \infty)$

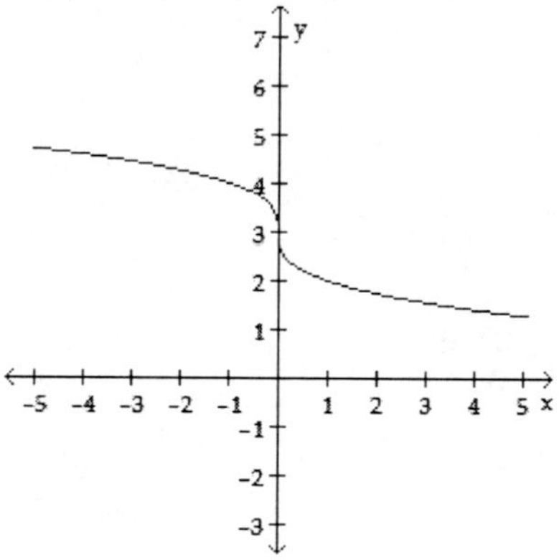

SECTION 7.2
Rational Exponents

31.

$$t^{1/2}=\sqrt{t}$$

32.

$$(5xy^3)^{1/4}=\sqrt[4]{5xy^3}$$

33.

$$25^{1/2}=\sqrt{25}$$
$$=5$$

34.

$$-36^{1/2}=-\sqrt{36}$$
$$=-6$$

35.

$$(-36)^{1/2}=\sqrt{-36}$$

is not a real number

36.

$$1^{1/5} = \sqrt[5]{1}$$
$$= 1$$

37.

$$\left(\frac{9}{x^2}\right)^{1/2} = \sqrt{\frac{9}{x^2}}$$
$$= \frac{\sqrt{9}}{\sqrt{x^2}}$$
$$= \frac{3}{x}$$

38.

$$(-8)^{1/3} = \sqrt[3]{-8}$$
$$= -2$$

39.

$$625^{1/4} = \sqrt[4]{625}$$
$$= 5$$

40.

$$\left(81c^4d^4\right)^{1/4} = \sqrt[4]{81c^4d^4}$$
$$= 3cd$$

41.

$$9^{3/2} = \left(\sqrt{9}\right)^3$$
$$= 3^3$$
$$= 27$$

42.

$$8^{-2/3} = \frac{1}{8^{2/3}}$$
$$= \frac{1}{\left(\sqrt[3]{8}\right)^2}$$
$$= \frac{1}{2^2}$$
$$= \frac{1}{4}$$

43.

$$-49^{5/2} = -\left(\sqrt{49}\right)^5$$
$$= -(7)^5$$
$$= -16{,}807$$

44.

$$\frac{1}{100^{-1/2}} = 100^{1/2}$$
$$= \sqrt{100}$$
$$= 10$$

45.

$$\left(\frac{4}{9}\right)^{-3/2} = \left(\frac{9}{4}\right)^{3/2}$$
$$= \left(\sqrt{\frac{9}{4}}\right)^3$$
$$= \left(\frac{3}{2}\right)^3$$
$$= \frac{27}{8}$$

46.

$$\frac{1}{25^{5/2}} = \frac{1}{\left(\sqrt{25}\right)^5}$$
$$= \frac{1}{5^5}$$
$$= \frac{1}{3{,}125}$$

47.

$$\left(25x^2y^4\right)^{3/2} = \left(\sqrt{25x^2y^4}\right)^3$$
$$= \left(5xy^2\right)^3$$
$$= 125x^3y^6$$

48.

$$\left(8u^6v^3\right)^{-2/3} = \frac{1}{\left(8u^6v^3\right)^{2/3}}$$
$$= \frac{1}{\left(\sqrt[3]{8u^6v^3}\right)^2}$$
$$= \frac{1}{\left(2u^2v\right)^2}$$
$$= \frac{1}{4u^4v^2}$$

49.

$$\begin{aligned}5^{1/4}5^{1/2} &= 5^{1/4+1/2}\\ &= 5^{1/4+2/4}\\ &= 5^{3/4}\end{aligned}$$

50.

$$\begin{aligned}a^{3/7}a^{-2/7} &= a^{3/7+(-2/7)}\\ &= a^{1/7}\end{aligned}$$

51.

$$\begin{aligned}\left(k^{4/5}\right)^{10} &= k^{(4/5)(10)}\\ &= k^{8}\end{aligned}$$

52.

$$\begin{aligned}\frac{3^{5/6}3^{1/3}}{3^{1/2}} &= \frac{3^{5/6+1/3}}{3^{1/2}}\\ &= \frac{3^{5/6+2/6}}{3^{1/2}}\\ &= \frac{3^{7/6}}{3^{1/2}}\\ &= 3^{7/6-1/2}\\ &= 3^{7/6-3/6}\\ &= 3^{4/6}\\ &= 3^{2/3}\end{aligned}$$

53.

$$\begin{aligned}u^{1/2}\left(u^{1/2}-u^{-1/2}\right) &= u^{1/2}u^{1/2}-u^{1/2}u^{-1/2}\\ &= u^{1/2+1/2}-u^{1/2+(-1/2)}\\ &= u^{2/2}-u^{0}\\ &= u^{1}-1\\ &= u-1\end{aligned}$$

54.

$$\begin{aligned}v^{2/3}\left(v^{1/3}+v^{4/3}\right) &= v^{2/3}v^{1/3}+v^{2/3}v^{4/3}\\ &= v^{2/3+1/3}+v^{2/3+4/3}\\ &= v^{3/3}+v^{6/3}\\ &= v^{1}+v^{2}\\ &= v+v^{2}\end{aligned}$$

55.

$$\begin{aligned}\sqrt[4]{a^2} &= a^{2/4}\\ &= a^{1/2}\\ &= \sqrt{a}\end{aligned}$$

56.

$$\begin{aligned}\sqrt[3]{\sqrt{c}} &= \sqrt[3]{c^{1/2}}\\ &= \left(c^{1/2}\right)^{1/3}\\ &= c^{1/6}\\ &= \sqrt[6]{c}\end{aligned}$$

57.

$$\begin{aligned}d &= 1.22a^{1/2}\\ &= 1.22(22{,}500)^{1/2}\\ &= 1.22\left(\sqrt{22{,}500}\right)\\ &= 1.22(150)\\ &= 183\text{ miles}\end{aligned}$$

58. Check (64, 64).

$$\begin{aligned}x^{2/3}+y^{2/3} &= 32\\ (64)^{2/3}+(64)^{2/3} &\stackrel{?}{=} 32\\ \left(\sqrt[3]{64}\right)^2+\left(\sqrt[3]{64}\right)^2 &\stackrel{?}{=} 32\\ 4^2+4^2 &\stackrel{?}{=} 32\\ 16+16 &\stackrel{?}{=} 32\\ 32 &= 32\end{aligned}$$

Check (–64, 64).

$$\begin{aligned}x^{2/3}+y^{2/3} &= 32\\ (-64)^{2/3}+(64)^{2/3} &\stackrel{?}{=} 32\\ \left(\sqrt[3]{-64}\right)^2+\left(\sqrt[3]{64}\right)^2 &\stackrel{?}{=} 32\\ (-4)^2+4^2 &\stackrel{?}{=} 32\\ 16+16 &\stackrel{?}{=} 32\\ 32 &= 32\end{aligned}$$

SECTION 7.3
Simplifying and Combining Radical Expressions

59.

$$\begin{aligned}\sqrt{80} &= \sqrt{16}\sqrt{5}\\ &= 4\sqrt{5}\end{aligned}$$

60.

$$\begin{aligned}\sqrt[3]{54} &= \sqrt[3]{27}\sqrt[3]{2}\\ &= 3\sqrt[3]{2}\end{aligned}$$

61.

$$\sqrt[4]{160} = \sqrt[4]{16}\sqrt[4]{10} = 2\sqrt[4]{10}$$

62.

$$\sqrt[5]{-96} = \sqrt[5]{-32}\sqrt[5]{3} = -2\sqrt[5]{3}$$

63.

$$\sqrt{8x^5} = \sqrt{4x^4}\sqrt{2x} = 2x^2\sqrt{2x}$$

64.

$$\sqrt[4]{r^{17}} = \sqrt[4]{r^{16}}\sqrt[4]{r} = r^4\sqrt[4]{r}$$

65.

$$\sqrt[3]{-27j^7k} = \sqrt[3]{-27j^6}\sqrt[3]{jk} = -3j^2\sqrt[3]{jk}$$

66.

$$\sqrt[3]{-16x^5y^4} = \sqrt[3]{-8x^3y^3}\sqrt[3]{2x^2y} = -2xy\sqrt[3]{2x^2y}$$

67.

$$\sqrt{\frac{m}{144n^{12}}} = \frac{\sqrt{m}}{\sqrt{144n^{12}}} = \frac{\sqrt{m}}{12n^6}$$

68.

$$\sqrt{\frac{17xy}{64a^4}} = \frac{\sqrt{17xy}}{\sqrt{64a^4}} = \frac{\sqrt{17xy}}{8a^2}$$

69.

$$\frac{\sqrt[5]{64x^8}}{\sqrt[5]{2x^3}} = \sqrt[5]{\frac{64x^8}{2x^3}} = \sqrt[5]{32x^5} = 2x$$

70.

$$\frac{\sqrt[5]{243x^{16}}}{\sqrt[5]{x}} = \sqrt[5]{\frac{243x^{16}}{x}} = \sqrt[5]{243x^{15}} = 3x^3$$

71.

$$\sqrt{2} + 2\sqrt{2} = 3\sqrt{2}$$

72.

$$6\sqrt{20} - \sqrt{5} = 6\sqrt{4}\sqrt{5} - \sqrt{5} = 6(2)\sqrt{5} - \sqrt{5} = 12\sqrt{5} - \sqrt{5} = 11\sqrt{5}$$

73.

$$2\sqrt[3]{3} - \sqrt[3]{24} = 2\sqrt[3]{3} - \sqrt[3]{8}\sqrt[3]{3} = 2\sqrt[3]{3} - 2\sqrt[3]{3} = 0$$

74.

$$-\sqrt[4]{32a^5} - 2\sqrt[4]{162a^5}$$
$$= -\sqrt[4]{16a^4}\sqrt[4]{2a} - 2\sqrt[4]{81a^4}\sqrt[4]{2a}$$
$$= -2a\sqrt[4]{2a} - 2(3a)\sqrt[4]{2a}$$
$$= -2a\sqrt[4]{2a} - 6a\sqrt[4]{2a}$$
$$= -8a\sqrt[4]{2a}$$

75.

$$2x\sqrt{8} + 2\sqrt{200x^2} + \sqrt{50x^2}$$
$$= 2x\sqrt{4}\sqrt{2} + 2\sqrt{100x^2}\sqrt{2} + \sqrt{25x^2}\sqrt{2}$$
$$= 2x(2)\sqrt{2} + 2(10x)\sqrt{2} + 5x\sqrt{2}$$
$$= 4x\sqrt{2} + 20x\sqrt{2} + 5x\sqrt{2}$$
$$= 29x\sqrt{2}$$

76.

$$\sqrt[3]{54x^3} - 3\sqrt[3]{16x^3} + 4\sqrt[3]{128x^3}$$
$$= \sqrt[3]{27x^3}\sqrt[3]{2} - 3\sqrt[3]{8x^3}\sqrt[3]{2} + 4\sqrt[3]{64x^3}\sqrt[3]{2}$$
$$= 3x\sqrt[3]{2} - 3(2x)\sqrt[3]{2} + 4(4x)\sqrt[3]{2}$$
$$= 3x\sqrt[3]{2} - 6x\sqrt[3]{2} + 16x\sqrt[3]{2}$$
$$= 13x\sqrt[3]{2}$$

77.

$$2\sqrt[4]{32t^3}-8\sqrt[4]{6t^3}+5\sqrt[4]{2t^3}$$
$$=2\sqrt[4]{16}\sqrt[4]{2t^3}-8\sqrt[4]{6t^3}+5\sqrt[4]{2t^3}$$
$$=2(2)\sqrt[4]{2t^3}-8\sqrt[4]{6t^3}+5\sqrt[4]{2t^3}$$
$$=4\sqrt[4]{2t^3}-8\sqrt[4]{6t^3}+5\sqrt[4]{2t^3}$$
$$=9\sqrt[4]{2t^3}-8\sqrt[4]{6t^3}$$

78.

$$10\sqrt[4]{16x^9}-8x^2\sqrt[4]{x}+5\sqrt[4]{x^5}$$
$$=10\sqrt[4]{16x^8}\sqrt[4]{x}-8x^2\sqrt[4]{x}+5\sqrt[4]{x^4}\sqrt[4]{x}$$
$$=10\left(2x^2\right)\sqrt[4]{x}-8x^2\sqrt[4]{x}+5x\sqrt[4]{x}$$
$$=20x^2\sqrt[4]{x}-8x^2\sqrt[4]{x}+5x\sqrt[4]{x}$$
$$=12x^2\sqrt[4]{x}-5x\sqrt[4]{x}$$

79.

a. You do not add the radicands together.
b. They are not like terms.
c. It should be $2\sqrt[3]{y^2}$.
d. You do not subtract the radicands.

80.

$$\sqrt{40}+\sqrt{32}+\sqrt{8}$$
$$=\sqrt{4}\sqrt{10}+\sqrt{16}\sqrt{2}+\sqrt{4}\sqrt{2}$$
$$=2\sqrt{10}+4\sqrt{2}+2\sqrt{2}$$
$$=\left(2\sqrt{10}+6\sqrt{2}\right)\text{ in.}$$
$$=14.8\text{ in.}$$

SECTION 7.4
Multiplying and Dividing Radical Expressions

81.

$$\sqrt{7}\sqrt{7}=\sqrt{49}$$
$$=7$$

82.

$$\left(2\sqrt{5}\right)\left(3\sqrt{2}\right)=6\sqrt{10}$$

83.

$$\left(-2\sqrt{8}\right)^2=\left(-2\sqrt{8}\right)\left(-2\sqrt{8}\right)$$
$$=4\sqrt{64}$$
$$=4(8)$$
$$=32$$

84.

$$2\sqrt{6}\sqrt{15}=2\sqrt{90}$$
$$=2\sqrt{9}\sqrt{10}$$
$$=2(3)\sqrt{10}$$
$$=6\sqrt{10}$$

85.

$$\sqrt{9x}\sqrt{x}=\sqrt{9x^2}$$
$$=3x$$

86.

$$\left(\sqrt[3]{x+1}\right)^3=x+1$$

87.

$$-\sqrt[3]{2x^2}\sqrt[3]{4x^8}=-\sqrt[3]{8x^{10}}$$
$$=-\sqrt[3]{8x^9}\sqrt[3]{x}$$
$$=-2x^3\sqrt[3]{x}$$

88.

$$\sqrt[5]{9}\bullet\sqrt[5]{27}=\sqrt[5]{243}$$
$$=3$$

89.

$$3\sqrt{7t}\left(2\sqrt{7t}+3\sqrt{3t^2}\right)$$
$$=6\sqrt{49t^2}+9\sqrt{21t^3}$$
$$=6\sqrt{49t^2}+9\sqrt{t^2}\sqrt{21t}$$
$$=6(7t)+9(t)\sqrt{21t}$$
$$=42t+9t\sqrt{21t}$$

90.

$$-\sqrt[4]{4x^5y^{11}}\sqrt[4]{8x^9y^3}=-\sqrt[4]{32x^{14}y^{14}}$$
$$=-\sqrt[4]{16x^{12}y^{12}}\sqrt[4]{2x^2y^2}$$
$$=-2x^3y^3\sqrt[4]{2x^2y^2}$$

91.

$$\left(\sqrt{3b}+\sqrt{3}\right)^2=\left(\sqrt{3b}+\sqrt{3}\right)\left(\sqrt{3b}+\sqrt{3}\right)$$
$$=3b+\sqrt{9b}+\sqrt{9b}+3$$
$$=3b+2\sqrt{9b}+3$$
$$=3b+2\sqrt{9}\sqrt{b}+3$$
$$=3b+2(3)\sqrt{b}+3$$
$$=3b+6\sqrt{b}+3$$

92.

$$\left(\sqrt[3]{3p}-2\sqrt[3]{2}\right)\left(\sqrt[3]{3p}+\sqrt[3]{2}\right)$$
$$=\sqrt[3]{9p^2}+\sqrt[3]{6p}-2\sqrt[3]{6p}-2\sqrt[3]{4}$$
$$=\sqrt[3]{9p^2}-\sqrt[3]{6p}-2\sqrt[3]{4}$$

93.

$$\frac{10}{\sqrt{3}}=\frac{10}{\sqrt{3}}\bullet\frac{\sqrt{3}}{\sqrt{3}}$$
$$=\frac{10\sqrt{3}}{\sqrt{9}}$$
$$=\frac{10\sqrt{3}}{3}$$

94.

$$\sqrt{\frac{3}{5xy}}=\frac{\sqrt{3}}{\sqrt{5xy}}$$
$$=\frac{\sqrt{3}}{\sqrt{5xy}}\bullet\frac{\sqrt{5xy}}{\sqrt{5xy}}$$
$$=\frac{\sqrt{15xy}}{\sqrt{25x^2y^2}}$$
$$=\frac{\sqrt{15xy}}{5xy}$$

95.

$$\frac{\sqrt[3]{6u}}{\sqrt[3]{u^5}}=\sqrt[3]{\frac{6u}{u^5}}$$
$$=\sqrt[3]{\frac{6}{u^4}}$$
$$=\sqrt[3]{\frac{6}{u^4}}\bullet\sqrt[3]{\frac{u^2}{u^2}}$$
$$=\sqrt[3]{\frac{6u^2}{u^6}}$$
$$=\frac{\sqrt[3]{6u^2}}{u^2}$$

96.

$$\frac{\sqrt[4]{a}}{\sqrt[4]{3b^2}}=\frac{\sqrt[4]{a}}{\sqrt[4]{3b^2}}\bullet\frac{\sqrt[4]{27b^2}}{\sqrt[4]{27b^2}}$$
$$=\frac{\sqrt[4]{27ab^2}}{\sqrt[4]{81b^4}}$$
$$=\frac{\sqrt[4]{27ab^2}}{3b}$$

97.

$$\frac{2}{\sqrt{2}-1}=\frac{2}{\sqrt{2}-1}\bullet\frac{\sqrt{2}+1}{\sqrt{2}+1}$$
$$=\frac{2\left(\sqrt{2}+1\right)}{\left(\sqrt{2}-1\right)\left(\sqrt{2}+1\right)}$$
$$=\frac{2\sqrt{2}+2}{2-1}$$
$$=\frac{2\sqrt{2}+2}{1}$$
$$=2\sqrt{2}+2$$
$$=2\left(\sqrt{2}+1\right)$$

98.

$$\frac{4\sqrt{x}-2\sqrt{z}}{\sqrt{z}+4\sqrt{x}}=\frac{4\sqrt{x}-2\sqrt{z}}{\sqrt{z}+4\sqrt{x}}\bullet\frac{\sqrt{z}-4\sqrt{x}}{\sqrt{z}-4\sqrt{x}}$$
$$=\frac{\left(4\sqrt{x}-2\sqrt{z}\right)\left(\sqrt{z}-4\sqrt{x}\right)}{\left(\sqrt{z}+4\sqrt{x}\right)\left(\sqrt{z}-4\sqrt{x}\right)}$$
$$=\frac{4\sqrt{xz}-16\sqrt{x^2}-2\sqrt{z^2}+8\sqrt{xz}}{\sqrt{z^2}-16\sqrt{x^2}}$$
$$=\frac{-16x-2z+12\sqrt{xz}}{z-16x}$$

99.

$$\frac{\sqrt{a}-\sqrt{b}}{\sqrt{a}}=\frac{\sqrt{a}-\sqrt{b}}{\sqrt{a}}\bullet\frac{\sqrt{a}+\sqrt{b}}{\sqrt{a}+\sqrt{b}}$$
$$=\frac{(\sqrt{a}-\sqrt{b})(\sqrt{a}+\sqrt{b})}{\sqrt{a}(\sqrt{a}+\sqrt{b})}$$
$$=\frac{\sqrt{a^2}-\sqrt{b^2}}{\sqrt{a^2}+\sqrt{ab}}$$
$$=\frac{a-b}{a+\sqrt{ab}}$$

100.

$$r=\sqrt[3]{\frac{3V}{4\pi}}$$
$$=\frac{\sqrt[3]{3V}}{\sqrt[3]{4\pi}}\bullet\frac{\sqrt[3]{2\pi^2}}{\sqrt[3]{2\pi^2}}$$
$$=\frac{\sqrt[3]{6\pi^2V}}{\sqrt[3]{8\pi^3}}$$
$$=\frac{\sqrt[3]{6\pi^2V}}{2\pi}$$

SECTION 7.5
Solving Radical Equations

101.

$$\sqrt{7x-10}-1=11$$
$$\sqrt{7x-10}=12$$
$$\left(\sqrt{7x-10}\right)^2=12^2$$
$$7x-10=144$$
$$7x=154$$
$$x=22$$

102.

$$u=\sqrt{25u-144}$$
$$u^2=\left(\sqrt{25u-144}\right)^2$$
$$u^2=25u-144$$
$$u^2-25u+144=0$$
$$(u-9)(u-16)=0$$
$$u-9=0 \quad\text{or}\quad u-16=0$$
$$u=9 \qquad\qquad u=16$$

103.

$$2\sqrt{y-3}=\sqrt{2y+1}$$
$$\left(2\sqrt{y-3}\right)^2=\left(\sqrt{2y+1}\right)^2$$
$$4(y-3)=2y+1$$
$$4y-12=2y+1$$
$$4y=2y+13$$
$$2y=13$$
$$y=\frac{13}{2}$$

104.

$$\sqrt{z+1}+\sqrt{z}=2$$
$$\sqrt{z+1}=2-\sqrt{z}$$
$$\left(\sqrt{z+1}\right)^2=\left(2-\sqrt{z}\right)^2$$
$$z+1=4-2\sqrt{z}-2\sqrt{z}+z$$
$$z+1=4-4\sqrt{z}+z$$
$$z+1-z-4=4-4\sqrt{z}+z-z-4$$
$$-3=-4\sqrt{z}$$
$$(-3)^2=\left(-4\sqrt{z}\right)^2$$
$$9=16z$$
$$\frac{9}{16}=z$$

105.

$$\sqrt[3]{x^3+56}-2=x$$
$$\sqrt[3]{x^3+56}=x+2$$
$$\left(\sqrt[3]{x^3+56}\right)^3=(x+2)^3$$
$$x^3+56=(x+2)(x+2)^2$$
$$x^3+56=(x+2)(x^2+4x+4)$$
$$x^3+56=x^3+4x^2+4x+2x^2+8x+8$$
$$x^3+56=x^3+6x^2+12x+8$$
$$x^3+56-x^3=x^3+6x^2+12x+8-x^3$$
$$56=6x^2+12x+8$$
$$56-56=6x^2+12x+8-56$$
$$0=6x^2+12x-48$$
$$\frac{0}{6}=\frac{6x^2}{6}+\frac{12x}{6}-\frac{48}{6}$$
$$0=x^2+2x-8$$
$$0=(x+4)(x-2)$$
$$x+4=0 \quad \text{or} \quad x-2=0$$
$$x=-4 \qquad\qquad x=2$$

106.

$$a=\sqrt{a^2+5a-35}$$
$$a^2=\left(\sqrt{a^2+5a-35}\right)^2$$
$$a^2=a^2+5a-35$$
$$0=5a-35$$
$$35=5a$$
$$7=a$$

107.

$$(x+2)^{1/2}-(4-x)^{1/2}=0$$
$$\sqrt{x+2}-\sqrt{4-x}=0$$
$$\sqrt{x+2}=\sqrt{4-x}$$
$$\left(\sqrt{x+2}\right)^2=\left(\sqrt{4-x}\right)^2$$
$$x+2=4-x$$
$$2x+2=4$$
$$2x=2$$
$$x=1$$

108.

$$\sqrt{b^2+b}=\sqrt{3-b^2}$$
$$\left(\sqrt{b^2+b}\right)^2=\left(\sqrt{3-b^2}\right)^2$$
$$b^2+b=3-b^2$$
$$b^2+b+b^2=3-b^2+b^2$$
$$2b^2+b=3$$
$$2b^2+b-3=0$$
$$(2b+3)(b-1)=0$$
$$2b+3=0 \quad \text{or} \quad b-1=0$$
$$2b=-3 \qquad\qquad b=1$$
$$b=-\frac{3}{2}$$

109.

$$\sqrt[4]{8x-8}+2=0$$
$$\sqrt[4]{8x-8}=-2$$
$$\left(\sqrt[4]{8x-8}\right)^4=(-2)^4$$
$$8x-8=16$$
$$8x=24$$
$$x=\not{3}$$

No solution

110.

$$\sqrt{2m+4}-\sqrt{m+3}=1$$
$$\sqrt{2m+4}=\sqrt{m+3}+1$$
$$\left(\sqrt{2m+4}\right)^2=\left(\sqrt{m+3}+1\right)^2$$
$$2m+4=m+3+2\sqrt{m+3}+1$$
$$2m+4=m+4+2\sqrt{m+3}$$
$$2m=m+2\sqrt{m+3}$$
$$m=2\sqrt{m+3}$$
$$m^2=\left(2\sqrt{m+3}\right)^2$$
$$m^2=4(m+3)$$
$$m^2=4m+12$$
$$m^2-4m-12=0$$
$$(m-6)(m+2)=0$$
$$m-6=0 \quad \text{or} \quad m+2=0$$
$$m=6 \qquad\qquad m=\not{-2}$$

111.

$$f(x) = g(x)$$
$$\sqrt{5x+1} = x+1$$
$$\left(\sqrt{5x+1}\right)^2 = (x+1)^2$$
$$5x+1 = x^2+2x+1$$
$$0 = x^2-3x$$
$$0 = x(x-3)$$
$$x = 0 \quad \text{or} \quad x-3=0$$
$$x = 0 \qquad x = 3$$
$$x = -\frac{1}{2}$$

112.

$$f(x) = \sqrt{2x^2-7x}$$
$$\sqrt{2x^2-7x} = 2$$
$$\left(\sqrt{2x^2-7x}\right)^2 = 2^2$$
$$2x^2-7x = 4$$
$$2x^2-7x-4 = 0$$
$$(2x+1)(x-4) = 0$$
$$2x+1=0 \quad \text{or} \quad x-4=0$$
$$2x = -1 \qquad x = 4$$
$$x = -\frac{1}{2}$$

113.

$$r = \sqrt{\frac{A}{P}} - 1$$
$$(r+1)^2 = \left(\sqrt{\frac{A}{P}}\right)^2$$
$$(r+1)^2 = \frac{A}{P}$$
$$P(r+1)^2 = P\left(\frac{A}{P}\right)$$
$$P(r+1)^2 = A$$
$$\frac{P(r+1)^2}{(r+1)^2} = \frac{A}{(r+1)^2}$$
$$P = \frac{A}{(r+1)^2}$$

114.

$$h = \sqrt[3]{\frac{12I}{b}}$$
$$(h)^3 = \left(\sqrt[3]{\frac{12I}{b}}\right)^3$$
$$h^3 = \frac{12I}{b}$$
$$b(h^3) = b\left(\frac{12I}{b}\right)$$
$$h^3b = 12I$$
$$\frac{h^3b}{12} = \frac{12I}{12}$$
$$\frac{h^3b}{12} = I$$

SECTION 7.6
Geometric Applications of Radicals

115. On the left right triangle, one leg is 8 and the other leg is $\frac{1}{2}(30) = 15$. The roof line, x, is the hypotenuse of the right triangle. Use the Pythagorean Theorem to find x.

$$a^2+b^2=c^2$$
$$8^2+15^2=c^2$$
$$64+225=c^2$$
$$289=c^2$$
$$c=\sqrt{289}$$
$$c=17$$

The roof line is 17 ft.

116. The hypotenuse is 125 and one leg is 117. Let x = the length of the other leg, which is one–half the d. Use the Pythagorean Theorem to find x.

$$a^2+b^2=c^2$$
$$x^2+117^2=125^2$$
$$x^2+13{,}689=15{,}625$$
$$x^2=1{,}936$$
$$x=\sqrt{1{,}936}$$
$$x=44 \text{ yd}$$

$$d=2x$$
$$=2(44)$$
$$=88 \text{ yd}$$

The distance the boat advances is 88 yards.

117. In an isosceles right triangle, the legs are equal. Let $a=b=7$. Find c.

$$a^2+b^2=c^2$$
$$7^2+7^2=c^2$$
$$49+49=c^2$$
$$98=c^2$$
$$c=\sqrt{98}$$
$$c=\sqrt{49}\sqrt{2}$$
$$c=7\sqrt{2} \text{ meters}$$
$$c\approx 9.90 \text{ meters}$$

118. The hypotenuse is $\sqrt{2}$ times the length of the equal sides. Let x = the length of the equal sides.

$$x\sqrt{2}=15$$
$$\frac{x\sqrt{2}}{\sqrt{2}}=\frac{15}{\sqrt{2}}$$
$$x=\frac{15}{\sqrt{2}}\bullet\frac{\sqrt{2}}{\sqrt{2}}$$
$$x=\frac{15\sqrt{2}}{2} \text{ yards}$$
$$x\approx 10.61 \text{ yards}$$

119. The hypotenuse is 12, which is twice the length of the shorter leg. The length of the longer leg is $\sqrt{3}$ times the length of the shorter leg. Let x = length of the shorter leg and y = length of the longer leg.

$$2x=12$$
$$\frac{2x}{2}=\frac{12}{2}$$
$$x=6 \text{ cm}$$

$$y=6\left(\sqrt{3}\right)$$
$$=6\sqrt{3}$$
$$\approx 10.39 \text{ cm}$$

120. The length of the longer leg is $\sqrt{3}$ times the length of the shorter leg. Let x = length of the shorter leg and y = length of the hypotenuse.

$$x\sqrt{3}=60$$
$$\frac{x\sqrt{3}}{\sqrt{3}}=\frac{60}{\sqrt{3}}$$
$$x=\frac{60}{\sqrt{3}}\bullet\frac{\sqrt{3}}{\sqrt{3}}$$
$$x=\frac{60\sqrt{3}}{3}$$
$$x=20\sqrt{3}$$
$$x\approx 34.64 \text{ feet}$$

$$y=2x$$
$$y=2\left(20\sqrt{3}\right)$$
$$y=40\sqrt{3}$$
$$y\approx 69.28 \text{ feet}$$

121. The hypotenuse is $\sqrt{2}$ times the length of a leg.

$$x=5\sqrt{2}$$
$$=7.07$$
$$y=5$$

122. The hypotenuse is 50, which is twice the length of the shorter leg. The length of the longer leg is $\sqrt{3}$ times the length of the shorter leg.
Let y = length of the shorter leg and x = length of the longer leg.

$$2y = 50$$
$$y = 25 \text{ cm}$$

$$x = 25\left(\sqrt{3}\right)$$
$$= 25\sqrt{3}$$
$$\approx 43.30 \text{ cm}$$

123.

$$d = \sqrt{(x_2 - x_1)^2 + (y_2 - y_1)^2}$$
$$= \sqrt{(6-1)^2 + (-9-3)^2}$$
$$= \sqrt{(5)^2 + (-12)^2}$$
$$= \sqrt{25+144}$$
$$= \sqrt{169}$$
$$= 13$$

124.

$$d = \sqrt{(x_2 - x_1)^2 + (y_2 - y_1)^2}$$
$$= \sqrt{[-2-(-4)]^2 + (8-6)^2}$$
$$= \sqrt{(2)^2 + (2)^2}$$
$$= \sqrt{4+4}$$
$$= \sqrt{8}$$
$$= \sqrt{4}\sqrt{2}$$
$$= 2\sqrt{2}$$

SECTION 7.7
Complex Numbers

125.

$$\sqrt{-25} = \sqrt{25}\sqrt{-1}$$
$$= 5i$$

126.

$$\sqrt{-18} = \sqrt{-1}\sqrt{9}\sqrt{2}$$
$$= 3i\sqrt{2}$$

127.

$$-\sqrt{-6} = -\sqrt{-1}\sqrt{6}$$
$$= -i\sqrt{6}$$

128.

$$\sqrt{-\frac{9}{64}} = \sqrt{-1}\sqrt{\frac{9}{64}}$$
$$= \frac{3}{8}i$$

129. Complex Numbers

Real Numbers	Imaginary Numbers

130. a. true
b. true
c. false
d. false

131. a. $3 - 6i$
b. $0 - 19i$

132. a. $-1 + 7i$
b. $0 + i$

133.

$$(3+4i)+(5-6i) = (3+5)+(4-6)i$$
$$= 8-2i$$

134.

$$\left(7-\sqrt{-9}\right)-\left(4+\sqrt{-4}\right)$$
$$= (7-3i)-(4+2i)$$
$$= (7-3i)+(-4-2i)$$
$$= (7-4)+(-3-2)i$$
$$= 3-5i$$

135.

$$3i(2-i) = 6i - 3i^2$$
$$= 6i - 3(-1)$$
$$= 6i + 3$$
$$= 3 + 6i$$

136.

$$\begin{aligned}(2-7i)(-3+4i) &= -6+8i+21i-28i^2\\ &= -6+29i-28(-1)\\ &= -6+29i+28\\ &= 22+29i\end{aligned}$$

137.

$$\begin{aligned}\sqrt{-3}\cdot\sqrt{-9} &= i\sqrt{3}\cdot 3i\\ &= 3i^2\sqrt{3}\\ &= 3(-1)\sqrt{3}\\ &= -3\sqrt{3}\\ &= -3\sqrt{3}+0i\end{aligned}$$

138.

$$\begin{aligned}(9i)^2 &= 81i^2\\ &= 81(-1)\\ &= -81\\ &= -81+0i\end{aligned}$$

139.

$$\begin{aligned}\frac{5+14i}{2+3i} &= \frac{5+14i}{2+3i}\cdot\frac{2-3i}{2-3i}\\ &= \frac{(5+14i)(2-3i)}{(2+3i)(2-3i)}\\ &= \frac{10-15i+28i-42i^2}{4-9i^2}\\ &= \frac{10+13i-42(-1)}{4-9(-1)}\\ &= \frac{10+13i+42}{4+9}\\ &= \frac{52+13i}{13}\\ &= \frac{52}{13}+\frac{13}{13}i\\ &= 4+i\end{aligned}$$

140.

$$\begin{aligned}\frac{3}{11i} &= \frac{3}{11i}\cdot\frac{i}{i}\\ &= \frac{3i}{11i^2}\\ &= \frac{3i}{11(-1)}\\ &= \frac{3i}{-11}\\ &= 0-\frac{3}{11}i\end{aligned}$$

141.

$$\begin{aligned}i^{42} &= i^{4\cdot 10+2}\\ &= \left(i^4\right)^{10}\cdot i^2\\ &= (1)^{10}\cdot i^2\\ &= 1\cdot i^2\\ &= i^2\\ &= -1\end{aligned}$$

142.

$$\begin{aligned}i^{97} &= i^{4\cdot 24+1}\\ &= \left(i^4\right)^{24}\cdot i^1\\ &= (1)^{24}\cdot i\\ &= 1\cdot i\\ &= i\end{aligned}$$

CHAPTER 7 TEST

1. a. The symbol $\sqrt{\ }$ is called a **radical** symbol.
   b. The **imaginary** number i is defined as $i=\sqrt{-1}$.
   c. Squaring both sides of an equation can introduce **extraneous** solutions.
   d. An **isosceles** right triangle is a right triangle with two legs of equal length.
   e. To **rationalize** the denominator of $\frac{4}{\sqrt{5}}$, we multiply the fraction by $\frac{\sqrt{5}}{\sqrt{5}}$.
   f. A **complex** number is any number that can be written in the form $a + bi$, where a and b are real numbers and $i=\sqrt{-1}$.

2. a. If $\sqrt[n]{a}$ and $\sqrt[n]{b}$ are real numbers, then $\sqrt[n]{ab}=\sqrt[n]{a}\sqrt[n]{b}$.
   b. If $\sqrt[n]{a}$ and $\sqrt[n]{b}$ are real numbers, then $\sqrt[n]{\frac{a}{b}}=\frac{\sqrt[n]{a}}{\sqrt[n]{b}},\ (b\neq 0)$.
   c. No real number raised to the fourth power is –16.

3. $f(x)=\sqrt{x-1}$
   D: $[1, \infty)$; R: $[0, \infty)$

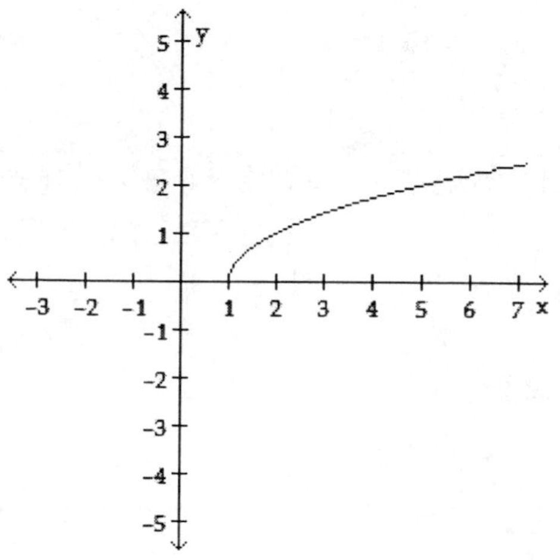

4.
$$v(d)=\sqrt{64.4d}$$
$$v(32.8)=\sqrt{64.4(32.8)}$$
$$=\sqrt{2{,}112.32}$$
$$=46 \text{ mph}$$

5. a. $f(-1)=-1$
   b. $f(8)=2$
   c. $x=1$
   d. D: $(-\infty, \infty)$, R: $(-\infty, \infty)$

6.
$$10x+50\geq 0$$
$$10x\geq -50$$
$$x\geq -5$$
$$D:[-5,\infty)$$

7.
$$\left(49x^4\right)^{1/2}=\sqrt{49x^4}$$
$$=7x^2$$

8.
$$-27^{2/3}=-\left(\sqrt[3]{27}\right)^2$$
$$=-(3)^2$$
$$=-9$$

9.
$$36^{-3/2}=\frac{1}{36^{3/2}}$$
$$=\frac{1}{\left(\sqrt{36}\right)^3}$$
$$=\frac{1}{6^3}$$
$$=\frac{1}{216}$$

10.
$$\left(-\frac{8}{125n^6}\right)^{-2/3}=\left(-\frac{125n^6}{8}\right)^{2/3}$$
$$=\left(-\sqrt[3]{\frac{125n^6}{8}}\right)^2$$
$$=\left(-\frac{5n^2}{2}\right)^2$$
$$=\frac{25n^4}{4}$$

11.

$$\frac{2^{5/3}2^{1/6}}{2^{1/2}} = \frac{2^{5/3+1/6}}{2^{1/2}}$$
$$= \frac{2^{10/6+1/6}}{2^{1/2}}$$
$$= \frac{2^{11/6}}{2^{1/2}}$$
$$= 2^{11/6-1/2}$$
$$= 2^{11/6-3/6}$$
$$= 2^{8/6}$$
$$= 2^{4/3}$$

12.

$$\left(a^{2/3}\right)^{1/6} = a^{(2/3)(1/6)}$$
$$= a^{2/18}$$
$$= a^{1/9}$$

13.

$$\sqrt{x^2} = |x|$$

14.

$$\sqrt{y^2 + 10y + 25} = \sqrt{(y+5)^2}$$
$$= |y+5|$$

15.

$$\sqrt[3]{-64x^3y^6} = -4xy^2$$

16.

$$\sqrt{\frac{4a^2}{9}} = \frac{2}{3}a$$

17.

$$\sqrt[5]{(t+8)^5} = t+8$$

18.

$$\sqrt{540x^3y^5} = \sqrt{36x^2y^4}\sqrt{15xy}$$
$$= 6xy^2\sqrt{15xy}$$

19.

$$\frac{\sqrt[3]{24x^{15}y^4}}{\sqrt[3]{y}} = \sqrt[3]{\frac{24x^{15}y^4}{y}}$$
$$= \sqrt[3]{24x^{15}y^3}$$
$$= \sqrt[3]{8x^{15}y^3}\sqrt[3]{3}$$
$$= 2x^5y\sqrt[3]{3}$$

20.

$$\sqrt[4]{32} = \sqrt[4]{16}\sqrt[4]{2}$$
$$= 2\sqrt[4]{2}$$

21.

$$2\sqrt{48y^5} - 3y\sqrt{12y^3}$$
$$= 2\sqrt{16y^4}\sqrt{3y} - 3y\sqrt{4y^2}\sqrt{3y}$$
$$= 2\left(4y^2\right)\sqrt{3y} - 3y(2y)\sqrt{3y}$$
$$= 8y^2\sqrt{3y} - 6y^2\sqrt{3y}$$
$$= 2y^2\sqrt{3y}$$

22.

$$2\sqrt[3]{40} - \sqrt[3]{5{,}000} + 4\sqrt[3]{625}$$
$$= 2\sqrt[3]{8}\sqrt[3]{5} - \sqrt[3]{1{,}000}\sqrt[3]{5} + 4\sqrt[3]{125}\sqrt[3]{5}$$
$$= 2(2)\sqrt[3]{5} - 10\sqrt[3]{5} + 4(5)\sqrt[3]{5}$$
$$= 4\sqrt[3]{5} - 10\sqrt[3]{5} + 20\sqrt[3]{5}$$
$$= 14\sqrt[3]{5}$$

23.

$$\sqrt[4]{243z^{13}} + z\sqrt[4]{48z^9} = \sqrt[4]{81z^{12}}\sqrt[4]{3z} + z\sqrt[4]{16z^8}\sqrt[4]{3z}$$
$$= 3z^3\sqrt[4]{3z} + z\left(2z^2\right)\sqrt[4]{3z}$$
$$= 3z^3\sqrt[4]{3z} + 2z^3\sqrt[4]{3z}$$
$$= 5z^3\sqrt[4]{3z}$$

24.

$$-2\sqrt{xy}\left(3\sqrt{x} + \sqrt{xy^3}\right) = -6\sqrt{x^2y} - 2\sqrt{x^2y^4}$$
$$= -6\sqrt{x^2}\sqrt{y} - 2\sqrt{x^2y^4}$$
$$-6x\sqrt{y} - 2xy^2$$

25.

$$\left(3\sqrt{2}+\sqrt{3}\right)\left(2\sqrt{2}-3\sqrt{3}\right)$$
$$=6\sqrt{4}-9\sqrt{6}+2\sqrt{6}-3\sqrt{9}$$
$$=6(2)-7\sqrt{6}-3(3)$$
$$=12-7\sqrt{6}-9$$
$$=3-7\sqrt{6}$$

26.

$$\left(\sqrt[3]{2a}+9\right)^2=\left(\sqrt[3]{2a}+9\right)\left(\sqrt[3]{2a}+9\right)$$
$$=\sqrt[3]{4a^2}+9\sqrt[3]{2a}+9\sqrt[3]{2a}+81$$
$$=\sqrt[3]{4a^2}+18\sqrt[3]{2a}+81$$

27.

$$\frac{8}{\sqrt{10}}=\frac{8}{\sqrt{10}}\bullet\frac{\sqrt{10}}{\sqrt{10}}$$
$$=\frac{8\sqrt{10}}{\sqrt{100}}$$
$$=\frac{8\sqrt{10}}{10}$$
$$=\frac{4\sqrt{10}}{5}$$

28. → $\frac{\sqrt{x}+\sqrt{y}}{\sqrt{x-y}}=\frac{x+2\sqrt{xy}+y}{x-y}$

$$\frac{3t-1}{\sqrt{3t}-1}=\frac{3t-1}{\sqrt{3t}-1}\bullet\frac{\left(\sqrt{3t}+1\right)}{\left(\sqrt{3t}+1\right)}$$
$$=\frac{(3t-1)\left(\sqrt{3t}+1\right)}{\left(\sqrt{3t}-1\right)\left(\sqrt{3t}+1\right)}$$
$$=\frac{3t\sqrt{3t}+3t-\sqrt{3t}-1}{3t-1}$$
$$=\frac{3t\left(\sqrt{3t}+1\right)-1\left(\sqrt{3t}+1\right)}{3t-1}$$
$$=\frac{(3t-1)\left(\sqrt{3t}+1\right)}{3t-1}$$
$$=\sqrt{3t}+1$$

29.

$$\sqrt[3]{\frac{9}{4a}}=\sqrt[3]{\frac{9}{4a}}\bullet\sqrt[3]{\frac{2a^2}{2a^2}}$$
$$=\frac{\sqrt[3]{18a^2}}{\sqrt[3]{8a^3}}$$
$$=\frac{\sqrt[3]{18a^2}}{2a}$$

30.

$$\frac{\sqrt{5}+3}{-4\sqrt{2}}=\frac{\sqrt{5}+3}{-4\sqrt{2}}\bullet\frac{\sqrt{5}-3}{\sqrt{5}-3}$$
$$=\frac{\left(\sqrt{5}+3\right)\left(\sqrt{5}-3\right)}{-4\sqrt{2}\left(\sqrt{5}-3\right)}$$
$$=\frac{5-9}{-4\sqrt{10}+12\sqrt{2}}$$
$$=\frac{-4}{-4\left(\sqrt{10}-3\sqrt{2}\right)}$$
$$=\frac{1}{\sqrt{10}-3\sqrt{2}}$$
$$=\frac{1}{\sqrt{2}\sqrt{5}-3\sqrt{2}}$$
$$=\frac{1}{\sqrt{2}\left(\sqrt{5}-3\right)}$$

31.

$$4\sqrt{x}=\sqrt{x+1}$$
$$\left(4\sqrt{x}\right)^2=\left(\sqrt{x+1}\right)^2$$
$$16x=x+1$$
$$15x=1$$
$$x=\frac{1}{15}$$

32.

$$\sqrt[3]{6n+4}-4=0$$
$$\sqrt[3]{6n+4}=4$$
$$\left(\sqrt[3]{6n+4}\right)^3=4^3$$
$$6n+4=64$$
$$6n=60$$
$$n=10$$

33.

$$1=\sqrt{u-3}+\sqrt{u}$$
$$1-\sqrt{u}=\sqrt{u-3}$$
$$\left(1-\sqrt{u}\right)^2=\left(\sqrt{u-3}\right)^2$$
$$1-\sqrt{u}-\sqrt{u}+u=u-3$$
$$1-2\sqrt{u}+u=u-3$$
$$1-2\sqrt{u}+u-1-u=u-3-1-u$$
$$-2\sqrt{u}=-4$$
$$\frac{-2\sqrt{u}}{-2}=\frac{-4}{-2}$$
$$\sqrt{u}=2$$
$$\left(\sqrt{u}\right)^2=(2)^2$$
$$u=\cancel{4}$$

No solution

34.

$$\left(2m^2-9\right)^{1/2}=m$$
$$\sqrt{2m^2-9}=m$$
$$\left(\sqrt{2m^2-9}\right)^2=(m)^2$$
$$2m^2-9=m^2$$
$$2m^2-9-m^2=m^2-m^2$$
$$m^2-9=0$$
$$(m+3)(m-3)=0$$
$$m+3=0 \quad \text{or} \quad m-3=0$$
$$m=\cancel{-3} \qquad m=3$$

35.

$$\sqrt{t-2}-t+2=0$$
$$\sqrt{t-2}=t-2$$
$$\sqrt{t-2}=t-2$$
$$\left(\sqrt{t-2}\right)^2=(t-2)^2$$
$$t-2=t^2-4t+4$$
$$0=t^2-5t+6$$
$$0=(t-2)(t-3)$$
$$t-2=0 \quad \text{or} \quad t-3=0$$
$$t=2 \qquad t=3$$

36.

$$\sqrt{x-8}+10=0$$
$$\sqrt{x-8}=-10$$
$$\left(\sqrt{x-8}\right)^2=(-10)^2$$
$$x-8=100$$
$$x=\cancel{108}$$

37.

$$f(x)=g(x)$$
$$\sqrt[4]{15-x}=\sqrt[4]{13-2x}$$
$$\left(\sqrt[4]{15-x}\right)^4=\left(\sqrt[4]{13-2x}\right)^4$$
$$15-x=13-2x$$
$$15+x=13$$
$$x=-2$$

38.

$$r=\sqrt[3]{\frac{GMt^2}{4\pi^2}}$$
$$(r)^3=\left(\sqrt[3]{\frac{GMt^2}{4\pi^2}}\right)^3$$
$$r^3=\frac{GMt^2}{4\pi^2}$$
$$4\pi^2\left(r^3\right)=4\pi^2\left(\frac{GMt^2}{4\pi^2}\right)$$
$$4\pi^2r^3=GMt^2$$
$$\frac{4\pi^2r^3}{Mt^2}=\frac{GMt^2}{Mt^2}$$
$$\frac{4\pi^2r^3}{Mt^2}=G$$

39. The length of the longer side is $\sqrt{3}$ time the length of the shorter side. The length of the hypotenuse is twice the length of the shorter side. First find the length of the shorter side and use that to find h, the length of the hypotenuse.

Let x = the length of the shorter side.

$$x\sqrt{3} = 8$$
$$\frac{x\sqrt{3}}{\sqrt{3}} = \frac{8}{\sqrt{3}}$$
$$x = \frac{8}{\sqrt{3}}$$
$$x = \frac{8}{\sqrt{3}} \cdot \frac{\sqrt{3}}{\sqrt{3}}$$
$$x = \frac{8\sqrt{3}}{3}$$
$$x \approx 4.62 \text{ cm}$$

$$h = 2x$$
$$= 2(4.62)$$
$$\approx 9.24 \text{ cm}$$

40. The length of the hypotenuse is $\sqrt{2}$ times the length of a leg. Let x and y equal the length of a leg.

$$x\sqrt{2} = 12.26$$
$$\frac{x\sqrt{2}}{\sqrt{2}} = \frac{12.26}{\sqrt{2}}$$
$$x = \frac{12.26}{\sqrt{2}}$$
$$x = \frac{12.26}{\sqrt{2}} \cdot \frac{\sqrt{2}}{\sqrt{2}}$$
$$x = \frac{12.26\sqrt{2}}{2}$$
$$x = 6.13\sqrt{2}$$
$x = 8.67$ ~~cm~~ in

$$y = 6.13\sqrt{2}$$
$y = 8.67$ ~~cm~~ in

41.

$$d = \sqrt{(x_2 - x_1)^2 + (y_2 - y_1)^2}$$
$$= \sqrt{[22 - (-2)]^2 + (12 - 5)^2}$$
$$= \sqrt{(24)^2 + (7)^2}$$
$$= \sqrt{576 + 49}$$
$$= \sqrt{625}$$
$$= 25$$

42. Use the Pythagorean Theorem to find h.

$$a^2 + b^2 = c^2$$
$$(h)^2 + (45)^2 = (53)^2$$
$$h^2 + 2{,}025 = 2{,}809$$
$$h^2 = 784$$
$$h = \sqrt{784}$$
$$h = 28 \text{ in.}$$

43.

$$\sqrt{-45} = \sqrt{-9}\sqrt{5}$$
$$= 3i\sqrt{5}$$

44.

$$i^{106} = i^{4\cdot 26+2}$$
$$= (i^4)^{26} \cdot i^2$$
$$= (1)^{26} \cdot i^2$$
$$= 1 \cdot i^2$$
$$= i^2$$
$$= -1$$

45.

$$(9 + 4i) + (-13 + 7i) = (9 - 13) + (4 + 7)i$$
$$= -4 + 11i$$

46.

$$\left(3-\sqrt{-9}\right)-\left(-1+\sqrt{-16}\right)$$
$$=(3-3i)-(-1+4i)$$
$$=(3-3i)+(1-4i)$$
$$=(3+1)+(-3-4)i$$
$$=4-7i$$

47.

$$15i(3-5i)=45i-75i^2$$
$$=45i-75(-1)$$
$$=45i+75$$
$$=75+45i$$

48.

$$(8+10i)(-7-i)=-56-8i-70i-10i^2$$
$$=-56-78i-10(-1)$$
$$=-56-78i+10$$
$$=-46-78i$$

49.

$$\frac{1}{i\sqrt{2}}=\frac{1}{i\sqrt{2}}\bullet\frac{i\sqrt{2}}{i\sqrt{2}}$$
$$=\frac{i\sqrt{2}}{2i^2}$$
$$=\frac{i\sqrt{2}}{2(-1)}$$
$$=\frac{i\sqrt{2}}{-2}$$
$$=0-\frac{\sqrt{2}}{2}i$$

50.

$$\frac{2+i}{3-i}=\frac{2+i}{3-i}\bullet\frac{3+i}{3+i}$$
$$=\frac{(2+i)(3+i)}{(3-i)(3+i)}$$
$$=\frac{6+2i+3i+i^2}{9-i^2}$$
$$=\frac{6+5i+(-1)}{9-(-1)}$$
$$=\frac{5+5i}{10}$$
$$=\frac{5}{10}+\frac{5}{10}i$$
$$=\frac{1}{2}+\frac{1}{2}i$$

CHAPTER 7 CUMULATIVE REVIEW

1. Natural numbers: 1
   Whole numbers: 1
   Integers: –5, 1
   Rational numbers: –5, 3.4, 1, $\frac{16}{5}$, $9.\overline{7}$
   Irrational numbers: $-\pi, \sqrt{19}$
   Real numbers: all of them

2.
$$\frac{|Ax_0 + By_0 + C|}{\sqrt{A^2 + B^2}} = \frac{|(6)(-2)+(-8)(-2)+(-5)|}{\sqrt{(6)^2 + (-8)^2}}$$
$$= \frac{|-12+16+(-5)|}{\sqrt{36+64}}$$
$$= \frac{|-1|}{\sqrt{100}}$$
$$= \frac{1}{10}$$

3.
$$\frac{2}{3}(b+3) = \frac{5}{4}b + \frac{17}{12}$$
$$12\left[\frac{2}{3}(b+3)\right] = 12\left[\frac{5}{4}b + \frac{17}{12}\right]$$
$$8(b+3) = 15b + 17$$
$$8b + 24 = 15b + 17$$
$$8b + 24 - 15b = 15b + 17 - 15b$$
$$-7b + 24 = 17$$
$$-7b + 24 - 24 = 17 - 24$$
$$-7b = -7$$
$$b = 1$$

4.
$$S = \frac{a - \ell r}{1 - r}$$
$$(1-r)(S) = (1-r)\left(\frac{a-\ell r}{1-r}\right)$$
$$S - Sr = a - \ell r$$
$$S - Sr + \ell r = a - \ell r + \ell r$$
$$S - Sr + \ell r = a$$
$$\ell r = a - S + Sr$$
$$\ell = \frac{a - S + Sr}{r}$$

5.
$$\frac{x}{2} + (-8) + (-15) + x + (-10) = 0$$
$$\frac{3x}{2} - 33 = 0$$
$$2\left(\frac{3x}{2} - 33\right) = 2(0)$$
$$3x - 66 = 0$$
$$3x = 66$$
$$x = 22$$

6.

	Amount ·	Strength =	Pure concentrate
10%	x	0.10	$0.10x$
18%	$10 - x$	0.18	$0.18(10 - x)$
Mixture	10	0.15	0.15(10)

$$0.10x + 0.18(10 - x) = 0.15(10)$$
$$0.10x + 1.8 - 0.18x = 1.5$$
$$-0.08x + 1.8 = 1.5$$
$$-0.08x + 1.8 - 1.8 = 1.5 - 1.8$$
$$-0.08x = -0.3$$
$$x = 3.75 = 3\frac{3}{4}$$
$$10 - x = 6.25 = 6\frac{1}{4}$$

$3\frac{3}{4}$ cups of 10% and $6\frac{1}{4}$ cups of 18%

7. (3, –1) and (–6, 2)
$$m = \frac{y_2 - y_1}{x_2 - x_1}$$
$$= \frac{2-(-1)}{-6-3}$$
$$= \frac{3}{-9}$$
$$= -\frac{1}{3}$$

8. $x = 3$

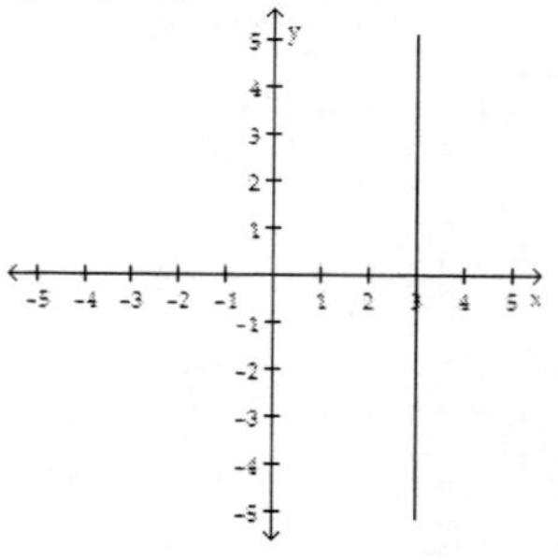

9. a. $-\dfrac{3}{5}$

b. (0, –3)

c. $y=-\dfrac{3}{5}x-3$

10. a. 3

b. Let $m=-\dfrac{1}{3}$.

Use point–slope formula.

$$y-y_1=m(x-x_1)$$

$$y-(-2)=-\frac{1}{3}[x-4]$$

$$y+2=-\frac{1}{3}x+\frac{4}{3}$$

$$y=-\frac{1}{3}x+\frac{4}{3}-2$$

$$y=-\frac{1}{3}x-\frac{2}{3}$$

11. The graph of $g(x)=(x+4)^2$ is the same as the graph of $f(x)=x^2$ except that it is shifted. **4** units to the **left**.

12. Yes, it does define a function.

13. D: (–∞, ∞); R: (–∞, –2]

14. $g(x)=1+x^3$

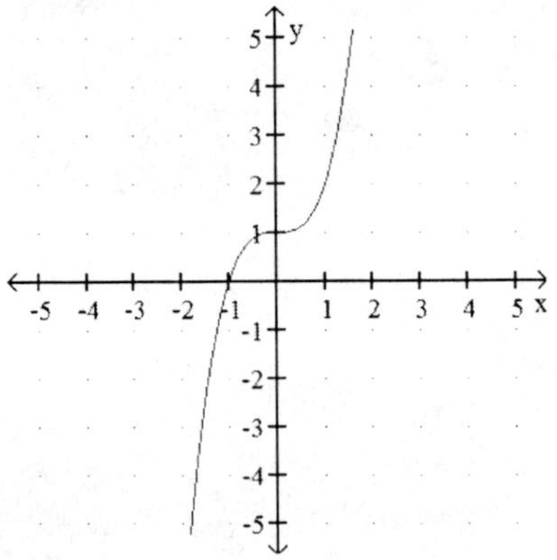

15. $\begin{cases}3x-y=-3\\ y=-2x-7\end{cases}$ (–2, –3)

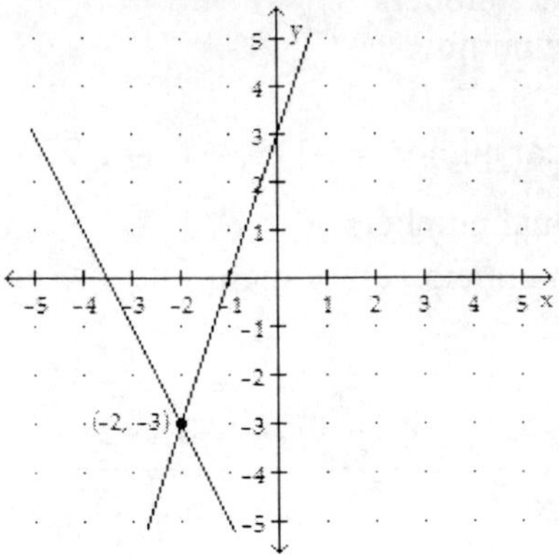

16.

$$\begin{cases}x=\dfrac{3}{2}y+5\\ 2x-3y=8\end{cases}$$

Stubstitute $x=\dfrac{3}{2}y+5$ into the 2nd equation.

$$2x-3y=8$$

$$2\left(\frac{3}{2}y+5\right)-3y=8$$

$$3y+10-3y=8$$

$$10\neq 8$$

no solution; ∅

inconsistent system

17.

$$\begin{vmatrix}-9 & 7\\ 4 & -2\end{vmatrix}=-9(-2)-7(4)$$

$$=18-28$$

$$=-10$$

18. $\begin{cases} 3x + y + 11z = 100 & (1) \\ x + y + 7z = 60 & (2) \\ 2x + z = 25 & (3) \end{cases}$

Subtract Equation 2 from Equation 1.

$$\begin{array}{r} 3x + y + 11z = 100 \\ \underline{-x - y - 7z = -60} \\ 2x + 4z = \ \ 40 \end{array}$$

Subtract Equation 3 from the new equation.

$$\begin{array}{r} 2x + 4z = \ \ 40 \\ \underline{-2x - z = -25} \\ 3z = 15 \\ z = 5 \end{array}$$

$$\begin{array}{rl} 2x + z = 25 & (3) \\ 2x + 5 = 25 & \\ 2x = 20 & \\ x = 10 & \\ x + y + 7z = 60 & (2) \\ 10 + y + 7(5) = 60 & \\ y + 45 = 60 & \\ y = 15 & \end{array}$$

19.

$$\begin{array}{rcr} 5(x+1) \le 4(x+3) & \text{and} & x + 12 < -3 \\ 5x + 5 \le 4x + 12 & & x + 12 - 12 < -3 - 12 \\ 5x + 5 - 5 \le 4x + 12 - 5 & & x < -15 \\ 5x \le 4x + 7 & & \\ 5x - 4x \le 4x + 7 - 4x & & \\ x \le 7 & & \end{array}$$

$(-\infty, -15)$

20.

$$|-1 - 2x| > 5$$

$$\begin{array}{rcr} -1 - 2x < -5 & \text{or} & -1 - 2x > 5 \\ -1 - 2x + 1 < -5 + 1 & & -1 - 2x + 1 > 5 + 1 \\ -2x < -4 & & -2x > 6 \\ x > 2 & & x < -3 \end{array}$$

$(-\infty, -3) \cup (2, \infty)$

21. $\begin{cases} x + 2y < 3 \\ 2x + 4y < 8 \end{cases}$

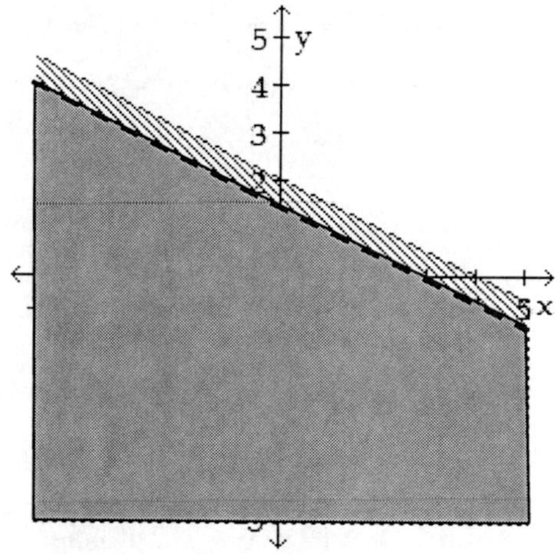

22.

$$\left(\frac{4a^{-2}b}{3ab^{-3}}\right)^3 = \left(\frac{4}{3}a^{-2-1}b^{1-(-3)}\right)^3$$
$$= \left(\frac{4}{3}a^{-3}b^4\right)^3$$
$$= \frac{64}{27}a^{-9}b^{12}$$
$$= \frac{64b^{12}}{27a^9}$$

23.

$$\left(6.1 \times 10^8\right)\left(3.9 \times 10^5\right) = \left(6.1 \times 3.9\right)\left(10^{8+5}\right)$$
$$= 23.79 \times 10^{13}$$
$$= 2.379 \times 10^{14}$$

24. a. 0
b. 16
c. −1 and 2
d. D: $(-\infty, \infty)$; R: $(-\infty, \infty)$

25.

$$f(x) + g(x) = \left(\frac{1}{5}x^6 + \frac{3}{4}x^2\right) + \left(\frac{2}{3}x^6 - \frac{1}{12}x^2\right)$$
$$= \frac{13}{15}x^6 + \frac{2}{3}x^2$$

26. $(-2x^2y^3 + 6xy + 5y^2) - (-4x^2y^3 - 7xy + 2y^2)$
$= -2x^2y^3 + 6xy + 5y^2 + 4x^2y^3 + 7xy - 2y^2$
$= 2x^2y^3 + 13xy + 3y^2$

27.

$$(3y+1)(2y^2+3y+2)$$
$$=3y(2y^2+3y+2)+1(2y^2+3y+2)$$
$$=6y^3+9y^2+6y+2y^2+3y+2$$
$$=6y^3+11y^2+9y+2$$

28.

$$(x+3)(x-3)+(2x-1)(x+2)$$
$$=x^2-9+2x^2+4x-x-2$$
$$=3x^2+3x-11$$

29.

$$(2y^5+5z)^2=(2y^5)^2+2(2y^5)(5z)+(5z)^2$$
$$=4y^{10}+20y^5z+25z^2$$

30.

$$(5s-t^3)^3$$
$$=(5s-t^3)(5s-t^3)^2$$
$$=(5s-t^3)\left[(5s)^2-2(5s)(t^3)+(t^3)^2\right]$$
$$=(5s-t^3)(25s^2-10st^3+t^6)$$
$$=125s^3-50s^2t^3+5st^6-25s^2t^3+10st^6-t^9$$
$$=125s^3-75s^2t^3+15st^6-t^9$$

31.

$$3c-cd+3d-c^2$$
$$=3c+3d-c^2-cd$$
$$=3(c+d)-c(c+d)$$
$$=(c+d)(3-c)$$

32.

$$x^3-8y^3=(x)^3-(2y)^3$$
$$=(x-2y)(x^2+x(2y)+(2y)^2)$$
$$=(x-2y)(x^2+2xy+4y^2)$$

33.

$$(a+b)^2-2(a+b)+1=[(a+b)-1]^2$$
$$=(a+b-1)^2$$

34.

$$x^4-17x^2+16=(x^2-16)(x^2-1)$$
$$=(x+4)(x-4)(x+1)(x-1)$$

35.

$$2z^3-200z=0$$
$$2z(z^2-100)=0$$
$$2z(z+10)(z-10)=0$$
$$2z=0,\quad z+10=0,\quad \text{or}\quad z-10=0$$
$$z=0 \qquad z=-10 \qquad z=10$$

36.

$$3m^2+10m=-3$$
$$3m^2+10m+3=0$$
$$(3m+1)(m+3)=0$$
$$3m+1=0 \quad \text{or} \quad m+3=0$$
$$3m=-1 \qquad m=-3$$
$$m=-\frac{1}{3}$$

37.

Let x = the width of the mat.
There is a part of the mat on each side of the picture so the width is $(12 + 2x)$ and the length is $(16 + 2x)$. The area of the mat is the total area – the area of the picture.

$$(12+2x)(16+2x)-12(16)=128$$
$$192+24x+32x+4x^2-192=128$$
$$4x^2+56x-128=0$$
$$x^2+14x-32=0$$
$$(x-2)(x+16)=0$$
$$x-2=0 \quad \text{or} \quad x+16=0$$
$$x=2 \qquad x=\cancel{-16}$$

The width of the mat is 2 inches.

38.

$$\frac{3x^2-10xy-8y^2}{4y^2-xy}=\frac{(3x+2y)(x-4y)}{y(4y-x)}$$
$$=\frac{-(3x+2y)\cancel{(4y-x)}}{y\cancel{(4y-x)}}$$
$$=\frac{-3x-2y}{y}$$

39.

$$\frac{2}{x^2-x-56}$$

$$x^2-x-56\neq 0$$

$$(x-8)(x+7)\neq 0$$

$$x-8\neq 0 \text{ and } x+7\neq 0$$

$$x\neq 8 \qquad x\neq -7$$

−7 and 8 make it undefined.

40. $f(x)=\dfrac{1}{x}$

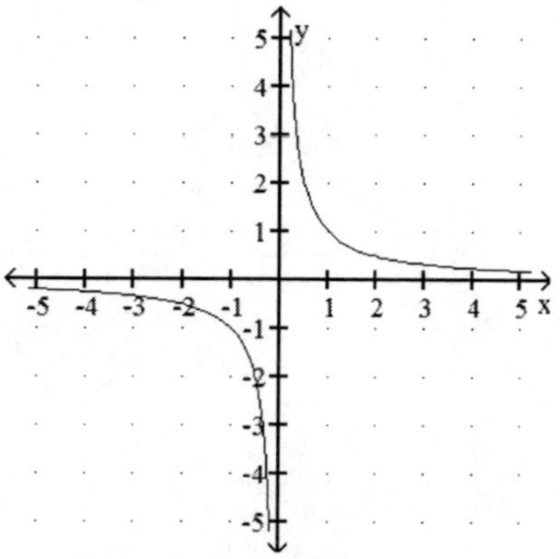

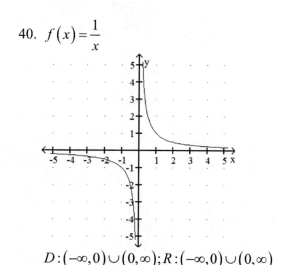

$D:(-\infty,0)\cup(0,\infty);R:(-\infty,0)\cup(0,\infty)$

41.

$$c(p)=\frac{60{,}000p}{100-p}$$

$$c(98)=\frac{60{,}000(98)}{100-98}$$

$$=\frac{5{,}880{,}000}{2}$$

$$=\$2{,}940{,}000$$

42.

$$f(t)\div g(t)=(10t-t^2)\div\frac{t^4-10t^3-2t^2+20t}{t^6}$$

$$=\frac{(10t-t^2)}{1}\cdot\frac{t^6}{t^4-10t^3-2t^2+20t}$$

$$=\frac{t(10-t)}{1}\cdot\frac{t^6}{t(t^3-10t^2-2t+20)}$$

$$=\frac{t(10-t)}{1}\cdot\frac{t^6}{t(t^2-2)(t-10)}$$

$$=-\frac{t^6}{t^2-2}$$

43.

$$(2x^2-9x-5)\cdot\frac{x}{2x^2+x}=\frac{\cancel{(2x+1)}(x-5)}{1}\cdot\frac{\cancel{x}}{\cancel{x}\cancel{(2x+1)}}$$

$$=x-5$$

44.

$$\frac{2x}{x^2-4}-\frac{1}{x^2-3x+2}+\frac{x+1}{x^2+x-2}$$

$$=\frac{2x}{(x+2)(x-2)}-\frac{1}{(x-1)(x-2)}+\frac{x+1}{(x+2)(x-1)}$$

$$=\left(\frac{x-1}{x-1}\right)\frac{2x}{(x+2)(x-2)}-\left(\frac{x+2}{x+2}\right)\frac{1}{(x-1)(x-2)}+\left(\frac{x-2}{x-2}\right)\frac{x+1}{(x+2)(x-1)}$$

$$=\frac{2x^2-2x}{(x+2)(x-2)(x-1)}-\frac{x+2}{(x+2)(x-2)(x-1)}+\frac{x^2+x-2x-2}{(x+2)(x-2)(x-1)}$$

$$=\frac{2x^2-2x-x-2+x^2-x-2}{(x+2)(x-2)(x-1)}$$

$$=\frac{3x^2-4x-4}{(x+2)(x-2)(x-1)}$$

$$=\frac{(3x+2)\cancel{(x-2)}}{(x+2)\cancel{(x-2)}(x-1)}$$

$$=\frac{3x+2}{(x+2)(x-1)}$$

45.

Let x = the number of hours it takes the two pipes to fill the tank working together.

One pipe's work in 1 hour = $\dfrac{1}{6}$

Other pipe's work in 1 hour = $\dfrac{1}{4}$

Amount of work together in 1 hour = $\dfrac{1}{x}$

$$\frac{1}{6}+\frac{1}{4}=\frac{1}{x}$$

$$12x\left(\frac{1}{6}+\frac{1}{4}\right)=12x\left(\frac{1}{x}\right)$$

$$2x+3x=12$$

$$5x=12$$

$$x=\frac{12}{5}$$

$$x=2\frac{2}{5}\text{ hours}$$

46.

$$r=\frac{k}{s}$$
$$40=\frac{k}{10}$$
$$10(40)=10\left(\frac{k}{10}\right)$$
$$400=k$$
$$r=\frac{400}{15}$$
$$r=\frac{80}{3}=26\frac{2}{3}$$

47. $f(x)=\sqrt{x}$;

$D:[0,\infty);R:[0,\infty)$

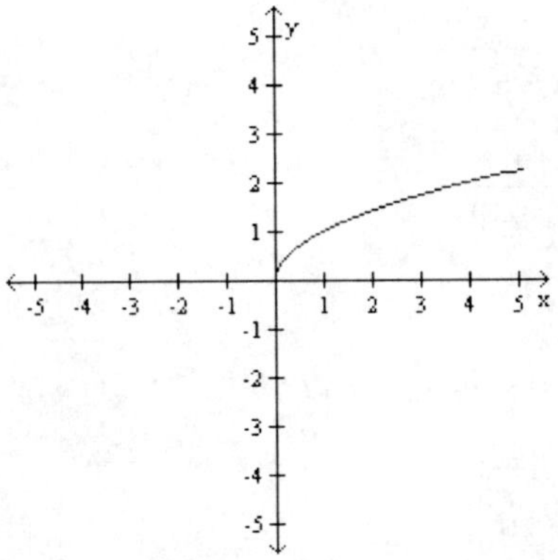

48.

$$f(x)=\sqrt[3]{-2x-4}$$
$$f(30)=\sqrt[3]{-2(30)-4}$$
$$=\sqrt[3]{-64}$$
$$=-4$$

49.

$$\sqrt{200x^4y^3z}=\sqrt{100x^4y^2}\sqrt{2yz}$$
$$=10x^2y\sqrt{2yz}$$

50.

$$\sqrt[3]{16}+\sqrt[3]{128}=\sqrt[3]{8}\sqrt[3]{2}+\sqrt[3]{64}\sqrt[3]{2}$$
$$=2\sqrt[3]{2}+4\sqrt[3]{2}$$
$$=6\sqrt[3]{2}$$

51.

$$\left(\sqrt{5z}+\sqrt{3}\right)\left(\sqrt{5z}+\sqrt{3}\right)$$
$$=\left(\sqrt{5z}+\sqrt{3}\right)^2$$
$$=\left(\sqrt{5z}\right)^2+2\left(\sqrt{5z}\right)\left(\sqrt{3}\right)+\left(\sqrt{3}\right)^2$$
$$=5z+2\sqrt{15z}+3$$

52.

$$\frac{\sqrt{3}}{\sqrt{50}}=\frac{\sqrt{3}}{\sqrt{25}\sqrt{2}}$$
$$=\frac{\sqrt{3}}{5\sqrt{2}}$$
$$=\frac{\sqrt{3}}{5\sqrt{2}}\left(\frac{\sqrt{2}}{\sqrt{2}}\right)$$
$$=\frac{\sqrt{6}}{5\sqrt{4}}$$
$$=\frac{\sqrt{6}}{10}$$

53.

$$\sqrt{-5x+24}=6-x$$
$$\left(\sqrt{-5x+24}\right)^2=(6-x)^2$$
$$-5x+24=36-6x-6x+x^2$$
$$-5x+24=36-12x+x^2$$
$$-5x+24+5x-24=36-12x+x^2+5x-24$$
$$0=x^2-7x+12$$
$$0=(x-3)(x-4)$$
$$x-3=0 \quad \text{or} \quad x-4=0$$
$$x=3 \qquad\qquad x=4$$

54.

$$\left(-\frac{8x^3}{27}\right)^{-1/3}=\left(-\frac{27}{8x^3}\right)^{1/3}$$
$$=\sqrt[3]{-\frac{27}{8x^3}}$$
$$=-\frac{3}{2x}$$

55.

$$a^2 + b^2 = c^2$$
$$10^2 + b^2 = 26^2$$
$$100 + b^2 = 676$$
$$b^2 = 576$$
$$b = \sqrt{576}$$
$$b = 24 \text{ cm}$$

56.

$$\frac{5-3i}{4+2i} = \frac{5-3i}{4+2i} \bullet \frac{(4-2i)}{(4-2i)}$$
$$= \frac{(5-3i)(4-2i)}{(4+2i)(4-2i)}$$
$$= \frac{20-10i-12i+6i^2}{16-4i^2}$$
$$= \frac{20-22i+6(-1)}{16-4(-1)}$$
$$= \frac{20-22i-6}{16+4}$$
$$= \frac{14-22i}{20}$$
$$= \frac{14}{20} - \frac{22}{20}i$$
$$= \frac{7}{10} - \frac{11}{10}i$$

SECTION 8.1

VOCABULARY

1. An equation of the form $ax^2 + bx + c = 0$, where $a \neq 0$, is called a **quadratic** equation.

3. When we add 16 to $x^2 + 8x$, we say that we have completed the **square** on $x^2 + 8x$.

CONCEPTS

5. For any nonnegative number c, if $x^2 = c$, then $x = \sqrt{c}$ or $x = -\sqrt{c}$.

7. a. $\left(\frac{12}{2}\right)^2 = 6^2 = 36$

   b. $\left(\frac{-5}{2}\right)^2 = \frac{25}{4}$

9. a. subtract 7 from both sides
   b. divide both sides by 4

11. Yes, it is a solution.

$$x^2 + 4x + 2 = 0$$
$$\left(-2+\sqrt{2}\right)^2 + 4\left(-2+\sqrt{2}\right) + 2 \stackrel{?}{=} 0$$
$$4 - 4\sqrt{2} + 2 - 8 + 4\sqrt{2} + 2 \stackrel{?}{=} 0$$
$$0 = 0$$

NOTATION

13. We read $8 \pm \sqrt{3}$ as "eight **plus or minus** the square root of 3."

GUIDED PRACTICE

15.
$$t^2 - 11 = 0$$
$$t^2 - 11 + 11 = 0 + 11$$
$$t^2 = 11$$
$$t = \sqrt{11} \quad \text{or} \quad t = -\sqrt{11}$$
$$t = \pm\sqrt{11}$$

17.
$$x^2 - 35 = 0$$
$$x^2 - 35 + 35 = 0 + 35$$
$$x^2 = 35$$
$$x = \sqrt{35} \quad \text{or} \quad x = -\sqrt{35}$$
$$x = \pm\sqrt{35}$$

19.
$$z^2 - 50 = 0$$
$$z^2 - 50 + 50 = 0 + 50$$
$$z^2 = 50$$
$$z = \pm\sqrt{50}$$
$$z = \pm\sqrt{25}\sqrt{2}$$
$$z = \pm 5\sqrt{2}$$
$$z = 5\sqrt{2} \quad \text{or} \quad z = -5\sqrt{2}$$

21.
$$3x^2 - 16 = 0$$
$$3x^2 = 16$$
$$x^2 = \frac{16}{3}$$
$$x = \sqrt{\frac{16}{3}} \quad \text{or} \quad x = -\sqrt{\frac{16}{3}}$$
$$x = \frac{4}{\sqrt{3}} \qquad x = -\frac{4}{\sqrt{3}}$$
$$x = \frac{4}{\sqrt{3}} \bullet \frac{\sqrt{3}}{\sqrt{3}} \qquad x = -\frac{4}{\sqrt{3}} \bullet \frac{\sqrt{3}}{\sqrt{3}}$$
$$x = \frac{4\sqrt{3}}{3} \qquad x = -\frac{4\sqrt{3}}{3}$$

23.
$$p^2 = -16$$
$$p = \sqrt{-16} \quad \text{or} \quad p = -\sqrt{-16}$$
$$p = 4i \qquad p = -4i$$
$$p = \pm 4i$$

25.
$$a^2 + 8 = 0$$
$$a^2 + 8 - 8 = 0 - 8$$
$$a^2 = -8$$
$$a = \pm\sqrt{-8}$$
$$a = \pm\sqrt{-1}\sqrt{4}\sqrt{2}$$
$$a = \pm 2i\sqrt{2}$$

27.

$$4m^2+81=0$$
$$4m^2=-81$$
$$m^2=-\frac{81}{4}$$
$$m=\sqrt{-\frac{81}{4}} \quad \text{or} \quad m=-\sqrt{-\frac{81}{4}}$$
$$m=\frac{9}{2}i \qquad m=-\frac{9}{2}i$$
$$m=\pm\frac{9}{2}i$$

29.

$$6b^2+144=0$$
$$6b^2=-144$$
$$b^2=-\frac{144}{6}$$
$$b=\sqrt{-\frac{144}{6}} \quad \text{or} \quad b=-\sqrt{-\frac{144}{6}}$$
$$b=\frac{12}{\sqrt{6}}i \qquad b=-\frac{12}{\sqrt{6}}i$$
$$b=\frac{12}{\sqrt{6}}\left(\frac{\sqrt{6}}{\sqrt{6}}\right)i \qquad b=-\frac{12}{\sqrt{6}}\left(\frac{\sqrt{6}}{\sqrt{6}}\right)i$$
$$b=\frac{12\sqrt{6}}{6}i \qquad b=-\frac{12\sqrt{6}}{6}i$$
$$b=2i\sqrt{6} \qquad b=-2i\sqrt{6}$$
$$b=\pm 2i\sqrt{6}$$

31.

$$(x+5)^2=9$$
$$\sqrt{(x+5)^2}=\sqrt{9}$$
$$x+5=\pm 3$$
$$x+5=3 \quad \text{or} \quad x+5=-3$$
$$x=-2 \qquad x=-8$$

33.

$$(t+4)^2=16$$
$$\sqrt{(t+4)^2}=\sqrt{16}$$
$$t+4=\pm 4$$
$$t+4=4 \quad \text{or} \quad t+4=-4$$
$$t=0 \qquad t=-8$$

35.

$$(x+5)^2=3$$
$$\sqrt{(x+5)^2}=\sqrt{3}$$
$$x+5=\sqrt{3} \quad \text{or} \quad x+5=-\sqrt{3}$$
$$x=-5+\sqrt{3} \qquad x=-5-\sqrt{3}$$
$$x=-5\pm\sqrt{3}$$

37.

$$(7a-2)^2=8$$
$$7a-2=\sqrt{8} \quad \text{or} \quad 7a-2=-\sqrt{8}$$
$$7a-2=\sqrt{4}\sqrt{2} \qquad 7a-2=-\sqrt{4}\sqrt{2}$$
$$7a-2=2\sqrt{2} \qquad 7a-2=-2\sqrt{2}$$
$$7a=2+2\sqrt{2} \qquad 7a=2-2\sqrt{2}$$
$$a=\frac{2+2\sqrt{2}}{7} \qquad a=\frac{2-2\sqrt{2}}{7}$$
$$a=\frac{2\pm 2\sqrt{2}}{7}$$

39.

$$\left(\frac{24}{2}\right)^2=(12)^2=144$$
$$x^2+24x+144=(x+12)^2$$

41.

$$\left(\frac{-7}{2}\right)^2=\frac{49}{4}$$
$$a^2-7a+\frac{49}{4}=\left(a-\frac{7}{2}\right)^2$$

43.

$$\frac{1}{2}\left(\frac{2}{3}\right)=\frac{1}{3}$$
$$\left(\frac{1}{3}\right)^2=\frac{1}{9}$$
$$x^2+\frac{2}{3}x+\frac{1}{9}=\left(x+\frac{1}{3}\right)^2$$

45.

$$\frac{1}{2}\left(-\frac{5}{6}\right)=-\frac{5}{12}$$

$$\left(-\frac{5}{12}\right)^2=\frac{25}{144}$$

$$m^2-\frac{5}{6}m+\frac{25}{144}=\left(m-\frac{5}{12}\right)^2$$

47.

$$x^2-4x-2=0$$
$$x^2-4x=2$$
$$x^2-4x+\left(\frac{-4}{2}\right)^2=2+\left(\frac{-4}{2}\right)^2$$
$$x^2-4x+(-2)^2=2+(-2)^2$$
$$x^2-4x+4=2+4$$
$$(x-2)^2=6$$
$$x-2=\pm\sqrt{6}$$
$$x=2\pm\sqrt{6}$$
$$x\approx 4.45 \quad \text{and} \quad x\approx -0.45$$

49.

$$x^2-12x+1=0$$
$$x^2-12x=-1$$
$$x^2-12x+\left(\frac{-12}{2}\right)^2=-1+\left(\frac{-12}{2}\right)^2$$
$$x^2-12x+(-6)^2=-1+(-6)^2$$
$$x^2-12x+36=-1+36$$
$$(x-6)^2=35$$
$$x-6=\pm\sqrt{35}$$
$$x=6\pm\sqrt{35}$$
$$x\approx 11.92 \quad \text{and} \quad x\approx 0.08$$

51.

$$t^2+20t+25=0$$
$$t^2+20t=-25$$
$$t^2+20t+\left(\frac{20}{2}\right)^2=-25+\left(\frac{20}{2}\right)^2$$
$$t^2+20t+(10)^2=-25+(10)^2$$
$$t^2+20t+100=-25+100$$
$$(t+10)^2=75$$
$$t+10=\pm\sqrt{75}$$
$$t+10=\pm\sqrt{25}\sqrt{3}$$
$$t+10=\pm 5\sqrt{3}$$
$$t=-10\pm 5\sqrt{3}$$
$$t\approx -1.34 \quad \text{and} \quad t\approx -18.66$$

53.

$$t^2+16t-16=0$$
$$t^2+16t=16$$
$$t^2+16t+\left(\frac{16}{2}\right)^2=16+\left(\frac{16}{2}\right)^2$$
$$t^2+16t+(8)^2=16+(8)^2$$
$$t^2+16t+64=16+64$$
$$(t+8)^2=80$$
$$t+8=\pm\sqrt{80}$$
$$t+8=\pm\sqrt{16}\sqrt{5}$$
$$t+8=\pm 4\sqrt{5}$$
$$t=-8\pm 4\sqrt{5}$$
$$t\approx 0.94 \quad \text{and} \quad t\approx -16.94$$

55.

$$2x^2 - x - 1 = 0$$
$$2x^2 - x = 1$$
$$\frac{2x^2}{2} - \frac{x}{2} = \frac{1}{2}$$
$$x^2 - \frac{1}{2}x = \frac{1}{2}$$
$$x^2 - \frac{1}{2}x + \left(\frac{1}{2} \bullet \frac{-1}{2}\right)^2 = \frac{1}{2} + \left(\frac{1}{2} \bullet \frac{-1}{2}\right)^2$$
$$x^2 - \frac{1}{2}x + \left(\frac{-1}{4}\right)^2 = \frac{1}{2} + \left(\frac{-1}{4}\right)^2$$
$$x^2 - \frac{1}{2}x + \frac{1}{16} = \frac{1}{2} + \frac{1}{16}$$
$$\left(x - \frac{1}{4}\right)^2 = \frac{8}{16} + \frac{1}{16}$$
$$\left(x - \frac{1}{4}\right)^2 = \frac{9}{16}$$
$$x - \frac{1}{4} = \pm\sqrt{\frac{9}{16}}$$
$$x - \frac{1}{4} = \pm\frac{3}{4}$$
$$x = \frac{1}{4} \pm \frac{3}{4}$$
$$x = \frac{1}{4} + \frac{3}{4} \quad \text{or} \quad x = \frac{1}{4} - \frac{3}{4}$$
$$x = \frac{4}{4} \qquad x = -\frac{2}{4}$$
$$x = 1 \qquad x = -\frac{1}{2}$$

57.

$$12t^2 - 5t - 3 = 0$$
$$12t^2 - 5t = 3$$
$$\frac{12t^2}{12} - \frac{5t}{12} = \frac{3}{12}$$
$$t^2 - \frac{5}{12}t = \frac{1}{4}$$
$$t^2 - \frac{5}{12}t + \left(\frac{1}{2} \bullet \frac{-5}{12}\right)^2 = \frac{1}{4} + \left(\frac{1}{2} \bullet \frac{-5}{12}\right)^2$$
$$t^2 - \frac{5}{12}t + \left(\frac{-5}{24}\right)^2 = \frac{1}{4} + \left(\frac{-5}{24}\right)^2$$
$$t^2 - \frac{5}{12}t + \frac{25}{576} = \frac{1}{4} + \frac{25}{576}$$
$$\left(t - \frac{5}{24}\right)^2 = \frac{144}{576} + \frac{25}{576}$$
$$\left(t - \frac{5}{24}\right)^2 = \frac{169}{576}$$
$$t - \frac{5}{24} = \pm\sqrt{\frac{169}{576}}$$
$$t - \frac{5}{24} = \pm\frac{13}{24}$$
$$t = \frac{5}{24} \pm \frac{13}{24}$$
$$t = \frac{5}{24} + \frac{13}{24} \quad \text{or} \quad t = \frac{5}{24} - \frac{13}{24}$$
$$t = \frac{18}{24} \qquad t = -\frac{8}{24}$$
$$t = \frac{3}{4} \qquad t = -\frac{1}{3}$$

59.

$$3x^2 - 12x + 1 = 0$$
$$\frac{3x^2}{3} - \frac{12x}{3} + \frac{1}{3} = \frac{0}{3}$$
$$x^2 - 4x + \frac{1}{3} = 0$$
$$x^2 - 4x = -\frac{1}{3}$$
$$x^2 - 4x + \left(\frac{1}{2} \bullet -4\right)^2 = -\frac{1}{3} + \left(\frac{1}{2} \bullet -4\right)^2$$
$$x^2 - 4x + (-2)^2 = -\frac{1}{3} + (-2)^2$$
$$x^2 - 4x + 4 = -\frac{1}{3} + 4$$
$$(x-2)^2 = -\frac{1}{3} + \frac{12}{3}$$
$$(x-2)^2 = \frac{11}{3}$$
$$x - 2 = \pm\sqrt{\frac{11}{3}}$$
$$x - 2 = \pm\frac{\sqrt{11}}{\sqrt{3}} \cdot \frac{\sqrt{3}}{\sqrt{3}}$$
$$x - 2 = \pm\frac{\sqrt{33}}{3}$$
$$x = 2 \pm \frac{\sqrt{33}}{3}$$
$$x = \frac{6}{3} \pm \frac{\sqrt{33}}{3}$$
$$x = \frac{6 \pm \sqrt{33}}{3}$$
$$x \approx 3.91 \quad \text{or} \quad x \approx 0.09$$

61.

$$2x^2 + 5x - 2 = 0$$
$$\frac{2x^2}{2} + \frac{5x}{2} - \frac{2}{2} = \frac{0}{2}$$
$$x^2 + \frac{5}{2}x - 1 = 0$$
$$x^2 + \frac{5}{2}x = 1$$
$$x^2 + \frac{5}{2}x + \left(\frac{1}{2} \bullet \frac{5}{2}\right)^2 = 1 + \left(\frac{1}{2} \bullet \frac{5}{2}\right)^2$$
$$x^2 + \frac{5}{2}x + \left(\frac{5}{4}\right)^2 = 1 + \left(\frac{5}{4}\right)^2$$
$$x^2 + \frac{5}{2}x + \frac{25}{16} = 1 + \frac{25}{16}$$
$$\left(x + \frac{5}{4}\right)^2 = \frac{16}{16} + \frac{25}{16}$$
$$\left(x + \frac{5}{4}\right)^2 = \frac{41}{16}$$
$$x + \frac{5}{4} = \pm\sqrt{\frac{41}{16}}$$
$$x + \frac{5}{4} = \pm\frac{\sqrt{41}}{4}$$
$$x = -\frac{5}{4} \pm \frac{\sqrt{41}}{4}$$
$$x = \frac{-5 \pm \sqrt{41}}{4}$$
$$x \approx 0.35 \quad \text{or} \quad x \approx -2.85$$

63.

$$p^2 + 2p + 2 = 0$$
$$p^2 + 2p = -2$$
$$p^2 + 2p + \left(\frac{2}{2}\right)^2 = -2 + \left(\frac{2}{2}\right)^2$$
$$p^2 + 2p + (1)^2 = -2 + (1)^2$$
$$p^2 + 2p + 1 = -2 + 1$$
$$(p+1)^2 = -1$$
$$p + 1 = \pm\sqrt{-1}$$
$$p + 1 = \pm i$$
$$p = -1 \pm i$$

65.

$$y^2+8y+18=0$$
$$y^2+8y=-18$$
$$y^2+8y+\left(\frac{8}{2}\right)^2=-18+\left(\frac{8}{2}\right)^2$$
$$y^2+8y+(4)^2=-18+(4)^2$$
$$y^2+8y+16=-18+16$$
$$(y+4)^2=-2$$
$$y+4=\pm\sqrt{-2}$$
$$y+4=\pm i\sqrt{2}$$
$$y=-4\pm i\sqrt{2}$$

67.

$$x^2+\frac{2}{3}x+7=0$$
$$x^2+\frac{2}{3}x=-7$$
$$x^2+\frac{2}{3}x+\left(\frac{1}{2}\bullet\frac{2}{3}\right)^2=-7+\left(\frac{1}{2}\bullet\frac{2}{3}\right)^2$$
$$x^2+\frac{2}{3}x+\left(\frac{1}{3}\right)^2=-7+\left(\frac{1}{3}\right)^2$$
$$x^2+\frac{2}{3}x+\frac{1}{9}=-7+\frac{1}{9}$$
$$\left(x+\frac{1}{3}\right)^2=-\frac{63}{9}+\frac{1}{9}$$
$$\left(x+\frac{1}{3}\right)^2=-\frac{62}{9}$$
$$x+\frac{1}{3}=\pm\sqrt{-\frac{62}{9}}$$
$$x+\frac{1}{3}=\pm\frac{i\sqrt{62}}{3}$$
$$x=-\frac{1}{3}\pm\frac{i\sqrt{62}}{3}$$

69.

$$a^2-\frac{1}{2}a+1=0$$
$$a^2-\frac{1}{2}a=-1$$
$$a^2-\frac{1}{2}a+\left(\frac{-1}{2}\bullet\frac{-1}{2}\right)^2=-1+\left(\frac{-1}{2}\bullet\frac{-1}{2}\right)^2$$
$$a^2-\frac{1}{2}a+\left(-\frac{1}{4}\right)^2=-1+\left(-\frac{1}{4}\right)^2$$
$$a^2-\frac{1}{2}a+\frac{1}{16}=-1+\frac{1}{16}$$
$$\left(a-\frac{1}{4}\right)^2=-\frac{16}{16}+\frac{1}{16}$$
$$\left(a-\frac{1}{4}\right)^2=-\frac{15}{16}$$
$$a-\frac{1}{4}=\pm\sqrt{-\frac{15}{16}}$$
$$a-\frac{1}{4}=\pm\frac{i\sqrt{15}}{4}$$
$$a=\frac{1}{4}\pm\frac{i\sqrt{15}}{4}$$

TRY IT YOURSELF

71.

$$(3x-1)^2=25$$
$$3x-1=\sqrt{25} \quad \text{or} \quad 3x-1=-\sqrt{25}$$
$$3x-1=5 \qquad 3x-1=-5$$
$$3x=6 \qquad 3x=-4$$
$$x=2 \qquad x=-\frac{4}{3}$$

73.

$$3x^2 - 6x = 1$$
$$\frac{3x^2}{3} - \frac{6x}{3} = \frac{1}{3}$$
$$x^2 - 2x = \frac{1}{3}$$
$$x^2 - 2x + \left(\frac{-2}{2}\right)^2 = \frac{1}{3} + \left(\frac{-2}{2}\right)^2$$
$$x^2 - 2x + 1 = \frac{1}{3} + 1$$
$$x^2 - 2x + 1 = \frac{1}{3} + \frac{3}{3}$$
$$(x-1)^2 = \frac{4}{3}$$
$$x - 1 = \pm\sqrt{\frac{4}{3}}$$
$$x - 1 = \pm\frac{2}{\sqrt{3}}$$
$$x - 1 = \pm\frac{2}{\sqrt{3}} \bullet \frac{\sqrt{3}}{\sqrt{3}}$$
$$x - 1 = \pm\frac{2\sqrt{3}}{3}$$
$$x = 1 \pm \frac{2\sqrt{3}}{3}$$
$$x = \frac{3}{3} \pm \frac{2\sqrt{3}}{3}$$
$$x = \frac{3 \pm 2\sqrt{3}}{3}$$
$x \approx 2.15$ ~~and~~ or $x \approx -0.15$

75.

$$x^2 + 8x + 6 = 0$$
$$x^2 + 8x = -6$$
$$x^2 + 8x + \left(\frac{8}{2}\right)^2 = -6 + \left(\frac{8}{2}\right)^2$$
$$x^2 + 8x + (4)^2 = -6 + (4)^2$$
$$x^2 + 8x + 16 = -6 + 16$$
$$(x+4)^2 = 10$$
$$x + 4 = \pm\sqrt{10}$$
$$x = -4 \pm \sqrt{10}$$
$x \approx -0.84$ ~~and~~ or $x \approx -7.16$

77.

$$6x^2 + 72 = 0$$
$$6x^2 = -72$$
$$\frac{6x^2}{6} = \frac{-72}{6}$$
$$x^2 = -12$$
$$x = \pm\sqrt{-12}$$
$$x = \pm i\sqrt{12}$$
$$x = \pm i\sqrt{4}\sqrt{3}$$
$$x = \pm 2i\sqrt{3}$$

79.

$$x^2 - 2x = 17$$
$$x^2 - 2x + \left(\frac{-2}{2}\right)^2 = 17 + \left(\frac{-2}{2}\right)^2$$
$$x^2 - 2x + (-1)^2 = 17 + (-1)^2$$
$$x^2 - 2x + 1 = 17 + 1$$
$$(x-1)^2 = 18$$
$$x - 1 = \pm\sqrt{18}$$
$$x - 1 = \pm\sqrt{9}\sqrt{2}$$
$$x - 1 = \pm 3\sqrt{2}$$
$$x = 1 \pm 3\sqrt{2}$$
$x \approx 5.24$ ~~and~~ or $x \approx -3.24$

81.

$$m^2 - 7m + 3 = 0$$
$$m^2 - 7m = -3$$
$$m^2 - 7m + \left(\frac{-7}{2}\right)^2 = -3 + \left(\frac{-7}{2}\right)^2$$
$$m^2 - 7m + \frac{49}{4} = -3 + \frac{49}{4}$$
$$m^2 - 7m + \frac{49}{4} = -\frac{12}{4} + \frac{49}{4}$$
$$\left(m - \frac{7}{2}\right)^2 = \frac{37}{4}$$
$$m - \frac{7}{2} = \pm\sqrt{\frac{37}{4}}$$
$$m - \frac{7}{2} = \pm\frac{\sqrt{37}}{2}$$
$$m = \frac{7}{2} \pm \frac{\sqrt{37}}{2}$$
$$m = \frac{7 \pm \sqrt{37}}{2}$$
$m \approx 6.54$ ~~and~~ or $m \approx 0.46$

83.

$$7h^2 = 35$$
$$\frac{7h^2}{7} = \frac{35}{7}$$
$$h^2 = 5$$
$$h = \pm\sqrt{5}$$
$h \approx 2.24$ ~~and~~ or $h \approx -2.24$

85.

$$\frac{7x + 1}{5} = -x^2$$
$$\frac{7x}{5} + \frac{1}{5} = -x^2$$
$$x^2 + \frac{7x}{5} = -\frac{1}{5}$$
$$x^2 + \frac{7}{5}x + \left(\frac{1}{2} \cdot \frac{7}{5}\right)^2 = -\frac{1}{5} + \left(\frac{1}{2} \cdot \frac{7}{5}\right)^2$$
$$x^2 + \frac{7}{5}x + \left(\frac{7}{10}\right)^2 = -\frac{1}{5} + \left(\frac{7}{10}\right)^2$$
$$x^2 + \frac{7}{5}x + \frac{49}{100} = -\frac{1}{5} + \frac{49}{100}$$
$$\left(x + \frac{7}{10}\right)^2 = -\frac{20}{100} + \frac{49}{100}$$
$$\left(x + \frac{7}{10}\right)^2 = \frac{29}{100}$$
$$x + \frac{7}{10} = \pm\sqrt{\frac{29}{100}}$$
$$x + \frac{7}{10} = \pm\frac{\sqrt{29}}{10}$$
$$x = -\frac{7}{10} \pm \frac{\sqrt{29}}{10}$$
$$x = \frac{-7 \pm \sqrt{29}}{10}$$
$x \approx -0.16$ ~~and~~ or $x \approx -1.24$

87.

$$t^2+t+3=0$$
$$t^2+t=-3$$
$$t^2+t+\left(\frac{1}{2}\right)^2=-3+\left(\frac{1}{2}\right)^2$$
$$t^2+t+\frac{1}{4}=-3+\frac{1}{4}$$
$$t^2+t+\frac{1}{4}=\frac{-12}{4}+\frac{1}{4}$$
$$\left(t+\frac{1}{2}\right)^2=-\frac{11}{4}$$
$$t+\frac{1}{2}=\pm\sqrt{-\frac{11}{4}}$$
$$t+\frac{1}{2}=\pm\frac{\sqrt{-11}}{2}$$
$$t+\frac{1}{2}=\pm\frac{\sqrt{11}}{2}i$$
$$t=-\frac{1}{2}\pm\frac{\sqrt{11}}{2}i$$

89.

$$(8x+5)^2=24$$
$$(8x+5)=\pm\sqrt{24}$$
$$8x+5=\pm\sqrt{4}\sqrt{6}$$
$$8x+5=\pm 2\sqrt{6}$$
$$8x=-5\pm 2\sqrt{6}$$
$$x=\frac{-5\pm 2\sqrt{6}}{8}$$
$$x\approx -0.01 \quad \text{and} \quad x\approx -1.24$$

91.

$$r^2-6r-27=0$$
$$r^2-6r=27$$
$$r^2-6r+\left(\frac{-6}{2}\right)^2=27+\left(\frac{-6}{2}\right)^2$$
$$r^2-6r+(-3)^2=27+(-3)^2$$
$$r^2-6r+9=27+9$$
$$(r-3)^2=36$$
$$r-3=\pm\sqrt{36}$$
$$r-3=\pm 6$$
$$r=3\pm 6$$
$$r=3+6 \quad \text{or} \quad r=3-6$$
$$r=9 \qquad r=-3$$

93.

$$4p^2+2p+3=0$$
$$4p^2+2p=-3$$
$$\frac{4p^2}{4}+\frac{2p}{4}=\frac{-3}{4}$$
$$p^2+\frac{1}{2}p=-\frac{3}{4}$$
$$p^2+\frac{1}{2}p+\left(\frac{1}{2}\bullet\frac{1}{2}\right)^2=-\frac{3}{4}+\left(\frac{1}{2}\bullet\frac{1}{2}\right)^2$$
$$p^2+\frac{1}{2}p+\left(\frac{1}{4}\right)^2=-\frac{3}{4}+\left(\frac{1}{4}\right)^2$$
$$p^2+\frac{1}{2}p+\frac{1}{16}=-\frac{3}{4}+\frac{1}{16}$$
$$\left(p+\frac{1}{4}\right)^2=-\frac{12}{16}+\frac{1}{16}$$
$$\left(p+\frac{1}{4}\right)^2=-\frac{11}{16}$$
$$p+\frac{1}{4}=\pm\sqrt{-\frac{11}{16}}$$
$$p+\frac{1}{4}=\pm\frac{\sqrt{-11}}{4}$$
$$p+\frac{1}{4}=\pm\frac{\sqrt{11}}{4}i$$
$$p=-\frac{1}{4}\pm\frac{\sqrt{11}}{4}i$$

LOOK ALIKES…

95. a.

$$x^2-24=0$$
$$x^2=24$$
$$x=\pm\sqrt{24}$$
$$x=\pm\sqrt{4}\sqrt{6}$$
$$x=\pm 2\sqrt{6}$$
$$x\approx\pm 4.90$$

b.

$$x^2+24=0$$
$$x^2=-24$$
$$x=\pm\sqrt{-24}$$
$$x=\pm\sqrt{-1}\sqrt{4}\sqrt{6}$$
$$x=\pm 2i\sqrt{6}$$

97. a.

$$2m^2 - 8m = 0$$
$$\frac{2m^2}{2} - \frac{8m}{2} = \frac{0}{2}$$
$$m^2 - 4m = 0$$
$$m^2 - 4m + \left(\frac{-4}{2}\right)^2 = 0 + \left(\frac{-4}{2}\right)^2$$
$$m^2 - 4m + 4 = 0 + 4$$
$$(m-2)^2 = 4$$
$$m - 2 = \pm 2$$
$$m - 2 = 2 \quad \text{and} \quad m - 2 = -2$$
$$m = 4 \quad \text{and} \quad m = 0$$

b.

$$2m^2 - 8m = 1$$
$$\frac{2m^2}{2} - \frac{8m}{2} = \frac{1}{2}$$
$$m^2 - 4m = \frac{1}{2}$$
$$m^2 - 4m + \left(\frac{-4}{2}\right)^2 = \frac{1}{2} + \left(\frac{-4}{2}\right)^2$$
$$m^2 - 4m + 4 = \frac{1}{2} + 4$$
$$(m-2)^2 = \frac{9}{2}$$
$$m - 2 = \pm\frac{3}{\sqrt{2}}$$
$$m - 2 = \pm\frac{3}{\sqrt{2}}\left(\frac{\sqrt{2}}{\sqrt{2}}\right)$$
$$m - 2 = \pm\frac{3\sqrt{2}}{2}$$
$$m = 2 \pm \frac{3\sqrt{2}}{2}$$
$$m = \frac{4}{2} \pm \frac{3\sqrt{2}}{2}$$
$$m = \frac{4 \pm 3\sqrt{2}}{2}$$
$$m \approx 4.12 \quad \text{and} \quad m \approx -0.12$$

99. a.

$$x^2 - 4x + 20 = 0$$
$$x^2 - 4x = -20$$
$$x^2 - 4x + \left(\frac{-4}{2}\right)^2 = -20 + \left(\frac{-4}{2}\right)^2$$
$$x^2 - 4x + 4 = -20 + 4$$
$$(x-2)^2 = -16$$
$$x - 2 = \pm\sqrt{-16}$$
$$x - 2 = \pm 4i$$
$$x = 2 \pm 4i$$

b.

$$x^2 - 4x - 20 = 0$$
$$x^2 - 4x = 20$$
$$x^2 - 4x + \left(\frac{-4}{2}\right)^2 = 20 + \left(\frac{-4}{2}\right)^2$$
$$x^2 - 4x + 4 = 20 + 4$$
$$(x-2)^2 = 24$$
$$x - 2 = \pm\sqrt{24}$$
$$x - 2 = \pm 2\sqrt{6}$$
$$x = 2 \pm 2\sqrt{6}$$
$$x \approx 6.90 \quad \text{and} \quad x \approx -2.90$$

101. a.

$$2r^2 - 4r + 3 = 0$$
$$2r^2 - 4r = -3$$
$$\frac{2r^2}{2} - \frac{4r}{2} = \frac{-3}{2}$$
$$r^2 - 2r = -\frac{3}{2}$$
$$r^2 - 2r + \left(\frac{-2}{2}\right)^2 = -\frac{3}{2} + \left(\frac{-2}{2}\right)^2$$
$$r^2 - 2r + 1 = -\frac{3}{2} + 1$$
$$(r-1)^2 = -\frac{1}{2}$$
$$r - 1 = \pm\sqrt{-\frac{1}{2}}$$
$$r - 1 = \pm i\frac{1}{\sqrt{2}}$$
$$r - 1 = \pm i\frac{1}{\sqrt{2}}\left(\frac{\sqrt{2}}{\sqrt{2}}\right)$$
$$r - 1 = \pm\frac{i\sqrt{2}}{2}$$
$$r = 1 \pm \frac{i\sqrt{2}}{2}$$
$$r = \frac{2}{2} \pm \frac{i\sqrt{2}}{2}$$
$$r = \frac{2 \pm i\sqrt{2}}{2}$$

b.

$$2r^2 - 4r - 3 = 0$$
$$2r^2 - 4r = 3$$
$$\frac{2r^2}{2} - \frac{4r}{2} = \frac{3}{2}$$
$$r^2 - 2r = \frac{3}{2}$$
$$r^2 - 2r + \left(\frac{-2}{2}\right)^2 = \frac{3}{2} + \left(\frac{-2}{2}\right)^2$$
$$r^2 - 2r + 1 = \frac{3}{2} + 1$$
$$(r-1)^2 = \frac{5}{2}$$
$$r - 1 = \pm\sqrt{\frac{5}{2}}$$
$$r - 1 = \pm\sqrt{\frac{5}{2}}\left(\frac{\sqrt{2}}{\sqrt{2}}\right)$$
$$r - 1 = \pm\frac{\sqrt{10}}{2}$$
$$r = 1 \pm \frac{\sqrt{10}}{2}$$
$$r = \frac{2}{2} \pm \frac{\sqrt{10}}{2}$$
$$r = \frac{2 \pm \sqrt{10}}{2}$$
$$r \approx 2.58 \quad \text{and} \quad r \approx -0.58$$

APPLICATIONS

103. MOVIE STUNTS

$$d = 16t^2$$
$$312 = 16t^2$$
$$\frac{312}{16} = \frac{16t^2}{16}$$
$$19.5 = t^2$$
$$t = \sqrt{19.5}$$
$$t = 4.4 \text{ sec}$$

105. ACCIDENTS

$$h = s - 16t^2$$
$$5 = 4(12) - 16t^2$$
$$5 = 48 - 16t^2$$
$$-43 = -16t^2$$
$$2.6875 = t^2$$
$$t = \sqrt{2.6875}$$
$$t = 1.6 \text{ sec}$$

107. AUTOMOBILE ENGINES

$$V = \pi r^2 h$$
$$47.75 = \pi r^2 (5.25)$$
$$\frac{47.75}{5.25\pi} = \frac{\pi r^2 (5.25)}{5.25\pi}$$
$$\frac{47.75}{16.5} = r^2$$
$$2.89 = r^2$$
$$r = \sqrt{2.89}$$
$$r = 1.70 \text{ in.}$$

109. PHYSICS

$$E = mc^2$$
$$\frac{E}{m} = \frac{mc^2}{m}$$
$$\frac{E}{m} = c^2$$
$$c^2 = \frac{E}{m}$$
$$c = \sqrt{\frac{E}{m}}$$
$$c = \frac{\sqrt{E}}{\sqrt{m}}\left(\frac{\sqrt{m}}{\sqrt{m}}\right)$$
$$c = \frac{\sqrt{Em}}{m}$$

WRITING

111. Answers will vary.

113. Answers will vary.

REVIEW

115.

$$\sqrt[3]{40a^3b^6} = \sqrt[3]{8a^3b^6}\sqrt[3]{5}$$
$$= 2ab^2\sqrt[3]{5}$$

117.

$$\sqrt[4]{\frac{16}{625}} = \sqrt[4]{\left(\frac{2}{5}\right)^4}$$
$$= \frac{2}{5}$$

CHALLENGE PROBLEMS

119. Take one-half of the coefficient of x and square it: $\left(\frac{\sqrt{3}}{2}\right)^2 = \frac{3}{4}$.

SECTION 8.2

VOCABULARY

1. The standard form of a **quadratic** equation is $ax^2 + bx + c = 0$.

CONCEPTS

3. a. $x^2 + 2x + 5 = 0$
   b. $3x^2 + 2x - 1 = 0$

5. a. true
   b. true
   c. false

7. a.

$$\frac{-2 \pm \sqrt{2^2 - 4(1)(-8)}}{2(1)} = \frac{-2 \pm \sqrt{4+32}}{2} = \frac{-2 \pm \sqrt{36}}{2} = \frac{-2 \pm 6}{2}$$

$$x = \frac{-2+6}{2} \quad \text{or} \quad x = \frac{-2-6}{2}$$

$$x = \frac{4}{2} \qquad x = \frac{-8}{2}$$

$$x = 2 \qquad x = -4$$

b.

$$\frac{-(-1) \pm \sqrt{(-1)^2 - 4(2)(-4)}}{2(2)} = \frac{1 \pm \sqrt{1+32}}{4} = \frac{1 \pm \sqrt{33}}{4}$$

9. a.

$$\frac{3 \pm 6\sqrt{2}}{3} = \frac{\cancel{3}(1 \pm 2\sqrt{2})}{\cancel{3}} = \frac{1 \pm 2\sqrt{2}}{1} = 1 \pm 2\sqrt{2}$$

b.

$$\frac{-12 \pm 4\sqrt{7}}{8} = \frac{\cancel{4}(-3 \pm \sqrt{7})}{\cancel{4} \cdot 2} = \frac{-3 \pm \sqrt{7}}{2}$$

NOTATION

11. a. The fraction bar wasn't drawn under both parts of the numerator.
    b. A ± sign wasn't written between b and the radical.

GUIDED PRACTICE

13. $x^2 - 3x + 2 = 0$

$a = 1$, $b = -3$, and $c = 2$

$$x = \frac{-b \pm \sqrt{b^2 - 4ac}}{2a} = \frac{-(-3) \pm \sqrt{(-3)^2 - 4(1)(2)}}{2(1)} = \frac{3 \pm \sqrt{9-8}}{2} = \frac{3 \pm \sqrt{1}}{2} = \frac{3 \pm 1}{2}$$

$$x = \frac{3+1}{2} \quad \text{or} \quad x = \frac{3-1}{2}$$

$$x = \frac{4}{2} \qquad x = \frac{2}{2}$$

$$x = 2 \qquad x = 1$$

15. $x^2 + 12x = -36$

$x^2 + 12x + 36 = 0$

$a = 1$, $b = 12$, and $c = 36$

$$x = \frac{-b \pm \sqrt{b^2 - 4ac}}{2a}$$
$$= \frac{-12 \pm \sqrt{(12)^2 - 4(1)(36)}}{2(1)}$$
$$= \frac{-12 \pm \sqrt{144 - 144}}{2}$$
$$= \frac{-12 \pm \sqrt{0}}{2}$$
$$= \frac{-12 \pm 0}{2}$$
$$x = \frac{-12+0}{2} \quad \text{or} \quad x = \frac{-12-0}{2}$$
$$x = \frac{-12}{2} \qquad x = \frac{-12}{2}$$
$$x = -6 \qquad x = -6$$

A repeated solution of -6

17. $2x^2 + x - 3 = 0$

$a = 2$, $b = 1$, and $c = -3$

$$x = \frac{-b \pm \sqrt{b^2 - 4ac}}{2a}$$
$$= \frac{-1 \pm \sqrt{(1)^2 - 4(2)(-3)}}{2(2)}$$
$$= \frac{-1 \pm \sqrt{1-(-24)}}{4}$$
$$= \frac{-1 \pm \sqrt{25}}{4}$$
$$= \frac{-1 \pm 5}{4}$$
$$x = \frac{-1+5}{4} \quad \text{or} \quad x = \frac{-1-5}{4}$$
$$x = \frac{4}{4} \qquad x = \frac{-6}{4}$$
$$x = 1 \qquad x = -\frac{3}{2}$$

19. $12t^2 - 5t - 2 = 0$

$a = 12$, $b = -5$, and $c = -2$

$$t = \frac{-b \pm \sqrt{b^2 - 4ac}}{2a}$$
$$= \frac{-(-5) \pm \sqrt{(-5)^2 - 4(12)(-2)}}{2(12)}$$
$$= \frac{5 \pm \sqrt{25 + 96}}{24}$$
$$= \frac{5 \pm \sqrt{121}}{24}$$
$$= \frac{5 \pm 11}{24}$$
$$t = \frac{5+11}{24} \quad \text{or} \quad t = \frac{5-11}{24}$$
$$t = \frac{16}{24} \qquad t = \frac{-6}{24}$$
$$t = \frac{2}{3} \qquad t = -\frac{1}{4}$$

21. $x^2 = x + 7$

$x^2 - x - 7 = 0$

$a = 1$, $b = -1$, and $c = -7$

$$x = \frac{-b \pm \sqrt{b^2 - 4ac}}{2a}$$
$$= \frac{-(-1) \pm \sqrt{(-1)^2 - 4(1)(-7)}}{2(1)}$$
$$= \frac{1 \pm \sqrt{1 + 28}}{2}$$
$$= \frac{1 \pm \sqrt{29}}{2}$$
$$x = \frac{1+\sqrt{29}}{2} \quad \text{or} \quad x = \frac{1-\sqrt{29}}{2}$$
$$x \approx 3.19 \qquad x \approx -2.19$$

23. $5x^2 + 5x = -1$

$5x^2 + 5x + 1 = 0$

$a = 5$, $b = 5$, and $c = 1$

$$x = \frac{-b \pm \sqrt{b^2 - 4ac}}{2a}$$

$$= \frac{-5 \pm \sqrt{(5)^2 - 4(5)(1)}}{2(5)}$$

$$= \frac{-5 \pm \sqrt{25 - 20}}{10}$$

$$x = \frac{-5 \pm \sqrt{5}}{10}$$

$$x = \frac{-5 + \sqrt{5}}{10} \quad \text{or} \quad x = \frac{-5 - \sqrt{5}}{10}$$

$$x \approx -0.28 \qquad x \approx -0.72$$

25. $3y^2 + 1 = -6y$

$3y^2 + 6y + 1 = 0$

$a = 3$, $b = 6$, and $c = 1$

$$y = \frac{-b \pm \sqrt{b^2 - 4ac}}{2a}$$

$$= \frac{-6 \pm \sqrt{(6)^2 - 4(3)(1)}}{2(3)}$$

$$= \frac{-6 \pm \sqrt{36 - 12}}{6}$$

$$= \frac{-6 \pm \sqrt{24}}{6}$$

$$= \frac{-6 \pm \sqrt{4}\sqrt{6}}{6}$$

$$= \frac{-6 \pm 2\sqrt{6}}{6}$$

$$= \frac{\cancel{2}\left(-3 \pm \sqrt{6}\right)}{\cancel{2} \cdot 3}$$

$$y = \frac{-3 \pm \sqrt{6}}{3}$$

$$y = \frac{-3 + \sqrt{6}}{3} \quad \text{or} \quad y = \frac{-3 - \sqrt{6}}{3}$$

$$y \approx -0.18 \qquad y \approx -1.82$$

27. $4m^2 = 4m + 19$

$4m^2 - 4m - 19 = 0$

$a = 4$, $b = -4$, and $c = -19$

$$m = \frac{-b \pm \sqrt{b^2 - 4ac}}{2a}$$

$$= \frac{-(-4) \pm \sqrt{(-4)^2 - 4(4)(-19)}}{2(4)}$$

$$= \frac{4 \pm \sqrt{16 + 304}}{8}$$

$$= \frac{4 \pm \sqrt{320}}{8}$$

$$= \frac{4 \pm \sqrt{64}\sqrt{5}}{8}$$

$$= \frac{4 \pm 8\sqrt{5}}{8}$$

$$= \frac{\cancel{4}\left(1 \pm 2\sqrt{5}\right)}{\cancel{4} \cdot 2}$$

$$m = \frac{1 \pm 2\sqrt{5}}{2}$$

$$m = \frac{1 + 2\sqrt{5}}{2} \quad \text{or} \quad m = \frac{1 - 2\sqrt{5}}{2}$$

$$m \approx 2.74 \qquad m \approx -1.74$$

29. $2x^2 + x + 1 = 0$

$a = 2$, $b = 1$, and $c = 1$

$$x = \frac{-b \pm \sqrt{b^2 - 4ac}}{2a}$$

$$= \frac{-1 \pm \sqrt{(1)^2 - 4(2)(1)}}{2(2)}$$

$$= \frac{-1 \pm \sqrt{1 - 8}}{4}$$

$$= \frac{-1 \pm \sqrt{-7}}{4}$$

$$= \frac{-1 \pm i\sqrt{7}}{4}$$

$$x = -\frac{1}{4} \pm \frac{\sqrt{7}}{4}i$$

31. $3x^2 - 2x + 1 = 0$

$a = 3, b = -2, \text{ and } c = 1$

$$x = \frac{-b \pm \sqrt{b^2 - 4ac}}{2a}$$
$$= \frac{-(-2) \pm \sqrt{(-2)^2 - 4(3)(1)}}{2(3)}$$
$$= \frac{2 \pm \sqrt{4-12}}{6}$$
$$= \frac{2 \pm \sqrt{-8}}{6}$$
$$= \frac{2 \pm \sqrt{-4}\sqrt{2}}{6}$$
$$= \frac{2 \pm 2i\sqrt{2}}{6}$$
$$= \frac{2}{6} \pm \frac{2\sqrt{2}}{6}i$$
$$x = \frac{1}{3} \pm \frac{\sqrt{2}}{3}i$$

33. $x^2 - 2x + 2 = 0$

$a = 1, b = -2, \text{ and } c = 2$

$$x = \frac{-b \pm \sqrt{b^2 - 4ac}}{2a}$$
$$= \frac{-(-2) \pm \sqrt{(-2)^2 - 4(1)(2)}}{2(1)}$$
$$= \frac{2 \pm \sqrt{4-8}}{2}$$
$$= \frac{2 \pm \sqrt{-4}}{2}$$
$$= \frac{2 \pm 2i}{2}$$
$$= \frac{2}{2} \pm \frac{2}{2}i$$
$$x = 1 \pm i$$

35. $4a^2 + 4a + 5 = 0$

$a = 4, b = 4, \text{ and } c = 5$

$$a = \frac{-b \pm \sqrt{b^2 - 4ac}}{2a}$$
$$= \frac{-4 \pm \sqrt{(4)^2 - 4(4)(5)}}{2(4)}$$
$$= \frac{-4 \pm \sqrt{16-80}}{8}$$
$$= \frac{-4 \pm \sqrt{-64}}{8}$$
$$= \frac{-4 \pm 8i}{8}$$
$$= -\frac{4}{8} \pm \frac{8}{8}i$$
$$a = -\frac{1}{2} \pm i$$

37. a. $-5x^2 + 9x - 2 = 0$

divide each term by –1

$5x^2 - 9x + 2 = 0$

b. $1.6t^2 + 2.4t - 0.9 = 0$

multiply each term to 10

$16t^2 + 24t - 9 = 0$

39. a. $45x^2 + 30x - 15 = 0$

divide each term by 15

$3x^2 + 2x - 1 = 0$

b. $\frac{1}{3}m^2 - \frac{1}{2}m - \frac{1}{3} = 0$

multiply each term by 6

$2m^2 - 3m - 2 = 0$

TRY IT YOURSELF

41.

$$x^2 - \frac{14}{15}x = \frac{8}{15}$$

$$15\left(x^2 - \frac{14}{15}x\right) = 15\left(\frac{8}{15}\right)$$

$$15x^2 - 14x = 8$$

$$15x^2 - 14x - 8 = 0$$

$a = 15$, $b = -14$, and $c = -8$

$$x = \frac{-b \pm \sqrt{b^2 - 4ac}}{2a}$$

$$= \frac{-(-14) \pm \sqrt{(-14)^2 - 4(15)(-8)}}{2(15)}$$

$$= \frac{14 \pm \sqrt{196 - (-480)}}{30}$$

$$= \frac{14 \pm \sqrt{676}}{30}$$

$$= \frac{14 \pm 26}{30}$$

$$x = \frac{14 + 26}{30} \quad \text{or} \quad x = \frac{14 - 26}{30}$$

$$x = \frac{40}{30} \qquad x = -\frac{12}{30}$$

$$x = \frac{4}{3} \qquad x = -\frac{2}{5}$$

43. $3x^2 - 4x = -2$

$3x^2 - 4x + 2 = 0$

$a = 3$, $b = -4$, and $c = 2$

$$x = \frac{-b \pm \sqrt{b^2 - 4ac}}{2a}$$

$$= \frac{-(-4) \pm \sqrt{(-4)^2 - 4(3)(2)}}{2(3)}$$

$$= \frac{4 \pm \sqrt{16 - 24}}{6}$$

$$= \frac{4 \pm \sqrt{-8}}{6}$$

$$= \frac{4 \pm \sqrt{-4}\sqrt{2}}{6}$$

$$= \frac{4 \pm 2i\sqrt{2}}{6}$$

$$= \frac{4}{6} \pm \frac{2\sqrt{2}}{6}i$$

$$x = \frac{2}{3} \pm \frac{\sqrt{2}}{3}i$$

45. $-16y^2 - 8y + 3 = 0$

$0 = 16y^2 + 8y - 3$

$a = 16$, $b = 8$, $c = -3$

$$y = \frac{-b \pm \sqrt{b^2 - 4ac}}{2a}$$

$$= \frac{-8 \pm \sqrt{(8)^2 - 4(16)(-3)}}{2(16)}$$

$$= \frac{-8 \pm \sqrt{64 - (-192)}}{32}$$

$$= \frac{-8 \pm \sqrt{256}}{32}$$

$$= \frac{-8 \pm 16}{32}$$

$$y = \frac{-8 + 16}{32} \quad \text{or} \quad y = \frac{-8 - 16}{32}$$

$$y = \frac{8}{32} \qquad y = \frac{-24}{32}$$

$$y = \frac{1}{4} \qquad y = -\frac{3}{4}$$

47. $2x^2 - 3x - 1 = 0$

$a = 2$, $b = -3$, and $c = -1$

$$x = \frac{-b \pm \sqrt{b^2 - 4ac}}{2a}$$
$$= \frac{-(-3) \pm \sqrt{(-3)^2 - 4(2)(-1)}}{2(2)}$$
$$= \frac{3 \pm \sqrt{9-(-8)}}{4}$$
$$x = \frac{3 \pm \sqrt{17}}{4}$$
$$x = \frac{3+\sqrt{17}}{4} \quad \text{or} \quad x = \frac{3-\sqrt{17}}{4}$$
$$x = 1.78 \qquad x = -0.28$$

49. $-x^2 + 10x = 18$

$0 = x^2 - 10x + 18$

$a = 1$, $b = -10$, and $c = 18$

$$x = \frac{-b \pm \sqrt{b^2 - 4ac}}{2a}$$
$$= \frac{-(-10) \pm \sqrt{(-10)^2 - 4(1)(18)}}{2(1)}$$
$$= \frac{10 \pm \sqrt{100-72}}{2}$$
$$= \frac{10 \pm \sqrt{28}}{2}$$
$$= \frac{10 \pm \sqrt{4}\sqrt{7}}{2}$$
$$= \frac{10 \pm 2\sqrt{7}}{2}$$
$$= \frac{2(5 \pm \sqrt{7})}{2}$$
$$x = 5 \pm \sqrt{7}$$
$$x = 5+\sqrt{7} \quad \text{or} \quad x = 5-\sqrt{7}$$
$$x = 2.35 \qquad x = 7.65$$

51. $x(x-6) = 391$

$x^2 - 6x = 391$

$x^2 - 6x - 391 = 0$

$a = 1$, $b = -6$, and $c = -391$

$$x = \frac{-b \pm \sqrt{b^2 - 4ac}}{2a}$$
$$= \frac{-(-6) \pm \sqrt{(-6)^2 - 4(1)(-391)}}{2(1)}$$
$$= \frac{6 \pm \sqrt{36-(-1{,}564)}}{2}$$
$$= \frac{6 \pm \sqrt{1{,}600}}{2}$$
$$= \frac{6 \pm 40}{2}$$
$$x = \frac{6+40}{2} \quad \text{or} \quad x = \frac{6-40}{2}$$
$$x = \frac{46}{2} \qquad x = \frac{-34}{2}$$
$$x = 23 \qquad x = -17$$

53. $x^2 + 5x - 5 = 0$

$a = 1$, $b = 5$, and $c = -5$

$$x = \frac{-b \pm \sqrt{b^2 - 4ac}}{2a}$$
$$= \frac{-5 \pm \sqrt{(5)^2 - 4(1)(-5)}}{2(1)}$$
$$= \frac{-5 \pm \sqrt{25+20}}{2}$$
$$= \frac{-5 \pm \sqrt{45}}{2}$$
$$= \frac{-5 \pm \sqrt{9}\sqrt{5}}{2}$$
$$x = \frac{-5 \pm 3\sqrt{5}}{2}$$
$$x = \frac{-5+3\sqrt{5}}{2} \quad \text{or} \quad x = \frac{-5-3\sqrt{5}}{2}$$
$$x = 0.85 \qquad x = -5.85$$

55. $9h^2 - 6h + 7 = 0$

$a = 9$, $b = -6$, and $c = 7$

$$h = \frac{-b \pm \sqrt{b^2 - 4ac}}{2a}$$

$$= \frac{-(-6) \pm \sqrt{(-6)^2 - 4(9)(7)}}{2(9)}$$

$$= \frac{6 \pm \sqrt{36 - 252}}{18}$$

$$= \frac{6 \pm \sqrt{-216}}{18}$$

$$= \frac{6 \pm i\sqrt{36}\sqrt{6}}{18}$$

$$h = \frac{6 \pm 6\sqrt{6}i}{18}$$

$$= \frac{6}{18} \pm \frac{6\sqrt{6}}{18}i$$

$$= \frac{1}{3} \pm \frac{\sqrt{6}}{3}i$$

57.

$$50x^2 + 30x - 10 = 0$$

$$\frac{50x^2}{10} + \frac{30x}{10} - \frac{10}{10} = 0$$

$$5x^2 + 3x - 1 = 0$$

$a = 5$, $b = 3$, and $c = -1$

$$x = \frac{-b \pm \sqrt{b^2 - 4ac}}{2a}$$

$$= \frac{-3 \pm \sqrt{(3)^2 - 4(5)(-1)}}{2(5)}$$

$$= \frac{-3 \pm \sqrt{9 - (-20)}}{10}$$

$$x = \frac{-3 \pm \sqrt{29}}{10}$$

$x = \dfrac{-3 - \sqrt{29}}{10}$ or $x = \dfrac{-3 - \sqrt{29}}{10}$

$x = 0.24$ $\qquad$ $x = -0.84$

59.

$$0.6x^2 + 0.03 - 0.4x = 0$$

$$0.6x^2 - 0.4x + 0.03 = 0$$

$$100(0.6x^2 - 0.4x + 0.03) = 100(0)$$

$$60x^2 - 40x + 3 = 0$$

$a = 60$, $b = -40$, and $c = 3$

$$x = \frac{-b \pm \sqrt{b^2 - 4ac}}{2a}$$

$$= \frac{-(-40) \pm \sqrt{(-40)^2 - 4(60)(3)}}{2(60)}$$

$$= \frac{40 \pm \sqrt{1,600 - 720}}{120}$$

$$= \frac{40 \pm \sqrt{880}}{120}$$

$$= \frac{40 \pm \sqrt{16}\sqrt{55}}{120}$$

$$= \frac{40 \pm 4\sqrt{55}}{120}$$

$$= \frac{\cancel{4}(10 \pm \sqrt{55})}{\cancel{4} \cdot 30}$$

$$x = \frac{10 \pm \sqrt{55}}{30}$$

$x = \dfrac{10 + \sqrt{55}}{30}$ or $x = \dfrac{10 - \sqrt{55}}{30}$

$x = 0.58$ $\qquad$ $x = 0.09$

61.

$$\frac{1}{8}x^2-\frac{1}{2}x+1=0$$
$$8\left(\frac{1}{8}x^2-\frac{1}{2}x+1\right)=8(0)$$
$$x^2-4x+8=0$$
$a=1$, $b=-4$, and $c=8$
$$x=\frac{-b\pm\sqrt{b^2-4ac}}{2a}$$
$$=\frac{-(-4)\pm\sqrt{(-4)^2-4(1)(8)}}{2(1)}$$
$$=\frac{4\pm\sqrt{16-32}}{2}$$
$$=\frac{4\pm\sqrt{-16}}{2}$$
$$=\frac{4\pm 4i}{2}$$
$$=\frac{4}{2}\pm\frac{4}{2}i$$
$$x=2\pm 2i$$

63.

$$\frac{a^2}{10}-\frac{3a}{5}+\frac{7}{5}=0$$
$$10\left(\frac{a^2}{10}-\frac{3a}{5}+\frac{7}{5}\right)=10(0)$$
$$a^2-6a+14=0$$
$a=1$, $b=-6$, and $c=14$
$$a=\frac{-b\pm\sqrt{b^2-4ac}}{2a}$$
$$=\frac{-(-6)\pm\sqrt{(-6)^2-4(1)(14)}}{2(1)}$$
$$=\frac{6\pm\sqrt{36-56}}{2}$$
$$=\frac{6\pm\sqrt{-20}}{2}$$
$$=\frac{6\pm\sqrt{-4}\sqrt{5}}{2}$$
$$=\frac{6\pm 2i\sqrt{5}}{2}$$
$$=\frac{6}{2}\pm\frac{2\sqrt{5}}{2}i$$
$$a=3\pm i\sqrt{5}$$

65.

$$\frac{x^2}{2}+\frac{5}{2}x=-1$$
$$2\left(\frac{x^2}{2}+\frac{5}{2}x\right)=2(-1)$$
$$x^2+5x=-2$$
$$x^2+5x+2=0$$
$a=1$, $b=5$, and $c=2$
$$x=\frac{-b\pm\sqrt{b^2-4ac}}{2a}$$
$$=\frac{-5\pm\sqrt{(5)^2-4(1)(2)}}{2(1)}$$
$$=\frac{-5\pm\sqrt{25-8}}{2}$$
$$x=\frac{-5\pm\sqrt{17}}{2}$$
$$x=\frac{-5+\sqrt{17}}{2} \quad \text{or} \quad x=\frac{-5-\sqrt{17}}{2}$$
$$x=-0.44 \qquad x=-4.56$$

67.

$$900x^2-8{,}100x=1{,}800$$
$$\frac{900x^2}{900}-\frac{8{,}100x}{900}=\frac{1{,}800}{900}$$
$$x^2-9x=2$$
$$x^2-9x-2=0$$
$a=1$, $b=-9$, and $c=-2$
$$x=\frac{-b\pm\sqrt{b^2-4ac}}{2a}$$
$$=\frac{-(-9)\pm\sqrt{(-9)^2-4(1)(-2)}}{2(1)}$$
$$=\frac{9\pm\sqrt{81-(-8)}}{2}$$
$$x=\frac{9\pm\sqrt{89}}{2}$$
$$x=\frac{9+\sqrt{89}}{2} \quad \text{or} \quad x=\frac{9-\sqrt{89}}{2}$$
$$x=9.22 \qquad x=-0.22$$

69.

$$\frac{1}{4}x^2 - \frac{1}{6}x - \frac{1}{6} = 0$$

$$12\left(\frac{1}{4}x^2 - \frac{1}{6}x - \frac{1}{6}\right) = 12(0)$$

$$3x^2 - 2x - 2 = 0$$

$a = 3$, $b = -2$, and $c = -2$

$$x = \frac{-b \pm \sqrt{b^2 - 4ac}}{2a}$$

$$= \frac{-(-2) \pm \sqrt{(-2)^2 - 4(3)(-2)}}{2(3)}$$

$$= \frac{2 \pm \sqrt{4 - (-24)}}{6}$$

$$= \frac{2 \pm \sqrt{28}}{6}$$

$$= \frac{2 \pm \sqrt{4}\sqrt{7}}{6}$$

$$= \frac{2 \pm 2\sqrt{7}}{6}$$

$$= \frac{1 \pm \sqrt{7}}{3}$$

$x = \frac{1 + \sqrt{7}}{3}$ or $x = \frac{1 - \sqrt{7}}{3}$

$x = 1.22$ $\quad$ $x = -0.55$

71.

$$f(x) = f(x)$$

$$0.7x^2 - 3.5x = 25$$

$$0.7x^2 - 3.5x - 25 = 0$$

$$10(0.7x^2 - 3.5x - 25) = 10(0)$$

$$7x^2 - 35x - 250 = 0$$

$a = 7$, $b = -35$, and $c = -250$

$$x = \frac{-b \pm \sqrt{b^2 - 4ac}}{2a}$$

$$= \frac{-(-35) \pm \sqrt{(-35)^2 - 4(7)(-250)}}{2(7)}$$

$$= \frac{35 \pm \sqrt{1,225 - (-7,000)}}{14}$$

$$= \frac{35 \pm \sqrt{8,225}}{14}$$

$$= \frac{35 \pm \sqrt{25}\sqrt{329}}{14}$$

$$= \frac{35 \pm 5\sqrt{329}}{14}$$

$x = \frac{35 + 5\sqrt{329}}{14}$ or $x = \frac{35 - 5\sqrt{329}}{14}$

$x = 8.98$ $\quad$ $x = -3.98$

73. a. $a^2 + 4a - 7 = 0$

$a = 1$, $b = 4$, and $c = -7$

$$a = \frac{-b \pm \sqrt{b^2 - 4ac}}{2a}$$
$$= \frac{-4 \pm \sqrt{(4)^2 - 4(1)(-7)}}{2(1)}$$
$$= \frac{-4 \pm \sqrt{16+28}}{2}$$
$$= \frac{-4 \pm \sqrt{44}}{2}$$
$$= \frac{-4 \pm \sqrt{4}\sqrt{11}}{2}$$
$$= \frac{-4 \pm 2\sqrt{11}}{2}$$
$$= \frac{2\left(-2 \pm \sqrt{11}\right)}{2}$$
$$a = -2 \pm \sqrt{11}$$

$a = -2 + \sqrt{11}$ or $a = -2 - \sqrt{11}$

$a = 1.32$ $\quad$ $a = -5.32$

b. $a^2 - 4a - 7 = 0$

$a = 1$, $b = -4$, and $c = -7$

$$a = \frac{-b \pm \sqrt{b^2 - 4ac}}{2a}$$
$$= \frac{-(-4) \pm \sqrt{(-4)^2 - 4(1)(-7)}}{2(1)}$$
$$= \frac{4 \pm \sqrt{16+28}}{2}$$
$$= \frac{4 \pm \sqrt{44}}{2}$$
$$= \frac{4 \pm \sqrt{4}\sqrt{11}}{2}$$
$$= \frac{4 \pm 2\sqrt{11}}{2}$$
$$= \frac{2\left(2 \pm \sqrt{11}\right)}{2}$$
$$a = 2 \pm \sqrt{11}$$

$a = 2 + \sqrt{11}$ or $a = 2 - \sqrt{11}$

$a = 5.32$ $\quad$ $a = -1.32$

75. a. $(x + 2)(x - 4) = 16$

$x^2 - 4x + 2x - 8 = 16$

$x^2 - 2x - 8 = 16$

$x^2 - 2x - 24 = 0$

$a = 1$, $b = -2$, and $c = -24$

$$x = \frac{-b \pm \sqrt{b^2 - 4ac}}{2a}$$
$$= \frac{-(-2) \pm \sqrt{(-2)^2 - 4(1)(-24)}}{2(1)}$$
$$= \frac{2 \pm \sqrt{4+96}}{2}$$
$$= \frac{2 \pm \sqrt{100}}{2}$$
$$x = \frac{2 \pm 10}{2}$$

$x = \frac{2+10}{2}$ or $x = \frac{2-10}{2}$

$x = \frac{12}{2}$ $\quad$ $x = \frac{-8}{2}$

$x = 6$ $\quad$ $x = -4$

b. $(x + 2)(x - 4) = -16$

$x^2 - 4x + 2x - 8 = -16$

$x^2 - 2x - 8 = -16$

$x^2 - 2x + 8 = 0$

$a = 1$, $b = -2$, and $c = 8$

$$x = \frac{-b \pm \sqrt{b^2 - 4ac}}{2a}$$
$$= \frac{-(-2) \pm \sqrt{(-2)^2 - 4(1)(8)}}{2(1)}$$
$$= \frac{2 \pm \sqrt{4-32}}{2}$$
$$= \frac{2 \pm \sqrt{-28}}{2}$$
$$= \frac{2 \pm i\sqrt{4}\sqrt{7}}{2}$$
$$= \frac{2 \pm 2i\sqrt{7}}{2}$$
$$= \frac{2\left(1 \pm i\sqrt{7}\right)}{2}$$
$$x = 1 \pm i\sqrt{7}$$

77. a. $x^2 - 42x + 441 = 0$

$a = 1$, $b = -42$, and $c = 441$

$$x = \frac{-b \pm \sqrt{b^2 - 4ac}}{2a}$$

$$= \frac{-(-42) \pm \sqrt{(-42)^2 - 4(1)(441)}}{2(1)}$$

$$= \frac{42 \pm \sqrt{1,764 - 1,764}}{2}$$

$$= \frac{42 \pm \sqrt{0}}{2}$$

$$= \frac{42 \pm 0}{2}$$

$$x = \frac{42+0}{2} \quad \text{or} \quad x = \frac{42-0}{2}$$

$$x = \frac{42}{2} \qquad x = \frac{42}{2}$$

$$x = 21 \qquad x = 21$$

A repeated solution of 21

b. $x^2 + 42x + 441 = 0$

$a = 1$, $b = 42$, and $c = 441$

$$x = \frac{-b \pm \sqrt{b^2 - 4ac}}{2a}$$

$$= \frac{-42 \pm \sqrt{(42)^2 - 4(1)(441)}}{2(1)}$$

$$= \frac{-42 \pm \sqrt{1,764 - 1,764}}{2}$$

$$= \frac{-42 \pm \sqrt{0}}{2}$$

$$= \frac{-42 \pm 0}{2}$$

$$x = \frac{-42+0}{2} \quad \text{or} \quad x = \frac{-42-0}{2}$$

$$x = \frac{-42}{2} \qquad x = \frac{-42}{2}$$

$$x = -21 \qquad x = -21$$

A repeated solution of –21

APPLICATIONS

79. CROSSWALKS

Let x = length of First Avenue crosswalk, and $x + 7$ = length of the Main Street crosswalk. Use the Pythagorean Theorem to find x.

$$x^2 + (x+7)^2 = 97^2$$

$$x^2 + x^2 + 14x + 49 = 97^2$$

$$2x^2 + 14x + 49 = 97^2$$

$$2x^2 + 14x - 9,360 = 0$$

$$\frac{2x^2}{2} + \frac{14x}{2} - \frac{9,360}{2} = 0$$

$$x^2 + 7x - 4,680 = 0$$

$a = 1$, $b = 7$, and $c = -4,680$

$$x = \frac{-b \pm \sqrt{b^2 - 4ac}}{2a}$$

$$= \frac{-7 \pm \sqrt{(7)^2 - 4(1)(-4,680)}}{2(1)}$$

$$= \frac{-7 \pm \sqrt{49 + 18,720}}{2}$$

$$= \frac{-7 \pm \sqrt{18,769}}{2}$$

$$= \frac{-7 \pm 137}{2}$$

$$x = \frac{-7+137}{2} \quad \text{or} \quad x = \frac{-7-137}{2}$$

$$= \frac{130}{2} \qquad = \frac{-144}{2}$$

$$= 65 \qquad = -72$$

Since the length cannot be negative, the length of First Avenue crosswalk must be 65 feet and the length of Main Street crosswalk is 65 + 7 = 72 feet. The total distance walked this way is 65 + 72 = 137 feet. The distance the shopper saves would be 137 – 97 = 40 feet.

81. RIGHT TRIANGLES

The hypotenuse is 2.5.
Let the shorter leg = x.
The longer leg = $x + 1.7$.
Use the Pythagorean Theorem.

$$a^2 + b^2 = c^2$$
$$(x)^2 + (x+1.7)^2 = (2.5)^2$$
$$x^2 + x^2 + 1.7x + 1.7x + 2.89 = 6.25$$
$$2x^2 + 3.4x + 2.89 = 6.25$$
$$2x^2 + 3.4x - 3.36 = 0$$
$$100(2x^2 + 3.4x - 3.36) = 100(0)$$
$$200x^2 + 340x - 336 = 0$$

$a = 200$, $b = 340$, and $c = -336$

$$x = \frac{-b \pm \sqrt{b^2 - 4ac}}{2a}$$
$$= \frac{-340 \pm \sqrt{(340)^2 - 4(200)(-336)}}{2(200)}$$
$$= \frac{-340 \pm \sqrt{115,600 - (-268,800)}}{400}$$
$$= \frac{-340 \pm \sqrt{384,400}}{400}$$
$$= \frac{-340 \pm 620}{400}$$

$$x = \frac{-340 + 620}{400} \quad \text{or} \quad x = \frac{-340 - 620}{400}$$
$$x = \frac{280}{400} \qquad x = \frac{-960}{400}$$
$$x = 0.7 \qquad x = -\cancel{2.4}$$

$x + 1.7 = 2.4$

The shorter leg is 0.7 units, the longer leg is 2.4 units, and the hypotenuse is 2.5 units.

83. IMAX SCREENS

Let the width = x and length = $x + 20$.
The area is 11,349.

$$(\text{width})(\text{length}) = \text{area}$$
$$x(x + 20) = 11,349$$
$$x^2 + 20x - 11,349 = 0$$

$a = 1$, $b = 20$, and $c = -11,349$

$$x = \frac{-b \pm \sqrt{b^2 - 4ac}}{2a}$$
$$= \frac{-20 \pm \sqrt{(20)^2 - 4(1)(-11,349)}}{2(1)}$$
$$= \frac{-20 \pm \sqrt{400 - (-45,396)}}{2}$$
$$= \frac{-20 \pm \sqrt{45,796}}{2}$$
$$= \frac{-20 \pm 214}{2}$$

$$x = \frac{-20 + 214}{2} \quad \text{or} \quad x = \frac{-20 - 214}{2}$$
$$x = \frac{194}{2} \qquad x = \frac{-234}{2}$$
$$x = 97 \qquad x = \cancel{-117}$$

$x + 20 = 117$

The dimensions are 97 ft by 117 ft.

85. PARKS

Let the width = x and length = $5x$.

Perimeter = $2x + 2(5x) = 2x + 10x = 12x$.

Area = $x(5x) = 5x^2$

Perimeter = Area + 4.75

$$12x = 5x^2 + 4.75$$
$$0 = 5x^2 - 12x + 4.75$$

$a = 5$, $b = -12$, and $c = 4.75$

$$x = \frac{-b \pm \sqrt{b^2 - 4ac}}{2a}$$
$$= \frac{-(-12) \pm \sqrt{(-12)^2 - 4(5)(4.75)}}{2(5)}$$
$$= \frac{12 \pm \sqrt{144 - 95}}{10}$$
$$= \frac{12 \pm \sqrt{49}}{10}$$
$$= \frac{12 \pm 7}{10}$$
$$x = \frac{12+7}{10} \quad \text{or} \quad x = \frac{12-7}{10}$$
$$x = \frac{19}{10} \qquad x = \frac{5}{10}$$
$$x = \cancel{1.9} \qquad x = 0.5$$

$$5x = 5(0.5)$$
$$= 2.5$$

$x = 1.9$ is not possible because it states that the width is less than 1 mile. The width must be 0.5 miles and the length must be 2.5 miles.

87. POLYGONS

$$275 = \frac{n(n-3)}{2}$$
$$2 \bullet 275 = 2 \bullet \frac{n(n-3)}{2}$$
$$550 = n(n-3)$$
$$550 = n^2 - 3n$$
$$0 = n^2 - 3n - 550$$

$a = 1$, $b = -3$, $c = -550$

$$x = \frac{-b \pm \sqrt{b^2 - 4ac}}{2a}$$
$$= \frac{-(-3) \pm \sqrt{(-3)^2 - 4(1)(-550)}}{2(1)}$$
$$= \frac{3 \pm \sqrt{9 - (-2,200)}}{2}$$
$$= \frac{3 \pm \sqrt{2,209}}{2}$$
$$= \frac{3 \pm 47}{2}$$
$$x = \frac{3+47}{2} \quad \text{or} \quad x = \frac{3-47}{2}$$
$$x = \frac{50}{2} \qquad x = \frac{-44}{2}$$
$$x = 25 \qquad x = -\cancel{22}$$

The polygon has 25 sides.

89. DANCES

Let x = number of increases in ticket price.
New price: $4 + 0.10x$
Number tickets sold: $300 - 5x$
New price · # tickets sold = new receipts

$$\begin{aligned}(4+0.10x)(300-5x)&=1,248\\ 1,200-20x+30x-0.5x^2&=1,248\\ 1,200+10x-0.5x^2&=1,248\\ 0&=0.5x^2-10x+48\\ 10(0)&=10(0.5x^2-10x+48)\\ 0&=5x^2-100x+480\end{aligned}$$

$a = 5$, $b = -100$, $c = 480$

$$\begin{aligned}x&=\frac{-b\pm\sqrt{b^2-4ac}}{2a}\\ &=\frac{-(-100)\pm\sqrt{(-100)^2-4(5)(480)}}{2(5)}\\ &=\frac{100\pm\sqrt{10,000-9,600}}{10}\\ &=\frac{100\pm\sqrt{400}}{10}\\ &=\frac{100\pm 20}{10}\end{aligned}$$

$$x=\frac{100+20}{10} \quad\text{or}\quad x=\frac{100-20}{10}$$

$$x=\frac{120}{10} \qquad x=\frac{80}{10}$$

$$x=12 \qquad x=8$$

$$4+0.10x=4+0.10(12) \qquad 4+0.10x=4+0.10(8)$$

$$=4+1.20 \qquad =4+0.80$$

$$=\$5.20 \qquad =\$4.80$$

When the ticket prices are $5.20 or $4.80, the receipts will be $1,248.

91. MAGAZINE SALES

Let x = number of new subscribers
New price: $20 + 0.01x$
Number subscribers: $3,000 + x$
New price · # subscribers = total profit

$$\begin{aligned}(20+0.01x)(3,000+x)&=120,000\\ 60,000+20x+30x+0.01x^2&=120,000\\ 60,000+50x+0.01x^2&=120,000\\ 0.01x^2+50x-60,000&=0\\ 100(0.01x^2+50x-60,000)&=100(0)\\ x^2+5,000x-6,000,000&=0\end{aligned}$$

$a = 1$, $b = 5,000$, $c = -6,000,000$

$$\begin{aligned}x&=\frac{-b\pm\sqrt{b^2-4ac}}{2a}\\ &=\frac{-5,000\pm\sqrt{(5,000)^2-4(1)(-6,000,000)}}{2(1)}\\ &=\frac{-5,000\pm\sqrt{25,000,000-(-24,000,000)}}{2}\\ &=\frac{-5,000\pm\sqrt{49,000,000}}{2}\\ &=\frac{-5,000\pm 7,000}{2}\end{aligned}$$

$$x=\frac{-5,000+7,000}{2} \quad\text{or}\quad x=\frac{-5,000-7,000}{2}$$

$$x=\frac{2,000}{2} \qquad x=\frac{-12,000}{2}$$

$$x=1,000 \qquad x=\cancel{-6,000}$$

$$3,000+x=3,000+1,000 = 4,000$$

There cannot be a number of subscribers.

4,000 subscribers will bring a profit of $120,000.

93. PATROL OFFICER

$$f(t) = 75{,}000$$
$$-24t^2 + 1{,}534t + 72{,}065 = 75{,}000$$
$$-24t^2 + 1{,}534t - 2{,}935 = 0$$
$$24t^2 - 1{,}534t + 2{,}935 = 0$$

$a = 24$, $b = -1{,}534$, $c = 2{,}935$

$$t = \frac{-b \pm \sqrt{b^2 - 4ac}}{2a}$$
$$= \frac{-(-1{,}534) \pm \sqrt{(-1{,}534)^2 - 4(24)(2{,}935)}}{2(24)}$$
$$= \frac{1{,}534 \pm \sqrt{2{,}353{,}156 - 281{,}760}}{48}$$
$$= \frac{1{,}534 \pm \sqrt{2{,}071{,}396}}{48}$$
$$\approx \frac{1{,}534 \pm 1{,}439}{48}$$
$$t \approx \frac{1{,}534 + 1{,}439}{48} \quad \text{or} \quad t \approx \frac{1{,}534 - 1{,}439}{48}$$
$$t \approx 62 \qquad\qquad t \approx 2$$

$t = 62$ would represent the year 2062 (2000 + 62 = 2062) which hasn't occurred yet.

$t = 2$ would represent the year 2002 (2002 + 2 = 2002). Thus, the model indicates that there were 75,000 female officers in 2002.

95. PICTURE FRAMING

Let x = the width of the matting. Then the total length of the picture plus the mat would be $(2x + 5)$ in. since there is the same width of matting on both sides of the picture, and the total width of the picture plus the mat would be $(2x + 4)$ in.
The area of the mat would be the total area minus the area of the picture, which is 4(5) or 20 in.^{2}

$$\text{area of mat} = \text{area of picture}$$
$$\text{total area - area of picture} = \text{area of picture}$$
$$(2x+5)(2x+4) - 20 = 20$$
$$4x^2 + 8x + 10x + 20 - 20 = 20$$
$$4x^2 + 18x = 20$$
$$4x^2 + 18x - 20 = 0$$

$a = 4$, $b = 18$, $c = -20$

$$x = \frac{-b \pm \sqrt{b^2 - 4ac}}{2a}$$
$$= \frac{-18 \pm \sqrt{(18)^2 - 4(4)(-20)}}{2(4)}$$
$$= \frac{-18 \pm \sqrt{324 + 320}}{8}$$
$$= \frac{-18 \pm \sqrt{644}}{8}$$
$$= \frac{-18 \pm 25.38}{8}$$
$$x = \frac{-18 + 25.38}{8} \quad \text{or} \quad x = \frac{-18 - 25.38}{8}$$
$$x \approx 0.92 \qquad\qquad x \approx \cancel{-5.42}$$

The mat is approximately 0.92 inches wide.

97. DIMENSIONS OF A RECTANGLE
Let the width = x ft. Then the length would be $(x + 4)$ ft..

$$(\text{width})(\text{length}) = \text{Area}$$
$$x(x+4) = 20$$
$$x^2 + 4x = 20$$
$$x^2 + 4x - 20 = 0$$

$a = 1$, $b = 4$, $c = -20$

$$x = \frac{-b \pm \sqrt{b^2 - 4ac}}{2a}$$
$$= \frac{-4 \pm \sqrt{(4)^2 - 4(1)(-20)}}{2(1)}$$
$$= \frac{-4 \pm \sqrt{16 + 80}}{2}$$
$$= \frac{-4 \pm \sqrt{96}}{2}$$
$$= \frac{-4 \pm 9.8}{2}$$
$$x = \frac{-4 + 9.8}{2} \quad \text{or} \quad x = \frac{-4 - 9.8}{2}$$
$$x \approx 2.9 \qquad x \approx \cancel{-6.9}$$

The width is 2.9 ft.
The length is 2.9 + 4 or 6.9 ft.

WRITING

99. Answers will vary.

REVIEW

101. $\sqrt{n} = n^{1/2}$

103. $\sqrt[4]{3b} = (3b)^{1/4}$

105. $t^{1/3} = \sqrt[3]{t}$

107. $(3t)^{1/4} = \sqrt[4]{3t}$

CHALLENGE PROBLEMS

109. $x^2 + 2\sqrt{2}x - 6 = 0$

$a = 1$, $b = 2\sqrt{2}$, $c = -6$

$$x = \frac{-b \pm \sqrt{b^2 - 4ac}}{2a}$$
$$= \frac{-2\sqrt{2} \pm \sqrt{(2\sqrt{2})^2 - 4(1)(-6)}}{2(1)}$$
$$= \frac{-2\sqrt{2} \pm \sqrt{8 - (-24)}}{2}$$
$$= \frac{-2\sqrt{2} \pm \sqrt{32}}{2}$$
$$= \frac{-2\sqrt{2} \pm \sqrt{16}\sqrt{2}}{2}$$
$$= \frac{-2\sqrt{2} \pm 4\sqrt{2}}{2}$$
$$x = \frac{-2\sqrt{2} + 4\sqrt{2}}{2} \quad \text{or} \quad x = \frac{-2\sqrt{2} - 4\sqrt{2}}{2}$$
$$x = \frac{2\sqrt{2}}{2} \qquad x = \frac{-6\sqrt{2}}{2}$$
$$x = \sqrt{2} \qquad x = -3\sqrt{2}$$

111. $x^2 - 3ix - 2 = 0$

$a = 1$, $b = -3i$, $c = -2$

$$x = \frac{-b \pm \sqrt{b^2 - 4ac}}{2a}$$
$$= \frac{-(-3i) \pm \sqrt{(-3i)^2 - 4(1)(-2)}}{2(1)}$$
$$= \frac{3i \pm \sqrt{9i^2 - (-8)}}{2}$$
$$= \frac{3i \pm \sqrt{-9 + 8}}{2}$$
$$= \frac{3i \pm \sqrt{-1}}{2}$$
$$= \frac{3i \pm i}{2}$$
$$x = \frac{3i + i}{2} \quad \text{or} \quad x = \frac{3i - i}{2}$$
$$x = \frac{4i}{2} \qquad x = \frac{2i}{2}$$
$$x = 2i \qquad x = i$$

SECTION 8.3

VOCABULARY

1. For the quadratic equation $ax^2+bx+c=0$, the **discriminant** is b^2-4ac.

CONCEPTS

3. If $b^2-4ac<0$, the solutions of the equation are two different imaginary numbers that are complex **conjugates**.

5. If b^2-4ac is a perfect square, the solutions of the equation are two different **rational** numbers.

7. a. $y=x^2$
   b. $y=\sqrt{x}$
   c. $y=x^{1/3}$
   d. $y=\dfrac{1}{x}$
   e. $y=x+1$

NOTATION

9.
$$\begin{aligned} b^2-\boxed{4ac}&=\boxed{5}^2-4(1)(\boxed{6})\\ &=25-\boxed{24}\\ &=1\end{aligned}$$
Since *a, b,* and *c* are rational numbers and the value of the discriminant is a perfect square, the solutions are two different **rational** numbers.

GUIDED PRACTICE

11.
$$4x^2-4x+1=0$$
$$\begin{aligned} b^2-4ac&=(-4)^2-4(4)(1)\\ &=16-16\\ &=0\end{aligned}$$
one repeated rational-number solution

13.
$$5x^2+x+2=0$$
$$\begin{aligned} b^2-4ac&=(1)^2-4(5)(2)\\ &=1-40\\ &=-39\end{aligned}$$
two imaginary -number solutions (complex conjugates)

15.
$$2x^2=4x-1$$
$$2x^2-4x+1=0$$
$$\begin{aligned} b^2-4ac&=(-4)^2-4(2)(1)\\ &=16-8\\ &=8\end{aligned}$$
two different irrational-number solutions

17.
$$x(2x-3)=20$$
$$2x^2-3x=20$$
$$2x^2-3x-20=0$$
$$\begin{aligned} b^2-4ac&=(-3)^2-4(2)(-20)\\ &=9-(-160)\\ &=169 \text{ (perfect square)}\end{aligned}$$
two different rational-number solutions

19.
$$3x^2-10=0$$
$$\begin{aligned} b^2-4ac&=0^2-4(3)(-10)\\ &=0+120\\ &=120\end{aligned}$$
two different irrational-number solutions

21.
$$x^2-\frac{14}{15}x=\frac{8}{15}$$
$$x^2-\frac{14}{15}x-\frac{8}{15}=0$$
$$\begin{aligned} b^2-4ac&=\left(-\frac{14}{15}\right)^2-4(1)\left(-\frac{8}{15}\right)\\ &=\frac{196}{225}+\frac{32}{15}\\ &=\frac{676}{225}\end{aligned}$$
two different rational-number solutions

23. Let $y = x^2$.

$$x^4 - 17x^2 + 16 = 0$$
$$\left(x^2\right)^2 - 17x^2 + 16 = 0$$
$$y^2 - 17y + 16 = 0$$
$$(y-16)(y-1) = 0$$
$$y - 16 = 0 \quad \text{or} \quad y - 1 = 0$$
$$y = 16 \qquad y = 1$$
$$x^2 = 16 \qquad x^2 = 1$$
$$x = \pm\sqrt{16} \qquad x = \pm\sqrt{1}$$
$$x = \pm 4 \qquad x = \pm 1$$
$$x = 4,\ -4,\ 1,\ -1$$

25. Let $y = x^2$.

$$x^4 + 5x^2 - 36 = 0$$
$$\left(x^2\right)^2 + 5x^2 - 36 = 0$$
$$y^2 + 5y - 36 = 0$$
$$(y+9)(y-4) = 0$$
$$y + 9 = 0 \quad \text{or} \quad y - 4 = 0$$
$$y = -9 \qquad y = 4$$
$$x^2 = -9 \qquad x^2 = 4$$
$$x = \pm\sqrt{-9} \qquad x = \pm\sqrt{2}$$
$$x = \pm 3i \qquad x = \pm 2$$
$$x = 3i,\ -3i,\ 2,\ -2$$

27. Let $y = \sqrt{x}$.

$$x - 13\sqrt{x} + 40 = 0$$
$$\left(\sqrt{x}\right)^2 - 13\sqrt{x} + 40 = 0$$
$$y^2 - 13y + 40 = 0$$
$$(y-8)(y-5) = 0$$
$$y - 8 = 0 \quad \text{or} \quad y - 5 = 0$$
$$y = 8 \qquad y = 5$$
$$\sqrt{x} = 8 \qquad \sqrt{x} = 5$$
$$\left(\sqrt{x}\right)^2 = (8)^2 \qquad \left(\sqrt{x}\right)^2 = (5)^2$$
$$x = 64 \qquad x = 25$$
$$x = 64,\ 25$$

29. Let $y = \sqrt{x}$.

$$2x + \sqrt{x} - 3 = 0$$
$$2\left(\sqrt{x}\right)^2 + \sqrt{x} - 3 = 0$$
$$2y^2 + y - 3 = 0$$
$$(2y+3)(y-1) = 0$$
$$2y + 3 = 0 \quad \text{or} \quad y - 1 = 0$$
$$y = -\frac{3}{2} \qquad y = 1$$
$$\sqrt{x} = -\frac{3}{2} \qquad \sqrt{x} = 1$$
$$\left(\sqrt{x}\right)^2 = \left(-\frac{3}{2}\right)^2 \qquad \left(\sqrt{x}\right)^2 = (1)^2$$
$$x = \frac{9}{4} \qquad x = 1$$

$\frac{9}{4}$ is extraneous;

$$x = 1$$

31. Let $y = a^{1/3}$.

$$a^{2/3} - 2a^{1/3} = 3$$
$$a^{2/3} - 2a^{1/3} - 3 = 0$$
$$\left(a^{1/3}\right)^2 - 2a^{1/3} - 3 = 0$$
$$y^2 - 2y - 3 = 0$$
$$(y+1)(y-3) = 0$$
$$y + 1 = 0 \quad \text{or} \quad y - 3 = 0$$
$$y = -1 \qquad y = 3$$
$$a^{1/3} = -1 \qquad a^{1/3} = 3$$
$$\left(a^{1/3}\right)^3 = (-1)^3 \qquad \left(a^{1/3}\right)^3 = (3)^3$$
$$a = -1 \qquad a = 27$$

33. Let $y = x^{1/3}$.

$$x^{2/3} + 2x^{1/3} - 8 = 0$$
$$\left(x^{1/3}\right)^2 + 2x^{1/3} - 8 = 0$$
$$y^2 + 2y - 8 = 0$$
$$(y-2)(y+4) = 0$$
$$y - 2 = 0 \quad \text{or} \quad y + 4 = 0$$
$$y = 2 \qquad y = -4$$
$$x^{1/3} = 2 \qquad x^{1/3} = -4$$
$$\left(x^{1/3}\right)^3 = (2)^3 \qquad \left(x^{1/3}\right)^3 = (-4)^3$$
$$x = 8 \qquad x = -64$$
$$x = 8,\ -64$$

35. Let $y = c + 1$.

$$(c+1)^2 - 4(c+1) + 3 = 0$$
$$y^2 - 4y + 3 = 0$$
$$(y-3)(y-1) = 0$$
$$y - 3 = 0 \qquad y - 1 = 0$$
$$y = 3 \qquad y = 1$$
$$c + 1 = 3 \qquad c + 1 = 1$$
$$c + 1 - 1 = 3 - 1 \qquad c + 1 - 1 = 1 - 1$$
$$c = 2 \qquad c = 0$$

37. Let $y = 2x + 1$.

$$2(2x+1)^2 - 7(2x+1) + 6 = 0$$
$$2y^2 - 7y + 6 = 0$$
$$(2y-3)(y-2) = 0$$
$$2y - 3 = 0 \quad \text{or} \quad y - 2 = 0$$
$$y = \frac{3}{2} \qquad y = 2$$
$$2x + 1 = \frac{3}{2} \qquad 2x + 1 = 2$$
$$2x = \frac{1}{2} \qquad 2x = 1$$
$$x = \frac{1}{4} \qquad x = \frac{1}{2}$$

39. Let $y = \frac{1}{m}$.

$$m^{-2} + m^{-1} - 6 = 0$$
$$\left(\frac{1}{m}\right)^2 + \left(\frac{1}{m}\right) - 6 = 0$$
$$y^2 + y - 6 = 0$$
$$(y+3)(y-2) = 0$$
$$y + 3 = 0 \quad \text{or} \quad y - 2 = 0$$
$$y = -3 \qquad y = 2$$
$$\frac{1}{m} = -3 \qquad \frac{1}{m} = 2$$
$$m\left(\frac{1}{m}\right) = m(-3) \qquad m\left(\frac{1}{m}\right) = m(2)$$
$$1 = -3m \qquad 1 = 2m$$
$$-\frac{1}{3} = m \qquad \frac{1}{2} = m$$
$$m = -\frac{1}{3}, \frac{1}{2}$$

41. Let $y = \frac{1}{x}$.

$$8x^{-2} - 10x^{-1} - 3 = 0$$
$$8\left(\frac{1}{x}\right)^2 - 10\left(\frac{1}{x}\right) - 3 = 0$$
$$8y^2 - 10y - 3 = 0$$
$$(2y-3)(4y+1) = 0$$
$$2y - 3 = 0 \quad \text{or} \quad 4y + 1 = 0$$
$$y = \frac{3}{2} \qquad y = -\frac{1}{4}$$
$$\frac{1}{x} = \frac{3}{2} \qquad \frac{1}{x} = -\frac{1}{4}$$
$$2 = 3x \qquad 4 = -x$$
$$\frac{2}{3} = x \qquad -4 = x$$
$$x = \frac{2}{3}, -4$$

43.

$$1 - \frac{5}{x} = \frac{10}{x^2}$$
$$x^2\left(1 - \frac{5}{x}\right) = x^2\left(\frac{10}{x^2}\right)$$
$$x^2 - 5x = 10$$
$$x^2 - 5x - 10 = 0$$
$$x = \frac{-b \pm \sqrt{b^2 - 4ac}}{2a}$$
$$= \frac{-(-5) \pm \sqrt{(-5)^2 - 4(1)(-10)}}{2(1)}$$
$$= \frac{5 \pm \sqrt{25 + 40}}{2}$$
$$x = \frac{5 \pm \sqrt{65}}{2}$$

45.

$$\frac{1}{2}+\frac{1}{b}=\frac{1}{b-7}$$

$$2b(b-7)\left(\frac{1}{2}+\frac{1}{b}\right)=2b(b-7)\left(\frac{1}{b-7}\right)$$

$$b(b-7)+2(b-7)=2b$$

$$b^2-7b+2b-14=2b$$

$$b^2-5b-14=2b$$

$$b^2-7b-14=0$$

$$b=\frac{-b\pm\sqrt{b^2-4ac}}{2a}$$

$$=\frac{-(-7)\pm\sqrt{(-7)^2-4(1)(-14)}}{2(1)}$$

$$=\frac{7\pm\sqrt{49+56}}{2}$$

$$b=\frac{7\pm\sqrt{105}}{2}$$

TRY IT YOURSELF

47. Let $y=\sqrt{x}$.

$$2x-\sqrt{x}=3$$

$$2x-\sqrt{x}-3=0$$

$$2\left(\sqrt{x}\right)^2-\sqrt{x}-3=0$$

$$2y^2-y-3=0$$

$$(2y-3)(y+1)=0$$

$$2y-3=0 \quad \text{or} \quad y+1=0$$

$$y=\frac{3}{2} \qquad y=-1$$

$$\sqrt{x}=\frac{3}{2} \qquad \sqrt{x}=-1$$

$$\left(\sqrt{x}\right)^2=\left(\frac{3}{2}\right)^2 \qquad \left(\sqrt{x}\right)^2=(-1)^2$$

$$x=\frac{9}{4} \qquad x=\not{1}$$

1 is extraneous

$$x=\frac{9}{4}$$

49. Let $y=\dfrac{1}{x}$.

$$x^{-2}+2x^{-1}-3=0$$

$$\left(\frac{1}{x}\right)^2+2\left(\frac{1}{x}\right)-3=0$$

$$y^2+2y-3=0$$

$$(y+3)(y-1)=0$$

$$y+3=0 \quad \text{or} \quad y-1=0$$

$$y=-3 \qquad y=1$$

$$\frac{1}{x}=-3 \qquad \frac{1}{x}=1$$

$$1=-3x \qquad 1=x$$

$$-\frac{1}{3}=x$$

$$x=-\frac{1}{3},\ 1$$

51. Let $y=x^2$.

$$x^4+19x^2+18=0$$

$$\left(x^2\right)^2+19x^2+18=0$$

$$y^2+19y+18=0$$

$$(y+18)(y+1)=0$$

$$y+18=0 \quad \text{or} \quad y+1=0$$

$$y=-18 \qquad y=-1$$

$$x^2=-18 \qquad x^2=-1$$

$$x=\pm\sqrt{-18} \qquad x=\pm\sqrt{-1}$$

$$x=\pm3i\sqrt{2} \qquad x=\pm i$$

$$x=3i\sqrt{2},\ -3i\sqrt{2},\ i,\ -i$$

53. Let $y = k - 7$.

$$(k-7)^2 + 6(k-7) + 10 = 0$$
$$y^2 + 6y + 10 = 0$$
$$y = \frac{-b \pm \sqrt{b^2 - 4ac}}{2a}$$
$$y = \frac{-6 \pm \sqrt{(6)^2 - 4(1)(10)}}{2(1)}$$
$$y = \frac{-6 \pm \sqrt{36 - 40}}{2}$$
$$y = \frac{-6 \pm \sqrt{-4}}{2}$$
$$y = \frac{-6 \pm 2i}{2}$$
$$y = \frac{2(-3 \pm i)}{2}$$
$$y = -3 \pm i$$
$$k - 7 = -3 \pm i$$
$$k - 7 + 7 = -3 \pm i + 7$$
$$k = 4 \pm i$$

55.

$$\frac{2}{x-1} + \frac{1}{x+1} = 3$$
$$(x-1)(x+1)\left(\frac{2}{x-1} + \frac{1}{x+1}\right) = 3(x-1)(x+1)$$
$$2(x+1) + (x-1) = 3(x^2 - 1)$$
$$2x + 2 + x - 1 = 3x^2 - 3$$
$$3x + 1 = 3x^2 - 3$$
$$0 = 3x^2 - 3x - 4$$
$$x = \frac{-b \pm \sqrt{b^2 - 4ac}}{2a}$$
$$x = \frac{-(-3) \pm \sqrt{(-3)^2 - 4(3)(-4)}}{2(3)}$$
$$x = \frac{3 \pm \sqrt{9 - (-48)}}{6}$$
$$x = \frac{3 \pm \sqrt{57}}{6}$$

57. Let $y = x^{1/2}$.

$$x - 6x^{1/2} = -8$$
$$x - 6x^{1/2} + 8 = 0$$
$$\left(x^{1/2}\right)^2 - 6x^{1/2} + 8 = 0$$
$$y^2 - 6y + 8 = 0$$
$$(y-2)(y-4) = 0$$
$$y - 2 = 0 \quad \text{or} \quad y - 4 = 0$$
$$y = 2 \qquad y = 4$$
$$x^{1/2} = 2 \qquad x^{1/2} = 4$$
$$\left(x^{1/2}\right)^2 = (2)^2 \qquad \left(x^{1/2}\right)^2 = (4)^2$$
$$x = 4 \qquad x = 16$$

59. Let $x = y^2 - 9$.

$$\left(y^2 - 9\right)^2 + 2\left(y^2 - 9\right) - 99 = 0$$
$$x^2 + 2x - 99 = 0$$
$$(x+11)(x-9) = 0$$
$$x + 11 = 0 \quad \text{or} \quad x - 9 = 0$$
$$x = -11 \qquad x = 9$$
$$y^2 - 9 = -11 \qquad y^2 - 9 = 9$$
$$y^2 = -2 \qquad y^2 = 18$$
$$y = \pm\sqrt{-2} \qquad y = \pm\sqrt{18}$$
$$y = \pm i\sqrt{2} \qquad y = \pm 3\sqrt{2}$$
$$y = i\sqrt{2},\ -i\sqrt{2},\ 3\sqrt{2},\ -3\sqrt{2}$$

61. Let $y = \dfrac{1}{x^2}$.

$$x^{-4} - 2x^{-2} + 1 = 0$$
$$\frac{1}{x^4} - 2\left(\frac{1}{x^2}\right) + 1 = 0$$
$$\left(\frac{1}{x^2}\right)^2 - 2\left(\frac{1}{x^2}\right) + 1 = 0$$
$$y^2 - 2y + 1 = 0$$
$$(y-1)(y-1) = 0$$
$$y - 1 = 0 \quad \text{or} \quad y - 1 = 0$$
$$y = 1 \qquad y = 1$$
$$\frac{1}{x^2} = \frac{1}{1} \qquad \frac{1}{x^2} = \frac{1}{1}$$
$$1 = x^2 \qquad 1 = x^2$$
$$x = \pm\sqrt{1} \qquad x = \pm\sqrt{1}$$
$$x = \pm 1 \qquad x = \pm 1$$
$$x = 1,\ -1,\ 1,\ -1$$

63. Let $y = t^2$.

$$t^4 + 3t^2 = 28$$
$$t^4 + 3t^2 - 28 = 0$$
$$\left(t^2\right)^2 + 3t^2 - 28 = 0$$
$$y^2 + 3y - 28 = 0$$
$$(y-4)(y+7) = 0$$
$$y - 4 = 0 \quad \text{or} \quad y + 7 = 0$$
$$y = 4 \qquad y = -7$$
$$t^2 = 4 \qquad t^2 = -7$$
$$t = \pm\sqrt{4} \qquad t = \pm\sqrt{-7}$$
$$t = \pm 2 \qquad t = \pm i\sqrt{7}$$
$$t = 2,\ -2,\ i\sqrt{7},\ -i\sqrt{7}$$

65. Let $y = x^{1/5}$.

$$2x^{2/5} - 5x^{1/5} = -3$$
$$2x^{2/5} - 5x^{1/5} + 3 = 0$$
$$2\left(x^{1/5}\right)^2 - 5x^{1/5} + 3 = 0$$
$$2y^2 - 5y + 3 = 0$$
$$(2y-3)(y-1) = 0$$
$$2y - 3 = 0 \quad \text{or} \quad y - 1 = 0$$
$$y = \frac{3}{2} \qquad y = 1$$
$$x^{1/5} = \frac{3}{2} \qquad x^{1/5} = 1$$
$$\left(x^{1/5}\right)^5 = \left(\frac{3}{2}\right)^5 \qquad \left(x^{1/5}\right)^5 = (1)^5$$
$$x = \frac{243}{32} \qquad x = 1$$

67. Let $y = \dfrac{3m+2}{m}$.

$$9\left(\frac{3m+2}{m}\right)^2 - 30\left(\frac{3m+2}{m}\right) + 25 = 0$$
$$9y^2 - 30y + 25 = 0$$
$$(3y-5)(3y-5) = 0$$
$$3y - 5 = 0 \quad \text{or} \quad 3y - 5 = 0$$
$$y = \frac{5}{3} \qquad y = \frac{5}{3}$$
$$\frac{3m+2}{m} = \frac{5}{3} \qquad \frac{3m+2}{m} = \frac{5}{3}$$
$$3(3m+2) = 5m \qquad 3(3m+2) = 5m$$
$$9m + 6 = 5m \qquad 9m + 6 = 5m$$
$$6 = -4m \qquad 6 = -4m$$
$$-\frac{3}{2} = m \qquad -\frac{3}{2} = m$$
$$m = -\frac{3}{2},\ -\frac{3}{2}$$

69.

$$\frac{3}{a-1} = 1 - \frac{2}{a}$$
$$a(a-1)\left(\frac{3}{a-1}\right) = a(a-1)\left(1 - \frac{2}{a}\right)$$
$$3a = a(a-1) - 2(a-1)$$
$$3a = a^2 - a - 2a + 2$$
$$3a = a^2 - 3a + 2$$
$$0 = a^2 - 6a + 2$$
$$a = \frac{-b \pm \sqrt{b^2 - 4ac}}{2a}$$
$$a = \frac{-(-6) \pm \sqrt{(-6)^2 - 4(1)(2)}}{2(1)}$$
$$= \frac{6 \pm \sqrt{36 - 8}}{2}$$
$$= \frac{6 \pm \sqrt{28}}{2}$$
$$= \frac{6 \pm \sqrt{4}\sqrt{7}}{2}$$
$$= \frac{6 \pm 2\sqrt{7}}{2}$$
$$= \frac{2\left(3 \pm \sqrt{7}\right)}{2}$$
$$a = 3 \pm \sqrt{7}$$

71. Let $y = 8-\sqrt{a}$.

$$\left(8-\sqrt{a}\right)^2+6\left(8-\sqrt{a}\right)-7=0$$
$$y^2+6y-7=0$$
$$(y+7)(y-1)=0$$
$$y+7=0 \quad \text{or} \quad y-1=0$$
$$y=-7 \qquad y=1$$
$$8-\sqrt{a}=-7 \qquad 8-\sqrt{a}=1$$
$$-\sqrt{a}=-15 \qquad -\sqrt{a}=-7$$
$$\sqrt{a}=15 \qquad \sqrt{a}=7$$
$$\left(\sqrt{a}\right)^2=(15)^2 \qquad \left(\sqrt{a}\right)^2=(7)^2$$
$$a=225 \qquad a=49$$
$$a=225,\ 49$$

73.

$$x+\frac{2}{x-2}=0$$
$$(x-2)\left(x+\frac{2}{x-2}\right)=(x-2)0$$
$$x(x-2)+2=0$$
$$x^2-2x+2=0$$
$$x=\frac{-b\pm\sqrt{b^2-4ac}}{2a}$$
$$x=\frac{-(-2)\pm\sqrt{(-2)^2-4(1)(2)}}{2(1)}$$
$$x=\frac{2\pm\sqrt{4-8}}{2}$$
$$x=\frac{2\pm\sqrt{-4}}{2}$$
$$x=\frac{2\pm 2i}{2}$$
$$x=\frac{2}{2}\pm\frac{2}{2}i$$
$$x=1\pm i$$

75. Let $y = \sqrt{x}$.

$$3x+5\sqrt{x}+2=0$$
$$3\left(\sqrt{x}\right)^2+5\sqrt{x}+2=0$$
$$3y^2+5y+2=0$$
$$(3y+2)(y+1)=0$$
$$3y+2=0 \quad \text{or} \quad y+1=0$$
$$y=-\frac{2}{3} \qquad y=-1$$
$$\sqrt{x}=-\frac{2}{3} \qquad \sqrt{x}=-1$$
$$\left(\sqrt{x}\right)^2=\left(-\frac{2}{3}\right)^2 \qquad \sqrt{x}=(-1)^2$$
$$x=\cancel{\frac{4}{9}} \qquad x=\cancel{1}$$

$\frac{4}{9}$ and 1 are extraneous

No Solution

77. Let $y = x^2$.

$$x^4-6x^2+5=0$$
$$\left(x^2\right)^2-6x^2+5=0$$
$$y^2-6y+5=0$$
$$(y-5)(y-1)=0$$
$$y-5=0 \quad \text{or} \quad y-1=0$$
$$y=5 \qquad y=1$$
$$x^2=5 \qquad x^2=1$$
$$x=\pm\sqrt{5} \qquad x=\pm\sqrt{1}$$
$$x=\pm\sqrt{5} \qquad x=\pm 1$$
$$x=\sqrt{5},\ -\sqrt{5},\ 1,\ -1$$

79. Let $y = \dfrac{1}{t+1}$.

$$8(t+1)^{-2} - 30(t+1)^{-1} + 7 = 0$$
$$8\left(\frac{1}{t+1}\right)^2 - 30\left(\frac{1}{t+1}\right) + 7 = 0$$
$$8y^2 - 30y + 7 = 0$$
$$(4y-1)(2y-7) = 0$$
$$4y-1=0 \quad \text{or} \quad 2y-7=0$$
$$y = \frac{1}{4} \qquad y = \frac{7}{2}$$
$$\frac{1}{t+1} = \frac{1}{4} \qquad \frac{1}{t+1} = \frac{7}{2}$$
$$4 = t+1 \qquad 7(t+1) = 2$$
$$3 = t \qquad 7t+7 = 2$$
$$t = 3 \qquad t = -\frac{5}{7}$$
$$t = 3,\ -\frac{5}{7}$$

81.

$$\frac{1}{x+2} + \frac{24}{x+3} = 13$$
$$(x+2)(x+3)\left(\frac{1}{x+2} + \frac{24}{x+3}\right) = (x+2)(x+3)(13)$$
$$x+3+24(x+2) = 13(x+2)(x+3)$$
$$x+3+24x+48 = 13(x^2+5x+6)$$
$$25x+51 = 13x^2+65x+78$$
$$0 = 13x^2+40x+27$$
$$0 = (13x+27)(x+1)$$
$$13x+27=0 \quad \text{or} \quad x+1=0$$
$$x = -\frac{27}{13} \qquad x = -1$$

83. a. Let $x = y^{1/3}$.

$$y^{2/3} + y^{1/3} - 20 = 0$$
$$\left(y^{1/3}\right)^2 + y^{1/3} - 20 = 0$$
$$x^2 + x - 20 = 0$$
$$(x-4)(x+5) = 0$$
$$x-4=0 \quad \text{or} \quad x+5=0$$
$$x = 4 \qquad x = -5$$
$$y^{1/3} = 4 \qquad y^{1/3} = -5$$
$$\left(y^{1/3}\right)^3 = (4)^3 \qquad \left(y^{1/3}\right)^3 = (-5)^3$$
$$y = 64 \qquad y = -125$$

b. Let $x = y^{-1}$.

$$y^{-2} + y^{-1} - 20 = 0$$
$$\left(y^{-1}\right)^2 + y^{-1} - 20 = 0$$
$$x^2 + x - 20 = 0$$
$$(x-4)(x+5) = 0$$
$$x-4=0 \quad \text{or} \quad x+5=0$$
$$x = 4 \qquad x = -5$$
$$y^{-1} = 4 \qquad y^{-1} = -5$$
$$\left(y^{-1}\right)^{-1} = (4)^{-1} \qquad \left(y^{-1}\right)^{-1} = (-5)^{-1}$$
$$y = \frac{1}{4} \qquad y = -\frac{1}{5}$$

85. a.

$$\frac{1}{x} = \frac{2x}{x+1}$$
$$x(x+1)\left(\frac{1}{x}\right) = x(x+1)\left(\frac{2x}{x+1}\right)$$
$$(x+1)(1) = x(2x)$$
$$x+1 = 2x^2$$
$$0 = 2x^2 - x - 1$$
$$0 = (2x+1)(x-1)$$
$$2x+1=0 \quad \text{or} \quad x-1=0$$
$$2x = -1 \qquad x = 1$$
$$x = -\frac{1}{2}$$

b.

$$\frac{1}{x} + \frac{2x}{x+1} = 1$$
$$x(x+1)\left(\frac{1}{x} + \frac{2x}{x+1}\right) = x(x+1)(1)$$
$$(x+1)(1) + x(2x) = x(x+1)$$
$$x+1+2x^2 = x^2+x$$
$$x^2+1 = 0$$
$$x^2 = -1$$
$$x = \pm\sqrt{-1}$$
$$x = \pm i$$

APPLICATIONS

87. CROWD CONTROL

Let x = the number of minutes it takes to empty the parking lot.

West exit's work in 1 minute = $\frac{1}{x}$

East exit's work in 1 minute = $\frac{1}{x+40}$

Amount of work together in 1 minute = $\frac{1}{60}$

$$\frac{1}{x}+\frac{1}{x+40}=\frac{1}{60}$$

$$60x(x+40)\left(\frac{1}{x}+\frac{1}{x+40}\right)=60x(x+40)\left(\frac{1}{60}\right)$$

$$60(x+40)+60x=x(x+40)$$

$$60x+2{,}400+60x=x^2+40x$$

$$120x+2{,}400=x^2+40x$$

$$0=x^2-80x-2{,}400$$

$$x=\frac{-b\pm\sqrt{b^2-4ac}}{2a}$$

$$x=\frac{-(-80)\pm\sqrt{(-80)^2-4(1)(-2400)}}{2(1)}$$

$$x=\frac{80\pm\sqrt{6{,}400-(-9{,}600)}}{2}$$

$$x=\frac{80\pm\sqrt{16{,}000}}{2}$$

$$x=\frac{80\pm 40\sqrt{10}}{2}$$

$$x=40\pm 20\sqrt{10}$$

$$x=40+20\sqrt{10} \text{ or } x=40-20\sqrt{10}$$

$$x=103 \qquad x=-23$$

It takes 103 minutes to clear the parking lot using only the west exit.

89. ASSEMBLY LINES

Let r = rate.

distance = (rate)(time)

Original Time:

$$300=rt$$

$$\frac{300}{r}=t$$

Faster Time:

$$300=(r+5)(t-3)$$

$$\frac{300}{r+5}=t-3$$

$$\frac{300}{r+5}+3=t$$

Substitute $\frac{300}{r}=t$ into the last equation and solve for r.

$$\frac{300}{r+5}+3=\frac{300}{r}$$

$$r(r+5)\left(\frac{300}{r+5}+3\right)=r(r+5)\left(\frac{300}{r}\right)$$

$$300r+3r(r+5)=300(r+5)$$

$$300r+3r^2+15r=300r+1{,}500$$

$$3r^2+15r-1{,}500=0$$

$$\frac{3r^2}{3}+\frac{15r}{3}-\frac{1{,}500}{3}=\frac{0}{3}$$

$$r^2+5r-500=0$$

$$(r+25)(r-20)=0$$

$$r+25=0 \quad \text{or} \quad r-20=0$$

$$r=-25 \qquad r=20$$

Since the rate cannot be negative, r = 20 feet per second.

91. ARCHITECTURE

$$\frac{\ell}{w}=\frac{w}{\ell-w}$$
$$\frac{\ell}{20}=\frac{20}{\ell-20}$$
$$\ell(\ell-20)=20(20)$$
$$\ell^2-20\ell=400$$
$$\ell^2-20\ell-400=0$$
$$\ell=\frac{-b\pm\sqrt{b^2-4ac}}{2a}$$
$$\ell=\frac{-(-20)\pm\sqrt{(-20)^2-4(1)(-400)}}{2(1)}$$
$$\ell=\frac{20\pm\sqrt{400-(-1600)}}{2}$$
$$\ell=\frac{20\pm\sqrt{2,000}}{2}$$
$$\ell=\frac{20\pm20\sqrt{5}}{2}$$
$$\ell=\frac{20}{2}\pm\frac{20}{2}\sqrt{5}$$
$$\ell=10\pm10\sqrt{5}$$
$$\ell=10+10\sqrt{5} \text{ or } \ell=10-10\sqrt{5}$$
$$\ell=32.4 \text{ ft.} \qquad \ell=\cancel{-12.4}$$

The length should be 32.4 ft.

WRITING

93. Answers will vary.

REVIEW

95. $x = 3$

97. $m = \frac{2}{3}$ and $b = (0, 0)$

$$y=mx+b$$
$$y=\frac{2}{3}x+0$$
$$y=\frac{2}{3}x$$

CHALLENGE PROBLEMS

99. Let $y = x^2$.

$$x^4-3x^2-2=0$$
$$(x^2)^2-3x^2-2=0$$
$$y^2-3y-2=0$$
$$a=1,\ b=-3,\ c=-2$$
$$y=\frac{-b\pm\sqrt{b^2-4ac}}{2a}$$
$$y=\frac{-(-3)\pm\sqrt{(-3)^2-4(1)(-2)}}{2(1)}$$
$$y=\frac{3\pm\sqrt{9-(-8)}}{2}$$
$$y=\frac{3\pm\sqrt{17}}{2}$$
$$y=\frac{3+\sqrt{17}}{2} \quad \text{or} \quad y=\frac{3-\sqrt{17}}{2}$$
$$x^2=\frac{3+\sqrt{17}}{2} \qquad x^2=\frac{3-\sqrt{17}}{2}$$
$$x=\pm\sqrt{\frac{3+\sqrt{17}}{2}} \qquad x=\pm\sqrt{\frac{3-\sqrt{17}}{2}}$$
$$x=\pm\frac{\sqrt{3+\sqrt{17}}}{\sqrt{2}}\cdot\frac{\sqrt{2}}{\sqrt{2}} \qquad x=\pm\frac{\sqrt{3-\sqrt{17}}}{\sqrt{2}}\cdot\frac{\sqrt{2}}{\sqrt{2}}$$
$$x=\pm\frac{\sqrt{6+2\sqrt{17}}}{2} \qquad x=\pm\frac{\sqrt{6-2\sqrt{17}}}{2}$$

Since $x=\pm\frac{\sqrt{6-2\sqrt{17}}}{2}$ is not real, the only solutions are $x=\pm\frac{\sqrt{6+2\sqrt{17}}}{2}$.

101.

$$x^3-x^2+16x-16=0$$
$$x^2(x-1)+16(x-1)=0$$
$$(x^2+16)(x-1)=0$$
$$x^2+16=0 \quad \text{or} \quad x-1=0$$
$$x^2=-16 \qquad x=1$$
$$\sqrt{x^2}=\pm\sqrt{-16}$$
$$x=\pm4i$$

SECTION 8.4

VOCABULARY

1. $f(x) = 2x^2 - 4x + 1$ is called a **quadratic** function. Its graph is a cup–shaped figure called a **parabola**.

3. The vertical line $x = 1$ divides the parabola into two halves. This line is called the **axis of symmetry**.

CONCEPTS

5. a. (1, 0) and (3, 0)
   b. (0, –3)
   c. (2, 1)
   d. $x = 2$
   e. D: $(-\infty, \infty)$; R: $(-\infty, 1]$

7.

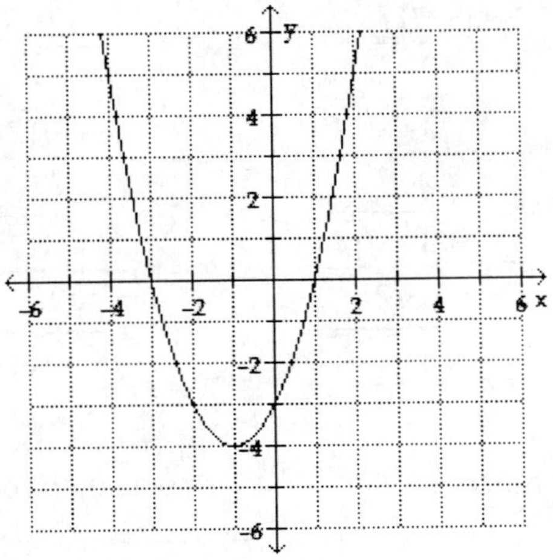

9. a. $f(x) = \boxed{2}(x^2 + 6x) + 11$
   b. $f(x) = \boxed{2}(x^2 + 6x + \boxed{9}) + 11 - \boxed{18}$

11. The x–intercepts are (–3, 0) and (5, 0), so the solutions of the function are $x = -3$ and $x = 5$.

NOTATION

13. $h = -1$

$f(x) = 2[x - (-1)]^2 + 6$

GUIDED PRACTICE

15.

x	$f(x) = x^2$
–3	$(-3)^2 = 9$
–2	$(-2)^2 = 4$
–1	$(-1)^2 = 1$
0	$(0)^2 = 0$
1	$(1)^2 = 1$
2	$(2)^2 = 4$
3	$(3)^2 = 9$

x	$g(x) = 2x^2$
–3	$2(-3)^2 = 18$
–2	$2(-2)^2 = 8$
–1	$2(-1)^2 = 2$
0	$2(0)^2 = 0$
1	$2(1)^2 = 2$
2	$2(2)^2 = 8$
3	$2(3)^2 = 18$

x	$s(x) = \frac{1}{2}x^2$
–3	$\frac{1}{2}(-3)^2 = \frac{9}{2}$
–2	$\frac{1}{2}(-2)^2 = 2$
–1	$\frac{1}{2}(-1)^2 = \frac{1}{2}$
0	$\frac{1}{2}(0)^2 = 0$
1	$\frac{1}{2}(1)^2 = \frac{1}{2}$
2	$\frac{1}{2}(2)^2 = 2$
3	$\frac{1}{2}(3)^2 = \frac{9}{2}$

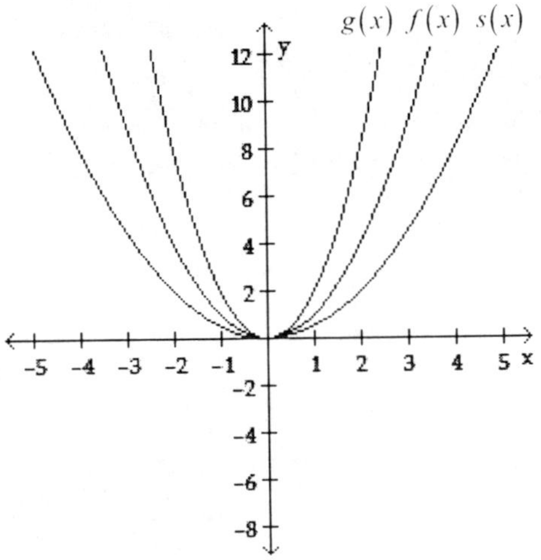

17.

x	$f(x) = 2x^2$
−3	$2(-3)^2 = 18$
−2	$2(-2)^2 = 8$
−1	$2(-1)^2 = 2$
0	$2(0)^2 = 0$
1	$2(1)^2 = 2$
2	$2(2)^2 = 8$
3	$2(3)^2 = 18$

x	$g(x) = -2x^2$
−3	$-2(-3)^2 = -18$
−2	$-2(-2)^2 = -8$
−1	$-2(-1)^2 = -2$
0	$-2(0)^2 = 0$
1	$-2(1)^2 = -2$
2	$-2(2)^2 = -8$
3	$-2(3)^2 = -18$

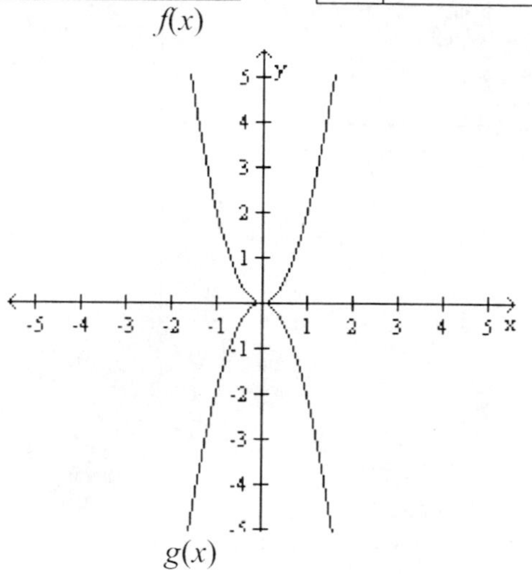

19.

x	$f(x) = 4x^2$
−3	$4(-3)^2 = 36$
−2	$4(-2)^2 = 16$
−1	$4(-1)^2 = 4$
0	$4(0)^2 = 0$
1	$4(1)^2 = 4$
2	$4(2)^2 = 16$
3	$4(3)^2 = 36$

The graph of $g(x) = 4x^2 + 3$ is shifted 3 units up, and the graph of $s(x) = 4x^2 - 2$ is shifted 2 units down.

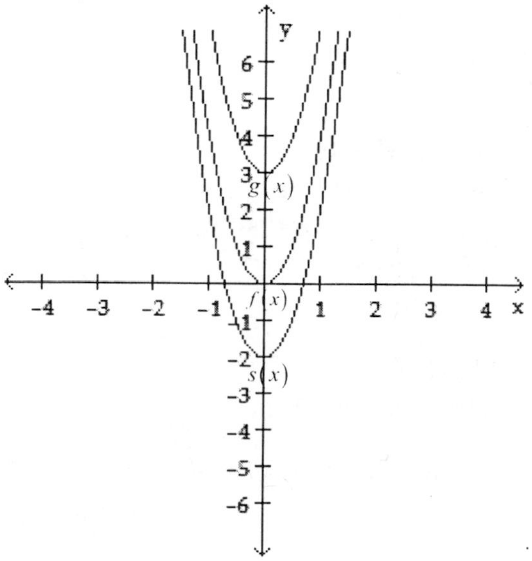

21.

x	$f(x) = 3x^2$
−3	$3(-3)^2 = 27$
−2	$3(-2)^2 = 12$
−1	$3(-1)^2 = 3$
0	$3(0)^2 = 0$
1	$3(1)^2 = 3$
2	$3(2)^2 = 12$
3	$3(3)^2 = 27$

The graph of $g(x) = 3(x+2)^2$ would be shifted 2 units left.

The graph of $s(x) = 3(x-3)^2$ would be shifted 3 units right.

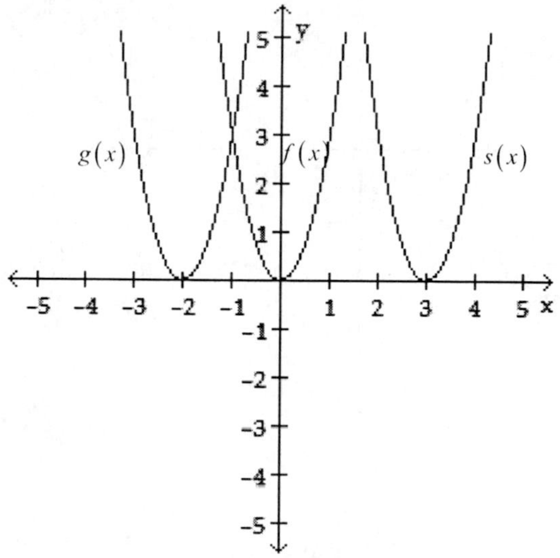

23. $f(x) = (x-1)^2 + 2$
The vertex is (1, 2).
The axis of symmetry is $x = 1$.
The parabola opens upward.

25. $f(x) = -2(x+3)^2 - 4$
The vertex is (−3, −4).
The axis of symmetry is $x = -3$.
The parabola opens downward.

27. $f(x) = -0.5(x-7.5)^2 + 8.5$
The vertex is (7.5, 8.5).
The axis of symmetry is $x = 7.5$.
The parabola opens downward.

29. $f(x) = 2x^2 - 4$

$f(x) = 2(x-0)^2 - 4$

The vertex is (0, –4).
The axis of symmetry is $x = 0$.
The parabola opens upward.

31. $f(x) = (x-3)^2 + 2$

The vertex is (3, 2).
The axis of symmetry is $x = 3$.
The parabola opens upward.

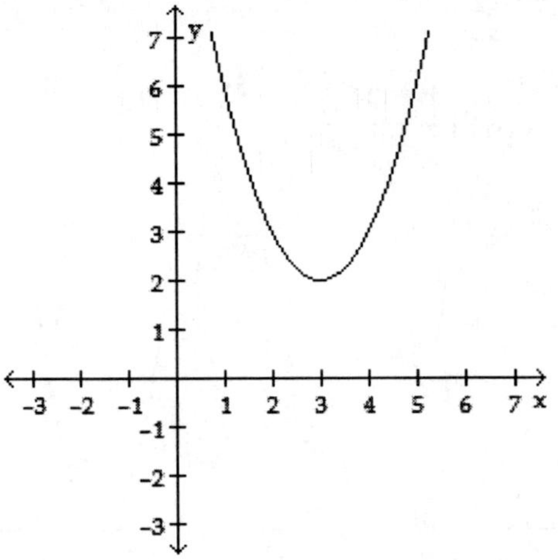

33. $f(x) = -(x-2)^2$

$f(x) = -(x-2)^2 + 0$

The vertex is (2, 0).
The axis of symmetry is $x = 2$.
The parabola opens downward.

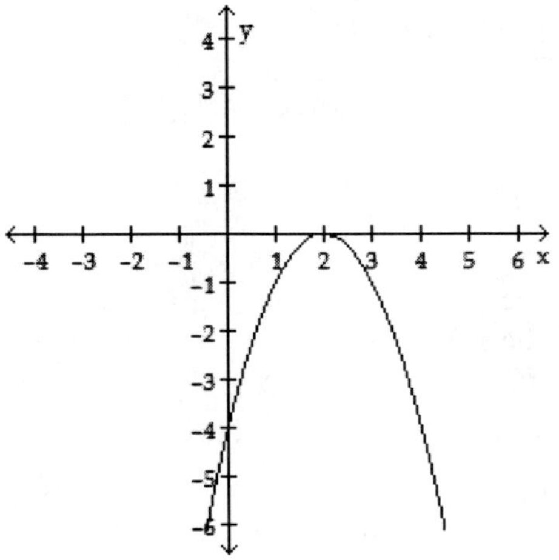

35. $f(x) = -2(x+3)^2 + 4$

The vertex is (–3, 4).
The axis of symmetry is $x = -3$.
The parabola opens downward.

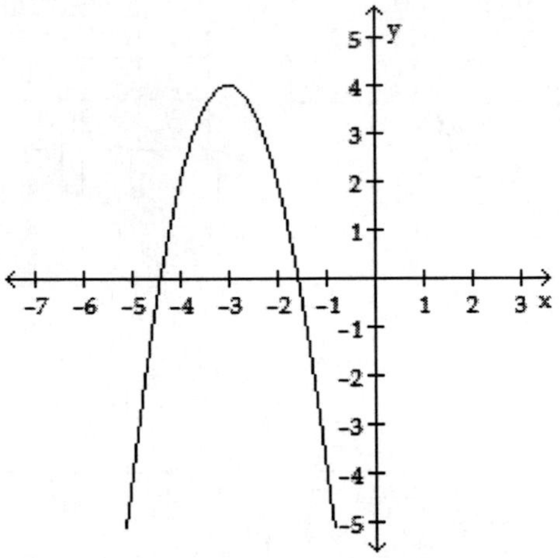

37. $f(x) = \frac{1}{2}(x+1)^2 - 3$

The vertex is (–1, –3).
The axis of symmetry is $x = -1$.
The parabola opens upward.

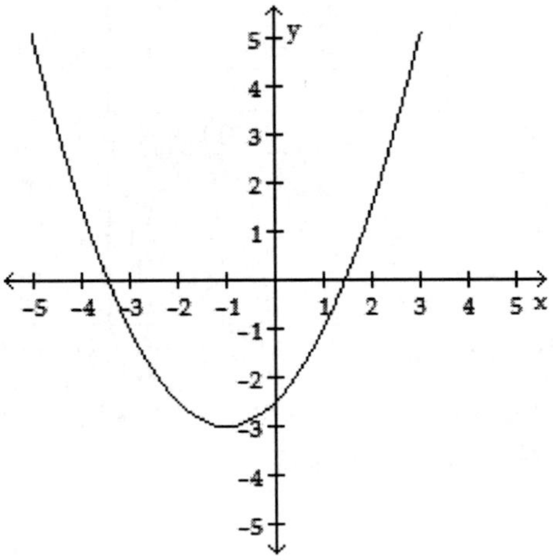

39. $f(x) = x^2 + 4x + 5$

$f(x) = \left(x^2 + 4x + \left(\frac{4}{2}\right)^2\right) + 5 - \left(\frac{4}{2}\right)^2$

$f(x) = (x^2 + 4x + 4) + 5 - 4$

$f(x) = (x+2)^2 + 1$

The vertex is (–2, 1).
The axis of symmetry is $x = -2$.
The parabola opens upward.

41. $f(x) = -x^2 + 6x - 15$

$$f(x) = -\left(x^2 - 6x + \left(\frac{-6}{2}\right)^2\right) - 15 + \left(\frac{-6}{2}\right)^2$$

$$f(x) = -(x^2 - 6x + 9) - 15 + 9$$

$$f(x) = -(x-3)^2 - 6$$

The vertex is (3, –6).
The axis of symmetry is $x = 3$.
The parabola opens downward.

43. $f(x) = x^2 + 2x - 3$

$$f(x) = (x^2 + 2x) - 3$$

$$f(x) = (x^2 + 2x + 1) - 3 - 1$$

$$f(x) = (x+1)^2 - 4$$

The vertex is (–1, –4).
The axis of symmetry is $x = -1$.
The parabola opens upward.

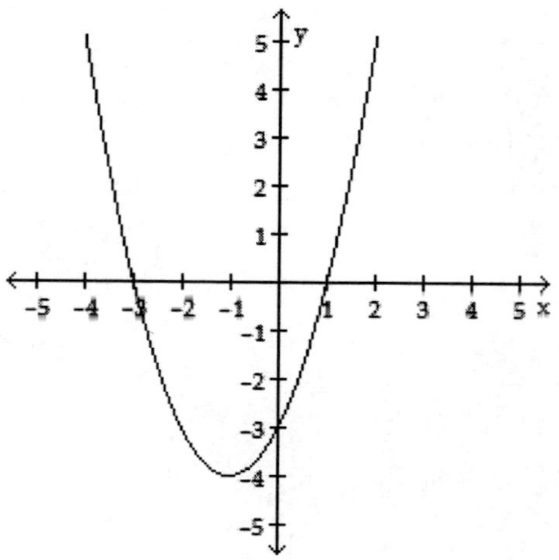

45. $f(x) = 4x^2 + 24x + 37$

$$f(x) = 4(x^2 + 6x) + 37$$

$$f(x) = 4(x^2 + 6x + 9) + 37 - 4(9)$$

$$f(x) = 4(x+3)^2 + 37 - 36$$

$$f(x) = 4(x+3)^2 + 1$$

The vertex is (–3, 1).
The axis of symmetry is $x = -3$.
The parabola opens upward.

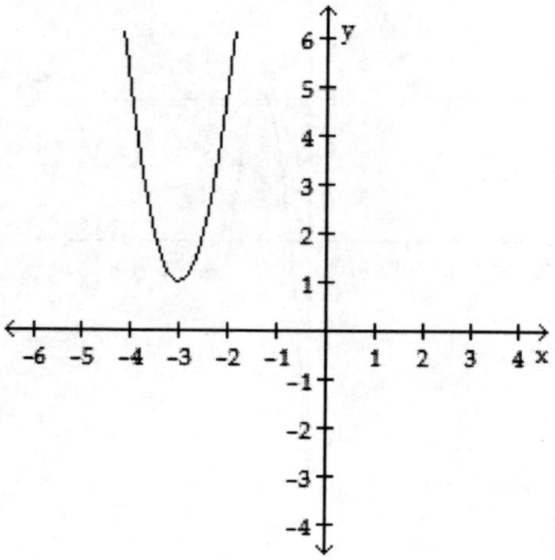

47. $f(x) = x^2 + x - 6$

$$f(x) = (x^2 + x) - 6$$

$$f(x) = \left(x^2 + x + \frac{1}{4}\right) - 6 - \frac{1}{4}$$

$$f(x) = \left(x + \frac{1}{2}\right)^2 - \frac{25}{4}$$

The vertex is $\left(-\frac{1}{2}, -\frac{25}{4}\right)$.

The axis of symmetry is $x = -\frac{1}{2}$.

The parabola opens upward.

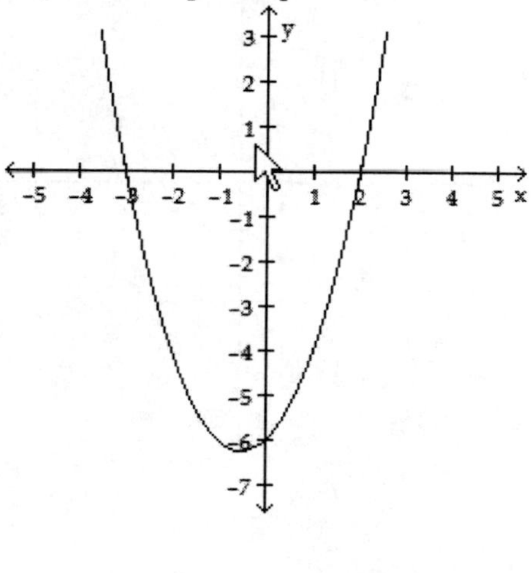

49. $f(x) = -4x^2 + 16x - 10$

$f(x) = -4(x^2 - 4x) - 10$

$f(x) = -4(x^2 - 4x + 4) - 10 - (-4)(4)$

$f(x) = -4(x-2)^2 - 10 + 16$

$f(x) = -4(x-2)^2 + 6$

The vertex is (2, 6).
The axis of symmetry is $x = 2$.
The parabola opens downward.

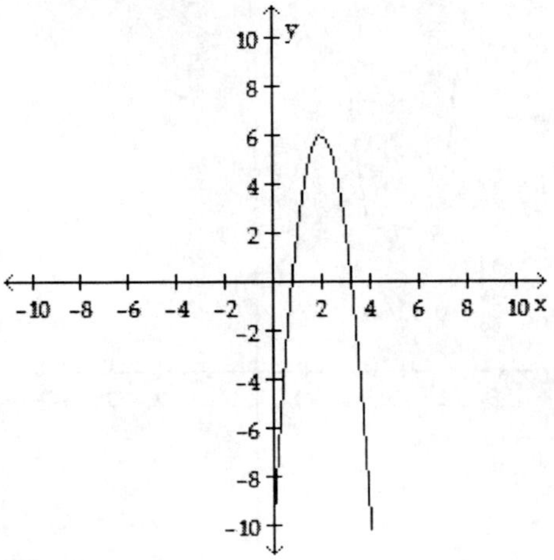

51. $f(x) = 2x^2 + 8x + 6$

$f(x) = 2(x^2 + 4x) + 6$

$f(x) = 2(x^2 + 4x + 4) + 6 - 2(4)$

$f(x) = 2(x+2)^2 + 6 - 8$

$f(x) = 2(x+2)^2 - 2$

The vertex is (–2, –2).
The axis of symmetry is $x = -2$.
The parabola opens upward.

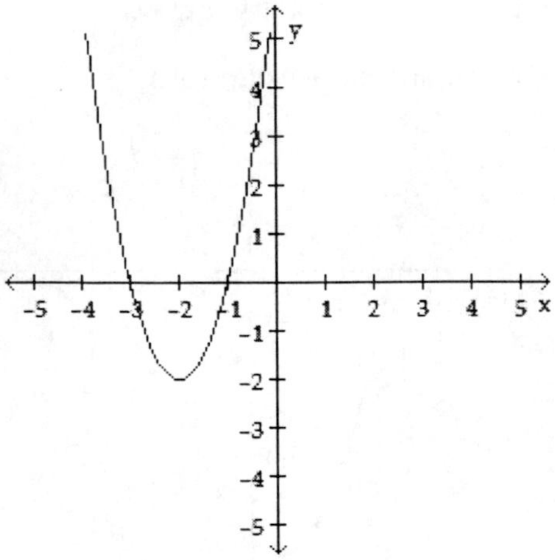

53. $f(x) = -x^2 - 8x - 17$

$f(x) = -(x^2 + 8x) - 17$

$f(x) = -(x^2 + 8x + 16) - 17 - (-1)(16)$

$f(x) = -(x+4)^2 - 17 + 16$

$f(x) = -(x+4)^2 - 1$

The vertex is (–4, –1).
The axis of symmetry is $x = -4$.
The parabola opens downward.

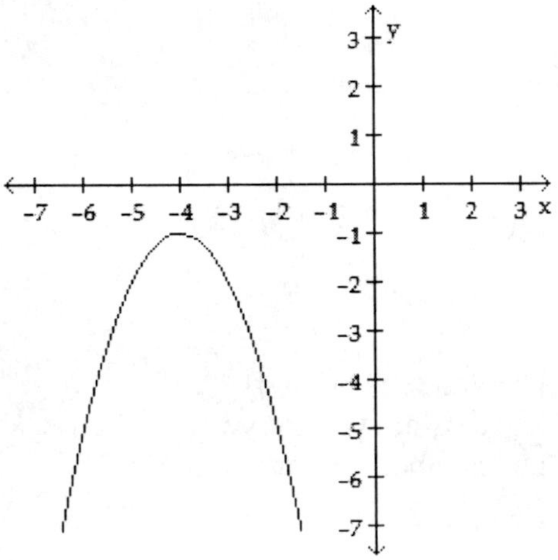

55. $f(x) = x^2 + 2x - 5$

$$x = \frac{-b}{2a} = \frac{-2}{2(1)} = \frac{-2}{2} = -1$$

$$f(-1) = (-1)^2 + 2(-1) - 5 = 1 - 2 - 5 = -6$$

The vertex is (–1, –6).

57. $f(x)=2x^2-3x+4$

$$x=\frac{-b}{2a}$$
$$=\frac{-(-3)}{2(2)}$$
$$=\frac{3}{4}$$

$$f\left(\frac{3}{4}\right)=2\left(\frac{3}{4}\right)^2-3\left(\frac{3}{4}\right)+4$$
$$=2\left(\frac{9}{16}\right)-\frac{9}{4}+4$$
$$=\frac{18}{16}-\frac{9}{4}+4$$
$$=\frac{18}{16}-\frac{36}{16}+\frac{64}{16}$$
$$=\frac{46}{16}$$
$$=\frac{23}{8}$$

The vertex is $\left(\frac{3}{4},\frac{23}{8}\right)$.

59. x–intercept: Let $f(x) = 0$.

$$0=x^2-2x-35$$
$$0=(x-7)(x+5)$$
$$x-7=0 \quad \text{or} \quad x+5=0$$
$$x=7 \qquad\qquad x=-5$$

x–intercepts are (7, 0) and (–5, 0)

y–intercept = $(0, c)$

$c = -35$

y–intercept is (0, –35)

61. x–intercept: Let $f(x) = 0$.

$$0=-2x^2+4x$$
$$0=-2x(x-2)$$
$$-2x=0 \quad \text{or} \quad x-2=0$$
$$x=0 \qquad\qquad x=2$$

x–intercepts are (0, 0) and (2, 0)

y–intercept = $(0, c)$

$c = 0$

y–intercept is (0, 0)

63. $f(x)=x^2+4x+4$

Step 1: Since $a = 1 > 0$, the parabola opens upward.

Step 2: Find the vertex and axis of symmetry.

$$x=\frac{-b}{2a}=\frac{-4}{2(1)}=\frac{-4}{2}=-2$$

$$y=(-2)^2+4(-2)+4=4-8+4=0$$

vertex: (–2, 0)

The axis of symmetry is $x = -2$.

Step 3: Find the x– and y–intercepts.

Since $c = 4$, the y–intercept is (0, 4).

To find the x–intercepts, let $y = 0$ and solve the equation for x.

$$0=x^2+4x+4$$
$$0=(x+2)(x+2)$$
$$x+2=0 \quad \text{or} \quad x+2=0$$
$$x=-2 \qquad\qquad x=-2$$

x–intercept: (–2, 0)

y–intercept: (0, 4)

Step 4: Find another point(s).

Let $x = -1$.

$$y=(-1)^2+4(-1)+4=1-4+4=1$$

(–1, 1)

Use symmetry to find the points (–3, 1) and (–4, 4).

Step 5: Plot the points and draw the parabola.

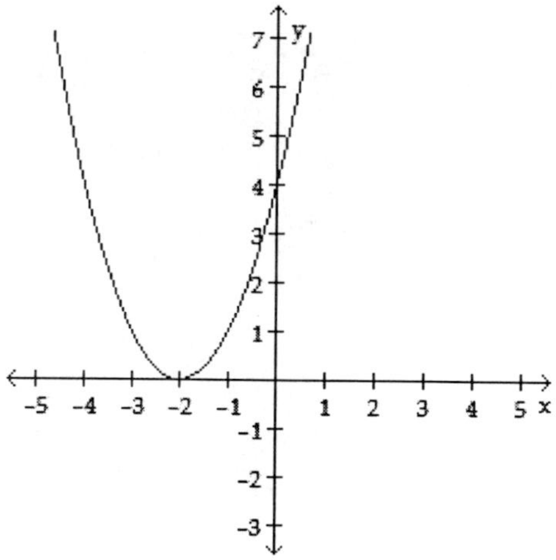

65. $f(x)=-x^2+2x-1$

Step 1: Since $a=-1<0$, the parabola opens downward.

Step 2: Find the vertex and axis of symmetry.

$$x=\frac{-b}{2a}=\frac{-2}{2(-1)}=\frac{-2}{-2}=1$$

$$y=-(1)^2+2(1)-1=-1+2-1=0$$

vertex: (1, 0)

The axis of symmetry is $x=1$.

Step 3: Find the x– and y–intercepts.

Since $c=-1$, the y–intercept is (0, –1).

To find the x–intercepts, let $y=0$ and solve the equation for x.

$$0=-x^2+2x-1$$

$$x^2-2x+1=0$$

$$(x-1)(x-1)=0$$

$$x-1=0 \quad \text{or} \quad x-1=0$$

$$x=1 \qquad\qquad x=1$$

x–intercept: (1, 0)

y–intercept: (0, –1)

Step 4: Find another point(s).

Let $x=-1$.

$$y=-(-1)^2+2(-1)-1=-1-2-1=-4$$

(–1, –4)

Use symmetry to find the points (2, –1) and (3, –4).

Step 5: Plot the points and draw the parabola.

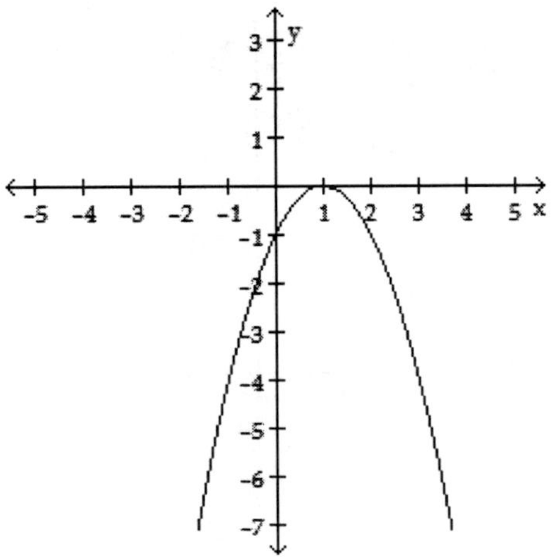

67. $f(x)=x^2-2x$

Step 1: Since $a=1>0$, the parabola opens upward.

Step 2: Find the vertex and axis of symmetry.

$$x=\frac{-b}{2a}=\frac{-(-2)}{2(1)}=\frac{2}{2}=1$$

$$y=(1)^2-2(1)=1-2=-1$$

vertex: (1, –1)

The axis of symmetry is $x=1$.

Step 3: Find the x– and y–intercepts.

Since $c=0$, the y–intercept is (0, 0).

To find the x–intercepts, let $y=0$ and solve the equation for x.

$$0=x^2-2x$$

$$0=x(x-2)$$

$$x=0 \quad \text{or} \quad x-2=0$$

$$x=0 \qquad\qquad x=2$$

x–intercepts: (0, 0) and (2, 0)

y–intercept: (0, 0)

Step 4: Find another point(s).

Let $x=-1$.

$$y=(-1)^2-2(-1)=1+2=3$$

(–1, 3)

Use symmetry to find the point (3, 3).

Step 5: Plot the points and draw the parabola.

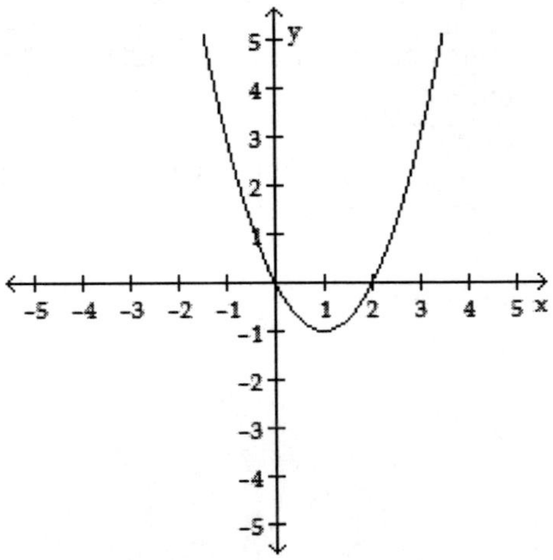

69. $f(x) = 2x^2 - 8x + 6$

Step 1: Since $a = 2 > 0$, the parabola opens upward.

Step 2: Find the vertex and axis of symmetry.

$$x = \frac{-b}{2a} = \frac{-(-8)}{2(2)} = \frac{8}{4} = 2$$

$$y = 2(2)^2 - 8(2) + 6 = 8 - 16 + 6 = -2$$

vertex: (2, –2)

The axis of symmetry is $x = 2$.

Step 3: Find the x– and y–intercepts.

Since $c = 6$, the y–intercept is (0, 6).

To find the x–intercepts, let $y = 0$ and solve the equation for x.

$$0 = 2x^2 - 8x + 6$$

$$0 = 2(x^2 - 4x + 3)$$

$$0 = (x-1)(x-3)$$

$$x - 1 = 0 \quad \text{or} \quad x - 3 = 0$$

$$x = 1 \qquad\qquad x = 3$$

x–intercepts: (1, 0) and (3, 0)

y–intercept: (0, 6)

Step 4: Use symmetry to find the point (4, 6)

Step 5: Plot the points and draw the parabola.

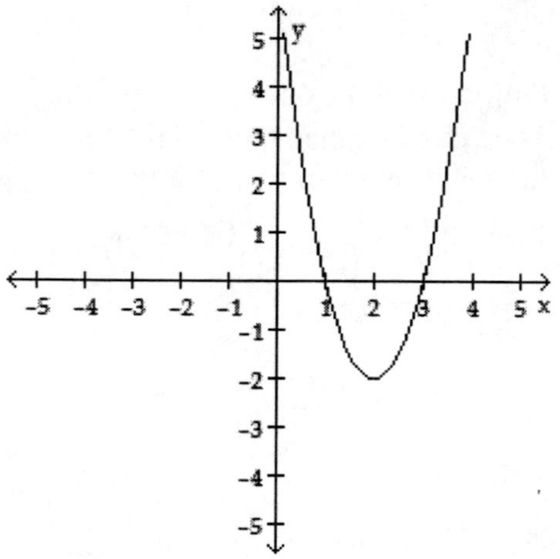

71. $f(x) = -6x^2 - 12x - 8$

Step 1: Since $a = -6 < 0$, the parabola opens downward.

Step 2: Find the vertex and axis of symmetry.

$$x = \frac{-b}{2a} = \frac{-(-12)}{2(-6)} = \frac{12}{-12} = -1$$

$$y = -6(-1)^2 - 12(-1) - 8$$

$$= -6 + 12 - 8$$

$$= -2$$

vertex: (–1, –2)

The axis of symmetry is $x = -1$.

Step 3: Find the x– and y–intercepts.

Since $c = -8$, the y–intercept is (0, –8).

Since the vertex is located below the x–axis and the parabola opens downward, there are no x–intercepts.

x–intercepts: none

y–intercept: (0, –8)

Step 4: Find another point(s).

Let $x = 1$.

$$y = -6(1)^2 - 12(1) - 8$$

$$= -6 - 12 - 8$$

$$= -26$$

(1, –26)

Use symmetry to find the points (–2, –8) and (–3, –26)

Step 5: Plot the points and draw the parabola.

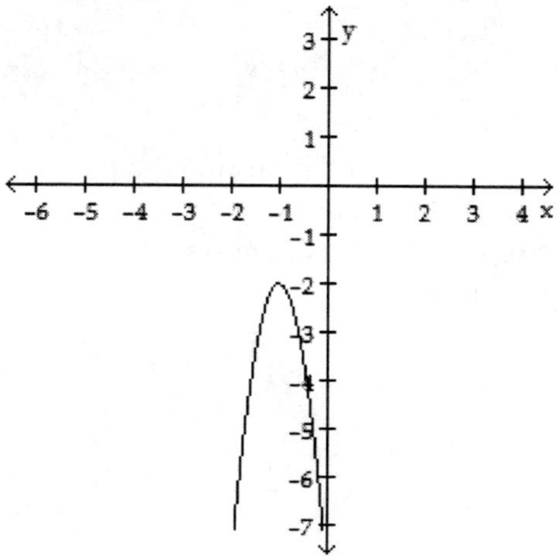

73. $f(x)=4x^2-12x+9$

Step 1: Since $a = 4 > 0$, the parabola opens upward.

Step 2: Find the vertex and axis of symmetry.

$$x=\frac{-b}{2a}=\frac{-(-12)}{2(4)}=\frac{12}{8}=\frac{3}{2}$$

$$y=4\left(\frac{3}{2}\right)^2-12\left(\frac{3}{2}\right)+9$$
$$=4\left(\frac{9}{4}\right)-12\left(\frac{3}{2}\right)+9$$
$$=9-18+9$$
$$=0$$

vertex: $\left(\frac{3}{2},0\right)$

The axis of symmetry is $x=\frac{3}{2}$.

Step 3: Find the x– and y–intercepts.
Since $c = 9$, the y–intercept is (0, 9).
To find the x–intercepts, let $y = 0$ and solve the equation for x.

$$0=4x^2-12x+9$$
$$0=(2x-3)(2x-3)$$
$$2x-3=0 \quad \text{or} \quad 2x-3=0$$
$$2x=3 \qquad 2x=3$$
$$x=\frac{3}{2} \qquad x=\frac{3}{2}$$

x–intercept: $\left(\frac{3}{2},0\right)$

y–intercept: (0, 9)

Step 4: Find another point(s).
Let $x = 1$.

$y=4(1)^2-12(1)+9=4-12+9=1$

(1, 1)

Use symmetry to find the points (2, 1) and (3, 9).

Step 5: Plot the points and draw the parabola.

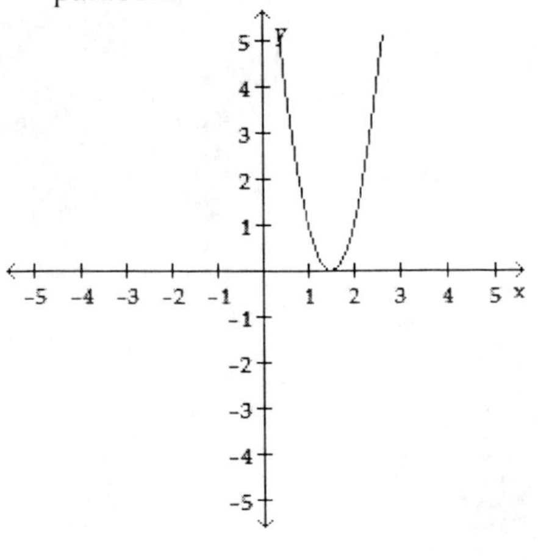

75. $f(x)=2x^2-x+1$

vertex: (0.25, 0.88)

77. $f(x)=-x^2+x+7$

vertex: (0.50, 7.25)

79. $x^2+x-6=0$

The x–intercepts are (–3, 0) and (2, 0), so the solutions of the equation are $x = -3$ and $x = 2$.

81. $0.5x^2-0.7x-3=0$

The x–intercepts are (–1.85, 0) and (3.25, 0), so the solutions of the equation are $x = -1.85$ and $x = 3.25$.

APPLICATIONS

83. CROSSWORD PUZZLES

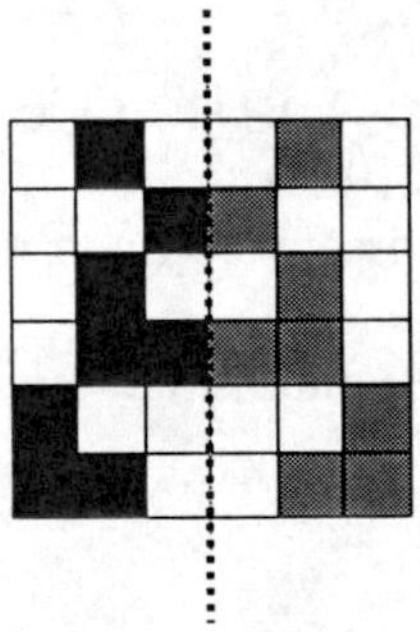

85. OPERATING COSTS

The vertex of the parabola represented by the function would be its minimum cost.

$$C(n)=2.2n^2-66n+655$$

$$n=\frac{-b}{2a}$$
$$=\frac{-(-66)}{2(2.2)}$$
$$=\frac{66}{4.4}$$
$$=15 \text{ minutes}$$

$$C(15)=2.2(15)^2-66(15)+655$$
$$=2.2(225)-66(15)+655$$
$$=495-990+655$$
$$=\$160$$

The cost of running the machine is a minimum at 15 minutes. The minimum cost is $160.

87. FIREWORKS

The vertex of the parabola would be its maximum height and time of explosion.

$$s = 120t - 16t^2$$
$$s = -16t^2 + 120t$$
$$t = \frac{-b}{2a} = \frac{-120}{2(-16)} = \frac{-120}{-32} = 3.75$$

$$s(3.75) = 120(3.75) - 16(3.75)^2 = 450 - 16(14.0625) = 450 - 225 = 225$$

The shell reaches its maximum height of 225 ft in 3.75 seconds.

89. POLICE PATROL OFFICER

Let the width = x and the length = y.
The sum of the four sides is 300 ft.

$$x + y + x + y = 300$$
$$2x + 2y = 300$$
$$2y = -2x + 300$$
$$y = -x + 150$$

To find a function for the area, multiply length times width.

$$\text{Area} = \text{width} \bullet \text{length} = x(y) = x(-x + 150)$$
$$A(x) = -x^2 + 150x$$

Use the vertex formula to find the width (x) and substitution to find the length (y).

$$A(x) = -x^2 + 150x$$
$$x = \frac{-b}{2a} = \frac{-150}{2(-1)} = \frac{-150}{-2} = 75 \text{ ft}$$
$$y = -x + 150 = -75 + 150 = 75$$

The dimensions are 75 ft by 75 ft.

To find the maximum area, find $A(75)$.

$$A(75) = -(75)^2 + 150(75) = -5{,}625 + 11{,}250 = 5{,}625 \text{ ft}^2$$

The maximum area is 5,625 square ft.

91. MILITARY HISTORY
The vertex of the parabola represented by the function would be its maximum strength.

$$N(x) = -0.0534x^2 + 0.337x + 0.97$$

$$\begin{aligned} x &= \frac{-b}{2a} \\ &= \frac{-0.337}{2(-0.0534)} \\ &= \frac{-0.337}{-0.1068} \\ &= 3.16 \end{aligned}$$

$$\begin{aligned} N(3.16) &= -0.0534(3.16)^2 + 0.337(3.16) + 0.97 \\ &= -0.0534(9.9856) + 0.337(3.16) + 0.97 \\ &= -0.533 + 1.065 + 0.97 \\ &= 1.502 \end{aligned}$$

It maximum strength would be in the 3rd year, which represents 1968. During that year, the army's personnel strength level was 1.5 million. At this time in history, the U.S. involvement in the Vietnam War was at its peak.

93. MAXIMIZING REVENUE
The vertex of the parabola represented by the function would be its maximum revenue.

$$R(x) = -\frac{x^2}{5} + 80x - 1{,}000$$

$$R(x) = -\frac{1}{5}x^2 + 80x - 1{,}000$$

$$R(x) = -0.2x^2 + 80x - 1{,}000$$

$$\begin{aligned} x &= \frac{-b}{2a} \\ &= \frac{-80}{2(-0.2)} \\ &= \frac{-80}{-0.4} \\ &= 200 \end{aligned}$$

$$\begin{aligned} R(200) &= -\frac{(200)^2}{5} + 80(200) - 1{,}000 \\ &= -\frac{40{,}000}{5} + 80(200) - 1{,}000 \\ &= -8{,}000 + 16{,}000 - 1{,}000 \\ &= 7{,}000 \end{aligned}$$

To obtain the maximum revenue, they must sell 200 stereos. The maximum revenue is $7,000.

WRITING

95. Answers will vary.

97. Answers will vary.

99. Answers will vary.

REVIEW

101.

$$\begin{aligned} \frac{\sqrt{3}}{\sqrt{50}} &= \frac{\sqrt{3}}{\sqrt{25}\sqrt{2}} \\ &= \frac{\sqrt{3}}{5\sqrt{2}} \\ &= \frac{\sqrt{3}}{5\sqrt{2}} \cdot \frac{\sqrt{2}}{\sqrt{2}} \\ &= \frac{\sqrt{6}}{5\sqrt{4}} \\ &= \frac{\sqrt{6}}{5(2)} \\ &= \frac{\sqrt{6}}{10} \end{aligned}$$

103.

$$\begin{aligned} 3\left(\sqrt{5b} - \sqrt{3}\right)^2 &= 3\left(\sqrt{5b} - \sqrt{3}\right)\left(\sqrt{5b} - \sqrt{3}\right) \\ &= 3\left(\sqrt{25b^2} - \sqrt{15b} - \sqrt{15b} + \sqrt{9}\right) \\ &= 3\left(5b - 2\sqrt{15b} + 3\right) \\ &= 15b - 6\sqrt{15b} + 9 \end{aligned}$$

CHALLENGE PROBLEMS

105.

$$f(x) = x - x^2$$

$$f(x) = -x^2 + x$$

$$\begin{aligned} \frac{-b}{2a} &= \frac{-1}{2(-1)} \\ &= \frac{1}{2} \end{aligned}$$

SECTION 8.5

VOCABUALRY

1. $x^2 + 3x - 18 < 0$ is an example of a **quadratic** inequality in one variable.

3. $y \le x^2 - 4x + 3$ is an example of a nonlinear inequality in **two** variables.

CONCEPTS

5. $(-\infty, -1)$, $(1, 4)$, $(4, \infty)$

7. a. yes
   b. no
   c. yes
   d. no

9. a. $(-3, 2)$
   b. $(-\infty,-1] \cup [1,\infty)$

11. a. solid
    b. yes

$$y \le x^2 + 2x + 1$$
$$0 \overset{?}{\le} 0^2 + 2(0) + 1$$
$$0 \overset{?}{\le} 0 + 0 + 1$$
$$0 \le 1$$

NOTATION

13. $x^2 - 6x - 7 \ge 0$

GUIDED PRACTICE

15.

$$x^2 - 5x + 4 < 0$$
$$x^2 - 5x + 4 = 0$$
$$(x-1)(x-4) = 0$$
$$x - 1 = 0 \quad \text{or} \quad x - 4 = 0$$
$$x = 1 \qquad x = 4$$
$$(1,4)$$

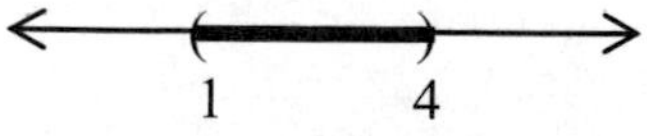

17.

$$x^2 - 8x + 15 > 0$$
$$x^2 - 8x + 15 = 0$$
$$(x-5)(x-3) = 0$$
$$x - 5 = 0 \quad \text{or} \quad x - 3 = 0$$
$$x = 5 \qquad x = 3$$
$$(-\infty,3) \cup (5,\infty)$$

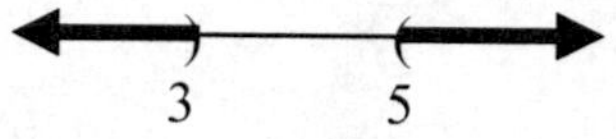

19.

$$x^2 - x \ge 42$$
$$x^2 - x - 42 \ge 0$$
$$x^2 - x - 42 = 0$$
$$(x+6)(x-7) = 0$$
$$x + 6 = 0 \quad \text{or} \quad x - 7 = 0$$
$$x = -6 \qquad x = 7$$
$$(-\infty,-6] \cup [7,\infty)$$

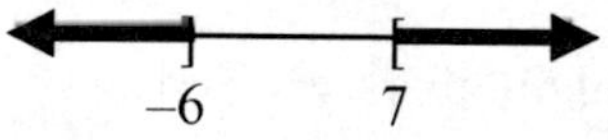

21.

$$x^2 + x \le 12$$
$$x^2 + x - 12 \le 0$$
$$x^2 + x - 12 = 0$$
$$(x-3)(x+4) = 0$$
$$x - 3 = 0 \quad \text{or} \quad x + 4 = 0$$
$$x = 3 \qquad x = -4$$
$$[-4,3]$$

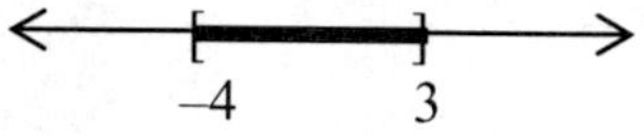

23.

$$\frac{1}{x} < 2$$

$$\frac{1}{x} = 2$$

$$\frac{1}{x} - 2 = 0$$

$$\frac{1}{x} - \frac{2x}{x} = 0$$

$$\frac{1-2x}{x} = 0$$

$$x\left(\frac{1-2x}{x}\right) = x(0)$$

$$1 - 2x = 0$$

$$-2x = -1$$

$$\frac{1}{2} = x$$

Set the denominator $= 0.$

$$x = 0$$

Critical numbers $= 0$ and $\frac{1}{2}$

$$(-\infty, 0) \cup \left(\frac{1}{2}, \infty\right)$$

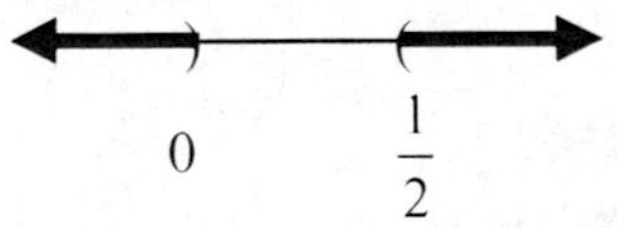

25.

$$\frac{5}{x} \geq -3$$

$$\frac{5}{x} = -3$$

$$\frac{5}{x} + 3 = 0$$

$$\frac{5}{x} + \frac{3x}{x} = 0$$

$$\frac{5+3x}{x} = 0$$

$$x\left(\frac{5+3x}{x}\right) = x(0)$$

$$5 + 3x = 0$$

$$3x = -5$$

$$x = -\frac{5}{3}$$

Set the denominator $= 0.$

$$x = 0$$

Critical numbers $= 0$ and $-\frac{5}{3}$

$$\left(-\infty, -\frac{5}{3}\right] \cup (0, \infty)$$

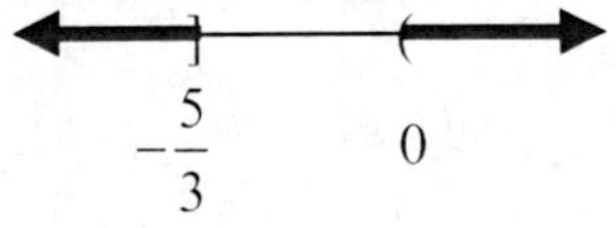

27.

$$\frac{x^2-x-12}{x-1}<0$$

$$\frac{x^2-x-12}{x-1}=0$$

$$(x-1)\left(\frac{x^2-x-12}{x-1}\right)=(x-1)(0)$$

$$x^2-x-12=0$$

$$(x-4)(x+3)=0$$

$$x-4=0 \quad \text{or} \quad x+3=0$$

$$x=4 \qquad\qquad x=-3$$

Set the denominator $=0$.

$$x-1=0$$

$$x=1$$

Critical numbers $=-3,\ 1,$ and 4

$$(-\infty,-3)\cup(1,4)$$

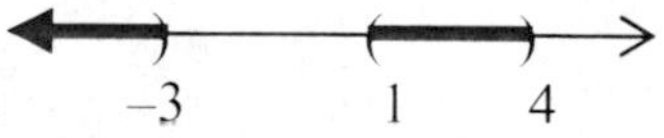

29.

$$\frac{6x^2-5x+1}{2x+1}>0$$

$$\frac{6x^2-5x+1}{2x+1}=0$$

$$(2x+1)\left(\frac{6x^2-5x+1}{2x+1}\right)=(2x+1)(0)$$

$$6x^2-5x+1=0$$

$$(2x-1)(3x-1)=0$$

$$2x-1=0 \quad \text{or} \quad 3x-1=0$$

$$x=\frac{1}{2} \qquad\qquad x=\frac{1}{3}$$

Set the denominator $=0$.

$$2x+1=0$$

$$x=-\frac{1}{2}$$

Critical numbers $=-\frac{1}{2},\ \frac{1}{3},$ and $\frac{1}{2}$

$$\left(-\frac{1}{2},\frac{1}{3}\right)\cup\left(\frac{1}{2},\infty\right)$$

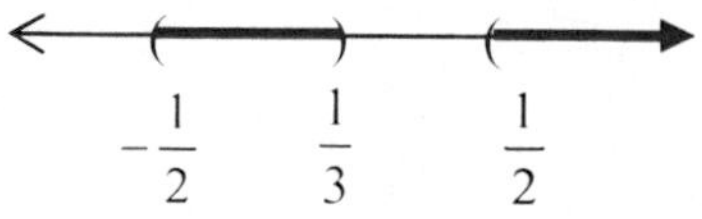

31.

$$\frac{3}{x-2}<\frac{4}{x}$$

$$\frac{3}{x-2}-\frac{4}{x}<0$$

$$\frac{3}{x-2}-\frac{4}{x}=0$$

$$\frac{3}{x-2}\cdot\frac{x}{x}-\frac{4}{x}\cdot\frac{x-2}{x-2}=0$$

$$\frac{3x-4(x-2)}{x(x-2)}=0$$

$$\frac{3x-4x+8}{x(x-2)}=0$$

$$\frac{-x+8}{x(x-2)}=0$$

$$x(x-2)\left(\frac{-x+8}{x(x-2)}\right)=x(x-2)(0)$$

$$-x+8=0$$

$$8=x$$

Set the denominator $=0$.

$$x(x-2)=0$$

$$x=0 \quad \text{or} \quad x-2=0$$

$$x=0 \qquad\qquad x=2$$

Critical numbers $=0,\ 2,$ and 8

$$(0,2)\cup(8,\infty)$$

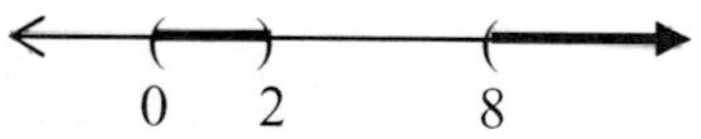

33.

$$\frac{7}{x-3} \geq \frac{2}{x+4}$$

$$\frac{7}{x-3} - \frac{2}{x+4} \geq 0$$

$$\frac{7}{x-3} - \frac{2}{x+4} = 0$$

$$\frac{7}{x-3} \bullet \frac{x+4}{x+4} - \frac{2}{x+4} \bullet \frac{x-3}{x-3} = 0$$

$$\frac{7(x+4)-2(x-3)}{(x-3)(x+4)} = 0$$

$$\frac{7x+28-2x+6}{(x-3)(x+4)} = 0$$

$$\frac{5x+34}{(x-3)(x+4)} = 0$$

$$(x-3)(x+4)\left(\frac{5x+34}{(x-3)(x+4)}\right) = (x-3)(x+4)(0)$$

$$5x+34 = 0$$

$$5x = -34$$

$$x = -\frac{34}{5}$$

Set the denominator $= 0$.

$$(x-3)(x+4) = 0$$

$$x-3=0 \quad \text{or} \quad x+4=0$$

$$x=3 \qquad x=-4$$

Critical numbers $= -\frac{34}{5}, -4$, and 3

$$\left[-\frac{34}{5}, -4\right) \cup (3, \infty)$$

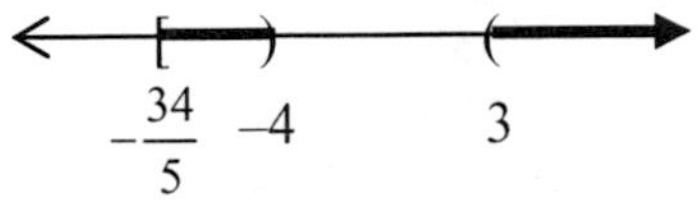

35. $y < x^2 + 1$

Complete a table of values and use (0, 0) as a check point.

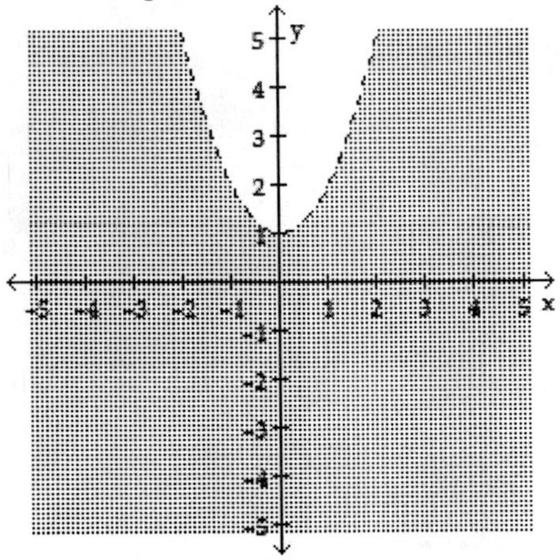

37. $y \leq x^2 + 5x + 6$

Complete a table of values and use (0, 0) as a check point.

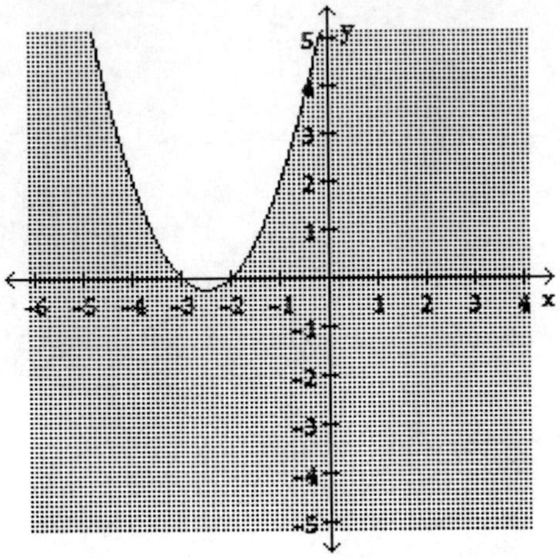

39. $y < |x+4|$

Complete a table of values and use (0, 0) as a check point.

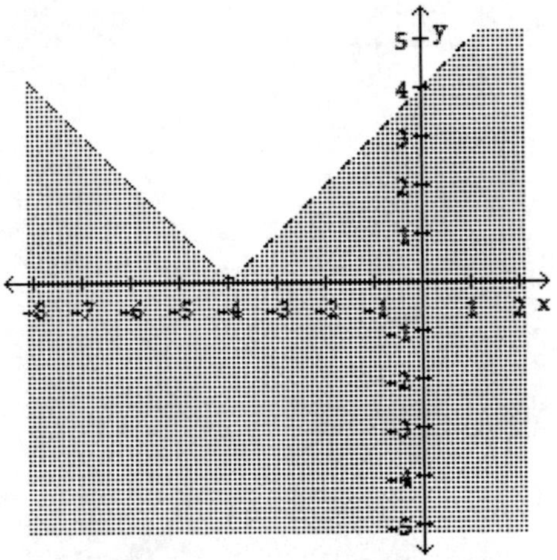

41. $y \leq -|x| + 2$

Complete a table of values and use (0, 0) as a check point.

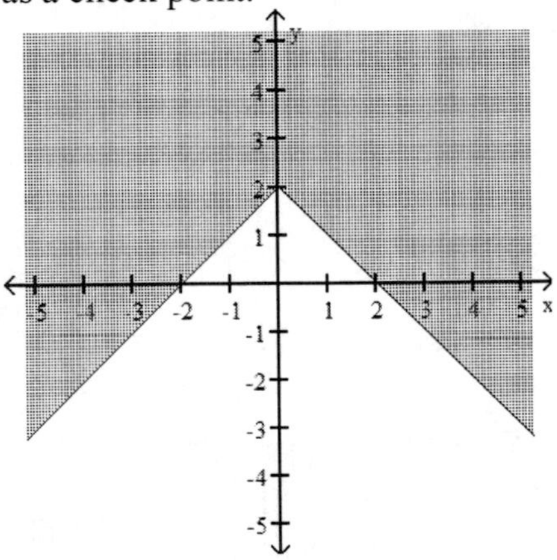

43. $x^2 - 2x - 3 < 0$

$(-1, 3)$

45. $\dfrac{x+3}{x-2} > 0$

$(-\infty, -3) \cup (2, \infty)$

TRY IT YOURSELF

47.

$$\frac{x}{x+4} \le \frac{1}{x+1}$$
$$\frac{x}{x+4} - \frac{1}{x+1} \le 0$$
$$\frac{x}{x+4} - \frac{1}{x+1} = 0$$
$$\frac{x}{x+4} \cdot \frac{x+1}{x+1} - \frac{1}{x+1} \cdot \frac{x+4}{x+4} = 0$$
$$\frac{x(x+1)-(x+4)}{(x+1)(x+4)} = 0$$
$$\frac{x^2+x-x-4}{(x+1)(x+4)} = 0$$
$$\frac{x^2-4}{(x+1)(x+4)} = 0$$
$$(x+1)(x+4)\left(\frac{x^2-4}{(x+1)(x+4)}\right) = (x+1)(x+4)(0)$$
$$x^2 - 4 = 0$$
$$(x+2)(x-2) = 0$$
$$x+2=0 \quad \text{or} \quad x-2=0$$
$$x=-2 \qquad\qquad x=2$$

Set the denominator $= 0$.

$$(x+1)(x+4) = 0$$
$$x+1=0 \quad \text{or} \quad x+4=0$$
$$x=-1 \qquad\qquad x=-4$$

Critical numbers $= -4, -2, -1,$ and 2

$(-4, -2] \cup (-1, 2]$

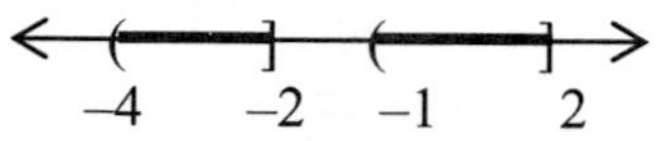

49.

$$x^2 \ge 9$$
$$x^2 - 9 \ge 0$$
$$x^2 - 9 = 0$$
$$(x+3)(x-3) = 0$$
$$x+3=0 \quad \text{or} \quad x-3=0$$
$$x=-3 \qquad\qquad x=3$$

$(-\infty, -3] \cup [3, \infty)$

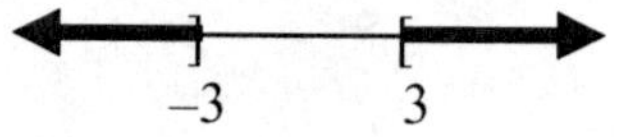

51.

$$x^2 + 6x \ge -9$$
$$x^2 + 6x + 9 \ge 0$$
$$x^2 + 6x + 9 = 0$$
$$(x+3)(x+3) = 0$$
$$x+3=0 \quad \text{or} \quad x+3=0$$
$$x=-3 \qquad\qquad x=-3$$

$(-\infty, \infty)$

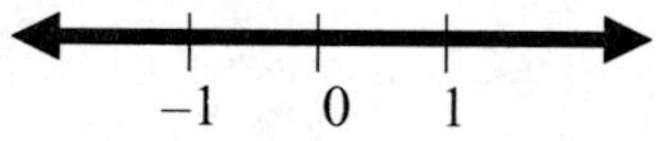

53.

$$\frac{x^2+x-2}{x-3} > 0$$
$$\frac{x^2+x-2}{x-3} = 0$$
$$(x-3)\left(\frac{x^2+x-2}{x-3}\right) = (x-3)(0)$$
$$x^2 + x - 2 = 0$$
$$(x-1)(x+2) = 0$$
$$x-1=0 \quad \text{or} \quad x+2=0$$
$$x=1 \qquad\qquad x=-2$$

Set the denominator $= 0$.

$$x - 3 = 0$$
$$x = 3$$

Critical numbers $= -2,\ 1,$ and 3

$(-2, 1) \cup (3, \infty)$

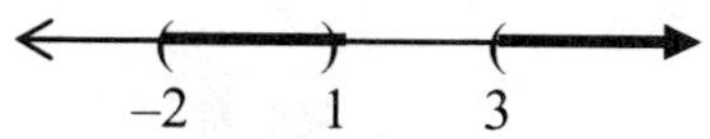

55.

$$2x^2 - 50 < 0$$
$$2(x^2 - 25) = 0$$
$$2(x+5)(x-5) = 0$$
$$x+5=0 \quad \text{or} \quad x-5=0$$
$$x=-5 \qquad x=5$$
$$(-5,5)$$

57.

$$\frac{2x-3}{3x+1} < 0$$
$$\frac{2x-3}{3x+1} = 0$$
$$(3x+1)\left(\frac{2x-3}{3x+1}\right) = (3x+1)0$$
$$2x-3=0$$
$$2x=3$$
$$x=\frac{3}{2}$$

Set the denominator $=0$.

$$3x+1=0$$
$$3x=-1$$
$$x=-\frac{1}{3}$$

Critical numbers $=-\frac{1}{3}$ and $\frac{3}{2}$

$$\left(-\frac{1}{3},\frac{3}{2}\right)$$

59.

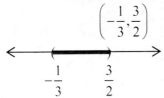

$$x^2 - 6x + 9 < 0$$
$$x^2 - 6x + 9 = 0$$
$$(x-3)(x-3) = 0$$
$$x-3=0 \quad \text{or} \quad x-3=0$$
$$x=3 \qquad x=3$$
$$\varnothing$$

No Solution

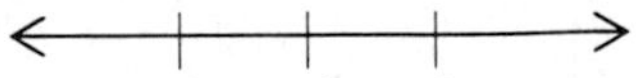

61.

$$\frac{5}{x+1} > \frac{3}{x-4}$$
$$\frac{5}{x+1} - \frac{3}{x-4} > 0$$
$$\frac{5}{x+1} - \frac{3}{x-4} = 0$$
$$\frac{x-4}{x-4}\cdot\frac{5}{x+1} - \frac{3}{x-4}\cdot\frac{x+1}{x+1} = 0$$
$$\frac{5(x-4)-3(x+1)}{(x+1)(x-4)} = 0$$
$$\frac{5x-20-3x-3}{(x+1)(x-4)} = 0$$
$$\frac{2x-23}{(x+1)(x-4)} = 0$$
$$(x+1)(x-4)\left(\frac{2x-23}{(x+1)(x-4)}\right) = (x+1)(x-4)(0)$$
$$2x-23=0$$
$$2x=23$$
$$x=\frac{23}{2}$$

Set the denominator $=0$.

$$(x+1)(x-4)=0$$
$$x+1=0 \quad \text{or} \quad x-4=0$$
$$x=-1 \qquad x=4$$

Critical numbers $=-1$, 4, and $\frac{23}{2}$

$$(-1,4)\cup\left(\frac{23}{2},\infty\right)$$

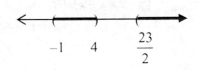

APPLICATIONS

63. BRIDGES

$$L = \frac{1}{9{,}000}x^2 + 5$$

$$\frac{1}{9{,}000}x^2 + 5 > 95$$

$$\frac{1}{9{,}000}x^2 + 5 = 95$$

$$\frac{1}{9{,}000}x^2 - 90 = 0$$

$$\frac{x^2}{9{,}000} - \frac{90}{1} \cdot \frac{9{,}000}{9{,}000} = 0$$

$$\frac{x^2 - 810{,}000}{9{,}000} = 0$$

$$9{,}000\left(\frac{x^2 - 810{,}000}{9{,}000}\right) = 9{,}000(0)$$

$$x^2 - 810{,}000 = 0$$

$$(x-900)(x+900) = 0$$

$$x - 900 = 0 \quad \text{or} \quad x + 900 = 0$$

$$x = 900 \qquad x = -900$$

$$(-2{,}100, -900) \cup (900, 2100)$$

WRITING

65. Answers will vary.

67. Answers will vary.

REVIEW

69. $x = ky$

71. $t = kxy$

CHALLENGE PROBLEMS

73. a.

$$x^2 - x - 12 > 0$$

$$x^2 - x - 12 = 0$$

$$(x-4)(x+3) = 0$$

$$x - 4 = 0 \quad \text{or} \quad x + 3 = 0$$

$$x = 4 \qquad x = -3$$

$$(-\infty, -3) \cup (4, \infty)$$

b. Answers will vary.

$$\frac{x-4}{x+3} > 0$$

CHAPTER 8 REVIEW

SECTION 8.1
The Square Root Property and Completing the Square

1.

$$x^2+9x+20=0$$
$$(x+4)(x+5)=0$$
$$x+4=0 \quad \text{or} \quad x+5=0$$
$$x=-4 \qquad x=-5$$

2.

$$6x^2+17x+5=0$$
$$(2x+5)(3x+1)=0$$
$$2x+5=0 \quad \text{or} \quad 3x+1=0$$
$$2x=-5 \qquad 3x=-1$$
$$x=-\frac{5}{2} \qquad x=-\frac{1}{3}$$

3.

$$x^2=28$$
$$x=\pm\sqrt{28}$$
$$x=\pm\sqrt{4}\sqrt{7}$$
$$x=\pm2\sqrt{7}$$

4.

$$(t+2)^2=36$$
$$t+2=\pm\sqrt{36}$$
$$t+2=\pm6$$
$$t+2=6 \quad \text{or} \quad t+2=-6$$
$$t=4 \qquad t=-8$$

5.

$$a^2+25=0$$
$$a^2=-25$$
$$a=\pm\sqrt{-25}$$
$$a=\pm5i$$

6.

$$5x^2-49=0$$
$$5x^2=49$$
$$x^2=\frac{49}{5}$$
$$x=\pm\sqrt{\frac{49}{5}}$$
$$x=\pm\frac{7}{\sqrt{5}}$$
$$x=\pm\frac{7}{\sqrt{5}}\bullet\frac{\sqrt{5}}{\sqrt{5}}$$
$$x=\pm\frac{7\sqrt{5}}{5}$$

7.

$$A=\pi r^2$$
$$\frac{A}{\pi}=\frac{\pi r^2}{\pi}$$
$$\frac{A}{\pi}=r^2$$
$$r^2=\frac{A}{\pi}$$
$$r=\sqrt{\frac{A}{\pi}}$$
$$r=\frac{\sqrt{A}}{\sqrt{\pi}}$$
$$r=\frac{\sqrt{A}}{\sqrt{\pi}}\bullet\frac{\sqrt{\pi}}{\sqrt{\pi}}$$
$$r=\frac{\sqrt{A\pi}}{\pi}$$

8. $\frac{1}{4}$ must be added to both sides because

$$\left(\frac{-1}{2}\right)^2=\frac{1}{4}$$
$$x^2-x+\frac{1}{4}=\left(x-\frac{1}{2}\right)^2$$

9.

$$x^2+6x+8=0$$
$$x^2+6x=-8$$
$$x^2+6x+\left(\frac{6}{2}\right)^2=-8+\left(\frac{6}{2}\right)^2$$
$$x^2+6x+(3)^2=-8+(3)^2$$
$$x^2+6x+9=-8+9$$
$$(x+3)^2=1$$
$$x+3=\pm\sqrt{1}$$
$$x+3=\pm 1$$
$$x+3=1 \quad \text{or} \quad x+3=-1$$
$$x=-2 \qquad x=-4$$

10.

$$2x^2-6x+3=0$$
$$2x^2-6x=-3$$
$$\frac{2x^2}{2}-\frac{6x}{2}=\frac{-3}{2}$$
$$x^2-3x=-\frac{3}{2}$$
$$x^2-3x+\left(\frac{-3}{2}\right)^2=-\frac{3}{2}+\left(\frac{-3}{2}\right)^2$$
$$x^2-3x+\frac{9}{4}=-\frac{3}{2}+\frac{9}{4}$$
$$\left(x-\frac{3}{2}\right)^2=-\frac{6}{4}+\frac{9}{4}$$
$$\left(x-\frac{3}{2}\right)^2=\frac{3}{4}$$
$$x-\frac{3}{2}=\pm\sqrt{\frac{3}{4}}$$
$$x-\frac{3}{2}=\pm\frac{\sqrt{3}}{\sqrt{4}}$$
$$x-\frac{3}{2}=\pm\frac{\sqrt{3}}{2}$$
$$x=\frac{3}{2}\pm\frac{\sqrt{3}}{2}$$
$$x=\frac{3\pm\sqrt{3}}{2}$$

11.

$$6a^2-12a=-1$$
$$\frac{6a^2}{6}-\frac{12a}{6}=\frac{-1}{6}$$
$$a^2-2a=-\frac{1}{6}$$
$$a^2-2a+\left(\frac{-2}{2}\right)^2=-\frac{1}{6}+\left(\frac{-2}{2}\right)^2$$
$$a^2-2a+1=-\frac{1}{6}+1$$
$$(a-1)^2=\frac{5}{6}$$
$$a-1=\pm\sqrt{\frac{5}{6}}$$
$$a-1=\pm\sqrt{\frac{5\cdot 6}{6\cdot 6}}$$
$$a-1=\pm\sqrt{\frac{30}{36}}$$
$$a-1=\pm\frac{\sqrt{30}}{6}$$
$$a=1\pm\frac{\sqrt{30}}{6}$$
$$a=\frac{6\pm\sqrt{30}}{6}$$

12.

$$x^2-2x=-13$$
$$x^2-2x+\left(\frac{-2}{2}\right)^2=-13+\left(\frac{-2}{2}\right)^2$$
$$x^2-2x+(-1)^2=-13+(-1)^2$$
$$x^2-2x+1=-13+1$$
$$(x-1)^2=-12$$
$$x-1=\pm\sqrt{-12}$$
$$x-1=\pm\sqrt{-4}\sqrt{3}$$
$$x-1=\pm 2i\sqrt{3}$$
$$x=1\pm 2i\sqrt{3}$$

13.

$$x^2 = 32$$
$$x = \pm\sqrt{32}$$
$$x = \pm\sqrt{16}\sqrt{2}$$
$$x = \pm 4\sqrt{2}$$

14.

$$(7x+51)^2 = 11$$
$$\sqrt{(7x+51)^2} = \pm\sqrt{11}$$
$$7x+51 = \pm\sqrt{11}$$
$$7x = -51 \pm \sqrt{11}$$
$$x = \frac{-51 \pm \sqrt{11}}{7}$$

15. Because 7 is an odd number and not divisible by 2, the computations involved in completed the square on $x^2 + 7x$ involve fractions. The computations involved in completing the square on $x^2 + 6x$ do not.

16.

$$d = 16t^2$$
$$605 = 16t^2$$
$$\frac{605}{16} = \frac{16t^2}{16}$$
$$37.8125 = t^2$$
$$t = \pm\sqrt{37.8125}$$
$$t = \pm 6$$

Since the time cannot be negative, the ball should be dropped 6 seconds before midnight.

SECTION 8.2

The Quadratic Formula

17.

$$2x^2 + 13x = 7$$
$$2x^2 + 13x - 7 = 0$$
$$x = \frac{-b \pm \sqrt{b^2 - 4ac}}{2a}$$
$$= \frac{-13 \pm \sqrt{(13)^2 - 4(2)(-7)}}{2(2)}$$
$$= \frac{-13 \pm \sqrt{169 - (-56)}}{4}$$
$$= \frac{-13 \pm \sqrt{225}}{4}$$
$$= \frac{-13 \pm 15}{4}$$
$$x = \frac{-13+15}{4} \quad \text{or} \quad x = \frac{-13-15}{4}$$
$$x = \frac{2}{4} \qquad x = \frac{-28}{4}$$
$$x = \frac{1}{2} \qquad x = -7$$

18.

$$-x^2 + 10x - 18 = 0$$
$$0 = x^2 - 10x + 18$$
$$x = \frac{-b \pm \sqrt{b^2 - 4ac}}{2a}$$
$$= \frac{-(-10) \pm \sqrt{(-10)^2 - 4(1)(18)}}{2(1)}$$
$$= \frac{10 \pm \sqrt{100 - 72}}{2}$$
$$= \frac{10 \pm \sqrt{28}}{2}$$
$$= \frac{10 \pm \sqrt{4}\sqrt{7}}{2}$$
$$= \frac{10 \pm 2\sqrt{7}}{2}$$
$$= \frac{10}{2} \pm \frac{2\sqrt{7}}{2}$$
$$x = 5 \pm \sqrt{7}$$
$$x = 5 + \sqrt{7} \quad \text{or} \quad x = 5 - \sqrt{7}$$
$$x \approx 7.65 \qquad x \approx 2.35$$

19.

$$x^2 - 10x = 0$$

$$x = \frac{-b \pm \sqrt{b^2 - 4ac}}{2a}$$

$$= \frac{-(-10) \pm \sqrt{(-10)^2 - 4(1)(0)}}{2(1)}$$

$$= \frac{10 \pm \sqrt{100 - 0}}{2}$$

$$= \frac{10 \pm \sqrt{100}}{2}$$

$$= \frac{10 \pm 10}{2}$$

$$x = \frac{10 + 10}{2} \quad \text{or} \quad x = \frac{10 - 10}{2}$$

$$x = \frac{20}{2} \qquad x = \frac{0}{2}$$

$$x = 10 \qquad x = 0$$

20.

$$3y^2 = 26y - 2$$

$$3y^2 - 26y + 2 = 0$$

$$y = \frac{-b \pm \sqrt{b^2 - 4ac}}{2a}$$

$$= \frac{-(-26) \pm \sqrt{(-26)^2 - 4(3)(2)}}{2(3)}$$

$$= \frac{26 \pm \sqrt{676 - 24}}{6}$$

$$= \frac{26 \pm \sqrt{652}}{6}$$

$$= \frac{26 \pm \sqrt{4}\sqrt{163}}{6}$$

$$= \frac{26 \pm 2\sqrt{163}}{6}$$

$$= \frac{\not{2}\left(13 \pm \sqrt{163}\right)}{\not{2} \cdot 3}$$

$$y = \frac{13 \pm \sqrt{163}}{3}$$

$$y = \frac{13 + \sqrt{163}}{3} \quad \text{or} \quad y = \frac{13 - \sqrt{163}}{3}$$

$$y \approx 8.59 \qquad y \approx 0.08$$

21.

$$\frac{1}{3}p^2 + \frac{1}{2}p + \frac{1}{2} = 0$$

$$6\left(\frac{1}{3}p^2 + \frac{1}{2}p + \frac{1}{2}\right) = 6(0)$$

$$2p^2 + 3p + 3 = 0$$

$$p = \frac{-b \pm \sqrt{b^2 - 4ac}}{2a}$$

$$= \frac{-3 \pm \sqrt{(3)^2 - 4(2)(3)}}{2(2)}$$

$$= \frac{-3 \pm \sqrt{9 - 24}}{4}$$

$$= \frac{-3 \pm \sqrt{-15}}{4}$$

$$= \frac{-3 \pm i\sqrt{15}}{4}$$

$$p = -\frac{3}{4} \pm \frac{\sqrt{15}}{4}i$$

22.

$$3{,}000t^2 - 4{,}000t = -2{,}000$$

$$\frac{3{,}000t^2}{1{,}000} - \frac{4{,}000t}{1{,}000} = \frac{-2{,}000}{1{,}000}$$

$$3t^2 - 4t = -2$$

$$3t^2 - 4t + 2 = 0$$

$$t = \frac{-b \pm \sqrt{b^2 - 4ac}}{2a}$$

$$= \frac{-(-4) \pm \sqrt{(-4)^2 - 4(3)(2)}}{2(3)}$$

$$= \frac{4 \pm \sqrt{16 - 24}}{6}$$

$$= \frac{4 \pm \sqrt{-8}}{6}$$

$$= \frac{4 \pm \sqrt{-4}\sqrt{2}}{6}$$

$$= \frac{4 \pm 2i\sqrt{2}}{6}$$

$$= \frac{4}{6} \pm \frac{2\sqrt{2}}{6}i$$

$$t = \frac{2}{3} \pm \frac{\sqrt{2}}{3}i$$

23.

$$0.5x^2 + 0.3x - 0.1 = 0$$
$$10(0.5x^2 + 0.3x - 0.1) = 10(0)$$
$$5x^2 + 3x - 1 = 0$$
$$x = \frac{-b \pm \sqrt{b^2 - 4ac}}{2a}$$
$$= \frac{-3 \pm \sqrt{(3)^2 - 4(5)(-1)}}{2(5)}$$
$$= \frac{-3 \pm \sqrt{9 - (-20)}}{10}$$
$$= \frac{-3 \pm \sqrt{29}}{10}$$
$$x = \frac{-3 + \sqrt{29}}{10} \quad \text{or} \quad x = \frac{-3 - \sqrt{29}}{10}$$
$$x \approx 0.24 \qquad x \approx -0.84$$

24.

$$x^2 - 3x - 27 = 0$$
$$x = \frac{-b \pm \sqrt{b^2 - 4ac}}{2a}$$
$$= \frac{-(-3) \pm \sqrt{(-3)^2 - 4(1)(-27)}}{2(1)}$$
$$= \frac{3 \pm \sqrt{9 + 108}}{2}$$
$$= \frac{3 \pm \sqrt{117}}{2}$$
$$= \frac{3 \pm 3\sqrt{13}}{2}$$
$$x = \frac{3 + 3\sqrt{13}}{2} \quad \text{or} \quad x = \frac{3 - 3\sqrt{13}}{2}$$
$$x \approx 6.91 \qquad x \approx -3.91$$

25.

$$x^2 + 3x - 7 = 1$$
$$x^2 + 3x - 8 = 0$$
$$x = \frac{-b \pm \sqrt{b^2 - 4ac}}{2a}$$
$$= \frac{-3 \pm \sqrt{(3)^2 - 4(1)(-8)}}{2(1)}$$
$$= \frac{-3 \pm \sqrt{9 + 32}}{2}$$
$$= \frac{-3 \pm \sqrt{41}}{2}$$

26.

$$-4x^2 - 2x = -3$$
$$0 = 4x^2 + 2x - 3$$
$$x = \frac{-b \pm \sqrt{b^2 - 4ac}}{2a}$$
$$= \frac{-2 \pm \sqrt{(2)^2 - 4(4)(-3)}}{2(4)}$$
$$= \frac{-2 \pm \sqrt{4 + 48}}{8}$$
$$= \frac{-2 \pm \sqrt{52}}{8}$$
$$= \frac{-2 \pm 2\sqrt{13}}{8}$$
$$= \frac{-1 \pm \sqrt{13}}{4}$$

27. –2 is not a factor of the numerator—it is a term. Only common factors of the numerator and denominator can be removed.

28. a. $(x + x + 2)$ ft = $(2x + 2)$ ft
b. $(x + x + 6)$ ft = $(2x + 6)$ ft

29. Let the border on top/bottom be x. Then the borders on the side would be $\frac{1}{2}x$. The length of the picture is $35 - 2(x)$ and the width of the picture is $23 - 2(\frac{1}{2}x)$.

$$(\text{length})(\text{width}) = \text{Area}$$

$$(35-2x)\left(23-2\bullet\frac{1}{2}x\right)=615$$

$$(35-2x)(23-x)=615$$

$$805-35x-46x+2x^2=615$$

$$2x^2-81x+805=615$$

$$2x^2-81x+190=0$$

$$x=\frac{-b\pm\sqrt{b^2-4ac}}{2a}$$

$$=\frac{-(-81)\pm\sqrt{(-81)^2-4(2)(190)}}{2(2)}$$

$$=\frac{81\pm\sqrt{6,561-1,520}}{4}$$

$$=\frac{81\pm\sqrt{5,041}}{4}$$

$$=\frac{81\pm71}{4}$$

$$x=\frac{81+71}{4} \quad \text{or} \quad x=\frac{81-71}{4}$$

$$x=\frac{152}{4} \qquad x=\frac{10}{4}$$

$$x=38 \qquad x=2.5$$

Since the total length is only 35 inches, the border cannot be 38 inches. So, the border on top and bottom is 2.5 inches and the border on the sides is ½ of 2.5, which is 1.25 inches.

30. The # of students would be $300 - 5x$. The price per student would be $20 + 0.50x$. (# of students)(price per student) = revenue

$$(300-5x)(20+0.5x)=6,240$$

$$6,000+150x-100x-2.5x^2=6,240$$

$$6,000+50x-2.5x^2=6,240$$

$$0=2.5x^2-50x+240$$

$$\frac{0}{2.5}=\frac{2.5x^2}{2.5}-\frac{50x}{2.5}+\frac{240}{2.5}$$

$$0=x^2-20x+96$$

$$x=\frac{-b\pm\sqrt{b^2-4ac}}{2a}$$

$$=\frac{-(-20)\pm\sqrt{(-20)^2-4(1)(96)}}{2(1)}$$

$$=\frac{20\pm\sqrt{400-384}}{2}$$

$$=\frac{20\pm\sqrt{16}}{2}$$

$$=\frac{20\pm4}{2}$$

$$x=\frac{20+4}{2} \quad \text{or} \quad x=\frac{20-4}{2}$$

$$x=\frac{24}{2} \qquad x=\frac{16}{2}$$

$$x=12 \qquad x=8$$

$$\begin{aligned}\text{price}&=20+0.5x & \text{price}&=20+0.5x\\ &=20+0.5(12) & &=20+0.5(8)\\ &=20+6 & &=20+4\\ &=\$26 & &=\$24\end{aligned}$$

31.

$$d = -16t^2 + 40t + 5$$
$$25 = -16t^2 + 40t + 5$$
$$16t^2 - 40t + 20 = 0$$
$$t = \frac{-b \pm \sqrt{b^2 - 4ac}}{2a}$$
$$= \frac{-(-40) \pm \sqrt{(-40)^2 - 4(16)(20)}}{2(16)}$$
$$= \frac{40 \pm \sqrt{1{,}600 - 1{,}280}}{32}$$
$$= \frac{40 \pm \sqrt{320}}{32}$$
$$= \frac{40 \pm \sqrt{64}\sqrt{5}}{32}$$
$$= \frac{40 \pm 8\sqrt{5}}{32}$$
$$= \frac{8(5 \pm \sqrt{5})}{32}$$
$$= \frac{5 \pm \sqrt{5}}{4}$$
$$t = \frac{5+\sqrt{5}}{4} \quad \text{or} \quad t = \frac{5-\sqrt{5}}{4}$$
$$t = 1.8 \qquad t = 0.7$$

He will be able to grab it in 0.7 seconds and 1.8 seconds.

32. Let x = the length of the shorter leg. Then $x + 23$ is the length of the longer leg.

$$(\text{short leg})^2 + (\text{long leg})^2 = (\text{hypotenuse})^2$$
$$x^2 + (x+23)^2 = 65^2$$
$$x^2 + x^2 + 46x + 529 = 4{,}225$$
$$2x^2 + 46x - 3{,}696 = 0$$
$$\frac{2x^2}{2} + \frac{46x}{2} - \frac{3{,}696}{2} = \frac{0}{2}$$
$$x^2 + 23x - 1{,}848 = 0$$
$$x = \frac{-b \pm \sqrt{b^2 - 4ac}}{2a}$$
$$x = \frac{-23 \pm \sqrt{23^2 - 4(1)(-1{,}848)}}{2(1)}$$
$$= \frac{-23 \pm \sqrt{529 + 7{,}392}}{2}$$
$$= \frac{-23 \pm \sqrt{7{,}921}}{2}$$
$$= \frac{-23 \pm 89}{2}$$
$$x = \frac{-23+89}{2} \quad \text{or} \quad x = \frac{-23-89}{2}$$
$$= \frac{66}{2} \qquad = \frac{-112}{2}$$
$$= 33 \qquad = -56$$

Since the length cannot be negative, then the length of the shorter leg is 33 in.
Short leg = x = 33 in.
Long leg = x + 23 = 33 + 23 = 56 in.

SECTION 8.3
The Discriminant and Equations That Can be Written in Quadratic Form

33. $3x^2 + 4x - 3 = 0$

$$b^2 - 4ac = (4)^2 - 4(3)(-3)$$
$$= 16 - (-36)$$
$$= 52$$

two different irrational-number solutions

34. $4x^2 - 5x + 7 = 0$

$$b^2 - 4ac = (-5)^2 - 4(4)(7)$$
$$= 25 - 112$$
$$= -87$$

two imaginary-number solutions that are complex conjugates

35. $3x^2 - 4x + \frac{4}{3} = 0$

$$3\left(3x^2 - 4x + \frac{4}{3}\right) = 3(0)$$
$$9x^2 - 12x + 4 = 0$$
$$b^2 - 4ac = (-12)^2 - 4(9)(4)$$
$$= 144 - 144$$
$$= 0$$

one repeated solution, a rational number

36. $m(2m - 3) = 20$

$$2m^2 - 3m = 20$$
$$2m^2 - 3m - 20 = 0$$
$$b^2 - 4ac = (-3)^2 - 4(2)(-20)$$
$$= 9 - (-160)$$
$$= 169$$

two different rational-number solutions

37. Let $y = \sqrt{x}$.

$$x - 13\sqrt{x} + 12 = 0$$
$$\left(\sqrt{x}\right)^2 - 13\sqrt{x} + 12 = 0$$
$$y^2 - 13y + 12 = 0$$
$$(y - 12)(y - 1) = 0$$
$$y - 12 = 0 \quad \text{and} \quad y - 1 = 0$$
$$y = 12 \qquad y = 1$$
$$\sqrt{x} = 12 \qquad \sqrt{x} = 1$$
$$\left(\sqrt{x}\right)^2 = (12)^2 \quad \left(\sqrt{x}\right)^2 = (1)^2$$
$$x = 144 \qquad x = 1$$

38. Let $y = a^{1/3}$.

$$a^{2/3} + a^{1/3} - 6 = 0$$
$$\left(a^{1/3}\right)^2 + a^{1/3} - 6 = 0$$
$$y^2 + y - 6 = 0$$
$$(y + 3)(y - 2) = 0$$
$$y + 3 = 0 \quad \text{or} \quad y - 2 = 0$$
$$y = -3 \qquad y = 2$$
$$a^{1/3} = -3 \qquad a^{1/3} = 2$$
$$\left(a^{1/3}\right)^3 = (-3)^3 \quad \left(a^{1/3}\right)^3 = (2)^3$$
$$a = -27 \qquad a = 8$$

39. Let $y = x^2$.

$$3x^4 + x^2 - 2 = 0$$
$$3\left(x^2\right)^2 + x^2 - 2 = 0$$
$$3y^2 + y - 2 = 0$$
$$(3y - 2)(y + 1) = 0$$
$$3y - 2 = 0 \quad \text{or} \quad y + 1 = 0$$
$$3y = 2$$
$$y = \frac{2}{3} \qquad y = -1$$
$$x^2 = \frac{2}{3} \qquad x^2 = -1$$
$$x = \pm\sqrt{\frac{2}{3}} \qquad x = \pm\sqrt{-1}$$
$$x = \pm\sqrt{\frac{2}{3}\cdot\frac{3}{3}} \qquad x = \pm i$$
$$x = \pm\frac{\sqrt{6}}{3}$$
$$x = \frac{\sqrt{6}}{3},\ -\frac{\sqrt{6}}{3},\ i,\ -i$$

40.

$$\frac{6}{x + 2} + \frac{6}{x + 1} = 5$$
$$(x + 2)(x + 1)\left(\frac{6}{x + 2} + \frac{6}{x + 1}\right) = 5(x + 2)(x + 1)$$
$$6(x + 1) + 6(x + 2) = 5\left(x^2 + 2x + x + 2\right)$$
$$6x + 6 + 6x + 12 = 5\left(x^2 + 3x + 2\right)$$
$$12x + 18 = 5x^2 + 15x + 10$$
$$0 = 5x^2 + 3x - 8$$
$$0 = (5x + 8)(x - 1)$$
$$5x + 8 = 0 \quad \text{or} \quad x - 1 = 0$$
$$5x = -8 \qquad x = 1$$
$$x = -\frac{8}{5}$$

41. Let $y = (x - 7)$.

$$(x-7)^2 + 6(x-7) + 10 = 0$$
$$y^2 + 6y + 10 = 0$$
$$y = \frac{-b \pm \sqrt{b^2 - 4ac}}{2a}$$
$$y = \frac{-6 \pm \sqrt{(6)^2 - 4(1)(10)}}{2(1)}$$
$$y = \frac{-6 \pm \sqrt{36 - 40}}{2}$$
$$y = \frac{-6 \pm \sqrt{-4}}{2}$$
$$y = \frac{-6 \pm 2i}{2}$$
$$y = \frac{-6}{2} \pm \frac{2}{2}i$$
$$y = -3 \pm i$$
$$x - 7 = -3 \pm i$$
$$x = -3 \pm i + 7$$
$$x = 4 \pm i$$

42. Let $y = \dfrac{1}{m^2}$.

$$m^{-4} - 2m^{-2} + 1 = 0$$
$$\frac{1}{m^4} - 2\left(\frac{1}{m^2}\right) + 1 = 0$$
$$\left(\frac{1}{m^2}\right)^2 - 2\left(\frac{1}{m^2}\right) + 1 = 0$$
$$y^2 - 2y + 1 = 0$$
$$(y-1)(y-1) = 0$$
$$y - 1 = 0 \quad \text{or} \quad y - 1 = 0$$
$$y = 1 \qquad y = 1$$
$$\frac{1}{m^2} = \frac{1}{1} \qquad \frac{1}{m^2} = \frac{1}{1}$$
$$m^2 = 1 \qquad m^2 = 1$$
$$m = \pm\sqrt{1} \qquad m = \pm\sqrt{1}$$
$$m = \pm 1 \qquad m = \pm 1$$
$$m = 1,\ -1,\ 1,\ -1$$

43. Let $y = \dfrac{x+1}{x}$.

$$4\left(\frac{x+1}{x}\right)^2 + 12\left(\frac{x+1}{x}\right) + 9 = 0$$
$$4y^2 + 12y + 9 = 0$$
$$(2y+3)(2y+3) = 0$$
$$2y + 3 = 0 \quad \text{or} \quad 2y + 3 = 0$$
$$2y = -3 \qquad 2y = -3$$
$$y = -\frac{3}{2} \qquad y = -\frac{3}{2}$$
$$\frac{x+1}{x} = -\frac{3}{2} \qquad \frac{x+1}{x} = -\frac{3}{2}$$
$$2(x+1) = -3x \qquad 2(x+1) = -3x$$
$$2x + 2 = -3x \qquad 2x + 2 = -3x$$
$$2 = -5x \qquad 2 = -5x$$
$$-\frac{2}{5} = x \qquad -\frac{2}{5} = x$$

44. Let $y = m^{1/5}$.

$$2m^{2/5} - 5m^{1/5} + 2 = 0$$
$$2\left(m^{1/5}\right)^2 - 5m^{1/5} + 2 = 0$$
$$2y^2 - 5y + 2 = 0$$
$$(2y-1)(y-2) = 0$$
$$2y - 1 = 0 \quad \text{or} \quad y - 2 = 0$$
$$2y = 1 \qquad y = 2$$
$$y = \frac{1}{2}$$
$$m^{1/5} = \frac{1}{2} \qquad m^{1/5} = 2$$
$$\left(m^{1/5}\right)^5 = \left(\frac{1}{2}\right)^5 \qquad \left(m^{1/5}\right)^5 = (2)^5$$
$$m = \frac{1}{32} \qquad m = 32$$

45. Let x = the number of minutes it takes the younger girl to do the yard work.

Younger girl's work in 1 minute = $\frac{1}{x}$

Older girl's work in 1 minute = $\frac{1}{x-20}$

Work together in 1 minute = $\frac{1}{45}$

Younger girl + older girl = work together

$$\frac{1}{x}+\frac{1}{x-20}=\frac{1}{45}$$

$$45x(x-20)\left(\frac{1}{x}+\frac{1}{x-20}\right)=45x(x-20)\left(\frac{1}{45}\right)$$

$$45(x-20)+45x=x(x-20)$$

$$45x-900+45x=x^2-20x$$

$$90x-900=x^2-20x$$

$$0=x^2-110x+900$$

$$x=\frac{-b\pm\sqrt{b^2-4ac}}{2a}$$

$$x=\frac{-(-110)\pm\sqrt{(-110)^2-4(1)(900)}}{2(1)}$$

$$x=\frac{110\pm\sqrt{12{,}100-3{,}600}}{2}$$

$$x=\frac{110\pm\sqrt{8{,}500}}{2}$$

$$x=\frac{110\pm 92.2}{2}$$

$$x=\frac{110+92.2}{2} \quad \text{or} \quad x=\frac{110-92.2}{2}$$

$$x=\frac{202}{2} \qquad x=\frac{17}{2}$$

$$x=101 \qquad x=8$$

$$x-20=101-20 \qquad x-20=8-20$$

$$=81 \qquad =-12$$

Since the older sister's time cannot be negative, her time must be about 81 min.

46. Recall $t=\frac{d}{r}$.

	distance	rate	time
original rate	150	r	$t=\frac{150}{r}$
increased rate	150	$r+20$	$t-2=\frac{150}{r+20}$

Solve the second time for t.

$$t-2=\frac{150}{r+20}$$

$$t=\frac{150}{r+20}+2$$

Set the times equal and solve for r.

$$\frac{150}{r}=\frac{150}{r+20}+2$$

$$r(r+20)\left(\frac{150}{r}\right)=r(r+20)\left(\frac{150}{r+20}+2\right)$$

$$150(r+20)=150(r)+2r(r+20)$$

$$150r+3{,}000=150r+2r^2+40r$$

$$150r+3{,}000-150r=150r+2r^2+40r-150r$$

$$3{,}000=2r^2+40r$$

$$0=2r^2+40r-3{,}000$$

$$0=\frac{2r^2}{2}+\frac{40r}{2}-\frac{3{,}000}{2}$$

$$0=r^2+20r-1{,}500$$

$$r=\frac{-b\pm\sqrt{b^2-4ac}}{2a}$$

$$=\frac{-20\pm\sqrt{20^2-4(1)(-1{,}500)}}{2(1)}$$

$$=\frac{-20\pm\sqrt{400+6{,}000}}{2}$$

$$=\frac{-20\pm\sqrt{6{,}400}}{2}$$

$$=\frac{-20\pm 80}{2}$$

$$r=\frac{-20+80}{2} \quad \text{or} \quad r=\frac{-20-80}{2}$$

$$r=\frac{60}{2} \qquad r=\frac{-100}{2}$$

$$r=30 \qquad r=-50$$

Since the rate cannot be negative, her original rate is 30 mph.

SECTION 8.4
Quadratic Functions and Their Graphs

47. Since 2005 – 1980 = 25, let $x = 25$.

$$A(x) = 0.03x^2 - 0.88x + 37.3$$
$$A(25) = 0.03x^2 - 0.88x + 37.3$$
$$= 0.03(25)^2 - 0.88(25) + 37.3$$
$$= 0.03(625) - 0.88(25) + 37.3$$
$$= 18.75 - 22 + 37.3$$
$$= 34.05$$
$$\approx 34.1 \text{ million}$$

48. The graph of the quadratic function $f(x) = a(x-h)^2 + k$ is a parabola with vertex at (h, k). The axis of symmetry is the line $x = h$. The parabola opens upward when $a > 0$ and downward when $a < 0$.

49.

x	$f(x) = 2x^2$
–2	$2(-2)^2 = 8$
–1	$2(-1)^2 = 2$
0	$2(0)^2 = 0$
1	$2(1)^2 = 2$
2	$2(2)^2 = 8$

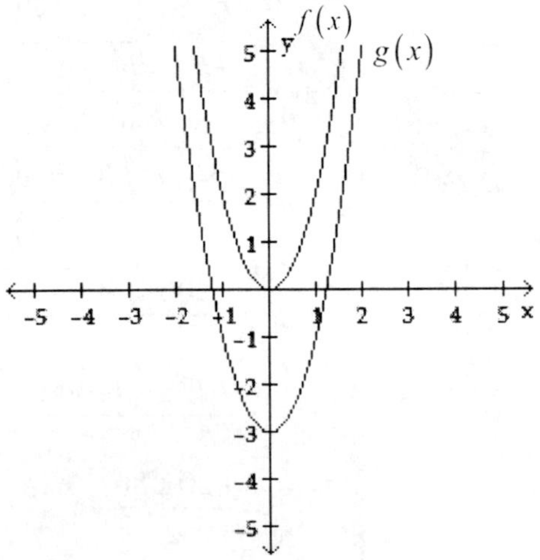

The graph of $g(x)$ is shifted down 3 units.

50.

x	$f(x) = -\frac{1}{4}x^2$
–2	$-\frac{1}{4}(-2)^2 = -1$
–1	$-\frac{1}{4}(-1)^2 = -\frac{1}{4}$
0	$-\frac{1}{4}(0)^2 = 0$
1	$-\frac{1}{4}(1)^2 = -\frac{1}{4}$
2	$-\frac{1}{4}(2)^2 = -1$

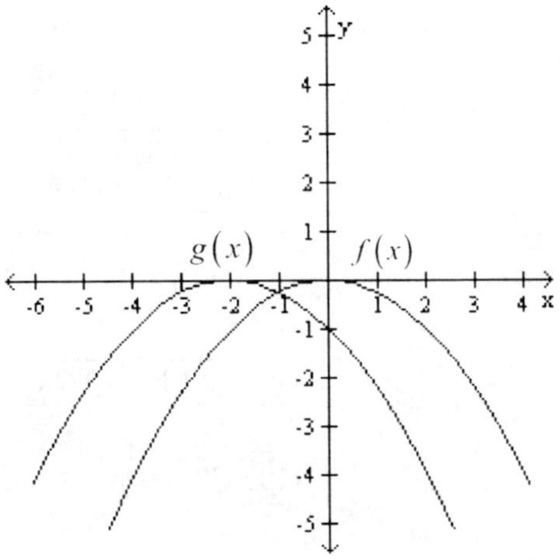

The graph of $g(x)$ is shifted left 2 units.

51. $f(x) = -2(x-1)^2 + 4$

Vertex: $(1, 4)$

Axis of symmetry: $x = 1$

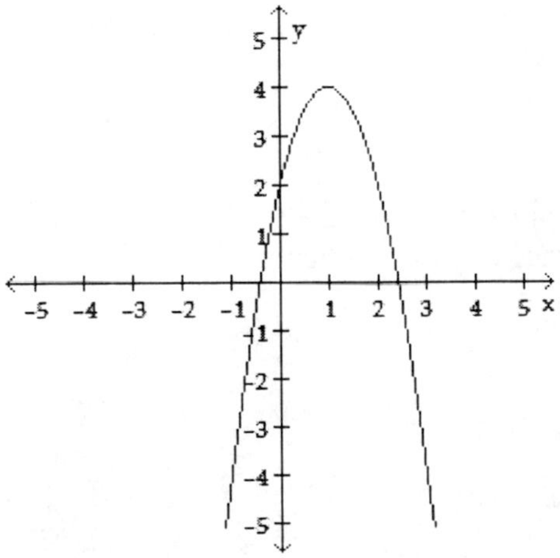

52. $f(x)=4x^2+16x+9$

$f(x)=4(x^2+4x)+9$

$f(x)=4(x^2+4x+4)+9-4(4)$

$f(x)=4(x+2)^2+9-16$

$f(x)=4(x+2)^2-7$

Vertex: $(-2,-7)$

Axis of symmetry: $x=-2$

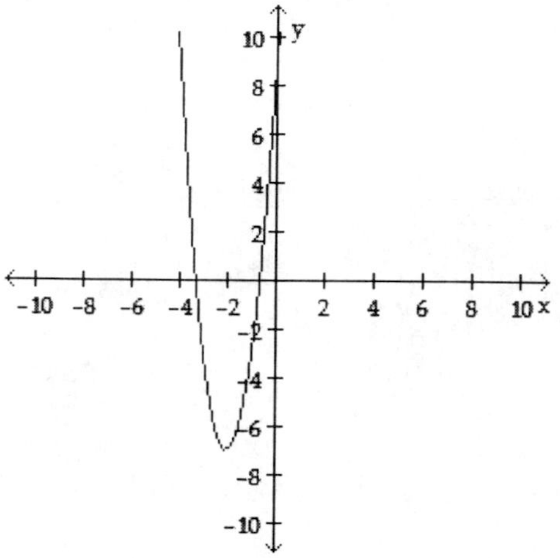

53. $f(x)=-2x^2+4x-8$

$x=\dfrac{-b}{2a}$

$=\dfrac{-4}{2(-2)}$

$=\dfrac{-4}{-4}$

$=1$

$f(1)=-2(1)^2+4(1)-8$

$=-2(1)+4(1)-8$

$=-2+4-8$

$=-6$

The vertex is $(1,-6)$.

54. $f(x)=x^2+x-2$

Step 1: Since $a=1>0$, the parabola opens upward.

Step 2: Find the vertex and axis of symmetry.

$x=\dfrac{-b}{2a}=\dfrac{-1}{2(1)}=-\dfrac{1}{2}$

$y=\left(-\dfrac{1}{2}\right)^2+\left(-\dfrac{1}{2}\right)-2$

$=\dfrac{1}{4}-\dfrac{1}{2}-2$

$=-\dfrac{9}{4}$

vertex: $\left(-\dfrac{1}{2},-\dfrac{9}{4}\right)$

The axis of symmetry is $x=-\dfrac{1}{2}$.

Step 3: Find the x– and y–intercepts. Since $c=-2$, the y–intercept is $(0,-2)$. To find the x–intercepts, let $y=0$ and solve the equation for x.

$0=x^2+x-2$

$0=(x+2)(x-1)$

$x+2=0$ or $x-1=0$

$x=-2$ $\quad$ $x=1$

x–intercepts: $(-2,0)$ and $(1,0)$

y–intercept: $(0,-2)$

Step 4: Use symmetry to find the point $(-1,-2)$.

Step 5: Plots the points and draw the parabola.

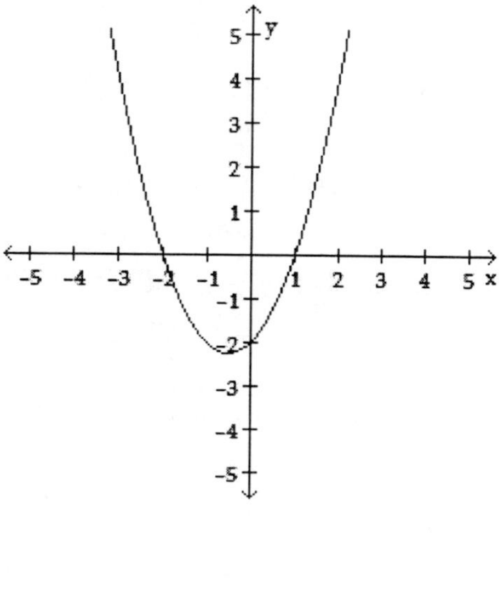

55. The vertex would represent the maximum of the function. Use the vertex formula to find the value of x and use substitution to find the number of farms.

$$\begin{aligned} x &= \frac{-b}{2a} \\ &= \frac{-155{,}652}{2(-1{,}526)} \\ &= \frac{-155{,}652}{-3{,}052} \\ &= 51 \end{aligned}$$

$$1870 + 51 = 1921$$

$$\begin{aligned} N(51) &= -1{,}526(51)^2 + 155{,}652(51) + 2{,}500{,}200 \\ &= -1{,}526(2{,}601) + 155{,}652(51) + 2{,}500{,}200 \\ &= -3{,}969{,}126 + 7{,}938{,}252 + 2{,}500{,}200 \\ &= 6{,}469{,}326 \text{ farms} \end{aligned}$$

56. The x–intercepts are (–2, 0) and $\left(\frac{1}{3}, 0\right)$, so the solutions to the equation are $x = -2$ and $x = \frac{1}{3}$.

SECTION 8.5
Quadratic and Other Nonlinear Inequalities

57.

$$\begin{aligned} x^2 + 2x - 35 &> 0 \\ x^2 + 2x - 35 &= 0 \\ (x+7)(x-5) &= 0 \end{aligned}$$

$$x + 7 = 0 \quad \text{or} \quad x - 5 = 0$$

$$x = -7 \qquad\qquad x = 5$$

Critical numbers = -7 and 5.

$$(-\infty, -7) \cup (5, \infty)$$

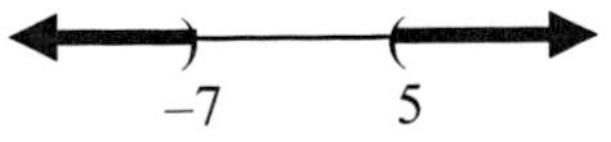

58.

$$\begin{aligned} x^2 &\le 81 \\ x^2 - 81 &\le 0 \\ x^2 - 81 &= 0 \\ (x+9)(x-9) &= 0 \end{aligned}$$

$$x + 9 = 0 \quad \text{or} \quad x - 9 = 0$$

$$x = -9 \qquad\qquad x = 9$$

Critical numbers = -9 and 9

$$[-9, 9]$$

59.

$$\begin{aligned} \frac{3}{x} &\le 5 \\ \frac{3}{x} &= 5 \\ \frac{3}{x} - 5 &= 0 \\ \frac{3}{x} - \frac{5x}{x} &= 0 \\ \frac{3-5x}{x} &= 0 \\ x\left(\frac{3-5x}{x}\right) &= x(0) \\ 3 - 5x &= 0 \\ -5x &= -3 \\ x &= \frac{3}{5} \end{aligned}$$

Set the denominator $= 0$.

$$x = 0$$

Critical numbers $= 0$ and $\frac{3}{5}$

$$(-\infty, 0) \cup \left[\frac{3}{5}, \infty\right)$$

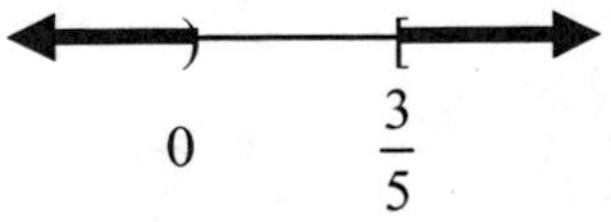

60.

$$\frac{2x^2 - x - 28}{x - 1} > 0$$

$$\frac{2x^2 - x - 28}{x - 1} = 0$$

$$(x-1)\left(\frac{2x^2 - x - 28}{x - 1}\right) = (x-1)(0)$$

$$2x^2 - x - 28 = 0$$

$$(x-4)(2x+7) = 0$$

$$x - 4 = 0 \quad \text{or} \quad 2x + 7 = 0$$

$$x = 4 \qquad\qquad x = -\frac{7}{2}$$

Set the denominator $= 0$.

$$x - 1 = 0$$

$$x = 1$$

Critical numbers $= -\frac{7}{2}$, 1, and 4

$$\left(-\frac{7}{2}, 1\right) \cup (4, \infty)$$

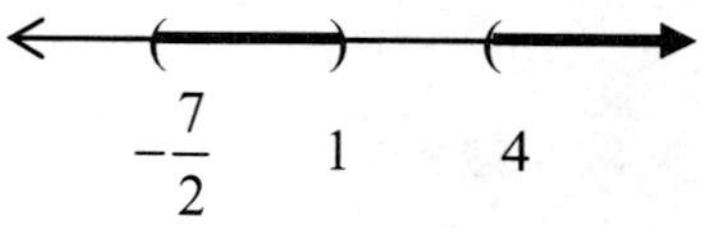

61. $\left[-4, \frac{2}{3}\right]$

62. $(-\infty, 0) \cup (1, \infty)$

63. $y < \frac{1}{2}x^2 - 1$

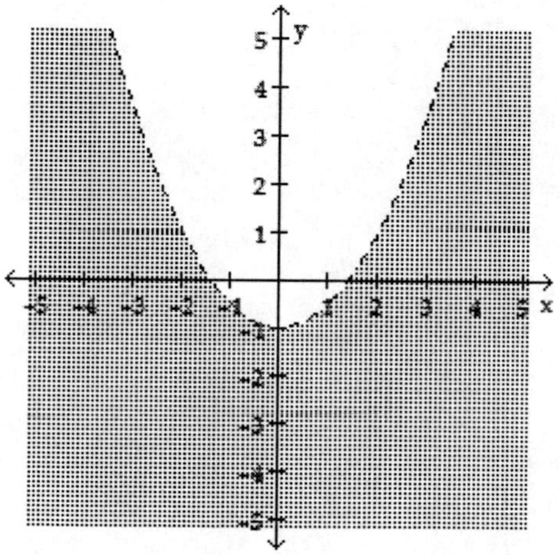

64. $y \geq -|x|$

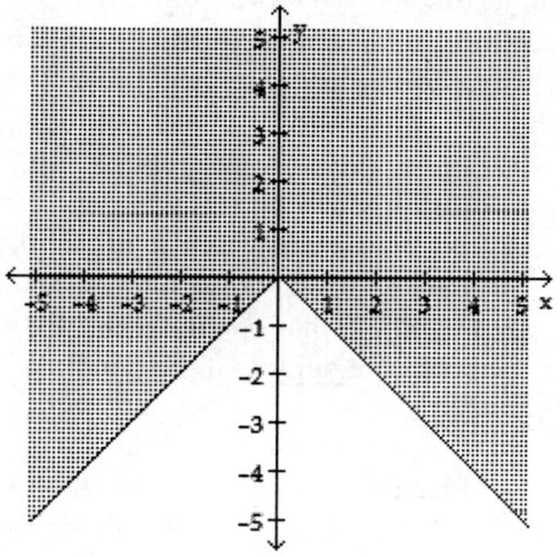

CHAPTER 8 TEST

1. a. An equation of the form $ax^2+bx+c=0$, where $a\neq 0$, is called a **quadratic** equation.
   b. When we add 81 to $x^2 + 18x$, we say that we have **completed** the **square** on $x^2 + 18x$.
   c. The lowest point on a parabola that opens upward, or the highest point on a parabola that opens downward, is called the **vertex** of the parabola.
   d. $\frac{x-5}{x^2-x-56}>0$ is an example of a _ **rational** inequality in one variable.
   e. $y\le x^2-4x+3$ is an example of a **nonlinear** inequality in two variables.

2.

$$x^2-63=0$$
$$x^2=63$$
$$x=\pm\sqrt{63}$$
$$x=\pm\sqrt{9}\sqrt{7}$$
$$x=\pm3\sqrt{7}$$
$$x\approx\pm7.94$$

3.

$$(a+7)^2=50$$
$$a+7=\pm\sqrt{50}$$
$$a+7=\pm\sqrt{25}\sqrt{2}$$
$$a+7=\pm5\sqrt{2}$$
$$a=-7\pm5\sqrt{2}$$

4.

$$m^2+4=0$$
$$m^2=-4$$
$$m=\pm\sqrt{-4}$$
$$m=\pm2i$$

5.

$$\left(\frac{11}{2}\right)^2=\frac{121}{4}$$
$$x^2+11x+\frac{121}{4}=\left(x+\frac{11}{2}\right)^2$$

6.

$$x^2+3x-2=0$$
$$x^2+3x=2$$
$$x^2+3x+\left(\frac{3}{2}\right)^2=2+\left(\frac{3}{2}\right)^2$$
$$x^2+3x+\frac{9}{4}=2+\frac{9}{4}$$
$$\left(x+\frac{3}{2}\right)^2=\frac{17}{4}$$
$$\sqrt{\left(x+\frac{3}{2}\right)^2}=\pm\sqrt{\frac{17}{4}}$$
$$x+\frac{3}{2}=\pm\frac{\sqrt{17}}{2}$$
$$x=-\frac{3}{2}\pm\frac{\sqrt{17}}{2}$$
$$x=-\frac{3}{2}+\frac{\sqrt{17}}{2} \quad\text{or}\quad x=-\frac{3}{2}-\frac{\sqrt{17}}{2}$$
$$x\approx0.56 \qquad\qquad x\approx-3.56$$

7.

$$2x^2+8x+12=0$$
$$\frac{2x^2}{2}+\frac{8x}{2}+\frac{12}{2}=\frac{0}{2}$$
$$x^2+4x+6=0$$
$$x^2+4x=-6$$
$$x^2+4x+\left(\frac{4}{2}\right)^2=-6+\left(\frac{4}{2}\right)^2$$
$$x^2+4x+4=-6+4$$
$$(x+2)^2=-2$$
$$\sqrt{(x+2)^2}=\pm\sqrt{-2}$$
$$x+2=\pm i\sqrt{2}$$
$$x=-2\pm i\sqrt{2}$$

8.

$$(3x-2)^2 = 18$$
$$\sqrt{(3x-2)^2} = \sqrt{18}$$
$$3x-2 = \pm\sqrt{9}\sqrt{2}$$
$$3x-2 = \pm 3\sqrt{2}$$
$$3x = 2 \pm 3\sqrt{2}$$
$$x = \frac{2 \pm 3\sqrt{2}}{3}$$

9.

$$4x^2 + 4x - 1 = 0$$
$$x = \frac{-b \pm \sqrt{b^2 - 4ac}}{2a}$$
$$= \frac{-4 \pm \sqrt{(4)^2 - 4(4)(-1)}}{2(4)}$$
$$= \frac{-4 \pm \sqrt{16+16}}{8}$$
$$= \frac{-4 \pm \sqrt{32}}{8}$$
$$= \frac{-4 \pm \sqrt{16}\sqrt{2}}{8}$$
$$= \frac{-4 \pm 4\sqrt{2}}{8}$$
$$= \frac{4(-1 \pm \sqrt{2})}{4 \cdot 2}$$
$$= \frac{-1 \pm \sqrt{2}}{2}$$

10.

$$\frac{t^2}{8} - \frac{t}{4} = \frac{1}{2}$$
$$8\left(\frac{t^2}{8} - \frac{t}{4}\right) = 8\left(\frac{1}{2}\right)$$
$$t^2 - 2t = 4$$
$$t^2 - 2t - 4 = 0$$
$$t = \frac{-b \pm \sqrt{b^2 - 4ac}}{2a}$$
$$= \frac{-(-2) \pm \sqrt{(-2)^2 - 4(1)(-4)}}{2(1)}$$
$$= \frac{2 \pm \sqrt{4-(-16)}}{2}$$
$$= \frac{2 \pm \sqrt{20}}{2}$$
$$= \frac{2 \pm \sqrt{4}\sqrt{5}}{2}$$
$$= \frac{2 \pm 2\sqrt{5}}{2}$$
$$= \frac{\cancel{2}(1 \pm \sqrt{5})}{\cancel{2}}$$
$$t = 1 \pm \sqrt{5}$$

11.

$$-t^2 + 4t - 13 = 0$$
$$0 = t^2 - 4t + 13$$
$$t = \frac{-b \pm \sqrt{b^2 - 4ac}}{2a}$$
$$= \frac{-(-4) \pm \sqrt{(-4)^2 - 4(1)(13)}}{2(1)}$$
$$= \frac{4 \pm \sqrt{16-52}}{2}$$
$$= \frac{4 \pm \sqrt{-36}}{2}$$
$$= \frac{4 \pm 6i}{2}$$
$$= \frac{4}{2} \pm \frac{6}{2}i$$
$$t = 2 \pm 3i$$

12.

$$0.01x^2 = -0.08x - 0.15$$
$$100(0.01x^2) = 100(-0.08x - 0.15)$$
$$x^2 = -8x - 15$$
$$x^2 + 8x + 15 = 0$$
$$x = \frac{-b \pm \sqrt{b^2 - 4ac}}{2a}$$
$$= \frac{-8 \pm \sqrt{(8)^2 - 4(1)(15)}}{2(1)}$$
$$= \frac{-8 \pm \sqrt{64 - 60}}{2}$$
$$= \frac{-8 \pm \sqrt{4}}{2}$$
$$= \frac{-8 \pm 2}{2}$$
$$x = \frac{-8+2}{2} \quad \text{or} \quad x = \frac{-8-2}{2}$$
$$x = \frac{-6}{2} \qquad x = \frac{-10}{2}$$
$$x = -3 \qquad x = -5$$

13.

$$m^2 - 94m = -2,209$$
$$m^2 - 94m + 2,209 = 0$$
$$m = \frac{-b \pm \sqrt{b^2 - 4ac}}{2a}$$
$$= \frac{-(-94) \pm \sqrt{(-94)^2 - 4(1)(2,209)}}{2(1)}$$
$$= \frac{94 \pm \sqrt{8,836 - 8,836}}{2}$$
$$= \frac{94 \pm \sqrt{0}}{2}$$
$$= \frac{94 \pm 0}{2}$$
$$x = \frac{94+0}{2} \quad \text{or} \quad x = \frac{94-0}{2}$$
$$x = \frac{94}{2} \qquad x = \frac{94}{2}$$
$$x = 47 \qquad x = 47$$

A repeated solution of 47

14.

$$3x^2 - 20 = 10$$
$$3x^2 = 30$$
$$\frac{3x^2}{3} = \frac{30}{3}$$
$$x^2 = 10$$
$$\sqrt{x^2} = \sqrt{10}$$
$$x = \pm\sqrt{10}$$

15. Let $x = \sqrt{y}$

$$2y - 3\sqrt{y} + 1 = 0$$
$$2(\sqrt{y})^2 - 3\sqrt{y} + 1 = 0$$
$$2x^2 - 3x + 1 = 0$$
$$(2x-1)(x-1) = 0$$
$$2x - 1 = 0 \quad \text{or} \quad x - 1 = 0$$
$$2x = 1 \qquad x - 1 + 1 = 0 + 1$$
$$x = \frac{1}{2} \qquad x = 1$$
$$\sqrt{y} = \frac{1}{2} \qquad \sqrt{y} = 1$$
$$(\sqrt{y})^2 = \left(\frac{1}{2}\right)^2 \qquad (\sqrt{y})^2 = (1)^2$$
$$y = \frac{1}{4} \qquad y = 1$$

16. Let $y=\frac{1}{m}$.

$$3=m^{-2}-2m^{-1}$$
$$0=m^{-2}-2m^{-1}-3$$
$$m^{-2}-2m^{-1}-3=0$$
$$\left(\frac{1}{m}\right)^2-2\left(\frac{1}{m}\right)-3=0$$
$$y^2-2y-3=0$$
$$(y-3)(y+1)=0$$
$$y-3=0 \quad \text{or} \quad y+1=0$$
$$y=3 \qquad y=-1$$

$$\frac{1}{m}=3 \qquad \frac{1}{m}=-1$$
$$m\left(\frac{1}{m}\right)=m(3) \qquad m\left(\frac{1}{m}\right)=m(-1)$$
$$1=3m \qquad 1=-m$$
$$\frac{1}{3}=m \qquad -1=m$$

17. Let $y=x^2$

$$x^4-x^2-12=0$$
$$\left(x^2\right)^2-x^2-12=0$$
$$y^2-y-12=0$$
$$(y-4)(y+3)=0$$
$$y-4=0 \quad \text{or} \quad y+3=0$$
$$y=4 \qquad y=-3$$
$$x^2=4 \qquad x^2=-3$$
$$x=\pm\sqrt{4} \qquad x=\pm\sqrt{-3}$$
$$x=\pm 2 \qquad x=\pm i\sqrt{3}$$
$$x=2,\ -2,\ i\sqrt{3},\ -i\sqrt{3}$$

18. Let $y=\frac{x+2}{3x}$.

$$4\left(\frac{x+2}{3x}\right)^2-4\left(\frac{x+2}{3x}\right)-3=0$$
$$4y^2-4y-3=0$$
$$(2y+1)(2y-3)=0$$
$$2y+1=0 \quad \text{or} \quad 2y-3=0$$
$$2y=-1 \qquad 2y=3$$
$$y=-\frac{1}{2} \qquad y=\frac{3}{2}$$
$$\frac{x+2}{3x}=-\frac{1}{2} \qquad \frac{x+2}{3x}=\frac{3}{2}$$
$$2(x+2)=-1(3x) \qquad 2(x+2)=3(3x)$$
$$2x+4=-3x \qquad 2x+4=9x$$
$$4=-5x \qquad 4=7x$$
$$-\frac{4}{5}=x \qquad \frac{4}{7}=x$$

19.

$$\frac{1}{n+2}=\frac{1}{3}-\frac{1}{n}$$
$$3n(n+2)\left(\frac{1}{n+2}\right)=3n(n+2)\left(\frac{1}{3}-\frac{1}{n}\right)$$
$$3n=n(n+2)-3(n+2)$$
$$3n=n^2+2n-3n-6$$
$$3n=n^2-n-6$$
$$0=n^2-4n-6$$
$$n=\frac{-b\pm\sqrt{b^2-4ac}}{2a}$$
$$=\frac{-(-4)\pm\sqrt{(-4)^2-4(1)(-6)}}{2(1)}$$
$$=\frac{4\pm\sqrt{16+24}}{2}$$
$$=\frac{4\pm\sqrt{40}}{2}$$
$$=\frac{4\pm\sqrt{4}\sqrt{10}}{2}$$
$$=\frac{4\pm 2\sqrt{10}}{2}$$
$$n=2\pm\sqrt{10}$$

20. Let $y = a^{1/3}$.

$$5a^{2/3} + 11a^{1/3} = -2$$
$$5a^{2/3} + 11a^{1/3} + 2 = 0$$
$$5\left(a^{1/3}\right)^2 + 11a^{1/3} + 2 = 0$$
$$5y^2 + 11y + 2 = 0$$
$$(5y+1)(y+2) = 0$$
$$5y + 1 = 0 \quad \text{or} \quad y + 2 = 0$$
$$5y = -1 \qquad y = -2$$
$$y = -\frac{1}{5}$$
$$a^{1/3} = -\frac{1}{5} \qquad a^{1/3} = -2$$
$$\left(a^{1/3}\right)^3 = \left(-\frac{1}{5}\right)^3 \qquad \left(a^{1/3}\right)^3 = (-2)^3$$
$$a = -\frac{1}{125} \qquad a = -8$$

21.

$$10x(x+1) = -3$$
$$10x^2 + 10x + 3 = 0$$
$$x = \frac{-b \pm \sqrt{b^2 - 4ac}}{2a}$$
$$= \frac{-10 \pm \sqrt{(10)^2 - 4(10)(3)}}{2(10)}$$
$$= \frac{-10 \pm \sqrt{100 - 120}}{20}$$
$$= \frac{-10 \pm \sqrt{-20}}{20}$$
$$= \frac{-10 \pm 2i\sqrt{5}}{20}$$
$$= \frac{-5 \pm i\sqrt{5}}{10}$$

22.

$$a^3 - 3a^2 + 8a - 24 = 0$$
$$\left(a^3 - 3a^2\right) + (8a - 24) = 0$$
$$a^2(a-3) + 8(a-3) = 0$$
$$\left(a^2 + 8\right)(a-3) = 0$$
$$a^2 + 8 = 0 \quad \text{or} \quad a - 3 = 0$$
$$a^2 = -8 \qquad a = 3$$
$$a = \pm\sqrt{-8}$$
$$a = \pm 2i\sqrt{2}$$

23.

$$E = mc^2$$
$$\frac{E}{m} = \frac{mc^2}{m}$$
$$\frac{E}{m} = c^2$$
$$c = \sqrt{\frac{E}{m}}$$
$$c = \frac{\sqrt{E}}{\sqrt{m}} \cdot \frac{\sqrt{m}}{\sqrt{m}}$$
$$c = \frac{\sqrt{Em}}{m}$$

24. a. $3x^2 + 5x + 17 = 0$

$$b^2 - 4ac = (5)^2 - 4(3)(17)$$
$$= 25 - 204$$
$$= -179$$

two different imaginary-number solutions that are complex conjugates

b. $9m^2 - 12m = -4$

$$9m^2 - 12m + 4 = 0$$
$$b^2 - 4ac = (-12)^2 - 4(9)(4)$$
$$= 144 - 144$$
$$= 0$$

one repeated solution, a rational number

25. Let the width = x.
Then length = $332x + 8$.

$$(\text{width})(\text{length}) = \text{Area}$$
$$x(332x+8) = 6,759$$
$$332x^2 + 8x - 6,759 = 0$$
$$x = \frac{-b \pm \sqrt{b^2 - 4ac}}{2a}$$
$$= \frac{-8 \pm \sqrt{(8)^2 - 4(332)(-6,759)}}{2(332)}$$
$$= \frac{-8 \pm \sqrt{64 - (-8,975,952)}}{664}$$
$$= \frac{-8 \pm \sqrt{8,976,016}}{664}$$
$$= \frac{-8 \pm 2,996}{664}$$
$$x = \frac{-8 + 2,996}{664} \quad \text{or} \quad x = \frac{-8 - 2,996}{664}$$
$$x = \frac{2,988}{664} \qquad x = \frac{-3,004}{664}$$
$$x = 4.5 \text{ ft} \qquad x = \cancel{-4.52}$$
$$332x + 8 = 332(4.5) + 8$$
$$= 1,502 \text{ ft}$$

The dimensions are 4.5 ft by 1,502 ft.

26. Let x = the number of minutes it takes the assistant to make the pastry dessert.

Assistant's work in 1 minute = $\frac{1}{x}$

Chef's work in 1 minute = $\frac{1}{x-8}$

Work together in 1 minute = $\frac{1}{25}$

Assistant + chef = work together

$$\frac{1}{x} + \frac{1}{x-8} = \frac{1}{25}$$
$$25x(x-8)\left(\frac{1}{x} + \frac{1}{x-8}\right) = 25x(x-8)\left(\frac{1}{25}\right)$$
$$25(x-8) + 25x = x(x-8)$$
$$25x - 200 + 25x = x^2 - 8x$$
$$50x - 200 = x^2 - 8x$$
$$0 = x^2 - 58x + 200$$
$$x = \frac{-b \pm \sqrt{b^2 - 4ac}}{2a}$$
$$x = \frac{-(-58) \pm \sqrt{(-58)^2 - 4(1)(200)}}{2(1)}$$
$$x = \frac{58 \pm \sqrt{3,364 - 800}}{2}$$
$$x = \frac{58 \pm \sqrt{2,564}}{2}$$
$$x = \frac{58 \pm 50.6}{2}$$
$$x = \frac{58 + 50.6}{2} \quad \text{or} \quad x = \frac{58 - 50.6}{2}$$
$$x = \frac{108.6}{2} \qquad x = \frac{7.4}{2}$$
$$x = 54.3 \qquad x = 3.7$$

$$x - 8 = 54.3 - 8 \qquad x - 8 = 3.7 - 8$$
$$= 46.3 \qquad = \cancel{-4.3}$$

It takes the assistant about 54 minutes and the chef about 46 minutes to make the pastry dessert.

27. Let x = shorter side of a triangle.
Then $x + 14$ = longer leg of a triangle.
Use Pythagorean Theorem to find x.

$$x^2+(x+14)^2=26^2$$
$$x^2+x^2+28x+196=676$$
$$2x^2+28x-480=0$$
$$\frac{2x^2}{2}+\frac{28x}{2}-\frac{480}{2}=\frac{0}{2}$$
$$x^2+14x-240=0$$
$$(x-10)(x+24)=0$$
$$x-10=0 \quad \text{or} \quad x+24=0$$
$$x=10 \qquad x=-24$$

The segment extending from ground to the top of the building is the sum of 2 shorter legs. Since the length of one shorter leg is 10 in., the length of the segment is 20 in.

28. Let x = width of border. Then the new length of the mirror with border is $2x + 18$ and the new width of the mirror with border is $2x + 24$.
Total area - area of picture = area of border

$$(2x+18)(2x+24)-(18)(24)=(18)(24)$$
$$4x^2+48x+36x+432-432=432$$
$$4x^2+84x-432=0$$
$$x^2+21x-108=0$$
$$x=\frac{-b\pm\sqrt{b^2-4ac}}{2a}$$
$$=\frac{-21\pm\sqrt{(21)^2-4(1)(-108)}}{2(1)}$$
$$=\frac{-21\pm\sqrt{441+432}}{2}$$
$$=\frac{-21\pm\sqrt{873}}{2}$$
$$x=\frac{-21+\sqrt{873}}{2} \quad \text{or} \quad x=\frac{-21-\sqrt{873}}{2}$$
$$x\approx 4.3 \qquad x\approx -25.3$$

The border should be approximately 4.3 in. wide.

29. Find when $E(t)=5{,}000$.

$$2.29t^2-75.72t+5{,}206.95=5{,}000$$
$$2.29t^2-75.72t+206.95=0$$
$$t=\frac{-b\pm\sqrt{b^2-4ac}}{2a}$$
$$t=\frac{-(-75.72)\pm\sqrt{(-75.72)^2-4(2.29)(206.95)}}{2(2.29)}$$
$$t=\frac{75.72\pm\sqrt{3837.8564}}{4.58}$$
$$t=\frac{75.72+\sqrt{3837.8564}}{4.58} \quad \text{or} \quad t=\frac{75.72-\sqrt{3837.8564}}{4.58}$$
$$t\approx 30 \qquad t\approx 3$$

Since 30 years from 1990 has occurred yet, then $t = 3$ which would be 1993.

30. The vertex is (0, 6) and the parabola is going down. The mathematical model that has those characteristics is **iii**.

31. $f(x)=-3(x-1)^2-2$ (1,2)
Vertex: (1, -2) → (1,2)
Axis of Symmetry: $x = 1$

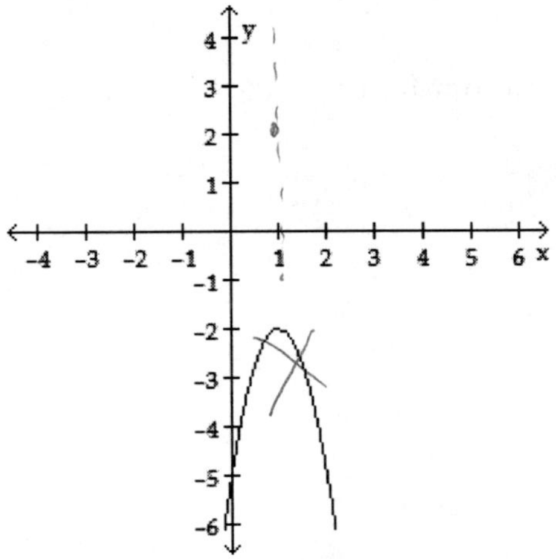

32. $f(x) = 5x^2 + 10x - 1$

$f(x) = 5(x^2 + 2x) - 1$

$f(x) = 5(x^2 + 2x + 1) - 1 - 5(1)$

$f(x) = 5(x+1)^2 - 1 - 5$

$f(x) = 5(x+1)^2 - 6$

Vertex: (–1, –6)
Axis of Symmetry: $x = -1$

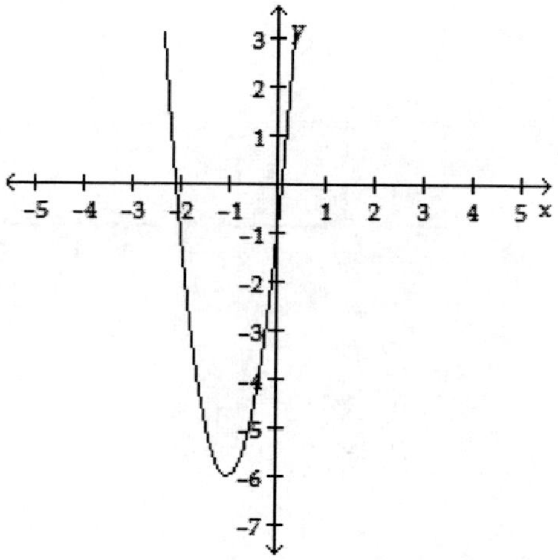

33. $f(x) = 2x^2 + x - 1$

Step 1: Since $a = 2 > 0$, the parabola opens upward.

Step 2: Find the vertex and axis of symmetry.

$$x = \frac{-b}{2a} = \frac{-1}{2(2)} = -\frac{1}{4}$$

$$y = 2\left(-\frac{1}{4}\right)^2 + \left(-\frac{1}{4}\right) - 1$$

$$= 2\left(\frac{1}{16}\right) - \frac{1}{4} - 1$$

$$= \frac{1}{8} - \frac{2}{8} - \frac{8}{8}$$

$$= -\frac{9}{8}$$

vertex: $\left(-\frac{1}{4}, -\frac{9}{8}\right)$

The axis of symmetry is $x = -\frac{1}{4}$.

Step 3: Find the x– and y–intercepts. Since $c = -1$, the y–intercept is (0, –1). To find the x–intercepts, let $y = 0$ and solve the equation for x.

$$0 = 2x^2 + x - 1$$

$$0 = (2x-1)(x+1)$$

$2x - 1 = 0$ or $x + 1 = 0$

$x = \frac{1}{2}$ $\qquad x = -1$

x–intercepts: $\left(\frac{1}{2}, 0\right)$ and (–1, 0)

y–intercept: (0, –1)

Step 4: Let $x = 1$ and $x = -2$ to find two more points..

$y = 2(1)^2 + (1) - 1 = 2 + 1 - 1 = 2$

$y = 2(-2)^2 + (-2) - 1 = 8 - 2 - 1 = 5$

(1, 2) and (–2, 5)

Step 5: Plot the points and draw the parabola.

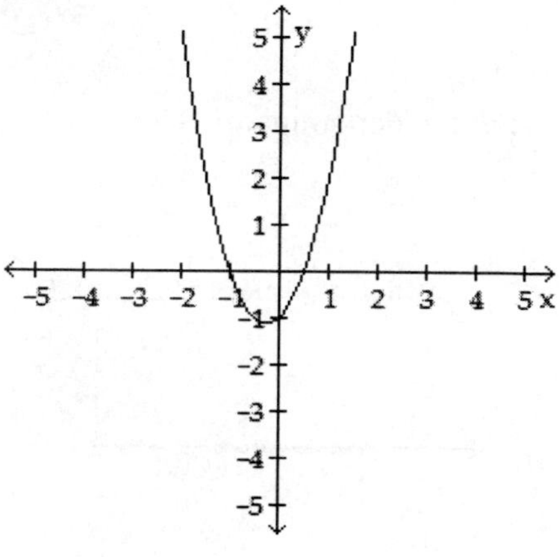

34. The vertex of the parabola represented by the equation would be the highest point. Use the vertex formula to find the vertex.

$$h = -16t^2 + 112t + 15$$

$$x = \frac{-b}{2a} = \frac{-112}{2(-16)} = \frac{-112}{-32} = 3.5$$

$$\begin{aligned} y &= -16(3.5)^2 + 112(3.5) + 15 \\ &= -16(12.25) + 112(3.5) + 15 \\ &= -196 + 392 + 15 \\ &= 211 \text{ ft} \end{aligned}$$

The flare explodes at 211 ft.

35.

$$x^2 - 2x > 8$$

$$x^2 - 2x - 8 > 0$$

$$x^2 - 2x - 8 = 0$$

$$(x-4)(x+2) = 0$$

$$x - 4 = 0 \quad \text{or} \quad x + 2 = 0$$

$$x = 4 \qquad x = -2$$

critical numbers = 4 and -2

$$(-\infty, -2) \cup (4, \infty)$$

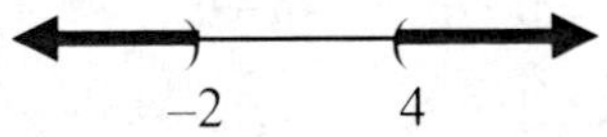

36.

$$\frac{x-2}{x+3} \le 0$$

$$\frac{x-2}{x+3} = 0$$

$$(x+3)\left(\frac{x-2}{x+3}\right) = (x+3)(0)$$

$$x - 2 = 0$$

$$x = 2$$

Set the denominator = 0

$$x + 3 = 0$$

$$x = -3$$

critical numbers = -3 and 2

$$(-3, 2]$$

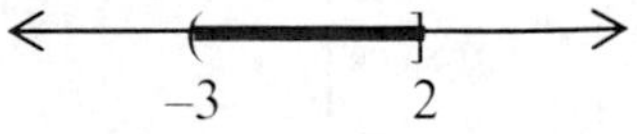

37. Since July is the 7th month, find $T(7)$.

$$T(m) = -1.1m^2 + 15.3m + 29.5$$

$$\begin{aligned} T(7) &= -1.1(7)^2 + 15.3(7) + 29.5 \\ &= -1.1(49) + 15.3(7) + 29.5 \\ &= -53.9 + 107.1 + 29.5 \\ &= 82.7^\circ \text{ F} \end{aligned}$$

38. $y \le -x^2 + 3$

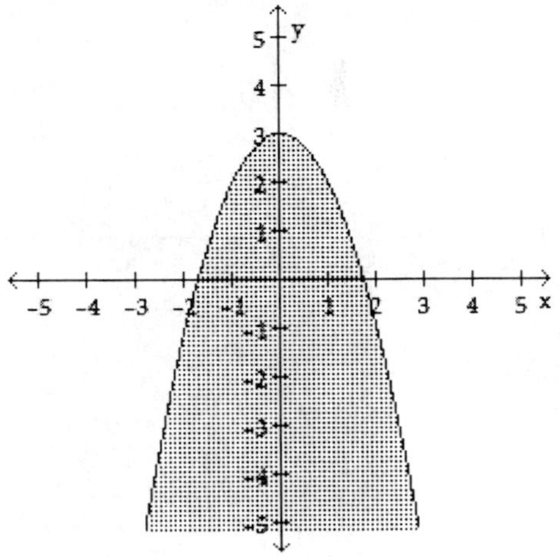

39. The x–intercepts are (3, 0) and (–2, 0), so the solutions to the equation are $x = 3$ and $x = -2$.

40. [–2, 3]

CHAPTER 8 CUMULATIVE REVIEW

1. $3(x+2)-2=-(5+x)+x$

$$\begin{aligned}3x+6-2&=-5-x+x\\3x+4&=-5\\3x+4-4&=-5-4\\3x&=-9\\x&=-3\end{aligned}$$

2. Let x = liters of 1% glucose solution.

$$\begin{aligned}0.01x+0.05(2)&=0.02(x+2)\\0.01x+0.1&=0.02x+0.04\\100(0.01x+0.1)&=100(0.02x+0.04)\\x+10&=2x+4\\x+10-4&=2x+4-4\\x+6&=2x\\x+6-x&=2x-x\\6&=x\end{aligned}$$

6 Liters of 1% glucose solution are needed

3. Use point–slope formula.

$$\begin{aligned}y-y_1&=m(x-x_1)\\y-(-4)&=3[x-(-2)]\\y+4&=3(x+2)\\y+4&=3x+6\\y&=3x+2\end{aligned}$$

4. Parallel lines have the same slope, so find the slope of the given equation by solving the equation for y.

$$\begin{aligned}2x+3y&=6\\3y&=-2x+6\\y&=-\frac{2}{3}x+2\\m&=-\frac{2}{3}\end{aligned}$$

Use the point–slope formula.

$$\begin{aligned}y-y_1&=m(x-x_1)\\y-(-2)&=-\frac{2}{3}(x-0)\\y+2&=-\frac{2}{3}x-0\\y&=-\frac{2}{3}x-2\end{aligned}$$

5. For the DEMAND, find the slope using the ordered pairs (2020, 2,850,000) and (2010, 2,300,000).

$$\begin{aligned}m&=\frac{y_2-y_1}{x_2-x_1}\\&=\frac{2{,}850{,}000-2{,}300{,}000}{2020-2010}\\&=\frac{550{,}000}{10}\\&=55{,}000\end{aligned}$$

For the SUPPLY, find the slope using the ordered pairs (2020, 1,800,000) and (2010, 1,950,000).

$$\begin{aligned}m&=\frac{y_2-y_1}{x_2-x_1}\\&=\frac{1{,}800{,}000-1{,}950{,}000}{2020-2010}\\&=\frac{-150{,}000}{10}\\&=-15{,}000\end{aligned}$$

6. a. Use the ordered pairs (10, 6.2) and (20, 5.4) to find the slope. Note that t is the number of years after 1960.

$$\begin{aligned}m&=\frac{y_2-y_1}{x_2-x_1}\\&=\frac{5.4-6.2}{20-10}\\&=\frac{-0.8}{10}\\&=-0.08\end{aligned}$$

Use (10, 6.2) and $m=-0.08$ to write an equation.

$$\begin{aligned}y-y_1&=m(x-x_1)\\y-6.2&=-0.08(x-10)\\y-6.2&=-0.08x+0.8\\y&=-0.08x+7\\P(t)&=-0.08t+7\end{aligned}$$

b. Let $t=50$.

$$\begin{aligned}P(50)&=-0.08(50)+7\\&=-4+7\\&=3\end{aligned}$$

In 2010, the percent of income used was 3%.

7. No, it is not a function. Answers may vary. Examples are (0, 3) and (0, –1).

8. a. domain: [0, 24]
   b. $f(6)=1.5$
   c. At noon, the low tide mark was –2.5 meters.
   d. 0, 2, 9, 17

9.

$$\begin{cases} x+y=-1 \\ y=2x-4 \end{cases}$$

The lines intersect at (1, –2).

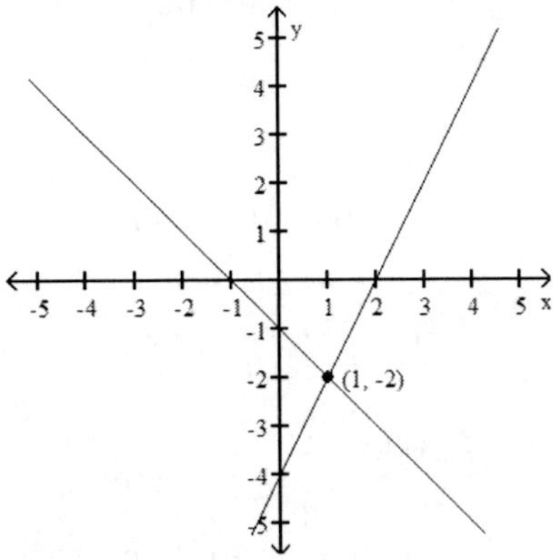

10.

$$\begin{cases} x-y=-5 \\ 3x-2y=-7 \end{cases}$$

Solve the first equation for x.

$$\begin{cases} x=y-5 \\ 3x-2y=-7 \end{cases}$$

Substitute $x = y - 5$ into the second equation and solve for y.

$$3x-2y=-7$$
$$3(y-5)-2y=-7$$
$$3y-15-2y=-7$$
$$y-15=-7$$
$$y-15+15=-7+15$$
$$y=8$$

Substitute $y = 8$ into the first equation and solve for x.

$$x-y=-5$$
$$x-8=-5$$
$$x-8+8=-5+8$$
$$x=3$$

The solution is (3, 8).

11.

$$\begin{cases} x-y+z=4 & (1) \\ x+2y-z=-1 & (2) \\ x+y-3z=-2 & (3) \end{cases}$$

Add Equation 1 and 2 to eliminate z.

$$\begin{array}{ll} x-\ \ y+z=\ 4 & (1) \\ \underline{x+2y-z=-1} & (2) \\ 2x+y\qquad\ =3 & (4) \end{array}$$

Multiply Equation 2 by -3 and add Equations 2 and 3 to eliminate z.

$$\begin{array}{ll} -3x-6y+3z=3 & (2) \\ \underline{x+\ \ y-3z=-2} & (3) \\ -2x-5y\qquad =1 & (5) \end{array}$$

Add Equations 4 and 5 to eliminate x.

$$\begin{array}{ll} 2x+\ \ y=3 & (4) \\ \underline{-2x-5y=1} & (5) \\ -4y=4 & (5) \end{array}$$
$$y=-1$$

Substitute $y=-1$ into Equation 4 to find x.

$$2x+y=3 \quad (4)$$
$$2x-1=3$$
$$2x=4$$
$$x=2$$

Substitute $x=2$ and $y=-1$ into Equation 1 to find z.

$$x-y+z=4 \quad (1)$$
$$2-(-1)+z=4$$
$$3+z=4$$
$$z=1$$

The solution is (2, -1, 1).

12.

$$\begin{vmatrix} -6 & -2 \\ 15 & 4 \end{vmatrix} = -6(4)-(-2)(15)$$
$$=-24+30$$
$$=6$$

13. Let x = the number of episodes of *That 70's Show*. Then *Friends* = $x + 36$ and *The King of Queens* = $x + 7$. The sum of all three is 643.

$$(x+36)+x+(x+7)=643$$
$$3x+43=643$$
$$3x=600$$
$$x=200$$
$$x+36=200+36=236$$
$$x+7=200+7=207$$

Friends had 236 episodes, *That 70's Show* had 200 episodes, and *The King of Queens* had 207 episodes.

14. Let x = the amount of sales per month. Then Package 1 is $\$2,500 + 0.05x$ and Package 2 is $\$3,500 + 0.03x$. Find x when package 1 is more than package 2.

$$2,500+0.05x>3,500+0.03x$$
$$2,500+0.05x-0.03x>3,500+0.03x-0.03x$$
$$2,500+0.02x>3,500$$
$$2,500+0.02x-2,500>3,500-2,500$$
$$0.02x>1,000$$
$$\frac{0.02x}{0.02}>\frac{1,000}{0.02}$$
$$x>50,000$$

Package 1 is more profitable when sales are more than \$50,000 per month.

15.

$$f(x)>g(x)$$
$$5(-2x+2)>20-x$$
$$-10x+10>20-x$$
$$-9x+10>20$$
$$-9x>10$$
$$x<-\frac{10}{9}$$
$$\left(-\infty,-\frac{10}{9}\right)$$

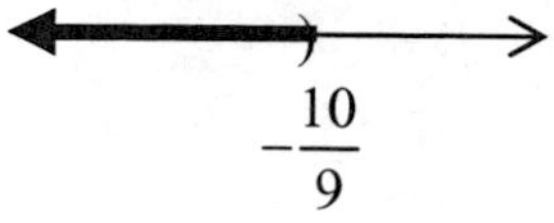

16.

$$5x-3\geq 2 \quad \text{and} \quad 6\geq 4x-3$$
$$5x-3+3\geq 2+3 \qquad 6+3\geq 4x-3+3$$
$$5x\geq 5 \qquad 9\geq 4x$$
$$\frac{5x}{5}\geq\frac{5}{5} \qquad \frac{9}{4}\geq\frac{4x}{4}$$
$$x\geq 1 \qquad \frac{9}{4}\geq x$$
$$x\leq\frac{9}{4}$$
$$\left[1,\frac{9}{4}\right]$$

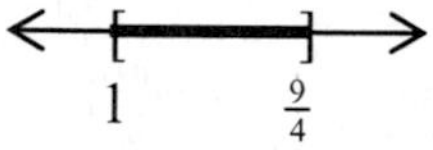

17.

$$|2x-5|\geq 25$$
$$2x-5\geq 25 \quad \text{or} \quad 2x-5\leq -25$$
$$2x\geq 30 \qquad 2x\leq -20$$
$$x\geq 15 \qquad x\leq -10$$
$$(-\infty,-10]\cup[15,\infty)$$

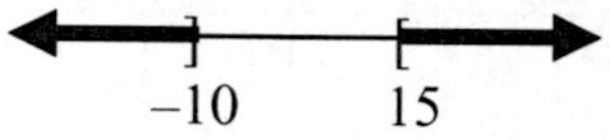

18.

$$\begin{cases} 3x+2y>6 \\ x+3y\leq 2 \end{cases}$$

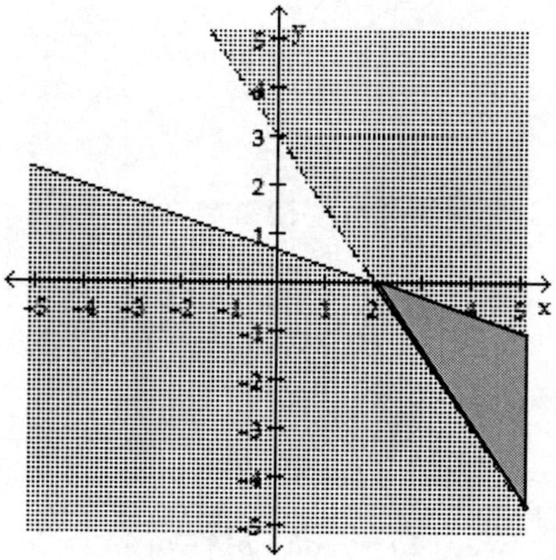

19.

$$\left(\frac{2a^2b^3c^{-4}}{5a^{-2}b^{-1}c^3}\right)^{-3}=\left(\frac{5a^{-2}b^{-1}c^3}{2a^2b^3c^{-4}}\right)^3$$
$$=\left(\frac{5}{2}a^{-2-2}b^{-1-3}c^{3-(-4)}\right)^3$$
$$=\left(\frac{5}{2}a^{-4}b^{-4}c^7\right)^3$$
$$=\left(\frac{5c^7}{2a^4b^4}\right)^3$$
$$=\frac{5^3c^{7\cdot3}}{2^3a^{4\cdot3}b^{4\cdot3}}$$
$$=\frac{125c^{21}}{8a^{12}b^{12}}$$

20. $$\frac{(1,280,000,000)(2,700,000)}{(240,000)}$$
$$=\frac{(1.28\times10^9)(2.7\times10^6)}{2.4\times10^5}$$
$$=\frac{(1.28\times2.7)(10^9\times10^6)}{2.4\times10^5}$$
$$=\frac{3.456\times10^{15}}{2.4\times10^5}$$
$$=\frac{3.456}{2.4}\times10^{15-5}$$
$$=1.44\times10^{10}$$
$$=14,400,000,000$$

21. $f(t)-g(t)=$

$$(-8.9t^3-2.4t)-(2.1t^3+0.8t^2-t)$$
$$=-8.9t^3-2.4t-2.1t^3-0.8t^2+t$$
$$=-11t^3-0.8t^2-1.4t$$

22.

$$(2a-b)(4a^2+2ab+b^2)$$
$$=2a(4a^2+2ab+b^2)-b(4a^2+2ab+b^2)$$
$$=8a^3+4a^2b+2ab^2-4a^2b-2ab^2-b^3$$
$$=8a^3-b^3$$

23.

$$(9mn^3-4)^2=(9mn^3)^2-2(9mn^3)(4)+(-4)^2$$
$$=81m^2n^6-72mn^3+16$$

24.

$$(2x^2-y^2)(2x^2+y^2)=(2x^2)^2-(-y^2)^2$$
$$=4x^4-y^4$$

25.

$$x^2+4y-xy-4x=x^2-xy-4x+4y$$
$$=x(x-y)-4(x-y)$$
$$=(x-y)(x-4)$$

26.

$$30a^4-4a^3-16a^2=2a^2(15a^2-2a-8)$$
$$=2a^2(3a+2)(5a-4)$$

27.

$$49s^6-84s^3n^2+36n^4=(7s^3-6n^2)^2$$

28.

$$x^2+10x+25-y^8=(x^2+10x+25)-y^8$$
$$=(x+5)^2-y^8$$
$$=(x+5-y^4)(x+5+y^4)$$

29.

$$x^4-16y^4=(x^2+4y^2)(x^2-4y^2)$$
$$=(x^2+4y^2)(x+2y)(x-2y)$$

30.

$$8x^6+125y^3$$
$$=(2x^2+5y)\left[(2x^2)^2-(2x^2)(5y)+(5y)^2\right]$$
$$=(2x^2+5y)(4x^4-10x^2y+25y^2)$$

31.

$$(m+4)(2m+3)-22=10m$$
$$2m^2+3m+8m+12-22=10m$$
$$2m^2+11m-10=10m$$
$$2m^2+m-10=0$$
$$(2m+5)(m-2)=0$$
$$2m+5=0 \quad \text{or} \quad m-2=0$$
$$m=-\frac{5}{2} \qquad m=2$$

32.

$$6a^3 - 2a = a^2$$
$$6a^3 - a^2 - 2a = 0$$
$$a(6a^2 - a - 2) = 0$$
$$a(2a+1)(3a-2) = 0$$
$$a = 0 \quad 2a+1 = 0 \quad \text{or} \quad 3a-2 = 0$$
$$a = 0 \qquad a = -\frac{1}{2} \qquad a = \frac{2}{3}$$

33. Find $h(t) = 0$.

$$-16t^2 + 96t = 0$$
$$-16t(t-6) = 0$$
$$-16t = 0 \quad \text{or} \quad t - 6 = 0$$
$$t = \cancel{0} \qquad t = 6$$

34. $f(x) = \dfrac{x^2 + 3x + 2}{x^2 - x - 56}$

$$x^2 - x - 56 \neq 0$$
$$(x-8)(x+7) = 0$$
$$x - 8 = 0 \quad \text{and} \quad x + 7 = 0$$
$$x = 8 \qquad x = -7$$

Domain: $(-\infty, -7) \cup (-7, 8) \cup (8, \infty)$

35. $f(x) = |x-1| + 2$

$D: (-\infty, \infty), R: [2, \infty)$

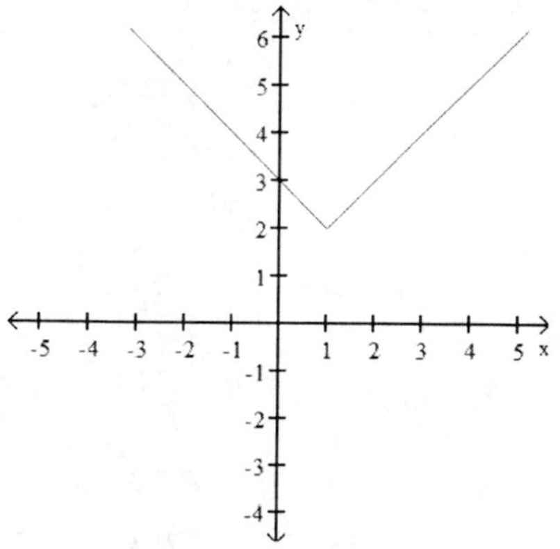

36. $f(x) = \dfrac{4}{x}$

$D: (0, \infty), R: (0, \infty)$

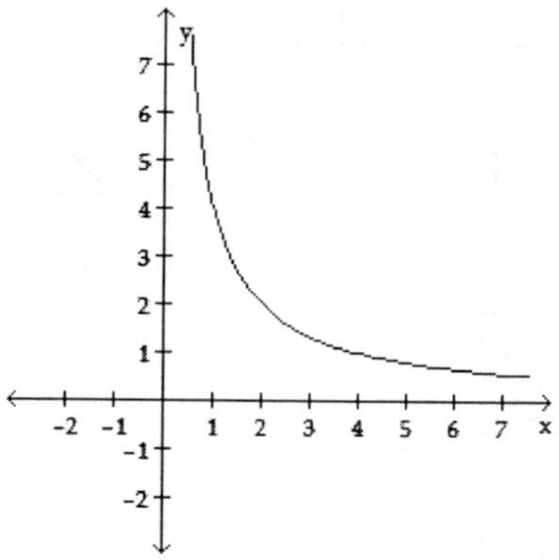

37.

$$\frac{6x^2 - 7x - 5}{2x^2 + 5x + 2} = \frac{(3x-5)\cancel{(2x+1)}}{(x+2)\cancel{(2x+1)}}$$
$$= \frac{3x-5}{x+2}$$

38.

$$\frac{x^3 + y^3}{x^3 - y^3} \div \frac{x^2 - xy + y^2}{x^2 + xy + y^2}$$
$$= \frac{x^3 + y^3}{x^3 - y^3} \bullet \frac{x^2 + xy + y^2}{x^2 - xy + y^2}$$
$$= \frac{(x+y)\cancel{(x^2 - xy + y^2)}\cancel{(x^2 + xy + y^2)}}{(x-y)\cancel{(x^2 + xy + y^2)}\cancel{(x^2 - xy + y^2)}}$$
$$= \frac{x+y}{x-y}$$

39.

$$\frac{1}{x+y}-\frac{1}{x-y}+\frac{2y}{x^2-y^2}$$
$$=\frac{1}{x+y}-\frac{1}{x-y}+\frac{2y}{(x+y)(x-y)}$$
$$=\frac{1}{x+y}\left(\frac{x-y}{x-y}\right)-\frac{1}{x-y}\left(\frac{x+y}{x+y}\right)+\frac{2y}{(x-y)(x+y)}$$
$$=\frac{x-y}{(x+y)(x-y)}-\frac{x+y}{(x+y)(x-y)}+\frac{2y}{(x+y)(x-y)}$$
$$=\frac{x-y-(x+y)+2y}{(x+y)(x-y)}$$
$$=\frac{x-y-x-y+2y}{(x+y)(x-y)}$$
$$=\frac{0}{(x+y)(x-y)}$$
$$=0$$

40.

$$\frac{\frac{1}{r^2+4r+4}}{\frac{r}{r+2}+\frac{r}{r+2}}=\frac{\frac{1}{(r+2)(r+2)}}{\frac{r}{r+2}+\frac{r}{r+2}}$$
$$=\frac{\frac{1}{(r+2)(r+2)}}{\frac{r}{r+2}+\frac{r}{r+2}}\bullet\frac{(r+2)(r+2)}{(r+2)(r+2)}$$
$$=\frac{\frac{(r+2)(r+2)}{(r+2)(r+2)}}{\frac{r(r+2)(r+2)}{r+2}+\frac{r(r+2)(r+2)}{r+2}}$$
$$=\frac{1}{r(r+2)+r(r+2)}$$
$$=\frac{1}{r^2+2r+r^2+2r}$$
$$=\frac{1}{2r^2+4r}$$
$$=\frac{1}{2r(r+2)}$$

41.

$$\frac{24x^6y^7-12x^5y^{12}+36xy}{48x^2y^3}$$
$$=\frac{24x^6y^7}{48x^2y^3}-\frac{12x^5y^{12}}{48x^2y^3}+\frac{36xy}{48x^2y^3}$$
$$=\frac{x^4y^4}{2}-\frac{x^3y^9}{4}+\frac{3}{4xy^2}$$

42.

$$\begin{array}{r} 5a^2-3a-4 \\ 3a-4\overline{)15a^3-29a^2+0a+16} \\ \underline{15a^3-20a^2} \\ -9a^2+0a \\ \underline{-9a^2+12a} \\ -12a+16 \\ \underline{-12a+16} \\ 0 \end{array}$$

43.

$$\frac{x-4}{x-3}+\frac{x-2}{x-3}=x-3$$
$$x-3\left(\frac{x-4}{x-3}+\frac{x-2}{x-3}\right)=(x-3)(x-3)$$
$$x-4+x-2=x^2-3x-3x+9$$
$$2x-6=x^2-6x+9$$
$$0=x^2-8x+15$$
$$0=(x-3)(x-5)$$
$$x-3=0 \quad \text{or} \quad x-5=0$$
$$x=\not{3} \qquad\qquad x=5$$

3 is extraneous

44.

$$\frac{1}{R}=\frac{1}{R_1}+\frac{1}{R_2}+\frac{1}{R_3}$$
$$RR_1R_2R_3\left(\frac{1}{R}\right)=RR_1R_2R_3\left(\frac{1}{R_1}+\frac{1}{R_2}+\frac{1}{R_3}\right)$$
$$R_1R_2R_3=RR_2R_3+RR_1R_3+RR_1R_2$$
$$R_1R_2R_3=R(R_2R_3+R_1R_3+R_1R_2)$$
$$\frac{R_1R_2R_3}{R_2R_3+R_1R_3+R_1R_2}=\frac{R(R_2R_3+R_1R_3+R_1R_2)}{R_2R_3+R_1R_3+R_1R_2}$$
$$\frac{R_1R_2R_3}{R_2R_3+R_1R_3+R_1R_2}=R$$

45. Let x = time it takes to fill the sink if both faucets are working together.

$$\frac{1}{30}+\frac{1}{45}=\frac{1}{x}$$
$$90x\left(\frac{1}{30}+\frac{1}{45}\right)=90x\left(\frac{1}{x}\right)$$
$$3x+2x=90$$
$$5x=90$$
$$x=18$$

It would take 18 seconds.

46. Let x = tons of sand.

$$\frac{7}{2}=\frac{x}{6}$$
$$42=2x$$
$$21=x$$

21 tons of sand are needed.

47. Let c = costs, t = # of trucks, and h = # of hours used. Find k when c = \$3,600, t = 8, and h = 12.

$$c=kth$$
$$3{,}600=k(8)(12)$$
$$3{,}600=k(96)$$
$$3{,}600=96k$$
$$37.5=k$$

Find c for k = 37.5, t = 20, and h = 12.

$$c=kth$$
$$c=(37.5)(20)(12)$$
$$c=\$9{,}000$$

48. $f(x)=\sqrt{x-2}$

$D:[2,\infty), R:[0,\infty)$

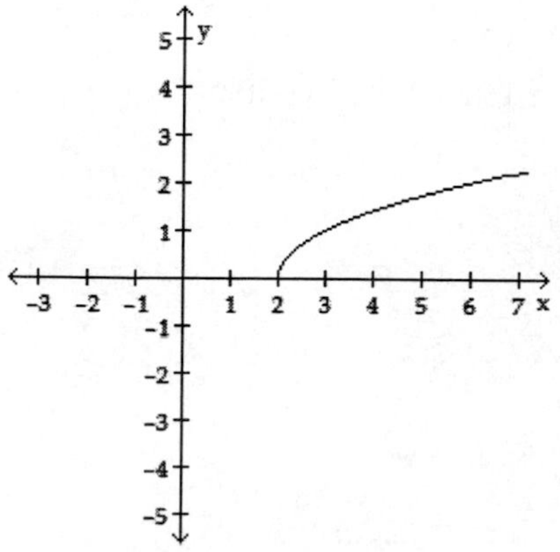

49. $C(L)=1.25\sqrt{L}$

$$C(6.5)=1.25\sqrt{6.5}$$
$$\approx 3.2$$

About 3.2 meters per second

50.

$$f(t)=g(t)$$
$$\sqrt[3]{x-5}=2\sqrt[3]{x+9}$$
$$\left(\sqrt[3]{x-5}\right)^3=\left(2\sqrt[3]{x+9}\right)^3$$
$$x-5=8(x+9)$$
$$x-5=8x+72$$
$$-7x-5=72$$
$$-7x=77$$
$$x=-11$$

51.

$$\sqrt[3]{-27x^3}=-3x$$

52.

$$\sqrt{48t^3}=\sqrt{16t^2}\sqrt{3t}$$
$$=4t\sqrt{3t}$$

53.

$$64^{-2/3}=\frac{1}{64^{2/3}}$$
$$=\frac{1}{\left(\sqrt[3]{64}\right)^2}$$
$$=\frac{1}{(4)^2}$$
$$=\frac{1}{16}$$

54.

$$\frac{x^{5/3}x^{1/2}}{x^{3/4}}=\frac{x^{5/3+1/2}}{x^{3/4}}$$
$$=\frac{x^{10/6+3/6}}{x^{3/4}}$$
$$=\frac{x^{13/6}}{x^{3/4}}$$
$$=x^{13/6-3/4}$$
$$=x^{26/12-9/12}$$
$$=x^{17/12}$$

55.

$$-3\sqrt[4]{32}-2\sqrt[4]{162}+5\sqrt[4]{48}$$
$$=-3\sqrt[4]{16}\sqrt[4]{2}-2\sqrt[4]{81}\sqrt[4]{2}+5\sqrt[4]{16}\sqrt[4]{3}$$
$$=-3(2)\sqrt[4]{2}-2(3)\sqrt[4]{2}+5(2)\sqrt[4]{3}$$
$$=-6\sqrt[4]{2}-6\sqrt[4]{2}+10\sqrt[4]{3}$$
$$=-12\sqrt[4]{2}+10\sqrt[4]{3}$$

56.

$$3\sqrt{2}\left(2\sqrt{3}-4\sqrt{12}\right)=6\sqrt{6}-12\sqrt{24}$$
$$=6\sqrt{6}-12\sqrt{4}\sqrt{6}$$
$$=6\sqrt{6}-12(2)\sqrt{6}$$
$$=6\sqrt{6}-24\sqrt{6}$$
$$=-18\sqrt{6}$$

57.

$$\frac{\sqrt{x}+2}{\sqrt{x}-1}=\frac{\sqrt{x}+2}{\sqrt{x}-1}\bullet\frac{\sqrt{x}+1}{\sqrt{x}+1}$$
$$=\frac{\left(\sqrt{x}+2\right)\left(\sqrt{x}+1\right)}{\left(\sqrt{x}-1\right)\left(\sqrt{x}+1\right)}$$
$$=\frac{\sqrt{x^2}+\sqrt{x}+2\sqrt{x}+2}{\sqrt{x^2}-1}$$
$$=\frac{x+3\sqrt{x}+2}{x-1}$$

58.

$$\frac{5}{\sqrt[3]{x}}=\frac{5}{\sqrt[3]{x}}\bullet\frac{\sqrt[3]{x^2}}{\sqrt[3]{x^2}}$$
$$=\frac{5\sqrt[3]{x^2}}{\sqrt[3]{x^3}}$$
$$=\frac{5\sqrt[3]{x^2}}{x}$$

59.

$$5\sqrt{x+2}=x+8$$
$$\left(5\sqrt{x+2}\right)^2=(x+8)^2$$
$$25(x+2)=x^2+8x+8x+64$$
$$25x+50=x^2+16x+64$$
$$0=x^2-9x+14$$
$$0=(x-2)(x-7)$$
$$x-2=0 \quad \text{or} \quad x-7=0$$
$$x=2 \qquad\qquad x=7$$

60.

$$\sqrt{x}+\sqrt{x+2}=2$$
$$\sqrt{x+2}=2-\sqrt{x}$$
$$\left(\sqrt{x+2}\right)^2=\left(2-\sqrt{x}\right)^2$$
$$x+2=4-2\sqrt{x}-2\sqrt{x}+\sqrt{x^2}$$
$$x+2=4-4\sqrt{x}+x$$
$$x+2-4-x=4-4\sqrt{x}+x-4-x$$
$$-2=-4\sqrt{x}$$
$$(-2)^2=\left(-4\sqrt{x}\right)^2$$
$$4=16x$$
$$\frac{4}{16}=x$$
$$\frac{1}{4}=x$$

61. The length of the hypotenuse in an isosceles right triangles is the length of the leg times $\sqrt{2}$.
The length of the hypotenuse is $3\sqrt{2}$ in.

62. The length of the longer leg is $\sqrt{3}$ times the length of the shorter leg. The length of the hypotenuse is 2 times the length of the shorter side.

Let x = length of the shorter leg.

longer leg:

$$x\sqrt{3} = 3$$
$$\frac{x\sqrt{3}}{\sqrt{3}} = \frac{3}{\sqrt{3}}$$
$$x = \frac{3}{\sqrt{3}} \bullet \frac{\sqrt{3}}{\sqrt{3}}$$
$$x = \frac{3\sqrt{3}}{3}$$
$$x = \sqrt{3} \text{ in.}$$

hypotenuse:

$$2x = 2\sqrt{3} \text{ in.}$$

63.

$$\begin{aligned} d &= \sqrt{(x_1 - x_2)^2 + (y_1 - y_2)^2} \\ &= \sqrt{[4-(-2)]^2 + (14-6)^2} \\ &= \sqrt{(6)^2 + (8)^2} \\ &= \sqrt{36+64} \\ &= \sqrt{100} \\ &= 10 \end{aligned}$$

64.

$$\begin{aligned} i^{43} &= i^{40} i^2 i \\ &= 1(-1)i \\ &= -i \end{aligned}$$

65.

$$\begin{aligned} &\left(-7+\sqrt{-81}\right)-\left(-2-\sqrt{-64}\right) \\ &= (-7+9i)-(-2-8i) \\ &= -7+9i+2+8i \\ &= -5+17i \end{aligned}$$

66.

$$\begin{aligned} \frac{5}{3-i} &= \frac{5}{3-i} \bullet \frac{3+i}{3+i} \\ &= \frac{5(3+i)}{(3-i)(3+i)} \\ &= \frac{15+5i}{9-i^2} \\ &= \frac{15+5i}{9-(-1)} \\ &= \frac{15+5i}{10} \\ &= \frac{15}{10} + \frac{5}{10}i \\ &= \frac{3}{2} + \frac{1}{2}i \end{aligned}$$

67.

$$\begin{aligned} (2+i)^2 &= (2+i)(2+i) \\ &= 4+2i+2i+i^2 \\ &= 4+4i+(-1) \\ &= 3+4i \end{aligned}$$

68.

$$\begin{aligned} \frac{-4}{6i^7} &= \frac{-4}{6i^7} \bullet \frac{i}{i} \\ &= \frac{-4i}{6i^8} \\ &= \frac{-4i}{6(1)} \\ &= -\frac{4}{6}i \\ &= -\frac{2}{3}i \\ &= 0 - \frac{2}{3}i \end{aligned}$$

69.

$$\begin{aligned} x^2 &= 28 \\ x &= \pm\sqrt{28} \\ x &= \pm\sqrt{4}\sqrt{7} \\ x &= \pm 2\sqrt{7} \end{aligned}$$

70.

$$(x-19)^2=-5$$
$$x-19=\sqrt{-5} \quad \text{or} \quad x-19=-\sqrt{-5}$$
$$x-19=i\sqrt{5} \qquad x-19=-i\sqrt{5}$$
$$x=19+i\sqrt{5} \qquad x=19-i\sqrt{5}$$
$$x=19\pm i\sqrt{5}$$

71.

$$2x^2-6x+3=0$$
$$2x^2-6x=-3$$
$$\frac{2x^2}{2}-\frac{6x}{2}=-\frac{3}{2}$$
$$x^2-3x=-\frac{3}{2}$$
$$x^2-3x+\left(\frac{-3}{2}\right)^2=-\frac{3}{2}+\left(\frac{-3}{2}\right)^2$$
$$x^2-3x+\frac{9}{4}=-\frac{3}{2}+\frac{9}{4}$$
$$\left(x-\frac{3}{2}\right)^2=\frac{3}{4}$$
$$x-\frac{3}{2}=\pm\sqrt{\frac{3}{4}}$$
$$x-\frac{3}{2}=\pm\frac{\sqrt{3}}{2}$$
$$x=\frac{3}{2}\pm\frac{\sqrt{3}}{2}$$
$$x=\frac{3\pm\sqrt{3}}{2}$$

72.

$$a^2-\frac{2}{5}a=-\frac{1}{5}$$
$$5\left(a^2-\frac{2}{5}a\right)=5\left(-\frac{1}{5}\right)$$
$$5a^2-2a=-1$$
$$5a^2-2a+1=0$$
$$a=\frac{-b\pm\sqrt{b^2-4ac}}{2a}$$
$$=\frac{-(-2)\pm\sqrt{(-2)^2-4(5)(1)}}{2(5)}$$
$$=\frac{2\pm\sqrt{4-20}}{10}$$
$$=\frac{2\pm\sqrt{-16}}{10}$$
$$=\frac{2\pm 4i}{10}$$
$$=\frac{2}{10}\pm\frac{4i}{10}$$
$$=\frac{1}{5}\pm\frac{2}{5}i$$

73. Let x = width of the uniform sidewalk.
To find the length or width, you must take the total length or width and subtract $2x$ since there is a sidewalk of width x on both ends of the garden.
The length is $(24 - 2x)$ and the width is $(16 - 2x)$. Solve the inequality to find x.

$$(\text{length})(\text{width}) \le 180$$
$$(24-2x)(16-2x) \le 180$$
$$384-48x-32x+4x^2 \le 180$$
$$4x^2-80x+384 \le 180$$
$$4x^2-80x+204 \le 0$$
$$4x^2-80x+204 = 0$$
$$\frac{4x^2}{4}-\frac{80x}{4}+\frac{204}{4}=\frac{0}{4}$$
$$x^2-20x+51=0$$
$$(x-3)(x-17)=0$$
$$x-3=0 \quad \text{or} \quad x-17=0$$
$$x=3 \qquad x=17$$

17 is not possible: it would make the length and width negative.

Length:
$$24-2(3)=24-6$$
$$=18\text{ ft}$$

Width:
$$16-2(3)=16-6$$
$$=10\text{ ft}$$

The dimension of the largest possible garden is 10 ft by 18 ft.

74. Let x = length of shorter leg.
Then $170 - x$ = length of longer leg.
Use the Pythagorean Theorem to find x.

$$x^2+(170-x)^2=130^2$$
$$x^2+28{,}900-170x-170x+x^2=16{,}900$$
$$2x^2-340x+28{,}900=16{,}900$$
$$2x^2-340x+12{,}000=0$$
$$\frac{2x^2}{2}-\frac{340x}{2}+\frac{12{,}000}{2}=0$$
$$x^2-170x+6{,}000=0$$
$$x=\frac{-(-170)\pm\sqrt{(-170)^2-4(1)(6{,}000)}}{2(1)}$$
$$x=\frac{170\pm\sqrt{28{,}900-24{,}000}}{2}$$
$$x=\frac{170\pm\sqrt{4{,}900}}{2}$$
$$x=\frac{170\pm 70}{2}$$
$$x=\frac{170+70}{2} \quad \text{or} \quad x=\frac{170-70}{2}$$
$$x=\frac{240}{2} \qquad x=\frac{100}{2}$$
$$x=120 \qquad x=50$$

The two segments are 50 m and 120 m.

75. Let $y = x^{1/3}$.
$$t^{2/3}-t^{1/3}=6$$
$$t^{2/3}-t^{1/3}-6=0$$
$$\left(t^{1/3}\right)^2-t^{1/3}-6=0$$
$$y^2-y-6=0$$
$$(y-3)(y+2)=0$$
$$y-3=0 \quad \text{or} \quad y+2=0$$
$$y=3 \qquad y=-2$$
$$x^{1/3}=3 \qquad x^{1/3}=-2$$
$$\left(x^{1/3}\right)^3=(3)^3 \qquad \left(x^{1/3}\right)^3=(-2)^3$$
$$x=27 \qquad x=-8$$
$$x=27,\ -8$$

76. Let $y=\frac{1}{x^2}$.

$$x^{-4}-2x^{-2}+1=0$$
$$\frac{1}{x^4}-2\left(\frac{1}{x^2}\right)+1=0$$
$$\left(\frac{1}{x^2}\right)^2-2\left(\frac{1}{x^2}\right)+1=0$$
$$y^2-2y+1=0$$
$$(y-1)(y-1)=0$$
$$y-1=0 \quad \text{or} \quad y-1=0$$
$$y=1 \qquad y=1$$
$$\frac{1}{x^2}=1 \qquad \frac{1}{x^2}=1$$
$$x^2=1 \qquad x^2=1$$
$$x=\pm\sqrt{1} \qquad x=\pm\sqrt{1}$$
$$x=\pm 1 \qquad x=\pm 1$$
$$x=1,\ -1,\ 1,\ -1$$

77. $f(x)=-x^2-4x$

Step 1: Since $a=-1<0$, the parabola opens downward.

Step 2: Find the vertex and axis of symmetry.

$$x=\frac{-b}{2a}=\frac{-(-4)}{2(-1)}=\frac{4}{-2}=-2$$
$$y=-(-2)^2-4(-2)=-4+8=4$$

vertex: (–2, 4)

The axis of symmetry is $x=-2$.

Step 3: Find the x– and y–intercepts.

Since $c=0$, the y–intercept is (0, 0).

To find the x–intercept, let $y=0$.

$$0=-x^2-4x$$
$$0=-x(x+4)$$
$$-x=0 \quad \text{or} \quad x+4=0$$
$$x=0 \qquad x=-4$$

x–intercepts: (0, 0) and (–4, 0)

y–intercepts: (0, 0)

Step 4: Find another point(s).

Let $x=-1$.

$$y=-(-1)^2-4(-1)=-1+4=3$$

(–1, 3)

Use symmetry to find the point (–3, 3)

Step 5: Plots the points and draw the parabola.

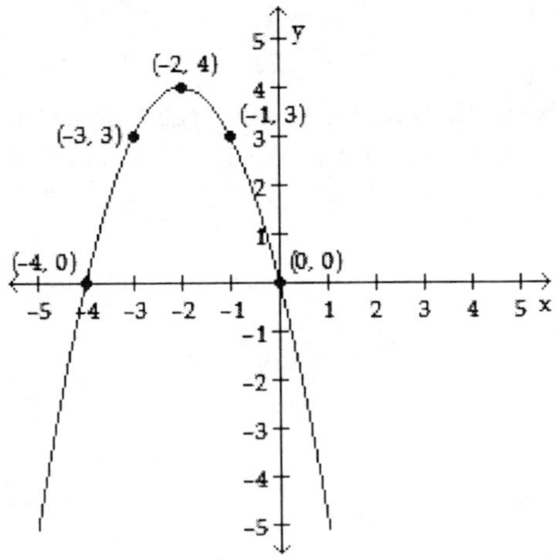

78.

$$x^2 - 81 < 0$$
$$x^2 - 81 = 0$$
$$(x-9)(x+9) = 0$$
$$x - 9 = 0 \quad \text{or} \quad x + 9 = 0$$
$$x = 9 \qquad\qquad x = -9$$
$$(-9, 9)$$

79.

$$\frac{1}{x+1} \geq \frac{x}{x+4}$$
$$\frac{1}{x+1} - \frac{x}{x+4} \geq 0$$
$$\frac{1}{x+1} - \frac{x}{x+4} = 0$$
$$\frac{1}{x+1} \bullet \frac{x+4}{x+4} - \frac{x}{x+4} \bullet \frac{x+1}{x+1} = 0$$
$$\frac{(x+4) - x(x+1)}{(x+1)(x+4)} = 0$$
$$\frac{x + 4 - x^2 - x}{(x+1)(x+4)} = 0$$
$$\frac{4 - x^2}{(x+1)(x+4)} = 0$$
$$(x+1)(x+4)\left(\frac{4 - x^2}{(x+1)(x+4)}\right) = (x+1)(x+4)(0)$$
$$4 - x^2 = 0$$
$$(2+x)(2-x) = 0$$
$$2 + x = 0 \quad \text{or} \quad 2 - x = 0$$
$$x = -2 \qquad\qquad 2 = x$$

Set the denominator $= 0$.

$$(x+1)(x+4) = 0$$
$$x + 1 = 0 \quad \text{or} \quad x + 4 = 0$$
$$x = -1 \qquad\qquad x = -4$$

Critical numbers $= -4, -2, -1,$ and 2

$$(-4, -2] \cup (-1, 2]$$

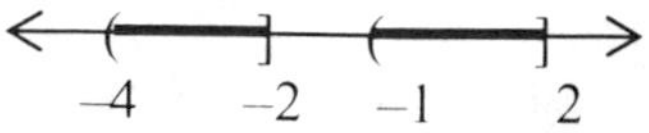

80. a. The x–intercept is $\left(-\frac{3}{4}, 0\right)$, so the solution to the equation is $x = -\frac{3}{4}$.

b. Use $x = -\frac{3}{4}$ as the critical number. The answer would be **no solution**.

SECTION 9.1

VOCABULARY

1. The **sum** of f and g, denoted as $f+g$, is defined by $(f+g)(x)=\boxed{f(x)+g(x)}$ and the **difference** of f and g, denoted as $f-g$, is defined by $(f-g)(x)=\boxed{f(x)-g(x)}$.

3. The **domain** of the function $f+g$ is the set of real numbers x that are in the domain of both f and g.

5. When we write $(f\circ g)(x)$ as $f(g(x))$, we have changed from $\circ$ notation to **nested** parentheses notation.

CONCEPTS

7. a. $(f\circ g)(3)=f(\boxed{g(3)})$

   b. To find $f(g(3))$, we first find $\boxed{g(3)}$ and then substitute that value for x in $f(x)$.

9.
$$\begin{aligned}f(-2)&=2-3(-2)^2\\&=2-3(4)\\&=2-12\\&=-10\\g(-10)&=-10+10\\&=0\end{aligned}$$

NOTATION

11.
$$\begin{aligned}(f\cdot g)(x)&=f(x)\cdot\boxed{g(x)}\\&=\boxed{(3x-1)}(2x+3)\\&=6x^2+\boxed{9x}-\boxed{2x}-3\\(f\cdot g)(x)&=6x^2+7x-3\end{aligned}$$

GUIDED PRACTICE

13. $f+g$
$$\begin{aligned}(f+g)(x)&=f(x)+g(x)\\&=2x+1+x-3\\&=3x-2\\D&:(-\infty,\infty)\end{aligned}$$

15. $g-f$
$$\begin{aligned}(g-f)(x)&=g(x)-f(x)\\&=(x-3)-(2x+1)\\&=x-3-2x-1\\&=-x-4\\D&:(-\infty,\infty)\end{aligned}$$

17. $f\cdot g$
$$\begin{aligned}(f\cdot g)(x)&=f(x)\cdot g(x)\\&=(2x+1)(x-3)\\&=2x^2-6x+x-3\\&=2x^2-5x-3\\D&:(-\infty,\infty)\end{aligned}$$

19. g/f
$$\begin{aligned}(g/f)(x)&=\frac{g(x)}{f(x)}\\&=\frac{x-3}{2x+1}\\D&:\left(-\infty,-\frac{1}{2}\right)\cup\left(-\frac{1}{2},\infty\right)\end{aligned}$$

21. $f+g$
$$\begin{aligned}(f+g)(x)&=f(x)+g(x)\\&=3x+4x\\&=7x\\D&:(-\infty,\infty)\end{aligned}$$

23. $g-f$
$$\begin{aligned}(g-f)(x)&=g(x)-f(x)\\&=4x-3x\\&=x\\D&:(-\infty,\infty)\end{aligned}$$

25. $f \cdot g$

$$(f \cdot g)(x) = f(x) \cdot g(x)$$
$$= (3x)(4x)$$
$$= 12x^2$$
$$D: (-\infty, \infty)$$

27. g/f

$$(g/f)(x) = \frac{g(x)}{f(x)}$$
$$= \frac{4x}{3x}$$
$$= \frac{4}{3}$$
$$D: (-\infty, 0) \cup (0, \infty)$$

29. $(f+g)(8)$

$$f(8) + g(8) = (2(8) - 5) + (8+1)$$
$$= (16-5) + (9)$$
$$= 11 + 9$$
$$= 20$$

31. $(f \cdot g)(0)$

$$f(0)g(0) = (2(0) - 5)(0+1)$$
$$= (0-5)(1)$$
$$= (-5)(1)$$
$$= -5$$

33. $(s \cdot t)(-2)$

$$s(-2)t(-2) = (3-(-2))((-2)^2 - (-2) - 6)$$
$$= (3+2)(4+2-6)$$
$$= (5)(0)$$
$$= 0$$

35. $(s/t)(1)$

$$s(1)/t(1) = \frac{3-(1)}{(1)^2 - (1) - 6}$$
$$= \frac{2}{1-1-6}$$
$$= \frac{2}{-6}$$
$$= -\frac{1}{3}$$

37. $(f \circ g)(2)$

$$(f \circ g)(2) = f(g(2))$$
$$= f(2^2 - 1)$$
$$= f(4-1)$$
$$= f(3)$$
$$= 2(3) + 1$$
$$= 6 + 1$$
$$= 7$$

39. $(g \circ f)(-3)$

$$(g \circ f)(-3) = g(f(-3))$$
$$= g(2(-3)+1)$$
$$= g(-6+1)$$
$$= g(-5)$$
$$= (-5)^2 - 1$$
$$= 25 - 1$$
$$= 24$$

41. $(f \circ g)\left(\frac{1}{2}\right)$

$$(f \circ g)\left(\frac{1}{2}\right) = f\left(g\left(\frac{1}{2}\right)\right)$$
$$= f\left(\left(\frac{1}{2}\right)^2 - 1\right)$$
$$= f\left(\frac{1}{4} - 1\right)$$
$$= f\left(-\frac{3}{4}\right)$$
$$= 2\left(-\frac{3}{4}\right) + 1$$
$$= \frac{-3}{2} + 1$$
$$= -\frac{1}{2}$$

43. $(g \circ f)(2x)$

$$\begin{aligned}(g \circ f)(2x) &= g(f(2x)) \\ &= g(2(2x)+1) \\ &= g(4x+1) \\ &= (4x+1)^2 - 1 \\ &= 16x^2 + 8x + 1 - 1 \\ &= 16x^2 + 8x\end{aligned}$$

45. a. $$\begin{aligned}(f+g)(-5) &= f(-5) + g(-5) \\ &= 3 + (-3) \\ &= 0\end{aligned}$$

b. $$\begin{aligned}(f-g)(3) &= f(3) - g(3) \\ &= 3 - (-3) \\ &= 6\end{aligned}$$

c. $$\begin{aligned}(f \cdot g)(-3) &= f(-3)g(-3) \\ &= 1(-3) \\ &= -3\end{aligned}$$

47. a. $$\begin{aligned}(g+f)(2) &= g(2) + f(2) \\ &= -3 + 2 \\ &= -1\end{aligned}$$

b. $$\begin{aligned}(g-f)(-5) &= g(-5) - f(-5) \\ &= 3 - (-3) \\ &= 6\end{aligned}$$

c. $$\begin{aligned}(g \cdot f)(1) &= g(1)f(1) \\ &= (-3)(2) \\ &= -6\end{aligned}$$

49. Let $f(x) = x^2$ and $g(x) = x + 15$.

$$\begin{aligned}h(x) &= (f \circ g)(x) \\ &= f(g(x)) \\ &= f(x+15) \\ &= (x+15)^2\end{aligned}$$

51. Let $f(x) = x + 9$ and $g(x) = x^5$.

$$\begin{aligned}h(x) &= (f \circ g)(x) \\ &= f(g(x)) \\ &= f(x^5) \\ &= x^5 + 9\end{aligned}$$

53. Let $f(x) = \sqrt{x}$ and $g(x) = 16x - 1$

$$\begin{aligned}h(x) &= (f \circ g)(x) \\ &= f(g(x)) \\ &= f(16x-1) \\ &= \sqrt{16x-1}\end{aligned}$$

55. Let $f(x) = \dfrac{1}{x}$ and $g(x) = x - 4$

$$\begin{aligned}h(x) &= (f \circ g)(x) \\ &= f(g(x)) \\ &= f(x-4) \\ &= \frac{1}{x-4}\end{aligned}$$

TRY IT YOURSELF

57. $(f \circ g)(4)$

$$\begin{aligned}(f \circ g)(4) &= f(g(4)) \\ &= f((4)^2 + 4) \\ &= f(16+4) \\ &= f(20) \\ &= 3(20) - 2 \\ &= 60 - 2 \\ &= 58\end{aligned}$$

59. $(g \circ f)(-3)$

$$\begin{aligned}(g \circ f)(-3) &= g(f(-3)) \\ &= g(3(-3)-2) \\ &= g(-9-2) \\ &= g(-11) \\ &= (-11)^2 - 11 \\ &= 121 - 11 \\ &= 110\end{aligned}$$

61. $(g \circ f)(0)$

$$\begin{aligned}(g \circ f)(0) &= g(f(0)) \\ &= g(3(0)-2) \\ &= g(0-2) \\ &= g(-2) \\ &= (-2)^2 - 2 \\ &= 4-2 \\ &= 2\end{aligned}$$

63. $(g \circ f)(x)$

$$\begin{aligned}(g \circ f)(x) &= g(f(x)) \\ &= g(3x-2) \\ &= (3x-2)^2 + 3x - 2 \\ &= 9x^2 - 12x + 4 + 3x - 2 \\ &= 9x^2 - 9x + 2\end{aligned}$$

65. $f-g$

$$\begin{aligned}(f-g)(x) &= f(x) - g(x) \\ &= (3x-2)-\left(2x^2+1\right) \\ &= 3x - 2 - 2x^2 - 1 \\ &= -2x^2 + 3x - 3\end{aligned}$$

$D:(-\infty,\infty)$

67. f/g

$$\begin{aligned}(f/g)(x) &= \frac{f(x)}{g(x)} \\ &= \frac{3x-2}{2x^2+1}\end{aligned}$$

$D:(-\infty,\infty)$

69. $(g \circ f)\left(\frac{1}{3}\right)$

$$\begin{aligned}(g \circ f)\left(\frac{1}{3}\right) &= g\left(f\left(\frac{1}{3}\right)\right) \\ &= g\left(\frac{1}{\frac{1}{3}}\right) \\ &= g\left(1 \div \frac{1}{3}\right) \\ &= g\left(1 \cdot \frac{3}{1}\right) \\ &= g(3) \\ &= \frac{1}{(3)^2} \\ &= \frac{1}{9}\end{aligned}$$

71. $(g \circ f)(8x)$

$$\begin{aligned}(g \circ f)(8x) &= g(f(8x)) \\ &= g\left(\frac{1}{8x}\right) \\ &= \frac{1}{\left(\frac{1}{8x}\right)^2} \\ &= \frac{1}{\frac{1}{64x^2}} \\ &= 1 \div \frac{1}{64x^2} \\ &= 1 \cdot \frac{64x^2}{1} \\ &= 64x^2\end{aligned}$$

73. $f-g$

$$\begin{aligned}(f-g)(x) &= f(x) - g(x) \\ &= \left(x^2-1\right)-\left(x^2-4\right) \\ &= x^2 - 1 - x^2 + 4 \\ &= 3\end{aligned}$$

$D:(-\infty,\infty)$

75. g/f

$$(g/f)(x)=\frac{g(x)}{f(x)}$$
$$=\frac{x^2-4}{x^2-1}$$
$$D:(-\infty,-1)\cup(-1,1)\cup(1,\infty)$$

77. $(h\circ k)(18)$

$$(h\circ k)(18)=h(k(18))$$
$$=h(18-5)$$
$$=h(13)$$
$$=\sqrt{13+3}$$
$$=\sqrt{16}$$
$$=4$$

79. $(k\circ h)(22)$

$$(k\circ h)(22)=k(h(22))$$
$$=k(\sqrt{22+3})$$
$$=k(\sqrt{25})$$
$$=k(5)$$
$$=5-5$$
$$=0$$

81. a.

$$(f+g)(1)=f(1)+g(1)$$
$$=3+4$$
$$=7$$

b.

$$(f-g)(5)=f(5)-g(5)$$
$$=8-0$$
$$=8$$

c.

$$(f\cdot g)(1)=f(1)g(1)$$
$$=3\cdot 4$$
$$=12$$

d.

$$(g/f)(5)=\frac{g(5)}{f(5)}$$
$$=\frac{0}{8}$$
$$=0$$

83.

$$(f\circ g)(x)=f(g(x))$$
$$=f(2x-5)$$
$$=(2x-5)+1$$
$$=2x-4$$

$$(g\circ f)(x)=g(f(x))$$
$$=g(x+1)$$
$$=2(x+1)-5$$
$$=2x+2-5$$
$$=2x-3$$

So, $(f\circ g)(x)\neq(g\circ f)(x)$.

APPLICATIONS

85. SAT SCORES

a. In 2004, the average combined math and reading score was 1,026.

$$(m+r)(4)=m(4)+r(4)$$
$$=518+508$$
$$=1{,}026$$

b. In 2004, the average difference in the math and reading scores was 10.

$$(m-r)(4)=m(4)-r(4)$$
$$=518-508$$
$$=10$$

c.

$$(m+r)(9)=m(9)+r(9)$$
$$=515+501$$
$$=1{,}016$$

d.

$$(m-r)(9)=m(9)-r(9)$$
$$=515-501$$
$$=14$$

87. METALLURGY

$$F(t) = -200t + 2{,}700$$

$$C(F) = \frac{5}{9}(F - 32)$$

$$\begin{aligned}(C \circ F)(t) &= C(F(t)) \\ &= C(-200t + 2{,}700) \\ &= \frac{5}{9}\left[(-200t + 2{,}700) - 32\right] \\ &= \frac{5}{9}(-200t + 2{,}668)\end{aligned}$$

$$C(t) = \frac{5}{9}(2{,}668 - 200t)$$

89. VACATION MILEAGE COSTS

a. On the first graph, 500 miles corresponds with about 25 gallons consumed. On the second graph, 25 gallons consumed corresponds with about $75.

b. Find $(C \circ G)(m)$.

$$\begin{aligned}(C \circ G)(m) &= C(G(m)) \\ &= C\left(\frac{m}{20}\right) \\ &= 3\left(\frac{m}{20}\right)\end{aligned}$$

$$C(m) = \frac{3m}{20}$$

$$C(m) = 0.15m$$

WRITING

91. Answers will vary.

93. Answers will vary.

REVIEW

95.

$$\begin{aligned}\frac{\dfrac{ac - ad - c + d}{a^3 - 1}}{\dfrac{c^2 - 2cd + d^2}{a^2 + a + 1}} &= \frac{ac - ad - c + d}{a^3 - 1} \div \frac{c^2 - 2cd + d^2}{a^2 + a + 1} \\ &= \frac{ac - ad - c + d}{a^3 - 1} \cdot \frac{a^2 + a + 1}{c^2 - 2cd + d^2} \\ &= \frac{a(c-d) - 1(c-d)}{(a-1)(a^2 + a + 1)} \cdot \frac{a^2 + a + 1}{(c-d)(c-d)} \\ &= \frac{(c-d)(a-1)}{(a-1)(a^2 + a + 1)} \cdot \frac{a^2 + a + 1}{(c-d)(c-d)} \\ &= \frac{1}{c-d}\end{aligned}$$

CHALLENGE PROBLEMS

97. If $f(x) = x^2$ and $g(x) = \boxed{2x + 5}$, then $(f \circ g)(x) = 4x^2 + 20x + 25$.

99.

$$f(x) = \frac{1}{12}x + \frac{7}{2};\quad g(x) = 2$$

$$\begin{aligned}f(x) + g(x) &= \frac{1}{12}x + \frac{7}{2} + 2 \\ &= \frac{1}{12}x + \frac{11}{2}\end{aligned}$$

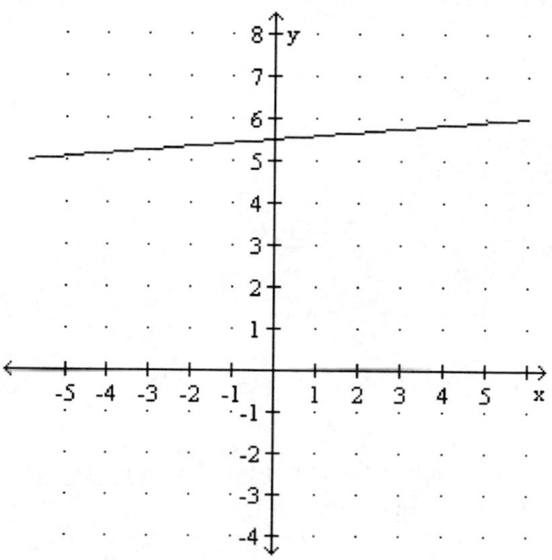

VOCABULARY

1. A function is called a **one–to–one** function if different inputs determine different outputs.

3. The functions f and f^{-1} are **inverses**.

CONCEPTS

5. If any horizontal line that intersects the graph of a function does so more than once, the function is not **one–to–one**.

7. If f is a one–to–one function, the domain of f is the **range** of f^{-1}, and the range of f is the **domain** of f^{-1}.

9. If f is a one-to-one function, and if $f(1)=6$, then $f^{-1}(6)=\boxed{1}$.

11. a. no
    b. no

13.

x	$f^{-1}(x)$
−2	**−4**
0	**0**
4	**8**

NOTATION

15.

$$\boxed{y}=2x-3$$
$$x=\boxed{2y}-3$$
$$x+\boxed{3}=2y$$
$$\frac{x+3}{2}=\boxed{y}$$

The inverse of $f(x)=2x-3$ is $\boxed{f^{-1}}(x)=\dfrac{x+3}{2}$.

17. The symbol f^{-1} is read as "the **inverse of** f" or "f **inverse**."

GUIDED PRACTICE

19. Yes, each output corresponds to exactly one input, so the function is one–to–one.

21. No. $2^4=(-2)^4=16$ so the output 16 corresponds to two different inputs, 2 and −2.

23. No. $-(3)^2+3(3)=-(0)+3(0)=0$ so the output 0 corresponds to two different inputs, 0 and −3.

25. No. The output 1 corresponds to more than one input, 1, 2, 3 and 4.

27. one–to–one

29. not one–to–one

31. not one–to–one

33. one-to-one

35.

$$f(x)=2x+4$$
$$y=2x+4$$
$$x=2y+4$$
$$x-4=2y$$
$$\frac{x-4}{2}=y$$
$$y=\frac{x-4}{2} \quad \text{or} \quad y=\frac{1}{2}x-2$$
$$f^{-1}(x)=\frac{x-4}{2} \quad \text{or} \quad f^{-1}(x)=\frac{1}{2}x-2$$

37.

$$f(x)=\frac{x}{5}+\frac{4}{5}$$
$$y=\frac{x}{5}+\frac{4}{5}$$
$$x=\frac{y}{5}+\frac{4}{5}$$
$$5(x)=5\left(\frac{y}{5}+\frac{4}{5}\right)$$
$$5x=y+4$$
$$5x-4=y$$
$$y=5x-4$$
$$f^{-1}(x)=5x-4$$

39.

$$f(x)=\frac{x-4}{5}$$
$$y=\frac{x-4}{5}$$
$$x=\frac{y-4}{5}$$
$$5(x)=5\left(\frac{y-4}{5}\right)$$
$$5x=y-4$$
$$5x+4=y$$
$$y=5x+4$$
$$f^{-1}(x)=5x+4$$

41.

$$f(x)=\frac{2}{x-3}$$
$$y=\frac{2}{x-3}$$
$$x=\frac{2}{y-3}$$
$$(y-3)(x)=(y-3)\left(\frac{2}{y-3}\right)$$
$$xy-3x=2$$
$$xy=2+3x$$
$$y=\frac{2+3x}{x}$$
$$y=\frac{2}{x}+\frac{3x}{x}$$
$$y=\frac{2}{x}+3$$
$$f^{-1}(x)=\frac{2}{x}+3$$

43.

$$f(x)=\frac{4}{x}$$
$$y=\frac{4}{x}$$
$$x=\frac{4}{y}$$
$$y(x)=y\left(\frac{4}{y}\right)$$
$$xy=4$$
$$y=\frac{4}{x}$$
$$f^{-1}(x)=\frac{4}{x}$$

45.

$$f(x)=x^3+8$$
$$y=x^3+8$$
$$x=y^3+8$$
$$x-8=y^3$$
$$y^3=x-8$$
$$\sqrt[3]{y^3}=\sqrt[3]{x-8}$$
$$y=\sqrt[3]{x-8}$$
$$f^{-1}(x)=\sqrt[3]{x-8}$$

47.

$$f(x)=\sqrt[3]{x}$$
$$y=\sqrt[3]{x}$$
$$x=\sqrt[3]{y}$$
$$(x)^3=\left(\sqrt[3]{y}\right)^3$$
$$x^3=y$$
$$y=x^3$$
$$f^{-1}(x)=x^3$$

49.

$$\begin{aligned} f(x) &= (x+10)^3 \\ y &= (x+10)^3 \\ x &= (y+10)^3 \\ \sqrt[3]{x} &= \sqrt[3]{(y+10)^3} \\ \sqrt[3]{x} &= y+10 \\ \sqrt[3]{x}-10 &= y \\ y &= \sqrt[3]{x}-10 \\ f^{-1}(x) &= \sqrt[3]{x}-10 \end{aligned}$$

51.

$$\begin{aligned} f(x) &= 2x^3-3 \\ y &= 2x^3-3 \\ x &= 2y^3-3 \\ x+3 &= 2y^3 \\ \frac{x+3}{2} &= y^3 \\ \sqrt[3]{\frac{x+3}{2}} &= \sqrt[3]{y^3} \\ \sqrt[3]{\frac{x+3}{2}} &= y \\ y &= \sqrt[3]{\frac{x+3}{2}} \\ f^{-1}(x) &= \sqrt[3]{\frac{x+3}{2}} \end{aligned}$$

53.

$$\begin{aligned} f(x) &= \frac{x^7}{2} \\ y &= \frac{x^7}{2} \\ x &= \frac{y^7}{2} \\ 2(x) &= 2\left(\frac{y^7}{2}\right) \\ 2x &= y^7 \\ \sqrt[7]{2x} &= \sqrt[7]{y^7} \\ \sqrt[7]{2x} &= y \\ y &= \sqrt[7]{2x} \\ f^{-1}(x) &= \sqrt[7]{2x} \end{aligned}$$

55.

$$\begin{aligned} (f\circ f^{-1})(x) &= f(f^{-1}(x)) \\ &= f\left(\frac{x-9}{2}\right) \\ &= 2\left(\frac{x-9}{2}\right)+9 \\ &= x-9+9 \\ &= x \end{aligned}$$

$$\begin{aligned} (f^{-1}\circ f)(x) &= f^{-1}(f(x)) \\ &= f^{-1}(2x+9) \\ &= \frac{2x+9-9}{2} \\ &= \frac{2x}{2} \\ &= x \end{aligned}$$

57.

$$\begin{aligned} (f\circ f^{-1})(x) &= f(f^{-1}(x)) \\ &= f\left(\frac{2}{x}+3\right) \\ &= \frac{2}{\left(\frac{2}{x}+3\right)-3} \\ &= \frac{2}{\frac{2}{x}} \\ &= \frac{\frac{2}{2}\bullet\frac{x}{x}}{x} \\ &= \frac{2x}{2} \\ &= x \end{aligned}$$

$$\begin{aligned} (f^{-1}\circ f)(x) &= f^{-1}(f(x)) \\ &= f^{-1}\left(\frac{2}{x-3}\right) \\ &= \frac{2}{\frac{2}{x-3}}+3 \\ &= \frac{2}{1}\bullet\frac{x-3}{2}+3 \\ &= \frac{2(x-3)}{2}+3 \\ &= x-3+3 \\ &= x \end{aligned}$$

59.

$$f(x)=2x$$
$$y=2x$$
$$x=2y$$
$$\frac{x}{2}=y$$
$$y=\frac{x}{2} \text{ or } y=\frac{1}{2}x$$
$$f^{-1}(x)=\frac{x}{2} \text{ or } f^{-1}(x)=\frac{1}{2}x$$

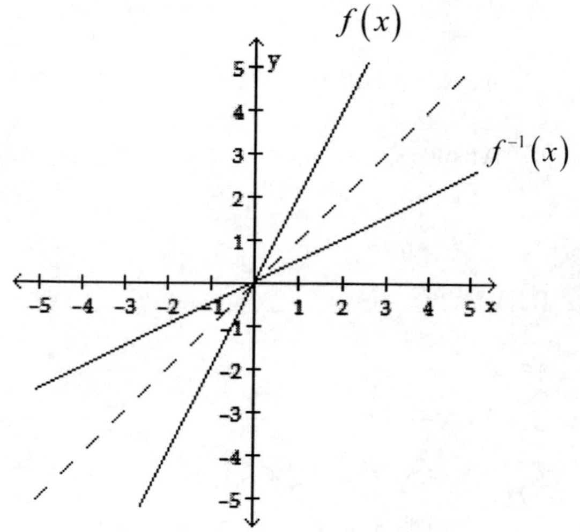

61.

$$f(x)=4x+3$$
$$y=4x+3$$
$$x=4y+3$$
$$x-3=4y$$
$$\frac{x-3}{4}=y$$
$$y=\frac{x-3}{4}$$
$$f^{-1}(x)=\frac{x-3}{4}$$

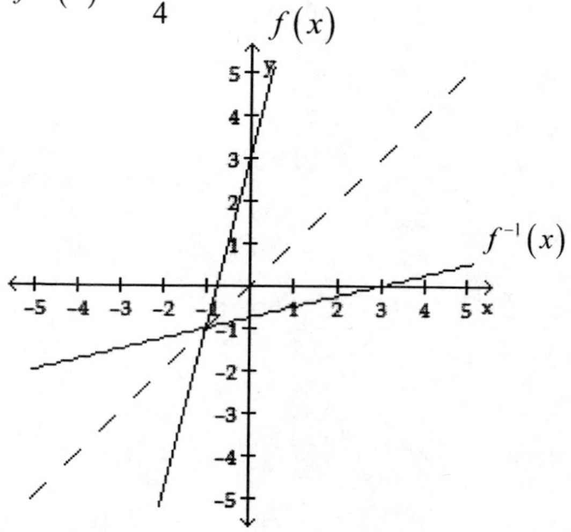

63.

$$f(x)=-\frac{2}{3}x+3$$
$$y=-\frac{2}{3}x+3$$
$$x=-\frac{2}{3}y+3$$
$$3(x)=3\left(-\frac{2}{3}y+3\right)$$
$$3x=-2y+9$$
$$3x-9=-2y$$
$$\frac{3x-9}{-2}=y$$
$$-\frac{3}{2}x+\frac{9}{2}=y$$
$$y=-\frac{3}{2}x+\frac{9}{2}$$
$$f^{-1}(x)=-\frac{3}{2}x+\frac{9}{2}$$

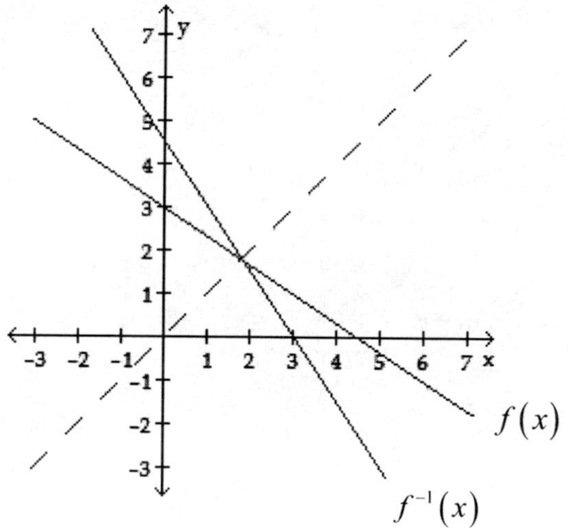

65.

$$f(x) = x^3$$
$$y = x^3$$
$$x = y^3$$
$$\sqrt[3]{x} = \sqrt[3]{y^3}$$
$$\sqrt[3]{x} = y$$
$$y = \sqrt[3]{x}$$
$$f^{-1}(x) = \sqrt[3]{x}$$

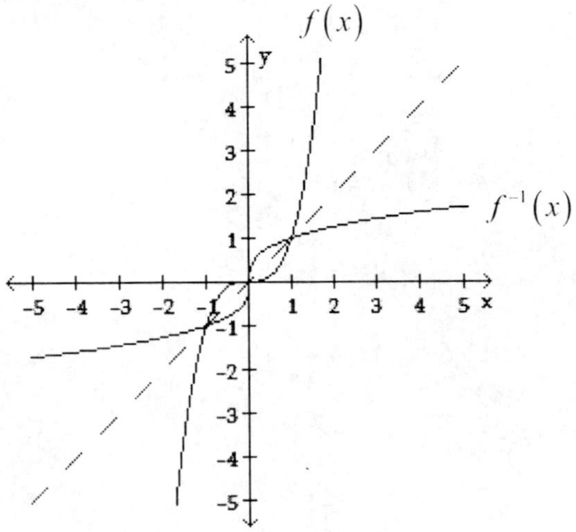

67.

$$f(x) = x^2 - 1$$
$$y = x^2 - 1$$
$$x = y^2 - 1$$
$$x + 1 = y^2$$
$$\sqrt{x+1} = \sqrt{y^2}$$
$$\sqrt{x+1} = y$$
$$y = \sqrt{x+1}$$
$$f^{-1}(x) = \sqrt{x+1}$$

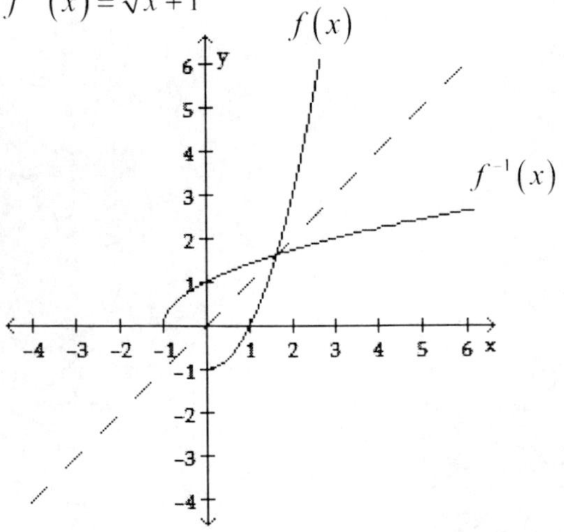

APPLICATIONS

69. INTERPERSONAL RELATIONSHIPS
   a. The graph is a function, but its inverse is not.
   b. No. Twice during this period, the person's anxiety level was at the maximum threshold value.

WRITING

71. Answers will vary.

73. Answers will vary.

75. Answers will vary.

REVIEW

77.

$$3 - \sqrt{-64} = 3 - 8i$$

79.

$$\begin{aligned}(3+4i)(2-3i) &= 6 - 9i + 8i - 12i^2 \\ &= 6 - i - 12(-1) \\ &= 6 - i + 12 \\ &= 18 - i\end{aligned}$$

81.

$$\begin{aligned}(6-8i)^2 &= (6-8i)(6-8i) \\ &= 36 - 48i - 48i + 64i^2 \\ &= 36 - 96i + 64(-1) \\ &= 36 - 96i - 64 \\ &= -28 - 96i\end{aligned}$$

CHALLENGE PROBLEMS

83.

$$f(x) = \frac{x+1}{x-1}$$

$$y = \frac{x+1}{x-1}$$

$$x = \frac{y+1}{y-1}$$

$$(y-1)(x) = (y-1)\left(\frac{y+1}{y-1}\right)$$

$$xy - x = y + 1$$

$$xy = y + 1 + x$$

$$xy - y = x + 1$$

$$y(x-1) = x + 1$$

$$y = \frac{x+1}{x-1}$$

$$f^{-1}(x) = \frac{x+1}{x-1}$$

85.

$$f^{-1}(f(4)) = f^{-1}(10)$$
$$= 4$$

$$f(f^{-1}(2)) = f(0)$$
$$= 2$$

SECTION 9.3

VOCABULARY

1. $f(x)=2^x$ and $f(x)=\left(\frac{1}{4}\right)^x$ are examples of **exponential** functions.

3. The graph of $f(x)=3^x$ approaches, but never touches, the negative portion of the x-axis. Thus, the x-axis is an **asymptote** of the graph.

5. a. exponential
   b. $(-\infty, \infty)$
   c. $(0, \infty)$
   d. (0, 1); none
   e. yes
   f. the x–axis ($y = 0$)
   g. increasing
   h. $y = 3$

CONCEPTS

7. a. $3^{-2}=\frac{1}{3^2}=\frac{1}{9}$
   b. $\left(\frac{1}{2}\right)^4=\frac{1^4}{2^4}=\frac{1}{16}$
   c. $\left(\frac{1}{5}\right)^{-2}=(5)^2=25$

9. a. iii
   b. iv
   c. ii
   d. i

11. a. D: $(-\infty,\infty)$, R: $(-3,\infty)$
    b. D: $(-\infty,\infty)$, R: $(1,\infty)$

13. $f(x)=5^x$

x	$f(x)$
-3	$5^{-3}=\frac{1}{5^3}$ $=\frac{1}{125}$
-2	$5^{-2}=\frac{1}{5^2}$ $=\frac{1}{25}$
-1	$5^{-1}=\frac{1}{5}$
0	$5^0=1$
1	$5^1=5$
2	$5^2=25$
3	$5^3=125$

15. a. ii.
    b. i.
    c. ii.
    d. ii.
    e. i.

NOTATION

17. $b > 0$ and $b \neq 1$

GUIDED PRACTICE

19. $f(x)=3^x$

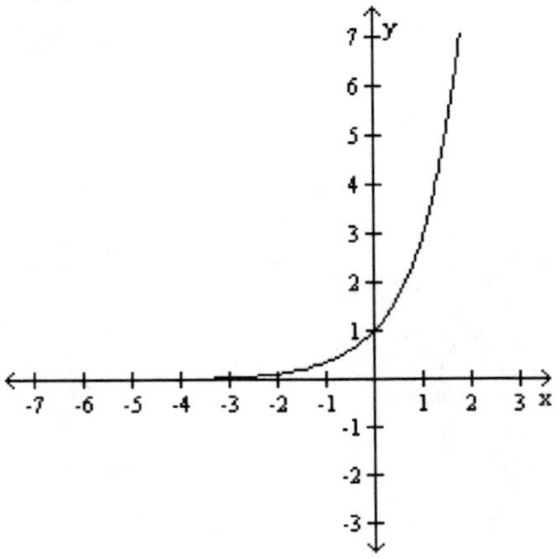

21. $f(x)=5^x$

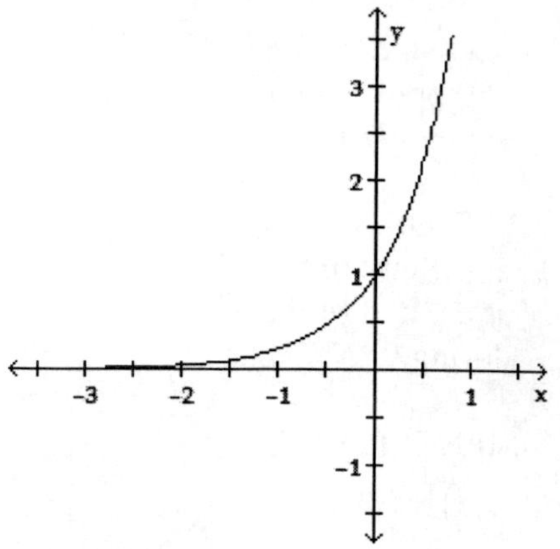

23. $f(x)=\left(\frac{1}{4}\right)^x$

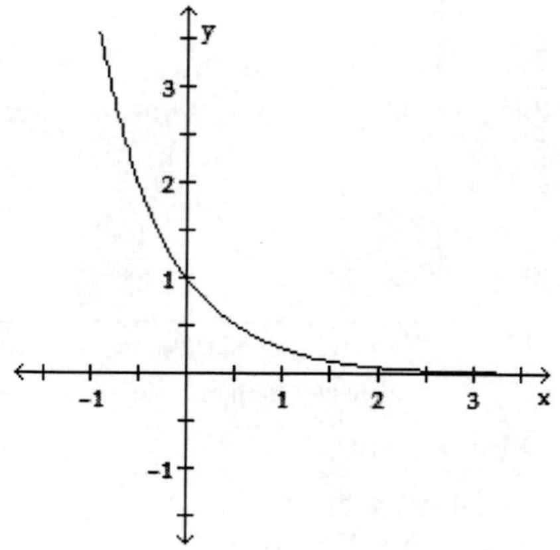

25. $f(x)=\left(\frac{1}{6}\right)^x$

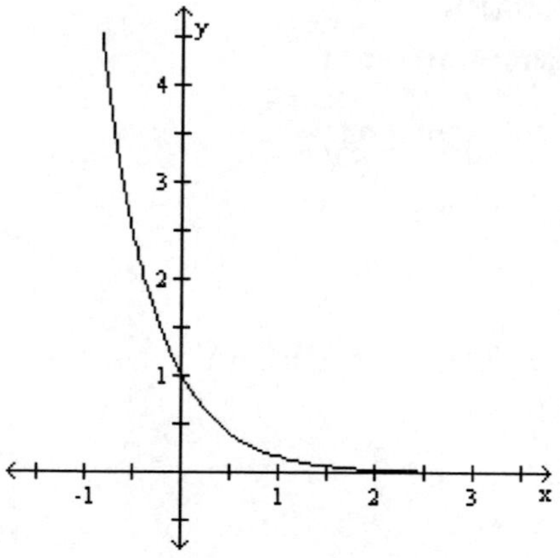

27. $g(x)=3^x-2$

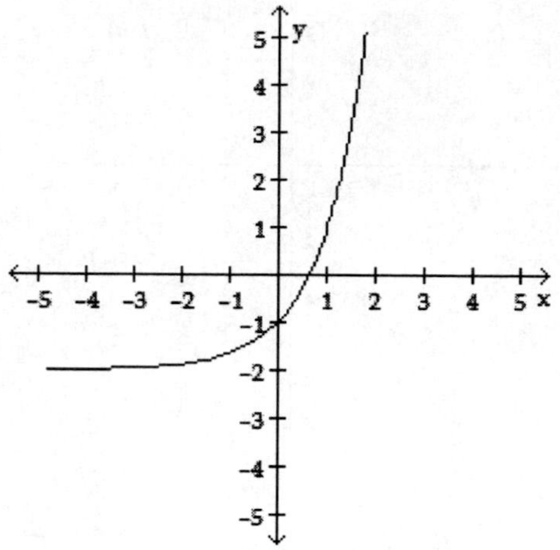

29. $g(x)=2^{x+1}$

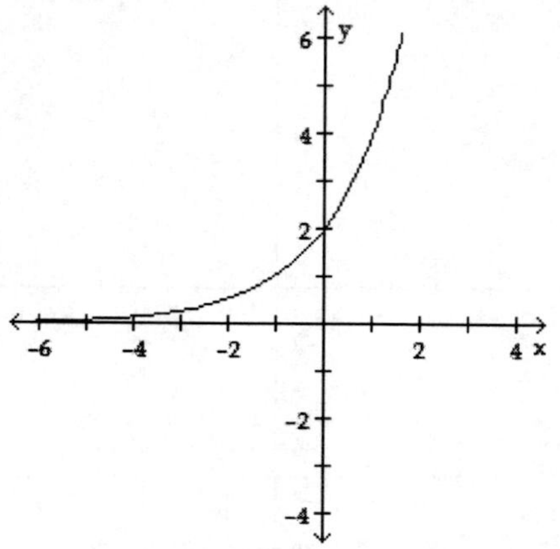

31. $g(x)=4^{x-1}+2$

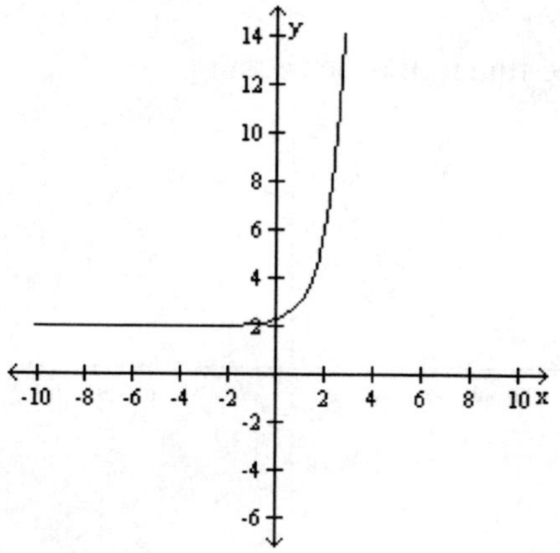

33. $g(x) = -2^x$

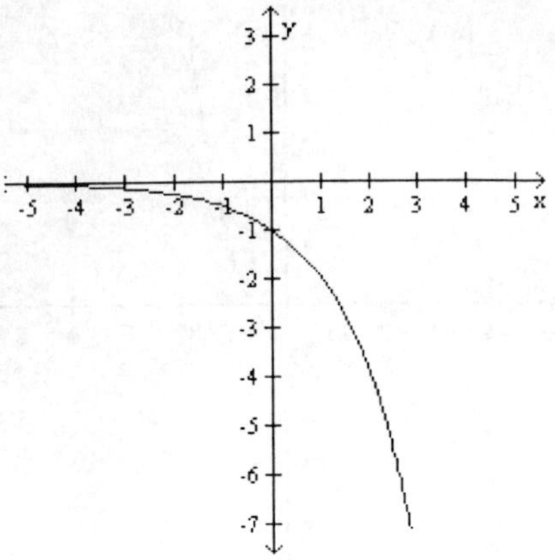

35. $f(x) = \frac{1}{2}(3^{x/2})$

The function is increasing.

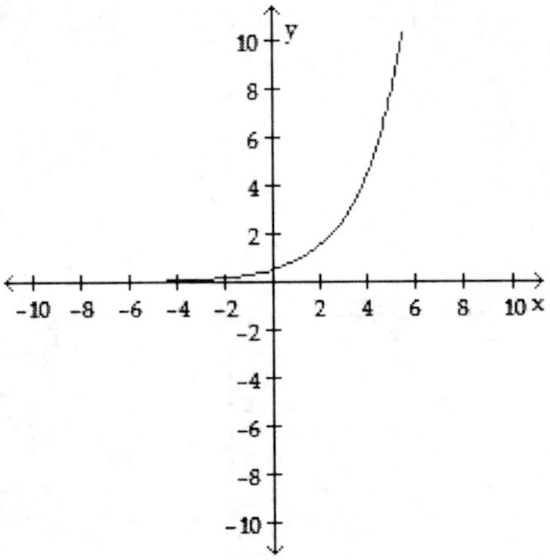

37. $y = 2(3^{-x/2})$

The function is decreasing.

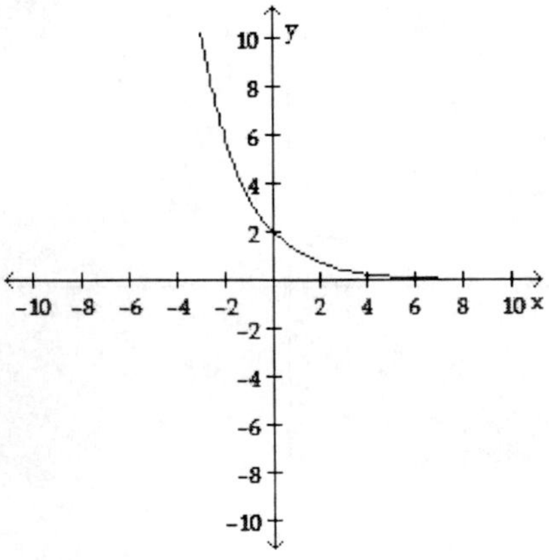

APPLICATIONS

39. CO_2 CONCENTRATION
   a. 280 ppm, 295 ppm, 370 ppm
   b. about 1970

41. VALUE OF A CAR
   a. at the end of the 2nd year
   b. at the end of the 4th year
   c. during the 7th year

43. COMPUTER VIRUSES
   a. $c(t) = 5(1.034)^t$

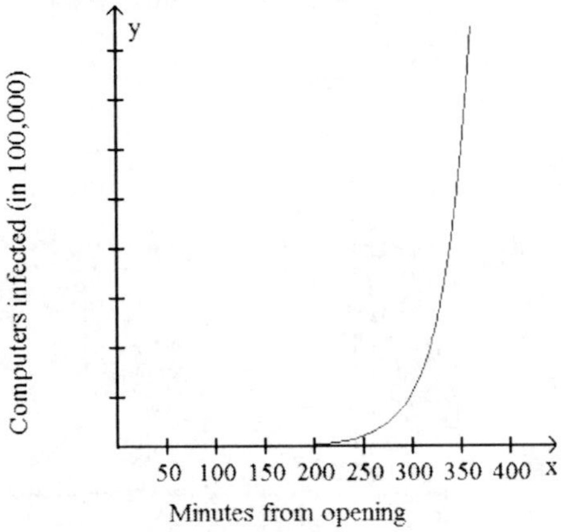

   b. Let $t = 480$.

$$c(480) = 5(1.034)^{480}$$
$$= 5(9{,}329{,}505.043)$$
$$= 46{,}647{,}545.22$$

45. GUITARS

$$f(n) = 650(0.94)^n$$
$$f(7) = 650(0.94)^7$$
$$= 650(0.648477594)$$
$$\approx 422 \text{ mm}$$

47. RADIOACTIVE DECAY

Let $t = 5$.

$$A = 500\left(\frac{2}{3}\right)^t$$
$$= 500\left(\frac{2}{3}\right)^{10}$$
$$= 500\left(\frac{1{,}024}{59{,}049}\right)$$
$$= 8.7 \text{ gm}$$

49. SOCIAL WORKER

$P(t)=35.8(1.06)^t$

For 2006, let $t = 0$.

$$\begin{aligned}P(0)&=35.8(1.06)^0\\&=35.8(1)\\&=35.8\end{aligned}$$

For 2007, let $t = 1$.

$$\begin{aligned}P(1)&=35.8(1.06)^1\\&=35.8(1.06)\\&=37.9\end{aligned}$$

For 2008, let $t = 2$.

$$\begin{aligned}P(2)&=35.8(1.06)^2\\&=35.8(1.1236)\\&=40.2\end{aligned}$$

For 2009, let $t = 3$.

$$\begin{aligned}P(3)&=35.8(1.06)^3\\&=35.8(1.191016)\\&=42.6\end{aligned}$$

51. POPULATION GROWTH

Let $t = 6\frac{9}{12}=6\frac{3}{4}=6.75$.

$$\begin{aligned}P&=3,745(0.93)^t\\&=3,745(0.93)^{6.75}\\&=3,745(0.612717)\\&\approx 2,295\end{aligned}$$

53. COMPOUND INTEREST

$P = 10,000$, $r = 0.08$, $k = 4$, $t = 10$

$$\begin{aligned}A&=P\left(1+\frac{r}{k}\right)^{kt}\\&=10,000\left(1+\frac{0.08}{4}\right)^{4(10)}\\&=10,000(1+0.02)^{40}\\&=10,000(1.02)^{40}\\&=10,000(2.20804)\\&=\$22,080.40\end{aligned}$$

55. COMPARING INTEREST RATES

Find the amount earned at $5\frac{1}{2}\%$.

$P = 1,000$, $r = 0.055$, $k = 4$, $t = 5$

$$\begin{aligned}A&=P\left(1+\frac{r}{k}\right)^{kt}\\&=1,000\left(1+\frac{0.055}{4}\right)^{4(5)}\\&=1,000(1.01375)^{20}\\&=1,000(1.314066)\\&=\$1,314.07\end{aligned}$$

Find the amount earned at 5%.

$P = 1,000$, $r = 0.05$, $k = 4$, $t = 5$

$$\begin{aligned}A&=P\left(1+\frac{r}{k}\right)^{kt}\\&=1,000\left(1+\frac{0.05}{4}\right)^{4(5)}\\&=1,000(1.0125)^{20}\\&=1,000(1.282037)\\&=\$1,282.04\end{aligned}$$

Find the difference in the two amounts.

$$\$1,314.07-1,282.04=\$32.03$$

57. COMPOUND INTEREST

$P = 1$, $r = 0.05$, $k = 1$, $t = 300$

$$\begin{aligned}A&=P\left(1+\frac{r}{k}\right)^{kt}\\&=1\left(1+\frac{0.05}{1}\right)^{1(300)}\\&=1(1+0.05)^{300}\\&=1(1.05)^{300}\\&=1(2,273,996.129)\\&=\$2,273,996.13\end{aligned}$$

WRITING

59. Answers will vary.

61. Answers will vary.

63. Answers will vary.

65. Answers will vary.

67. Answers will vary.

REVIEW

69. The sum of same–side interior angles is 180°.

$$3x+(2x-20)=180$$
$$5x-20=180$$
$$5x=200$$
$$x=40^{\circ}$$

71. Angle 2 is equal to $3x$ because they are alternate interior angles.

$$3x=3(40)$$
$$=120^{\circ}$$

CHALLENGE PROBLEMS

73. $f(x)=3^x$

$f(1.5)\approx 5.2$

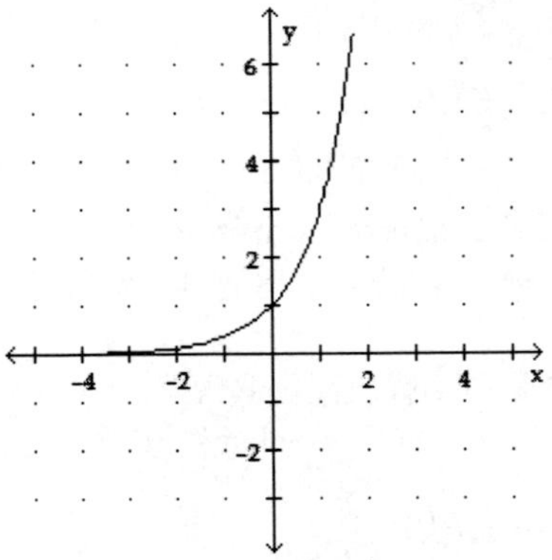

75. For $b=\dfrac{1}{50}$, the graph would be correct.

77. $f(x)=2^{|x|}$

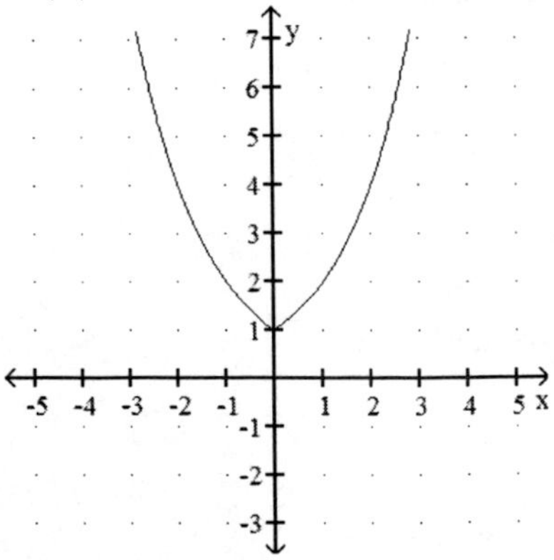

SECTION 9.4

VOCABULARY

1. $f(x)=\log_2 x$ and $g(x)=\log x$ are examples of **logarithmic** functions.

3. The graph of $f(x)=\log_2 x$ approaches, but never touches, the negative portion of the y-axis. Thus the y-axis is an **asymptote** of the graph.

CONCEPTS

5. a. logarithmic
   b. D: $(0, \infty)$; R: $(-\infty, \infty)$
   c. none, (1, 0)
   d. yes
   e. the y–axis ($x = 0$)
   f. increasing
   g. 1

7. a. ii.
   b. iii.
   c. i.
   d. iv.

9. $\log_6 36 = 2$ means $6^2 = 36$.

11. $\log_b x$ is the **exponent** to which b is raised to get x.

13. The inverse of an exponential function is called a **logarithmic** function.

15. $f(x) = \log x$

x	$f(x)$
100	2
$\frac{1}{100}$	–2

17. $f(x)=\log_6 x$

input	output
6	1
–6	undefined
0	Undefined

19. a. $f(x)=\log x$

x	$f(x)$
0.5	–0.30
1	0
2	0.30
4	0.60
6	0.78
8	0.90
10	1

b.

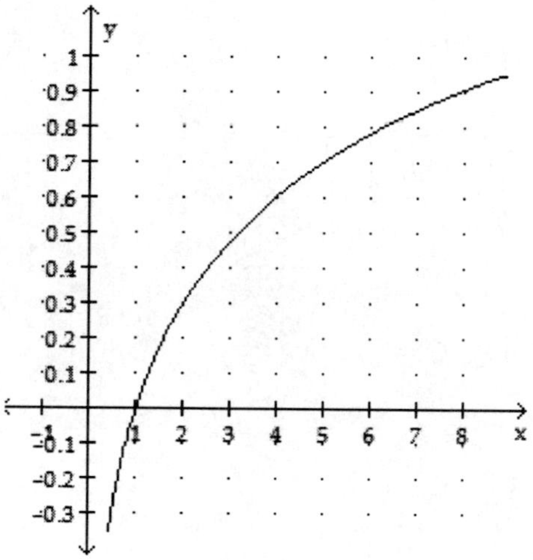

NOTATION

21. a. $\log x = \log_{\boxed{10}} x$
    b. $\log_{10} 10^x = \boxed{x}$

GUIDED PRACTICE

23.
$\log_3 81 = 4$
$3^4 = 81$

25.
$\log_{10} 10 = 1$
$10^1 = 10$

27.
$\log_4 \frac{1}{64} = -3$
$4^{-3} = \frac{1}{64}$

29.

$$\log_5 \sqrt{5} = \frac{1}{2}$$
$$5^{1/2} = \sqrt{5}$$

31.

$$\log 0.1 = -1$$
$$10^{-1} = 0.1$$

33.

$$x = \log_8 64$$
$$8^x = 64$$

35.

$$t = \log_b T_1$$
$$b^t = T_1$$

37.

$$\log_n C = -42$$
$$n^{-42} = C$$

39.

$$8^2 = 64$$
$$\log_8 64 = 2$$

41.

$$4^{-2} = \frac{1}{16}$$
$$\log_4 \frac{1}{16} = -2$$

43.

$$\left(\frac{1}{2}\right)^{-5} = 32$$
$$\log_{1/2} 32 = -5$$

45.

$$x^y = z$$
$$\log_x z = y$$

47.

$$y^t = 8.6$$
$$\log_y 8.6 = t$$

49.

$$7^{4.3} = B + 1$$
$$\log_7 (B+1) = 4.3$$

51.

$$\log_x 81 = 2$$
$$x^2 = 81$$
$$x^2 = 9^2$$
$$x = 9$$

53.

$$\log_8 x = 2$$
$$8^2 = x$$
$$64 = x$$

55.

$$\log_5 125 = x$$
$$5^x = 125$$
$$5^x = 5^3$$
$$x = 3$$

57.

$$\log_5 x = -2$$
$$5^{-2} = x$$
$$\frac{1}{5^2} = x$$
$$\frac{1}{25} = x$$

59.

$$\log_{36} x = -\frac{1}{2}$$
$$36^{-1/2} = x$$
$$\frac{1}{36^{1/2}} = x$$
$$\frac{1}{\sqrt{36}} = x$$
$$\frac{1}{6} = x$$

61.

$$\log_x 0.01 = -2$$
$$x^{-2} = 0.01$$
$$\left(x^{-2}\right)^{-1/2} = (0.01)^{-1/2}$$
$$x = \left(\frac{1}{100}\right)^{-1/2}$$
$$x = (100)^{1/2}$$
$$x = \sqrt{100}$$
$$x = 10$$

63.

$$\log_{27} 9 = x$$
$$27^x = 9$$
$$3^{3x} = 3^2$$
$$3x = 2$$
$$x = \frac{2}{3}$$

65.

$$\log_x 5^3 = 3$$
$$x^3 = 5^3$$
$$x^3 = 125$$
$$\sqrt[3]{x^3} = \sqrt[3]{125}$$
$$x = 5$$

67.

$$\log_{100} x = \frac{3}{2}$$
$$100^{3/2} = x$$
$$\left(\sqrt{100}\right)^3 = x$$
$$(10)^3 = x$$
$$1{,}000 = x$$

69.

$$\log_x \frac{1}{64} = -3$$
$$x^{-3} = \frac{1}{64}$$
$$\left(x^{-3}\right)^{-1/3} = \left(\frac{1}{64}\right)^{-1/3}$$
$$x = (64)^{1/3}$$
$$x = \sqrt[3]{64}$$
$$x = 4$$

71.

$$\log_8 x = 0$$
$$8^0 = x$$
$$1 = x$$

73.

$$\log_x \frac{\sqrt{3}}{3} = \frac{1}{2}$$
$$x^{1/2} = \frac{\sqrt{3}}{3}$$
$$\left(x^{1/2}\right)^2 = \left(\frac{\sqrt{3}}{3}\right)^2$$
$$x = \frac{\sqrt{9}}{9}$$
$$x = \frac{3}{9}$$
$$x = \frac{1}{3}$$

75.

$$\log_2 8 = x$$
$$2^x = 8$$
$$2^x = 2^3$$
$$x = 3$$

77.

$$\log_4 16 = x$$
$$4^x = 16$$
$$4^x = 4^2$$
$$x = 2$$

79.

$$\log 1{,}000{,}000 = x$$
$$10^x = 1{,}000{,}000$$
$$10^x = 10^6$$
$$x = 6$$

81.

$$\log \frac{1}{10} = x$$
$$10^x = \frac{1}{10}$$
$$10^x = 10^{-1}$$
$$x = -1$$

83.

$$\log_{1/2}\frac{1}{32}=x$$
$$\left(\frac{1}{2}\right)^x=\frac{1}{32}$$
$$2^{-x}=2^{-5}$$
$$-x=-5$$
$$x=5$$

85.

$$\log_9 3=x$$
$$9^x=3$$
$$3^{2x}=3^1$$
$$2x=1$$
$$x=\frac{1}{2}$$

87.

$\log 3.25=0.5119$

89.

$\log 0.00467=-2.3307$

91.

$\log x=3.7813$

$x=6{,}043.6597$

93.

$\log x=-0.7630$

$x=0.1726$

95.

$\log x=-0.5$

$x=0.3162$

97.

$\log x=-1.71$

$x=0.0195$

99. $f(x)=\log_3 x$

increasing

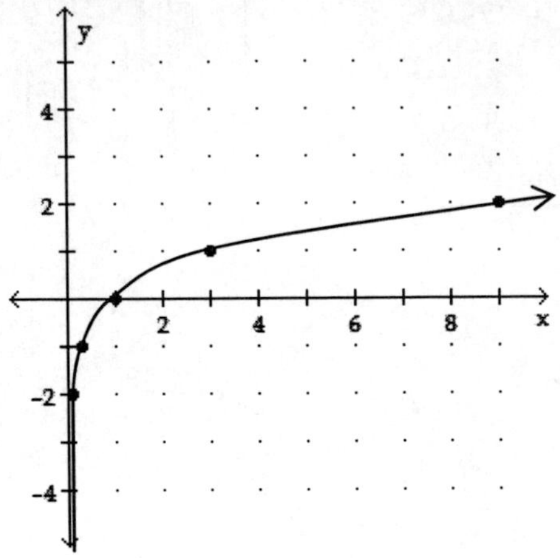

101. $y=\log_{1/2} x$

decreasing

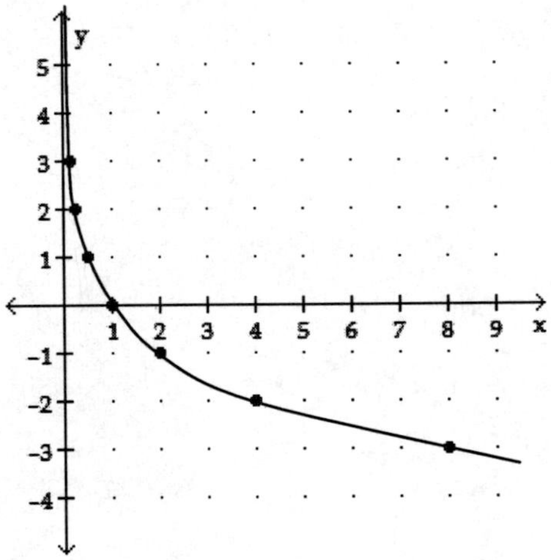

103. $f(x)=3+\log_3 x$

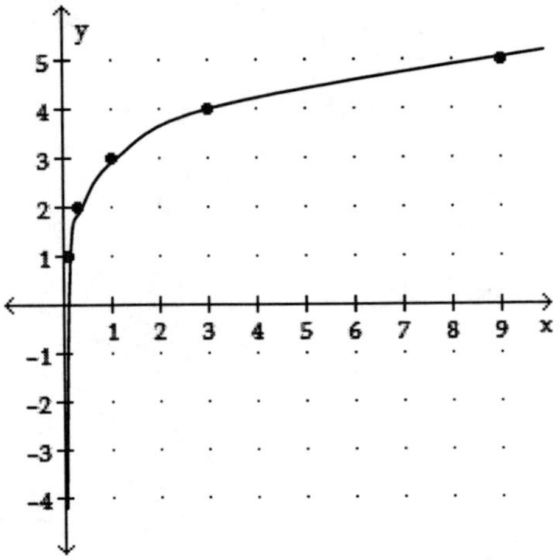

105. $y=\log_{1/2}(x-2)$

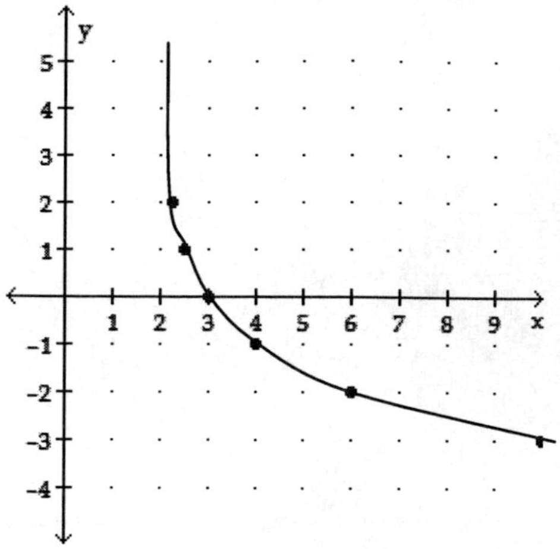

107.

$f(x)=6^x$

$f^{-1}(x)=\log_6 x$

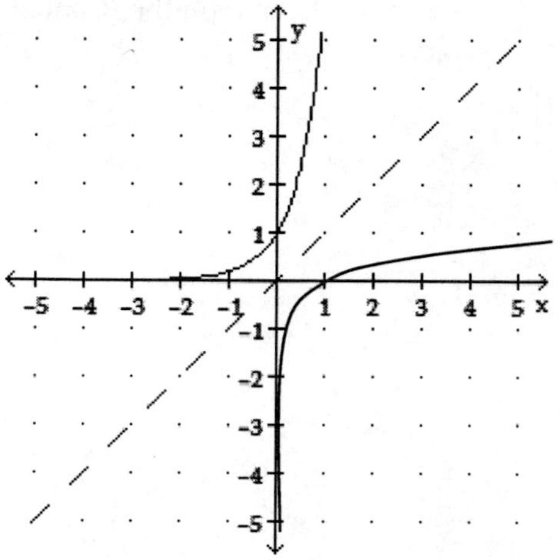

109. $f(x)=5^x$ and $f^{-1}(x)=\log_5 x$

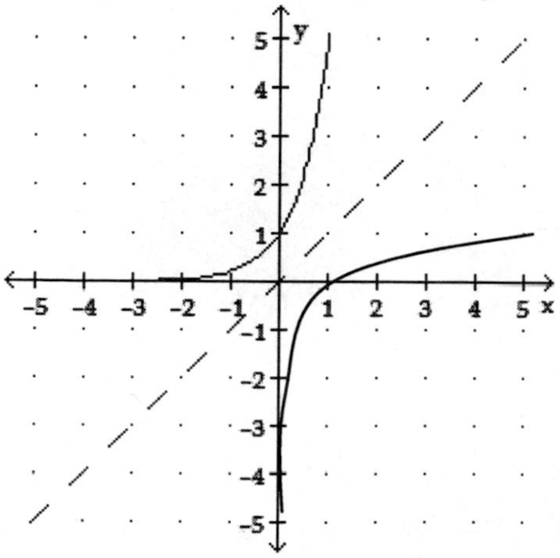

APPLICATIONS

111. db GAIN

$E_O = 30$ and $E_I = 0.1$

$$\begin{aligned}\text{db gain} &= 20\left(\log\frac{E_O}{E_I}\right)\\ &= 20\left(\log\frac{30}{0.1}\right)\\ &= 20(\log 300)\\ &= 20(2.477)\\ &= 49.5\text{ db}\end{aligned}$$

113. EARTHQUAKES

$A = 5{,}000$ and $P = 0.2$

$$\begin{aligned}R &= \log\frac{A}{P}\\ &= \log\frac{5{,}000}{0.2}\\ &= \log 25{,}000\\ &\approx 4.4\end{aligned}$$

115. SOCIAL WORKER

Let $m = 6$.

$$\begin{aligned}c(m) &= 500\log(m+1)\\ c(6) &= 500\log(6+1)\\ &= 500\log 7\\ &\approx 422\text{ cases}\end{aligned}$$

117. STOCKING LAKES

Let $t = 2.5$.

$$\begin{aligned}f(t) &= 75+45\log(t+1)\\ f(2.5) &= 75+45\log(2.5+1)\\ &= 75+45\log 3.5\\ &\approx 99\text{ sunfish}\end{aligned}$$

119. CHILDREN'S HEIGHT

$$\begin{aligned}h(A) &= 29+48.8\log(A+1)\\ h(9) &= 29+48.8\log(9+1)\\ &= 29+48.8\log(10)\\ &= 29+48.8\\ &= 77.8\end{aligned}$$

77.8% of his adult height.

121. INVESTING

$P = 1{,}000$, $A = 20{,}000$, and $r = 0.12$

$$n = \frac{\log\left(\frac{Ar}{P}+1\right)}{\log(1+r)}$$

$$= \frac{\log\left(\frac{20{,}000 \bullet 0.12}{1{,}000}+1\right)}{\log(1+0.12)}$$

$$= \frac{\log(2.4+1)}{\log(1+0.12)}$$

$$= \frac{\log 3.4}{\log 1.12}$$

$$\approx \frac{0.5315}{0.0492}$$

$$\approx 10.8 \text{ years}$$

WRITING

123. Answers will vary.

125. Answers will vary.

REVIEW

127.

$$\sqrt[3]{6x+4} = 4$$
$$\left(\sqrt[3]{6x+4}\right)^3 = (4)^3$$
$$6x+4 = 64$$
$$6x = 60$$
$$x = 10$$

129.

$$\sqrt{a+1} - 1 = 3a$$
$$\sqrt{a+1} = 3a+1$$
$$\left(\sqrt{a+1}\right)^2 = (3a+1)^2$$
$$a+1 = 9a^2 + 6a + 1$$
$$0 = 9a^2 + 5a$$
$$0 = a(9a+5)$$
$$a = 0 \quad \text{or} \quad 9a+5 = 0$$
$$a = 0 \qquad a = -\cancel{\tfrac{5}{9}}$$

CHALLENGE PROBLEMS

131. The domain would be $(-\infty,-1)\cup(1,\infty)$

133.

$$R = \log\frac{A}{P}$$
$$10^R = \frac{A}{P}$$

1985:

$$10^{8.1} = \frac{A}{P}$$

1989:

$$10^{7.1} = \frac{A}{P}$$

If the period remains constant, the amplitude of an earthquake must change by a factor of 10 to increase its severity by 1 point on the Richter scale.

SECTION 9.5

VOCABULARY

1. $f(x)=e^x$ is called the natural **exponential** function. The base is ***e***.

3. If a bank pays interest infinitely many times a year, we say that the interest is compounded **continuously**.

CONCEPTS

5. a. the natural exponential function
   b. D: $(-\infty, \infty)$; R: $(0, \infty)$
   c. (0, 1); none
   d. yes
   e. the x–axis ($y = 0$)
   f. increasing
   g. e

7. $e = 2.718281828459\ldots$

9. If n gets larger and larger, the value of $\left(1+\frac{1}{n}\right)^n$ approaches the value of ***e***.

11. To find $\ln e^2$, we ask, "To what power must we raise ***e*** to get e^2?" Since the answer is the 2nd power, $\ln e^2 = 2$.

13. $\sqrt{2} \approx 1.41$; $e \approx 2.718$; $\pi \approx 3.14$

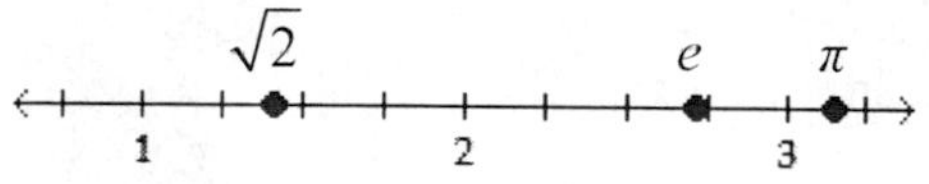

15. a. The y–coordinate of the point on the graph having an x–coordinate of 1 is 2.7182818… The name of that number is e.
    b. As x decreases, the values of $f(x)$ decreases. The value of $f(x)$ will never be 0 or negative.

17. $f^{-1}(x)=e^x$

NOTATION

19. $P = 1{,}000$, $r = 0.09$, and $t = 10$

$$A = \boxed{1{,}000}e^{(0.09)(\boxed{10})}$$
$$= 1{,}000e^{\boxed{0.9}}$$
$$\approx \boxed{2{,}459.6}$$

21. We read $\ln x$ letter-by-letter as "$\boxed{\ell}$…$\boxed{n}$… of x."

23. To evaluate a base-10 logarithm with a calculator, use the **LOG** key. To evaluate the base-e logarithm, use the **LN** key.

GUIDED PRACTICE

25. $f(x)=e^x$

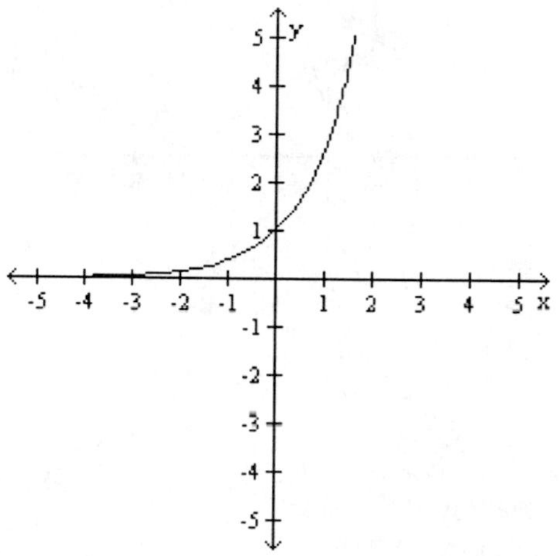

27. $f(x)=e^x+1$

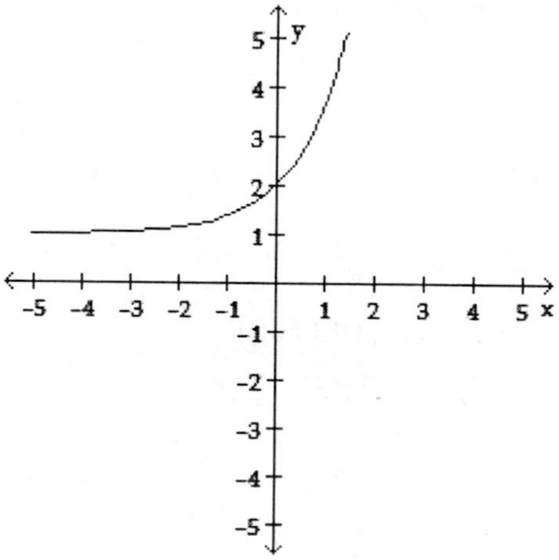

29. $y = e^{x+3}$

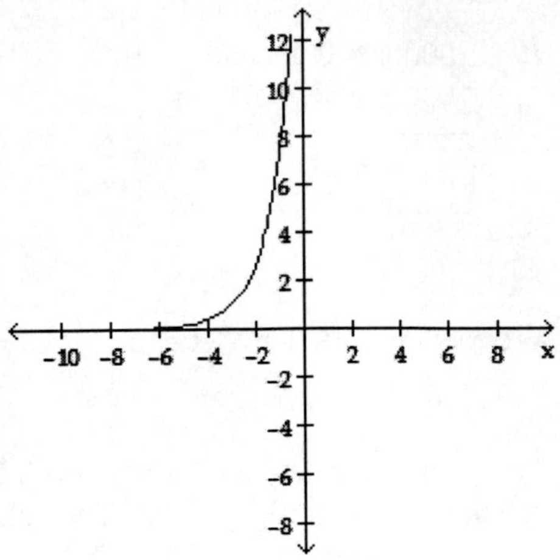

31. $f(x) = 2e^x$

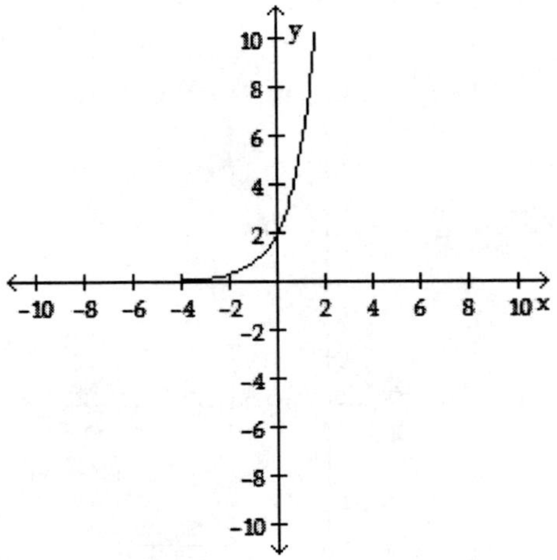

33. $A = Pe^{rt}$
$$A = (5{,}000)e^{0.08(20)}$$
$$= (5{,}000)e^{1.6}$$
$$= 5{,}000(4.953)$$
$$= 24{,}765.16$$

35. $A = Pe^{rt}$
$$A = 20{,}000e^{0.105(50)}$$
$$= 20{,}000e^{5.25}$$
$$= 20{,}000(190.566)$$
$$= 3{,}811{,}325.37$$

37. $A = Pe^{rt}$
$$A = 15{,}895e^{-0.02(16)}$$
$$= 15{,}895e^{-0.32}$$
$$= 15{,}895(0.726)$$
$$= 11{,}542.14$$

39. $A = Pe^{rt}$
$$A = 565e^{-0.005(8)}$$
$$= 565e^{-0.04}$$
$$= 565(0.961)$$
$$= 542.85$$

41. $\ln e^5 = 5$

43. $\ln e^6 = 6$

45. $\ln \frac{1}{e} = \ln e^{-1} = -1$

47. $\ln \sqrt[4]{e} = \ln e^{1/4} = \frac{1}{4}$

49. $\ln \sqrt[3]{e^2} = \ln e^{2/3} = \frac{2}{3}$

51. $\ln e^{-7} = -7$

53. ln 35.15 = 3.5596

55. ln 0.00465 = –5.3709

57. ln 1.72 = 0.5423

59. ln (–0.1) = undefined

61.
$$\ln x = 1.4023$$
$$e^{1.4023} = x$$
$$4.0645 = x$$

63.
$$\ln x = 4.24$$
$$e^{4.24} = x$$
$$69.4079 = x$$

65.

$$\ln x = -3.71$$
$$e^{-3.71} = x$$
$$0.0245 = x$$

67.

$$1.001 = \ln x$$
$$x = e^{1.001}$$
$$x = 2.7210$$

69. $f(x) = \ln\left(\frac{1}{2}x\right)$

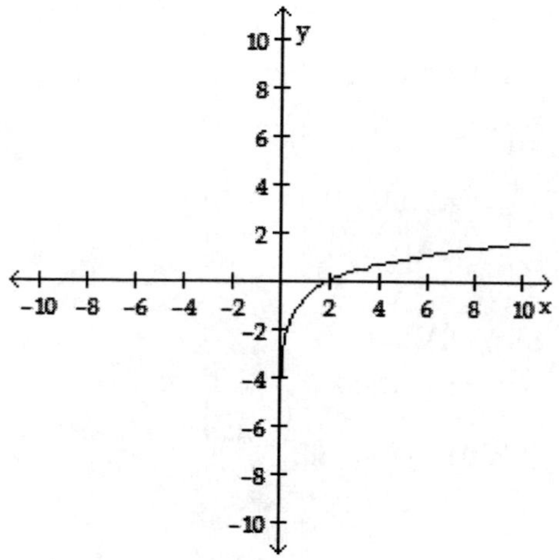

71. $f(x) = \ln(-x)$

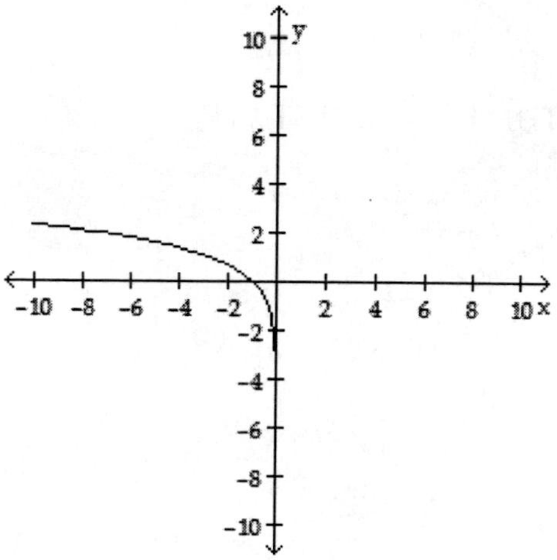

APPLICATIONS

73. CONTINUOUS COMPOUND INTEREST

$P = 5{,}000$, $r = 0.082$, and $t = 12$

$$\begin{aligned} A &= Pe^{rt} \\ &= 5{,}000e^{(0.082)(12)} \\ &= 5{,}000e^{0.984} \\ &\approx 5{,}000(2.675135411) \\ &\approx \$13{,}375.68 \end{aligned}$$

75. COMPARISON OF COMPOUNDING METHODS

Annual Compounding:

$PV = 5{,}000$, $i = 0.085$, and $n = 5$

$$\begin{aligned} FV &= PV(1+i)^n \\ &= 5{,}000(1+0.085)^5 \\ &= 5{,}000(1.085)^5 \\ &= 5{,}000(1.5037) \\ &= \$7{,}518.28 \end{aligned}$$

Continuous Compounding:

$P = 5{,}000$, $r = 0.085$, and $t = 5$

$$\begin{aligned} A &= Pe^{rt} \\ &= 5{,}000e^{(0.085)(5)} \\ &= 5{,}000e^{0.425} \\ &\approx 5{,}000(1.52959042) \\ &\approx \$7{,}647.95 \end{aligned}$$

77. DETERMINING INITIAL DEPOSIT

$A = 11{,}180$, $r = 0.07$, and $t = 7$

$$A = Pe^{rt}$$
$$11{,}180 = Pe^{(0.07)(7)}$$
$$11{,}180 = Pe^{0.49}$$
$$\frac{11{,}180}{e^{0.49}} = \frac{Pe^{0.49}}{e^{0.49}}$$
$$\frac{11{,}180}{e^{0.49}} = P$$
$$P \approx \frac{11{,}180}{1.63231622}$$
$$P \approx \$6{,}849.16$$

79. The 20th CENTURY

$$A(t)=123e^{0.0117t}$$

$$1937-A(7)=123e^{0.0117(7)}$$
$$=133 \text{ million}$$

$$1941-A(11)=123e^{0.0117(11)}$$
$$=140 \text{ million}$$

$$1955-A(25)=123e^{0.0117(25)}$$
$$=165 \text{ million}$$

$$1969-A(39)=123e^{0.0117(39)}$$
$$=194 \text{ million}$$

$$1974-A(44)=123e^{0.0117(44)}$$
$$=206 \text{ million}$$

$$1986-A(56)=123e^{0.0117(56)}$$
$$=237 \text{ million}$$

$$1997-A(67)=123e^{0.0117(67)}$$
$$=269 \text{ million}$$

81. HIGHS AND LOWS

Kuwait:

$P = 2{,}789{,}132$, $r = 0.03501$, and $t = 15$

$$A = Pe^{rt}$$
$$= 2{,}789{,}132e^{(0.03501)(15)}$$
$$= 2{,}789{,}132e^{0.52515}$$
$$\approx 4{,}715{,}620$$

Bulgaria:

$P = 7{,}148{,}785$, $r = -0.00768$, and $t = 15$

$$A = Pe^{rt}$$
$$= 7{,}148{,}785e^{(-0.00768)(15)}$$
$$= 7{,}148{,}785e^{-0.1152}$$
$$\approx 6{,}370{,}911$$

83. EPIDEMICS

$$P(t)=2e^{0.27t}$$
$$P(12)=2e^{0.27(12)}$$
$$=2e^{3.24}$$
$$\approx 2(25.53372175)$$
$$\approx 51$$

85. ANTS

Let $t = 40$.

$$a(t)=1.36\left(\frac{e}{2.5}\right)^t$$
$$a(40)=1.36(1.087)^{40}$$
$$=1.36(28.4563)$$
$$\approx 39$$

87. SOCIAL WORKERS

The trainee had to assemble 14 chairs before meeting company standards.

89. DISINFECTANTS

$$A(t)=2{,}000{,}000e^{-0.588t}$$
$$A(5)=2{,}000{,}000e^{-0.588(5)}$$
$$=2{,}000{,}000e^{-2.94}$$
$$=105{,}731$$

91. SKYDIVING

$$f(t)=50\left(1-e^{-0.2t}\right)$$
$$f(20)=50\left(1-e^{-0.2(20)}\right)$$
$$=50\left(1-e^{-4}\right)$$
$$\approx 50(1-0.018315639)$$
$$\approx 50(0.981684361)$$
$$\approx 49 \text{ mps}$$

93. THE TARHEEL STATE

Let $r = 0.043$.

$$t=\frac{\ln 2}{r}$$
$$=\frac{\ln 2}{0.043}$$
$$\approx\frac{0.6931}{0.043}$$
$$\approx 16 \text{ years}$$

95. THE EQUALITY STATE

Let $r = 0.0213$.

$$t=\frac{\ln 2}{r}$$
$$=\frac{\ln 2}{0.0213}$$
$$\approx 33 \text{ years}$$

33 years from 2009 is the year 2042.

97. POPULATION GROWTH

Let $r = 0.12$.

$$t = \frac{\ln 3}{r}$$
$$= \frac{\ln 3}{0.12}$$
$$\approx \frac{1.0986}{0.12}$$
$$\approx 9.2 \text{ years}$$

99. FORENSIC MEDICINE

Let $T_s = 70$.

$$t = \frac{1}{0.25}\ln\frac{98.6 - T_s}{82 - T_s}$$
$$= \frac{1}{0.25}\ln\frac{98.6 - 70}{82 - 70}$$
$$= \frac{1}{0.25}\ln\frac{28.6}{12}$$
$$= \frac{1}{0.25}\ln 2.383$$
$$\approx \frac{1}{0.25}(0.8684)$$
$$\approx 3.5 \text{ hours}$$

101. CROSS COUNTRY SKIING

Let $s = 7.5$.

$$H(s) = -47.73 + 107.38\ln s$$
$$H(7.5) = -47.73 + 107.38\ln(7.5)$$
$$= -47.73 + 216.36$$
$$\approx 169 \text{ beats per minute}$$

103. THE PACE OF LIFE

Use $s(p) = 0.05 + 0.37\ln p$.

For New York City:

$$s(8{,}392{,}000) = 0.05 + 0.37\ln 8{,}392{,}000$$
$$= 0.05 + 5.8988$$
$$\approx 5.9 \text{ ft.sec}$$

For Atlanta:

$$s(541{,}000) = 0.05 + 0.37\ln 541{,}000$$
$$= 0.05 + 488$$
$$\approx 4.9 \text{ ft.sec}$$

The average pedestrian in NYC walks about 1 foot per second faster than the average pedestrian in Atlanta.

WRITING

105. Answers will vary.

107. Answers will vary.

109. Answers will vary.

111. Answers will vary.

113. Answers will vary.

REVIEW

115. $\sqrt{240x^5} = \sqrt{16x^4}\sqrt{15x}$
$$= 4x^2\sqrt{15x}$$

117.

$$4\sqrt{48y^3} - 3y\sqrt{12y} = 4\sqrt{16y^2}\sqrt{3y} - 3y\sqrt{4}\sqrt{3y}$$
$$= 4(4y)\sqrt{3y} - 3y(2)\sqrt{3y}$$
$$= 16y\sqrt{3y} - 6y\sqrt{3y}$$
$$= 10y\sqrt{3y}$$

CHALLENGE PROBLEMS

119. False.

$$e^e \overset{?}{>} e^3$$
$$e^{2.71829} \overset{?}{>} e^3$$
$$15.15 \not> 20.09$$

121.

$$e^{t+5} = ke^t$$
$$\frac{e^{t+5}}{e^t} = \frac{ke^t}{e^t}$$
$$e^{t+5-t} = k$$
$$e^5 = k$$
$$k = e^5$$

123.

$$P = P_0e^{rt}$$
$$P = P_0e^{r\left(\frac{\ln 2}{r}\right)}$$
$$P = P_0e^{\ln 2}$$
$$P = P_0(2)$$
$$P = 2P_0$$

125. $y = f(x) = \dfrac{1}{1+e^{-2x}}$

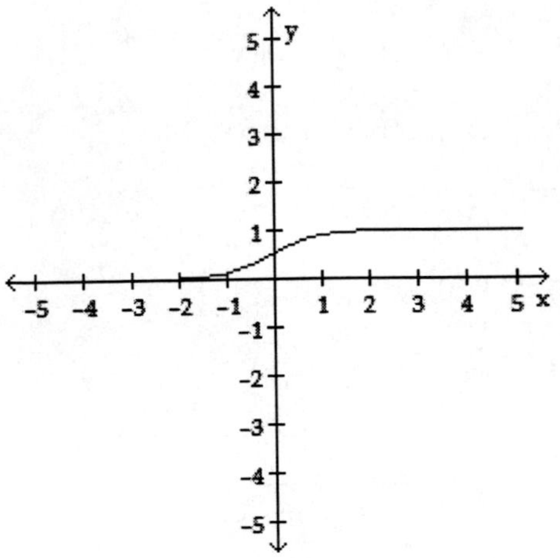

SECTION 9.6

VOCABULARY

1. The logarithm of a **product**, such as $\log_3 4x$, equals the sum of the logarithms of the factors.

3. The logarithm of a **power**, such as $\log_4 5^3$, equals the power times the logarithm of the number.

CONCEPTS

5. a. $\log_b 1 = \boxed{0}$
   b. $\log_b b = \boxed{1}$
   c. $\log_b b^x = \boxed{x}$
   d. $b^{\log_b x} = \boxed{x}$

7.
$$\begin{aligned}\log[(2.5)(3.7)] &= \log 2.5 + \log 3.7\\ \log(9.25) &= 0.3979 + 0.5682\\ 0.9661 &= 0.9661\end{aligned}$$

9.
$$\begin{aligned}\ln\frac{11.3}{6.1} &= \ln 11.3 - \ln 6.1\\ \ln 1.8525 &= 2.4248 - 1.8082\\ 0.617 &= 0.617\end{aligned}$$

11. c. apply the product rule

13. b. apply the power rule

NOTATION

15.
$$\begin{aligned}\log_8 8a^3 &= \log_8 \boxed{8} + \log_8 \boxed{a^3}\\ &= \log_8 8 + \boxed{3}\log_8 a\\ &= \boxed{1} + 3\log_8 a\end{aligned}$$

17. True

GUIDED PRACTICE

19. $\log_6 1 = 0$

21. $\log_4 4^7 = 7$

23. $5^{\log_5 10} = 10$

25. $\log_5 5^2 = 2$

27. $\ln e = 1$

29. $\log_3 3^7 = 7$

31.
$$\begin{aligned}\log_2(4\cdot 5) &= \log_2 4 + \log_2 5\\ &= \log_2 2^2 + \log_2 5\\ &= 2 + \log_2 5\end{aligned}$$

33.
$$\log 25y = \log 25 + \log y$$

35.
$$\begin{aligned}\log 100pq &= \log 100 + \log p + \log q\\ &= \log 10^2 + \log p + \log q\\ &= 2 + \log p + \log q\end{aligned}$$

37.
$$\log 5xyz = \log 5 + \log x + \log y + \log z$$

39.
$$\begin{aligned}\log\frac{100}{9} &= \log 100 - \log 9\\ &= \log 10^2 - \log 9\\ &= 2 - \log 9\end{aligned}$$

41.
$$\begin{aligned}\log_6\frac{x}{36} &= \log_6 x - \log_6 36\\ &= \log_6 x - \log_6 6^2\\ &= \log_6 x - 2\end{aligned}$$

43.
$$\begin{aligned}\log\frac{7c}{2} &= \log 7c - \log 2\\ &= \log 7 + \log c - \log 2\end{aligned}$$

45.
$$\begin{aligned}\log\frac{10x}{y} &= \log 10x - \log y\\ &= \log 10 + \log x - \log y\\ &= 1 + \log x - \log y\end{aligned}$$

47.

$$\ln\frac{exy}{z} = \ln exy - \ln z = \ln e + \ln x + \ln y - \ln z = 1 + \ln x + \ln y - \ln z$$

49.

$$\log_8 \frac{1}{8m} = \log_8 1 - \log_8 8m = \log_8 1 - \log_8 8 - \log_8 m = \log_8 1 - 1 - \log_8 m = 0 - 1 - \log_8 m = -1 - \log_8 m$$

51.

$$\ln y^7 = 7\ln y$$

53.

$$\log\sqrt{5} = \log 5^{1/2} = \frac{1}{2}\log 5$$

55.

$$\log e^{-3} = -3\log e$$

57.

$$\log_7\left(\sqrt[5]{100}\right)^3 = \log_7(100)^{3/5} = \frac{3}{5}\log_7 100$$

59.

$$\log xyz^2 = \log x + \log y + \log z^2 = \log x + \log y + 2\log z$$

61.

$$\log_2 \frac{2\sqrt[3]{x}}{y} = \log_2 \frac{2x^{1/3}}{y} = \log_2\left(2x^{1/3}\right) - \log_2 y = \log_2 2 + \log_2 x^{1/3} - \log_2 y = 1 + \frac{1}{3}\log_2 x - \log_2 y$$

63.

$$\log x^3 y^2 = \log x^3 + \log y^2 = 3\log x + 2\log y$$

65.

$$\log_b \sqrt{xy} = \log_b (xy)^{1/2} = \frac{1}{2}\log_b(xy) = \frac{1}{2}\left(\log_b x + \log_b y\right)$$

67.

$$\log_a \frac{\sqrt[3]{x}}{\sqrt[4]{yz}} = \log_a \frac{x^{1/3}}{(yz)^{1/4}} = \log_a x^{1/3} - \log_a (yz)^{1/4} = \frac{1}{3}\log_a x - \frac{1}{4}\log_a(yz) = \frac{1}{3}\log_a x - \frac{1}{4}\left(\log_a y + \log_a z\right) = \frac{1}{3}\log_a x - \frac{1}{4}\log_a y - \frac{1}{4}\log_a z$$

69.

$$\ln x^{20}\sqrt{z} = \ln x^{20} + \ln\sqrt{z} = \ln x^{20} + \ln z^{1/2} = 20\ln x + \frac{1}{2}\ln z$$

71.

$$\log_5\left(\frac{1}{t^3}\right)^d = \log_5\left(t^{-3}\right)^d = \log_5 t^{-3d} = -3d\log_5 t$$

73.

$$\ln\sqrt{ex} = \ln(ex)^{1/2} = \frac{1}{2}\ln(ex) = \frac{1}{2}\ln e + \frac{1}{2}\ln x = \frac{1}{2}(1) + \frac{1}{2}\ln x = \frac{1}{2} + \frac{1}{2}\ln x$$

75.

$$\log_2(x+1) + 9\log_2 x = \log_2(x+1) + \log_2 x^9 = \log_2 x^9(x+1)$$

77.

$$\log_3 x + \log_3(x+2) - \log_3 8$$
$$= \log_3 x(x+2) - \log_3 8$$
$$= \log_3 \frac{x(x+2)}{8}$$

79.

$$-3\log_b x - 2\log_b y + \frac{1}{2}\log_b z$$
$$= \frac{1}{2}\log_b z - 3\log_b x - 2\log_b y$$
$$= \log_b z^{1/2} - \log_b x^3 - \log_b y^2$$
$$= \log_b z^{1/2} - \left(\log_b x^3 + \log_b y^2\right)$$
$$= \log_b z^{1/2} - \left(\log_b x^3 y^2\right)$$
$$= \log_b \frac{z^{1/2}}{x^3 y^2}$$
$$= \log_b \frac{\sqrt{z}}{x^3 y^2}$$

81.

$$\frac{1}{3}\left[\log_b\left(M^2-9\right) - \log_b(M+3)\right]$$
$$= \log_b\left(M^2-9\right)^{1/3} - \log_b(M+3)^{1/3}$$
$$= \log_b \frac{\left(M^2-9\right)^{1/3}}{(M+3)^{1/3}}$$
$$= \log_b\left(\frac{M^2-9}{M+3}\right)^{1/3}$$
$$= \log_b\left(\frac{(M+3)(M-3)}{M+3}\right)^{1/3}$$
$$= \log_b(M-3)^{1/3}$$
$$= \log_b \sqrt[3]{M-3}$$

83.

$$\ln\left(\frac{x}{z}+x\right) - \ln\left(\frac{y}{z}+y\right) = \ln\left(\frac{\frac{x}{z}+x}{\frac{y}{z}+y}\right)$$
$$= \ln\left(\frac{\frac{x}{z}+x}{\frac{y}{z}+y}\bullet\frac{z}{z}\right)$$
$$= \ln\left(\frac{x+xz}{y+yz}\right)$$
$$= \ln\left(\frac{x\cancel{(1+z)}}{y\cancel{(1+z)}}\right)$$
$$= \ln\frac{x}{y}$$

85.

$$\frac{1}{2}\log_6\left(x^2+1\right) - \log_6\left(x^2+2\right)$$
$$= \log_6\left(x^2+1\right)^{1/2} - \log_6\left(x^2+2\right)$$
$$= \log_6 \frac{\left(x^2+1\right)^{1/2}}{x^2+2}$$
$$= \log_6 \frac{\sqrt{x^2+1}}{x^2+2}$$

87.

$$\log_b 28 = \log_b(4\bullet 7)$$
$$= \log_b 4 + \log_b 7$$
$$= 0.6021 + 0.8451$$
$$= 1.4472$$

89.

$$\log_b \frac{4}{63} = \log_b 4 - \log_b 63$$
$$= \log_b 4 - \log_b(7\bullet 9)$$
$$= \log_b 4 - \left(\log_b 7 + \log_b 9\right)$$
$$= \log_b 4 - \log_b 7 - \log_b 9$$
$$= 0.6021 - 0.8451 - 0.9542$$
$$= -1.1972$$

91.

$$\begin{aligned}\log_b \frac{63}{4} &= \log_b 63 - \log_b 4\\ &= \log_b (7\bullet 9) - \log_b 4\\ &= \log_b 7 + \log_b 9 - \log_b 4\\ &= 0.8451 + 0.9542 - 0.6021\\ &= 1.1972\end{aligned}$$

93.

$$\begin{aligned}\log_b 64 &= \log_b 4^3\\ &= 3\log_b 4\\ &= 3(0.6021)\\ &= 1.8063\end{aligned}$$

95.

$$\begin{aligned}\log_3 7 &= \frac{\log 7}{\log 3}\\ &\approx \frac{0.84509}{0.47712}\\ &\approx 1.7712\end{aligned}$$

97.

$$\begin{aligned}\log_{1/3} 3 &= \frac{\log 3}{\log \frac{1}{3}}\\ &\approx \frac{0.47712}{-0.47712}\\ &\approx -1.0000\end{aligned}$$

99.

$$\begin{aligned}\log_3 8 &= \frac{\log 8}{\log 3}\\ &\approx \frac{0.90309}{0.47712}\\ &\approx 1.8928\end{aligned}$$

101.

$$\begin{aligned}\log_{\sqrt{2}} \sqrt{5} &= \frac{\log\sqrt{5}}{\log\sqrt{2}}\\ &\approx \frac{0.349485}{0.150514}\\ &\approx 2.3219\end{aligned}$$

103. a. and c. are equivalent
$\log(9\cdot 3) = \log 9 + \log 3$

105. a. and b. are equivalent
$\log_2 11^4 = 4\log_2 11$

APPLICATIONS

107. pH OF A SOLUTION

$$\begin{aligned}\text{pH} &= -\log\left[\text{H}^+\right]\\ &= -\log\left[1.7\times 10^{-5}\right]\\ &= -\left(\log 1.7 + \log 10^{-5}\right)\\ &= -\left(\log 1.7 - 5\log 10\right)\\ &= -\log 1.7 + 5\log 10\\ &= -0.2304 + 5\bullet 1\\ &= -0.2304 + 5\\ &= 4.8\end{aligned}$$

109. FORMULAS

a.

$$\begin{aligned}B &= 10(\log I - \log I_0)\\ &= 10\left(\log\frac{I}{I_0}\right)\\ &= \log\left(\frac{I}{I_0}\right)^{10}\end{aligned}$$

b.

$$\begin{aligned}T &= \frac{1}{k}(\ln C_2 - \ln C_1)\\ &= \frac{1}{k}\left(\ln\frac{C_2}{C_1}\right)\\ &= \ln\sqrt[k]{\frac{C_2}{C_1}}\end{aligned}$$

WRITING

111. Answers will vary.

113. Answers will vary.

115. Answers will vary.

117. Answers will vary.

REVIEW

119.

$$m = \frac{y_2 - y_1}{x_2 - x_1}$$
$$= \frac{-4-3}{4-(-2)}$$
$$= \frac{-7}{6}$$
$$= -\frac{7}{6}$$

121.

$$\left(\frac{x_1+x_2}{2}, \frac{y_1+y_2}{2}\right) = \left(\frac{-2+4}{2}, \frac{3+(-4)}{2}\right)$$
$$= \left(\frac{2}{2}, \frac{-1}{2}\right)$$
$$= \left(1, -\frac{1}{2}\right)$$

CHALLENGE PROBLEMS

123. Answers will vary.

125.

$$\log_{b^2} x = \frac{1}{2}\log_b x$$
$$\log_{b^2} x = \log_b x^{1/2}$$
$$\log_{b^2} x = \log_{b^2} x$$

SECTION 9.7

VOCABULARY

1. An equation with a variable in its exponent, such as $3^{2x} = 8$, is called an **exponential** equation.

CONCEPTS

3. a. If two exponential expressions with the same base are equal, their exponents are **equal**.
$b^x = b^y$ is equivalent to $\boxed{x = y}$

b. If the logarithms base–b of two numbers are equal, the numbers are **equal**.
$\log_b x = \log_b y$ is equivalent to $\boxed{x = y}$

5. If $6^{4x} = 6^{-2}$, then $4x = \boxed{-2}$.

7. To solve $5^x = 2$, we can take the **logarithm** of both sides of the equation to get $\log 5^x = \log 2$.

9. If $e^{x+2} = 4$, then $\ln e^{x+2} = \boxed{\ln 4}$.

11. No, it is not a solution.

$$\log_5(x+1) = 2$$
$$\log_5(4+1) \stackrel{?}{=} 2$$
$$\log_5(5) \stackrel{?}{=} 2$$
$$5^2 \stackrel{?}{=} 5$$
$$25 \neq 5$$

13. a. Divide both sides by ln 3.

b. $\dfrac{\ln 5}{\ln 3}$

c. 1.4650

15. a. $\dfrac{\log 8}{\log 5} \approx 1.2920$

b. $\dfrac{3\ln 12}{\ln 4 - \ln 2} \approx 10.7549$

17. a. $\text{pH} = -\log\left[\text{H}^+\right]$

b. $A = \boxed{A_0 2^{-t/h}}$

c. $P = \boxed{P_0 e^{kt}}$

NOTATION

19.

$$2^x = 7$$
$$\boxed{\log} 2^x = \log 7$$
$$x \boxed{\log 2} = \log 7$$
$$x = \frac{\log 7}{\log 2}$$
$$x \approx \boxed{2.8074}$$

GUIDED PRACTICE

21.

$$6^{x-2} = 36$$
$$6^{x-2} = 6^2$$
$$x - 2 = 2$$
$$x = 4$$

23.

$$5^{4x} = \frac{1}{125}$$
$$5^{4x} = \frac{1}{5^3}$$
$$5^{4x} = 5^{-3}$$
$$4x = -3$$
$$x = -\frac{3}{4}$$
$$x = -0.75$$

25.

$$2^{x^2-2x} = 8$$
$$2^{x^2-2x} = 2^3$$
$$x^2 - 2x = 3$$
$$x^2 - 2x - 3 = 0$$
$$(x-3)(x+1) = 0$$
$$x - 3 = 0 \quad \text{or} \quad x + 1 = 0$$
$$x = 3 \qquad\qquad x = -1$$

27.

$$3^{x^2+4x} = \frac{1}{81}$$
$$3^{x^2+4x} = \frac{1}{3^4}$$
$$3^{x^2+4x} = 3^{-4}$$
$$x^2 + 4x = -4$$
$$x^2 + 4x + 4 = 0$$
$$(x+2)(x+2) = 0$$
$$x+2=0 \quad \text{or} \quad x+2=0$$
$$x=-2 \qquad\qquad x=-2$$

29.

$$4^x = 5$$
$$\log 4^x = \log 5$$
$$x\log 4 = \log 5$$
$$\frac{x\log 4}{\log 4} = \frac{\log 5}{\log 4}$$
$$x \approx 1.1610$$

31.

$$13^{x-1} = 2$$
$$\log 13^{x-1} = \log 2$$
$$(x-1)\log 13 = \log 2$$
$$x\log 13 - \log 13 = \log 2$$
$$x\log 13 = \log 2 + \log 13$$
$$\frac{x\log 13}{\log 13} = \frac{\log 2 + \log 13}{\log 13}$$
$$x \approx 1.2702$$

33.

$$2^{x+1} = 3^x$$
$$\log 2^{x+1} = \log 3^x$$
$$(x+1)\log 2 = x\log 3$$
$$x\log 2 + \log 2 = x\log 3$$
$$\log 2 = x\log 3 - x\log 2$$
$$\log 2 = x(\log 3 - \log 2)$$
$$\frac{\log 2}{\log 3 - \log 2} = \frac{x(\log 3 - \log 2)}{\log 3 - \log 2}$$
$$x = \frac{\log 2}{\log 3 - \log 2}$$
$$x \approx 1.7095$$

35.

$$5^{x-3} = 3^{2x}$$
$$\log 5^{x-3} = \log 3^{2x}$$
$$(x-3)\log 5 = 2x\log 3$$
$$x\log 5 - 3\log 5 = 2x\log 3$$
$$-3\log 5 = 2x\log 3 - x\log 5$$
$$-3\log 5 = x(2\log 3 - \log 5)$$
$$\frac{-3\log 5}{2\log 3 - \log 5} = \frac{x(2\log 3 - \log 5)}{2\log 3 - \log 5}$$
$$x = \frac{-3\log 5}{2\log 3 - \log 5}$$
$$x \approx -8.2144$$

37.

$$e^{2.9x} = 4.5$$
$$\ln e^{2.9x} = \ln 4.5$$
$$2.9x\ln e = \ln 4.5$$
$$2.9x \cdot 1 = \ln 4.5$$
$$2.9x = \ln 4.5$$
$$x = \frac{\ln 4.5}{2.9}$$
$$x \approx 0.5186$$

39.

$$e^{-0.2t} = 14.2$$
$$\ln e^{-0.2t} = \ln 14.2$$
$$-0.2t\ln e = \ln 14.2$$
$$-0.2t \bullet 1 = \ln 14.2$$
$$-0.2t = \ln 14.2$$
$$t = \frac{\ln 14.2}{-0.2}$$
$$t \approx -13.2662$$

41.

$$\log 2x = 4$$
$$10^4 = 2x$$
$$10{,}000 = 2x$$
$$5{,}000 = x$$

43.

$$\log_3(x-3) = 2$$
$$3^2 = x-3$$
$$9 = x-3$$
$$12 = x$$

45.

$$\log(7-x)=2$$
$$10^2=7-x$$
$$100=7-x$$
$$93=-x$$
$$-93=x$$

47.

$$\log\frac{1}{8}x=-2$$
$$10^{-2}=\frac{1}{8}x$$
$$\frac{1}{100}=\frac{1}{8}x$$
$$800\left(\frac{1}{100}\right)=800\left(\frac{1}{8}x\right)$$
$$8=100x$$
$$\frac{8}{100}=x$$
$$0.08=x$$

49.

$$\log(3-2x)=\log(x+24)$$
$$3-2x=x+24$$
$$3-3x=24$$
$$-3x=21$$
$$x=-7$$

51.

$$\ln(3x+1)=\ln(x+7)$$
$$3x+1=x+7$$
$$2x+1=7$$
$$2x=6$$
$$x=3$$

53.

$$\log x+\log(x-48)=2$$
$$\log_{10}\left[x(x-48)\right]=2$$
$$10^2=x(x-48)$$
$$100=x^2-48x$$
$$0=x^2-48x-100$$
$$0=(x-50)(x+2)$$
$$x-50=0 \quad \text{or} \quad x+2=0$$
$$x=50 \qquad x=\cancel{-2}$$

A negative number does not have a logarithm, so the only answer is 50.

55.

$$\log_5(4x-1)+\log_5 x=1$$
$$\log_5\left[x(4x-1)\right]=1$$
$$5^1=x(4x-1)$$
$$5=4x^2-x$$
$$0=4x^2-x-5$$
$$0=(4x-5)(x+1)$$
$$4x-5=0 \quad \text{or} \quad x+1=0$$
$$4x=5 \qquad x=\cancel{-1}$$
$$x=\frac{5}{4}$$

A negative number does not have a logarithm, so the only answer is $\frac{5}{4}$.

57.

$$\log 5-\log x=1$$
$$\log\frac{5}{x}=1$$
$$10^1=\frac{5}{x}$$
$$10=\frac{5}{x}$$
$$10x=5$$
$$x=0.5$$

59.

$$\log_3 4x-\log_3 7=2$$
$$\log_3\frac{4x}{7}=2$$
$$3^2=\frac{4x}{7}$$
$$9=\frac{4x}{7}$$
$$63=4x$$
$$15.75=x$$

61.

$$\log 2x=\log 4$$
$$2x=4$$
$$x=2$$

63.

$$\ln x = 1$$
$$e^1 = x$$
$$x = e$$
$$x \approx 2.7183$$

65.

$$7^{x^2} = 10$$
$$\log 7^{x^2} = \log 10$$
$$x^2 \log 7 = \log 10$$
$$x^2 = \frac{\log 10}{\log 7}$$
$$x = \pm\sqrt{\frac{\log 10}{\log 7}}$$
$$x \approx \pm 1.0878$$

67.

$$\log(x+90) + \log x = 3$$
$$\log_{10}\left[x(x+90)\right] = 3$$
$$10^3 = x(x+90)$$
$$1{,}000 = x^2 + 90x$$
$$0 = x^2 + 90x - 1{,}000$$
$$0 = (x+100)(x-10)$$
$$x + 100 = 0 \quad \text{or} \quad x - 10 = 0$$
$$x = \cancel{-100} \qquad x = 10$$

Since A negative number does not have a logarithm, the only possible answer is 10.

69.

$$3^{x-6} = 81$$
$$3^{x-6} = 3^4$$
$$x - 6 = 4$$
$$x = 10$$

71.

$$\log\frac{4x+1}{2x+9} = 0$$
$$10^0 = \frac{4x+1}{2x+9}$$
$$1 = \frac{4x+1}{2x+9}$$
$$2x + 9 = 4x + 1$$
$$9 = 2x + 1$$
$$8 = 2x$$
$$4 = x$$

73.

$$15 = 9^{x+2}$$
$$\log 15 = \log 9^{x+2}$$
$$\log 15 = (x+2)\log 9$$
$$\log 15 = x\log 9 + 2\log 9$$
$$\log 15 - 2\log 9 = x\log 9$$
$$\frac{\log 15 - 2\log 9}{\log 9} = x$$
$$x \approx -0.7675$$

75.

$$\log x^2 = 2$$
$$10^2 = x^2$$
$$100 = x^2$$
$$x^2 = 100$$
$$x = \pm\sqrt{100}$$
$$x = \pm 10$$
$$x = 10, -10$$

77.

$$\log(x-6) - \log(x-2) = \log\frac{5}{x}$$
$$\log\left(\frac{x-6}{x-2}\right) = \log\frac{5}{x}$$
$$\frac{x-6}{x-2} = \frac{5}{x}$$
$$x(x-2)\left(\frac{x-6}{x-2}\right) = x(x-2)\left(\frac{5}{x}\right)$$
$$x(x-6) = 5(x-2)$$
$$x^2 - 6x = 5x - 10$$
$$x^2 - 6x - 5x + 10 = 0$$
$$x^2 - 11x + 10 = 0$$
$$(x-1)(x-10) = 0$$
$$x - 1 = 0 \quad \text{or} \quad x - 10 = 0$$
$$x = \cancel{1} \qquad x = 10$$

When $x = 1$, the two numbers on the left side would be negative, which is not possible. The only possible answer for x is 10.

79.

$$\log_3 x = \log_3\left(\frac{1}{x}\right) + 4$$

$$\log_3 x - \log_3\left(\frac{1}{x}\right) = 4$$

$$\log_3 \frac{x}{\frac{1}{x}} = 4$$

$$\log_3 x^2 = 4$$

$$3^4 = x^2$$

$$81 = x^2$$

$$x^2 = 81$$

$$x = \pm\sqrt{81}$$

$$x = \pm 9$$

$$x = 9, \cancel{-9}$$

A negative number does not have a logarithm, so the only answer is $x = 9$.

81.

$$2\log_2 x = 3 + \log_2(x-2)$$

$$2\log_2 x - \log_2(x-2) = 3$$

$$\log_2 x^2 - \log_2(x-2) = 3$$

$$\log_2 \frac{x^2}{x-2} = 3$$

$$2^3 = \frac{x^2}{x-2}$$

$$8 = \frac{x^2}{x-2}$$

$$(x-2)(8) = (x-2)\left(\frac{x^2}{x-2}\right)$$

$$8x - 16 = x^2$$

$$0 = x^2 - 8x + 16$$

$$0 = (x-4)(x-4)$$

$$x - 4 = 0 \quad \text{or} \quad x - 4 = 0$$

$$x = 4 \qquad\qquad x = 4$$

83.

$$\log(7y+1) = 2\log(y+3) - \log 2$$

$$\log(7y+1) = \log(y+3)^2 - \log 2$$

$$\log(7y+1) = \log\frac{(y+3)^2}{2}$$

$$7y + 1 = \frac{(y+3)^2}{2}$$

$$2(7y+1) = 2\left(\frac{(y+3)^2}{2}\right)$$

$$14y + 2 = (y+3)^2$$

$$14y + 2 = y^2 + 6y + 9$$

$$0 = y^2 + 6y + 9 - 14y - 2$$

$$0 = y^2 - 8y + 7$$

$$0 = (y-1)(y-7)$$

$$y - 1 = 0 \quad \text{or} \quad y - 7 = 0$$

$$y = 1 \qquad\qquad y = 7$$

85.

$$e^{3x} = 9$$

$$\ln e^{3x} = \ln 9$$

$$3x \ln e = \ln 9$$

$$3x \cdot 1 = \ln 9$$

$$3x = \ln 9$$

$$x = \frac{\ln 9}{3}$$

$$x \approx 0.7324$$

87.

$$\frac{\log(5x+6)}{2} = \log x$$

$$2\left(\frac{\log(5x+6)}{2}\right) = 2(\log x)$$

$$\log(5x+6) = \log x^2$$

$$5x + 6 = x^2$$

$$0 = x^2 - 5x - 6$$

$$0 = (x-6)(x+1)$$

$$x - 6 = 0 \quad \text{or} \quad x + 1 = 0$$

$$x = 6 \qquad\qquad x = \cancel{-1}$$

A negative number does not have a logarithm, so the only answer is $x = 6$.

89. a.

$$\log 5x = 1.7$$
$$10^{1.7} = 5x$$
$$x = \frac{10^{1.7}}{5}$$
$$x \approx 10.0237$$

b.

$$\ln 5x = 1.7$$
$$e^{1.7} = 5x$$
$$x = \frac{e^{1.7}}{5}$$
$$x \approx 1.0948$$

91. a.

$$4^{3x-5} = 90$$
$$\log 4^{3x-5} = \log 90$$
$$(3x-5)\log 4 = \log 90$$
$$3x\log 4 - 5\log 4 = \log 90$$
$$3x\log 4 = \log 90 + 5\log 4$$
$$\frac{3x\log 4}{3\log 4} = \frac{\log 90 + 5\log 4}{3\log 4}$$
$$x = \frac{\log 90 + 5\log 4}{3\log 4}$$
$$x \approx 2.7486$$

b.

$$e^{3x-5} = 90$$
$$\ln e^{3x-5} = \ln 90$$
$$(3x-5)\ln e = \ln 90$$
$$3x - 5 = \ln 90$$
$$3x = \ln 90 + 5$$
$$x = \frac{\ln 90 + 5}{3}$$
$$x \approx 3.1666$$

93. a.

$$\log_2(x+5) - \log_2 4x = \log_2 x$$
$$\log_2 \frac{x+5}{4x} = \log_2 x$$
$$\frac{x+5}{4x} = x$$
$$4x\left(\frac{x+5}{4x}\right) = 4x(x)$$
$$x + 5 = 4x^2$$
$$0 = 4x^2 - x - 5$$
$$0 = (4x-5)(x+1)$$
$$4x - 5 = 0 \quad \text{and} \quad x + 1 = 0$$
$$4x = 5 \qquad x = \not{-1}$$
$$x = \frac{5}{4}$$

b.

$$\ln(x+5) - \ln 4x = \ln x$$
$$\ln \frac{x+5}{4x} = \ln x$$
$$\frac{x+5}{4x} = x$$
$$4x\left(\frac{x+5}{4x}\right) = 4x(x)$$
$$x + 5 = 4x^2$$
$$0 = 4x^2 - x - 5$$
$$0 = (4x-5)(x+1)$$
$$4x - 5 = 0 \quad \text{and} \quad x + 1 = 0$$
$$4x = 5 \qquad x = \not{-1}$$
$$x = \frac{5}{4}$$

95. a.

$$\left(\frac{2}{3}\right)^{6-x} = \frac{8}{27}$$

$$\left(\frac{2}{3}\right)^{6-x} = \left(\frac{2}{3}\right)^{3}$$

$$6 - x = 3$$

$$-x = -3$$

$$x = 3$$

b.

$$\left(\frac{2}{3}\right)^{6-x} = \frac{16}{81}$$

$$\left(\frac{2}{3}\right)^{6-x} = \left(\frac{2}{3}\right)^{4}$$

$$6 - x = 4$$

$$-x = -2$$

$$x = 2$$

97.

$$2^{x+1} = 7$$

$$2^{x+1} - 7 = 0$$

$$y = 2^{x+1} - 7$$

The x–intercept is (1.8, 0) so the solution is $x \approx 1.8$.

99.

$$\log x + \log(x - 15) = 2$$

$$\log_{10}\left[x(x-15)\right] = 2$$

$$10^2 = x(x-15)$$

$$100 = x^2 - 15x$$

$$0 = x^2 - 15x - 100$$

$$0 = (x-20)(x+5)$$

$$x - 20 = 0 \quad \text{or} \quad x + 5 = 0$$

$$x = 20 \qquad x = \cancel{-5}$$

A negative number does not have a logarithm, so the only possible answer is 20.

APPLICATIONS

101. HYDROGEN ION CONCENTRATION

$$\text{pH} = -\log\left[\text{H}^+\right]$$

$$13.2 = -\log\left[\text{H}^+\right]$$

$$-13.2 = \log\left[\text{H}^+\right]$$

$$\left[\text{H}^+\right] = 10^{-13.2}$$

$$\left[\text{H}^+\right] \approx 6.3 \times 10^{-14} \text{ grams-ions per liter}$$

103. TRITIUM DECAY

If 25% of the tritium has decayed, then 75%, or 0.75, of the tritium remains. (100% – 25% = 75%)

Let $h = 12.4$ and $A = 0.75A_0$.

$$A = A_0 2^{-t/h}$$

$$0.75A_0 = A_0 2^{-t/12.4}$$

$$\frac{0.75A_0}{A_0} = \frac{A_0 2^{-t/12.4}}{A_0}$$

$$0.75 = 2^{-t/12.4}$$

$$\log 0.75 = \log 2^{-t/12.4}$$

$$\log 0.75 = -\frac{t}{12.4}\log 2$$

$$-12.4(\log 0.75) = -12.4\left(-\frac{t}{12.4}\log 2\right)$$

$$-12.4(\log 0.75) = t\log 2$$

$$\frac{-12.4(\log 0.75)}{\log 2} = \frac{t\log 2}{\log 2}$$

$$\frac{-12.4(\log 0.75)}{\log 2} = t$$

$$t = \frac{-12.4(\log 0.75)}{\log 2}$$

$$t \approx 5.1 \text{ years}$$

105. THORIUM DECAY
If 80% of the thorium has decayed, then 20%, or 0.20, of the thorium remains.
(100% – 80% = 20%)

Let $h = 18.4$ and $A = 0.20A_0$.

$$A = A_0 2^{-t/h}$$
$$0.20A_0 = A_0 2^{-t/18.4}$$
$$\frac{0.20A_0}{A_0} = \frac{A_0 2^{-t/18.4}}{A_0}$$
$$0.20 = 2^{-t/18.4}$$
$$\log 0.20 = \log 2^{-t/18.4}$$
$$\log 0.20 = -\frac{t}{18.4}\log 2$$
$$-18.4(\log 0.20) = -18.4\left(-\frac{t}{18.4}\log 2\right)$$
$$-18.4(\log 0.20) = t\log 2$$
$$\frac{-18.4(\log 0.20)}{\log 2} = \frac{t\log 2}{\log 2}$$
$$\frac{-18.4(\log 0.20)}{\log 2} = t$$
$$t = \frac{-18.4(\log 0.20)}{\log 2}$$
$$t \approx 42.7 \text{ days}$$

107. CARBON–14 DATING
60% of the carbon–14 remains and the half–life is 5,700 years (from Example 10).

Let $h = 5{,}700$ and $A = 0.60A_0$.

$$A = A_0 2^{-t/h}$$
$$0.60A_0 = A_0 2^{-t/5{,}700}$$
$$\frac{0.60A_0}{A_0} = \frac{A_0 2^{-t/5{,}700}}{A_0}$$
$$0.60 = 2^{-t/5{,}700}$$
$$\log 0.60 = \log 2^{-t/5{,}700}$$
$$\log 0.60 = -\frac{t}{5{,}700}\log 2$$
$$-5{,}700(\log 0.60) = -5{,}700\left(-\frac{t}{5{,}700}\log 2\right)$$
$$-5{,}700(\log 0.60) = t\log 2$$
$$\frac{-5{,}700(\log 0.60)}{\log 2} = \frac{t\log 2}{\log 2}$$
$$\frac{-5{,}700(\log 0.60)}{\log 2} = t$$
$$t = \frac{-5{,}700(\log 0.60)}{\log 2}$$
$$t \approx 4{,}200 \text{ years}$$

109. COMPOUND INTEREST
Let $P = 800$, $P_0 = 500$, $r = 0.085$, and $k = 2$.

$$P = P_0\left(1+\frac{r}{k}\right)^{kt}$$
$$800 = 500\left(1+\frac{0.085}{2}\right)^{2t}$$
$$800 = 500(1+0.0425)^{2t}$$
$$800 = 500(1.0425)^{2t}$$
$$\frac{800}{500} = \frac{500(1.0425)^{2t}}{500}$$
$$1.6 = (1.0425)^{2t}$$
$$\log 1.6 = \log 1.0425^{2t}$$
$$\log 1.6 = 2t\log 1.0425$$
$$\frac{\log 1.6}{\log 1.0425} = 2t$$
$$2t = \frac{\log 1.6}{\log 1.0425}$$
$$t = \frac{\log 1.6}{2\log 1.0425}$$
$$t \approx 5.6 \text{ years}$$

111. COMPOUND INTEREST
Let $P = 2{,}100$, $P_0 = 1{,}300$, $r = 0.09$, and $k = 4$.

$$P = P_0\left(1+\frac{r}{k}\right)^{kt}$$
$$2{,}100 = 1{,}300\left(1+\frac{0.09}{4}\right)^{4t}$$
$$2{,}100 = 1{,}300(1+0.0225)^{4t}$$
$$2{,}100 = 1{,}300(1.0225)^{4t}$$
$$\frac{2{,}100}{1{,}300} = \frac{1{,}300(1.0225)^{4t}}{1{,}300}$$
$$1.6154 = (1.0225)^{4t}$$
$$\log 1.6154 = \log(1.0225)^{4t}$$
$$\log 1.6154 = 4t\log 1.0225$$
$$\frac{\log 1.6154}{\log 1.0225} = 4t$$
$$4t = \frac{\log 1.6154}{\log 1.0225}$$
$$t = \frac{\log 1.6154}{4\log 1.0225}$$
$$t \approx 5.4 \text{ years}$$

113. RULE OF SEVENTY
This formula works because $\ln 2 \approx 0.7$.

115. RODENT CONTROL
Let P_0=30,000, P = 2(30,000)=60,000, and t = 5.

$$P = P_0e^{kt}$$
$$60,000 = 30,000e^{k5}$$
$$60,000 = 30,000e^{5k}$$
$$\frac{60,000}{30,000} = \frac{30,000e^{5k}}{30,000}$$
$$2 = e^{5k}$$
$$5k = \ln 2$$
$$k = \frac{\ln 2}{5}$$

Let P_0 = 30,000, P = 1,000,000, and $k = \frac{\ln 2}{5}$.

$$P = P_0e^{kt}$$
$$1,000,000 = 30,000e^{\left(\frac{\ln 2}{5}\right)t}$$
$$\frac{1,000,000}{30,000} = \frac{30,000e^{\left(\frac{\ln 2}{5}\right)t}}{30,000}$$
$$\frac{100}{3} = e^{\left(\frac{\ln 2}{5}\right)t}$$
$$\left(\frac{\ln 2}{5}\right)t = \ln\frac{100}{3}$$
$$\left(\frac{5}{\ln 2}\right)\left(\frac{\ln 2}{5}\right)t = \left(\frac{5}{\ln 2}\right)\left(\ln\frac{100}{3}\right)$$
$$t \approx 25.3 \text{ years}$$

117. BACTERIA CULTURE
Let P_0 = 1, P = 2(1) = 2, and t = 24.

$$P = P_0e^{kt}$$
$$2 = 1e^{k24}$$
$$2 = e^{24k}$$
$$\ln 2 = \ln e^{24k}$$
$$\ln 2 = 24k$$
$$24k = \ln 2$$
$$k = \frac{\ln 2}{24}$$

Let P_0 = 1, t = 36, and $k = \frac{\ln 2}{24}$.

$$P = P_0e^{kt}$$
$$P = 1e^{\left(\frac{\ln 2}{24}\right)(36)}$$
$$P = e^{\frac{3\ln 2}{2}}$$
$$P \approx 2.828 \text{ times larger}$$

119. NEWTON'S LAW OF COOLING
Let t = 3 and T = 90.

$$T = 60 + 40e^{kt}$$
$$90 = 60 + 40e^{3k}$$
$$90 - 60 = 60 + 40e^{3k} - 60$$
$$30 = 40e^{3k}$$
$$\frac{30}{40} = \frac{40e^{3k}}{40}$$
$$0.75 = e^{3k}$$
$$\ln 0.75 = \ln e^{3k}$$
$$\ln 0.75 = 3k$$
$$\frac{\ln 0.75}{3} = \frac{3k}{3}$$
$$\frac{\ln 0.75}{3} = k$$
$$k = \frac{\ln 0.75}{3}$$
$$k = \frac{1}{3}\ln 0.75$$
$$k \approx -0.0959$$

WRITING

121. Answers will vary.

123. Answers will vary.

REVIEW

125. Use the Pythagorean Theorem.

$$a^2 + b^2 = c^2$$
$$\left(\sqrt{7}\right)^2 + b^2 = (12)^2$$
$$7 + b^2 = 144$$
$$b^2 = 137$$
$$b = \sqrt{137} \text{ in.}$$

CHALLENGE PROBLEMS

127.

$$\log_3 x + \log_3(x+2) = 2$$
$$\log_3 x(x+2) = 2$$
$$\log_3(x^2 + 2x) = 2$$
$$3^2 = x^2 + 2x$$
$$9 = x^2 + 2x$$
$$0 = x^2 + 2x - 9$$
$$x = \frac{-b \pm \sqrt{b^2 - 4ac}}{2a}$$
$$x = \frac{-2 \pm \sqrt{2^2 - 4(1)(-9)}}{2(1)}$$
$$x = \frac{-2 \pm \sqrt{4+36}}{2}$$
$$x = \frac{-2 \pm \sqrt{40}}{2}$$
$$x = \frac{-2 \pm 2\sqrt{10}}{2}$$
$$x = -1 + \sqrt{10} \quad \text{or} \quad x = \cancel{-1 - \sqrt{10}}$$
$$x \approx 2.1623$$

129.

$$\frac{\log_2(6x-8)}{\log_2 x} = 2$$
$$(\log_2 x)\left(\frac{\log_2(6x-8)}{\log_2 x}\right) = (\log_2 x)(2)$$
$$\log_2(6x-8) = 2\log_2 x$$
$$\log_2(6x-8) = \log_2 x^2$$
$$6x - 8 = x^2$$
$$0 = x^2 - 6x + 8$$
$$0 = (x-2)(x-4)$$
$$x - 2 = 0 \quad \text{or} \quad x - 4 = 0$$
$$x = 2 \qquad\qquad x = 4$$

CHAPTER 9 REVIEW

SECTION 9.1
Algebra and Composition of Functions

1. $f+g$

$$\begin{aligned}(f+g)(x)&=f(x)+g(x)\\&=(2x)+(x+1)\\&=3x+1\end{aligned}$$

$D:(-\infty,\infty)$

2. $f-g$

$$\begin{aligned}(f-g)(x)&=f(x)-g(x)\\&=(2x)-(x+1)\\&=2x-x-1\\&=x-1\end{aligned}$$

$D:(-\infty,\infty)$

3. $f\cdot g$

$$\begin{aligned}(f\bullet g)(x)&=f(x)\bullet g(x)\\&=(2x)(x+1)\\&=2x^2+2x\end{aligned}$$

$D:(-\infty,\infty)$

4. f/g

$$\begin{aligned}(f/g)(x)&=\frac{f(x)}{g(x)}\\&=\frac{2x}{x+1}\end{aligned}$$

$D:(-\infty,-1)\cup(-1,\infty)$

5.

$$\begin{aligned}(f\circ g)(-1)&=f(g(-1))\\&=f(2(-1)+1)\\&=f(-2+1)\\&=f(-1)\\&=(-1)^2+2\\&=1+2\\&=3\end{aligned}$$

6.

$$\begin{aligned}(g\circ f)(0)&=g(f(0))\\&=g(0^2+2)\\&=g(0+2)\\&=g(2)\\&=2(2)+1\\&=4+1\\&=5\end{aligned}$$

7.

$$\begin{aligned}(f\circ g)(x)&=f(g(x))\\&=f(2x+1)\\&=(2x+1)^2+2\\&=4x^2+2x+2x+1+2\\&=4x^2+4x+3\end{aligned}$$

8.

$$\begin{aligned}(g\circ f)(x)&=g(f(x))\\&=g(x^2+2)\\&=2(x^2+2)+1\\&=2x^2+4+1\\&=2x^2+5\end{aligned}$$

9.

a.

$$\begin{aligned}(f+g)(2)&=f(2)+g(2)\\&=2+(-2)\\&=0\end{aligned}$$

b.

$$\begin{aligned}(f\bullet g)(-4)&=f(-4)\bullet g(-4)\\&=4(-2)\\&=-8\end{aligned}$$

c.

$$\begin{aligned}(f\circ g)(4)&=f(g(4))\\&=f(4)\\&=0\end{aligned}$$

d.

$$\begin{aligned}(g\circ f)(6)&=g(f(6))\\&=g(-2)\\&=-1\end{aligned}$$

10.

$$(C \circ f)(m) = C(f(m))$$
$$= C\left(\frac{m}{8}\right)$$
$$= 2.85\left(\frac{m}{8}\right)$$
$$C(m) = \frac{2.85m}{8}$$

SECTION 9.2
Inverse Functions

11. No. $f(1) = 4$ and $f(-1) = 4$

12. Yes.

13. Yes.

14. No. The output –5 corresponds with more than one input, 0 and 4.

15. Yes.

16. No. It does not pass the horizontal line test.

17.

x	$f^{-1}(x)$
–6	–6
–3	–1
12	7
3	20

18.

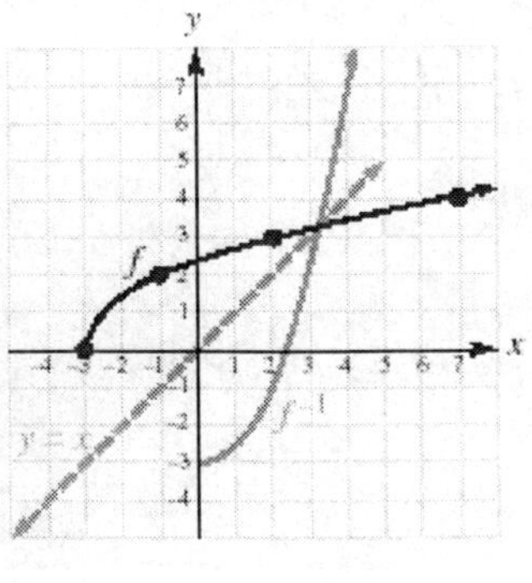

19.

$$f(x) = 6x - 3$$
$$y = 6x - 3$$
$$x = 6y - 3$$
$$x + 3 = 6y$$
$$\frac{x+3}{6} = y$$
$$y = \frac{x+3}{6}$$
$$f^{-1}(x) = \frac{x+3}{6}$$

20.

$$f(x) = \frac{4}{x-1}$$
$$y = \frac{4}{x-1}$$
$$x = \frac{4}{y-1}$$
$$(y-1)(x) = (y-1)\left(\frac{4}{y-1}\right)$$
$$xy - x = 4$$
$$xy = 4 + x$$
$$y = \frac{4+x}{x}$$
$$y = \frac{4}{x} + \frac{x}{x}$$
$$y = \frac{4}{x} + 1$$
$$f^{-1}(x) = \frac{4}{x} + 1$$

21.

$$f(x) = (x+2)^3$$
$$y = (x+2)^3$$
$$x = (y+2)^3$$
$$\sqrt[3]{x} = \sqrt[3]{(y+2)^3}$$
$$\sqrt[3]{x} = y + 2$$
$$\sqrt[3]{x} - 2 = y$$
$$y = \sqrt[3]{x} - 2$$
$$f^{-1}(x) = \sqrt[3]{x} - 2$$

22.

$$f(x) = \frac{x}{6} - \frac{1}{6}$$
$$y = \frac{x}{6} - \frac{1}{6}$$
$$x = \frac{y}{6} - \frac{1}{6}$$
$$x + \frac{1}{6} = \frac{y}{6}$$
$$6\left(x + \frac{1}{6}\right) = 6\left(\frac{y}{6}\right)$$
$$6x + 1 = y$$
$$y = 6x + 1$$
$$f^{-1}(x) = 6x + 1$$

23.

$$f(x) = \sqrt[3]{x-1}$$
$$y = \sqrt[3]{x-1}$$
$$x = \sqrt[3]{y-1}$$
$$(x)^3 = \left(\sqrt[3]{y-1}\right)^3$$
$$x^3 = y - 1$$
$$x^3 + 1 = y$$
$$y = x^3 + 1$$
$$f^{-1}(x) = x^3 + 1$$

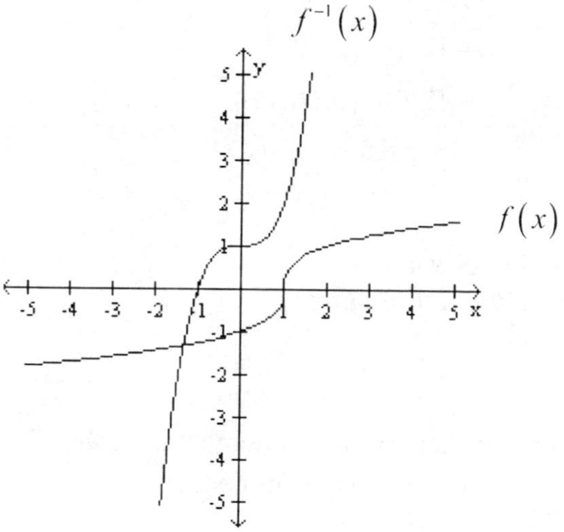

24.

$$(f \circ f^{-1})(x) = f(f^{-1}(x))$$
$$= f\left(-\frac{x-5}{4}\right)$$
$$= 5 - 4\left(-\frac{x-5}{4}\right)$$
$$= 5 + x - 5$$
$$= x$$
$$(f^{-1} \circ f)(x) = f^{-1}(f(x))$$
$$= f^{-1}(5-4x)$$
$$= -\frac{(5-4x)-5}{4}$$
$$= -\frac{5-4x-5}{4}$$
$$= -\frac{-4x}{4}$$
$$= x$$

SECTION 9.3
Exponential Functions

25. a. $n(x) = 2^x$; $s(t) = 1.08^t$
    b. $0.9(1.42)^{14} \approx 121.9774$

26. a. exponential decay
    b. exponential growth

27. $f(x) = 3^x$
    D: $(-\infty, \infty)$, R: $(0, \infty)$

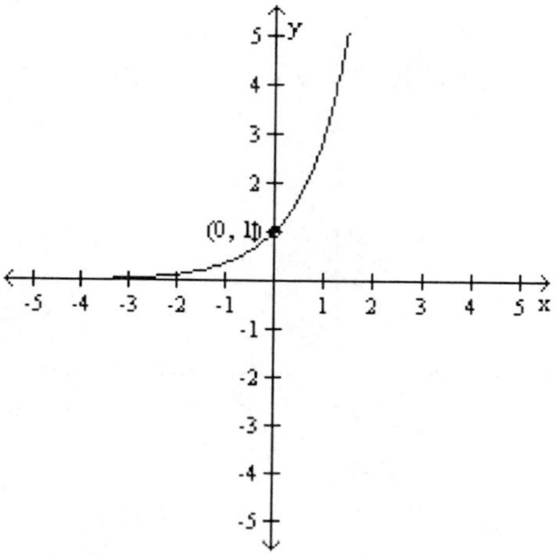

28. $f(x)=\left(\frac{1}{3}\right)^x$

D: $(-\infty, \infty)$, R: $(0, \infty)$

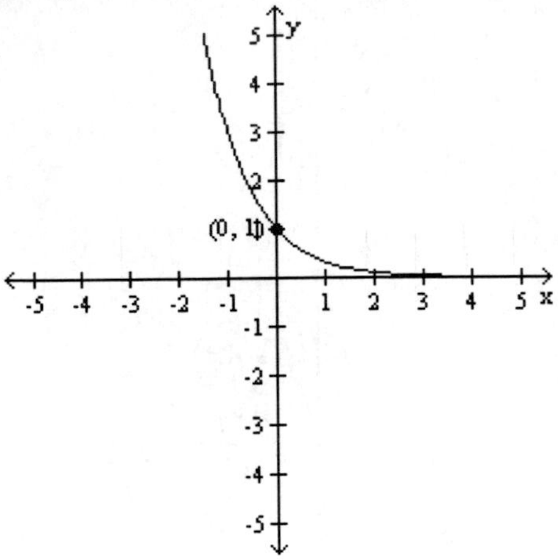

29. $f(x)=\left(\frac{1}{2}\right)^x-2$

D: $(-\infty, \infty)$, R: $(-2, \infty)$

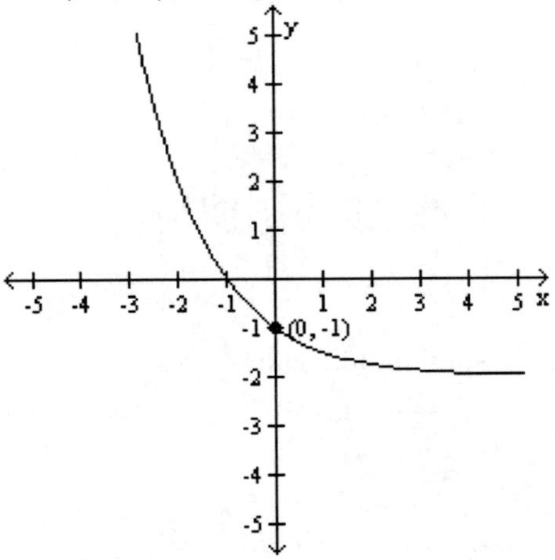

30. $f(x)=3^{x-1}$

D: $(-\infty, \infty)$, R: $(0, \infty)$

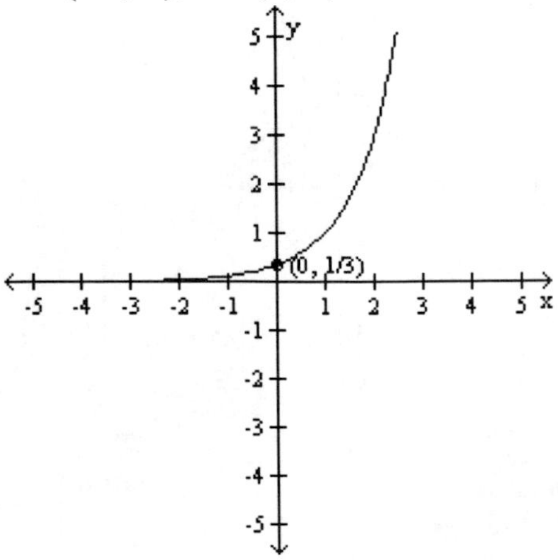

31. the x–axis ($y = 0$)

32. a. Let $t = 0$.

$$c(t)=128{,}000(1.08)^t$$
$$c(0)=128{,}000(1.08)^0$$
$$=128{,}000(1)$$
$$=128{,}000 \text{ tons}$$

b. Let $t = 100$.

$$c(t)=128{,}000(1.08)^t$$
$$c(0)=128{,}000(1.08)^{100}$$
$$\approx 128{,}000(2199.76)$$
$$\approx 281{,}569{,}441 \text{ tons}$$

33. Let $P = 10{,}500$, $r = 0.09$, $k = 4$, and $t = 60$

$$A=10{,}500\left(1+\frac{0.09}{4}\right)^{4(60)}$$
$$=10{,}500(1+0.0225)^{240}$$
$$=10{,}500(1.0225)^{240}$$
$$=10{,}500(208.543186)$$
$$=\$2{,}189{,}703.45$$

34. Let $t = 5$.

$$V(t)=12{,}000\left(10^{-0.155t}\right)$$
$$V(5)=12{,}000\left(10^{-0.155(5)}\right)$$
$$=12{,}000\left(10^{-0.775}\right)$$
$$=12{,}000(0.16788)$$
$$\approx \$2{,}015$$

SECTION 9.4
Logarithmic Functions

35. D: $(0, \infty)$; R: $(-\infty, \infty)$

36. Since there is no real number such that $10^? = 0$, log 0 is undefined.

37. $4^3 = 64$

38. $\log_7 \frac{1}{7} = -1$

39.

$$\log_3 9 = x$$
$$3^x = 9$$
$$3^x = 3^2$$
$$x = 2$$

40.

$$\log_9 \frac{1}{81} = x$$
$$9^x = \frac{1}{81}$$
$$9^x = \frac{1}{9^2}$$
$$9^x = 9^{-2}$$
$$x = -2$$

41.

$$\log_{1/2} 1 = x$$
$$\left(\frac{1}{2}\right)^x = 1$$
$$x = 0$$

42.

$$\log_5(-25) = x$$
$$5^x = -25$$

not possible

43.

$$\log_6 \sqrt{6} = x$$
$$6^x = \sqrt{6}$$
$$6^x = 6^{1/2}$$
$$x = \frac{1}{2}$$

44.

$$\log 1{,}000 = x$$
$$\log_{10} 1{,}000 = x$$
$$10^x = 1{,}000$$
$$10^x = 10^3$$
$$x = 3$$

45.

$$\log_2 x = 5$$
$$2^5 = x$$
$$32 = x$$

46.

$$\log_3 x = -4$$
$$3^{-4} = x$$
$$\frac{1}{3^4} = x$$
$$\frac{1}{81} = x$$

47.

$$\log_x 16 = 2$$
$$x^2 = 16$$
$$\sqrt{x^2} = \sqrt{16}$$
$$x = 4$$

48.

$$\log_x \frac{1}{100} = -2$$
$$x^{-2} = \frac{1}{100}$$
$$\left(x^{-2}\right)^{-1/2} = \left(\frac{1}{100}\right)^{-1/2}$$
$$x = (100)^{1/2}$$
$$x = \sqrt{100}$$
$$x = 10$$

49.

$$\log_9 3 = x$$
$$9^x = 3$$
$$\left(3^2\right)^x = 3^1$$
$$3^{2x} = 3^1$$
$$2x = 1$$
$$x = \frac{1}{2}$$

50.

$$\log_{27} 3 = x$$
$$27^x = 3$$
$$\left(3^3\right)^x = 3$$
$$(3)^{3x} = 3^1$$
$$3x = 1$$
$$x = \frac{1}{3}$$

51.

$\log 4.51 = x$

$0.6542 = x$

52.

$\log x = 1.43$

$10^{1.43} = x$

$x = 26.9153$

53. $f(x) = \log_4 x$ and $g(x) = 4^x$

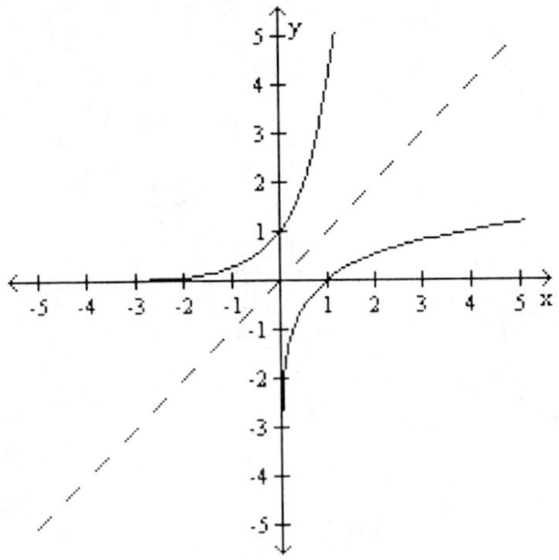

54. $f(x) = \log_{1/3} x$ and $g(x) = \left(\frac{1}{3}\right)^x$

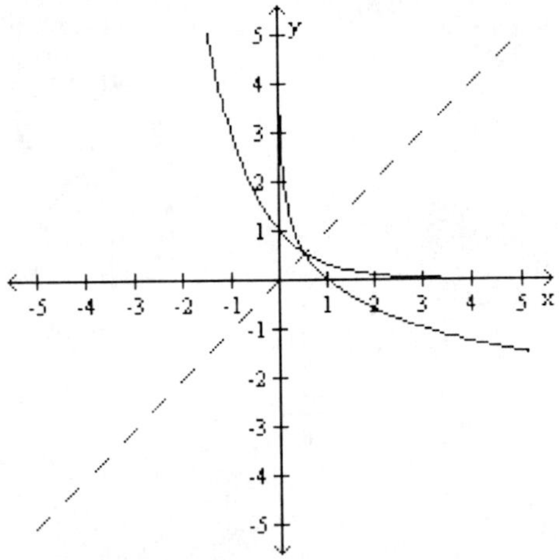

55. $f(x) = \log(x-2)$

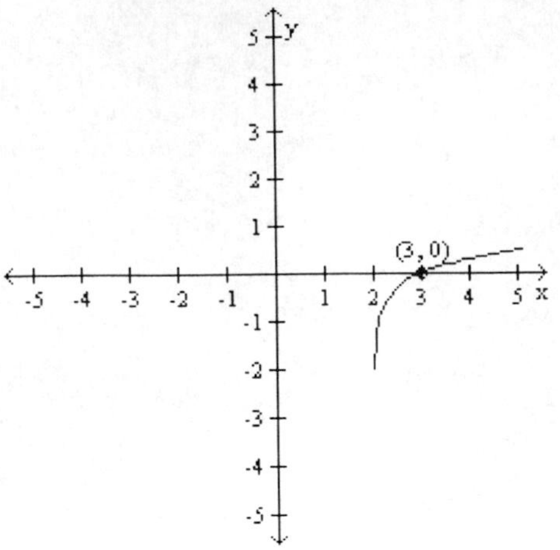

56. $f(x) = 3 + \log x$

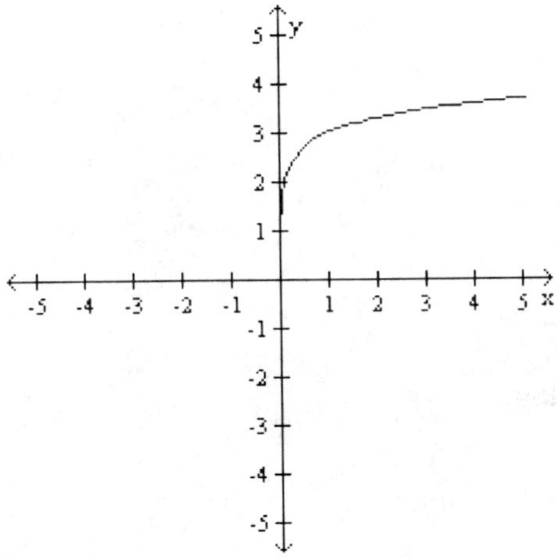

57. Let $E_O = 18$ and $E_I = 0.04$.

$$\begin{aligned} \text{db gain} &= 20\log\frac{E_O}{E_I} \\ &= 20\log\frac{18}{0.04} \\ &= 20\log 450 \\ &\approx 20(2.65321) \\ &\approx 53 \end{aligned}$$

58. Let $P = 0.3$ and $A = 7{,}500$.

$$\begin{aligned} R &= \log\frac{A}{P} \\ &= \log\frac{7{,}500}{0.3} \\ &= \log 25{,}000 \\ &\approx 4.4 \end{aligned}$$

59.a. Let $n = 1$.

$$h(1) = 52 + 25\log 1$$
$$= 52 + 25(0)$$
$$= 52 + 0$$
$$= 52 \text{ cm}$$

b. Let $n = 8$.

$$h(8) = 52 + 25\log 8$$
$$= 52 + 25(0.9031)$$
$$\approx 74.6 \text{ cm}$$

60. Let $A = 13$.

$$P(13) = 61.8 + 34.9\log(13-4)$$
$$= 61.8 + 34.9\log 9$$
$$\approx 95\%$$

SECTION 9.5
Base-e Exponential and Logarithmic Functions

61. a. $e \approx 2.72$

b. ln 15 ≈ 2.7081 means $e^{\boxed{2.7081}} \approx \boxed{15}$

62. 16.6704

63. $f(x) = e^x + 1$

D: $(-\infty, \infty)$, R: $(1, \infty)$

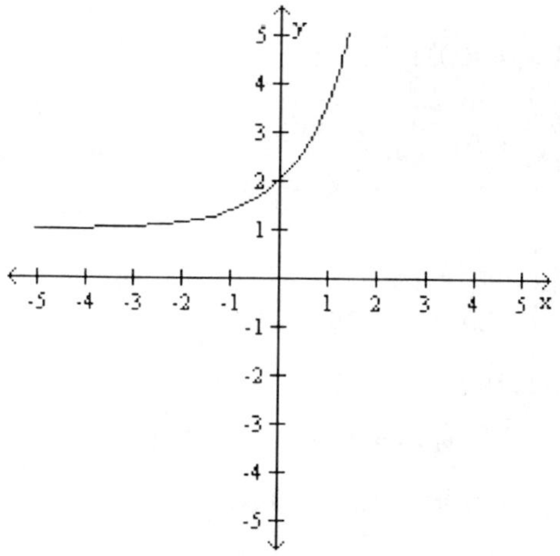

64. $f(x) = e^{x-3}$

D: $(-\infty, \infty)$, R: $(0, \infty)$

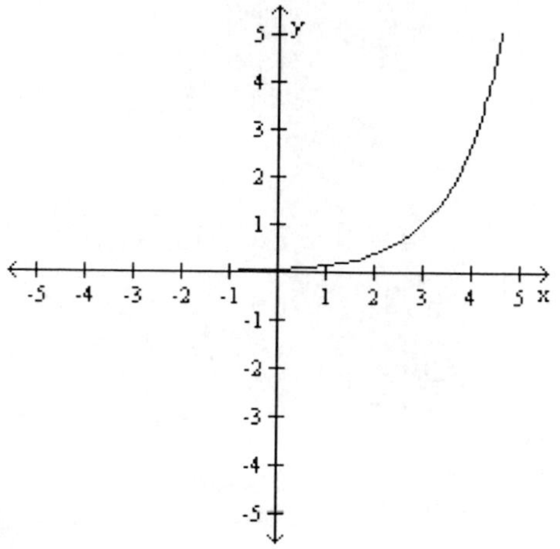

65. Let $P_0 = 10{,}500$, $k = 0.09$, and $t = 60$.

$$P = P_0 e^{kt}$$
$$= 10{,}500e^{0.09(60)}$$
$$= 10{,}500e^{5.4}$$
$$= 2{,}324{,}767.37$$

66. Let $P_0 = 142.5$ billion, $k = 0.066$, and $t = 6$.

$$P = P_0 e^{kt}$$
$$= 142.5e^{0.066(6)}$$
$$= 142.5e^{0.396}$$
$$\approx \$211.7 \text{ billion}$$

67. For 1980, let $t = 0$.

$$r(t) = 13.9e^{-0.035t}$$
$$r(0) = 13.9e^{-0.035(0)}$$
$$= 13.9e^{0}$$
$$= 13.9\%$$

For 1985, let $t = 5$.

$$r(t) = 13.9e^{-0.035t}$$
$$r(5) = 13.9e^{-0.035(5)}$$
$$= 13.9e^{-0.175}$$
$$= 11.67\%$$

For 1990, let $t = 10$.

$$r(t) = 13.9e^{-0.035t}$$
$$r(10) = 13.9e^{-0.035(10)}$$
$$= 13.9e^{-0.35}$$
$$= 9.8\%$$

68. The exponent on the base e is negative.

69.

$\ln e = 1$

70.

$\ln e^2 = 2$

71.

$$\ln \frac{1}{e^5} = \ln e^{-5}$$
$$= -5$$

72.

$$\ln \sqrt{e} = \ln e^{1/2}$$
$$= \frac{1}{2}$$

73.

$\ln(-e) =$ undefined

74.

$\ln 0 =$ undefined

75.

$\ln 1 = 0$

76.

$\ln e^{-7} = -7$

77.

$\ln 452 = 6.1137$

78.

$\ln 0.85 = -0.1625$

79.

$$\ln x = 2.336$$
$$e^{\ln x} = e^{2.336}$$
$$x = 10.3398$$

80.

$$\ln x = -8.8$$
$$e^{\ln x} = e^{-8.8}$$
$$x = 0.0002$$

81. They have different bases:

$\log x = \log_{10} x$ and $\ln x = \log_e x$

82. $f^{-1}(x) = e^x$

83. $f(x) = 1 + \ln x$

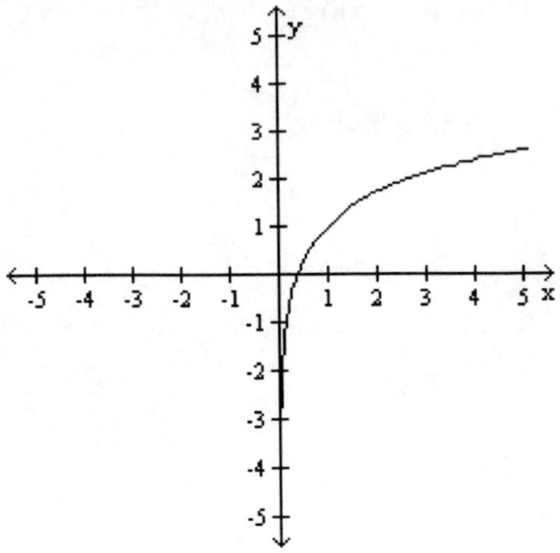

84. $f(x) = \ln(x+1)$

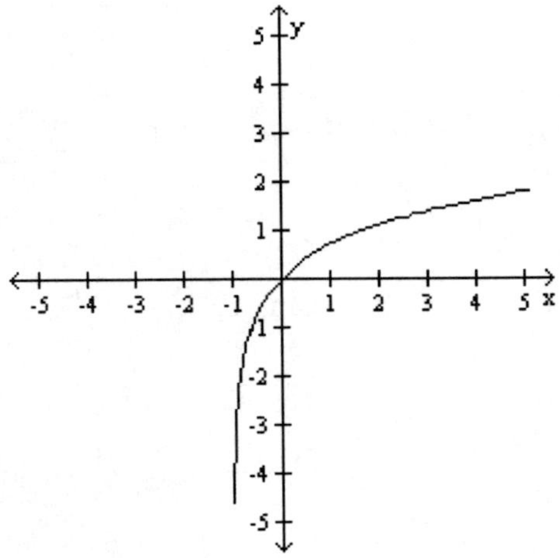

85. Let $r = 0.01118$.

$$t = \frac{\ln 2}{0.01118}$$
$$\approx 62$$

86. Let $a = 19$.

$$H(a) = 13 + 20.03 \ln a$$
$$H(19) = 13 + 20.03 \ln 19$$
$$\approx 13 + 20.03(2.94444)$$
$$\approx 72 \text{ in. or } 6 \text{ ft.}$$

SECTION 9.6
Properties of Logarithms

87.

$\log_2 1 = 0$

88.

$\log_9 9 = 1$

89.

$$\log 10^3 = 3$$

90.

$$7^{\log_7 4} = 4$$

91.

$$\begin{aligned}\log_3 27x &= \log_3 27 + \log_3 x \\ &= 3 + \log_3 x\end{aligned}$$

92.

$$\begin{aligned}\log\frac{100}{x} &= \log 100 - \log x \\ &= 2 - \log x\end{aligned}$$

93.

$$\begin{aligned}\log_5 \sqrt{27} &= \log_5 27^{1/2} \\ &= \frac{1}{2}\log_5 27\end{aligned}$$

94.

$$\begin{aligned}\log_b 10ab &= \log_b 10 + \log_b a + \log_b b \\ &= \log_b 10 + \log_b a + 1\end{aligned}$$

95.

$$\begin{aligned}\log_b \frac{x^2y^3}{z} &= \log_b x^2 + \log_b y^3 - \log_b z \\ &= 2\log_b x + 3\log_b y - \log_b z\end{aligned}$$

96.

$$\begin{aligned}\ln\sqrt{\frac{x}{yz^2}} &= \ln\left(\frac{x}{yz^2}\right)^{1/2} \\ &= \frac{1}{2}\ln\frac{x}{yz^2} \\ &= \frac{1}{2}\left(\ln x - \ln y - \ln z^2\right) \\ &= \frac{1}{2}\left(\ln x - \ln y - 2\ln z\right)\end{aligned}$$

97.

$$\begin{aligned}3\log_2 x - 5\log_2 y + 7\log_2 z &= \log_2 x^3 - \log_2 y^5 + \log_2 z^7 \\ &= \log_2 \frac{x^3z^7}{y^5}\end{aligned}$$

98.

$$\begin{aligned}&-3\log_b y - 7\log_b z + \frac{1}{2}\log_b (x+2) \\ &= -\log_b y^3 - \log_b z^7 + \log_b (x+2)^{1/2} \\ &= -\log_b y^3 - \log_b z^7 + \log_b \sqrt{x+2} \\ &= \log \frac{\sqrt{x+2}}{y^3z^7}\end{aligned}$$

99.

$$\begin{aligned}&\log_b (a^2 - 25) - \log_b (a+5) \\ &= \log_b \frac{a^2-25}{a+5} \\ &= \log_b \frac{(a+5)(a-5)}{a+5} \\ &= \log_b (a-5)\end{aligned}$$

100.

$$\begin{aligned}3\log_8 x + 4\log_8 x &= \log_8 x^3 + \log_8 x^4 \\ &= \log_8 \left(x^3x^4\right) \\ &= \log_8 x^7\end{aligned}$$

101.

$$\begin{aligned}\log_b 40 &= \log_b (5\bullet 8) \\ &= \log_b 5 + \log_b 8 \\ &= 1.1609 + 1.5000 \\ &= 2.6609\end{aligned}$$

102.

$$\begin{aligned}\log_b 64 &= \log_b (8\bullet 8) \\ &= \log_b 8 + \log_b 8 \\ &= 1.5000 + 1.5000 \\ &= 3.0000\end{aligned}$$

103.

$$\begin{aligned}\log_5 17 &= \frac{\log 17}{\log 5} \\ &\approx \frac{1.23045}{0.69897} \\ &\approx 1.7604\end{aligned}$$

104.

$$\text{pH} = -\log\left[\text{H}^+\right]$$
$$\text{pH} = -\log\left[7.9\times 10^{-4}\right]$$
$$\text{pH} = -(-3.1)$$
$$\text{pH} = 3.1$$

SECTION 9.7
Exponential and Logarithmic Equations

105.

$$5^{x+6} = 25$$
$$5^{x+6} = 5^2$$
$$x + 6 = 2$$
$$x = -4$$

106.

$$2^{x^2+4x} = \frac{1}{8}$$
$$2^{x^2+4x} = 2^{-3}$$
$$x^2 + 4x = -3$$
$$x^2 + 4x + 3 = 0$$
$$(x+1)(x+3) = 0$$
$$x + 1 = 0 \quad \text{or} \quad x + 3 = 0$$
$$x = -1 \qquad x = -3$$

107.

$$3^x = 7$$
$$\log 3^x = \log 7$$
$$x \log 3 = \log 7$$
$$x = \frac{\log 7}{\log 3}$$
$$x \approx 1.7712$$

108.

$$2^x = 3^{x-4}$$
$$\log 2^x = \log 3^{x-4}$$
$$x \log 2 = (x-4)\log 3$$
$$x \log 2 = x \log 3 - 4\log 3$$
$$x \log 2 - x \log 3 = -4\log 3$$
$$x(\log 2 - \log 3) = -4\log 3$$
$$x = \frac{-4\log 3}{\log 2 - \log 3}$$
$$x \approx 10.8380$$

109.

$$e^x = 7$$
$$\ln e^x = \ln 7$$
$$x = \ln 7$$
$$x = 1.9459$$

110.

$$e^{-0.4t} = 25$$
$$\ln e^{-0.4t} = \ln 25$$
$$-0.4t = \ln 25$$
$$t = \frac{\ln 25}{-0.4}$$
$$t = -8.0472$$

111.

$$\left(\frac{2}{5}\right)^{3x-4} = \frac{8}{125}$$
$$\left(\frac{2}{5}\right)^{3x-4} = \left(\frac{2}{5}\right)^3$$
$$3x - 4 = 3$$
$$3x = 7$$
$$x = \frac{7}{3}$$

112.

$$9^{x^2} = 33$$
$$\log 9^{x^2} = \log 33$$
$$x^2 \log 9 = \log 33$$
$$x^2 = \frac{\log 33}{\log 9}$$
$$x = \sqrt{\frac{\log 33}{\log 9}}$$
$$x \approx 1.2615$$

113.

$$\log(x-4) = 2$$
$$10^2 = x - 4$$
$$100 = x - 4$$
$$104 = x$$

114.

$$\ln(2x-3)=\ln 15$$
$$e^{\ln(2x-3)}=e^{\ln 15}$$
$$2x-3=15$$
$$2x=18$$
$$x=9$$

115.

$$\log x+\log(29-x)=2$$
$$\log[x(29-x)]=2$$
$$10^2=x(29-x)$$
$$100=29x-x^2$$
$$x^2-29x+100=0$$
$$(x-25)(x-4)=0$$
$$x-25=0 \quad \text{or} \quad x-4=0$$
$$x=25 \qquad x=4$$

116.

$$\log_2 x+\log_2(x-2)=3$$
$$\log_2 x(x-2)=3$$
$$2^3=x(x-2)$$
$$8=x^2-2x$$
$$0=x^2-2x-8$$
$$0=(x-4)(x+2)$$
$$x-4=0 \quad \text{or} \quad x+2=0$$
$$x=4 \qquad x=\cancel{-2}$$

117.

$$\frac{\log(7x-12)}{\log x}=2$$
$$\log x\left(\frac{\log(7x-12)}{\log x}\right)=2\log x$$
$$\log(7x-12)=\log x^2$$
$$7x-12=x^2$$
$$0=x^2-7x+12$$
$$0=(x-3)(x-4)$$
$$x-3=0 \quad \text{or} \quad x-4=0$$
$$x=3 \qquad x=4$$

118.

$$\log_2(x+2)+\log_2(x-1)=2$$
$$\log_2[(x+2)(x-1)]=2$$
$$2^2=(x+2)(x-1)$$
$$4=x^2-x+2x-2$$
$$4=x^2+x-2$$
$$0=x^2+x-6$$
$$0=(x+3)(x-2)$$
$$x+3=0 \quad \text{or} \quad x-2=0$$
$$x=\cancel{-3} \qquad x=2$$

119.

$$\log x+\log(x-5)=\log 6$$
$$\log x(x-5)=\log 6$$
$$x(x-5)=6$$
$$x^2-5x=6$$
$$x^2-5x-6=0$$
$$(x-6)(x+1)=0$$
$$x-6=0 \quad \text{or} \quad x+1=0$$
$$x=6 \qquad x=\cancel{-1}$$

120.

$$\log 3-\log(x-1)=-1$$
$$\log\frac{3}{x-1}=-1$$
$$10^{-1}=\frac{3}{x-1}$$
$$\frac{1}{10}=\frac{3}{x-1}$$
$$10(x-1)\left(\frac{1}{10}\right)=10(x-1)\left(\frac{3}{x-1}\right)$$
$$x-1=30$$
$$x=31$$

121.

$$\frac{\log 8}{\log 15}\neq\log 8-\log 15$$
$$0.7679\neq-0.2730$$

122. Let $h = 5{,}700$ and $A = \frac{2}{3}A_0$.

$$A = A_0 2^{-t/h}$$
$$\frac{2}{3}A_0 = A_0 2^{-t/5{,}700}$$
$$\frac{2A_0}{3A_0} = \frac{A_0 2^{-t/5{,}700}}{A_0}$$
$$\frac{2}{3} = 2^{-t/5{,}700}$$
$$\log\frac{2}{3} = \log 2^{-t/5{,}700}$$
$$\log\frac{2}{3} = -\frac{t}{5{,}700}\log 2$$
$$-5{,}700\left(\log\frac{2}{3}\right) = -5{,}700\left(-\frac{t}{5{,}700}\log 2\right)$$
$$-5{,}700\left(\log\frac{2}{3}\right) = t\log 2$$
$$\frac{-5{,}700\left(\log\frac{2}{3}\right)}{\log 2} = \frac{t\log 2}{\log 2}$$
$$\frac{-5{,}700\left(\log\frac{2}{3}\right)}{\log 2} = t$$
$$t = \frac{-5{,}700\left(\log\frac{2}{3}\right)}{\log 2}$$
$$t \approx 3{,}300 \text{ years}$$

123. Let $P_0 = 800$, $P = 3(800) = 2{,}400$, and $t = 14$.

$$P = P_0 e^{kt}$$
$$2{,}400 = 800e^{k(14)}$$
$$3 = e^{14k}$$
$$\ln 3 = \ln e^{14k}$$
$$\ln 3 = 14k$$
$$14k = \ln 3$$
$$k = \frac{\ln 3}{14}$$

Let $P_0 = 800$, $P = 1{,}000{,}000$, and $k = \frac{\ln 3}{14}$.

$$P = P_0 e^{kt}$$
$$1{,}000{,}000 = 800e^{\left(\frac{\ln 3}{14}\right)t}$$
$$1{,}250 = e^{\left(\frac{\ln 3}{14}\right)t}$$
$$\ln 1{,}250 = \ln e^{\left(\frac{\ln 3}{14}\right)t}$$
$$\ln 1{,}250 = \left(\frac{\ln 3}{14}\right)t$$
$$\left(\frac{14}{\ln 3}\right)(\ln 1{,}250) = \left(\frac{14}{\ln 3}\right)\left(\frac{\ln 3}{14}\right)t$$
$$\frac{14\ln 1{,}250}{\ln 3} = t$$
$$t = \frac{14\ln 1{,}250}{\ln 3}$$
$$t \approx 91 \text{ days}$$

124. $x = 2$ and $x = 5$

For $x = 2$:

$$\log x = 1 - \log(7 - x)$$
$$\log 2 \stackrel{?}{=} 1 - \log(7 - 2)$$
$$\log 2 \stackrel{?}{=} 1 - \log 5$$
$$0.3010 = 0.3010$$

For $x = 5$:

$$\log x = 1 - \log(7 - x)$$
$$\log 5 \stackrel{?}{=} 1 - \log(7 - 5)$$
$$\log 5 \stackrel{?}{=} 1 - \log 2$$
$$0.6990 = 0.6990$$

CHAPTER 9 TEST

1. a. A **composite** function is denoted by $f \circ g$.
   b. $f(x) = e^x$ is the **natural** exponential function.
   c. In **continuous** compound interest, the number of compoundings is infinitely large.
   d. The functions $f(x) = \log_{10} x$ and $f(x) = 10^x$ are **inverse** functions.
   e. $f(x) = \log_4 x$ is **logarithmic** a function.

2. a. *f* composed with *g* of *x*
   b. *g* of *f* of eight
   c. *f* inverse of *x*

3. $f+g$

$$
\begin{aligned}
(f+g)(x) &= f(x)+g(x) \\
&= x+9+4x^2-3x+2 \\
&= 4x^2-2x+11
\end{aligned}
$$

$D:(-\infty,\infty)$

4. g/f

$$
\begin{aligned}
(g/f)(x) &= \frac{g(x)}{f(x)} \\
&= \frac{4x^2-3x+2}{x+9}
\end{aligned}
$$

$D:(-\infty,-9)\cup(-9,\infty)$

5.

$$
\begin{aligned}
(g\circ f)(-3) &= g(f(-3)) \\
&= g\left(2(-3)^2+3\right) \\
&= g(2(9)+3) \\
&= g(18+3) \\
&= g(21) \\
&= 4(21)-8 \\
&= 84-8 \\
&= 76
\end{aligned}
$$

6.

$$
\begin{aligned}
(f\circ g)(x) &= f(g(x)) \\
&= f(4x-8) \\
&= 2(4x-8)^2+3 \\
&= 2\left(16x^2-32x-32x+64\right)+3 \\
&= 2\left(16x^2-64x+64\right)+3 \\
&= 32x^2-128x+128+3 \\
&= 32x^2-128x+131
\end{aligned}
$$

7. a.

$$
\begin{aligned}
(f\bullet g)(9) &= f(9)\bullet g(9) \\
&= -1\bullet 16 \\
&= -16
\end{aligned}
$$

b.

$$
\begin{aligned}
(f\circ g)(-3) &= f(g(-3)) \\
&= f(10) \\
&= 17
\end{aligned}
$$

8.

a. $$
\begin{aligned}
(g/f)(-4) &= \frac{g(-4)}{f(-4)} \\
&= \frac{0}{4} \\
&= 0
\end{aligned}
$$

b. $$
\begin{aligned}
(f\circ g)(1) &= f(g(1)) \\
&= f(-2) \\
&= 3
\end{aligned}
$$

c. $$
\begin{aligned}
(f+g)(2) &= f(2)+g(2) \\
&= 4+(-3) \\
&= 1
\end{aligned}
$$

d. $$
\begin{aligned}
(f\cdot g)(0) &= (f)(0)\cdot(g)(0) \\
&= 2(-1) \\
&= -2
\end{aligned}
$$

e. $$
\begin{aligned}
(g-f)(1) &= (g)(1)-(f)(1) \\
&= -2-3 \\
&= -5
\end{aligned}
$$

9. a. No
   b. Yes
   c. Yes
   d. No

10.

$$f(x) = -\frac{1}{3}x$$
$$y = -\frac{1}{3}x$$
$$x = -\frac{1}{3}y$$
$$3(x) = 3\left(-\frac{1}{3}y\right)$$
$$3x = -y$$
$$-3x = y$$
$$y = -3x$$
$$f^{-1}(x) = -3x$$

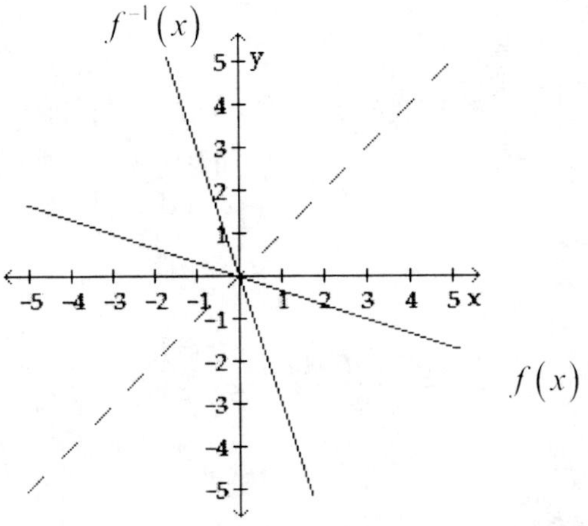

11. one-to-one function

$$f(x) = \frac{1}{3}x + 2$$
$$y = \frac{1}{3}x + 2$$
$$x = \frac{1}{3}y + 2$$
$$3(x) = 3\left(\frac{1}{3}y + 2\right)$$
$$3x = y + 6$$
$$3x - 6 = y$$
$$y = 3x - 6$$
$$f^{-1}(x) = 3x - 6$$

12.

$$f(x) = (x-15)^3$$
$$y = (x-15)^3$$
$$x = (y-15)^3$$
$$\sqrt[3]{x} = \sqrt[3]{(y-15)^3}$$
$$\sqrt[3]{x} = y - 15$$
$$\sqrt[3]{x} + 15 = y$$
$$y = \sqrt[3]{x} + 15$$
$$f^{-1}(x) = \sqrt[3]{x} + 15$$

13.

$$(f \circ f^{-1})(x) = f(f^{-1}(x))$$
$$= f\left(\frac{x-4}{4}\right)$$
$$= 4\left(\frac{x-4}{4}\right) + 4$$
$$= x - 4 + 4$$
$$= x$$
$$(f^{-1} \circ f)(x) = f^{-1}(f(x))$$
$$= f^{-1}(4x+4)$$
$$= \frac{4x+4-4}{4}$$
$$= \frac{4x}{4}$$
$$= x$$

14. a. Yes.
    b. Yes.
    c. $f^{-1}(260) = 80$; when the temperature of the tire tread is 260°, the vehicle is traveling 80 mph.

15. $f(x) = 2^x + 1$

D: $(-\infty, \infty)$, R: $(1, \infty)$

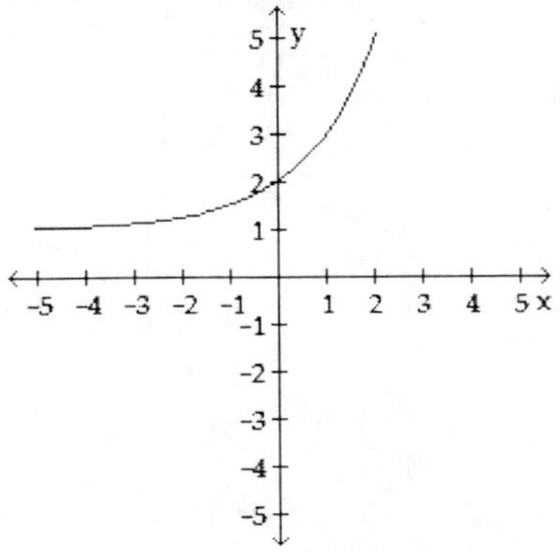

16. $f(x) = 3^{-x}$

D: $(-\infty, \infty)$, R: $(0, \infty)$

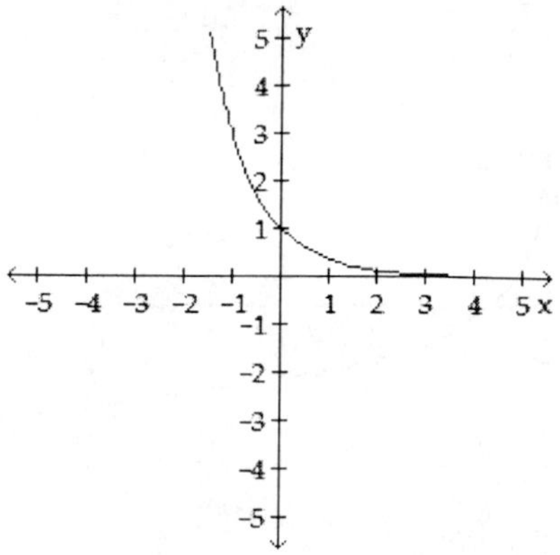

17. Let $t = 6$ and $A_0 = 3$.

$$A = A_0(2)^{-t}$$
$$= 3(2)^{-6}$$
$$= 3(0.015625)$$
$$= 0.046875 \text{ g}$$

18. Let $P = 1{,}000$, $r = 0.06$, $k = 2$, and $t = 1$.

$$A = P\left(1 + \frac{r}{k}\right)^{kt}$$
$$= 1{,}000\left(1 + \frac{0.06}{2}\right)^{2(1)}$$
$$= 1{,}000(1 + 0.03)^2$$
$$= 1{,}000(1.03)^2$$
$$= 1{,}000(1.0609)$$
$$= \$1{,}060.90$$

19. a. $f(x) = e^x$

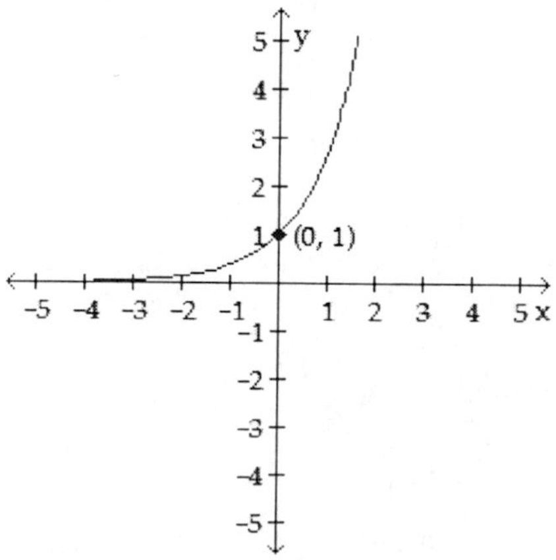

b. D: $(-\infty, \infty)$; R: $(0, \infty)$

c. $f^{-1}(x) = \ln x$

20. Let $P_0 = 1{,}173{,}108{,}018$, $r = 0.01376$, and $t = 10$.

$$P = P_0 e^{kt}$$
$$= (1{,}173{,}108{,}018)e^{0.01376(10)}$$
$$= (1{,}173{,}108{,}018)e^{0.1376}$$
$$\approx 1{,}346{,}161{,}000$$

21. $H(a) = 1{,}425(1.052)^{-a}$

$$H(55) = 1{,}425(1.052)^{-55}$$
$$\approx 88 \text{ micrograms per day}$$

22

$$\log_6 \frac{1}{36} = -2$$
$$6^{-2} = \frac{1}{36}$$

23 a. D: $(0, \infty)$, R: $(-\infty, \infty)$

b. $f^{-1}(x) = 10^x$

24. logarithmic growth

25.

$$\log_5 25 = x$$
$$5^x = 25$$
$$5^x = 5^2$$
$$x = 2$$

26.

$$\log_9 \frac{1}{81}$$
$$9^x = \frac{1}{81}$$
$$9^x = 9^{-2}$$
$$x = -2$$

27. $\log(-100)$ is undefined since $10^? \neq -100$

28.

$$\ln \frac{1}{e^6} = \ln e^{-6}$$
$$= -6$$

29.

$$\begin{aligned}\log_4 2 &= x\\ 4^x &= 2\\ 2^{2x} &= 2^1\\ 2x &= 1\\ x &= \frac{1}{2}\end{aligned}$$

30.

$$\begin{aligned}\log_{1/3} 1 &= x\\ \frac{1}{3}^x &= 1\\ 3^{-x} &= 1\\ x &= 0\end{aligned}$$

31.

$$\begin{aligned}\log_x 32 &= 5\\ x^5 &= 32\\ \sqrt[5]{x^5} &= \sqrt[5]{32}\\ x &= 2\end{aligned}$$

32.

$$\begin{aligned}\log_8 x &= \frac{4}{3}\\ 8^{4/3} &= x\\ \left(\sqrt[3]{8}\right)^4 &= x\\ (2)^4 &= x\\ 16 &= x\end{aligned}$$

33.

$$\begin{aligned}\log_3 x &= -3\\ 3^{-3} &= x\\ \frac{1}{3^3} &= x\\ \frac{1}{27} &= x\end{aligned}$$

34.

$$\begin{aligned}\ln x &= 1\\ e^{\ln x} &= e^1\\ x &= e\end{aligned}$$

35. $f(x) = -\log_3 x$

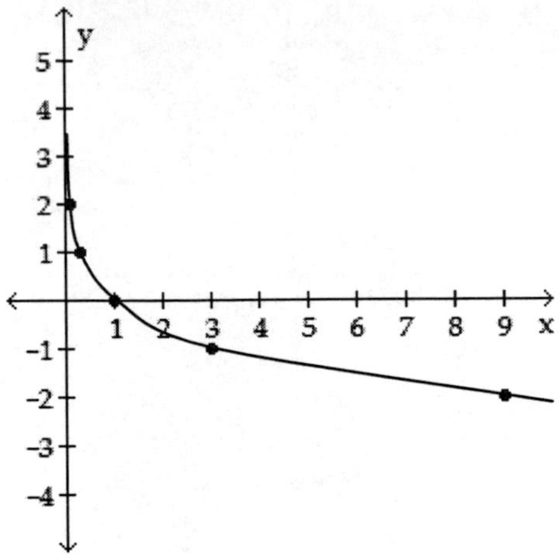

36. $f(x) = \ln x$

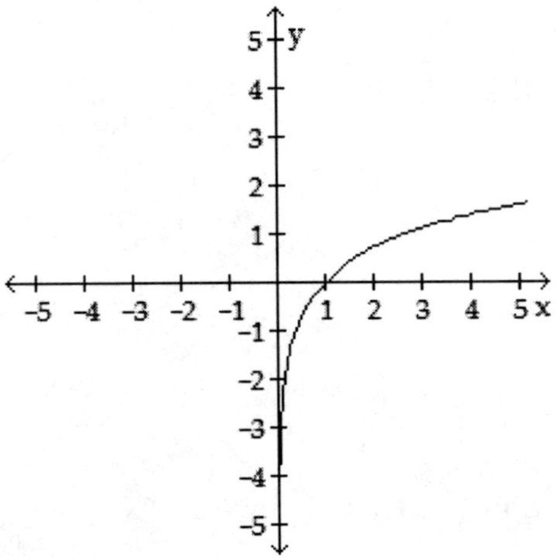

37. Let $H^+ = 3.7 \times 10^{-7}$.

$$\begin{aligned}\text{pH} &= -\log\left[H^+\right]\\ &= -\log\left(3.7\times10^{-7}\right)\\ &\approx -(-6.4)\\ &\approx 6.4\end{aligned}$$

38. Let $E_O = 60$ and $E_I = 0.3$.

$$\begin{aligned}\text{db gain} &= 20\log\left(\frac{E_O}{E_I}\right)\\ &= 20\log\left(\frac{60}{0.3}\right)\\ &= 20\log 200\\ &\approx 20(2.3)\\ &\approx 46\end{aligned}$$

39.

$$\log x = -1.06$$
$$10^{-1.06} = x$$
$$x = 10^{-1.06}$$
$$x \approx 0.0871$$

40.

$$\log_7 3 = \frac{\log 3}{\log 7}$$
$$\approx 0.5646$$

41.

$$\log_b a^2bc^3 = \log_b a^2 + \log_b b + \log_b c^3$$
$$= 2\log_b a + 1 + 3\log_b c$$

42.

$$\frac{1}{2}\ln(a+2) + \ln b - 3\ln c$$
$$= \ln(a+2)^{1/2} + \ln b - \ln c^3$$
$$= \ln\sqrt{a+2} + \ln b - \ln c^3$$
$$= \ln b\sqrt{a+2} - \ln c^3$$
$$= \ln\frac{b\sqrt{a+2}}{c^3}$$

43.

$$5^x = 3$$
$$\log 5^x = \log 3$$
$$x\log 5 = \log 3$$
$$x = \frac{\log 3}{\log 5}$$
$$x \approx 0.6826$$

44.

$$3^{x-1} = 27$$
$$3^{x-1} = 3^3$$
$$x - 1 = 3$$
$$x = 4$$

45.

$$\left(\frac{3}{2}\right)^{6x+2} = \frac{27}{8}$$
$$\left(\frac{3}{2}\right)^{6x+2} = \left(\frac{3}{2}\right)^3$$
$$6x + 2 = 3$$
$$6x = 1$$
$$x = \frac{1}{6}$$

46.

$$e^{0.08t} = 4$$
$$\ln e^{0.08t} = \ln 4$$
$$0.08t = \ln 4$$
$$t = \frac{\ln 4}{0.08}$$
$$t \approx 17.3287$$

47.

$$2\log x = \log 25$$
$$\log x^2 = \log 25$$
$$x^2 = 25$$
$$x = \pm\sqrt{25}$$
$$x = 5 \quad \text{or} \quad x = -5$$

48.

$$\log_2(x+2) - \log_2(x-5) = 3$$
$$\log_2\frac{x+2}{x-5} = 3$$
$$\frac{x+2}{x-5} = 2^3$$
$$\frac{x+2}{x-5} = 8$$
$$(x-5)\left(\frac{x+2}{x-5}\right) = (x-5)(8)$$
$$x + 2 = 8x - 40$$
$$-7x + 2 = -40$$
$$-7x = -42$$
$$x = 6$$

49.

$$\ln(5x+2) = \ln(2x+5)$$
$$5x+2 = 2x+5$$
$$3x+2 = 5$$
$$3x = 3$$
$$x = 1$$

50.

$$\log x + \log(x-9) = 1$$
$$\log x(x-9) = 1$$
$$10^1 = x(x-9)$$
$$10 = x^2 - 9x$$
$$0 = x^2 - 9x - 10$$
$$0 = (x-10)(x+1)$$
$$x - 10 = 0 \quad \text{or} \quad x + 1 = 0$$
$$x = 10 \qquad x = \cancel{-1}$$

The log cannot be negative, so the only solution is 10.

51.

$$x \approx 5$$
$$\frac{1}{2}\ln(x-1) = \ln 2$$
$$\frac{1}{2}\ln(5-1) \stackrel{?}{=} \ln 2$$
$$\frac{1}{2}\ln 4 \stackrel{?}{=} \ln 2$$
$$\frac{1}{2}(1.3863) \stackrel{?}{=} 0.6931$$
$$0.6931 = 0.6931$$

52. Let $P_0 = 5$, $P = 4(5) = 20$, and $t = 6$.

$$P = P_0 e^{kt}$$
$$20 = 5e^{k(6)}$$
$$4 = e^{6k}$$
$$\ln 4 = \ln e^{6k}$$
$$\ln 4 = 6k$$
$$6k = \ln 4$$
$$k = \frac{\ln 4}{6}$$

Let $P_0 = 5$, $P = 500$, and $k = \dfrac{\ln 4}{6}$.

$$P = P_0 e^{kt}$$
$$500 = 5e^{\left(\frac{\ln 4}{6}\right)t}$$
$$100 = e^{\left(\frac{\ln 4}{6}\right)t}$$
$$\ln 100 = \ln e^{\left(\frac{\ln 4}{6}\right)t}$$
$$\ln 100 = \left(\frac{\ln 4}{6}\right)t$$
$$\left(\frac{6}{\ln 4}\right)(\ln 100) = \left(\frac{6}{\ln 4}\right)\left(\frac{\ln 4}{6}\right)t$$
$$\frac{6\ln 100}{\ln 4} = t$$
$$t = \frac{6\ln 100}{\ln 4}$$
$$t \approx 20 \text{ minutes}$$

CHAPTER 9 CUMULATIVE REVIEW

1. $P = 2l + 2w;\ A = lw$

2. $A = \pi r^2;\ C = 2\pi r = \pi d$

3. $A = \dfrac{1}{2}bh$

4. $V = lwh;\ V = s^3$

5. amount = percent · base

6. Total value = amount · price

7. $I = Prt$

8. $d = rt$

9. $\left(\dfrac{x_1 + x_2}{2}, \dfrac{y_1 + y_2}{2}\right)$

10. a. $m = \dfrac{y_2 - y_1}{x_2 - x_1}$

    b. $\text{Slope} = \dfrac{\text{rise}}{\text{run}}$

11. $y = mx + b$

12. $y - y_1 = m(x - x_1)$

13. $y = b,\ x = a$

14. $\begin{vmatrix} a & b \\ c & d \end{vmatrix} = ad - bc$

15. a. $y = kx$

    b. $y = \dfrac{k}{x}$

16. $a^2 + b^2 = c^2$

17. $d = \sqrt{(x_2 - x_1)^2 + (y_2 - y_1)^2}$

18. $x = \dfrac{-b \pm \sqrt{b^2 - 4ac}}{2a}$

19. $f(x) = a(x - h)^2 + k$

20. $(f \circ g)(x) = f(g(x))$

21. $A = Pe^{rt}$

22. $\log_b x = \dfrac{\log_a x}{\log_a b}$

23. Parallel lines have the **same** slope. The slopes of perpendicular lines are **negative reciprocals**.

24. The addition (elimination) method

25. a. The solution set of a compound inequality containing the word *and* consists of all the numbers that make **both** inequalities true.

    b. The solution set of a compound inequality containing the word *or* consists of all the numbers that make **one** or the other or **both** inequalities true.

26. a. $X = k$ or $X = -k$

    b. $X = Y$ or $X = -Y$

    c. $X > k$ or $X < -k$

    d. $-k < X < k$

27. a. $x^1 = x$

    b. $x^m x^n = x^{m+n}$

    c. $(x^m)^n = x^{mn}$

    d. $(xy)^n = x^n y^n$

    e. $x^0 = 1$

    f. $\left(\dfrac{x}{y}\right)^n = \dfrac{x^n}{y^n}$

    g. $\dfrac{x^m}{x^n} = x^{m-n}$

    h. $x^{-n} = \dfrac{1}{x^n}$

    i. $\dfrac{1}{x^{-n}} = x^n$

    j. $\left(\dfrac{x}{y}\right)^{-n} = \left(\dfrac{y}{x}\right)^n$

    k. $x^{1/n} = \sqrt[n]{x}$

    l. $x^{m/n} = \left(\sqrt[n]{x}\right)^m$

28. Pick a test point on one side of the boundary line. In $y \geq x$, replace x and y with the coordinates of that point. If the inequality is satisfied, shade the side that contains that point. If the inequality is not satisfied, shade the other side.

29. a. $(x+y)^2 = x^2 + 2xy + y^2$
   b. $(x-y)^2 = x^2 - 2xy + y^2$
   c. $(x+y)(x-y) = x^2 - y^2$

30. a. $x^2 - y^2 = (x+y)(x-y)$
   b. $x^3 - y^3 = (x-y)(x^2 + xy + y^2)$
   c. $x^3 + y^3 = (x+y)(x^2 - xy + y^2)$

31. An equation of the form $ax^2 + bx + c = 0$ where $a \neq 0$.

32. $x + 1 = 0$ or $x - 7 = 0$

33. A factor equal to 1 is being removed: $\frac{3a-2}{3a-2} = 1$.

34. A fraction is undefined when the **denominator** is 0.

35. a. $\frac{A}{B} \cdot \frac{C}{D} = \frac{AC}{BD}$
   b. $\frac{A}{B} \div \frac{C}{D} = \frac{AD}{BC}$

36. a. $\frac{A}{D} + \frac{B}{D} = \frac{A+B}{D}$
   b. $\frac{A}{D} - \frac{B}{D} = \frac{A-B}{D}$

37. a. To build up the fractions so they have the same denominator
   b. To simplify a complex fraction
   c. To rationalize the denominator

38. If $\frac{a}{b} = \frac{c}{d}$, the $ad = $ **bc**. The **cross products** are equal.

39. Multiply both sides of the equation by the LCD, which is $x(x-2)$

40. a. $\sqrt[n]{ab} = \sqrt[n]{a}\sqrt[n]{b}$
   b. $\sqrt[n]{\frac{a}{b}} = \frac{\sqrt[n]{a}}{\sqrt[n]{b}}$

41. square both sides

42. a. $\pi = 3.14$
   b. $\sqrt{2} = 1.41$
   c. $e = 2.72$

43. a. $\log_b 1 = 0$
   b. $\log_b b = 1$
   c. $\log_b b^x = x$
   d. $b^{\log_b x} = x$
   e. $\log_b MN = \log_b M + \log_b N$
   f. $\log_b \frac{M}{N} = \log_b M - \log_b N$
   g. $\log_b M^p = p \log_b M$

44. a. take the log on both sides
   b. write the left side as a single logarithm: $\log x(x-3) = 1$

45. If a and b are real numbers, $a - b = a + (-b)$.

46. For any real number x,
$$\begin{cases} \text{if } x \geq 0, \text{ then } |x| = \boxed{x} \\ \text{if } x < 0, \text{ then } |x| = \boxed{-x} \end{cases}$$

47. a. The number b is a square root of a if $\boxed{b^2} = a$.
   b. If x can be any real number, then $\sqrt{x^2} = \boxed{|x|}$.
   c. If x is any real number, then $\sqrt[3]{x^3} = \boxed{x}$.

48. a. $i = \boxed{\sqrt{-1}}$
   b. $i^2 = \boxed{-1}$

SECTION 10.1

VOCABULARY

1. The curves formed by the intersection of a plane with an infinite right–circular cone are called **conic sections**.

3. A **circle** is the set of all points in a plane that are a fixed distance from a fixed point called its center. The fixed distance is called the **radius**.

CONCEPTS

5. a. $(x-h)^2+(y-k)^2=r^2$
   b. $x^2+y^2=r^2$

7. a. $(2, -1)$; $r=4$
   b. $(x-2)^2+(y-(-1))^2=4^2$
   $(x-2)^2+(y+1)^2=16$

9. a. $y=a(x-h)^2+k$
   b. $x=a(y-k)^2+h$

11. a. circle
    b. parabola
    c. parabola
    d. circle

NOTATION

13. $h=6$, $k=-2$, and $r=3$

GUIDED PRACTICE

15. $x^2+y^2=9$
    center (0, 0) and $r=3$

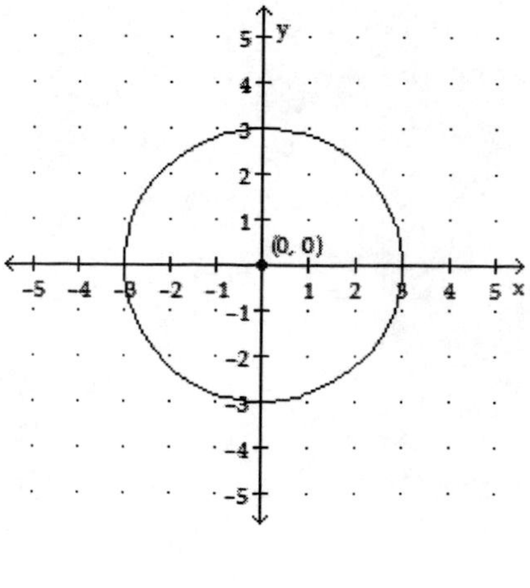

17. $x^2+(y+3)^2=1$
    center (0, –3) and $r=1$

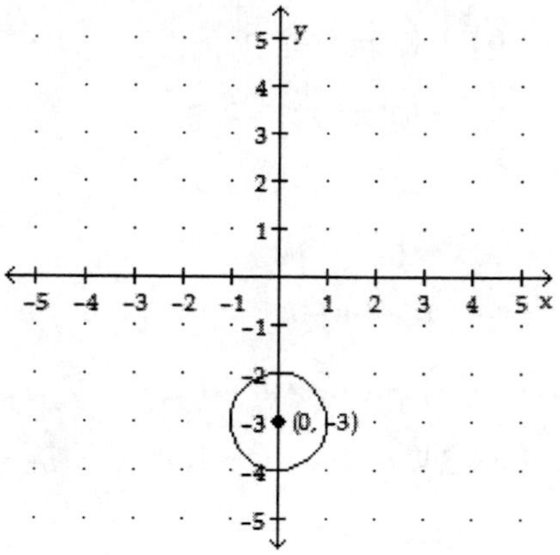

19. $(x+3)^2+(y-1)^2=16$
    center (–3, 1) and $r=4$

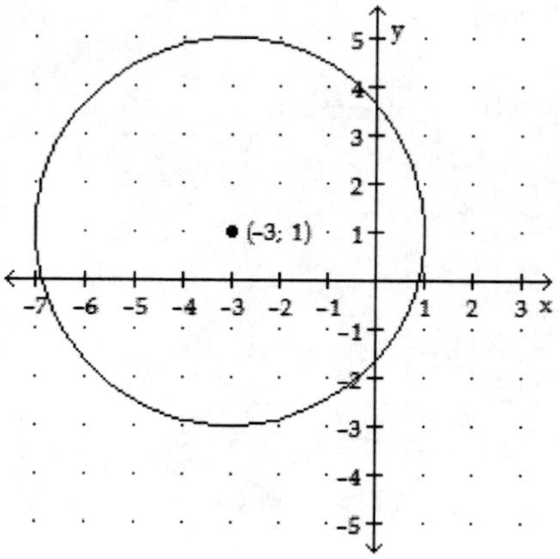

21. $x^2+y^2=6$
    center (0, 0) and $r=\sqrt{6}\approx 2.4$

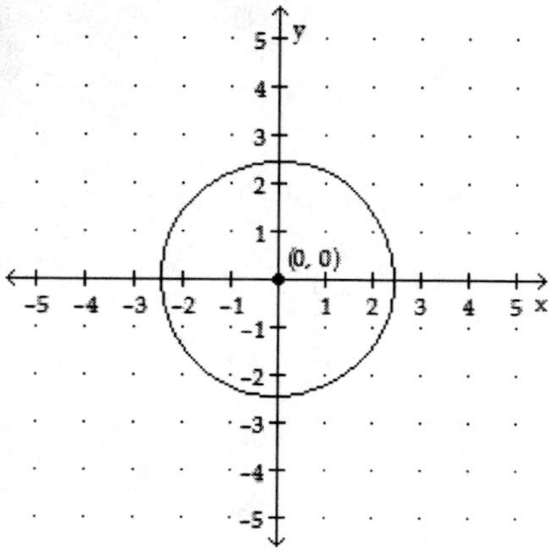

23. $h=0$, $k=0$, $r=1$
    $(x-h)^2+(y-k)^2=r^2$
    $(x-0)^2+(y-0)^2=1^2$
    $x^2+y^2=1$

25. $h=6, k=8, r=5$

$$(x-h)^2+(y-k)^2=r^2$$
$$(x-6)^2+(y-8)^2=5^2$$
$$(x-6)^2+(y-8)^2=25$$

27. $h=-2, k=6, r=12$

$$(x-h)^2+(y-k)^2=r^2$$
$$(x-(-2))^2+(y-6)^2=12^2$$
$$(x+2)^2+(y-6)^2=144$$

29. $h=0, k=0, r=\frac{1}{4}$

$$(x-h)^2+(y-k)^2=r^2$$
$$(x-0)^2+(y-0)^2=\left(\frac{1}{4}\right)^2.$$
$$x^2+y^2=\frac{1}{16}$$

31. $h=\frac{2}{3}, k=-\frac{7}{8}, r=\sqrt{2}$

$$(x-h)^2+(y-k)^2=r^2$$
$$\left(x-\frac{2}{3}\right)^2+\left(y-\left(-\frac{7}{8}\right)\right)^2=\left(\sqrt{2}\right)^2.$$
$$\left(x-\frac{2}{3}\right)^2+\left(y+\frac{7}{8}\right)^2=2$$

33. $h=0,\ k=0,\ d=4\sqrt{2}$ so $r=\frac{4\sqrt{2}}{2}=2\sqrt{2}$

$$(x-h)^2+(y-k)^2=r^2$$
$$(x-0)^2+(y-0)^2=\left(2\sqrt{2}\right)^2$$
$$x^2+y^2=4(2)$$
$$x^2+y^2=8$$

35.

$$x^2+y^2-2x+4y=-1$$
$$\left(x^2-2x\right)+\left(y^2+4y\right)=-1$$
$$\left(x^2-2x+1\right)+\left(y^2+4y+4\right)=-1+1+4$$
$$(x-1)^2+(y+2)^2=4$$

Center at $(1,\ -2)$; radius = 2

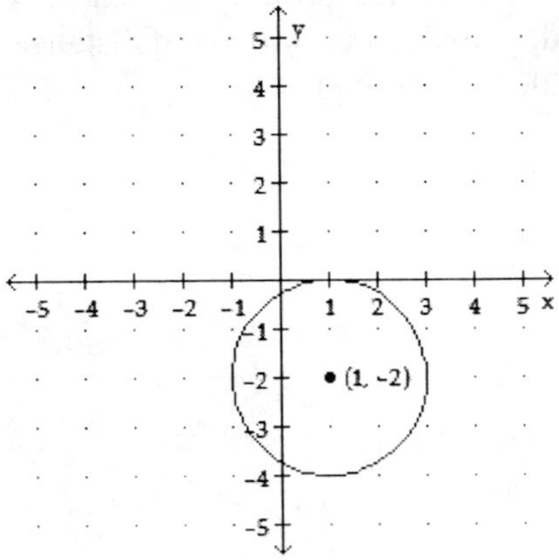

37.

$$x^2+y^2+4x+2y=4$$
$$\left(x^2+4x\right)+\left(y^2+2y\right)=4$$
$$\left(x^2+4x+4\right)+\left(y^2+2y+1\right)=4+4+1$$
$$(x+2)^2+(y+1)^2=9$$

Center at $(-2,\ -1)$; radius = 3

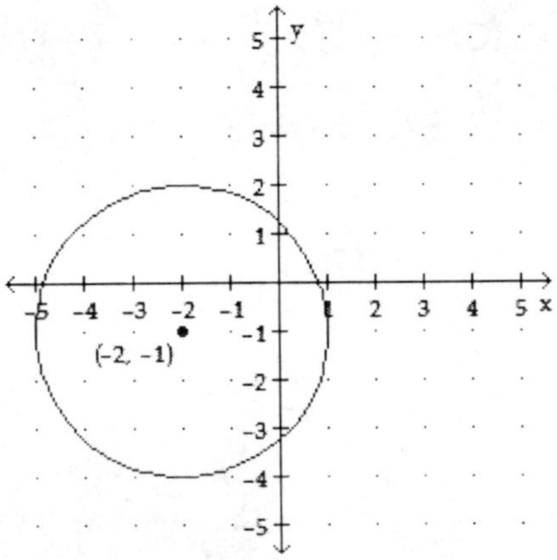

39.

$$y = 2x^2 - 4x + 5$$
$$y = 2(x^2 - 2x) + 5$$
$$y = 2(x^2 - 2x + 1) + 5 - 2(1)$$
$$y = 2(x-1)^2 + 5 - 2$$
$$y = 2(x-1)^2 + 3$$
$$a = 2,\ h = 1,\ k = 3$$
vertex at $(1, 3)$

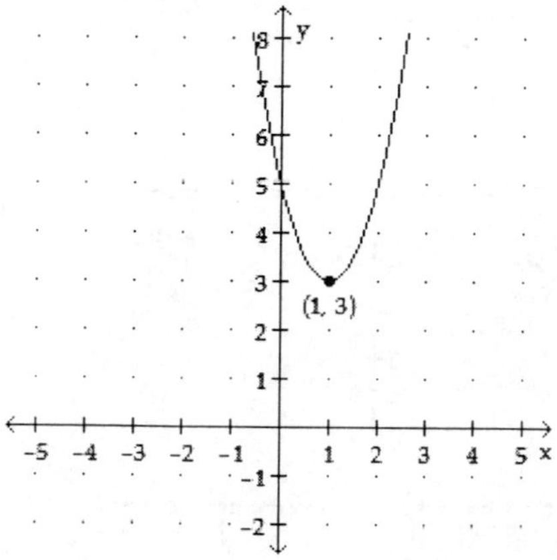

41.

$$y = -x^2 - 2x + 3$$
$$y = -(x^2 + 2x) + 3$$
$$y = -(x^2 + 2x + 1) + 3 - (-1)(1)$$
$$y = -(x+1)^2 + 3 + 1$$
$$y = -(x+1)^2 + 4$$
$$a = -1,\ h = -1,\ k = 4$$
vertex at $(-1, 4)$

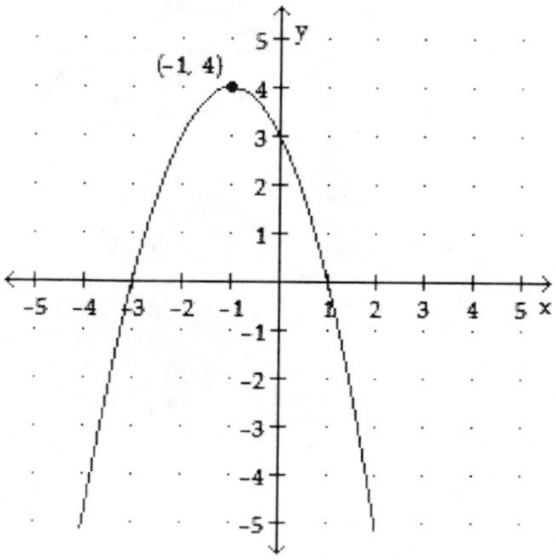

43.

$$x = y^2$$
$$a = 1,\ h = 0,\ k = 0$$
vertex at $(0, 0)$

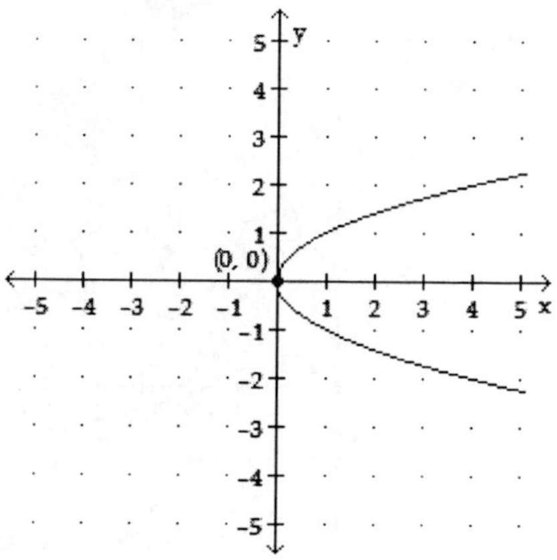

45.

$$x = 2(y+1)^2 + 3$$
$$a = 2,\ h = 3,\ k = -1$$
vertex at $(3, -1)$

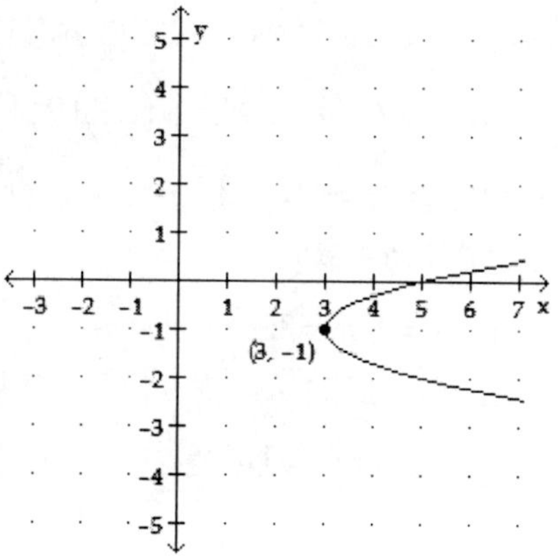

47.

$x = y^2 - 2y + 5$

$x = y^2 - 2y + 1 + 5 - 1$

$x = (y-1)^2 + 4$

$a = 1, h = 4, k = 1$

vertex at (4, 1)

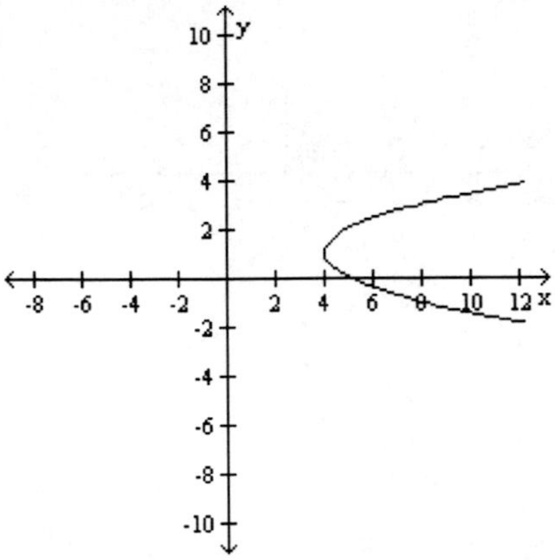

49.

$x = -3y^2 + 18y - 25$

$x = -3(y^2 - 6y) - 25$

$x = -3(y^2 - 6y + 9) - 25 - (-3)(9)$

$x = -3(y-3)^2 - 25 + 27$

$x = -3(y-3)^2 + 2$

$a = -3,\ h = 2, k = 3$

vertex at $(2,\ 3)$

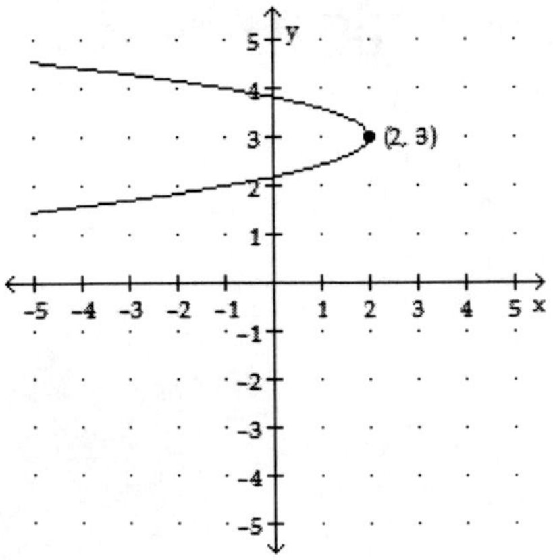

51. $x^2 + y^2 = 7$

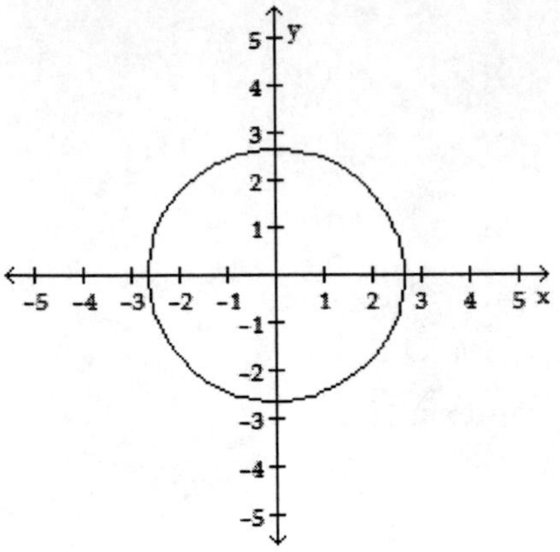

53. $(x+1)^2 + y^2 = 16$

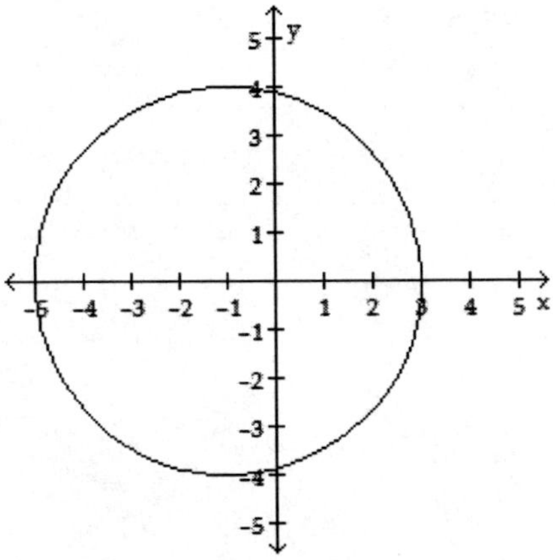

55. $x = 2y^2$

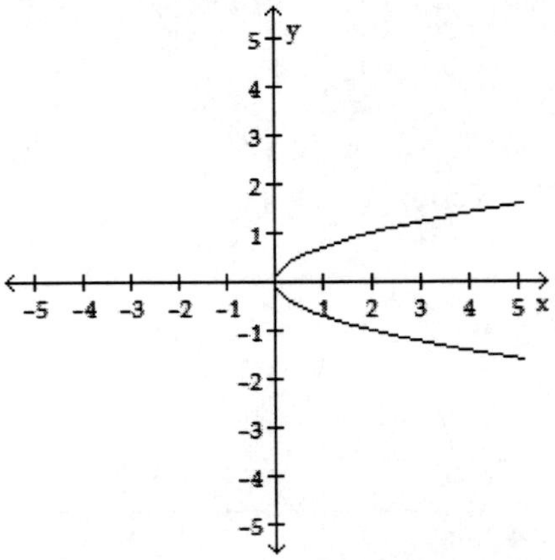

57. $x^2 - 2x + y = 6$

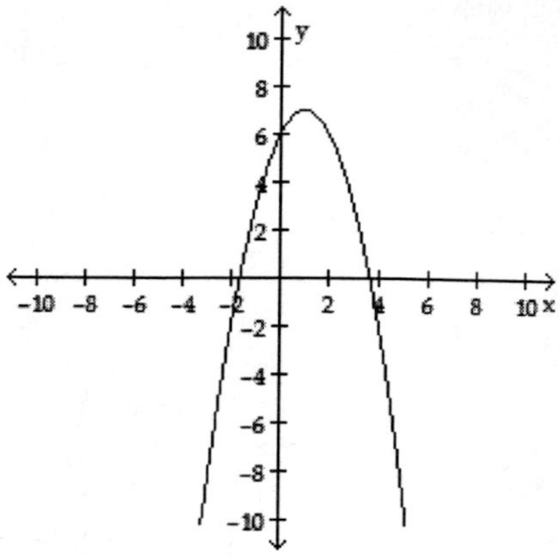

TRY IT YOURSELF

59.

$$x = y^2 - 6y + 4$$
$$x = y^2 - 6y + 9 + 4 - 9$$
$$x = (y-3)^2 - 5$$
$$a = 1,\ h = -5,\ k = 3$$

vertex at $(-5, 3)$

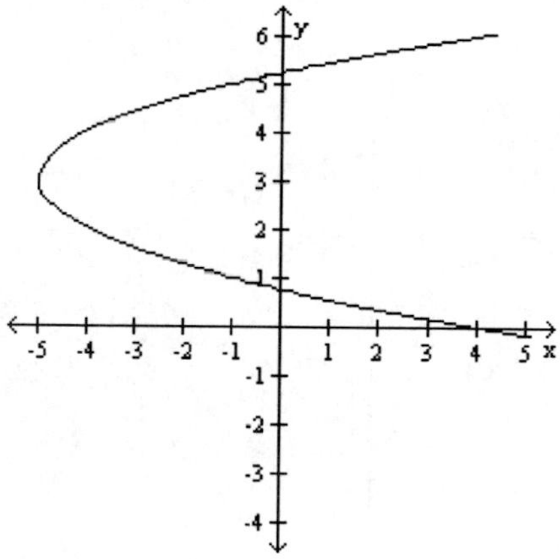

61. $(x-2)^2 + y^2 = 25$

center (2, 0) and $r = 5$

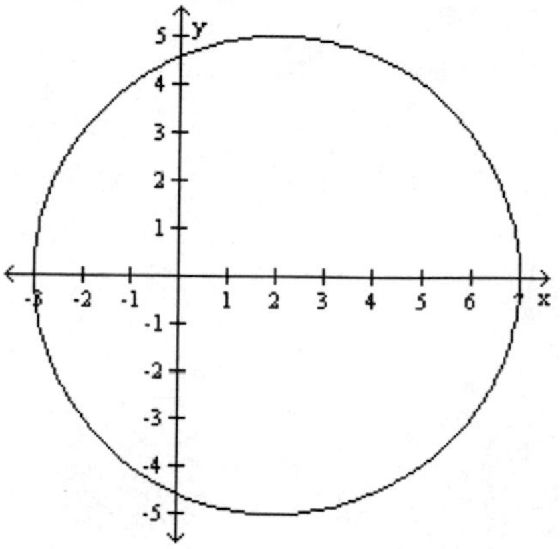

63.

$$x^2 + y^2 - 6x + 8y + 18 = 0$$
$$(x^2 - 6x) + (y^2 + 8y) = -18$$
$$(x^2 - 6x + 9) + (y^2 + 8y + 16) = -18 + 9 + 16$$
$$(x-3)^2 + (y+4)^2 = 7$$

Center at $(3, -4)$; radius $= \sqrt{7}$

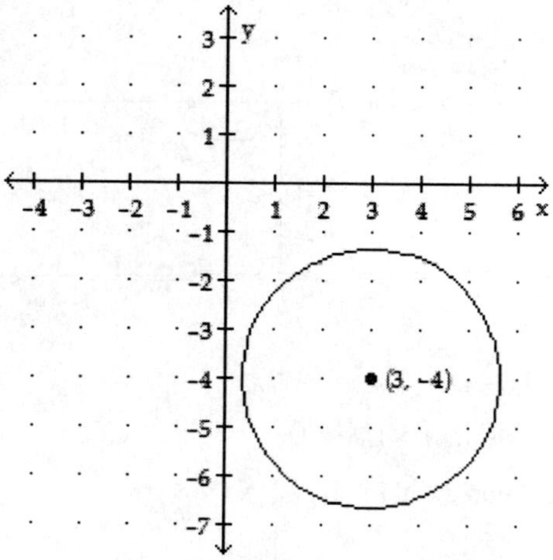

65.

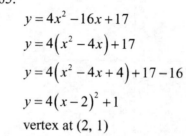

$$y = 4x^2 - 16x + 17$$
$$y = 4(x^2 - 4x) + 17$$
$$y = 4(x^2 - 4x + 4) + 17 - 16$$
$$y = 4(x-2)^2 + 1$$

vertex at (2, 1)

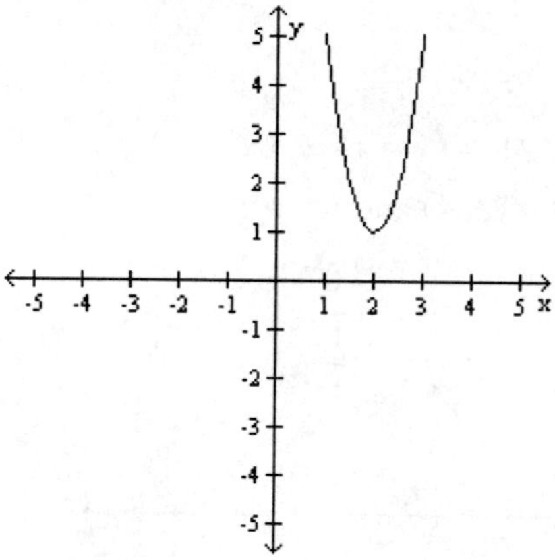

67. $(x-1)^2+(y-3)^2=15$

center (1, 3) and $r=\sqrt{15}\approx 3.9$

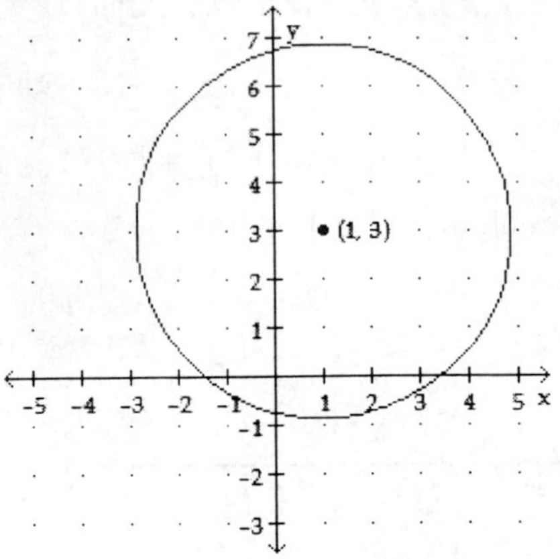

69.

$x=-y^2+1$

$a=-1,\ h=1,\ k=0$

vertex at $(1,\ 0)$

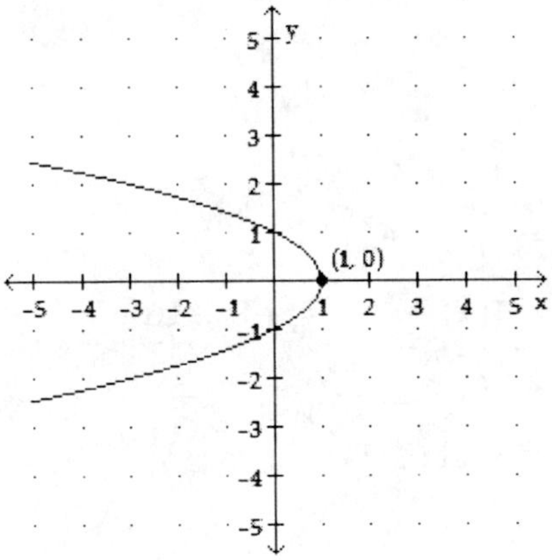

71. $(x-2)^2+(y-4)^2=36$

center (2, 4) and $r=6$

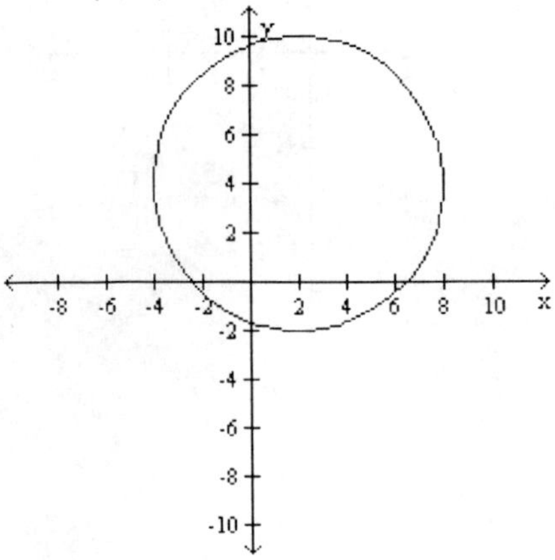

73.

$x=-\dfrac{1}{4}y^2$

$a=-\dfrac{1}{4},\ h=0,\ k=0$

vertex at $(0,\ 0)$

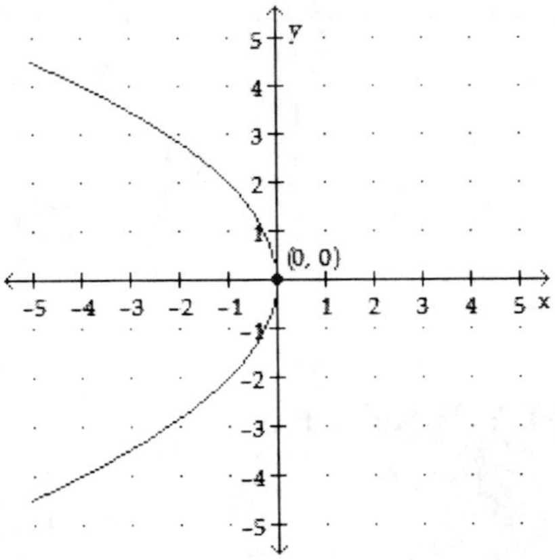

75.

$x=-6(y-1)^2+3$

$a=-6,\ h=3,\ k=1$

vertex at $(3,\ 1)$

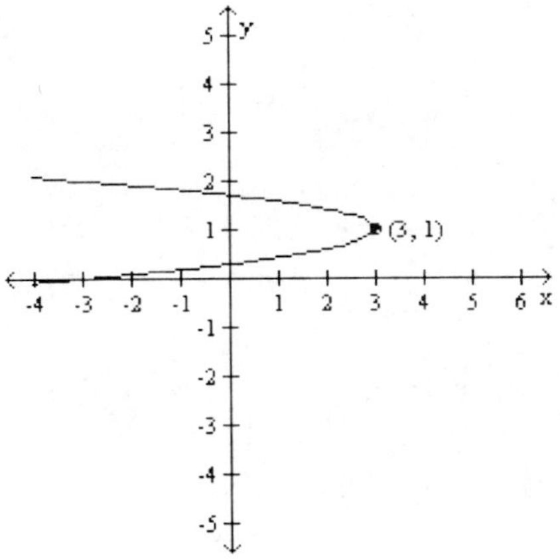

77.

$$x^2+y^2+2x-8=0$$
$$\left(x^2+2x\right)+y^2=0+8$$
$$\left(x^2+2x+1\right)+y^2=8+1$$
$$\left(x+1\right)^2+y^2=9$$

Center at $\left(-1,\ 0\right)$; radius = 3

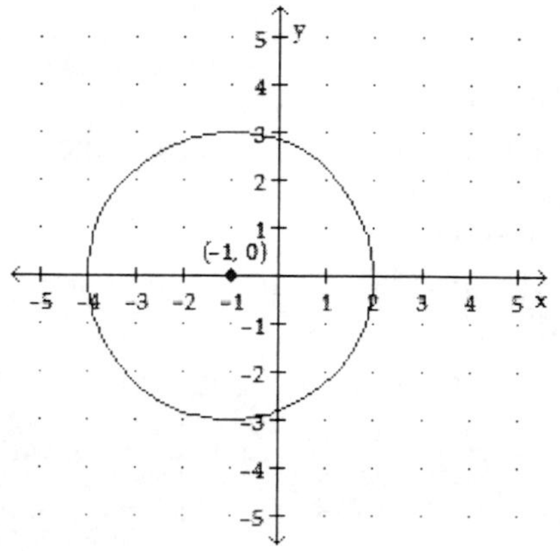

79.

$$x=\frac{1}{2}y^2+2y$$
$$x=\frac{1}{2}\left(y^2+4y\right)$$
$$x=\frac{1}{2}\left(y^2+4y+4\right)-\frac{1}{2}(4)$$
$$x=\frac{1}{2}\left(y+2\right)^2-2$$
$$a=\frac{1}{2},\ h=-2,\ k=-2$$

vertex at $\left(-2,-2\right)$

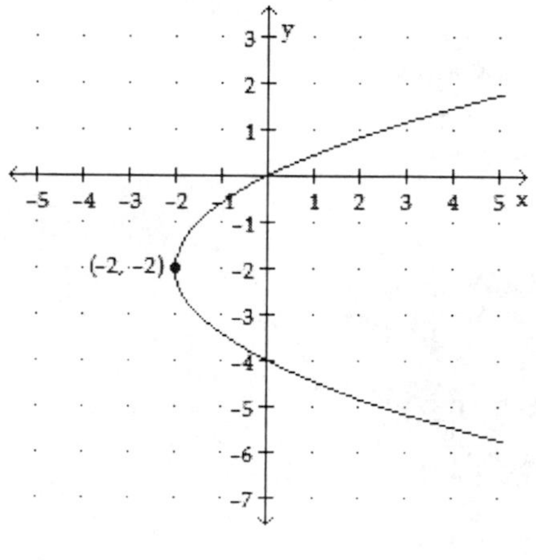

81.

$$y=-4(x+5)^2+5$$
$$a=-4,\ h=-5,\ k=5$$

vertex at (–5, 5)

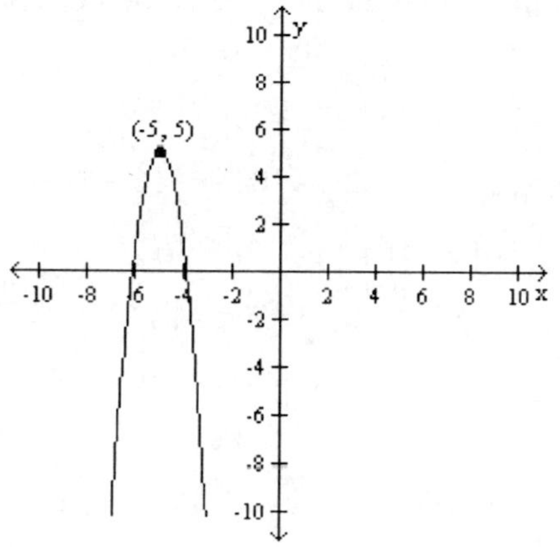

83. a. $\left(x-4\right)^2+\left(y+7\right)^2=28$

center: (4, –7), r = $2\sqrt{7}$

b. $\left(x+4\right)^2+\left(y-7\right)^2=28$

center: (–4, 7), r = $2\sqrt{7}$

85. a. $y=8\left(x-3\right)^2+6$

V(3, 6); opens upward

b. $x=8\left(y-3\right)^2+6$

V(6, 3); opens to the right

APPLICATIONS

87. BROADCAST RANGES

Find the standard equation for each coverage and graph to see if they overlap.

$$x^2+y^2-8x-20y+16=0$$
$$(x^2-8x)+(y^2-20y)=-16$$
$$(x^2-8x+16)+(y^2-20y+100)=-16+16+100$$
$$(x-4)^2+(y-10)^2=100$$

center at $(4, 10)$; radius = 10

$$x^2+y^2+2x+4y-11=0$$
$$(x^2+2x)+(y^2+4y)=11$$
$$(x^2+2x+1)+(y^2+4y+4)=11+1+4$$
$$(x+1)^2+(y+2)^2=16$$

center at $(-1, -2)$; radius = 4

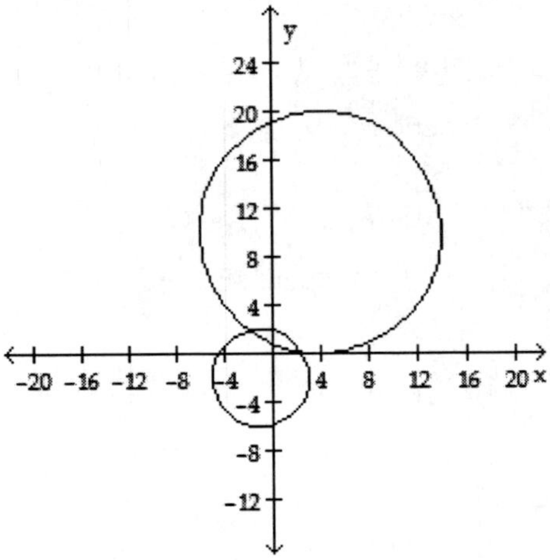

Since the coverage areas overlap, they cannot be licensed for the same frequency.

89. CIVIL ENGINEER

Find the center and radius of the circle.

$$x^2+y^2-10x-12y+52=0$$
$$(x^2-10x)+(y^2-12y)=-52$$
$$(x^2-10x+25)+(y^2-12y+36)=-52+25+36$$
$$(x-5)^2+(y-6)^2=9$$

center at $(5, 6)$; radius = 3

To find the locations of the intersections, add the radius of 3 to the x- and y-coordinates of the center.

a. The intersection with State (x-value) is $5 + 3 = 8$ miles.
b. The intersection with Highway 60 (y-value) is $6 + 3 = 9$ miles.

91. PROJECTILES

Its landing position is an x–intercept, so let $y = 0$, and solve the equation for x.

$$y=30x-x^2$$
$$0=30x-x^2$$
$$0=x(30-x)$$
$$x=0 \quad \text{or} \quad 30-x=0$$
$$x=0 \qquad\qquad 30=x$$

If it lands 30 feet away from its starting point, it is 35 – 30 = 5 ft from the castle.

93. COMETS

Find the vertex of the comet's orbit.

$$2y^2-9x=18$$
$$-9x=-2y^2+18$$
$$x=\frac{2}{9}y^2-2$$
$$k=0 \text{ and } h=-2$$

vertex = $(-2, 0)$

Find the distance between the sun (0, 0) and the comet (–2, 0) using the distance formula.

$$d=\sqrt{(x_2-x_1)^2+(y_2-y_1)^2}$$
$$=\sqrt{(-2-0)^2+(0-0)^2}$$
$$=\sqrt{4}$$
$$=2 \text{ AU}$$

WRITING

95. Answers will vary.

97. Answers will vary.

REVIEW

99.

$$|3x-4|=11$$
$$3x-4=11 \quad \text{or} \quad 3x-4=-11$$
$$3x=15 \qquad\qquad 3x=-7$$
$$x=5 \qquad\qquad x=-\frac{7}{3}$$

101.

$$|3x+4| = |5x-2|$$

$$\begin{aligned} 3x+4 &= 5x-2 \quad \text{or} \quad & 3x+4 &= -(5x-2) \\ 3x+4 &= 5x-2 & 3x+4 &= -5x+2 \\ -2x+4 &= -2 & 8x+4 &= 2 \\ -2x &= -6 & 8x &= -2 \\ x &= 3 & x &= -\frac{1}{4} \end{aligned}$$

CHALLENGE PROBLEMS

103. Yes. Answers will vary.

105. To find the center of the circle, find the midpoint of the points (–2, –6) and (8, 10).

$$\begin{aligned} \left(\frac{x_1+x_2}{2}, \frac{y_1+y_2}{2}\right) &= \left(\frac{-2+8}{2}, \frac{-6+10}{2}\right) \\ &= \left(\frac{6}{2}, \frac{4}{2}\right) \\ &= (3,2) \end{aligned}$$

To find the radius, use the distance formula to find the distance from (8, 10) and the center (3, 2).

$$\begin{aligned} d &= \sqrt{(x_2-x_1)^2+(y_2-y_1)^2} \\ &= \sqrt{(3-8)^2+(2-10)^2} \\ &= \sqrt{(-5)^2+(-8)^2} \\ &= \sqrt{25+64} \\ &= \sqrt{89} \end{aligned}$$

Write the equation of the circle with center (3, 2) and radius $\sqrt{89}$.

$$(x-3)^2+(y-2)^2 = \left(\sqrt{89}\right)^2$$

$$(x-3)^2+(y-2)^2 = 89$$

SECTION 10.2

GUIDED PRACTICE

VOCABULARY

1. The curve graphed below is an **ellipse**.

3. In the graph above, F_1 and F_2 are the **foci** of the ellipse. Each one is called a **focus** of the ellipse.

5. The line segment joining the vertices of an ellipse is called the **major** axis of the ellipse.

CONCEPTS

7. $\frac{x^2}{a^2}+\frac{y^2}{b^2}=1$

9. x–intercepts: $(a, 0)$ and $(-a, 0)$
   y–intercepts: $(0, b)$ and $(0, -b)$

11. a. $(-2, 1)$; $a = 2$ and $b = 5$
    b. vertical
    c. $\frac{(x+2)^2}{4}+\frac{(y-1)^2}{25}=1$

13.
$$4(x-1)^2+64(y+5)^2=64$$
$$\frac{4(x-1)^2}{64}+\frac{64(y+5)^2}{64}=\frac{64}{64}$$
$$\frac{(x-1)^2}{16}+\frac{(y+5)^2}{1}=1$$

NOTATION

15. $h=-8$, $k=6$, $a=\sqrt{100}=10$, $b=\sqrt{144}=12$

GUIDED PRACTICE

17. $\frac{x^2}{25}+\frac{y^2}{4}=1$

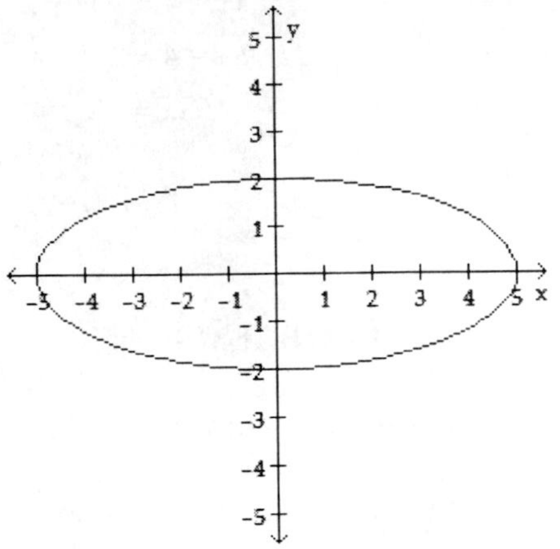

19. $\frac{x^2}{4}+\frac{y^2}{9}=1$

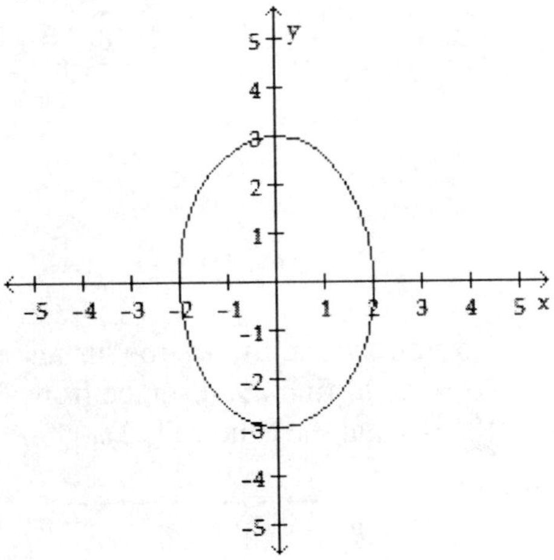

21.
$$x^2+9y^2=9$$
$$\frac{x^2}{9}+\frac{9y^2}{9}=\frac{9}{9}$$
$$\frac{x^2}{9}+\frac{y^2}{1}=1$$

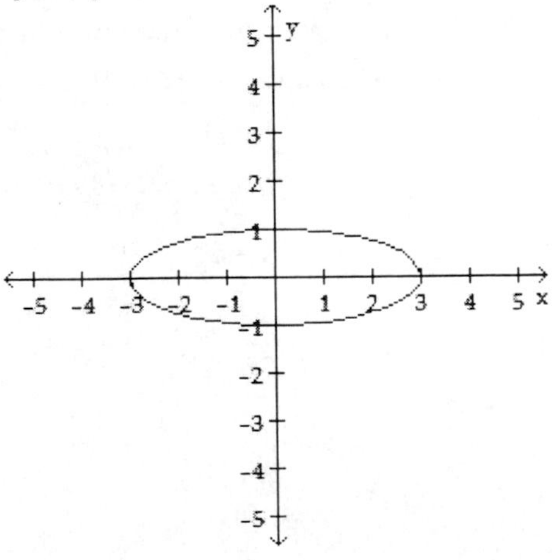

23.

$$16x^2 + 4y^2 = 64$$

$$\frac{16x^2}{64} + \frac{4y^2}{64} = \frac{64}{64}$$

$$\frac{x^2}{4} + \frac{y^2}{16} = 1$$

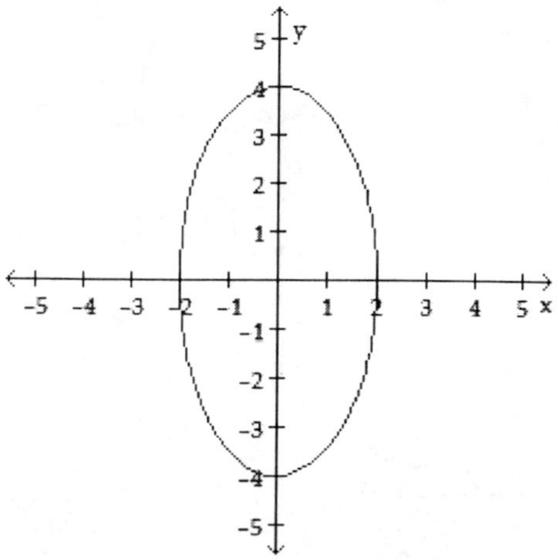

27. $\dfrac{(x+2)^2}{64} + \dfrac{(y-2)^2}{100} = 1$

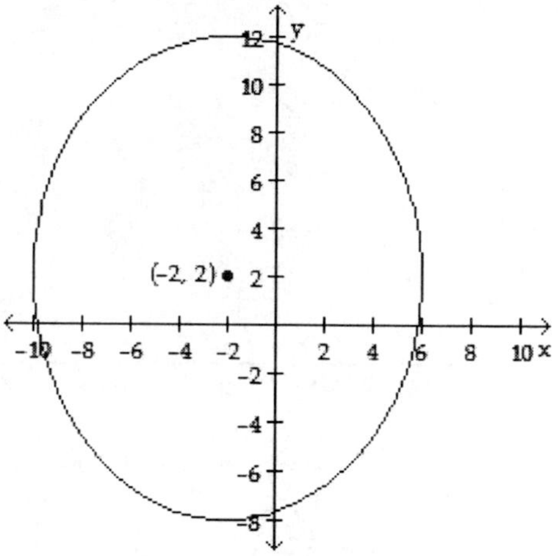

25. $\dfrac{(x-2)^2}{9} + \dfrac{(y-1)^2}{4} = 1$

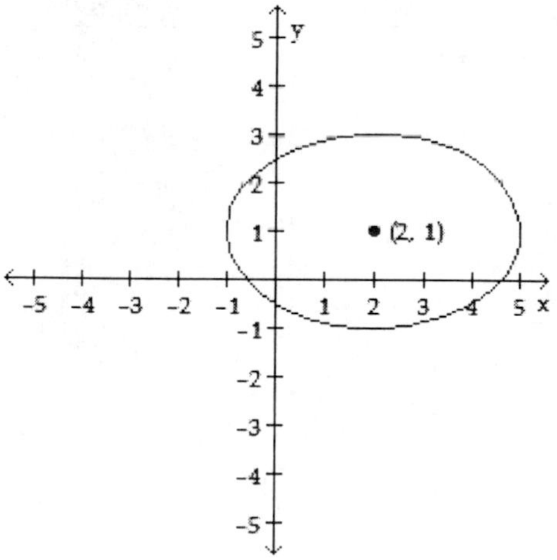

29.

$$(x+1)^2 + 4(y+2)^2 = 4$$

$$\frac{(x+1)^2}{4} + \frac{4(y+2)^2}{4} = \frac{4}{4}$$

$$\frac{(x+1)^2}{4} + \frac{(y+2)^2}{1} = 1$$

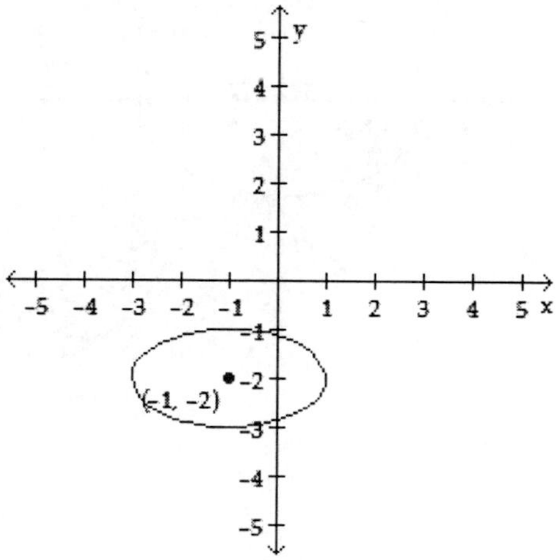

31.

$$16(x-2)^2+4(y+4)^2=256$$

$$\frac{16(x-2)^2}{256}+\frac{4(y+4)^2}{256}=\frac{256}{256}$$

$$\frac{(x-2)^2}{16}+\frac{(y+4)^2}{64}=1$$

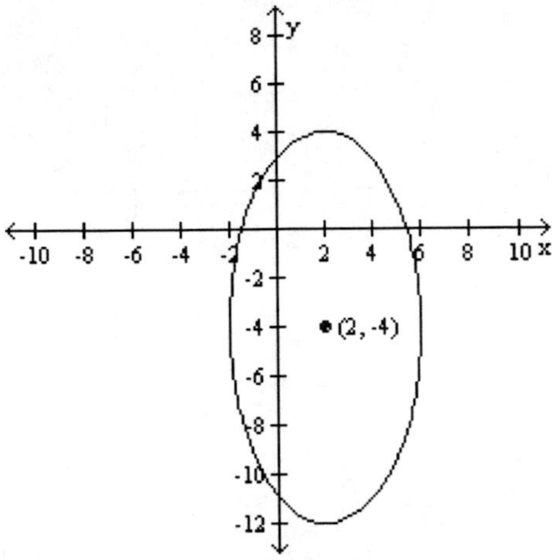

33.

$$\frac{x^2}{9}+\frac{y^2}{4}=1$$

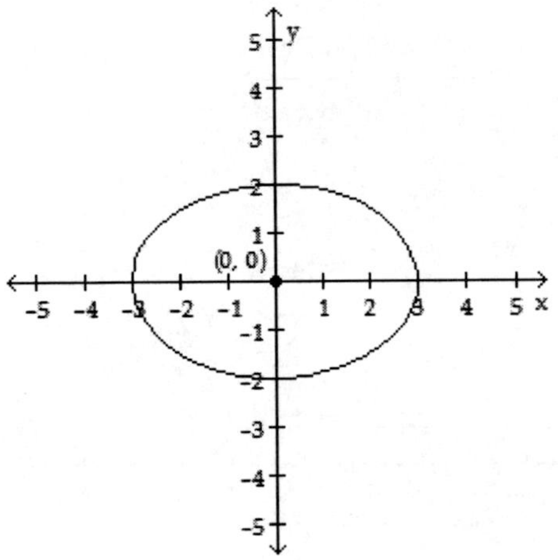

35.

$$\frac{x^2}{4}+\frac{(y-1)^2}{9}=1$$

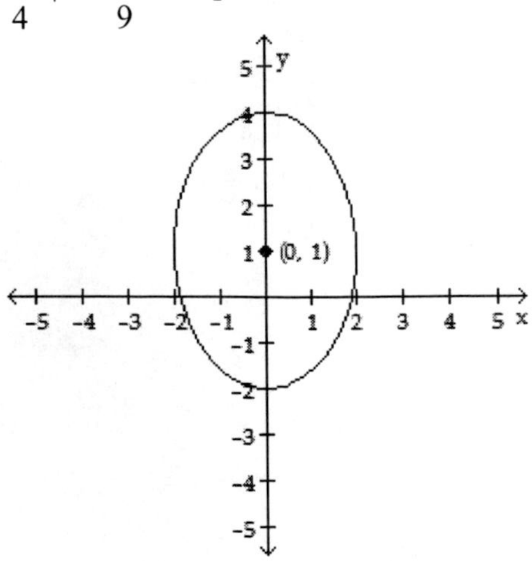

TRY IT YOURSELF

37. $(x+1)^2+(y-2)^2=16$

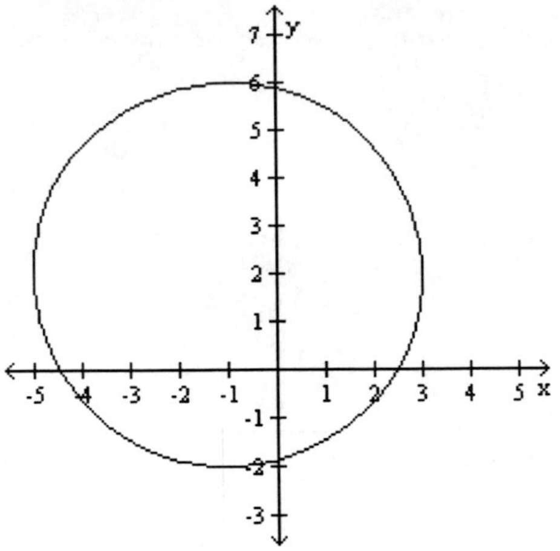

39. $\frac{x^2}{16}+\frac{y^2}{1}=1$

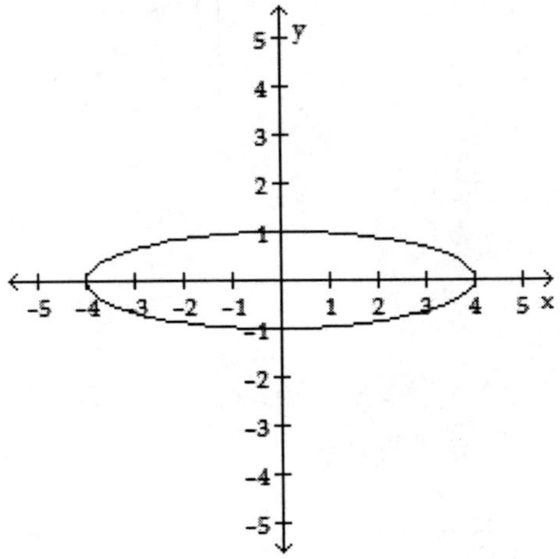

41. $x=\frac{1}{2}(y-1)^2-2$

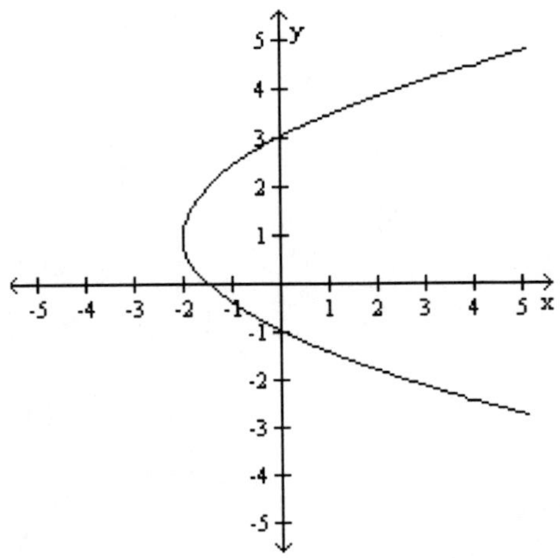

43. $x^2 + y^2 - 25 = 0$

$x^2 + y^2 = 25$

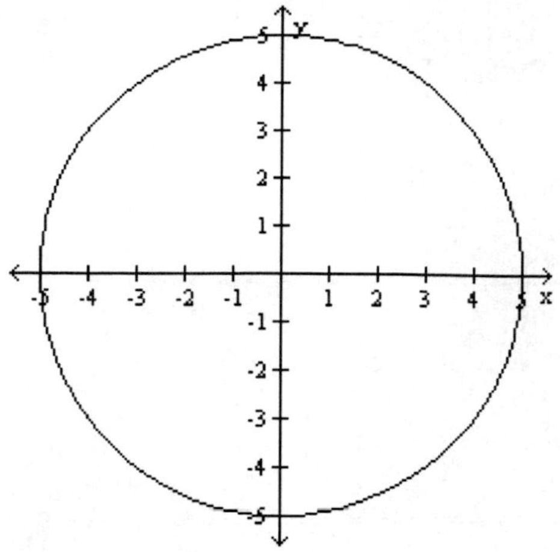

45.

$$x^2 = 100 - 4y^2$$
$$x^2 + 4y^2 = 100$$
$$\frac{x^2}{100} + \frac{4y^2}{100} = \frac{100}{100}$$
$$\frac{x^2}{100} + \frac{y^2}{25} = 1$$

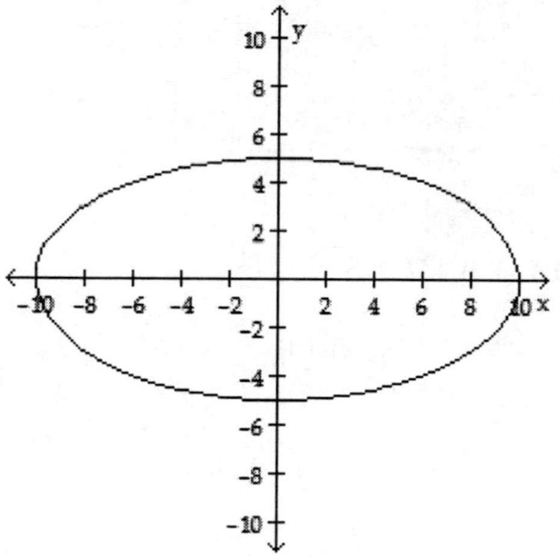

47.

$$y = -3x^2 - 24x - 43$$
$$y = -3(x^2 + 8x) - 43$$
$$y = -3(x^2 + 8x + 16) - 43 + 48$$
$$y = -3(x+4)^2 + 5$$

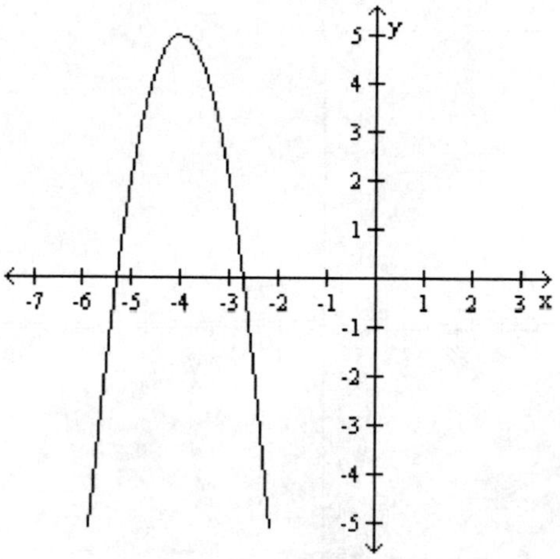

49.

$$x^2 + y^2 - 2x + 4y - 4 = 0$$
$$(x^2 - 2x) + (y^2 + 4y) = 4$$
$$(x^2 - 2x + 1) + (y^2 + 4y + 4) = 4 + 1 + 4$$
$$(x-1)^2 + (y+2)^2 = 9$$

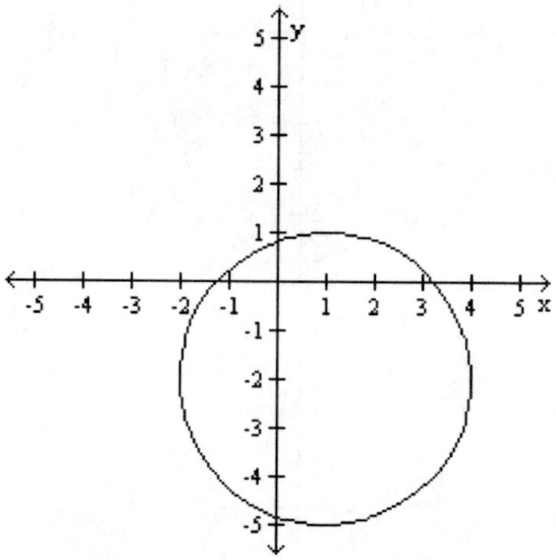

51.

$$9(x-1)^2+4(y+2)^2=36$$
$$9(x-1)^2+4(y+2)^2=36$$
$$\frac{9(x-1)^2}{36}+\frac{4(y+2)^2}{36}=\frac{36}{36}$$
$$\frac{(x-1)^2}{4}+\frac{(y+2)^2}{9}=1$$

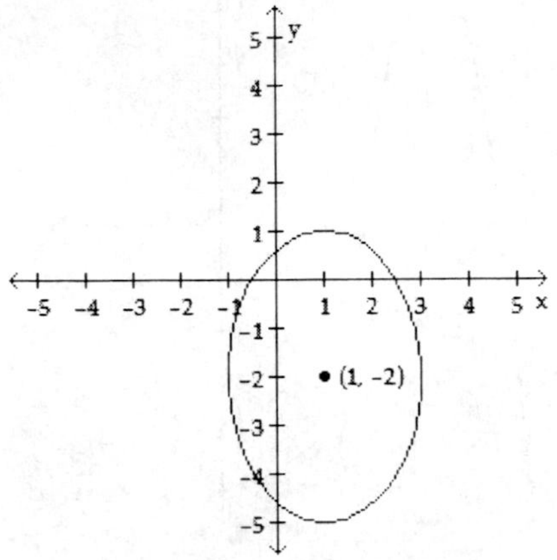

53. a. $\frac{x^2}{9}+\frac{y^2}{25}=1$

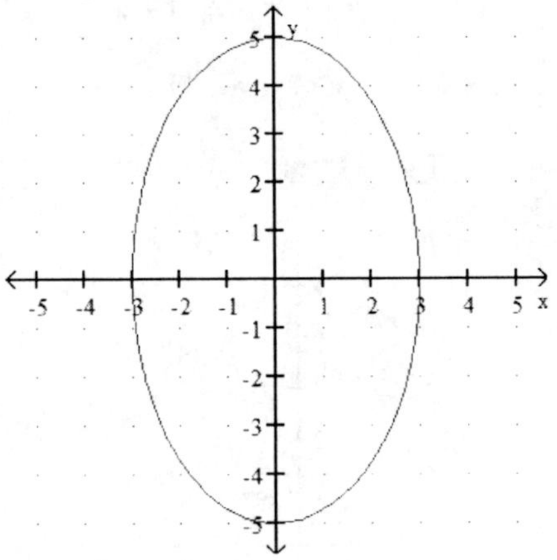

b. $\frac{x^2}{25}+\frac{y^2}{9}=1$

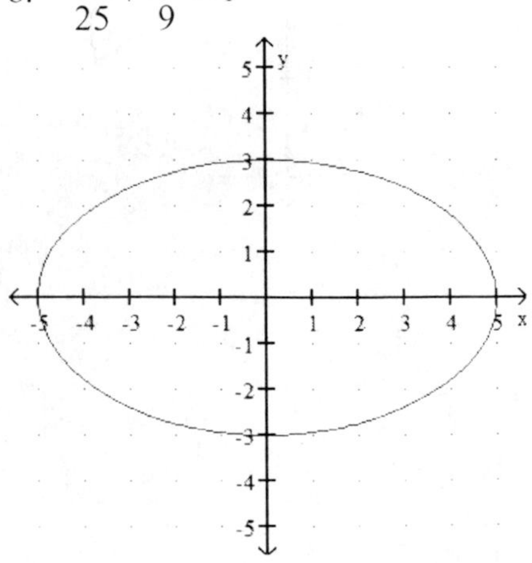

55. a. $\frac{(x-3)^2}{100}+\frac{(y-2)^2}{36}=1$

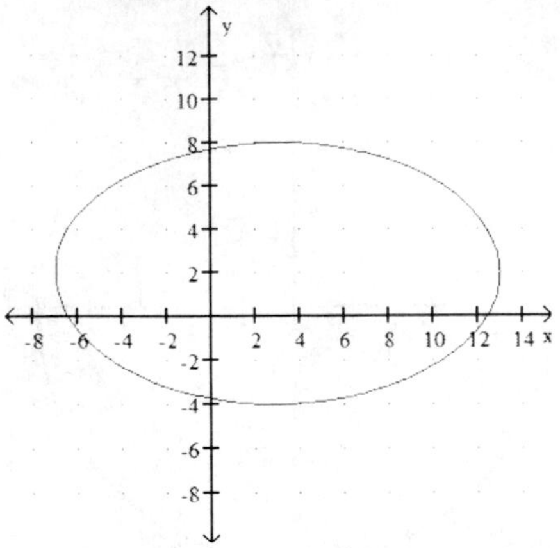

b. $\frac{(x+3)^2}{100}+\frac{(y+2)^2}{36}=1$

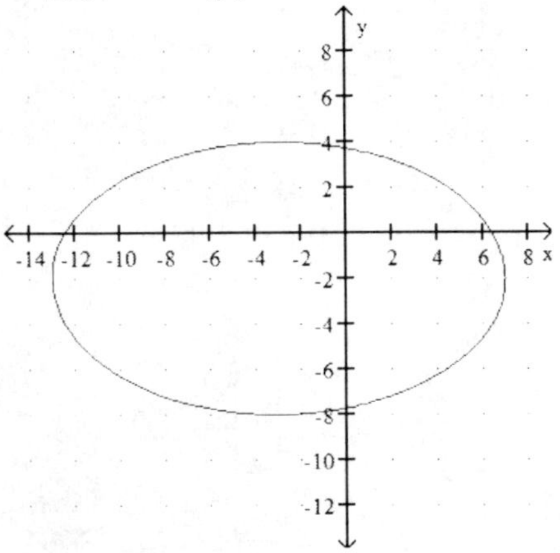

APPLICATIONS

57. CIVIL ENGINEER

a. $a=\pm 20,\ b\pm 10$

$$\frac{x^2}{400}+\frac{y^2}{100}=1$$

b. Let $x = 10$.

$$\frac{10^2}{400}+\frac{y^2}{100}=1$$
$$\frac{100}{400}+\frac{y^2}{100}=1$$
$$0.25+\frac{y^2}{100}=1$$
$$\frac{y^2}{100}=0.75$$
$$y^2=75$$
$$y=5\sqrt{3}$$
$$y\approx 8.7 \text{ ft}$$

59. KOI PONDS

$a = 110/2 - 15 = 40$

$b = 100/2 - 15 = 35$

$a = \pm 40,\ b \pm 35$

$$\frac{x^2}{1,600} + \frac{y^2}{1,125} = 1$$

61. AREA OF AN ELLIPSE

$$9x^2 + 16y^2 = 144$$

$$\frac{9x^2}{144} + \frac{16y^2}{144} = \frac{144}{144}$$

$$\frac{x^2}{16} + \frac{y^2}{9} = 1$$

$a = 4$ and $b = 3$

$$\begin{aligned} A &= \pi ab \\ &= \pi(4)(3) \\ &= 12\pi \text{ sq. units} \\ &\approx 37.7 \text{ sq. units} \end{aligned}$$

WRITING

63. Answers will vary.

65. Answers will vary.

REVIEW

67.

$$\begin{aligned} 3x^{-2}y^2\left(4x^2 + 3y^{-2}\right) &= 12x^0y^2 + 9x^{-2}y^0 \\ &= 12(1)y^2 + 9x^{-2}(1) \\ &= 12y^2 + \frac{9}{x^2} \end{aligned}$$

69.

$$\begin{aligned} \frac{x^{-2} + y^{-2}}{x^{-2} - y^{-2}} &= \frac{\frac{1}{x^2} + \frac{1}{y^2}}{\frac{1}{x^2} - \frac{1}{y^2}} \\ &= \frac{\frac{1}{x^2} + \frac{1}{y^2}}{\frac{1}{x^2} - \frac{1}{y^2}} \bullet \frac{x^2y^2}{x^2y^2} \\ &= \frac{\frac{x^2y^2}{x^2} + \frac{x^2y^2}{y^2}}{\frac{x^2y^2}{x^2} - \frac{x^2y^2}{y^2}} \\ &= \frac{y^2 + x^2}{y^2 - x^2} \\ &= \frac{y^2 + x^2}{(y+x)(y-x)} \end{aligned}$$

CHALLENGE PROBLEMS

71. It forms a circle.

73.

$$9x^2 + 4y^2 - 18x + 16y = 11$$

$$\left(9x^2 - 18x\right) + \left(4y^2 + 16y\right) = 11$$

$$9\left(x^2 - 2x\right) + 4\left(y^2 + 4y\right) = 11$$

$$9\left(x^2 - 2x + 1\right) + 4\left(y^2 + 4y + 4\right) = 11 + 9(1) + 4(4)$$

$$9(x-1)^2 + 4(y+2)^2 = 36$$

$$\frac{9(x-1)^2}{36} + \frac{4(y+2)^2}{36} = \frac{36}{36}$$

$$\frac{(x-1)^2}{4} + \frac{(y+2)^2}{9} = 1$$

SECTION 10.3

VOCABULARY

1. The two–branch curve graphed below is a **hyperbola**.

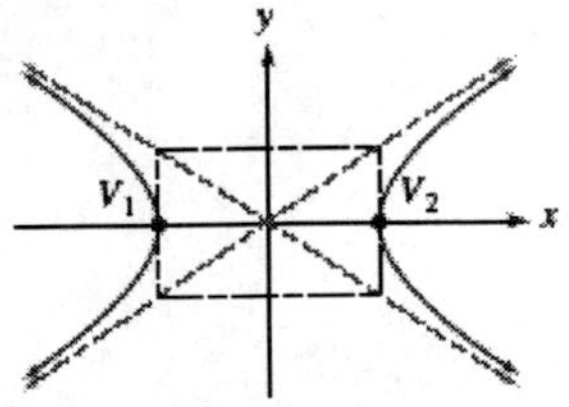

3. In the graph above, V_1 and V_2 are called the **vertices** of the hyperbola.

5. The extended **diagonals** of the central rectangle are asymptotes of the hyperbola.

CONCEPTS

7. $\dfrac{x^2}{a^2}-\dfrac{y^2}{b^2}=1$

9. $\dfrac{(x-h)^2}{a^2}-\dfrac{(y-k)^2}{b^2}=1$

11. a. $(-1,-2)$; $a=3$, $b=1$

    b. $\dfrac{(y+2)^2}{9}-\dfrac{(x+1)^2}{1}=1$

13.

$$100(x+1)^2-25(y-5)^2=100$$
$$\frac{100(x+1)^2}{100}-\frac{25(y-5)^2}{100}=\frac{100}{100}$$
$$\frac{(x+1)^2}{1}-\frac{(y-5)^2}{4}=1$$

NOTATION

15. $h=5$, $k=-11$, $a=5$, $b=6$

GUIDED PRACTICE

17. $\dfrac{x^2}{9}-\dfrac{y^2}{4}=1$

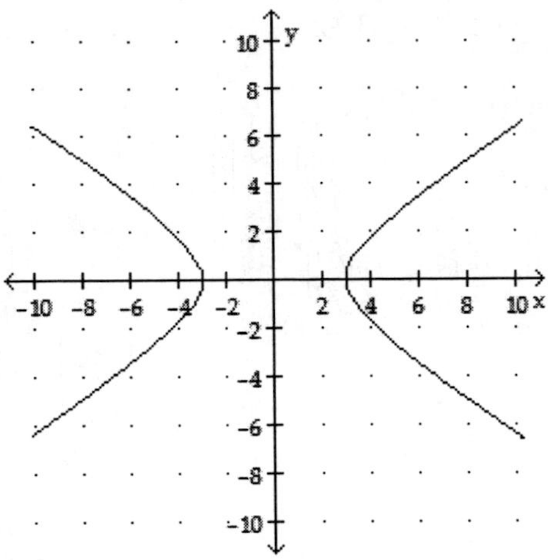

19. $\dfrac{y^2}{4}-\dfrac{x^2}{9}=1$

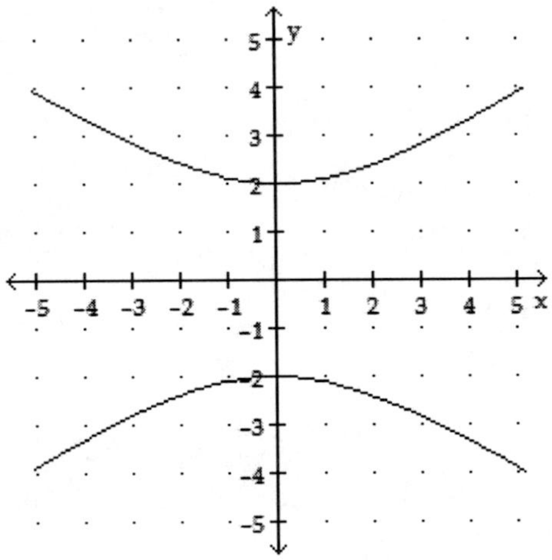

21.

$$y^2 - 4x^2 = 16$$

$$\frac{y^2}{16} - \frac{4x^2}{16} = \frac{16}{16}$$

$$\frac{y^2}{16} - \frac{x^2}{4} = 1$$

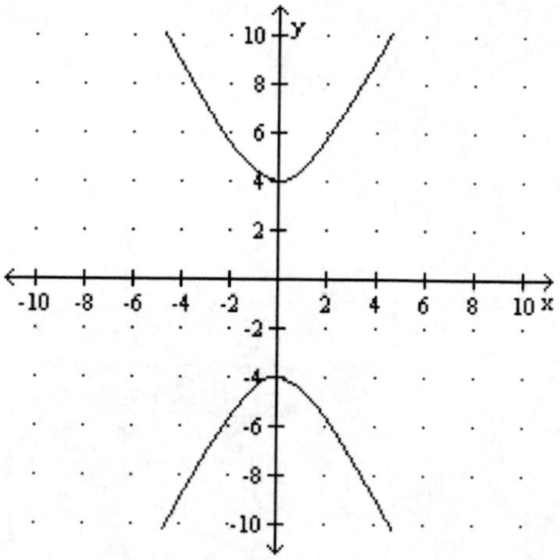

23.

$$25x^2 - y^2 = 25$$

$$\frac{25x^2}{25} - \frac{y^2}{25} = \frac{25}{25}$$

$$\frac{x^2}{1} - \frac{y^2}{25} = 1$$

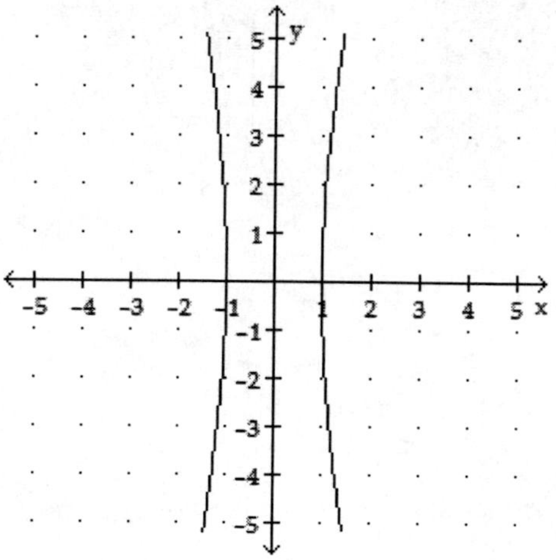

25. $\dfrac{(x-2)^2}{9} - \dfrac{y^2}{16} = 1$

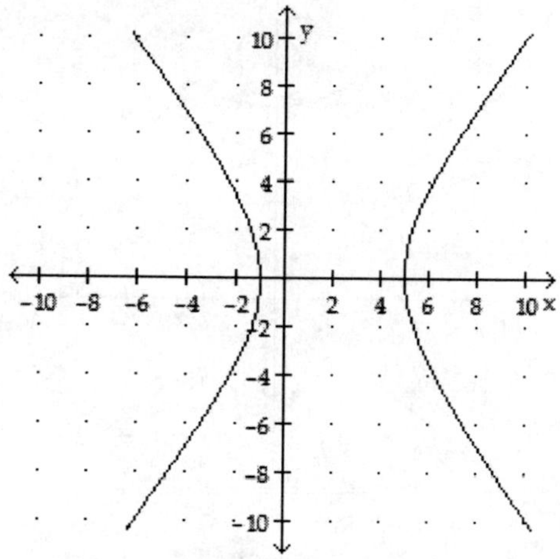

27. $\dfrac{(y+1)^2}{1} - \dfrac{(x-2)^2}{4} = 1$

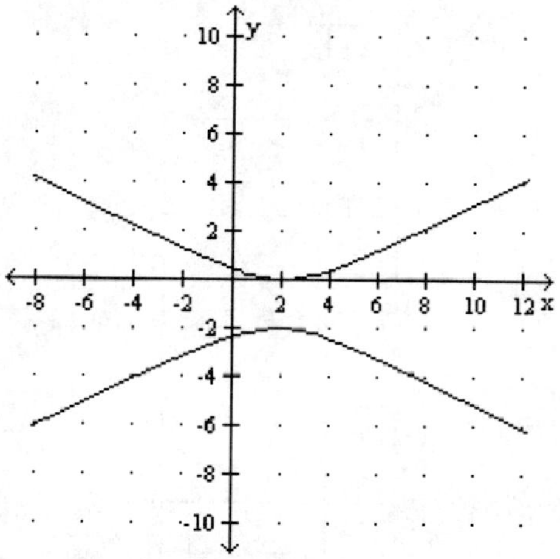

29. $\dfrac{(x+1)^2}{9} - \dfrac{(y+1)^2}{9} = 1$

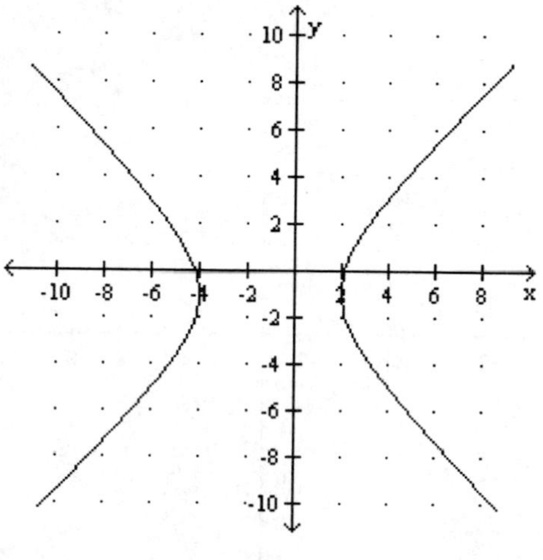

31. $\dfrac{(y-3)^2}{25} - \dfrac{x^2}{25} = 1$

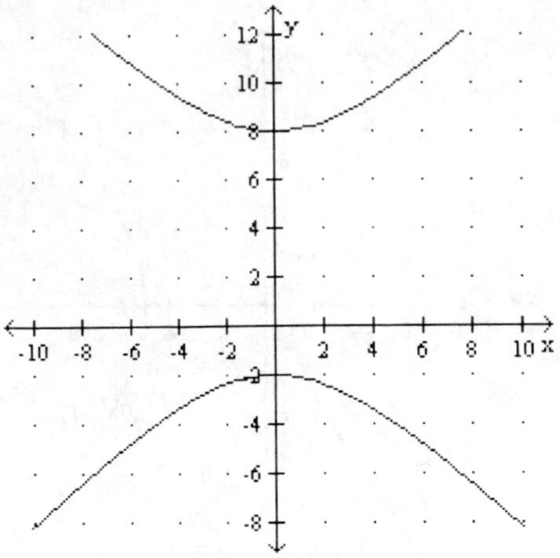

33.

$xy = 8$

$y = \dfrac{8}{x}$

x	$y = \dfrac{8}{x}$
1	$\dfrac{8}{1} = 8$
2	$\dfrac{8}{2} = 4$
4	$\dfrac{8}{4} = 2$
−4	$\dfrac{8}{-4} = -2$
−2	$\dfrac{8}{-2} = -4$
−1	$\dfrac{8}{-1} = -8$

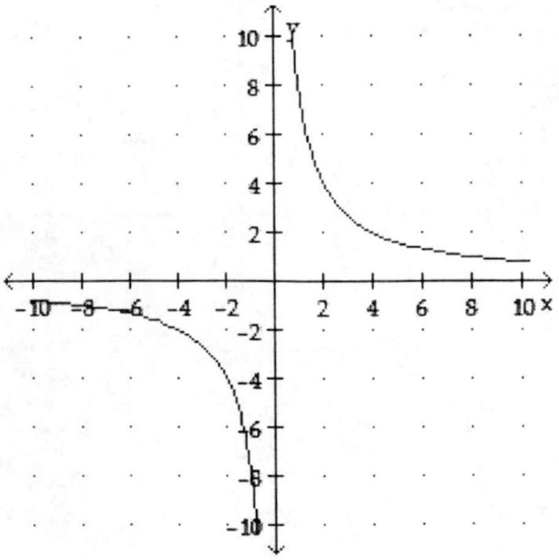

35.

$xy = -10$

$y = \dfrac{-10}{x}$

x	$y = \dfrac{-10}{x}$
1	$\dfrac{-10}{1} = -10$
2	$\dfrac{-10}{2} = -5$
5	$\dfrac{-10}{5} = -2$
−5	$\dfrac{-10}{-5} = 2$
−2	$\dfrac{-10}{-2} = 5$
−1	$\dfrac{-10}{-1} = 10$

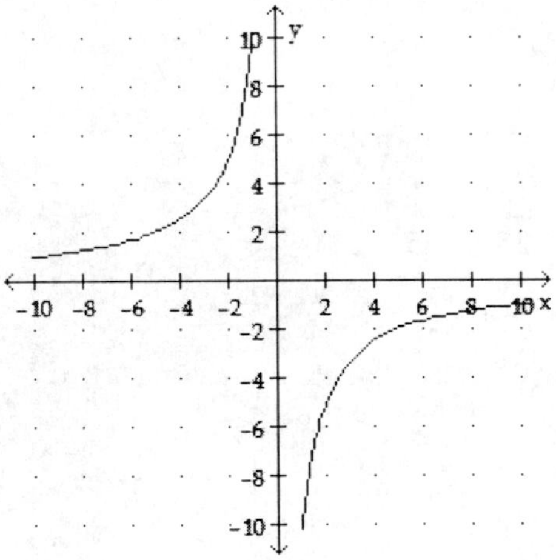

37. $\dfrac{x^2}{9} - \dfrac{y^2}{4} = 1$

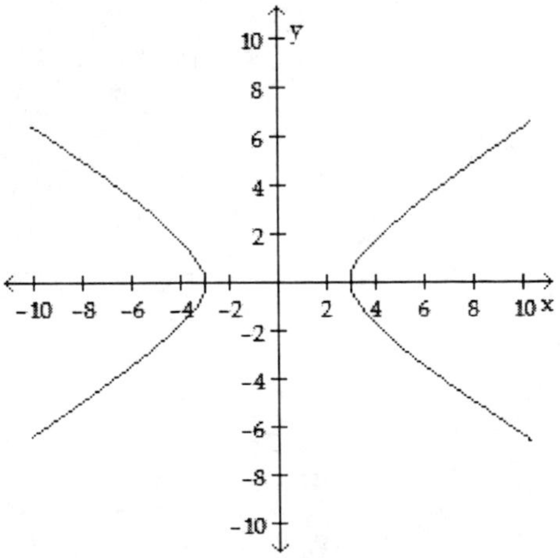

39. $\frac{x^2}{4}-\frac{(y-1)^2}{9}=1$

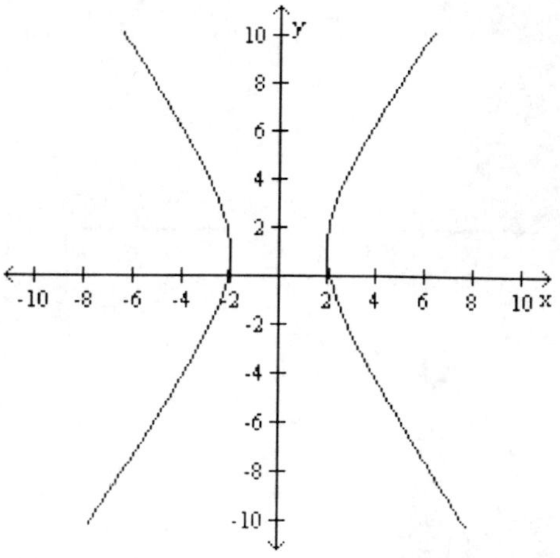

43.

$$9x^2-49y^2=441$$

$$\frac{9x^2}{441}-\frac{49y^2}{441}=\frac{441}{441}$$

$$\frac{x^2}{49}-\frac{y^2}{9}=1$$

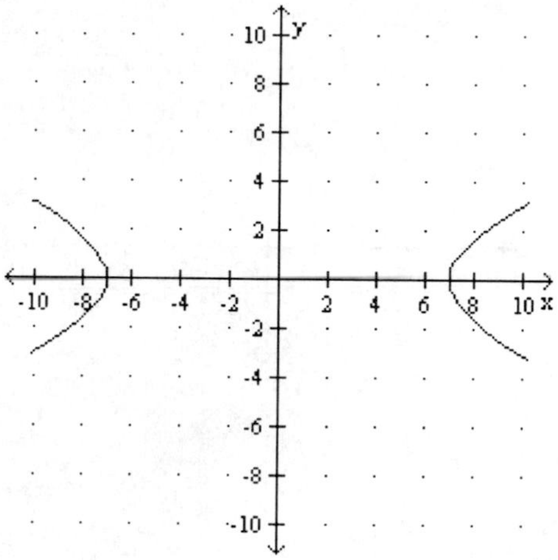

TRY IT YOURSELF

41. $(x+1)^2+(y-2)^2=16$

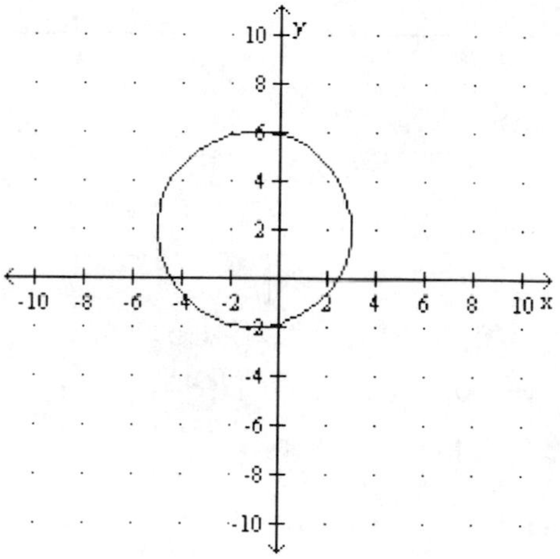

45.

$$4(x+1)^2+9(y+1)^2=36$$

$$\frac{4(x+1)^2}{36}+\frac{9(y+1)^2}{36}=\frac{36}{36}$$

$$\frac{(x+1)^2}{9}+\frac{(y+1)^2}{4}=1$$

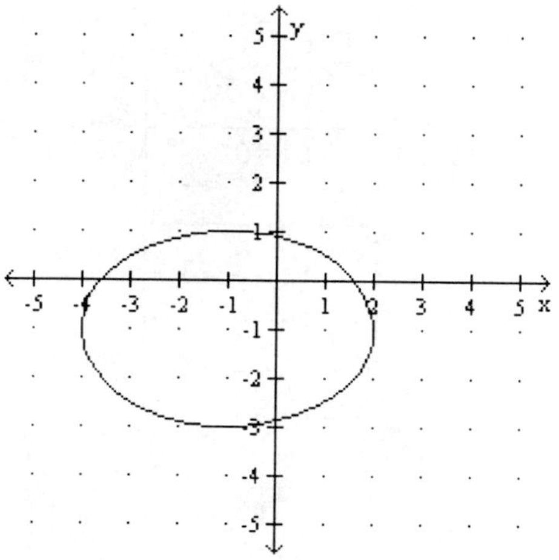

47.

$$4(x+3)^2-(y-1)^2=4$$

$$\frac{4(x+3)^2}{4}-\frac{(y-1)^2}{4}=\frac{4}{4}$$

$$\frac{(x+3)^2}{1}-\frac{(y-1)^2}{4}=1$$

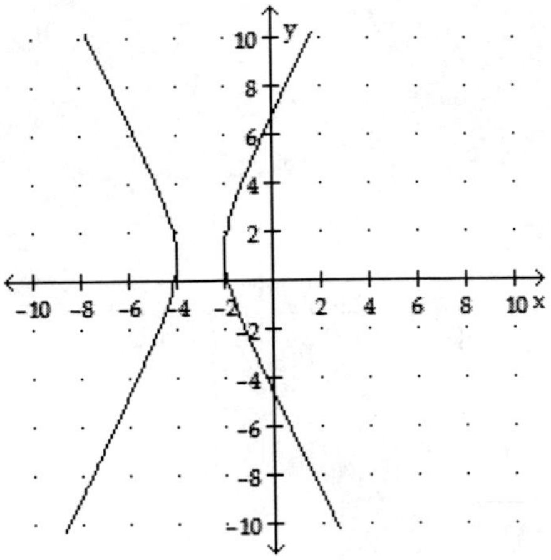

49.

$$xy=-6$$

$$y=\frac{-6}{x}$$

x	$y=\frac{-6}{x}$
1	$\frac{-6}{1}=-6$
2	$\frac{-6}{2}=-3$
3	$\frac{-6}{3}=-2$
−3	$\frac{-6}{-3}=2$
−2	$\frac{-6}{-2}=3$
−1	$\frac{-6}{-1}=6$

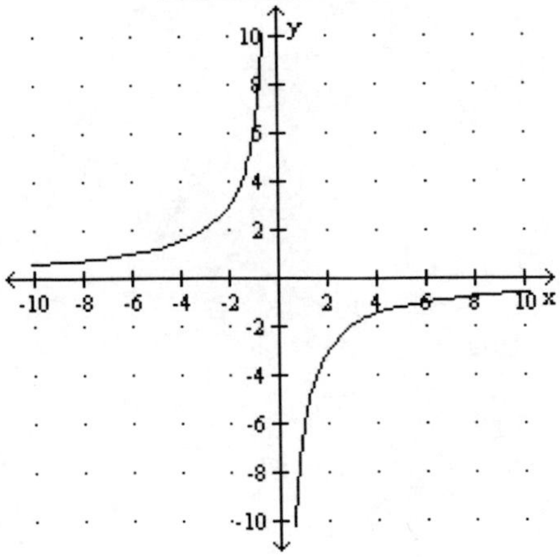

51. $x=\frac{1}{2}(y-1)^2-2$

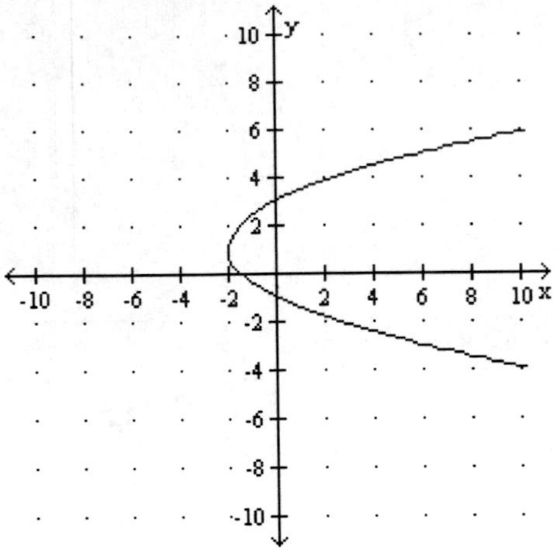

53. $\frac{y^2}{25}-\frac{(x-2)^2}{4}=1$

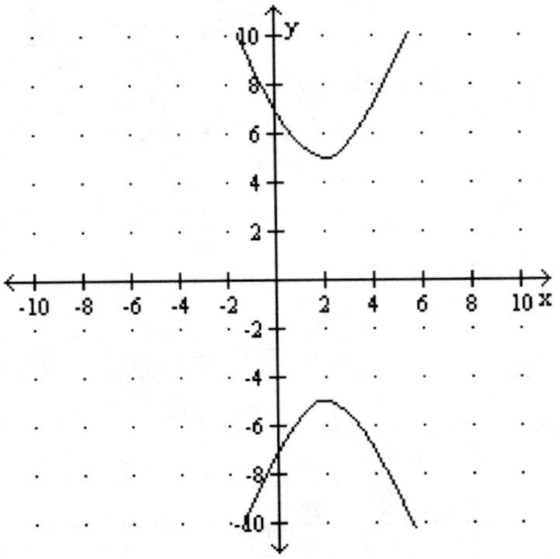

55.

$$y=-x^2+6x-4$$

$$y=-(x^2-6x)-4$$

$$y=-(x^2-6x+9)-4+9$$

$$y=-(x-3)^2+5$$

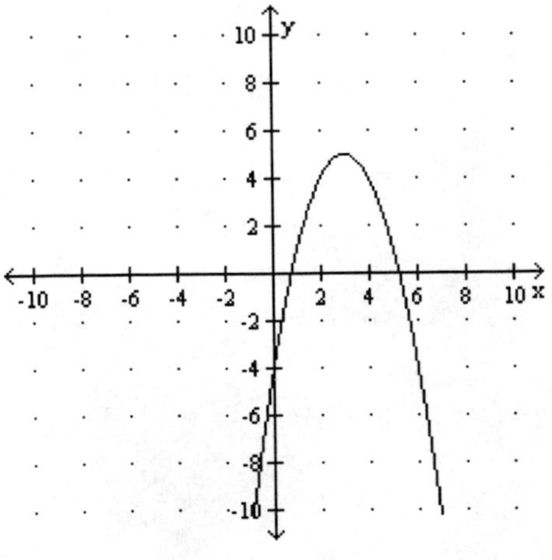

57. $\frac{x^2}{1}+\frac{y^2}{36}=1$

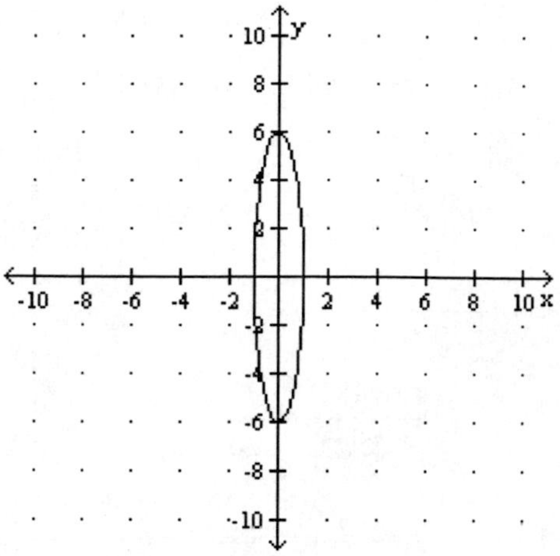

59.

$$x^2+y^2+4x-6y-23=0$$
$$\left(x^2+4x\right)+\left(y^2-6y\right)=23$$
$$\left(x^2+4x+4\right)+\left(y^2-6y+9\right)=23+4+9$$
$$(x+2)^2+(y-3)^2=36$$

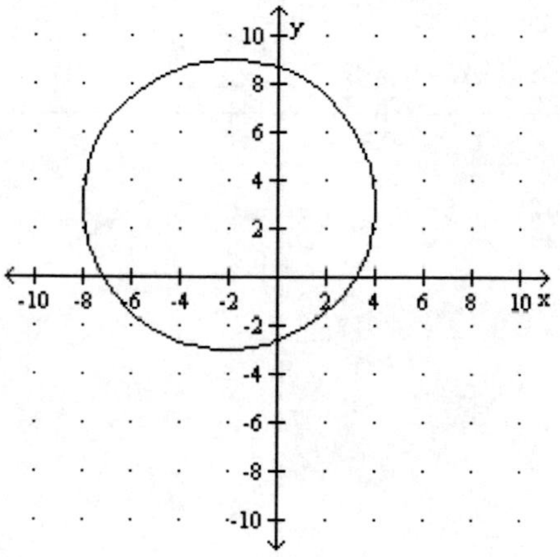

APPLICATIONS

61. ALPHA PARTICLES
Find the coordinates of the vertex.

$$9y^2-x^2=81$$
$$\frac{9y^2}{81}-\frac{x^2}{81}=\frac{81}{81}$$
$$\frac{y^2}{9}-\frac{x^2}{81}=1$$
$$a=3 \text{ and } b=9$$

The vertex is (3, 0), so the particle comes within 3 units of the nucleus.

63. CIVIL ENGINEER
Let $x=5$ and solve for y to find half of the width of the hyperbola.

$$y^2-x^2=25$$
$$y^2-5^2=25$$
$$y^2-25=25$$
$$y^2=50$$
$$y=\sqrt{50}$$
$$y=5\sqrt{2}$$

The width of the hyperbola would be $2y$.

$$2y=2\left(5\sqrt{2}\right)$$
$$=10\sqrt{2} \text{ miles}$$
$$\approx 14.1 \text{ miles}$$

65. NUCLEAR POWER
The tower is in the shape of a hyperbola.

WRITING

67. Answers will vary.

69. Answers will vary.

REVIEW

71.

$$\log_8 x=2$$
$$8^2=x$$
$$64=x$$

73.

$$\log_{1/2}\frac{1}{8}=x$$
$$\left(\frac{1}{2}\right)^x=\frac{1}{8}$$
$$2^{-x}=2^{-3}$$
$$-x=-3$$
$$x=3$$

75.

$$\log_x \frac{9}{4} = 2$$
$$x^2 = \frac{9}{4}$$
$$x = \sqrt{\frac{9}{4}}$$
$$x = \frac{3}{2}$$

77.

$$\log_x 1{,}000 = 3$$
$$x^3 = 1{,}000$$
$$x^3 = 10^3$$
$$x = 10$$

CHALLENGE PROBLEMS

79.

$$x^2 - y^2 - 2x + 4y = 12$$
$$x^2 - 2x - y^2 + 4y = 12$$
$$(x^2 - 2x) - (y^2 - 4y) = 12$$
$$(x^2 - 2x + 1) - (y^2 - 4y + 4) = 12 + 1 - 4$$
$$(x-1)^2 - (y-2)^2 = 9$$
$$\frac{(x-1)^2}{9} - \frac{(y-2)^2}{9} = 1$$

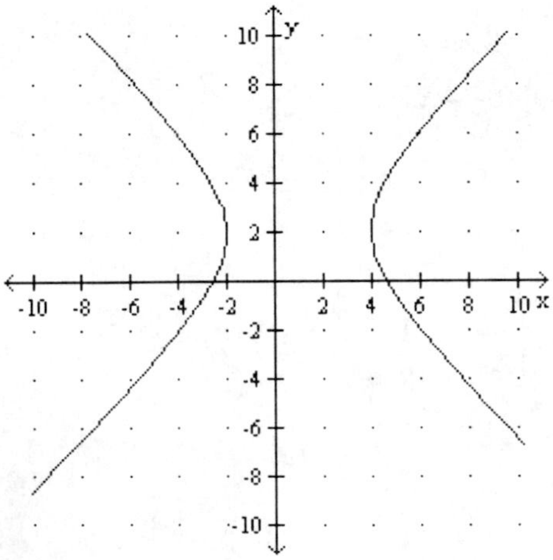

81.

$$36x^2 - 25y^2 - 72x - 100y = 964$$
$$(36x^2 - 72x) - (25y^2 + 100y) = 964$$
$$36(x^2 - 2x) - 25(y^2 + 4y) = 964$$
$$36(x^2 - 2x + 1) - 25(y^2 + 4y + 4) = 964 + 36(1) - 25(4)$$
$$36(x-1)^2 - 25(y+2)^2 = 900$$
$$\frac{36(x-1)^2}{900} - \frac{25(y+2)^2}{900} = \frac{900}{900}$$
$$\frac{(x-1)^2}{25} - \frac{(y+2)^2}{36} = 1$$

83.

$$16x^2 - 25y^2 = 1$$
$$\frac{x^2}{\frac{1}{16}} - \frac{y^2}{\frac{1}{25}} = 1$$

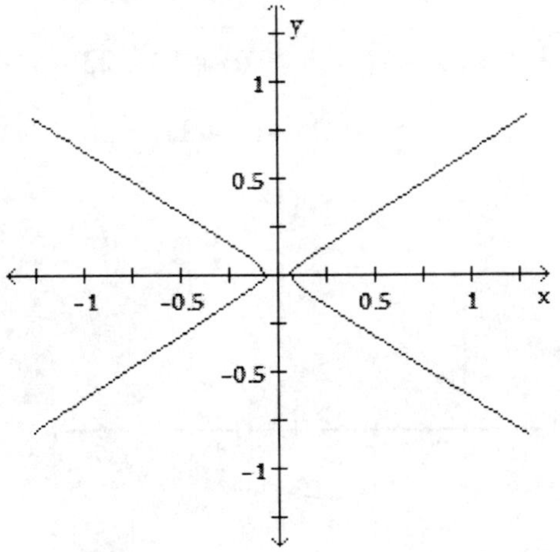

SECTION 10.4

VOCABULARY

1. $\begin{cases} 4x^2 + 6y^2 = 24 \\ 9x^2 - y^2 = 9 \end{cases}$ is a **system** of two nonlinear equations.

3. When solving a system by graphing, it is often difficult to determine the coordinates of the points of **intersection** of the graphs.

5. A **secant** is a line that intersects a circle at two points.

CONCEPTS

7. a. A line can intersect an ellipse in at most **two** points.
   b. An ellipse can intersect a parabola in at most **four** points.
   c. An ellipse can intersect a circle in at most **four** points.
   d. A hyperbola can intersect a circle in at most **four** points.

9. They intersect at (–3, 2), (3, 2), (–3, –2), and (3, –2).

11. a. –4
    b. –2

NOTATION

13.

$$x^2 + y^2 = 5$$
$$x^2 + \left(\boxed{2x}\right)^2 = 5$$
$$x^2 + 4x^2 = \boxed{5}$$
$$\boxed{5}x^2 = 5$$
$$x^2 = \boxed{1}$$
$$x = 1 \quad \text{or} \quad x = -1$$

If $x = 1$, then

$$y = 2\left(\boxed{1}\right) = 2$$

If $x = -1$, then

$$y = 2\left(\boxed{-1}\right) = -2$$

The solutions are $(1,2)$ and $\left(-1, \boxed{-2}\right)$.

GUIDED PRACTICE

15. $\begin{cases} x^2 + y^2 = 9 \\ y - x = 3 \end{cases}$

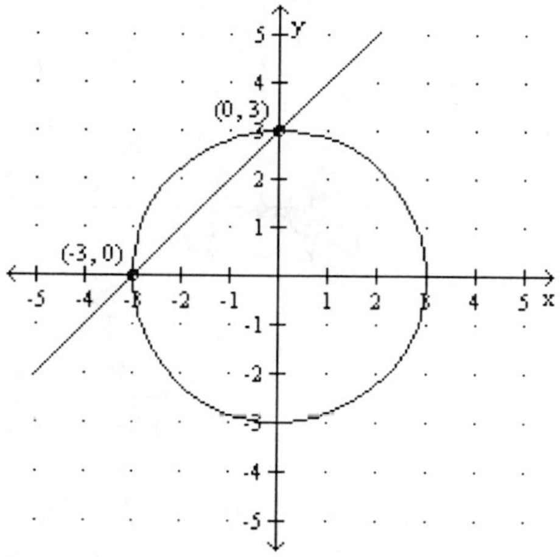

The solutions are (0, 3) and (–3, 0).

17. $\begin{cases} 9x^2 + 16y^2 = 144 \\ 9x^2 - 16y^2 = 144 \end{cases} \Rightarrow \begin{cases} \dfrac{x^2}{16} + \dfrac{y^2}{9} = 1 \\ \dfrac{x^2}{16} - \dfrac{y^2}{9} = 1 \end{cases}$

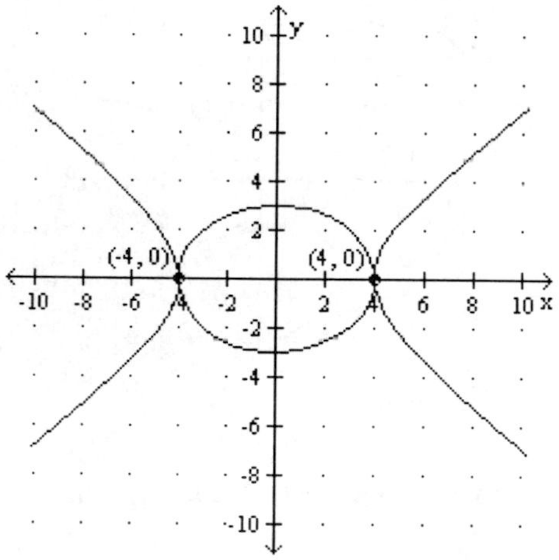

The solutions are (–4, 0) and (4, 0).

19. $\begin{cases} y = x^2 - 4x \\ x^2 + y = 0 \end{cases} \Rightarrow \begin{cases} y = x^2 - 4x \\ y = -x^2 \end{cases}$

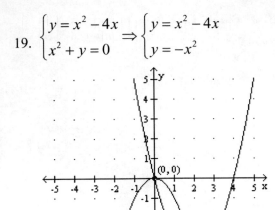

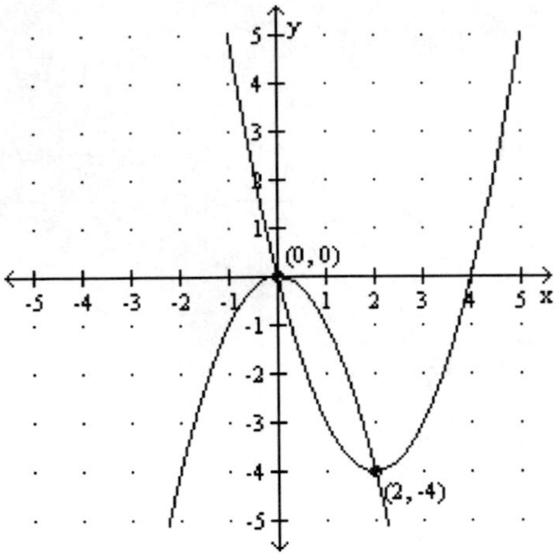

The solutions are (0, 0) and (2, –4).

21. $\begin{cases} x^2 + 4y^2 = 4 \\ x = 2y^2 - 2 \end{cases} \Rightarrow \begin{cases} \dfrac{x^2}{4} + \dfrac{y^2}{1} = 1 \\ x = 2y^2 - 2 \end{cases}$

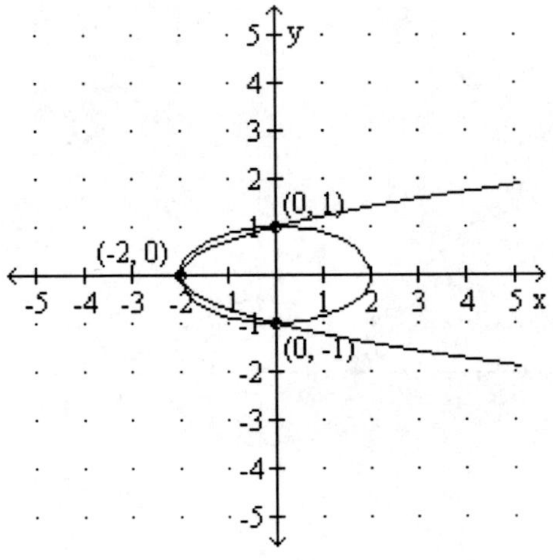

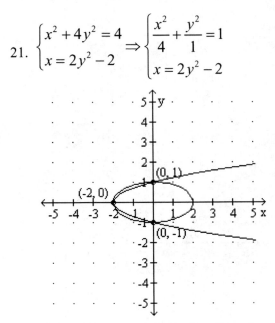

The solutions are (–2, 0), (0, –1), and (0, 1).

23.

$\begin{cases} x^2 + y^2 = 5 \\ x + y = 3 \end{cases}$

Solve the second equation for *y*.

$x + y = 3$

$y = -x + 3$

Substitute $y = -x + 3$ into the first equation and solve for *x*.

$$x^2 + y^2 = 5$$
$$x^2 + (-x + 3)^2 = 5$$
$$x^2 + (x^2 - 3x - 3x + 9) = 5$$
$$x^2 + (x^2 - 6x + 9) = 5$$
$$2x^2 - 6x + 9 = 5$$
$$2x^2 - 6x + 4 = 0$$
$$2(x^2 - 3x + 2) = 0$$
$$2(x - 1)(x - 2) = 0$$
$$x - 1 = 0 \quad \text{or} \quad x - 2 = 0$$
$$x = 1 \qquad\qquad x = 2$$

Substitute $x = 1$ into $x + y = 2$ and solve for *y*.

$(1) + y = 3 \qquad (2) + y = 3$

$y = 2 \qquad\qquad y = 1$

The solutions are $(1, 2)$ and $(2, 1)$.

25.

$$\begin{cases} y = x^2 + 6x + 7 \\ 2x + y = -5 \end{cases}$$

Substitute $y = x^2 + 6x + 7$ into the second equation and solve for x.

$$2x + y = -5$$
$$2x + (x^2 + 6x + 7) = -5$$
$$x^2 + 8x + 7 = -5$$
$$x^2 + 8x + 12 = 0$$
$$(x + 2)(x + 6) = 0$$
$$x + 2 = 0 \quad \text{and} \quad x + 6 = 0$$
$$x = -2 \qquad x = -6$$

Substitute $x = -2$ and $x = -6$ into the second equation and solve for y.

$$2x + y = -5 \qquad 2x + y = -5$$
$$2(-2) + y = -5 \qquad 2(-6) + y = -5$$
$$-4 + y = -5 \qquad -12 + y = -5$$
$$y = -1 \qquad y = 7$$

The solutions are (–2, –1) and (–6, 7).

27.

$$\begin{cases} x^2 + y^2 = 13 \\ y = x^2 - 1 \end{cases}$$

Substitute $y = x^2 - 1$ into the first equation and solve for x.

$$x^2 + y^2 = 13$$
$$x^2 + (x^2 - 1)^2 = 13$$
$$x^2 + (x^4 - x^2 - x^2 + 1) = 13$$
$$x^2 + (x^4 - 2x^2 + 1) = 13$$
$$x^4 - x^2 + 1 = 13$$
$$x^4 - x^2 - 12 = 0$$
$$(x^2 - 4)(x^2 + 3) = 0$$
$$x^2 - 4 = 0 \quad \text{or} \quad x^2 + 3 = 0$$
$$x^2 = 4 \qquad x^2 = -3$$
$$x = \sqrt{4} \qquad x = \sqrt{-3}$$
$$x = \pm 2 \qquad \text{not real}$$

Substitute $x = 2$ and $x = -2$ into $y = x^2 - 1$ and solve for y.

$$y = (2)^2 - 1 \qquad y = (-2)^2 - 1$$
$$y = 4 - 1 \qquad y = 4 - 1$$
$$y = 3 \qquad y = 3$$

The solutions are $(2,3)$ and $(-2,3)$.

29.

$$\begin{cases} x^2 + y^2 = 30 \\ y = x^2 \end{cases}$$

Substitute $y = x^2$ into the first equation and solve for x.

$$x^2 + y^2 = 30$$
$$x^2 + (x^2)^2 = 30$$
$$x^2 + x^4 = 30$$
$$x^4 + x^2 - 30 = 0$$
$$(x^2 + 6)(x^2 - 5) = 0$$
$$x^2 + 6 = 0 \quad \text{and} \quad x^2 - 5 = 0$$
$$x^2 = -6 \qquad x^2 = 5$$
$$x = \pm\sqrt{-6} \qquad x = \pm\sqrt{5}$$
$$\text{not real}$$

Substitute $x = \pm\sqrt{5}$ into the second equation and solve for y.

$$y = x^2 \qquad y = x^2$$
$$y = (\sqrt{5})^2 \qquad y = (-\sqrt{5})^2$$
$$y = 5 \qquad y = 5$$

The solutions are $(\sqrt{5}, 5)$ and $(-\sqrt{5}, 5)$.

31.

$$\begin{cases} x^2 + y^2 = 20 \\ x^2 - y^2 = -12 \end{cases}$$

Add the equations and use elimination to solve for x.

$$\begin{array}{r} x^2 + y^2 = 20 \\ \underline{x^2 - y^2 = -12} \\ 2x^2 = 8 \end{array}$$
$$x^2 = 4$$
$$x = \sqrt{4}$$
$$x = \pm 2$$

Substitute $x = 2$ and $x = -2$ into $x^2 + y^2 = 20$ and solve for y.

$$x^2 + y^2 = 20 \qquad x^2 + y^2 = 20$$
$$(2)^2 + y^2 = 20 \qquad (-2)^2 + y^2 = 20$$
$$4 + y^2 = 20 \qquad 4 + y^2 = 20$$
$$y^2 = 16 \qquad y^2 = 16$$
$$y = \sqrt{16} \qquad y = \sqrt{16}$$
$$y = \pm 4 \qquad y = \pm 4$$

The solutions are $(2,4)$, $(2,-4)$, $(-2,4)$ and $(-2,-4)$.

33.

$$\begin{cases} 9x^2 - 7y^2 = 81 \\ x^2 + y^2 = 9 \end{cases}$$

Multiply the second equation by -9 use elimination to solve for y.

$$\begin{aligned} 9x^2 - 7y^2 &= 81 \\ -9x^2 - 9y^2 &= -81 \\ \hline -16y^2 &= 0 \\ y^2 &= 0 \\ y &= \sqrt{0} \\ y &= 0 \end{aligned}$$

Substitute $y = 0$ into $x^2 + y^2 = 9$ and solve for x.

$$\begin{aligned} x^2 + y^2 &= 9 \\ x^2 + 0^2 &= 9 \\ x^2 &= 9 \\ x &= \sqrt{9} \\ x &= \pm 3 \end{aligned}$$

The solutions are $(3,0)$ and $(-3,0)$.

35.

$$\begin{cases} 2x^2 + y^2 = 6 \\ x^2 - y^2 = 3 \end{cases}$$

Add the two equations and use elimination to solve for x.

$$\begin{aligned} 2x^2 + y^2 &= 6 \\ x^2 - y^2 &= 3 \\ \hline 3x^2 &= 9 \\ x^2 &= 3 \\ x &= \pm\sqrt{3} \end{aligned}$$

Substitute $x = \sqrt{3}$ and $x = -\sqrt{3}$ into $x^2 - y^2 = 3$ and solve for y.

$$\begin{aligned} x^2 - y^2 &= 3 & x^2 - y^2 &= 3 \\ \left(\sqrt{3}\right)^2 - y^2 &= 3 & \left(-\sqrt{3}\right)^2 - y^2 &= 3 \\ 3 - y^2 &= 3 & 3 - y^2 &= 3 \\ -y^2 &= 0 & -y^2 &= 0 \\ y^2 &= 0 & y^2 &= 0 \\ y &= \sqrt{0} & y &= \sqrt{0} \\ y &= 0 & y &= 0 \end{aligned}$$

The solutions are $\left(\sqrt{3},0\right)$ and $\left(-\sqrt{3},0\right)$.

37.

$$\begin{cases} x^2 - y^2 = -5 \\ 3x^2 + 2y^2 = 30 \end{cases}$$

Multiply the first equation by 2 and use elimination to solve for x.

$$\begin{aligned} 2x^2 - 2y^2 &= -10 \\ 3x^2 + 2y^2 &= 30 \\ \hline 5x^2 &= 20 \\ x^2 &= 4 \\ x &= \sqrt{4} \\ x &= \pm 2 \end{aligned}$$

Substitute $x = 2$ and $x = -2$ into $x^2 - y^2 = -5$ and solve for y.

$$\begin{aligned} x^2 - y^2 &= -5 & x^2 - y^2 &= -5 \\ (2)^2 - y^2 &= -5 & (-2)^2 - y^2 &= -5 \\ 4 - y^2 &= -5 & 4 - y^2 &= -5 \\ -y^2 &= -9 & -y^2 &= -9 \\ y^2 &= 9 & y^2 &= 9 \\ y &= \pm\sqrt{9} & y &= \pm\sqrt{9} \\ y &= \pm 3 & y &= \pm 3 \end{aligned}$$

The solutions are $(2, 3)$, $(2,-3)$, $(-2, 3)$, and $(-2,-3)$.

39.

$$\begin{cases} x^2 - 6x - y = -5 \\ x^2 - 6x + y = -5 \end{cases}$$

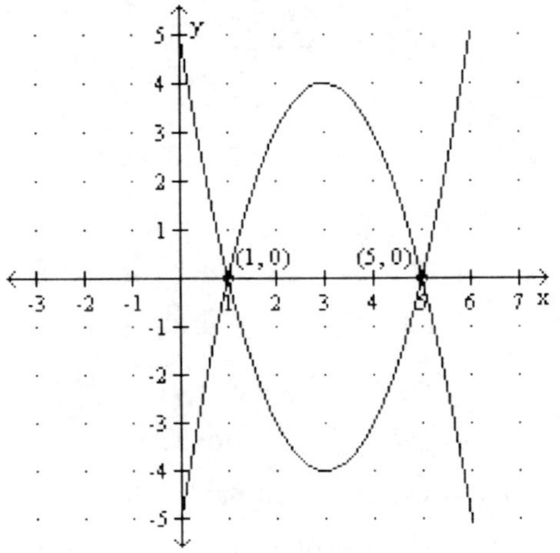

The solutions are (1, 0) and (5, 0).

TRY IT YOURSELF

41. $\begin{cases} 2x^2 - 3y^2 = 5 \\ 3x^2 + 4y^2 = 16 \end{cases}$

Multiply the first equation by 4 and the second equation by 3. Add the equations and use elimination to solve for x.

$$\begin{aligned} 8x^2 - 12y^2 &= 20 \\ 9x^2 + 12y^2 &= 48 \\ \hline 17x^2 &= 68 \\ x^2 &= 4 \\ x &= \pm 2 \end{aligned}$$

Substitute $x = 2$ and $x = -2$ into the first equation and solve for y.

$$\begin{aligned} 2x^2 - 3y^2 &= 5 \\ 2(2)^2 - 3y^2 &= 5 \\ 2(4) - 3y^2 &= 5 \\ 8 - 3y^2 &= 5 \\ -3y^2 &= -3 \\ y^2 &= 1 \\ y &= \pm 1 \end{aligned} \qquad \begin{aligned} 2x^2 - 3y^2 &= 5 \\ 2(-2)^2 - 3y^2 &= 5 \\ 2(4) - 3y^2 &= 5 \\ 8 - 3y^2 &= 5 \\ -3y^2 &= -3 \\ y^2 &= 1 \\ y &= \pm 1 \end{aligned}$$

The solutions are (2, 1), (2, –1), (–2, 1), and (–2, –1).

43. $\begin{cases} y = x^2 - 4 \\ x^2 - y^2 = -16 \end{cases}$

Write the first equation in standard form and use elimination.

$$\begin{aligned} -x^2 + y &= -4 \\ x^2 - y^2 &= -16 \\ \hline -y^2 + y &= -20 \\ y^2 - y - 20 &= 0 \\ (y+4)(y-5) &= 0 \end{aligned}$$

$$y + 4 = 0 \quad \text{or} \quad y - 5 = 0$$
$$y = -4 \qquad\qquad y = 5$$

Substitute $y = -4$ and $y = 5$ into $y = x^2 - 4$ and solve for x.

$$\begin{aligned} -4 &= x^2 - 4 \\ x^2 - 4 &= -4 \\ x^2 &= 0 \\ x &= \sqrt{0} \\ x &= 0 \end{aligned} \qquad \begin{aligned} 5 &= x^2 - 4 \\ x^2 - 4 &= 5 \\ x^2 &= 9 \\ x &= \sqrt{9} \\ x &= \pm 3 \end{aligned}$$

The solutions are $(0,-4), (3,5)$, and $(-3,5)$.

45.

$\begin{cases} 3y^2 = xy \\ 2x^2 + xy - 84 = 0 \end{cases}$

Solve the first equation for x.

$$\begin{aligned} 3y^2 &= xy \\ \frac{3y^2}{y} &= x \\ 3y &= x \end{aligned}$$

Substitute $3y = x$ into the second equation and solve for y.

$$\begin{aligned} 2(3y)^2 + (3y)y - 84 &= 0 \\ 2(9y^2) + 3y^2 - 84 &= 0 \\ 18y^2 + 3y^2 - 84 &= 0 \\ 21y^2 - 84 &= 0 \\ 21y^2 &= 84 \\ y^2 &= 4 \\ y &= \sqrt{4} \\ y &= \pm 2 \end{aligned}$$

Substitute $y = 2$, and $y = -2$ into the first equation to solve for x.

$$\begin{aligned} 3y^2 &= xy \\ 3(2)^2 &= x(2) \\ 12 &= 2x \\ 6 &= x \end{aligned} \qquad \begin{aligned} 3y^2 &= xy \\ 3(-2)^2 &= x(-2) \\ 12 &= -2x \\ -6 &= x \end{aligned}$$

Set the denominator of $\frac{3y^2}{y} = 0$ and you also get $y = 0$. Substitute $y = 0$ into the second equation and solve for x.

$$\begin{aligned} 2x^2 + xy - 84 &= 0 \\ 2x^2 + x(0) - 84 &= 0 \\ 2x^2 - 84 &= 0 \\ 2(x^2 - 42) &= 0 \\ x^2 &= 42 \\ x &= \pm\sqrt{42} \end{aligned}$$

The solutions are $(6,2), (-6,-2), (\sqrt{42},0)$ and $(-\sqrt{42},0)$.

47.

$$\begin{cases} y^2 = 40 - x^2 \\ y = x^2 - 10 \end{cases}$$

Write both equations in standard form and use elimination.

$$\begin{array}{r} x^2 + y^2 = 40 \\ \underline{-x^2 + y = -10} \\ y^2 + y = 30 \end{array}$$

$$y^2 + y - 30 = 0$$

$$(y+6)(y-5) = 0$$

$$y + 6 = 0 \quad \text{or} \quad y - 5 = 0$$

$$y = -6 \qquad y = 5$$

Substitute $y = -6$ and $y = 5$ into $y = x^2 - 10$ and solve for x.

$$-6 = x^2 - 10 \qquad 5 = x^2 - 10$$

$$x^2 - 10 = -6 \qquad x^2 - 10 = 5$$

$$x^2 = 4 \qquad x^2 = 15$$

$$x = \sqrt{4} \qquad x = \sqrt{15}$$

$$x = \pm 2 \qquad x = \pm\sqrt{15}$$

The solutions are $(2,-6),(-2,-6),(\sqrt{15},5)$, and $(-\sqrt{15},5)$.

49.

$$\begin{cases} 3x - y = -3 \\ 25y^2 - 9x^2 = 225 \end{cases}$$

Solve the first equation for y.

$$\begin{cases} y = 3x + 3 \\ 25y^2 - 9x^2 = 225 \end{cases}$$

Substitute $y = 3x + 3$ into the second equation and solve for x.

$$25y^2 - 9x^2 = 225$$

$$25(3x+3)^2 - 9x^2 = 225$$

$$25(9x^2 + 18x + 9) - 9x^2 = 225$$

$$225x^2 + 450x + 225 - 9x^2 = 225$$

$$216x^2 + 450x = 0$$

$$18x(12x + 25) = 0$$

$$18x = 0 \quad \text{and} \quad 12x + 25 = 0$$

$$x = 0 \qquad x = -\frac{25}{12}$$

Substitute $x = 0$ and $x = -\frac{25}{12}$ into the first equation and solve for y.

$$y = 3x + 3 \qquad y = 3x + 3$$

$$y = 3(0) + 3 \qquad y = 3\left(-\frac{25}{12}\right) + 3$$

$$y = 0 + 3 \qquad y = -\frac{25}{4} + 3$$

$$y = 3 \qquad y = -\frac{13}{4}$$

The solutions are (0, 3) and $\left(-\frac{25}{12}, -\frac{13}{4}\right)$.

51.

$$\begin{cases} x^2 - y = 0 \\ x^2 - 4x + y = 0 \end{cases}$$

Add the two equations and solve for x.

$$\begin{array}{r} x^2 \qquad - y = 0 \\ \underline{x^2 - 4x + y = 0} \end{array}$$

$$2x^2 - 4x = 0$$

$$2x(x-2) = 0$$

$$2x = 0 \quad \text{and} \quad x - 2 = 0$$

$$x = 0 \qquad\qquad x = 2$$

Substitute $x = 0$ and $x = 2$ into the first equation and solve for y.

$$\begin{array}{rr} x^2 - y = 0 & x^2 - y = 0 \\ 0^2 - y = 0 & 2^2 - y = 0 \\ -y = 0 & 4 - y = 0 \\ y = 0 & 4 = y \end{array}$$

The solutions are (0, 0) and (2, 4).

53.

$$\begin{cases} x^2 - 2y^2 = 6 \\ x^2 + 2y^2 = 2 \end{cases}$$

Add the equations and solve for x.

$$\begin{array}{r} x^2 - 2y^2 = 6 \\ \underline{x^2 + 2y^2 = 2} \\ 2x^2 = 8 \\ x^2 = 4 \\ x = \pm 2 \end{array}$$

Substitute $x = 2$ and $x = -2$ into the first equation and solve for y.

$$\begin{array}{rr} x^2 - 2y^2 = 6 & x^2 - 2y^2 = 6 \\ (2)^2 - 2y^2 = 6 & (-2)^2 - 2y^2 = 6 \\ 4 - 2y^2 = 6 & 4 - 2y^2 = 6 \\ -2y^2 = 2 & -2y^2 = 2 \\ y^2 = -1 & y^2 = -1 \\ y = \sqrt{-1} & y = \sqrt{-1} \\ y = i & y = i \end{array}$$

not real   not real

No solution; ∅

55.

$$\begin{cases} y = x^2 - 4 \\ 6x - y = 13 \end{cases}$$

Substitute $y = x^2 - 4$ into the second equation and solve for x.

$$6x - y = 13$$

$$6x - (x^2 - 4) = 13$$

$$6x - x^2 + 4 = 13$$

$$0 = x^2 - 6x + 9$$

$$0 = (x-3)(x-3)$$

$$x - 3 = 0 \quad \text{and} \quad x - 3 = 0$$

$$x = 3 \qquad\qquad x = 3$$

Substitute $x = 3$ into the second equation and solve for y.

$$6x - y = 13$$

$$6(3) - y = 13$$

$$18 - y = 13$$

$$-y = -5$$

$$y = 5$$

The solution is (3, 5).

57. $\begin{cases} x^2 + y^2 = 4 \\ 9x^2 + y^2 = 9 \end{cases}$

Multiply the first equation by -1 and add the equations to solve for x.

$$\begin{aligned} -x^2 - y^2 &= -4 \\ 9x^2 + y^2 &= 9 \\ \hline 8x^2 &= 5 \\ x^2 &= \frac{5}{8} \\ x &= \pm\sqrt{\frac{5}{8}} \\ x &= \pm\sqrt{\frac{5\cdot 2}{8\cdot 2}} \\ x &= \pm\frac{\sqrt{10}}{4} \end{aligned}$$

Substitute $x = \pm\frac{\sqrt{10}}{4}$ into the first equation and solve for y.

$$\begin{aligned} x^2 + y^2 &= 4 \\ \left(\frac{\sqrt{10}}{4}\right)^2 + y^2 &= 4 \\ \frac{10}{16} + y^2 &= 4 \\ \frac{5}{8} + y^2 &= 4 \\ y^2 &= 4 - \frac{5}{8} \\ y^2 &= \frac{27}{8} \\ y &= \pm\sqrt{\frac{27}{8}} \\ y &= \pm\sqrt{\frac{27\cdot 2}{8\cdot 2}} \\ y &= \pm\frac{\sqrt{54}}{\sqrt{16}} \\ y &= \pm\frac{3\sqrt{6}}{4} \end{aligned}$$

The solutions are $\left(\frac{\sqrt{10}}{4}, \frac{3\sqrt{6}}{4}\right)$, $\left(-\frac{\sqrt{10}}{4}, \frac{3\sqrt{6}}{4}\right)$, $\left(\frac{\sqrt{10}}{4}, -\frac{3\sqrt{6}}{4}\right)$, and $\left(-\frac{\sqrt{10}}{4}, -\frac{3\sqrt{6}}{4}\right)$.

59. $\begin{cases} xy = \frac{1}{6} \\ y + x = 5xy \end{cases}$

Solve the first equation for y.

$$\begin{aligned} xy &= \frac{1}{6} \\ \frac{1}{x}(xy) &= \frac{1}{x}\left(\frac{1}{6}\right) \\ y &= \frac{1}{6x} \end{aligned}$$

Substitute $y = \frac{1}{6x}$ into the second equation and solve for x.

$$\begin{aligned} y + x &= 5xy \\ \frac{1}{6x} + x &= 5x\left(\frac{1}{6x}\right) \\ \frac{1}{6x} + x &= \frac{5}{6} \\ 6x\left(\frac{1}{6x} + x\right) &= 6x\left(\frac{5}{6}\right) \\ 1 + 6x^2 &= 5x \\ 6x^2 - 5x + 1 &= 0 \\ (2x-1)(3x-1) &= 0 \end{aligned}$$

$$2x - 1 = 0 \quad \text{or} \quad 3x - 1 = 0$$

$$x = \frac{1}{2} \qquad\qquad x = \frac{1}{3}$$

Substitute $x = \frac{1}{2}$ and $x = \frac{1}{3}$ into $xy = \frac{1}{6}$ and solve for y.

$$\begin{aligned} xy &= \frac{1}{6} & xy &= \frac{1}{6} \\ \frac{1}{2}y &= \frac{1}{6} & \frac{1}{3}y &= \frac{1}{6} \\ 6\left(\frac{1}{2}y\right) &= 6\left(\frac{1}{6}\right) & 6\left(\frac{1}{3}y\right) &= 6\left(\frac{1}{6}\right) \\ 3y &= 1 & 2y &= 1 \\ y &= \frac{1}{3} & y &= \frac{1}{2} \end{aligned}$$

The solutions are $\left(\frac{1}{2}, \frac{1}{3}\right)$ and $\left(\frac{1}{3}, \frac{1}{2}\right)$.

61. $\begin{cases} x^2 = 4 - y \\ y = x^2 + 2 \end{cases}$

Substitute $y = x^2 + 2$ into the first equation and solve for x.

$$x^2 = 4 - y$$
$$x^2 = 4 - (x^2 + 2)$$
$$x^2 = 4 - x^2 - 2$$
$$2x^2 = 2$$
$$x^2 = 1$$
$$x = \pm 1$$

Substitute $x = 1$ and $x = -1$ into the second equation and solve for y.

$y = x^2 + 2$	$y = x^2 + 2$
$y = (1)^2 + 2$	$y = (-1)^2 + 2$
$y = 1 + 2$	$y = 1 + 2$
$y = 3$	$y = 3$

The solutions are (1, 3) and (–1, 3).

63. $\begin{cases} x^2 - y^2 = 4 \\ x + y = 4 \end{cases}$

Solve the second equation for x.

$$\begin{cases} x^2 - y^2 = 4 \\ x = -y + 4 \end{cases}$$

Substitute $x = -y + 4$ into the first equation and solve for y.

$$x^2 - y^2 = 4$$
$$(-y + 4)^2 - y^2 = 4$$
$$y^2 - 8y + 16 - y^2 = 4$$
$$-8y + 16 = 4$$
$$-8y = -12$$
$$y = \frac{3}{2}$$

Substitute $y = \frac{3}{2}$ into the second equation and solve for x.

$$x + y = 4$$
$$x + \frac{3}{2} = 4$$
$$x = \frac{5}{2}$$

The solution is $\left(\frac{5}{2}, \frac{3}{2}\right)$.

APPLICATIONS

65. INTEGER PROBLEM

Let the integers be x and y.

$$\begin{cases} xy = 32 \\ x + y = 12 \end{cases}$$

Solve the first equation for y.

$$xy = 32$$
$$y = \frac{32}{x}$$

Substitute $y = \frac{32}{x}$ into $x + y = 12$ and solve for x.

$$x + y = 12$$
$$x + \frac{32}{x} = 12$$
$$x\left(x + \frac{32}{x}\right) = x(12)$$
$$x^2 + 32 = 12x$$
$$x^2 - 12x + 32 = 0$$
$$(x - 4)(x - 8) = 0$$
$$x - 4 = 0 \quad \text{or} \quad x - 8 = 0$$
$$x = 4 \qquad\qquad x = 8$$

Substitute $x = 4$ and $x = 8$ into the first equation to solve for y.

$xy = 32$	$xy = 32$
$4y = 32$	$8y = 32$
$y = 8$	$y = 4$

The two integers are 4 and 8.

67. ARCHERY

$$\begin{cases} y = -\frac{1}{6}x^2 + 2x \\ y = \frac{1}{3}x \end{cases}$$

Substitute $y = \frac{1}{3}x$ into the first equation and solve for x.

$$y = -\frac{1}{6}x^2 + 2x$$
$$\frac{1}{3}x = -\frac{1}{6}x^2 + 2x$$
$$6\left(\frac{1}{3}x\right) = 6\left(-\frac{1}{6}x^2 + 2x\right)$$
$$2x = -x^2 + 12x$$
$$x^2 - 10x = 0$$
$$x(x-10) = 0$$
$$x = 0 \quad \text{and} \quad x - 10 = 0$$
$$x = 0 \qquad\qquad x = 10$$

Substitute $x = 10$ into the second equation.

$$y = \frac{1}{3}x$$
$$y = \frac{1}{3}(10)$$
$$y = \frac{10}{3}$$

The coordinates of the point of impact are $\left(10, \frac{10}{3}\right)$.

To find the distance of the gun from the point of impact, you must use the Pythagorean Theorem. 10 is one leg (the distance on the x–axis), $\frac{10}{3}$ is the other leg (the height from the x–axis of the hill at the point of impact), and the hypotenuse is the distance on the hill from the archer to the point of impact.

$$a^2 + b^2 = c^2$$
$$10^2 + \left(\frac{10}{3}\right)^2 = c^2$$
$$100 + \frac{100}{9} = c^2$$
$$\frac{1{,}000}{9} = c^2$$
$$c = \sqrt{\frac{1{,}000}{9}}$$
$$c = \frac{10\sqrt{10}}{3} \text{ miles}$$

69. FENCING PASTURES

Let x = length and y = width.
Area = length · width
The perimeter is 2 widths + 1 length since the other length is riverbank.

$$\begin{cases} xy = 8{,}000 \\ x + 2y = 260 \end{cases}$$

Solve the first equation for y.

$$xy = 63$$
$$y = \frac{8{,}000}{x}$$

Substitute $y = \frac{8{,}000}{x}$ into the second equation and solve for x.

$$x + 2y = 260$$
$$x + 2\left(\frac{8{,}000}{x}\right) = 260$$
$$x + \frac{16{,}000}{x} = 260$$
$$x\left(x + \frac{16{,}000}{x}\right) = x(260)$$
$$x^2 + 16{,}000 = 260x$$
$$x^2 - 260x + 16{,}000 = 0$$
$$(x-100)(x-160) = 0$$
$$x - 100 = 0 \quad \text{or} \quad x - 160 = 0$$
$$x = 100 \qquad\qquad x = 160$$

Substitute $x = 100$ and $x = 160$ into the first equation and solve for y.

$$xy = 8{,}000 \qquad\qquad xy = 8{,}000$$
$$100y = 8{,}000 \qquad\qquad 160y = 8{,}000$$
$$y = 80 \qquad\qquad y = 50$$

The dimensions are 80 ft by 100 ft or 50 ft by 160 ft.

71. INVESTING

Let x = the amount invested and y = rate of investment.

Grant's investment:

$$xy = 225$$

Jeff's investment:

$$(x + 500)(y - 0.01) = 240$$

$$\begin{cases} xy = 225 \\ (x+500)(y-0.01) = 240 \end{cases}$$

Solve the first equation for y and simplify the second equation

$$\begin{cases} y = \dfrac{225}{x} \\ xy - 0.01x + 500y - 5 = 240 \end{cases}$$

$$\begin{cases} y = \dfrac{225}{x} \\ xy - 0.01x + 500y = 245 \end{cases}$$

$$xy - 0.01x + 500y = 245$$

$$x\left(\frac{225}{x}\right) - 0.01x + 500\left(\frac{225}{x}\right) = 245$$

$$225 - 0.01x + \frac{112{,}500}{x} = 245$$

$$-0.01x + \frac{112{,}500}{x} = 20$$

$$x\left(-0.01x + \frac{112{,}500}{x}\right) = x(20)$$

$$-0.01x^2 + 112{,}500 = 20x$$

$$-0.01x^2 - 20x + 112{,}500 = 0$$

$$-100(-0.01x^2 - 20x + 112{,}500) = -100(0)$$

$$x^2 + 2{,}000x - 11{,}250{,}000 = 0$$

$$(x + 4{,}500)(x - 2{,}500) = 0$$

$$x + 4{,}500 = 0 \quad \text{or} \quad x - 2{,}500 = 0$$

$$x = \cancel{-4{,}500} \qquad x = 2{,}500$$

$$y = \frac{225}{x}$$

$$y = \frac{225}{2{,}50}$$

$$y = 0.09$$

$$y = 9\%$$

Since the amount invested cannot be negative, Grant invested \$2,500 at 9%.

WRITING

73. Answers will vary.

REVIEW

75.

$$\log 5x = 4$$

$$10^4 = 5x$$

$$10{,}000 = 5x$$

$$2{,}000 = x$$

77.

$$\frac{\log(8x - 7)}{\log x} = 2$$

$$\log(8x - 7) = 2\log x$$

$$\log(8x - 7) = \log x^2$$

$$8x - 7 = x^2$$

$$0 = x^2 - 8x + 7$$

$$0 = (x - 7)(x - 1)$$

$$x - 7 = 0 \quad \text{or} \quad x - 1 = 0$$

$$x = 7 \qquad x = \cancel{1}$$

$x \neq 1$ because it make the denominator undefined

CHALLENGE PROBLEMS

79. a. The system may have 0, 1, 2, 3, or 4 solutions.
    b. The system may have 0, 1, 2, 3, or 4 solutions.

81.

$$\begin{cases} \dfrac{1}{x}+\dfrac{3}{y}=4 \\ \dfrac{2}{x}-\dfrac{1}{y}=7 \end{cases}$$

Multiply the second equation by 3 and add the two equations to eliminate y.

$$\frac{1}{x}+\frac{3}{y}=4$$
$$\frac{6}{x}-\frac{3}{y}=21$$
$$\frac{7}{x}=25$$
$$x\left(\frac{7}{x}\right)=x(25)$$
$$7=25x$$
$$\frac{7}{25}=x$$

Substitute $x=\dfrac{7}{25}$ into the first equation and solve for y.

$$\frac{1}{\frac{7}{25}}+\frac{3}{y}=4$$
$$\frac{25}{7}+\frac{3}{y}=4$$
$$7y\left(\frac{25}{7}+\frac{3}{y}\right)=7y(4)$$
$$25y+21=28y$$
$$21=3y$$
$$7=y$$

The solution is $\left(\dfrac{7}{25},7\right)$.

CHAPTER 10 REVIEW

SECTION 10.1
The Circle and the Parabola

1. $x^2 + y^2 = 16$

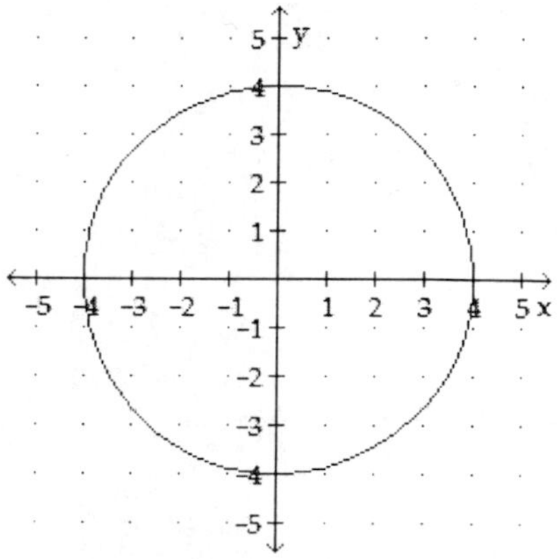

2. $(x-4)^2 + (y+3)^2 = 4$

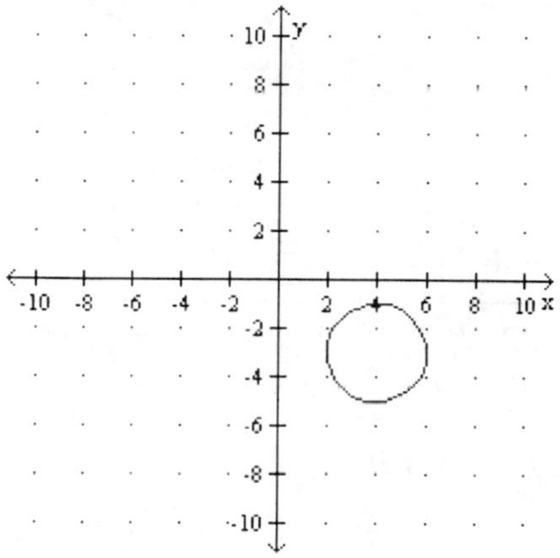

3.

$$x^2 + y^2 + 4x - 2y = 4$$
$$\left(x^2 + 4x\right) + \left(y^2 - 2y\right) = 4$$
$$\left(x^2 + 4x + 4\right) + \left(y^2 - 2y + 1\right) = 4 + 4 + 1$$
$$(x+2)^2 + (y-1)^2 = 9$$

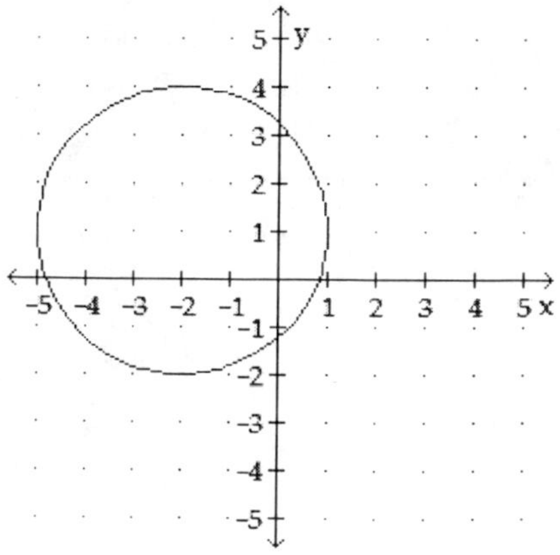

4.

$$(x-9)^2 + (y-9)^2 = (9)^2$$
$$(x-9)^2 + (y-9)^2 = 81$$

5. center: $(-6, 0)$
radius: $\sqrt{24} = 2\sqrt{6}$

6. A circle is the set of all points in a plane that are a fixed distance from a point called its **center**. The fixed distance is called the **radius** of the circle.

7. $x = y^2$
vertex: $(0, 0)$

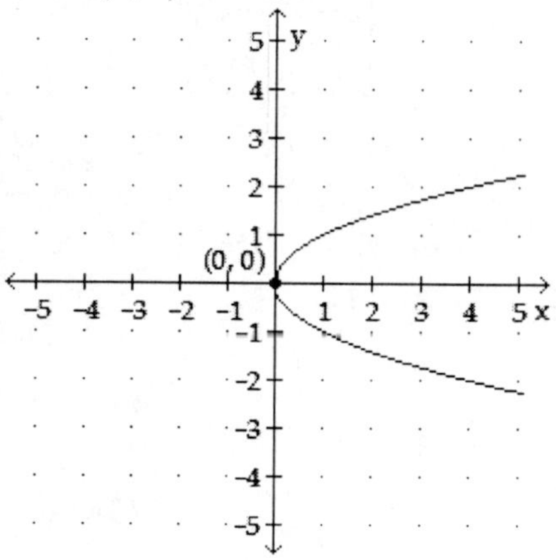

8. $x = 2(y+1)^2 - 2$
vertex: $(-2, -1)$

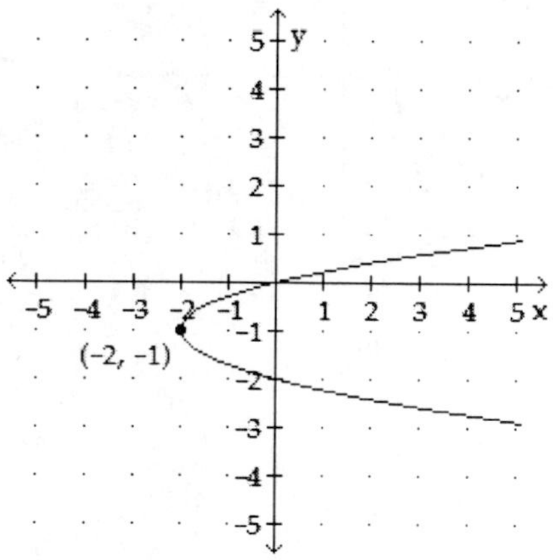

9.

$$x=-3y^2+12y-7$$
$$x=-3(y^2-4y+4)-7+12$$
$$x=-3(y-2)^2+5$$
vertex: $(5,\ 2)$

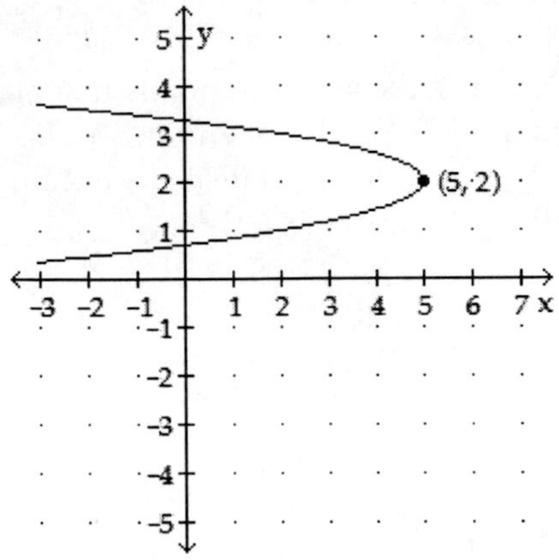

10.

$$y=x^2+8x+11$$
$$y=(x^2+8x+16)+11-16$$
$$y=(x+4)^2-5$$

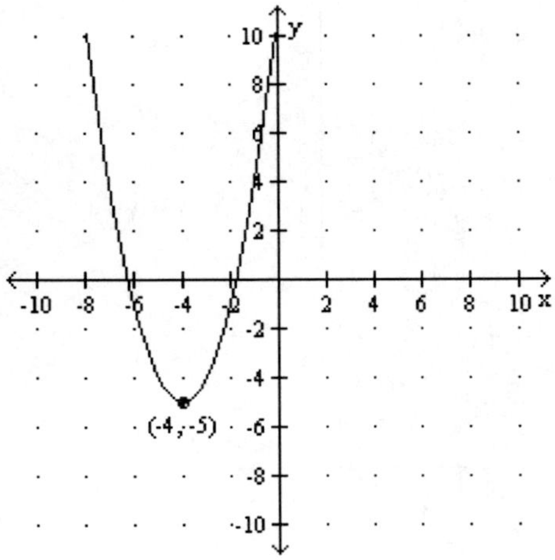

11. $(2, -2)$ and $(8, -3)$

12. When $x = 22$, $y = 0$.

$$y=-\frac{5}{121}(x-11)^2+5$$
$$0\overset{?}{=}-\frac{5}{121}(22-11)^2+5$$
$$0\overset{?}{=}-\frac{5}{121}(11)^2+5$$
$$0\overset{?}{=}-\frac{5}{121}(121)+5$$
$$0\overset{?}{=}-5+5$$
$$0=0$$

SECTION 10.2
The Ellipse

13.

$$\frac{x^2}{16}+\frac{y^2}{9}=1$$

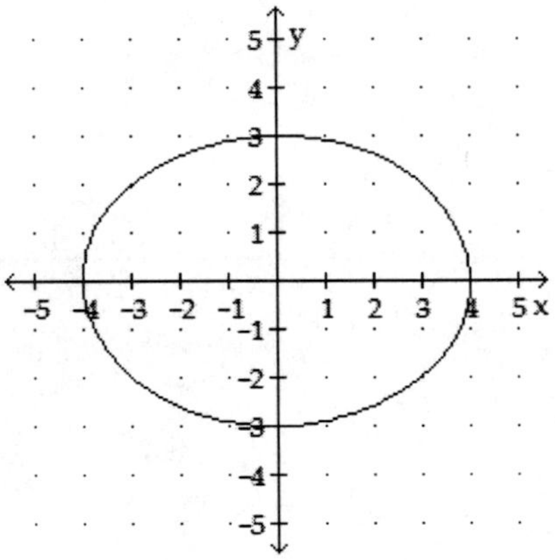

14.

$$\frac{(x-2)^2}{4}+\frac{(y-1)^2}{25}=1$$

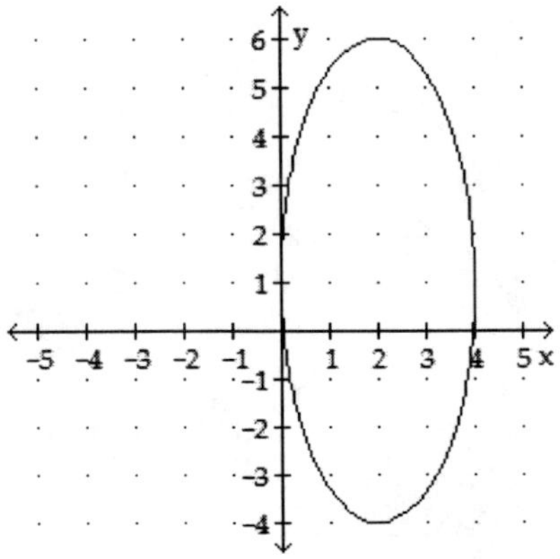

15.

$$4(x+1)^2 + 9(y-1)^2 = 36$$

$$\frac{4(x+1)^2}{36} + \frac{9(y-1)^2}{36} = \frac{36}{36}$$

$$\frac{(x+1)^2}{9} + \frac{(y-1)^2}{4} = 1$$

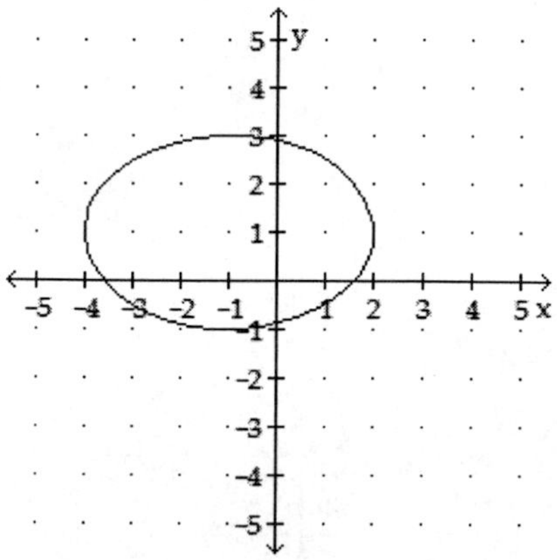

16.

$$\frac{x^2}{144} + \frac{y^2}{1} = 1$$

$$\frac{x^2}{12^2} + \frac{y^2}{1^2} = 1$$

17. a. circle
    b. ellipse
    c. parabola
    d. ellipse

18.

$$\frac{x^2}{5^2} + \frac{y^2}{3^2} = 1$$

$$\frac{x^2}{25} + \frac{y^2}{9} = 1$$

19. An **ellipse** is the set of all points in a plane for which the sum of the distances from two fixed points is a constant. Each of the fixed points is called a **focus**.

20. Answers may vary.

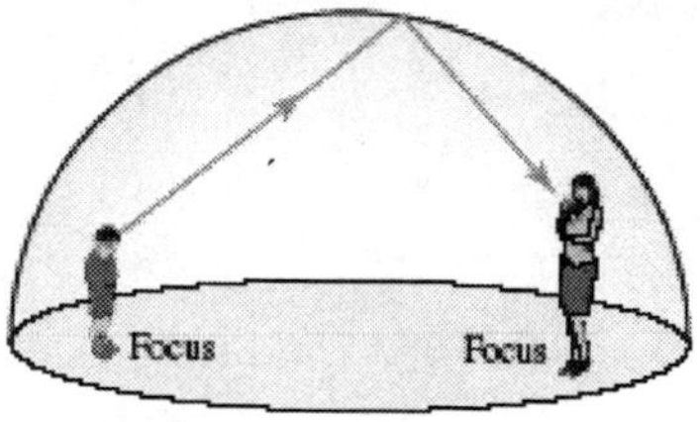

SECTION 10.3
The Hyperbola

21.

$$\frac{y^2}{9} - \frac{x^2}{1} = 1$$

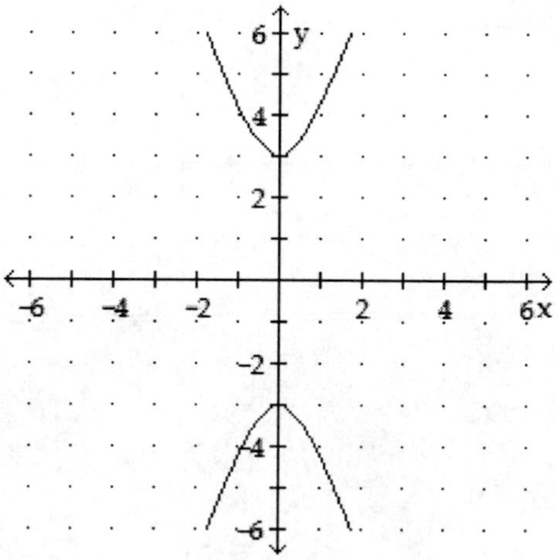

22.

$$9(x-1)^2 - 4(y+1)^2 = 36$$

$$\frac{9(x-1)^2}{36} - \frac{4(y+1)^2}{36} = \frac{36}{36}$$

$$\frac{(x-1)^2}{4} - \frac{(y+1)^2}{9} = 1$$

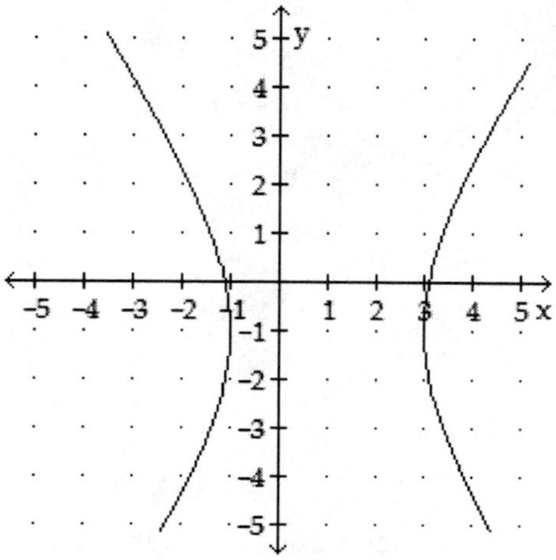

23. $\dfrac{(y-2)^2}{25} - \dfrac{(x+1)^2}{25} = 1$

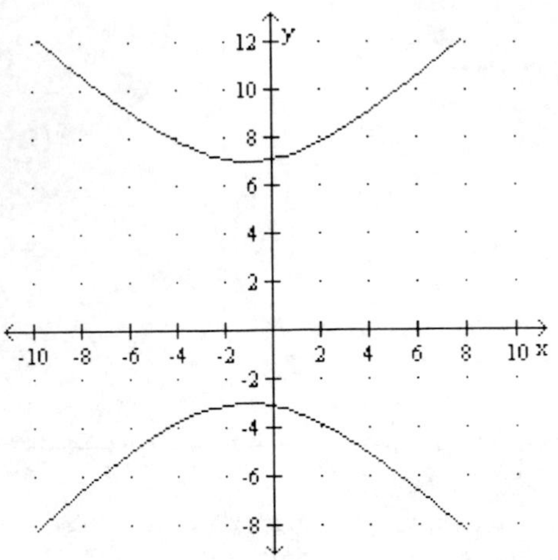

24. $xy = 9$

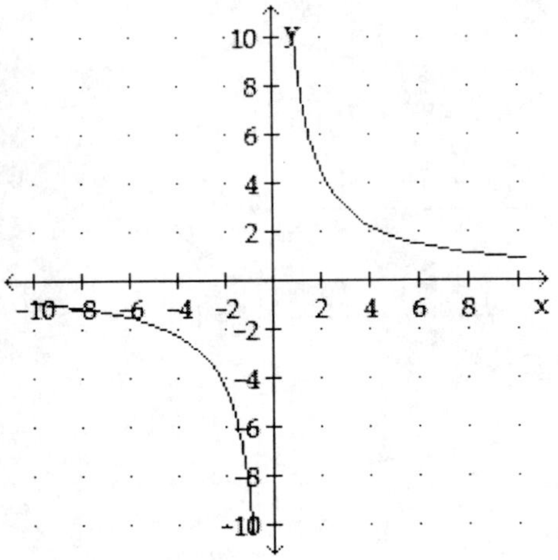

25.

$$x^2 - 4y^2 = 4$$
$$\frac{x^2}{4} - \frac{4y^2}{4} = \frac{4}{4}$$
$$\frac{x^2}{4} - \frac{y^2}{1} = 1$$

$a = 2$, so they were $2^2 = 4$ units apart.

26. a. ellipse
    b. hyperbola
    c. parabola
    d. circle

SECTION 10.4
Solving Systems of Nonlinear Equations

27. Yes, it is a solution.

$$x^2 + y^2 = 20 \qquad\qquad x^2 - y^2 = 2$$
$$\left(-\sqrt{11}\right)^2 + (-3)^2 \stackrel{?}{=} 20 \qquad \left(-\sqrt{11}\right)^2 - (-3)^2 \stackrel{?}{=} 2$$
$$11 + 9 \stackrel{?}{=} 20 \qquad\qquad 11 - 9 \stackrel{?}{=} 2$$
$$20 = 20 \qquad\qquad 2 = 2$$

28. The graphs intersect at (0, 3) and (0, –3).

29. $\begin{cases} xy = 4 \\ y = 2x - 2 \end{cases}$

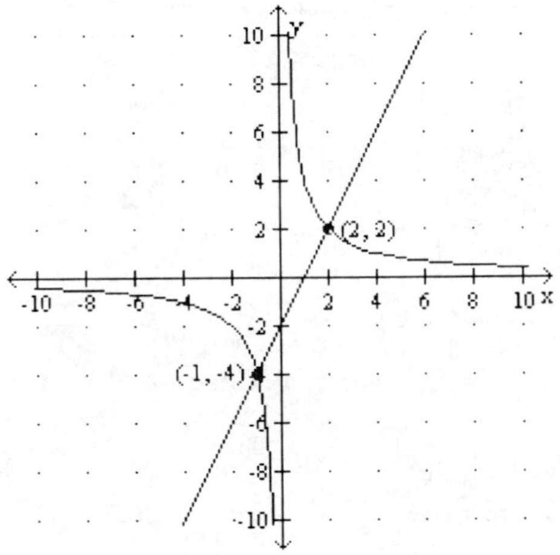

The solutions are (2, 2) and (–1, –4).

30. a. 2
    b. 4
    c. 4
    d. 4

31. Let $x = 0$ in both equations to find the y–coordinate.

$$x^2 + y^2 = 1 \qquad\qquad 4y^2 - x^2 = 4$$
$$0^2 + y^2 = 1 \qquad\qquad 4y^2 - 0^2 = 4$$
$$y^2 = 1 \qquad\qquad 4y^2 = 4$$
$$y = \sqrt{1} \qquad\qquad y^2 = 1$$
$$y = \pm 1 \qquad\qquad y = \sqrt{1}$$
$$y = \pm 1$$

The solutions are (0, 1) and (0, –1).

32. Solve the second equation for y: $y = 3x - 1$

33. $\begin{cases} y^2 - x^2 = 16 \\ y + 4 = x^2 \end{cases}$

Substitute $y + 4 = x^2$ into the first equation and solve for y.

$$y^2 - x^2 = 16$$
$$y^2 - (y+4) = 16$$
$$y^2 - y - 4 = 16$$
$$y^2 - y - 20 = 0$$
$$(y-5)(y+4) = 0$$
$$y - 5 = 0 \quad \text{or} \quad y + 4 = 0$$
$$y = 5 \qquad\qquad y = -4$$

Substitute $y = 5$ and $y = -4$ into the second equation and solve for x.

$$y + 4 = x^2 \qquad y + 4 = x^2$$
$$5 + 4 = x^2 \qquad -4 + 4 = x^2$$
$$9 = x^2 \qquad 0 = x^2$$
$$x = \sqrt{9} \qquad x = \sqrt{0}$$
$$x = \pm 3 \qquad x = 0$$

The solutions are $(3, 5), (-3, 5)$, and $(0, -4)$.

34. $\begin{cases} y = -x^2 + 2 \\ x^2 - y - 2 = 0 \end{cases}$

Substitute $y = -x^2 + 2$ into the second equation and solve for x.

$$x^2 - y - 2 = 0$$
$$x^2 - (-x^2 + 2) - 2 = 0$$
$$x^2 + x^2 - 2 - 2 = 0$$
$$2x^2 - 4 = 0$$
$$2(x^2 - 2) = 0$$
$$x^2 - 2 = 0$$
$$x^2 = 2$$
$$x = \pm\sqrt{2}$$

Substitute $x = \sqrt{2}$ and $x = -\sqrt{2}$ into the first equation and solve for y.

$$y = -x^2 + 2 \qquad y = -x^2 + 2$$
$$y = -\left(\sqrt{2}\right)^2 + 2 \qquad y = -\left(-\sqrt{2}\right)^2 + 2$$
$$y = -2 + 2 \qquad y = -2 + 2$$
$$y = 0 \qquad y = 0$$

The solutions are $\left(\sqrt{2}, 0\right)$ and $\left(-\sqrt{2}, 0\right)$.

35.

$\begin{cases} x^2 + 2y^2 = 12 \\ 2x - y = 2 \end{cases}$

Solve the second equation for y.

$$2x - y = 2$$
$$-y = -2x + 2$$
$$y = 2x - 2$$

Substitute $y = 2x - 2$ into the first equation and solve for x.

$$x^2 + 2y^2 = 12$$
$$x^2 + 2(2x-2)^2 = 12$$
$$x^2 + 2(4x^2 - 4x - 4x + 4) = 12$$
$$x^2 + 2(4x^2 - 8x + 4) = 12$$
$$x^2 + 8x^2 - 16x + 8 = 12$$
$$9x^2 - 16x - 4 = 0$$
$$(9x+2)(x-2) = 12$$
$$9x + 2 = 0 \quad \text{or} \quad x - 2 = 0$$
$$9x = -2 \qquad\qquad x = 2$$
$$x = -\frac{2}{9}$$

Substitute $x = -\frac{2}{9}$ and $x = 2$ into the second equation and solve for y.

$$y = 2x - 2 \qquad y = 2x - 2$$
$$y = 2\left(-\frac{2}{9}\right) - 2 \qquad y = 2(2) - 2$$
$$y = -\frac{4}{9} - 2 \qquad y = 4 - 2$$
$$y = -\frac{22}{9} \qquad y = 2$$

The solutions are $\left(-\frac{2}{9}, -\frac{22}{9}\right)$ and $(2, 2)$.

36.

$$\begin{cases} 3x^2 + y^2 = 52 \\ x^2 - y^2 = 12 \end{cases}$$

Add the equations and solve for x.

$$\begin{aligned} 3x^2 + y^2 &= 52 \\ x^2 - y^2 &= 12 \\ \hline 4x^2 &= 64 \\ x^2 &= 16 \\ x &= \sqrt{16} \\ x &= \pm 4 \end{aligned}$$

Substitute $x = 4$ and $x = -4$ into the second equation and solve for y.

$$\begin{aligned} x^2 - y^2 &= 12 \\ (4)^2 - y^2 &= 12 \\ 16 - y^2 &= 12 \\ -y^2 &= -4 \\ y^2 &= 4 \\ y &= \sqrt{4} \\ y &= \pm 2 \end{aligned} \qquad \begin{aligned} x^2 - y^2 &= 12 \\ (-4)^2 - y^2 &= 12 \\ 16 - y^2 &= 12 \\ -y^2 &= -4 \\ y^2 &= 4 \\ y &= \sqrt{4} \\ y &= \pm 2 \end{aligned}$$

The solutions are $(4, 2), (4, -2)$, $(-4, 2)$, and $(-4, -2)$

37.

$$\begin{cases} \dfrac{x^2}{16} + \dfrac{y^2}{12} = 1 \\ \dfrac{x^2}{1} - \dfrac{y^2}{3} = 1 \end{cases}$$

Multiply the first equation by 48 and the second equation to 3 to eliminate fractions.

$$\begin{cases} 3x^2 + 4y^2 = 48 \\ 3x^2 - y^2 = 3 \end{cases}$$

Multiply the second equation by 4.

Add the equations and solve for x.

$$\begin{aligned} 3x^2 + 4y^2 &= 48 \\ 12x^2 - 4y^2 &= 12 \\ \hline 15x^2 &= 60 \\ x^2 &= 4 \\ x &= \sqrt{4} \\ x &= \pm 2 \end{aligned}$$

Substitute $x = 2$ and $x = -2$ into the second equation and solve for y.

$$\begin{aligned} 3x^2 - y^2 &= 3 \\ 3(2)^2 - y^2 &= 3 \\ 3(4) - y^2 &= 3 \\ 12 - y^2 &= 3 \\ -y^2 &= -9 \\ y^2 &= 9 \\ y &= \sqrt{9} \\ y &= \pm 3 \end{aligned} \qquad \begin{aligned} 3x^2 - y^2 &= 3 \\ 3(-2)^2 - y^2 &= 3 \\ 3(4) - y^2 &= 3 \\ 12 - y^2 &= 3 \\ -y^2 &= -9 \\ y^2 &= 9 \\ y &= \sqrt{9} \\ y &= \pm 3 \end{aligned}$$

The solutions are $(2, 3), (2, -3)$, $(-2, 3)$, and $(-2, -3)$

38.

$$\begin{cases} xy = 4 \\ \frac{x^2}{1} + \frac{y^2}{2} = 9 \end{cases}$$

Solve the first equation for y and multiply the second equation by 2 to eliminate fractions.

$$\begin{cases} y = \frac{4}{x} \\ 2x^2 + y^2 = 18 \end{cases}$$

Substitute $y = \frac{4}{x}$ into the second equation and solve for x.

$$2x^2 + y^2 = 18$$
$$2x^2 + \left(\frac{4}{x}\right)^2 = 18$$
$$2x^2 + \frac{16}{x^2} = 18$$
$$x^2\left(2x^2 + \frac{16}{x^2}\right) = x^2(18)$$
$$2x^4 + 16 = 18x^2$$
$$2x^4 - 18x^2 + 16 = 0$$
$$2(x^4 - 9x^2 + 8) = 0$$
$$2(x^2 - 1)(x^2 - 8) = 0$$
$$x^2 - 1 = 0 \quad \text{or} \quad x^2 - 8 = 0$$
$$x^2 = 1 \qquad x^2 = 8$$
$$x = \sqrt{1} \qquad x = \sqrt{8}$$
$$x = \pm 1 \qquad x = \pm 2\sqrt{2}$$

Substitute all four values into the first equation and solve for y.

$$xy = 4 \qquad xy = 4$$
$$1y = 4 \qquad -1y = 4$$
$$y = 4 \qquad y = -4$$

$$xy = 4 \qquad xy = 4$$
$$2\sqrt{2} \cdot y = 4 \qquad -2\sqrt{2} \cdot y = 4$$
$$y = \frac{4}{2\sqrt{2}} \qquad y = \frac{4}{-2\sqrt{2}}$$
$$y = \frac{2}{\sqrt{2}} \qquad y = -\frac{2}{\sqrt{2}}$$
$$y = \frac{2\sqrt{2}}{2} \qquad y = -\frac{2\sqrt{2}}{2}$$
$$y = \sqrt{2} \qquad y = -\sqrt{2}$$

The solutions are $(1, 4), (-1, -4)$, $(2\sqrt{2}, \sqrt{2})$, and $(-2\sqrt{2}, -\sqrt{2})$

39. $\begin{cases} y = -x^2 + 1 \\ x + y = 5 \end{cases}$

Substitute $y = -x^2 + 1$ into the second equation and solve for x.

$$x + y = 5$$
$$x + (-x^2 + 1) = 5$$
$$-x^2 + x + 1 = 5$$
$$0 = x^2 - x + 4$$

Since the quadratic equation $x^2 - x + 4$ cannot be factored using real numbers, there is no solution to this system.

40. $\begin{cases} x = y^2 - 3 \\ x = y^2 - 3y \end{cases}$

Substitute $x = y^2 - 3$ into the second equation for x.

$$x = y^2 - 3y$$
$$y^2 - 3 = y^2 - 3y$$
$$-3 = -3y$$
$$1 = y$$

Substitute $y = 1$ into the first equation and solve for x.

$$x = y^2 - 3$$
$$x = 1^2 - 3$$
$$x = 1 - 3$$
$$x = -2$$

The solution is (–2, 1).

CHAPTER 10 TEST

1. a. The curves formed by the intersection of a plane with an infinite right-circular cone are called **conic** sections.

   b. A circle is the set of all points in a plane that are a fixed distance from a point called its **center**. The fixed distance is called the **radius** of the circle.

   c. The standard form for the equation of a **hyperbola** centered at the origin that opens left and right is $\frac{x^2}{a^2} - \frac{y^2}{b^2} = 1$.

   d. $\begin{cases} y = x^2 + x - 4 \\ x^2 + y^2 = 36 \end{cases}$ is a(n) **nonlinear** system of equations

   e. The standard form for the equation of an **ellipse** centered at the origin is $\frac{x^2}{a^2} + \frac{y^2}{b^2} = 1$.

2. center: (0, 0); radius: $\sqrt{100} = 10$

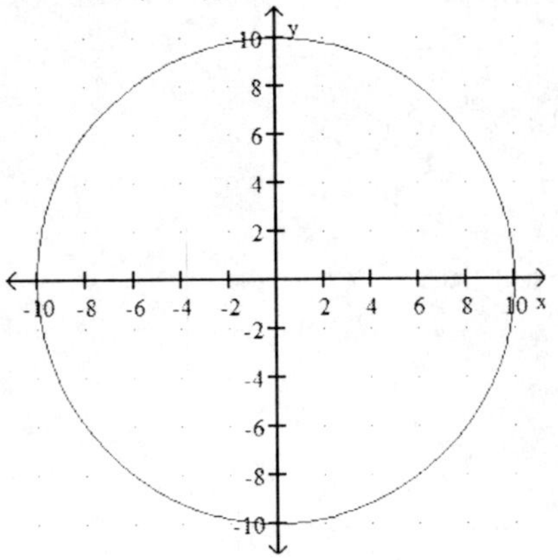

3.

$$x^2 + y^2 + 4x - 6y = 5$$
$$(x^2 + 4x) + (y^2 - 6y) = 5$$
$$(x^2 + 4x + 4) + (y^2 - 6y + 9) = 5 + 4 + 9$$
$$(x+2)^2 + (y-3)^2 = 18$$

center: $(-2, 3)$

radius: $\sqrt{18} = 3\sqrt{2}$

4. The center is (4, 3) and the radius is 3. The equation of the circle is

$$(x-4)^2 + (y-3)^2 = 3^2$$
$$(x-4)^2 + (y-3)^2 = 9$$

5.

$$r^2 = \frac{441}{16}$$
$$r = \sqrt{\frac{441}{16}}$$
$$r = \frac{21}{4}$$

diameter $= 2r$

$$= 2\left(\frac{21}{4}\right)$$
$$= \frac{21}{2} \text{ in.}$$
$$= 10.5 \text{ in.}$$

6.

$$x = y^2 + 8y + 10$$
$$x = \left(y^2 + 8y + \boxed{16}\right) + 10 - \boxed{16}$$
$$x = \left(y + \boxed{4}\right)^2 - 6$$

7. $(x+2)^2 + (y-1)^2 = 9$

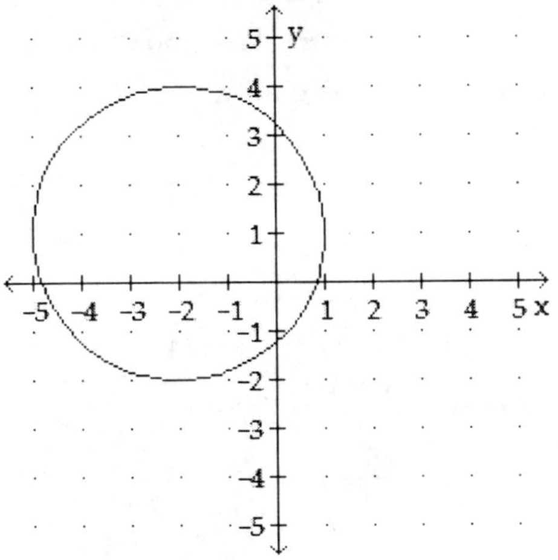

8.

$$x = y^2 - 4y + 3$$
$$x = (y^2 - 4y) + 3$$
$$x = (y^2 - 4y + 4) + 3 - 4$$
$$x = (y - 2)^2 - 1$$

vertex: $(-1, 2)$

axis of symmetry: $y = 2$

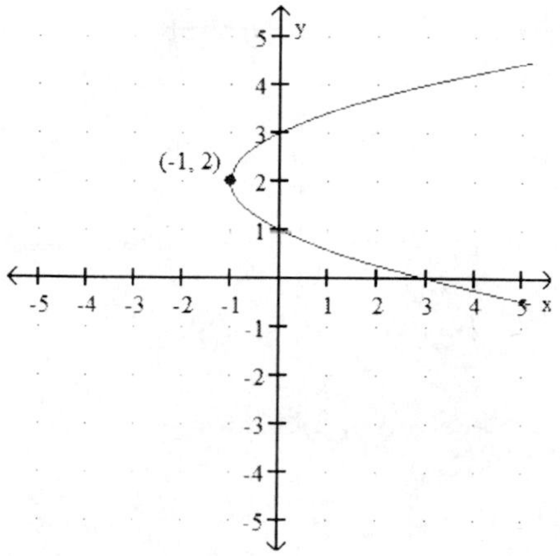

9.

$$y = -2x^2 - 4x + 5$$
$$y = -2(x^2 + 2x) + 5$$
$$y = -2(x^2 + 2x + 1) + 5 - (-2)(1)$$
$$y = -2(x + 1)^2 + 5 + 2$$
$$y = -2(x + 1)^2 + 7$$

vertex: $(-1, 7)$

axis of symmetry: $x = -1$

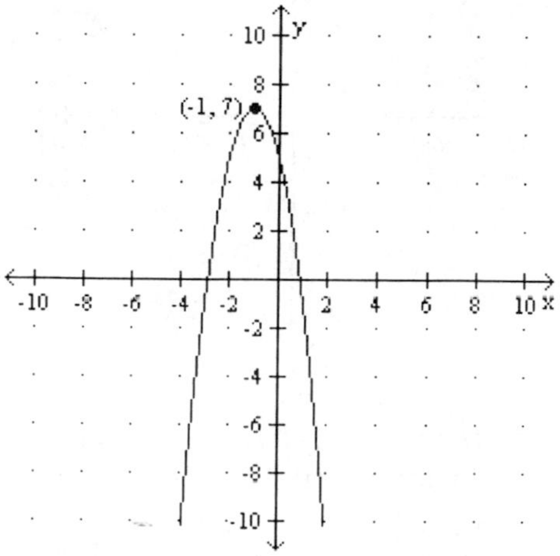

10.

$$xy = 4$$
$$y = \frac{4}{x}$$

x	$y = \frac{4}{x}$
1	$\frac{4}{1} = 4$
2	$\frac{4}{2} = 2$
4	$\frac{4}{4} = 1$
-4	$\frac{4}{-4} = -1$
-2	$\frac{4}{-2} = -2$
-1	$\frac{4}{-1} = -4$

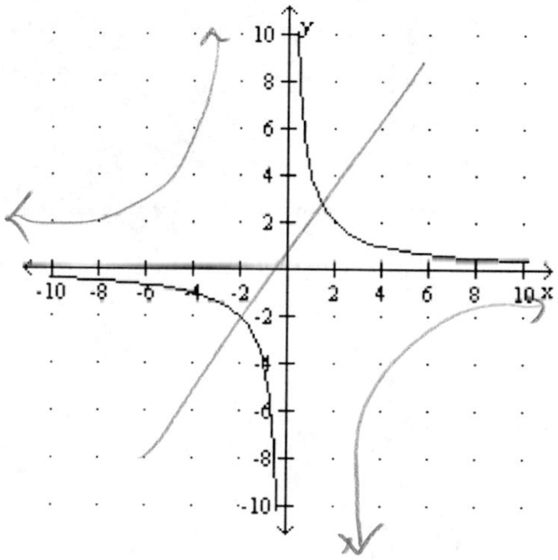

11.

$$9x^2 + 4y^2 = 36$$

$$\frac{9x^2}{36} + \frac{4y^2}{36} = \frac{36}{36}$$

$$\frac{x^2}{4} + \frac{y^2}{9} = 1$$

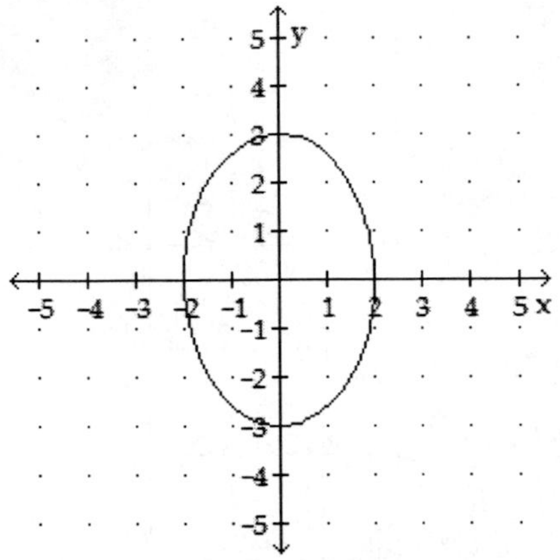

12. $\dfrac{(x-2)^2}{9} - \dfrac{y^2}{1} = 1$

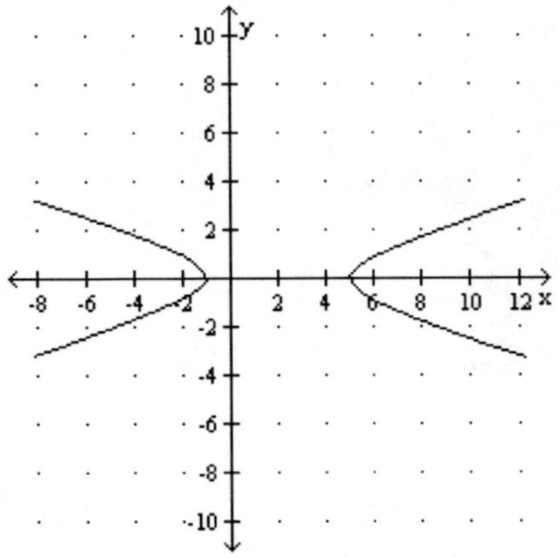

13. $\dfrac{(x-3)^2}{49} + \dfrac{(y+2)^2}{16} = 1$

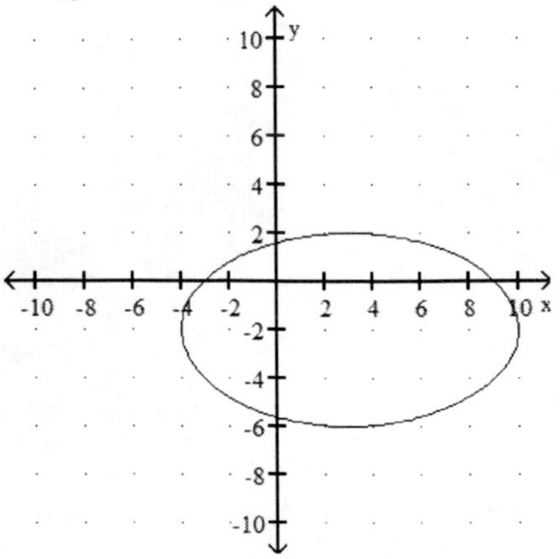

14.

$$x^2 + y^2 = 7$$

$$x^2 + y^2 = \left(\sqrt{7}\right)^2$$

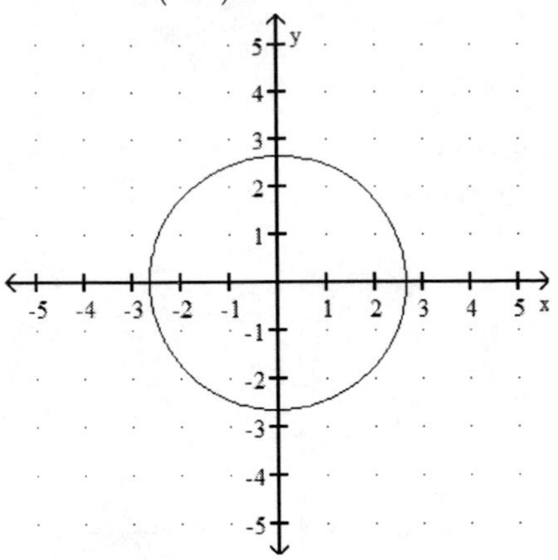

15. $x = -\dfrac{1}{2}y^2$

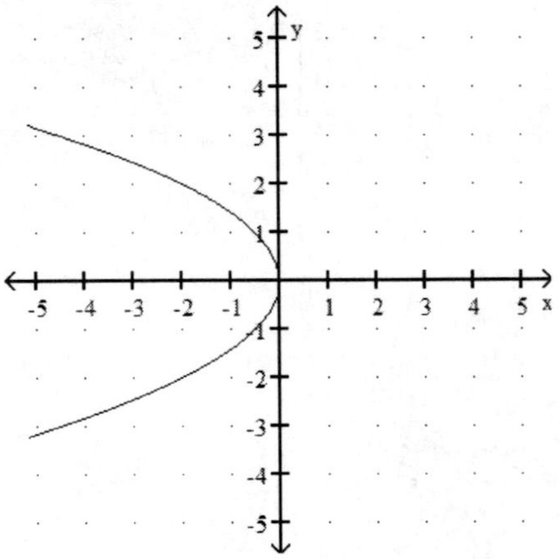

16. $\dfrac{y^2}{25} - \dfrac{x^2}{9} = 1$

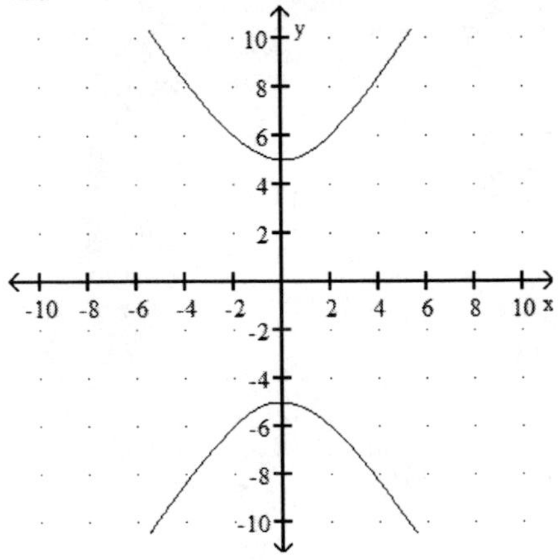

17. The center is (1, –2), $a = 4$, and $b = 3$. The equation of the ellipse is

$$\frac{(x-1)^2}{16} + \frac{(y+2)^2}{9} = 1.$$

18. If the width is 10, then $y = 5$.

$$x = \frac{1}{10}y^2$$
$$x = \frac{1}{10}(5)^2$$
$$x = \frac{1}{10}(25)$$
$$x = 2.5$$

19. The ellipse is centered at the origin and extends on the x-axis to –28 and 28 (half of 60 – 4 for the 2 inches on each side) and extends on the y-axis to 16 and –16 (half of 36 – 4 for the 2 inches on each side).

$$\frac{x^2}{28^2} + \frac{y^2}{16^2} = 1$$
$$\frac{x^2}{784} + \frac{y^2}{256} = 1$$

20.

$$(x+1)^2 - (y-1)^2 = 4$$

center: $(-1,1)$

radius: 2

horizonal dimensions: 4 units

vertical dimensions: 4 units

21. The center is (0, 0). The y-value is 4 and the x-value is 6. It opens up.

$$\frac{y^2}{9} - \frac{x^2}{36} = 1$$

22. a. ellipse
b. hyperbola
c. circle
d. parabola

23. $\begin{cases} x^2 + y^2 = 25 \\ y - x = 1 \end{cases}$

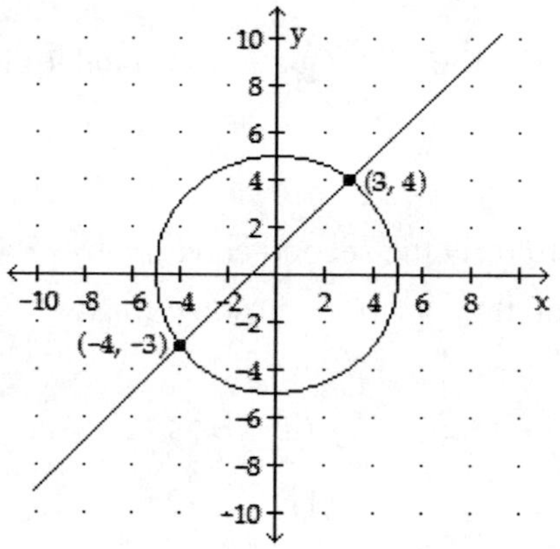

24.

$$\begin{cases} 2x - y = -2 \\ x^2 + y^2 = 16 + 4y \end{cases}$$

Solve the first equation for y.

$$\begin{cases} y = 2x + 2 \\ x^2 + y^2 = 16 + 4y \end{cases}$$

Substitute $y = 2x + 2$ into the second equation and solve for x.

$$x^2 + y^2 = 16 + 4y$$
$$x^2 + (2x+2)^2 = 16 + 4(2x+2)$$
$$x^2 + (4x^2 + 4x + 4x + 4) = 16 + 8x + 8$$
$$5x^2 + 8x + 4 = 8x + 24$$
$$5x^2 - 20 = 0$$
$$5(x^2 - 4) = 0$$
$$5(x-2)(x+2) = 0$$
$$x - 2 = 0 \quad \text{or} \quad x + 2 = 0$$
$$x = 2 \qquad\qquad x = -2$$

Substitute $x = 2$ and $x = -2$ into the first equation and solve for y.

$y = 2x + 2$	$y = 2x + 2$
$y = 2(2) + 2$	$y = 2(-2) + 2$
$y = 4 + 2$	$y = -4 + 2$
$y = 6$	$y = -2$

The solutions are $(2,6)$ and $(-2,-2)$.

25.

$$\begin{cases} 5x^2 - y^2 - 3 = 0 \\ x^2 + 2y^2 = 5 \end{cases}$$

Write the first equation in standard form

$$\begin{cases} 5x^2 - y^2 = 3 \\ x^2 + 2y^2 = 5 \end{cases}$$

Multiply the second equation by -5 and add the equations to solve for y.

$$\begin{aligned} 5x^2 - \quad y^2 &= 3 \\ -5x^2 - 10y^2 &= -25 \\ \hline -11y^2 &= -22 \\ y^2 &= 2 \\ y &= \pm\sqrt{2} \end{aligned}$$

Substitute $y = \sqrt{2}$ and $y = -\sqrt{2}$ into the second equation and solve for x.

$$\begin{aligned} x^2 + 2y^2 &= 5 & \qquad x^2 + 2y^2 &= 5 \\ x^2 + 2\left(\sqrt{2}\right)^2 &= 5 & x^2 + 2\left(-\sqrt{2}\right)^2 &= 5 \\ x^2 + 2(2) &= 5 & x^2 + 2(2) &= 5 \\ x^2 + 4 &= 5 & x^2 + 4 &= 5 \\ x^2 &= 1 & x^2 &= 1 \\ x &= \sqrt{1} & x &= \sqrt{1} \\ x &= \pm 1 & x &= \pm 1 \end{aligned}$$

The solutions are $\left(1, \sqrt{2}\right), \left(-1, \sqrt{2}\right)$, $\left(1, -\sqrt{2}\right)$, and $\left(-1, -\sqrt{2}\right)$.

26.

$$\begin{cases} xy = -\dfrac{9}{2} \\ 3x + 2y = 6 \end{cases}$$

Solve the first equation for y.

$$\begin{cases} y = -\dfrac{9}{2x} \\ 3x + 2y = 6 \end{cases}$$

Substitute $y = -\dfrac{9}{2x}$ into the second equation and solve for x.

$$\begin{aligned} 3x + 2y &= 6 \\ 3x + 2\left(-\frac{9}{2x}\right) &= 6 \\ 3x - \frac{9}{x} &= 6 \\ x\left(3x - \frac{9}{x}\right) &= x(6) \\ 3x^2 - 9 &= 6x \\ 3x^2 - 6x - 9 &= 0 \\ 3\left(x^2 - 2x - 3\right) &= 0 \\ 3(x-3)(x+1) &= 0 \end{aligned}$$

$$x - 3 = 0 \quad \text{or} \quad x + 1 = 0$$
$$x = 3 \qquad\qquad x = -1$$

Substitute $x = 3$ and $x = -1$ into the first equation and solve for y.

$$\begin{aligned} xy &= -\frac{9}{2} & \qquad xy &= -\frac{9}{2} \\ 3y &= -\frac{9}{2} & -1y &= -\frac{9}{2} \\ \frac{1}{3}(3y) &= \frac{1}{3}\left(-\frac{9}{2}\right) & -1(-1y) &= -1\left(-\frac{9}{2}\right) \\ y &= -\frac{3}{2} & y &= \frac{9}{2} \end{aligned}$$

The solutions are $\left(3, -\dfrac{3}{2}\right)$ and $\left(-1, \dfrac{9}{2}\right)$.

27. $\begin{cases} y = x + 1 \\ x^2 - y^2 = 1 \end{cases}$

Substitute $y = x + 1$ into the second equation and solve for x.

$$x^2 - y^2 = 1$$
$$x^2 - (x+1)^2 = 1$$
$$x^2 - (x^2 + 2x + 1)^2 = 1$$
$$x^2 - x^2 - 2x - 1 = 1$$
$$-2x = 2$$
$$x = -1$$

Substitute $x = -1$ into the first equation and solve for y.

$$y = x + 1$$
$$y = -1 + 1$$
$$y = 0$$

The solution is $(-1, 0)$.

28. $\begin{cases} x^2 + 3y^2 = 6 \\ x^2 + y = 8 \end{cases}$

Solve the second equation for x^2

$$\begin{cases} x^2 + 3y^2 = 6 \\ x^2 = -y + 8 \end{cases}$$

Substitute $x^2 = -y + 8$ into the first equation and solve for y,

$$-y + 8 + 3y^2 = 6$$
$$3y^2 - y + 2 = 0$$

No Solution; $\varnothing$

The quadratic equation is not factorable so the solutions are not real

SECTION 11.1

VOCABULARY

1. The two-term polynomial expression $a + b$ is called a **binomial**.

3. We can use the **binomial** theorem to raise binomials to positive-integer powers without doing the actual multiplication.

5. $n!$ (read as "n **factorial**") is the product of consecutively **decreasing** natural numbers from n to 1.

CONCEPTS

7. The binomial expansion of $(m+n)^6$ has **one** more term than the power of the binomial.

9. The first term of the expansion of $(r + s)^{20}$ is $\boxed{r^{20}}$ and the last term is $\boxed{s^{20}}$.

11. The coefficients of the terms of the expansion of $(c + d)^{20}$ begin with $\boxed{1}$, increase through some values, and then decrease through those same values, back to $\boxed{1}$.

13. $n\cdot\boxed{(n-1)!} = n!$

15. $0! = \boxed{1}$

17. The coefficient of the fourth term of the expansion of $(a + b)^9$ is 9! Divided by $\underline{3!(9-3)!}$.

19. The exponent on the a in the fifth term of the expansion of $(a + b)^6$ is $\boxed{2}$ and the exponent on b is $\boxed{4}$.

21. $(x + y)^3$

$$= x^{\boxed{3}} + \frac{\boxed{3!}}{1!(3-1)!}x^2\boxed{y} + \frac{\boxed{3!}}{\boxed{2}!(3-2)!}xy^{\boxed{2}} + y^{\boxed{3}}$$

NOTATION

23. $n! = n\cdot(\boxed{n-1})(n-2)\ldots3\cdot2\cdot1$

GUIDED PRACTICE

25.
$$(a+b)^3 = a^3 + 3a^2b + 3ab^2 + b^3$$

27.
$$(m-p)^5 = m^5 - 5m^4p + 10m^3p^2 - 10m^2p^3 + 5mp^4 - p^5$$

29.
$$\begin{aligned}3! &= 3\cdot2\cdot1\\ &= 6\end{aligned}$$

31.
$$\begin{aligned}5! &= 5\cdot4\cdot3\cdot2\cdot1\\ &= 120\end{aligned}$$

33.
$$\begin{aligned}3! + 4! &= 3\cdot2\cdot1 + 4\cdot3\cdot2\cdot1\\ &= 6 + 24\\ &= 30\end{aligned}$$

35.
$$\begin{aligned}3!(4!) &= (3\cdot2\cdot1)(4\cdot3\cdot2\cdot1)\\ &= (6)(24)\\ &= 144\end{aligned}$$

37.
$$\begin{aligned}8(7!) &= 8(7\cdot6\cdot5\cdot4\cdot3\cdot2\cdot1)\\ &= 8(5{,}040)\\ &= 40{,}320\end{aligned}$$

39.
$$\begin{aligned}\frac{49!}{47!} &= \frac{49\cdot48\cdot\cancel{47!}}{\cancel{47!}}\\ &= \frac{49\cdot48}{1}\\ &= 2{,}352\end{aligned}$$

41.

$$\frac{9!}{11!}=\frac{\cancel{9}\cdot\cancel{8}\cdot\cancel{7}\cdot\cancel{6}\cdot\cancel{5}\cdot\cancel{4}\cdot\cancel{3}\cdot\cancel{2}\cdot\cancel{1}}{11\cdot10\cdot\cancel{9}\cdot\cancel{8}\cdot\cancel{7}\cdot\cancel{6}\cdot\cancel{5}\cdot\cancel{4}\cdot\cancel{3}\cdot\cancel{2}\cdot\cancel{1}}$$
$$=\frac{1}{11\cdot10}$$
$$=\frac{1}{110}$$

43.

$$\frac{9!}{7!0!}=\frac{9\cdot8\cdot\cancel{7!}}{\cancel{7!}0!}$$
$$=\frac{9\cdot8}{1}$$
$$=72$$

45.

$$\frac{5!}{1!(5-1)!}=\frac{5!}{1!4!}$$
$$=\frac{5\cdot\cancel{4!}}{1\cdot\cancel{4!}}$$
$$=5$$

47.

$$\frac{5!}{3!(5-3)!}=\frac{5!}{3!2!}$$
$$=\frac{5\cdot4\cdot\cancel{3!}}{\cancel{3!}(2\cdot1)}$$
$$=\frac{20}{2}$$
$$=10$$

49.

$$\frac{5!(8-5)!}{4!7!}=\frac{\cancel{5!}\,\cancel{3!}}{(4\cdot\cancel{3!})(7\cdot6\cdot\cancel{5!})}$$
$$=\frac{1}{4\cdot7\cdot6}$$
$$=\frac{1}{168}$$

51.

$$\frac{7!}{5!(7-5)!}=\frac{7!}{5!2!}$$
$$=\frac{7\cdot6\cdot\cancel{5!}}{\cancel{5!}(2\cdot1)}$$
$$=\frac{42}{2}$$
$$=21$$

53.

$$11!=39{,}916{,}800$$

55.

$$20!=2.432902008\times10^{18}$$

57.

$$(m+n)^4$$
$$=m^4+\frac{4!}{1!(4-1)!}m^3n+\frac{4!}{2!(4-2)!}m^2n^2+\frac{4!}{3!(4-3)!}mn^3+n^4$$
$$=m^4+\frac{4\cdot\cancel{3!}}{1!\cancel{3!}}m^3n+\frac{4\cdot3\cdot\cancel{2!}}{2!\cancel{2!}}m^2n^2+\frac{4\cdot\cancel{3!}}{\cancel{3!}1!}mn^3+n^4$$
$$=m^4+\frac{4}{1}m^3n+\frac{12}{2}m^2n^2+\frac{4}{1}mn^3+n^4$$
$$=m^4+4m^3n+6m^2n^2+4mn^3+n^4$$

59.

$$(c-d)^5$$
$$=c^5+\frac{5!}{1!(5-1)!}c^4(-d)+\frac{5!}{2!(5-2)!}c^3(-d)^2+\frac{5!}{3!(5-3)!}c^2(-d)^3+\frac{5!}{4!(5-4)!}c(-d)^4+(-d)^5$$
$$=c^5+\frac{5\cdot\cancel{4!}}{1!\cancel{4!}}c^4(-d)+\frac{5\cdot4\cdot\cancel{3!}}{2!\cancel{3!}}c^3(-d)^2+\frac{5\cdot4\cdot\cancel{3!}}{\cancel{3!}2!}c^2(-d)^3+\frac{5\cdot\cancel{4!}}{\cancel{4!}1!}c(-d)^4+(-d)^5$$
$$=c^5-\frac{5}{1}c^4d+\frac{20}{2}c^3d^2-\frac{20}{2}c^2d^3+\frac{5}{1}cd^4-d^5$$
$$=c^5-5c^4d+10c^3d^2-10c^2d^3+5cd^4-d^5$$

61.

$$(a-b)^9=(a+(-b))^9$$

$$=a^9+\frac{9!}{1!(9-1)!}a^8(-b)+\frac{9!}{2!(9-2)!}a^7(-b)^2$$
$$+\frac{9!}{3!(9-3)!}a^6(-b)^3+\frac{9!}{4!(9-4)!}a^5(-b)^4$$
$$+\frac{9!}{5!(9-5)!}a^4(-b)^5+\frac{9!}{6!(9-6)!}a^3(-b)^6$$
$$+\frac{9!}{7!(9-7)!}a^2(-b)^7+\frac{9!}{8!(9-8)!}a(-b)^8+(-b)^9$$
$$=a^9+\frac{9!}{1!8!}a^8(-b)+\frac{9!}{2!7!}a^7(-b)^2+\frac{9!}{3!6!}a^6(-b)^3$$
$$+\frac{9!}{4!5!}a^5(-b)^4+\frac{9!}{5!4!}a^4(-b)^5+\frac{9!}{6!3!}a^3(-b)^6$$
$$+\frac{9!}{7!2!}a^2(-b)^7+\frac{9!}{8!1!}a(-b)^8+(-b)^9$$
$$=a^9-\frac{9}{1}a^8b+\frac{72}{2}a^7b^2-\frac{504}{6}a^6b^3+\frac{3,024}{24}a^5b^4$$
$$-\frac{3,024}{24}a^4b^5+\frac{504}{6}a^3b^6-\frac{72}{2}a^2b^7+\frac{9}{1}ab^8-b^9$$
$$=a^9-9a^8b+36a^7b^2-84a^6b^3+126a^5b^4-126a^4b^5$$
$$+84a^3b^6-36a^2b^7+9ab^8-b^9$$

63.

$$(s+t)^6$$
$$=s^6+\frac{6!}{1!(6-1)!}s^5t+\frac{6!}{2!(6-2)!}s^4t^2$$
$$+\frac{6!}{3!(6-3)!}s^3t^3+\frac{6!}{4!(6-4)!}s^2t^4$$
$$+\frac{6!}{5!(6-5)!}st^5+t^6$$
$$=s^6+\frac{6!}{1!5!}s^5t+\frac{6!}{2!4!}s^4t^2$$
$$+\frac{6!}{3!3!}s^3t^3+\frac{6!}{4!2!}s^2t^4$$
$$+\frac{6!}{5!1!}st^5+t^6$$
$$=s^6+\frac{6}{1}s^5t+\frac{30}{2}s^4t^2+\frac{120}{6}s^3t^3+\frac{30}{2}s^2t^4$$
$$+\frac{6}{1}st^5+t^6$$
$$=s^6+6s^5t+15s^4t^2+20s^3t^3+15s^2t^4+6st^5+t^6$$

65.

$$(2x+y)^3$$
$$=(2x)^3+\frac{3!}{1!(3-1)!}(2x)^2y+\frac{3!}{2!(3-2)!}(2x)y^2+y^3$$
$$=8x^3+\frac{3!}{1!2!}(4x^2)y+\frac{3!}{2!1!}(2x)y^2+y^3$$
$$=8x^3+\frac{3}{1}(4x^2)y+\frac{3}{1}(2x)y^2+y^3$$
$$=8x^3+12x^2y+6xy^2+y^3$$

67.

$$(2t-3)^5$$
$$=(2t)^5+\frac{5!}{1!(5-1)!}(2t)^4(-3)+\frac{5!}{2!(5-2)!}(2t)^3(-3)^2$$
$$+\frac{5!}{3!(5-3)!}(2t)^2(-3)^3+\frac{5!}{4!(5-4)!}(2t)(-3)^4+(-3)^5$$
$$=(2t)^5+\frac{5\cdot\cancel{4!}}{1!\cancel{4!}}(2t)^4(-3)+\frac{5\cdot4\cdot\cancel{3!}}{2!\cancel{3!}}(2t)^3(-3)^2$$
$$+\frac{5\cdot4\cdot\cancel{3!}}{\cancel{3!}2!}(2t)^2(-3)^3+\frac{5\cdot\cancel{4!}}{\cancel{4!}1!}(2t)(-3)^4+(-3)^5$$
$$=32t^5+\frac{5}{1}(16t^4)(-3)+\frac{20}{2}(8t^3)(9)+\frac{20}{2}(4t^2)(-27)$$
$$+\frac{5}{1}(2t)(81)-243$$
$$=32t^5-240t^4+720t^3-1,080t^2+810t-243$$

69.

$$(5m-2n)^4$$
$$=(5m)^4+\frac{4!}{1!(4-1)!}(5m)^3(-2n)$$
$$+\frac{4!}{2!(4-2)!}(5m)^2(-2n)^2+\frac{4!}{3!(4-3)!}(5m)(-2n)^3$$
$$+(-2n)^4$$
$$=625m^4+\frac{4\cdot\cancel{3!}}{1!\cancel{3!}}(125m^3)(-2n)$$
$$+\frac{4\cdot3\cdot\cancel{2!}}{2!\cancel{2!}}(25m^2)(4n^2)+\frac{4\cdot\cancel{3!}}{\cancel{3!}1!}(5m)(-8n^3)+16n^4$$
$$=625m^4+\frac{4}{1}(-250m^3n)+\frac{12}{2}(100m^2n^2)$$
$$+\frac{4}{1}(-40mn^3)+16n^4$$
$$=625m^4-1,000m^3n+600m^2n^2-160mn^3+16n^4$$

71.

$$\left(\frac{x}{3}+\frac{y}{2}\right)^3$$

$$=\left(\frac{x}{3}\right)^3+\frac{3!}{1!(3-1)!}\left(\frac{x}{3}\right)^2\left(\frac{y}{2}\right)+\frac{3!}{2!(3-2)!}\left(\frac{x}{3}\right)\left(\frac{y}{2}\right)^2+\left(\frac{y}{2}\right)^3$$

$$=\frac{x^3}{27}+\frac{3!}{1!2!}\left(\frac{x^2}{9}\right)\left(\frac{y}{2}\right)+\frac{3!}{2!1!}\left(\frac{x}{3}\right)\left(\frac{y^2}{4}\right)+\frac{y^3}{8}$$

$$=\frac{x^3}{27}+\frac{3}{1}\left(\frac{x^2y}{18}\right)+\frac{3}{1}\left(\frac{xy^2}{12}\right)+\frac{y^3}{8}$$

$$=\frac{x^3}{27}+\frac{x^2y}{6}+\frac{xy^2}{4}+\frac{y^3}{8}$$

73.

$$\left(\frac{x}{3}-\frac{y}{2}\right)^4$$

$$=\left(\frac{x}{3}\right)^4+\frac{4!}{1!(4-1)!}\left(\frac{x}{3}\right)^3\left(-\frac{y}{2}\right)+\frac{4!}{2!(4-2)!}\left(\frac{x}{3}\right)^2\left(-\frac{y}{2}\right)^2+\frac{4!}{3!(4-3)!}\left(\frac{x}{3}\right)\left(-\frac{y}{2}\right)^3+\left(-\frac{y}{2}\right)^4$$

$$=\frac{x^4}{81}+\frac{4\cdot\cancel{3!}}{1!\cancel{3!}}\left(\frac{x^3}{27}\right)\left(-\frac{y}{2}\right)+\frac{4\cdot3\cdot\cancel{2!}}{2!\cancel{2!}}\left(\frac{x^2}{9}\right)\left(\frac{y^2}{4}\right)+\frac{4\cdot\cancel{3!}}{\cancel{3!}1!}\left(\frac{x}{3}\right)\left(-\frac{y^3}{8}\right)+\frac{y^4}{16}$$

$$=\frac{x^4}{81}+\frac{4}{1}\left(-\frac{x^3y}{54}\right)+\frac{12}{2}\left(\frac{x^2y^2}{36}\right)+\frac{4}{1}\left(-\frac{xy^3}{24}\right)+\frac{y^4}{16}$$

$$=\frac{x^4}{81}-\frac{2x^3y}{27}+\frac{x^2y^2}{6}-\frac{xy^3}{6}+\frac{y^4}{16}$$

75.

$$\left(c^2-d^2\right)^5$$

$$=\left(c^2\right)^5+\frac{5!}{1!(5-1)!}\left(c^2\right)^4\left(-d^2\right)+\frac{5!}{2!(5-2)!}\left(c^2\right)^3\left(-d^2\right)^2+\frac{5!}{3!(5-3)!}\left(c^2\right)^2\left(-d^2\right)^3+\frac{5!}{4!(5-4)!}\left(c^2\right)\left(-d^2\right)^4+\left(-d^2\right)^5$$

$$=c^{10}+\frac{5\cdot\cancel{4!}}{1!\cancel{4!}}\left(c^8\right)\left(-d^2\right)+\frac{5\cdot4\cdot\cancel{3!}}{2!\cancel{3!}}\left(c^6\right)\left(d^4\right)+\frac{5\cdot4\cdot\cancel{3!}}{\cancel{3!}2!}\left(c^4\right)\left(-d^6\right)+\frac{5\cdot\cancel{4!}}{\cancel{4!}1!}\left(c^2\right)\left(d^8\right)-d^{10}$$

$$=c^{10}+\frac{5}{1}\left(-c^8d^2\right)+\frac{20}{2}\left(c^6d^4\right)+\frac{20}{2}\left(-c^4d^6\right)+\frac{5}{1}\left(c^2d^8\right)-d^{10}$$

$$=c^{10}-5c^8d^2+10c^6d^4-10c^4d^6+5c^2d^8-d^{10}$$

77. The 3rd term of $(x+y)^8$

$$\frac{8!}{2!(8-2)!}x^6y^2=\frac{8\cdot7\cdot\cancel{6!}}{2!\cancel{6!}}x^6y^2=\frac{56}{2}x^6y^2=28x^6y^2$$

79. The 5th term of $(r+s)^6$

$$\frac{6!}{4!(6-4)!}r^2s^4=\frac{6\cdot5\cdot\cancel{4!}}{\cancel{4!}2!}r^2s^4=\frac{30}{2}r^2s^4=15r^2s^4$$

81. The 3rd term of $(x-1)^{13}$

$$\frac{13!}{2!(13-2)!}x^{11}(-1)^2=\frac{13\cdot12\cdot\cancel{11!}}{2!\cancel{11!}}x^{11}(1)=\frac{156}{2}x^{11}=78x^{11}$$

83. The 2nd term of $(x - 3y)^4$

$$\frac{4!}{1!(4-1)!}x^3(-3y) = \frac{4\cdot\cancel{3!}}{1!\cancel{3!}}(-3x^3y)$$
$$= \frac{4}{1}(-3x^3y)$$
$$= -12x^3y$$

85. The 5th term of $(2x - 3y)^5$

$$\frac{5!}{4!(5-4)!}(2x)(-3y)^4 = \frac{5\cdot\cancel{4!}}{\cancel{4!}1!}(2x)(81y^4)$$
$$= \frac{5}{1}(162xy^4)$$
$$= 810xy^4$$

87. The 2nd term of $\left(\frac{c}{2} - \frac{d}{3}\right)^4$

$$\frac{4!}{1!(4-1)!}\left(\frac{c}{2}\right)^3\left(-\frac{d}{3}\right) = \frac{4\cdot\cancel{3!}}{1!\cancel{3!}}\left(\frac{c^3}{8}\right)\left(-\frac{d}{3}\right)$$
$$= \frac{4}{1}\left(-\frac{c^3d}{24}\right)$$
$$= -\frac{c^3d}{6}$$
$$= -\frac{1}{6}c^3d$$

89. The 4th term of $(2t - 5)^7$

$$\frac{7!}{3!(7-3)!}(2t)^4(-5)^3 = \frac{7\cdot6\cdot5\cdot\cancel{4!}}{3!\cancel{4!}}(16t^4)(-125)$$
$$= \frac{210}{6}(-2{,}000t^4)$$
$$= 35(-2{,}000t^4)$$
$$= -70{,}000t^4$$

91. The 2nd term of $(a^2 + b^2)^6$

$$\frac{6!}{1!(6-1)!}(a^2)^5(b^2) = \frac{6\cdot\cancel{5!}}{1!\cancel{5!}}(a^{10})(b^2)$$
$$= \frac{6}{1}(a^{10}b^2)$$
$$= 6a^{10}b^2$$

WRITING

93. Answers will vary.

95. Answers will vary.

REVIEW

97.

$$2\log x + \frac{1}{2}\log y = \log x^2 + \log y^{1/2}$$
$$= \log x^2y^{1/2}$$

99.

$$\ln(xy + y^2) - \ln(xz + yz) + \ln z$$
$$= \ln z(xy + y^2) - \ln(xz + yz)$$
$$= \ln\frac{z(xy + y^2)}{(xz + yz)}$$
$$= \ln\frac{\cancel{z}y\cancel{(x+y)}}{\cancel{z}\cancel{(x+y)}}$$
$$= \ln y$$

CHALLENGE PROBLEMS

101. The constant term is the coefficient of $x^{10}x^{-10} = x^0 = 1$. It is the 10th term in the expansion.

$$\frac{10!}{5!(10-5)!}(x)^5\left(\frac{1}{x}\right)^5 = \frac{10\cdot9\cdot8\cdot7\cdot6\cdot\cancel{5!}}{\cancel{5!}5!}(x)^5(x^{-1})^5$$
$$= \frac{30{,}240}{120}(a^5)(a^{-5})$$
$$= 252x^{5+(-5)}$$
$$= 252x^0$$
$$= 252(1)$$
$$= 252$$

103. a.

$$\frac{n!}{0!(n-0)!} = \frac{\cancel{n!}}{(1)\cancel{n!}}$$
$$= 1$$

b.

$$\frac{n!}{n!(n-n)!} = \frac{\cancel{n!}}{\cancel{n!}0!}$$
$$= \frac{1}{1}$$
$$= 1$$

SECTION 11.2

VOCABULARY

1. A **sequence** is a function whose domain is the set of natural numbers.

3. Each term of an **arithmetic** sequence is found by adding the same number to the previous term.

5. If a single number is inserted between a and b to form an arithmetic sequence, the number is called the arithmetic **mean** between a and b.

CONCEPTS

7. 1, 7, 13

9. a. $a_n = a_1 + (n-1)d$
   b. `

NOTATION

11. The notation a_n represents the ***n*th** term of a sequence.

13. The symbol Σ is the Greek letter **sigma**.

15. We read $\sum_{k=1}^{10} 3k$ as "the **summation** of $3k$ as k **runs** from 1 to 10."

PRACTICE

17. $a_n = 4n - 1$
$$a_1 = 4(1) - 1 = 3$$
$$a_2 = 4(2) - 1 = 7$$
$$a_3 = 4(3) - 1 = 11$$
$$a_4 = 4(4) - 1 = 15$$
$$a_5 = 4(5) - 1 = 19$$
$$a_{40} = 4(40) - 1 = 159$$

19. $a_n = -3n + 1$
$$a_1 = -3(1) + 1 = -2$$
$$a_2 = -3(2) + 1 = -5$$
$$a_3 = -3(3) + 1 = -8$$
$$a_4 = -3(4) + 1 = -11$$
$$a_5 = -3(5) + 1 = -14$$
$$a_{30} = -3(30) + 1 = -89$$

21 $a_n = -n^2$
$$a_1 = -1^2 = -1$$
$$a_2 = -2^2 = -4$$
$$a_3 = -3^2 = -9$$
$$a_4 = -4^2 = -16$$
$$a_5 = -5^2 = -25$$
$$a_{20} = -20^2 = -400$$

23. $a_n = \frac{n-1}{n}$
$$a_1 = \frac{1-1}{1} = \frac{0}{1} = 0$$
$$a_2 = \frac{2-1}{2} = \frac{1}{2}$$
$$a_3 = \frac{3-1}{3} = \frac{2}{3}$$
$$a_4 = \frac{4-1}{4} = \frac{3}{4}$$
$$a_5 = \frac{5-1}{5} = \frac{4}{5}$$
$$a_{12} = \frac{12-1}{12} = \frac{11}{12}$$

25. $a_n = \frac{(-1)^n}{3^n}$
$$a_1 = \frac{(-1)^1}{3^1} = \frac{-1}{3} = -\frac{1}{3}$$
$$a_2 = \frac{(-1)^2}{3^2} = \frac{1}{9}$$
$$a_3 = \frac{(-1)^3}{3^3} = \frac{-1}{27} = -\frac{1}{27}$$
$$a_4 = \frac{(-1)^4}{3^4} = \frac{1}{81}$$

27. $a_n = (-1)^n(n+6)$

$a_1 = (-1)^1(1+6) = -1(7) = -7$

$a_2 = (-1)^2(2+6) = 1(8) = 8$

$a_3 = (-1)^3(3+6) = -1(9) = -9$

$a_4 = (-1)^4(4+6) = 1(10) = 10$

29. $a_n = a_1 + (n-1)d$

$a_n = 3 + 2(n-1)$

$a_1 = 3 + 2(1-1) = 3 + 0 = 3$

$a_2 = 3 + 2(2-1) = 3 + 2 = 5$

$a_3 = 3 + 2(3-1) = 3 + 4 = 7$

$a_4 = 3 + 2(4-1) = 3 + 6 = 9$

$a_5 = 3 + 2(5-1) = 3 + 8 = 11$

$a_{10} = 3 + 2(10-1) = 3 + 18 = 21$

31. $a_n = a_1 + (n-1)d$

$a_n = -5 - 3(n-1)$

$a_1 = -5 - 3(1-1) = -5 + 0 = -5$

$a_2 = -5 - 3(2-1) = -5 - 3 = -8$

$a_3 = -5 - 3(3-1) = -5 - 6 = -11$

$a_4 = -5 - 3(4-1) = -5 - 9 = -14$

$a_5 = -5 - 3(5-1) = -5 - 12 = -17$

$a_{15} = -5 - 3(15-1) = -5 - 42 = -47$

33. $a_n = a_1 + (n-1)d$

$a_n = 7 + 12(n-1)$

$a_1 = 7 + 12(1-1) = 7 + 0 = 7$

$a_2 = 7 + 12(2-1) = 7 + 12 = 19$

$a_3 = 7 + 12(3-1) = 7 + 24 = 31$

$a_4 = 7 + 12(4-1) = 7 + 36 = 43$

$a_5 = 7 + 12(5-1) = 7 + 48 = 55$

$a_{30} = 7 + 12(30-1) = 7 + 348 = 355$

35. $a_n = a_1 + (n-1)d$

$a_n = -7 - 2(n-1)$

$a_1 = -7 - 2(1-1) = -7 - 0 = -7$

$a_2 = -7 - 2(2-1) = -7 - 2 = -9$

$a_3 = -7 - 2(3-1) = -7 - 4 = -11$

$a_4 = -7 - 2(4-1) = -7 - 6 = -13$

$a_5 = -7 - 2(5-1) = -7 - 8 = -15$

$a_{15} = -7 - 2(15-1) = -7 - 28 = -35$

37.

$a_1 = 1, \quad d = 4 - 1 = 3, \quad n = 30$

$a_n = 1 + 3(30-1)$

$= 1 + 3(29)$

$= 88$

39.

$a_1 = -5, \quad d = -1 - (-5) = 4, \quad n = 17$

$a_n = -5 + 4(17-1)$

$= -5 + 4(16)$

$= 59$

41. $a_n = a_1 + (n-1)d$

Find d.

$29 = 5 + d(5-1)$

$29 = 5 + d(4)$

$24 = 4d$

$6 = d$

$a_1 = 5 + 6(1-1) = 5 + 0 = 5$

$a_2 = 5 + 6(2-1) = 5 + 6 = 11$

$a_3 = 5 + 6(3-1) = 5 + 12 = 17$

$a_4 = 5 + 6(4-1) = 5 + 18 = 23$

$a_5 = 5 + 6(5-1) = 5 + 24 = 29$

The first five terms are 5, 11, 17, 23, and 29.

43. $a_n = a_1 + (n-1)d$

Find d.

$-39 = -4 + d(6-1)$

$-39 = -4 + d(5)$

$-35 = 5d$

$-7 = d$

$a_1 = -4 - 7(1-1) = -4 - 0 = -4$

$a_2 = -4 - 7(2-1) = -4 - 7 = -11$

$a_3 = -4 - 7(3-1) = -4 - 14 = -18$

$a_4 = -4 - 7(4-1) = -4 - 21 = -25$

$a_5 = -4 - 7(5-1) = -4 - 28 = -32$

The first five terms are –4, –11, –18, –25, and –32.

45. Let $a_1 = 2$ and $a_4 = 11$

$a_2 = 2 + d, a_3 = 2 + 2d$

$a_4 = a_1 + (4-1)d$

$11 = 2 + 3d$

$9 = 3d$

$3 = d$

$a_2 = 2 + 3(2-1) = 2 + 3 = 5$

$a_3 = 2 + 3(3-1) = 2 + 6 = 8$

47 Let $a_1 = 10$ and $a_6 = 20$

$a_2 = 10 + d, a_3 = 10 + 2d, a_4 = 10 + 3d, a_5 = 10 + 4d$

$a_6 = a_1 + (6-1)d$

$20 = 10 + 5d$

$10 = 5d$

$2 = d$

$a_2 = 10 + d = 10 + 2 = 12$

$a_3 = 10 + 2d = 10 + 2(2) = 14$

$a_4 = 10 + 3d = 10 + 3(2) = 16$

$a_5 = 10 + 4d = 10 + 4(2) = 18$

49. Let $a_1 = 20$ and $a_5 = 30$

$a_2 = 20 + d, a_3 = 20 + 2d, a_4 = 20 + 3d$

$a_5 = a_1 + (5-1)d$

$30 = 20 + 4d$

$10 = 4d$

$\frac{5}{2} = d$

$a_2 = 20 + d = 20 + \frac{5}{2} = \frac{45}{2}$

$a_3 = 20 + 2d = 20 + 2\left(\frac{5}{2}\right) = 25$

$a_4 = 20 + 3d = 20 + 3\left(\frac{5}{2}\right) = \frac{55}{2}$

51. Let $a_1 = -4.5$ and $a_3 = 7$

$a_2 = -4.5 + d$

$a_3 = a_1 + (3-1)d$

$7 = -4.5 + 2d$

$11.5 = 2d$

$5.75 = d$

$a_2 = -4.5 + d = -4.5 + 5.75 = 1.25 = \frac{5}{4}$

53. Find a_{35}.

$a_1 = 5, d = 4, n = 35$

$a_n = a_1 + (n-1)d$

$a_{35} = 5 + 4(35-1) = 5 + 4(34) = 141$

$S_n = \frac{n(a_1 + a_n)}{2}$

$S_{35} = \frac{35(5 + 141)}{2}$

$= \frac{35(146)}{2}$

$= 2,555$

55. Find a_{40}.

$$a_1 = -5, d = 4, n = 40$$
$$a_n = a_1 + (n-1)d$$
$$a_{40} = -5 + 4(40-1) = -5 + 4(39) = 151$$

$$S_n = \frac{n(a_1 + a_n)}{2}$$
$$S_{40} = \frac{40(-5+151)}{2}$$
$$= 2,920$$

57.

$$\sum_{k=1}^{4}(3k) = 3(1) + 3(2) + 3(3) + 3(4)$$
$$= 3 + 6 + 9 + 12$$

59.

$$\sum_{k=2}^{4} k^2 = 2^2 + 3^2 + 4^2$$
$$= 4 + 9 + 16$$

61.

$$\sum_{k=1}^{4}(6k) = 6(1) + 6(2) + 6(3) + 6(4)$$
$$= 6 + 12 + 18 + 24$$
$$= 60$$

63.

$$\sum_{k=3}^{4} k^3 = 3^3 + 4^3$$
$$= 27 + 64$$
$$= 91$$

65.

$$\sum_{k=3}^{4}(k^2 + 3) = (3^2 + 3) + (4^2 + 3)$$
$$= (9+3) + (16+3)$$
$$= (12) + (19)$$
$$= 31$$

67.

$$\sum_{k=4}^{4}(2k + 4) = 2(4) + 4$$
$$= 8 + 4$$
$$= 12$$

69.

$$\sum_{k=2}^{5}(5k) = 5(2) + 5(3) + 5(4) + 5(5)$$
$$= 10 + 15 + 20 + 25$$
$$= 70$$

71.

$$\sum_{k=4}^{6}(4k - 1) = (4 \cdot 4 - 1) + (4 \cdot 5 - 1) + (4 \cdot 6 - 1)$$
$$= 15 + 19 + 23$$
$$= 57$$

TRY IT YOURSELF

73. Let $n = 44$ and $a_n = 556$, $a_1 = 40$

$$556 = 40 + d(44-1)$$
$$556 = 40 + d(43)$$
$$516 = 43d$$
$$12 = d$$

75.

$$a_2 = 7, \quad a_3 = 12, \quad d = 12 - 7 = 5, \quad n = 12$$
$$7 = a_1 + 5(2-1)$$
$$7 = a_1 + 5(1)$$
$$7 = a_1 + 5$$
$$2 = a_1$$

$$a_{12} = 2 + 5(12-1)$$
$$= 2 + 5(11)$$
$$= 57$$

$$S_n = \frac{n(a_1 + a_n)}{2}$$
$$S_{17} = \frac{12(2+57)}{2}$$
$$= \frac{12(59)}{2}$$
$$= 354$$

77. $a_n = a_1 + (n-1)d$

Find a_1.

$-83 = a_1 + 7(6-1)$

$-83 = a_1 + 7(5)$

$-83 = a_1 + 35$

$-118 = a_1$

$a_1 = -118 + 7(1-1) = -118 + 0 = -118$

$a_2 = -118 + 7(2-1) = -118 + 7 = -111$

$a_3 = -118 + 7(3-1) = -118 + 14 = -104$

$a_4 = -118 + 7(4-1) = -118 + 21 = -97$

$a_5 = -118 + 7(5-1) = -118 + 28 = -90$

The first five terms are –118, –111, –104, –97, and –90.

79. $a_n = a_1 + (n-1)d$

Find a_1.

$10 = a_1 - 3(9-1)$

$10 = a_1 - 3(8)$

$10 = a_1 - 24$

$34 = a_1$

$a_1 = 34 - 3(1-1) = 34 - 0 = 34$

$a_2 = 34 - 3(2-1) = 34 - 3 = 31$

$a_3 = 34 - 3(3-1) = 34 - 6 = 28$

$a_4 = 34 - 3(4-1) = 34 - 9 = 25$

$a_5 = 34 - 3(5-1) = 34 - 12 = 22$

The first five terms are 34, 31, 28, 25, and 22.

81. Find a_{200}.

$a_1 = 5, d = 7, n = 200$

$a_n = a_1 + (n-1)d$

$a_{200} = 5 + 7(200-1) = 5 + 7(199) = 1,398$

83.

$a_1 = 1,\ a_n = 50,\ n = 50$

$S_n = \dfrac{n(a_1 + a_n)}{2}$

$= \dfrac{50(1+50)}{2}$

$= \dfrac{50(51)}{2}$

$= 1,275$

85. Find d

–9 – (–4), so $d = -5$.

Find a_1 if $a_n = -4$ and $n = 2$.

$-4 = a_1 - 5(2-1)$

$-4 = a_1 - 5(1)$

$-4 = a_1 - 5$

$1 = a_1$

Find a_{37}

$a_{37} = 1 - 5(37-1)$

$= 1 - 5(36)$

$= 1 - 180$

$= -179$

87. Let $d = 11$, $n = 27$ and $a_n = 263$.

$263 = a_1 + 11(27-1)$

$263 = a_1 + 11(26)$

$263 = a_1 + 286$

$-23 = a_1$

89. Find a_{15}.

$a_1 = \frac{1}{2}, d = -\frac{1}{4}, n = 15$

$a_n = a_1 + (n-1)d$

$a_{15} = \frac{1}{2} - \frac{1}{4}(15-1) = \frac{1}{2} - \frac{1}{4}(14) = -3$

91.

$a_1 = 1,\ a_n = 99,\ n = 50$

$S_n = \dfrac{n(a_1 + a_n)}{2}$

$= \dfrac{50(1+99)}{2}$

$= \dfrac{50(100)}{2}$

$= 2,500$

APPLICATIONS

93. SAVING MONEY

Let $a_1 = 60$ and $d = 50$.

Let $n =$ the month. Find a_n to find the amount in the account each month with $a_1 = 60$ and $d = 50$.

$$a_n = a_1 + d(n-1)$$
$$a_1 = 60$$
$$a_2 = 60 + 50(2-1) = 60 + 50(1) = 110$$
$$a_3 = 60 + 50(3-1) = 60 + 50(2) = 160$$
$$a_4 = 60 + 50(4-1) = 60 + 50(3) = 210$$
$$a_5 = 60 + 50(5-1) = 60 + 50(4) = 260$$
$$a_6 = 60 + 50(6-1) = 60 + 50(5) = 310$$

To find the amount after 10 years, or $10(12) = 120$ months, plus the first deposit, let $n = 120 + 1 = 121$ and find a_{121}.

$$a_{121} = 60 + 50(121-1) = 60 + 50(120) = 6{,}060$$

She will have a savings of \$6,060 after 10 years.

95. DESIGNING PATIOS

$$a_1 = 1, \quad a_n = 150, \quad n = 150$$
$$S_n = \frac{n(a_1 + a_n)}{2} = \frac{150(1+150)}{2} = \frac{150(151)}{2} = 11{,}325$$

11,325 bricks are needed.

97. HOLIDAY SONGS

$$S_n = \frac{n(a_1 + a_n)}{2} = \frac{12(1+12)}{2} = \frac{12(13)}{2} = \frac{156}{2} = 78$$

WRITING

99. Answers will vary.

101. Answers will vary.

REVIEW

103.

$$\log_2 \frac{2x}{y} = \log_2 2 + \log_2 x - \log_2 y = 1 + \log_2 x - \log_2 y$$

105.

$$\log x^3 y^2 = \log x^3 + \log y^2 = 3\log x + 2\log y$$

CHALLENGE PROBLEMS

107.

$$1 + 4 + 9 + 16 + 25 = 1^2 + 2^2 + 3^2 + 4^2 + 5^2 = \sum_{k=1}^{5} k^2$$

109. Let $x = 9$. Then,

$x - 2 = 9 - 2 = 7$

$2x + 4 = 2(9) + 4 = 22$

$5x - 8 = 5(9) - 8 = 37$

It forms an arithmetic sequence with a difference of 15.

SECTION 11.3

VOCABULARY

1. Each term of a **geometric** sequence is found by multiplying the previous term by the same number.

3. If a single number is inserted between a and b to form a geometric sequence, the number is called the geometric **mean** between a and b.

CONCEPTS

5.
$$a_1 = 16$$
$$a_2 = 16\left(\frac{1}{4}\right) = 4$$
$$a_3 = 16\left(\frac{1}{4}\right)^2 = 1$$
16, 4, 1

7. $a_n = a_1 r^{n-1}$

9. a. yes
b. no, $3 \not< 1$
c. no, $6 \not< 1$
d. yes

NOTATION

11. An infinite geometric sequence is of the form $a_1, a_1r, \boxed{a_1r^2}, a_1r^3, \boxed{a_1r^4}, \ldots$

13. To find the common ratio of a geometric sequence, we use the formula $r = \dfrac{a_{\boxed{n+1}}}{a_{\boxed{n}}}$.

PRACTICE

15. $a_1 = 3, r = 2$
$$a_1 = 3$$
$$a_2 = a_1r = 3(2) = 6$$
$$a_3 = a_1r^2 = 3(2)^2 = 12$$
$$a_4 = a_1r^3 = 3(2)^3 = 24$$
$$a_5 = a_1r^4 = 3(2)^4 = 48$$
$$a_9 = a_1r^8 = 3(2)^8 = 768$$

17. $a_1 = -5, r = \frac{1}{5}$
$$a_1 = -5$$
$$a_2 = a_1r = -5\left(\frac{1}{5}\right) = -1$$
$$a_3 = a_1r^2 = -5\left(\frac{1}{5}\right)^2 = -\frac{1}{5}$$
$$a_4 = a_1r^3 = -5\left(\frac{1}{5}\right)^3 = -\frac{1}{25}$$
$$a_5 = a_1r^4 = -5\left(\frac{1}{5}\right)^4 = -\frac{1}{125}$$
$$a_8 = a_1r^7 = -5\left(\frac{1}{5}\right)^7 = -\frac{1}{15{,}625}$$

19. $a_1 = 2, r = 3, n = 7$
$$a_n = a_1r^{n-1}$$
$$a_7 = 2(3)^{7-1}$$
$$= 2(3)^6$$
$$= 2(729)$$
$$= 1{,}458$$

21.

$$a_1=\frac{1}{2}, a_2=-\frac{5}{2}, n=6$$

$$r=\frac{a_2}{a_1}=\frac{-\frac{5}{2}}{\frac{1}{2}}=-5$$

$$a_n=a_1r^{n-1}$$

$$a_6=\frac{1}{2}(-5)^{6-1}$$

$$=\frac{1}{2}(-5)^5$$

$$=\frac{1}{2}(-3{,}125)$$

$$=-\frac{3{,}125}{2}$$

23. $a_1=2$, $r>0$, third term is 18

Use $a_n=ar^{n-1}$ to find r.

$$a=2,\ a_3=18 \text{ and } n=3$$

$$18=2r^{3-1}$$

$$18=2r^2$$

$$9=r^2$$

$$3=r \text{ since } r>0$$

$$a_1=2$$

$$a_2=a_1r=2(3)=6$$

$$a_3=a_1r^2=2(3)^2=18$$

$$a_4=a_1r^3=2(3)^3=54$$

$$a_5=a_1r^4=2(3)^4=162$$

2, 6, 18, 54, 162

25. $a_1=3$, fourth term is 24

Use $a_n=ar^{n-1}$ to find r.

$$a=3,\ a_4=24 \text{ and } n=4$$

$$24=3r^{4-1}$$

$$24=3r^3$$

$$8=r^3$$

$$2=r$$

$$a_1=3$$

$$a_2=a_1r=3(2)=6$$

$$a_3=a_1r^2=3(2)^2=12$$

$$a_4=a_1r^3=3(2)^3=24$$

$$a_5=a_1r^4=3(2)^4=48$$

3, 6, 12, 24, 48

27. $a_1=2$, $a_5=162$ for three geometric means between 2 and 162

Use $a_n=ar^{n-1}$ to find r.

$$a_1=2,\ a_5=162 \text{ and } n=5$$

$$162=2(r)^{5-1}$$

$$162=2r^4$$

$$81=r^4$$

$$3=r$$

$$a_1=2$$

$$a_2=2(3)=6$$

$$a_3=2(3)^2=18$$

$$a_4=2(3)^3=54$$

$$a_5=2(3)^4=162$$

6, 18, 54

29. $a_1 = -4$, $a_6 = -12{,}500$ for four geometric means between –4 and –12,500

Use $a_n = ar^{n-1}$ to find r.

$$a_1 = -4,\ a_6 = -12{,}500 \text{ and } n = 6$$
$$-12{,}500 = -4(r)^{6-1}$$
$$-12{,}500 = -4r^5$$
$$3{,}125 = r^5$$
$$5 = r$$
$$a_1 = -4$$
$$a_2 = -4(5) = -20$$
$$a_3 = -4(5)^2 = -100$$
$$a_4 = -4(5)^3 = -500$$
$$a_5 = -4(5)^4 = -2{,}500$$
$$a_6 = -4(5)^5 = -12{,}500$$
$$-20, -100, -500, -2{,}500$$

31. $a_1 = 2$, $a_3 = 128$ for the geometric mean between 2 and 128

Use $a_n = ar^{n-1}$ to find r.

$$a_1 = 2,\ a_3 = 128 \text{ and } n = 3$$
$$128 = 2(r)^{3-1}$$
$$128 = 2r^2$$
$$64 = r^2$$
$$\pm 8 = r$$
$$a_1 = 2$$
$$a_2 = 2(\pm 8) = \pm 16$$

The geometric mean is 16 or –16.

33. $a_1 = 10$, $a_3 = 20$ for the geometric mean between 10 and 20

Use $a_n = ar^{n-1}$ to find r.

$$a_1 = 10,\ a_3 = 20 \text{ and } n = 3$$
$$20 = 10(r)^{3-1}$$
$$20 = 10r^2$$
$$2 = r^2$$
$$\pm\sqrt{2} = r$$
$$a_1 = 10$$
$$a_2 = 10(\pm\sqrt{2}) = \pm 10\sqrt{2}$$

The geometric mean is $\pm 10\sqrt{2}$.

35. Find r.

$$r = \frac{a_2}{a_1} = \frac{6}{2} = 3$$

Find the sum with $a_1 = 2$, $r = 3$, and $n = 6$.

$$S_n = \frac{a_1(1-r^n)}{1-r} = \frac{2(1-3^6)}{1-3} = \frac{2(1-729)}{-2} = \frac{2(-728)}{-2} = 728$$

37. Find r.

$$r = \frac{a_2}{a_1} = \frac{-6}{2} = -3$$

Find the sum with $a_1 = 2$, $r = -3$, and $n = 5$.

$$S_n = \frac{a_1(1-r^n)}{1-r} = \frac{2(1-(-3)^5)}{1-(-3)} = \frac{2(1+243)}{4} = \frac{2(244)}{4} = 122$$

39. Find r.

$$r = \frac{a_2}{a_1} = \frac{-6}{3} = -2$$

Find the sum with $a_1 = 3$, $r = -2$, and $n = 8$.

$$S_n = \frac{a_1(1-r^n)}{1-r} = \frac{3(1-(-2)^8)}{1-(-2)} = \frac{3(1-256)}{3} = \frac{3(-255)}{3} = -255$$

41. Find r.

$$r=\frac{a_2}{a_1}=\frac{6}{3}=2$$

Find the sum with $a_1=3$, $r=2$, and $n=7$.

$$\begin{aligned}S_n&=\frac{a_1(1-r^n)}{1-r}\\&=\frac{3(1-2^7)}{1-2}\\&=\frac{3(1-128)}{-1}\\&=\frac{3(-127)}{-1}\\&=381\end{aligned}$$

43. Find r.

$$r=\frac{a_2}{a_1}=\frac{4}{8}=\frac{1}{2}$$

$|r|=\left|\frac{1}{2}\right|=\frac{1}{2}<1$ so the sum does exist

Find the sum with $a_1=8$ and $r=\frac{1}{2}$.

$$\begin{aligned}S&=\frac{a_1}{1-r}\\&=\frac{8}{1-\frac{1}{2}}\\&=\frac{8}{\frac{1}{2}}\\&=8\bullet\frac{2}{1}\\&=16\end{aligned}$$

45. Find r.

$$r=\frac{a_2}{a_1}=\frac{18}{54}=\frac{1}{3}$$

$|r|=\left|\frac{1}{3}\right|=\frac{1}{3}<1$ so the sum does exist

Find the sum with $a_1=54$ and $r=\frac{1}{3}$.

$$\begin{aligned}S&=\frac{a_1}{1-r}\\&=\frac{54}{1-\frac{1}{3}}\\&=\frac{54}{\frac{2}{3}}\\&=54\bullet\frac{3}{2}\\&=81\end{aligned}$$

47. Find r.

$$r=\frac{a_2}{a_1}=\frac{-9}{-\frac{27}{2}}=\frac{2}{3}$$

$|r|=\left|\frac{2}{3}\right|=\frac{2}{3}<1$ so the sum does exist

Find the sum with $a_1=-\frac{27}{2}$ and $r=\frac{2}{3}$.

$$\begin{aligned}S&=\frac{a_1}{1-r}\\&=\frac{-\frac{27}{2}}{1-\frac{2}{3}}\\&=\frac{-\frac{27}{2}}{\frac{1}{3}}\\&=-\frac{27}{2}\bullet 3\\&=-\frac{81}{2}\end{aligned}$$

49. Find r.

$$r = \frac{a_2}{a_1} = \frac{6}{\frac{9}{2}} = \frac{4}{3}$$

Since $|r| = \left|\frac{4}{3}\right| = \frac{4}{3} \geq 1$, the sum of the terms of the sequence, S, does not exist

51. Find r.

$$r = \frac{a_2}{a_1} = \frac{-6}{12} = -\frac{1}{2}$$

$$|r| = \left|-\frac{1}{2}\right| = \frac{1}{2} < 1 \text{ so the sum does exist}$$

Find the sum with $a_1 = 12$ and $r = -\frac{1}{2}$.

$$\begin{aligned} S &= \frac{a_1}{1-r} \\ &= \frac{12}{1-\left(-\frac{1}{2}\right)} \\ &= \frac{12}{\frac{3}{2}} \\ &= 12 \bullet \frac{2}{3} \\ &= 8 \end{aligned}$$

53. Find r.

$$r = \frac{a_2}{a_1} = \frac{15}{-45} = -\frac{1}{3}$$

$$|r| = \left|-\frac{1}{3}\right| = \frac{1}{3} < 1 \text{ so the sum does exist}$$

Find the sum with $a_1 = -45$ and $r = -\frac{1}{3}$.

$$\begin{aligned} S &= \frac{a_1}{1-r} \\ &= \frac{-45}{1-\left(-\frac{1}{3}\right)} \\ &= \frac{-45}{\frac{4}{3}} \\ &= -45 \bullet \frac{3}{4} \\ &= -\frac{135}{4} \end{aligned}$$

55.

$$0.\bar{1} = 0.111\ldots = \frac{1}{10} + \frac{1}{100} + \frac{1}{1{,}000} + \ldots$$

where $a_1 = \frac{1}{10}$ and $r = \frac{1}{10}$.

$$|r| = \left|\frac{1}{10}\right| = \frac{1}{10} < 1 \text{ so the sum does exist.}$$

Find the sum with $a_1 = \frac{1}{10}$ and $r = \frac{1}{10}$.

$$\begin{aligned} S &= \frac{a_1}{1-r} \\ &= \frac{\frac{1}{10}}{1-\frac{1}{10}} \\ &= \frac{\frac{1}{10}}{\frac{9}{10}} \\ &= \frac{1}{10} \bullet \frac{10}{9} \\ &= \frac{1}{9} \end{aligned}$$

$$0.\bar{1} = \frac{1}{9}$$

57.

$$0.\overline{3} = 0.333... = \frac{3}{10} + \frac{3}{100} + \frac{3}{1,000} + ...$$

where $a_1 = \frac{3}{10}$ and $r = \frac{1}{10}$.

$|r| = \left|\frac{1}{10}\right| = \frac{1}{10} < 1$ so the sum does exist.

Find the sum with $a_1 = \frac{3}{10}$ and $r = \frac{1}{10}$.

$$\begin{aligned} S &= \frac{a_1}{1-r} \\ &= \frac{\frac{3}{10}}{1-\frac{1}{10}} \\ &= \frac{\frac{3}{10}}{\frac{9}{10}} \\ &= \frac{3}{10} \bullet \frac{10}{9} \\ &= \frac{1}{3} \end{aligned}$$

$$0.\overline{3} = \frac{1}{3}$$

59.

$$0.\overline{12} = 0.121212... = \frac{12}{100} + \frac{12}{10,000} + \frac{12}{1,000,000} + ...$$

where $a_1 = \frac{12}{100}$ and $r = \frac{1}{100}$.

$|r| = \left|\frac{1}{100}\right| = \frac{1}{100} < 1$ so the sum does exist.

Find the sum with $a_1 = \frac{12}{100}$ and $r = \frac{1}{100}$.

$$\begin{aligned} S &= \frac{a_1}{1-r} \\ &= \frac{\frac{12}{100}}{1-\frac{1}{100}} \\ &= \frac{\frac{12}{100}}{\frac{99}{100}} \\ &= \frac{12}{100} \bullet \frac{100}{99} \\ &= \frac{12}{99} \\ &= \frac{4}{33} \\ &= \frac{4}{33} \end{aligned}$$

61.

$$0.\overline{75} = 0.757575... = \frac{75}{100} + \frac{75}{10,000} + \frac{75}{1,000,000} + ...$$

where $a_1 = \frac{75}{100}$ and $r = \frac{1}{100}$.

$|r| = \left|\frac{1}{100}\right| = \frac{1}{100} < 1$ so the sum does exist.

Find the sum with $a_1 = \frac{75}{100}$ and $r = \frac{1}{100}$.

$$\begin{aligned} S &= \frac{a_1}{1-r} \\ &= \frac{\frac{75}{100}}{1-\frac{1}{100}} \\ &= \frac{\frac{75}{100}}{\frac{99}{100}} \\ &= \frac{75}{100} \cdot \frac{100}{99} \\ &= \frac{75}{99} \\ &= \frac{25}{33} \\ 0.\overline{75} &= \frac{25}{33} \end{aligned}$$

TRY IT YOURSELF

63. $a_1 = -8$, $a_6 = -1,944$

Use $a_n = ar^{n-1}$ to find a_1.

$$a_1 = -8,\ a_6 = -1,944 \text{ and } n = 6$$

$$\begin{aligned} -1,944 &= -8(r)^{6-1} \\ -1,944 &= -8r^5 \\ 243 &= r^5 \\ 3 &= r \end{aligned}$$

65. $a_1 = -64$, $r < 0$, fifth term is -4

Use $a_n = ar^{n-1}$ to find r.

$$a = -64,\ a_5 = -4 \text{ and } n = 5$$

$$\begin{aligned} -4 &= -64r^{5-1} \\ -4 &= -64r^4 \\ \frac{1}{16} &= r^4 \\ -\frac{1}{2} &= r \text{ since } r < 0 \end{aligned}$$

$$a_1 = -64$$

$$a_2 = a_1 r = -64\left(-\frac{1}{2}\right) = 32$$

$$a_3 = a_1 r^2 = -64\left(-\frac{1}{2}\right)^2 = -16$$

$$a_4 = a_1 r^3 = -64\left(-\frac{1}{2}\right)^3 = 8$$

$$a_5 = a_1 r^4 = -64\left(-\frac{1}{2}\right)^4 = -4$$

$$-64, 32, -16, 8, -4$$

67. $a_1 = -50$, $a_3 = 10$ for the geometric mean between -50 and 10

Use $a_n = ar^{n-1}$ to find r.

$$a_1 = -50,\ a_3 = 10 \text{ and } n = 3$$
$$10 = -50(r)^{3-1}$$
$$10 = -50r^2$$
$$-\frac{1}{5} = r^2$$

There is no geometric mean because no real number squared is $-\frac{1}{5}$. When you square a real number, it is always positive.

69. $a_1 = 7$, $r = 2$

Use $a_n = ar^{n-1}$ to find a_{10}.

$$r = 2,\ a_1 = 7 \text{ and } n = 10$$
$$a_{10} = 7(2)^{10-1}$$
$$a_{10} = 7(2)^9$$
$$a_{10} = 7(512)$$
$$a_{10} = 3{,}584$$

71. $a_1 = -64$, sixth term is -2

Use $a_n = ar^{n-1}$ to find r.

$$a_1 = -64,\ a_6 = -2 \text{ and } n = 6$$
$$-2 = -64r^{6-1}$$
$$-2 = -64r^5$$
$$\frac{1}{32} = r^5$$
$$\frac{1}{2} = r$$
$$a_1 = -64$$
$$a_2 = a_1 r = -64\left(\frac{1}{2}\right) = -32$$
$$a_3 = a_1 r^2 = -64\left(\frac{1}{2}\right)^2 = -16$$
$$a_4 = a_1 r^3 = -64\left(\frac{1}{2}\right)^3 = -8$$
$$-64, -32, -16, -8$$

73. Find r.

$$r = \frac{a_2}{a_1} = \frac{\frac{3}{4}}{3} = \frac{1}{4}$$

$|r| = \left|\frac{1}{4}\right| = \frac{1}{4} < 1$ so the sum does exist

Find the sum with $a_1 = 3$ and $r = \frac{1}{4}$.

$$S = \frac{a_1}{1-r}$$
$$= \frac{3}{1-\frac{1}{4}}$$
$$= \frac{3}{\frac{3}{4}}$$
$$= 3 \cdot \frac{4}{3}$$
$$= 4$$

75. $r = -3$, $a_8 = -81$

Use $a_n = ar^{n-1}$ to find a_1.

$$r = -3,\ a_8 = -81 \text{ and } n = 8$$
$$-81 = a_1(-3)^{8-1}$$
$$-81 = a_1(-3)^7$$
$$-81 = a_1(-2{,}187)$$
$$\frac{-81}{-2{,}187} = a_1$$
$$\frac{1}{27} = a_1$$

77. Find the sum with $a_1 = 3$, $r = 2$, and $n = 5$.

$$S_n = \frac{a_1(1-r^n)}{1-r}$$
$$= \frac{3(1-2^5)}{1-2}$$
$$= \frac{3(1-32)}{-1}$$
$$= \frac{3(-31)}{-1}$$
$$= 93$$

APPLICATIONS

79. DECLINING SAVINGS
If he spends 12% of the funds each year, then 88% of the funds remain. The amount of money in the savings box after 15 years is represented by the 16th term of a geometric series, where $a_1 = 10{,}000$, $n = 16$, and $r = 0.88$.

$$a_n = a_1 r^{n-1}$$
$$a_{16} = (10{,}000)(0.88)^{16-1}$$
$$a_{16} = (10{,}000)(0.88)^{15}$$
$$a_{16} = (10{,}000)(0.1469738)$$
$$a_{16} = \$1{,}469.74$$

He will have $1,469.74 after 15 years.

81. REAL ESTATE SALES AGENT
If the house appreciates 8% each year, then it will be worth 108% of the preceding years' value (100% + 8% interest). The value of the house after 10 years is represented by the 11th term of a geometric series, where $a_1 = 250{,}000$, $n = 11$, and $r = 1.08$.

$$a_n = a_1 r^{n-1}$$
$$a_{11} = (250{,}000)(1.08)^{11-1}$$
$$a_{11} = (250{,}000)(1.08)^{10}$$
$$a_{11} = (250{,}000)(2.158924997)$$
$$a_{11} = \$539{,}731.25$$

The house will be worth approximately $539,731 in 10 years.

83. INSCRIBED SQUARES
The area of each new is ½ the area of the previous square. The area of the 12th square is represented by the 12th term of a geometric series, where $a_1 = 1$, $r = ½$, and $n = 12$.

$$a_n = a_1 r^{n-1}$$
$$a_{12} = (1)\left(\frac{1}{2}\right)^{12-1}$$
$$a_{12} = (1)\left(\frac{1}{2}\right)^{11}$$
$$a_{12} = (1)(0.00048828)$$
$$a_{12} \approx 0.0005$$

85. BOUNCING BALLS
The total distance the ball travels is the sum of two motions, falling and rebounding. The distance the ball falls is given by the sum $10 + \frac{1}{2}\bullet 10 + \frac{1}{2}\left(\frac{1}{2}\bullet 10\right) + \ldots$ or $10 + 5 + \frac{5}{2} + \ldots$. The distance the ball rebounds begins with the first bounce up which is 5 m, and then is one–half the distance afterwards or $5 + \frac{5}{2} + \frac{5}{4} + \ldots$.
Since these are infinite geometric series, use the formula $S = \frac{a_1}{1-r}$ to find the sum.

Falling:

$a_1 = 10$ and $r = \frac{1}{2}$

$$S = \frac{a_1}{1-r} = \frac{10}{1-\frac{1}{2}} = \frac{10}{\frac{1}{2}} = 10\bullet\frac{2}{1} = 20$$

Rebounding:

$a_1 = 5$ and $r = \frac{1}{2}$

$$S = \frac{a_1}{1-r} = \frac{5}{1-\frac{1}{2}} = \frac{5}{\frac{1}{2}} = 5\bullet\frac{2}{1} = 10$$

The distance the ball travels is the sum of the distances falling and rebounding:

20 + 10 = 30 m.
The ball travels a total of 30 m.

87. PEST CONTROL
To find the sum of the given infinite geometric series use the formula $S = \frac{a_1}{1-r}$ with $a_1 = 1{,}000$ and $r = 0.8$.

$$S = \frac{a_1}{1-r} = \frac{1{,}000}{1-0.8} = \frac{1{,}000}{0.2} = 5{,}000$$

The long–term population is 5,000.

WRITING

89. Answers will vary.

91. Answers will vary.

REVIEW

93.

$$x^2 - 5x - 6 \le 0$$
$$(x-6)(x+1) = 0$$
$$x - 6 = 0 \text{ or } x + 1 = 0$$
$$x = 6 \qquad x = -1$$

The critical numbers are 6 and –1. Pick numbers in all three parts and substitute to determine which areas to shade.

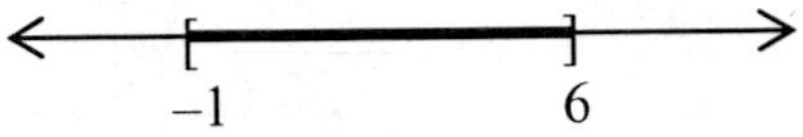

[–1, 6]

95.

$$\frac{x-4}{x+3} > 0$$
$$\frac{x-4}{x+3} = 0$$
$$(x+3)\left(\frac{x-4}{x+3}\right) = (x+3)(0)$$
$$x - 4 = 0$$
$$x = 4$$

Set the denominator $= 0$.

$$x + 3 = 0$$
$$x = -3$$

Critical numbers $= -3$ and 4

$$(-\infty, -3) \cup (4, \infty)$$

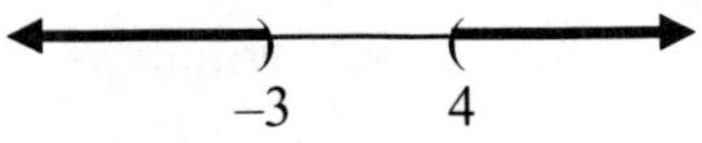

CHALLENGE PROBLEMS

97.

$$f(x) = 1 + x + x^2 + x^3 + x^4 + \ldots$$
$$f\left(\frac{1}{2}\right) = 1 + \frac{1}{2} + \left(\frac{1}{2}\right)^2 + \left(\frac{1}{2}\right)^3 + \left(\frac{1}{2}\right)^4 + \ldots$$
$$f\left(\frac{1}{2}\right) = 1 + \text{ infinite geometric series } \left(a = \frac{1}{2}, r = \frac{1}{2}\right)$$
$$f\left(\frac{1}{2}\right) = 1 + \frac{\frac{1}{2}}{1 - \frac{1}{2}}$$
$$f\left(\frac{1}{2}\right) = 1 + \frac{\frac{1}{2}}{\frac{1}{2}}$$
$$f\left(\frac{1}{2}\right) = 1 + 1$$
$$f\left(\frac{1}{2}\right) = 2$$

$$f\left(-\frac{1}{2}\right) = 1 + \left(-\frac{1}{2}\right) + \left(-\frac{1}{2}\right)^2 + \left(-\frac{1}{2}\right)^3 + \left(-\frac{1}{2}\right)^4 + \ldots$$
$$f\left(-\frac{1}{2}\right) = 1 + \text{ infinite geometric series } \left(a = -\frac{1}{2}, r = -\frac{1}{2}\right)$$
$$f\left(-\frac{1}{2}\right) = 1 + \frac{-\frac{1}{2}}{1 - \left(-\frac{1}{2}\right)}$$
$$f\left(-\frac{1}{2}\right) = 1 + \frac{-\frac{1}{2}}{\frac{3}{2}}$$
$$f\left(-\frac{1}{2}\right) = 1 - \frac{1}{3}$$
$$f\left(-\frac{1}{2}\right) = \frac{2}{3}$$

99. the arithmetic mean

CHAPTER 11 REVIEW

SECTION 11.1
The Binomial Theorem

1. The coefficients for the expansion of $(a + b)^5$ are 1, 5, 10, 10, 5, 1.

$$
\begin{array}{ccccccccccccc}
 & & & & & & 1 & & & & & & \\
 & & & & & \boxed{1} & & 1 & & & & & \\
 & & & & 1 & & 2 & & 1 & & & & \\
 & & & 1 & & \boxed{3} & & 3 & & \boxed{1} & & & \\
 & & \boxed{1} & & 4 & & 6 & & 4 & & 1 & & \\
 & 1 & & 5 & & \boxed{10} & & 10 & & 5 & & 1 & \\
1 & & 6 & & 15 & & 20 & & \boxed{15} & & 6 & & 1
\end{array}
$$

2. a. 13 terms
   b. 12
   c. The first term is a^{12} and the last term is b^{12}.
   d. The exponents of a decrease and the exponents of b increase.

3.
$$
\begin{aligned}
(4!)(3!) &= (4\cdot3\cdot2\cdot1)(3\cdot2\cdot1)\\
&= (24)(6)\\
&= 144
\end{aligned}
$$

4.
$$
\begin{aligned}
\frac{5!}{3!} &= \frac{5\cdot4\cdot\cancel{3!}}{\cancel{3!}}\\
&= 20
\end{aligned}
$$

5.
$$
\begin{aligned}
\frac{6!}{2!(6-2)!} &= \frac{6!}{2!4!}\\
&= \frac{6\cdot5\cdot\cancel{4!}}{2\cdot1\cdot\cancel{4!}}\\
&= \frac{30}{2}\\
&= 15
\end{aligned}
$$

6.
$$
\begin{aligned}
\frac{12!}{3!(12-3)!} &= \frac{12!}{3!9!}\\
&= \frac{12\cdot11\cdot10\cdot\cancel{9!}}{3\cdot2\cdot1\cdot\cancel{9!}}\\
&= \frac{1,320}{6}\\
&= 220
\end{aligned}
$$

7.
$$
\begin{aligned}
(n-n)! &= 0!\\
&= 1
\end{aligned}
$$

8.
$$
\begin{aligned}
\frac{8!}{7!} &= \frac{8\cdot\cancel{7!}}{\cancel{7!}}\\
&= 8
\end{aligned}
$$

9.
$$
\begin{aligned}
&(x+y)^5\\
&= x^5 + \frac{5!}{1!(5-1)!}x^4y + \frac{5!}{2!(5-2)!}x^3y^2\\
&\quad + \frac{5!}{3!(5-3)!}x^2y^3 + \frac{5!}{4!(5-4)!}xy^4 + y^5\\
&= x^5 + \frac{5\cdot\cancel{4!}}{1!\cancel{4!}}x^4y + \frac{5\cdot4\cdot\cancel{3!}}{2!\cancel{3!}}x^3y^2\\
&\quad + \frac{5\cdot4\cdot\cancel{3!}}{\cancel{3!}2!}x^2y^3 + \frac{5\cdot\cancel{4!}}{\cancel{4!}1!}xy^4 + y^5\\
&= x^5 + \frac{5}{1}x^4y + \frac{20}{2}x^3y^2 + \frac{20}{2}x^2y^3 + \frac{5}{1}xy^4 + y^5\\
&= x^5 + 5x^4y + 10x^3y^2 + 10x^2y^3 + 5xy^4 + y^5
\end{aligned}
$$

10.

$$(x-y)^9=(x+(-y))^9$$

$$=x^9+\frac{9!}{1!(9-1)!}x^8(-y)+\frac{9!}{2!(9-2)!}x^7(-y)^2$$

$$+\frac{9!}{3!(9-3)!}x^6(-y)^3+\frac{9!}{4!(9-4)!}x^5(-y)^4$$

$$+\frac{9!}{5!(9-5)!}x^4(-y)^5+\frac{9!}{6!(9-6)!}x^3(-y)^6$$

$$+\frac{9!}{7!(9-7)!}x^2(-y)^7+\frac{9!}{8!(9-8)!}x(-y)^8+(-y)^9$$

$$=x^9+\frac{9!}{1!8!}x^8(-y)+\frac{9!}{2!7!}x^7(-y)^2+\frac{9!}{3!6!}x^6(-y)^3$$

$$+\frac{9!}{4!5!}x^5(-y)^4+\frac{9!}{5!4!}x^4(-y)^5+\frac{9!}{6!3!}x^3(-y)^6$$

$$+\frac{9!}{7!2!}x^2(-y)^7+\frac{9!}{8!1!}x(-y)^8+(-y)^9$$

$$=x^9-\frac{9}{1}x^8y+\frac{72}{2}x^7y^2-\frac{504}{6}x^6y^3+\frac{3{,}024}{24}x^5y^4$$

$$-\frac{3{,}024}{24}x^4y^5+\frac{504}{6}x^3y^6-\frac{72}{2}x^2y^7+\frac{9}{1}xy^8-y^9$$

$$=x^9-9x^8y+36x^7y^2-84x^6y^3+126x^5y^4-126x^4y^5$$

$$+84x^3y^6-36x^2y^7+9xy^8-y^9$$

11.

$$(4x-y)^3$$

$$=(4x)^3+\frac{3!}{1!(3-1)!}(4x)^2(-y)+\frac{3!}{2!(3-2)!}(4x)(-y)^2+(-y)^3$$

$$=64x^3+\frac{3!}{1!2!}(16x^2)(-y)+\frac{3!}{2!1!}(4x)(y^2)-y^3$$

$$=64x^3-\frac{3}{1}(16x^2)y+\frac{3}{1}(4x)y^2-y^3$$

$$=64x^3-48x^2y+12xy^2-y^3$$

12.

$$\left(\frac{c}{2}+\frac{d}{3}\right)^4$$

$$=\left(\frac{c}{2}\right)^4+\frac{4!}{1!(4-1)!}\left(\frac{c}{2}\right)^3\left(\frac{d}{3}\right)$$

$$+\frac{4!}{2!(4-2)!}\left(\frac{c}{2}\right)^2\left(\frac{d}{3}\right)^2+\frac{4!}{3!(4-3)!}\left(\frac{c}{2}\right)\left(\frac{d}{3}\right)^3$$

$$+\left(\frac{d}{3}\right)^4$$

$$=\frac{c^4}{16}+\frac{4\cdot\cancel{3!}}{1!\cancel{3!}}\left(\frac{c^3}{8}\right)\left(\frac{d}{3}\right)+\frac{4\cdot3\cdot\cancel{2!}}{2!\cancel{2!}}\left(\frac{c^2}{4}\right)\left(\frac{d^2}{9}\right)$$

$$+\frac{4\cdot\cancel{3!}}{\cancel{3!}1!}\left(\frac{c}{2}\right)\left(-\frac{d^3}{27}\right)+\frac{d^4}{81}$$

$$=\frac{c^4}{16}+\frac{4}{1}\left(\frac{c^3d}{24}\right)+\frac{12}{2}\left(\frac{c^2d^2}{36}\right)$$

$$+\frac{4}{1}\left(\frac{cd^3}{54}\right)+\frac{d^4}{81}$$

$$=\frac{c^4}{16}+\frac{c^3d}{6}+\frac{c^2d^2}{6}+\frac{2cd^3}{27}+\frac{d^4}{81}$$

13. The 3rd term of $(x+y)^4$

$$\frac{4!}{2!(4-2)!}x^2y^2=\frac{4\cdot3\cdot\cancel{2!}}{\cancel{2!}2!}x^2y^2$$

$$=\frac{12}{2}x^2y^2$$

$$=6x^2y^2$$

14. The 4th term of $(x-y)^6$

$$\frac{6!}{3!(6-3)!}x^3(-y)^3=\frac{6\cdot5\cdot4\cdot\cancel{3!}}{\cancel{3!}3\cdot2\cdot1}x^3(-y)^3$$

$$=-\frac{120}{6}x^3y^3$$

$$=-20x^3y^3$$

15. The 2nd term of $(3x-4y)^3$

$$\frac{3!}{1!(3-1)!}(3x)^2(-4y)=\frac{3\cdot\cancel{2!}}{1!\cancel{2!}}(9x^2)(-4y)$$

$$=\frac{3}{1}(-36x^2y)$$

$$=-108x^2y$$

16. The 5th term of $\left(u^2 - v^2\right)^5$

$$\frac{5!}{4!(5-4)!}\left(u^2\right)\left(-v^3\right)^4 = \frac{5\cdot\cancel{4!}}{\cancel{4!}1!}\left(u^2\right)\left(v^{12}\right)$$
$$= \frac{5}{1}\left(u^2v^{12}\right)$$
$$= 5u^2v^{12}$$

SECTION 11.2
Arithmetic Sequences and Series

17. $a_n = 2n - 4$

$$a_1 = 2(1) - 4 = -2$$
$$a_2 = 2(2) - 4 = 0$$
$$a_3 = 2(3) - 4 = 2$$
$$a_4 = 2(4) - 4 = 4$$

The first four terms are –2, 0, 2, and 4.

18. $a_n = \dfrac{(-1)^n}{n+1}$

$$a_1 = \frac{(-1)^1}{1+1} = -\frac{1}{2}$$
$$a_2 = \frac{(-1)^2}{2+1} = \frac{1}{3}$$
$$a_3 = \frac{(-1)^3}{3+1} = -\frac{1}{4}$$
$$a_4 = \frac{(-1)^4}{4+1} = \frac{1}{5}$$
$$a_5 = \frac{(-1)^5}{5+1} = -\frac{1}{6}$$

The first five terms are $-\frac{1}{2}, \frac{1}{3}, -\frac{1}{4}, \frac{1}{5}, -\frac{1}{6}$.

19. $a_n = 100 - \dfrac{n}{2}$

$$a_{50} = 100 - \frac{50}{2}$$
$$= 100 - 25$$
$$= 75$$

20. Find a_8 if $a_1 = 7$ and $d = 5$.

$$a_8 = a_1 + (n-1)(d)$$
$$a_8 = 7 + (8-1)(5)$$
$$a_8 = 7 + (7)(5)$$
$$a_8 = 7 + 35$$
$$a_8 = 42$$

21. Find d

242 – 212 = 30, so $2d = 30$ and $d = 15$.
Find a_1 if $a_n = 212$ and $n = 7$.

$$212 = a_1 + 15(7-1)$$
$$212 = a_1 + 15(6)$$
$$212 = a_1 + 90$$
$$122 = a_1$$
$$a_1 = 122 + 15(1-1) = 122 + 0 = 122$$
$$a_2 = 122 + 15(2-1) = 122 + 15 = 137$$
$$a_3 = 122 + 15(3-1) = 122 + 30 = 152$$
$$a_4 = 122 + 15(4-1) = 122 + 45 = 167$$
$$a_5 = 122 + 15(5-1) = 122 + 16 = 182$$

22. Find d

–6 – 6 = –12, so $d = -12$.
Find a_{101} if $a_1 = 6$, $d = -12$ and $n = 101$.

$$a_n = a_1 + d(n-1)$$
$$a_{101} = 6 - 12(101-1)$$
$$= 6 - 12(100)$$
$$= 6 - 1{,}200$$
$$= -1{,}194$$

23. Let $n = 23$ and $a_n = -625$, $a_1 = -515$

$$-625 = -515 + d(23-1)$$
$$-625 = -515 + d(22)$$
$$-110 = 22d$$
$$-5 = d$$

24. Let $a_1 = 8$ and $a_4 = 25$.

$a_2 = 2 + d, a_3 = 2 + 2d, a_4 = 2 + 3d$

$a_4 = a_1 + (4-1)d$

$25 = 8 + 3d$

$17 = 3d$

$\frac{17}{3} = d$

$a_2 = 8 + d = 8 + \frac{17}{3} = \frac{41}{3}$

$a_3 = 8 + 2d = 8 + 2\left(\frac{17}{3}\right) = \frac{58}{3}$

25.

$a_1 = 9,\ d = 6\frac{1}{2} - 9 = -2\frac{1}{2} = -\frac{5}{2},\quad n = 10$

$a_{10} = 9 + -\frac{5}{2}(10-1)$

$= 9 + -\frac{5}{2}(9)$

$= -\frac{27}{2}$

$= -13.5$

$S_n = \frac{n(a_1 + a_n)}{2}$

$S_{10} = \frac{10(9-13.5)}{2}$

$= \frac{10(-4.5)}{2}$

$= -22.5$

$= -\frac{45}{2}$

26. Find d

$22 - 6 = 16$, so $4d = 16$ and $d = 4$.

Find a_1 if $a_n = 6$ and $n = 2$.

$6 = a_1 + 4(2-1)$

$6 = a_1 + 4(1)$

$6 = a_1 + 4$

$2 = a_1$

Find a_{28}.

$a_1 = 2,\quad d = 4,\quad n = 28$

$a_{28} = 2 + 4(28-1)$

$= 2 + 4(27)$

$= 110$

$S_n = \frac{n(a_1 + a_n)}{2}$

$S_{28} = \frac{28(2+110)}{2}$

$= \frac{28(112)}{2}$

$= 1{,}568$

27.

$\sum_{k=4}^{6} \frac{1}{2}k = \frac{1}{2}(4) + \frac{1}{2}(5) + \frac{1}{2}(6)$

$= 2 + \frac{5}{2} + 3$

$= \frac{15}{2}$

28.

$\sum_{k=2}^{5} 7k^2 = 7(2^2) + 7(3^2) + 7(4^2) + 7(5^2)$

$= 7(4) + 7(9) + 7(16) + 7(25)$

$= 28 + 63 + 112 + 175$

$= 378$

29.

$\sum_{k=1}^{4} (3k-4) = (3\cdot1-4) + (3\cdot2-4) + (3\cdot3-4) + (3\cdot4-4)$

$= -1 + 2 + 5 + 8$

$= 14$

30.

$$\sum_{k=10}^{10} 36k = 36(10)$$
$$= 360$$

31.

$$a_1 = 1,\ a_n = 200,\ n = 200,$$
$$S_n = \frac{n(a_1 + a_n)}{2}$$
$$S_{100} = \frac{200(1+200)}{2}$$
$$= \frac{200(201)}{2}$$
$$= 20{,}100$$

32. $a_1 = 10$, $a_2 = 12$, so $d = 2$.
Let $n = 30$ and find a_{30} to find the number of seats on the last row.

$$a_{30} = 10 + 2(30-1)$$
$$= 10 + 2(29)$$
$$= 68$$

Find the sum of all of the seats.

$$S_n = \frac{n(a_1 + a_n)}{2}$$
$$S_{30} = \frac{30(10+68)}{2}$$
$$= \frac{30(78)}{2}$$
$$= 1{,}170$$

SECTION 11.3
Geometric Sequences and Series

33. $a_1 = \frac{1}{8}, r = 2$

$$a_n = a_1 r^{n-1}$$
$$a_6 = a_1 r^{6-1}$$
$$= \frac{1}{8}(2)^{6-1}$$
$$= \frac{1}{8}(2)^5$$
$$= \frac{1}{8}(32)$$
$$= 4$$

34. 4th term is 3, 5th term is $\frac{3}{2}$

Use $r = \frac{a_{n+1}}{a_n}$ to find r.

$$a_4 = 3 \text{ and } a_5 = \frac{3}{2}$$
$$r = \frac{\frac{3}{2}}{3}$$
$$r = \frac{3}{2} \cdot \frac{1}{3}$$
$$r = \frac{1}{2}$$

Use $a_n = a_1 r^{n-1}$ to find a_1.

$$r = \frac{1}{2},\ a_4 = 3 \text{ and } n = 4$$
$$3 = a_1\left(\frac{1}{2}\right)^{4-1}$$
$$3 = a_1\left(\frac{1}{2}\right)^3$$
$$3 = \frac{1}{8}a_1$$
$$24 = a_1$$

$$a_1 = 24$$
$$a_2 = a_1 r = 24\left(\frac{1}{2}\right) = 12$$
$$a_3 = a_1 r^2 = 24\left(\frac{1}{2}\right)^2 = 6$$
$$a_4 = a_1 r^3 = 24\left(\frac{1}{2}\right)^3 = 3$$
$$a_5 = a_1 r^4 = 24\left(\frac{1}{2}\right)^4 = \frac{3}{2}$$

$$24,\ 12,\ 6,\ 3,\ \frac{3}{2}$$

35. $r = -3$, $a_9 = 243$

Use $a_n = ar^{n-1}$ to find a_1.

$$r = -3,\ a_9 = 243 \text{ and } n = 9$$

$$243 = a_1(-3)^{9-1}$$

$$243 = a_1(-3)^8$$

$$243 = a_1(6,561)$$

$$\frac{243}{6,561} = a_1$$

$$\frac{1}{27} = a_1$$

36. $a_1 = -6$, $a_4 = 384$ for two geometric means between –6 and 384

Use $a_n = ar^{n-1}$ to find r.

$$a_1 = -6,\ a_4 = 384 \text{ and } n = 4$$

$$384 = -6(r)^{4-1}$$

$$384 = -6r^3$$

$$-64 = r^3$$

$$-4 = r$$

$$a_1 = -6$$

$$a_2 = -6(-4) = 24$$

$$a_3 = -6(-4)^2 = -96$$

$$a_4 = -6(-4)^3 = 384$$

The geometric means are 24 and – 96.

37. Find r.

$$r = \frac{a_2}{a_1} = \frac{54}{162} = \frac{1}{3}$$

Find the sum of $a_1 = 162$, $r = \frac{1}{3}$, and $n = 7$.

$$S_n = \frac{a_1(1-r^n)}{1-r}$$

$$= \frac{162\left(1-\left(\frac{1}{3}\right)^7\right)}{1-\frac{1}{3}}$$

$$= \frac{162\left(1-\frac{1}{2,187}\right)}{\frac{2}{3}}$$

$$= \frac{162\left(\frac{2,186}{2,187}\right)}{\frac{2}{3}}$$

$$= \frac{\frac{4,372}{27}}{\frac{2}{3}}$$

$$= \frac{4,372}{27} \bullet \frac{3}{2}$$

$$= \frac{2,186}{9}$$

38. Find r.

$$r = \frac{a_2}{a_1} = \frac{-\frac{1}{4}}{\frac{1}{8}} = -2$$

Find the sum of $a_1 = \frac{1}{8}$, $r = -2$, and $n = 8$.

$$\begin{aligned} S_n &= \frac{a_1(1-r^n)}{1-r} \\ &= \frac{\frac{1}{8}(1-(-2)^8)}{1-(-2)} \\ &= \frac{\frac{1}{8}(1-256)}{3} \\ &= \frac{\frac{1}{8}(-255)}{3} \\ &= \frac{-255}{8} \bullet \frac{1}{3} \\ &= \frac{-255}{24} \\ &= -\frac{85}{8} \end{aligned}$$

39. If he uses 25% of the birdseed each month, then 75% of the birdseed remains. The amount of in the bag after 12 months is represented by the 13th term of a geometric series, where $a_1 = 50$, $n = 13$, and $r = 0.75$.

$$\begin{aligned} a_n &= a_1 r^{n-1} \\ a_{13} &= (50)(0.75)^{13-1} \\ a_{13} &= (50)(0.75)^{12} \\ a_{13} &= (50)(0.031676352) \\ a_{13} &\approx 1.6 \text{ lb} \end{aligned}$$

He will have about 1.6 lbs of birdseed left.

40. Find r.

$$r = \frac{a_2}{a_1} = \frac{20}{25} = \frac{4}{5}$$

Find the sum of the infinite geometric sequence with $a_1 = 25$ and $r = \frac{4}{5}$.

$$\begin{aligned} S &= \frac{a_1}{1-r} \\ &= \frac{25}{1-\frac{4}{5}} \\ &= \frac{25}{\frac{1}{5}} \\ &= 25 \bullet \frac{5}{1} \\ &= 125 \end{aligned}$$

41.

$$0.\overline{05} = 0.050505\ldots = \frac{5}{100} + \frac{5}{10{,}000} + \frac{5}{1{,}000{,}000} + \ldots$$

where $a_1 = \frac{5}{100}$ and $r = \frac{1}{100}$.

Find the sum with $a_1 = \frac{5}{100}$ and $r = \frac{1}{100}$.

$$\begin{aligned} S &= \frac{a_1}{1-r} \\ &= \frac{\frac{5}{100}}{1-\frac{1}{100}} \\ &= \frac{\frac{5}{100}}{\frac{99}{100}} \\ &= \frac{5}{100} \bullet \frac{100}{99} \\ &= \frac{5}{99} \end{aligned}$$

$$0.\overline{05} = \frac{5}{99}$$

42. The total distance the ball travels is the sum of two motions, falling and rebounding. The distance the ball falls is given by the sum $10+\frac{9}{10}\bullet 10+\frac{9}{10}\left(\frac{9}{10}\bullet 10\right)+\ldots$ or $10+9+\frac{81}{10}+\ldots$. The distance the ball rebounds begins with the first bounce up which is 9 ft, and then is nine-tenths the distance afterwards or $9+\frac{81}{10}+\frac{729}{100}+\ldots$. Since these are infinite geometric series, use the formula $S=\frac{a_1}{1-r}$ to find the sum.

Falling:

$a_1 = 10$ and $r = \frac{9}{10}$

$$\begin{aligned} S &= \frac{a_1}{1-r} \\ &= \frac{10}{1-\frac{9}{10}} \\ &= \frac{10}{\frac{1}{10}} \\ &= 10\bullet\frac{10}{1} \\ &= 100 \end{aligned}$$

Rebounding:

$a_1 = 9$ and $r = \frac{9}{10}$

$$\begin{aligned} S &= \frac{a_1}{1-r} \\ &= \frac{9}{1-\frac{9}{10}} \\ &= \frac{9}{\frac{1}{10}} \\ &= 9\bullet\frac{10}{1} \\ &= 90 \end{aligned}$$

The distance the ball travels is the sum of the distances falling and rebounding:

100 + 90 = 190 ft.

The ball travels a total of 190 ft.

CHAPTER 11 TEST

1. a. The array of numbers that gives the coefficients of the terms of a binomial expansion is called **Pascal's** triangle.
   b. In the expansion $a^3 - 3a^2b + 3ab^2 - b^3$, the signs **alternate** between + and –.
   c. Each term of an **arithmetic** sequence is found by adding the same number to the previous term.
   d. The sum of the terms of an arithmetic sequence is called an arithmetic **series**.
   e. Each term of a **geometric** sequence is found by multiplying the previous term by the same number.

2. $a_n = -6n + 8$

$$a_1 = -6(1) + 8 = 2$$
$$a_2 = -6(2) + 8 = -4$$
$$a_3 = -6(3) + 8 = -10$$
$$a_4 = -6(4) + 8 = -16$$

The first four terms are 2, –4, –10, and –16.

3. $a_n = \dfrac{(-1)^n}{n^3}$

$$a_1 = \frac{(-1)^1}{1^3} = \frac{-1}{1} = -1$$
$$a_2 = \frac{(-1)^2}{2^3} = \frac{1}{8}$$
$$a_3 = \frac{(-1)^3}{3^3} = -\frac{1}{27}$$
$$a_4 = \frac{(-1)^4}{4^3} = \frac{1}{64}$$
$$a_5 = \frac{(-1)^5}{5^3} = -\frac{1}{125}$$

4.

$$\frac{10!}{6!(10-6)!} = \frac{10\cdot 9\cdot 8\cdot 7\cdot \cancel{6!}}{\cancel{6!}4!} = \frac{5{,}040}{24} = 210$$

5.

$$(a-b)^6 = (a+(-b))^6$$
$$= a^6 + \frac{6!}{1!(6-1)!}a^5(-b) + \frac{6!}{2!(6-2)!}a^4(-b)^2 + \frac{6!}{3!(6-3)!}a^3(-b)^3 + \frac{6!}{4!(6-4)!}a^2(-b)^4 + \frac{6!}{5!(6-5)!}a(-b)^5 + (-b)^6$$
$$= a^6 + \frac{6!}{1!5!}a^5(-b) + \frac{6!}{2!4!}a^4(-b)^2 + \frac{6!}{3!3!}a^3(-b)^3 + \frac{6!}{4!2!}a^2(-b)^4 + \frac{6!}{5!1!}a(-b)^5 + (-b)^6$$
$$= a^6 - \frac{6}{1}a^5b + \frac{30}{2}a^4b^2 - \frac{120}{6}a^3b^3 + \frac{30}{2}a^2b^4 - \frac{6}{1}ab^5 + b^6$$
$$= a^6 - 6a^5b + 15a^4b^2 - 20a^3b^3 + 15a^2b^4 - 6ab^5 + b^6$$

6. The 3rd term of $(x^2 + 2y)^4$

$$\frac{4!}{2!(4-2)!}(x^2)^2(2y)^2 = \frac{4\cdot 3\cdot \cancel{2!}}{2!\cancel{2!}}(x^4)(4y^2) = \frac{12}{2}(4x^4y^2) = 24x^4y^2$$

7. Find d.
   $10 - 3 = 7$, so $d = 7$.
   Find a_{10} if $a_1 = 3$, $d = 7$ and $n = 10$.

$$a_n = a_1 + d(n-1)$$
$$a_{10} = 3 + 7(10-1) = 3 + 7(9) = 3 + 63 = 66$$

8.

$$a_1 = -2,\ d = 3-(-2) = 5,\ n = 12$$
$$a_{12} = -2+5(12-1)$$
$$= -2+5(11)$$
$$= -2+55$$
$$= 53$$

$$S_n = \frac{n(a_1+a_n)}{2}$$
$$S_{12} = \frac{12(-2+53)}{2}$$
$$= \frac{12(51)}{2}$$
$$= 306$$

9.

Let $a_1 = 2$ and $a_4 = 98$.

$$a_4 = a_1 + (4-1)d$$
$$98 = 2+3d$$
$$96 = 3d$$
$$32 = d$$

$$a_n = a_1 + (n-1)d$$
$$a_2 = 2+(2-1)(32) = 2+1(32) = 34$$
$$a_3 = 2+(3-1)(32) = 2+2(32) = 66$$

The arithmetic means are 34 and 66.

10.

$$\left(5-\frac{5}{4}\right) = (17-2)d$$
$$\frac{15}{4} = 15d$$
$$\frac{1}{15}\bullet\frac{15}{4} = \frac{1}{15}\bullet 15d$$
$$\frac{1}{4} = d$$

11. Find d

$$(-75-(-11)) = (20-4)d$$
$$-64 = 16d$$
$$-4 = d$$

Find a_1 if $a_n = -11$, $d = -4$ and $n = 4$.

$$-11 = a_1 - 4(4-1)$$
$$-11 = a_1 - 4(3)$$
$$-11 = a_1 - 12$$
$$1 = a_1$$

Find a_{27}.

$$a_1 = 1,\quad d = 4,\quad n = 27$$
$$a_{27} = 1-4(27-1)$$
$$= 1-4(26)$$
$$= -103$$

$$S_n = \frac{n(a_1+a_n)}{2}$$
$$S_{27} = \frac{27(1-103)}{2}$$
$$= \frac{27(-102)}{2}$$
$$= -1{,}377$$

12. The sum of all 25 rows minus the sum of the first 15 rows equals the number of pipe left in the stack after removing the first 15 rows. For S_{25}, let $a_1 = 1$, $a_{25} = 25$, and $n = 25$. For S_{15}, let $a_1 = 1$, $a_{15} = 15$, and $n = 15$.

$$S_n = \frac{n(a_1+a_n)}{2}$$
$$S_{25} - S_{15} = \frac{25(1+25)}{2} - \frac{15(1+15)}{2}$$
$$= \frac{25(26)}{2} - \frac{15(16)}{2}$$
$$= 325-120$$
$$= 205$$

There are 205 pipes remaining in the stack.

13. Find d if $a_1 = 16$, $a_2 = 48$, and $n = 2$.

$$a_n = a_1 + d(n-1)$$
$$48 = 16 + d(2-1)$$
$$48 = 16 + d$$
$$32 = d$$

Find a_{10} to find out how far the object fell during the 10th second.

$$a_1 = 16,\quad d = 32,\quad n = 10$$
$$a_{10} = 16 + 32(10-1)$$
$$= 16 + 32(9)$$
$$= 304$$

Find the sum of the distances fell during the first 10 seconds.

$$S_n = \frac{n(a_1 + a_n)}{2}$$
$$S_{10} = \frac{10(16+304)}{2}$$
$$= \frac{10(320)}{2}$$
$$= 1{,}600$$

14.

$$\sum_{k=1}^{3}(2k-3) = (2\cdot1-3)+(2\cdot2-3)+(2\cdot3-3)$$
$$= -1+1+3$$
$$= 3$$

15.

Use $r = \dfrac{a_{n+1}}{a_n}$ to find r.

$$a_2 = -\frac{1}{3} \text{ and } a_1 = -\frac{1}{9}$$
$$r = \frac{-\frac{1}{3}}{-\frac{1}{9}}$$
$$r = -\frac{1}{3}\bullet -\frac{9}{1}$$
$$r = 3$$

Use $a_n = a_1 r^{n-1}$ to find a_7.

$$r = 3,\ a_1 = -\frac{1}{9} \text{ and } n = 7$$
$$a_7 = -\frac{1}{9}(3)^{7-1}$$
$$a_7 = -\frac{1}{9}(3)^6$$
$$a_7 = -\frac{1}{9}(729)$$
$$a_7 = -81$$

16. Find r.

$$r = \frac{a_2}{a_1} = \frac{\frac{1}{9}}{\frac{1}{27}} = \frac{1}{9}\bullet\frac{27}{1} = 3$$

Find the sum of $a_1 = \dfrac{1}{27}$, $r = 3$, and $n = 6$.

$$S_n = \frac{a_1(1-r^n)}{1-r}$$
$$= \frac{\frac{1}{27}(1-3^6)}{1-3}$$
$$= \frac{\frac{1}{27}(1-729)}{-2}$$
$$= \frac{\frac{1}{27}(-728)}{-2}$$
$$= \frac{-728}{27}\bullet\frac{1}{-2}$$
$$= \frac{364}{27}$$

17. Let $r = -\frac{2}{3}$ and $a_4 = -\frac{16}{9}$

$$\begin{aligned} a_n &= a_1r^{n-1} \\ a_4 &= a_1r^{4-1} \\ a_4 &= a_1r^3 \\ -\frac{16}{9} &= a_1\left(-\frac{2}{3}\right)^3 \\ -\frac{16}{9} &= -\frac{8}{27}a_1 \\ -\frac{27}{8}\bullet-\frac{16}{9} &= -\frac{27}{8}\bullet-\frac{8}{27}a_1 \\ 6 &= a_1 \end{aligned}$$

18. $a_1 = 3$, $a_4 = 648$ for two geometric means between 3 and 648

Use $a_n = ar^{n-1}$ to find r.

$$\begin{aligned} &a_1 = 3,\ a_4 = 648 \text{ and } n = 4 \\ &648 = 3(r)^{4-1} \\ &648 = 3r^3 \\ &216 = r^3 \\ &\sqrt[3]{216} = \sqrt[3]{r^3} \\ &6 = r \\ &a_1 = 3 \\ &a_2 = 3(6) = 18 \\ &a_3 = 3(6)^2 = 108 \\ &a_4 = 3(6)^3 = 648 \end{aligned}$$

The geometric means are 18 and 108.

19. Find r.

$$r = \frac{a_2}{a_1} = \frac{3}{9} = \frac{1}{3}$$

Find the sum of the infinite geometric series with $a_1 = 9$ and $r = \frac{1}{3}$.

$$\begin{aligned} S &= \frac{a_1}{1-r} \\ &= \frac{9}{1-\frac{1}{3}} \\ &= \frac{9}{\frac{2}{3}} \\ &= 9\bullet\frac{3}{2} \\ &= \frac{27}{2} \end{aligned}$$

20. DEPRECIATION

If the yacht depreciates 8% of its value each year, then 92% of its value remains at the end of the first year. The value of the boat in 10 years is represented by the 11th term of a geometric series, where $a_1 =$ 1,500,000, $n = 11$, and $r = 0.92$.

$$\begin{aligned} a_n &= a_1r^{n-1} \\ a_{11} &= (1{,}500{,}000)(0.92)^{11-1} \\ a_{11} &= (1{,}500{,}000)(0.92)^{10} \\ a_{11} &\approx (1{,}500{,}000)(0.43) \\ a_{11} &\approx \$651{,}583 \end{aligned}$$

The boat is worth \$651,583 after 10 years.

21. PENDULUMS

Let $a_1 = 60$, and $r = \frac{97}{100}$.

$$\begin{aligned} S &= \frac{a_1}{1-r} \\ &= \frac{60}{1-\frac{97}{100}} \\ &= \frac{60}{\frac{3}{100}} \\ &= 2{,}000 \text{ in.} \end{aligned}$$

22.

$$0.\overline{7} = 0.777... = \frac{7}{10} + \frac{7}{100} + \frac{7}{1,000} + ...$$

where $a_1 = \frac{7}{10}$ and $r = \frac{1}{10}$.

$|r| = \left|\frac{1}{10}\right| = \frac{1}{10} < 1$ so the sum does exist.

Find the sum with $a_1 = \frac{7}{10}$ and $r = \frac{1}{10}$.

$$S = \frac{a_1}{1-r}$$

$$= \frac{\frac{7}{10}}{1-\frac{1}{10}}$$

$$= \frac{\frac{7}{10}}{\frac{9}{10}}$$

$$= \frac{7}{10} \bullet \frac{10}{9}$$

$$= \frac{7}{9}$$

$$0.\overline{7} = \frac{7}{9}$$

CHAPTER 11 CUMULATIVE REVIEW

1. a. 0

   b. $-\frac{4}{3}, 5.6, 0, -23$

   c. $\pi, \sqrt{2}, e$

   d. $-\frac{4}{3}, \pi, 5.6, \sqrt{2}, 0, -23, e$

2.

$$6[x-(2-x)] = -4(8x+3)$$
$$6[x-2+x] = -32x-12$$
$$6[2x-2] = -32x-12$$
$$12x-12 = -32x-12$$
$$12x-12+32x = -32x-12+32x$$
$$44x-12 = -12$$
$$44x-12+12 = -12+12$$
$$44x = 0$$
$$\frac{44x}{44} = \frac{0}{44}$$
$$x = 0$$

3.

$$A = \frac{1}{2}h(b_1+b_2)$$
$$2 \cdot A = 2 \cdot \frac{1}{2}h(b_1+b_2)$$
$$2A = h(b_1+b_2)$$
$$2A = hb_1 + hb_2$$
$$2A - hb_1 = hb_1 + hb_2 - hb_1$$
$$2A - hb_1 = hb_2$$
$$\frac{2A-hb_1}{h} = \frac{hb_2}{h}$$
$$\frac{2A-hb_1}{h} = b_2$$

4.

Let x = measure of angle C.
Then $\angle B = 5+5x$ and
$\angle A = 5+(5+5x) = 10+5x$. The sum of the angles of the triangle is 180°.

$$\angle A + \angle B + \angle C = 180$$
$$(10+5x)+(5+5x)+(x) = 180$$
$$11x+15 = 180$$
$$11x = 165$$
$$x = 15$$
$$\angle C = 15^\circ$$
$$\angle B = 5+5(15) = 80^\circ$$
$$\angle A = 10+5(15) = 85^\circ$$

5. Let x = the amount of money she has to invest. Then x + 3,000 = the needed amount of money to earn the greater interest.

	principal	rate	Interest
original	x	0.075	$0.075x$
larger amount	$x + 3{,}000$	0.11	$0.11(x + 3{,}000)$

11% investment = 2(7.5% investment)

$$0.11(x+3{,}000) = 2(0.075x)$$
$$0.11x + 330 = 0.15x$$
$$330 = 0.04x$$
$$8{,}250 = x$$

She had $8,250 to invest originally.

6. Use the points (1,600, 68) and (2,800, 78).

$$\text{rate of change} = \frac{y_2 - y_1}{x_2 - x_1} = \frac{78-68}{2{,}800-1{,}600} = \frac{10}{1{,}200} = \frac{1}{120} \text{ db/rpm}$$

7. a. Find the slope of each line.

$$3x-4y=12$$
$$-4y=-3x+12$$
$$y=\frac{3}{4}x-3$$
$$m=\frac{3}{4}$$

$$y=\frac{3}{4}x-5$$
$$m=\frac{3}{4}$$

Since the slopes for both lines are the same, the lines must be parallel.

b. Find the slope of each line.

$$y=3x+4$$
$$m=3$$

$$x=-3y+4$$
$$x-4=-3y$$
$$y=-\frac{1}{3}x+\frac{4}{3}$$
$$m=-\frac{1}{3}$$

Since the slopes for both lines are the opposite reciprocals, the lines must be perpendicular.

8. SALVAGE VALUE

Write ordered pairs in the form (age, value). Use (0, 28,000) for the purchase value and (6, 7,600) for the predicted value.

Find the rate of change (m).

$$m=\frac{y_2-y_1}{x_2-x_1}$$
$$=\frac{7{,}600-28{,}000}{6-0}$$
$$=\frac{-20{,}400}{6}$$
$$=-3{,}400$$

Use the point (0, 28,000) and $m=-3{,}400$.

$$y-y_1=m(x-x_1)$$
$$y-28{,}000=-3{,}400(x-0)$$
$$y-28{,}000=-3{,}400x+0$$
$$y=-3{,}400x+28{,}000$$

9. $y=mx+b$; $m=-2$ and $b=5$

$y=-2x+5$

10. Find m.

$$m=\frac{y_2-y_1}{x_2-x_1}$$
$$=\frac{4-(-5)}{-5-8}$$
$$=\frac{9}{-13}$$
$$=-\frac{9}{13}$$

Use $m=-\frac{9}{13}$ and $(8,-5)$.

$$y-y_1=m(x-x_1)$$
$$y-(-5)=-\frac{9}{13}(x-8)$$
$$y+5=-\frac{9}{13}x+\frac{72}{13}$$
$$y=-\frac{9}{13}x+\frac{7}{13}$$

11. $f(x)=3x^5-2x^2+1$

$$f(-1)=3(-1)^5-2(-1)^2+1$$
$$=3(-1)-2(1)+1$$
$$=-3-2+1$$
$$=-4$$

$$f(a)=3a^5-2a^2+1$$

12. a. Find the slope using (95, 10.3) and (100, 9).

$$m=\frac{y_2-y_1}{x_2-x_1}$$
$$=\frac{10.3-9}{95-100}$$
$$=\frac{1.3}{-5}$$
$$=-0.26$$

Use $m=-0.26$ and $(100,\ 9)$.

$$y-y_1=m(x-x_1)$$
$$y-9=-0.26(x-100)$$
$$y-9=-0.26x+26$$
$$y=-0.26x+35$$
$$P(t)=-0.26t+35$$

b. Let $t=130$.

$$P(130)=-0.26(130)+35$$
$$=1.2\%$$

13. It doesn't pass the vertical line test. The graph passes through (0, 2) and (0, –2).

14. a. $h(-3) = 4$
    b. $h(4) = 3$
    c. $x = 0$ and $x = 2$
    d. $x = 1$

15. $g(x) = (x-6)^2$

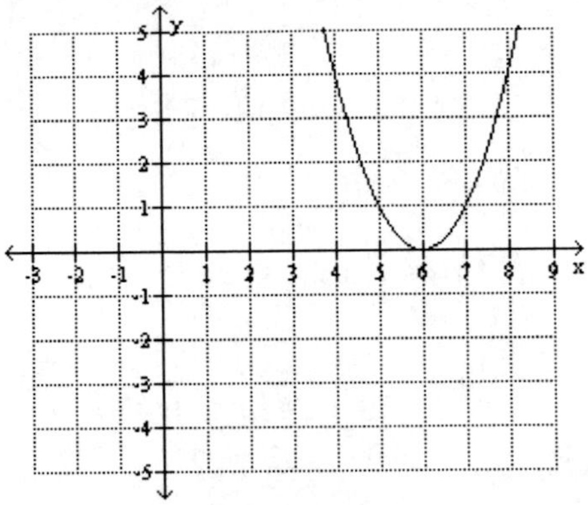

D: the set of real numbers $(-\infty, \infty)$
R: the set of nonnegative real numbers $[0, \infty)$

16. $\begin{cases} 2x - y = 6 \\ y = -x \end{cases}$

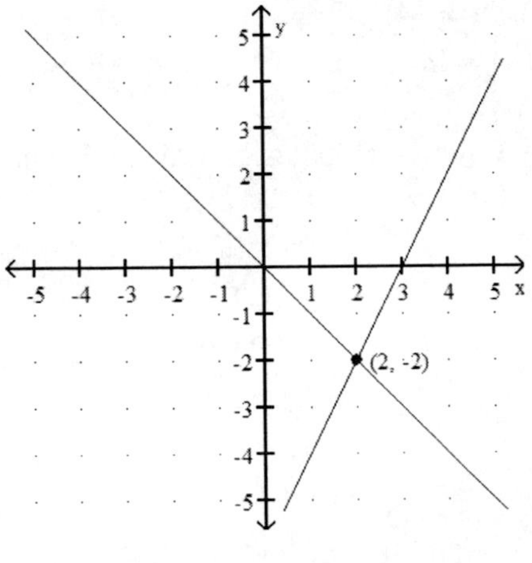

The solution is (2, –2).

17.
$\begin{cases} 2x - y = -21 \\ 4x + 5y = 7 \end{cases}$

Solve the first equation for y.

$$2x - y = -21$$
$$-y = -2x - 21$$
$$y = 2x + 21$$

Substitute $y = 2x + 2y$ into the second equation and solve for x.

$$4x + 5y = 7$$
$$4x + 5(2x + 21) = 7$$
$$4x + 10x + 105 = 7$$
$$14x = -98$$
$$x = -7$$
$$y = 2x + 21$$
$$y = 2(-7) + 21$$
$$y = -14 + 21$$
$$y = 7$$

The solution is $(-7, 7)$.

18.

$$\begin{cases} 4y+5x-7=0 \\ \frac{10}{7}x-\frac{4}{9}y=\frac{17}{21} \end{cases}$$

Write the equations in standard form. Multiply the second equation by 63 to eliminate fractions.

$$\begin{cases} 5x+4y=7 \\ 90x-28y=51 \end{cases}$$

Multiply the first equation by 7 and add the equations to solve for x.

$$\begin{aligned} 35x+28y&=49 \\ \underline{90x-28y}&\underline{=51} \\ 125x&=100 \\ x&=\frac{4}{5} \end{aligned}$$

Substitute $x=\frac{4}{5}$ into the first equation and solve for y.

$$\begin{aligned} 5x+4y&=7 \\ 5\left(\frac{4}{5}\right)+4y&=7 \\ 4+4y&=7 \\ 4y&=3 \\ y&=\frac{3}{4} \end{aligned}$$

The solution is $\left(\frac{4}{5},\frac{3}{4}\right)$.

19.

$$\begin{cases} b+2c=7-a \\ a+c=8-2b \\ 2a+b+c=9 \end{cases}$$

Write the equations in standard form.

$$\begin{cases} a+b+2c=7 & (1) \\ a+2b+c=8 & (2) \\ 2a+b+c=9 & (3) \end{cases}$$

Multiply Equation 2 by -2 and add Equations 1 and 2.

$$\begin{aligned} a+\ b+2c&=7 && (1) \\ \underline{-2a-4b-2c}&\underline{=-16} && (2) \\ -a-3b&=-9 && (4) \end{aligned}$$

Multiply Equation 2 by -1 and add Equations 2 and 3.

$$\begin{aligned} -a-2b-c&=-8 && (2) \\ \underline{2a+b+c}&\underline{=9} && (3) \\ a-b&=1 && (5) \end{aligned}$$

Add Equations 4 and 5.

$$\begin{aligned} -a-3b&=-9 && (4) \\ \underline{a-b}&\underline{=1} && (5) \\ -4b&=-8 \\ b&=2 \end{aligned}$$

Substitute $b=2$ into Equation 5.

$$\begin{aligned} a-b&=1 && (5) \\ a-2&=1 \\ a&=3 \end{aligned}$$

Substitute $b=2$ and $a=3$ into Equation 1.

$$\begin{aligned} a+b+2c&=7 && (1) \\ 3+2+2c&=7 \\ 5+2c&=7 \\ 2c&=2 \\ c&=1 \end{aligned}$$

The solution is $(3,2,1)$.

20. $\begin{cases} 2x + y = 1 \\ x + 2y = -4 \end{cases}$

$$\left[\begin{array}{cc|c} 2 & 1 & 1 \\ 1 & 2 & -4 \end{array}\right]$$

$R_1 \leftrightarrow R_2$

$$\left[\begin{array}{cc|c} 1 & 2 & -4 \\ 2 & 1 & 1 \end{array}\right]$$

$-2R_1 + R_2$

$$\left[\begin{array}{cc|c} 1 & 2 & -4 \\ 0 & -3 & 9 \end{array}\right]$$

$-\frac{1}{3}R_2$

$$\left[\begin{array}{cc|c} 1 & 2 & -4 \\ 0 & 1 & -3 \end{array}\right]$$

This matrix represents the system

$$\begin{cases} x + 2y = -4 \\ y = -3 \end{cases}$$

$x - 6 = -4$

$x = 2$

The solution is $(2, -3)$.

21.

Write the first equation in standard form.

$$\begin{cases} 2x + 2y = -1 \\ 3x + 4y = 0 \end{cases}$$

$$x = \frac{D_x}{D} = \frac{\begin{vmatrix} -1 & 2 \\ 0 & 4 \end{vmatrix}}{\begin{vmatrix} 2 & 2 \\ 3 & 4 \end{vmatrix}} = \frac{-1(4) - 2(0)}{2(4) - 2(3)} = \frac{-4 - 0}{8 - 6} = \frac{-4}{2} = -2$$

$$y = \frac{D_y}{D} = \frac{\begin{vmatrix} 2 & -1 \\ 3 & 0 \end{vmatrix}}{\begin{vmatrix} 2 & 2 \\ 3 & 4 \end{vmatrix}} = \frac{2(0) - (-1)(3)}{2(4) - 2(3)} = \frac{0 + 3}{8 - 6} = \frac{3}{2}$$

The solution is $\left(-2, \frac{3}{2}\right)$.

22. MIXING COFFEE

Let x = # of lb of regular coffee and y = # of lb of Brazilian coffee.

	Regular coffee	Brazilian coffee	Mixture
# of pounds	x	y	40
Value	$4x$	$11.50y$	6(40)

$$\begin{cases} x + y = 40 \\ 4x + 11.50y = 6(40) \end{cases}$$

$$\begin{cases} x + y = 40 \\ 4x + 11.50y = 240 \end{cases}$$

Multiply the 1st equation by -4.

Add equations to eliminate x.

$$-4x - 4y = -160$$
$$4x + 11.5y = 240$$
$$7.5y = 80$$
$$y = 10\frac{2}{3}$$
$$x + y = 40$$
$$x + 10\frac{2}{3} = 40$$
$$x = 29\frac{1}{3}$$

$29\frac{1}{3}$ pounds of regular coffee and $10\frac{2}{3}$ pounds of Brazilian coffee are needed.

23. AVIATION

Let s = the speed of the plane in still air and w = the speed of the wind. Then the speed of the plane flying with the wind is $s + w$ and the speed of the plane flying against the wind is $s - w$. Using the formula $\boldsymbol{d = rt}$, we find that $2(s + w)$ represents the distance traveled with the wind and $2.5(s - w)$ represents the distance traveled against the wind.

$$\begin{cases} 2(s+w) = 1{,}000 \\ 2.5(s-w) = 1{,}000 \end{cases}$$

Distribute.

$$\begin{cases} 2s + 2w = 1{,}000 \\ 2.5s - 2.5w = 1{,}000 \end{cases}$$

Multiply the 1st equation by 2.5 and the 2nd equation by 2 and add equations to eliminate the w.

$$\begin{aligned} 5s + 5w &= 2{,}500 \\ \underline{5s - 5w} &\underline{= 2{,}000} \\ 10s &= 4{,}500 \\ s &= 450 \end{aligned}$$

$$\begin{aligned} 2s + 2w &= 1{,}000 \\ 2(450) + 2w &= 1{,}000 \\ 900 + 2w &= 1{,}000 \\ 2w &= 100 \\ w &= 50 \end{aligned}$$

The speed of the plane in still air is 450 mph and the speed of the wind is 50 mph.

24. SPORTS SOCKS

Let x = number of ankle socks, y = number of low cut socks, and z = number of crew socks.

$$\begin{cases} x + y + z = 500 \\ 2x + 3y + 4z = 1{,}650 \\ 3x + 5y + 6z = 2{,}550 \end{cases}$$

Multiply the first equation by –2 and add Equations 1 and 2.

$$\begin{aligned} -2x - 2y - 2z &= -1{,}000 \\ \underline{2x + 3y + 4z} &\underline{= \quad 1{,}650} \\ y + 2z &= \quad 650 \end{aligned}$$

Multiply the first equation by –3 and add Equations 1 and 3.

$$\begin{aligned} -3x - 3y - 3z &= -1{,}500 \\ \underline{3x + 5y + 6z} &\underline{= \quad 2{,}550} \\ 2y + 3z &= \quad 1{,}050 \end{aligned}$$

Multiply the first new equation by –2 and it to the other new equation..

$$\begin{aligned} -2y - 4z &= -1{,}300 \\ \underline{2y + 3z} &\underline{= \quad 1{,}050} \\ -z &= \ -250 \\ z &= \quad 250 \end{aligned}$$

Substitute $z = 250$ into the first new equation to solve for y.

$$\begin{aligned} y + 2z &= 650 \\ y + 2(250) &= 650 \\ y + 500 &= 650 \\ y &= 150 \end{aligned}$$

Substitute $y = 150$ and $z = 250$ into the frist equation and solve for x.

$$\begin{aligned} x + y + z &= 500 \\ x + 150 + 250 &= 500 \\ x + 400 &= 500 \\ x &= 100 \end{aligned}$$

They make 100 ankle socks, 150 low cut, and 250 crew socks each day.

25.

$$\begin{aligned} 4.5x - 1 < -10 \quad &\text{or} \quad 6 - 2x \ge 12 \\ 4.5x < -9 \quad &\qquad\quad -2x \ge 6 \\ x < -2 \quad &\qquad\qquad x \le -3 \end{aligned}$$

$$(-\infty, -2)$$

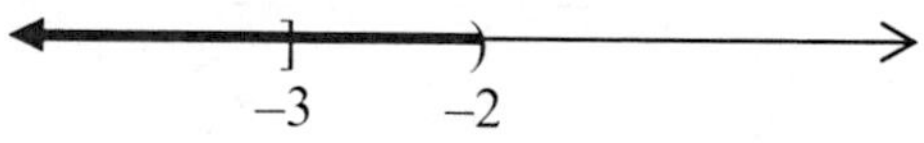

26.

$$|5-3x|-14 \le 0$$
$$|5-3x| \le 14$$
$$-14 \le 5-3x \le 14$$
$$-14-5 \le 5-3x-5 \le 14-5$$
$$-19 \le -3x \le 9$$
$$\frac{19}{3} \ge x \ge -3$$
$$-3 \le x \le \frac{19}{3}$$
$$\left[-3, \frac{19}{3}\right]$$

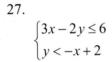

27.

$$\begin{cases} 3x-2y \le 6 \\ y < -x+2 \end{cases}$$

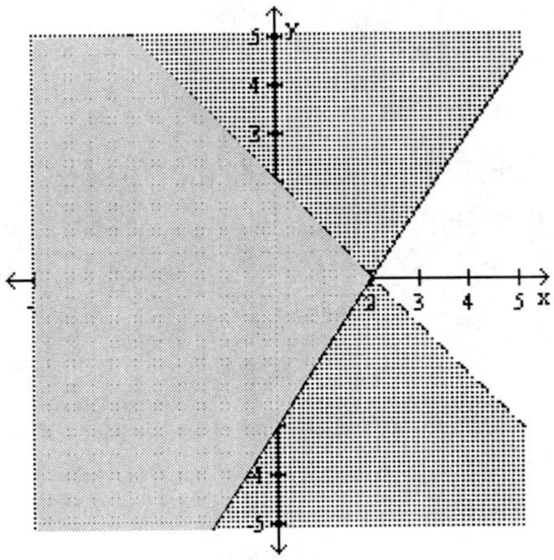

28. WORK SCHEDULES

job	pay per hour	number of hr worked	amount earned
bookstore	\$12	x	$\$12x$
airport	\$22.5	$25-x$	$22.5(25-x)$

$$12x+22.5(25-x) \ge 450$$
$$12x+562.5-22.5x \ge 450$$
$$-10.5x+562.5 \ge 450$$
$$-10.5x+562.5-562.5 \ge 450-562.5$$
$$-10.5x \ge -112.5$$
$$\frac{-10.5x}{-10.5} \le \frac{-112.5}{-10.5}$$
$$x \le 10.7$$

She can work **10 hours** at the bookstore.

29.

$$(x^2)^5 y^7 y^3 x^{-2} y^0 = x^{2\cdot5} y^7 y^3 x^{-2} y^0$$
$$= x^{10} y^7 y^3 x^{-2} y^0$$
$$= x^{10+(-2)} y^{7+3+0}$$
$$= x^8 y^{10}$$

30.

$$\left(\frac{3x^5y^2}{6x^5y^{-2}}\right)^{-4} = \left(\frac{6x^5y^{-2}}{3x^5y^2}\right)^4$$
$$= \left(2x^{5-5}y^{-2-2}\right)^4$$
$$= \left(2x^0y^{-4}\right)^4$$
$$= \left(2\cdot 1y^{-4}\right)^4$$
$$= 2^4 y^{-4\cdot4}$$
$$= 16y^{-16}$$
$$= \frac{16}{y^{16}}$$

31. 1.73×10^{14}; 4.6×10^{-8}

32.

$$\frac{(0.00024)(96{,}000{,}000)}{(640{,}000{,}000)(0.025)} = \frac{(2.4\times10^{-4})(9.6\times10^7)}{(6.4\times10^8)(2.5\times10^{-2})}$$
$$= \frac{(2.4\times9.6)(10^{-4+7})}{(6.4\times2.5)(10^{8+(-2)})}$$
$$= \frac{23.04\times10^3}{16.0\times10^6}$$
$$= \frac{23.04}{16.0}\times10^{3-6}$$
$$= 1.44\times10^{-3}$$
$$= 0.00144$$

33. PRODUCE

The square base has 6 canteloupes on each side so let $c = 6$.

$$n(c) = \frac{1}{3}c^3 + \frac{1}{2}c^2 + \frac{1}{6}c$$
$$n(6) = \frac{1}{3}(6)^3 + \frac{1}{2}(6)^2 + \frac{1}{6}(6)$$
$$= \frac{1}{3}(216) + \frac{1}{2}(36) + \frac{1}{6}(6)$$
$$= 72+18+1$$
$$= 91$$

They need 91 cantaloupes.

34. $\frac{9}{4}rt^2 - \frac{5}{3}rt - \frac{1}{2}rt^2 + \frac{5}{6}rt$
$= \frac{7}{4}rt^2 - \frac{5}{6}rt$

35.
$(-2x^2y^3 + 6xy + 5y^2) - (-4x^2y^3 - 7xy + 2y^2)$
$= -2x^2y^3 + 6xy + 5y^2 + 4x^2y^3 + 7xy - 2y^2$
$= 2x^2y^3 + 13xy + 3y^2$

36.
$(x - 3y)(x^2 + 3xy + 9y^2)$
$= x(x^2 + 3xy + 9y^2) - 3y(x^2 + 3xy + 9y^2)$
$= x^3 + 3x^2y + 9xy^2 - 3x^2y - 9xy^2 - 27y^3$
$= x^3 - 27y^3$

37.
$(2m^5 - 7)(3m^5 - 1) = 6m^{10} - 2m^5 - 21m^5 + 7$
$= 6m^{10} - 23m^5 + 7$

38.
$(9ab^2 - 4)^2 = (9ab^2 - 4)(9ab^2 - 4)$
$= 81a^2b^4 - 36ab^2 - 36ab^2 + 16$
$= 81a^2b^4 - 72ab^2 + 16$

39.
$3x^3y - 4x^2y^2 - 6x^2y + 8xy^2$
$= xy(3x^2 - 4xy - 6x + 8y)$
$= xy(3x^2 - 4xy) - (6x - 8y)$
$= xy[x(3x - 4y) - 2(3x - 4y)]$
$= xy(3x - 4y)(x - 2)$

40.
$b^3 - 4b^2 - 3b + 12 = b^2(b - 4) - 3(b - 4)$
$= (b - 4)(b^2 - 3)$

41.
$12y^2 + 23y + 10 = (3y + 2)(4y + 5)$

42.
$256x^4y^4 - z^8 = (16x^2y^2 + z^4)(16x^2y^2 - z^4)$
$= (16x^2y^2 + z^4)(4xy + z^2)(4xy - z^2)$

43.
$27t^3 + u^3 = (3t + u)((3t)^2 + 3tu + u^2)$
$= (3t + u)(9t^2 + 3tu + u^2)$

44.
$a^4b^2 - 20a^2b^2 + 64b^2$
$= b^2(a^4 - 20a^2 + 64)$
$= b^2(a^2 - 4)(a^2 - 16)$
$= b^2(a + 2)(a - 2)(a + 4)(a - 4)$

45.
$\frac{A\lambda}{2} + 1 = 2d + 3\lambda$
$\frac{A\lambda}{2} - 3\lambda = 2d - 1$
$2\left(\frac{A\lambda}{2} - 3\lambda\right) = 2(2d - 1)$
$A\lambda - 6\lambda = 4d - 2$
$\lambda(A - 6) = 4d - 2$
$\frac{\lambda(A - 6)}{A - 6} = \frac{4d - 2}{A - 6}$
$\lambda = \frac{4d - 2}{A - 6}$

46. Let $t = -\frac{5}{7}$.
$7t^2 - 2t - 5 = 0$
$7\left(-\frac{5}{7}\right)^2 - 2\left(-\frac{5}{7}\right) - 5 \stackrel{?}{=} 0$
$\frac{25}{7} + \frac{10}{7} - 5 \stackrel{?}{=} 0$
$0 = 0$
Yes, it is a solution.

47. Let $u = (x + 7)$.
$(x + 7)^2 = -2(x + 7) - 1$
$(x + 7)^2 + 2(x + 7) + 1 = 0$
$u^2 + 2u + 1 = 0$
$(u + 1)(u + 1) = 0$
$u + 1 = 0$ or $u + 1 = 0$
$u = -1$ $\quad u = -1$
$x + 7 = -1$ $\quad x + 7 = -1$
$x = -8$ $\quad x = -8$

48

$$x^3+8x^2=9x$$
$$x^3+8x^2-9x=0$$
$$x\left(x^2+8x-9\right)=0$$
$$x(x-1)(x+9)=0$$
$$x=0,\ x-1=0,\ \text{or}\ x+9=0$$
$$x=1 \qquad x=-9$$

49. PAINTING

Let the width = w.

Then the length = $5w + 1$.

$$(\text{width})(\text{length}) = \text{Area}$$
$$w(5w+1)=84$$
$$5w^2+w=84$$
$$5w^2+w-84=0$$
$$(5w+21)(w-4)=0$$
$$5w+21=0 \quad \text{or} \quad w-4=0$$
$$w=-\frac{21}{5} \qquad w=4$$

Since length and width cannot be negative, the width is 4 inches and the length is 5(4) + 1 = 21 inches.

50.

$$h(t)=-16t^2+64t+80$$
$$0=-16t^2+64t+80$$
$$0=-16\left(t^2-4t-5\right)$$
$$0=-16(t-5)(t+1)$$
$$t-5=0 \quad \text{or} \quad t+1=0$$
$$t=5 \qquad t=\cancel{-1}$$

It will take 5 seconds.

51. $f(x)=\dfrac{2x^2-3x-2}{x^2+2x-24}$

Set the denominator equal to 0.

$$x^2+2x-24=0$$
$$(x+6)(x-4)=0$$
$$x+6=0 \ \text{or}\ x-4=0$$
$$x=-6 \qquad x=4$$

The domain is $(-\infty,-6)\cup(-6,4)\cup(4,\infty)$.

52.

$$\frac{6x^2+13x+6}{6x^2+5x-6}=\frac{\cancel{(2x+3)}(3x+2)}{\cancel{(2x+3)}(3x-2)}$$
$$=\frac{3x+2}{3x-2}$$

53.

$$\frac{p^3-q^3}{q^2-p^2}\bullet\frac{q^2+pq}{p^3+p^2q+pq^2}$$
$$=\frac{\cancel{(p-q)}\cancel{(p^2+pq+q^2)}}{-\cancel{(p-q)}\cancel{(q+p)}}\bullet\frac{q\cancel{(q+p)}}{p\cancel{(p^2+pq+q^2)}}$$
$$=-\frac{q}{p}$$

54.

$$\frac{2}{a-2}+\frac{3}{a+2}-\frac{a-1}{a^2-4}$$
$$=\frac{2}{a-2}+\frac{3}{a+2}-\frac{a-1}{(a+2)(a-2)}$$
$$=\frac{2}{a-2}\left(\frac{a+2}{a+2}\right)+\frac{3}{a+2}\left(\frac{a-2}{a-2}\right)-\frac{a-1}{(a+2)(a-2)}$$
$$=\frac{2a+4}{(a+2)(a-2)}+\frac{3a-6}{(a+2)(a-2)}-\frac{a-1}{(a+2)(a-2)}$$
$$=\frac{2a+4+3a-6-(a-1)}{(a+2)(a-2)}$$
$$=\frac{2a+4+3a-6-a+1}{(a+2)(a-2)}$$
$$=\frac{4a-1}{(a+2)(a-2)}$$

55.

$$\frac{\frac{y}{x}-\frac{x}{y}}{\frac{1}{x}+\frac{1}{y}}=\frac{xy\left(\frac{y}{x}-\frac{x}{y}\right)}{xy\left(\frac{1}{x}+\frac{1}{y}\right)}$$
$$=\frac{y^2-x^2}{y+x}$$
$$=\frac{\cancel{(y+x)}(y-x)}{\cancel{y+x}}$$
$$=y-x$$

56.

$$(16x^4+3x^2+13x+3)\div(4x+3)$$

$$\begin{array}{r} 4x^3-3x^2+3x+1 \\ 4x+3\overline{)16x^4+0x^3+3x^2+13x+3} \\ \underline{16x^4+12x^3} \qquad\qquad\qquad \\ -12x^3+3x^2 \qquad\quad \\ \underline{-12x^3-9x^2} \qquad\quad \\ 12x^2+13x \quad \\ \underline{12x^2+9x} \quad \\ 4x+3 \\ \underline{4x+3} \\ 0 \end{array}$$

57.

$$\frac{1}{a+5}=\frac{1}{3a+6}-\frac{a+2}{a^2+7a+10}$$

$$\frac{1}{a+5}=\frac{1}{3(a+2)}-\frac{a+2}{(a+5)(a+2)}$$

$$3(a+5)(a+2)\left(\frac{1}{a+5}\right)=3(a+5)(a+2)\left(\frac{1}{3(a+2)}-\frac{a+2}{(a+5)(a+2)}\right)$$

$$3(a+2)=a+5-3(a+2)$$

$$3a+6=a+5-3a-6$$

$$3a+6=-2a-1$$

$$3a+6+2a=-2a-1+2a$$

$$5a+6=-1$$

$$5a+6-6=-1-6$$

$$5a=-7$$

$$a=-\frac{7}{5}$$

58.

$$\frac{1}{R}=\frac{1}{R_1}+\frac{1}{R_2}+\frac{1}{R_3}$$

$$(RR_1R_2R_3)\left(\frac{1}{R}\right)=(RR_1R_2R_3)\left(\frac{1}{R_1}+\frac{1}{R_2}+\frac{1}{R_3}\right)$$

$$R_1R_2R_3=RR_2R_3+RR_1R_3+RR_1R_2$$

$$R_1R_2R_3=R(R_2R_3+R_1R_3+R_1R_2)$$

$$\frac{R_1R_2R_3}{R_2R_3+R_1R_3+R_1R_2}=\frac{R(R_2R_3+R_1R_3+R_1R_2)}{R_2R_3+R_1R_3+R_1R_2}$$

$$\frac{R_1R_2R_3}{R_2R_3+R_1R_3+R_1R_2}=R$$

$$R=\frac{R_1R_2R_3}{R_2R_3+R_1R_3+R_1R_2}$$

59. Let x = the amount of time it takes when the machines work together.

$$\frac{1}{6}+\frac{1}{4}=\frac{1}{x}$$

$$12x\left(\frac{1}{6}+\frac{1}{4}\right)=12x\left(\frac{1}{x}\right)$$

$$2x+3x=12$$

$$5x=12$$

$$x=\frac{12}{5}$$

$x=2.4$ hours

60. Use a ratio.

$$\frac{24}{x}=\frac{8}{31}$$

$$24(31)=8(x)$$

$$744=8x$$

$$93=x$$

61. Let I = intensity of the light and d = distance from the source.

$$I=\frac{k}{d^2}$$

$$18=\frac{k}{4^2}$$

$$18=\frac{k}{16}$$

$$16(18)=16\left(\frac{k}{16}\right)$$

$$288=k$$

Find I if $d=12$ and $k=288$.

$$I=\frac{k}{d^2}$$

$$I=\frac{288}{12^2}$$

$$I=\frac{288}{144}$$

$I=2$ lumens

62. $f(x)=\sqrt{x}+2$

$D:[0,\infty);\ R:[2,\infty)$

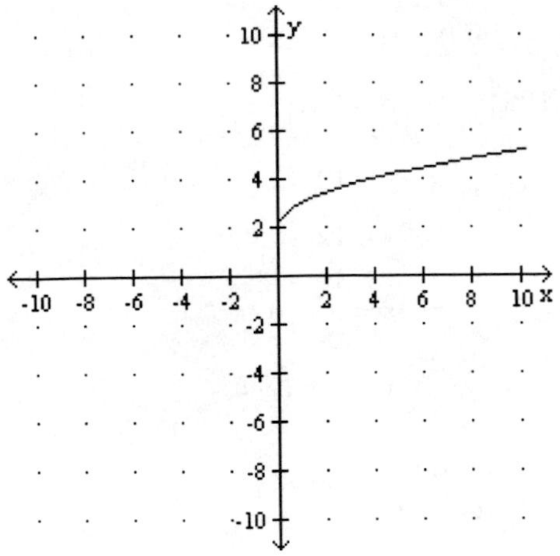

63. $f(x)=\sqrt{x+3}$

$$\sqrt{x+3}=0$$
$$\left(\sqrt{x+3}\right)^2=(0)^2$$
$$x+3=0$$
$$x=-3$$

D: $[-3,\ \infty)$

64.

$$L(g)=\sqrt[3]{\frac{g}{7.5}}$$
$$L(480)=\sqrt[3]{\frac{480}{7.5}}$$
$$=\sqrt[3]{64}$$
$$=4\text{ ft}$$

65.

$$\left(\frac{25}{49}\right)^{-3/2}=\left(\frac{49}{25}\right)^{3/2}$$
$$=\left(\sqrt{\frac{49}{25}}\right)^3$$
$$=\left(\frac{7}{5}\right)^3$$
$$=\frac{343}{125}$$

66.

$$\sqrt{112a^3b^5}=\sqrt{16a^2b^4}\sqrt{7ab}$$
$$=4ab^2\sqrt{7ab}$$

67.

$$\sqrt{98}+\sqrt{8}-\sqrt{32}=\sqrt{49}\sqrt{2}+\sqrt{4}\sqrt{2}-\sqrt{16}\sqrt{2}$$
$$=7\sqrt{2}+2\sqrt{2}-4\sqrt{2}$$
$$=5\sqrt{2}$$

68.

$$12\sqrt[3]{648x^4}+3\sqrt[3]{81x^4}=12\sqrt[3]{216x^3}\sqrt[3]{3x}+3\sqrt[3]{27x^3}\sqrt[3]{3x}$$
$$=12(6x)\left(\sqrt[3]{3x}\right)+3(3x)\left(\sqrt[3]{3x}\right)$$
$$=72x\sqrt[3]{3x}+9x\sqrt[3]{3x}$$
$$=81x\sqrt[3]{3x}$$

69.

$$\left(2\sqrt{7}+1\right)\left(\sqrt{7}-1\right)=2\sqrt{49}-2\sqrt{7}+\sqrt{7}-1$$
$$=14-\sqrt{7}-1$$
$$=13-\sqrt{7}$$

70.

$$3\left(\sqrt{5x}-\sqrt{3}\right)^2=3\left(\sqrt{5x}-\sqrt{3}\right)\left(\sqrt{5x}-\sqrt{3}\right)$$
$$=3\left(\sqrt{25x^2}-\sqrt{15x}-\sqrt{15x}+\sqrt{9}\right)$$
$$=3\left(5x-2\sqrt{15x}+3\right)$$
$$=15x-6\sqrt{15x}+9$$

71.

$$\frac{\sqrt[3]{4}}{\sqrt[3]{b}}=\sqrt[3]{\frac{4}{b}}$$
$$=\sqrt[3]{\frac{4}{b}\bullet\frac{b^2}{b^2}}$$
$$=\sqrt[3]{\frac{4b^2}{b^3}}$$
$$=\frac{\sqrt[3]{4b^2}}{b}$$

72.

$$\frac{3t-1}{\sqrt{3t}+1}=\frac{3t-1}{\sqrt{3t}+1}\bullet\frac{\sqrt{3t}-1}{\sqrt{3t}-1}$$
$$=\frac{3t\sqrt{3t}-3t-\sqrt{3t}+1}{\sqrt{9t^2}-1}$$
$$=\frac{3t\left(\sqrt{3t}-1\right)-\left(\sqrt{3t}-1\right)}{3t-1}$$
$$=\frac{\left(\sqrt{3t}-1\right)\cancel{(3t-1)}}{\cancel{3t-1}}$$
$$=\sqrt{3t}-1$$

73.

$$\begin{aligned}2x &= \sqrt{16x-12}\\ (2x)^2 &= \left(\sqrt{16x-12}\right)^2\\ 4x^2 &= 16x-12\\ 4x^2-16x+12 &= 0\\ 4\left(x^2-4x+3\right) &= 0\\ 4(x-1)(x-3) &= 0\end{aligned}$$

$$x-1=0 \quad \text{or} \quad x-3=0$$
$$x=1 \qquad\qquad x=3$$

74.

$$\begin{aligned}\sqrt[3]{12m+4} &= 4\\ \left(\sqrt[3]{12m+4}\right)^3 &= 4^3\\ 12m+4 &= 64\\ 12m+4-4 &= 64-4\\ 12m &= 60\\ m &= 5\end{aligned}$$

75.

$$\begin{aligned}\sqrt{x+3}-\sqrt{3} &= \sqrt{x}\\ \left(\sqrt{x+3}-\sqrt{3}\right)^2 &= \left(\sqrt{x}\right)^2\\ x+3-2\sqrt{3(x+3)}+3 &= x\\ x+6-2\sqrt{3x+9} &= x\\ -2\sqrt{3x+9} &= -6\\ \frac{-2\sqrt{3x+9}}{-2} &= \frac{-6}{-2}\\ \sqrt{3x+9} &= 3\\ \left(\sqrt{3x+9}\right)^2 &= 3^2\\ 3x+9 &= 9\\ 3x &= 0\\ x &= 0\end{aligned}$$

76. Use the Pythagorean Theorem. The legs of the triangle formed are both 16 (sides of the square) and the waist size is the hypotenuse, c.

$$\begin{aligned}a^2+b^2 &= c^2\\ 16^2+16^2 &= c^2\\ 256+256 &= c^2\\ 512 &= c^2\\ \sqrt{512} &= \sqrt{c^2}\\ 22.6 &= c\end{aligned}$$

Subtract 1 inch from 22.6 to allow for the pin, and the largest waist size the diaper can wrap around is about 21.6 inches.

77.

$$\begin{aligned}\sqrt{-25} &= \sqrt{25}\sqrt{-1}\\ &= 5i\end{aligned}$$

78.

$$\begin{aligned}i^{42} &= i^{40}\bullet i^2\\ &= \left(i^4\right)^{10}\bullet i^2\\ &= (1)^{10}\bullet(-1)\\ &= 1\bullet(-1)\\ &= -1\end{aligned}$$

79.

$$\begin{aligned}(-7+9i)-(-2-8i) &= -7+9i+2+8i\\ &= -5+17i\end{aligned}$$

80.

$$\begin{aligned}\frac{2-5i}{2+5i} &= \frac{2-5i}{2+5i}\bullet\frac{2-5i}{2-5i}\\ &= \frac{4-10i-10i+25i^2}{4-25i^2}\\ &= \frac{4-20i+25(-1)}{4-25(-1)}\\ &= \frac{4-20i-25}{4+25}\\ &= \frac{-21-20i}{29}\\ &= -\frac{21}{29}-\frac{20}{29}i\end{aligned}$$

81.

$$t^2 = 24$$
$$\sqrt{t^2} = \pm\sqrt{24}$$
$$t = \pm\sqrt{4}\sqrt{6}$$
$$t = \pm 2\sqrt{6}$$

82.

$$m^2 + 10m - 7 = 0$$
$$m^2 + 10m = 7$$
$$m^2 + 10m + 25 = 7 + 25$$
$$(m+5)^2 = 32$$
$$\sqrt{(m+5)^2} = \pm\sqrt{32}$$
$$m + 5 = \pm\sqrt{16}\sqrt{2}$$
$$m + 5 = \pm 4\sqrt{2}$$
$$m = -5 \pm 4\sqrt{2}$$

83.

$$4w^2 + 6w + 1 = 0$$
$$w = \frac{-b \pm \sqrt{b^2 - 4ac}}{2a}$$
$$w = \frac{-6 \pm \sqrt{6^2 - 4(4)(1)}}{2(4)}$$
$$w = \frac{-6 \pm \sqrt{36 - 16}}{8}$$
$$w = \frac{-6 \pm \sqrt{20}}{8}$$
$$w = \frac{-6 \pm 2\sqrt{5}}{8}$$
$$w = -\frac{6}{8} \pm \frac{2\sqrt{5}}{8}$$
$$w = -\frac{3}{4} \pm \frac{\sqrt{5}}{4}$$
$$w = \frac{-3 \pm \sqrt{5}}{4}$$

84.

$$3x^2 - 4x = -2$$
$$3x^2 - 4x + 2 = 0$$
$$x = \frac{-b \pm \sqrt{b^2 - 4ac}}{2a}$$
$$x = \frac{-(-4) \pm \sqrt{(-4)^2 - 4(3)(2)}}{2(3)}$$
$$x = \frac{4 \pm \sqrt{16 - 24}}{6}$$
$$x = \frac{4 \pm \sqrt{-8}}{6}$$
$$x = \frac{4 \pm 2i\sqrt{2}}{6}$$
$$x = \frac{4}{6} \pm \frac{2\sqrt{2}}{6}i$$
$$x = \frac{2}{3} \pm \frac{\sqrt{2}}{3}i$$

85. Let $u = 2x + 1$.

$$2(2x+1)^2 - 7(2x+1) + 6 = 0$$
$$2u^2 - 7u + 6 = 0$$
$$(2u - 3)(u - 2) = 0$$
$$2u - 3 = 0 \quad \text{or} \quad u - 2 = 0$$
$$u = \frac{3}{2} \qquad u = 2$$
$$2x + 1 = \frac{3}{2} \qquad 2x + 1 = 2$$
$$2x = \frac{1}{2} \qquad 2x = 1$$
$$x = \frac{1}{4} \qquad x = \frac{1}{2}$$

86.

$$x^4 + 19x^2 + 18 = 0$$
$$(x^2 + 1)(x^2 + 18) = 0$$
$$x^2 + 1 = 0 \quad \text{or} \quad x^2 + 18 = 0$$
$$x^2 = -1 \qquad x^2 = -18$$
$$x = \sqrt{-1} \qquad x = \sqrt{-18}$$
$$x = \pm i \qquad x = \pm 3i\sqrt{2}$$
$$x = -i, i, 3i\sqrt{2}, -3i\sqrt{2}$$

87. a. a quadratic function
b. at about 85% and 120% of the suggested inflation

88. $f(x) = -6x^2 - 12x - 8$

Step 1: Since $a = 1 < 0$, the parabola opens downward.

Step 2: Find the vertex and axis of symmetry.

$$x = \frac{-b}{2a} = \frac{-(-12)}{2(-6)} = \frac{12}{-12} = -1$$

$$f(-1) = -6(-1)^2 - 12(-1) - 8$$
$$= -6 + 12 - 8$$
$$= -2$$

vertex: $(-1, -2)$

The axis of symmetry is $x = -1$.

Step 3: Find the x- and y-intercepts. Since $c = -8$, the y-intercept is $(0, -8)$. The vertex is located below the x-axis and it opens downward, so there are no x-intercepts.

Step 4: Use symmetry to find the point $(-4, -8)$.

Step 5: Plots the points and draw the parabola.

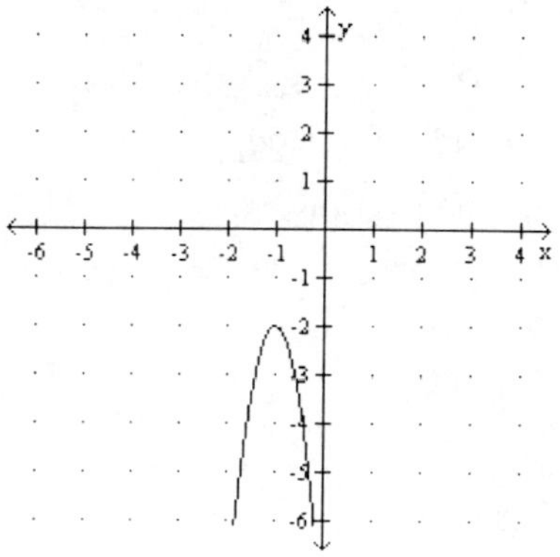

89.

$$x^2 - 8x \le -15$$
$$x^2 - 8x + 15 \le 0$$
$$(x-3)(x-5) = 0$$
$$x - 3 = 0 \quad \text{or} \quad x - 5 = 0$$
$$x = 3 \qquad\qquad x = 5$$
$$[3, 5]$$

90.

$$(f \circ g)(x) = f(g(x))$$
$$= f(2x+1)$$
$$= (2x+1)^2 - 2$$
$$= 4x^2 + 4x + 1 - 2$$
$$= 4x^2 + 4x - 1$$

91.

$$f(x) = 2x^3 - 1$$
$$y = 2x^3 - 1$$
$$x = 2y^3 - 1$$
$$x + 1 = 2y^3$$
$$\frac{x+1}{2} = y^3$$
$$\sqrt[3]{\frac{x+1}{2}} = y$$
$$y = \sqrt[3]{\frac{x+1}{2}}$$
$$f^{-1}(x) = \sqrt[3]{\frac{x+1}{2}}$$

92. $f(x) = \left(\frac{1}{2}\right)^x \quad D: (-\infty, \infty); \; R: (0, \infty)$

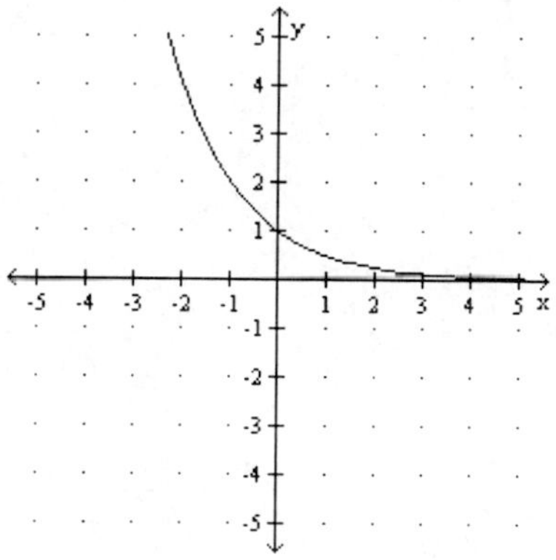

93.

$$\log 1{,}000 = x$$
$$\log_{10} 1{,}000 = x$$
$$10^x = 1{,}000$$
$$10^x = 10^3$$
$$x = 3$$

94.

$$\log_8 64 = x$$
$$8^x = 64$$
$$8^x = 8^2$$
$$x = 2$$

95.

$$\log_3 x = -3$$
$$3^{-3} = x$$
$$\frac{1}{3^3} = x$$
$$\frac{1}{27} = x$$

96.

$$\log_x 25 = 2$$
$$x^2 = 25$$
$$x^2 = 5^2$$
$$x = 5$$

97.

$$\ln e = x$$
$$\log_e e = x$$
$$e^x = e^1$$
$$x = 1$$

98.

$$\ln \frac{1}{e} = x$$
$$\log_e \frac{1}{e} = x$$
$$e^x = \frac{1}{e}$$
$$e^x = e^{-1}$$
$$x = -1$$

99. $f(x) = e^x$

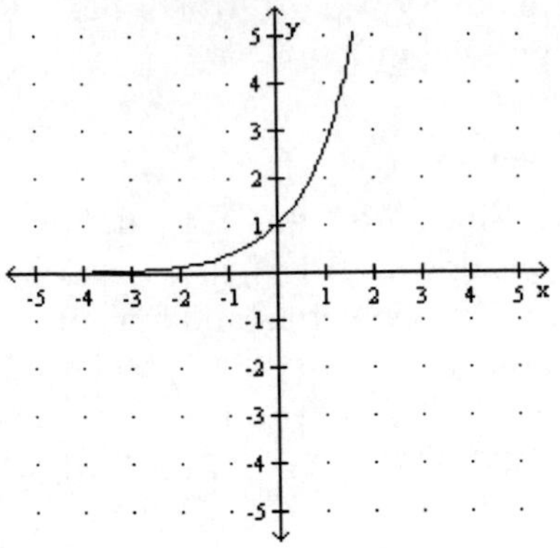

100. POPULATION GROWTH
$P = 114$ million, $r = 0.01102$, and $t = 25$

$$A = Pe^{rt}$$
$$= 114e^{(0.01102)(25)}$$
$$= 114e^{0.2755}$$
$$\approx 150 \text{ million}$$

101. $\log 0$ is undefined

102.

$$\log_6 \frac{36}{x^3} = \log_6 36 - \log_6 x^3$$
$$= \log_6 6^2 - \log_6 x^3$$
$$= 2\log_6 6 - 3\log_6 x$$
$$= 2(1) - 3\log_6 x$$
$$= 2 - 3\log_6 x$$

103.

$$\frac{1}{2}\ln x + \ln y - \ln z = \ln x^{1/2} + \ln y - \ln z$$
$$= \ln \frac{x^{1/2}y}{z}$$
$$= \ln \frac{\sqrt{x} \bullet y}{z}$$
$$= \ln \frac{y\sqrt{x}}{z}$$

104. BACTERIA GROWTH

Let $P_0 = 200$, $P = 600$, and $t = 4$.

$$P = P_0 e^{kt}$$
$$600 = 200e^{k\cdot 4}$$
$$600 = 200e^{4k}$$
$$3 = e^{4k}$$
$$\ln 3 = \ln e^{4k}$$
$$\ln 3 = 4k \ln e$$
$$\ln 3 = 4k$$
$$k = \frac{\ln 3}{4}$$

Let $P_0 = 200$, $P = 8{,}000$, and $k = \dfrac{\ln 3}{4}$.

$$P = P_0 e^{kt}$$
$$8{,}000 = 200e^{\left(\frac{\ln 3}{4}\right)t}$$
$$40 = e^{\frac{\ln 3}{4}t}$$
$$\ln 40 = \ln e^{\frac{\ln 3}{4}t}$$
$$\ln 40 = \frac{\ln 3}{4}t$$
$$\left(\frac{4}{\ln 3}\right)(\ln 40) = \left(\frac{4}{\ln 3}\right)\left(\frac{\ln 3}{4}t\right)$$
$$\frac{4\ln 40}{\ln 3} = t$$
$$t = 13.4 \text{ hours}$$

105.

$$5^{4x} = \frac{1}{125}$$
$$5^{4x} = 5^{-3}$$
$$4x = -3$$
$$x = -\frac{3}{4}$$

106.

$$2^{x+2} = 3^x$$
$$\log 2^{x+2} = \log 3^x$$
$$(x+2)\log 2 = x\log 3$$
$$x\log 2 + 2\log 2 = x\log 3$$
$$2\log 2 = x\log 3 - x\log 2$$
$$2\log 2 = x(\log 3 - \log 2)$$
$$\frac{2\log 2}{\log 3 - \log 2} = x$$
$$x = \frac{2\log 2}{\log 3 - \log 2}$$
$$x \approx 3.4190$$

107.

$$\log x + \log(x+9) = 1$$
$$\log(x^2 + 9x) = 1$$
$$10^1 = x^2 + 9x$$
$$0 = x^2 + 9x - 10$$
$$0 = (x+10)(x-1)$$
$$x+10 = 0 \quad \text{or} \quad x-1 = 0$$
$$x = \cancel{-10} \qquad x = 1$$

108.

$$\log_3 x = \log_3\left(\frac{1}{x}\right) + 4$$
$$\log_3 x - \log_3\left(\frac{1}{x}\right) = 4$$
$$\log_3 \frac{x}{\frac{1}{x}} = 4$$
$$\log_3 x^2 = 4$$
$$3^4 = x^2$$
$$81 = x^2$$
$$9 = x$$

109. a. ii.
b. iv.
c. i.
d. iii.

110. center $(1, -3)$, radius $= 2$

$$(x-1)^2+(y+3)^2=2^2$$

$$(x-1)^2+(y+3)^2=4$$

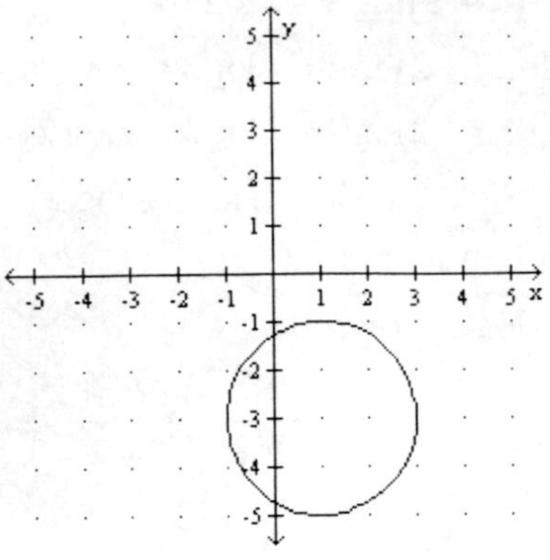

111.

$$y^2+4x-6y=-1$$

$$4x=-y^2+6y-1$$

$$x=-\frac{1}{4}y^2+\frac{3}{2}y-\frac{1}{4}$$

$$x=-\frac{1}{4}\left(y^2-6y\right)-\frac{1}{4}$$

$$x=-\frac{1}{4}\left(y^2-6y+9\right)-\frac{1}{4}+\frac{1}{4}(9)$$

$$x=-\frac{1}{4}(y-3)^2+2$$

vertex: $(2,3)$

axis of symmetry: $y=3$

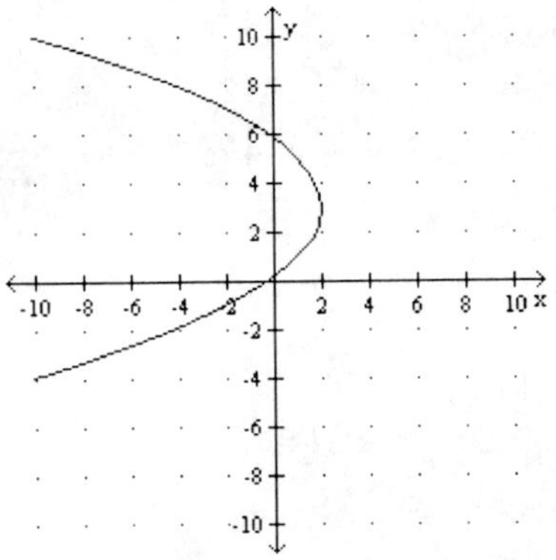

112. $$\frac{(x+1)^2}{4}+\frac{(y-3)^2}{16}=1$$

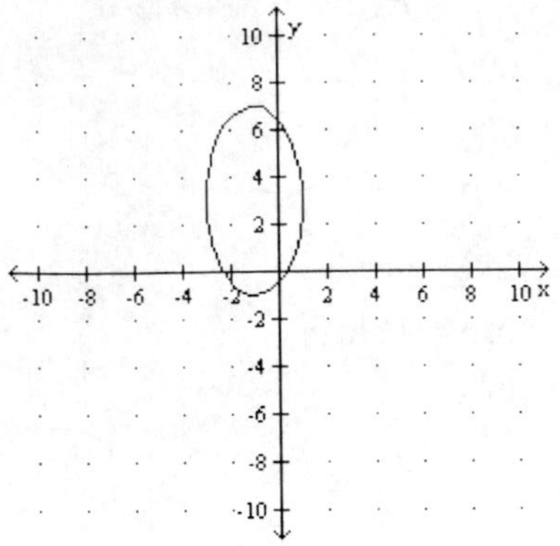

113.

$$(x-2)^2-9y^2=9$$

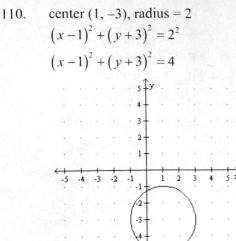

$$\frac{(x-2)^2}{9}-\frac{9y^2}{9}=\frac{9}{9}$$

$$\frac{(x-2)^2}{9}-\frac{y^2}{1}=1$$

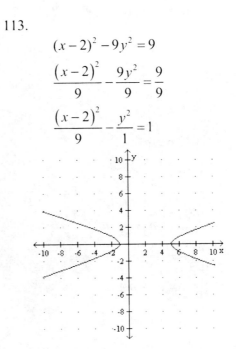

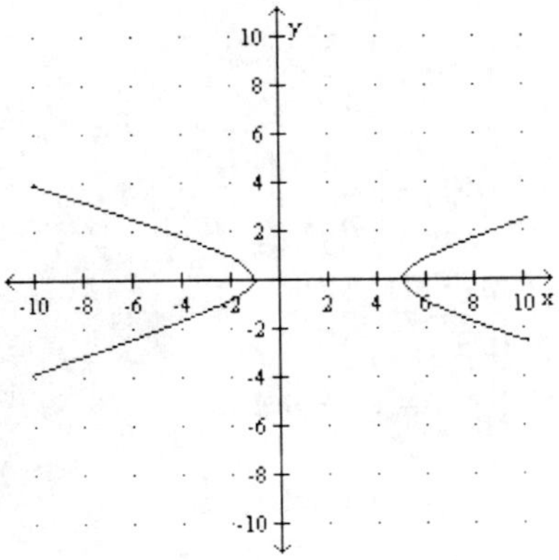

114.

$$(3a-b)^4$$

$$=(3a)^4+\frac{4!}{1!(4-1)!}(3a)^3(-b)$$

$$+\frac{4!}{2!(4-2)!}(3a)^2(-b)^2+\frac{4!}{3!(4-3)!}(3a)(-b)^3$$

$$+(-b)^4$$

$$=81a^4+\frac{4\cdot\cancel{3!}}{1!\cancel{3!}}\left(27a^3\right)(-b)+\frac{4\cdot3\cdot\cancel{2!}}{2!\cancel{2!}}\left(9a^2\right)\left(b^2\right)$$

$$+\frac{4\cdot\cancel{3!}}{\cancel{3!}1!}(3a)\left(-b^3\right)+b^4$$

$$=81a^4-108a^3b+54a^2b^2-12ab^3+b^4$$

115.

$$\frac{12!}{10!(12-10)!}=\frac{12\cdot 11\cdot \cancel{10!}}{\cancel{10!}2!}$$
$$=\frac{132}{2}$$
$$=66$$

116. Let $n = 20$, $a_1 = -11$, and $d = 6$.

$$a_n = a_1 + d(n-1)$$
$$a_{20} = -11 + 6(20-1)$$
$$a_{20} = -11 + 6(19)$$
$$a_{20} = -11 + 114$$
$$a_{20} = 103$$

117. Let $n = 20$, $a_1 = 6$, and $d = 3$.

$$a_n = a_1 + d(n-1)$$
$$a_{20} = 6 + 3(20-1)$$
$$a_{20} = 6 + 3(19)$$
$$a_{20} = 6 + 57$$
$$a_{20} = 63$$

Find S if $a_{20} = 63$ and $a_1 = 6$.

$$S_n = \frac{n(a_1 + a_n)}{2}$$
$$S_{20} = \frac{20(6+63)}{2}$$
$$= \frac{20(69)}{2}$$
$$= 690$$

118.

$$\sum_{k=3}^{5}(2k+1) = (2\cdot 3+1)+(2\cdot 4+1)+(2\cdot 5+1)$$
$$=7+9+11$$
$$=27$$

119. Let $a_1 = \frac{1}{27}$ and $r = 3$.

$$a_n = a_1 r^{n-1}$$
$$a_7 = \left(\frac{1}{27}\right)(3)^{7-1}$$
$$a_7 = \left(\frac{1}{27}\right)(3)^{6}$$
$$a_7 = \left(\frac{1}{27}\right)(729)$$
$$a_7 = 27$$

120. If the boat depreciates 12% of its value each year, then 88% of its value remains. The value of the boat in 9 years is represented by the 10th term of a geometric series, where $a_1 = 9{,}000$, $n = 10$, and $r = 0.88$.

$$a_n = a_1 r^{n-1}$$
$$a_{10} = (9{,}000)(0.88)^{10-1}$$
$$a_{10} = (9{,}000)(0.88)^{9}$$
$$a_{10} = (9{,}000)(0.316478)$$
$$a_{10} = \$2{,}848.31$$

The boat is worth \$2,848.31 after 9 years.

121. Find r.

$$r = \frac{a_2}{a_1} = \frac{\frac{1}{32}}{\frac{1}{64}} = \frac{1}{32} \bullet \frac{64}{1} = 2$$

Find the sum with $a_1 = \frac{1}{64}$ $n = 10$, and $r = 2$.

$$S_n = \frac{a_1(1-r^n)}{1-r}$$
$$= \frac{\frac{1}{64}(1-2^{10})}{1-2}$$
$$= \frac{\frac{1}{64}(1-1{,}024)}{-1}$$
$$= \frac{\frac{1}{64}(-1{,}023)}{-1}$$
$$= \frac{-\frac{1{,}023}{64}}{-1}$$
$$= \frac{1{,}023}{64}$$

122. Find r.

$$r = \frac{a_2}{a_1} = \frac{3}{9} = \frac{1}{3}$$

Find the sum with $a_1 = 9$ and $r = \frac{1}{3}$.

$$S = \frac{a_1}{1-r}$$
$$= \frac{9}{1-\frac{1}{3}}$$
$$= \frac{9}{\frac{2}{3}}$$
$$= \frac{9}{1} \bullet \frac{3}{2}$$
$$= \frac{27}{2}$$

CPSIA information can be obtained
at www.ICGtesting.com
Printed in the USA
FFOW03n0716301216
30861FF

9 781111 987589